AA002398

2019 31st International Symposium on Power Semiconductor Devices and ICs (ISPSD 2019)

Shanghai, China
19 – 23 May 2019

IEEE Catalog Number: CFP19ISP-POD

ISBN: 978-1-7281-0582-6

**Copyright © 2019 by the Institute of Electrical and Electronics Engineers, Inc.
All Rights Reserved**

Copyright and Reprint Permissions: Abstracting is permitted with credit to the source. Libraries are permitted to photocopy beyond the limit of U.S. copyright law for private use of patrons those articles in this volume that carry a code at the bottom of the first page, provided the per-copy fee indicated in the code is paid through Copyright Clearance Center, 222 Rosewood Drive, Danvers, MA 01923.

For other copying, reprint or republication permission, write to IEEE Copyrights Manager, IEEE Service Center, 445 Hoes Lane, Piscataway, NJ 08854. All rights reserved.

*** This is a print representation of what appears in the IEEE Digital Library. Some format issues inherent in the e-media version may also appear in this print version.

IEEE Catalog Number: CFP19ISP-POD
ISBN (Print-On-Demand):   978-1-7281-0582-6
ISBN (Online):   978-1-7281-0581-9
ISSN: 1063-6854

Additional Copies of This Publication Are Available From:

Curran Associates, Inc
57 Morehouse Lane
Red Hook, NY 12571 USA
Phone: (845) 758-0400
Fax: (845) 758-2633
E-mail: curran@proceedings.com
Web: www.proceedings.com

TABLE OF CONTENTS

Monday - May 20, 2019

8:50 – 10:10
Plenary 1
Chairs: Kuang Sheng, *Zhejiang University*
Oliver Häberlen, *Infineon*

8:50

PL1-1 Powering 5G Era Computing Platforms – The Road Toward Integrated Power Delivery 1
Peng Zou[1], Qiang Xie[1], Wei Song[1], Qimeng Jiang[2], Yuliang Lu[2], Boning Huang[2]
[1]*Hisilicon Semiconductor Co. Ltd.;* [2]*Huawei Technologies Co., Ltd.*

9:30

PL1-2 Magnetics in the GaN/SiC Power Electronics World 7
Alexander Gerfer
Würth Elektronik eiSos GmbH & Co. KG

10:40 – 12:10
Plenary 2
Chairs: Kevin Chen, *The Hong Kong University of Science and Technology*
Nando Kaminski, *Universität Bremen*

10:40

PL2-1 Foreseeable Industrial Changes of Automotive and Transportation Driven by Semiconductors 11
Tsuguo Nobe
Nagoya University

11:25

PL2-2 Power Challenges caused by IOT Edge Nodes: Securing and Sensing Our World 17
James Spehar, Adam Fuks, Marc Vauclair, Maurice Meijer, Joost van Beek, Bin Shao
NXP Semiconductors

13:30 – 15:35
Session 1 – Advanced SiC MOSFETs
Chairs: Jon Zhang, *PowerAmerica*
Kung-Yen Lee, *National Taiwan University*

13:30

**1-1 Experimental Demonstration on Superior Switching Characteristics of
1.2 kV SiC SWITCH-MOS** 23
Ruito Aiba[1], Masataka Okawa[1], Taiga Kanamori[1], Hiroshi Yano[1], Noriyuki Iwamuro[1],
Yusuke Kobayashi[2], Shinsuke Harada[2]
[1]*University of Tsukuba;* [2]*National Institute of Advanced Industrial Science and Technology*

13:55

1-2 Superior Switching Characteristics of SiC-MOSFET Embedding SBD 27
Takaaki Tominaga, Shiro Hino, Yohei Mitsui, Junichi Nakashima, Koutarou Kawahara,
Shingo Tomohisa, Naruhisa Miura
Mitsubishi Electric Corporation

14:20

1-3 **High-Temperature Performance of 1.2 kV-Class SiC Super Junction MOSFET** 31
Yusuke Kobayashi[1,2], Shinya Kyogoku[1,3], Tadao Morimoto[1], Teruaki Kumazawa[1,4],
Yusuke Yamashiro[1,5], Manabu Takei[1,2], Shinsuke Harada[1]
[1]National Institute of Advanced Industrial Science and Technology; [2]Fuji Electric Co., Ltd.;
[3]Toshiba Corporation; [4]Toyota Motor Corporation; [5]Mitsubishi Electric Corporation

14:45

1-4 **Suppression of Bipolar Degradation in Deep-P Encapsulated 4H-SiC Trench**
MOSFETs up to Ultra-High Current Density 35
Yasuhiro Ebihara, Junichi Uehara, Aicko Ichimura, Shuhei Mitani, Masato Noborio,
Yuichi Takeuchi, Kazuhiro Tsuruta
DENSO Corporation

15:10

1-5 **Breaking the Theoretical Limit of 6.5 kV-Class 4H-SiC Super-Junction (SJ)**
MOSFETs by Trench-Filling Epitaxial Growth (Late News) 39
Ryoji Kosugi, Shiyang Ji, Kazuhiro Mochizuki, Kohei Adachi, Satoshi Segawa,
Yasuyuki Kawada, Yoshiyuki Yonezawa, Hajime Okumura
National Institute of Advanced Industrial Science and Technology

16:00 – 17:40
Session 2 – IGBTs
Chairs: Giovanni Breglio, *University of Naples Federico II*
Marina Antoniou, *University of Warwick*

16:00

2-1 **3300V Scaled IGBTs Driven by 5V Gate Voltage** 43
Takuya Saraya[1], Kazuo Itou[1], Toshihiko Takakura[1], Munetoshi Fukui[1], Shinichi Suzuki[1], Kiyoshi Takeuchi[1],
Masanori Tsukuda[2], Yohichiroh Numasawa[3], Katsumi Satoh[4], Tomoko Matsudai[5], Wataru Saito[5],
Kuniyuki Kakushima[6], Takuya Hoshii[6], Kazuyoshi Furukawa[6], Masahiro Watanabe[6], Naoyuki Shigyo[6],
Hitoshi Wakabayashi[6], Kazuo Tsutsui[6], Hiroshi Iwai[6], Atsushi Ogura[3], Shin-ichi Nishizawa[7],
Ichiro Omura[8], Hiromichi Ohashi[6], Toshiro Hiramoto[1]
[1]The University of Tokyo; [2]Green Electronics Research Institute; [3]Meiji University; [4]Mitsubishi Electric
Corporation; [5]Toshiba Electronic Devices & Storage Corporation; [6]Tokyo Institute of Technology;
[7]Kyushu University; [8]Kyushu Institute of Technology

16:25

2-2 **A Critical View of IGBT Buffer Designs for 200 °C Operation** 47
Elizabeth Buitrago[1], Athanasios Mesemanolis[1], Maxi Andenna[1], Charalampos Papadopoulos[1],
Chiara Corvasce[1], Jan Vobecky[1], Munaf Rahimo[2]
[1]ABB Semiconductors; [2]MTAL GmbH

16:50

2-3 **A Novel 3300V Trench IGBT with Hole Extraction Structure for Low Power Loss** 51
Xin Peng[1], Zehong Li[1], Yishang Zhao[1], Yang Yang[1], Min Ren[1], Jinping Zhang[1], Wei Gao[1],
Bo Zhang[1], Zhaoji Li[1], Defu Sun[2], Yingxin Song[2], Kuncun Zhu[2], Donghua Li[2]
[1]University of Electronic Science and Technology of China; [2]Jinan Semiconductor Institute

17:15

2-4 **Effects of the HV-BIGT Design Elements on the High-Frequency Oscillation Instability**
during Short Circuit Transients 55
P. Diaz Reigosa[1], C. Papadopoulos[2], F. Iannuzzo[3], C. Corvasce[2], M. Rahimo[4]
[1]University of Applied Sciences Northwestern Switzerland; [2]ABB Switzerland Ltd.;
[3]Aalborg University, Denmark; [4]MTAL GmbH

Tuesday – May 21, 2019

8:30 – 10:10
Session 3 – GaN Device Technology
Chairs: Shu Yang, *Zhejiang University*
         Hideyuki Okita, *Panasonic Corp.*

8:30

3-1 **Estimation of Impact Ionization Coefficient in GaN by Photomulitiplication Measurement Utilizing Franz-Keldysh Effect** 59
Takuya Maeda[1], Tetsuo Narita[2], Hiroyuki Ueda[2], Masakazu Kanechika[2], Tsutomu Uesugi[2], Tetsu Kachi[3], Tsunenobu Kimoto[1], Masahiro Horita[1,3], Jun Suda[1,3]
[1]Kyoto University; [2]Toyota Central R&D Labs., Inc.; [3]Nagoya University

8:55

3-2 **Photon-Enhanced Conductivity Modulation and Surge Current Capability in Vertical GaN Power Rectifiers** 63
Shaowen Han, Shu Yang, Yongkai Li, Yinxiang Liu, Kuang Sheng
Zhejiang University

9:20

3-3 **High Performance GaN-on-Si Power Devices with Ultralow Specific On-Resistance using Novel Strain Method Fabricated on 200 mm CMOS-Compatible Process Platform** 67
Roy K.-Y. Wong[1], H.-C. Chiu[1], J.H. Zhang[1], C. Zhou[1,2], Thomas Zhao[1], Y.B. Wu[1], Henry Liao[1], Simon He[1], A.B. Zhang[1], Y.B. Zou[1], Seiya Li[1], Martin Zhang[1], Macro Wu[1], John Lee[1], P.W. Chen[1], Andy Xie[1], Jeff Zhang[1]
[1]Innoscience Technology; [2]University of Electronic Science and Technology of China

9:45

3-4 **High-Performance Normally-Off Tri-Gate GaN Power MOSFETs** 71
Minghua Zhu, Jun Ma, Luca Nela, Elison Matioli
École Polytechnique Fédérale de Lausanne

10:30 – 12:10
Session 4 – Gate Driver ICs and Invited Paper
Chairs: Jing Zhu, *Southeast University*
         Maarten Swanenberg, *NXP Semiconductors*

10:30

4-1 **Sub-Nanosecond Delay CMOS Active Gate Driver for Closed-Loop dv/dt Control of GaN Transistors** 75
Plinio Bau[1,2], Marc Cousineau[1], Bernardo Cougo[2], Frederic Richardeau[1], Sebastien Vinnac[1], Didier Flumian[1], Nicolas Rouger[1]
[1]LAPLACE, Université de Toulouse; [2]Institute of Technology Antoine de Saint Exupéry

10:55

4-2 **Building Blocks for Future Dual-Channel GaN Gate Drivers: Arbitrary Waveform Driver, Bootstrap Voltage Supply, and Level Shifter** 79
Dawei Liu[1], Harry C.P. Dymond[1], Jianjing Wang[1], Bernard H. Stark[1], Simon J. Hollis[2]
[1]University of Bristol; [2]Xilinx, Inc.

11:20

4-3 **An Integrated Gate Driver for E-mode GaN HEMTs with Active Clamping for Reverse Conduction Detection** 83
Wei Jia Zhang[1], Yahui Leng[2], Jingshu Yu[1], Yu Shen Lu[1], Chu Yao Cheng[1], Wai Tung Ng[1]
[1]University of Toronto, Canada; [2]Zhejiang University

11:45

4-4 **Driving GaN Power Transistors – Circuit Design Challenges and Opportunities (Invited)** 87
D. Brian Ma
University of Texas at Dallas

13:30 – 15:10
Session 5 – Low Voltage Discrete Devices and Late News
Chairs: Naoto Fujishima, *Fuji Electric Co., Ltd.*
Mark Gajda, *Nexperia*

13:30

5-1 **Alpha-Particle Shielding Effect of Thick Copper Plating Film on Power MOSFETs** 91
Tatsuya Nishiwaki, Kohei Oasa, Kentaro Ichinoseki, Kikuo Aida, Tatsuya Ohguro,
Yoshiharu Takada, Hideharu Kojima, Yusuke Kawaguchi
Toshiba Electronic Devices & Storage Corporation

13:55

5-2 **A New 200V Dual Trench MOSFET with Stepped Oxide for Ultra Low R_{DS}(on)** 95
Chanho Park, Misbah Azam, Gabriel Dengel, Ayman Shibib, Kyle Terrill
Vishay Intertechnology Inc.

14:20

5-3 **100-V Class Two-Step-Oxide Field-Plate Trench MOSFET to Achieve Optimum**
RESURF Effect and Ultralow On-Resistance 99
Kenya Kobayashi, Hiroaki Kato, Toshifumi Nishiguchi, Saya Shimomura, Tetsuya Ohno, Tatsuya Nishiwaki,
Kikuo Aida, Kentaro Ichinoseki, Kohei Oasa, Yusuke Kawaguchi
Toshiba Electronic Devices & Storage Corporation

14:45

5-4 **Vertical 1.2kV SiC Power MOSFETs with High-k/Metal Gate Stack (Late News)** 103
Stephan Wirths, Yulieth Christina Arango, Alyssa Prasmusinto, Giovanni Alfieri,
Enea Bianda, Andrei Mihaila, Lukas Kranz, Marco Bellini, Lars Knoll
ABB Switzerland Ltd.

15:30 – 17:30
Poster Session 1 – IC Design, SiC, Packaging
Chairs: Konji Hara, *Hitachi*
Takeharu Kuroiwa, *Mitsubishi Electric Corporation*

P1-1 **Feasibility and Limitation of DC/DC Multilevel Converter Power ICs using**
Standard CMOS Transistors 107
Runtao Ning[1], Yuanfeng Zhou[2], Aritra Kundu[2], Z. John Shen[2]
[1]*Alpha and Omega Semiconductor;* [2]*Illinois Institute of Technology*

P1-2 **Integrated Current Sensing in GaN Power ICs** 111
Stefan Moench[1], Richard Reiner[1], Patrick Waltereit[1], Rüdiger Quay[1], Oliver Ambacher[1], Ingmar Kallfass[2]
[1]*Fraunhofer Institute for Applied Solid State Physics;* [2]*University of Stuttgart*

P1-3 **500°C SiC-Based Driver IC for SiC Power MOSFETs** 115
Saleh Kargarrazi[1], Hossein Elahipanah[2], Zikang Tong[1], Debbie Senesky[1], Carl-Mikael Zetterling[2]
[1]*Stanford University;* [2]*KTH Royal Institute of Technology*

P1-4 **A Total Ionizing Dose Detecting Circuit based on Off-State Leakage Current of**
NLDMOS in Power IC 119
Ping Luo, Rongxun Ling, Xiao Zhou, Pengkai Jiang, Yucao Wu
University of Electronic Science and Technology of China

P1-5 A Novel Cross-Over CMR Transformer Technology for Magnetic
Isolation Gate Driver Applications 123
Yunzhong Luo, Jian Fang, Erli Zhang, Ming Li, Bo Zhang
University of Electronic Science and Technology of China

P1-6 A High-Reliability Half-Bridge GaN FET Gate Driver with Advanced
Floating Bias Control Techniques 127
Xin Ming, Xuan Zhang, Zhi-wen Zhang, Li Hu, Yao Qin, Xu-dong Feng, Qi Zhou, Zhuo Wang, Bo Zhang
University of Electronic Science and Technology of China

P1-7 A Predictive Gate Driver Suitable for Half-Bridge Applications 131
Zekun Zhou[1], Yandong Yuan[1], Yunkun Wang[1], Bo Zhang[1], Yue Shi[2]
[1]University of Electronic Science and Technology of China; [2]Chengdu University of Information Technology

P1-8 SiC MOSFET with Integrated Zener Diode as an Asymmetric Bidirectional Voltage
Clamp between the Gate and Source for Overvoltage Protection 135
Cheng-Tyng Yen, Fu-Jen Hsu, Kuo-Ting Chu, Chien-Chung Hung, Lurng-Shehng Lee, Chwan-Ying Lee
Hestia Power Inc.

P1-9 Breakthrough in Channel Mobility Limit of Nitrided Gate Insulator for
SiC DMOSFET with Novel High-Temperature N_2 Annealing 139
S. Asaba[1], T. Ito[1], S. Fukatsu[1], Y. Nakabayashi[1], T. Shimizu[1], M. Furukawa[2], T. Suzuki[2], R. Iijima[1]
[1]Toshiba Corporation; [2]Toshiba Electronic Devices & Storage Corporation

P1-10 Trench Field Plate Engineering for High Efficient Edge Termination of 1200 V-Class SiC Devices 143
Yong Liu[1], Wentao Yang[1], Hao Feng[1], Yuichi Onozawa[2], Setsuko Wakimoto[2],
Naoto Fujishima[2], Johnny K.O. Sin[1]
[1]The Hong Kong University of Science and Technology; [2]Fuji Electric Co., Ltd.

P1-11 Power Cycling Test on 3.3kV SiC MOSFETs and the effects of Bipolar
Degradation on the Temperature Estimation by V_{SD}-Method 147
Felix Hoffmann[1], Victor Soler[2], Andrei Mihaila[3], Nando Kaminski[1]
[1]University of Bremen; [2]CNM-CSIC; [3]ABB Switzerland Ltd.

P1-12 175V, > 5.4 MV/cm, 50 mΩ.cm² at 250°C Diamond MOSFET and its Reverse Conduction 151
Cédric Masante[1], Julien Pernot[1], Juliette Letellier[1], David Eon[1], Nicolas Rouger[2]
[1]Université Grenoble Alpes; [2]LAPLACE, Université de Toulouse

P1-13 Practical One-Step Solution of Smoothly Tapered Junction Termination Extension for
High Voltage SiC Gate Turn-off Thyristor 155
Hu Long, Qing Guo, Kuang Sheng
Zhejiang University

P1-14 Comparison of New Octagonal Cell Topology for 1.2 kV 4H-SiC JBSFETs with
Linear and Hexagonal Topologies: Analysis and Experimental Results 159
Kijeong Han, Aditi Agarwal, B. Jayant Baliga
North Carolina State University

P1-15 Differential Variable Base Charge Pumping (Δ-CP) for SiO_2/SiC Interface Characterization 163
P. Moens[1], A. Constant[1], A. Stockman[1,3], J. Franchi[1], F. Allerstam[1]
[1]ON Semiconductor; [2]Gent University

P1-16 Experimental and Numerical Investigations of Short-Circuit Failure
Mechanisms for State-of-the-Art 1.2kV SiC Trench MOSFETs 167
Masataka Okawa[1], Ruito Aiba[1], Taiga Kanamori[1], Hiroshi Yano[1], Noriyuki Iwamuro[1], Shinsuke Harada[2]
[1]University of Tsukuba; [2]National Institute of Advanced Industrial Science and Technology

P1-17 Fabrication of a High-Voltage SiC Avalanche Diode with a Superior Voltage Clamp Property 171
Masayuki Yamamoto[1,2], Kunio Koseki[1], Yasunori Tanaka[1]
[1]National Institute of Advanced Industrial Science and Technology; [2]University of Yamanashi

P1-18 Operation of Ultra-High Voltage (>10kV) SiC IGBTs at Elevated Temperatures: Benefits & Constraints 175
Amit K. Tiwari[1], Florin Udrea[1], Neophytos Lophitis[2], Marina Antoniou[3]
[1]University of Cambridge; [2]Coventry University; [3]University of Warwick

P1-19 3kV SiC Charge-Balanced Diodes Breaking Unipolar Limit 179
Reza Ghandi, Alexander Bolotnikov, David Lilienfeld, Stacey Kennerly, Raju Ravisekhar
GE Global Research Center

P1-20 Failure Mechanism Analysis of SiC MOSFETs in Unclamped Inductive Switching Conditions 183
Na Ren[1], Kang L. Wang[1], Jiupeng Wu[2], Hongyi Xu[2], Kuang Sheng[2]
[1]University of California, Los Angeles; [2]Zhejiang University

P1-21 Investigation of UIS Capability for -600V Class Vertical SiC p-Channel MOSFET 187
Kailun Yao, Hiroshi Yano, Noriyuki Iwamuro
University of Tsukuba

P1-22 Ultra-High Speed 7mohm, 650V SiC Half-Bridge Module with Robust Short Circuit Capability for EV Inverters 191
Anup Bhalla, Melvin Nava, Mike Zhu, Frank Sudario, Deborah Sumaoang, Peter Alexandrov, Xueqing Li, Peter Losee
United Silicon Carbide Inc.

P1-23 Determination of the Transient Threshold Voltage Hysteresis in SiC MOSFETs after Positive and Negative Gate Bias 195
Christian Unger, Martin Pfost
TU Dortmund

P1-24 Investigations of p-Shielded SiC Trench IGBT with Considerations on IE Effect, Oxide Protection and Dynamic Degradation 199
Jin Wei[1], Meng Zhang[2], Huaping Jiang[3], Baikui Li[2], Zheyang Zheng[1], Kevin J. Chen[1]
[1]The Hong Kong University of Science and Technology; [2]Shenzhen University; [3]Chongqing University

P1-25 Design and Experimental Study of 1.2kV 4H-SiC Merged PiN Schottky Diode 203
Jiupeng Wu, Na Ren, Kuang Sheng
Zhejiang University

P1-26 Characterization of BTI in SiC MOSFETs using Third Quadrant Characteristics 207
Jose Ortiz Gonzalez, Olayiwola Alatise, Philip Mawby
University of Warwick

P1-27 Planar 1.2kV SiC MOSFETs with Retrograde Channel Profile for Enhanced Ruggedness 211
L. Knoll, A. Mihaila, S. Wirths, Y. Arango, A. Prasmusinto, E. Bianda, L. Kranz, M. Bellini, G. Romano, C. Papadopoulos
ABB Switzerland Ltd.

P1-28 An Experimentally Verified 3.3 kV SiC MOSFET Model Suitable for High-Current Modules Design 215
A. Borghese[1], M. Riccio[1], L. Maresca[1], G. Breglio[1], A. Irace[1], G. Romano[2], E. Bianda[2], A. Mihaila[2], M. Bellini[2], L. Knoll[2], S. Wirths[2]
[1]University of Naples Federico II; [2]ABB Switzerland Ltd.

P1-29 Time-Resolved Short Circuit Failure Analysis of SiC MOSFETs 219
Thomas Ziemann, Alexander Tsibizov, Bhagyalakshmi Kakarla, Lorenz Bort, Ulrike Grossner
ETH Zürich

P1-30 Design Considerations for High Voltage SiC Power Devices: An Experimental Investigation into Channel Pinching of 10kV SiC Junction Barrier Schottky (JBS) Diodes 223
Justin Lynch, Nick Yun, Woongje Sung
SUNY Polytechnic Institute

P1-31 Experimental Investigations of SiC MOSFETs under Short-Circuit Operations 227
Lei Cao, Zijian Gao, Qing Guo, Kuang Sheng
Zhejiang University

P1-32 **Switching Loss Model of SiC MOSFET Promoting High Frequency Applications** 231

Xuan Li[1], Xu Li[1], Liping Yang[1], Alex Q. Huang[2], Pengku Liu[2], Xiaochuan Deng[1], Bo Zhang[1]
[1]University of Electronic Science and Technology of China; [2]University of Texas at Austin

P1-33 **New Compact Automotive SiC-Sixpack Converter System with Stacked 3D-Gate Driver** 235

S. Buetow, R. Herzer, N. Burani, R. Bittner, M. Kujath
Semikron Elektronik GmbH & Co. KG

P1-34 **A Novel All-in-One Digital-Analog Heterogeneous Integrated Intelligent Power Module** 239

Y.T. Lin, K.S. Kao, C.M. Tzeng, H.H. Lin, W.K. Han, J.Y. Chang, T.C. Chang
Industrial Technology Research Institute

P1-35 **An Integrated Packaging Structure of Press Pack for High Power IGBTs** 243

Erping Deng[1,2], Bin Ren[1], Anqi Li[1], Yanhao Wang[1], Zhibin Zhao[1], Yongzhang Huang[1]
[1]North China Electric Power University; [2]Chemnitz University of Technology

P1-36 **The Development of a 1200V/400A SiC Hybrid Module** 247

Puqi Ning[1], Tianshu Yuan[2], Han Cao[2], Lei Li[2], Yuhui Kang[2]
[1]Institute of Electrical Engineering of the Chinese Academy of Science;
[2]University of Chinese Academy of Sciences

P1-37 **Analog Basis, Low-Cost Inverter Output Current Sensing with Tiny**
PCB Coil Implemented Inside IPM 251

Battuvshin Bayarkhuu[1], Bat-Otgon Bat-Ochir[2], Kazunori Hasegawa[1], Masanori Tsukuda[1],
Bayasgalan Dugarjav[2], Ichiro Omura[1]
[1]Kyushu Institute of Technology; [2]National University of Mongolia

Wednesday - May 22, 2019

8:30 – 10:10
Session 6 – SiC MOSFET Ruggedness and Stability
Chairs: Sei-Hyung Ryu, *Wolfspeed*
          Andrei Mihaila, *ABB Switzerland Ltd.*

8:30
6-1 **Short-Circuit Ruggedness Analysis of SiC JMOS and DMOS** 255

Fu-Jen Hsu, Cheng-Tyng Yen, Chien-Chung Hung, Kuo-Ting Chu, Lurng-Shehng Lee, Chwan-Ying Lee
Hestia Power Inc.

8:55
6-2 **Analysis of Degradation Phenomena in Bipolar Degradation Screening Process for SiC-MOSFETs** 259

Takashi Ishigaki[1], Tatsunori Murata[1], Koyo Kinoshita[1], Takahiro Morikawa[1], Tetsuo Oda[1],
Ryuusei Fujita[2], Kumiko Konishi[2], Yuki Mori[2], Akio Shima[2]
[1]Hitachi Power Semiconductor Device, Ltd.; [2]Hitachi, Ltd.

9:20
6-3 **Repetitive Short Circuit Energy Dependent V_{TH} Instability of 1.2kV SiC Power MOSFETs** 263

Jiahui Sun, Jin Wei, Zheyang Zheng, Yuru Wang, Kevin J. Chen
The Hong Kong University of Science and Technology

9:45
6-4 **Short-Circuit Robustness of 4600 V SiC DMOSFETs** 267

Siddarth Sundaresan, Vamsi Mulpuri, Sumit Jadav, Stoyan Jeliazkov, Ranbir Singh
GeneSiC Semiconductor

10:30 – 12:10

Session 7 – Monolithic Power ICs

Chairs: Shuichi Nagai, *Panasonic Corp.*
Budong Albert You, *Silergy Corp.*

10:30

7-1 **Development of GaN Power IC Platform and All GaN DC-DC Buck Converter IC** 271
Ruize Sun[1,2,3], Y.C. Liang[2,3], Yee-Chia Yeo[2], Cezhou Zhao[4], Wanjun Chen[1], Bo Zhang[1]
[1]University of Electronic Science and Technology of China; [2]National University of Singapore;
[3]National University of Singapore (Suzhou) Research Institute; [4]Xi'an Jiaotong-Liverpool University

10:55

7-2 **Integrated High-Speed Over-Current Protection Circuit for GaN Power Transistors** 275
Han Xu, Gaofei Tang, Jin Wei, Kevin J. Chen
The Hong Kong University of Science and Technology

11:20

7-3 **Gate Control Circuit for the LIGBT to Improve the Freewheeling Characteristics in Monolithic IC** 279
Siyuan Yu[1], Jing Zhu[1], Yangyang Lu[1], Weifeng Sun[1], Bowei Yang[1], Ding Yan[1],
Chuanyi Cheng[1], Yan Gu[2], Sen Zhang[2], Yunwu Zhang[3]
[1]Southeast University; [2]CSMC Technologies Corporation; [3]Wuxi i-Driver Electronics Co., Ltd

11:45

7-4 A New 1200 V HVIC Free of Latch-Off Failures for High Power Applications 283
Kinam Song, Junho Lee, Seunghyun Hong, Zhou Kong Shan
ON Semiconductor

WITHDRAWN

13:30 – 15:35

Session 8 – GaN Reliability

Chairs: Jun Suda, *Nagoya University*
JL Tom Tsai, *TSMC*

13:30

8-1 **Threshold Voltage Instability Mechanisms in p-GaN Gate AlGaN/GaN HEMTs** 287
Arno Stockman[1,2], Eleonora Canato[3], Matteo Meneghini[3], Gaudenzio Meneghesso[3],
Peter Moens[1], Benoit Bakeroot[2]
[1]ON Semiconductor; [2]CMST, imec and Ghent University; [3]University of Padova

13:55

8-2 **Dynamic Threshold Voltage in p-GaN Gate HEMT** 291
Jin Wei[1], Han Xu[1], Ruiliang Xie[1], Meng Zhang[2], Hanxing Wang[1], Yuru Wang[1],
Kailun Zhong[1], Mengyuan Hua[1], Jiabei He[1], Kevin J. Chen[1]
[1]The Hong Kong University of Science and Technology; [2]Shenzhen University

14:20

8-3 **Temperature-Dependent Gate Degradation of p-GaN Gate HEMTs under**
Static and Dynamic Positive Gate Stress 295
Jiabei He, Jin Wei, Song Yang, Mengyuan Hua, Kailun Zhong, Kevin J. Chen
The Hong Kong University of Science and Technology

14:45

8-4 **New Circuit Topology for System-Level Reliability of GaN** 299
Ming-Cheng Lin, Wen-Che Chang, Haw-Yun Wu, Gabriel Petrus Lansbergen, Man-Ho Kwan,
Jiun-Lei Yu, Cheng-Pao Wu, Chun-Lin Tsai, Hsiao-Chin Tuan, Alex Kalnitsky
Taiwan Semiconductor Manufacturing Company

15:10

8-5 **100 A Vertical GaN Trench MOSFETs with a Current Distribution Layer (Late News)** 303
Tohru Oka, Tsutomu Ina, Yukihisa Ueno, Junya Nishii
Toyoda Gosei Co. Ltd.

15:55 – 17:55
Poster Session 2 – Low Voltage, High Voltage, GaN
Chairs: Wataru Saito, *Toshiba Electronic Devices & Storage Corporation*
Tsung-Yi Huang, *Richtek*

P2-1   **High Accurate IGBT/IEGT Compact Modeling for Prediction of Power Efficiency and EMI Noise** 307
Takeshi Mizoguchi, Yoko Sakiyama, Naoto Tsukamoto, Wataru Saito
Toshiba Electronic Devices & Storage Corporation

P2-2   **Impact of Three-Dimensional Current Flow on Accurate TCAD Simulation for Trench-Gate IGBTs** 311
Masahiro Watanabe[1], Naoyuki Shigyo[1], Takuya Hoshii[1], Kazuyoshi Furukawa[1], Kuniyuki Kakushima[1],
Katsumi Satoh[2], Tomoko Matsudai[3], Takuya Saraya[4], Toshihiko Takakura[4], Kazuo Itou[4], Munetoshi Fukui[4],
Shinichi Suzuki[4], Kiyoshi Takeuchi[4], Iriya Muneta[1], Hitoshi Wakabayashi[1], Akira Nakajima[5],
Shin-ichi Nishizawa[6], Kazuo Tsutsui[1], Toshiro Hiramoto[4], Hiromichi Ohashi[1], Hiroshi Iwai[1]
[1]Tokyo Institute of Technology; [2]Mitsubishi Electric Corporation; [3]Toshiba Electronic Devices & Storage Corporation; [4]The University of Tokyo; [5]National Institute of Advanced Industrial Science and Technology; [6]Kyushu University

P2-3   **Self-Sustained Oscillation of Superjunction MOSFET Intrinsic Diode**
during Reverse Recovery Transient 315
Peng Xue, Luca Maresca, Michele Riccio, Giovanni Breglio, Andrea Irace
University of Naples Federico II

P2-4   **Static and Dynamic Figures of Merits (FOM) for Superjunction MOSFETs** 319
H. Kang, F. Udrea
University of Cambridge

P2-5   **New LOCOS Trench Oxide IGBT Enables 25% Higher Current Density in 4.5Kv/1500A Module** 323
L. Ngwendson[1], I. Deviny[1], C. Zhu[1], I. Saddiqui[1], J. Hutchings[1], C. Kong[1], Y. Wang[1], H. Luo[2]
[1]Dynex Semiconductors Ltd.; [2]Zhuzhou CRRC Times Electric Co., Ltd.

P2-6   **Broadband TCAD Mixed-Mode Simulation Framework for Predictive Modeling**
of Fast Dynamic Switching Events 327
Dan Popescu, Maximilian Treiber
Infineon Technologies AG

P2-7   **A Novel CSTBT with Hole Barrier for High dV/dt Controllability and Low EMI Noise** 331
Xiaorui Xu, Wanjun Chen, Chao Liu, Yuan Wang, Nan Chen, Fangzhou Wang, Qi Shi,
Kenan Zhang, Qi Zhou, Zhaoji Li, Bo Zhang
University of Electronic Science and Technology of China

P2-8   **Influence of External Gate Resistance on UIS Capability in Superjunction MOSFET** 335
Masaaki Honda, Mizue Yamaji, Daisuke Arai, Noriaki Suzuki, Takeshi Asada,
Wataru Hirasawa, Takeshi Yamaguchi, Yuji Watanabe
Shindengen Electric Manufacturing Co., Ltd.

P2-9   **Self-Turn-on-Free 5V Gate Driving for 1200V Scaled IGBT** 339
Masanori Tsukuda[1,2], Masaki Sudo[2], Kazunori Hasegawa[2], Seiya Abe[2], Takuya Saraya[3],
Toshihiko Takakura[3], Munetoshi Fukui[3], Kazuo Itou[3], Shinichi Suzuki[3], Kiyoshi Takeuchi[3],
Tamotsu Ninomiya[1], Toshiro Hiramoto[3], Ichiro Omura[2]
[1]Green Electronics Research Institute; [2]Kyushu Institute of Technology; [3]The University of Tokyo

P2-10   **Insulated Gate Bipolar Transistors based on Pure Boron Collectors** 343
Ahmed Elsayed, Jan Frederik Dick, Joerg Schulze
University of Stuttgart

P2-11   **4.5kV Insulated Gate Triggered Thyristor (IGTT) with High di/dt**
Characteristics for Pulse Power Applications 347
Chao Liu, Wanjun Chen, Yijun Shi, Bin Qiao, Qian Jiang, Yun Xia, Qijun Zhou,
Xiaorui Xu, Qi Zhou, Zhaoji Li, Bo Zhang
University of Electronic Science and Technology of China

P2-12 **Silicon RC-Snubber for 900 V Applications Utilising Non-Stoichiometric Silicon Nitride** 351
N. Boettcher[1], T. Heckel[1], T. Erlbacher[1], K. Pelaic[2]
[1]Fraunhofer Institute for Integrated Systems and Device Technology IISB; [2]Friedrich-Alexander University

P2-13 **Condition Monitoring of High Voltage IGBT Devices based on Controllable RF Oscillations** 355
Miaosong Gu[1], Xiang Cui[1], Xinling Tang[2], Cheng Peng[1], Xuebao Li[1], Rui Jin[2], Zhibin Zhao[1]
[1]North China Electric Power University; [2]Global Energy Interconnection Research Institute

P2-14 **An Injection Enhanced LIGBT on Thin SOI Layer with Low ON-state Voltage** 359
Gaoqiang Deng, Xiaorong Luo, Diao Fan, Tao Sun, Bo Zhang
University of Electronic Science and Technology of China

P2-15 **Design Method and Mechanism Study of LDMOS to Conquer Stress Induced Degradation of Leakage Current and HTRB Reliability** 363
Kanako Komatsu[1], Tomoko Kinoshita[1], Saori Shioda[1], Toshihiro Sakamoto[1], Koji Kimura[1], Koji Yonemura[1], Fumitomo Matsuoka[1], Keita Takahashi[2], Akihiro Urata[2], Shoichi Sakaguchi[2], Takahito Nagamatsu[2]
[1]Toshiba Electronic Devices & Storage Corporation; [2]Japan Semiconductor Corporation

P2-16 **Low On-state Voltage and Latch-up Immunity Thin SOI LIGBT with Multi-Segmented Trench Gates** 367
Chao Yang, Xiaorong Luo, Tao Sun, Dongfa Ouyang, Anbang Zhang, Zhaoji Li, Bo Zhang
University of Electronic Science and Technology of China

P2-17 **Full-Chip Simulation Analysis of Power MOSFET's during Unclamped Inductive Switching with Physics-Base Device Models** 371
Tsuyoshi Kachi, Katsumi Eikyu, Takashi Saito
Renesas Electronics Corp.

P2-18 **Experimental Study on the Effect of Recessed Gates in Drain STI Regions of nLDMOSFETs** 375
Takahiro Mori, Hirokazu Sayama, Takashi Ipposhi, Koji Iizuka
Renesas Electronics Corporation

P2-19 **Fast-Switching Lateral IGBT with Trench/Planar Gate and Integrated Schottky Barrier Diode (SBD)** 379
Licheng Sun, Baoxing Duan, Yandong Wang, Yintang Yang
Xidian University

P2-20 **Diode Reverse Recovery Characteristics of a Shielded-Gate Trench Power MOSFET** 383
Zia Hossain[1], Raghuram Mullapudi[2], Harshad Surdi[3]
[1]ON Semiconductor; [2]Rochester Institute of Technology; [3]Arizona State University

P2-21 **Channel-Off Avalanche Instability in SOI Lateral IGBT at Low Temperature: Mechanism and Optimization Schemes** 387
Jie Ma[1], Long Zhang[1], Jing Zhu[1], Shilin Cao[1], Ankang Li[1], Yanqin Zou[1], Weifeng Sun[1], Yan Gu[2], Sen Zhang[2], Yunwu Zhang[3], Zhuo Yang[4]
[1]Southeast University; [2]CSMC Technologies Corporation; [3]Wuxi i-driver Electronics Co., Ltd; [4]Wuxi Nce Power Co., Ltd

P2-22 **The Lowest On-Resistance and Robust 130nm BCDMOS Technology Implementation Utilizing HFP and DPN for Mobile PMIC Applications** 391
Daehoon Kim, Kuemju Lee, Jaeeuk Kim, Junghun Choi, Jaehee Lee, Inwook Cho
SK Hynix System IC Inc.

P2-23 **Experimental Study on the Electrical Properties of Lateral IGBT under the Mechanical Strain** 395
Wangran Wu, Yaohui Wang, Long Zhang, Guangan Yang, Siyang Liu, Jing Zhu, Weifeng Sun
Southeast University

P2-24 **Circuit Dependent Plasma Charging Effect Robustness in 0.16 um BCD Technology Platform** 399
Michele Basso, Antonino Martino, Simone Bertaiola, Damiano Riccardi, Paola Galbiati
STMicroelectronics

P2-25 A Novel Self-Regulated Potential SOI LIGBT with Low ON-State Voltage and Turn-Off Loss 403
Yun Xia, Wanjun Chen, Wuhao Gao, Bin Qiao, Chao Liu, Yijun Shi, Yajie Xin,
Fangzhou Wang, Yu Shi, Ruize Sun, Qi Zhou, Zhaoji Li, Bo Zhang
University of Electronic Science and Technology of China

P2-26 Performance and Reliability Co-Design of LDMOS-SCR for Self-Protected
High Voltage Applications On-Chip 407
Nagothu Karmel Kranthi[1], B. Sampath Kumar[1], Akram Salman[2], Gianluca Boselli[2], Mayank Shrivastava[1]
[1]Indian Institute of Science Bangalore, India; [2]Texas Instruments Inc.

P2-27 Revealing the Positive Bias Temperature Instability in Normally-OFF AlGaN/GaN
MIS-HFETs by Constant-Capacitance DLTS 411
Sen Huang[1], Xinhua Wang[1], Xinyu Liu[1], Xuanwu Kang[1], Jie Fan[1], Shuo Yang[1], Haibo Yin[1], Ke Wei[1],
Yingkui Zheng[1], Xiaolei Wang[1], Wenwu Wang[1], Jingyuan Shi[1], Hongwei Gao[2], Qian Sun[2], Kevin J. Chen[3]
*[1]Institute of Microelectronics of Chinese Academy of Sciences and University of CAS; [2]Suzhou Institute of
Nano-Tech and Nano-Bionics, CAS; [3]The Hong Kong University of Science and Technology*

P2-28 Demonstration of Electron/Hole Injections in the Gate of p-GaN/AlGaN/GaN Power
Transistors and their Effect on Device Dynamic Performance 415
Xi Tang[1,2,3], Baikui Li[1], Jun Zhang[2], Hui Li[3], Jisheng Han[2], Nam-Trung Nguyen[2],
Sima Dimitrijev[2], Jiannong Wang[3]
[1]Shenzhen University; [2]Griffith University; [3]The Hong Kong University of Science and Technology

P2-29 Trading off between Threshold Voltage and Subthreshold Slope in
AlGaN/GaN HEMTs with a p-GaN Gate 419
Benoit Bakeroot[1], Steve Stoffels[2], Niels Posthuma[2], Dirk Wellekens[2], Stefaan Decoutere[2]
[1]imec & Ghent University; [2]imec

P2-30 Observation of Self-Recoverable Gate Degradation in p-GaN AlGaN/GaN HEMTs after
Long-Term Forward Gate Stress: The Trapping & Detrapping Dynamics of Hole/Electron 423
Yuanyuan Shi, Qi Zhou, Wei Xiong, Xi Liu, Xin Ming, Zhaoji Li, Wanjun Chen, Bo Zhang
University of Electronic Science and Technology of China

P2-31 Recess-Free Normally-Off GaN MIS-HEMT Fabricated on Ultra-Thin-Barrier
AlGaN/GaN Heterostructure 427
Ping-Cheng Han, Chia-Hsun Wu, Yu-Hsuan Ho, Zong-Zheng Yan, Edward Yi Chang
National Chiao Tung University

P2-32 Over Kilovolt GaN Vertical Super-Junction Trench MOSFET:
Approach for Device Design and Optimization 431
Peng Huang, Qi Zhou, Kuangli Chen, Xiaoqi Han, Dong Wei, Yuanyuan Shi, Wanjun Chen, Bo Zhang
University of Electronic Science and Technology of China

P2-33 Identifying the Location of Hole-Induced Gate Degradation in LPCVD-SiN$_x$/GaN
MIS-FETs under High Reverse-Bias Stress 435
Zheyang Zheng[1], Mengyuan Hua[2], Jin Wei[1], Zhaofu Zhang[1], Kevin J. Chen[1]
[1]The Hong Kong University of Science and Technology; [2]Southern University of Science and Technology

P2-34 Surge Current Capability of GaN E-HEMTs in Reverse Conduction Mode 439
Yinxiang Liu, Shaowen Han, Shu Yang, Kuang Sheng
Zhejiang University

P2-35 Impact of Carrier Accumulation on the Transient Behavior of p-Gate GaN HEMTs 443
Thorsten Oeder, Martin Pfost
TU Dortmund

P2-36 Integrated GaN MIS-HEMT with Multi-Channel Heterojunction SBD Structures 447
Sheng Li[1], Siyang Liu[1], Chi Zhang[1], Jiaxing Wei[1], Long Zhang[1], Weifeng Sun[1], Youhua Zhu[2],
Tingting Zhang[2], Dongsheng Wang[2], Yinxia Sun[2], Yiheng Li[2], Tinggang Zhu[2]
[1]Southeast University; [2]CorEnergy Semiconductor Co., Ltd

P2-37 Damage Accumulation in GaN GITs Exposed to Repetitive Short-Circuit 451
F. D'Aniello[1], A. Fayyaz[1], A. Castellazzi[1], T. Oeder[2], M. Pfost[2]
[1]University of Nottingham; [2]TU Dortmund

P2-38 **Soft-Switching Losses in GaN and SiC Power Transistors based on New Calorimetric Measurements** 455
Julian Weimer, Ingmar Kallfass
University of Stuttgart

P2-39 **Charge-Modulated Schottky Barrier Lowering Effect in GaN Double-Channel Lateral Power SBDs with Gated Anode** 459
Jiacheng Lei, Jin Wei, Gaofei Tang, Zhaofu Zhang, Qingkai Qian, Mengyuan Hua, Zheyang Zheng, Yuru Wang, Kevin J. Chen
The Hong Kong University of Science and Technology

P2-40 **Characterization of Dynamic I_{OFF} in Schottky-Type p-GaN Gate HEMTs** 463
Yuru Wang, Jin Wei, Song Yang, Jiacheng Lei, Mengyuan Hua, Kevin J. Chen
The Hong Kong University of Science and Technology

P2-41 **Effects of Substrate Termination on Reverse-Bias Stress Reliability of Normally-Off Lateral GaN-on-Si MIS-FETs** 467
Mengyuan Hua[1,2], Song Yang[2], Zheyang Zheng[2], Jin Wei[2], Zhaofu Zhang[2], Kevin J. Chen[2]
[1]Southern University of Science and Technology; [2]The Hong Kong University of Science and Technology

P2-42 **Novel 2000 V Normally-Off MOS-HEMTs using AlN/GaN Superlattice Channel** 471
Ming Xiao[1], Weihang Zhang[1], Yuhao Zhang[2], Hong Zhou[1], Kui Dang[1], Jincheng Zhang[1], Yue Hao[1]
[1]Xidian University; [2]Virginia Polytechnic Institute and State University

Thursday - May 23, 2019

8:30 – 10:10
Session 9 – High Voltage Devices: Si, Ga_2O_3, SiC
Chairs: Jun Zeng, *MaxPower Semiconductor*
David Sheridan, *Alpha & Omega Semiconductor*

8:30

9-1 **Analysis the Complex Tradeoff Among E_{on}-V_{CEsat}-SCSOA and EMI Noise through the Single Chip Evaluation Method** 475
Koichi Nishi, Tetsuo Takahashi, Atsushi Narazaki
Mitsubishi Electric Corporation

8:55

9-2 **An Ultra-Low Qrr Cell-Distributed Schottky Contacts SJ-MOSFET with Integrated Isolated NMOS** 479
Shaohong Li, Ajiang Li, Tian Tian, Jing Zhu, Long Zhang, Weifeng Sun, Guichuang Zhu, Yanqin Zou, Xin Tong, Yangyang Lu, Jiaxing Wei, Ran Ye
Southeast University

9:20

9-3 **1.6 kV Vertical Ga_2O_3 FinFETs with Source-Connected Field Plates and Normally-Off Operation** 483
Zongyang Hu[1], Kazuki Nomoto[1], Wenshen Li[1], Riena Jinno[1,2], Tohru Nakamura[3], Debdeep Jena[1], Huili Xing[1]
[1]Cornell University; [2]Kyoto University; [3]Hosei University

9:45

9-4 **Influence of Carrier Lifetime on Silicon Carbide Power Devices for Pulsed Power Application** 487
Kun Zhou, Yingxin Cui, Lianghui Li, Yunfei Gu, Lin Zhang, Shuairong Deng, Zhiqiang Li, Juntao Li
China Academy of Engineering Physics

10:30 – 12:10
Session 10 – Packaging: Performance, Modeling and Reliability
Chairs: Tomoyuki Miyoshi, *Hitachi*
Bassem Mouawad, *University of Nottingham*

10:30

10-1 **An 8.5kV Sacrificial Bypass Thyristor with Unprecedented Rupture Resilience** 491
Tobias Wikström, Bjørn Ødegård, Remo Baumann
ABB Switzerland Ltd.

10:55

10-2 **Power Cycling Capability of Silicon Low-Voltage MOSFETs under Different Operation Conditions** 495
Christian Schwabe, Peter Seidel, Josef Lutz
Chemnitz University of Technology

11:20

10-3 **Die-Attach on Copper by Pressureless Silver Sintering in Formic Acid** 499
Meiyu Wang[1], Yijing Xie[1], Yunhui Mei[1], Xin Li[1], Guo-Quan Lu[2]
[1]Tianjin University; [2]Virginia Polytechnic Institute and State University

11:45

10-4 **Mutual Inductance Influence to Switching Speed and TDR Measurements for**
Separating Self- and Mutual Inductances in the Package 503
H. Iida, K. Hasegawa, I. Omura
Kyushu Institute of Technology

13:30 – 14:45
Session 11 – Integrated Power Technologies
Chairs: Hiroki Fujii, *Samsung Electronics*
Ronghua Zhu, *NXP Semiconductors*

13:30

11-1 **Experiments of a Novel Low On-Resistance LDMOS with 3-D Floating Vertical Field Plate** 507
Guangsheng Zhang[1], Wentong Zhang[1,2], Junqing He[2], Xuhan Zhu[2], Sen Zhang[1], Jingchuan Zhao[1],
Zhili Zhang[1], Ming Qiao[2], Xin Zhou[2], Zhaoji Li[2], Bo Zhang[2]
[1]CSMC Technologies Corporation; [2]University of Electronic Science and Technology of China

13:55

11-2 **A Laterally Monolithic-Integrated Multi-Cascode for Applications with 600V and**
more based on 20V-FINFETs in 90nm Technology 511
Rolf Weis[1], Marko Lemke[1], Marco Müller[1], Ralf Rudolf[1], Knut Stahrenberg[1],
Thomas Bertrams[1], Martin Bartels[1], Ahmed Mahmoud[2], Nicolas Nagel[2]
[1]Infineon Technologies Dresden GmbH & Co. KG; [2]Infineon Technologies AG

14:20

11-3 **Cu Double Side Plating Technology for High Performance and Reliable Si Power Devices** 515
Hitoshi Kobayashi, Tatsuya Ohguro, Tetsuya Kai, Takako Motai, Masaaki Ogawa, Mie Matsuo,
Kenichi Oohashi, Shinsuke Kozumi, Yoshiharu Takada, Hideharu Kojima, Shingo Masuko,
Naoki Yonezawa, Akira Komatsu, Tatsuya Nishiwaki, Takuma Hara, Mari Takahashi, Akira Ezaki,
Kenichi Ohtsuka, Seiji Inumiya, Kyoichi Suguro
Toshiba Electronic Devices & Storage Corporation

GENERAL CHAIR'S MESSAGE

On behalf of the Conference Committee, it is my great honor and pleasure to welcome you to the 31st IEEE International Symposium on Power Semiconductor Devices and ICs (ISPSD 2019). It will be held between 19th and 23rd of May, 2019, in the Shanghai Marriott Hotel Parkview, Shanghai, one of the most cosmopolitan, diverse, dynamic, vibrant and fascinating cities in China.

ISPSD brings together the world's foremost experts and leading companies on power devices and integrated circuit technology. Since the first meeting held in Tokyo in 1988, ISPSD has become the premier international forum for technical discussions in all aspects of power semiconductor devices and integrated circuits. The conference location has rotated each year among Japan, North America, Europe and other regions since 2015. ISPSD 2019 will be the first time the conference is held in mainland China. The economic growth in China in the last decades has led to tremendous progress in power electronics industry. In particular, the rapid development of high-speed train, hybrid/electric vehicle and renewable energy has been accelerating the advancements of power semiconductor devices and ICs in China.

Shanghai, located in the Yangtze River Delta and the middle portion of the East China coast, is the country's largest city, and a global hub for finance, technology, industry and transportation. Shanghai is a special city featuring futuristic skyline, historic buildings and numerous museums, such as the Oriental Pearl TV Tower, the Yu Garden and China Art Museum. We trust you and your family will enjoy your stay in Shanghai.

I sincerely thank the Organizing Committee, the Technical Program Committee and the Advisory Committee for their contributions and supports to ISPSD 2019.

Finally, I would like to extend a warm welcome to all of you attending ISPSD 2019. I wish you an informative, valuable and memorable experience in Shanghai.

Sincerely Yours,

Prof. Kuang Sheng
General Chair, ISPSD 2019

ORGANIZATION

ORGANIZING COMMITTEE

General Chair
Kuang Sheng, *Zhejiang University (OA)* [*]

TPC Chair
Kevin Chen, *HKUST (OA)* [*]

Past General Chair
Johnny Sin, *HKUST (OA)* [*]

Vice General Chair
Bo Zhang, *UESTC (OA)* [*]
Oliver Haeberlen, *Infineon (EU)* [*]
Kimimori Hamada, *Toyota (JP)* [*]

Publicity Chair
Wai Tung Ng, *University of Toronto (NA)* [*]

Local Arrangement Chair
Simon Zhang, *TSMC (OA)* [*]

Short Course Chair
Tanya Trajkovic, *Camutronics (EU)* [*]

Publication Chair
Weifeng Sun, *Southeast University (OA)* [*]

Exhibition Chair
Lin Liang, *HUST (OA)* [*]

Webmaster
Xiaorong Luo, *UESTC (OA)* [*]

Treasurer
Qing Guo, *Zhejiang University (OA)* [*]

Secretary
Shu Yang, *Zhejiang University (OA)* [*]

[*] *EU: Europe* [*] *JP: Japan* [*] *NA: North America* [*] *OA: Other Area*

ADVISORY COMMITTEE

Gehan Amaratunga, *University of Cambridge, (EU)* [*]
Tat-Sing Paul Chow, *Rensselaer Polytechnic Institute, (NA)* [*]
Mohamed Darwish, *MaxPower Semiconductor, (NA)* [*]
Don Disney, *Infineon Tecnologies, (NA)* [*]
Dan Kinzer, *Navitas Semiconductor, (NA)* [*]
Leo Lorenz, *ECPE, (EU)* [*]
Gourab Majumdar, *Mitsubishi Electric, (JP)* [*]
Peter Moens, *ON Semiconductors, (NA)* [*]
Mutsuhiro Mori, *Hitachi Power Semiconductor Device, Ltd., (JP)* [*]
Hiromichi Ohashi, *NPERC-J, (JP)* [*]
Yasukazu Seki, *Fuji Electric Co., Ltd., (JP)* [*]
John Shen, *Illinois Institute of Technology, (NA)* [*]
M. Ayman Shibib, *Vishay Siliconix, (NA)* [*]
Johnny Sin, *Hong Kong University of Science and Technology, (OA)* [*]
Jan Šonský, *NXP Semiconductors, (EU)* [*]
Yoshitaka Sugawara, *Ibaraki University, (JP)* [*]
Richard K. Williams, *Adventive Technology, (NA)* [*]
Toshiaki Yachi, *Tokyo University of Science, (JP)* [*]

TECHNICAL PROGRAM COMMITTEE

Chair
Kevin J. Chen, *The Hong Kong University of Science and Technology (OA)* [*]

Category 1: High Voltage Power Devices (HV)
Category Chair
Breglio Giovanni, *University of Naples Federico II (EU)* [*]

Members
Marina Antoniou, *University of Cambridge (EU)* [*]
Corvasce Chiara, *ABB (EU)* [*]
Shigeto Honda, *Mitsubishi Electric (JP)* [*]
Thomas Laska, *Infineon Technologies (EU)* [*]
Xiaorong Luo, *University of Electronic Science and Technology of China (OA)* [*]
Yasuhiko Onishi, *Fuji Electric (JP)* [*]
Wataru Saito, *Kyushu University (JP)* [*]
Yi Tang, *Starpower Semiconductor (OA)* [*]
Jun Zeng, *MaxPower Semiconductor (NA)* [*]
Shuai (Simon) Zhang, *TSMC (OA)* [*]

Category 2: Low Voltage Devices and Power IC Device Technology (LVT)
Category Chair
Naoto Fujishima, *Fuji Electric (JP)* [*]

Members
Riccardo Depetro, *STMicroelectronics (EU)* [*]
Hiroki Fujii, *Samsung Electronics (OA)* [*]
Mark Gajda, *Nexperia (EU)* [*]
David Tsung-Yi Huang, *Richtek (OA)* [*]
Kenya Kobayashi, *Toshiba Electronic Devices & Storage Corporation (JP)* [*]
Sang-Gi Lee, *Dongbu HiTek's (OA)* [*]
Takahiro Mori, *Renesas Semiconductor Manufacturing (JP)* [*]
Amit Paul, *ON Semiconductor (NA)* [*]
Purakh Raj Verma, *UMC (OA)* [*]
Ronghua Zhu, *NXP Semiconductors (NA)* [*]

Category 3: Power IC Design (ICD)
Category Chair
Nicolas Rouger, *CNRS (EU)* [*]

Members
Katsumi Eikyu, *Renesas Electronics (JP)* [*]
Kenji Hara, *Hitachi (JP)* [*]
Shuichi Nagai, *Panasonic (JP)* [*]
Wai Tung Ng, *University of Toronto (NA)* [*]
Maarten Swanenberg, *NXP Semiconductors (EU)* [*]
Budong (Albert) You, *Silergy Corp. (OA)* [*]
Alessandro Zafarana, *ON Semiconductor (EU)* [*]
Jing Zhu, *Southeast University (OA)* [*]

Category 4: GaN and Nitride Base Compound Materials (GaN)
Category Chair
Tom Tsai, *TSMC (OA)* [*]

Members
Oliver Haeberlen, *Infineon Technologies (EU)* [*]
Alex Huang, *University of Texas at Austin (NA)* [*]
Yang Liu, *Sun Yat-sen University (OA)* [*]
Peter Moens, *ON Semiconductor (EU)* [*]
Hideyuki Okita, *Panasonic (JP)* [*]
Tomas Palacios, *Massachusetts Institute of Technology (NA)* [*]
Sameer Pendharkar, *Texas Instruments (NA)* [*]
Jun Suda, *Nagoya University (JP)* [*]
Shu Yang, *Zhejiang University (OA)* [*]

Category 5: SiC and Other Materials (SiC)
Category Chair
Jon Zhang, *Power America (NA)* [*]

Members
Philippe Godignon, *CNM institute (EU)* [*]
Chih-Fang Huang, *National Tsing Hua University (OA)* [*]
Takeharu Kuroiwa, *Mitsubishi Electric (JP)* [*]
Chwan Ying Lee, *Hestia-Power Inc. (OA)* [*]
Kung-Yen Lee, *National Taiwan University (OA)* [*]
Kevin Matocha, *Littelfuse (NA)* [*]
Andrei Petru Mihaila, *ABB (EU)* [*]
Sei-Hyung Ryu, *Wolfspeed (NA)* [*]
David Sheridan, *Alpha & Omega Semiconductor (NA)* [*]
Sid Sundaresan, *GeneSiC (NA)* [*]
Victor Veliadis, *North Carolina State University (NA)* [*]
Yoshiyuki Yonezawa, *AIST (JP)* [*]

Category 6: Module and Package Technologies (PK)
Category Chair
Ichiro Omura, *Kyushu Institute of Technology (JP)* [*]

Members
Sven Berberich, *Semikron (EU)* [*]
Sameh Khalil, *Infineon Technologies (NA)* [*]
Tomoyuki Miyoshi, *Hitachi (JP)* [*]
Bassem Mouawad, *Nottingham University (EU)* [*]
Zhenqing Zhao, *Delta Power Electronic Research Center (OA)* [*]

[*]*EU: Europe* [*]*JP: Japan* [*]*NA: North America* [*]*OA: Other Area*

AWARDS

THE OHMI BEST PAPER AWARD

The Best Paper Award was renamed to "The Ohmi Best Paper Award" in honor of the late Prof. Ohmi's outstanding contributions to the ISPSD. The Ohmi Best Paper Award will be granted to the author(s) of a paper determined to be the best overall in the ISPSD2019.

ISPSD2018 THE OHMI BEST PAPER AWARD
Deep-P Encapsulated 4H-SiC Trench MOSFETs with Ultra Low $R_{on}Q_{gd}$

Yasuhiro Ebihara, Aiko Ichimura, Shuhei Mitani, Masato Noborio, Yuichi Takeuchi, Shoji Mizuno, Toshimasa Yamamoto, Kazuhiro Tsuruta
Denso Corporation, Japan

Abstract: Deep-P encapsulated 4H-SiC trench MOSFET was proposed. The fabricated MOSFET with a blocking voltage of 1800V demonstrated an ultra low $R_{on}Q_{gd}$ of 133 nCmΩ. The structure optimization was carried out for switching-loss reduction. The improved switching characteristics were obtained by the balanced JFET resistance and the gate-drain capacitance.

Yasuhiro Ebihara received his M.S. degree from University of Tsukuba in 2012. In 2012, he joined DENSO CORPORATION in Aichi, Japan, where he has been engaged in the research and development of SiC power devices.

Aiko Ichimura received her B.S. and M.S. degrees from Kwansei Gakuin University in 2012 and 2014, respectively. In 2014, she joined DENSO CORPORATION in Aichi, Japan, where she has been engaged in the research and development of SiC power devices.

Shuhei Mitani received his M.E. from Kyoto Institute of Technology in 2006. Since 2004, He has been developing total FEOL processes for SiC power devices. He joined DENSO CORPORATION in 2013. His current main interest is the development of miniaturization technology for SiC trench MOSFET.

Masato Noborio received his B.E., M.E., and Ph.D. degrees, based on his work on SiC MOS interface characterization, device physics, and lateral power devices from Kyoto University, Kyoto, Japan, in 2004, 2006, and 2009, respectively. He joined DENSO CORPORATION, Aichi, Japan, in April 2009. His current research interests include the development of SiC power devices.

Yuichi Takeuchi received his B.S. and M.S. degrees in 1988 and 1990 from Nagoya University, Nagoya, Japan. He joined DENSO CORPORATION in 1990. He was engaged in development of Si power devices. From 1992, he has been engaged in the research and development of SiC power devices.

Shoji Mizuno received his B.E. and M.E. degrees in 1985 and 1987 from Nagoya University, Nagoya, Japan. He joined DENSO CORPORATION in 1987. He was engaged in development of semiconductor devices, especially SOI BCD devices and Si power devices for automotive applications. From 2010, he has been engaged in the research and development of SiC power devices.

Toshimasa Yamamoto received his B.E. and M.E. degrees in 1986 and 1988 from Osaka University, Osaka, Japan. He joined DENSO CORPORATION in 1988. He was engaged in the research and development of MEMS devices like accelerometers and an air flow meters. From 2007, he has been engaged in the research and development of SiC power devices.

Kazuhiro Tsuruta received his B.E. degree in electrical engineering from Osaka University in 1987. He joined DENSO CORPORATION in 1987. He was engaged in the research and development of Si devices. From 2004, he has been engaged in the research and development of SiC power devices & modules.

CHARITAT AWARD (YOUNG RESEARCHER AWARD)

A young researcher (age less than 30 at the time of the conference) who is both first author and presenter of a paper will be nominated to the award.

ISPSD 2018 CHARITAT AWARD
1 kV/1.3 mΩ·cm² Vertical GaN-on-GaN Schottky Barrier Diodes
with High Switching Performance

Shu Yang, Shaowen Han, Rui Li, Kuang Sheng,
Zhejiang University, China

Abstract: In this work, we developed 1 kV/1.3 mΩ·cm² vertical GaN Schottky barrier diodes (SBDs) on bulk GaN substrate. The vertical GaN SBDs exhibit a nearly ideal Schottky contact with an ideality factor of 1.04, a forward current density of 2000 A/cm², and a high current swing over 13 orders of magnitude. By virtue of a planar nitridation-based termination technique, the excess leakage current at the junction edge can be effectively suppressed and an enhanced breakdown voltage of ~1 kV is realized in the vertical SBDs. In addition to the fast reverse recovery characteristics, the vertical GaN-on-GaN SBDs also deliver a current-collapse-free performance with no dynamic ON-resistance degradation at ~500 ns after switching from a high-voltage OFF-state.

Shu Yang received her B.S. degree in Microelectronics from Fudan University, China, and her Ph.D. degree in Electronic and Computer Engineering from the Hong Kong University of Science and Technology (HKUST) with SENG PhD Research Excellence Award. She was a visiting assistant professor at HKUST and a postdoctoral research associate at the University of Cambridge. She is currently a faculty member at the College of Electrical Engineering, Zhejiang University, China. Her research interests include characterization, fabrication and application-relevant study of GaN-based power devices.

A Smart Gate Driver IC for GaN Power Transistors

Jingshu Yu, Wei Jia Zhang, Andrew Shorten, Rophina Li, Wai Tung Ng,
University of Toronto, Canada

Abstract: In this paper, an integrated smart gate driver IC with segmented output stage topology, programmable sense-FET, current sensing circuits and an on-chip stacked-based CPU for flexible digital control is presented. This IC is fabricated using TSMC's 0.18 µm BCD GEN2 process for driving a d-mode GaN power HEMT in cascode configuration. Using a segmentation technique, this IC can dynamically adjust the gate driving strength during switching transition to achieve slope control and EMI reduction. Programmable sense-FET and current sensing circuit monitor the load current for peak-current regulation. The embedded CPU can update all digital configuration bits on-the-fly. In dynamic driving mode, current spike at turn-on transition is reduced by 83% without sacrificing the switching speed. Current sensing circuit can detect peak current value and response within 5 ns. The pre-stored driving patterns can be loaded to the driving circuit in 1 µs under active driving mode.

Jingshu Yu received her B.Eng. degree in Electrical Engineering from Zhejiang University, Hangzhou, China, in 2011. She then joined the University of Toronto, Toronto, Canada, where she received her M.A.Sc. degree in 2014 under the supervision of Prof. Wai Tung Ng. She is currently working toward her Ph.D. degree with the Smart Power Integration and Semiconductor Devices Research Group at the same university. Her research interests include smart gate driving techniques, EMI reduction and power stage design for GaN power applications.

ISPSD HALL OF FAME

The purpose of the ISPSD Hall of Fame (IHF) is to honor individuals who have made high impact contributions in advancing power semiconductor technology and/or in sustaining the success of ISPSD. Starting this year, the IHF replaces the traditional "Contributory Awards" and "Pioneer Awards". The AdCom has selected the following 32 distinguished colleagues as the first inductees into the ISPSD Hall of Fame:

Michael S. Adler	for contributions to modern power semiconductor technology, and his leadership role in organizing ISPSD conferences
G.A.J. Amaratunga	for contributions to modern power semiconductor technology, and his leadership role in organizing ISPSD conferences
B. Jayant Baliga	for contributions to modern power semiconductor technology, and his leadership role in organizing ISPSD conferences
Xingbi Chen	for contributions to superjunction power semiconductor devices
Tat-Sing Paul Chow	for contributions to silicon and wide bandgap power semiconductor devices, and his leadership role in organizing ISPSD conferences
Mohamed Darwish	for contributions to the advancement of power semiconductor technology, and his leadership role in organizing ISPSD conferences
Taylor R. Efland	for contributions to power IC technology, and his leadership role in organizing ISPSD conferences
Wolfgang Fichtner	for contributions to MOS gated thyristors and TCAD modeling tools, and his leadership role in organizing ISPSD conferences
Min-Koo Han	for contributions to modern power semiconductor technology, and his leadership role in organizing ISPSD conferences
Phil Hower	for contributions to power device safe operating area study and power IC technology
A. A. Jaecklin	for contributions to modern power semiconductor technology, and his leadership role in organizing ISPSD conferences
Daniel Kinzer	for contributions to power MOSFET technology, and his leadership role in organizing ISPSD conferences
Leo Lorenz	for contributions to modern power semiconductor technology, and his leadership role in organizing ISPSD conferences
Gourab Majumdar	for contributions to IGBT and intelligent power module technology, and his leadership role in organizing ISPSD conferences
Jose Millan	for contributions to modern power semiconductor technology, and his leadership role in organizing ISPSD conferences
Peter Moens	for contributions to integrated power technology and GaN power device and reliability, and his leadership role in organizing ISPSD conferences
Akio Nakagawa	for contributions to IGBT and power IC technology
Hiromichi Ohashi	for contributions to modern power semiconductor technology, and his leadership role in organizing ISPSD conferences
Tadahiro Ohmi	for contributions to modern power semiconductor technology, and his leadership role in organizing ISPSD conferences

Masahiro Okamura	for contributions to modern power semiconductor technology, and his leadership role in organizing ISPSD conferences
James Plummer	for contributions to MOS-bipolar power devices and power ICs, and for inspiring and training a new generation of device researchers
C. A. T. Salama	for contributions to power IC technology, and his leadership role in organizing ISPSD conferences
Yasukazu Seki	for contributions to IGBT technology, and his leadership role in organizing ISPSD conferences
M. A. Shibib	for contributions to modern power semiconductor technology, and his leadership role in organizing ISPSD conferences
Dieter Silber	for contributions to modern power semiconductor technology, and his leadership role in organizing ISPSD conferences
Paolo Spirito	for contributions to modern power semiconductor technology, and his leadership role in organizing ISPSD conferences
Yoshitaka Sugawara	for contributions to modern power semiconductor technology, and his leadership role in organizing ISPSD conferences
Yoshiyuki Uchida	for contributions to modern power semiconductor technology, and his leadership role in organizing ISPSD conferences
Harry Vaes	for contributions to RESURF technology
Carl F. Wheatley	for contributions to IGBT and radiation-hard power device technology
Richard K. Williams	for contributions to trench power MOSFET and power IC technology, and his leadership role in organizing ISPSD conferences
Toshiaki Yachi	for contributions to modern power semiconductor technology, and his leadership role in organizing ISPSD conferences

Deceased members in BOLD

PARTNERSHIP ORGANIZATIONS

PLATINUM PARTNERS

GOLD PARTNERS

SILVER PARTNERS

SHORT COURSE

	Agenda for Short Course May 19th, 2019 Sunday
09:00-10:00	**Superjunction Concept, History and State of the Art** Prof. Florin Udrea, *University of Cambridge*
10:00-10:15	*Coffee break*
10:15-11:15	**The Current Status of SiC MOSFET Technology** Dr. Andrei Mihaila, *ABB*
11:15-13:30	*Coffee break and lunch*
13:30-14:30	**Si Trench IGBTs for Automotive** Dr. Haihui Luo, *CRRC Corporation Limited*
14:30-15:30	**Full Mark Validation of 600+V GaN Power Devices** Dr. Yifeng Wu, *Transphorm Inc.*
15:30-15:45	*Coffee break*
15:45-16:45	**Application Reliability of a Wide Bandagap (WBG) Semiconductor Power Electronics Switch** Prof. Krishna Shenai, *University of Chicago*

SC1. Superjunction Concept, History and State of the Art
Prof. Florin Udrea, *University of Cambridge, UK*

Speaker Biography: Professor Florin Udrea FREng is a professor in semiconductor engineering and the head of the High Voltage Microelectronics and sensors group at Cambridge University. Florin published over 500 papers in journals and international conferences and holds more than 100 patents in power semiconductor devices and sensors. Florin co-founded five companies, Cambridge Semiconductor (Camsemi) in power ICs – sold to Power Integrations (US), Cambridge CMOS Sensors (CCS) in smart sensors – sold to ams (Austria), Cambridge Microelectronics (Camutronics) in Power Devices, Cambridge GaN Device in GaN technology and Flusso in Flow and temperature sensors. For his 'outstanding personal contribution to British Engineering' he has been awarded the Silver Medal from the Royal Academy of Engineering in 2012. In 2015 Prof. Florin Udrea was elected a Fellow of Royal Academy of Engineering. In 2018 Prof Florin Udrea was awarded the Pro Vice-Chancellor award from Cambridge University, the international Nanosmat medal and the Mullard award from the Royal Society.

Abstract: Superjunction has arguably been the most creative and important concept in the power device field since the introduction of the Insulated Gate Bipolar Transistor (IGBT) in 1980s. It is the only concept known today that has challenged and ultimately proved wrong the well-known theoretical study on the limit of silicon in high voltage devices. The short-course will deal with the history, device and process development, and the future prospects of superjunction technologies. It covers fundamental physics, technological and application challenges as well as aspects of design and modelling of unipolar devices, such as CoolMOS. New figures of merit (FOMs) and the ultimate limit of the superjunction concept in Silicon and SiC will be shown. The course will finish with the application of the superjunction concept to other structures or materials, such as terminations, Superjunction IGBTs, Silicon Carbide FETs and GaN HEMTs.

SC2. The Current Status of SiC MOSFET Technology
Dr. Andrei Mihaila, *ABB, Switzerland*

Speaker Biography: Dr Andrei Mihaila graduated in 1999 with an MPhil from University "Politehnica" in Bucharest. In 2002, he received his Phd from University of Cambridge, Department of Engineering with the thesis "Silicon Carbide high power Field Effect Transistor switches". Between 2006 and 2008, he was with Cambridge Semiconductors (UK). Dr Mihaila joined ABB Corporate Research Centre (CH) in 2010, in the semiconductor group. Since then, he has been extensively involved in developing ABB's next generation of SiC diodes and MOSFETs. He is a TPC member of ISPSD and ECSCRM conferences. Recently, he's been appointed in the Editorial Board of the IEEE Transactions on Electron Devices (T-ED) as an Editor in the area "Solid-State Power".

Abstract: For decades, Si has been the semiconductor of choice for power electronics applications. Recently, several other materials such as SiC, GaN, diamond or Ga_2O_3, have challenged Si dominant position. Among this list, SiC has established itself as the most mature technology today. This short course will review the current status of SiC MOSFET technology. Aspects related to the active and edge termination area design will be discussed upon. How SiC MOSFETs perform statically and dynamically will also be reviewed, with a focus on their behaviour under fault conditions operation. The reliability of SiC MOSFETs will be reviewed. Finally, some examples of SiC MOSFET utilization in several applications will be presented. The short course is intended for an intermediate-advanced audience.

SC3. Si Trench IGBTs for Automotive
Dr. Haihui Luo, *Zhuzhou CRRC Times Electric Co. Ltd, China*

Speaker Biography: Dr. Haihui Luo received his Masters degree in materials science from Yanshan University, and his Ph.D. degree in condensed matter physics from Institute of Semiconductors, CAS, in 2006 and 2009, respectively. He is currently employed with Semiconductor Business Unit of Zhuzhou CRRC Times Electric Co. Ltd. as vice general manager in charge of technology.

From 2010 to 2012 he was the key talent appointed to work in Dynex, UK, where he was involved in research and development for high voltage, high power density planar gate IGBT and trench gate IGBT technology. From 2013 to 2015, he organized the establishment of IGBT and diode wafer process technology platform from 650V to 6500V based on Zhuzhou 8-inch process line, which was founded by Zhuzhou CRRC Times Electric Co. Ltd. Since 2016, he has been responsible of IGBT R&D and manufacturing business, including IGBT chip and module.

Abstract: Hybrid and electric vehicle growth has hit unpresented levels across the globe. At present, propulsion system in automotive applications is being widely addressed by IGBTs realized in smaller discrete, over-molded packages or in large power modules. This short course will discuss IGBT technologies from the perspectives of high power density, controllability and reliability, which are key concerns especially in automotive applications with complex drive cycles and power dissipation patterns. Continuous developments of the IGBTs at the chip level are discussed in thin wafer technology, fine pattern technology and their dv/dt and di/dt controllability. Besides, Reverse-Conducting, Super-Junction concepts and wideband gap materials are also discussed. In addition, in order to improve the performance, reliability and cost-saving, the advanced packaging technologies and new materials for xEV power modules are presented as well.

The tutorial is intended for intermediate level audiences interested in advanced IGBT technologies for automotive applications with an in-depth overview of the chip and package developments.

SC4. Full Market Validation of 600+V GaN Power Devices: From Design, Manufacturability, Performance and Reliability to System Economics, User Satisfaction and PPM Field Failure Rate

Dr. Yifeng Wu, *Transphorm Inc., USA*

Speaker Biography: Dr. Yifeng Wu received the B.E. degree from Tsinghua University in 1985, the M.S. degree in Mechanical Eng. and the Ph.D. degree in Electrical Eng. from UC Santa Barbara in 1994 and 1997, respectively. He served as a lead scientist at Witech/Cree for 11 years. He joined Transphorm in 2008 where he is now Sr. VP of Engineering.

Since Ph.D. project from 1995, Dr. Wu has been at the forefront of GaN electronics with contributions from basic device discovery to cutting-edge device designs, from millimeter-wave power HEMTs to kV high-efficiency power switches. He demonstrated the first GaN microwave power transistor [EDL 1996] and multiple times extended the record of the highest power density of a solid-state transistor [EDL/IEDM/DRC 1998-2006]. He led an engineering team at Transphorm and succeeded in qualifying industry's first 600V GaN-on-Si power device products [WiPDA 2013] and overcame stringent reliability challenges to establish automotive qualification [2017]. His team also demonstrated true kV-class GaN transistors exceeding vertical SiC efficiency, extending the limit for horizontal devices [TPE 2014]. Dr. Wu holds 109 US patents and has authored 200 publications with >12,000 citations in Google Scholar.

Abstract: The development of GaN-on-Si power devices has come a long way and 600+V transistor products are now in mass production for high-efficiency applications such as titanium-grade power supplies. The success in technology development after a decade-long effort and the continued cost reduction will expand the GaN power device market at 35% compounded growth to exceed $1 billion by 2027 as predicted by IHS Markit. This tutorial offers a comprehensive look at how 600+V GaN switch products complete a full cycle of technology validation: from design, manufacturability, performance and reliability to system economics, user satisfaction and ppm field failure rate. Topics to cover include GaN's physical advantages as a power semiconductor, device choices as a robust product, device properties and performance, competitiveness over Si and SiC counterparts, circuit design considerations, operation analysis in hard and resonant-switching topologies, as well as reliability study and quality control for ppm-level failure rates.

SC5. Application Reliability of a Wide Bandagap (WBG) Semiconductor Power Electronics Switch

Prof. Krishna Shenai, *University of Chicago, USA*

Speaker Biography: Krishna Shenai earned his B. Tech. (electronics) degree from IIT-Madras in 1979, MS (EE) degree from the University of Maryland – College Park, Maryland (USA) in 1981, and PhD (EE) degree from Stanford University, Stanford, California (USA) in 1986. For nearly 40 years, Dr. Shenai and his students have made seminal contributions to silicon and wide bandgap (WBG) power electronics technologies that have shaped the world-wide industry. He is a Fellow of IEEE, a Fellow of American Association for the Advancement of Science (AAAS), a Fellow of the American Physical Society (APS), and a member of the Academy of Engineers of Serbia (AES). Dr. Shenai currently serves as a Distinguished Lecturer of IEEE Power Electronics Society (PELS) and as an Editor of IEEE J. Electron Devices Society (EDS). He has authored over 400 peer-reviewed archived papers in top international conference digests and journals, 10 books, 9 book chapters, and holds 12 issued US patents.

Abstract: The field-reliability of a power electronics switching device is among the least understood topics today. There are no established guidelines available for the design and manufacture of high-performance low-cost power devices in order to guarantee a prescribed meantime-between-failure (MTBF) in a power electronics switching converter. This is particularly important when developing next-generation compact power systems based on advanced silicon and emerging wide bandgap (WBG) semiconductor power devices. This talk will discuss the current approach followed in industry for assessing the field-reliability of power devices, present extensive experimental results on the reliability of compact silicon-based high-end computer power supplies with MTBF of 1 million hours, and outline a novel design and manufacturing approach for WBG power devices that emphasizes on "physics-based" component failure mechanisms.

SHORT COURSE AND TUTORIAL

	Agenda for Tutorial May 19th, 2019 Sunday
14:00~15:30	**Basic Principles and Technology Development of Si Power Devices** Prof. Xiaorong Luo, *University of Electronic Science and Technology of China*
15:30~15:45	*Coffee break*
15:45~17:15	**Smart Power Devices and ICs of SiC and GaN** Prof. Paul Chow, *Rensselaer Ploytechnic Institute*

Tutorial 1: Basic Principles and Technology Development of Si Devices
Prof. Xiaorong Luo, *University of Electronic Science and Technology of China, China*

Speaker Biography: Xiaorong Luo received the M.S. and Ph.D. degree in 2001 and 2007, respectively. She joined University of Electronic Science and Technology of China in 2001, working on power semiconductor devices and ICs. She was with the University of Cambridge as a postdoctoral researcher from 2009 to 2010. She has published about 100 papers, including about 40 papers published in the IEEE TPE, IEEE EDL, and IEEE TED. She holds 4 US Patents and 40 China invention patents. She has got the following honors and awards: National Scientific and Technological Progress Award, the 2nd Prize; Sichuan province Science and Technology Progress Award, the 1st Prize; Natural Science Award of the Ministry of Education, the 2nd Prize; New Century Excellent Talents Award Program from Ministry of Education of China; She is the winner of the National Defence Science Funds for Distinguished Young Scholar. She is the ISPSD's TPC Member and TC Member of IEEE EDS Power Devices and ICs. She is the vice Director of the Department of Microelectronic and Solid-state electronic.

Abstract: Power semiconductor devices are the essential components determining the efficiency, size, and cost of electronic systems for energy conditioning. High breakdown voltage (BV), low static and dynamic power loss, high reliability, wide safe operation area (SOA) and downsizing are important design targets of power devices. In this talk, the main categories, operation principles, development history and trend of power devices are presented. Among them, MOSFET and IGBT are the most popular and widespread application in both discrete device and integrated devices in the medium-power and high-power range, owing to their good trade-off among high BV, high power, low loss and acceptable frequency. Therefore, the novel structures, new mechanisms and fabrication process, as well as state-of-the-art performances of Si-based MOSFETs and IGBTs to realize the design targets above will be demonstrated in detail.

Tutorial 2: Smart Power Devices and ICs of SiC and GaN
Prof. Paul Chow, *Rensselaer Ploytechnic Institute, USA*

Speaker Biography: Prof. T. Paul Chow received his Ph.D. in Electrical Engineering from RPI in 1982. He was a member of the technical staff at GE Corporate Research and Development from 1977 to 1989. Since 1989, he has been with RPI, where he is now professor of the Electrical, Computer and Systems Engineering Department. He has been working in the power semiconductor device area since 1982. His present research activities include high-voltage silicon, GaAs and wide (particularly SiC and GaN) and ultra-wide (diamond and AlN) bandgap semiconductor power devices and ICs. He has published over 150 papers in scientific journals, has contributed eight chapters in technical textbooks, and has procured over fifteen patents. He is a Fellow of the IEEE.

Abstract: The design, layout, simulations, fabrication and characterization of high-voltage vertical and lateral SiC and GaN power transistors are presented. The basic device structures and their refinements, with respect to their potential and experimental performance, are evaluated and compared to their Si or GaAs counterparts. The integrated and unit processing technologies (such as MOS, implant activation, epi regrowth, trench etching) needed for the fabrication and manufacturing of these power devices are examined with their compatibility with commercial silicon foundry capabilities. The robustness and reliability of present commercial discrete power transistors (like threshold shift, current collapse) are assessed. Examples of demonstrated power and/or optoelectronic integrated circuits with SiC and GaN technologies are shown to indicate their potentials. Future directions and challenges for the widespread adoption of these emerging power device technologies are identified.

Proceedings of the 31st International Symposium on Power Semiconductor Devices & ICs
May 19-23, 2019, Shanghai, China

Powering 5G Era Computing Platforms – the Road toward Integrated Power Delivery

Peng Zou
Platform & Key Technologies Development Dept.
Hisilicon Semiconductor Co. Ltd.
Shenzhen, China
Peng.zou@huawei.com

Qiang Xie
Platform & Key Technologies Development Dept.
Hisilicon Semiconductor Co. Ltd.
Shenzhen, China
frank.xie@hisilicon.com

Wei Song
Platform & Key Technologies Development Dept.
Hisilicon Semiconductor Co. Ltd.
Shenzhen, China
wei.song@huawei.com

Qimeng Jiang
Advanced Power Conversion Lab
Network Energy Product Line
Huawei Technologies Co., Ltd.
Shenzhen, China
jiangqimeng@huawei.com

Yuliang Lu
Secondary Power Supply Product
Network Energy Product Line
Huawei Technologies Co., Ltd.
Shenzhen, China
scott.lu@huawei.com

Boning Huang
Advanced Power Conversion Lab
Network Energy Product Line
Huawei Technologies Co., Ltd.
Shenzhen, China
huangboning@huawei.com

Abstract—**The projected core density in future processor chip creates pressing demand to improve computing energy efficiency. The challenge lies in delivering more accurate on-die supply voltage. The conventional discrete voltage regulator based power delivery becomes very inadequate. Breakthroughs in high frequency power stage design, integrated magnetic inductors, and silicon capacitors enables IVR based integrated power delivery architecture for future processor's energy efficiency need.**

Keywords—*IVR, integrated voltage regulator, power delivery, integrated magnetics, integrated inductors, near-threshold, power density, integrated magnetic thin film inductor, 3DIC*

I. INTRODUCTION

5G rollout is widely anticipated to transform the technology and economic landscape. The data rate will be increased to 1~10Gbps from 4G network's peak data rate of 150Mbps, network round-trip latency will be reduced to 1 millisecond compared to 4G's 10 millisecond round trip time [1] [2], and the network will support one million connected devices per square kilometer. Applications such as high-definition video, autonomous vehicles, augmented reality, virtual reality, and hundreds of billions connected Internet of Things (IoT) would become viable. The exponentially increased data volume, more responsive data processing and communication through such high bandwidth network foretell the staggering demand to scale up computing throughput of data centers, fog computing, edge computing, and even terminal devices. It becomes even more crucial to significantly improve energy efficiency to sustain such high data rates, large amount of data processing, and massive connectivity of 5G network [3].

Power delivery increasingly becomes a performance limiting factor as transistor scaling approaches to the end. This paper discussed why current platform power delivery becomes inadequate for the desired computing energy efficiency in 5G era, and why it needs integrating power conversion (or integrated voltage regulator) closely with processor to realize integrated power delivery. Then we review key ingredients and challenges of the emerging integrated voltage regulator (IVR) technology. After more than 20 years endeavor and many breakthroughs in power stage, integrated magnetics, and low parasitic output

capacitors have been achieved. We are at the cusp of integrated power delivery becoming ubiquitous in future computing platforms. Lastly, this paper shares the authors' thoughts on the implications to future roadmap of power electronic devices and systems as computing platforms adopt integrated power conversion.

II. POWER : A LEADING CONSTRAINT FOR FUTURE ARCHITECTURE

In the past several decades, computing performance achieved exponential growth as semiconductor industry closely follows Moore's law to double transistor density generation by generation [4], and the projection of "Dennard scaling" [5] predicting that power consumption per transistor would drop as geometry dimensions shrink to keep the power per mm^2 of silicon near constant. Since computational capability per given silicon area gets increased, computing energy efficiency would be improved. Dennard scaling slowed significantly since 2007 and became almost unnoticeable by 2012 [6] [7].

Fig. 1 shows the significantly worsened silicon power density problem in the most recent decade [7].

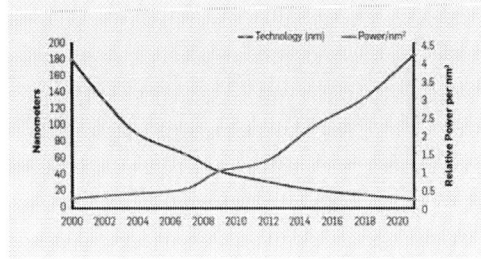

Fig. 1. Transistor scaling and chip power density trend [7]

A. Power Delivery Limits the Future of Muticore Era

Around 65nm CMOS process node, scaling down supply voltage brought the significantly diminished achievable clock speed as the threshold voltage can no longer be scaled down [8], as shown in (1) .

$$f_{max} \propto (V - V_{threshold})^2/V \qquad (1)$$

978-1-7281-0582-6/19 $31.00 © 2019 IEEE

To deal with power density constraint, current architecture utilized increased parallelism with core count growth by 1.4X to 2X in each generation instead of greatly scaling up clock speed [9]. According to IEEE International Roadmap for Devices and Systems (IRDS) 2017 roadmap, number of CPU and GPU cores in a given silicon area will increase more than 10 fold to meet the projected computing throughput in 5G era, as shown in Fig. 2 [10].

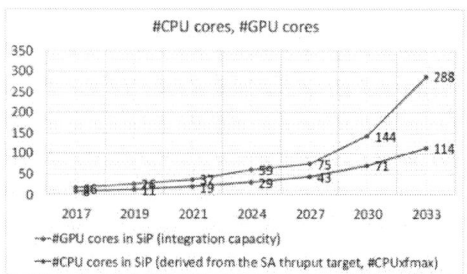

Fig. 2. IRDS projection of number of CPU and GPU core in an 80mm² area

3D integration scheme such as die stacking [11] would be expected to integrate so many cores over a fixed chip area. Without improving CPU and GPU cores' energy efficiency, stacking such a large number of CPU and GPU cores brings a question to its sustainability, as thermal design power (TDP) may be beyond silicon's physical limitation. The limited PCB and package routing resources also cast doubt on viability of power delivery network (PDN).

Studies showed that more than 50% of cores have to be turned off, i.e. going "dark", at around 8nm or smaller CMOS node if there is no improvement in core energy efficiency [12]. This is due to practical TDP constraint of processor chips. More granular core level dynamic power and thermal management are needed to balance core level hotspots, and improve performance of 2D and 3D microprocessors to some extent [13]. More dramatic scheme was proposed to improve energy efficiency per instruction by 10X using near-threshold computing [14].

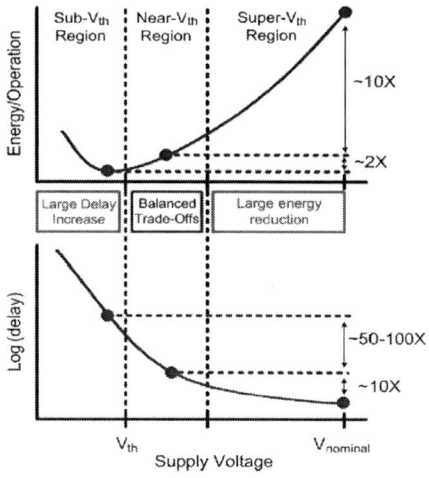

Fig. 3. Energy and delay in different supply voltage operating regions by Dreslinski et al. [14]

Large sensitivity of delay versus supply voltage near threshold voltage (~300mV) (Fig. 3) hinders the potential to harvest high computing energy efficiency. To accommodate large delay variation, the implementation of core logic would have to slow down the clock rate to ensure sufficient timing margin. This leads to reduced computing throughput and negated energy efficiency gain from lowering supply voltage. The accuracy of voltage, i.e. minimized voltage variation, at the on-die power grid has far greater impact on energy efficiency than DC-DC converter's conversion efficiency [15].

In a 3D stacked multi-core processor, we can expect the variation of optimal supply voltages due to different process corners. To maximize computing energy efficiency, different voltages are expected for cores or core clusters at different locations, leading to the need of a large number of dedicated supply rails.

B. Power Delivery Network's Impact

A typical processor PDN model is shown in Fig. 4. The logic operations of processor generate fluctuation of loading current to PDN. The higher is computing throughput, the larger is the amplitude of current fluctuation. The voltage variation at the on-die power grid due to PDN impedance can be described in (2). Even though motherboard VRM tightly regulates its output node at bulk capacitor, the on-die power grid can still experience significant voltage fluctuation due to large ΔI amplitude. The voltage noise caused by PDN impedance is the most significant part of supply voltage uncertainty for processor, and is accounted for tens of millivolts extra voltage guard band.

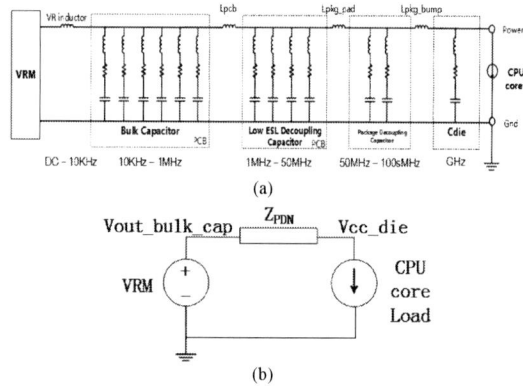

Fig. 4. Typical processor PDN model (a) components and corresponding operation frequencies (b) high level abstraction model

$$\Delta Vcc\_die = Z_{PDN}\Delta I \qquad (2)$$

Previous study on current spectrum signature of processor showed prominent peaks of spectrum energy at frequencies above 10MHz [16]. Conventional power integrity design relies on passive decoupling to lower Z_{PDN} above 100s KHz. Under the constraints of fixed area, Z_{PDN} is expected to be relatively constant in future generations. The increase of processor dynamic current ΔI would worsen ΔVcc_die as number of cores scaling up over the fixed PDN. It becomes prohibitive for energy efficiency to deal with delay variation due to 100s millivolt PDN noise, considering the desire to lower supply voltage toward 300mV threshold voltage. Placing switching voltage regulator with high regulation bandwidth in very close proximity of CPU cores can bypass majority PDN network, shown in Fig. 5.

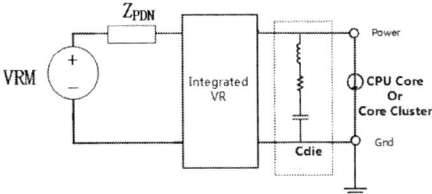

Fig. 5. High bandwidth VR close to CPU cores to bypass PDN bottleneck

C. Emerging Power Delivery Architecture

The limit of thermal design power likely becomes the primary gating factor for powering future computing platforms. The peak power delivered to a single processor chip may have to be kept at relatively constant level even the number of cores continue to increase generation over generation. More attention should be on minimizing on-die supply voltage variation for computing energy efficiency.

It is well recognized that integrated voltage regulators in close proximity with CPU cores for low fluctuation supply voltage is essential for high energy efficiency of future computing platform [8] [14] [15] [17]. Moving to this integrated power delivery scheme bypasses PDN bottleneck, generates tightly regulated on-die voltage supply for high energy efficiency, and enables granular dynamic power and thermal management at core level.

III. INTEGRATED VOLTAGE REGULATOR TECHNOLOGY

The miniaturization of switching voltage regulator for integrating onto processor chip has been a long sought breakthrough by power electronics community since at least 15 years ago. The operation frequency of such integrated voltage regulator was projected over 50MHz, as shown in Fig. 6 [18].

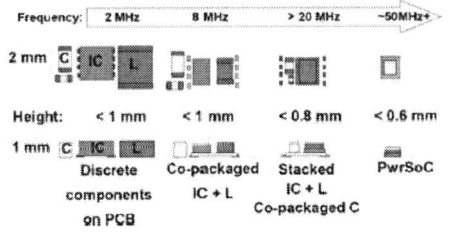

Fig. 6. Envisioned voltage regulator size evolution versus frequency by O'Mathuna et al. [18]

Intel was the first to productize fully integrated voltage regulator technology (FIVR) with their Core and Xeon processors [19] [21]. Air core inductors made of package routings in organic substrate in the range of 1nH~6.7nH and with footprint around 2.3mm² were used. The inductance density is only within the range of 0.4nH/mm² ~ 3.2nH/mm² [20]. The operation frequency of Intel's FIVR is reported around 140MHz due to small inductance. Even though the products demonstrated the improvements of system energy efficiency from FIVR technology [21], such high operation frequency of switching voltage regulator presented a

significant challenge to achieve comparable conversion efficiency as conventional low frequency POL voltage regulator based power delivery. Due to its reliance on routings in thick organic package substrate, FIVR solution is not suitable for 3D die stacking schemes. A hypothetical scheme with alternatingly stacking integrated VR die and processor die is shown in Fig. 7.

Nevertheless, FIVR showed the feasibility of integrated switching voltage regulator circuitry with processor cores. Increasing inductance density over package air core inductors provides possibility of integration into 3D-IC and improving conversion efficiency. Therefore, magnetic based micro inductors, especially magnetic thin film inductors, becomes essential technology for future high throughput computing platforms.

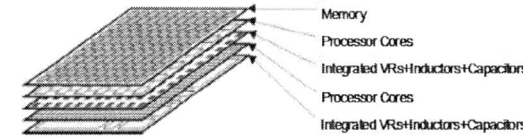

Fig. 7. A 3D stacking scheme for integrating voltage regulators with processor cores, modified from original drawing in [11]

A. Integrated Magnetics

Nanocrystalline and amorphous ferromagnetic alloys, e.g. NiFe and CoZrTa etc., have become the primary material of choice in developing micro inductor structures over the past decade. These ferromagnetic materials have high saturation flux density between 1T and 2T, and can have relative permeability greater than 300. Majority of research work focused on reducing micro inductor's AC loss, defined by quality factor in (3). R_{ac} in most reported data is usually characterized through small signal measurement. The survey of existing work, in Fig. 8 has showed the feasibility of achieving low AC loss with quality factor above 60 for frequency 100MHz and below [18].

$$Q=2\pi f L/R_{ac} \qquad (3)$$

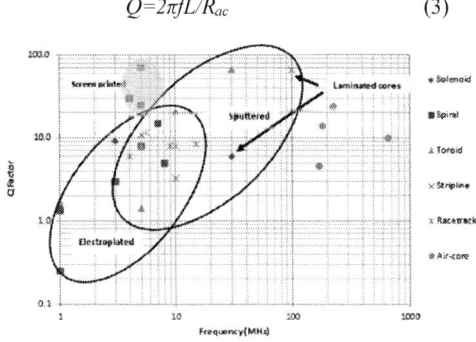

Fig. 8. Survey of micro inductor AC performance by O'Mathuna et al. [18]

Unlike low-frequency inductor, eddy current loss in ferromagnetic alloy becomes a dominant loss mechanism, especially when material resistivity is only 100µΩ cm or less. The classical eddy current loss in unit volume is described as,

$$P_{eddy}=\pi^2 t^2 B_m^2 f^2/6\rho \qquad (4)$$

Where B_m is the amplitude of AC flux density, t is the thickness of magnetic material, and ρ is the resistivity of magnetic material. Laminating magnetic thin film has been proven a necessary way to mitigate eddy current loss. When taking into account more balanced requirements among inductance density, DC loss, AC loss and saturation current, recent work showed that it would be still possible to achieve quality factor around 15 at 100MHz [24][25]. Leary et al. defined a power ratio in (5), which can be viewed as large-signal quality-factor of inductors with magnetic material characterized by empirical Steinmetz equation.

Power ratio = Stored Power/Power Loss

$$= B_m{}^{2-\beta}(f/1000)^{1-\alpha}/k\mu_0\mu_r \qquad (5)$$

The challenge of micro inductor lies in the fundamental tradeoff between miniaturizing structure with high permeability material and minimizing inductor AC loss. Besides classical eddy current loss, losses in magnetic materials consist of static and dynamic hysteresis losses. Static hysteresis is characterized by coercivity, and soft magnetic material sometimes refers to coercivity less than 5Oe. Due to high switching frequency, such static loss per cycle should be greatly minimized. Most sputtered nanocrystalline and amorphous films can rapidly lower coercivity to 0.5Oe or below by reduced grain size well below ferromagnetic moment exchange length (~10nm) [26]. Dynamic hysteresis loss or anomalous loss is believed to stem from domain wall movement, and is detrimental to high frequency inductor's loss. To minimize this loss in high frequency power applications, film deposition can induce anisotropy transverse to the field direction (hard axis) to restrict domain wall movement and promote low loss rotational magnetization flux conduction [22][23].

Controlling anisotropy in film processing and proper structure design, micro inductor's magnetic loss mechanism can be limited to classical eddy current only. Operation frequency and ferromagnetic alloy thickness can then be optimized for the specific application to achieve reasonable efficiency performance. For integrated voltage regulator, magnetic loss likely accounts for 20% or more of total converter loss. System level optimization can be explored for the overall power delivery architecture to lower voltage conversion ratio of high frequency integrated voltage regulator portion, thus the amplitude of AC flux density can be lowered to exponentially reduce magnetic loss.

B. Power FET Choice for Integrated Voltage Regulator

To have sufficiently low switching loss around 100MHz, conventional 5V FET becomes unsuitable. CMOS can be scaled to very small feature size and provide interesting options. Table I shows the figure of merits (Ron*Qg) for core devices in different process nodes. It is shown that, the scaling rule lowers the FOM value (indicating high frequency operation capability). However, the voltage handling capability is somewhat reduced due to thinner gate oxide and smaller gate length. Therefore, in order to achieve fast/large power delivery capability, cascode configuration (as shown in Fig. 9) of core devices was usually adopted to elevate the operation voltage [27][28]. Obviously, in the same voltage rating, e.g. 1.8V, cascode-connection of the advanced-node core devices power stage is capable of delivering much better high-frequency performance compared with 180nm single core devices. In the cascode configuration, cascode devices PM2/NM2 should be carefully biased and designed to prevent the reliability issues (e.g., HCI effect), especially in the switch transient period.

Table I. Figure of Merit (FOM) for CMOS core devices

Process node	NFOM	PFOM	Voltage rating
180nm	1.59mohm·nC	6.68 mohm·nC	1.8V
65nm	0.56 mohm·nC	1.74 mohm·nC	1.2V
40nm	0.44 mohm·nC	1.09 mohm·nC	1.1V
28nm	0.32 mohm·nC	0.33 mohm·nC	0.9V
40nm cascode	0.885mohm·nC	2.42 mohm·nC	1.8V

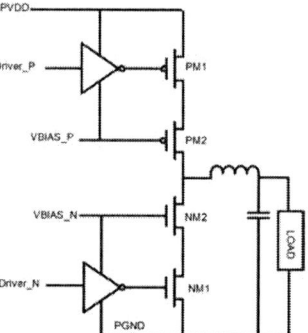

Fig. 9. Cascode configuration for IVR power stage

Intel's FIVR has proven the performance and reliability of cascode power stage using 22nm FinFET and 14nm FinFET devices at frequency over 100MHz. The concerns of integrating VR in the same die as CPU and GPU cores have been resolved by product applications. Future cascode power stage in 7nm FinFET or smaller CMOS technology can be expected to be integrated closely with CPU and GPU cores. However, at least two conversion stages are needed between 12V or 5V input voltage to core supply voltage, because high frequency cascode power stage can only handle input voltage of 1.8V or lower. As future computing platforms continue scales down core supply voltage, properly chosen 2-stage conversion architecture can achieve superior conversion efficiency than single stage design [29].

C. Output Capacitor Choice

As micro inductors of 10nH or less are expected for high-frequency IVR application, the stray inductance (ESL) of output capacitor can become the dominant factor of output voltage ripple amplitude, as described in (6).

$$\Delta V_{out\_ripple} = \Delta I_L(1/8C_of + ESR) + ESL(V_{in(max)}/L) \qquad (6)$$

The output voltage ripple amplitude is part of supply voltage variation which needs to be minimized. Conventional ceramic capacitor with ESL between 100~200pH is an inadequate choice. Form factor of ceramic capacitor is also not suitable for future 3D die stacking scheme as in Fig. 7. Silicon deep trench capacitors already achieved capacitance density of $500nF/mm^2$ and greater with ESL of 10pH or less [30]. It is feasible to fabricate magnetic thin film inductors, silicon capacitors, and IVR circuits on the same wafer. This can be a very attractive solution for 5G era's 3D processor power delivery.

IV. HISILICON IVR PMIC TEST CHIP

A 100MHz IVR PMIC was implemented in 40nm CMOS Technology and integrated with on-die magnetic thin film inductors, as shown in

Fig. 10. The inductor's height is well below 100μm.

Table II shows the key parameters of this multiphase IVR PMIC design.

In this particular test setup, IVR powered up 4 big ARM cores, and ran AnTuTu and GeekBench benchmarks. The output capacitor was ARM cores' parasitic capacitance of 56nF, thus it demonstrated an integrated power delivery scheme with IVR completely bypassed package and PCB PDN. For comparable work load, integrated power delivery scheme used lower supply voltage than conventional low frequency motherboard VR with PCB and package portions of PDN, as shown in Fig. 12. Its high bandwidth responds to load transient in 20ns, as shown in Fig. 13.

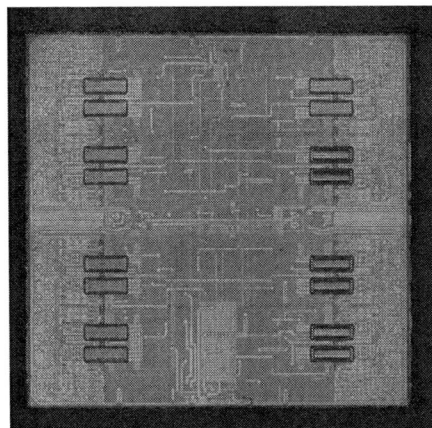

(a)

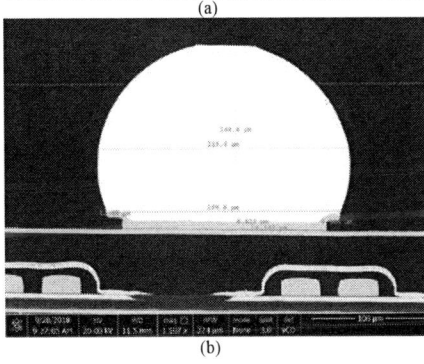

(b)

Fig. 10. Multi-phase IVR PMIC chip with on-die magnetic thin film stripline inductors (a) topview (b) cross-section view at one inductor structure with solder ball.

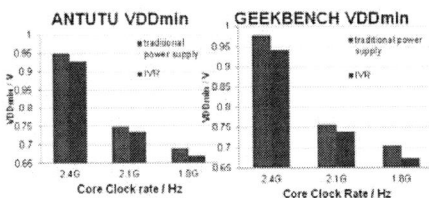

Fig. 11. Benchmark energy efficiency comparison between IVR based power delivery and conventional PMIC.

The limited comparison showed the promising potential of IVR based integrated power delivery scheme to improve processor's energy efficiency.

Table II. Key Parameters of IVR PMIC

Inductor Technology	On-die Magnetic Core Strip-line Inductor
Core Material	Laminated CoZrTaB
Core coercivity @hard axis	0.27Oe
Inductor Quality Factor (@Fsw)	9~13
Fsw(MHz)	100
L(nH)	6~8
Inductor DCR(mΩ)	60
C(nF)	56
Inductor saturation current	0.6A when single phase only, >2A coupled
Vin(V)	1.8
Vout(V)	0.85
Peak Efficiency	82%
Iout_max(A)	20 for 16 phases
Power Density (with Inductor and FET)	~6A/mm^2

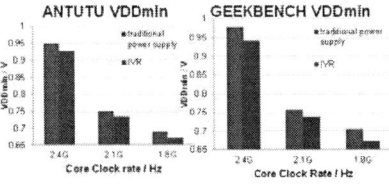

Fig. 12. Benchmark energy efficiency comparison between IVR based power delivery and conventional PMIC

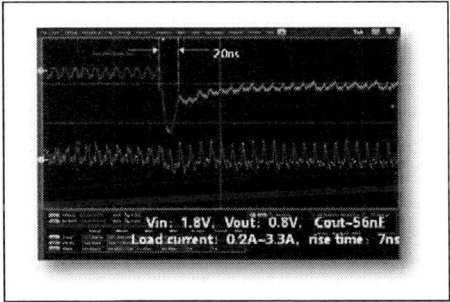

Fig. 13. Fast load transient response from 100MHz IVR.

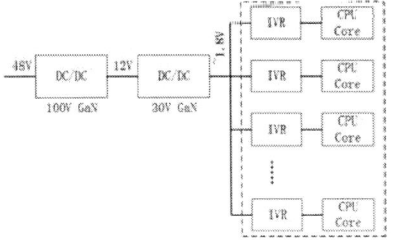

Fig. 14. Data center server power delivery diagram with IVRs

V. IMPLICATIONS FROM INTEGRATED POWER DELIVERY

Fig. 14 shows a typical data center server power delivery networks, in which the IVRs are the last stage to deliver power directly to the CPU/GPU cores. With a high-frequency high-bandwidth IVR stages integrated on processor chip close to CPU/GPU cores, the input DC-DC conversion stages can consider topologies realizing higher efficiency. Since load transient performance is handled by IVRs, converters for 48V bus down conversion and 12V down conversion no longer need design tradeoff between efficiency and transient response. Switched capacitor DC-DC converter, resonant switched capacitor DC-DC converter, or soft switching DC-DC converter design can be good choices for 48V and 12V converters.

For 48-V bus DC-DC application, 100-V GaN device becomes a promising candidate, which can achieve much higher efficiency at operation frequency above 500KHz (need for volume reduction) [31][32]. With the possibility of deploying soft switching circuit techniques, over stress to power FET can be reduce to improve reliability. For 12-V bus dc-dc application, both conventional MOSFET and 30V GaN device can be viable candidates.

VI. CONCLUSIONS

To support the substantially increased computing performance, the number of cores in processor would continue scale up generation over generation. Minimizing on-die supply variation becomes critical for improving computing energy efficiency of CPU/GPU cores, and sustaining such large number of cores under practical processor chip TDP constraint. Computing platforms adopting IVR based integrated power delivery architecture seems the only viable path. More emphasis is needed on the effect voltage accuracy (less variation) to energy efficiency.

Breakthroughs have been achieved in key ingredients of IVR, such as high frequency power stage, integrated magnetics, and silicon capacitors. IVR silicon realizations have been demonstrated, and showed the potential to deliver improved computing energy efficiency.

REFERENCES

[1] A. Gupta, A. R. Kumar Jha, "A Survey of 5G Network: Architecture and Emerging Technologies," *IEEE Access*, vol. 3, pp. 1206-1232, 2015

[2] M. Agiwal, A. Roy, N. Saxena, "Next Generation 5G Wireless Networks: A Comprehensive Survey," *IEEE Communications Surveys & Tutorials*, Vo.. 18, Issue 3, pp. 1617-1655, 2016

[3] GSMA Intelligence, "Understanding 5G: Perspectives on future tecnological advancements in mobile," White paper, 2014

[4] G. Moore, "Cramming more components onto integrated circuits," *Electronics*, Vol. 38, No. 8, pp. 56-59, Apr. 19, 1965

[5] R. H. Dennard, R. F. H. Gaensslen, V. L. Rideout, E. Bassous, A. R. LeBlanc, "Design of ion-implanted MOSFETs with very small physical dimensions," *IEEE Journal of Solid State Circuits*, Vol. 9, Issue 5, pp. 256-268, Oct. 1974

[6] M. Bohr, "A 30 Year Retrospective on Dennard's MOSFET Scaling Paper," *IEEE Solid-State Circuits Society Newsletter*, Vol. 12, Issue 1, pp. 11-13, 2007

[7] J. Hennessy, D. Patterson, "A New Golden Age of Computer Architecture," *Communications of the ACM*, Vol. 62 No. 2, pp. 48-60, February 2019

[8] R. Dennard, "Past Progress and Future Challenges in LSI Technology from DRAM and Scaling to Ultra-Low-Power CMOS," *IEEE Solid-State Circuits Magazine*, pp. 29-38, Spring 2015

[9] W. Huang, K. Rajamani, M. R. Stan, K. Skadron, "Scaling with Design Constraints: Predicting the Future of Big Chips," *IEEE Micro*, Vol. 31, Issue 4, pp. 16-29, 2011

[10] "International Roadmap for Devices and Systems 2017 Edition More Moore," *IEEE IRDS Executive Committee*, 2017

[11] R. S. Patti, "Three-Dimensional Integrated Circuits and the Future of System-on-Chip Designs," *Proceedings of the IEEE*, Vol. 94, Issue 6, pp. 1214-1224, 2006

[12] H. Esmaeilzadeh, E. Blem, R. S. Amant, K. Sankaralingam, D. Burger, "Dark Silicon and the End of Multicore Scaling," *38th Annual International Symposium on Computer Architecture*, pp. 365-376, 2011

[13] J. Kong, S. W. Chuang, K. Skadron, "Recent Thermal Management Techniques for Microprocessors," *ACM Computing Surveys*, Vol. 44, No. 3, Article 13, pp. 13:1-13:42, 2012

[14] R. G. Dreslinski, M. Wieckowski, D. Blaauw, D. Sylvester, T. Mudge, "Near-Threshold Computing: Reclaiming Moore's Law Through Energy Efficient Integrated Circuits," *Proceedings of the IEEE*, Vol. 98, No. 2, February 2010

[15] J. Torrellas, "Extreme-Scale Computer Architecture: Energy Efficiency from the Ground Up," *IEEE Design, Automation & Test in Europe*, pp. 1-5, 2014

[16] R. Weekly, S. Chun, F. O'Connell, "Characterization of Current Signatures for Microprocessors," *IEEE Conference Proceeding for Electrical Performance of Electronic Packaging*, pp. 95-98, 2004

[17] J. Torrellas, N. S. Kim, R. Teodorescu, "Parameter Variation at Near Threshold Voltage: The Power Efficiency versus Resilience Tradeoff," Technical Report, Dept. of Computer Science, University of Illinois, March 2012

[18] C. O'Mathuna, N. Wang, S. Kulkarni, S. Roy, "Review of Integrated Magnetics for Power Supply on Chip (PwrSoC)," *IEEE Trans. on Power Electronics*, Vol. 27, No. 11, pp. 4799-4816, November 2012

[19] E. A. Burton, G. Schrom, F. Paillet, W. J. Lambert, K. Radhakrishnan, "FIVR – Fully Integrated Voltage Regulators on 4th Generation Intel Core SoCs," *IEEE Applied Power Electronics Conference and Exposition (APEC)*, pp. 432-439, 2014

[20] W. J. Lambert, M. J. Hill, K. Radhakrishnan, L. Wojewoda, A. E. Augustine, "Package Inductors for Intel Fully Integrated Voltage Regulators," *IEEE Trans. On Components, Packaging, and Manufaturing Technology*, pp. 528-534, 2014

[21] A. Varma, B. Bowhill, J. Crop, C. Gough, B. Griffith, D. Kingsley, K. Sistla, "Power Management in the Intel Xeon E5 v3," *IEEE ACM International Symposium on Low Power Electronics and Design (ISLPED)*, pp. 371-376, 2015

[22] A. M. Leary, P. R. Ohodnicki, M. E. McHenry, "Soft Magnetic Materials in High-Frequency, High-Power Conversion Applications," *Journal of the Minerals, Metals, and Materials Society*, Vol. 64, No. 7, pp. 772-781, 2012

[23] V. Korenivski, "GHz Magnetic Film Inductors," *Journal of Magnetism and Magnetic Materials*, Vol. 215-216, pp. 800-806, June 2000

[24] N. Sturcken, R. Davies, H. Wu, M. Lekas, K. Shepard, K. W. Cheng, et al., "Magnetic Thin-Film Inductors for Monolithic Integration with CMOS," *IEEE International Electron Devices Meeting (IEDM)*, pp. 114.1-11.4.4, 2015

[25] P. Zou, S. Wei, X. Qiang, C. Liang, C. Yue, H. Chen, et al., "A 100Mhz IVR PMIC with On-Silicon Magnetic Thin Film Inductors," *2018 International Power Supply-on-Chip (PwrSoC) Workshop*, plenary talk, Hsinchu, October 2018

[26] G. Herzer, "Grain Size Dependence of Coercivity and Permeability in Nanocrystalline Ferromagnets," *IEEE Trans. On Magnetics*, Vol. 26, Issue 5, pp. 1397-1402, 1990

[27] P. Hazucha, S. T. Moon, G. Schrom, F. Paillet, D. S. Gardner, S. Rajapandian, et al., "A Linear Regulator with Fast Digital Control for Biasing Integrated DC-DC Converters", *IEEE International Solid State Circuits Confernce (ISSCC)*, pp. 2180-2189, 2006

[28] G. Schrom *et al.*, "A 100MHz Eight-Phase Buck Converter Delivering 12A in 25mm2 Using Air-Core Inductors," *APEC 07 - Twenty-Second Annual IEEE Applied Power Electronics Conference and Exposition, Anaheim, CA, , pp. 727-730, 2007

[29] Y. Ren, M. Xu, Y. Meng, F. C. Lee, "12V VR Efficiency Improvement Based on Two-stage Approach and a Novel Gate Driver," *IEEE 36th Power Electronics Specialist Conference*, pp. 2635-2641, 2005

[30] Murata silicon capacitor catalogs and technical reference, www.murata.com

[31] D. C. Reusch, "High Frequency, High Power Density Integrated Point of Load and Bus converters," PhD Dissertation of the Virginia Polytechnic Institute and State University, 2012

[32] E. Aklimi, "Magnetics and GaN for Integrated CMOS Voltage Regulators," PhD Dissertation of the Columbia university, 2016

Proceedings of the 31st International Symposium on Power Semiconductor Devices & ICs
May 19-23, 2019, Shanghai, China

Magnetics in the GaN/SiC Power Electronics World

Dipl. Ing. (FH) Alexander Gerfer
Würth Elektronik eiSos GmbH&CoKG
Max-Eyth-Str 1
74638 Waldenburg, Germany
alexander.gerfer@we-online.com

Abstract—**The power electronics world made a big step in higher efficiency by introducing Gallium-Nitride (GaN) and Silicon-Carbide (SiC) electronic switches. The fast switching time and high voltage capabilities with low losses bring new opportunities for the optimization of magnetic components. For the majority of power electronics designers, the impression is that core material manufacturers are not innovative and the industry is making slow progress with the development of more efficient materials. The keynote and this paper will demonstrate that progress is indeed being made, as better materials become available while expanding on their limitations.**

Keywords — magnetics, core loss, GaN, SiC, transformer, inductor, airgap.

I. INTRODUCTION

For the majority of electronic engineers, the subject of inductors and magnetics is much disliked. The required level of knowledge of how magnetic components work, how to characterize them and how different phenomena interact can be severely lacking. A number of factors contribute to this including minimal coverage of this topic at tertiary level educational institutes, lack of sufficient parameter characterization, industrial standards lagging behind technological advancement in how magnetic devices are used today, and crucially the implementation of new and innovative design approaches and methodologies. This paper will cover what magnetic components are, before discussing current limitations and methods for optimizing magnetic design. The paper finishes with a discussion of new design approaches.

II. CORE MATERIAL

Cores can be composed of different materials and can be selected to fabricate smaller parts with fewer winding turns, ensuring lower winding loses. The core material can also be chosen depending on the operating frequency range. However, it must be remembered that each material has intrinsic characteristics that result in different performance characteristics at different frequencies. Up to the frequency where AC losses begin to dominate, the component is primarily reactive with relatively low losses.

As the frequency increases, the material becomes less susceptible and AC losses begin to dominate the impedance, resulting in a loss of reactance and a sharp increase in losses. Past this point, the material loses all magnetic susceptibility and no longer performs as a magnetic material and therefore, as an inductor.

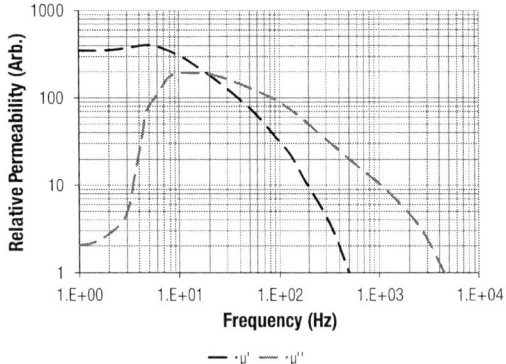

Fig. 1. Typical complex permeability profile of a ferrite material

This is clearly seen in the concept of complex permeability (Figure 1) which allows us to separate the ideal inductive component (μ') and the frequency dependent resistive component (μ'') which represent the core losses. We see that when the permeability is high and the losses are low, useful inductors and transformers can be built. Once the losses start to increase the material becomes very useful for EMI suppression because the combined impedances form a frequency dependence resistance, which converts high frequency noise into heat. Figure 2 and 3 show the useful frequency ranges of core materials materials.

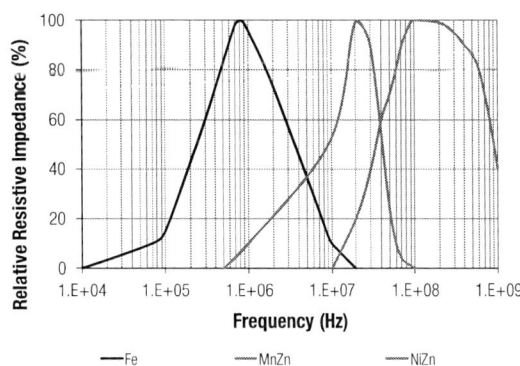

Fig. 2. Relative frequency dependent resistive impedance of different core materials

978-1-7281-0582-6/19 $31.00 © 2019 IEEE

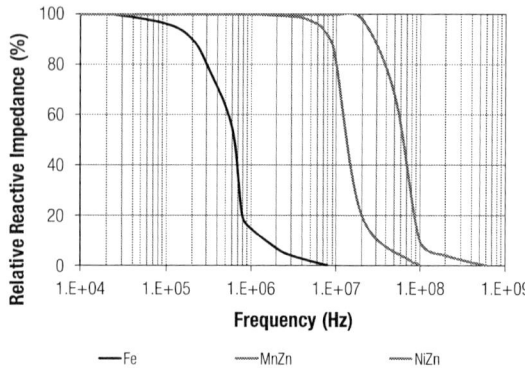

Fig. 3. Relative frequency dependent reactive impedance of different core materials

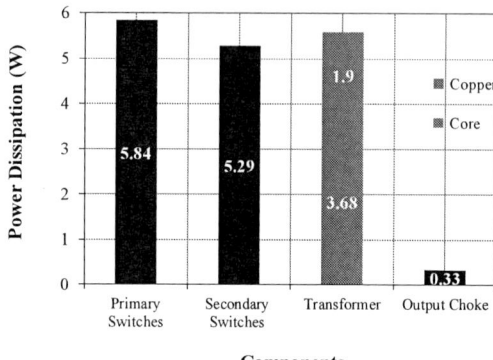

Fig. 4. Loss distribution with conventional electronic switches (I. Jitaro, APEC 2018)

III. PERMEABILITY AND LIMITATIONS

Each core material composition has intrinsic advantages that influence key parameters. For a given inductance value, higher permeability materials result in the possibility of reducing the winding turns or to build smaller volume components. However, increasing permeability has the negative impact of limiting the maximum operating frequency range while lowering the Curie, and therefore operating temperature. As far back as 1947, J.L. Snoek postulated that, due to gyromagnetic resonance, there are frequency and permeability limits to ferrite materials. We know it today as Snoek's limit where the product of frequency and permeability is fixed. As the frequency increases, the permeability decreases.

Different material compositions also effect the overall frequency band, the saturation flux density, and insulation resistance. The complex nature of magnetic materials means that one parameter cannot be changed without influencing almost all other parameters. A deep understanding of the physics involved in the material science, electromagnetic phenomena and the application all contribute to the research and development of new core materials. Manufacturers are actively working to further optimize and improve existing material grades such as NiZn, NiCu, alloys, iron powder and MnZn.

IV. MAGNETICS ARE BACK IN FOCUS

As previously stated, existing electronic switches result in the designer having to live with and mitigate high losses in switches, as well in the magnetics (Figure 4).

GaN and SiC switches have dramatically changed the capabilities in the scope of operating frequency and therefore the losses (Figure 5) and size of components. This has brought magnetics back into focus for design engineers.

Most designs start with the standard core shapes and sizes, which are often optimized for cost and ease of manufacturing (Figure 6). Are these really the best design choice to ensure optimal magnetic field characteristics and do such designs offer the best use of the volume?

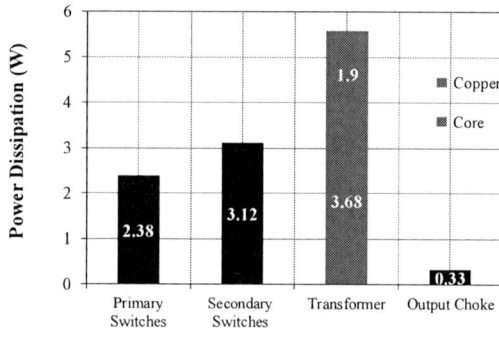

Fig. 5. Loss distribution with GaN electronic switches (I. Jitaro, APEC 2018)

To be useful the GaN switches needed to change the way they were packaged. No more plastic over molding with leads and bond wires, but direct die attachment for minimal inductance. Can magnetics learn from this changed paradigm?

Fig. 6. A typical selection of core shapes available [1]

V. OPTIMIZE MAGNETIC DESIGN

The optimization of magnetics for GaN and SiC switches is based on different approaches.

1) Introducing magnetic materials with higher saturation characteristics while maintaining low losses across a wide frequency range. This is done to mitigate the effect of switching harmonics, which could result in significantly higher and unnecessary core losses. On the other hand, many high frequency designs are core loss limited need and the concentration is on reducing eddy current losses. The new switches can also operate at high temperatures, which is another possible area of development.

2) Introducing novel core material shapes that move away from conventionally used bobbins and cores. Careful analysis of the whole system has led to more optimized shapes. I. Jitaru (Figure 7) showed a very interesting example at APEC 2018 [2].

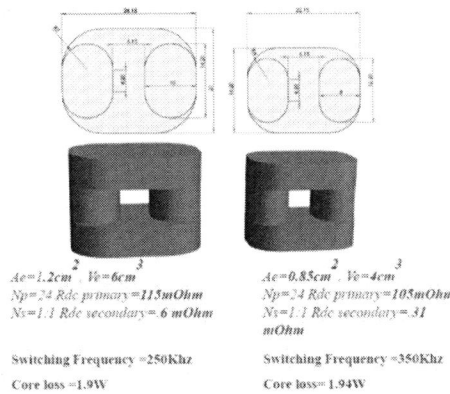

Ae=1.2cm², Ve=6cm³
Np=24 Rdc primary=115mOhm
Ns=1:1 Rdc secondary=.6 mOhm

Switching Frequency =250Khz
Core loss =1.9W

Ae=0.85cm², Ve=4cm³
Np=24 Rdc primary=105mOhm
Ns=1:1 Rdc secondary=.31 mOhm

Switching Frequency =350Khz
Core loss= 1.94W

Fig. 7.   New core material design and shape [2]

3) For better saturation behavior and less impact on winding structure, it is recommended to use distributed air gap design approaches. Nevertheless, the designer should keep in mind that the highest magnetic field strength is in the air gap and be thorough in the development process to ensure any negative effects of distributed air gaps, such as couplings, are reduced (Figure 8).

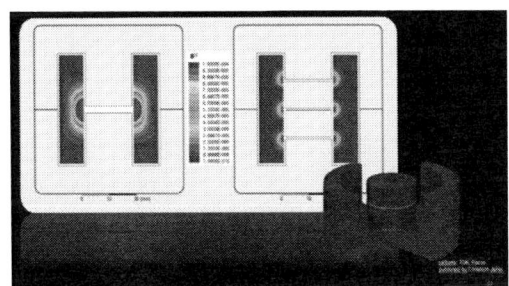

Fig. 8.   Distributed airgap design by TDK [3]

4) Topologies that build on the strengths of each component and mitigate their weakness. Resonant based converters with zero voltage and current switching (ZVS, ZCS) are prime examples. Though these do not need the fast switching speeds of GaN, they benefit from the lower Rdson and smaller package sizes of new switches.

VI. DESIGN THINKING AND AI

The most interesting question: With all the knowledge we have about magnetics and the behavior of magnetic fields, the materials and experience from many designs – are we able to forget all of this and start from scratch, brand new, without limitation in thinking limiting our designs?!

My advice would be to explore the design thinking method. This process involves a methodical, logical process whereby design is focused and framed in a context that works towards resolving specific criteria using solution-focused strategies. This could lead to the development of and experimentation with unconventional core shapes and winding forms that are not confined or blocked by conventional design practices and thinking. Bringing into focus how the magnetic field builds up, what would happen if you would let it do as it would naturally do?! Is it really optimal to squeeze the field lines in a core shape like E-I and hope it is efficient while maintaining low losses?

Framing the question differently – how would AI design transformers or inductors? Would AI, with its unconfined deep-learning logic, present an EE-core as the optimized solution for an inductor core design?

Fig. 9.   Portrait of Edmond Belamy, 2018, created by GAN (Generative Adversarial Network) [4]

REFERENCES

[1]   https://www.we-online.com/, accessed 27/02/2019.

[2]   Ionel Dan Jitaru, "Modern Soft Switching Technologies," Applied Power Electronics Conference and Exposition (APEC), 2018, San Antonio, USA, unpublished.

[3]   TDK, "Distributed air-gap cores improve performance of power electronics", Applied Power Electronics Conference and Exposition (APEC), 2018, San Antonio, USA, unpublished.

[4]   https://www.christies.com/features/A-collaboration-between-two-artists-one-human-one-a-machine-9332-1.aspx, accessed 27/02/19.

(This page is intentionally left blank.)

Proceedings of the 31st International Symposium on Power Semiconductor Devices & ICs
May 19-23, 2019, Shanghai, China

Foreseeable Industrial Changes of Automotive and Transportation Driven by Semiconductors

Tsuguo Nobe
Institute of Innovation for Future Society
Nagoya University
Aichi, Japan
tsuguon@gmail.com

Abstract—**This manuscript discusses the changing roles of Semiconductors for the development of Autonomous Driving within the scope of Vehicle IoT and Deep Learning. Also the development of Autonomous Driving will change the outlook of transportation industries with shorter product life cycle and different modes of business profitability. Along with these changes of technologies and market, the semiconductor business model will also require changes.**

Keywords—Semiconductor, ECU, ICT, Internet, IoT, Connected, Vehicle IoT, Deep Learning, Data Center, Cloud, CASE, EV, Autonomous Driving, Level 3, Level4, Operational Design Domain, 3D Map, Driving Algorism, Self-driving, Robot Taxi, Ride-hailing, Sharing, Mobility Service, MaaS, "Sell-off Business" and "Operational Business".

I. INTRODUCTION

A Vehicle became an ingot of semiconductors and improved the driving safety. Now more than 90% of accidents are said to be caused by human errors of perception, judgment and operation. In order to further improve the safety of vehicle, it is expected that computers will assist or replace the human driving toward Autonomous Driving, whose realization will be materialized by the exponential advancement of ICT/IoT and Deep Learning. As for a recent trend of development of Autonomous Driving, Level 4, where human does not intervene at all on the move, precedes the semi-autonomous driving called Level 3, where human driver intervenes while driving. Level 4 vehicles will mostly be supplied for Mobile Service Providers which will be further grown into the core of MaaS. In the face of these drastic changes of the market, the automobile industry and related semiconductor business are forced to change their business model.

II. CHANGING RELATIONSHIP BETWEEN CARS AND ICT

Since the 1970s, semiconductors began to be implemented in cars and have contributed to the advancement of environmental friendliness and safety performance. For example, the Electric Fuel Injection system (EFI), which appeared around 1970, electronically controlled the amount and timing of fuel supply to engines and contributed to improve fuel efficiency to meet the automobile exhaust gas regulations with lesser excess emission of greenhouse effect gas. After that, semiconductors rapidly penetrated into AT control, CVT control, anti-lock brake system (ABS) and etc…

Since 2000, especially in Japan, Car-navigation systems had well penetrated in the market and so-called "Galapagos Mobile Phones", which allowed data communication with HTTP protocol had penetrated as well significantly ahead of the world trends.

When those Car-navigation systems and "Galapagos Mobile Phones" were connected via Bluetooth or USB, various types of data such as car locations, settings of wiper and State of Charge of EV batteries could be uploaded and analyzed in data centers and it became possible to provide information to Car-navigation systems with valuable information for driving assistance and safe driving. These combinations of Car-navigation systems and "Galapagos Mobile Phones" had already composed the total system called IoT of cars (Vehicle IoT) even before the word of IoT appeared in the market. Those are also identified as "Connected" and the relationship between cars and ICT/IoT expanded from early 2000s onwards to the present (Figure 1).

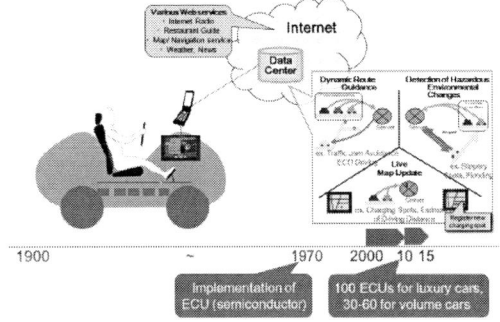

Figure 1: Relation between Cars and ICT (past and present)

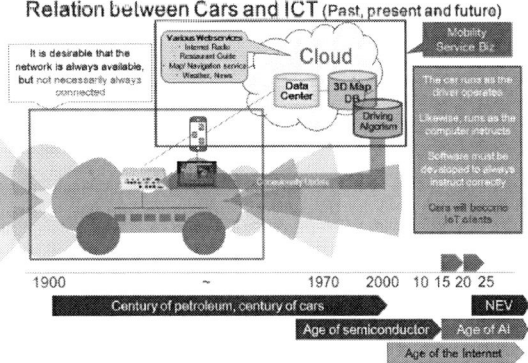

Figure 2: Relation between Cars and ICT (present and future)

Around 2010, around 30 ECUs for a volume zone car and more than 100 ECUs, recently almost 200 ECUs, for a luxury cars came to be embedded. With these, lots of semiconductors

978-1-7281-0582-6/19 $31.00 © 2019 IEEE

are now supporting safety and over 90% of accidents are said caused by human errors of recognition, judgment and operations. If we can substitute ever growing computer and communication capabilities for human driving activities, further reduction of car accidents will be made possible. This is "the automation of driving". (Figure 2)

III. FULLY AUTOMATED VEHICLES (LEVEL 4) WILL BE COMMECIALLIZED EARLIER THAN EXPECTED.

Figure 3 depicts difficulty level of autonomous driving development according to the Operational Design Domain, based on the surrounding environmental conditions, the types of road and the applied speed on the road.

For example, on an expressway, activities, such as keeping the lane, maintaining the distance from the previous car, successful control at junction parts and safe overtaking as needed, will enable cars to continue driving. (Of course, the system needs to drive more diligently than a human driver).

When it comes to ordinary roads, the difficulty of autonomous driving rapidly increases. It runs while intermittently performing various recognitions and judgments based on more complex traffic situations and rules. There are intersections and traffic lights, and some people cross pedestrian crossings or jaywalk, where a person standing on the sidewalk may suddenly cross the area other than the pedestrian crossing. It is difficult to distinguish between bicycles and motorcycles. When making a left or right turn, you must predict the position and speed of the oncoming vehicle and complete it safely before the relevant traffic light turns red.

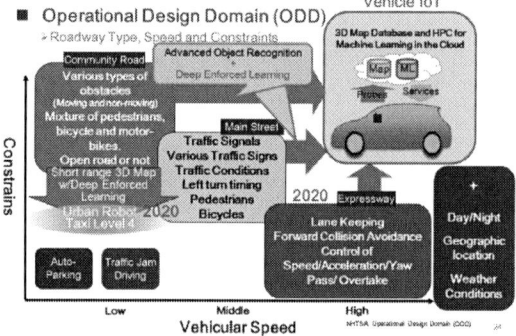

Figure 3: Classification of Autonomous Driving

In addition, there may be no white lines on the low-speed and narrow community roads and it is not clear whether the road is dead-end or not. Also, a high degree of regionally different rules and cultural dependency among countries mat make the judgment of autonomous driving very difficult.

In round 2014, based on this recognition, it was assumed that the timing of commercialization of autonomous driving would be around 2020 with level 3 on expressways and around or beyond 2025 on main streets and community roads.

Meanwhile, in the world of ICT/IoT, Deep Learning has exponentially advanced since 2012, and it came to be recognized that the commercialization of Level 4 will be expedited. Drivable range will be limited to about 1 mile to 10 miles of radius though, it was expected that advanced image recognition would enable the development of high-definition and dynamic three-dimensional map and Deep Reinforcement Learning would help to reveal edge-case scenarios in order to complete rule-based autonomous-driving algorithms on every corner of the roads. Then, in 2016, major OEMs announced that the estimated commercialization year of driverless self-driving robot taxis would be in around 2021.

IV. THE DEVELOPMENT OF AUTONOMOUS DRIVING WILL PROCEED IN TWO DIFFERENT DIRECTIONS

In parallel with the expedited feasibility of Level 4 in limited areas, the recognition that "Level 3 is difficult to commercialize" has spread among automobile companies. Unlike levels 4 and 5 which the computer operates from the start to the goal, at level 3, a "transfer of driving authority" occurs between the computer and a human driver.

From the 2016 Federal Automated Vehicle Policy of NHTSA, the condition of Level 3 has been changed to "(human driver) must always be ready to return to driving" and, conversely, if an accident occurs due to the failure of transferring authority to a person, it became necessary for car companies or mobility service operators to prove that the driver was not ready to return to driving. Furthermore, with the advancement of technology, it is also pointed out that the range that can be run by autonomous driving mode is expanded, and the driver "over-trusts" the autonomous driving system and "loses situational awareness" that he or she is driving and it becomes more difficult to return to driving.

Due to this background, since the summer of 2016, major automotive companies have announced almost the similar plans to accelerate the Level 4 market introduction rather than Level 3. Each company will realize fully automated vehicles i) without one handle, accelerator and brake, ii) with EV, iii) to carry not only people but also goods, iv) not to sell to individuals but offer them to mobility service operators, v) even themselves might become mobility operators and vi) with commercialization target by 2021.

Thus, the engineering and market development trends of autonomous driving are going in two directions: "Level 3" where privately-owned cars are partially automated especially on expressways, and "Level 4" where mobility operators provide new services like robot taxis with self-driving cars within geo-fenced areas of about one to ten miles of radius.

V. POSITIVE ATTITUDE OF THE US-CHINA GOVERNMENT

Competition for commercialization of "mobility service" is overheating among major global automotive companies. Behind this is the government's proactive measures in the US and China.

In the first place, as the US, which is the world's most advanced in the world in computer and communication technologies, there is a prospect among not only within both industries but also among Congress that it can once again become the world's leader in the automobile industry by combining cars and ICT.

On the other hand, in China, while selling 28.88 million cars in 2017, only about one-third of households currently own cars. If car sales volume continues to increase, there may occur worsening problems such as energy security, air pollution, traffic congestion, etc., and the shift form Internal Combustion Engine to EV and from ownership to sharing would be urgently required. And the government is also actively promoting autonomous driving in the face of aging population.

In other words, both countries are working hard to realize CASE <Connected, Automated, Sharing Services and Electrification>, and European automobile companies are also making significant investments targeting the US and China, the world's two largest markets.

VI. IMPORTANCE OF VEHICLE IoT AND DEEP LEARNING

For example, when a person drives and intends to make a right-turn (or left-turn on a left-hand steering wheel) at the signal two turns ahead, the second signal is instantaneously recognized without any difficulty (Fig. 4). At the same time, however, it is important to pay attention to the immediate signal more, and if red, stop before the stop line. When this is done with a computer, it is extremely difficult to find a signal from a camera image while the car also moves itself. There may be a car tail lamp ahead and lights of all colors in the surrounding area. Furthermore, with the camera alone, it is difficult to identify the correct immediate signal of the direction". Even if radar and LiDAR are used in combination, they require enormous computing power and errors are likely to occur.

Figure 4: How do you drive a car?

Here, what has become an eminent and staple solution since around 2014 is the method of using a three-dimensional map. Also, it became possible with the help of Deep Learning. As attributes of the three-dimensional map, signals, street lights, traffic signs present on roads or in the space, etc. are assigned as singular points and are given as XYZ coordinates of three-dimensional information. It is possible to locate one's position by superimposing the results of sensed data from vehicles with multiple singular points on such a three-dimensional map and simultaneously improve the accuracy of recognition and judgment of the environmental situations with reducing the burden of spatial processing and increasing the accuracy at the same time.

In addition, in order to avoid collisions, we will complement and fuse various sensors (cameras, radars, riders, etc.) in a complex manner (so-called sensor fusion) to analyze and grasp environmental information accurately. Sensor Fusion recognition is used not only for automatic driving but also for ADAS such as lane departure warning, forward collision warning, blind spot recognition, and advanced ACC. Here too, it is effective to apply deep learning on the cloud

connected with Vehicle IoT to improve the recognition accuracy of each sensor.

Up to this point is the story of the "visible world". People are driving without seeing only the "visible world" as a matter of course, but even in the "invisible world" ahead, they are always careful to be able to respond to unexpected situations from a certain level of experience doing.

Vehicle IoT is used also to cope with such a situation as beyond "Line-of-Sight". Upload sensor data from multiple cars driving ahead in the past, such as 1 minute ago, 1 hour ago, 1 day ago, etc., perform statistical processing and big data analysis in the data center, and associate attributes with quasi-immobile body (associated with position; i.e. update the traffic congestion and white line deterioration over time, the decrease of lanes ahead due to construction, the existence of an accident car, and further information such as the position of traffic lights and traffic signs on a three-dimensional map, and the following cars refer to it. With these, Vehicle IoT achieves plan-ahead autonomous driving within the area of "non-Line-of-Sight".

Thus, for the autonomous driving, three-dimensional maps and generation of traveling algorithms on the cloud, distribution and downloading them via OTA (Over-the-Air), plan-ahead autonomous driving, and for MaaS to be described below, Vehicle IoT is a prerequisite to construct the new era of mobility and transportation.

VII. EMERGENCE OF MAAS

Without having the property of taxis, a raid hailing company like Uber created a taxi business on the Cloud by connecting people who want to get a ride, and people who want to offer a ride with SNS-like functions via combination of data centers and smartphones. Also, since it is on the Cloud, with some regional customization, services and business penetrated internationally within a snap.

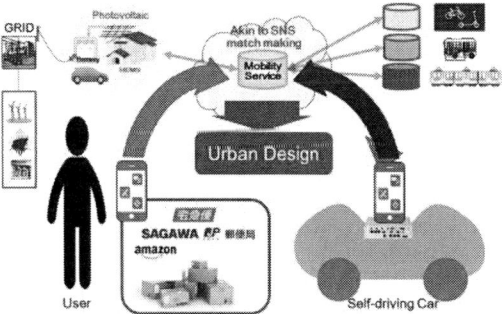

Figure 5: Emergence of MaaS

It is also possible to improve service quality for users, drivers and vehicles via data center programming. The same functionality is also required on the data center of Mobility Service Providers when delivering taxis and microbuses that have become driverless in the future. In addition, Mobility Service Providers can collect environmental information, traffic utilization and traffic congestion information, make them into an API and provide them as valuable information to third party service providers.

The data center of Mobility Service Providers will also gather information of railways, buses, shared bicycles, electric scooters and all types of transportation means are combined to offer faster, cheaper and comfortable means of transport from any starting points to any destinations. This is the MaaS.

It provides not only an utterly new style of transportation in urban areas, but also a solution of issues such as "increasing traffic accidents by elderly people" and "diminishing life line due to abolition of unprofitable bus routes" in rural and depopulated areas. Drivers of courier companies will be able to respond flexibly to changing requests of delivery and collection and it will also be possible to reduce the workload, which will lead to the elimination of the "truck driver shortage".

In addition, it is estimated that such Self-driving vehicles will mostly be EVs, and it is considered that such EV carrying a battery of about 100 KWh will be a moving electricity storage in the era of renewable energy. Furthermore, Mobility Service Providers can also be the main players of future Urban Design as they grasp the flow of people, goods and energy and will be able to delineate and shape the user needs. (Figure 5).

Since data used to build such services include personal information, as collected with the consent of individuals based on contracts, raw data from sold products and services can only be collected and analyzed by sellers and providers. Basically, as the more data there is, the more accurate the analyzed information will be, the user scale is the key success factor and the company with more users and with the ability to process their data gains the international competitiveness.

VIII. MssS will change the Future Structure of Automotive Industry

According to Strategy Analytics, the percentage of users who use car sharing services at least once a week is 49% in China compared to 20% in both Europe and the US. (Figure 6). The ratio of current vehicle ownership in China is roughly one in three households, and if there is a ride hailing company like DiDi, it is understandable that the market for matchmaking between people who want a ride and people who want to offer a ride has rapidly expanded with using smartphones. In fact, as of November 2017, the number of registered DiDi users is reported as 440 million, and it is said that there are 30,000 ride requests per minute during rush hour.

What is suggested from here is the possibility that the relationship between vehicles and users will be changed from "owning" to "sharing" in emerging countries. Under these circumstances, it is necessary for car companies to develop and manufacture vehicles targeted for hailing/sharing in the near future and sell to Mobility Service Providers in order to expand the revenue, or otherwise, automobile companies need to be Mobile Service Providers by themselves.

Of course, when it comes to sharing, the development, manufacturing, and sales procedures also need to be changed significantly. Currently, the owned cars are manufactured and sold as new vehicles for three years beyond the start of sales after a three-year development period. On top of that, car companies' sales only occur just once during the three years of new-vehicle period, in the form of sales to the dealership, not the customers.

When it comes to sharing, Mobility Service Providers capitalize the vehicles and make revenue as the "mobility

service" compensation, which is currently estimated in the US to be about $ 1.5 per mile, subject to verification of feasibility and marketability. The utilization rate of shared cars will be more than 10 times of that of owned cars, which is currently said to be about 4%, and run 100,000 miles within two years. If revenue occurs from the mobility of people, goods and energy at the rate of 50% of driven miles, it is discussed that a car that costs only $ 25,000, if sold, will earn about $ 75,000 in two years. Mobility operators will have the cost increase from EV and autonomous vehicle operation , and expenses for electricity, maintenance, insurance etc. However, due to the change of cost structure, the profit ratio is less than 10% of the current companies. It is pointed out that at least more than twice.

Penetration of ride-hailing in the US, EU and China

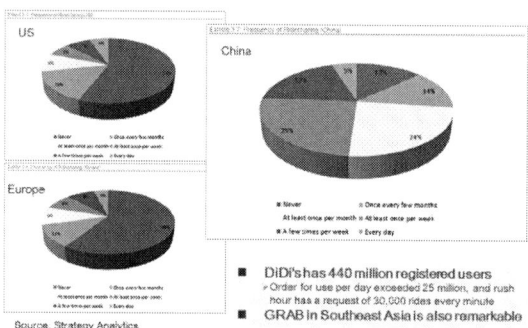

Figure 6: Market Penetration of Ride-hailing Services

In addition, when the life cycle will become about 2 years which is same as the smartphone, it will be shipped with ample hardware and software headroom in the car, and it will be possible to catch up with the advancement of technology with adding the service daily with the software update after the shipment, so the variety Service solutions are expected to be born and market will expand as well.

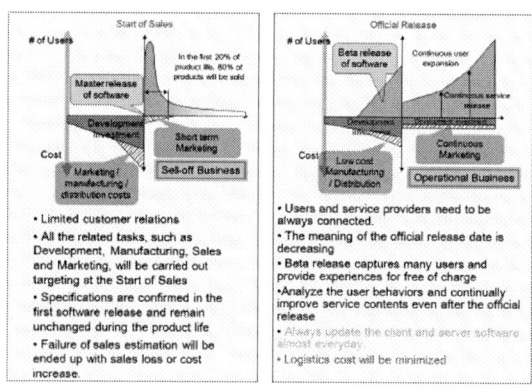

Figure 7: Differences of Business Models

The daily business content will be fundamentally different between "Sell-off Business" and "Operational Business". In the "Sell-off Business", product development, production and logistics will be organized toward the start of sales. Failure of sales forecast and production adjustment will lead to business loss and cost increase if products become short or surplus. Furthermore, the popularity of new cars continuously declines since the launch.

"Operational Business" is very different from "Sell-off Business", and it is important to continuously expand the popularity among users after the launch. Even after the launch (service release), Operational Business requires the continued analysis of end users, consecutively update the software to improve the service almost daily, and along with the continuous marketing activities, occasionally upgrade versions to keep the user's expectations higher and satisfied (Figure 7). Conversely, once the user base declines, business will shrink sharply.

Furthermore, the cost structures of the "Sell-off Business" of vehicles and the "Operational Business" of Mobility Services differ significantly. While hardware-based "sell-off Business" has heavy fixed and variable costs, software-based "Operational Business" has light fixed and variable costs. As a result, in the "Operational Business", the breakeven point is reached when a relatively small number of users are acquired and subsequent user expansion brings enormous profits, but completely opposite for "Sell-off Business".

Another characteristic of the Internet business is that the top 10% of heavy users use 90% of network resources (costs). With this equation, services can be widely provided for free to obtain more users as broadly as possible to analyze the market needs from immense data obtained and improve daily services. As a result, the loyalty of the top 10% of premium user is further enhanced and the total cost could be covered by the premium fee. This is a concept known as Freemium, but it works well also with the concept of Unit Economy and will be an important key to the success of the Mobility Service.

(This page is intentionally left blank.)

978-1-7281-0582-6/19 $31.00 © 2019 IEEE

Proceedings of the 31st International Symposium on Power Semiconductor Devices & ICs
May 19-23, 2019, Shanghai, China

Power Challenges Caused by IOT Edge Nodes: Securing and Sensing Our World

James Spehar
Secure Interface and Power,
NXP, USA
Chandler AZ, USA
jim.spehar@nxp.com

Adam Fuks
Secure Interface and Power
NXP, USA
San Jose CA, USA
adam.fuks@nxp.com

Marc Vauclair
CTO, Security Concepts
NXP, Belgium
Leuven, Belgium
adam.fuks@nxp.com

Maurice Meijer
Secure Interface and Power
NXP Semiconductors
Eindhoven, The Netherlands
j.t.m.van.beek@nxp.com

Joost van Beek
Secure Interface and Power
NXP Semiconductors
Eindhoven, The Netherlands
maurice.meijer@nxp.com

Bin Shao
Secure Interface and Power
NXP Semiconductors
Chandler AZ, USA
bin.shao_1@nxp.com

Abstract—This paper discusses power challenges caused by an explosion in data driven application at the Edge of IOT. How technology, power management techniques, machine learning (AI) and energy recycling are evolving to do more computing at the Edge of IOT to solve Power, Data, Communication and Latency issues. The paper will show how innovation is needed to bring different topics together to provide an optimal solution for the power challenges ahead. A use case with biometric security at the edge will be discussed

Keywords—IOT, Internet of Things, AI, Sensors, Microcontrollers, Power Management, Biometrics, Security, Edge, Cloud

I. INTRODUCTION

Recent years have seen an explosion in data-driven applications. This explosion is powered by the technologies which allow low-power always-on sensing, ever more localized processing of data, and low-power connectivity.
There are two classes of devices which are seeing a particularly sharp rise in prominence. The two classes of devices are End-Nodes, which are predominantly focused on duty-cycled always-on sensing, and Edge-Nodes, focused on computing and interpreting the mass volumes of data prior to sharing it with large internet-connected compute resources. It is projected that the number of devices with intelligence will grow from 40 billion in 2020 to 75 billion in 2025. Shown in Figure 1 is an example of these "intelligent devices" that move data to the cloud "Data Centers" for further processing. This change in the way data is stored and the large amount of data will create power, storage and communication issues. This shift in paradigm, has created a need to make these intelligent devices at "Edge of IOT" smarter, to reduce data traffic and power efficient by using advanced technology nodes, directed power techniques, energy recycling, and artificial intelligence.

II. CURRENT STATUS AND CHALLENGES OF IOT

A. IOT IS TRANSFORMING OUR WORLD

IOT is evolving into multiple End Nodes. In Figure 2, Smart Home, Smart Health, Smart Transportation, Smart

Transportation, Smart Cities and Connected Consumer are a few examples. As we move from the Cloud to the Edge Computing, the goal is to control the Economics, Robustness, Safety and Data Protection by making Edge computing smart. What makes these IOT Nodes Smart is how AI is used to solve: (1) time sensitive applications not hampered by network latency, (2) On-Board Machine Learning will provide Precise & fast detection, classification, adaptation, (3) Reduce power by reducing Data Center Traffic with data centers only process & store relevant data, (4) Reduce Network Cost by Local processing shields Cloud from large part of raw data, (5) Safeguard Privacy by Transmit semantic data rather than raw data, and (6) Increase Security resilience to offline conditions.

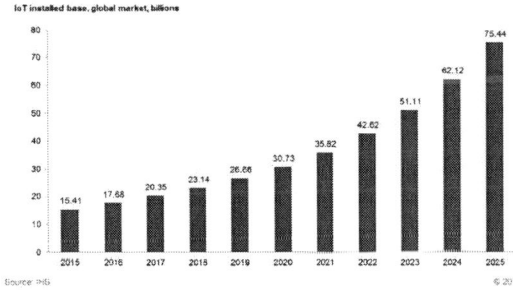

Fig. 1. IOT Installed Base Market, billions. Source HIS

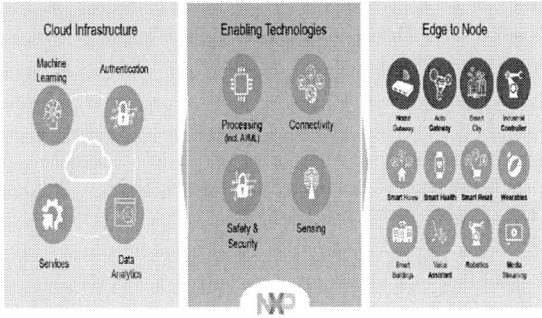

Fig.2. IOT, Edge Nodes, Enabling Technologies, Cloud Infrastructure. Source NXP

B. WHAT IS AN IOT EDGE DEVICE

Figure 3 is a simplified view of IOT. We need these Edge devices to Sense, Process, Connect and Act. This will advance solutions which make lives easier, better and safer. The Edge comprises a Gateway, Devices, Sensor and Actuators. There are five main functions in IOT: Sensors & Actuators, Devices, Gateway, Cloud and Servers.

- Sensors & Actuators "Sense and Execute" are used to convert physical parameters into measurable parameters like optical and electrical signals. IOT needs sensors to be smart. Examples of Sensors are microphones, accelerometers, proximity, pressure, humidity and biometric. One of the key functions is the power management on sensors & actuators portion. The power management needs may change depending if the node is mains-powered, or battery-powered. It can be a driver for functional partitioning e.g. edge vs. cloud.
- Devices "Process" collect, process and store data in a more distributed fashion and closer to endpoints results in hastened response times, reducing latency and a conservation of network resources[?]. Process technology has been a major contributor to making devices smaller and more powerful.
- Gateways "Connect" are needed to connect Edge Devices to the Cloud for computing and data storage. They are used to move data to a network
- Cloud "Process" is a term used for delivering on demand services over the internet. For example, data centers used for computing and data storage.
- Systems "Act" are use the information from the cloud computing to provide a service. For example, A command to Alexa from Amazon may cause a purchase or return of an answer. In many cases "Act" may also refer to driving an actuator such as an audio speaker, stepper motor, or robot arm.

This paper will focus on two aspects of the IOT system: Sensors and Devices.

C. TECHNOLOGY: MAKING IOT FASTER

There are the challenges being presented with making every aspect of services or products. A brief look back in time shows there have been major waves of innovation. In Figure 4, different waves were enabled by different technologies. Transistors enabled seemingly endless computing and storage. Bipolar devices enabled the Mainframe, DRAM enabled the PC, wireless connectivity enabled the cellphone, and IOT/AI is enabled by almost endless computing and storage. Each wave has an order of magnitude growth over the previous wave. The future may hold a factor of ten or greater for IOT over the Cell Phone Revolution which started in mid-1990's

Are we ready for this explosive growth as a society? What is the impact to resources like power, wafer capacity, storage and communication? What do we do with all this analog sensor data at the Edge?

An SRC presentation by Hansen, Yeh and Zhirnov[1] identifies significant trends that are driving information technology and what roadblocks and challenges the industry is facing. There are four trends (1) Computing energy, (2) Global Storage-Communication cross-over, (3) Dramatic global Storage requirement increase, (4) Analog data deluge.

(1) The Computing Energy has seen large leaps forward and has followed Moore's Law. The amount of Joules/bit required for computation has decreased by a factor of 1000 from 1991 (100fJ/b in 1um) to 2014 (0.1fJ/b in 22nm) due to process technology. If the Landauer's limit is $\sim 3 \times 10^{-21}$ J/b, then the potential exists to still see a reduction of 30,000. These advances are allowing us to do more data collection and computation at the edge, sending only useful information to the Cloud[1].

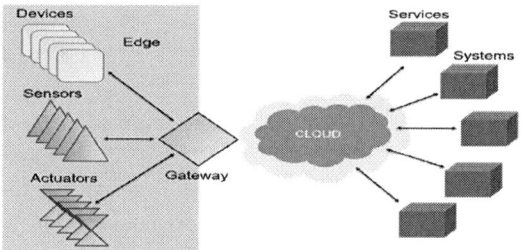

Fig. 3. IOT System, Cloud and Edge.
Source:https://searchnetworking.techtarget.com/definition/edge-device

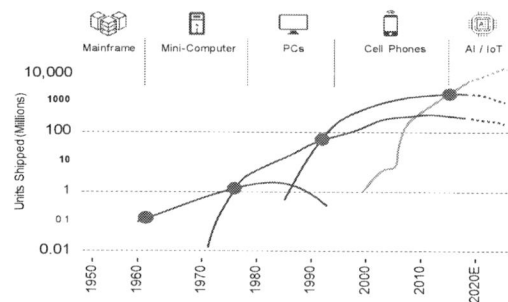

Fig. 4. AI/IOT Will Create New Industry Leaders. Source Jeffries

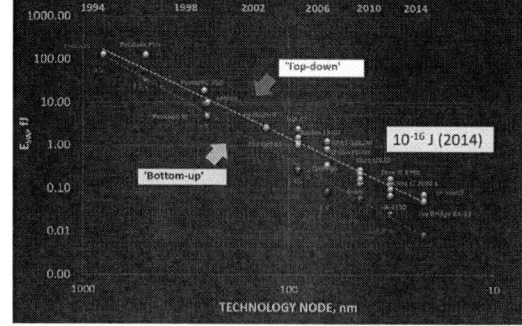

Fig. 5. Computing energy: Energy per bit in MPU[1]

(2) Storage and Communication cross over will occur in 2020. Data over load will increase latency between IOT Edge and System response[1].

(3) Storage requirements[1] are going to outstrip available silicon by a factor of 10.

(4) Analog Data Deluge what do we do with data from trillions of sensors (45 Trillion Sensor by 2032)[1]. How can we move these Smart decisions into the Smart Nodes[1].

    a. This situation will be getting even more serious in the near future, when data from our lives as well as from IoT sensors may create a **data deluge** that will obscure valuable information when we need it most[1].

    b. Extracting key information in the predicted data deluge and applying it in an appropriate way is key to harnessing the data revolution[1].

In addition, smaller geometries create further challenges. The reduction in geometries on advanced technology nodes have created the ability to do extreme amounts of computing at the edge. Unfortunately, these smaller geometry nodes are leaky and consume more power during operation. As a result, battery operated devices require more clever power techniques to address these.

Another innovation in technology, lower resistance MOS devices: The R_{DSON} of power devices, has been reduced significantly over the years which leads to increased efficiencies for buck, boost, LDO and switching regulators. These power devices are seeing advances by reducing resistance to 1mohm/mm[2]. This is a factor of 2-3 decrease in the last 10 years.

D. POWER TECHNIQUES: MAKING IOT EFFICIENT

Low power operation is one of the main requirements for IoT edge devices. This is especially true when devices are battery operated or are relying on some form of energy scavenging. Lower power consumption can enable longer operating lifetime of the IoT edge device, or smaller form-factor devices by reducing the size of the energy source. There exist many approaches to reduce power. This section discusses a few of the main techniques to be employed.

Duty-Cycled Operation: Static power consumption is typically a large contributor to the total power for IoT devices with low activity profiles. The main causes of static power are active chip infrastructure components e.g. clock generation, power-consuming analog/mixed-signal blocks, or caused by device leakage. In such case, a main approach towards low power operation is to make use of duty-cycling and following the so-called 'run fast to sleep' method. An example can be found in sensor hub MCUs[2]. Such MCU disables the sensors after the needed data has been read out. Processing of the stored data is done in a burst mode fashion, to allow MCUs dynamic power to dominate during that period (thus mitigating the proportional overhead costs of power-expensive infra-structure). After the data processing is completed, the data is stored or transmitted to a host device. The IoT device is then put to sleep, until the next data collection cycle starts. We follow a similar approach for connectivity functionality (e.g. Bluetooth Low-Energy) where cost of radio communication is fairly high. Radio operation is performed in regular short bursts with high power, while the radio is put to sleep in between bursts. In such applications, duty-cycled (or burst-mode) operation can reach sweet spots of power efficient operation. Low power modes shall be employed during inactive time. A power delivery system for such devices needs to have low quiescent current for the long sleep duration, but also the capability of good transient response during the active portions. Further to this, a fast ramp-up is highly desirable.

Near Threshold Computing (NTC): With NTC, the digital core operates at a reduced supply voltage just above the threshold voltage of transistors. In this way, significant energy savings can be achieved with NTC, but at the expense of a lower operating speed[3] (see Figure 6). Run-time supply voltage control or body biasing can be utilized to compensate for process and temperature variations[4,5]. Challenges of NTC are to avoid the large speed degradation and to ensure efficient power supply delivery. This calls for innovative system architectures and circuit solutions. For a IoT edge device, the overall power shall be optimized based on the speed requirements and static power consumed.

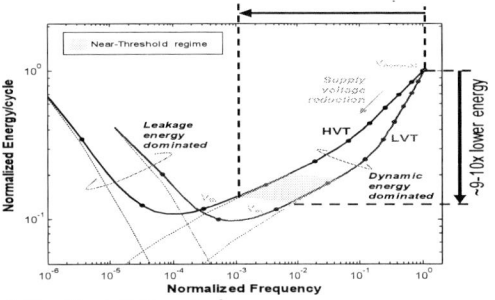

Fig.6. Near Threshold Computing[3]

Fine Grained Control: Fine-grained control[6] can achieve further system power reductions. Design-time optimization includes static approaches such as multi-V_{th} design, and multiple clock- and power domains. Run-time power-performance management can be achieved by employing, among others, introduction of various power modes, dynamic voltage and frequency scaling (DVFS), power gating, clock gating, body biasing, low energy quick wakeup circuits. This is illustrated in Figure 7.

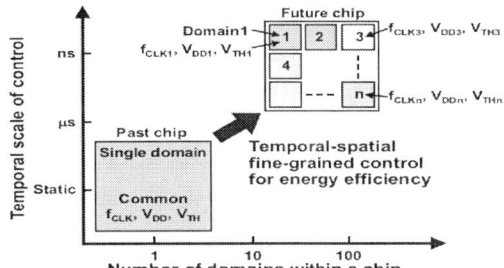

f_{CLK}: Clock frequency, V_{DD}: Supply voltage, V_{TH}: Threshold voltage

Fig. 7. Fine Grained Control[5]

Ultra-Low-Power Always On Sensing: The main purpose of always sensing is to enable ultra-low-power operation for the IoT edge device while offering wake-up functionality. Once a wake-up event is received, the IoT edge device will be waken up to perform its tasks. Until that time, the IoT edge devices operates at its minimum power. The challenge is to optimize power consumption for always-on sensing while ensuring fast wake-up to full functionality of the IoT edge node.

E. ENERGY HARVESTING: MAKING IOT GREEN

Another class of products where either, power is in short supply (e.g. energy-harvesting devices), or static power overheads are very low, a more efficient tradeoff is processing information slowly over long periods. This change avoids the sharp peaks in consumption, but also allows for a simpler circuit design where performance targets are reduced if a task can take longer to complete. In these sort of products, both active and dynamic power need optimization to be efficient with the power available. Furthermore, robustness to voltage dips or intermittent supply may be necessary. However, energy harvesting only has specific applications.

F. ENERGY RECYCLING: MAKING IOT GREEN

Voltage stacking for charge recycling is a technique that involves connecting power domains in series rather than in parallel. This approach has limited applications, but can be used to help with power recycling (e.g. memory leakage)

Antenna based standards are becoming used for energy harvesting. These communication standards like NFC, BLE, BTC, mm Wave, optics are possibilities for power and charging edge node during communication.

G. ARTIFICIAL INTELLIGENCE: MAKING IOT SMART

Smarter devices require more compute power which can increase overall energy consumption. This increase can manifest itself as heavy computation for longer periods, or short duty-cycled high peaks. Alternatively, some devices may consume a steady continuous low current. The activity duty-cycle, but also system overheads can affect the choice of how to most efficiently power a device.

Let's begin by examining some basic notions which help increase power efficiency in IOT devices and then explore the impact of AI on these.

AI could be used to optimize power where computation may be more prolonged. This could be an edge device where a whole network's data may be analyzed, sorted, and potentially examined for local reaction, or for sending over the internet to the cloud for further processing or logging.

Computing at the Edge: AI computation may be fairly intensive, but since it is inherently designed to be noise-immune, it offers interesting possibilities. AI computation enables applications from sensor aggregation nodes to edge devices. If our understanding of human biology is correct, a human being has no more resolution in sensing than 8-bits at each individual node. The human body is also somewhat immune to noisy data or inaccurate thinking[7]. The human brain becomes a reference for the potential for AI and innovation is possible.

Advances in technology nodes are allowing cheaper local computing. This allows a wider range of complex problems to be solved closer to the sensing nodes. To facilitate the computation of complex problems, a larger volume of data needs to be analyzed. AI shows promise in performing this larger analysis and data reduction, such that data need not need leave the IC. This can be a major power saving, especially considering that toggling IOs may cost an order of magnitude more of power than on-chip processing.

Increasing AI capabilities of IOT devices, while maintaining a low power footprint, means that we need to dig further into approximate, intermittent computing. As such, it is important is to be able to cope with noisy or unstable power supply (perhaps even intermittent). The use of AI capabilities such as neural networks show a lot of promise for improved efficiency because they are somewhat immune to noisy data (whether as input or during computation). The trend going forward is to change focus from the fast-and-accurate computing of the past to the parallel-and-approximate computing of the future. This fundamentally changes the possibilities of not only design choices (e.g. traditional adders/multipliers vs approximate, cheap arithmetical building blocks), but also opens the door to innovative implementations of approximate computing (e.g. Analog current summations etc.) as well as non-CMOS methods.

Communication from the Edge: Inter-device communication between IOT devices is another major contributor to power consumption. Data transmitted from one device to another impacts the application's power consumption. This is true, not only in the sender device but also the receiver device. Furthermore, quantity of data in transfer impacts memory storage sizing and thus leakage too. The further data has to flow from its source, the more compounded the increase in energy is. For this reason, local computation as close to the source of the data is important. Data traffic reduction can only happen by filtering of redundant data by levels of importance/relevance. AI can play a major role in this filtering. Whereas traditional models for processing data are based on pre-determined characteristics (or mathematical models), to produce precise results, AI does allow a form of 'intuition' and approximate context recognition. In fact, this is not dissimilar to human perception of data. We do not look at all details, and we allow ourselves to make an initial observation which may be wrong, as long as, within a short space of time, we correct our perception based on wider context.

Optimizing Power at the Edge: This all means that, apart from the aforementioned consumption impact, the AI's 'smart' approximation capability is not only part of the power consumption problem, it can also be part of the energy reduction solution by the Edge optimizing amount of data to and from the cloud. In addition to the data reduction, there are other areas where AI can help. For example, in smarter power cycling (e.g. learning when to wake up, or learning how accurate to be in sensing). Context-aware power delivery offers an opportunity to find dynamically changing sweet spots for power efficiency. Traditionally, power delivery has been rigidly pre-programmed. In some cases, reactive cycling can be very useful for coping with power increase steps. Nonetheless, in order for AI to be an effective tool for power reduction, there needs to be a focus on not only approximate computing, but also robust, allowed-to-fail, computing. Having the opportunity to sometimes fail and correct can significantly reduce power consumption through a more aggressive removal of margins.

H. SENSORS: MAKING IOT AWARE

IoT nodes often form part of a cyber-physical feedback system depicted in Figure 8 enabling self-regulated, near-autonomous operation of large and complex systems. Many examples exist in Smart Building (indoor climate control, adaptive lighting), Automotive (autonomous driving), e-Health (e-pill, patient monitoring and medicine delivery), and Industry4.0 (self-optimizing logistic chains, remote monitoring and maintenance of manufacturing lines). The deployment of many IoT nodes uniquely allows for networking of sensor and actuator conglomerates that are spread out over space and detecting and manipulating various physical attributes (e.g. distance, temperature, pressure, light intensity, air flow, magnetic or electric field). The sensor data is processed, events are classified, actuator settings are updated using a combination of edge processors and remote cloud servers capable of running advanced self-learning, data- and sensor fusion algorithms.

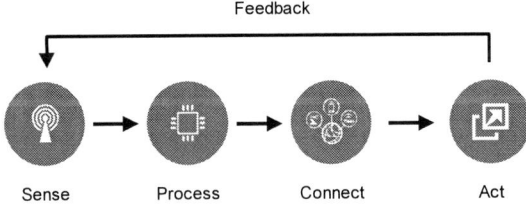

Feedback

Sense   Process   Connect   Act

Fig. 8. The IoT node as part of a cyber-physical feedback loop as applied in Industrial, Smart Home, e-Health, and Automotive systems.
Source: NXP

In many of cases, a human-being will be part of the IoT feedback loop depicted in Fig. 8, as is the case in gaming, automotive driver assistance, and many use-cases involving drones and co-bots. The amount of accessible data enabled by cheap and numerous IoT devices available to the end-user can be substantial. To process and steer this flow of information requires sophisticated and intuitive Human-Machine Interfaces (HMI) both in terms of sensing (e.g. haptics, vision), as well as actuation (voice activation, gesture recognition). Ultimately leading to multi-modal augmented reality HMI involving various human senses. This need for sophisticated HMI's is also in accordance with the evolution of HMIs that has taken place over many decades, as is depicted in Figure 9. Most notably the recent large-scale deployment of voice control underscores this trend and has been enabled to large extend by IoT edge nodes (such Smart Speakers, Smart Watches) providing data

to advanced voice algorithms running on remote cloud servers. This trend will continue to more advanced user-interfaces and human centered services involving person localization, navigation, gesture detection and intuitive visual, auditory, and haptic feedback.

Sensors for IoT will need to be deployed in large numbers, so need to be low cost, easy to install and maintain, and low power. Furthermore, their output needs to be digitized to allow for digital signal processing, as is required for networking, feature extraction and event classification. In many ways, these requirements are identical to the development of integrated circuits (ICs). Therefore, IoT sensors and actuators are, whenever possible, designed and batch manufactured in a similar fashion to CMOS ICs. This also explains the success of silicon-based sensors such as MEMS inertial sensors, MEMS microphones, CMOS mm-wave radar, Hall magnetic sensors, PTAT temperature sensors, and IR photo-diodes. It is therefore expected that these types of sensors will be adopted in IoT edge nodes.

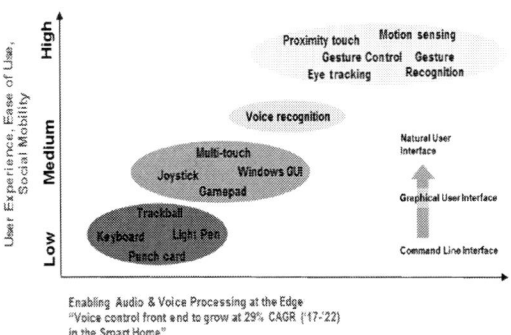

Fig. 9. Evolution of intuitive Human-Machine interface for IoT edge nodes.

I. BIOMETRICS: A CASE STUDY FOR THE EDGE.

The following is a use case where computing at the edge has significant benefit for multiple reasons. A biometric edge device has significant advantages over Local and Cloud processing.

Local processing (Figure 10: Local Biometric Processing): as a low-end IOT node because it misses both ability to communicate with the cloud or the computing power of Edge device. (1) The limited processing and memory available in low end IoT nodes makes it difficult to process complex biometrics like face recognition and speaker authentication. Moreover, the biometrics are limited to user verification (i.e. the system checks on whether the user is the one he pretends to be against one biometric profile). Fingerprint recognition is an example of where Local processing may be adequate. (2) The security of the local processing is also limited if no secure subsystem is added or no embedded secure element. And if such secure subsystems are added, they even more limited in processing and memory capabilities[8].

978-1-7281-0582-6/19 $31.00 © 2019 IEEE   21

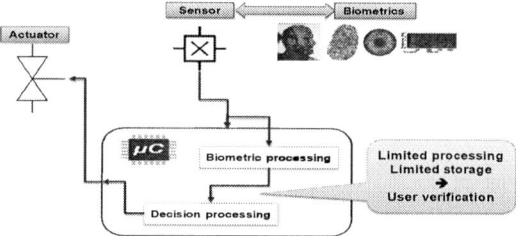

Fig. 10. Local Biometric Processing

Cloud processing (Figure 11: Cloud Biometric Processing) suffers from several key issues with regards to biometrics. (1) The network connection to the cloud must be available permanently and must have a high bandwidth because of the volume of data needed for some biometrics processing and for the billions of devices that will be connected to such a cloud service. (2) The bare sensor data is available to the cloud service for all users worldwide of the service ➔ this implies a big issue with privacy preserving regulations and expectations. (3) The security is in the cloud and the local device connected to the cloud is exposed to remote attackers (they come in through the same connection as the one to the cloud that is permanently available). This implies that the processing power required by the cloud will be "huge" along with a strain on cloud due to communication bandwidth, cloud processing and storage.

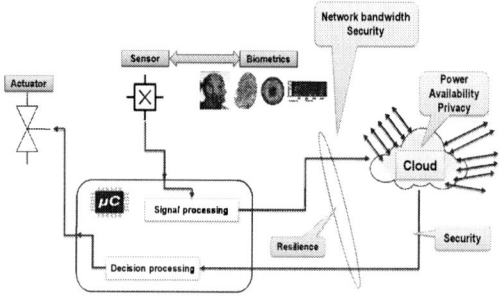

Fig. 11: Cloud Biometric Processing

Edge processing (Figure 12: Edge Biometric Processing) is the best solution for biometrics for the following reasons. This solution has the best of all worlds. It reduces power, storage, computing, communication bandwidth and power on the cloud. Other reasons are:

(1) More resilient: The architecture does not depend on the permanent availability of the cloud service and the connection to the cloud service.

(2) Less power consumption: The equation is that if you do the biometric matching in the cloud, you need the power for the gateway to the cloud anyway; this power in the cloud is an additional factor while if you do it in the local edge gateway that is powered for many other local reasons, the cost of the biometric matching is marginal.

(3) Privacy preserving: The personal data remains beyond a perimeter under the control of the user.

(4) Improved security: The local management of the connection is isolated from the outside world. If the gateway/edge is doing its job properly but having one gateway well protected is easier than having all IoT devices

protected in a Smart Home; not all of them will have the resources to protect themselves against remote attacks. The decision process based on the biometric matching remains local to the house and cannot be influenced from outside of the house. The wall between the edge and the cloud can be a strong one.

(5) User identification: Moving from Local processing to Edge processing, you can encompass user identification (e.g. identifying a user amongst a set of users).

(6) More complex biometrics: We can envision face recognition but also maybe speaker authentication and identification with anti-spoofing techniques.

(7) Improved multi-factor authentication: It is known that due to spoofing attacks, one biometric is not enough and that the future solutions will be based on multi-factor authentication. The Edge can manage this multiplicity.

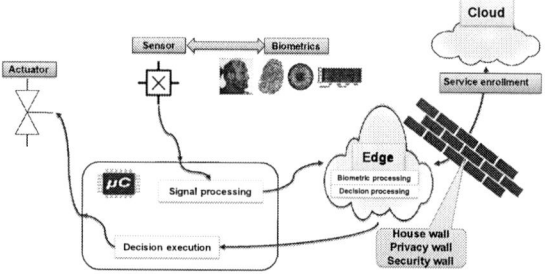

Fig. 12: Edge Biometric Processing

III. CONCLUSIONS

The importance of doing more at the Edge of IOT is creating revolution on how we sense and secure our world. Computing moving to the Edge is solving power, security, data and communication issues in the cloud. Sensors are being expected to do more to enhance HMI. The applications are endless. There is still much innovation needed around Power at the edge needed to manage the power needed by 75B IOT devices in 2025. AI and clever power techniques which reduce device on time, leakage and decrease latency.

J. REFERENCES

1. Ken Hansen, David Yeh and Victor Zhirnov, The Case for a Decadal Plan for Semiconductors(2030 SRC research goals).

2. Fuks, A. "Sensor-hub sweet-spot analysis for ultra-low-power always-on operation," *2015 Symposium on VLSI Circuits (VLSI Circuits)*, Kyoto, 2015, pp. C154-C155.

3. Meijer, M., et.al. "Ultra-Low-Power Digital Design with Body Biasing for Low Area and Performance-Efficient Operation," JOLPE, Vol.6, No. 4, 2011, pp. 1-12

4. Dreslinski, R. Wieckowski, M. Blaauw, D. Sylvester, D. Mudge, T. "Near Threshold Computing: Overcoming Performance Degradation from Aggressive Voltage Scaling." IEEE, 2019

5. Makoto Takamiya, "Energy efficient design and energy harvesting for energy autonomous systems", VLSI Design, Automation and Test (VLSI-DAT), 2015

6. Le, Long N. Jones, D., "Guided-Processing Outperforms Duty-Cycling for Energy-Efficient Systems", https://arxiv.org/abs/1705.00615

7. https://www.theatlantic.com/technology/archive/2012/11/noam-chomsky-on-where-artificial-intelligence-went-wrong/261637/

8. "From the Internet of Things to the Internet of Trust" by NXP.,https://www.nxp.com/docs/en/white-paper/NXP-FROM-IOT-TO-IOTRUST-WP.pdf

Proceedings of the 31st International Symposium on Power Semiconductor Devices & ICs
May 19-23, 2019, Shanghai, China

Experimental Demonstration on Superior Switching Characteristics of 1.2 kV SiC SWITCH-MOS

Ruito Aiba, Masataka Okawa*, Taiga Kanamori,
Hiroshi Yano*, and Noriyuki Iwamuro*
College of Engineering Sciences
*Graduate School of Pure and Applied Sciences
University of Tsukuba
Tsukuba, Japan
s1510957@s.tsukuba.ac.jp
iwamuro.noriyuki.fb@u.tsukuba.ac.jp

Yusuke Kobayashi, Shinsuke Harada
National Institute of Advanced Industrial Science and
Technology (AIST)
Tsukuba, Japan

Abstract—**A 1.2kV silicon carbide (SiC) SBD-wall-integrated trench MOSFET (SWITCH-MOS) had been proposed and fabricated in order to solve body-PiN-diode related problems such as bipolar degradation and reverse recovery loss. In this paper, switching characteristics of turn-on and turn-off of the SWITCH-MOS are investigated and discussed in comparison with conventional trench MOSFET structures with p+ region at trench gate bottom (IE-UMOSFET) and without the p+ region (named "device A" in this paper). The SWITCH-MOS shows an extremely small turn-on loss and turn-off loss because of the high *dV/dt* and little reverse recovery current. In addition, it is found that the SWITCH-MOS exhibits smaller turn-off loss than "device A" at almost the same *dJ/dt*, which means that drain surge voltage was suppressed effectively while keeping the smaller turn-off loss.**

Keywords—Bipolar degradation, SWITCH-MOS, Switching loss, Reverse transfer capacitance, Reverse recovery current

I. INTRODUCTION

In the high power applications, silicon (Si) power devices are currently the most commonly used switching devices. However, Si power device technology is relatively mature, and it is not easy to achieve innovative breakthroughs using this technology. Silicon carbide metal-oxide-semiconductor filed-effect transistor (SiC-MOSFET) attracts more attention as a candidate replacing Si power devices owing to the excellent material properties such as its high breakdown electric field and high thermal capability. Moreover, SiC-MOSFET is expected to eliminate external Schottky Barrier Diode (SBD) by means of the parasitic body-PiN-diode, resulting in reduction of the total chip area and high power density. However, in the case of body-PiN-diode of SiC-MOSFET, the increase in the forward voltage known as the bipolar degradation is a serious problem. Embedding SBD in parallel with SiC-MOSFET is one of the proposed approaches to solve the problem [1-7], but causes increasing a specific on-resistance owing to reducing a channel density and introducing extra parasitic capacitance, which read to increase switching loss.

For the purpose of both suppressing bipolar degradation and reducing total chip area, a 1.2kV SiC SBD-wall-integrated trench MOSFET (SWITCH-MOS) was proposed and fabricated [8]. The SiC SWITCH-MOS is a device structure of trench-SBD-integrated trench MOSFET with a small cell pitch, so this device shows the low on resistance. This device was designed to improve its switching characteristic by reducing reverse recovery loss in its turn-on operation. Furthermore, it is so beneficial because MOSFET and SBD share not only the forward conduction layer but also the edge termination regions, resulting in a significant reduction of SiC wafer area. In this paper, switching characteristics of turn-on and turn-off of the SWITCH-MOS are investigated and discussed through the analysis of switching waveforms and the reverse transfer capacitance (Crss) characteristics in comparison with conventional trench MOSFET structures with p+ region at trench gate bottom (IE-UMOSFET [9]) and without the p+ region (named "device A" in this paper).

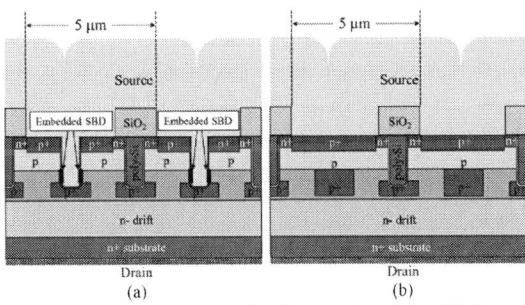

Fig. 1 Schematic cross section of (a) the SWITCH-MOS and (b) the IE-UMOSFET.

Table I. Measured results of static and dynamic characteristics of the SWITCH-MOS, the IE-UMOSFET and the "device A".

	SWITCH-MOS	IE-UMOSFET	device A
Die size	3.0×3.0 mm^2		4.4×3.0 mm^2
Breakdown voltage (V, @I_d = 100 μm)	1576	1605	1746
Thresholad voltage (V, @R.T.)	3.7	3.2	4.6
Specific on resistance (mΩ cm^2, @V_{gs} = 20 V, R.T.)	3.3	3.2	2.8
Turn-on loss (mJ/cm^2, @175°C)	18.5	26.2	49.5
Turn-off loss (mJ/cm^2, @175°C)	6.3	7.1	17.2
Crss (pF/cm^2) (@V_{ds}=800 V, 100 kHz)	519	53	490

978-1-7281-0582-6/19 $31.00 © 2019 IEEE

II. EXPERIMENTAL RESULTS AND DISCUTTION

Figure 1 shows schematic cross sections of the SWITCH-MOS and the IE-UMOSFET with a cell pitch of 5 μm. Its die size was designed of 3×3 mm². The SWITCH-MOS and the IE-UMOSFET have an exactly same structure and dimensions except for the presence of embedded trench SBD. Static and switching characteristics of the SWITCH-MOS, the IE-UMOSFET, and the "device A" are measured and summarized in Table 1. The SWITCH-MOS shows almost the same static characteristics with the IE-UMOSFET and the "device A". Figure 2 shows measured forward I-V and breakdown characteristics of the SWITCH-MOS. Switching characteristics are measured using a test bench as shown in Fig.3. Drain currents for these measurements were set of 25 A in the SWITCH-MOS and the IE-UMOSFET and 50 A in the "device A" to equalize current density J ($J\sim 460$ A/cm²).

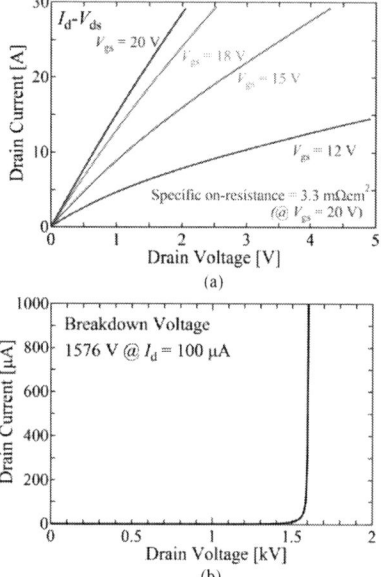

Fig. 2. Measured waveforms of (a) I_d-V_{ds} and (b) breakdown voltage in the SWITCH-MOS.

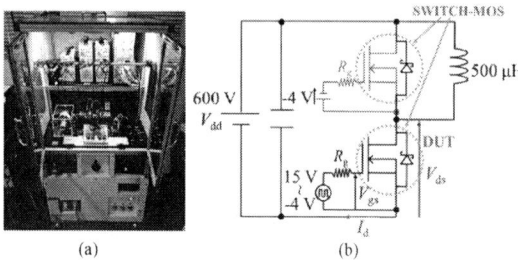

Fig. 3. (a) Test bench and (b) equivalent circuit for switching characteristics measurements. Internal diodes of MOSFETs in the upper arm (SBDs in the SWITCH-MOS, pn diodes in the IE-UMOSFET and the "device A") were used as free-wheeling diodes.

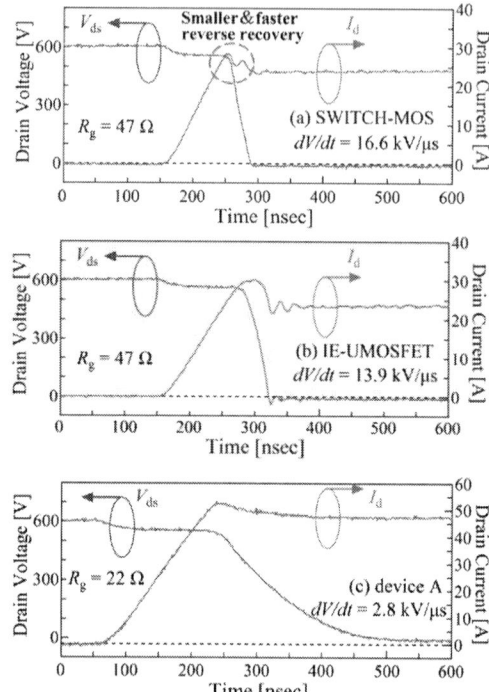

Fig. 4. Comparison of the measured turn-on waveforms (@175°C). In (c), gate resistance R_g was set to 22 Ω in order to adjust the difference of active area. (V_{gs} = +15 V/-4 V)

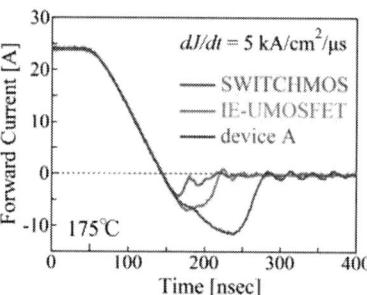

Fig. 5. Comparison of reverse recovery waveforms of the body diode as a free-wheeling diode at the same dJ/dt of 5 kA/cm²/μsec.

Figure 4 shows a comparison of turn-on waveforms at a high temperature of 175°C. It is noted that the SWITCH-MOS exhibits a fast turn-on characteristics and smaller turn-on loss by about 29% than that of the IE-UMOSFET whereas the dV/dt was almost the same value as high as 14.0-16.0 kV/μsec. Further, the SWITCH-MOS shows an extremely smaller turn-on loss by about 63% than that of the "device A".

One main reason is little recovery current which can be seen in the drain current waveform of the SWITCH-MOS. Figure 5 shows a comparison of reverse recovery waveforms of the body diode as a free-wheeling diode at the same dJ/dt of 5 kA/cm²/μsec. Owing to unipolar operation of the embedded SBD, reverse recovery current in the SWITCH-MOS is sufficiently suppressed even in a high temperature of 175°C. Figure 6 shows a comparison of forward I-V characteristics of the built-in diodes in the SWITCH-MOS and the IE-UMOSFET. It is found from this figure that a forward voltage drop in the body-PiN-diode of the IE-UMOSFET decreases by about 0.4 V (@25 A) in 175°C owing to the enhanced minority carrier injection and lifetime, resulting in high reverse recovery current.

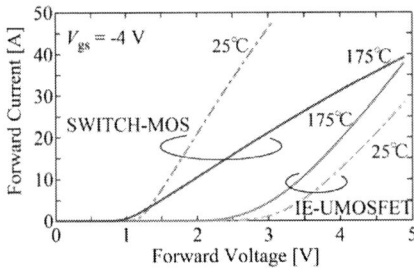

Fig. 6. Comparison of I-V characteristics of the SWITCH-MOS and the IE-UMOSFET built-in diodes. A forward voltage drop in the body-PiN-diode of the IE-UMOSFET is 4.7 V (@25 A) in 25 °C and 4.3 V (@25 A) in 175°C.

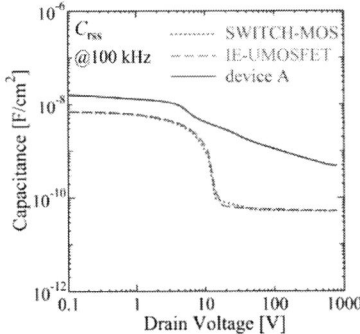

Fig. 7. Comparison of Crss characteristics of the SWITCH-MOS, the IE-UMOSFET and device A

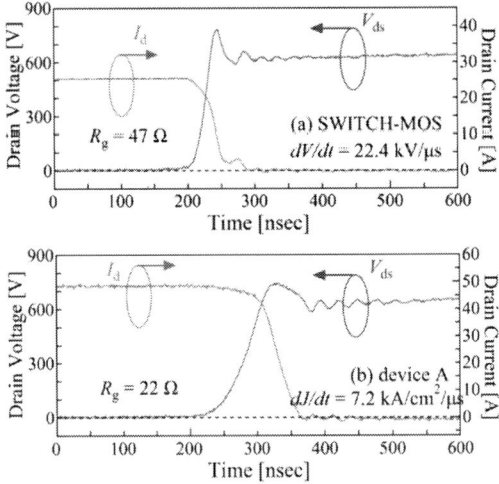

Fig. 8. Comparison of measured turn-off waveforms between the SWITCH-MOS and the "device A" (@175°C). In (b), gate resistance R_g was set to 22 Ω to adjust the difference of active area. (V_{gs} = +15 V/-4 V)

The other one is the extremely high dV/dt. Recently, improved switching characteristics of SiC trench MOSFETs with much smaller Crss were reported [10, 11]. Figure 7 shows the comparison of measured Crss characteristics of these devices. The Crss values of the SWITCH-MOS and the IE-UMOSFET structure are about 52 pF/cm² at drain voltage of 800V, which are approximately one-tenth of "device A". There are p+ regions at trench gate bottom which are connected to the source electrode in the SWITCH-MOS and the IE-UMOSFET, resulting in smaller facing area between the gate and drain electrodes; therefore, the fast turn-on characteristics are attained in the SWITCH-MOS and the IE-UMOSFET. In case of turn-off switching, this smaller Crss leads to very high dV/dt of about 22 kV/μsec in turn-off waveforms resulting in fast switching characteristics and less turn-off loss (see Table 1 and Fig. 8).

Figure 9 shows the comparison of turn-off loss dependence on dJ/dt between the SWITCH-MOS and the "device A". In order to apply SiC MOSFETs for many power electronics systems, such as a motor drive inverter and a DC-DC converter, it is necessary to control switching speed and suppress a drain surge in turn-off drain voltage waveforms, so turn-off loss dependence of dJ/dt is very important. As can be seen in Fig. 10, the turn-off loss of the SWITCH-MOS is smaller than that of the "device A" at the same dJ/dt. This is because of an extremely high dV/dt owing to a small Crss in the SWITCH-MOS. Figure 10 shows a result of turn-off loss analysis of the SWITCH-MOS and the "device A" at approximately 7 and 10 kA/cm²/μsec as dJ/dt. In this analysis, a turn-off waveform was divided into the two regions, "region A" and "region B" as shown in Fig.10. The "region A" corresponds to first half from time zero to a point when the drain current falls to 90% of load current of 25 A (50 A in the "device A") and the "region B" to second half after the point. It is clear from this figure that the SWITCH-MOS exhibits much smaller dissipated loss in "region A" whereas the dissipated loss in "region B" is almost the same each other. This means that the SWITCH-MOS can successfully reduce its turn-off loss in the first half of the turn-off operation while keeping the same value of dJ/dt with the "device A" in the second half. It should be noted that the trade-off characteristics between the turn-off loss and dJ/dt was improved by the smaller Crss in the SWITCH-MOS, which means that drain surge voltage was successfully suppressed effectively while keeping the smaller dissipated turn-off loss as shown in Fig. 11.

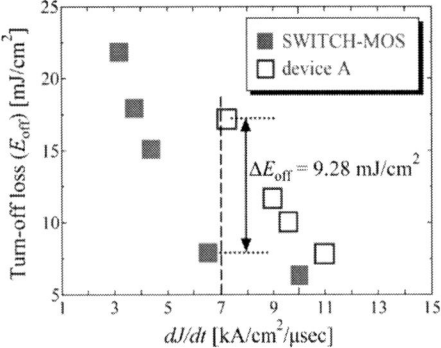

Fig. 9. Comparison of turn-off loss dependence on dJ/dt (@175°C). E_{off} can be successfully reduced by 9.28 mJ/cm² at almost the same dJ/dt of 7 kA/cm²/μsec. Values of R_g were changed to control the dJ/dt in this measurement.

978-1-7281-0582-6/19 $31.00 © 2019 IEEE

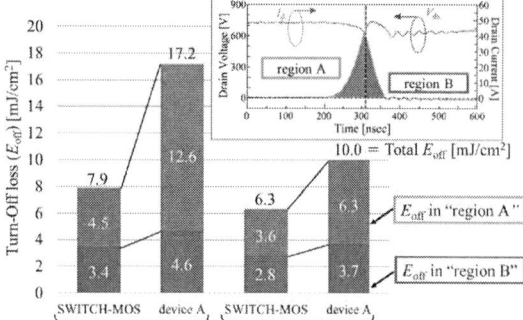

Fig. 10. Turn-off loss components of the SWITCH-MOS and the "device A" at the almost the same dJ/dt of 7 and 10 kA/cm^2/μsec. Turn-off waveform was divided into the two parts, "region A" and "region B" in this paper.

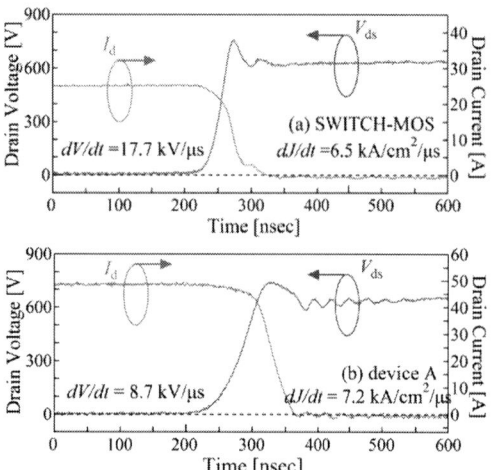

Fig. 11. Comparison of measured turn-off waveforms between the SWITCH-MOS and the "device A" at almost the same dJ/dt of 7 kA/cm^2/μsec. (@175°C)

III. SUMMARY

In this paper, switching characteristics of turn-on and turn-off of the SWITCH-MOS are investigated and discussed through the analysis of switching waveforms and the reverse transfer capacitance (Crss) characteristics comparison with conventional trench MOSFET structures, the IE-UMOSFET and the "device A". The SWITCH-MOS exhibits fast switching characteristics and smaller turn-on loss by about 29% than that of the IE-UMOSFET and by about 63% than that of the "device A" because of fast switching speed by small Crss and little reverse recovery current owing to embedded SBD unipolar recovery operation. In turn-off operation, the SWITCH-MOS shows very fast switching characteristics and less turn-off loss. Furthermore, the SWITCH-MOS has a superior turn-off loss dependence of dJ/dt; therefore, by utilizing the SWITCH-MOS, the extremely higher frequency inverter can be realized for possible applications.

ACKNOWLEDGMENT

A part of this paper has been implemented under a joint research project of Tsukuba Power Electronics Constellations (TPEC). Then, co-author of Yusuke Kobayashi is assigned from Fuji Electric Co., Ltd.

REFERENCES

[1] W. Sung and B. J. Baliga, "Monothithically Integrated 4H-SiC MOSFET and JBS Diode (JBSFET) Using a Single Ohmic/Schottky Process Scheme," *IEEE Electron Devices Lett.* vol. 37, no. 12, 2016, pp. 1605-1608.

[2] H. Jiang, J. Wei, X. Dai, M. Ke, C. Zheng, and I. Deviny, "Silicon Carbide split-gate MOSFET with merged Schottky barrier diode and reduced switching loss, in *Proc. Int. Symp. Power Semiconductors and ICs,* Jun. 2016, pp.59-63.

[3] K. Kawahara, S.Hino, K.Sadamatsu, Y.Nakao, Y.Yamashiro, Y.Yamamoto, T.Iwamatsu, S.Nakata, S.Tomohisa, and S.Yamakawa, "6.5 kV Schottky-barrier-diode-embedded SiC-MOSFET for compact full-unipolar module," in *Proc. Int. Symp. Power Semiconductors and ICs,* May 2017, pp.41-44

[4] S. Hino, H. Hatta, K. Sadadamatsu, Y. Nagahisa, S. Yamamoto, T. Iwamatsu, Y. Yamamoto, M. Imaizumi, S. Nakata, and S. Yamakawa, "Demonstration of SiC-MOSFET Embedding Schottky Barrier Diode for Inactivation of Parasitic Body Diode", *Material Science Forum,* vol.897, pp.477–482, (2017).

[5] F. J. Hsu, C. T. Yen, C. C. Hung, H. T. Hung, C. Y. Lee, L. S. Lee, Y. F. Huang, T. Z. Chen, P. J. Chuang, "High efficiency high reliability SiC MOSFET with monolithically integrated Schottky rectifier", *Proc. Int. Symp. Power Semiconductors and ICs,* May 2017, pp. 45–48.

[6] H. Jiang, J. Wei, X. Dai, C. Zheng, M. Ke, X. Deng, Y. Sharma, I. Deviny, and P. Mawby, "SiC MOSFET with built-in SBD for reduction of reverse recovery charge and switching loss in 10-kV applications," in *Proc. Int. Symp. Power Semiconductors and ICs,* May 2017, pp.49-52.

[7] A Kanale, K.Han, B.J.Baliga, and S.Bhattachaya, "Superior short circuit performance of 1.2 kV SiC JBSFETs compared to 1.2 kV SiC MOSFETs, in *Abstract of European Conference on Silicon Carbide and Related Materials 2018,* Sept. 2018, ID1364.

[8] Y. Kobayashi, N. Ohse, T. Morimoto, M. Kato, T. Kojima, M. Miyazato, M. Takei, H. Kimura, and S. Harada, "Body PIN diode inactivation eith low on-resistance achieved by a 1.2 kV-class 4H-SiC SWITCH-MOS," in *IEEE IEDM Tech., Dig.,* 2017, 9.1.1, pp.211-214.

[9] S. Harada, K. Kobayashi, K. Kinoshita, N. Ose, T. Kojima, M. Iwaya, H. Shiomi, H. Kitai, S. Kyogoku, K. Ariyoshi, Y. Onishi, and H. Kimura, "1200 V SiC IE-UMOSFET with Low On-Resistance and High Threshold Voltage," *Material Science Forum,* vol.897, pp.497-500, (2017).

[10] S. Kyogoku, K. Tanaka, K. Ariyoshi, R. Iijima, Y. Kobayashi, and S. Harada, "Role of Trench Bottom Shielding Region on Switching Characteristics of 4H-SiC Double-Trench MOSFETs," *Material Science Forum,* vol.924, pp.748-751, (2018).

[11] Y. Ebihara, A. Ichimura, A. Mitani, M. Noborio, Y. Takeuchi, S. Mizuno, T.Yamamoto, and K. Tsuruta, "Deep-P Encapsulated 4H-SiC Trench MOSFETs With Ultra Low $RonQgd$", in *Proc. Int. Symp. Power Semiconductors and ICs,* May 2018, pp. 44-48.

Proceedings of the 31st International Symposium on Power Semiconductor Devices & ICs
May 19-23, 2019, Shanghai, China

Superior Switching Characteristics of SiC-MOSFET Embedding SBD

Takaaki Tominaga, Shiro Hino, Yohei Mitsui, Junichi Nakashima, Koutarou Kawahara, Shingo Tomohisa, and Naruhisa Miura
Advanced Technology R&D Center, Mitsubishi Electric Corporation, Amagasaki, HYOGO, JAPAN,
Tominaga.Takaaki@dy.MitsubishiElectric.co.jp

Abstract—**Superior switching characteristics of SiC-MOSFET embedding SBD is demonstrated compared with conventional MOSFET and MOSFET with external SBD. Inactivation of parasitic body diode by embedding SBD enables a suppression of recovery charge during turn-on process, which results in a reduction of turn-on loss. Furthermore, elimination of external SBD reduces total chip size, or output capacitance charge, which results in a reduction of output capacitance loss.**

Keywords—SiC-MOSFET, SBD, switching characteristics

I. INTRODUCTION

Silicon carbide is a promising candidate for future power electronics due to its superior properties such as large band gap and high thermal conductivity compared with silicon. Although the parasitic body diode of SiC-MOSFET ideally plays a role for reverse conduction as free-wheeling diode, it is widely recognized that current conduction of pn diode of SiC causes reliability degradation due to expansion of stacking faults by recombination processes of electron-hole pairs, so called bipolar degradation. In order to avoid the reverse current conduction through the parasitic body diode, anti-paralleled external Schottky barrier diodes (SBD) are generally introduced. However, since the voltage drop through the external SBD needs to be lower than the built-in potential of the pn junction in order to suppress the body diode conduction of the paralleled MOSFET, the size of an external SBD chip is 1.5 times larger than that of a 3.3 kV MOSFET, assuming rated current density of 100 A/cm² for the MOSFET and specific resistance of 22 mΩcm² for the

SBD at 175 °C [1]. By embedding an SBD into each unit cell of a MOSFET, the chip size of only 1.15 times larger than that of a conventional MOSFET has been achieved without any activation of parasitic body diode [1, 2]. However, switching characteristics of SiC-MOSFET embedding SBD have been seldom reported, although they are also essential elements for the application of power device.

In this paper, we have investigated the switching characteristics of SiC-MOSFET embedding SBD, and revealed the superior performance to conventional MOSFET including body diode and MOSFET with external SBD.

II. EXPERIMENTAL

Schematic behaviors of (a) a MOSFET embedding SBD and (b) a conventional MOSFET coupled with an external SBD are depicted in Fig. 1. The applied voltage of each pn junction is slightly lower than its built-in potential ($V_{built-in}$), which means the parasitic body diode is not activated. As shown in Fig. 1 (a), voltage drops at Schottky contact (V_K) and at the JFET region below SBD contact due to its specific resistance (R_S) are applied to the pn junction. This behavior is clearly different from that of the conventional MOSFET coupled with external SBD, in which the whole voltage drop of the external SBD is applied to the pn junction of the conventional MOSFET as shown in Fig. 1 (b), resulting in the necessity of large external SBD.

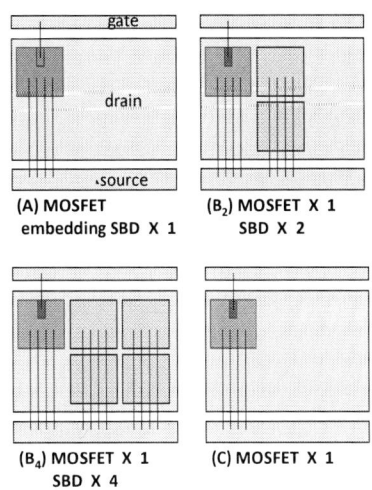

Fig. 2. Evaluated device combinations,
(A) a MOSFET embedding SBD, (B₂) a MOSFET with 2 SBDs, (B₄) a MOSFET with 4 SBDs, and (C) a MOSFET.

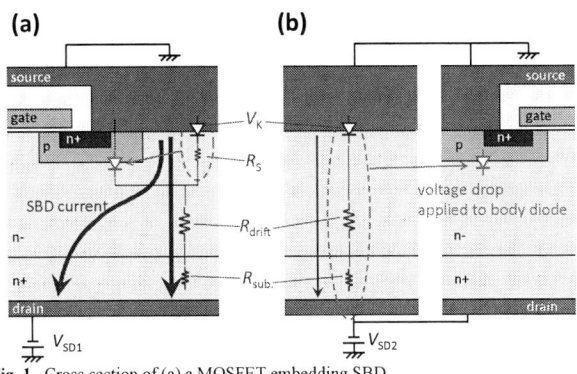

Fig. 1. Cross section of (a) a MOSFET embedding SBD,
(b) a conventional MOSFET coupled with an external SBD.

978-1-7281-0582-6/19 $31.00 © 2019 IEEE   27

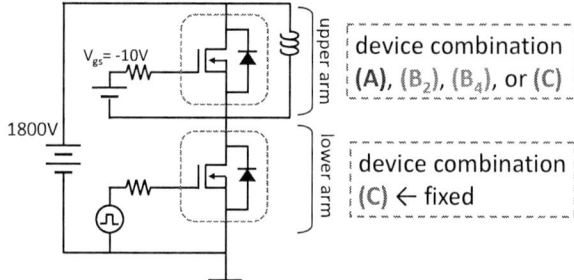

Fig. 3. Configuration of measurement circuit.

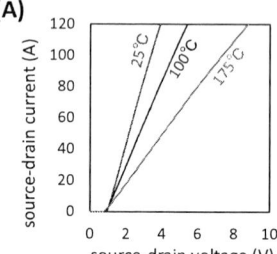

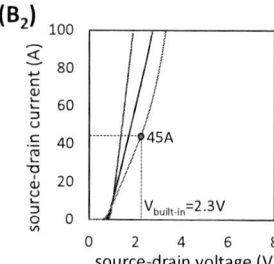

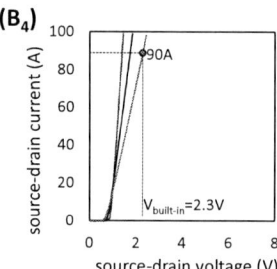

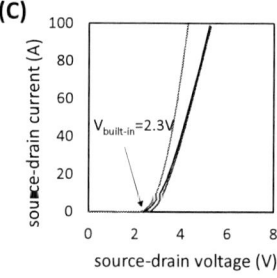

Fig. 4. Diode characteristics of each device combination at off-state at 25 ℃ (blue), 100 ℃ (black), and 175 ℃ (red).

In this study, we investigated four device combinations as shown in Fig. 2, which are (A) a MOSFET embedding SBD, (B₂) a MOSFET with two external SBDs, (B₄) a MOSFET with four external SBDs, and (C) a MOSFET. The chip size of each device is almost same and the rated voltage is 3.3 kV.

We evaluated the influence of the diode characteristics of each device combination on the switching performance with circuit connections shown in Fig. 3. The transistor of the upper arm is kept off during the measurement and each device combination works as the free-wheeling diode. The switching transistor of the lower arm is always (C) a conventional MOSFET in order to distinguish the difference of the diode characteristics of the upper arm.

III. RESULTS AND DISCUSSIONS

The diode characteristics of each device combination at off-state were evaluated at 25, 100, 175 ℃ in Fig. 4. As described in Fig. 1, the current conduction through the parasitic body diode of the conventional MOSFET occurs when the applied voltage to source-drain exceeds $V_{\text{built-in}}$, while the parasitic body diode of the MOSFET embedding SBD is not activated at $V_{\text{built-in}}$. The maximum current (I_{UCmax}) without parasitic body diode activation of (A), (B₂), and (B₄) at 175 ℃ were estimated as >120 A, 45 A, and 90 A, respectively.

Fig. 5 shows the turn-on waveforms of upper arm and lower arm. The large reverse recovery current of the upper arm diode (B₄) and (C) is observed, which derives from the large total recovery charge (Q_{tot}). The large recovery current due to Q_{tot} of the upper arm, which is around 3.6 µC for (C) as an example, causes the excessive current of the lower arm transistor, resulting in the increase of turn-on loss (E_{on}).

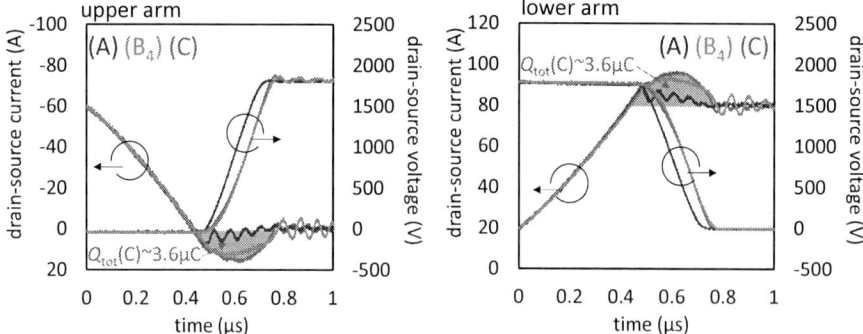

Fig. 5. Turn-on waveforms of upper arm and lower arm at 80 A at 175 ℃.

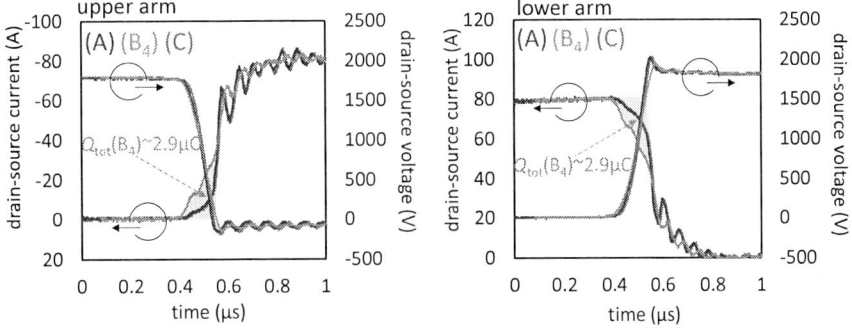

Fig. 6. Turn-off waveforms of upper arm and lower arm at 80 A at 175 °C.

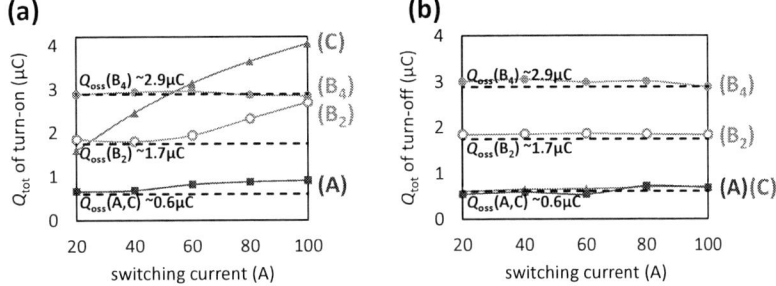

Fig. 7. Dependence of total recovery charge (Q_{tot}) for turn-on (a) and turn-off (b) on switching current at 175 °C (solid line), and calculated output capacitance charge (broken line).

The turn-off characteristics were also investigated as shown in Fig. 6. The displacement current due to Q_{tot} of the upper arm, which is around 2.9 µC for (B$_4$) as an example, reduces the switching current of the lower arm transistor, resulting in the decreased turn-off loss (E_{off}).

We evaluated Q_{tot} of turn-on and turn-off characteristics as a function of switching current for each device combination as shown in Fig. 7. If the switching current of turn-on is small enough to inactivate the body diode, Q_{tot} is same as output capacitance charge (Q_{oss}) of the upper arm. In contrast, excessive charge over Q_{oss} means the existence of recovery charge (Q_{rec}) due to minority carrier injection by body diode activation. For example, Q_{tot} of (B$_2$) exceeds the estimated Q_{oss} of (B$_2$) at a certain switching current, while Q_{tot} of (B$_4$) or (A) doesn't. On the other hand, as shown in Fig. 7 (b), Q_{tot} is same as Q_{oss} independent of switching current due to inexistence of Q_{rec} during turn-off procedure.

E_{on} and E_{off} of the lower arm transistor (C) with the upper arm diode (A), (B$_4$), (C) at 175 °C are evaluated as a function of switching current in Fig. 8. E_{on} of the lower arm transistor with (B$_4$) upper arm is larger than that with (A) upper arm due to the larger Q_{oss}, which coincides with smaller E_{off} of (B$_4$) upper arm compared with (A) upper arm. On the other hand, (C) upper arm shows larger E_{on} than that of (A) upper arm because of Q_{rec}, while E_{off} shows almost the same value. It should be also noted that large Q_{rec} could induce ringing oscillation [3], and large Q_{oss} increases output capacitance loss (E_{oss}), which stems from charging and discharging process of output capacitance of lower arm device.

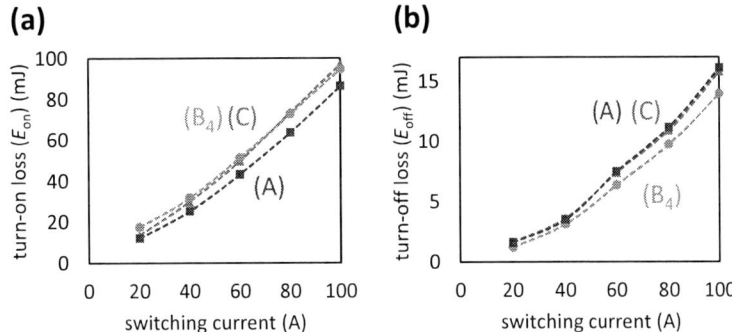

Fig. 8. Turn-on loss (E_{on}) and turn-off loss (E_{off}) of lower arm transistor (C) at 175 °C.

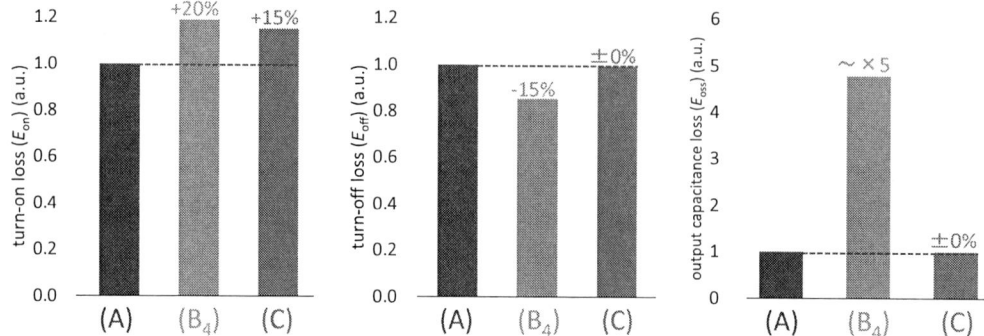

Fig. 9. Comparison of measured turn-on loss (E_{on}) and turn-off loss (E_{off}) at 60 A at 175 °C, and calculated output capacitance loss (E_{oss}).

TABLE I. Summary of advantages and disadvantages of SiC-MOSFET embedding SBD.

	(A) MOSFET embedding SBD	(C) MOSFET (w/o SBD)	○ advantage / × disadvantage of MOSFET embedding SBD
chip size (a. u.)	1	1	-
I_{UCmax} @175°C	> 120A	0A	○ free form bipolar degradation
Q_{rec}	0 (I_{sd} < 120A)	Large	○ smaller E_{on}
Q_{oss} (a. u.)	1	1	-

	(A) MOSFET embedding SBD	(B_n) MOSFET + SBD x n (n=2, 4)	○ advantage / × disadvantage of MOSFET embedding SBD
chip size (a. u.)	1	1 + n	○ drastically smaller chip
I_{UCmax} @175°C	> 120A	n x 22.5A	-
Q_{rec}	0 (I_{sd} < 120A)	0 (I_{sd} < n x 22.5A)	-
Q_{oss} (a. u.)	1	1 + n	○ smaller E_{on} × larger E_{off} ○ smaller E_{oss}

These results clarify the impact of both large Q_{oss} due to large external SBD and Q_{rec} due to body diode conduction on the increase of E_{on}, and thus the superior switching performance is achieved by SiC-MOSFET embedding SBD.

Fig. 9 shows the comparison of measured E_{on} and E_{off} for each device combination of upper arm at switching current of 60 A at 175 °C as well as calculated E_{oss} for each device combination of lower arm. E_{on} with (A) upper arm is smaller than that with (B_4) upper arm due to smaller Q_{oss}. E_{on} with (A) upper arm is smaller than that with (C) upper arm due to elimination of Q_{rec}, while E_{off} is almost same. E_{oss} with (B_4) lower arm is around five times larger than that with (A) or (C) lower arm due to larger Q_{oss}.

Table I summarizes the superior performance of SiC-MOSFET embedding SBD described in this paper as well as reported advantages such as reduction of total chip size and suppression of bipolar degradation [1, 2]. (A) a MOSFET embedding SBD is better than (C) a MOSFET due to smaller E_{on} and suppressed bipolar degradation, and is better than (B_n) a MOSFET with external SBDs due to smaller chip size, smaller E_{on}, and smaller E_{oss}.

IV. CONCLUSIONS

We have investigated the switching characteristics of SiC-MOSFET embedding SBD, and revealed a superior performance due to suppressed recovery charge of minority carrier injection compared with conventional MOSFET and reduced total chip size compared with conventional MOSFET coupled with external SBDs.

Despite the 1.1 times larger on-resistance than that of conventional 3.3 kV MOSFET, SiC-MOSFET embedding SBD would be the most promising candidate especially for future railcar traction systems owing to its superior features, such as improved switching characteristics, reduced total chip size, and suppressed bipolar degradation.

REFERENCES

[1] S. Hino, *et al.*, "Demonstration of SiC-MOSFET Embedding Schottky Barrier Diode for Inactivation of Parasitic Body Diode", Mater. Sci. Forum, vol. 897, p. 477-482, 2017.

[2] K. Kawahara, *et al.*, "6.5 kV Schottky-Barrier-Diode-Embedded SiC-MOSFET for Compact Full-Unipolar Module", Proceedings of ISPSD'17, p. 41-44, 2017.

[3] J. Nakashima, *et al.*, "6.5-kV Full SiC Power Module (HV100) with SBD-embedded SiC-MOSFETs", Proceedings of PCIM'18, p. 441-447, 2018.

Proceedings of the 31st International Symposium on Power Semiconductor Devices & ICs
May 19-23, 2019, Shanghai, China

High-temperature Performance of 1.2 kV-class SiC Super Junction MOSFET

Yusuke Kobayashi, Shinya Kyogoku, Tadao Morimoto, Teruaki Kumazawa, Yusuke Yamashiro, Manabu Takei, and Shinsuke Harada

National Institute of Advanced Industrial Science and Technology (AIST)
Advanced Power Electronics Research Center
Tsukuba, Ibaraki, Japan
Email: yusuke-kobayashi@aist.go.jp, kobayashi-yusuk@fujielectric.com, s-harada@aist.go.jp

Abstract— SiC SJ MOSFETs had exhibited soft recovery characteristics at room temperature in comparison with that of non-SJ MOSFETs. In this study, the static and dynamic characteristics of the SiC SJ MOSFET at high temperature have been first demonstrated to clarify the mechanism of the soft recovery of the body-diode. The R_{onA} at high temperature was drastically decreased by the SJ structure due to small drift resistance with suppressed injection level. The hard recovery characteristics are suppressed by the low injection level and large output capacitance (C_{oss}) compared with those in the Si SJ MOSFETs, indicating the great potential of SiC SJ MOSFET in various applications.

Keywords—SiC, MOSFET, Super Junction, High temperature, reverse recovery.

I.   INTRODUCTION

SiC power devices have attractively low power loss because of its higher critical electric field than that of Si power devices. Recently, 1.2 kV-class SiC metal oxide-silicon field effect transistors (MOSFETs) have achieved low specific on-resistance (R_{onA}) by both shrinking the cell pitch and improving the channel mobility by a trench-gate [1]. In addition, the electric field at the trench bottom has been successfully suppressed by applying buried p+ layers in the Implantation and Epitaxial Trench MOSFET (IE-UMOSFET) [2]. Further improvement requires the reduction of the drift resistance, which is the largest component of the R_{onA} in recent trench MOSFETs [3]. Super-junction (SJ) structures have effectively decreased the drift resistance without degrading the blocking voltage [4] and have demonstrated low on-resistance in SiC at room temperature [5–6]. One known issue in Si SJ MOSFETs is their hard recovery characteristics of the body diode because the injected hole disappears at low drain voltage due to the short distance from the p-column [7]. Nevertheless, SiC SJ MOSFETs have soft recovery characteristics at room temperature in comparison with that of non-SJ MOSFETs [8]. It is expected that the displacement current of the output capacitance dominates the reverse recovery current. In this study, to clarify insight of the soft recovery characteristics, the high-temperature performance of the static and dynamic characteristics in a 1.2 kV-class SiC SJ MOSFET are investigated in terms of reverse recovery of the body diode.

II.   FABRICATION

Figure 1 shows the schematic diagrams of the fabricated 1.2 kV-class SiC SJ MOSFETs. The n-drift layer of 4.4 μm was grown on a 4-degree off-axis 4H-SiC substrate. The SJ structure was formed on the drift layer using seven times multi-epitaxial growth with a thickness of 0.65 μm each. The p-column was formed by a total of eight times aluminum implantation to the drift-layer and multi-epitaxial layers. The structure, named narrow SJ pitch, has p- and n-columns with half-width and high doping concentration in comparison with that of the standard SJ structure. An IE-UMOSFET without SJ structure was also formed as a comparison. Figure 2 shows a cross-sectional scanning electron microscope (SEM) micrograph of the fabricated SJ devices. It is obvious that the SJ structures have been successfully formed by the multi-epitaxial growth method employed in this study. The fabricated device with a chip size of 3 mm x 3 mm was packaged to TO-247. The 18 A rated current of the package corresponds to a current density of about 330 A/cm^2.

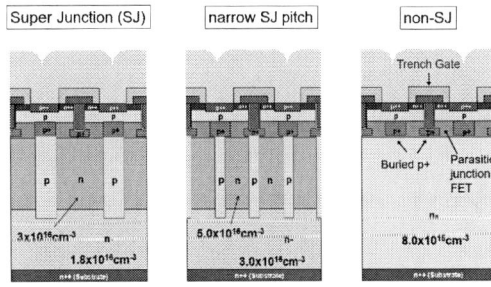

Fig. 1 Schematic diagrams of fabricated devices

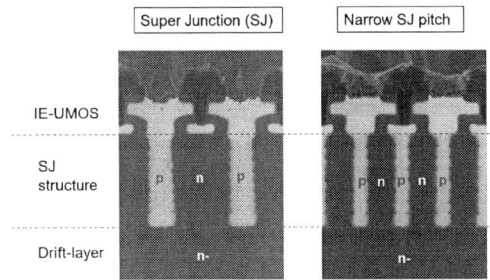

Fig. 2 SEM micrographs of fabricated SJ MOSFETs

978-1-7281-0582-6/19 $31.00 © 2019 IEEE

III. RESULTS

Figure 3 and 4 show the I-V characteristics at room temperature (RT) and at 175 °C, and the evaluated temperature dependence of the R_{onA} and threshold voltage (V_{th}). The V_{th} is defined as the gate voltage at a drain current of 18 mA and drain voltage of 20 V. The R_{onA} is successfully reduced by the SJ structure over the entire temperature range. In spite of the small difference at RT, the R_{onA} of the SJ devices become much lower than that of the non-SJ devices with increasing temperature, indicating the suppressed R_{onA} increase with temperature in SJ devices. The drift resistance becomes dominant to the R_{onA} at high temperature due to the higher temperature coefficient than that of other resistances (e.g., channel, substrate, and contact). Therefore, the low drift resistance of the SJ devices offers a critical advantage in reducing the R_{onA} at high temperature. The temperature dependence of the R_{onA} is least with a narrow SJ pitch because the high doping concentration of the n-layer further decreases the temperature coefficient. On the other hand, the V_{th} decrease at high temperature is similar for all devices since it is determined by the MOS channel properties. The blocking voltages of these structures are approximately 1.5 kV, as shown in Figure 5. Figure 6 shows the output capacitance (C_{oss}). The C_{oss} is increased by the SJ structure and is further increased in the narrow SJ pitch due to the large p-n junction area with a high doping concentration of the n-column. The total charge in the SJ and narrow SJ pitch are about twice and thrice of the non-SJ, respectively. The C_{oss} is rapidly decreased by the pinch-off of the parasitic junction FET and SJ structure, resulting in similar values among the three devices after full depletion of the SJ structure.

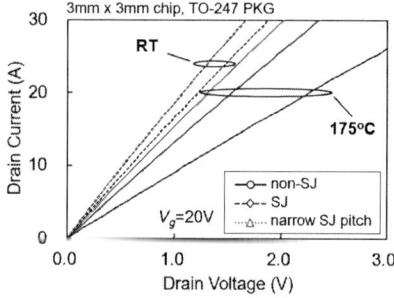

Fig. 3 I-V characteristics at RT and 175 °C

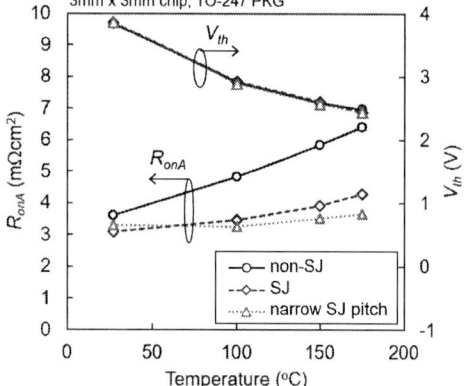

Fig. 4 Temperature dependency of R_{onA} and V_{th}

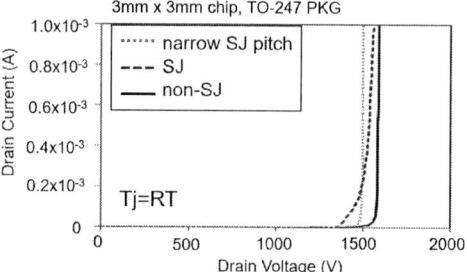

Fig. 5 Blocking voltage waveforms

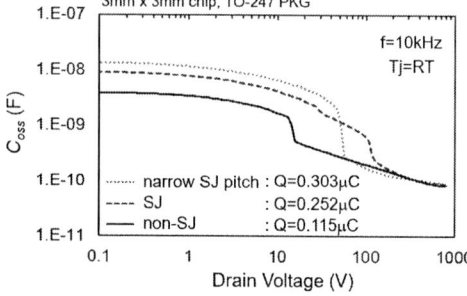

Fig. 6 Output capacitance waveforms at 10 kHz

Figure 7 and 8 show the forward characteristics of the body diode at RT and 175 °C, and the evaluated temperature dependence of the differential conductance (dI/dV). With increasing temperature, the dI/dV decreases in SJ devices and is constant in non-SJ devices. The dI/dV at high temperature is increased by increasing the injection level. Inversely, it is decreased by decreasing the mobility of the n-layer. Therefore, the dI/dV is determined by the balance of the temperature dependences between the mobility and injection level. The injection level is also determined by the doping concentration and carrier lifetime. The decrease of dI/dV at high temperature in SJ devices indicates the dominant influence of the mobility decrease in the n-layer rather than the injection level increase. A lower injection level in SJ devices than in non-SJ devices may be due to a higher doping concentration of the n-layer with a shorter lifetime by ion implantation damage during the p-column fabrication.

Next, the injection level of the body diode in the SiC SJ is evaluated from the reverse recovery current. The reverse recovery charge (Q_{rr}) is defined as the sum of the charges of C_{oss} and injected carriers, and is divided by the temperature and drain current (Id) dependences since only Q_{rr} by the injected carrier is influenced by them. The reverse recovery characteristics of the body diode were measured using a simple chopper circuit shown in Figure 9. The gate voltage and resistance are +20 V/-5 V and 100 Ω, respectively. Figure 10 shows the reverse recovery current for Id of 18 A and 7 A measured at RT and 175 °C. The Q_{rr} is calculated considering the current oscillation in which the ideal current flows in the center of the oscillation if the noise is ignored. At RT, the difference of the Q_{rr} between 18 A and 7 A is only less than 10% in both SJ and non-SJ devices, indicating that Q_{rr} nearly consists of only C_{oss} due to the low injection level. By increasing the temperature to 175 °C in the non-SJ device, the Q_{rr} at 18 A is increased to approximately twice its original value, while the Q_{rr} at 7 A is increased with only about 15%. On the other hand, in the SJ device, the Q_{rr} increase at both 18 A and 7 A is only 6%. This means that the injection level in the SJ device is kept low when increasing the temperature to 175 °C, which is consistent with the forward characteristics.

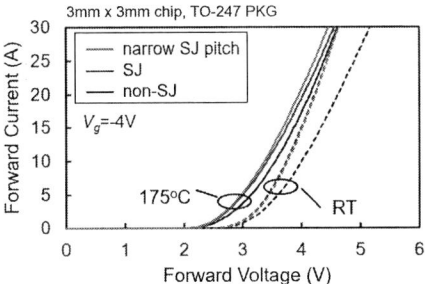

Fig. 7 Forward characteristics at RT and 175 °C

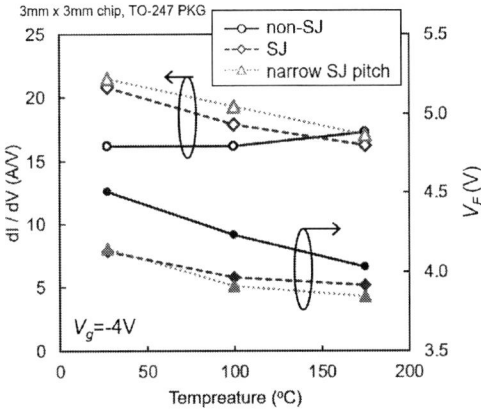

Fig. 8 Temperature dependency dI/dV and V_F

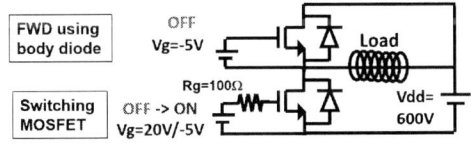

Fig. 9 Measurement circuit of reverse recovery

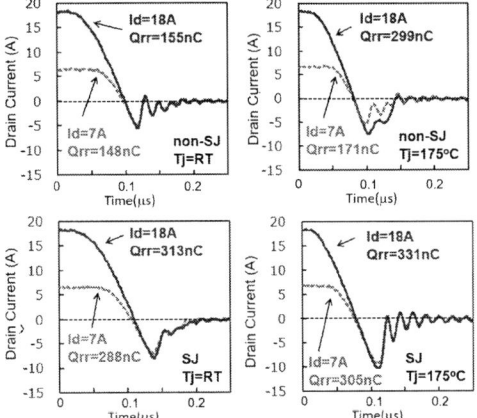

Fig. 10 Reverse recovery at a current of 18 A and 7 A

Figures 11 and 12 show the reverse recovery waveforms in the non-SJ, SJ, and narrow SJ pitch at RT and 175 °C, respectively. Figures 13 and 14 show the evaluated temperature dependencies of the Q_{rr} and the maximum recovery di/dt (max di/dt), respectively. The waveforms shift to an earlier period with an increasing temperature because the threshold voltage decreases. The Q_{rr} of SJ devices is larger than that of non-SJ devices and the Q_{rr} of the narrow SJ pitch is further larger than that of SJ devices. The temperature dependence of Q_{rr} in both SJ devices is small in comparison with that of non-SJ devices, indicating that the injection level is independent of the SJ structure and is kept low even at high temperature. The small max di/dt is preferred because the surge voltage is increased by increasing the max di/dt, especially in circuits with large parasitic inductance. The SiC SJ devices achieves small max di/dt at RT, which means soft recovery, in comparison with that of non-SJ devices, because of the large C_{oss} of the SJ device maintained until a high V_D, as shown in Figure 6. The max di/dt of SiC SJ devices is increased with increasing temperature. This is due to the rapid disappearance of the increased injection carrier with the temperature at a low drain voltage in the SJ structure, in spite of low injection level in SiC SJ. The narrow SJ pitch has a larger di/dt than SJ because the narrow n-column accelerated the disappearance of the injection carrier. Although the max di/dt of the SiC SJ is increased by increasing the temperature, the low injection level and high C_{oss} contribute to avoiding an extremely large di/dt by the SJ structure.

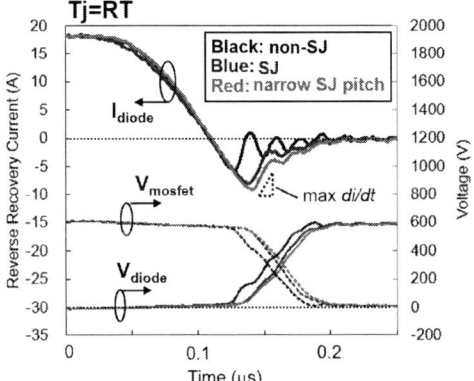

Fig. 11 Reverse recovery characteristics at RT

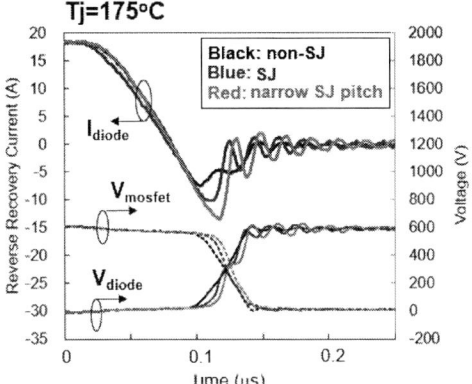

Fig. 12 Reverse recovery characteristics at 175 °C

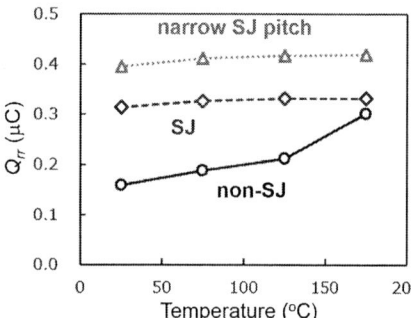

Fig. 13 Temperature dependence of Q_{rr}

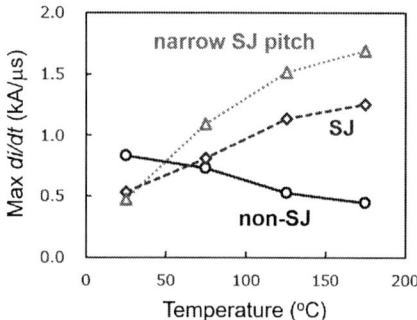

Fig. 14 Temperature dependence of max di/dt

Figure 15 compares the reverse recovery characteristic at 175 °C of the SiC SJ device with that of a commercial 650 V-class Si SJ device with a similar rated current. The drain voltage for the switching test was set to half of the blocking voltage classes. The same non-SJ MOSFET was utilized as the switching MOSFET to compare the body diode characteristics only. It is clarified that the Q_{rr} of the SiC SJ is drastically smaller than that of the Si SJ. Furthermore, the max di/dt and surge voltage of the SiC SJ are also smaller than those of the Si SJ. These results indicate the advantage of the low injection level of SiC in improving the reverse recovery characteristics of the SJ device. Furthermore, it is considered that the smaller chip size due to the low R_{onA} of the SiC device may also contribute to the Q_{rr} (and C_{oss}) reduction in spite of the larger C_{oss} per area than that of the Si device.

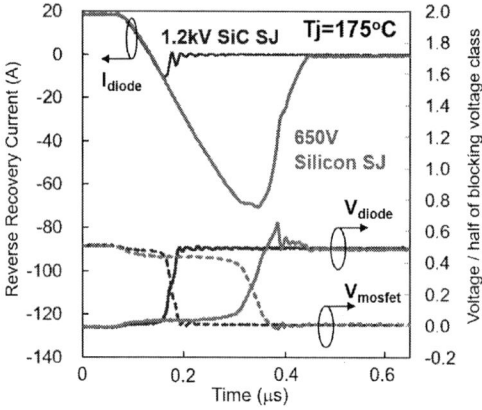

Fig. 15 Reverse recovery of the Si and SiC SJ devices

IV. CONCLUSION

The static and dynamic characteristics of the SiC SJ devices at high temperature were analyzed. At high temperature, the R_{onA} of the SiC SJ was drastically reduced by the small temperature coefficient of the n-layer and low drift resistance. The low injection level of body-diode in SiC SJ even in high temperature was substantiated by forward characteristics and reverse recovery current. The high C_{oss} and low injection level of the SiC SJ are great advantageous in suppressing the hard recovery characteristics in comparison with Si SJ.

ACKNOWLEDGMENT

This paper has been implemented under a joint research project of Tsukuba Power Electronics Constellations (TPEC).

The author, Y. Kobayashi, and co-author, M. Takei, are assigned from Fuji Electric Co., Ltd., the co-author, S. Kyogoku, is assigned from Toshiba Co., Ltd., the co-author, T. Kumazawa, is assigned from Toyota Co., Ltd., and the co-author, Y. Yamashiro, is assigned from Mitsubishi Electric Co., Ltd.

REFERENCES

[1] T. Nakamura, Y. Nakano, M. Aketa, R. Nakamura, S. Mitani, H. Sakairi, and Y. Yokotsuji, "High performance SiC trench devices with ultra-lon ron", IEEE International Electron Devices Meeting (IEDM), 26.5.1, Dec. 2011.

[2] S. Harada, Y. Kobayashi, A. Kinoshita, N. Ohse, T. Kojima, M. Iwaya, H. Shiomi, H. Kitai, S. Kyogoku, K. Ariyoshi, Y. Ohnishi, and H. Kimura, "1200 V SiC IE-UMOSFET with low on-resistance and high threshold voltage", Materials Science Forum, Vol. 897, pp. 497–500, 2017.

[3] Y, Kobayashi, N. Ohse T. Morimoto, T. Kojia, M Takei, H. Kimura, and S. Harada, "Low on-resistance SiC trench MOSFET with suppressed short channel effect by halo implantation", ICSCRM, FR.D2 1, Sep. 2017.

[4] T. Fujihira, "Theory of semiconductor superjunction devices", J. Appl. Phys., Vol. 36, pp. 6254–6262, 1997.

[5] R. Kosugi, Y. Sakuma, K. Kojima, S. Itoh, A. Nagata, T. Yatsuo, Y. Tanaka, and H. Okumura, "First experimental demonstration of SiC super-junction (SJ) structure by multi-epitaxial growth method", ISPSD, pp. 346–349, Jun 2014.

[6] T. Masuda, R. Kosugi, and T. Hiyoshi, "0.97 mΩcm²/820 V 4H-SiC super junction V-Groove trench MOSFET", Materials Science Forum, Vol. 897, pp. 483–488, 2016.

[7] W. Saito, I. Omura, S. Aida, S. Koduki, M. Izumisawa, and T. Ogura, "600V semi-superconjunction MOSFET", ISPSD, pp. 45–48, April 2003.

[8] S. Harada, Y. Kobayashi, S. Kyogoku, T. Morimoto, T. Tanaka, M. Takei, and H. Okumura, "First demonstration of dynamic characteristics for SiC superjunction MOSFET realized using multi-epitaxial growth method", IEEE International Electron Devices Meeting (IEDM), 8.2, Dec. 2018.

Proceedings of the 31st International Symposium on Power Semiconductor Devices & ICs
May 19-23, 2019, Shanghai, China

Suppression of Bipolar Degradation in Deep-P Encapsulated 4H-SiC Trench MOSFETs up to Ultra-High Current Density

Yasuhiro Ebihara, Junichi Uehara, Aiko Ichimura, Shuhei Mitani, Masato Noborio, Yuichi Takeuchi,
and Kazuhiro Tsuruta
DENSO CORPORATION
Nisshin, Aichi, 470-0111, Japan
Yasuhiro_ebihara@denso.co.jp

Abstract— **The bipolar degradation of Deep-p encapsulated 4H-SiC trench MOSFETs was investigated by using the pulse-current conduction system up to ultra-high current density. The fabricated MOSFETs exhibit the low hole injection compared with the PN diodes and conventional MOSFETs, resulted in the fabricated MOSFETs suppressed the bipolar degradation up to a current density of 3000 A/cm².**

Keywords; SiC-MOSFET, bipolar degradation, body-diode,

I. INTRODUCTION

SiC MOSFETs are attractive for low-loss, voltage-controlled, and normally-off power devices. The performances and reliabilities of SiC MOSFETs have been enhanced by reducing specific on-resistance and using a technique to relax electric-field crowding at the gate oxide. Among many types of SiC MOSFETs, 4H-SiC trench MOSFETs have great potential owing to its high MOS-channel density and high channel mobility on the trench sidewall [1,2]. The novel structure called Deep-p encapsulated MOSFET was presented and demonstrated superior Figure-of-merit ($R_{on}Q_{gd}$), which means low conduction and switching losses [3,4].

In the inverter and boost converter for motor generator, free-wheel diodes are often added externally to the MOSFETs to suppress the voltage spikes, especially for the 1200 V class system. In these cases, SiC SBDs are generally used due to its low conduction loss and low reverse-recovery loss. However, there is a more cost-saving alternative which is to use the body diode of the MOSFET itself.

The major problem of the body diode in SiC-MOSFETs is the body diode degradation [5,6]. When the body diode is conducting, hole injection has occurred from the top p-type layer into the n-type drift layer, which causes the hole-electron recombination at the BPD (Basal Plane Dislocation) at various locations. The BPD gains energy from the recombination, and consequently the stacking faults grow into the drift region. The stacking faults act as a resistance, which results in the increase of conduction loss. Therefore, the bipolar degradation at the body-diode conduction should be solved.

In addition, the diode conduction test up to ultra-high currents are necessary if large chip sizes are to be investigated. However, the forward voltage (V_f) of the diode is generally high, which leads to high heat generation. Therefore, better cooling test systems are required to fulfill the high-current requirement.

In this work, the bipolar degradation of Deep-p encapsulated MOSFETs were investigated up to ultra-high current density.

II. DEVICE STRUCTURE AND FABRICATION

Figures 1 (a), (b), and (c) show the schematic cross-sectional illustrations of PN diode, conventional MOSFET, and Deep-p encapsulated MOSFET, respectively. The p-type layer called Deep-p region is formed in all areas (except terminal region) in the PN diode while the Deep-p regions of conventional MOSFETs and Deep-p encapsulated MOSFETs are formed parallel and orthogonal to the trench side wall, respectively. The Deep-p layer and the P$^+$ layer of the PN diode were formed by using the Al$^+$ implantation. The concentration and thickness of the drift layer are selected to have a blocking voltage of over 1200 V. The structures and the fabrication process of conventional MOSFETs and the Deep-p encapsulated MOSFETs are the same in ref [4,7]. All structures used commercial wafers of the same specifications to focus on the structural differences.

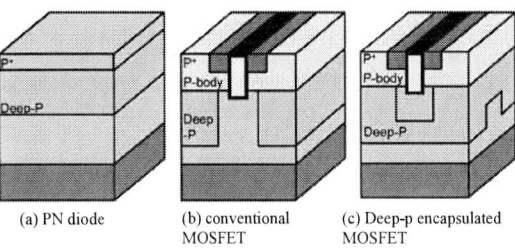

(a) PN diode (b) conventional MOSFET (c) Deep-p encapsulated MOSFET

Fig.1 Schematic cross-sectional illustrations of (a) PN diode,
(b) conventional MOSFET, and (c) Deep-p encapsulated MOSFET.

978-1-7281-0582-6/19 $31.00 © 2019 IEEE

III. TEST CIRCUIT AND CONDITIONS

Figure 2 (a) shows the test circuit. Figures 2 (b) and (c) show the typical pulse waveforms of small current and high current tests, respectively. The DUT current is controlled by changing the gate voltage (V_g) of the IGBT module. For high-current tests, a current controlling IGBT of 3300 A/1500 V was used, and similarly a 200 A/200 V IGBT for small current tests. Compared with the waveforms at a DUT current of 100 A in Fig. 2 (b), the waveform at a DUT current of 1500 A is interrupted due to stray components, resulting in relatively small slope as shown in Fig. 2 (c). The total stress time at high currents is defined to be the integrated value within the range of 90% of peak current.

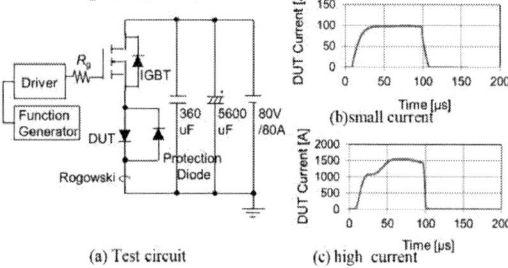

Fig.2 (a) Test circuit and the typical wafeforms of (b) small current and (c) high current.

To investigate the degradation mechanism at high current-density, the relation between transient temperature and current was studied first. Figures 3 (a), (b), and (c) show the test circuit, typical waveforms of transient measurements, and the current density dependence of transient temperatures, respectively. In order to estimate the transient temperature, the forward voltage at small currents is measured soon after the DUT current turns off as shown in Fig. 3 (b). The transient temperature is then estimated by using the temperature dependence of the body-diode forward voltage at a gate voltage of 0 V. The transient temperature at a current density of 3000 A/cm^2 is around 80 degrees Celsius as shown in Fig. 3 (c), which means that the DUT is well heated in one pulse. The forward voltage of the body diode was not degraded because only one pulse was applied at each current density. The forward voltage value was extracted 50 usec after the main pulse turns off to avoid the influence of noise. From the transient temperature response, the desired test temperature can be set by adjusting the steady-state temperature. For example, to achieve a total test temperature of 130 degrees Celsius at a current density of 3000 A/cm^2, the steady-state temperature is set around 50 degrees Celsius by appropriately changing the pulse duty.

Figure 4 (a) shows the test sequence. The forward voltage was initially measured. Thereafter, the DUT was tested at each current stress. After the diode conduction, the forward voltage of the diode was re-measured. If the forward voltage of the body diode did not degrade, the DUT was tested at a next elevated current. When the forward voltage has degraded, the diode conduction test was finished. Figure 4 (b) shows the test condition of each structure. The PN diodes were packaged into a single-side module. The conventional MOSFET and the Deep-p encapsulated MOSFET were packaged into a power card [8]

to enable double-side cooling, which is necessary because the chip size is much larger than that of the PN diode.

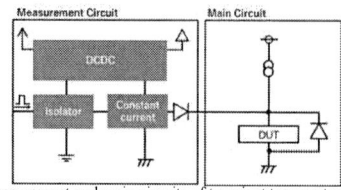

(a) Measurement and main circuits of transient temperature measurement

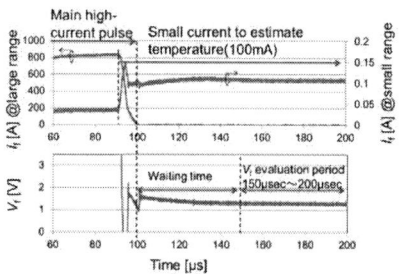

(b) Typical waveforms of transient temperature measurement.

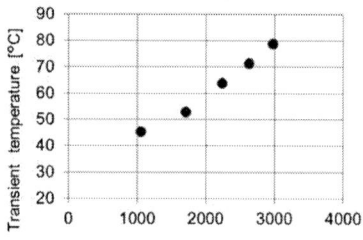

(c) Current density dependence of the transient temperature.

Fig.3 (a) Test circuit of the transient temperature, (b) typical waveforms and (c) current density dependence of the transient temperature.

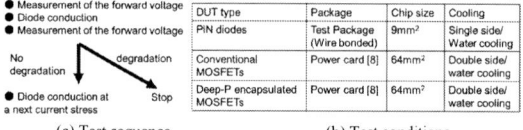

	DUT type	Package	Chip size	Cooling
	PIN diodes	Test Package (Wire bonded)	9mm²	Single side/ Water cooling
	Conventional MOSFETs	Power card [8]	64mm²	Double side/ water cooling
	Deep-P encapsulated MOSFETs	Power card [8]	64mm²	Double side/ water cooling

(a) Test sequence   (b) Test conditions

Figure 4. (a) Test sequence and (b) test condition. Water temperature was set to 20 degrees Celcius.

IV. RESULTS AND DISCUSSIONS

A. Test Results

a) PN Diodes

In the stress time of 0.06 sec, there were no degradation of the forward voltage up to a current density of 4000 A/cm^2 as shown in the Fig. 5. Consequently, the stacking faults were not observed in the PL image, which indicates that the stress time is insufficient to initiate the forward degradation. On the other hand, the forward voltages were degraded in a stress time of 6 sec, 60 sec, 600 sec, and 3600 sec, and stacking faults were observed at each stress time. In the 6 sec case, the current density when forward degradation occurred is much larger than those at 60 sec, 600 sec, and 3600 sec. As the stress time becomes longer, the forward degradation seems to saturate. It is pointed out in ref [9,10] that the forward degradation has a threshold hole density depending on the temperature and concentration of the electron and hole density, although there is still some debate regarding the absolute values of those density. The stress time is fixed at a 60 sec for subsequent tests to reduce the measurement time. The forward degradation strength of the PN diode is estimated to be around 500-800 A/cm^2.

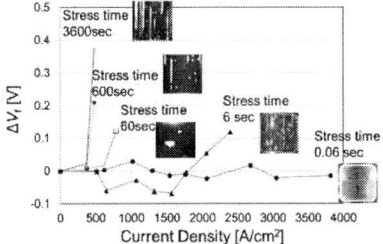

Figure 5. Forward voltage degradation as a function of current density and PL mapping image of PN diodes. V_f was defined as a forward current of 563 A/cm^2 and a temperature of RT.

b) Conventional MOSFETs

Figure 6 shows the forward degradation at a gate voltage of 0 V and -5 V as a function of the current density. Corresponding PL mapping images are also shown. The forward voltage degraded at a current density of 1000-1100 A/cm^2 for both gate voltages, although the degree of forward degradation at the gate voltage of -5 V is about three times larger than that at 0 V. Since the forward degradation threshold is determined by the hole concentration, this result suggests that the hole concentrations at both gate voltages do not differ at high current density. As shown in the PL mapping image, the number of the stacking faults at a gate voltage of -5 V is more than that at 0 V. Because the forward degradation values depend on the number of defects in the substrate and drift layer, it is assumed that the differences in the degradation are contributed by the defect numbers in the drift layer and/or substrate. Figure 7 shows the forward voltage-current curve. For the high current condition, the difference between the curve of $V_g = 0$ V and $V_g = -5$ V becomes close, which agrees with the degradation results.

c) Deep-p encapsulated MOSFETs

Figure 8 shows the forward degradation at gate voltages of 0 V and -5 V as a function of the current density. Corresponding PL mapping images are also shown. In contrast to the results of PN diodes and conventional MOSFETs, the forward degradation does not occur up to a current density of 3000 A/cm^2, and stacking faults were not observed, which indicated that the hole injection in the Deep-p encapsulated MOSFETs are much smaller than those of the PN diodes and conventional MOSFETs.

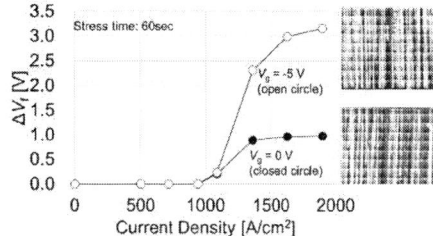

Figure 6. Forward voltage degradation of $V_g = 0$ V and $V_g = -5$ V as a function of current density and PL mapping image of conventional MOSFETs. V_f was defined as a current of 285 A/cm^2, a temperature of RT, and a gate voltage of -5 V. The PL image of only active area is shown in this figure.

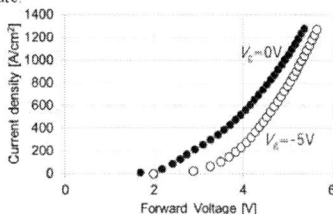

Figure 7. Measured V_f-I_f curve of V_g=0V and -5V.

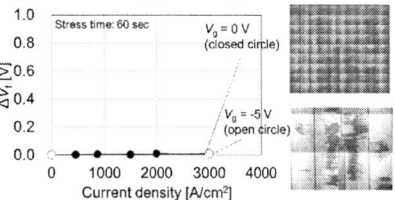

Figure 8. Forward voltage degradation of $V_g = 0$ V and $V_g = -5$ V as a function of current density and PL mapping image of Deep-p encapsulated MOSFETs. V_f was defined as a current of 263 A/cm^2, a temperature of RT, and a gate voltage of -5 V. The PL images of only active area is shown in this figure.

B. Analysis by using the TCAD

In order to investigate the injected hole concentration for each structure, device simulations were performed. The implanted regions of each structure are created by using the process simulator [11].

The forward voltage-current curves are initially fitted to the measured curve by including and changing the concentration of the hole traps in the Al$^+$ implanted layer. In this simulation, the single hole trap level is set to the E_t-E_v of 1.3 eV corresponding to the HK3 center [12,13], and the hole traps are set proportional to the Al concentration. Figure 9 shows the measured and simulated forward voltage-current curve with and without the hole traps for each structure. By considering the hole traps, the simulated curve becomes close to the measured one. Figures 10 (a) and (b) show the hole density as a function of the current density with and without hole traps in the Al$^+$ implanted layer, respectively.

The hole density of the Deep-p encapsulated MOSFETs become much larger than that of the PN diodes and conventional MOSFETs without the hole traps, which deviates from the experimental results. In contrast, in the case with hole traps, the hole density of Deep-p encapsulated MOSFETs become much smaller than that of the PN diodes and conventional MOSFETs. These results agree with the experimental results.

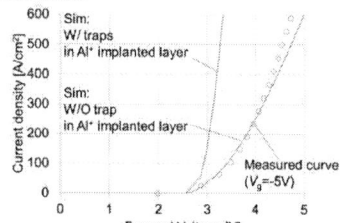

Figure 9. Measured and simulated V_f-I_f curve with or without traps.

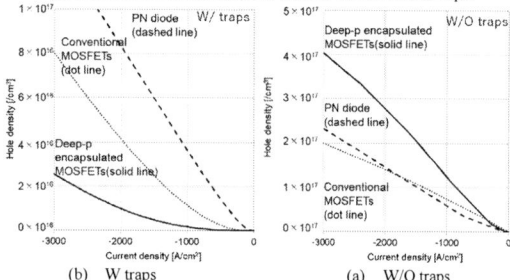

(b) W traps   (a) W/O traps

Figure 10. Hole density profile as a function of current density. Hole density of 1 um under the drift/substrate interface are plotted.

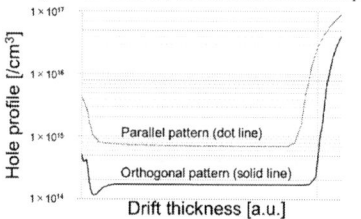

Figure 11. hole density profile in the drift layer at a current density of 2368 A/cm² and a V_g of -5 V, RT.

To investigate further, only geometrical effects are studied by using the parallel pattern and orthogonal pattern of Deep-p as shown in Figs. 1 (b) and (c), respectively. In order to focus only on geometrical effects, the trench cell pitch are selected to the Deep-p encapsulated MOSFETs. Thus, only the geometry of the Deep-p is changed. The hole current density beneath the Deep-p layer was 9.1 A/cm² and 104.7 A/cm² in the orthogonal and parallel pattern, respectively. Therefore, as shown in Fig. 11, the hole are less injected into the drift layer in the orthogonal pattern while the hole density of the parallel pattern is about 10 times larger than that of its orthogonal counterpart. This result indicates that the geometry of the Deep-p pattern is also important to suppress the hole injection into the drift layer in the body diode conduction.

V. SUMMARY

The bipolar degradation of the PN diodes, conventional MOSFETs, and Deep-p encapsulated MOSFETs were investigated up to the ultra-high current density using in-house developed high current testing system. The strength of the PN diodes, conventional MOSFETs, and Deep-p encapsulated MOSFETs were 500-800 A/cm², 1000-1100 A/cm², and over 3000 A/cm², respectively. The device simulations were performed and the injected hole density estimated for each structure. By including the traps and geometrical effects, it was revealed that the injected hole density of the Deep-p encapsulated MOSFETs was much smaller than those of the PN diodes and conventional MOSFETs.

ACKNOWLEDGMENT

The authors would want to give special thanks to K. Sakamoto and T. Kondo for their helps on fabrication of the high current system and the experiments.

This work is supported by Toyota Motor Corporation and Toyota Central R&D Labs., Inc..

REFERENCES

[1] J.A. Cooper Jr., M.R. Melloch, R. Singh, A. Agarwal, J.W. Palmour, "Status and prospects for SiC power MOSFETs" IEEE Trans. Electron Devices vol.49, p.658-64, 2002.

[2] H. Yano, H. Nakao, T. Hatayama, Y. Uraoka, and T. Fuyuki, "Increased Channel Mobility in 4H-SiC UMOSFETs Using On-Axis Substrates" Mater. Sci. Forum, Vols.556-557, pp.807-810, 2007.

[3] A. Ichimura, Y. Ebihara, S. Mitani, M. Noborio, Y. Takeuchi, S. Mizuno, T. Yamamoto, and K. Tsuruta, "4H-SiC Trench MOSFET with Ultra-Low On-Resistance by using Miniaturization Technology", Mater. Sci. Forum, Vol. 924, pp. 707-710, 2018.

[4] Y. Ebihara, A. Ichimura, S. Mitani, M. Noborio, Y. Takeuchi, S. Mizuno, T. Yamamoto, and K. Tsuruta, "Deep-P Encapsulated 4H-SiC Trench MOSFETs With Ultra Low $R_{on}Q_{gd}$", Proceedings of ISPSD, pp. 89-92, 2018.

[5] J. P. Bergman, H. Lendenmann, P. A. Nilsson, U. Lindefelt and P. Skytt, "Crystal Defects as Source of Anomalous Forward Voltage Increase of 4H-SiC Diodes", Mater. Sci. Forum 353-356, 299 (2001).

[6] M. Skowronski and S. Ha, "Degradation of hexagonal silicon-carbide-based bipolar devices", J. Appl. Phys. 99, 011101 (2006).

[7] https://www.denso.com/jp/ja/products-and-services/industrial-products/sic

[8] N. Hirao, K. Mamitsu, and T. Okumura, "Structural Development of Double-sided Cooling Power Modules", DENSO TECHINICAL REV. Vol. 16, pp. 30-37, 2011.

[9] T. Tawara, S. Matsunaga, T. Fujimoto, M. Ryo, M. Miyazato, T. Miyazawa, K. Takenaka, M. Miyajima, A. Otsuki, Y. Yonezawa1, T. Kato1, H. Okumura, T. Kimoto, and H. Tsuchida, " Injected carrier concentration dependence of the expansion of single Shockley-type stacking faults in 4H-SiC PiN diodes", J. Appl. Phys. 123, 025707 (2018).

[10] A. Iijima and T. Kimoto, "Theoretical and Experimental Investigation of Critical Condition for Expansion/Contraction of a Single Shockley Stacking Fault in 4H-SiC", Ext. Abstr. (ECSCRM 2018), MO.04.02. (2018)

[11] SenTaurus at [http://www.synopsys.com/].

[12] K. Danno and T. Kimoto, "Deep level transient spectroscopy on as-grown and electron-irradiated p-type 4H-SiC epilayers", J. Appl. Phys. 101, 103704 (2007).

[13] K. Kawahara, J. Suda, G. Pensl and T. Kimoto, "Reduction of deep levels generated by ion implantation into n- and p-type 4H-SiC", J. Appl. Phys. 108, 033706 (2010).

Proceedings of the 31st International Symposium on Power Semiconductor Devices & ICs
May 19-23, 2019, Shanghai, China

Breaking the Theoretical Limit of 6.5 kV-Class 4H-SiC Super-Junction (SJ) MOSFETs by Trench-Filling Epitaxial Growth

Ryoji Kosugi*, Shiyang Ji, Kazuhiro Mochizuki, Kohei Adachi, Satoshi Segawa, Yasuyuki Kawada, Yoshiyuki Yonezawa, and Hajime Okumura

Advanced Power Electronics Research Center (ADPERC)/National Institute of Advanced Industrial Science and Technology (AIST)/Japan
Email: r-kosugi@aist.go.jp

Abstract— A super-junction (SJ) device has been developed to improve the trade-off relationship between the breakdown voltage (V_{BD}) and specific on-resistance (R_{onA}). A multi-epitaxial (ME) growth method had been used for fundamental demonstrations, but this method needs a lot of repetitions of epitaxial growth and implantation in the case of SiC material. A trench-filling epitaxial (TFE) growth method is expected as a promising alternative, especially for high-voltage devices. In this study, we have established critical fabrication processes for a thick (> 20 μm) and high-aspect-ratio SJ structure. The measured R_{onA} of a 7.8 kV SJ MOSFET was 17.8 mΩ·cm², which is less than half the R_{onA} of the state-of-the-art 6.5 kV-class SiC MOSFET with an n-type drift layer. Improvement of trade-off relationship exceeding the 4H-SiC theoretical limit was experimentally demonstrated for the first time.

I. Introduction

A super-junction (SJ) device, in which alternating p- and n-type columns are located in a drift layer, has been developed to improve the trade-off relationship between breakdown voltage (V_{BD}) and specific on-resistance (R_{onA}) in unipolar power devices [1]. This improvement becomes prominent especially for high voltage applications, where a fabrication method of SJ structure is critical issue. Multi-epitaxial (ME) growth method had been used for the first fundamental demonstrating SiC SJ diodes [2, 3] and MOSFETs [4−6]. However, the ME growth method needs a lot of repetitions of epitaxial growth and ion implantation, especially for thick SJ structures, because thermal diffusion coefficients of dopants in SiC are very small. In order to reduce the number of repetitions, a fabrication method using Al implantation at a high acceleration voltage over 20 MeV has been reported [7, 8].

A trench-filling epitaxial (TFE) growth method is expected as an alternative candidate for realizing mass production of SJ devices. Recently, it has become possible to homo-epitaxially fill several μm- [9, 10] to 50 μm-deep [11, 12] 4H-SiC trenches. In addition, understanding of the TFE growth mechanism has been significantly advanced [13−15]. Therefore, establishment of the TFE growth method for SiC SJ devices corresponding to a wide breakdown-voltage region is expected.

In this study, we have developed critical process techniques to fabricate SJ wafers for 6.5 kV-class SJ

MOSFETs, such as deep-trench formation, TFE growth with p-type doping control, and flattening process of TFE grown wafers. Formation of deep (> 20 μm) trenches followed by void-free TFE growth is indispensable for 6.5 kV-class SJ devices. One of the important points in this work is to clarify whether vertical pn junctions fabricated by the TFE growth function normally or not. First, two types of test elemental groups (TEGs) were used for evaluations of blocking characteristics and specific resistivity of the SJ drift layer (R_{drift}). Then, 6.5 kV-class SJ MOSFETs were fabricated on the SJ wafer whose dopant concentration was well controlled.

II. Experiment

Fabrication of partial SJ structures by deep trench-filling epitaxial (TFE) growth method

A partial SJ structure was used for device fabrication, instead of a full SJ structure. In the partial SJ structure, an n-type buffer layer is connected to the bottom of the SJ structure. With the introduction of a buffer layer whose thickness is optimized, a thin SJ structure can be designed with a narrow pn column pitch while high V_{BD} and low R_{onA} are maintained. It is known that narrowing the pn column pitch enables us to increase the TFE growth rate [10] and to increase the donor concentration of n-type columns [1]. High-aspect-ratio trenches are preferred for the fabrication of the partial SJ structure in SiC. Design of the SJ structure and the buffer layer was optimized from the standpoints of high V_{BD} and low R_{onA}, as well as manufacturability including process margin. This study was targeted at halving the R_{onA} of the state-of-the-art 6.5 kV-class SiC MOSFET [16].

Figure 1 shows schematic of fabrication process of the partial SJ structure and SJ-MOSFETs. Typical concentrations/thicknesses of the pn columns and the buffer layer were 1.5×10^{16} cm^{-3}/22 μm and 2.0×10^{15} cm^{-3}/41 μm, respectively. Fabrication of the SJ structure began with a deep stripe-trench formation in an n-type epilayer by using inductively coupled plasma (ICP) etching in a SF$_6$/O$_2$ gas ambient. The trench pitch was 5 μm with the aspect ratio of 9−10. In order to suppress the tilt of the film grown on the mesa top during TFE growth, the stripe-trenches were formed along [11-20] direction as strictly as possible [17]. These trenches are separated both in the longitudinal

978-1-7281-0582-6/19 $31.00 © 2019 IEEE

direction and in the vertical direction in units of exposure shots in the photolithography process.

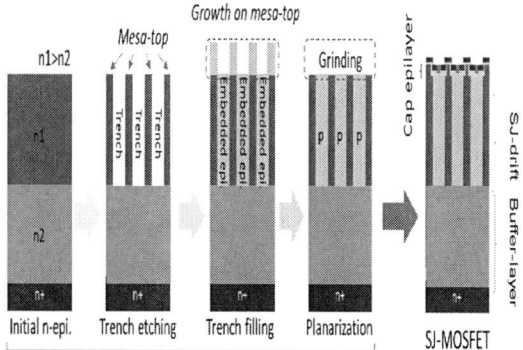

Figure 1. Schematic of a fabrication process of partial SJ structure by trench-filling epitaxial (TFE) growth method. The partial-SJ structure consists of the SJ-drift part and n-type buffer layer (n1>n2).

A hot-wall chemical vapor deposition system that was equipped with the SiC source gases of silane (SiH_4), propane (C_3H_8) and H_2 carrier gas was used for TFE growth. Trimethyl-aluminum (TMA) was used as an aluminum source to provide p-type dopants for charge balance control. HCl gas was added as an etchant gas to suppress the mesa top growth and thus avoid fast trench closure [11].

A growth rate of the epilayer on the mesa top relative to that on the trench bottom can be controlled by the flow-rate ratio of etchant gas to source gas [12]. By adjusting the ratio of $HCl/SiH_4 = 50$, a quasi-selective growth becomes possible, as shown in Fig. 2.

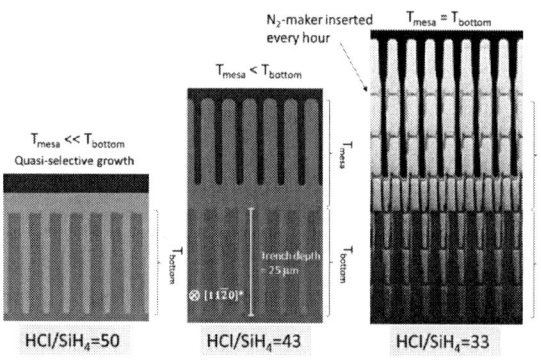

Figure 2. Cross-sectional SEM images of SJ structures grown under different TFE growth conditions in terms of HCl/SiH_4 ratio.

TFE growth was performed for 6 hours under following conditions. Surface temperature and total pressure were 1650°C and 60 kPa, respectively, and the gas flow rates for SiH_4, C_3H_8, HCl and H_2 were 42 sccm, 14 sccm, 1.8 slm and 40 slm, respectively. A typical TFE growth rate was 5 µm/h.

After TFE growth, the wafer surface was flattened by thinning the TFE grown wafer to the initial epiwafer thickness using grinding and polishing. Finally, a thin n-type epilayer was grown on the SJ structures to avoid an overlapping of doping profiles of the MOS part and SJ part. PN junction and resistivity TEGs shown in Fig. 3 were used for evaluations of V_{BD} and R_{drift}, respectively, prior to SJ MOSFETs fabrication.

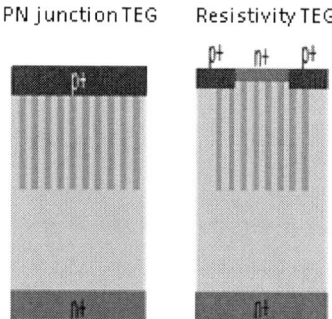

Figure 3. Schematic section views of pn junction TEG for evaluation of breakdown voltage and resistivity TEG for evaluation of specific resistance of drift layer including n+ substrate.

III. EXPERIMENTAL RESULTS AND DISCUSSIONS

Resistivity and breakdown voltage of pn junction TEGs on SJ wafer

R_{drift} including an n+ substrate estimated from I-V characteristics of resistivity TEGs on a 3-inch wafer was in the range of 13.0−13.7 mΩ·cm^2 (Ave. = 13.3 mΩ·cm^2) at room temperature (RT), which is consistent with TCAD simulation. R_{drift} increased to 42.9−45.0 mΩ·cm^2 (Ave. = 43.9 mΩ·cm^2) at 175°C. Figure 4 shows forward characteristics of the pn junction TEGs. The PN junction TEGs demonstrated clear rectifying property, indicating the pn junction functions normally.

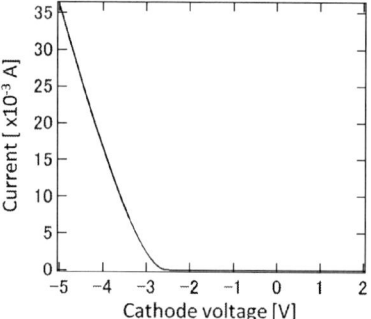

Figure 4. Forward characteristics of the pn junction TEG.

However, large leakage current in the reverse characteristics was observed, as shown by the blue lines in Fig. 5. Leakage current increases from low cathode voltage and saturates when cathode voltage of about 30 V is applied. The pn junction TEGs formed on the same wafer without SJ structures did not show such large leakage, which suggests an SJ structure-driven leakage path.

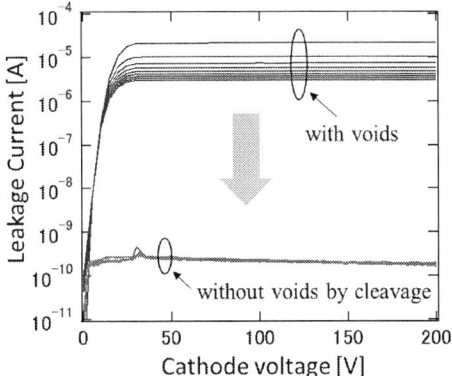

Figure 5. Reverse characteristics of the pn junction TEGs with voids at the longitudinal end of trenches (blue lines) and without voids (red lines). Voids are removed by cleavage.

Emission analysis of the pn junction TEG reverse-biased at 200V revealed that the cause of leakage current was unintended voids formed near the trench edges during TFE growth, as shown in Fig. 6. This result suggests an existence of leakage path on the edge region of each stripe-trench, although the formation mechanism and detailed structure of these voids are not clear yet [14]. In fact, we confirmed the leakage current was decreased by four orders of magnitude by removing those voids, for example by cleavage, as shown by the red lines in Fig. 5.

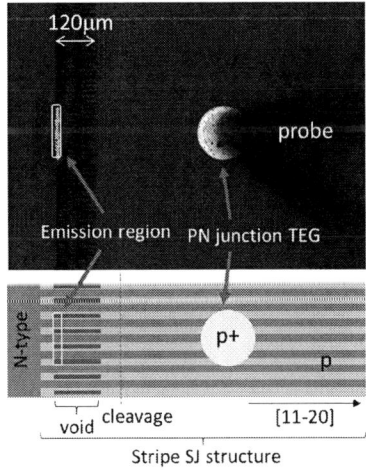

Figure 6. Emission image of pn junction TEG under reverse bias of 200V. Emission region corresponds to the longitudinal end of stripe-trenches along the [11-20] direction.

The measured V_{BD} of pn junction TEGs with JTE terminations was less than 2.5 kV, suggesting that the charge compensation within the partial SJ structures was not achieved. TCAD simulation revealed that the acceptor concentration in the p-type columns was as low as 0.8×10^{16} cm^{-3}. We therefore checked the amount of incorporated Al atoms in the embedded epilayer directly by SIMS analysis of

the SJ region. The actual Al concentration was deduced by considering the areal ratio of n and p-pillar regions observed in SEM images. As shown in Fig. 7, incorporated Al concentration grown under the TFE growth conditions (blue symbols) is remarkably decreased compared to that grown under the standard condition without HCl gas (black symbols). Furthermore, the incorporated Al concentration in the case of trenched substrates (red symbols) shows further decrease compared to that in the case of flat substrates. This was the reason why the Al concentration in the p-type column, became low. However, in the range of at least 0.5 to 3×10^{16} cm^{-3}, linearity of the incorporated Al concentration with respect to TMA flow rate was confirmed. The SJ wafers for MOSFETs fabrication were thus prepared according to the calibration curves in Fig. 7.

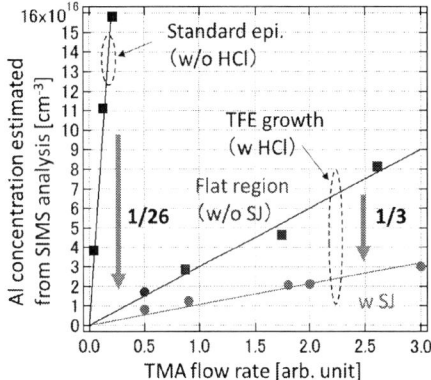

Figure 7. TMA flow rate dependence of the incorporated Al concentration in the case of flat substrates grown under standard epitaxial conditions (black squares) and in the case of flat substrates (blue squares and circle) and trenched substrates (red circles) grown under TFE growth conditions.

Fabrication of 6.5 kV-class SJ MOSFETs

6.5 kV-class SJ MOSFETs were fabricated on the SJ wafer, where the Al concentration in the SJ region was designed to be 1.5×10^{16} cm^{-3} using the calibration curves. A cell pitch of SJ MOSFETs was 10 µm, corresponding to twice the pn column pitch. Gate oxide formation was performed at 1450°C in dry O$_2$ following by post-oxidation annealing in nitric oxide gas ambient at the same temperature. Figure 8 and 9 show forward and blocking characteristics, respectively.

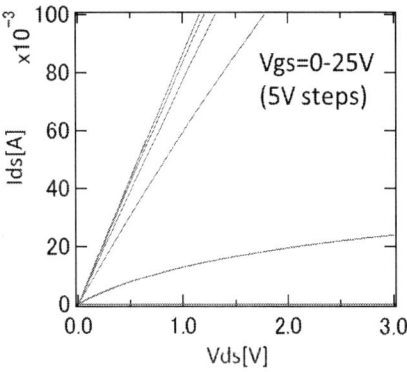

Figure 8. Ids-Vds of SJ MOSFET at RT.

The measured R_{onA} at RT and V_{BD} were 17.8 mΩcm^2 and 7.8 kV, respectively. Typical R_{drift} value at RT was 13.3 m$\Omega \cdot$cm^2 as mentioned above, indicating other resistance components considering of channel and JFET resistances were 4.5 m$\Omega \cdot$cm^2. Temperature dependence of R_{onA} of another device fabricated on the same wafer revealed that the R_{onA} was 18.2 m$\Omega \cdot$cm^2 at RT and 48.7 m$\Omega \cdot$cm^2 at 175°C. The R_{onA} values are less than half the values of the state-of-the-art 6.5 kV-class SiC MOSFET with an n-type drift layer [16]. As a result, the SiC SJ MOSFET exceeding the unipolar limit of 4H-SiC were successfully fabricated for the first time, as shown in Fig. 10.

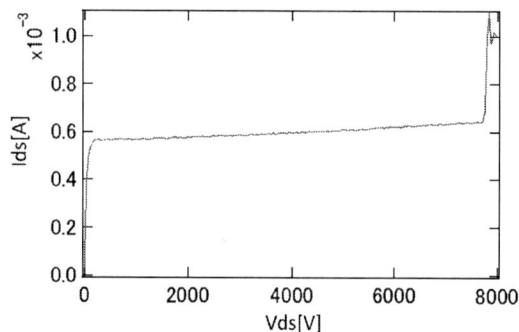

Figure 9. Blocking characteristics of SJ MOSFET (not removing voids by cleavage).

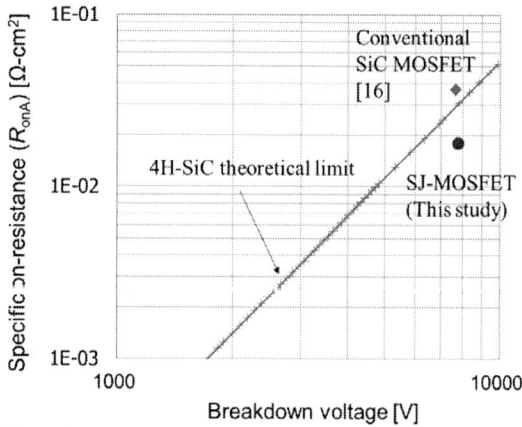

Figure 10. Trade-off relationship between specific resistivity of drift layer and breakdown voltage. Blue circle corresponds to this study.

IV. CONCLUSIONS

We have developed elemental fabrication processes for deep trench-filling epitaxial (TFE) growth method, and confirmed for the first time that vertical pn junctions

fabricated by the deep TFE growth function normally. The measured V_{BD} and R_{onA} of the SJ MOSFETs were 7.8 kV and 17.8 m$\Omega \cdot$cm^2, respectively. These results show that the trade-off relationship between V_{BD} and R_{onA} of 4H-SiC is improved by the introduction of a pn column structure into the drift layer. We believe that SiC SJ devices open a new application field for high-voltage unipolar power devices, which is not realized with Si-IGBTs.

ACKNOWLEDGMENT

This work was supported by the Council for Science, Technology and Innovation (CSTI), the Cross-Ministerial Strategic Innovation Promotion Program (SIP), and the "Next-generation power electronics/Consistent R&D of next-generation SiC power electronics" funded by NEDO.

REFERENCES

[1] T. Fujihira, Jpn. J. Appl. Phys. 36, (1997) pp. 6254.

[2] R. Kosugi, Y. Sakuma, K. Kojima, S. Itoh, A. Nagata, T. Yatsuo, Y. Tanaka, H. Okumura, Mat. Sci. Forum 778–780, (2014) pp. 845.

[3] R. Kosugi, Y. Sakuma, K. Kojima, S. Itoh, A. Nagata, T. Yatsuo, Y. Tanaka, and H. Okumura, Proc. Int. Symp. Power Semicond. Devices & IC's, Waikoloa, (2014) pp. 346.

[4] T. Masuda, R. Kosugi, and T. Hiyoshi, Mat. Sci. Forum 897, (2017) pp. 483.

[5] T. Masuda, Y. Saito, T. Kumazawa, T. Hatayama, and S. Harada, Tech. Dig. Int. Electron Devices Meeting, San Francisco, (2018) p. 177.

[6] S. Harada, Y. Kobayashi, S. Kyogoku, T. Morimoto, T. Tanaka, M. Takei, and H. Okumura, Tech. Dig. Int. Electron Devices Meeting, San Francisco, (2018) pp. 181.

[7] K. Mochizuki, R. Kosugi, Y. Yonezawa, and H. Okumura, to be published in Mat. Sci. Forum.

[8] R. Ghandi, P. Losee, A. Bolotnikov, and D. Lilienfeld, Mat. Sci. Forum vol. 924, (2018) pp. 573.

[9] R. Kosugi, Y. Sakuma, K. Kojima, S. Itoh, A. Nagata, T. Yatsuo, Y. Tanaka, and H. Okumura, Mat. Sci. Forum 740-742, (2013) pp. 785.

[10] K. Kojima, A. Nagata, S. Ito, Y. Sakuma, R. Kosugi, and Y. Tanaka, Materials Science Forum 740–742, (2013) pp. 79.

[11] S.Y. Ji, K. Kojima, R. Kosugi, S. Saito, Y. Sakuma, Y. Tanaka, S. Yoshida, H. Himi, and H. Okumura, Applied Physics Express, vol.8, (2015) pp. 065502-1/065502-4.

[12] S. Ji, R. Kosugi, K. Kojima, K. Mochizuki, S. Saito, A. Nagata, Y. Matsukawa, Y. Yonezawa, and H. Okumura, Appl. Phys. Express, vo 10, (2017) pp. 05550.

[13] K. Mochizuki, S. Ji., R. Kosugi, K. Kojima, Y. Yonezawa, and H. Okumura, Appl. Phys. Express 9, (2015) pp. 035601.

[14] K. Mochizuki, S. Ji., R. Kosugi, Y. Yonezawa, and H. Okumura, Tech. Dig. Int. Electron Devices Meeting, San Francisco, (2017) p. 788.

[15] K. Mochizuki, S. Ji., R. Kosugi, Y. Yonezawa, and H. Okumura, Proc. Int. Conf. Simulation of Semicond. Processes & Devices, Austin, (2018) pp. 331.

[16] K. Kawahara, S. Hino, K. Sadamatsu, Y. Nakao, Y. Yamashiro, Y, Yamamoto, T. Iwamatsu, S. Tomohisa, and S. Yamakawa, Proc. The 29th Int. Symp. Power Semicond. Devices & IC's, Sapporo, (2017) pp. 41.

[17] R. Kosugi, S. Ji, K. Mochizuki, H. Kouketsu, H. Kawada, H. Fujisawa, K. Kojima, Y. Yonezawa, and H. Okumura, Jpn. J. Appl. Phys. 56, (2017) pp. 04CR05.

Proceedings of the 31st International Symposium on Power Semiconductor Devices & ICs
May 19-23, 2019, Shanghai, China

3300V Scaled IGBTs Driven by 5V Gate Voltage

Takuya Saraya[1], Kazuo Itou[1], Toshihiko Takakura[1], Munetoshi Fukui[1], Shinichi Suzuki[1], Kiyoshi Takeuchi[1], Masanori Tsukuda[2], Yohichiroh Numasawa[3], Katsumi Satoh[4], Tomoko Matsudai[5], Wataru Saito[5], Kuniyuki Kakushima[6], Takuya Hoshii[6], Kazuyoshi Furukawa[6], Masahiro Watanabe[6], Naoyuki Shigyo[6], Hitoshi Wakabayashi[6], Kazuo Tsutsui[6], Hiroshi Iwai[6], Atsushi Ogura[3], Shin-ichi Nishizawa[7], Ichiro Omura[8], Hiromichi Ohashi[6], and Toshiro Hiramoto[1]

Email: saraya@nano.iis.u-tokyo.ac.jp
[1]The University of Tokyo, Tokyo, Japan,
[2]Green Electronics Research Institute, Kitakyushu, Japan, [3]Meiji University, Kawasaki, Japan, [4]Mitsubishi Electric Corp., Fukuoka, Japan, [5]Toshiba Electronic Devices & Storage Corp., Tokyo, Japan, [6]Tokyo Inst. of Technology, Yokohama, Japan, [7]Kyushu University, Kasuga, Japan, [8]Kyushu Inst. of Technology, Kitakyushu, Japan

Abstract— **In this work, 5V gate drive 3300V IGBTs, designed based on a scaling principle, have been demonstrated. Turn-off characteristics without noticeable degradation in the gate voltage waveforms were confirmed. Turn-off tail current of the scaled devices significantly decreased than conventional 15V-driven devices. As a result of both V_{ce} and turn-off loss reduction, 35% improvement in E_{off} vs V_{cesat} relationship was achieved.**

Keywords— IGBT, Scaling, Injection Enhancement

I. INTRODUCTION

While many candidates for next generation power switching devices, e.g. SiC, GaN, Ga_2O_3, are developed [1], Si-IGBTs (Insulated Gate Bipolar Transistors) are still the main stream of power devices, and their performance improvement is strongly required to keep high efficiency, especially in high voltage region [2-4]. For a long time, gate voltage has been fixed to 15V for conventional IGBTs [5]. Reducing the voltage to 5V will realize not only much smaller size and lower cost gate drivers, but also more flexible and intelligent power control schemes [6] thanks to CMOS compatibility.

Recently, a scaling concept for Si-IGBTs has been proposed [7,8] (Fig.1 and Table 1), which realizes low voltage gate control while keeping or even improving the performance, thanks to an Injection Enhancement (IE) effect.

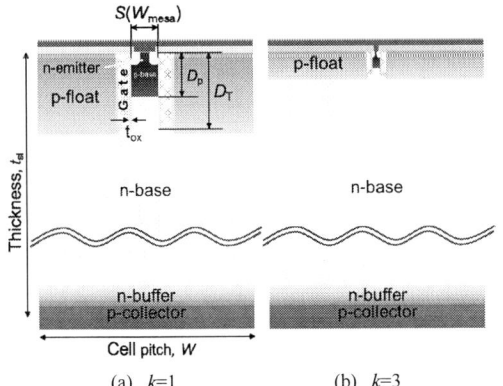

(a) k=1　　　(b) k=3
Fig. 1. Scaling principle of trench gate IGBTs.

That is, similarly to the CMOS scaling, all the geometrical dimensions, both horizontal and vertical, as well as gate voltage, are scaled down proportionately, while keeping the cell pitch W constant. Applying this concept, we have

Table 1. Structural parameters of scaled IGBTs.

	k=1	k=3
Cell pitch, W	1	1
Mesa width, S	1	1/3
Trench depth, D_T	1	1/3
p-base depth, D_P	1	1/3
Trench extrusion, $D_T - D_P$	1	1/3
Gate oxide thickness, t_{ox}	1	1/3
Gate voltage, V_g	1	1/3

demonstrated 5V driven 1200V-10A class scaled IGBTs (scaling factor k=3) with improved on-state voltage (V_{cesat}) vs turn-off loss (E_{off}) tradeoff [9]. But, the applicability of 5V gate drive to higher blocking voltage (BV) has not been proven yet. The purpose of this work is to show that 3.3kV IGBT can be turned-off by 5V gate drive and hence the gate drive voltage scaling is applicable to a higher BV range.

II. DEVICE DESIGN AND FABRICATION

A. Design and Simulation

Starting from the 1200V design, modifications and optimizations were made to achieve 3300V BV. To achieve low V_{cesat} with the increased BV, design of the p-float region was identified to be a key parameter. It should be optimized to balance switching stability and IE effect. N-emitter/p-body structures were also modified to improve latch-up resistance [8]. An n-buffer [10] (Field Stop [11]) structure was adopted to obtain 3300V BV with thin and low dose wafer.

Simulated collector-to-emitter current density (J_{ce}) vs collector-to-emitter voltage (V_{ce}) characteristics for k=1 and k=3 for the same p-collector doping and corresponding on-state carrier distributions at J_{ce}=80A/cm^2 are shown in Figs. 2 and 3. The carrier lifetime in n-base region (τ_e) is defined as 10μsec for both k=1 and k=3. The higher J_{ce} for k=3 than k=1 in Fig.2 was obtained. From the carrier distribution in Fig.3, it is expected that the scaling will increase the carrier concentration selectively at the front side. That is, it will decrease hole sinking efficiency and increase electron injection efficiency of the trench gated emitter (IE effect). As is well known, the back side carrier concentration can be easily controlled by the p-collector doping. But the increased carrier concentration of the back side by high p-collector doping makes worse the switching loss. An advantage of the

978-1-7281-0582-6/19 $31.00 © 2019 IEEE　　43

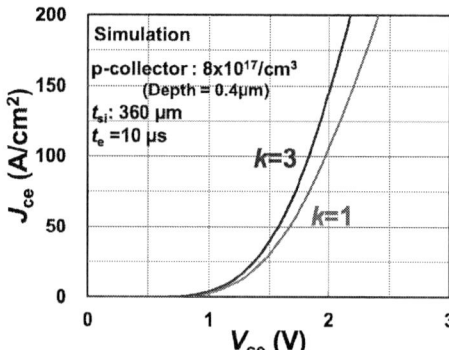

Fig. 2. Simulated on-state characteristics of k=1 and k=3 trench gate IGBTs.

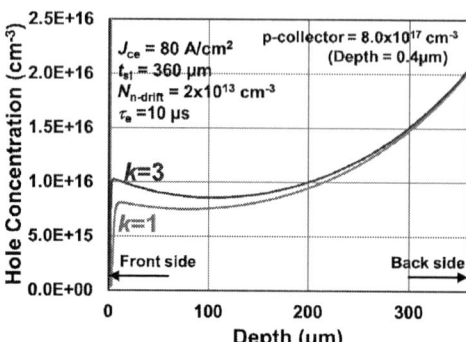

Fig. 3. Simulated carrier distribution on k=1 and k=3 IGBTs.

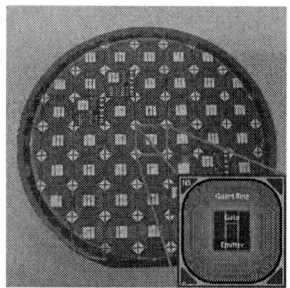

Fig. 4. Chip photograph of IGBT wafer fabricated in facilities of the Univ. of Tokyo.

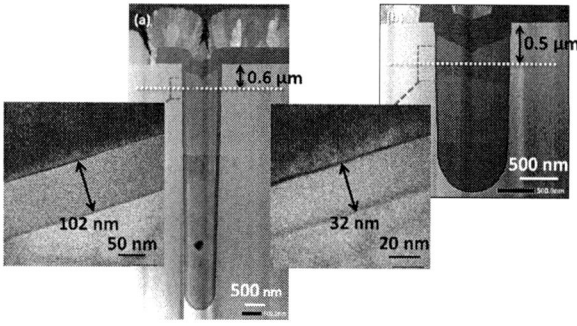

Fig. 5. Cross sectional TEM images at the trench gate of IGBT device. (a) k=1 IGBT with 6μm trench depth and 3μm mesa width. (b) k=3 IGBT with 2μm trench depth and 1μm mesa width.

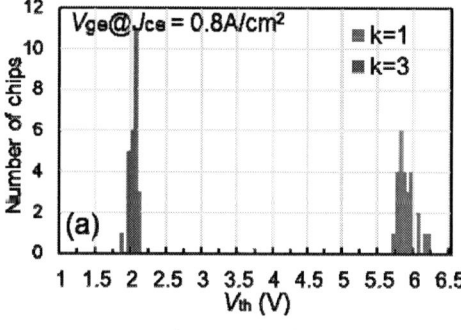

Fig. 6. (a) Threshold voltage and (b) On-state voltage distribution of k=1 and k=3 IGBTs.

k=3 device is that a flatter carrier profile is obtained, which is desirable for achieving both low on-resistance and low switching loss. It is expected that the scaling of IGBTs will lead to reduced V_{cesat} without turn-off loss degradation.

B. Fabrication Process

Table 1 shows structural parameters of conventional IGBT (k=1) and scaled IGBT (k=3). Both IGBTs were fabricated on 3-inch wafers (Fig. 4). P-collector concentration was also varied to reveal trade-off of V_{cesat} and switching losses. The chip size is 81mm² and active emitter area is 5mm². The width and depth of the gourd ring (surface termination) structure were not scaled in this work. Therefore, the size of k=1 and k=3 chips are same. Fig.5 shows cross sectional TEM images of the trench gate regions. The trench depth, the distance between the two trenches (i.e. mesa width S) and gate oxide thickness are reduced to 1/3 for k=3, though the trench width is constant (1μm) to avoid complication of the fabrication process.

III. DEVICE CHARACTERISTICS

Figure 6 shows histograms of (a) threshold voltage (V_{th}) and (b) V_{cesat} for both k=1 and k=3 IGBTs. (a) Threshold voltages were reduced by around 1/3 from k=1 to k=3 by reducing gate oxide thickness and adjusting the p-base doping concentration. According to the reduction of gate oxide thickness, the variability of V_{th} in k=3 is less than that of k=1. (b) V_{cesat} distribution also shows good uniformity across the wafer for the same p-collector doping.

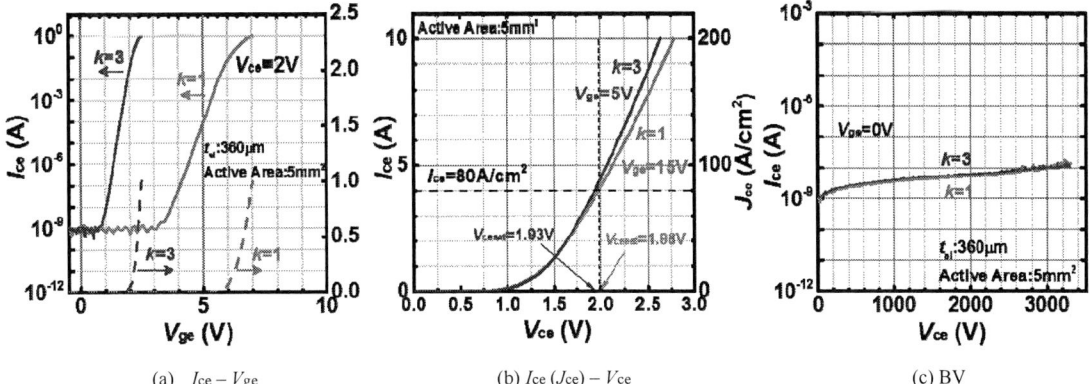

(a) $I_{ce} - V_{ge}$ (b) $I_{ce}(J_{ce}) - V_{ce}$ (c) BV

Fig. 7. Typical static characteristics of 3300V k=1 and k=3 IGBTs.

A. Static Characteristics

Fig.7 compares typical (a) collector current (I_{ce}) vs gate voltage (V_{ge}), (b) I_{ce} (J_{ce}) vs V_{ce}, and (c) BV characteristics. (a) Threshold voltage for k=3 was adjusted to be 1/3 that for k=1. Subthreshold slope and transconductance for k=3 were improved by the gate oxide scaling. (b) I_{ce}-V_{ce} slopes of k=1 and k=3 IGBTs were compared, keeping the p-collector doping the same. I_{ce}-V_{ce} slope of k=3 was steeper than that of k=1, which can be attributed to the IE effect. Despite the reduced gate voltage by 1/3, the V_{cesat} (defined as V_{ce} at J_{ce}=80A/cm^2) improvement from k=1 to 3 of 0.05V was obtained. (c) Both k=1 and k=3 IGBTs withstood 3300V, while keeping the leakage current small. No significant change by the scaling in I-V characteristics was observed.

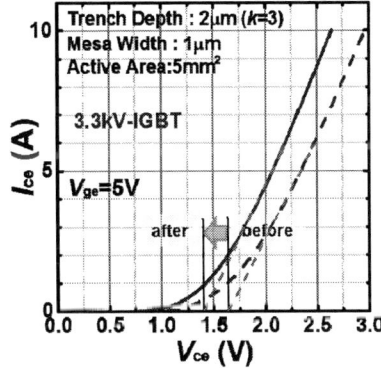

Fig. 8. I_{ce} - V_{ce} characteristics of 3300V IGBTs (k=3)

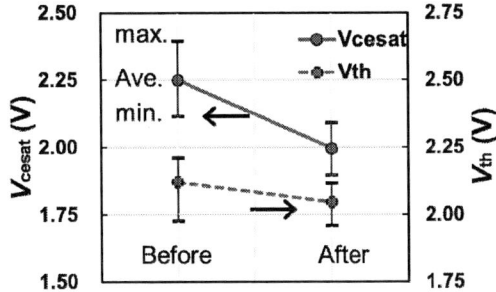

Fig. 9. Vth and V_{cesat} distribution before and after p-float modification on k=3 IGBTs.

This means the shrinking of the trench structures did not affect the critical electric filed and BV.

It was found that the p-float design should be modified from 1200V to 3300V BV. Not only I_{ce}-V_{ce} slope but also offset voltage on V_{ce} were degraded by increasing the thickness and lowering doping concentration of the n-base. By optimizing the p-float doping concentration and resistive grounding, V_{cesat} was significantly improved by reducing offset voltage on V_{ce} (Fig.8). Variability of V_{cesat} and Vth were also reduced by modifying the p-flaot design (Fig. 9). That is, the optimization of the p-float design improved both IE effect and uniformity.

B. Switching Characteristics

Switching characteristics were measured using the circuit in Fig.10. Switching voltage of 1700V was applied for both k=1 and k=3 devices. The gate voltage applied was ±15V for k=1 and ±5V for k=3, respectively. Fig.11 shows typical switching waveforms for a switched voltage of 1700V and current density of 80A/cm^2. The scaled k=3 IGBT was successfully turned off by ±5V gate voltage, without noticeable degradation in the gate voltage waveform. A magnified view of the turn-off waveforms are shown in Fig.12, because the switching loss in IGBT is mainly determined by the turn-off loss. The turn-off tail current of the k=3 device significantly decreased than k=1 for the same p-collector dose. This indicates that the amount of accumulated charge near the bottom side is less in k=3 than k=1, in spite of the improved V_{cesat} in Fig.7(b).

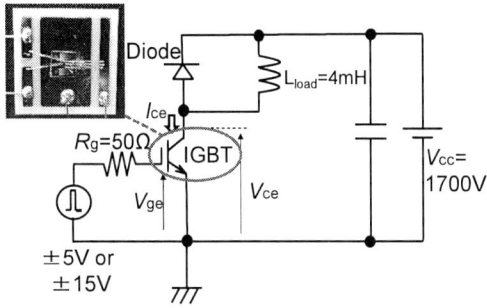

Fig. 10. Measurement circuit for switching losses.

978-1-7281-0582-6/19 $31.00 © 2019 IEEE

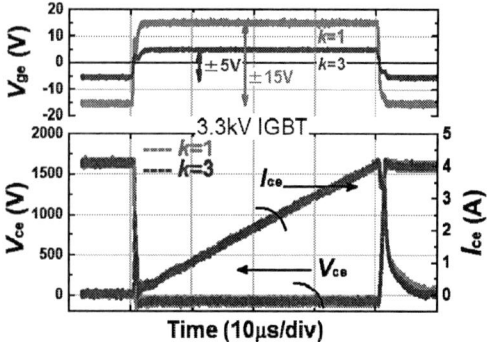

Fig. 11. Switching waveform of 3300V IGBTs under 1700V-80A/cm² switching condition.

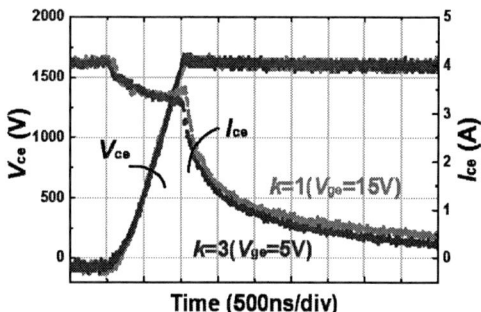

Fig. 12. Turn-off waveform of 3300V IGBTs under 1700V-80A/cm² switching condition.

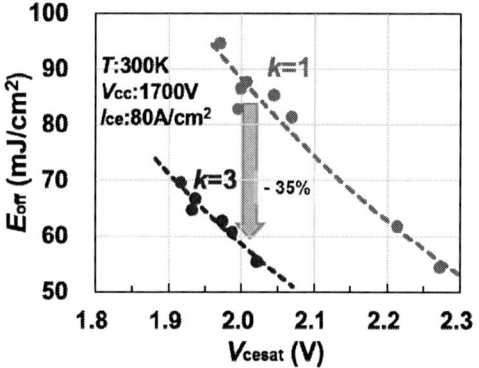

Fig. 13. E_{off} - V_{cesat} trade-off curve comparing k=1 and k=3 IGBTs.

Fig.13 shows tradeoff between E_{off} and V_{cesat} for both k=1 and k=3 IGBTs. Here, V_{cesat} is controlled by changing the p-collector dose. E_{off} vs V_{cesat} plots for k=1 and k=3 fall on different curves, which are clearly separated. As a result of both the V_{ce} reduction, and also turn-off loss reduction, 35% improvement in E_{off} was achieved for the same V_{cesat}. That is, clear improvement of the performance by IGBT scaling with 5V gate drive was confirmed even though the BV became higher.

IV. CONCLUSIONS

3300V-5A class scaled IGBTs with 5V gate dive were fabricated, and their static and switching characteristics were verified. V_{cesat} was improved by 0.05V from k=1 to 3 for the same p-collector dose. It was confirmed that 3300V IGBTs can be turned-off by 5V gate drive without noticeable gate voltage waveform degradation. Scaled IGBTs show 35% reduction of turn-off loss for same on-state voltage with high blocking voltage. These results demonstrate that 5V gate drive is applicable even to 3300V IGBTs, which makes more efficient and intelligent gate control viable even in the high voltage range.

ACKNOWLEDGMENT

The authors are grateful to M. Sakamoto and T. Tanaka of Sumitomo Heavy Industries for technical supporting on the laser anneal.

This work is based on results obtained from a project commissioned by the New Energy and Industrial Technology Development Organization (NEDO).

REFERENCES

[1] T. P. Chow, I. Omura, M. Higashiwaki, H. Kawarada, and V. Pala, "Smart Power Devices and ICs Using GaAs and Wide and Extreme Bandgap Semiconductors", IEEE Transaction on Electron Device, vol.64, No.3, pp.856-873, 2017.

[2] A. Kopta, et al., "Next Generation IGBT and Package Technologies for High Voltage Applications", IEEE Transaction on Electron Device, vol.64, No.3, pp.753-759, 2017.

[3] M. Mori, T. Miyoshi, T. Furukawa, Y. Takeuchi, Y. Hotta, and M. Shiraishi, "An Innovative Silicon Power Device (i-Si) through Time and Space Control of a Stored Carrier (TASC)", Proc. 30th International Symposium on Power Semiconductor Devices and ICs, pp.520-523, Chicago, USA, May 2018.

[4] Y. Toyota et al., "Novel 3.3kV advanced trench HiGT with low loss and low dv/dt noise", Proc. 25th International Symposium on Power Semiconductor Devices and ICs, pp.29-32, Kanazawa, Japan, May 2013.

[5] N. Iwamuro, and T. Laska, "IGBT History, State-of-the-Art, and Future Prospects", IEEE Transaction on Electron Device, vol.64, No.3, pp.741-752, 2017.

[6] K. Miyazaki et al., "General-Purpose Clocked Gate Driver IC with Programmable 63-Level Drivability to Optimize Overshoot and Energy Loss in Switching by a Simulated Annealing Algorithm", IEEE Transactions on Industry Applications, Vol.53, No.3, pp.2350-2357, 2017.

[7] M. Tanaka and I. Omura, "IGBT scaling principle toward CMOS compatible wafer processes", Solid-State Electronics, No.80, pp.118-123, 2013.

[8] K. Kakushima et al., "Experimental Verification of a 3D Scaling Principle for Low Vce(sat) IGBT", International Electron Devices Meeting pp.268-271, San Francisco, USA, December. 2016.

[9] T. Saraya et al., "Demonstration of 1200V Scaled IGBTs Driven by 5V Gate Voltage with Superiorly Low Switching Loss", International Electron Devices Meeting, pp.189-192, San Francisco, USA, December, 2018.

[10] T. Matsudai, K. Kinoshita, and A. Nakagawa, "New 600 V trench gate punch-through IGBT concept with very thin wafer and low efficiency p-emitter, having an on-state voltage drop lower than diode", IPEC-Tokyo, pp.292–296, Tokyo, Japan, Apr. 2000.

[11] T. Laska, M. Munzer, F. Pfirsch, C. Schaeffer, and T. Schmidt, "The Field Stop IGBT (FS IGBT) – A New Power Device Concept with a Great Improvement Potential", Proc. 12th International Symposium on Power Semiconductor Devices and ICs, pp.355-358, Toulouse, France, May 2000.

978-1-7281-0582-6/19 $31.00 © 2019 IEEE

Proceedings of the 31st International Symposium on Power Semiconductor Devices & ICs
May 19-23, 2019, Shanghai, China

A Critical View of IGBT Buffer Designs for 200 °C Operation

Elizabeth Buitrago[a], Athanasios Mesemanolis[a], Maxi Andenna[a], Charalampos Papadopoulos[a], Chiara Corvasce[a], Jan Vobecky[a], Munaf Rahimo[b]

[a]ABB Semiconductors, Lenzburg, Switzerland
[b] MTAL GmbH, Gänsbrunnen, Switzerland
elizabeth.buitrago@ch.abb.com

Abstract—Different state-of-the-art buffer concepts compatible with ≥ 200 mm diameter, thin wafer technology (< 200 μm) have been optimized in enhanced planar insulated gate bipolar transistor (IGBT) technology for 1200 V and 1700 V, 150 A rated, 13.6 x 13.6 mm² devices and evaluated with the goal of reducing leakage currents to enable a stable operation up to T_j = 200 °C. With the maximum punch through (MPT) ultra-thin field stop design in particular, we can produce devices with the lowest leakage currents up to T_j = 200 °C. These devices were found to provide rugged RBSOA and soft switching, can withstand short circuit pulses as long as 10 μs and be thermally stable at T_j = 200 °C.

I. INTRODUCTION

Low voltage (V_{ce} < 2000 V) IGBT designers have proposed different N+ field-stop concepts throughout the years to deal with the need for thin N-drift layers to produce devices with the lowest possible static on-state and dynamic turn-off losses while maintaining sufficiently high blocking voltage capabilities [1]. Yet performance and cost must be carefully weighted to produce the most efficient device for the target application. In this investigation, low leakage currents for high temperature functionality up to T_j = 200 °C while maintaining other high-performance characteristics was targeted.

The simple addition of an ultra-shallow (< 1 μm) but high doping concentration phosphorus peak into the soft-punch-through (SPT) buffer IGBT based on deep diffused (DD) wafer technology has already been shown to significantly reduce the leakage current at elevated operation temperatures up to 200 °C [2]. This type of implementation still enables full front-side cathode processing before thinning to the desired thickness. Nonetheless, design freedom is limited as grinding consumes the buffer in a less controllable fashion. Moreover, not only is the field stop profile fixed by the silicon supplier, but manufacture of DD wafers with large diameters (≥ 150 mm) is restricted and costly. To overcome such constraints, we have benchmarked several buffer designs that allow for full front-side MOS (metal oxide semiconductor) processing without any fabrication limitations for 200 mm wafer diameter LV IGBT compatibility. These concepts have been further optimized for leakage reduction and high temperature IGBT operation. Fig. 1 shows the spreading resistance profiles (SRP) showing the activated carrier concentration for the different buffers investigated.

1. **Maximum punch through (MPT)** buffer (Fig.1a) defined by its V_{pt}/V_{bd} ratio at 70 - 99%, where V_{pt} is the punch through voltage and V_{bd} is the breakdown voltage [3]. This buffer

(and anode) can been fully produced by low/medium energy ion implantation of phosphorous and boron. Dopant activation can then be achieved by consecutive or serial laser annealing (LA). LA is a powerful technique in which heat dissipates within the first few microns from the annealing surface so that the entire fully processed front-side thinned wafer does not experience the high temperatures needed for dopant activation.

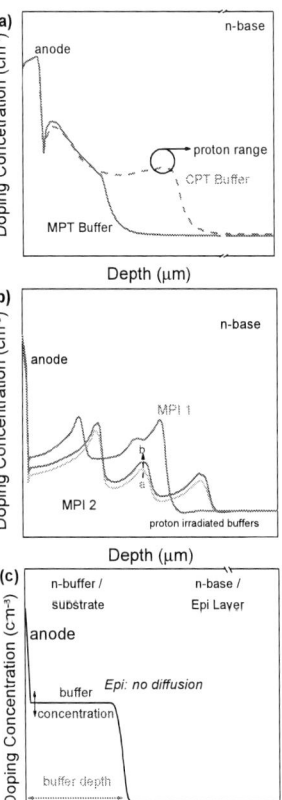

Fig. 1. Doping profiles for a) MPT and CPT type buffers, b) MPI buffer and c) Epi SPT buffer concept, where the thinned n-type substrate serves as the buffer region and grown epitaxial layer serves as the n-type drift region.

2. **Controlled punch through (CPT)** buffer (Fig.1a) consisting of a double layer profile: a detached but relatively shallow proton peak (< 10 μm) to stop the electric field farther away from the anode and a phosphorous layer next to the anode to reduce the leakage current [4]. Proton activation can be performed at low temperatures < 400 °C while the phosphorous layer (and anode) can be activated by LA in a separate step.

3. **Multiple proton implantation (MPI)** buffer (Fig.1b) produced by three different, medium to high energy (> 1 MeV) hydrogen sequential implantations and low temperature annealing (< 450° C) [5]. The annealing temperature and time has been optimized to maximize a stable hydrogen related donor formation. This design offers high flexibility to shape the field stop to produce a device with the proper static and switching characteristics depending on the desired application.

4. **Epi SPT** buffer (Fig.1c) design consisting of an epitaxial layer as the n-type drift region and substrate field stop that can be adjusted by thinning after cathode processing [6].

II. EXPERIMENTAL RESULTS

A. Static performance at high temperatures

Figs. 2 and 3 show technology curves (switching vs. conduction losses) for the 1200 V and 1700 V devices respectively, measured at 175 °C. A constant device total thickness was maintained regardless of the type of buffer implemented per voltage except for the Epi SPT, where reduced losses were further achieved by decreasing the device thickness without compromising other high performance characteristics (with the right combination of substrate and epi layer resistivities and thicknesses). In all buffer concepts, turn-off softness can be adjusted by tuning the anode injection efficiency by LA. In particular, as can be seen in Fig. 2, two different MPT and EPI SPT buffer devices with different positions on the technology curve have been produced.

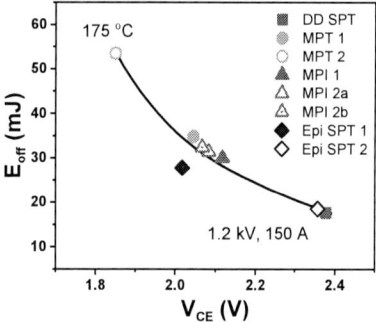

Fig. 2. Trade-off curve measured at 175 °C (L_s = 200 nH, same gate drive conditions) for IGBTs with different buffer concepts. The buffer depth and device thickness for the Epi SPT buffer1 is reduced and sits in technology curve with reduced losses.

As previously stated, the different buffer designs were optimized to produce devices with the lowest leakage current for high temperature operation up to 200 °C. For these measurements, each IGBT substrate is mounted on a temperature-controlled board with resistive heaters. Nonetheless, it's clear that the measured (thermocouples) heat sink temperature is not the same as the actual junction temperature. For this reason, leakage

current was measured for 1200 and 1700 V chips at 200°C during a very short 10 ms pulse to avoid self-heating (Figs. 4 and 5). Thermal resistance between the substrate and the heat sink is minimized by using a thermally conductive foil.

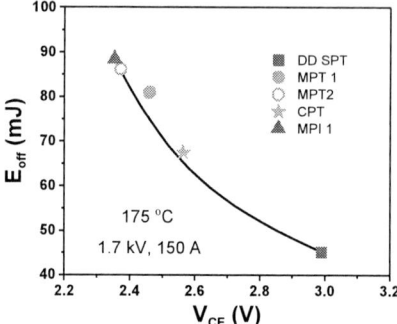

Fig. 3. Trade-off curve measured at 175 °C (L_s = 200 nH, same gate drive conditions) for IGBTs with different buffer concepts.

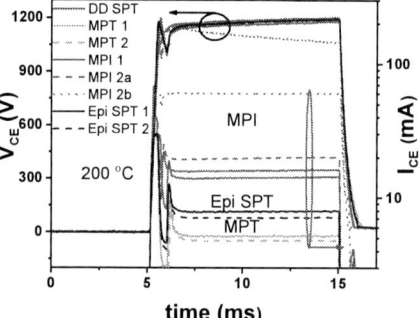

Fig. 4. Leakage current measured at 200 °C while applying a blocking voltage of 1200V for 10 ms to prevent self-heating for all buffer designs.

The MPT and Epi SPT buffers have the lowest leakage currents below the reference DD SPT and well below the MPI 2 design for the full temperature range tested (Fig. 6). The leakage current of the MPI design (Fig. 6) can be improved by increasing the overall peak doping concentration (holes coming from anode face higher potential energy barrier as they diffuse through buffer and n-base (Fig. 1b doping profiles) when going from buffer profile MPI 2a to 2b. Furthermore, by decreasing the depth and increasing the doping concentration of the deepest peak the leakage current was significantly reduced to levels closer to the reference DD SPT IGBT without affecting blocking. A deep buffer design is typically desirable for softness and to be able to withstand harsh turn-off conditions. The deepest peak serves to slow down the expansion of the space-charge during turn off so that plasma can be extracted in a smoother way [5]. Softness nonetheless can also be improved by anode activation via LA as it was done for the MPI 1 buffer here with which an over x4 leakage reduction (vs. MPI 2a) was possible. All three MPI buffer designs lie closely together in the technology curve as can be seen in Fig. 2. It is therefore possible to optimize the buffer depth for the reduction of leakage current for high temperature applications and blocking without compromising in softness and turn off losses in the MPI design.

For leakage current reduction optimization of the MPT (Figs. 5 and 6) and CPT (Fig. 6) buffer designs it is important to produce shallow (< 3 μm) but highly doped phosphorous layer buffer profiles (Fig. 1a). This is to ensure blocking while still maintaining a sufficient separation between buffer and anode for bipolar gain adjustment and positioning in the technology curve. Maximum leakage reduction is possible as the hole concentration in the drift region efficiently decreases as holes coming from the anode need to diffuse across this highly N+ doped area.

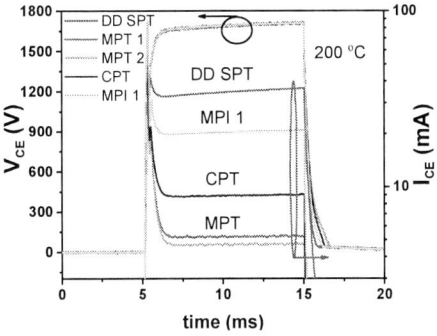

Fig. 5. Leakage current measured at 200 °C while applying a blocking voltage of 1700V for 10 ms to prevent self-heating for different buffer designs.

As the anode can also be activated in the same step, the bipolar gain can be adjusted so that the device has similar on-state losses and keep the same position in the technology curve. The high doping concentration of the buffer peak may prevent the electric field and prevent it from reaching the collector during blocking.

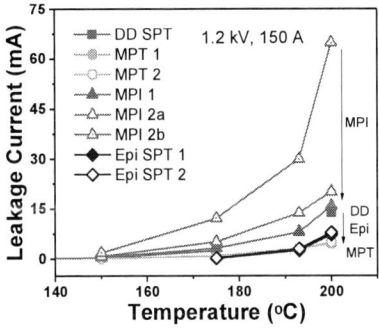

Fig. 6. Leakage current measured at different temperatures from 150 °C to 200 °C for 1200 V devices.

B. Switchig and RBSOA Capability up to 200°C

For the elevated temperature measurements (200°C), the different IGBT/diode substrates are mounted on a heating plate to maintain the desired device temperature. Fig. 7 shows the turn-off waveforms for two different MPT devices and the CPT buffer IGBT measured at 200 °C (1700V, 150 A) under nominal conditions. A favorable place in the technology curve (Fig. 3) and soft-switching is possible to be achieved with both state-of-the art buffer concepts at 1200 (not shown) and 1700 V at elevated temperatures. Furthermore, these devices were also shown to produce rugged switching under RBSOA conditions as can be seen in Fig. 8.

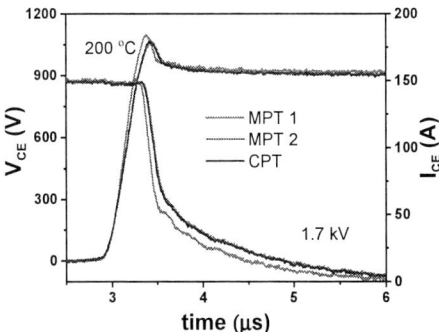

Fig. 7. Turn-off waveforms under nominal conditions (900 V, 150 A, 200 nH, 8.2 Ohm) at 200°C for CPT and MPT buffer IGBTs rated to 1700V and 150A.

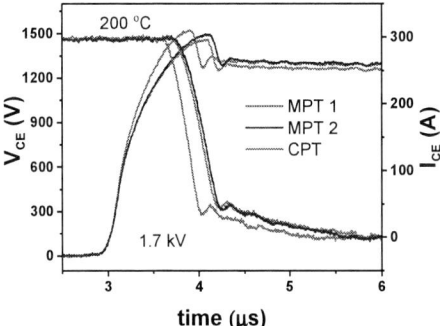

Fig. 8. Turn-off waveforms at 2x nominal conditions (1300 V, 300 A, Ls = 200 nH, Rg = 8.2 Ohm) measured at 200 °C for 1700V / 150A IGBTs with MPT and CPT buffers.

C. Short-Circuit capability up to 200°C

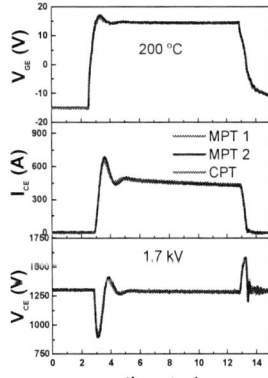

Fig. 9. Short-circuit waveforms (1300 V, 200 nH, 8.2 Ohm) measured at 200 °C for IGBTs (1700V, 150A) with MPT and CPT buffers. These devices pass the short-circuit pulses up to 10 μs at 200 °C.

Short-circuit (SC) capability was investigated at 200 °C. Thermal runaway due to excess energy during short circuit pulse is the typical failure mechanism observed during this test [5][6]. Power cannot be controlled at the testbench due to a slow feedback system. There is also no active cooling implemented therefore the generated heat cannot be easily dissipated leading to a severe

978-1-7281-0582-6/19 $31.00 © 2019 IEEE   49

increase of the junction temperature up to more than double the value at the heating plate for a 10 μs pulse [7]. The MPT and CPT devices are still able to withstand short circuit pulses as long as 10 μs without thermal runaway (Fig. 9 for 1700 V).

D. HTRB stability

Since short pulse measurements are not a criterion of thermal stability, high temperature reverse bias HTRB tests were performed. A voltage of 1200 or 1700 V is applied for 2 minutes to see if the device self-stabilizes over time with the heating system without active cooling or is unstable due to thermal runaway. These measurements were performed at 200 °C and up to 205 °C for 120 and up to 180 seconds for 1200 V IGBTs (Fig.10 and 11). The MPI and DD SPT devices with leakage currents >10 mA are thermally unstable at 200 °C. The MPT and Epi SPT devices nonetheless with leakage currents < 10 mA are thermally stable at 200 °C and the MPT buffer provides even thermal stability at 205 °C (Fig. 11).

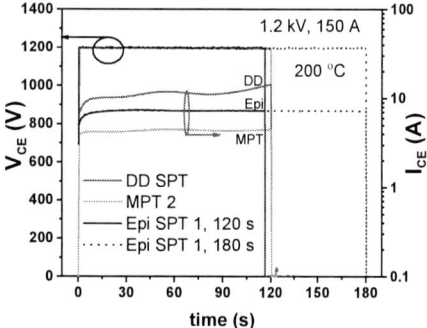

Fig. 10. Leakage current stability over time at 1200 V. Devices stabilize over time as long as the leakage current at 200 °C is below 10 mA.

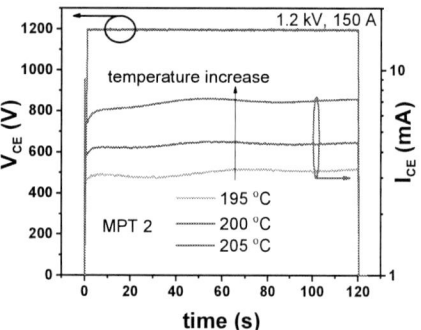

Fig. 11. Leakage current stability over time at 1200 V at increasing temperatures up to 205 °C for MPT buffer IGBT.

For 1700 V devices the MPT buffer shows again the lowest leakage current and thermal stability (Figs.5 and 12). The CPT buffer shows also low leakage current at 200 °C, but it does not stabilize over time. Only the buffer concepts produced with high temperature drive-in dopants alone provide devices with stable 200 °C operation as the buffer depth can be kept at a minimum. These devices can generally be stabilized as long as the leakage current stays below 10 mA. Above this value the test system cannot react fast enough and the device thermally runs away.

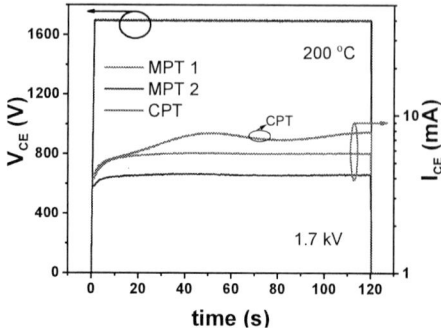

Fig. 12. Leakage current stability over time at 1700 V and 200°C (1700 V, 150 A devices). CPT buffer device thermally runs away at 200 °C.

I. CONCLUSIONS

Different buffer designs with respect to thermal stability at $T_j = 200$ °C have been compared. For 1200 V IGBTs the MPT and Epi SPT buffers allow for the lowest leakage currents below not just the reference DD SPT devices, but well below the MPI design for the full temperature range tested up to 200 °C and provide thermal stability even at 205 °C. The MPI and DD SPT devices with leakage currents >10 mA are thermally unstable at 200 °C. For 1700 V IGBTs, the MPT buffers also provide the lowest leakage currents and thermal stability at 200°C. The CPT buffer shows also low leakage currents but it suffers thermal runaway over time. The leakage current of the MPI design can be significantly reduced (> 4x) by reducing the deepest hydrogen buffer peak (and increasing its concentration) while softness can be maintained by modifying the anode injection efficiency through LA. The MPT, Epi SPT and CPT buffers allow for soft turn-off at 200°C up to 2x nominal current. These devices can also withstand short-circuit current pulses up to 10 μs at 200 °C. All this makes them a powerful candidate when extremely high temperature capabilities are needed.

ACKNOWLEDGMENTS

We thank J. Lauri, B. Karadaghi, W. Janisch and R. Jabrany for process support.

REFERENCES

[1] N. Iwamuro, T. Laska, "IGBT History, State-of-the-art, and Future Prospects", IEEE Transactions on Electron Devices, 64, pp.741 – 752, 2017.

[2] E. Buitrago et al., "An advanced soft punch through buffer design for thin wafer IGBTs targeting lower losses and higher operating temperatures up to 200 °C," ISPSD, Chicago, 2018, pp. 499-502.

[3] M. Rahimo et al., (2014). U.S. Patent No. US8829571B2. Washington, DC: U.S. Patent and Trademark Office.

[4] J. Vobecky et al., "Exploring the Silicon Design Limits of Thin Wafer IGBT Technology: The Controlled Punch Through (CPT) IGBT," ISPSD, Orlando, 2008, pp. 76-79.

[5] F. Niedernostheide et al., "Tailoring of field-stop layers in power devices by hydrogen-related donor formation," ISPSD, Prague, 2016, pp. 351-354.

[6] M. Andenna et al., "A New Soft-Punch-Through Buffer Concept for 600V – 1200V IGBTs." ISPS, Prague, 2018.

[7] Schlapbach U. et al., "1200V IGBTs operating at 200°C? An investigation on the potentials and the design constraints", Proc. ISPSD'07, Jeju Island, 2007.

Proceedings of the 31st International Symposium on Power Semiconductor Devices & ICs
May 19-23, 2019, Shanghai, China

A Novel 3300V Trench IGBT with Hole Extraction Structure for Low Power Loss

Xin Peng, Zehong Li, Yishang Zhao, Yang Yang, Min Ren,
Jinping Zhang, Wei Gao, Bo Zhang, Zhaoji Li

State Key Laboratory of Electronic Thin Films and Integrated
Devices, University of Electronic Science and Technology of
China, Chengdu, China
Email: lizh@uestc.edu.cn

Defu Sun, Yingxin Song, Kuncun Zhu, Donghua Li

Jinan Semiconductor Institute, Jinan, China

Abstract—**A novel 3300V Trench IGBT with Hole Extraction Structure (HE-IGBT) is proposed in this paper. The hole extraction is controlled by the JFET structure in the Adjustable Hole Path (AHP) region. The AHP region is in the floating potential during on-state and extracts holes swiftly during the turn-off transition. Meanwhile, the AHP region effectively reduces electric fields under trench gates and shields IGBT gates during switching, contributing to the higher BV and lower Miller capacitance. Based on the theoretical analysis and numerical simulations, the breakdown voltage of HE-IGBT is 10.7% higher than that of the Separate Floating P-layer IGBT (SFP-IGBT). Furthermore, compared with SFP-IGBT, the proposed device decreases Miller capacitance and switching loss by 67% and 30%, respectively.**

Keywords—IGBT, Miller capacitance, adjustable hole path, switching loss

I. Introduction

High voltage IGBTs play an important role in motor drives and power converters. And the floating p-layer structure is widely applied in high voltage trench IGBTs to achieve highly reliable breakdown voltage (BV), robust short-circuit capability and low power loss. However, the floating p-layer region will reduce its effects on weakening electric fields under trench gates, generate dV/dt noise and increase the switching loss. The Separate Floating p-layer structure has been proposed to solve the dV/dt noise but ignores the increased turn-off loss induced by accumulation of holes in the floating p-layer region [1-5]. Meanwhile, lower Miller capacitance structures have been developed to reduce the power loss [6-9]. Hole path has been proposed to realize the low switching loss but the uncontrollable hole path will increase the on-state voltage, contributing to the higher conduction loss [10].

In this work, an Adjustable Hole Path (AHP) is proposed to extract holes and reduce power loss, which is more compatible with the front-side fabrication process than extracting electron from collector electrode. And the AHP region can effectively reduce electric fields under trench gates and shield IGBT gates during switching, contributing to the higher BV and lower Miller capacitance.

II. Device Structure and Mechanism

The Separate Floating P-layer IGBT (SFP-IGBT) and proposed Hole Extraction IGBT (HE-IGBT) are shown in Fig. 1. The location and doping of AHP region are in consistency with SFP region, which is deeper than the P-base

This work was supported in part by the Science & Technology Program of Guangdong Province, China (Granted No. 2018B010142001).

region to support higher BV. As shown in Fig. 1(b), a n-type JFET gate region, a lightly doped channel and a JFET source region connected to the emitter constitute the JFET structure in the AHP region to control the behavior of holes.

The trench gate of HE-IGBT is connected to the JFET gate through polysilicon. In addition, the JFET gate is surrounded by thick oxide to reduce leakage currents. The channel depth T_{JC} and channel doping concentration N_{JC} of JFET are key parameters to determine characteristics of the AHP region.

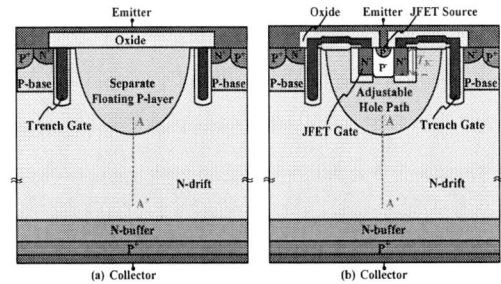

Fig. 1. Schematic cross section of (a) SFP-IGBT and (b) the proposed HE-IGBT. T_{JC} is the channel depth of JFET.

As the equivalent circuit of HE-IGBT shown in Fig. 2(a), the AHP region is in a floating potential and equivalent to the SFP region during on-state, because the JFET channel is pinched off by high gate voltages. Therefore, the hole concentration in the AHP region is equivalent to that in the SFP region.

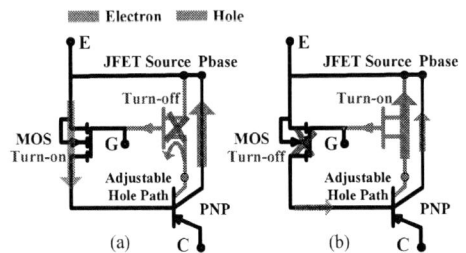

Fig. 2. Equivalent circuit of HE-IGBT during (a) on-state and (b) turn-off. When MOS channel is turned on, the JFET channel is turned off and vice versa.

During the turn-off transition shown in Fig. 2(b), the JFET gate follows the decreasing IGBT gate voltage and the JFET channel is formed accordingly. At this time, holes in the AHP region are transported to the emitter swiftly, shortening the transport path and saving turn-off time. On the

978-1-7281-0582-6/19 $31.00 © 2019 IEEE

contrary, holes in the SFP region can only be transported to the emitter through P-base in the SFP-IGBT.

Meanwhile, the AHP region is in ground potential during the switching transient and effectively shields IGBT gates, resulting in the lower Miller capacitance. Hence, the switching time and switching loss are reduced. On the other hand, the low potential of AHP region during blocking state reduces electric fields under trench gates, contributing to the higher BV.

III. RESULTS AND DISCUSSION

Characteristics of the 3300V SFP-IGBT and HE-IGBT are simulated by TCAD. The cell pitch is 26μm and width of the JFET channel is 2μm. Moreover, the depth and width of a trench gate are 6μm and 2μm, respectively. And the doping concentration and thickness of the N-drift region are $1.7 \times 10^{13} cm^{-3}$ and 340μm, respectively. The static characteristics and power loss of two structures are discussed below.

A. Static Characteristics

As shown in Fig. 3, the BV of SFP-IGBT increases with larger doping concentration of floating p-layer N_{FP} because larger N_{FP} increases the depletion region between the SFP and the N-drift region. The SFP-GND-IGBT means the SFP region is in ground potential, thus its BV is insensitive to the larger N_{FP} shown in Fig. 3.

When the SFP region is in ground potential, it results in the wider depletion region and larger BV compared with the floating p-layer. For HE-IGBT, the AHP region is in ground potential shown in the inset of Fig. 4 and BV is increased by 10.7% during blocking state when N_{FP} is $2 \times 10^{17} cm^{-3}$.

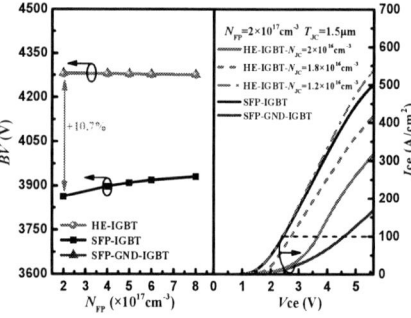

Fig. 3. Comparison of breakdown voltage and on-state voltage between SFP-IGBT and HE-IGBT.

Although SFP region in ground potential is beneficial for larger BV, it will lead to larger on-state voltage as well. For HE-IGBT, N_{JC} is the JFET channel doping concentration and it determines the on-state voltage. Fig. 3 shows that larger N_{JC} results in lower on-state voltage Vcesat. When N_{JC} is $1.2 \times 10^{16} cm^{-3}$, the Vcesat of HE-IGBT is equal to that of SFP-IGBT because holes are stored in the AHP region.

Fig. 4 reveals the carrier distribution in the SFP and AHP regions with different N_{JC} during on-state. It is clearly shown that when N_{JC} is $1.2 \times 10^{16} cm^{-3}$, carrier distributions in the AHP region is equal to that in the SFP region of SFP-IGBT, because the JFET channel is completely pinched off at the positive gate voltage. On the contrary, the JFET channel is not pinched off and extracts holes to the emitter at larger N_{JC},

leading to the carrier distribution decreasing in the N-drift region and larger Vcesat. Thus, Parameter N_{JC} for HE-IGBT determines carrier distributions in the AHP region and on-state voltage.

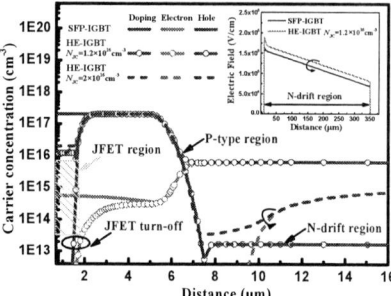

Fig. 4. Distributions of carrier concentration during on-state and electric fields during forward blocking state along line AA' at $N_{FP}=2 \times 10^{17} cm^{-3}$.

B. Miller Capacitance

According to (1), reducing the Miller capacitance C_{res} is an effective way to reduce the turn-on loss E_{on} and turn-off loss E_{off}[11-12]. And k_1, k_2, k_3 and k_4 is the coefficient related to the bus voltage Vcc, load current I_{Load} and threshold voltage.

$$\begin{cases} E_{on} \propto R_g \left(k_1 \cdot C_{ge} + k_2 \cdot C_{res} \right) \\ E_{off} \propto R_g \left(k_3 \cdot C_{ge} + k_4 \cdot C_{res} \right) \end{cases} \quad (1)$$

The C_{res} and C_{res}/C_{ies} ratio between two different structures at different Vce are shown in Fig. 5. The C_{res} of HE-IGBT is reduced by 67% at Vce=25V using the AC small-signal analysis, which is beneficial to reduce the switching loss as illustrated in (1). This is because the JFET channel always forms at $V_{GE}=0V$ and the AHP region is in the ground potential all the time even at higher Vce. Hence lower potential AHP region shields the effect of trench gate on the collector, which reduces the Miller capacitance compared with the floating potential of SFP region.

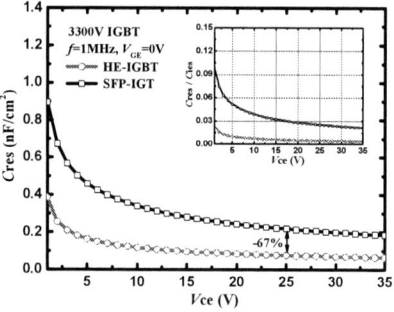

Fig. 5. Simulation results of Miller capacitance and C_{res}/C_{ies} ratio at different Vce, where C_{ies} is the input capacitance.

The inset of Fig. 5 gives the C_{res}/C_{ies} ratio at different Vce. Although the connection between trench gate and JFET gate will increase the C_{ies} slightly because of the depletion capacitance of PN junction, the lower proportion of C_{res} over C_{ies} of HE-IGBT is achieved by lower Miller capacitance, which is good for suppressing the faulty turn-on of trench gate when the dV/dt is too high.

C. Adjustable Hole Path during Switching

Fig. 6 shows the switching circuit for transient analysis simulation, where L_{s1} and L_{s2} are the stray inductance. The load current is 100A and the area of the active region is 1cm^2.

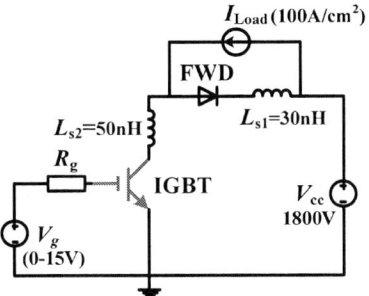

Fig. 6. Circuit schematic for transient switching analysis. BV of the FWD is in 3300V-class.

Based on the above circuit, the simulated turn-on waveforms of IGBTs are illustrated in Fig. 7. The current overshoot together with the voltage spike of trench gate in voltage plateau phase is due to reverse recovery characteristics of FWD shown in Fig. 6.

During the turn-on transient of HE-IGBT, the hole path is gradually pinched off by JFET structure with gate voltage increasing. The potential of AHP region is low and the shielding effect of AHP region exists until the hole path is blocked. The time of voltage plateau is in proportion to the Miller capacitance value and the larger Rg results in the longer plateau time.

Because Miller capacitance of HE-IGBT is smaller than that of SFP-IGBT, the voltage plateau time of HE-IGBT is shorter at the same Rg. As a result, the turn-on time of the HE-IGBT is shortened by 28% than that of the SFP-IGBT when Rg is 25 ohms.

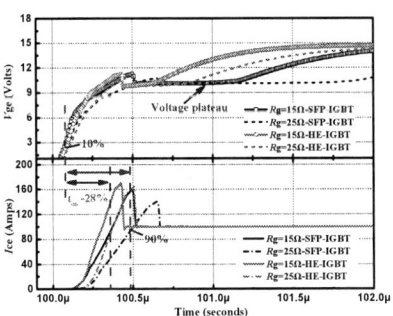

Fig. 7. Turn-on waveforms with different gate resistance Rg. The voltage plateau period is related to the Miller capacitance.

Fig. 8 is the switching waveforms of current and variation of hole carriers proportion in emitter when Rg is 5 ohms. Because gate resistor Rg is smaller than that in Fig.7, the gate current is large enough and Miller capacitance can be fully charged or discharged in a short time. Consequently, the smaller Miller capacitance of HE-IGBT is not dominant in shortening the turn-on and turn-off time. However, the

lower Miller capacitance can always be achieved as the JFET channel for hole extraction forms during switching operation.

As shown in Fig. 8, hole carriers proportion is gradually in dominant during the switching process in the emitter electrode. During the turn-off period, hole currents constitute the whole emitter current in the SFP-IGBT as well as HE-IGBT. Hence, providing an extra hole extraction path in the emitter is an effective way to reduce the turn-off loss besides the lower Miller capacitance.

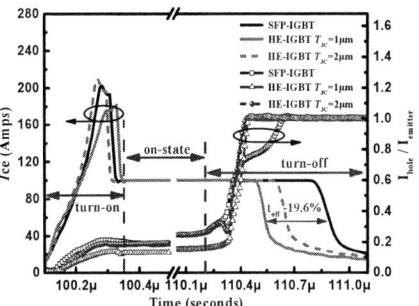

Fig. 8. Switching waveforms of current and variation of hole carriers proportion in emitter during the switching. Over-current of Ice is due to the reverse recovery of FWD.

A hole extraction path of HE-IGBT is formed during turn-off period and therefore shortens the turn-off time by 19.6% in comparison with that of the SFP-IGBT shown in Fig. 8. When channel depth T_{JC} is 1μm, the hole extraction path is formed and hole carriers are extracted to the emitter more swiftly than the one when T_{JC} is 2μm, which is beneficial from the lower PN junction capacitance of JFET structure.

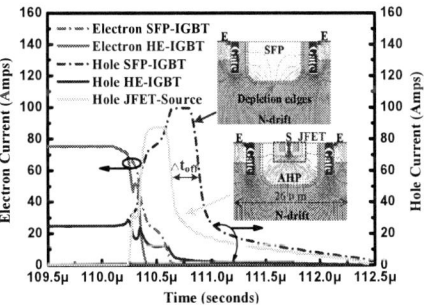

Fig. 9. Simulation results of electron and hole currents at emitter when T_{JC} is 2μm. The insets are the current density distribution (blank line) of SFP-IGBT and HE-IGBT during turn-off.

The variation of electron and hole currents at emitter during turn-off can be found in Fig. 9. In SFP region of SFP-IGBT, the electron current decreases rapidly with gate voltage decreasing and stored hole carriers can only be transported to the emitter through the adjacent P-base region as illustrated in the inset.

On the contrary, the AHP region introduced in the emitter of HE-IGBT provides an additional hole path. And the potential of AHP region is controlled by the JFET structure to extract hole carriers from the AHP region to the emitter via the source of JFET as shown in the inset of Fig. 9.

978-1-7281-0582-6/19 $31.00 © 2019 IEEE

Because of the lower Miller capacitance of HE-IGBT, the decreasing of electron currents is faster than that of SFP-IGBT. Moreover, hole current flowing through the P-base of HE-IGBT reduces with the electron current decreasing. And the majority of hole currents flow to the AHP region after the hole extraction path forms, because the AHP region along with the SFP region is deeper than the P-base. As a result, the turn-off time of HE-IGBT with an adjustable hole path is shorter than that of SFP-IGBT during turn-off transient.

D. Switching Loss

From above results and discussion, the hole extraction path of HE-IGBT is controlled by the gate voltage. When HE-IGBT is turned on, the hole path is pinched off to prevent the hole extraction, resulting in the lower Vcesat compared with that of the SFP region in ground potential shown in Fig. 10.

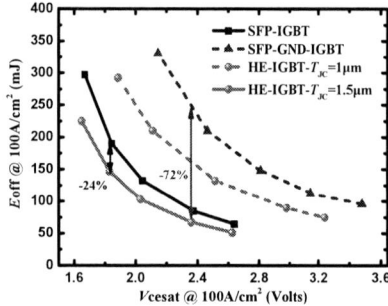

Fig. 10. Trade-off relationship between Eoff and Vcesat for different structure (@Rg=5Ω, N_{JC}=5×10^{15}cm^{-3}).

When T_{JC} is 1μm, the JFET channel is not fully pinched off during on-state, resulting in the higher Vcesat. Thus the trade-off relationship between Vcesat and E_{off} is better with the deeper T_{JC}. Moreover, the E_{off} of HE-IGBT is reduced by 24% and 72% compared with that of SFP-IGBT and SFP-GND-IGBT, respectively.

TABLE I. Comparison of static and dynamic characteristics between the 3300V SFP-IGBT, SFP-GND-IGBT and HE-IGBT (@Rg=25Ω, T_{JC}=1.5μm, N_{JC}=5×10^{15}cm^{-3}).

	SFP-IGBT	SFP-GND-IGBT	HE-IGBT
BV(V)	3863	4281	4277
Vcesat(V)	2.38	4.4	2.36
C_{res}(nF/cm^2)	0.22	0.08	0.08
E_{on}(mJ)	65.8	63.7	45.6
E_{off}(mJ)	101.6	60.8	70.7
E_{total}(mJ)	167.4	124.5	116.3

Static and dynamic characteristics of three different structures are shown in Table I. Although the E_{off} of SFP-GND-IGBT is lower, the on-state voltage is too high,

resulting in the higher on-state loss. Meanwhile, compared with the SFP-IGBT, the E_{on} and E_{total} of HE-IGBT with lower Miller capacitance and adjustable hole path are reduced by 30.6% and 30%, respectively.

IV. CONCLUSION

In this work, a novel HE-IGBT with higher BV, lower Miller capacitance and lower power loss is investigated by simulation. Compared with the SFP-IGBT, the AHP region effectively reduces electric fields under trench gates and increases BV by 10.7%. Moreover, the AHP region shields IGBT gates with the formation of an extra hole extraction path during switching and consequently reduces Miller capacitance by 67%. Hence, total switching loss of HE-IGBT is decreased by 30%. The proposed HE-IGBT is compatible with the front-side fabrication process and provides a new way to achieve a lower switching loss.

REFERENCES

[1] S. Watanabe, M. Mori, T. Arai, K. Ishibashi, Y. Toyoda and T. Oda, et al., "1.7kv Trench IGBT with deep and separate floating p-layer designed for low loss, low EMI noise, and high reliability," in *Proc. 23th ISPSD*, May 2011, pp. 48-51.

[2] Y. Toyota, S. Watanabe, T. Arai, M. Wakagi, M. Mori and M. Shinagawa, et al., "Novel 3.3-kV advanced trench HiGT with low loss and low dv/dt noise," in *Proc. 25th ISPSD*, May 2013, pp. 29-32.

[3] M. Sawada, Y. Sakurai, K. Ohi, Y. Ikura, Y. Onozawa and T. Yamazaki, et al., "1800A/3.3kV IGBT module using advanced trench HiGT structure and module design optimization," in *Proc. PCIM Europe 2014*, May 2014, pp. 346-353.

[4] C. Corvasce, M. Andenna, S. Matthias, L. Storasta, A. Kopta and M. Rahimo, et al., "3300V HiPak2 modules with enhanced trench (TSPT+) IGBTs and field charge extraction diodes rated up to 1800A," in *Proc. PCIM Europe 2016*, May 2016, pp. 417-424.

[5] Y. Ikura, Y. Onozawa and A. Nakagawa, "IGBT structure with electrically separated floating-p region improving turn-on dVak/dt controllability," in *Proc. 30th ISPSD*, May 2018, pp. 168-171.

[6] H. Feng, W. Yang, Y. Onozawa, T. Yoshimura, A. Tamenori and J. K. O. Sin, "A new fin p-body insulated gate bipolar transistor with low miller capacitance," *IEEE Electron Device Letters*, vol. 36, no. 6, June 2015, pp. 591-593.

[7] H. Feng, W. Yang, Y. Onozawa, T. Yoshimura, A. Tamenori and J. K. O. Sin, "A 1200 V-class fin p-body IGBT with ultra-narrow-mesas for low conduction loss," in *Proc. 28th ISPSD*, June 2016, pp. 203-206.

[8] M. Sawada, K. Ohi, Y. Ikura, Y. Onozawa, M. Otsuki and Y. Nabetani, "Trench shielded gate concept for improved switching performance with the low miller capacitance," in *Proc. 28th ISPSD*, June 2016, pp. 207-210.

[9] M. Shiraishi, T. Furukawa, S. Watanabe, T. Arai and M. Mori, "Side gate HiGT with low dv/dt noise and low loss," in *Proc. 28th ISPSD*, June 2016, pp. 199-202.

[10] M. Sawada, Y. Sakurai, K. Ohi, Y. Ikura, Y. Onozawa and T. Yamazaki, et al., "Hole path concept for low switching loss and low EMI noise with high IE-effect," in *Proc. 29th ISPSD*, May 2017, pp. 65-6.

[11] V. K. Khanna, The Insulated Gate Bipolar Transistor IGBT Theory and Design, Wiley-IEEE Press, 2003, pp.69-70.

[12] B. J. Baliga, Fundamentals of Power Semiconductor Devices, Springer Science, 2008, pp.440-443.

Proceedings of the 31st International Symposium on Power Semiconductor Devices & ICs
May 19-23, 2019, Shanghai, China

Effects of the HV-BIGT Design Elements on the High-Frequency Oscillation Instability during Short Circuit Transients

P. Diaz Reigosa[a], C. Papadopoulos[b], F. Iannuzzo[c], C. Corvasce[b] and M. Rahimo[d]

[a] Applied University of Sciences, Windisch, Klosterzelgstrasse 2, CH-5210, Switzerland
[b] ABB Switzerland Ltd. Semiconductors, CH-5600, Lenzburg, Switzerland
[c] Department of Energy Technology, Aalborg University, Pontoppidanstraede 111 DK-9220, Denmark
[d] MTAL GmbH, Gänsbrunnen, Switzerland

Abstract—**The design elements of the Bi-mode Insulated Gate Transistor BIGT show that the combination of the high hole injection levels supplied from the collector together with the presence of a localized lifetime control at the MOS cells have brought improvements on the short circuit capability, strongly minimizing the high-frequency oscillations observed in IGBTs. The BIGT concept and the traditional IGBT structures have been compared under short circuit conditions to investigate the charge-field interactions at the MOS cells, triggering the oscillation mechanism. The effect of the lifetime control and the irradiation method on the short circuit capability is investigated.**

Index Terms—**IGBT, BIGT, bipolar gain, short circuit, gate oscillations, parametric oscillation, robustness, Kirk Effect, TCAD.**

I. INTRODUCTION

The modern IGBT technologies based on the Soft Punch Through or Field Stop lowly doped buffer concepts combined with the low hole injection from the p-type collector have provided very low on-state and switching losses when compared to previous generations [1]. However, the targeted low bipolar gain is not optimum for the performance of the IGBT during short circuit transients. It has been observed in [2], [3] that the IGBT shows a gate oscillation instability during the short circuit event, which in most of the cases leads to the device failure. This instability is mainly dependent on the amount of excess charge and the strength of the electric field near the emitter, which in turn is dependent on the collector and buffer design [4], [5]. IGBTs with low bipolar gains show a weak electric field at the emitter during the short circuit event, which is more critical at low V_{CE}. Under this condition, the electric field is strongly influenced by the amount of the excess charge, thus, the electric field fluctuates and leads to a time-varying Miller capacitance. The time-varying Miller capacitance combined with the stray inductance from the gate circuit creates a parametric oscillation, as demonstrated in [3]. To avoid this instability, IGBTs with high hole injections are preferred but this comes at the expense of higher turn-off losses and higher leakage currents [4].

The Bi-mode-Insulated-Gate-Transistor BIGT is also based on the Soft Punch Through buffer but employing n$^+$ shorts at the collector side to integrate the diode mode within the same cell [6]. the BIGT is designed with higher p-type collector doping concentrations compared to the IGBT collector, to compensate the n$^+$ short regions which do not inject any

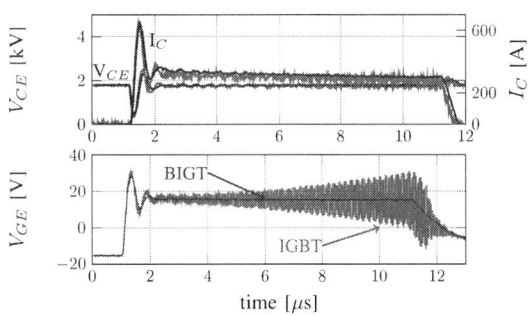

Fig. 1. Short-circuit experiments of a 3.3 kV planar IGBT and a 3.3-kV planar BIGT tested under the same conditions: V_{CE} = 1800 V, $R_{g,on}$ = 2.2 Ω, L_σ = 530 nH and T_C = 25°C.

holes, and hence, obtain similar injection efficiency, on-state voltage drop, turn-off losses and leakage current as the IGBT [7]. Additionally, the BIGT employs a Local p-well Lifetime (LpL) control technique utilizing a particle implantation which reduces the diode reverse recovery, which will not be part of the discussion in this paper. Also, the adjustment of the reverse recovery losses is achieved with a uniform local lifetime control employing proton irradiation. These techniques adjust the amount of excess charge and the strength of the electric field near the emitter, and hence, the electric field is less influenced by the amount of the excess charge, resulting in a stable Miller capacitance. The BIGT is an attractive power device concept, which has been already implemented in very demanding high power applications such as High Voltage Direct Currents (HVDCs). Hence, a detailed understanding of the device behavior, especially under fault conditions is needed to further improve its SOA.

II. SHORT-CIRCUIT EXPERIMENTS

In an IGBT, the low hole injection levels needed to achieve a good switching performance triggers high-frequency oscillations under the event of a short circuit, as presented in Fig. 1. For the sake of comparison, the performance of the BIGT under a short circuit test with the same external circuit is also presented in the same figure. Both devices are designed to

978-1-7281-0582-6/19 $31.00 © 2019 IEEE

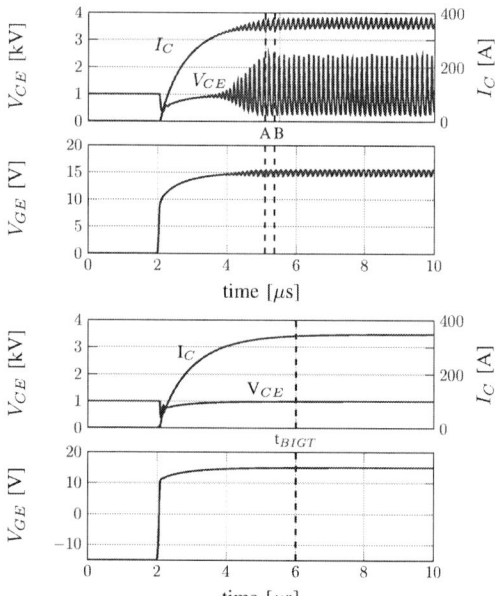

Fig. 2. Short-circuit simulation of a 3.3-kV planar IGBT (upper waveforms) and a 3.3-kV planar BIGT (lower waveforms). Conditions: $V_{CE} = 1$ kV, $R_{g,on} = 2$ Ω, $L_{\sigma} = 800$ nH, $L_{gate} = 40$ nH, $L_e = 10$ nH and $T_C = 25°C$. The time instant at which the device is evaluated is highlighted.

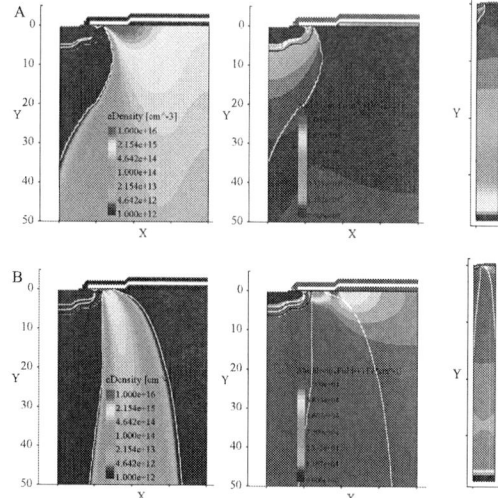

Fig. 3. The SPT 3.3-kV planar IGBT at low (A) and high (B) V_{CE} during the oscillations in Fig. 2. Left: electron density; middle: electric field; right: electric field through the whole cell.

have same saturation current and turn-off losses, as can be seen from the matching short circuit waveforms. However, the design elements of the BIGT help to eliminate the field-charge interactions, and therefore, the high-frequency oscillations do not occur, as will be clarified in the following section.

III. MECHANISMS DURING THE SHORT-CIRCUIT OSCILLATIONS

TCAD simulations have been performed, where a 3.3-kV IGBT model and 3.3-kV BIGT model have been coupled with the same external circuit to reproduce the oscillations, as can be observed in Fig. 2. It is important to include a gate stray inductance and a relatively low gate resistor in the external circuit, otherwise, the instability creating the short-circuit oscillations could not be reproduced. The simulations prove that the IGBT shows high-frequency oscillations under the short circuit transient while the BIGT does not show such instability. The field-charge interactions occurring in the IGBT have been plotted when the V_{CE} is low and high along one oscillation cycle, as presented in Fig. 3. The time instant where the device has been evaluated is highlighted in Fig. 2. During the oscillation cycle, the electric field fluctuates causing charge-storage effects at the MOS cell when V_{CE} is low (A in Fig. 3) and no charge accumulation when V_{CE} is high (B in Fig. 3). This phenomenon creates a Miller capacitance that continuously changes with time, leading to a nonlinear amplification mechanism, which is called parametric oscillation as discussed in [3].

IV. EFFECT OF LIFETIME CONTROL

A. The BIGT structure

Reducing the lifetime value by proton irradiation has two benefits for the short circuit robustness of the BIGT. The carrier lifetime is reduced at the surface of the BIGT, which leads to the reduction of the excess carrier concentration, and therefore, the electric field in this region is less influenced by the mobile carriers. The second effect is related to the proton irradiation, which generates n-type donors, and this is accompanied by an overall increase in the background doping. This results in a higher electric field at the surface of the BIGT dominating over the excess carrier concentration. Both effects minimize the field-charge interactions at the surface of the BIGT during the short-circuit event and contribute to a stable Miller capacitance.

The high electric field at the MOS cells is observed when the BIGT is simulated under short circuit conditions in Fig. 4. The same BIGT has been simulated under the same short circuit conditions but without lifetime control, showing a lower electric field at the MOS cells (see Fig. 5). For the sake of comparison, two vertical cuts have been performed to show the electric field shape across the pilot-IGBT region (i.e., C_1) and the n^+ short region (i.e., C_2), as presented in Fig. 6. When the BIGT is evaluated in the n^+ areas (i.e, C_2), the electric field becomes weaker at the MOS cells, which may trigger the conditions for having oscillations. However, the electric field in the pilot-IGBT region remains strong because of the high hole injection from the back side, and therefore, this structure does not show oscillations during short circuit. Evaluating the BIGT without lifetime control across the the n^+ region, the electric field at the front side slightly changes sign (i.e, rotates)

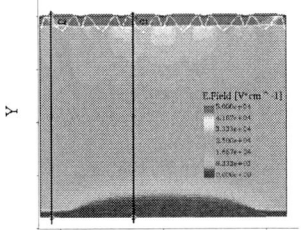
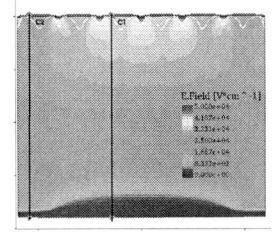
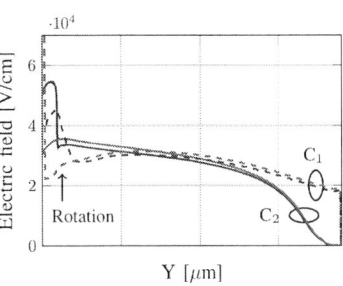

Fig. 4. Electric field of the BIGT with lifetime control at the conditions in Fig. 2.

Fig. 5. Electric field of the BIGT without lifetime control at the conditions in Fig. 2.

Fig. 6. Comparison between BIGT with lifetime (blue) and without lifetime control (red).

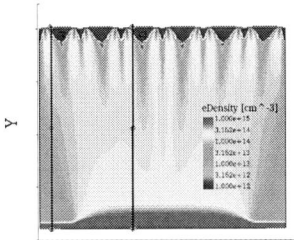
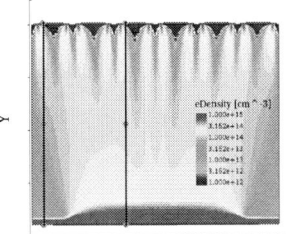
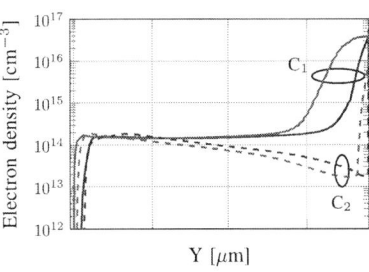

Fig. 7. Electron density of the BIGT with lifetime control at the conditions in Fig. 2.

Fig. 8. Electron density distribution of the BIGT without lifetime control at the conditions in Fig. 2.

Fig. 9. Comparison between BIGT with lifetime (blue) and without lifetime control (red).

as highlighted in Fig. 6. This instability is counteracted by implementing the lifetime control.

During the short circuit of the BIGT, the pilot-IGBT region significantly modifies the BIGT charge distribution compared to a traditional IGBT. The pilot region is designed with a strong hole injection to counteract the regions where hole injection does not occur, e.g. the n^+ shorts. This effect can be observed in Fig. 7, where the excess carrier distribution is the highest in the pilot region. When the lifetime control is removed from the BIGT structure, a charge accumulation effect between the MOS cells can be observed, which is a sign of instability for triggering oscillations during the short circuit event. The charge distribution differences between the pilot-IGBT and the n^+ regions are plotted in Fig. 9.

B. The IGBT structure

In BIGTs, the combination of high anode injection levels and the front side local lifetime control have brought improvements on the short circuit SOA capability. For the IGBT, the increase of electric field at the front side by increasing the hole injection comes at the expense of increased turn-off losses. Another way to increase the gradient of the dE/dy at the front side is to apply the lifetime control technique that is used for the BIGT. Fig. 10 shows that, while the traditional IGBT oscillates during short circuit, the IGBT with lifetime control does not oscillate. The electric field has been increased at

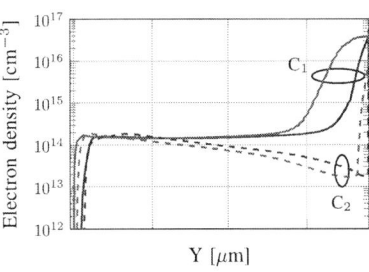

Fig. 10. Simulation of the short-circuit waveforms of a 3.3-kV planar IGBT including lifetime control (blue) and without lifetime control (red). The time instant at which the two devices are evaluated in Fig. 11 is highlighted.

the surface of the IGBT by inserting the well-defined lifetime control (see Fig. 11). The carrier distribution is adjusted in a way that prevents field-charge interactions, hence, the Miller capacitance is stable.

C. Influence of carrier lifetime control by proton irradiation

As discussed previously in section IV-A, proton irradiation significantly minimizes the field-charge interactions at the surface of the BIGT during the short-circuit event. First,

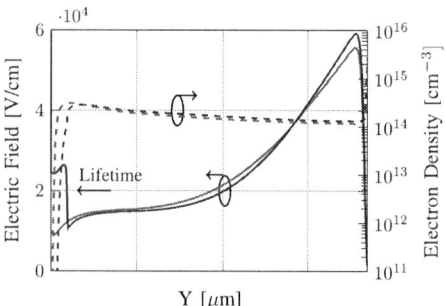

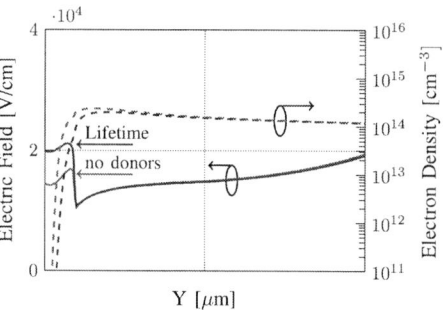

Fig. 11. Simulated electric field (solid line) and electron carrier concentration (dashed line) for the planar IGBT with and without lifetime control during short circuit at t= 3.8 μs (see Fig. 10).

Fig. 13. Simulated electric field (solid line) and electron carrier concentration (dashed line) for the planar IGBT irradiated with protons and without donors at t= 3 μs (see Fig. 12)

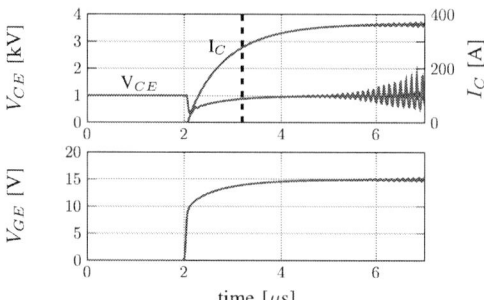

Fig. 12. Simulation of the short-circuit waveforms of a 3.3-kV planar IGBT irradiated with protons and without donors. The time instant at which the two devices are evaluated in Fig. 13 is highlighted.

V. CONCLUSIONS

The IGBT structure based on the Soft Punch Through lowly doped buffer concept combined with the low hole injection from the p-type collector leads to short-circuit oscillation instabilities that can be mitigated by introducing the design elements of the BIGT. The collector structure of a BIGT with the pilot-IGBT region and the n^+ region significantly modifies the BIGT charge distribution compared to an IGBT, where a strong hole injection is observed in the pilot region. Additionally, the lifetime reduction by proton irradiation at the MOS cells, decreases the excess carrier concentration and increases the electric field strength, mitigating field-charge interactions and leading to a stable Miller capacitance. During short circuit conditions, both design elements from the BIGT (i.e., high hole injection and lifetime control) help to mitigate the oscillation instability and the design trade-off improvements of the BIGT can be better understood.

the carrier lifetime reduction shows a lower excess carrier concentration, which has the advantage that the electric field is less influenced by the excess carriers. The second advantage is related to the generation of n-type donors from the proton irradiation. The n-type donor contributes to the increase of the drift doping concentration, and therefore, the electric field is less influenced by the excess carrier concentration. To prove that the oscillation instability can be mitigated, the IGBT is simulated by using proton irradiation in Fig. 10. The differences in the electric field strength and the electron density have been plotted in Fig. 11.

To figure out which one among the excess carrier reduction or the donor contribution is more effective in mitigating the oscillation, the IGBT has been simulated by using proton irradiation first and then removing the donor contribution. The simulation results can be observed in Fig. 12, where the waveforms are superimposed, showing that the oscillations are triggered without the n-type donors. The electric field becomes weaker and therefore it is influenced by the excess carriers, as shown in Fig. 13. This effect is similar as reducing the lifetime carrier by helium irradiation, which does not produce n-type donors. The increase of the background doping at the surface of the IGBT has been proven to be effective for mitigating the oscillations, as discussed in [3].

REFERENCES

[1] . M. Rahimo et al, "Novel soft-punch-through (spt) 1700v IGBT sets benchmark on technology curve." in *Proc. of the PCIM*, May 2001, pp. 1–8.

[2] P. D. Reigosa, F. Iannuzzo, and M. Rahimo, "TCAD analysis of short-circuit oscillations in IGBTs," in *Proc. of ISPSD*, May 2017, pp. 151–154.

[3] P. D. Reigosa, F. Iannuzzo, M. Rahimo, C. Corvasce, and F. Blaabjerg, "Improving the short-circuit reliability in IGBTs: How to mitigate oscillations," *IEEE Transactions on Power Electronics*, vol. 33, no. 7, pp. 5603–5612, July 2018.

[4] P. D. Reigosa, F. Iannuzzo, M. Rahimo, C. Corvasce, and F. Blaabjerg, "Increasing emitter efficiency in 3.3-kv enhanced trench IGBTs for higher short-circuit capability," in *Proc. of APEC*, March 2018, pp. 1722–1728.

[5] V. van Treek, H. Schulze, F. Niedernostheide, C. Sandow, R. Baburske, and F. Pfirsch, "Influence of doping profiles and chip temperature on short-circuit oscillations of IGBTs," in *Proc. of ISPSD*, May 2018, pp. 108–111.

[6] L. Storasta, M. Rahimo, C. Corvasce, and A. Kopta, "Resolving design trade-offs with the BIGT concept," in *Proc. of the PCIM*, May 2014, pp. 1–8.

[7] L. Storasta, S. Matthias, A. Kopta, and M. Rahimo, "Bipolar transistor gain influence on the high temperature thermal stability of HV-BiGTs," in *Proc. of the ISPSD*, June 2012, pp. 157–160.

978-1-7281-0582-6/19 $31.00 © 2019 IEEE

Proceedings of the 31st International Symposium on Power Semiconductor Devices & ICs
May 19-23, 2019, Shanghai, China

Estimation of Impact Ionization Coefficient in GaN by Photomulitiplication Measurement Utilizing Franz-Keldysh Effect

Takuya Maeda*[1], Tetsuo Narita[2], Hiroyuki Ueda[2], Masakazu Kanechika[2], Tsutomu Uesugi[2], Tetsu Kachi[3],
Tsunenobu Kimoto[1], Masahiro Horita[1, 3], and Jun Suda[1, 3]
[1]Kyoto University, Kyoto, 615-8510, Japan,
[2]Toyota Central R&D Labs., Inc., Aichi, 480-1118, Japan,
[3]Nagoya University, Aichi, 464-8603, Japan
*Email: maeda@semicon.kuee.kyoto-u.ac.jp

Abstract—Avalanche multiplication characteristics in GaN p-n junction diodes (PNDs) under high reverse bias conditions were investigated. The GaN-on-GaN PNDs with double-side-depleted shallow bevel termination, which showed low reverse leakage current and excellent avalanche capability, were used for the measurements. Under sub-bandgap light illumination, the photocurrents induced by Franz-Keldysh (FK) effect, which can be well reproduced by the theoretical calculations of the optical absorption, and their avalanche multiplications were observed. The multiplication factors were extracted as the ratios of the experimental photocurrents to the calculated FK-induced photocurrent. Under an assumption of equal impact ionization coefficients of electrons and holes, the electric-field dependence of an impact ionization coefficient in GaN were estimated.

Keywords—GaN, p-n junction diode, avalanche breakdown, impact ionization coefficient, Franz-Keldysh effect

I. INTRODUCTION

Owing to its high critical electric field (~3 MV/cm), GaN has attracted much attention as a material for the next-generation power devices [1-5]. Impact ionization coefficient is one of the most important material parameters, which determines an breakdown voltage and a safe-operation-area of a power device. To obtain impact ionization coefficients, photomultiplication measurement is employed [6-8], since impact ionization coefficients cannot be directly measured. For the measurement, above-bandgap light ($h\nu > E_g$), which is absorbed at a semiconductor surface, is generally used.

To obtain both electrons and holes impact ionization coefficients (α_n, α_p), electron-initiated and hole-initiated multiplication factors (M_n, M_p) should be obtained by the top-side and back-side injections as shown in Fig. 1(a). However, it is difficult to measure M_p in GaN devices fabricated on an n-type GaN substrate by back-side illumination, since the penetration depth of above-bandgap light is small (< 1 μm) and the hole diffusion length in n-GaN are short (< 1 μm). Precise removal of n-type GaN substrate is required, which is very difficult.

Recently, Cao *et al.* reported on the impact ionization coefficients of electrons and holes in GaN [6]. For the measurements, the GaN p-n junction diodes (PNDs) with thin pseudomorphic InGaN photo-absorption layers on the cathode sides of the drift layers combined with sub-bandgap light illumination ($E_{g,InGaN} < h\nu < E_{g,GaN}$) were used to obtain M_p. Although the devices were fabricated on GaN bulk substrate, large reverse leakage currents were observed, which may be originated from the inserted InGaN layers.

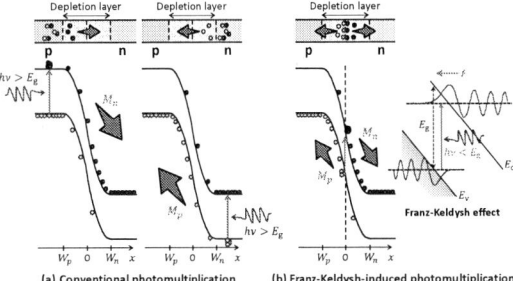

Fig. 1. Conceptual diagrams of (a) conventional and (b) Franz-Keldysh-induced photomultiplication measurements proposed in this study.

In this study, we propose the novel photomultiplication measurements for GaN PNDs using sub-bandgap light. Figure 1(b) shows the diagrams of the conventional and proposed photomultiplication measurements. In our previous study, we have reported that the optical absorption due to Franz-Keldysh (FK) effect [9, 10] occurs at high electric field region in reverse-biased GaN PNDs [11] and Schottky barrier diodes [12], and the voltage dependence of the FK-induced photocurrent well agreed with the theoretical calculation of optical absorption considering the FK effect [13]. Utilizing the sub-bandgap light absorption due to the FK effect, the electron-hole pairs can be generated at the high electric field region selectively.

To measure a multiplication factor in precise, a device with low leakage current and high avalanche capability is needed. Recently, we have reported GaN PNDs with double-side-depleted shallow bevel termination, in which the acceptor concentrations in p-layers and the donor concentrations in n-layers are almost comparable [5]. These devices showed excellent avalanche capabilities, low reverse leakage currents, positive temperature dependences of breakdown voltages, and uniform luminescence in the entire p-n junction at the breakdown. By combination of these PNDs and the FK-induced photomultiplication method, the reliable multiplication factors can be obtained.

In this study, we prepared GaN PNDs with excellent avalanche capability and measured the photocurrent under sub-bandgap illumination up to breakdown voltages. By comparing measured photocurrents with the theoretical FK-induced photocurrents, the multiplication factors were obtained. By assuming that impact ionization coefficients of electrons and holes are equal ($\alpha_p = \alpha_n$), we estimated the impact ionization coefficient in GaN.

978-1-7281-0582-6/19 $31.00 © 2019 IEEE

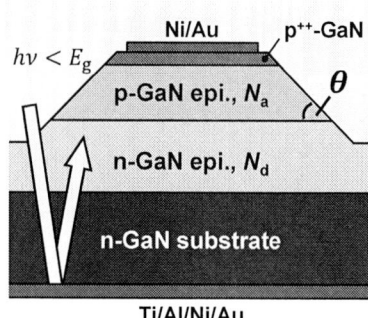

Fig. 2. Schematic cross-section of a GaN PND with double-side-depleted shallow bevel termination. The mesa height and the mesa angle are 3.5 μm and 12°, respectively. The monochromatic light (λ = 405 nm) was irradiated to the devices.

TABLE I. THE THICKNESSES AND DOPING CONCENTRATIONS IN THE EPITAXIAL LAYERS OF THE GAN PNDs.

	d_p (μm)	[Mg] (cm^{-3})	d_n (μm)	[Si] (cm^{-3})	$\frac{[Mg][Si]}{[Mg]+[Si]}$ (cm^{-3})	$\frac{N_a N_d}{N_a+N_d}$ (cm^{-3})
PN1	3.3	1.2×10^{17}	3.3	7.6×10^{16}	4.7×10^{16}	4.7×10^{16}
PN2	2.5	1.4×10^{17}	2.5	1.2×10^{17}	6.5×10^{16}	6.4×10^{16}
PN3	2.0	2.7×10^{17}	2.0	2.1×10^{17}	1.2×10^{17}	1.2×10^{17}
PN4	1.5	3.9×10^{17}	1.5	4.1×10^{17}	2.0×10^{17}	1.9×10^{17}

II. EXPERIMENTS

Figure 2 shows the device structure of the GaN PND with double-side-depleted shallow bevel termination [5]. We prepared four GaN PNDs with various epitaxial structures (PN1-4). The epitaxial layers were grown by metal organic vapor phase epitaxy (MOVPE) on GaN(0001) freestanding substrates grown by hydride vapor phase epitaxy (HVPE). When an acceptor concentration in a p-layer and a donor concentration in an n-layer are almost comparable, the depletion layer extends to the both p- and n-layers and electric field crowding at the mesa surface is effectively suppressed, confirmed by the TCAD simulation [5, 14]. Mg concentrations in the p-layers and Si in the n-layers are distributed along the depth directions, confirmed by the secondary ion mass spectrometry (SIMS). The obtained thicknesses and doping concentrations are shown in Table I. Owing to the optimized growth conditions [15], the residual carbon concentrations are very low ([C]~3×10^{15} cm^{-3}). It should be noted that [Mg][Si]/([Mg]+[Si]) is almost equal to the $N_a N_d/(N_a+N_d)$ obtained from the capacitance-voltage (C-V) measurements for each device. The shallow beveled-mesa structures were formed by Cl$_2$-based inductive coupled plasma-reactive ion etching (ICP-RIE) with a thick photoresist mask [8, 16]. The mesa height and mesa angle were 3.5 μm and 12°, respectively. Ni/Au was deposited to the epitaxial layers as the anode electrodes, and Ti/Al/Ti/Au was deposited to the backside of the substrates as cathode electrodes. The fabrication process and their avalanche breakdown characteristics are described in [5] in detail.

For the photocurrent measurements, a Hg lamp combined with a band-pass filter of 405 nm (bandwidth: 3 nm, irradiated photon flux: 4×10^{17} cm^{-2}s^{-1}) [17] and a Keysight B1505 parameter analyzer were used. The devices were dipped into Fluorinert to avoid an air spark.

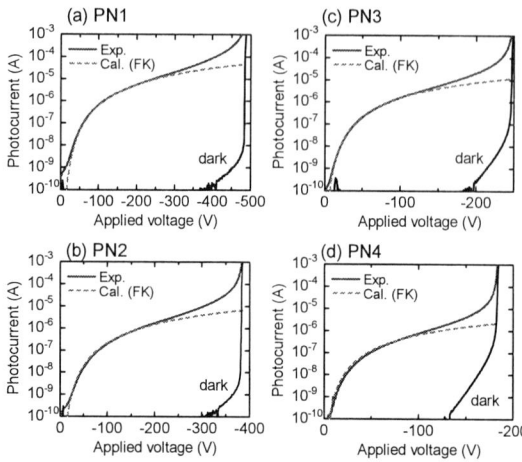

Fig. 3. Reverse-voltage dependences of the photocurrents in (a) PN1, (b) PN2, (c) PN3, and (d) PN4. The solid lines and the red broken lines are the measured photocurrents and the calculated photocurrents induced by the Franz-Keldyh effect, respectively.

III. RESULTS AND DISCUSSION

Figure 3 shows the reverse-voltage dependence of the photocurrents in PN1-PN4 under monochromatic light illumination. The wavelength of the light (405 nm) is longer than the GaN absorption edge (~365 nm, 3.4 eV); thus, the light penetrated into a GaN layer, and reached the p-n junction from the back via the reflection at the backside electrodes [11, 12]. Under the reverse bias condition, the sub-bandgap light absorption induced by FK effect occurs at the p-n junction (high electric field region), resulting in the photocurrent. The calculated curves with consideration of the theoretical optical absorption due to the FK effect [13] in the depletion layers are also shown as the red broken lines, which well reproduced the measured values up to about half of the breakdown voltages for all the devices, where avalanche multiplication can be negligible. The reduced effective mass of GaN parallel to the c-axis ($\mu = 0.16 m_0$) [18] was used in the calculation. The calculation method of the FK-induced photocurrent for a GaN PND is shown in Fig. 4 and described in [11] in detail.

Above half of the breakdown voltages, the measured photocurrents far exceeded the calculated photocurrents, reflecting the avalanche multiplications.

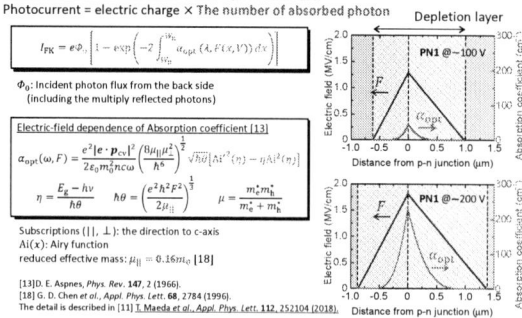

Fig. 4. Calculation method of the photocurrent induced by the FK effect. The reduced effective mass is a key parameter, since effective masses determine the tails of the electrons and holes wavefunctions.

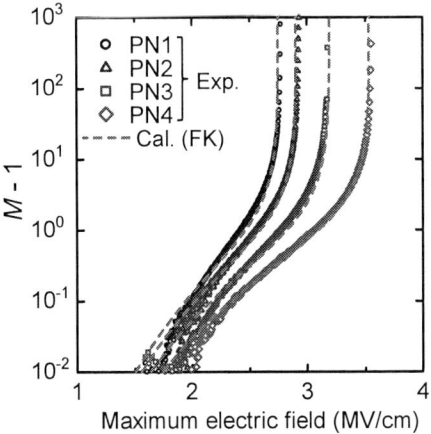

Fig. 5. Electric-field dependence of the multiplication factors in PN1-4. The dots are the experimetal values extracted as $M = I_{m}/I_{FK}$. The red broken lines are the simulated curves using the obtained Chynoweth's empirical formula for the impact ionization coefficint in GaN in this study.

Figure 5 shows the multiplication factors (M) extracted as the ratios of the measured photocurrents (I_m) to the calculated FK-induced photocurrents (I_{FK}) plotted as $M-1$ vs. the maximum electric field in PN1-4. A multiplication factor at higher electric field region was obtained for a device with a higher doping concentration, since a narrower depletion layer needs a higher electric field to cause the same avalanche multiplication.

An avalanche multiplication factor depends on a position where electron-hole pairs are generated, since impact ionization coefficients of electrons and holes are not equal. A multiplication factor when electron-hole pairs are generated at x_0 in a depletion layer of a p-n junction can be written as

$$1 - \frac{1}{M} = \left(1 - \frac{1}{M_p}\right) + \left(1 - \frac{1}{M_n}\right), \tag{1}$$

$$1 - \frac{1}{M_p} = \int_{W_p}^{x_0} \alpha_p \exp\left[-\int_x^{x_0}(\alpha_p - \alpha_n)dx'\right]dx,$$

$$1 - \frac{1}{M_p} = \int_{x_0}^{W_n} \alpha_n \exp\left[-\int_{x_0}^{x}(\alpha_n - \alpha_p)dx'\right]dx.$$

Here, M_p and M_p are hole-initiated and electron-initiated multiplication factors from x_0 to the depletion layer edges in a p-region and an n-region, respectively. The case when the electron-hole pairs are generated at the p-n junction interface ($x_0 = 0$) is described in Fig. 1(b).

In this study, electron-hole pairs are generated in the middle of the depletion layer. Obtained multiplication factors (M) contain the M_p and M_n as shown in Fig. 1(b). Although impact ionization coefficients of electrons and holes are not equal, we assume equal impact ionization coefficients of electrons and holes ($\alpha_n = \alpha_p \equiv \alpha$) for simplicity. Under this assumption, an avalanche multiplication factor does not depend on the position where electron-hole pairs are generated and can be written as

$$1 - \frac{1}{M} = \int_{W_p}^{W_n} \alpha \, dx. \tag{2}$$

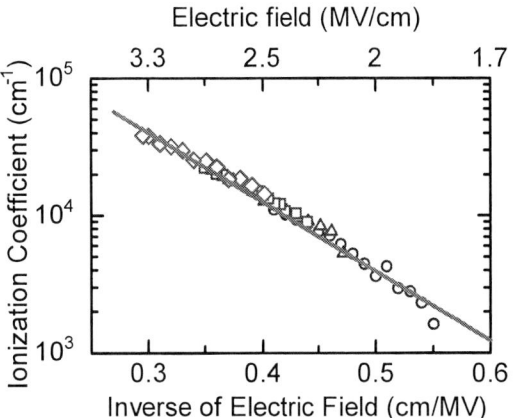

Fig. 6. The extracted impact ionzation coefficients from the multiplication factors in PN1-4. The obtained values for all the devices lie on the same linear line in the $\alpha - E^{-1}$ plot and were fitted by Chynoweth's empirical formula: $\alpha = 1.30 \times 10^6 \cdot \exp[-(1.18 \times 10^7/E)]$.

By transforming the integral in the right side as $dx \rightarrow dE$, we can obtain

$$1 - \frac{1}{M} = \frac{\varepsilon_s}{e}\frac{N_d + N_a}{N_d N_a}\int_0^{E_m} \alpha \, dE. \tag{3}$$

The electric field distribution in the GaN PNDs with double-side-depleted shallow termination is

$$E(x) = \begin{cases} \dfrac{eN_a}{\varepsilon_s}(x - W_p), & W_p \le x \le 0 \\ \dfrac{eN_d}{\varepsilon_s}(W_n - x), & 0 \le x \le W_n. \end{cases} \tag{4}$$

Thus, dE/dx for p-region and that for n-region are different. $E_m = E(0)$ is a maximum electric field. By differentiating the both sides of (3), we can obtain

$$\alpha(E_m) = \frac{e}{\varepsilon_s}\frac{N_d N_a}{N_d + N_a}\frac{1}{M^2}\frac{dM}{dE_m}. \tag{5}$$

Using (5), the impact ionization coefficient (α) for GaN can be estimated from $M(E_m)$. Figure 6 shows the extracted impact ionization coefficients in PN1-4 as the $\alpha - E^{-1}$ plot. The obtained values for all the devices lie on the same linear line in the range of 2.0-3.3 MV/cm and were fitted by Chynoweth's empirical formula [19];

$$\alpha(E) = 1.30 \times 10^6 \cdot \exp\left[-\left(\frac{1.18 \times 10^7}{E}\right)\right]. \tag{6}$$

The multiplication factors in PN1-4 were simulated using the obtained (6), and the simulated results are shown in Fig. 5 as the red broken lines. The simulated curves well reproduce the experimental results. These results suggest that the obtained impact ionization coefficient α is consistent for all the devices in this study.

In the future, α_n and α_p in GaN will be determined by using this measurement method for two different epitaxial structures; p$^+$/n$^-$ and p$^-$/n$^+$ junctions.

IV. CONCLUSION

In this study, the voltage dependence of the photocurrents in GaN-on-GaN PNDs with double-side-depleted shallow bevel termination under sub-bandgap light illumination were investigated. The FK-induced photocurrents and their avalanche multiplications were clearly observed, and the multiplication factors were successfully extracted as the ratios of the measured values to the calculated FK-induced photocurrents. Under the assumption that the impact ionization coefficients of electrons and holes are equal, the ionization coefficient in GaN was estimated by numerical analyses of the multiplication factors in the range of 2.0-3.3 MV/cm. The Chynoweth's empirical formula for the impact ionization coefficient in GaN were presented as $\alpha = 1.30 \times 10^6 \cdot \exp[-(1.18 \times 10^7/E)]$. This FK-induced photomultiplication measurement is useful for investigation of an avalanche multiplication in GaN.

ACKNOWLEDGMENT

This work was supported by the Council for Science, Technology and Innovation CSTI), the Cross-ministerial Strategic Innovation Promotion Program (SIP), and Next-Generation power electronics (funding agency: NEDO).

REFERENCES

[1] I. C. Kizilyalli, T. Pruty, O. Aktas, IEEE Electron Device Lett. vol. 36, no. 10 (2015).

[2] K. Nomoto, Z. Hu, B. Song, M. Zhu, M. Qi, R. Yan, V. Protasenko, E. Imhoff, J. Kuo, N. Kaneda, T. Mishima, T. Nakamura, D. Jena, and H. G. Xing, IEDM Tech. Digest, 2015, pp. 9.7.1-9.7.4.

[3] D. Shibata, R. Kajitani, M. Ogawa, K. Tanaka, S. Tamura, T. Hatsuda, M. Ishida, and T. Ueda, IEDM Tech. Digest, 2016, pp. 10.1.1-10.1.4.

[4] Y. Zhang, M. Sun, D. Piedra, J. Hu, Z. Liu, Y. Lin, X. Gao, K. Shepard, and T. Palacios, IEDM Tech. Digest, 2017, pp. 9.2.1-9.2.4.

[5] T. Maeda, T. Narita, H. Ueda, M. Kanechika, T. Uesugi, T. Kachi, T. Kimoto, M. Horita, and J. Suda, IEDM Tech. Digest, 2018, pp. 30.1.1-30.1.4.

[6] L. Cao, J. Wang, G. Harden, H. Ye, R. Stillwell, A. J. Hoffman, and P. Fay, Appl. Phys. Lett. vol. 112, no. 262103 (2018).

[7] G. E. Bulman, L. W. Cook, and G. E. Stillman, Solid-Stae Electron. vol. 25, no. 12 (1982).

[8] H. Niwa, J. Suda, and T. Kimoto, IEEE Trans. Electron Devices, vol. 62, no. 10 (2015).

[9] V. W. Franz, Z. Naturforsch. vol. 13 a, no. 489 (1958).

[10] L. V. Keldysh, Sov. Phys. JETP vol. 34, no. 1138 (1958).

[11] T. Maeda, T. Narita, M. Kanechika, T. Uesugi, T. Kachi, T. Kimoto, M. Horita, and J. Suda, Appl. Phys. Lett. vol. 112, no. 252104 (2018).

[12] T. Maeda, M. Okada, M. Ueno, Y. Yamamoto, M. Horita, and J. Suda, Appl. Phys. Express vol. 9, no. 091002 (2016).

[13] D. E. Aspnes, Phys. Rev. vol 147, no. 2 (1966).

[14] T. Maeda, T. Narita, H. Ueda, M. Kanechika, T. Uesugi, T. Kachi, T. Kimoto, M. Horita, and J. Suda, "Design and Fabrication of GaN p-n Junction Diodes with Negative-Beveled Mesa Termination", IEEE Electron Device Lett. under revison.

[15] T. Narita, N. Ikarashi, K. Tomita, K. Kataoka, and T. Kachi, J. Appl. Phys. vol. 124, 165706 (2018).

[16] F. Yan, C. Qin, J. H. Zhao, and M. Weiner, Mater. Sci. Forum vol. 389, no. 1305 (2002).

[17] T. Maeda, X. Chi, M. Horita, T. Kimoto, and J. Suda, Appl. Phys. Express vol. 11, no. 091302 (2018).

[18] G. D. Chen, M. Smith, J. Y. Lin, H. X. Jiang, S. H. Wei, M. A. Khan, and C. J. Sun, Appl. Phys. Lett. vol. 68, no. 2784 (1996).

[19] A. G. Chynoweth, Phys. Rev. vol 109, no. 5 (1958).

Proceedings of the 31st International Symposium on Power Semiconductor Devices & ICs
May 19-23, 2019, Shanghai, China

Photon-Enhanced Conductivity Modulation and Surge Current Capability in Vertical GaN Power Rectifiers

Shaowen Han, Shu Yang[*], Yongkai Li, Yinxiang Liu, Kuang Sheng

College of Electrical Engineering, Zhejiang University, Hangzhou, China

[*]Email: eesyang@zju.edu.cn

Abstract—In this work, we investigate the temperature-dependent forward conduction performance, the reverse recovery behavior and surge current capability of the vertical GaN Schottky barrier diode (SBD) and PN diode (PND). Compared with the SBD, the vertical GaN PND can deliver photon-enhanced conductivity modulation (PECM), temperature-independent differential ON-resistance, and higher surge current capability. It is experimentally verified that the PECM and zero reverse recovery can be simultaneously realized in the vertical GaN PND.

Keywords—Conductivity modulation, electroluminescence, GaN, power rectifiers, reverse recovery, surge current

I. INTRODUCTION

With the availability of high-quality bulk GaN substrates, vertical GaN-on-GaN devices with record power figure-of-merit [1, 2] and current-collapse-free performance [3] have been recently demonstrated. The GaN-on-GaN power devices with a vertical architecture can inherently deliver a high conduction current with small footprint and low thermal resistance [4]. In particular, a pulsed current up to 400 A (or 2.5 kA/cm²) has been realized in a 16-mm² vertical GaN-on-GaN PN diode (PND) [5], with a short ON-state pulse width and low duty cycle to minimize the self-heating effect.

In bipolar power rectifiers based on indirect bandgap semiconductors (e.g. Si and SiC), the injection and storage of minority carriers (or the conductivity modulation) usually result in a trade-off between the forward conduction and the switching speed/loss [6]. By contrast, the *direct-bandgap* GaN, in spite of its short *intrinsic* minority-carrier lifetime, enables intrinsic/extrinsic photon recycling (i.e., re-absorption of the radiative recombination) [7, 8] which could lead to optically-enhanced p-GaN activation and photoconduction modulation in vertical GaN PNDs [9, 10]. Therefore, it is of significance to conduct a comprehensive investigation on the photon-enhanced conductivity modulation and switching performance in the direct-bandgap GaN power devices.

Power devices would undergo a high surge current under the circumstances of current overshoot or oscillation [11]. Repetitive avalanche breakdown ruggedness with reverse surge current has been demonstrated using unclamped inductive switching (UIS) test in fully vertical GaN-on-GaN and quasi-vertical GaN-on-Si PNDs [12, 13]. On the other hand, as for vertical GaN-on-GaN power rectifiers that are

capable of delivering high forward current density, it is of particular interest to investigate the forward surge current capability and to reveal the impact of the hole injection.

In this work, the photon-enhanced conductivity modulation (PECM) in a vertical GaN bipolar PND is investigated. The temperature dependence of the forward conduction characteristics in the vertical GaN Schottky barrier diode (SBD) and PND has been compared and analyzed. With board-level tests, the reverse recovery behavior and surge current capability of the vertical GaN SBD and PND are also characterized. The influences of the hole injection on the forward conduction, switching performance and surge current capability in the direct-bandgap GaN power rectifiers are revealed and discussed.

II. DEVICE STRUCTURE AND FABRICATION

Fig. 1 shows the schematic cross sections and key process steps of the vertical GaN-on-GaN SBD and PND. The epitaxial structures were grown by metal-organic chemical vapor deposition on free-standing n⁺-GaN substrates. The net doping concentration in the n⁻-GaN drift layer is ~1×10¹⁶ cm⁻³ and ~6×10¹⁵ cm⁻³ in the SBD and PND, respectively.

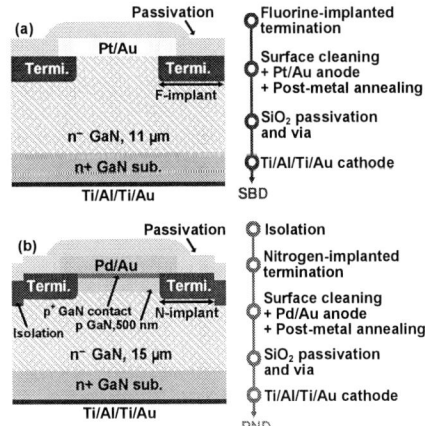

Fig. 1 Schematic cross sections and key process steps of vertical GaN (a) SBD and (b) PND, respectively.

978-1-7281-0582-6/19 $31.00 © 2019 IEEE

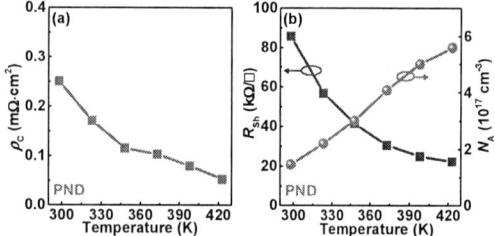

Fig. 2 (a) P-GaN contact resistivity (ρ_C) as a function of temperature. (b) Sheet resistance (R_{Sh}) and effective acceptor concentration (N_A) in p-GaN as a function of temperature, assuming a hole mobility of 10 cm²/V·s.

The vertical GaN SBD consists of Pt/Au anode deposited on the pre-cleaned surface [14], Ti/Al/Ti/Au cathode on the backside, and fluorine (F)-implanted termination [15], in which the negatively charged F⁻ in GaN can alleviate the electric field concentration near the main junction.

The vertical GaN PND consists of Pd/Au anode, Ti/Al/Ti/Au cathode, and nitrogen-implanted termination that can convert p-GaN into an insulating region around the device periphery [16].

III. STATIC ELECTRICAL CHARACTERISTICS

The contact resistivity (ρ_C) and sheet resistance (R_{Sh}) of the p-GaN layer have been characterized using the transfer length method (TLM). At elevated temperatures (T), the reduction of ρ_C and R_{Sh} of the p-GaN layer are mainly attributed to thermally-enhanced p-GaN activation, in which the effective

acceptor concentration (N_A) increases from ~1.5×10¹⁷ cm⁻³ at 25 °C to ~5.6×10¹⁷ cm⁻³ at 150 °C, as shown in Fig. 2.

The vertical GaN SBD shows a current density of 2 kA/cm² at 3 V, a near-unity ideality factor ($\eta \sim 1.02$), and a differential specific ON-resistance ($R_{ON,sp}$) of 1.08 mΩ·cm² which saturates at 1 V (Fig. 3(a)). On the other hand, the vertical GaN PND exhibits a continuous reduction in differential $R_{ON,sp}$ at a higher forward bias/current (i.e., 0.81 mΩ·cm² at 2 kA/cm² and 0.46 mΩ·cm² at 6 kA/cm²), as shown in Fig. 3(b). Meanwhile, the measured $R_{ON,sp}$ of the PND is considerably lower than the calculated value (2.2 mΩ·cm²), suggesting the conductivity modulation. In the electroluminescence (EL) measurements of the vertical PND (Fig. 3(c)), the peaks at ~2.2 eV and ~2.9 eV are usually correlated to deep acceptor traps with a broad energy distribution and Mg-V$_N$ complexes in p-GaN [17, 18]. The radiative recombination of electron-hole pair in forward-biased PND can be re-absorbed by the deep-level states, which could lead to optically enhanced p-GaN activation and hole injection into n⁻-GaN [8]. As a consequence, the enhanced hole injection could induce excess electron injection into the n⁻-GaN drift layer from the n⁺-GaN, resulting in photon-enhanced conductivity modulation.

Fig. 4 shows the forward conduction characteristics of the vertical GaN SBD and PND at elevated temperatures. The reduced turn-on voltage (V_{ON}) in both SBD and PND at higher temperature primarily originates from the bandgap narrowing and thermally-enhanced carrier diffusion. The vertical GaN SBD shows an increased differential $R_{ON,sp}$ at higher

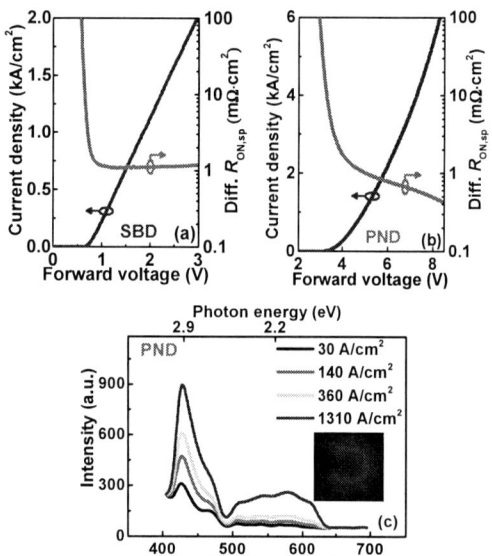

Fig. 3 Forward I-V characteristics and differential $R_{ON,sp}$ of the vertical GaN (a) SBD and (b) PND. (c) EL intensity vs. wavelength in PND at different current levels. The wavelength range is limited by the EL equipment. Inset: Photo showing the luminescence of PND at a current level of ~1000 A/cm².

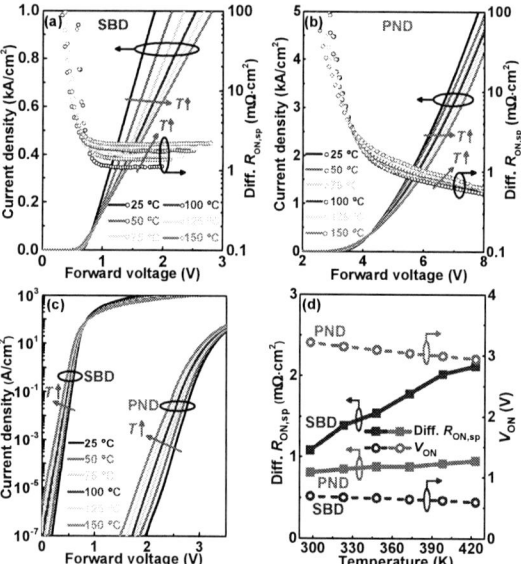

Fig. 4 Forward I-V characteristics in linear scale and extracted differential $R_{ON,sp}$ of the vertical GaN (a) SBD and (b) PND at varying T. (c) Forward I-V characteristics in semi-log scale of the vertical GaN SBD and PND at varying T. (d) Differential $R_{ON,sp}$ and V_{ON} vs. T. V_{ON} is extracted at 10 A/cm², while differential $R_{ON,sp}$ is extracted at 500 A/cm² for SBD and at 2000 A/cm² for PND, considering the inherent higher current capability of PND.

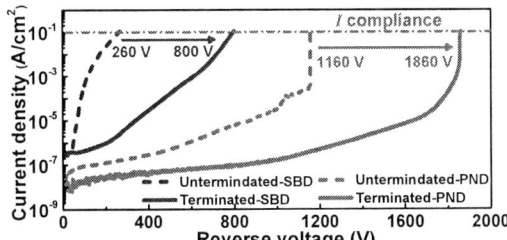

Fig. 5 Reverse *I-V* characteristics of the vertical GaN SBD and PND with and without termination.

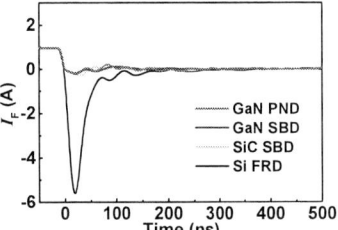

Fig. 6 Reverse recovery characteristics of the vertical GaN PND and SBD developed in this work when switching from a forward current of 1 A (or 1100 A/cm²) to a reverse bias of 400 V. The reverse recovery performance of the commercial 600 V/1 A SiC SBD and 700 V/1 A Si FRD is shown for reference. Note commercial SiC PND is not currently available.

temperature, arising from the enhanced phonon scattering and reduced electron mobility in the n⁻-GaN drift layer. By contrast, the vertical GaN PND exhibits a *T*-independent differential $R_{ON,sp}$, as the thermally-enhanced p-GaN activation enables the reduction in the contact and sheet resistance of the p-GaN layer which has been verified in Fig. 2(b).

As shown in Fig. 5, the ion implantation-based termination techniques can effectively boost the breakdown voltage from 260 V to 800 V for SBD, and from 1160 V to 1860 V for PND, respectively.

IV. REVERSE RECOVERY PERFORMANCE

By using a double-pulse tester (DPT) [3], the reverse recovery performance of the packaged vertical GaN SBD and PND, as well as that of the commercial SiC SBD and Si fast recovery diode (FRD) has been evaluated and compared. As shown in Fig. 6, when switching from 1 A to a reverse bias of 400 V with d*I*F/d*t* of 100 A/μs, the vertical GaN PND exhibits zero reverse recovery performance which is comparable with the vertical GaN and SiC SBDs, thanks to the short intrinsic carrier lifetime of single nanoseconds [19] at the reverse bias.

To investigate the impact of the forward bias-induced illumination on the transient electrical characteristics, the transient reverse leakage currents at −1200 V after switching from an ON-state at various forward biases (V_F) have been monitored, as shown in Fig. 7. The transient reverse leakage current after switching from ON-state increases with V_F, and is slightly higher than the dark current. The reverse leakage at −1200 V can maintain at ~1 nA, after switching from a relatively high V_F of 7 V. The following decay in the transient reverse leakage, in the range of tens of seconds, suggests the persistent photoconductivity effect which can be well fitted by a stretched-exponential law [20].

$$I(t) = I_0 + I(t = 0)\exp[-(t/\tau)^\beta], \qquad 0 < \beta < 1 \quad (1)$$

Where I_0, $I(t = 0)$, τ and β is the dark current, buildup level at the moment when illumination is turned off, time constant and decay exponent.

In summary, the vertical GaN PND can simultaneously realize PECM and zero reverse recovery performance, whereas the generated photons at forward bias only induce insignificant persistent photoconductivity.

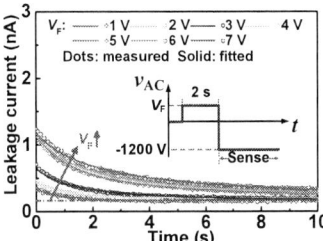

Fig. 7. Transient behaviors of the reverse leakage in vertical GaN PND at −1200 V after switching from ON-state at various V_F.

V. SURGE CURRENT CAPABILITY

The surge current capability of the vertical GaN SBD and PND has been evaluated using the board-level tests. Fig. 8 shows the circuit schematics for surge current evaluation, in which the peak current (I_{peak}) can be varied by adjusting V_C according to

$$I_{peak} = V_C \cdot \sqrt{\frac{C}{L}} \quad (2)$$

With increased I_{peak}, the vertical GaN PND exhibits relatively small peak voltage (V_{peak}) increase and can sustain a higher I_{peak} (10 A or 11 kA/cm²), compared with the SBD (Fig. 9). The enhancement in the surge current capability of the vertical GaN PND is attributed to the hole injection and *T*-independent differential $R_{ON,sp}$ (Fig. 4(d)), whereas the positive temperature coefficient of differential $R_{ON,sp}$ in the SBD results in more severe self-heating and lower surge current.

Table I summarizes the performances and key mechanisms in the vertical unipolar and bipolar GaN power rectifiers.

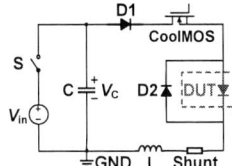

Fig. 8 Schematic circuit for surge current evaluation.

978-1-7281-0582-6/19 $31.00 © 2019 IEEE

Table I. Summary of vertical GaN-on-GaN SBD and PND.

Devices	Illumination at +V	Differential R_{ON}	Reveres recovery	Surge current
Vertical GaN-on-GaN SBD (800 V/1.08 mΩ·cm² @ 0.5 kA/cm²)	N/A	Saturates at certain +V; T-dependent	Zero (W/o minority carrier storage)	Low
Vertical GaN-on-GaN PND (1860 V/0.81 mΩ·cm² @ 2 kA/cm²)	Yes	Drops with increasing +V; T-independent (PECM; thermally enhanced p-GaN activation)	Zero (Minority carrier storage w/ ultrashort intrinsic lifetime)	High (Hole injection and photon recycling)

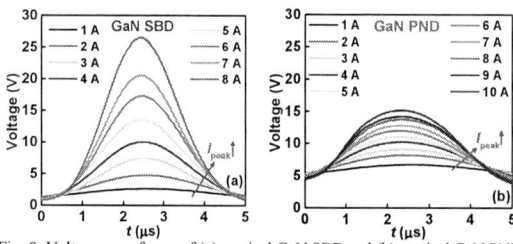

Fig. 9. Voltage waveforms of (a) vertical GaN SBD and (b) vertical GaN PND during surge current test with varying I_{peak}. The device area is 0.09 mm². Note that the large-area vertical GaN diodes have slightly higher $R_{ON,sp}$, which has also been observed in previous reports [10].

VI. CONCLUSIONS

The temperature-dependent forward conduction, reverse recovery performance and surge current capability of the vertical GaN SBD and PND have been evaluated, whereby the impacts of the hole injection and thermally-enhanced p-GaN activation have been analyzed and discussed. Compared with the SBD, the vertical GaN PND can simultaneously deliver photon-enhanced conductivity modulation, temperature-independent differential $R_{ON,sp}$, fast reverse recovery and high surge current capability.

ACKNOWLEDGMENT

This work was supported by the National Key Research and Development Program of China (No. 2017YFB0404100).

REFERENCES

[1] J. Wang, L. Cao, J. Xie, E. Beam, R. McCarthy, C. Youtsey, and P. Fay, "High voltage vertical pn diodes with ion-implanted edge termination and sputtered SiNx passivation on GaN substrates," in *IEDM Tech. Dig.*, Dec. 2017, pp. 9.6. 1-9.6. 4.

[2] H. Ohta, K. Hayashi, F. Horikiri, M. Yoshino, T. Nakamura, and T. Mishima, "5.0 kV breakdown-voltage vertical GaN p–n junction diodes," *Jpn. J. Appl. Phys.*, vol. 57, no. 4S, p. 04FG09, Feb. 2018.

[3] S. Han, S. Yang, R. Li, X. Wu, and K. Sheng, "Current-collapse-free and fast reverse recovery performance in vertical GaN-on-GaN Schottky barrier diode," *IEEE Trans. Power Electron.*, 2018, early access. DOI: 10.1109/TPEL.2018.2876444.

[4] N. Killat, M. Montes, J. Pomeroy, T. Paskova, K. Evans, J. Leach, X. Li, Ü. Ozgur, H. Morkoc, and K. Chabak, "Thermal properties of AlGaN/GaN HFETs on bulk GaN substrates," *IEEE Electron Device Lett.*, vol. 33, no. 3, pp. 366-368, Mar. 2012.

[5] I. C. Kizilyalli, A. P. Edwards, H. Nie, P. Bui-Quang, D. Disney, and D. Bour, "400-A (pulsed) vertical GaN p-n diode with breakdown voltage of 700 V," *IEEE Electron Device Lett.*, vol. 35, no. 6, pp. 654-656, Jun. 2014.

[6] D. T. Morisette and J. A. Cooper, "Theoretical comparison of SiC PiN and Schottky diodes based on power dissipation considerations," *IEEE Trans. Electron Devices*, vol. 49, no. 9, pp. 1657-1664, Sep. 2002.

[7] K. Mochizuki, K. Nomoto, Y. Hatakeyama, H. Katayose, T. Mishima, N. Kaneda, T. Tsuchiya, A. Terano, T. Ishigaki, and R. Tsuchiya, "Photon-recycling GaN pn diodes demonstrating temperature-independent, extremely low on-resistance," in *IEDM Tech. Dig.*, Dec. 2011, pp. 26.3. 1-26.3. 4.

[8] K. Mochizuki, "Vertical GaN bipolar devices: Gaining competitive advantage from photon recycling," *Phys. Status Solidi A*, vol. 214, no. 3, p. 1600489, Sep. 2017.

[9] K. Mochizuki, T. Mishima, A. Terano, N. Kaneda, T. Ishigaki, and T. Tsuchiya, "Numerical analysis of forward-current/voltage characteristics of vertical GaN Schottky-barrier diodes and p-n diodes on free-standing GaN substrates," *IEEE Trans. Electron Devices*, vol. 58, no. 7, pp. 1979-1985, Jul. 2011.

[10] K. Nomoto, B. Song, Z. Hu, M. Zhu, M. Qi, N. Kaneda, T. Mishima, T. Nakamura, D. Jena, and H. G. Xing, "1.7-kV and 0.55-mΩ·cm² GaN p-n diodes on bulk GaN substrates with avalanche capability," *IEEE Electron Device Lett.*, vol. 37, no. 2, pp. 161-164, Feb. 2016.

[11] L. Knoll, A. Mihaila, F. Bauer, V. Sundaramoorthy, E. Bianda, R. Minamisawa, L. Kranz, M. Bellini, U. Vemulapati, H. Bartolf, S. Kicin, S. Skibin, C. Papadopoulos, and M. Rahimo, "Robust 3.3kV silicon carbide MOSFETs with surge and short circuit capability," in *Proc. Int. Symp. Power Semiconductor Devices IC's (ISPSD)*, May/Jun. 2017, pp. 243-246.

[12] O. Aktas and I. C. Kizilyalli, "Avalanche capability of vertical GaN p-n junctions on bulk GaN substrates," *IEEE Electron Device Lett.*, vol. 36, no. 9, pp. 890-892, Sep. 2015.

[13] X. Zou, X. Zhang, X. Lu, C. W. Tang, and K. M. Lau, "Breakdown ruggedness of quasi-vertical GaN-based p-i-n diodes on Si substrates," *IEEE Electron Device Letters*, vol. 37, no. 9, pp. 1158-1161, Sep. 2016.

[14] S. Han, S. Yang, and K. Sheng, "High-voltage and high-I_{ON}/I_{OFF} vertical GaN-on-GaN Schottky barrier diode with nitridation-based termination," *IEEE Electron Device Lett.*, vol. 39, no. 4, pp. 572-575, Apr. 2018.

[15] S. Han, S. Yang, and K. Sheng, "Fluorine-implanted termination for vertical GaN Schottky rectifier with high blocking voltage and low forward voltage drop," under review.

[16] J. R. Dickerson, A. A. Allerman, B. N. Bryant, A. J. Fischer, M. P. King, M. W. Moseley, A. M. Armstrong, R. J. Kaplar, I. C. Kizilyalli, O. Aktas and J. J. Wierer, "Vertical GaN power diodes with a bilayer edge termination," *IEEE Trans. Electron Devices*, vol. 63, no. 1, pp. 419-425, Jan. 2016.

[17] U. Kaufmann, M. Kunzer, H. Obloh, M. Maier, C. Manz, A. Ramakrishnan, and B. Santic, "Origin of defect-related photoluminescence bands in doped and nominally undoped GaN," *Phys. Rev. B*, vol. 59, no. 8, p. 5561, Feb. 1999.

[18] I. Shalish, L. Kronik, G. Segal, Y. Rosenwaks, Y. Shapira, U. Tisch, and J. Salzman, "Yellow luminescence and related deep levels in unintentionally doped GaN films," *Phys. Rev. B*, vol. 59, no. 15, p. 9748, Apr. 1999.

[19] D. L. Mauch, F. J. Zutavern, J. J. Delhotal, M. P. King, J. C. Neely, I. C. Kizilyalli, and R. J. Kaplar, "Ultrafast reverse recovery time measurement for wide-bandgap diodes," *IEEE Trans. Power Electron.*, vol. 32, no. 12, pp. 9333-9341, Dec. 2017.

[20] B. Li, Q. Jiang, S. Liu, C. Liu, and K. J. Chen, "Degradation of transient OFF-state leakage current in AlGaN/GaN HEMTs induced by ON-state gate overdrive," *Phys. Status Solidi C*, vol. 11, no. 3-4, pp. 928-931, Feb. 2014.

Proceedings of the 31st International Symposium on Power Semiconductor Devices & ICs
May 19-23, 2019, Shanghai, China

High Performance GaN-on-Si Power Devices with Ultralow Specific On-resistance Using Novel Strain Method Fabricated on 200 mm CMOS-Compatible Process Platform

Roy K.-Y. Wong [a,*], H. C. Chiu [a], J. H. Zhang [a], C. Zhou [a, b], Thomas Zhao [a], Y. B. Wu [a], Henry Liao [a], Simon He [a], A. B. Zhang [a],
Y.B. Zou [a], Seiya Li [a], Martin Zhang [a], Macro Wu [a], John Lee [a], P.W. Chen [a], Andy Xie [a] & Jeff Zhang [a]

[a]*R&D Division, Innoscience Technology, Zhuhai, China*

[*]Tel: (+86)-07563819888 ext. 8051, Email: kingyuen@innoscience.com

[b]*State Key Laboratory of Electronic Thin Films and Integrated Devices University of Electronic Science and Technology of China,
Chengdu, China*, Email: czhou@uestc.edu.cn

Abstract—A novel strain engineering is reported to realize enhancement-mode high electron mobility transistors (HEMTs) with ultralow specific on-resistance ($R_{on,sp}$) fabricated on 200 mm CMOS-compatible process platform. In this scheme, a strain enhancement layer deposited on the access region of HEMT by a low cost CVD process is demonstrated to reduce $R_{on,sp}$. As comparing to 100 V-rated HEMT without the strain layer, HEMT with the strain layer features comparable V_{th} of +1.5 V but significant $R_{on,sp}$ reduction of 24 % and 28 % at ambient temperatures of 25 °C and 150 °C, respectively. The proposed device exhibits the off-state drain-to-source breakdown voltage increases about 12 %. System verification shows that high efficiency of 92.7 % which is higher than the state-of-art Si MOSFET about 3 % with stable burn-in performance. This work achieves a record low $R_{on,sp}$ in 100 V rating commercially available GaN and Si power transistors.

Keywords—Gallium nitride, HEMTs, Power transistors

I. INTRODUCTION

GaN-on-Si lateral high electron mobility transistors (HEMTs) have been successfully used for producing high efficiency converter applications with compact size, owing to their superior device properties such as low on-resistance (R_{on}), high breakdown voltage and high switching frequency. Over the past decade industrial companies and academic institutes made rapid progress in device design, epitaxial growth, processing technology, packaging technology, reliability enhancement and gate driving techniques, GaN-on-Si lateral heterojunction power devices have being commercialized [1-2]. However, one of the major challenges to popularize GaN-on-Si power devices is to make GaN-on-Si power devices cost-competitive to conventional Si power MOSFETs [3]. Large volume production with high yield can be achieved by CMOS compatible process to reduce the manufacturing cost. In addition, increase in wafer size and reduction of specific on-state resistance ($R_{on,sp} = R_{on} \times$ Die Area) can significantly increase the number of chips per wafer resulting in cost efficient. To achieve high performance E-mode HEMTs with small device size, the 2DEG density needs to be lower in the gate region to realize normally-off operation with positive threshold voltage (V_{th}) and higher in the access region to reduce the access resistance (R_{access}). One of the existing methods to reach this purpose is to perform a selective area

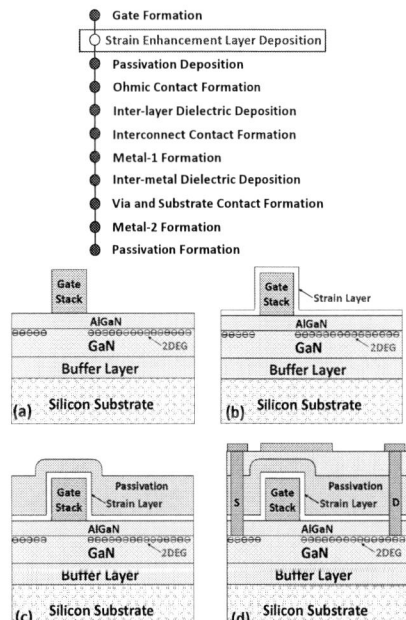

Fig. 1: Process flow for the fabrication of the proposed strain enhancement layer capped E-mode HEMTs. (a) The gate stack is defined by lithography and dry etching. (b) The strain enhancement layer is deposited by CVD process after surface cleaning. The stress in the capped layer induces sheet resistance reduction in the access region. (c) The passivation layer is deposited by CVD process. (d) Ohmic and field-plate are formed before the backend processes.

growth of thicker AlGaN in the access region [4]. Nevertheless, such method might increase the manufacturing cost. In order to reduce the cost for mass production, a strain enhancement layer deposited on the access region of HEMTs by a low cost CVD process is firstly demonstrated to reduce R_{access} without impact on V_{th} of E-mode HEMTs. This novel strain method is utilized to realize E-mode HEMTs with ultralow $R_{on,sp}$ fabricated on 200 mm CMOS-compatible process platform. The comprehensive characterization and analysis have been performed.

978-1-7281-0582-6/19 $31.00 © 2019 IEEE

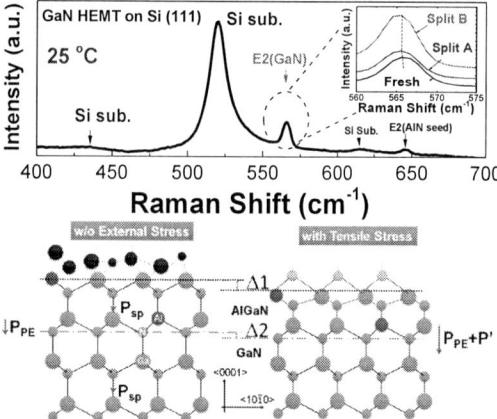

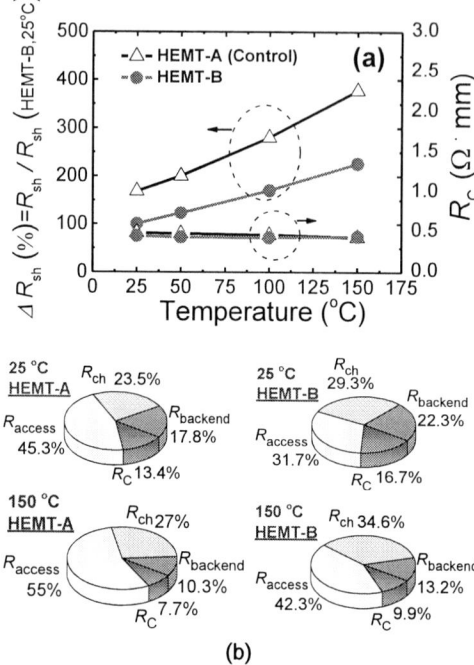

Fig. 2: Raman spectra of HEMTs in various strain layer splits were recorded at 532 nm excitation. E_2 phonon peak was used to study the stress presented in the GaN layer. Extracted tensile stresses measured from fresh (only AlGaN), thin strain layer (split-A), and thick strain layer (split-B) are 0.5, 0.58, and 0.64 GPa, respectively. Schematic shows extra piezoelectric polarization (P') is induced by the strain layer and then causes 2DEG density increasing.

Fig. 4: (a) Extracted sheet resistance ratios ($\Delta R_{sh} = R_{sh}/R_{sh}$ (HEMT-B, 25 °C)) and R_c of the TLM as a function of ambient temperature. (b) On-resistance component distributions of 100 V E-HEMT at 25 °C and 150 °C.

thick strain layer (split-B), respectively. Extracted tensile stresses measured from fresh (only AlGaN), thin strain layer (split-A), and thick strain layer (split-B) are 0.5, 0.58, and 0.64 GPa, respectively [5]. The stress modulation by the strain enhancement layer induces additional piezoelectric polarizations and then causes 2DEG density increasing.

III. DEVICE RESULTS AND DISCUSSION

Fig. 3 shows the measured on-state DC comparison of 100V E-HEMTs without (HEMT-A) and with (HEMT-B) the strain enhancement layer measured at the ambient temperatures of (a) 25 °C and (b) 150 °C. As comparing to 100V HEMT-A without the strain layer, HEMT-B with the strain layer features comparable V_{th} (+1.5V) but significant $R_{on,sp}$ reduction of 24 % (from 0.29 mΩ·cm^2 to 0.22 mΩ·cm^2) and 28 % (from 0.53 mΩ·cm^2 to 0.38 mΩ·cm^2) at the ambient temperatures of 25 °C and 150 °C, respectively. This on-resistance improvement is analyzed by the temperature dependence of sheet resistance (R_{sh}) and contact resistance (R_c) extracted by transfer length method (TLM) patterns [6]. The R_{sh} is improved about 66 % by utilized the strain enhancement layer scheme. The rate of change of R_{sh} is larger than that of R_c, as they are affected by different thermal factors. The increase in R_{sh} at higher temperature is consistent with the decreasing 2DEG mobility observed in the high-temperature transport experiment, which is attributed to the enhanced phonon scattering at higher temperature; however R_c is dominated by tunnel effects [6]. Fig. 4b depicts the on-resistance (R_{on}) is dominated by R_{access} in 100V E-HEMT. At higher ambient temperature, R_{access} becomes more dominated in R_{on} as the rate of change of R_{sh} is larger than that of R_c.

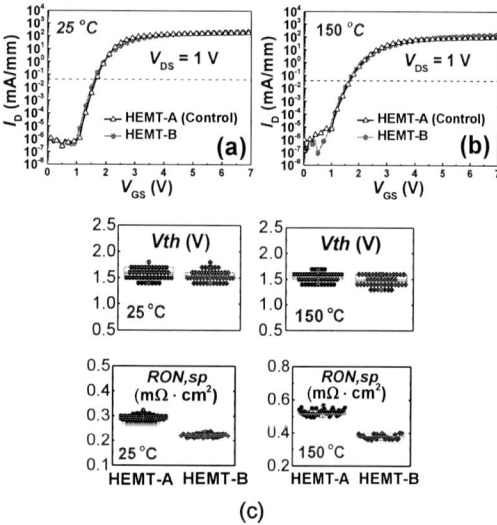

Fig. 3: DC comparison of 100 V E-HEMTs without (HEMT-A) and with (HEMT-B) the strain enhancement layer measured at the ambient temperatures of (a) 25 °C and (b) 150 °C. (c) Both devices show comparable V_{th} (+1.5 V) but significant $R_{on,sp}$ reduction in HEMT-B scheme about 24 % and 28 % at 25 °C and 150 °C, respectively.

II. DEVICE PROCESS AND STRAIN LAYER ANALYSIS

Device process flow for the fabrication of the proposed strain enhancement layer capped E-mode HEMTs is depicted in Fig. 1. The strain enhancement layer is deposited by CVD process after gate stack defined. The stress in the capped layer induces sheet resistance reduction in the access region. The process temperature range of the strain layer deposition is from 400 °C to 550 °C. Fig. 2(a) shows Raman spectra of HEMTs in various strain layer splits. E_2 phonon peak (567.5 cm^{-1}) was used to study the stress presented in the GaN layers [5]. The E_2 modes at 565.8, 565.5, and 565.3 cm^{-1} were obtained for fresh (only AlGaN), thin strain layer (split A) and

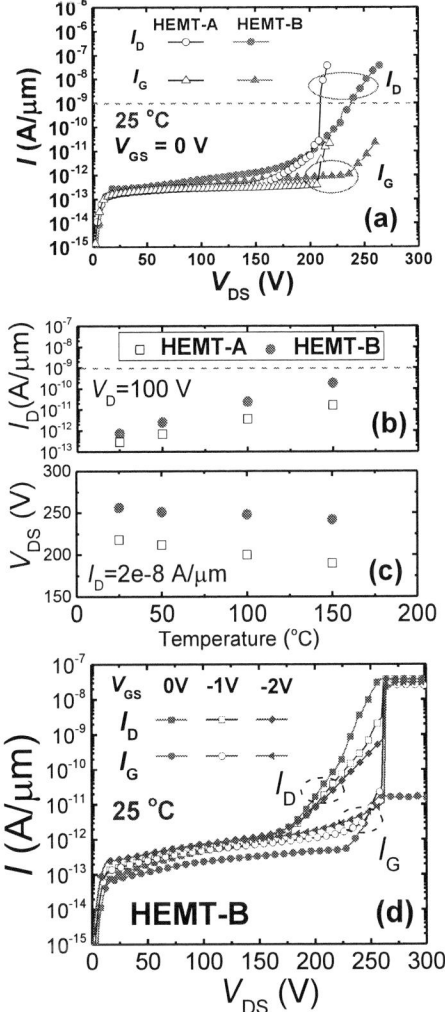

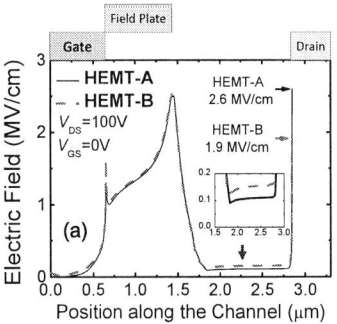

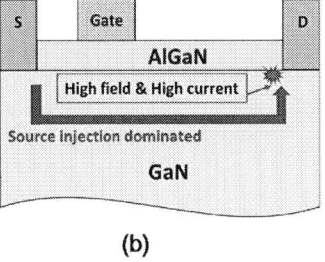

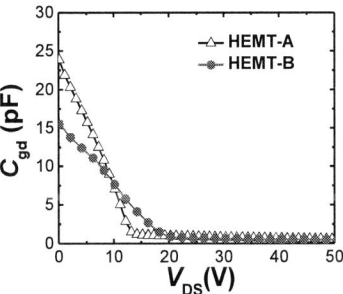

Fig. 5: (a) Measured off-state BV of 100 V HEMT-B is enhanced from 210 V to 237 V at I_D of 10^{-9} A/μm by applying the strain enhancement layer. (b) Off-state I_D against ambient temperatures at V_D of 100 V. (c) Hard-BV against ambient temperatures at I_D of 2×10^{-8} A/μm. (d) Off-state BV increasing as V_{GS} applied from 0 V to -2 V.

Fig. 6: (a) Simulated electric field along channel biased at off-state (V_{DS} = 100 V, V_{GS}= 0 V). At the drain edge, HEMT-B shows lower electric field strength as partial electric field shared in the drift region. (b) Off-state BV is dominated by the electric field at the drain edge and the source injected electrons flow to drain.

Fig. 7: Measured C_{gd} of 100 V E-HEMTs with R_{on} of 27 mΩ. FOM ($R_{on}{\times}Q_{gd}$) of HEMT-A and HEMT-B is 5.6 and 5.32 mΩ·nC, respectively.

At the off-state biasing, HEMT-B shows higher BV than HEMT-A as plotted in Fig. 5. Measured off-state BV of HEMT-B is enhanced from 210 V to 237 V at I_D of 10^{-9} A/μm by applying strain enhancement layer. Off-state I_D against ambient temperature at V_D of 100 V and hard-BV against ambient temperature at I_D of 2×10^{-8} A/μm are shown in Fig. 5b and 5c. The temperature dependence of off-state performance of devices with and without the strain enhancement layer scheme are comparable. Fig. 5d shows off-state BV increasing as V_{GS} applied from 0 to -2 V. The off-state breakdown mechanism is dominated by the electric field at the drain edge and the source injected electrons flowing to drain [7]. HEMT-B shows lower electric field strength at the drain edge as partial electric field is shared in the drift region. Fig. 6 shows the simulated electric field along channel biased at off-state (V_{DS} = 100 V, V_{GS}= 0 V). At the drain edge, HEMT-B shows lower electric field strength as partial electric field shared in the drift region. (Fig. 6b) off-state BV is dominated by the electric field at the drain edge and the source injected electrons flow to drain. Fig. 7 shows switching figure-of-merit (FOM) Measured C_{gd} of 100 V E-HEMTs with R_{on} of 27 mΩ. FOM ($R_{on}{\times}Q_{gd}$) of HEMT-A and HEMT-B is 5.6 and 5.32 mΩ·nC, respectively. It is contributed by HEMT-B features lower $R_{on,sp}$ and then smaller device size to achieve the same R_{on}.

In the system evaluation, Fig. 8 plots DC-DC converter efficiency against the output current (I_o) of 100 V HEMT-B in comparison with the state-of-art Si MOSFET. HEMT-B achieves high efficiency of 92.73 % at I_o of 1.5 A (heavy load) which is higher than the state-of-art Si MOSFET about 3 %. Stable performance observed in system burn-in (Fig. 8 insert). At the heavy load condition, low case temperature of 71.04 ℃ is measured by infrared detector as 100V HEMT-B achieves lower power losses such as output conductance loss (E_{oss}),

978-1-7281-0582-6/19 \$31.00 © 2019 IEEE   69

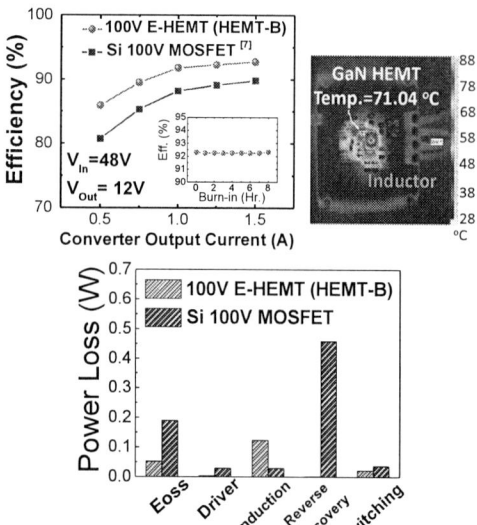

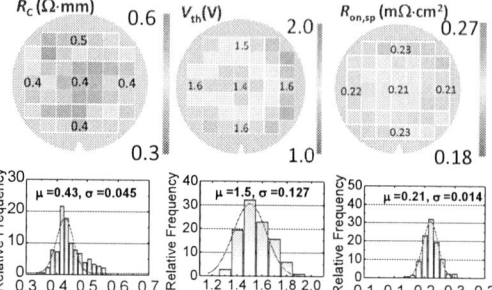

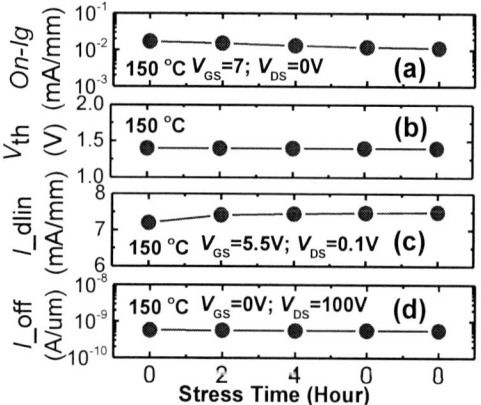

Fig. 8: DC-DC converter efficiency against output current (I_o) of 100 V HEMT-B in comparison with the state-of-art Si MOSFET. HEMT-B achieves high efficiency of 92.73 % at I_o of 1.5 A. Stable performance observed in system burn-in. At heavy load, low case temperature of 71.04 °C is measured by infrared detector as low power loss.

Fig. 10: Wafer mappings of HEMT-B fabricated on 200 mm GaN-on-Si show excellent electrical parameters uniformity and the statistic plots of 5 lots (20 pcs) exhibits high reproductivity.

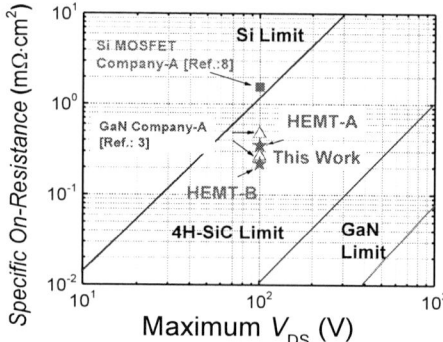

Fig. 11: Specific-on-resistance against maximum V_{DS} of commercially available GaN HEMTs and the state-of-art Si MOSFET. This work shows excellent performance.

CONCLUSION

Ultra-low specific on-resistance and high switching speed E-HEMTs fabricated on low cost 200 mm CMOS-compatible GaN-on-Si platform are demonstrated. The comprehensive characterization and analysis have been performed. Robust device performances are verified by energy system.

Fig. 9: HEMT-B shows positive results after HTGB stress ($V_{GS} = 7$ V, $V_{DS} = 0$ V, 150 °C and 8 hours).

driver loss, conduction loss, reverse recovery loss and switching loss (Fig. 8b). In addition, HEMT-B shows positive and stable performance after high temperature gate bias (HTGB) stress.

For the process stability, key DC parameters (R_c, V_{th} and $R_{on,sp}$) of HEMT-B fabricated on 200 mm CMOS-compatible GaN-on-Si process are shown in Fig. 10. Excellent wafer uniformity and high reproductivity (statistic plots of 5 lots (20 pcs)) are achieved. This scheme provides a method to make the GaN-on-Si power devices cost-competitive to conventional Si power MOSFETs. Fig. 11 shows $R_{on,sp}$ against maximum V_{DS} of commercially available GaN HEMTs and the state-of-art Si MOSFET. This work shows excellent performance and achieves a record low $R_{on,sp}$ in 100 V rating commercially available GaN and Si power transistors.

REFERENCES

[1] Kevin J. Chen, Oliver Häberlen, Alex Lidow, Chun Lin Tsai, Tetsuzo Ueda, Yasuhiro Uemoto and Yifeng Wu, "GaN-on-Si Power Technology Devices and Applications," IEEE Trans. on Electron Devices, vol. 64, pp. 779 - 795, Mar. 2017.

[2] Roy K.-Y. Wong, et al., "A Next Generation CMOS-Compatible GaN-on-Si Transistors for High Efficiency Energy Systems," in Proc. IEEE IEDM, pp. 229 - 232, Dec. 2015.

[3] Alex Lidow, "Gallium Nitride Integration: Going Where Silicon Power Can't Go," IEEE Power Electronics Magazine, vol. 5, Issue: 3, pp. 70 - 72, Sept. 2018.

[4] L. He, F. Yang, Y. No, Y. Zheng, L. Li, Z. Wu, B. Zhang and Y. Liu, "A Novel Normally-off GaN MISFET with an In-situ AlN Space Layer using Selective Area Growth," in IEEE ISPSD, pp.111 - 114, Jun. 2016.

[5] S. Dun, et al "Micro-Raman spectroscopy observation of field-induced strain relaxation in AlGaN/GaN heterojunction field-effect transistors," Phys. Status Solidi A, pp.1 - 5, Mar. 2012.

[6] Roy K.-Y. Wong, Wanjun Chen and Kevin J. Chen, "Characterization and Analysis of the Temperature-Dependent ON-Resistance in AlGaN/GaN Lateral Field-Effect Rectifiers," IEEE Trans. on Electron Devices, vol. 57, pp. 1924 - 1929, Aug. 2010.

[7] M. J. Wang, et al "Source Injection Induced Off-State Breakdown and Its Improvement by Enhanced Back Barrier with Fluorine Ion Implantation in AlGaN/GaN HEMTs," in Proc. IEDM, Dec. 2008.

[8] Infineon, Si MOSFET (BSZ096N10LS5).

Proceedings of the 31st International Symposium on Power Semiconductor Devices & ICs
May 19-23, 2019, Shanghai, China

High-performance normally-off tri-gate GaN power MOSFETs

Minghua Zhu, Jun Ma, Luca Nela and Elison Matioli

Power and Wide-band-gap Electronics Research Laboratory (POWERLAB)
École Polytechnique Fédérale de Lausanne (EPFL)
CH-1015 Lausanne, Switzerland
minghua.zhu@epfl.ch, elison.matioli@epfl.ch

Abstract— In this work, we present the investigation of the combination of gate recess and tri-gate structures to achieve high performance normally-off GaN-on-Si MOSFETs with high positive threshold voltage (V_{TH}), low specific on resistance ($R_{ON,SP}$) and high output current (I_D^{max}). The excellent channel control capability offered by tri-gate structure led to a reduced OFF-state leakage current (I_{OFF}), higher ON/OFF ratio, smaller sub-threshold slope (SS) compared to similar planar and recessed-gate devices. With gate to drain length (L_{GD}) of 20 µm, a hard V_{BR} of 2050 V were achieved, along with a low $R_{ON,SP}$ of 2.42 mΩ·cm², which corresponds to a state-of-the-art figure of merit (FOM) of 1.73 GW/cm². These results unveil the extraordinary prospects of tri-gate technology for future power electronics applications.

Keywords— Gallium Nitride, normally-off, MOSFET, tri-gate, recess, high breakdown, low leakage

I. INTRODUCTION

GaN transistors are very promising for future high-frequency power electronics converters with low conduction and switching losses and high blocking voltages [1]–[3]. Due to the presence of a polarization-induced two-dimensional electron gas (2DEG) [4], it is currently challenging to demonstrate concurrently a sufficiently positive V_{TH} with low R_{ON}, along with high breakdown voltage in GaN (MOS)HEMTs for normally-off (E-mode) operation, which is highly demanded by most power electronics applications [5]–[7].

Several techniques were reported in the literature to achieve normally-off operation, such as p-GaN gate [8], fluorine-based plasma treatment [9], [10], and recessing the barrier [11]–[14] under the gate region, either fully or partially [18], however these methods typically degrade R_{ON} and lower the I_D^{max}. Tri-gate structure can offer a way to control the V_{TH}, to enhance gate control [15]–[17] and to yield large V_{BR} [18]–[20]. Nonetheless, only a relatively limited positive V_{TH} can be achieved by relying solely on tri-gates, as shown in [21]. Much larger V_{TH} can be achieved by combining gate recess with tri-gate structures, however, to this date, only mild positive V_{TH} of 0.5 V has been demonstrated [15].

In this work, high performance normally-off GaN MOSFETs are demonstrated with high V_{BR}. A short gate recess was integrated with an optimized tri-gate geometry, fabricated with

This work was supported in part by the European Research Council under the European Union's H2020 Program/ERC Grant Agreement 67925 and in part by the Swiss National Science Foundation under Assistant Professor (AP) Energy Grant PYAPP2_166901.

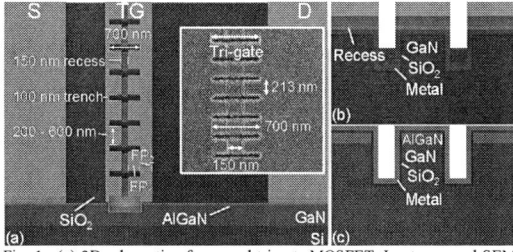

Fig. 1. (a) 3D schematic of recessed tri-gate MOSFET. Inset: zoomed SEM images of tri-gate regions. Cross-sectional views of recessed tri-gate MOSFET of (b) recessed regions and (c) tri-gate regions.

low-damage etching and a judicious surface treatment to minimize electron scattering in the channel. These devices presented V_{TH} of 1.4 V (defined at 1 µA/mm), high I_D^{max} of 622 mA/mm, improved SS of 95 mV/dec and large ON/OFF ratio beyond 10^9. In addition, the tri-gate region converts part of the gate electrode into a gate-connected FP (Fig. 1(a)), which resulted in an exceptional V_{BR} of up to 2050 V for L_{GD} of 20 µm with I_{OFF} below 9 µA/mm.

II. DEVICE STRUCTURE AND FABRICATION

Fig.1 depicts the 3D schematic (Fig. 1(a)) and cross-sectional schematics (Fig. 1(b-c)) of the fabricated normally-off tri-gate GaN MOSFETs based on GaN-on-Si wafers. The device fabrication started with mesa and tri-gate regions definition by e-beam lithography, and followed by Cl₂-based ICP etch. A 150 nm-wide gate recess (Fig. 1 (a)) was defined by e-beam lithography, followed by a 20 nm-deep low-damage slow-etch-rate Cl₂-based ICP etch, which resulted in a very precise control of the etch depth. The etched surfaces were treated with 5 cycles of O₂ plasma/HCl (37%) for 1 min each, followed by a 500°C annealing for 5 min. The low-damage slow-rate gate recess combined with cycled O₂ plasma/HCl treatment is a critical process for smoother etched surface, which minimizes electron scattering in the short recessed channel and results in E-mode transistors with reproducible V_{TH} and low R_{ON}. A metal stack composed of Ti (20 nm)/Al (120 nm)/Ti (40 nm)/ Ni (60 nm)/ Au (50 nm) was deposited in both source and drain regions, followed by rapid thermal annealing (RTA) at 780°C under N₂ atmosphere. The 25 nm-thick SiO₂

978-1-7281-0582-6/19 $31.00 © 2019 IEEE

gate dielectric was deposited by atomic layer deposition (ALD) at 300°C, immediately after a surface treatment in 37% HCl for 1 min and 500°C bake for 5 min. Finally, gate metal was formed by 50 nm Ni/ 150 nm Au. All device characteristics, such as I_D, R_{ON}, I_{OFF}, were normalized by 80 μm (width of all the devices), and the error bar was determined from 8 separate same kind devices.

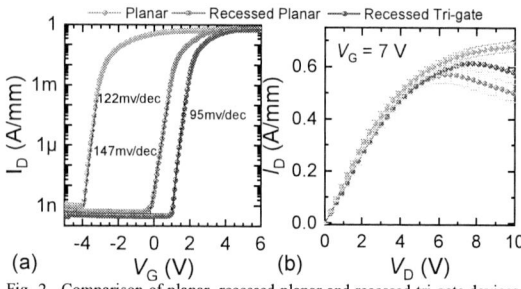

Fig. 2. Comparison of planar, recessed planar and recessed tri-gate devices. (a) Transfer at V_{DS} = 5 V and (b) Output characters of the three device with V_G = 7 V. The L_{GS}, L_G and L_{GD} were 1.5, 2 and 15 μm, respectively, and FF was 0.66.

III. RESULTS AND DISCUSSION

The comparison of the DC transfer and output characteristics of planar, recessed and recessed tri-gate devices is shown in Fig. 2(a) and (b). A noteworthy shift of V_{TH} was observed from the planar devices (–3.6 V), to +0.3 V for the recessed planar devices and +1.4 V for the recessed tri-gate (V_{TH} was defined at 1 μA/mm) (Fig. 2(a)). The normally-off tri-gate MOSFET presented a steeper SS and a lower I_{OFF} compared with recessed planar devices, revealing a better gate control by the tri-gate. The recessed tri-gate exhibited an ON/OFF ratio beyond 10^9, an enhanced SS of 95.5 ± 3.1 mV/dec and I_{OFF} as small as 300 pA/mm at V_G = 0, compared to 30 nA/mm for the recessed device. Moreover, these devices presented high I_D^{max} (at V_G = 7 V) of 622 ± 16 mA/mm compared with 581 ± 34.05 mA/mm for recessed planar, which was only slightly smaller than that of the normally-on planar device (672 ± 19 mA/mm) (Fig. 2(b)). The R_{ON} of planar, recessed planar and recessed tri-gate, extracted from I_D-V_D sweeps in the linear region, were 6.82 ± 0.28 Ω·mm, 7.37 ± 0.44 Ω·mm, and 7.32 ± 0.25 Ω·mm at V_G = 7 V, respectively (Fig. 2(b)).

The V_{TH} and SS for the recessed planar devices shifted with the variation of V_D, as observed in Fig. 3(a), which was not the case in planar and recessed tri-gate devices. This was mainly due to short-channel effects on the narrow recessed regions (100 to 150 nm) of the recessed gate transistors, which lacks effective gate control. The improved channel control of the tri-gate structure was effective in reducing the short-channel effects of the narrow recessed-gate MOSFETs [15], which makes possible to increase the I_D^{max} and reduce R_{ON} without increasing the recess length. Fig. 3(b) shows transfer characteristics of recessed tri-gate devices with

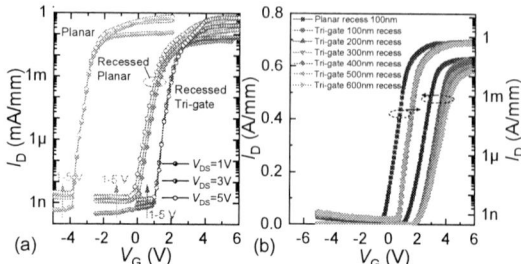

Fig. 3. (a) Comparison of the normally-off recessed tri-gate with planar and recessed devices under different V_{DS} (1V to 5V). (b) Comparison of planar recessed transistors with recessed tri-gate under different recess length (100nm to 600 nm).

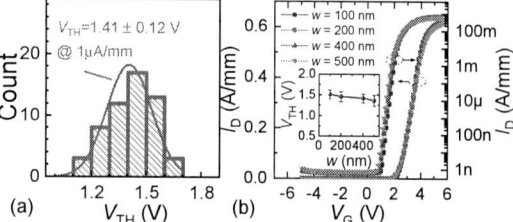

Fig. 4. (a) Distribution of V_{TH} of 56 recessed tri-gate devices. (b) Transfer characteristics of recessed tri-gate with different nanowire width. Inset: The extracted V_{TH} dependence with tri-gate nanowire width (w).

different recess length (100 nm to 600 nm). We observed that the 100 nm-wide recessed planar device showed near normally-off behavior, but when combining the 100 nm-wide recess with tri-gate structures led to a much higher V_{TH} compared to planar devices. Moreover, the threshold voltage of recessed tri-gate devices remained unchanged with the variation of recess length up to 600 nm. Nevertheless, the maximum saturation current decreased with increasing recess length. Therefore, the enhanced gate control of recessed tri-gate allows to significantly reduce the gate recess length to minimize current degradation, without any detriment from short channel effects.

Fig. 4(a) shows the narrow V_{TH} distribution among measured 56 recessed tri-gate devices, with a V_{TH} of +1.41 ± 0.12V, confirming the excellent process uniformity. To optimize the tri-gate geometry, we fabricated devices with fixed spacing s of 100nm and NW width w of 200, 400, 500, and 600 nm. The transfer characteristics of recessed tri-gate transistors with different w is shown in Fig. 4(b). The inset of Fig. 4(b) revealed that V_{TH} is not strongly dependent on the tri-gate width, which indicates the absence of 2DEG in the recessed region. A reduction of R_{ON} and an increase in I_D^{max} were observed when reducing the filling factor (FF) (Fig. 5(a)). The equivalent circuit (Fig. 5(b)) of the recessed tri-gate MOSFETs included 2 parallel parts: top (recessed + planar) and trench portions (sidewall and bottom portions). For small FF, the main conduction contribution is from the sidewall region, whereas by increasing FF, N_{NW} is reduced and the contribution from the top regions become dominant.

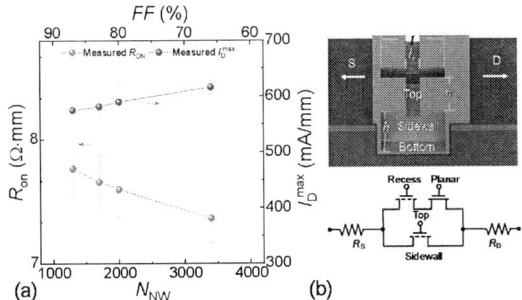

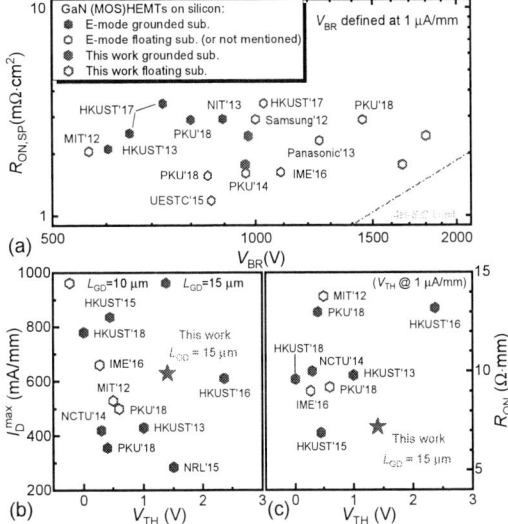

Fig. 5. (a) R_{ON} and I_D^{max} of the recessed tri-gate versus N_{NW} and FF. (b) Schematic and equivalent circuit of the recessed tri-gate region.

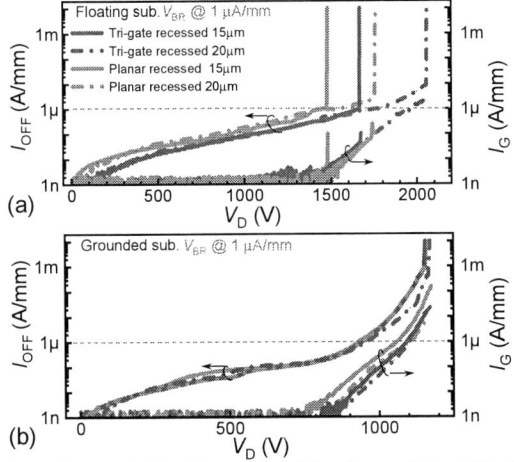

Fig. 6. Three terminal breakdown characteristics of recessed tri-gate ($V_G = 0$ V) and recessed planar ($V_G = -1$ V) MOSFETs with $L_{GD} = 15$ μm and 20 μm, for (a) floating and (b) grounded substrate.

Fig. 7. Benchmarking of (a) $R_{ON,SP}$ versus V_{BR} of the fabricated recessed tri-gate devices against GaN-on-Si E-mode transistors, by defining V_{BR} at $I_{OFF} \leq$ 1μA/mm with grounded and floating substrates. (b) I_D^{max} and (c) R_{ON} versus V_{TH} (defined at 1μA/mm). For a fair comparison, L_{GD} smaller than 10 μm, and literature results with unspecified $R_{ON,SP}$ or I_{OFF} were not included.

normally-off transistors benchmarked against state-of-the-art E-mode GaN-on-Si transistors in the literature. Fig. 7(a) was benchmarked with both floating and grounded substrate of $R_{ON,SP}$ versus V_{BR}, which outperform other device with both methods. We also benchmarked with I_D^{max} versus V_{TH} (Fig. 7(b)) and R_{ON} versus V_{TH} (Fig. 7(c)), our device exhibited high I_D^{max}, low R_{ON} and large V_{TH} of 1.4 V (Fig. 7(b-c)), with $R_{ON,SP}$ of 1.76 and 2.42 mΩ·cm^2 for L_{GD} of 15 μm and 20 μm, respectively.

The three terminal breakdown voltage of the devices was measured with floating (Fig. 6(a)) and grounded substrate (Fig. 6(b)), with $V_G = 0$ V for the recessed tri-gate transistors and $V_G = -1$ V for the recessed planar devices. With floating substrate, the hard V_{BR} of the recessed tri-gate with L_{GD} of 15 μm and 20 μm were 1650 V (at 1 μA/mm) and 2050 V (at 9 μA/mm), respectively, compared with 1480 V and 1750 V, respectively, in recessed planar devices. The soft V_{BR} at I_{OFF} of 1μA/mm of the recessed tri-gate with L_{GD} of 20 μm was 1800 V compared with 1600 V of the recessed planar devices. The improvement in V_{BR} of recessed tri-gate devices was mainly due to the integrated FP$_1$ and FP$_2$ in the tri-gate regions (Fig.1(a)) [17]. Moreover, the OFF-state gate leakage of recessed tri-gate was also much smaller than recessed planar devices, leading to a higher soft V_{BR}. With grounded substrate, a large V_{BR} of 960 V was observed, for both recessed planar and recessed tri-gate devices with $L_{GD} = 15$ μm or 20 μm, which was mainly limited by the 4.2 μm buffer thickness. The hard V_{BR} of the grounded substrate can be as high as 1100 V.

Fig. 7 shows the performance of our recessed tri-gate

IV. CONCLUSION

In this work, we present the investigation of the combination of gate recess and tri-gate structures to achieve high performance normally-off GaN-on-Si MOSFETs presenting V_{TH} of 1.4 V at 1μA/mm, high I_D^{max}, low $R_{ON,SP}$ of 2.42 mΩ·cm^2 and high V_{BR} of 2050 V, corresponding to a record Baliga FOM of 1.73 GW/cm^2. The results show the extraordinary prospects of recessed tri-gate devices for power electronics applications.

ACKNOWLEDGMENT

We would like to thank the staff in CMi and ICMP cleanrooms at EPFL for technical support and advice.

REFERENCES

[1] M. Tao *et al.*, "Characterization of 880 V Normally Off GaN MOSHEMT on Silicon Substrate Fabricated With a Plasma-Free, Self-Terminated Gate Recess Process," *IEEE Trans. Electron Devices*, vol. 65, no. 4, pp. 1453–1457, Apr. 2018.

[2] C. Lee, W. Lin, Y. Lee, and J. Huang, "Characterizations of Enhancement-Mode Double Heterostructure GaN HEMTs With

Gate Field Plates," *IEEE Trans. Electron Devices*, vol. 65, no. 2, pp. 488–492, Feb. 2018.

[3] P. Fiorenza, G. Greco, F. Iucolano, A. Patti, and F. Roccaforte, "Channel Mobility in GaN Hybrid MOS-HEMT Using SiO2as Gate Insulator," *IEEE Trans. Electron Devices*, vol. 64, no. 7, pp. 2893–2899, Jul. 2017.

[4] M. Ishida, T. Ueda, T. Tanaka, and D. Ueda, "GaN on Si Technologies for Power Switching Devices," *IEEE Trans. Electron Devices*, vol. 60, no. 10, pp. 3053–3059, Oct. 2013.

[5] J. Wei *et al.*, "Low On-Resistance Normally-Off GaN Double-Channel Metal–Oxide–Semiconductor High-Electron-Mobility Transistor," *IEEE Electron Device Lett.*, vol. 36, no. 12, pp. 1287–1290, Dec. 2015.

[6] L. Zhang *et al.*, "AlGaN-Channel Gate Injection Transistor on Silicon Substrate With Adjustable 4–7-V Threshold Voltage and 1.3-kV Breakdown Voltage," *IEEE Electron Device Lett.*, vol. 39, no. 7, pp. 1026–1029, Jul. 2018.

[7] J. Zhang *et al.*, "High-mobility normally-off Al2O3/AlGaN/GaN MISFET with damage-free recessed-gate structure," *IEEE Electron Device Lett.*, pp. 1–1, 2018.

[8] Y. Zhou *et al.*, "p-GaN Gate Enhancement-Mode HEMT Through a High Tolerance Self-Terminated Etching Process," *IEEE J. Electron Devices Soc.*, vol. 5, no. 5, pp. 340–346, Sep. 2017.

[9] Y. Zhang, M. Sun, S. J. Joglekar, and T. Palacios, "High threshold voltage in GaN MOS-HEMTs modulated by fluorine plasma and gate oxide," in *71st Device Research Conference*, 2013, pp. 141–142.

[10] L. Yang *et al.*, "High-Performance Enhancement-mode AlGaN/GaN high electron mobility transistors combined with TiN-based Source Contact Ledge and Two-Step Fluorine Treatment," *IEEE Electron Device Lett.*, pp. 1–1, 2018.

[11] Z. Xu *et al.*, "Fabrication of Normally Off AlGaN/GaN MOSFET Using a Self-Terminating Gate Recess Etching Technique," *IEEE Electron Device Lett.*, vol. 34, no. 7, pp. 855–857, Jul. 2013.

[12] M. Hua *et al.*, "Integration of LPCVD-SiNx gate dielectric with recessed-gate E-mode GaN MIS-FETs: Toward high performance, high stability and long TDDB lifetime," in *2016 IEEE International Electron Devices Meeting (IEDM)*, 2016, pp. 10.4.1-10.4.4.

[13] W. B. Lanford, T. Tanaka, Y. Otoki, and I. Adesida, "Recessed-gate enhancement-mode GaN HEMT with high threshold voltage," *Electron. Lett.*, vol. 41, no. 7, pp. 449–450, Mar. 2005.

[14] J. He, M. Hua, Z. Zhang, and K. J. Chen, "Performance andVTHStability in E-Mode GaN Fully Recessed MIS-FETs and Partially Recessed MIS-HEMTs With LPCVD-SiNx/PECVD-SiNxGate Dielectric Stack," *IEEE Trans. Electron Devices*, vol. 65, no. 8, pp. 3185–3191, Aug. 2018.

[15] B. Lu, E. Matioli, and T. Palacios, "Tri-Gate Normally-Off GaN Power MISFET," *IEEE Electron Device Lett.*, vol. 33, no. 3, pp. 360–362, Mar. 2012.

[16] S. Liu *et al.*, "Enhancement-Mode Operation of Nanochannel Array (NCA) AlGaN/GaN HEMTs," *IEEE Electron Device Lett.*, vol. 33, no. 3, pp. 354–356, Mar. 2012.

[17] J. Ma and E. Matioli, "High Performance Tri-Gate GaN Power MOSHEMTs on Silicon Substrate," *IEEE Electron Device Lett.*, vol. 38, no. 3, pp. 367–370, Mar. 2017.

[18] J. Ma and E. Matioli, "Slanted Tri-Gates for High-Voltage GaN Power Devices," *IEEE Electron Device Lett.*, vol. 38, no. 9, pp. 1305–1308, Sep. 2017.

[19] J. Ma, M. Zhu, and E. Matioli, "900 V Reverse-Blocking GaN-on-Si MOSHEMTs With a Hybrid Tri-Anode Schottky Drain," *IEEE Electron Device Lett.*, vol. 38, no. 12, pp. 1704–1707, Dec. 2017.

[20] J. Ma and E. Matioli, "2 kV slanted tri-gate GaN-on-Si Schottky barrier diodes with ultra-low leakage current," *Appl. Phys. Lett.*, vol. 112, no. 5, p. 052101, Jan. 2018.

[21] L. Nela, M. Zhu, J. Ma, and E. Matioli, "High-Performance Nanowire-Based E-Mode Power GaN MOSHEMTs With Large Work-Function Gate Metal," *IEEE Electron Device Lett.*, vol. 40, no. 3, pp. 439–442, Mar. 2019.

Proceedings of the 31st International Symposium on Power Semiconductor Devices & ICs
May 19-23, 2019, Shanghai, China

Sub-Nanosecond Delay CMOS Active Gate Driver for Closed-Loop dv/dt Control of GaN Transistors

Plinio Bau[1,2], Marc Cousineau[1], Bernardo Cougo[2], Frederic Richardeau[1], Sebastien Vinnac[1], Didier Flumian[1], Nicolas Rouger[1]

[1]LAPLACE, Université de Toulouse, CNRS, Toulouse, France
[2]Institute of Technology Antoine de Saint Exupéry
Email: plinio.bau@irt-saintexupery.com   nicolas.rouger@laplace.univ-tlse.fr

Abstract— This paper presents an AGD (active gate driver) implemented with a low voltage CMOS technology to control the dv/dt sequence of low voltage (100V) and high voltage (650V) GaN power transistors. Such an AGD can control and reduce the dv/dt of fast switching GaN devices with a reduced impact on switching losses. In the case of both low voltage and high voltage GaN fast switching transistors, such an AGD must have a total response time lower than 1ns. Therefore, introducing a feedback loop to control the dv/dt requires a specific design with a very high bandwidth (550MHz). Moreover, probing the v_{DS} voltage and its derivative is quite challenging, as the voltage level is higher than the low voltage gate driver supply. The purpose of this work is to optimize a low voltage CMOS AGD with fully integrated functions, and implement such a solution in GaN-based power converters.

Keywords—Active gate driver, GaN, switching analysis, dv/dt, EMI, power electronics, ASIC for power ic.

I. INTRODUCTION

Different solutions have been previously demonstrated to control separately the dv/dt and di/dt sequences with silicon, SiC and GaN FETs in the view of improving the loss versus EMI tradeoff. Different strategies to improve this tradeoff are: a variable impedance output stage, an open-loop control with previously programmed, an adjusted impedance sequence [1-4] or a closed-loop control to reduce the gate current during the dv/dt sequence [5,6]. The open-loop solutions rely on pre-optimization and fine tuning of the variable impedance switching sequence, which can be sensitive to parameter dispersion [1-4]. Closed-loop controls are either based on discrete components [5,6] or integrated solutions [3]. The discrete solutions typically have a large response time, above few nanoseconds or tens of nanoseconds. Our approach consists to get a full CMOS integration for all the required functions. The size of the high voltage capacitor required to sense the dv_{DS}/dt has been reduced to a few pico-Farads, which can be integrated on-chip in our work. This method can be used both for turn-on and turn-off transients. However, the first developments concern only the active control of the turn-on transient. The turn-off sequence is typically controlled by the output load current in the case of high switching speed (low switching losses), and the Zero Voltage Switching (ZVS) condition is more critical than the turn-off switching speed, in the case of synchronous buck converter [7]. However, it should be mentioned that the proposed active gate driver can be easily modified to actively control the turn-off of a fast switching GaN transistors using the exact same principle as the one detailed hereafter.

II. PRINCIPLE OF THE ACTIVE GATE DRIVER

Fig. 1 shows the transistor level schematics of the proposed CMOS AGD. Additionally, to the required output buffer (M5-M6), a high voltage sense capacitor C_S and a two stage current mirror M1-M2 and M3-M4 are integrated. During the dv/dt sequence, the dv/dt sense current i_{CS} is amplified and participates to reduce the gate current i_G during the Miller plateau (Fig. 2). Consequently, the dv/dt is reduced, while keeping a fast di/dt sequence unchanged. The trade-off between dv/dt and the turn-on energy loss E_{ON} is then optimized, comparatively to a change of the gate resistor R_G, as already demonstrated in [6].

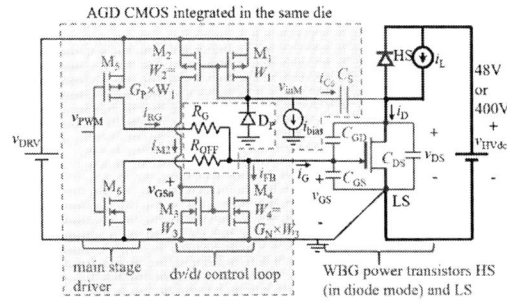

Fig. 1.   CMOS AGD topology in a transistor level schematic.

As presented earlier, the transistors M1-M6 are 5V transistors which are particularly suited to drive efficiently EPC GaN transistors, and can also drive GaN systems transistors, albeit with non-optimal driving voltages. The transistors M5 and M6 are designed with a +/-3A source/sink current rating, with integrated pre-amplification and short-circuit protection. D$_P$ is a protection diode to prevent the current mirror input voltage v_{inM} from achieving values outside the safe operating region of the transistor M1. i_{bias} is a current source improving the response time of the feedback loop, which will be further discussed in section IV.

Fig. 2 shows the qualitative current and voltage waveforms during a turn-on of the GaN power device. This time interval can be divided into the di/dt (between t_1 and t_2) and the dv/dt (between t_2 and t_3) sequences. The time duration t_D-t_2 is the feedback loop delay, further analyzed in section IV.

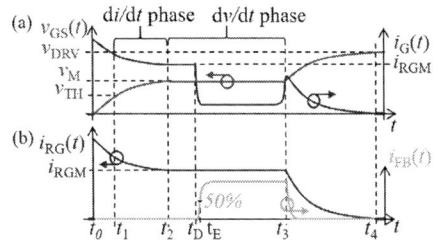

Fig. 2.   Qualitative waveforms for gate voltage and currents with the AGD during turn-on (closed-loop). (a) Current and voltage at the power gate and (b) currents are being supplied and sunk simultaneously by the AGD.

978-1-7281-0582-6/19 $31.00 © 2019 IEEE   75

The main idea of this method is to emulate a virtual capacitor added between the gate and drain nodes of the power device only during the dv/dt phase. The value of such an equivalent capacitor can be controlled by the gain G of the feedback loop (gain of the cascaded current mirrors M1-M2 and M3-M4). One should note that this gain G results from the first stage PMOS transistors and the second stage NMOS transistors $G = G_P \times G_N$. The emulated capacitance effect occurs when a feedback current i_{FB} is sinking from the gate of the power device during the dv/dt phase. This feedback current is generated by the current mirror gain G and the C_S sense capacitor value. This added emulated capacitance equal to GC_S can be easily computed using a simplified model for the power transistor where the dv/dt of the power device is expressed by (1):

$$\frac{dv_{DS}}{dt} = -\frac{v_{DRV}-v_M}{R_G C_{GD}} + \frac{i_{FB}}{C_{GD}} = -\frac{v_{DRV}-v_M}{R_G(C_{GD}+GC_S)} \quad (1)$$

where v_{DRV} is the power supply voltage of the driver and v_M is the Miller plateau voltage, shown in Fig. 2.

This principle has been previously and successfully applied to the control of Silicon IGBT [5] and high voltage GaN FET [6] for which the turn-on time is significantly high. However, this closed-loop approach is particularly difficult to achieve when GaN FET are switching within a few nanoseconds. This work aims to demonstrate experimentally the feasibility and the limitations of such an AGD with both low voltage and high voltage GaN FETS, thanks to a full CMOS integration.

TABLE I.
COMPARISON OF PERFORMANCE FOR THE SAME
METHOD OF DV/DT CLOSED-LOOP CONTROL

	Published by [5]	*Published by [6]*	*This work*
Reduction in | dv/dt |	1.8 to 0.5V/ns	27 to 8V/ns	45 to 6.6V/ns
v_{HVdc}	600V	300V	400V and 48V
v_{DS} fall time during turn-on	333ns	11ns	20ns and 4ns
Embedded in an ASIC	No	No	Yes

III. IMPLEMENTATION

A. CMOS AGD

Fig. 3 shows a microscope photograph of the fabricated AGD in AMS 180nm CMOS technology. This first prototype has been designed as a test chip with several current mirror designs and embedded high voltage capacitors. Even though quite simple, the design of the current mirrors is subject to important compromises such as the feedback loop gain splitting between the C_S value, the first and second current mirror stage gain values, the large maximum output current in M4, the dynamic bias current i_{bias} and the M2-M3 biasing during the active driving. Key constraints are current consumptions, total silicon area (embedded capacitor + current mirrors) and bandwidth. These design considerations will be illustrated in section IV. After a quick characterization in probe station [8], the AGDs are packaged in QFN24 6mm×6mm. The packaged AGDs are then implemented to drive commercially available low voltage and high voltage GaN FETs.

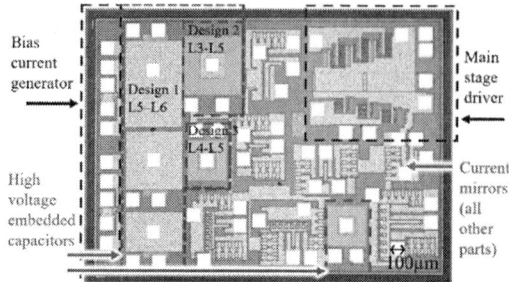

Fig. 3. Optical microscope photograph of the CMOS AGD test chip built in 0.18μm technology.

B. Implementation in 48V and 400V DC bus voltages

Two different boards have been developed to demonstrate the active control of GaN transistors in 48V and 400V applications. In both experiments, only the low side (LS) GaN transistor is driven by our CMOS AGD, while the high side (HS) GaN transistor has its gate and source shorted. The 4-layer PCB (48V) and 6-layer PCB (400V) have been designed to minimize the gate and power loops, hence only voltage probes are used. In this first implementation, an external C_S ceramic capacitor is used, 2pF/250V for 48V (J0603D2R0BXPAJ) and 1pF/500V for 400V (MC0805N1R0C501CT).

Fig. 4 (a) shows the system implementation in a 48V-36A power commutation cell with EPC2001 eGaN™FET. Fig. 4 (b) shows a similar implementation in a 400V-30A application, with GaN systems e-mode GS66508T transistors. Additional components are limited to decoupling capacitors, protection and configuration buffers and gate resistors.

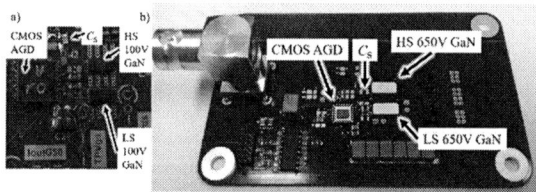

Fig. 4. AGD in a half bridge configuration for test in double pulse for (a) 4-layer board with 48V DC bus voltage with 100V GaN EPC 2001C device. (b) 6-layer board with 400V DC bus voltage with 650V GaN systems GS66508T devices.

C. Experimental Results

In the case of 48V bus voltage, the dv/dt is attenuated from 15V/ns to 6.3V/ns with the CMOS AGD (Fig. 5). Simulation results shows the switching losses are increased from 1.93μJ to 2.34μJ during turn-on. Cadence™ simulation results fitting experimental data shows the switching losses are reduced by 15.7% comparing to the increase of the gate resistance R_G, while keeping the same attenuation in dv/dt. For the 400V application, the dv/dt is attenuated from 45V/ns to 6.6V/ns, (Fig. 6). Both cases use $R_G = 4.4Ω$. This choice of R_G has been made for the EPC2001C because characterization tests in open-loop with lower values showed no more significant increase in peak dv/dt with lower R_G values. For 400V tests, the same value is used for comparative purposes.

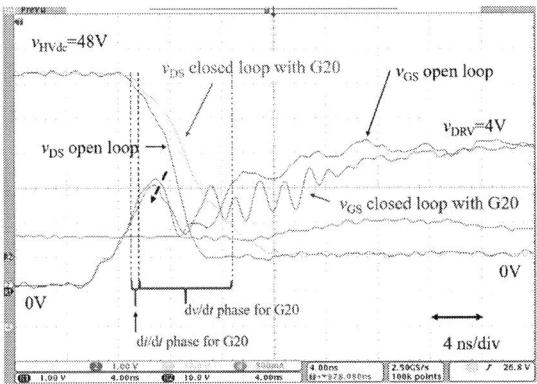

Fig. 5. Experimental results with the CMOS AGD in the 48V DC bus voltage. With the closed-loop, the dv/dt is attenuated from 15V/ns to 6.3V/ns.

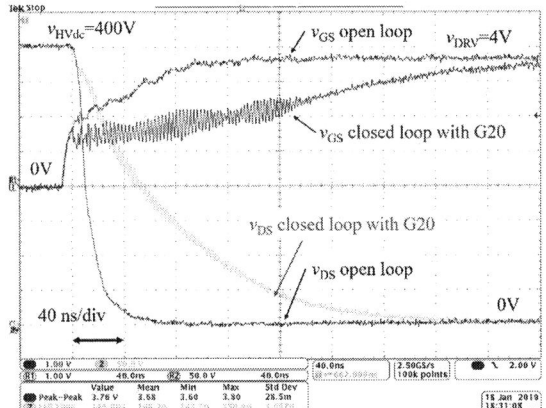

Fig. 6. Experimental results with the CMOS AGD in the 400V DC bus voltage. With the closed-loop, the dv/dt is attenuated from 45V/ns to 6.6V/ns.

IV. RESPONSE TIME OF THE FEEDBACK LOOP

A. Reducing the delay time by pre-biasing the current mirrors

One way to reduce the delay of the feedback loop (t_D-t_2) is to pre-bias the current mirrors M1-M2 and M3-M4 by pre-charging their capacitors with a low biasing current i_{bias}. Doing so, the pre-biased transistors, biased in their linear region, are able to provide a very fast response for any external transient. Table I shows the results obtained with Cadence™-Spectre™ transient simulations to determine the value of the delay defined in Fig. 2 for different biasing currents i_{bias}. Different total gain values (5, 10, 20 and 50) for the current mirrors are simulated and implemented within the prototype. One has to note that all transistors M1-M4 are designed differently for each case of total gain G value. If different gain modifies the dv/dt in closed-loop and consequently the input current i_{Cs}, it also implies different parasitic capacitance values which impact the overall response time. In these simulations, for comparison purpose, a constant dv/dt equal to 15V/ns with a sense capacitor of 2pF is considered. The design must be done carefully, splitting the total gain G into two different gain G_P and G_N to minimizing the parasitic capacitors of the transistors, leading to an optimized size reducing the several time constants of the feedback loop.

TABLE II
SIMULATED RESPONSE TIME FOR DIFFERENT
VALUES OF BIASING CURRENT WITH THREE DIFFERENT GAINS

bias current	G10 pre-biased		G20 pre-biased		G50 pre-biased	
	Time delay	50% settling time	Time delay	50% settling time	Time delay	50% settling time
i_{bias}	t_D-t2	t_E- t_D	t_D-t2	t_E- t_D	t_D-t2	t_E- t_D
0	218ps	214ps	244ps	279ps	351ps	503ps
20μA	80ps	256ps	83ps	337ps	102ps	608ps
2mA	27ps	251ps	28ps*	330ps*	31ps	589ps

* shown in Fig. 7

The time intervals t_D–t_2 and t_E–t_D are determined on Fig. 7 showing a simulation of the response time with the current mirror $G = 20$ pre-biased with a 2mA i_{bias} current. The input current goes from 0 to 26mA that correspond to a dv/dt equal to either 26V/ns with $C_S = 1$pF or 13V/ns with $C_S = 2$pF.

After tens of picoseconds the feedback loop provides a current at the output (i_{FB}) and therefore is already acting and affecting the dv/dt.

The delay decreases as the bias current i_{bias} increases. This improvement is at the expense of an additional power consumption. Indeed, using a pre-bias current, extra losses in transistors M1 to M4 are generated, which can reach high levels due to the high value of the gain G_P. It should be noted that, this biasing current can be provided only during the switching. Then, a trade-off between extra losses and bandwidth improvement has to be considered.

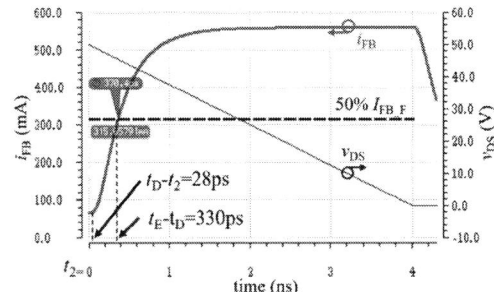

Fig. 7. Simulation of the transient response time with $G = 20$ ($v_{DRV} = 4$V, $i_{bias} = 2$mA, $i_{cs} = 30$mA).

B. Bandwidth analysis

If a non null current i_{bias} is used to pre bias the two current mirrors, a small signal analysis can be performed to determine the bandwidth of the feedback loop. Fig. 8 shows the equivalent circuit required for the study in the frequency domain. The input signal is the current coming from the capacitor C_S during a turn-on event assuming that a constant dv/dt occurs during the transient. The output of the second stage is connected to the resistance R_G and the gate of the power device behaving like a voltage-dependent capacitor C_{ISS} during the Miller's plateau. Small signal models of transistors M1 to M4 are used in Fig. 8:

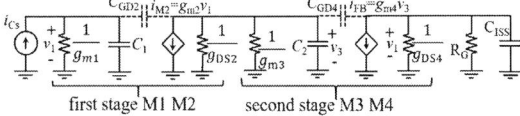

Fig. 8. Small signal analysis of the AGD current mirrors during dv/dt phase.

The current gain transfer function is expressed by (2):

978-1-7281-0582-6/19 $31.00 © 2019 IEEE

$$\frac{I_{FB}(s)}{I_{CS}(s)} = G_P G_N \frac{1 - \frac{C_{GD2}}{g_{m2}}s}{1 + \tau_{21}s} \cdot \frac{1 - \frac{C_{GD4}}{g_{m2}}s}{1 + \tau_{43}s} \quad (2)$$

where G_P and G_N are respectively the gains of the current mirrors M1-M2 and M3-M4, g_{mi} and g_{DSi} respectively the transconductances and the drain-to-source admittances of transistors M1 to M4, τ_{21} and τ_{43} the time constants of both current mirrors.

The mirror time constants are calculated with the following expressions:

$$\tau_{21} = \frac{C_1}{g_{m1}}; \ \tau_{43} = \frac{C_2}{g_{DS2} + g_{m3}} \quad (3)$$

with

$$\begin{cases} C_1 = C_{DP} + C_{DS1} + C_{GS1} + C_{GS2} + C_{GD2}(1 + \frac{g_{m2}}{g_{DS2} + g_{m3}}) \\ C_2 = C_{DS2} + C_{DS3} + C_{GS3} + C_{GS4} + C_{GD4}(1 + \frac{g_{m4}}{g_{DS4} + \frac{1}{R_G}}) \end{cases} \quad (4)$$

where C_{DP} is the capacitive contribution of the reverse biased protection diode D_P.

A Cadence™ AC simulation (Fig. 9) demonstrates a bandwidth close to 550MHz (-3dB gain) for the two-step current mirror with a current gain $G = 20$ and considering an input current made of a biasing current $i_{bias} = 2mA$ increased by a 12mA current generated by a constant dv/dt equal to 6v/ns in a sensing capacitor $C_S = 2pF$.

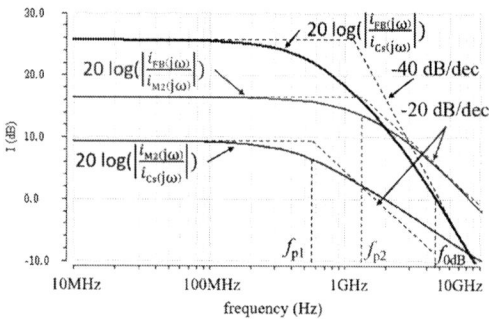

Fig 9. Cadence™ simulation of the two-step current mirror transfer function ($i_{bias} = 2mA$, $dv/dt = 6V/ns$, $G = 20$, CMOS 0.18µm technology).

With this large bandwidth, the step response of the cascaded mirrors provides a settle time equal to $(t_E - t_D)_{50\%} = 220ps$ that is close to the result provide on table II.

V. EMBEDDED CAPACITOR FOR DV/DT SENSING

In order to propose a fully integrated solution, high voltage pF range C_S capacitors are integrated on chip. The different designs involve different metal and oxide layers offered by the CMOS technology. The design compromises are the capacitor density, the breakdown voltage and the common mode parasitic capacitor. The breakdown voltages of three different designs are measured between 500V and 3.5kV (Fig. 10). The measurements have been done in a probe station under vacuum. For low voltage applications (48V), 5.5pF are measured with a surface of 0.13mm² and a destructive breakdown at 450V. For medium voltage applications (400V), 1.3pF to 2.9pF are measured for a surface of 0.19mm² to 0.35mm² respectively. The destructive breakdown for those capacitors occurs at 3.5kV. These measurements confirm the viability of a full CMOS integration of our solution.

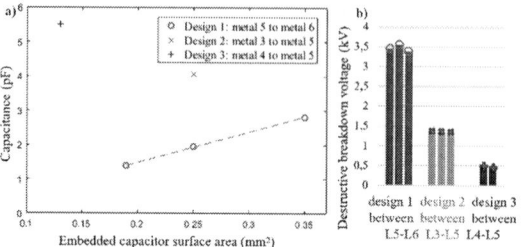

Fig. 10. High voltage embedded capacitor integration measurements for the proposed method (a) capacitance (b) breakdown voltage (L3 to L6 are the metal layer of the technology).

VI. CONCLUSION

A solution to reduce the dv/dt value while saving switching loss is presented and fully integrated in CMOS technology. 15.7% E_{ON} switching energy is saved compared to a simple reduction in dv/dt by changing the gate resistance. A high bandwidth is demonstrated both theoretically and experimentally for the feedback loop, showing sub-nanosecond delays. The key benefits of our technique are: a fully CMOS integrated solution, a reduced overcurrent during turn-on, an improved EMI vs switching-loss tradeoff, simple and fast analog circuits to implement a closed-loop technique and the possibility to integrate pF range high voltage capacitor used as a dv/dt sensor.

ACKNOWLEDGMENT

The authors would like to thank the French Ministry of Economy and Finance DGE (Directorate General for Entreprise), which contributed to the equipment acquisition (in the frame of 'FilSiC' research project) and Sorin Dinculescu and Marie-Laure Locatelli for their contributions to the capacitors tests at Laplace lab.

REFERENCES

[1] M. Takamiya, K. Miyazaki, H. Obara, T. Sai, K. Wada and T. Sakurai, "Power electronics 2.0: IoT-connected and AI-controlled power electronics operating optimally for each user," *2017 29th International Symposium on Power Semiconductor Devices and IC's (ISPSD)*, Sapporo, 2017, pp. 29-32.

[2] M. Rose, Y. Wen, R. Fernandez, R. Van Otten, H. J. Bergveld and O. Trescases, "A GaN HEMT driver IC with programmable slew rate and monolithic negative gate-drive supply and digital current-mode control," *2015 IEEE 27th International Symposium on Power Semiconductor Devices & IC's (ISPSD)*, Hong Kong, 2015, pp. 361-364.

[3] J. Yu, W. J. Zhang, A. Shorten, R. Li and W. T. Ng, "A smart gate driver IC for GaN power transistors," *2018 IEEE 30th International Symposium on Power Semiconductor Devices and ICs (ISPSD)*, Chicago, IL, 2018, pp. 84-87.

[4] H. C. P. Dymond *et al.*, "A 6.7-GHz Active Gate Driver for GaN FETs to Combat Overshoot, Ringing, and EMI," in *IEEE Trans. on Power Elect.*, vol. 33, no. 1, pp. 581-594, Jan. 2018.

[5] Shihong Park and T. M. Jahns, "Flexible dv/dt and di/dt control method for insulated gate power switches," *Conference Record of the 2001 IEEE Industry Applications Conference. 36th IAS Annual Meeting (Cat. No.01CH37248)*, Chicago, IL, USA, 2001, pp. 1038-1045 vol.2.

[6] B. Sun, R. Burgos, X. Zhang and D. Boroyevich, "Active dv/dt control of 600V GaN transistors," *2016 IEEE Energy Conversion Congress and Exposition (ECCE)*, Milwaukee, WI, 2016, pp. 1-8.

[7] R. Grezaud, F. Ayel, N. Rouger and J. Crebier, "A Gate Driver With Integrated Deadtime Controller," in *IEEE Trans. on Power Elect.*, vol. 31, no. 12, pp. 8409-8421, Dec. 2016.

[8] P. P. Bau, M. Cousineau, B. Cougo, F. Richardeau, D. Colin and N. Rouger, "A CMOS gate driver with ultra-fast dV/dt embedded control dedicated to optimum EMI and turn-on losses management for GaN power transistors," *2018 14th Conference on Ph.D. Research in Microelectronics and Electronics (PRIME)*, Prague, 2018, pp. 105-10.

Proceedings of the 31st International Symposium on Power Semiconductor Devices & ICs
May 19-23, 2019, Shanghai, China

Building blocks for future dual-channel GaN gate drivers: Arbitrary waveform driver, bootstrap voltage supply, and level shifter

Dawei Liu, Harry C. P. Dymond, Jianjing Wang, and
Bernard H. Stark
Electrical and Electronic Engineering
University of Bristol
Bristol, UK
dawei.liu;harry.dymond;jianjing.wang;bernard.stark@bristol.ac.uk

Simon J. Hollis
Xilinx, Inc.
San Jose, CA
USA
harryhollis@cantab.net

Abstract—Capitalising on the high-speed switching capability of 650 V GaN FETs in power-electronic bridge-legs is challenging. Whilst active gate driving has previously been shown to help overcome adverse switching behaviour, the best results are likely to be achieved through a combination of uncompromised circuit layout and active gate driving. A fully integrated dual-channel driver would minimise external circuitry and allow power devices to be placed as close together as possible. This would facilitate simultaneous minimization of parasitic inductances in the gate-drive and power-circuit loops. Other benefits would include ease of use, lower BOM cost, and providing a step towards full integration of driver and power stage. This paper presents three circuit blocks vital to the implementation of a fully integrated dual-channel gate driver – A 100 ps resolution, digitally-controlled active gate driver IC, a sub-ns propagation delay level shifter with 200 V/ns slew-rate immunity, and a regulated bootstrap supply that maintains its output voltage regardless of any switch-node undershoot during switching events. Measurement results show the efficacy of the high-resolution active gate driver in a GaN bridge leg, and the sub-ns propagation delay of the level shifter, both fabricated in a 50 V CMOS process. Simulation results demonstrate the slew-immunity of the level shifter, and operation of the bootstrap supply. It is also inferred how to increase the voltage rating of the level-shifter and bootstrap without adversely affecting performance.

Keywords—Active gate driver, bootstrap power supply, dual-channel gate driver, level shifter, GaN

I. INTRODUCTION

The half-bridge power stage is an essential circuit block in power electronics [1]. The layout of low-EMI 650 V GaN half-bridges is challenging, and a range of solutions have been proposed. Active gate drivers, for example, have been reported to increase power efficiency and suppress EMI for a given layout, for IGBTs [2], Silicon MOSFETs [3], and SiC JFETs [4], and recently, the first inroads have been made into the active gate driving of GaN FETs [5], albeit open-loop. Since GaN devices are so small for a given current rating, the layout itself could be significantly improved by integrating the high-and low-side drivers into a dual-channel driver. This would make possible a circuit layout with reduced gate-loop and power-loop parasitic inductances. This, in turn, improves

This work was supported by the UK Engineering and Physical Sciences Research Council (EPSRC) under Grants EP/K021273/1 and EP/R029504/1.

Fig.1. A possible future dual-channel digital active gate driver for GaN FETs: The sub-circuits addressed in this paper are shown in bold.

circuit performance, minimising overshoots and ringing, and allows better utilisation of the high-speed features of GaN FETs. It also permits the integration of functions that involve both power devices, such as dead-time optimisation[6].

This work aims to enable future dual-channel active driver ICs for mains-voltage GaN bridge legs, as shown in Fig.1, to provide safe and quiet switching at 100 V/ns. The paper focuses on three sub-circuits of a dual-channel gate driver that are critical to permit switch-node slew rates of 100 V/ns: A new single-channel arbitrary waveform gate driver with 10 GHz update rate; a level shifter with sub-nanosecond propagation delay, 200 V/ns slew rate immunity, and potential for use in mains-voltage applications; and an accurate high-side bootstrap power supply.

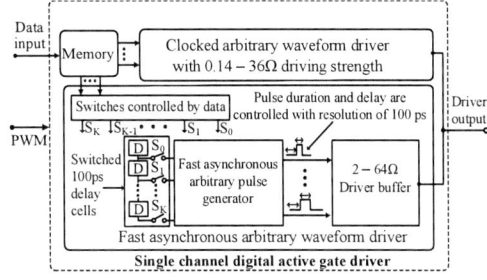

Fig.2. Architecture of the single-channel active gate driver.

978-1-7281-0582-6/19 $31.00 © 2019 IEEE

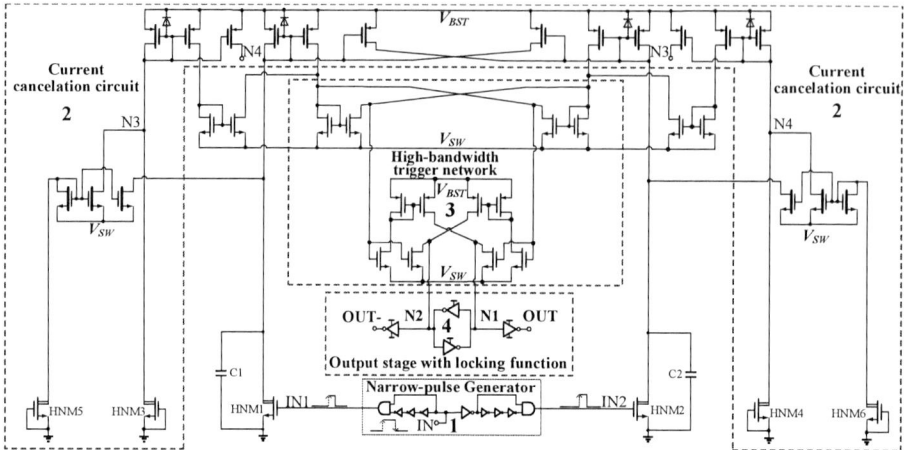

Fig.3. Schematic of level shifter, with sub-nanosecond delay, and 200 V/ns slew rate tolerance.

II. DESIGN OF THREE PROPOSED SUB-CIRCUITS

A. Single-channel active gate driver

The single-channel active gate driver with 100 ps driving resolution is shown in Fig. 2. The output stage is split into a slower clocked part (top) and a fast asynchronous part (bottom). The driver includes a memory whose stored data defines the driver's output impedance sequences that occur on input PWM signal transitions. The clocked arbitrary driver has 8-bit driving resolution and $0.14\ \Omega$ - $36\ \Omega$ driving strength. The driving strength is synchronously updated at a rate determined by the internal clock. The faster asynchronous arbitrary waveform driver includes an arbitrary pulse generator and a fast driver buffer. The pulse generator produces driving pulses with a 100 ps resolution, and the 6-bit fast driver buffer delivers these pulses with driving strengths of $2\ \Omega$ - $64\ \Omega$.

On each PWM edge, a gate-drive sequence of 8 internal clock cycles' duration is read from the memory by the internal control logic. The setting in the 8th clock period is maintained until the next PWM transition, and there are independent resistance sequences for low-to-high and high-to-low transitions.

B. Floating level shifter with sub-nanosecond propagation delay and high slew rate immunity

A level shifter with short propagation delay and high slew rate immunity is a vital component for high-side GaN FET gate drivers [7]. The level shifter circuit in this work is shown in Fig. 3. The design technique is detailed in [8]. This level shifter includes 4 main blocks:

1. An input narrow-pulse generator that detects the input *IN* edges and generates the narrow trigger pulses.

2. A current cancellation circuit that prevents currents flowing through parasitic capacitors C_1 and C_2 from triggering the level shifter during positive or negative transients.

3. A high-bandwidth network that mirrors the input triggering current generated by HNM1 and HNM2 to trigger the output stage.

4. An output stage that correctly locks the output state, according to input signal *IN*.

This level shifter has been verified through a 50 V CMOS process. To apply it to a higher voltage rating bridge-leg converter, an appropriate process should be selected. In this case, devices in blocks 2, 3, and 4 would still be low-voltage devices, built in a higher isolation N-well. The propagation delay is affected only by the features of HNM1-HNM6. The 'intrinsic' delay [10] of these devices is increased with higher voltage rating, but the overall propagation delay will remain low. The slew rate tolerance is derived from the symmetric architecture with good layout matching, and we therefore anticipate it to be unaffected by the use of higher-voltage rated devices when implementing a 600 V version.

C. Accurate high-side bootstrap power supply

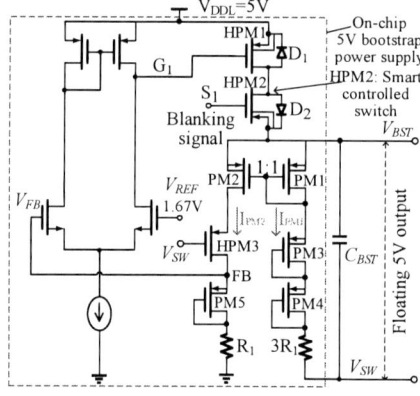

Fig.4. 5 V high-side bootstrap power supply generated from low-side 5V power supply.

The circuit of the high-side bootstrap power supply is shown in Fig. 4, and its operation illustrated in Fig. 5. The bootstrap capacitor C_{BST} is charged during the charging phase when the switch-node voltage is low (i.e., when Q_1 of Fig. 1 is on), but only after the deadtime is over, in order to protect from overcharging. To this end, HPM2 is gated with a blanking signal, Fig. 5. During this charging phase, the bootstrap voltage is regulated to 5 V, by HPM1, whose gate is controlled through the negative feedback that forces V_{FB} and V_{REF} to have the same voltage of 1.67 V. PM1 to PM5

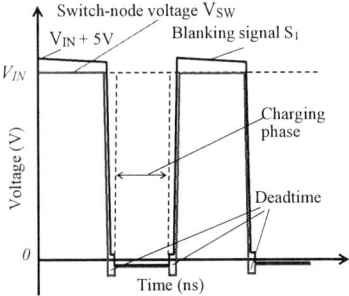

Fig.5. Control of bootstrap charging using blanking signal S_1: Switch off HPM2 outside of charging phase to protect bootstrap circuit. Switch on only after deadtime to avoid overshoot on bootstrap voltage.

have the same size and each PMOS device's bulk and source are connected together to eliminate body effect and to guarantee they have the same threshold voltage. Current mirrors composed by PM1 and PM2 make I_{PM1} and I_{PM2} equal, and gate-to-source voltages of PM1 to PM5 are equal as a result.

$$I_{PM1} = \frac{(V_{BST} - V_{SW}) - 3 * V_{GS(PM1)}}{3 * R1} \tag{1}$$

$$I_{PM2} = \frac{V_{REF} - V_{GS(PM4)}}{R1} \tag{2}$$

As $I_{PM1} = I_{PM2}$, it can be concluded:

$$V_{BST} - V_{SW} = 3 * V_{REF} = 5V \tag{3}$$

From equation (3), we conclude the voltage across C_{BST} is accurately maintained at 5 V regardless of the value of switch node voltage V_{SW} during the charging phase. Outside of the charging phase, the blanking signal S_1 is set to equal V_{BST} to turn off HPM2, and HPM2 blocks the high voltage when V_{SW} is equal to the bridge-leg DC link V_{IN}. HPM3 protects the source of PM4 when V_{SW} equals V_{IN}.

The charging phase operates with $V_{DDL} = 5V$ regardless of process choice. For a 600 V implementation, PM1-PM5 need to be placed in the 600 V isolation well. The parasitic diode D2 must block 600 V. HPM3 also needs to be a 600 V device to protect node *FB*, which sits at around 1.67 V.

III. MEASUREMENT AND SIMULATION RESULTS

The single-channel active gate driver is fabricated on an AMS 50 V HV CMOS process, occupying 5 mm². It is

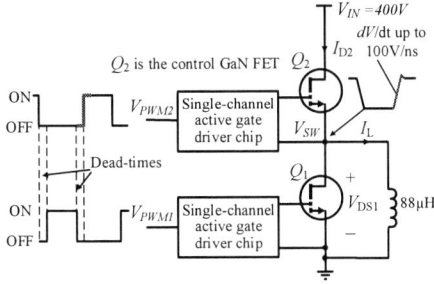

Fig.6. Measurement setup: Bridge-leg Buck converter with two GS66508P GaN FETs, each driven by one single-channel active gate driver IC.

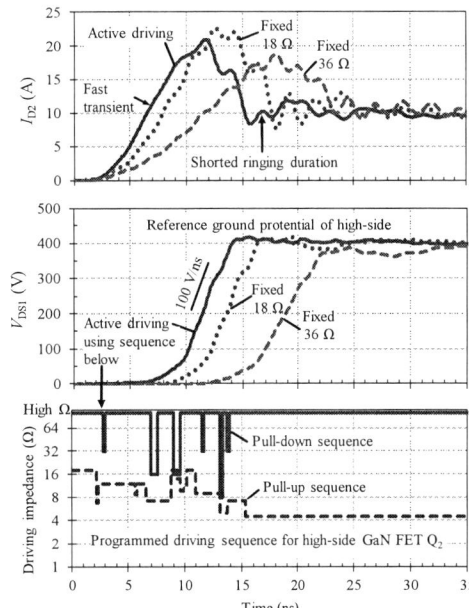

Fig.7. Demonstration of high-side driver actively controlling the upper device to shape the drain current, whilst its ground reference VDS1 is slewing at 100 V/ns. The resulting current switches faster with a significantly reduced ringing duration.

demonstrated experimentally on the high-side of the bridge leg Buck converter of Fig. 6, to validate its performance under fast slewing. Measured results for turn-on are given in Fig. 7. The drain current I_{D2} is measured using an Infinity Sensor [9] that has just 0.2 nH insertion inductance. The measurement results are shown to correctly output 100-ps-resolution, pre-programmed driving sequences, whilst the switch-node V_{SW} is slewing at 100 V/ns. This is, in part, enabled by use of TI's digital isolators (ISO78x) with a CMTI of 100 V/ns. Active driving is seen to reduce the duration of current ringing by more than half, as well as peak overshoot, whilst simultaneously speeding up the switching transient, when compared to driving with a fixed 18 Ω gate resistance. This improvement in the current overshoot behaviour is not possible by simply adding gate resistance; this is illustrated by the result using a fixed 36 Ω gate resistance: Here the switching losses are high, and yet the ringing is still present.

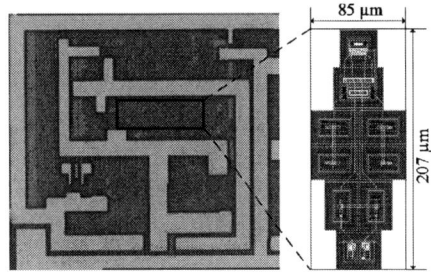

Fig.8. Die photo of level shifter.

The proposed signal level-shifting circuit is also fabricated with AMS 50 V HV CMOS process. Its micrograph photo is shown in Fig. 8. Fig. 9 shows the measurement result of propagation delay that is below 750 ps with V_{SW} up to 45 V.

978-1-7281-0582-6/19 $31.00 © 2019 IEEE

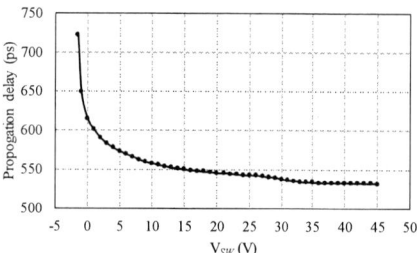

Fig.9. Measured propagation delay of the level shifter.

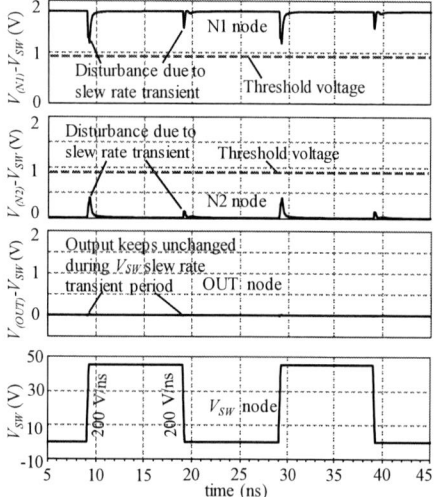

Fig. 10: Post-layout simulation showing that the level shifter is immune to switch-node slew-rates up to 200 V/ns.

Post-layout simulation results in Fig. 10 show that the level shifter is immune to 200 V/ns slewing of the switch node voltage.

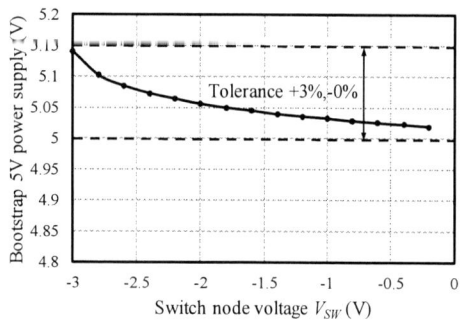

Fig.11. Bootstrap 5V power supply is independent of negative switch node V_{SW} voltage during charging phase (low-side GaN FET is on and reverse conducting).

The performance of the bootstrap power supply is demonstrated by simulation in Fig. 11. The supply maintains an output of between 5.14 V and 5.05 V during the charging phase, independent of the high-side reference potential, which is typically around -0.5 V during synchronous conduction, and -2 V during the dead time.

IV. CONCLUSION

Three key circuit blocks required to realise a fully integrated dual-channel active gate driver for GaN FETs have been presented: a single-channel active gate driver, a level shifter, and an accurate 5 V bootstrap power supply. The single-channel active gate driver and level shifter are fabricated with the AMS 50 V HV CMOS process and verified through measurements. The simulation results of the charging phase characteristic of the bootstrap power supply demonstrate reliable operation in the face of switch-node voltage undershoots. For the implementation for mains-voltage bridge-legs with 400 V DC links, an alternative process is required to give the required blocking capability to the level shifter and bootstrap supply [11]. It is noted which devices would have to support the additional voltage, and how circuit operation would be affected by these changes.

The presented circuit blocks could also be used to realise a fully-integrated power stage that includes the gate drivers and both power devices. The arbitrary waveform gate driver is particularly attractive here as it offers system designers full flexibility to tune the behaviour of the gate driver whilst eliminating the most challenging aspects of circuit layout.

REFERENCES

[1] D. Kinzer, "GaN Power IC Technology: Past, Present, and Future," in *Proc. 29th International Symposium Power Semiconductor Devices & ICs (ISPSD)*, pp. 19–24, June 2017.

[2] Makoto Takamiya, Koutaro Miyazaki, Hidemine Obara, Toru Sai, Keiji Wada, Takayasu Sakurai, "Power electronics 2.0: IoT-connected and AI-controlled power electronics operating optimally for each user", *Power Semiconductor Devices and IC's (ISPSD) 2017 29th International Symposium on*, pp. 29-32, June.

[3] Shorten, A.; Fomani, A.A.; Ng, W.T.; Nishio, H.; Takahashi, Y.; , "Reduction of conducted electromagnetic interference in SMPS using programmable gate driving strength," in *2011 IEEE 23rd International Symposium Power Semiconductor Devices and ICs (ISPSD)*, 2011 pp.364-367.

[4] R. Grezaud, F. Ayel, N. Rouger, and J. Chebier, "An adaptive output impedance gate drive for safer and more efficient control of wide bandgap devices," in *IEEE Workshop on Wide Bandgap Power Devices and Applications*, Oct. 2013, pp. 68-71.

[5] H. C. P. Dymond, J. Wang, D. Liu, J. J. O. Dalton, N. McNeill, D. Pamunuwa, S. J. Hollis, B. H. Stark, "A 6.7-GHz Active Gate Driver for GaN FETs to Combat Overshoot, Ringing, and EMI, *IEEE Transactions on Power Electronics*, vol. 33, no. 1, pp. 581-594, Jan. 2018.

[6] R. Grezaud, F. Ayel, N. Rouger, and J.-C. Crebier, "A Gate Driver with Integrated Dead-Time Controller," *IEEE Transactions on Power Electronics*, vol. 31, no. 12, pp. 8409–8421, Dec. 2016.

[7] X. Ming *et al.*, "A high-voltage half-bridge gate drive circuit for GaN devices with high-speed low-power and high-noise-immunity level shifter," in *2018 IEEE 30th International Symposium on Power Semiconductor Devices and ICs (ISPSD)*, 2018, pp. 355–358.

[8] D. Liu, S. J. Hollis, and B. H. Stark, "A New Design Technique for Sub-Nanosecond Delay and 200 V/ns Power Supply Slew-Tolerant Floating Voltage Level Shifters for GaN SMPS," *IEEE Transactions on Circuits and Systems I: Regular Papers*, in progress.

[9] J. Wang *et al.*, "Infinity Sensor: Temperature Sensing in GaN Power Devices using Peak di/dt," in *2018 IEEE Energy Conversion Congress and Exposition (ECCE)*, 2018, pp. 884–890.

[10] D. Liu, S. J. Hollis, H. C. P. Dymond, N. McNeill, and B. H. Stark, "Design of 370-ps Delay Floating-Voltage Level Shifters With 30-V/ns Power Supply Slew Tolerance," *IEEE Transactions on Circuits and Systems II: Express Briefs*, vol. 63, no. 7, pp. 688–692, Jul. 2016.

[11] Y. Lu et al., "A 600V high-side gate drive circuit with ultra-low propagation delay for enhancement mode GaN devices," in *2018 IEEE 30th International Symposium on Power Semiconductor Devices and ICs (ISPSD)*, 2018, pp. 80–83.

Proceedings of the 31st International Symposium on Power Semiconductor Devices & ICs
May 19-23, 2019, Shanghai, China

An Integrated Gate Driver for E-mode GaN HEMTs with Active Clamping for Reverse Conduction Detection

Wei Jia Zhang
The Edward S. Rogers Sr. Department of
Electrical and Computer Engineering
University of Toronto
Toronto, Canada
weijia@vrg.utoronto.ca

Yahui Leng
Institute of VLSI Design
Zhejiang University
Hangzhou, China
lengyahui@zju.edu.cn

Jingshu Yu
The Edward S. Rogers Sr. Department of
Electrical and Computer Engineering
University of Toronto
Toronto, Canada
yujingshu@vrg.utoronto.ca

Yu Shen Lu
The Edward S. Rogers Sr. Department of
Electrical and Computer Engineering
University of Toronto
Toronto, Canada
yushen.lu@mail.utoronto.ca

Chu Yao Cheng
The Edward S. Rogers Sr. Department of
Electrical and Computer Engineering
University of Toronto
Toronto, Canada
chuyao.cheng@mail.utoronto.ca

Wai Tung Ng
The Edward S. Rogers Sr. Department of
Electrical and Computer Engineering
University of Toronto
Toronto, Canada
ngwt@vrg.utoronto.ca

Abstract—**Detection of reverse-conduction in GaN-based switched-mode power converters is essential to reduce power loss and to protect the circuit from severe undershoot voltage. In this paper, an active clamping SenseFET circuit is proposed to measure the duration that the low-side GaN HEMT is in reverse-conduction and with protection from the high voltage at the switching node. The output of the SenseFET clamping circuit is processed by a custom designed gate driver IC. This IC is fabricated using TSMC's 0.18 μm BCD GEN2 process for driving e-mode GaN power HEMTs with an on-chip closed-loop dead-time correction circuit. The one-step correction mode can optimize the dead-times for both the turn-on and turn-off edges in one switching cycle for switching frequencies of up to 10 MHz with 0.32 ns precision. This allow the power converters to maintain optimal conversion efficiency over the full output current range.**

Keywords—*e-mode GaN driver; SenseFET; active driving; reverse-conduction reduction; dead-time correction*

I. INTRODUCTION

Gallium nitride (GaN) power transistors are attracting much attention as an alternate candidate for switched-mode power converters. Comparing to their silicon counterparts, GaN high electron mobility transistors (HEMTs) exhibit lower specific on-resistance, higher breakdown voltage and lower parasitic capacitance, leading to lower conduction and switching losses [1]. They also exhibit higher switching frequency, further reducing the size and cost of the output inductors and capacitors. There are two types of GaN power transistors, enhancement-mode (e-mode, normally-off) and depletion-mode (d-mode, normally on). D-mode GaN power transistors require a negative gate voltage to turn off and are impractical for integrated gate drivers. The d-mode GaN HEMT can be co-packaged with a low voltage silicon power MOSFET integrated with silicon drivers in a cascode configuration [2] to re-gain normal gate control. The emergence of e-mode GaN HEMTs enables their wide adoption in various power conversion applications. The driving method for e-mode GaN HEMTs is similar to that for silicon-based power MOSFETs. However, there are challenges such as the limited maximum gate voltage (7 V), the lack of intrinsic body diode and large undershoot voltage [3]. To mitigate these issues, precise dead-time detection and fast correction are essential to ensure optimal performance.

Dead-times are necessary in switching output stage to avoid shoot-through current between the high-side (HS) and the low-side (LS) power transistors. However, excessively long dead-times can lead to unwanted reverse-conduction and power loss [4]. Sensing the duration of the reverse conduction is especially difficult for high voltage e-mode GaN HEMTs due to their fast switching speed. High precision timing circuits are required to correct the dead-times as the load current changes [5]. Traditional CMOS-based sensing circuits (e.g. standard logic gates) are not suitable for GaN based converters as they can only handle limited voltage ranges. In addition, severe undershoots (up to −3 V) could damage the sensing circuit.

In this paper, a gate driver IC for e-mode GaN power output stages capable of detecting the presence of reverse-conduction with a resolution of 0.64 ns, a dead-time adjustment resolution of 0.32 ns [6] and on-chip closed-loop control is presented. Together with a 70V level shifter and digital control loop, the integrated HS and LS gate drivers can continuously optimize the dead-times to achieve optimal efficiency. In addition, a novel reverse-conduction sensing circuit that can tolerate the large voltage swings at the switching node will also be described. Section II discusses the implementation of this proposed design, explains the active clamping mechanism and demonstrates two dead-time correction modes. Experimental results are presented in Section III. Finally, Section IV summarizes this work.

II. PROPOSED DRIVER IC

A. Design Approach and System

Fig. 1 is the micrograph of the proposed gate driver IC. The chip was fabricated using TSMC's 0.18 μm BCD GEN2 process. The system level block diagram is shown in Fig. 2. The chip includes a dead-time generator, a time to digital converter (TDC), an on-chip state machine, an internal bootstrap capacitor, two gate drivers, a reverse conduction detection circuit and various level-shifters in four different voltage islands including digital 1.8 V, analog 3.3 V, power 5 V and 70 V high voltage wells. The SENSE pin is connected to the reverse-conduction active clamping circuit. The drain of the SenseFET is connected to the switching node (SW). Its gate is tied to a constant voltage.

978-1-7281-0582-6/19 $31.00 © 2019 IEEE

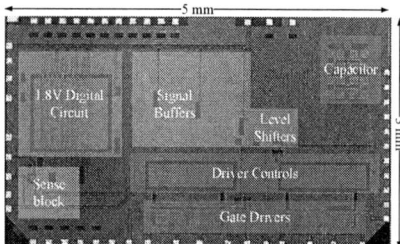

Fig. 1. Micrograph of the gate driver IC fabricated using TSMC's 0.18 μm BCD GEN2 process (5 mm × 3 mm) [6].

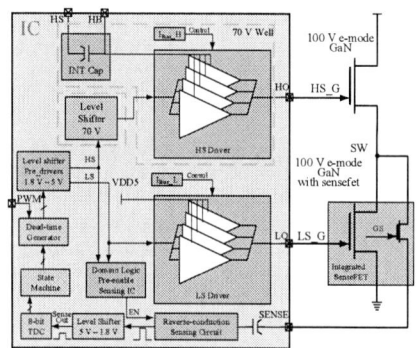

Fig. 2. System level block diagram of the proposed gate driver IC with a novel reverse-conduction sensing circuit [6].

B. Active Clamping Mechanism

An equivalent capacitive divider model between the switching node and the sense node (SENSE) is illustrated in Fig. 3. The GaN SenseFET is fabricated on the same chip as the main power HEMT to reduce parasitic inductance between packages. The SenseFET could also be an external GaN HEMT or a silicon n-type power MOSFETs.

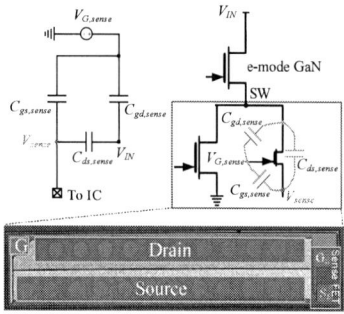

Fig. 3. GaN power HEMT with an integrated SenseFET (GaNPower 15 A SenseFET) and a capacitive divider model for the clamping circuit.

The constant gate bias voltage, $V_{G,sense}$ improves the tracking accuracy while protecting the sense node from the high voltage at the switching node. During the LS conduction period and dead-time periods, the sense node voltage (V_{sense}) follows the switching node. During the HS conduction period, V_{sense} is clamped to a pre-determined voltage, as described by Equation (1). Since the gate capacitance (C_{GS}) usually dominates in a power HEMT, V_{sense} is roughly clamped to the constant gate bias voltage, $V_{G,sense}$.

$$V_{sense} \approx V_{G,sense}/(1+C_{DS}/C_{GS}) + V_{IN}/(1+C_{GS}/C_{DS}) \quad (1)$$

C. Dead-time Detection

After receiving a negative pulse from the clamping circuit, the sensing circuit can detect reverse conduction duration up to 40 ns with a resolution of 0.5 ns [6]. The output pulse width of the sensing block is digitized by a time to digital converter consisting of a digital delay lock loop (DLL) and a 128 to 7-bit decoder as shown in Fig. 4. This digitized reserve conduction duration can be processed and corrected by an on-chip state machine.

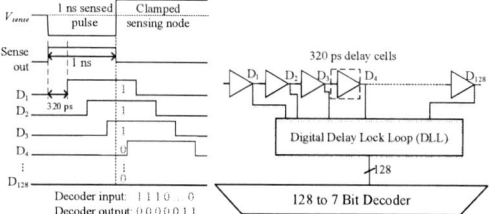

Fig. 4. Example of reverse conduction detection sequence for a 1 ns long pulse [6].

D. Dead-time Correction

Fig. 5 shows the simulated rise/fall times (10 to 90 % or 90 to 10 %) of the switching node for a 45 to 12 V synchronized buck converter implemented using 100 V e-mode GaN HEMTs with a switching frequency of 1 MHz. Both the rise and fall times at the switching node varies with the loading condition. It is important to keep track of the dead-times to avoid shoot-through and reduce reverse-conduction losses.

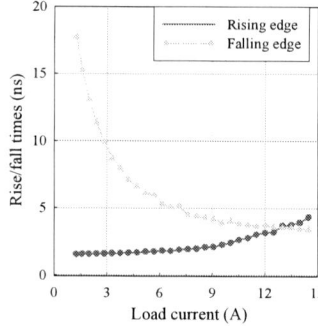

Fig. 5. Estimation of the optimum dead-times for a synchronized buck converter 100V using e-mode GaN power HEMTs under a wide range of load current. Simulated at V_{IN} = 45V, V_{OUT} = 12V and f_{sw} = 1 MHz.

Two different dead-time correction methods (one-step correction and programmable correction) are illustrated in the flow chart in Fig. 6. The state-machine can first be programmed with a target reverse-conduction time (D_t) which is set to 0 by default. The state machine is designed to correct the sensed reverse-conduction to be near D_t. The one-step mode can complete the correction and set the new dead-time within one switch cycle for switching frequencies up to 10 MHz. A gradual correction mode can be achieved by programming the adjustment speed (as a fraction of the one-step correct speed in X %) to allow a slower but more stable correction when load current varies drastically.

978-1-7281-0582-6/19 $31.00 © 2019 IEEE   84

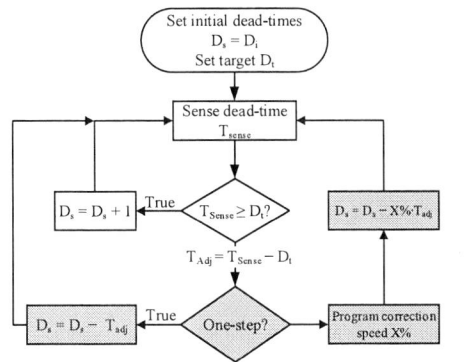

Fig. 6.   Flow chart representation of the one-step correction mode and programmable correction mode.

III. EXPERIMENTAL RESULTS

Two experimental test-benches for the gate driver IC are used: one with an external silicon SenseFET and another with an integrated GaN SenseFET, as shown in Fig. 7. The results are compared in terms of clamping performance, sensing accuracy and closed-loop reliability.

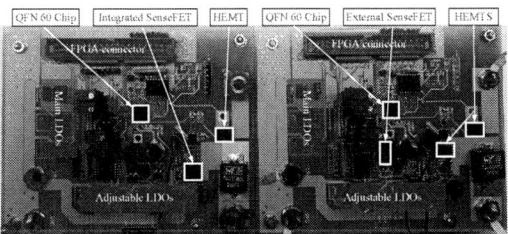

Fig. 7.   Two experimental test-benches for the gate driver IC with a silicon SenseFET or a GaN SenseFET.

A. Active Clamping Verification

The effectiveness of the clamping circuit is verified using either an external silicon SenseFET or an integrated GaN SenseFET. Fig. 8 (a) shows that for a constant gate voltage at 1.5 V, the sensing voltage increases gradually with increasing input voltage. Based on Equation (2), obtained from simplifying Equation (1), the clamping voltage is proportional to the product of V_{IN} and C_{DS}. The C_{DS} of the silicon SenseFET is a drain voltage dependent parameter which decreases with increasing V_{IN}. As a result, the sensing voltage of silicon SenseFET gradually increases to around 1.8 V. The sensing output for the GaN SenseFET stays between 1.5 and 2 V. In Fig. 8 (b), the sensing voltage increases linearly with the gate voltage for both SenseFETs. In contrast, the silicon SenseFET has a lower clamping voltage than the GaN SenseFET due to its larger gate capacitance.

$$V_{sense} \approx (V_{G,sense} \cdot C_{GS} + V_{IN} \cdot C_{DS})/(C_{GS} + C_{DS}) \quad (2)$$

B. Comparison between SeneFETs

The tracking accuracy was examined for both test-benches. The measurements of the negative pulse width were taken at a level of -100 mV for both the switching and sense nodes. Due to the larger capacitance of the silicon SenseFET, its tracking time has a larger variance compare to the actual

reverse-conduction pulse width measured at the switching node. The additional parasitic inductance resulted from the silicon packaging also contributes to the tracking offset. An example of the testing waveform is shown in Fig. 9 .

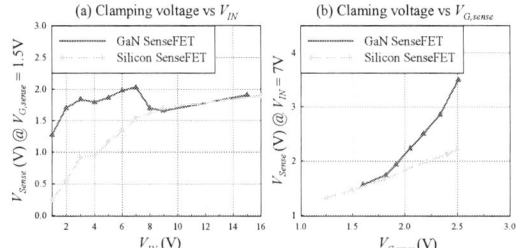

Fig. 8.   Clamping voltage verification between silicon n-MOS and integrated GaN SenseFET.

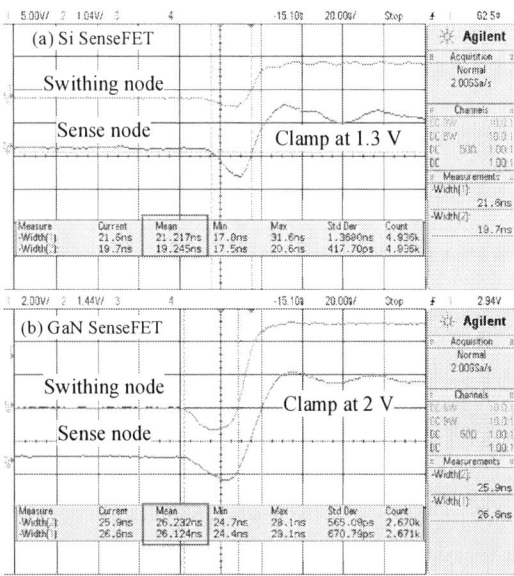

Fig. 9.   Tracking accuracy comparison between the two test-benches using (a) an external silicon SenseFET or (b) an integrated GaN SenseFET with $V_{IN} = 5$V and $V_{G,sense} = 2$ V.

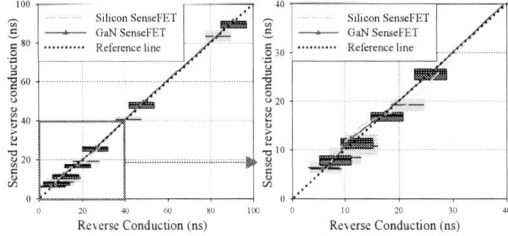

Fig. 10.   Sensed reverse-conduction duration (with error bars) as a function of the actual reversed conduction duration at switching node.

The tracking accuracy is summarized in Fig. 10. Since the reverse conduction measured at the switching node exhibits a standard deviation of around 0.5 ns, the tracking signals also carry this standard deviation. Both the silicon and GaN SenseFETs test-benches demonstrate good agreement with the actual reverse conduction time and with similar standard deviation. However, the average sensed reverse conduction duration from the silicon SenseFET circuit shows slightly larger discrepancy. On the other hand, the average sensed reverse conduction duration GaN SenseFET circuit shows better fitting to the expected value.

C. Programmable Correction Mode

With the external silicon SenseFET, the additional parasitic inductance and its slower switching speed contributes to tracking errors, leading to large over-shoots using the one-step correction mode, as shown in Fig. 11. This issue can be solved by using a gradual correction mode and a pre-set target dead-time as a safety margin.

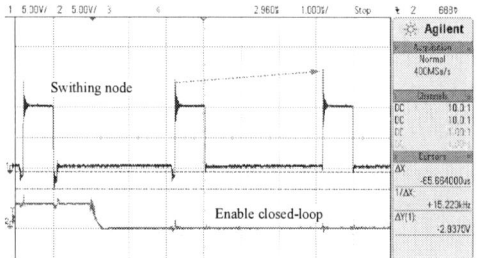

Fig. 11.   Test-bench with silicon SenseFET shows over-correction under closed-loop condition due to errors from sensing circuit.

As shown in Fig. 12, the GaN SenseFET testbench does not suffer from large overshoots when operating in the one-step correction mode. The gradual correction mode, shown at the falling edge, is also demonstrated. The reverse-conduction is suppressed in both modes and the undershoot voltage has also been eliminated.

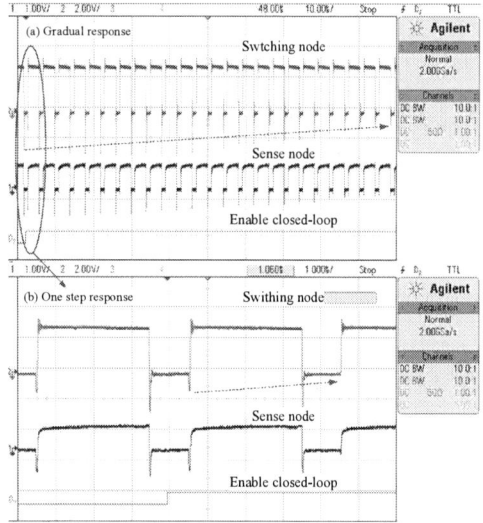

Fig. 12.   Waveforms for the one-step correction at the rising edge and the gradual correction at the falling edge for the integrated GaN SenseFET test-bench operating in closed-loop.

D. System Efficiency

The primary purpose of the dead-time correction control is to maintain optimal dead-times to reduce shoot-through current and reverse-conduction times. This allows the switched-mode power converters to maintain optimal efficiency across various load conditions, as shown in Fig. 13.

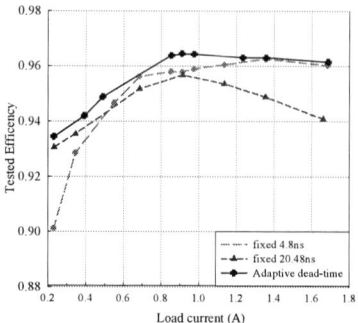

Fig. 13.   Measured power conversion efficiency for different dead-times, at V_{IN} = 20 V, V_{OUT} = 12 V, f_{sw} = 1 MHz with all losses included.

IV.   CONCLUSIONS

This paper demonstrates an innovative reverse-conduction detection method for the LS power switch using an active clamping sensing circuit. This detection circuit can be implemented with an integrated GaN SenseFET or an external silicon MOSFET. The clamping mechanism is explained with an equivalent capacitive divider model. After the duration of the reverse conduction is determined, a closed-loop dead-time correction circuit can effectively optimize the power conversion efficiency of the converter over a wide range of loading conditions. The suppression of reverse conduction in the LS power switch also reduces the occurrence of the negative undershoots at the switching node.

ACKNOWLEDGMENT

The authors would like to thank NSERC Canada for financial support, TSMC for providing access to their 0.18 μm BCD GEN2 technology and IC fabrication and GaNPower Inc. for their generous offer on GaN HEMT samples.

REFERENCES

[1]   D. Kinzer, "GaN power IC technology: Past, present, and future," in *2017 29th International Symposium on Power Semiconductor Devices and IC's (ISPSD)*, 28 May-1 June 2017 2017, pp. 19-24.

[2]   M. Rose, Y. Wen, R. Fernandes, R. V. Otten, H. J. Bergveld, and O. Trescases, "A GaN HEMT driver IC with programmable slew rate and monolithic negative gate-drive supply and digital current-mode control," in *2015 IEEE 27th International Symposium on Power Semiconductor Devices & IC's (ISPSD)*, 10-14 May 2015 2015, pp. 361-364.

[3]   R. Yan, S. Tang, J. Xi, L. He, and K. Sun, "A GaN HEMTs half-bridge driver with bandgap reference comparator clamping for high-frequency DC-DC converter," in *IECON 2017 - 43rd Annual Conference of the IEEE Industrial Electronics Society*, 29 Oct.-1 Nov. 2017 2017, pp. 539-545.

[4]   Z. Chen, Y. T. Wong, T. S. Yim, and W. H. Ki, "A 12A 50V half-bridge gate driver for enhancement-mode GaN HEMTs with digital dead-time correction," in *2015 IEEE International Symposium on Circuits and Systems (ISCAS)*, 24-27 May 2015 2015, pp. 1750-1753.

[5]   J. J. O. Dalton *et al.*, "Stretching in Time of GaN Active Gate Driving Profiles to Adapt to Changing Load Current," in *2018 IEEE Energy Conversion Congress and Exposition (ECCE)*, 23-27 Sept. 2018 2018, pp. 3497-3502.

[6]   W. J. Zhang, Y. Leng, J. Yu, X. Jiang, C. Cheng, and W. T. Ng, "A Gate Driver IC For Enhancement Mode GaN Power GaN Transistors with Precise Dead-time Correction," presented at the IEEE 14th International Seminar on Power Semiconductors (ISPS), Prague, Czech Republic, 2018.

Proceedings of the 31st International Symposium on Power Semiconductor Devices & ICs
May 19-23, 2019, Shanghai, China

Driving GaN Power Transistors
Circuit Design Challenges and Opportunities

D. Brian Ma
Department of Electrical and Computer Engineering
The University of Texas at Dallas
Richardson, TX 75080, USA
E-mail: brian.ma@utdallas.edu

Abstract—**In order to take full advantage of fast and efficient switching performance of emerging GaN power devices, it is important for circuit designers to understand the unique features of power GaN devices and their limitations. It is also important to appreciate new opportunities and challenges that these new devices have brought to the society. Motivated by this, this paper reviews some critical design issues in developing modern GaN based power ICs, with focus on reliability, noise, efficiency, speed and cost. State-of-art design solutions are also briefly reviewed.**

Keywords—GaN, gate driver, device breakdown, aging, EMI, deadtime control, level shifting

I. Introduction

Today Gallium Nitride (GaN) power devices have seen rapid growth in many power applications. Despite of the fact that the first release of commercial GaN power devices was only eight years ago, the global GaN power market is projected to reach around $423M by 2023, with an impressive compound annual growth rate of 93% [1]. With far superior figure of merits $(R_{DS\_ON} \times Q_G)$ [2-4], it is highly expected that GaN technology would replace conventional silicon counterparts especially in the high-end power circuit market. Applications include automotive electronics, data centers, telecommunication, industrial applications, and so on.

One significant advantage of GaN power switches is the capability of performing fast switching without compromising power efficiency, owing to a few attractive physics attributes. For instance, the electric breakdown field of a GaN device is usually much higher than that of silicon. This allows for much finer drift region, leading to reduced capacitance. Furthermore, the mobility of GaN device is high. This, together with high saturation velocity, contributes to low on-resistance [2-4]. Accordingly, in order to deliver the same amount of current in a power circuit, a GaN device needs less die area than its silicon equivalent, which further reduces the capacitance. As a result, it enables the device to operate at high switching frequency (f_{SW}) in a very efficient way. From the design perspective, high f_{SW} facilitates a drastic reduction on the sizes of power passive components such as inductors, transformers and capacitors in power stages. As these components get smaller, it leads to a series of benefits such as high power density, low EMI production, small PCB footprint and low cost.

However, to implement such a high f_{SW} operation in a GaN-based power circuit reliably and effectively, conventional silicon power circuits may not be directly applicable for a quick 'plug-and-play'. This is because GaN power devices possess other unique characteristics in addition to high switching speed. If not designed properly, severe performance and reliability issues would occur. Unfortunately, similar to all other new research subjects, there

has been very few literatures available in the area. Serving as a trial reference, this paper presents several important design challenges that were encountered by my research team in past few years. The intention is to use this as a stepping stone that lead to better and more effective solutions in this emerging research area.

It should be noted that there are primarily two system integration approaches in the area today. To minimize the parasitics and maximize the switching speed, it would be ideal to achieve monolithic integration of GaN switches and gate driver circuitries on the same die. *Navitas* has demonstrated the success on this effort recently [5]. However, due to lack of mature GaN technology, this approach limits the choice of available GaN devices. In addition, achievable functions and performance of these on-die gate drivers are still very limited. For these reasons, majority of commercial GaN power modules still opt for system-in-package, which integrates GaN power device with silicon gate driver on a surface-mount package [6, 7], or even board-level approaches. Hence, the scope of this paper will also focus on this mainstream aspect.

The rest of this paper is organized as follows. Section II provides a general review of a GaN gate driver. Functions of the key components in a GaN driver as well as their impacts on switching behavior and performance will be addressed. Section III primarily reviews major design challenges, followed by some recent design solutions. Finally, this research work will be concluded in Section IV.

II. Fundamentals of GaN Based Gate Drivers

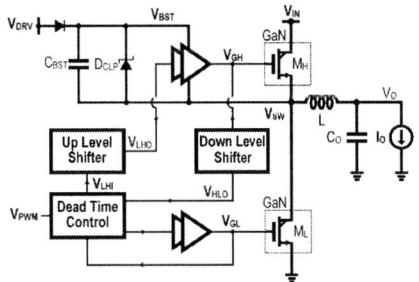

Fig. 1. Gate driver in a GaN based half-bridge power converter.

To illustrate the key circuit elements in GaN gate drivers, Fig. 1 shows the driver circuit schematic in a half-bridge converter topology, which is the most popular and commonly used converter topology [7-9]. As illustrated, two GaN transistors M_H and M_L serve as the converter's high-side and low-side switches respectively. While the gate-source breakdown voltage is relatively low in GaN devices, in order to allow the high-side switch M_H to sustain high voltage V_{IN}, bootstrap driver is usually used. Improperly designed bootstrap driver would cause device breakdown and excessive power loss. In the meantime, the gate drivers have

978-1-7281-0582-6/19 $31.00 © 2019 IEEE   87

significant influence on the gate voltage slew rates of V_{GH} and V_{GL}, which play a critical role on EMI noise, switching power loss and switching speed. The performance bottleneck of switching speed, on the other hand, is the propagation delays of level shifters and deadtime controller. The level shifters serve as the interfaces between low-voltage (control logic) and high-voltage (gate drive) domains. To reduce such delay, circuit topology and biasing current should be cautiously chosen, which often lead to a classic power-speed trade-off. To avoid large shoot-through current and switching noise, a deadtime is usually introduced between the turn-on periods of M_H and M_L. As f_{SW} increases, controlling the deadtime become very difficult, which is sensitive to the supply V_{IN} and load condition I_O.

III. CHALLENGING DESIGN ISSUES

A. Design for Reliability

1) C_{BST} Overcharge Breakdown

In a conventional power MOSFET, its body diode provides a path whenever reverse conduction is needed. This is extremely important during the deadtime control mentioned earlier. Use the gate driver in Fig. 1 as an example. The body diode of M_L clamps the switching node V_{SW} around $-0.7V$ during the reverse conduction. In general, $V_{DRV}-V_{SW}$ is the voltage across C_{BST}, which becomes gate drive voltage when M_H conducts. Hence, the presence of the body diode in M_L facilitates a very stable charge voltage across C_{BST} with little risk of overcharge. However, a GaN transistor has no body region or doping in the drift region, and thus it has no body diode. As a result, its mechanism of reverse conduction becomes quite different. When it occurs, the freewheeling inductor current forces the gate-drain voltage to exceed its threshold voltage. As a result, V_{SW} is equal to V_{SD} of M_L, when M_L conducts reversely during the deadtime periods. Unlike a voltage drop across body diode, V_{SD} is highly sensitive to load current (I_O) condition, and can easily fall below $-2V$ at high I_O, forcing C_{BST} to be overcharged ($>7V$). If it surpasses the V_{GS} breakdown voltage, it causes device breakdown and damages control logic circuits.

A Zener diode clamping technique can be used to protect the BST rail by sinking excessive charge to ground [8, 9]. However, this technique is poorly suited for high f_{SW} operation as the power loss is proportional to f_{SW}. In [10], an adaptive charging scheme called adaptive BST balancing was proposed. As shown in Fig. 2, the BST rail charge time t_{charge} is modulated adaptively to I_O or V_{IN}. To avoid overcharge, t_{charge} would not be initialized until the V_{SW} zero-crossing sensor determines that the charge voltage has enter the safe zone ($V_{SNS}>V_{ref}$).

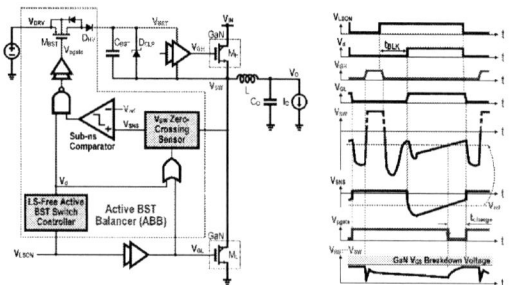

Fig. 2. Adaptive BST balancing scheme and circuit [10].

2) On-line Self-Prognosis on Aging & Failure

Current collapse, due to hot electron injection and charge trapping, has been widely considered as a major cause of GaN switching aging and premature failure [11]. It degrades channel conductivity and increases on-resistance R_{DS_ON}. On the other hand, for a GaN-on-silicon implementation, to reduce lattice mismatch, an AlGaN buffer layer is often inserted in the process, which increases junction-to-ambient thermal resistance $R_{\theta JA}$. Overall, the increased R_{DS_ON} and $R_{\theta JA}$ lead to higher heat generation, elevating junction temperature T_J. According to Arrhenius' Law, as T_J increases, the mean time to failure (MTTF) drops exponentially [12]. For even worse, the elevated T_J deteriorates the i-collapse effect with even higher R_{DS_ON}, significantly reducing device lifetime.

To monitor the device aging, R_{DS_ON} is highly desirable as the precursor, because its change is directly related to i-collapse effect. A similar but simplified approach was reported in [13] using V_{CE} as the actual precursor for IGBTs. However, the condition monitoring can only be accomplished offline. Because operation conditions between off- and on-line can change greatly, it can cause large errors. An online condition monitoring approach was reported [14] by using pole variation in loop gain for power MOSFETs. However, the implementation of pole location precursor is highly sophisticated and noise-sensitive. Also, both pole location and R_{DS_ON} are T_J-sensitive precursors, meaning that self-heating and ambient temperature fluctuation can confuse the aging monitoring and raise FMR largely. Online T_J effect calibration has to be implemented in R_{DS_ON} condition monitoring. For such, an I_{GSS}-inspired T_J sensor is developed on the fact that gate-leakage I_{GSS} of a GaN switch is both T_J-sensitive and aging-independent [15]. With sensed T_J, the in-situ condition monitor removes the impact of T_J on R_{DS_ON} through the T_J dependence remover effectively. This also allows to take more proactive measure to slow down the aging process whenever possible, through adaptive temperature frequency scaling (TFS) [15].

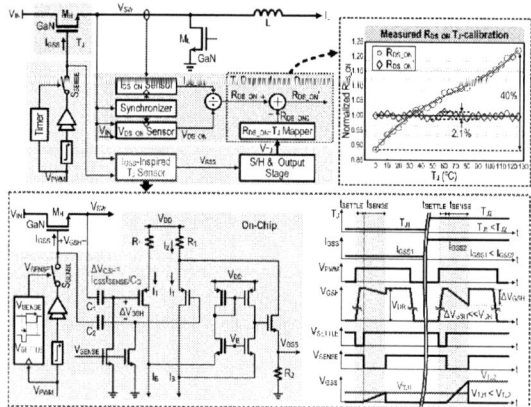

Fig. 3. T_J-independent in-situ condition monitoring [15].

B. Design for EMI Issues

1) Spread-Spectrum Modulation

When power switches operate at high f_{SW}, switching transitions need to be completed in very short period of time, leading to large di/dt and dv/dt changes, which directly connect to electromagnetic interference (EMI) emission.

A bulky input filter can reduce EMI, but it greatly increases size and cost. Several techniques [9, 10, 16, 17] are reported to mitigate EMI. Frequency hopping using discrete frequencies is proposed [16], but cannot spread the frequency spectrum evenly to lower the peak noise. Alternatively, a series resistor is typically added at the gate of the GaN FET to slow down the transition [9,10]. However, the switching loss is dramatically increased. To mitigate this, adjustable driving strength is proposed in [17]. Unfortunately, the sensing and driver delays confine its use to low f_{SW} applications, where the switch node rising time is several tens or hundreds of ns.

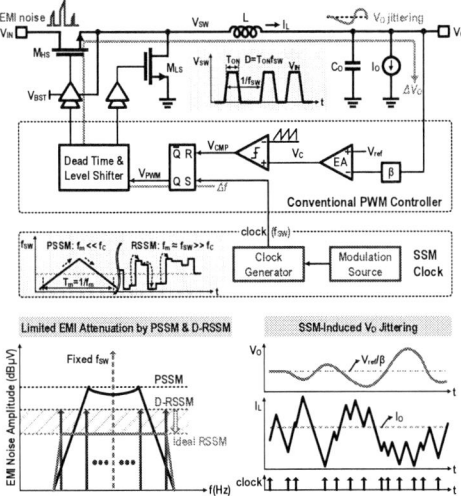

Fig. 4. Design challenges on SSM techniques [20].

Currently, spread-spectrum modulation (SSM) techniques are regarded as the most effective methods for EMI suppression. As depicted in Fig. 4, periodic SSM (PSSM) is straightforward and easy to implement. However, its EMI suppression is the least effective [18]. Randomized SSM (RSSM) can outperform PSSM, with lower peak EMI and near-uniform noise spreading, but its performance highly relies on the random clock design. In [19], an N-bit digital random clock was reported to achieve discrete RSSM (D-RSSM). However, the bit number N has to be large in order to achieve satisfying EMI attenuation, significantly increasing circuit complexity, chip area and power consumption. [20] proposes a Markov chain based random clock to achieve analog f_{SW} modulation. It conducts SSM continuously and spreads spurious noise at f_{SW0} and its harmonics uniformly, achieving the desirable C-RSSM.

2) Closed-Loop EMI Regulation

It is well known that there is a trade-off between EMI and switching loss in a power converter. To achieve optimized performance between the two aspects, it is critical to identify the Miller Plateau (MP) start point accurately. Conventionally, the first order derivative of V_{GSH}, I_{HS} or V_{DSH} is sensed with a differentiator to identify the MP, despite of I_O and V_{IN} [21, 22]. However, the delay of the sensing loop confines its application to low f_{SW} domain. In [9], the MP is predicted based on a pre-defined reference. It minimizes the detection delay, but it is not adaptive to I_O and V_{IN}, and clearly inaccurate. In [23], a closed-loop adaptive MP sensing technique is presented. The technique utilizes the low-side switch M_L reverse conduction behavior to detect the MP

starting point of the high-side switch M_H. This ease the design stress on the closed-loop propagation delay. Once the MP starting point is identified, a closed-loop control adaptively adjusts the driving strength in the gate driver to achieve low di/dt before the MP for low EMI generation and high di/dt after the MP to reduce the switching noise, leading to an optimal operation scenario between noise and efficiency.

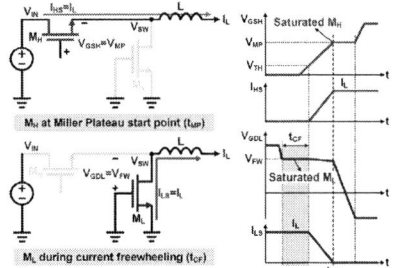

Fig. 5. Principle of emulated Miller Plateau tracking scheme [23].

C. Design for Performance

1) Deadtime (t_{dead}) Control

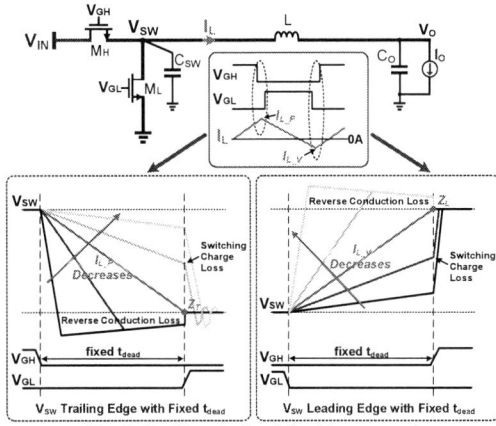

Fig. 6. Problems in fixed deadtime control.

As stated in Section II, in order to prevent disastrous shoot-through current in GaN power switches M_H & M_L, dead time (t_{dead}) control is essential. However, conventional fixed t_{dead} control is problematic due to the trailing edge slope of V_{SW} being inversely proportional to I_O, as shown in Fig. 6. At heavy I_O, the charge at C_{SW} is discharged quickly with a high inductor current I_L, resulting in a high falling slope at V_{SW}. But with a fixed t_{dead}, V_{SW} drops below 0V before t_{dead} expires, and M_L becomes reversely conducted to freewheel I_L, leading to a high reverse conduction loss. At light I_O, on the contrary, the charge at C_{SW} is discharged with a low level of I_L, leading to a slow falling slope. After the duration of a fixed t_{dead}, M_L is turned on instantly to pull V_{SW} to ground, leading to an excessive switching charge loss. In the meantime, as V_{SW} falling time is also proportional to V_{IN} level, a fixed t_{dead} causes either reverse conduction loss at low V_{IN}, or switching loss at high V_{IN}. At a high f_{SW} up to tens of MHz, the power loss significantly increases with fixed t_{dead} control.

To overcome these challenges, digital t_{dead} control is proposed which uses digital cells to generate adaptive t_{dead}, however, a high circuit complexity is involved [24]. A near-optimal t_{dead} control is proposed [25], but it suffers long delay in sensing loop limiting its use only to low f_{SW} applications.

978-1-7281-0582-6/19 $31.00 © 2019 IEEE

In [10], a V_{SW} dual-edge dead time (t_{dead}) modulator was designed, which senses I_O and V_{IN} and then generate modulated delays for V_{SW} leading and trailing edges to adjust the instant deadtime, realizing zero-voltage switching at GaN power switches.

2) High-Speed Level Shifting

Conventional high-voltage gate drivers usually carry propagation delay t_{delay} of up to several 10s of ns in the level shifter (LS), which becomes a critical problem as the f_{SW} reaches the 10MHz regime. In [8], large t_{delay} and long deadtime in the GaN FET driver limit its f_{SW} to 1 MHz. To reduce the t_{delay}, which is dominated by the high-voltage LS, a sub-ns delay BST GaN FET driver with dynamic level shifter (LS) was reported [9]. As shown in Fig. 7, during on-period (DT), a large dynamic current I_d is first mirrored to the output of the LS. This I_d is sustained till capacitor C_d charges up, achieving a fast level shifting at the output, V_{SWHb}. When V_{SWHb} reaches the BST supply level, and turns on the high-side GaN FET, only a small static current I_S holds the on-state until V_{SWH} goes low to minimize the power dissipation. When the operation environment exhibits high level of noise interference, in order to achieve reliable level shifting, ground isolation can be implemented in constructing high dv/dt immune level shifting, which can drastically improve system's common mode transient immunity (CMTI) [26].

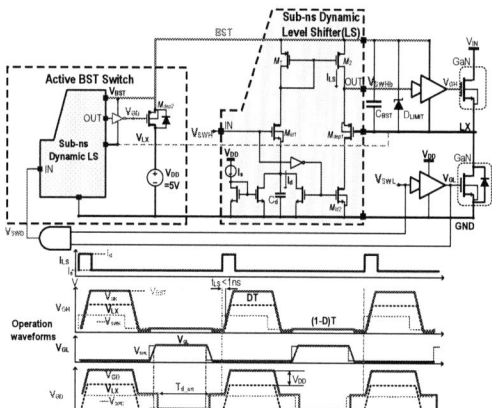

Fig. 7. Sub-ns delay GaN gate driver with dynamic level shifter [9].

IV. CONCLUSIONS

As GaN technology sees ever-increasing applications in power electronics area, there are growing demands on effective circuits and control techniques that could fully exploit the advantages of these devices. This paper reviews major circuit design challenges and existing solutions in the aspects of reliability, EMI noise and other performance relevant issues.

REFERENCES

[1] Yole Development, "Power GaN 2018: Epitaxy, Devices, Applications and Technology Trends", Yole Market & Technology Report, Nov. 2018.

[2] B. J. Baliga, "Gallium Nitride Devices for Power Electronic Applications", vol. 28, no. 7, Semicond. Sci. & Technol., pp. 1-8, 2013.

[3] A. Lidow, et. al., "GaN Transistors for Efficient Power Conversion," 2nd Ed, John Wiley & Sons, West Sussex, UK, 2015.

[4] E. A. Jones, F. Wang, and D. Costinett, "Review of Commercial GaN Power Devices and GaN-Based Converter Design Challenges", IEEE

J. on Emerging & Selected Topics in Power Elecs., vol 4, no. 3, pp. 707-719, Sept. 2016.

[5] D. Kinzer, "Breaking Speed limits with GaN power ICs", IEEE Appl. Power Electron. Conf., plenary talk, Mar. 2016.

[6] F. Luo, et. al., "Design considerations for GaN HEMT multichip halfbridge module for high-frequency power converters," IEEE Appl. Power Electron. Conf., pp. 537-544, Mar. 2014,

[7] Texas Instruments, *LMG5200 80V GaN Half Bridge Power Stage*, Available: http://www.ti.com/product/LMG5200.

[8] J. Delaine et al., "High frequency DC-DC converter using GaN device," IEEE Appl. Power Electron. Conf., pp. 1754-1761, Mar. 2012.

[9] M. K. Song et al., "A 20V 8.4W 20MHz four-phase GaN DC-DC converter with fully on-chip dual-SR bootstrapped GaN FET driver achieving 4ns constant propagation delay and 1ns switching rise time," ISSCC Dig. Tech. Papers, pp. 302-303, Feb. 2015.

[10] X. Ke, et al., "A 3-to-40V 10-to-30MHz Automotive-Use GaN Driver with Active BST Balancing and V_{SW} Dual-Edge Dead-Time Modulation Achieving 8.3% Efficiency Improvement and 3.4ns Constant Propagation Delay," ISSCC Dig. Tech. Papers, pp. 302-304, Feb. 2016.

[11] S. R. Bahl, et al., "Application Reliability Validation of GaN Power Devices," IEDM, pp. 544-547, Dec. 2016.

[12] B. M. Paine, et al., "Lifetesting GaN HEMTs with Multiple Degradation Mechanisms," IEEE Trans. Device and Materials Reliability, vol. 15, no. 4, pp. 486-494, Aug. 2015.

[13] V. Smet, et al., "Ageing and Failure Modes of IGBT Modules in High-Temperature Power Cycling," IEEE Trans. Industrial Electronics, vol. 58, no. 10, pp. 4931-4941, Oct. 2011.

[14] S. Dusmez, et al., "In Situ Condition Monitoring of High-Voltage Discrete Power MOSFET in Boost Converter Through Software Frequency Response Analysis," IEEE Trans. Industrial Electronics, vol. 63, no. 12, pp. 7693-7702, Dec. 2016.

[15] Y. Chen, D. Ma, "A 10MHz i-Collapse Failure Self-Prognostic GaN Power Converter with TJ-Independent In-Situ Condition Monitoring and Proactive Temperature Frequency Scaling," ISSCC Dig. Tech. Papers, pp. 248-249, San Francisco, CA, Feb. 2019.

[16] C. Tao, et al., "Spurious-noise-free buck regulator for direct powering of analog/RF loads using PWM control with random frequency hopping and random phase chopping," ISSCC Dig. Tech. Papers, pp. 396-398, Feb. 2011.

[17] M. Rose, et al., "Adaptive dv/dt and di/dt control for isolated gate power devices", *Energy Conversion Congress and Exposition*, pp. 927-934, Sep. 2010.

[18] K. K. Tse, et al., "A Comparative Study of Carrier-Frequency Modulation Techniques for Conducted EMI Suppression in PWM Converters," IEEE Trans. Industrial Electronics, vol. 49, no. 3, pp. 618-627, June. 2002.

[19] W. H. Yang, et al., "An Enhanced-Security Buck DC-DC Converter with True-Random-Number-Based Pseudo Hysteresis Controller for Internet-of-Everything (IoE) Devices," ISSCC Dig. Tech. Papers, pp. 126-128, Feb. 2018.

[20] Y. Chen, D. Ma, "An 8.3MHz GaN Power Converter Using Markov Continuous RSSM for 35dBμV Conducted EMI Attenuation and One-Cycle TON Rebalancing for 27.6dB VO Jittering Suppression," ISSCC Dig. Tech. Papers, pp. 250-251, San Francisco, CA, Feb. 2019.

[21] N. Idir, et al., "Active Gate Voltage Control of Turn-on di/dt and Turn-off dv/dt in Insulated Gate Transistors," IEEE Transactions on Power Electronics, vol. 21, pp. 849-855 Jul. 2006.

[22] Z. Wang et al., "A di/dt Feedback-Based Active Gate Driver for Smart Switching and Fast Overcurrent Protection of IGBT Modules," IEEE Transactions on Power Electronics, vol. 29, pp. 3720-3732 Jul. 2014.

[23] Y. Chen, X. Ke, D. Ma, "A 10MHz 5-to-40V EMI-Regulated GaN Power Driver with Closed-Loop Adaptive Miller Plateau Sensing", IEEE VLSI Symp. on Circuits, pp. C120-C121, June 2017.

[24] J. Wittmann, et al., "An 18 V Input 10 MHz Buck Converter With 125 ps Mixed-Signal Dead Time Control," in IEEE J. of Solid-State Circuits, vol. 51, no. 7, pp. 1705-1715, July 2016.

[25] S. Lee. et al., "Robust and Efficient Synchronous Buck Converter with Near-Optimal Dead-Time Control", IEEE ISSCC Dig. Tech. Papers, pp. 392-394, Feb. 2011.

[26] X. Ke, D. Ma, "A 3-to-40V VIN 10-to-50MHz 12W Isolated GaN Driver with Self-Excited t_{dead} Minimizer Achieving 0.2ns/0.3ns t_{dead}, 7.9% minimum Duty Ratio and 50V/ns CMTI," IEEE ISSCC Dig. Tech. Papers,, pp. 386-387, San Francisco, CA, Feb. 2018.

Proceedings of the 31st International Symposium on Power Semiconductor Devices & ICs
May 19-23, 2019, Shanghai, China

Alpha-Particle Shielding Effect of Thick Copper Plating Film on Power MOSFETs

Tatsuya Nishiwaki, Kohei Oasa, Kentaro Ichinoseki, Kikuo Aida,
Tatsuya Ohguro, Yoshiharu Takada, Hideharu Kojima and Yusuke Kawaguchi
Toshiba Electronic Devices & Storage Corporation, Ishikawa, Japan,
E-mail:tatsuya.nishiwaki@toshiba.co.jp

Abstract—We investigated characteristics impact of alpha-particles irradiation to power MOSFETs and demonstrated alpha-particle shielding effect of thick copper plating film on a power MOSFET. We used americium-241 alpha-source for irradiating alpha-particles to the surface of the power MOSFET die. Irradiation under gate-source bias caused threshold voltage (Vth) decrease due to generated trapped holes in gate oxide, however, no Vth shift occurred under drain–source bias. We found that sensitivity to alpha-particles depends on gate structures of trench or planar. Recovery behavior of Vth shift by gate bias or thermal annealing support the hole trapping model in the gate oxide. We evaluated alpha-particle shielding effect of thick copper plating film. We demonstrated more than 15 micrometer thick copper plating on the power MOSFET shielded alpha-particles successfully.

Keywords—alpha-particle, power MOSFET, copper plating film, threshold voltage shift, hole trapping

I. INTRODUCTION

Continuous cell pitch shrinking of trench power MOSFET has achieved significant chip On-resistance reduction in last three decades [1], [2]. Furthermore, charge coupling technologies such as superjunction [3] and trench Field Plate (FP) MOSFET [4] drastically reduced drift layer resistance and overcame silicon limit. In addition, copper clip source connection technology decreased parasitic package resistance [5], which is especially important for low voltage devices. These technologies enabled us to realize extremely low on-resistance power MOSFET.

Recently, threshold voltage (Vth) shift of power MOSFET due to alpha-particles which radiated from the lead solder of copper clip was reported [6]. The mechanism of the Vth shift is illustrated in Fig. 1 and 2. The lead solder on the source metal slightly radiates alpha-particles because the lead solder contains small amount of radioactive elements such as lead-210 (^{210}Pb) which become polonium-210 (^{210}Po) alpha-source [7]. These alpha-particles penetrates gate oxide in some probability and generates electron-hole pairs. Compare to electron, field effect mobility of hole is estimated to be very low [8]. Therefore, when the gate is biased, electrons are swept out and holes are left behind and gradually trapped in the gate oxide. These trapped holes in the gate oxide decrease Vth.

Historically, the effect of alpha-particles to semiconductor devices was recognized as one of the cause of "soft error" in dynamic memory devices at the end of the 1970s [9]. In this case, the main source of the alpha-particles was identified as radioactive decay of uranium and thorium in packaging material. In addition, it is reported that the sensitivity of the alpha-particles (error rate) increase as increasing density of memory cells. Considering analogy with memory devices, as the density of the cell increase in power MOSFET, the impact of alpha-particle become remarkable as well.

The practical influence of the Vth shift of power MOSFET by alpha-particles was also investigated in [6]. In fact, no significant difference was observed in TDDB, avalanche capability and Safe Operating Area (SOA). These results are because holes are trapped in very small region in active area and impact is limited to sub-threshold characteristics degradation. In addition, the Vth shift recovery by positive gate bias was also reported [6]. However, avoiding these alpha-particles effects is more desirable. One solution is using low-alpha (low alpha-particle emission) solder which is proposed in [6]. The other solution we propose is shielding alpha-particles. As well known, alpha-particles are very interactive with substances, therefore, shielding alpha-particles is relatively easy, for instance by using a paper.

In this article, we used thick copper (Cu) plating film on the power MOSFET [10] for shielding alpha-particles and demonstrated its effect.

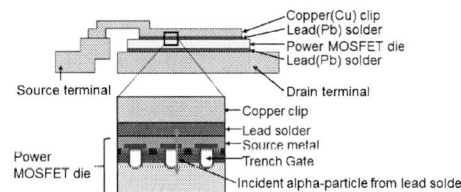

Fig. 1. A power MOSFET die mounted on copper clip package. Alpha-particles radiated from lead solder on the source metal penetrate the power MOSFET die.

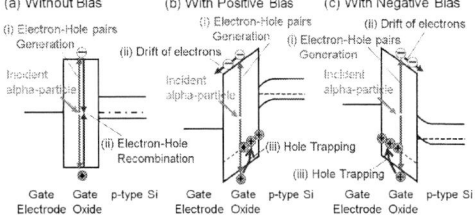

Fig. 2. Vth shift model of alpha-particle irradiation under positive or negative gate bias. Incident alpha-particles to gate oxide generates electron-hole pairs. Holes separated from electron due to gate electric field are trapped in the gate oxide.

978-1-7281-0582-6/19 $31.00 © 2019 IEEE

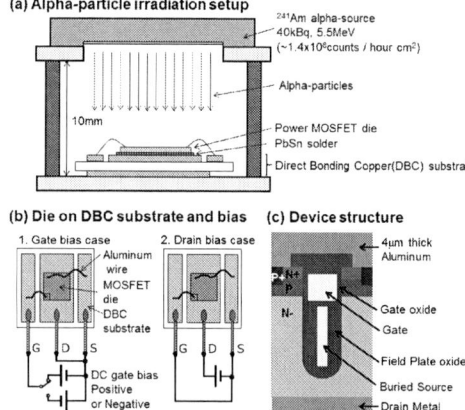

(a) Alpha-particle irradiation setup

241Am alpha-source
40kBq, 5.5MeV
(~1.4x10⁶counts / hour cm²)

Alpha-particles

Power MOSFET die
PbSn solder
Direct Bonding Copper(DBC) substrate

10mm

(b) Die on DBC substrate and bias **(c) Device structure**

1. Gate bias case   2. Drain bias case

Aluminum wire
MOSFET die
DBC substrate

G D S   G D S

DC gate bias
Positive or Negative

4μm thick Aluminum

Gate oxide
Gate
Field Plate oxide
Buried Source
Drain Metal

Fig. 3. (a) Alpha-particle irradiation setup, (b) Sample mounted on Direct Bonding Copper substrate and (c) Schematic device structure of the trench FP MOSFET. Alpha-particles emitted from ²⁴¹Am source was irradiated to the bare MOSFET chip with DC gate or drain bias.

II. Experimental Method

In order to investigate behavior to alpha-particle penetration, we irradiated alpha-particles to a bare power MOSFET die by using an americium-241(²⁴¹Am) artificial alpha-source as shown in Fig. 3. The energy of alpha-particles emitted from ²⁴¹Am is about 5.5MeV. This energy is almost the same as alpha-particles energy of 5.3MeV from ²¹⁰Po in lead solder. The radioactivity of exploited ²⁴¹Am alpha-source was 40kBq, which approximately corresponded to 1.4x10⁸ counts / hour cm². The tested power MOSFET die was mounted on Direct Bonding Copper (DBC) substrate as illustrated in Fig. 3(b). During alpha-irradiation, gate–source was biased with positive or negative DC voltage and drain-source was shorted. Besides, irradiation under drain–source reverse bias without gate bias was tested. We mainly evaluated trench FP MOSFETs. In addition, for investigating alpha-particle sensitivity dependence on gate structures, we evaluated planar gate power MOSFETs for comparison. In order to reveal trapped hole state, we studied recovery behavior of Vth shift by thermal annealing or gate bias. After these comprehensive studies of device behavior, we evaluated alpha-particle shielding effect of thick copper plating film on the surface of the trench FP MOSFET. We fabricated 5 to 20μm Cu plated samples and irradiated alpha-particles to these surface, then evaluated degree of Vth shift under gate bias.

III. Results and Discussion

A. Alpha-particle irradiation with gate bias

We irradiated alpha-particles to trench FP MOSFET under gate bias. Fig. 4 shows gate voltage (V_G) – drain current density (J_D) characteristics of before and after alpha-particle irradiation. Shoulder like shape due to negative Vth shift in low current regime was observed both positive and negative gate bias, however no significant Vth shift was observed without gate bias. These results supported the model of electron-hole separation and hole trapping due to gate electric field as illustrated in Fig. 2.

The shoulder like shape of V_G-J_D curve suggested localized current leakage resulted from local Vth decrease. Fig. 5 shows photo emission analysis of drain current leak points. Each leak spots in active area implies random incidence of alpha-particles.

We also checked impact of alpha-irradiation under drain-source reverse bias. In contrast with the gate bias case, no significant change was observed both V_G-J_D and drain voltage (V_D) –J_D characteristics as shown in Fig. 6. As previously stated, alpha-particles generate electron-hole pairs in the device. In the case of drain-source bias, generation of trapped holes in the oxide is negligible because electric field in the gate oxide is very low. Instead, electron-hole pairs in depleted layer might be spatially separated by drain-source electric field. However, these carriers are swept out quickly by drain-source electric field.

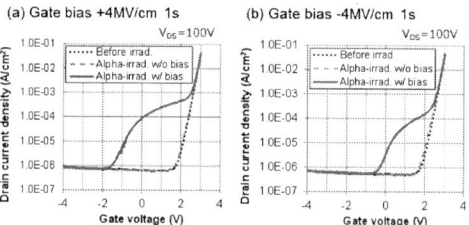

Fig. 4. Gate voltage – drain current density characteristics before and after alpha-particle irradiation without gate bias and with gate bias. The gate bias was (a) +4MV/cm 1sec. and (b) -4MV/cm 1sec. Negative Vth shift appeared both positive and negative gate bias. In contrast, no significant change was observed without gate bias.

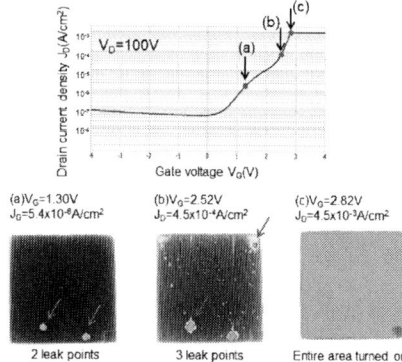

Fig. 5. Photo emission analysis of drain current leak points of after alpha-particle irradiated sample. Each spot in the active area indicated an incident point of an alpha-particle.

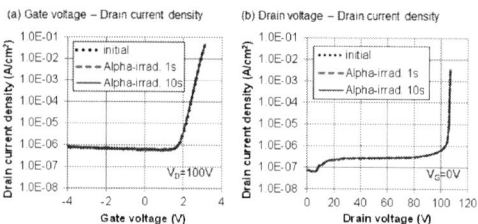

Fig. 6. Gate voltage – drain current density characteristics before and after alpha-particle irradiation under drain-source bias. Applied voltage between drain and source was 100V. No significant change was observed in alpha-particle irradiation under drain–source bias.

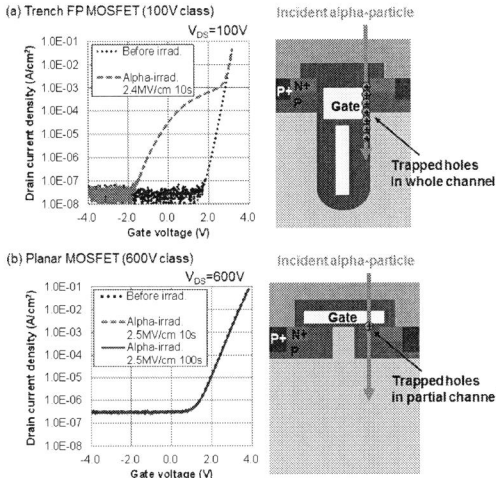

Fig. 7. Gate voltage – drain current density characteristics comparison between (a) trench and (b) planar gate structure after alpha-particle irradiation. No degradation of planar gate after 100s irradiation indicated that the trench structure is more sensitive to alpha-particle irradiation than planar structure. This difference comes from incident angle of alpha-particles to the the channel.

B. Alpha-particle sensitivety of different gate structures

Sensitivity of alpha-particle should depend on the gate structure because the Vth shift is caused by alpha-particles penetration to the gate oxide. We compared alpha-particle irradiation behavior of different gate structure: trench gate and planar gate as shown in Fig. 7. In the case of the trench gate, Vth shifted immediately after 10 seconds irradiation, however, in the case of the planar structure, no characteristics change was observed even at 100 seconds irradiation. We considered that this sensitivity difference comes from incident angle of alpha-particles to the channel as illustrated in Fig. 7. In the case of the trench gate, the incident alpha-particles in parallel to the channel create hole trapped region throughout the channel, which leads channel leakage. In contrast, in the case of the planar gate, incident alpha-particles in perpendicular to the channel partially create hole trapped region in channel; thus no Vth shift occurs. Similar results and a model were also reported in heavy ion irradiation to trench power MOSFET [11].

C. Vth shift recovery behavior

In order to investigate trapped hole state in the gate oxide, we estimated Vth shift recovery with thermal annealing and gate bias. We prepared Vth shifted samples by alpha-particles irradiation with positive gate bias for examining Vth recovery behavior.

Vth shift recovered gradually by annealing. From Arrhenius plot shown in Fig. 8, we obtained activation energy of the Vth recovery of 0.47eV, which correspond to trapping energy of the holes. Furthermore, the Vth shift recovered by gate bias as shown in Fig. 9. The recovery under positive gate bias started over +4.0MV/cm and became obvious in +6.0MV/cm. This result agreed with previous report [6]. Compare to positive gate bias case, the recovery under negative gate bias is slow in the same electric field of 6.0MV/cm. This is explained by different behavior of trapped holes near the channel in each gate bias direction as depicted in Fig. 10.

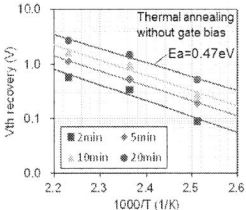

Fig. 8. Arrhenius plot of Vth shift recovery by thermal annealing. Vth was defined at drain current density $4.5 \times 10^{-7} A/cm^2$. We annealed three samples at 125, 150, 175 degrees C. Vth of each sample was measured at 2, 5, 10, 20 minutes. Activation energy of Ea=0.47eV was obtained.

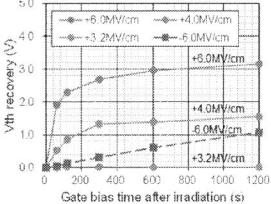

Fig. 9. Recovery of Vth shift under positive or negative gate electric field. Positive bias Vth recovery appeared over +4.0MV/cm. Vth recovery in negative gate bias is slower than that in positive gate bias.

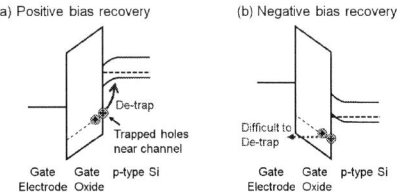

Fig. 10. Vth Recovery model under (a) positive gate bias and (b) negative bias. Trapped holes near channel are de-trapped by positive gate bias. In contrast, de-trapping is difficult under negative gate bias.

The trapped holes near the channel are significant for considering Vth influence. In the case of positive gate bias, the trapped holes near the channel de-trap to the channel side by gate electric field. In contrast, in the case of negative gate bias, these holes hardly escape to gate electrode side due to the long distance from gate electrode.

D. Alpha-particle shield effect of thick Cu plating film

After the comprehensive study of the alpha-irradiated power MOSFET behavior, we evaluated alpha-particle shielding effect of thick copper plating film on the trench FP MOSFET. We calculated penetration depth of 5.5MeV alpha-particles from ^{241}Am into Cu, aluminum (Al) and silicon (Si) using NIST standard reference database [12]. The penetration depth of 5.5MeV alpha-particle into Cu was calculated as 11.8μm as plotted in Fig. 11. As this calculation indicated, Cu is suitable material for shielding alpha-particle because the density of Cu is higher than typical source metal of Al. Therefore, we expected that over 11.8μm thick Cu film on the device is capable of shielding alpha-particles as illustrated in Fig. 12. In order to evaluate shielding effect of copper film, we fabricated 5, 10, 15 and 20μm thick Cu plated trench FP MOSFET samples. As shown in Fig. 13, 5μm thick Cu plating sample showed Vth shift as well as conventional sample in Fig. 4; and 10μm thick Cu plating sample showed slight Vth shift. In contrast, no Vth degradation was observed in 15 and 20 µm thick Cu plating samples. These results are reasonable for the alpha-particle penetration depth of 11.8µm. This results also indicated more than 15µm is enough Cu thickness for shielding 5.3MeV alpha-particles radiated from ^{210}Po in lead solder. Furthermore, the possible maximum alpha-particle energy is 8.78MeV (^{212}Po) in thorium decay chain [13]; thus maximum penetration depth in Cu is estimated to 23µm. Therefore, more than 23µm thick Cu plating film must completely shield alpha-particles.

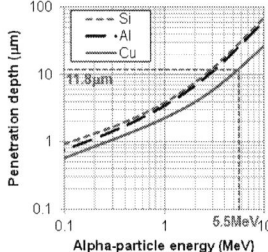

Fig. 11. Calculated penetration depth of Alpha-particle into copper (Cu), aluminum (Al) and silicon (Si). The penetration depth of 5.5MeV alpha-particles into Cu was calculated as 11.8µm.

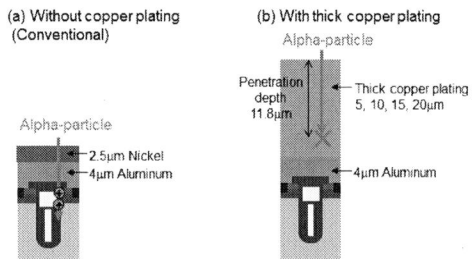

Fig. 12. The concept of alpha-particle shielding by thick copper plating film on the device. Copper plating film thicker than alpha-particle penetration depth is expected to shield alpha-particles.

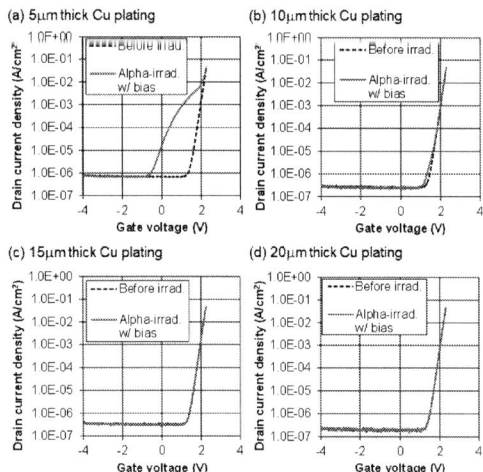

Fig. 13. Cu plating thickness dependence of alpha-particle shielding effect. These results indicated that 15 and 20µm thick copper suppressed penetration of 5.5MeV alpha-particles to the gate oxide. These are reasonable results for alpha-particles penetration depth of 11.8µm.

IV. CONCLUSIONS

We investigated Vth shift phenomena of power MOSFET by alpha-particles irradiation using ^{241}Am alpha-source. Incident alpha-particles to gate oxide create electron-hole pairs. Under gate bias condition, spatially separated holes trapped in the gate oxide and these holes decrease Vth. The study of Vth recovery behavior by annealing and gate bias supports this degradation model. The sensitivity of Vth degradation to alpha-particle incidence depends on gate structures. Compare to planar gate structure, trench gate structure is more sensitive due to alpha-particles penetration in parallel to the channel. In order to avoid subthreshold characteristics degradation by alpha-particles, we proposed shielding alpha-particles by thick Cu plating films on the devices. We demonstrated more than 15µm thick Cu plating shielded alpha-particles radiated from ^{241}Am which have almost equivalent energy to alpha-particle from lead solder.

ACKNOWLEDGMENT

The authors would like to thank Mr. Toshikazu Fukuda and Mr. Daisuke Minohara for cooperating alpha-particles irradiation experiments, Dr. Wataru Saito, Mr. Hiroaki Yamashita, and Mr. Kenya Kobayashi for significant discussion on this work.

REFERENCES

[1] W. Saito, "Power device trends for high-power density operation of power electronics system", Jpn. J. Appl. Phys., 53, 04EP02, 2014.

[2] R. K. Williams, M. N. Darwish, R. A. Blanchard, R. Siemieniec, P. Rutter and Y. Kawaguchi, "The Trench Power MOSFET: Part I – History, Technology, and Prospects", IEEE Trans. Electron Devices, 64, no. 3, pp. 674-691, 2017.

[3] G. Deboy, N. Marz, J. P. Stengl, H. Strack, J. Tihanyi, H. Weber, "A new generation of high voltage MOSFETs breaks the limit line of silicon", IEDM Tech. Dig., pp. 683-685, 1998.

[4] M.A. Gajda, S.W. Hodgkiss, L.A. Mounfield, N.T. Irwin, G.E.J. Koops, R. van Dalen, "Industrialisation of Resurf Stepped Oxide Technology for Power Transistors", Proc. of ISPSD, pp. 109-112, 2006.

[5] R. K. Williams, M. N. Darwish, R. A. Blanchard, R. Siemieniec, P. Rutter and Y. Kawaguchi, "The Trench Power MOSFET: Part II – Application Specific VDMOS, LDMOS, Packaging, and Reliability", IEEE Trans. Electron Devices, 64, no, 3, pp, 692-712, 2017.

[6] G. Schindler, K.-H. Bach, P. Nelle, M. Deckers, A. Knapp, K. Ermisch, C. Feuerbaum and W. v. Emden, "Impact of Alpha-Radiation on Power MOSFETs", Proc. of IRPS, 5C-2-1 - 5C-2-5, 2016.

[7] M. S. Gordon, K. P. Rodbell, D. F. Heidel, C. E. Murray, H. H. K. Tang, B. D. MacNally and W. K. Warburton, "Alpha-Particle Emission Energy Spectra From Materials Used for Solder Bumps", IEEE Trans. Nucl. Sci., 57, No. 6, pp. 3251-3256, 2010.

[8] R. C. Hughes, "Hole mobility and transport in thin SiO$_2$ films", Appl. Phys. Lett. 26, pp. 436-438, 1975.

[9] T. C. May, and M. H. Woods, "Alpha-Particle-Induced Soft Errors in Dynamic Memories", IEEE Trans. Electron Devices, 26, no. 1, pp. 2-9, 1979.

[10] Y. S. Chung, T. Willett, V. Macary, S. Merchant and B. Baird, "Energy Capablity of Power Devices with Cu Layer Integration", Proc. of ISPSD, pp. 63-66, 1999.

[11] J. A. Felix, M. R. Shaneyfelt, J. R. Schwank, S. M. Dalton, P. E. Dodd and J. B. Witcher, "Enhanced Degradation in Power MOSFET Devices Due to Heavy Ion Irradiation", IEEE Trans. Nucl. Sci., 54, No. 6, pp. 2181-2189, 2007.

[12] NIST standard reference database: "https://www.nist.gov/pml/stopping-power-range-tables-electrons-protons-and-helium-ions".

[13] M. Gedion, F. Wrobel, F. Saigné and R. D. Schrimpf, "Uranium and Thorium Contribution to Soft Error Rate in Advanced Technologies", IEEE Trans. Nucl. Sci., 58, No. 3, pp. 1098-1103, 2011.

Proceedings of the 31st International Symposium on Power Semiconductor Devices & ICs
May 19-23, 2019, Shanghai, China

A New 200 V Dual Trench MOSFET With Stepped Oxide for Ultra Low $R_{DS(on)}$

Chanho Park, Misbah Azam, *Gabriel Dengel, Ayman Shibib, and Kyle Terrill
Device Technology, R&D, Vishay Intertechnology Inc.,
2585 Junction Avenue, San Jose, CA 95134-1923, E-mail: Chanho.Park@vishay.com
*Fraunhoferstrasse 1, D-25524 Itzehoe, Germany

Abstract— **A new 200 V rating power trench MOSFET having dual trench structures with stepped oxide in the drift region, used here for the first time has been proposed and implemented to obtain higher breakdown voltages with lower drain-source on-state resistance, $R_{DS(on)}$. A more efficient reverse blocking voltage capability to reduce the drift region resistance by increasing doping concentration of the drift region has been demonstrated by this approach.**

We obtained Rsp=215 mΩ·mm² at the breakdown voltage of 225 V from the first fabricated devices, and Rsp=196 mΩ·mm² at 224 V from the simulated devices, which positions them as the best-in-class on-state drain-source specific resistance, Rsp, with comparable voltage rating devices.

Keywords—stepped; oxide; dual; trench; MOSFET; gate; charge; balance; medium; voltage; power; device; on-state; specific; resistance

I. INTRODUCTION

High efficient switching power devices are much more required in these days for new applications such as electric vehicles (EV), artificial intelligence (AI), and all kinds of big data, their servers and networks. In order to obtain near ideal breakdown voltages and high performance power devices, there have been numerous structures proposed and implemented since 1990's. Super-Junction (SJ) structures [1]-[4] for high voltages, Split Gate (SG) MOSFETs [5]-[9] for low and medium voltages and Dual Trench (DT) MOSFETs [10][12] for medium voltages are several examples of those efforts.

The Dual Trench MOSFET concept was first introduced as an oxide-bypassed VDMOS [11] in 2001 with a planar gate for 200 - 700 V devices, and as a tunable oxide-bypassed trench gate MOSFET [10] in 2003 for 80 - 100 V devices. In this paper, we present new 200 V rated MOSFETs having dual trench structures with stepped oxide in the drift region, used here for the first time to obtain higher breakdown voltages with lower on-state drain-source resistance, $R_{DS(on)}$.

The on-state drain-source resistance, $R_{DS(on)}$, is composed of several components as shown in TABLE I and Fig. 1 for typical 200V rating devices. We need to take particular note of the drift region resistance, R_{drift}, as a way of reducing $R_{DS(on)}$ since it accounts for 88 % of total resistance for the 200 V rating devices. As a consequence of this approach, we need a more efficient reverse blocking voltage capability to reduce R_{drift} by increasing doping concentration in the drift

region. It would be noted that the channel resistance is about 6 % for 200 V devices, while 20 - 30 V devices, the channel resistance portion is dominant over 50 - 60 %, and about 30 % for the 60 V rating devices [8].

TABLE I.   ON-STATE RESISTANCE, $R_{DS(on)}$, COMPONENTS FOR TYPICAL 200V RATING DEVICES

Components	Resistance (mΩ)	Portion
R_{drift}	10.98	87.8 %
$R_{channel}$	0.78	6.3 %
$R_{package}$	0.6	4.8 %
$R_{substrate}$	0.14	1.1 %
R_{total}	12.5	100 %

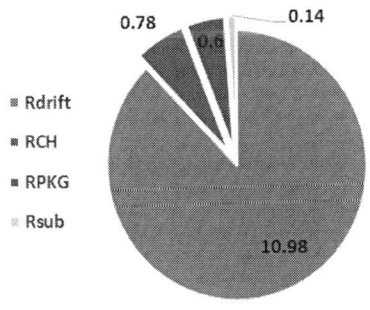

Fig. 1. On-state resistance portions for typical 200 V rating devices

II. DEVICE SRUCTURE AND FABRICATION

Fig. 2 (a) shows a conventional Dual Trench (DT) MOSFET structure and (b) the new Stepped Oxide Dual Trench (SODT) structure [12]. While the conventional dual trench structure has core MOSFET trenches and deep charge balance trenches having uniform shield oxide thickness, the new stepped oxide dual trench structure has core MOSFET trenches same as conventional DT but deep charge balance trenches having stepped oxide thickness. There are two important structural parameters affecting breakdown voltages and on-state resistance in the SODT, i.e. thin oxide thickness (Xso) in the stepped oxide and the step position (Yso) from the top of the silicon in the stepped oxide as shown in Fig. 1 (b). The effects of these parameters will be investigated in the section III.

Fig. 3 shows several key process steps to illustrate the stepped oxide forming method [13]: (a) the core cell structure after silicon deep trench etching with intended trench depth, (b) after shield oxide growth, (c) shield oxide partial etching with intended thickness, 1st shield poly deposition, CMP and etch back with intended depth, (d) 2nd shield poly deposition to fill the upper part of the trench, and CMP.

Fig. 4 demonstrates the cross sectional SEM picture of the fabricated Stepped Oxide Dual Trench (SODT) structure.

Fig. 4. Cross sectional SEM picture of the fabricated Stepped Oxide Dual Trench (SODT) structure

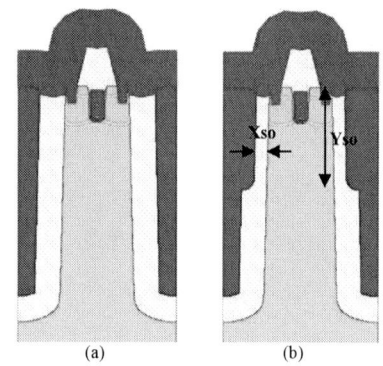

(a) (b)

Fig. 2. Schematic device structures for (a) Conventional Dual Trench (DT), (b) New Steppes Oxide Dual Trench (SODT)

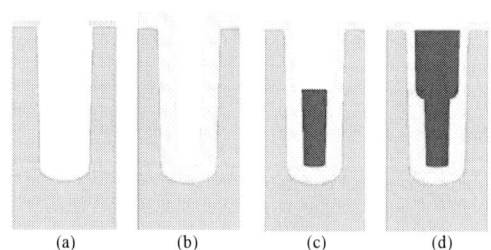

(a) (b) (c) (d)

Fig. 3. Key process steps to form a stepped oxide: (a) trench etch, (b) shield oxide growing, (c) first shield poly deposition, etch, shield oxide etch in the upper part as designed , (d) second shield poly deposition to fill the upper part of the trench and CMP

III. RESULTS AND DISCUSSION

We investigated the electric field distribution and potential distribution by 2-D simulation. Fig. 5 shows the side by side comparison between the conventional DT and SODT for electric field intensity and potential distribution, respectively. We can clearly see more uniformly distributed electric fields inside silicon beside the deep trenches in SODT than those in conventional DT.

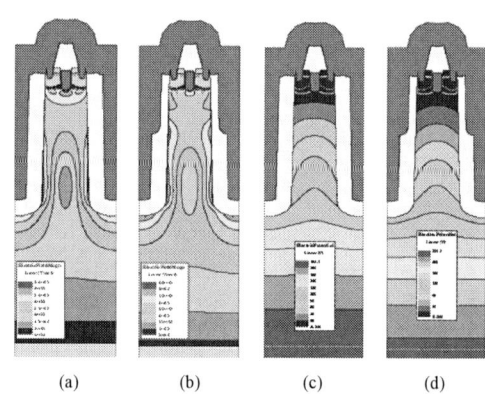

(a) (b) (c) (d)

Fig. 5: 2D-Electric field distributions (a) conventional dual trench (DT), (b) stepped oxide dual trench (SODT) MOSFETs, and potential distributions (c) conventional dual trench (DT), (d) stepped oxide dual trench (SODT) MOSFETs

Fig. 6 and Fig. 7 show the electric field distributions and the potential distributions for the Stepped Oxide Dual Trench (SODT) and the conventional Dual Trench (DT) structures along the trench side wall and at the center of mesa under the gate, respectively. We can confirm again that SODT shows more uniform field distributions and sustain higher potentials.

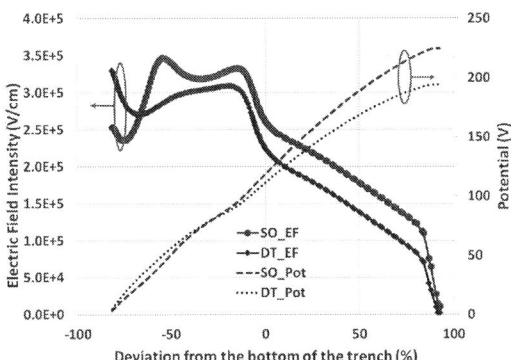

Fig. 6. Simulation results of the electric field distributions and the potential distributions along the trench side wall for the Stepped Oxide (SO) and the conventional Dual Trench (DT) structures

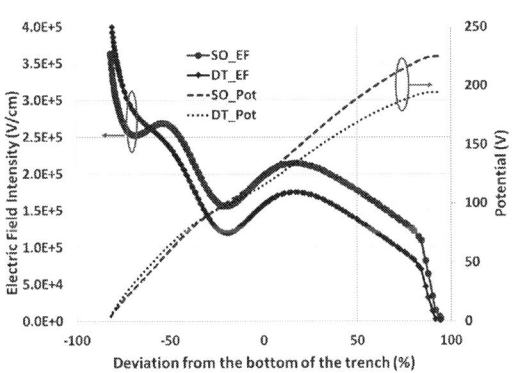

Fig. 7. Simulation results of the electric field distributions and the potential distributions at the center of mesa under the gate for the Stepped Oxide (SO) and the conventional Dual Trench (DT) structures

Fig. 8 typify the effects of the step position (Yso) on the breakdown voltages for four different epi resistivity cases by simulation. We can find the optimal step position for each epi condition separately. Furthermore we can get even higher breakdown voltages with much lower resistivity at the same time, which validates our approach using stepped oxide in the drift region. Please note that lower resistivity epi has relatively narrow window for the stable breakdown voltages.

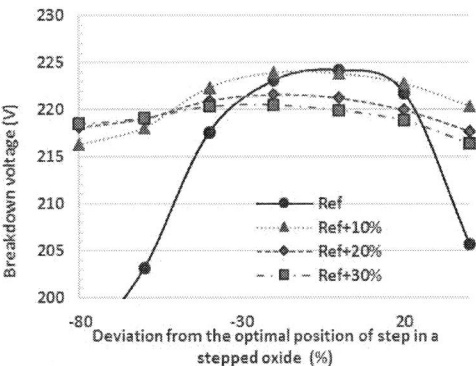

Fig. 8. Breakdown voltages as a function of the position of step (Yso) for the four different epi resistivity cases

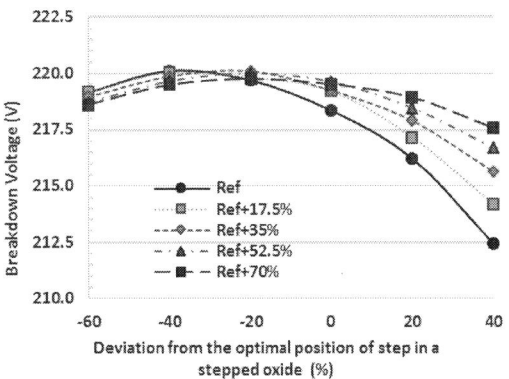

Fig. 9. Breakdown voltages as a function of the position of step (Yso) for the five different thin oxide thicknesses (Xso) in a stepped oxide structure

We found the effects of the step position (Yso) on the on-state specific resistance for four different epi resistivity cases, which was independent of the step position in the stepped oxide, as expected.

We investigated also the effects of the thin oxide thicknesses (Xso) in the stepped oxide. Fig. 9 represents the effects of the five different thin oxide thicknesses (Xso) on the breakdown voltages by simulation. Relatively thinner shield oxide shows higher breakdown voltages only at the 40% position of the optimal step position. Please note that thicker shield oxide shows wider window of stable breakdown voltages, the highest value of the breakdown voltage is slightly lower than those of thinner shield oxide cases.

Fig. 10 shows that the surface response of the deviation from the optimal position in a stepped oxide and the thin oxide thickness of the stepped oxide on the breakdown voltages. We can figure out more comprehensively from the picture where we can choose the optimal point for each case.

We obtained Rsp=215 mΩ·mm^2 at the breakdown voltage of 225 V from the first fabricated devices, and Rsp=196 mΩ·mm^2 at 224 V from the simulated devices, which positions them as the best-in-class $R_{DS(on)}$ with comparable voltage rating devices as shown in Fig. 11. The differences between the fabricated device and the simulation result partly comes from differences in the epitaxial layer conditions, since the first fabricated device has not been fully optimized. We are expecting much lower $R_{DS(on)}$ close to the simulation results from the second silicon devices.

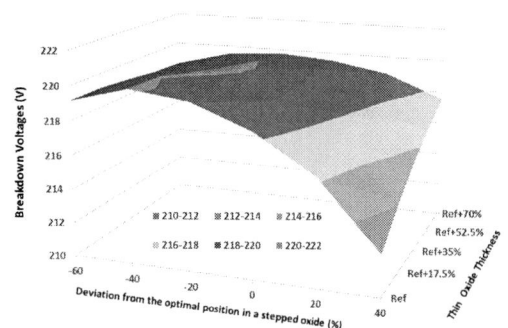

Fig. 10. Surface response of the deviation from the optimal position and the thin oxide thickness in the stepped oxide on the breakdown voltages

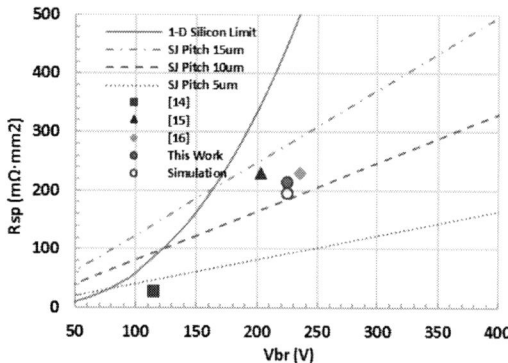

Fig. 11. Specific resista nce as a function of the breakdown voltage for 1-dimensional silicon limit, super-junction (SJ) limits for the pitches of 5, 10, and 15 µm, some references and this work

IV. CONCLUSION

We proposed and demonstrated a new 200 V rating dual trench MOSFETs having stepped oxide in the drift region. It shows the best-in class specific resistance of 225 mΩ·mm^2 from the first silicon and 196 mΩ·mm^2 from the

simulation. Comprehensive simulation study confirmed that these stepped oxide dual trench MOSFETs have more uniform electric field distributions in the drift region, show higher potentials, and can be a viable candidate for the high efficient switching power devices in the medium voltage ranges.

REFERENCES

[1] L. Lorenz, M. Marz, and G. Deboy, "COOMOS – An important milestone towards a new power MOSFET generation", Proc. of Power Conversion, pp. 151-160, 1998

[2] G. Deboy, M. Marz, J. P. Stengl, H. Weber, "A new generation of high voltage MOSFGETs breaks the limit line of silicon", IEDM Tech. Dig., pp. 683-685, 1998

[3] P. M. Shenoy, A. Bhalla, and G. M. Dolny, "Analysis of the effect of charge imbalance on the static and dynamic characteristics of the super junction MOSFET", ISPSD Proceedings, pp. 99-102, 1999

[4] W. Saito, I. Omura, S. Aida, S. Koduki, M. Izumisawa, H. Yoshioka, and T. Ogura, "High Breakdown Voltage (>1000 V) Semi-Superjunction MOSFETs Using 600-V Class Superjunction MOSFET Process", IEEE T-ED, Vol. 52, No. 10, pp 2317-2322, 2005

[5] B. J. Baliga, "Power Semiconductor Devices Having Improved High Frequency Switching and Breakdown Chanracteristics", US Patent 5,998, 833, 1999

[6] Z. Hossain, B. Burra, J. Sellers, B. Pratt, P. Venkatraman, G. Loechelt, and A. Salih, "Process & Design Impact on BVDSS Stability of a Shield Gate Trench Power MOSFET", ISPSD Proceedings, pp. 378-381, 2014

[7] G. E. K. Koops, E. A. Hijzen, R. J. E. Hueting, M. A. A. in 't Zandt, "Resurf Stepped Oxide (RSO) MOSFET for 85V having a record low specific on-resistance", ISPSD Proceedings, pp. 185-188, 2004

[8] P. Goarin, G. E. J. Koops, R. van Dalen, C. L. Cam, and J. Saby, "Split-gate Resurf Oxide (RSO) MOSFETs for 25V applications with record low gate-to-drain charge", ISPSD, pp. 61-64, 2007

[9] C. Park, S. Havanur, A. Shibib, and K. Terrill, "60V Rating Split Gate Trench MOSFETs Havng Best-in-Class Specific Resistance and Figure-of-Merit", ISPSD Proceedings, pp. 387-390, 2016

[10] X. Yang, Y. C. Liang, G. S. Samudra, and Y. Liu, "Tunable Oxide-Bypassed Trench Gate MOSFET: Breaking the Ideal Superjunction MOSFET Performance Line at Equal Column Width", IEEE Electron Device Letters, Vol. 24, No. 11, pp. 704-706, 2003

[11] Y. C. Liang, K. P. Gan, and G. S. Samudra, "Oxide-Bypassed VDMOS (OBVDMOS): An Alternative to Superjunction High Voltage MOS Power Devices", IEEE Electron Device Letters, Vol. 22, No. 8, pp. 407-409, 2001

[12] C. Park, A. Shibib, and K. Terrill, "Semiconductor device with non-uniform trench oxide layer", US Patent, US 9,673,314 B2, 2017

[13] C. Park, A. Shibib, and K. Terrill, "Semiconductor device with non-uniform trench oxide layer", US Patent, US 9,978,859 B2, 2018

[14] K. Kobayashi, T. Nishiguchi, S. Katoh, T. Kawano, and Y. Kawaguchi, " 100V Class Multiple Stepped Oxide Field Plate Trench MOSFET (MSO-FP-MOSFET) Aimed to Ultimate Structure Realization", ISPSD Proceedings, pp. 141-144, 2015

[15] Y. Hattori, K. Nakashima, M. Kuwahara, T. Yoshida, S. Yamauchi, and H. Yamaguchi, "Design of a 200V Super Junction MOSFET with n-buffer regions and its Fabrication by Trench Filling", ISPSD Proceedings, pp. 189-192, 2004

[16] T. Shibata, Y. Noda, S. Yamauchi, S. Nogami, T. Yamaoka, Y. Hattori, and H. Yamaguchi, "200V Trench Filling Type Super Junction MOSFET with Orthogonal Gate Structure", ISPSD Proceedings, pp. 37-40. 2007

Proceedings of the 31st International Symposium on Power Semiconductor Devices & ICs
May 19-23, 2019, Shanghai, China

100-V Class Two-step-oxide Field-Plate Trench MOSFET to Achieve Optimum RESURF Effect and Ultralow On-resistance

Kenya Kobayashi, Hiroaki Kato, Toshifumi Nishiguchi, Saya Shimomura, Tetsuya Ohno, Tatsuya Nishiwaki, Kikuo Aida,
Kentaro Ichinoseki, Kohei Oasa and Yusuke Kawaguchi
Advanced Discrete Development Center, Toshiba Electronic Devices & Storage Corporation, Ishikawa, Japan
Email: kenya.kobayashi@toshiba.co.jp

Abstract—**We propose a 100-V class two-step-oxide Field-Plate MOSFET (2-step FP-MOSFET), which is formed by two steps of thick-oxide to simplify the structure and fabrication process. By optimizing design parameters, we reveal the 2-step FP-MOSFET can achieve sufficient RESURF (Reduced Surface Field) effect and an ultralow specific on-resistance ($R_{ON}A$). Measurement results showed breakdown voltage of 109.9 V and the $R_{ON}A$ of 27.7 mohm·mm² with good process controllability. Moreover, as figure-of-merit of the 2-step FP-MOSFET, $R_{ON}·Q_g$ and $R_{ON}·Q_{sw}$ were reduced by 27.1% and 4.7%, respectively, compared with conventional one. Power loss estimation is also discussed by simple calculation.**

Keywords—field plate, RESURF, shielded gate, oxide slope, figure-of-merit, power loss.

I. INTRODUCTION

Field-plate trench MOSFETs (FP-MOSFETs) have been continuously developed for high efficiency and low energy consumption power electronics [1]–[9]. In particular, 100-V class MOSFETs are expected to be applied to 48-V input power converters and 48-V battery automobile systems. In the FP-MOSFET, vertical field plates inside trench have RESURF (Reduced Surface Field) effect in mesa region with high impurity concentration, so that tradeoff between breakdown voltage (V_B) and specific on-resistance ($R_{ON}A$) is drastically improved. We have reported that multiple stepped oxide (MSO) FP-MOSFET can achieve an ultralow $R_{ON}A$ [10]. It was close to an ideal gradient field-plate structure in [11], however, in order to realize the structure, many additional process steps such as deposition and etching were needed.

In this study, we propose an advanced device structure, named 2-step FP-MOSFET, which is formed by two steps of thick-oxide and two steps of poly-silicon field-plate to simplify the structure and fabrication process. By optimizing plural design parameters, we reveal the 2-step FP-MOSFET can achieve sufficient RESURF effect and an ultralow $R_{ON}A$. Moreover, we describe figure-of-merit (FOM) and power loss in comparison with conventional FP-MOSFET.

II. DEVICE STRUCTURE AND DESIGN PARAMETERS

The device structures and significant design parameters for conventional, MSO and 2-step FP-MOSFETs are shown in Fig. 1(a)–(c). Those FP-MOSFETs apply the source field-plate structure, so called the shielded-gate structure [7], to promise very low gate-drain charge (Q_{gd}), instead of the gate

field-plate structure in previous work [10]. The most part of the V_B is determined by poly-silicon field-plate length L_{FP}, and the remaining part is shared by p-base/n-drift junction and trench bottom region. Field-plate oxide thickness $t_{OX,t}$ and $t_{OX,b}$ correspond to a position of top and bottom of the L_{FP}, respectively. Oxide-slope K is defined by

$$K = \frac{L_{FP}}{\frac{L_{FP}}{\tan\alpha} - t_{OX,t}},$$ (1)

where α is trench angle. In particular, the $t_{OX,b}$ and drift layer concentration (N_D) are important to obtain an appropriate charge balance. An optimum charge density (Q_{Opt}) in the mesa region against both sides' field-plates, is given by

$$Q_{Opt} = \frac{2E_y\varepsilon_{Si}}{q},$$ (2)

where E_y is average of the vertical electric field at the V_B, ε_{Si} is permittivity of silicon and q is elementary charge. To obtain the V_B over 100 V, the E_y is estimated to 2.5E5 ~ 3.5E5 V/cm by TCAD simulation. The relationship between the mesa width (W_{Mesa}) and the N_D is expressed by

$$N_D = \frac{Q_{Opt}}{W_{Mesa}}.$$ (3)

When the W_{Mesa} is 1.0 μm, the N_D is calculated to 3.2E16 ~ 4.5E16 /cm³.

Fig. 2(a)–(d) shows simulated 2-D potential contours and 1-D vertical electric field distributions along trench sidewall for conventional, MSO and 2-step FP-MOSFETs, at N_D = 3.0E16 /cm³. Both of FP-MOSFETs indicate mostly uniform electric field distribution and obtain a sufficient V_B (112 ~ 117 V), on the contrary, the conventional one degrades V_B.

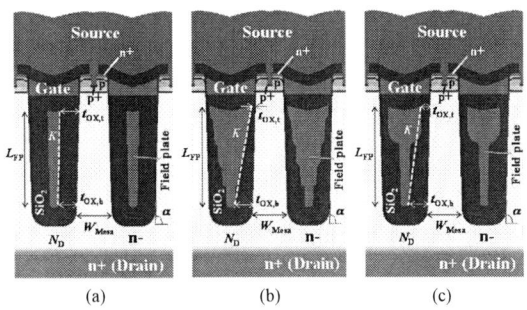

Fig. 1. Cross-sectional structures and significant design parameters of three kinds of FP-MOSFETs in this study. (a) Conventional FP-MOSFET, (b) MSO FP-MOSFET [10], and (c) 2-step FPMOSFET.

978-1-7281-0582-6/19 $31.00 © 2019 IEEE

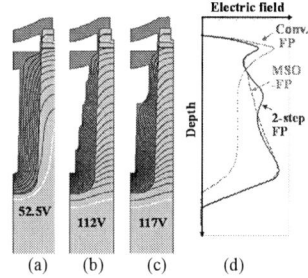
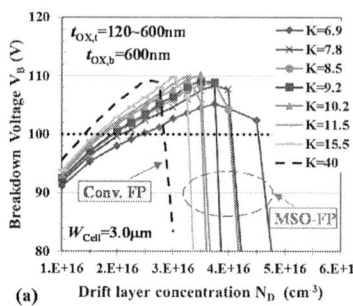
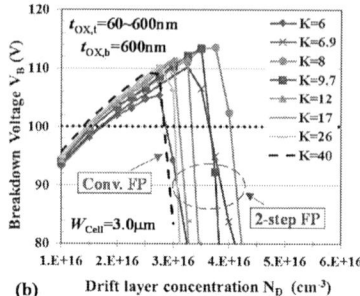

Fig. 2. Simulated 2-D potential contours for (a) conventional, (b) MSO and (c) 2-step FP-MOS FETs. (d) 1-D vertical electric field distributions along trench sidewall. (N_D = 3.0E16 /cm³.)

Fig. 3. Simulated N_D dependences of V_B for (a) MSO FP-MOFETs (K = 6.9 ~ 15.5) and (b) 2-step FP-MOSFETs (K = 6 ~ 26). Conventional FP-MOSFETs (K = 40) are shown in each graph.

III. OPTIMUM DESIGN BY TCAD SIMULATION

As investigation of optimum design, we simulated N_D dependence of the V_B of both MSO (K = 6.9 ~ 15.5) and 2-step (K = 6 ~ 26) FP-MOSFETs, as shown in Fig. 3. The conventional FP-MOSFETs (K = 40) are shown in each graph. The $t_{OX,b}$ is fixed to 600 nm, which can obtain approximately 110 V. It is found that peak V_B ($V_{B,peak}$) is changed by the N_D and the K. The tendency is different in three kinds of FP-MOSFETs.

By redrawing the results in Fig. 3 into Fig. 4, it is found that the $V_{B,peak}$ over 110 V can obtain at K = 10.2 in the MSO FP-MOSFETs and at wide range of K = 6.9 ~ 26 in the 2-step FP-MOSFETs (Fig. 4(a)). This is because there is existence of the point of inflection in the electric field distribution of the 2-step FP-MOSFET (Fig. 2(d)), therefore the V_B increases compared with the MSO FP-MOSFET even if the N_D is high. In addition, when "N_D at $V_{B,peak}$" is around 3.5E16 ~ 3.75E16 /cm³, minimum $R_{ON}A$ can achieve at K = 8 ~ 9.7 in the 2-step FP-MOSFET (Fig. 4(b), "$R_{ON}A$ at $V_{B,peak}$"). This N_D is 1.5 times higher and the $R_{ON}A$ is 21% lower than those of the conventional FP-MOSFET.

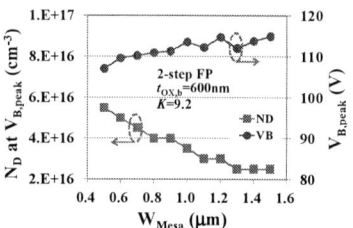

Fig. 5. Simulated W_{Mesa} dependences of $V_{B,peak}$ and "N_D at $V_{B,peak}$" for 2-step FP-MOSFETs. ($t_{OX,b}$ = 600 nm, K = 9.2)

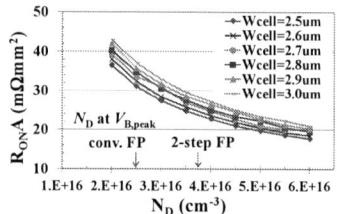

Fig. 6. Simulated N_D and W_{Cell} dependences of $R_{ON}A$ for 2-step (K=9.2) FP-MOSFETs. "N_D at $V_{B,peak}$" for conventional and 2-step FP-MOSFETs are shown in x-axis.

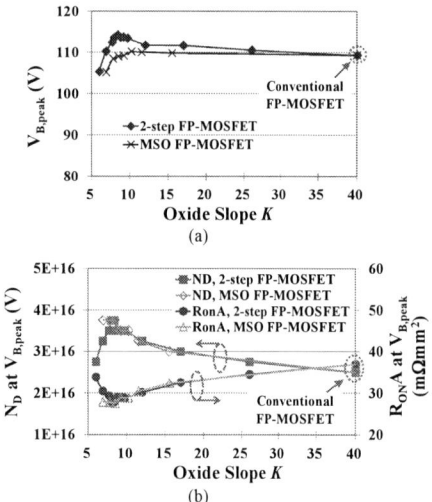

Fig. 4. Simulated oxide-slope K dependences of (a) peak breakdown voltage $V_{B,peak}$, and (b) "N_D at $V_{B,peak}$" and "$R_{ON}A$ at $V_{B,peak}$" for conventional, MSO and 2-step FP MOSFETs ($t_{OX,b}$ = 600 nm).

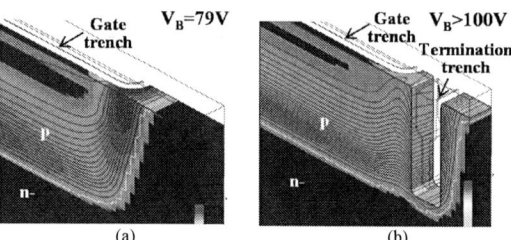

Fig. 7. Simulated 3-D potential contours (black lines) and impact-ionization generation rate (color). (a) Without and (b) with termination trench.

Fig. 5 shows W_{Mesa} dependences of the $V_{B,peak}$ and "N_D at $V_{B,peak}$" for K = 9.2. By the results, we chose W_{Mesa} = 1.0 μm in a view point of ease of the fabrication process. As shown in Fig. 6, the $R_{ON}A$ is monotonically decreased by the N_D increase and narrower W_{Mesa}.

In the FP-MOSFET chip design, the V_B in termination area has to be enough high and stable [12]. As shown in Fig. 7, it is confirmed that the termination trench layout in orthogonal direction of the gate trench is effective to improve the V_B in the 2-step FP-MOSFET by 3-D TCAD simulation.

978-1-7281-0582-6/19 $31.00 © 2019 IEEE

IV. FABRICATION PROCESS

Representative process steps for the 2-step FP-MOSFET are shown in Fig. 8. (a) The trench of around 5.5-μm depth is formed by reactive ion etching (RIE), and followed by first thick oxidation. (b) By using sacrificial layer inside the trench, the thick oxide is etched down to approximately half-depth of the trench, and followed by second thick oxidation. (c) Poly-silicon field-plate is formed by chemical vapor deposition (CVD) and RIE. (d) First interlayer oxide is filled in the trench by CVD and etched back to appropriate depth. (e) Gate oxidation and gate poly-silicon CVD are performed continuously. (f) The gate poly-silicon is etched. After that, following process steps such as p-base, n⁺-source, p⁺-body, second interlayer oxide, contact, surface metallization, wafer thinning and back-side metallization are performed.

TEM (Transmission Electron Microscope) photograph of a unit cell structure of fabricated 2-step FP-MOSFET is shown in Fig 9. To relax a strong stress inside the trench during the gate poly-silicon forming, U-shaped gate structure is applied.

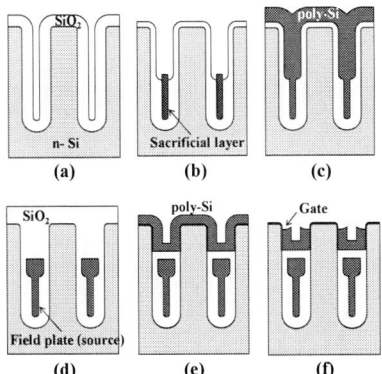

Fig. 8. Representative process steps for 2-step FP-MOSFET. (a) Trench etching and first thick oxidation. (b) Sacrificial layer formation, oxide half-etching in upper region, and second thick oxidation. (c) Field-plate poly-silicon CVD. (d) First interlayer oxide CVD and etching. (e) Gate oxidation and gate poly-silicon CVD. (f) Gate poly-silicon etching.

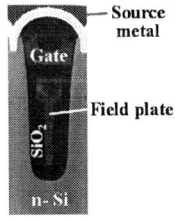

Fig. 9. TEM photograph of fabricated 2-step FP-MOSFET (unit cell).

V. CHARACTERIZATION OF DEVELOPED FP-MOSFET

A. Breakdown Voltage and On-resistance

Deviation of the V_B in three case of the trench depth that are deep (5.9 μm), medium (5.5 μm) and shallow (5.1 μm), is shown in Fig. 10. In both N_D variation of +/-10% (Fig. 10(a)) and $t_{OX,b}$ variation of +/-10% (Fig. 10(b)), it was confirmed that the process controllability was very good. Average of V_B at drain current I_D = 10 mA was 109.9 V and the standard deviation was 0.50 V, under the center design and the process condition.

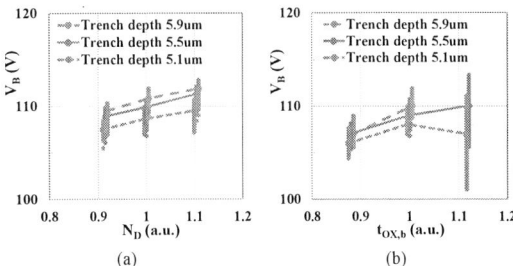

Fig. 10. Measured V_B deviation regarding process variations of (a) N_D and (b) $t_{OX,b}$ of 2-step FP-MOSFETs, which was additionally evaluated in each trench depth (deep, medium and shallow).

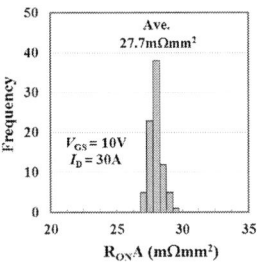

Fig. 11. Measured $R_{ON}A$ distribution of 2-step FP-MOSFETs fabricated by center design and process conditions.

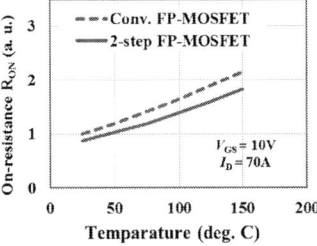

Fig. 12. Measured temperature dependence of R_{ON} in SOP-8 package for conventional FP-MOSFET and 2-step FP-MOSFET.

Subsequently, R_{ON} packaged in SOP-8 was measured, under the condition of gate voltage V_{GS} = 10 V and I_D = 30 A. The $R_{ON}A$ obtained by deduction of the package resistance indicated average value of 27.7 mΩ·mm² and good deviation of 0.46 mΩ·mm², as shown in Fig. 11. This $R_{ON}A$ of the 2-step FP-MOSFET is ultralow in ever reported 100-V class device and improved by 16.6% compared with that of the conventional FP-MOSFET. Moreover, as shown in Fig. 12, temperature coefficient of $R_{ON}A$ (from 25 to 150 degrees C) was 1.83 and it was superior to 2.15 of the conventional FP-MOSFET. This is because the 2-step FP-MOSFET has lower drift resistance compared with the conventional FP-MOSFET.

B. Figure-of-Merit

When the 2-step FP-MOSFET is applied in high efficiency switching circuit, gate charge properties, i.e., gate-source charge (Q_{gs}), gate-drain charge (Q_{gd}), total gate charge (Q_g) and output charge (Q_{oss}), are very important. Moreover, as an indicator of the switching property, Q_{sw} is defined by sum of the Q_{gs} after threshold voltage and the Q_{gd}. Fig. 13 compares figure-of-merit (*FOM*) of the conventional and the 2-step FP-MOSFET, under the conditions of V_{DS} = 50 V and I_D = 35 A as the gate charge measurement.

It was confirmed that $R_{ON} \cdot Q_g$ and $R_{ON} \cdot Q_{sw}$ were reduced by 27.1% and 4.7%, respectively, compared with those of the conventional device. The effect of the Q_g reduction includes modification of the gate-source insulating film structure in the fabricated device. On the other hand, $R_{ON} \cdot Q_{oss}$ was increased. The Q_{oss} is calculated by voltage integral of drain-source capacitance C_{ds}. In the 2-step FP-MOSFET, the N_D was increased to improve the $R_{ON}A$, so that depletion capacitance component of the C_{ds} was increased and it affected to the Q_{oss}. Thus, improvement of the tradeoff between $R_{ON}A$ and Q_{oss} is further challenge to realize ultimate power MOSFET.

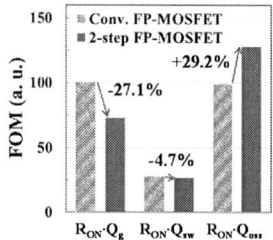

Fig. 13. Comparison of figure-of-merit (*FOM*) for conventional and 2-step FP-MOSFETs. Measurement conditions: $V_{GS} = 10$ V and $I_D = 30$ A for R_{ON}, and $V_{DS} = 50$ V and $I_D = 35$ A for gate charge.

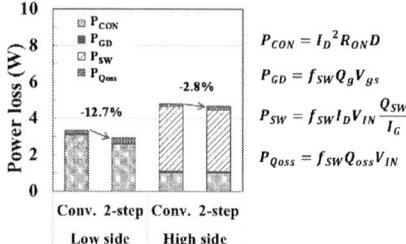

Fig. 14. Comparison of power loss estimation for conventional and 2-step FP-MOSFETs. For both low-side and high-side in a half-bridge circuit are calculated in assumption of $V_{IN} = 48$ V, $I_D = 35$ A, $V_{gs} = 10$ V, $I_g = 1$ mA and $f_{sw} = 100$ kHz.

C. Power Loss Estimation

As simple estimation of power loss, we took account of a half-bridge circuit and calculated each component of the power loss: conduction loss (P_{CON}), gate drive loss (P_{GD}), switching loss (P_{SW}) and output charge loss (P_{Qoss}) by general formula [13]. Fig. 14 shows comparison of the power loss estimation for the conventional and the 2-step FP-MOSFETs. In assumption of the low-side, same die size was applied to both MOSFETs so that the 2-step FP-MOSFET had low-R_{ON} advantage. For the high-side, same R_{ON} MOSFETs were used, thus die size of the 2-step FP-MOSFET became smaller. In addition, we assumed 48-V input, 35-A output, 100 kHz operation, and 75%/25% duty ratio for low-side/high-side. In the low-side switching, the power loss of the 2-step MOSFET was calculated as 2.91 W, which was 12.7% lower than that of the conventional device. In the high-side, although the reduction was only 2.8%, it showed the effect of $R_{ON} \cdot Q_{sw}$. The developed 2-step FP-MOSFET has advantage, especially in using as high current switching, due to the ultralow $R_{ON}A$.

VI. CONCLUSIONS

We proposed the advanced trench MOSFET, which has two steps of field-plate. By designing plural parameters, the 2-step FP-MOSFET achieved optimum RESURF effect and 1.5 times higher N_D compared with the conventional one. In the fabricated device, we confirmed superior tradeoff, V_B of 109.9 V and $R_{ON}A$ of 27.7 m$\Omega \cdot$mm^2. The measurement results also showed small deviation and good process controllability. Moreover, $R_{ON} \cdot Q_g$ and $R_{ON} \cdot Q_{sw}$ were reduced by 27.1% and 4.7%. The power loss was estimated to improve by 12.7% in the assumed half-bridge as low-side operation. The developed 2-step FP-MOSFET has advantage, especially in using as high current switching, due to the ultralow $R_{ON}A$.

ACKNOWLEDGMENT

The authors are grateful to T. Kawano for the first process design, Y. Oomuro for 3-D TCAD, T. Matsuda and Y. Himori for process engineering, T. Hishinuma and A. Tsuyuguchi for device measurement, W. Saito for useful discussion, and S. Jimbo and H. Kamijo for their encouragement.

REFERENCES

[1] Y. Baba, N. Matsuda, S. Yanagiya, S. Hiraki and S. Yasuda, "A study on a high blocking voltage UMOS-FET with a double gate structure," *Proc. of ISPSD '92*, pp. 300-302, 1992.

[2] M. Kodama, E. Hayashi, Y. Nishibe and T. Uesugi, "Temperature characteristics of a new 100V rated power MOSFET, VLMOS (vertical LOCOS MOS)," *Proc. of ISPSD '04*, pp.463-466, 2004.

[3] G. E. J. Koops, E. A. Hijzen, R. J. E Hueting and M. A. A. in't Zandt, "Resurf stepped oxide (RSO) MOSFET for 85V having a record-low specific on-resistance," *Proc. of ISPSD '04*, pp.185-188, 2004.

[4] A. Schlögl, F. Hirler, J. Ropohl, U. Hiller, M. Rösch, N. Soufi-Amlashi and R. Siemieniec, "A new robust power MOSFET family in the voltage range 80 V-150 V with superior low R_{DSon}, excellent switching properties and improved body diode," *European Conf. Power Electronics and Applications*, pp. 1-10, 2005.

[5] M. A. Gajda, S. W. Hodgskiss, L. A. Mounfield and N. T. Irwin, "Industrialisation of resurf stepped oxide technology for power transistors," *Proc. of ISPSD '06*, pp.109-112, 2006.

[6] P. Moens, F. Bauwens, B. Desoete, J. Baele, K. Vershinin, H. Ziad, E.M. Shankara Narayanan and M. Tack, "Record-low on-resistance for 0.35 μm based integrated XtreMOS™ Transistors," *Proc. of ISPSD '07*, pp. 57-60, 2007.

[7] P. Goarin, G. E. J. Koops, R. van Dalen, C. L. Cam and J. Saby, "Split-gate resurf stepped oxide (RSO) MOSFETs for 25 V applications with record low gate-to-drain charge," *Proc. of ISPSD '07*, pp. 61-64, 2007.

[8] K. Kobayashi, T. Yamaguchi, S. Tokuda, S. Tsuboi and H. Ninomiya "An 18V n-channel UMOSFET with super low on-resistance by using vertical resurf structure," *the Papers of Joint Technical Meeting, IEE Japan*, EDD-12-068, SPC-12-141, pp. 49-53, 2012. (In Japanese)

[9] C. Park, S. Havanur, A. Shibib and K. Terrill, "60 V rating split gate trench MOSFETs having best-in-class specific resistance and figure-of-merit," *Proc. of ISPSD '16*, pp. 387-390, 2016.

[10] K. Kobayashi, T. Nishiguchi, S. Katoh, T. Kawano and Y. Kawaguchi, "100 V class multiple stepped oxide field plate trench MOSFET aimed to ultimate structure realization," *Proc. of ISPSD '15*, pp. 141-144, 2015.

[11] Y. Chen, Y. C. Liang and G. S. Samudra, "Theoretical analyses of oxide-bypassed superjunction power metal oxide semiconductor field effect transistor devices," *Japanese Journal of Applied Physics*, vol. 44, no. 2, pp. 847-856, February 2005.

[12] Z. Hossain, G. Sabui, J. Sellers, B. Pratt and A. Salih, "3-D TCAD simulation to optimize the trench termination design for higher and robust BV$_{DSS}$," *Proc. of ISPSD '16*, pp. 391-394, 2016.

[13] R. Sodhi, S. Brown and D. Kinzer, "Integrated design environment for DC/DC converter FET optimization," *Proc. of ISPSD '99*, pp. 241-244, 1999.

Proceedings of the 31st International Symposium on Power Semiconductor Devices & ICs
May 19-23, 2019, Shanghai, China

Vertical 1.2kV SiC Power MOSFETs with High-k/Metal Gate Stack

Stephan Wirths, Yulieth Christina Arango, Alyssa Prasmusinto, Giovanni Alfieri, Enea Bianda,
Andrei Mihaila, Lukas Kranz, Marco Bellini and Lars Knoll
ABB Switzerland Ltd., Corporate Research,
5405 Baden-Dättwil, Switzerland,
stephan.wirths@ch.abb.com.

Abstract—We demonstrate the first integration of high-k/metal gate stacks in vertical 1.2kV SiC power MOSFETs including static and dynamic characterization as well as safe operation area (SOA). The high-k/4H-SiC MOS interface exhibits a remarkably low interface defect state density and improved threshold voltage stability compared to conventional gate stacks based on SiO_2. Moreover, we achieved an impressive performance boost in terms of on-resistance due to this low-defective interface and increased gate capacitance. Compared to vertical devices with SiO_2/poly Si gate stacks these devices exhibit a negligible hysteresis.

Keywords—Wide band gap, high-k, 1.2kV MOSFETs, SiC, SiC power MOSFETs

I. INTRODUCTION

SiC-based power semiconductor devices, e.g. vertical 1.2kV power MOSFETs, are promising candidates to replace Si IGBTs in a number of applications, e.g. motor drives, EV chargers or for renewable energy conversion. This is due to their higher on-state current densities at high frequencies for a given power dissipation limit of the package. Here, key is to reduce the on-resistance (R_{on}) and at the same time to guarantee a reliable switching behavior of all devices involved, especially when parallel operation is required. In order to improve the conduction in the inversion channel optimizing the poor quality of the SiO_2/SiC interface is a common strategy. Recently, several interface passivation techniques such as doped oxides [1] or B diffusion into thermal oxides [2] were investigated. Although fairly high inversion channel mobility values were reported high leakage currents, V_{th} instabilities or low critical fields prevent their integration into production. The most commonly used process in commercially available devices is the introduction of N near the interface via NO post oxidation annealing (POA), which reduces the interface defect density (D_{it}) [3] but enhances hole and electron trapping [4]. In this paper, however, we will present an approach that does not require sophisticated defect passivation techniques using group III or V elements, which can cause degradation of the oxide. We demonstrate the integration of high-k/metal gate stacks in 1.2kV vertical SiC power MOSFETs, which exhibit D_{it} values below $1 \cdot 10^{11}$ cm^{-2}eV^{-1}. The high-k/4H-SiC MOS-structures show negligible hysteresis and frequency-dependent flat band voltage (V_{FB}) shifts. Moreover, vertical SiC MOSFETs exhibit nearly no hysteresis and reduced V_{FB} shift in a V_G-range of -6V to 10V at temperatures up to 125°C.

II. CONCEPT

One advantage of a high-k dielectric is that a thicker gate dielectric film can be used while maintaining the same

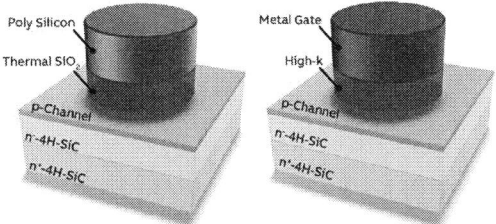

Figure 1: Schematics of (left) a state-of-the-art gate stack using SiO_2 as gate oxide and (right) a high-k/metal gate stack on 4H-SiC.

effective capacitance. That is, carrier injection into the semiconductor due to tunnel processes induced by high electric fields across the insulator can be reduced at the same channel resistance. This in turn results in an improved reliability (e.g. reduced risk of dielectric breakdown) and reduced short channel effects. Former studies on aluminum oxynitride (AlON) and HfAlON high-k dielectrics revealed reliability improvements [5,6]. On the other hand, using the same gate dielectric thickness as in a SiO_2/poly-Si gate stack the enlarged gate capacitance of a high-k dielectric/metal gate stack results in higher on-currents assuming similar inversion channel mobility. Additionally, a gate dielectric with a higher k value is beneficial in terms of reverse blocking capability, since it leads to a lower field in the gate dielectric. Figure 1 illustrates the schematics of the two gate stacks that are compared throughout this paper; a SiO_2/poly Si gate stack (cf Figure 1a) and a high-k/metal gate stack (cf Figure 1b). We investigated and compared the interface quality and electronic transport properties using MOS capacitors (MOSCAPs) and lateral MOSFETs, respectively, as well as the performance of vertical planar power MOSFETs. For these studies, we used 4° off axis 4H SiC (0001) wafers with lowly nitrogen-doped epitaxial layers deposited on high conductivity n-type substrates. The MOSCAP samples were wet-chemically cleaned using H_2SO_4:H_2O_2 and diluted HF prior to the experiments and we employ N_2O POA for the SiO_2/poly Si gate stacks. Metallic contacts (30nm Ti and 400nm Al) were formed using shadow masks and the backside contacts consist of Ni silicide/Al, Ti and Ag. Regarding the lateral MOSFET fabrication (schematic is given in the inset of Figure 5a), we used Al and P as channel and source implantation species, respectively. The Ni silicidation process is following the gate stack deposition and etching. The final devices exhibit gate widths of 50 μm and gate lengths between 20 μm and 240 μm. We performed Capacitance-Voltage (CV) measurements and Constant-Capacitance Deep Level Transient Spectroscopy (CC-DLTS) as well as temperature-dependent current-voltage (IV) measurements for the electrical characterization of MOSCAPs and lateral MOSFETs, respectively. Finally, planar 1.2kV vertical 4H-SiC power

978-1-7281-0582-6/19 $31.00 © 2019 IEEE   103

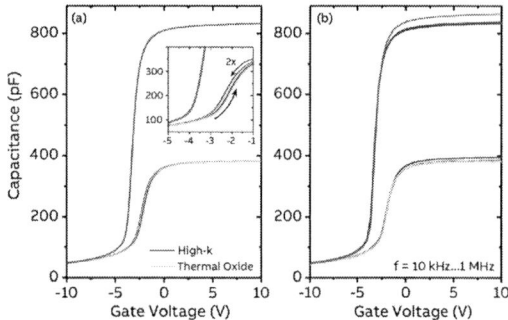

Figure 2: Room temperature CV characterization for the high-k and SiO₂ reference samples. (a) Bidirectional (2x) CV as well as (b) frequency-dependent (10 kHz to 1 MHz) CV sweeps.

MOSFETs were fabricated with a pitch of 14µm, a constant body diode region and channel lengths of 150nm and 250nm with a retrograded channel doping profile. To minimize electric fields at the gate oxide in between the cells, i.e. avoid a JFET implantation step, we optimized charge and depth distributions in the channel. Again, we processed reference vertical power devices using the identical design but a gate dielectric based on SiO₂ and N₂O POA. Static on-state measurements were performed using a pulsed I-V curve tracer with a duty cycle of 1% and a pulse duration of 50ms to avoid self-heating. For dynamic characterization the devices were soldered on test substrates, wire bonded and measured in a double pulse tester with a system inductance (L_σ) of 60nH.

III. RESULTS AND DISCUSSION

A. SiC/Dielectric Interface Optimization

Figure 2a and 2b show the room-temperature CV characterization of the high-k MOSCAP and the reference sample (SiO₂ and N₂O POA). The former exhibits the steepest CV characteristics as well as significantly higher C_{ox} at identical dielectric layer thickness due to the higher k value. The bidirectional CV sweeps (see Figure 2a) revealed a significantly reduced hysteresis for the high-k sample. In fact, the flatband voltage shift (V_{FB} shift) is negligible and, thus, indicating a decreased number or carriers trapped in the oxide. In addition, the frequency-dependent (10 kHz to 1 MHz) V_{FB} shift and stretch-out (Figure 2b) were not measurable. That is, less carriers of the inversion region are trapped in interface traps located close to the oxide. Consequently, the bands inside the semiconductor are strongly bent and, thus, result in steep CV curves (no Fermi level pinning at the dielectric-semiconductor interface). Moreover, we evaluated the gate leakage characteristic of several devices with high-k layers as displayed in Figure 3. Whereas we observed nearly constant leakage current levels up to $V_G = 25V$ the leakage current increases exponentially due to Fowler-Nordheim tunneling above 25V. The dielectric breakdown occurs above $V_G = 55V$ and, thus, demonstrating its suitability for device integration, i.e. gate voltages up to 25V can be applied without noteworthy increase of gate leakage current. We extracted the D_{it} using Constant-Capacitance Deep Level Transient Spectroscopy (CC-DLTS) as shown in Figure 4. Throughout the investigated energy range, we extracted considerably lower D_{it} values for the

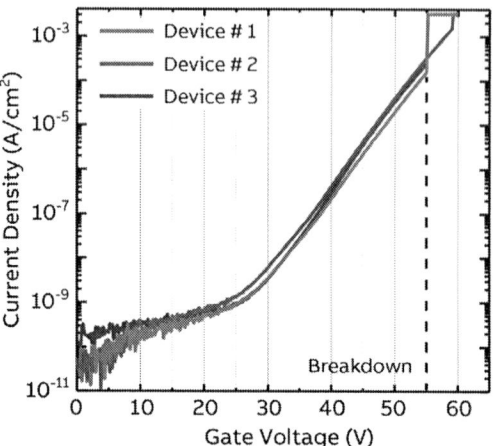

Figure 3: Gate leakage measurements of MOSCAPs using the high-k dielectric on SiC.

high-k sample (approx. $8 \cdot 10^{10}$ cm⁻²eV⁻¹ $- 2 \cdot 10^{12}$ cm⁻²eV⁻¹) compared to SiO₂-based gate stacks. Correspondingly bidirectional transfer sweeps ($V_G = -15V...+15V$ @ $V_D = 0.1V$) of lateral MOSFETs at room temperature up to 200°C notably reflect this improved SiC/dielectric interface quality, see Figure 5. Virtually no hysteresis effect and significantly reduced V_{TH}-shifts were observed for the devices with high-k/metal gate stacks. Whereas we determined V_{TH}-shifts due to hysteresis of 300-400mV (at 25°C and 125°C) as well as 800mV at 200°C for devices based on SiO₂ gate oxides, ΔV_{TH} for the high-k devices is merely measurable up to 125°C (approx. 10mV). At 200°C ΔV_{TH} amounts to 350mV. This reduction in V_{TH}-shift can be explained by the decreased density of interface states [7]. Additionally, the inset of Figure 5b clearly demonstrates 2.5x higher current levels at identical gate overdrive voltages for the high-k/metal gate stack compared to thermally oxidized devices and N₂O POA due to the higher dielectric constant.

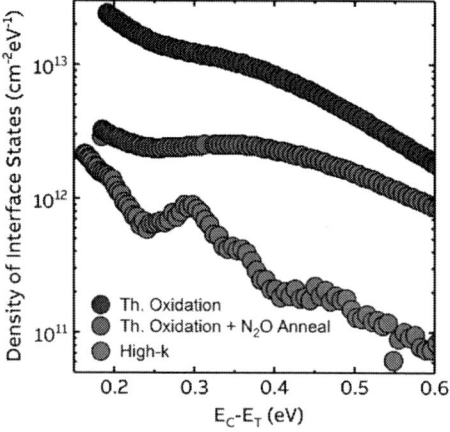

Figure 4: Density of interface states (D_{it}) for three different gate stacks extracted from constant capacitance deep level transient spectroscopy (CC-DLTS).

978-1-7281-0582-6/19 $31.00 © 2019 IEEE

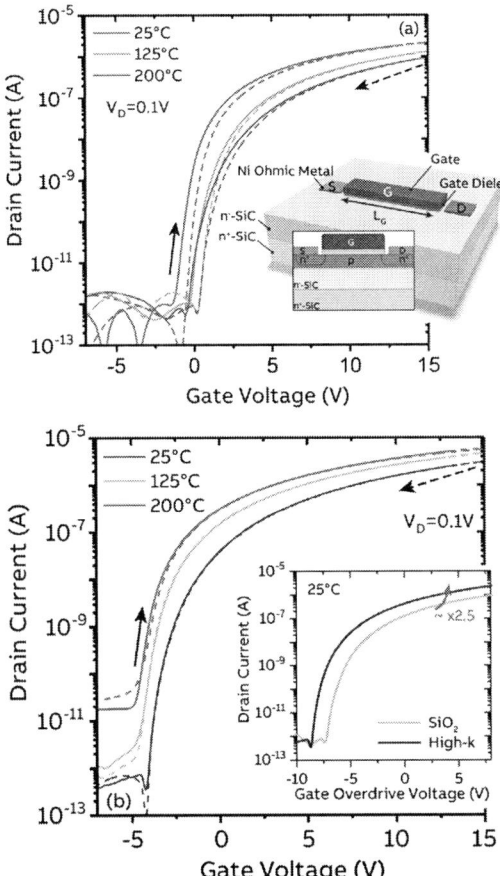

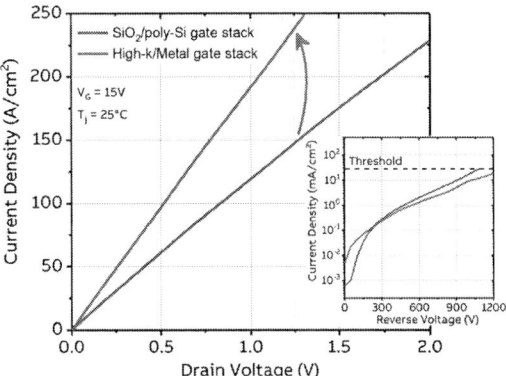

Figure 5: Temperature-dependent bidirectional transfer characteristics of lateral SiC MOSFETs using (a) SiO$_2$ and high-k dielectric (b). Solid and dashed lines correspond to forward and reverse measurement directions, respectively. The insets show a schematic of the lateral MOSFET structure (a) as well as (b) drain current vs. gate overdrive voltage for both gate stacks at room temperature.

Figure 6: Comparison of on-state performance (output characteristic @ Vg = 15V) at room temperature of 1.2 kV SiC power MOSFETs with high-k/metal and SiO$_2$/poly Si gate stacks. The inset shows the reverse blocking capability for both devices at V$_G$ = -10V.

k/metal gate stack are presented in Figure 7. We swept the DC gate voltage from -2V, -4V, -6V, -8V and -10V up to +10V. For sweeps from -6V to +10V we hardly measure any hysteresis or V$_{th}$-shift at room temperature and even at 125°C the effect is only slightly enlarged. However, at larger sweeps, i.e. -10V to +10V, we observed a hysteresis of approx. 0.2V. These results represent a substantial improvement compared to recently published data [7].

C. Dynamic Switching of Verical Power MOSFETs

Figure 8 shows the turn on and turn off waveforms of a MOSFET with high-k/metal gate under overvoltage and overcurrent conditions. The devices switched I$_D$ = 100A representing 4x I$_{NOM}$ at V$_D$ = 800V (V$_G$ = -10...+15V) and T$_j$ = 125°C using R$_G$ = 33Ω, respectively, and a stray inductance of 60 nH. A SiC JBS rectifier has been used as the free-wheeling diode. The Reverse Bias Safe Operating Area (RBSOA) capability of a high-k/metal gate vertical power MOSFET is plotted in Figure 9. The MOSFETs successfully turn-off 675A/cm^2 at V$_D$ = 900V without failures observed during the testing representing more than 3x I$_{NOM}$.

B. Static Characterization of Vertical Power MOSFETs

Figure 6 shows the static characterization, i.e. on-state performance and reverse blocking capability (see inset), of vertical SiC planar power MOSFET with high-k/metal gate and SiO$_2$/poly-Si gate stacks. The design and main processing steps such as active area, termination, doping profiles etc. of both devices are identical; the only differences during fabrication were the gate stack processing and integration. As can be seen, the current density at V$_{DS}$ = 1V (@ V$_G$ = 15V) is tremendously improved by more than 60%. As a consequence the specific on-resistance (R$_{DS,on}$) is reduced by more than 35%. Notably, this on-state performance was achieved using a comparatively large pitch of 14 μm and without wafer thinning. Hence, this boost in on-state performance is unambiguously due to the improved conductivity in the inversion channel, i.e. reduced channel resistance. Moreover, we observed an enhanced reverse blocking capability of the high-k/metal gate stack devices. Bidirectional transfer sweeps (at room temperature and at 125°C) of a vertical planar power MOSFET with high-

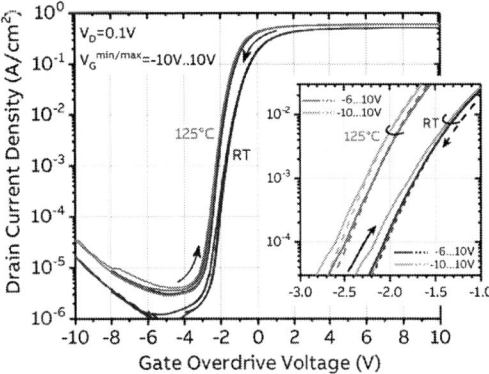

Figure 7: Bidirectional transfer characteristics of vertical 1.2kV power MOSFETs at V$_D$ = 0.1V for V$_G$ swings from max -10V to +10V at room temperature and 125°C. The inset shows the zoom-in for V$_G$ swings from -6V to +10V and -10V to +10V.

978-1-7281-0582-6/19 $31.00 © 2019 IEEE

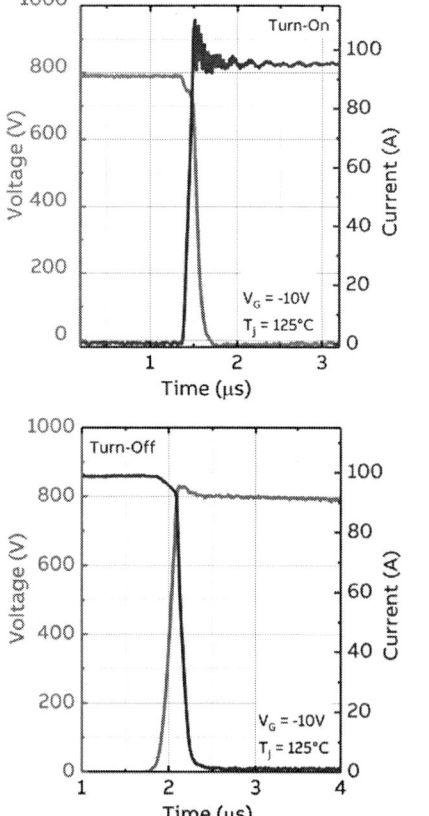

IV. CONCLUSIONS

We presented the first integration of a high-k/metal gate stack in planar vertical 1.2kV SiC power MOSFETs. Investigated MOSCAPs reveal exceptional SiC/high-k interface quality. In particular, bidirectional CV sweeps show negligible hysteresis and frequency-dependent V_{FB} shift. Moreover, we extracted D_{it} values below $10^{11}/cm^2eV^1$. Lateral as well as vertical power MOSFETs do not suffer from V_{th} shifts and hysteresis during gate voltage sweeps from -6V to +10V. For vertical power MOSFETs with the developed high-k/metal gate stack, we observed a reduction in specific on-resistance by more than 35% compared to SiO_2/poly-Si control devices. These vertical devices switched 100A at $V_D = 800V$ and exhibit a wide RBSOA.

These results demonstrate the huge potential of high-k/metal gate stacks for power semiconductor devices based on wide band gap materials. Particularly, in terms of specific on-resistance a higher k dielectric with the presented high interfacial quality lead to a tremendously improved inversion channel conductivity, enhanced reproducibility, predictability and reliability.

REFERENCES

[1] D. Okamoto H. Yano, K. Hirata, T. Hatayama, and T. Fuyuki, "Improved Inversion Channel Mobility in 4H-SiC MOSFETs on Si Face Utilizing Phosphorus-Doped Gate Oxide", IEEE EDL, vol. 31 (7), pp. 710-712, 2010.

[2] A. Modic, G. Liu, A. C. Ahyi, Y. Zhou, P. Xu et al., "High channel mobility 4H-SiC MOSFETs by antimony counter-doping", IEEE EDL, vol 35 (9), pp. 894-896, 2014.

[3] G. Liu, B. R. Tuttle and S. Dhar, "Silicon carbide: A unique platform for metal-oxide-semiconductor physics", Appl. Phys. Rev. 2, 021307, 2018.

[4] J. Rozen, S. Dhar, S. K. Dixit, V. V. Afanas`ev, F. O. Roberts, H. L. Dang, S. Wang, S. T. Pantelides, J. R. Williams, and L. C. Feldman, "Increase in oxide hole trap density associated with nitrogen incorporation at the SiO$_2$/SiC interface", JAP, vol. 103, 124513, 2008.

[5] T. Hosoi, S. Azumo, Y. Kashiwagi, S. Hosaka, R. Nakamura, S. Mitani, Y. Nakano, H. Asahara, T. Nakamura, T. Kimoto, T. Shimura, and H. Watanabe, "Performance and reliability improvement in SiC power MOSFETs by implementing AlON high-k gate dielectrics," in Tech. Dig. 2012 IEEE Int. Electron Devices Meeting (IEDM), 7.4.

[6] T. Hosoi, S. Azumo, Y. Kashiwagi, S. Hosaka, K. Yamamoto, M. Aketa, H. Asahara, T. Nakamura, T. Kimoto, T. Shimura, and H. Watanabe, "Reliability-aware design of metal/high-k gate stack for high-performance SiC power MOSFET", in 2017 IEEE 29th International Symposium on Power Semiconductor Devices and IC's (ISPSD), 247-250, 2017.

[7] D. Peters, T. Aichinger, T. Basler, G. Rescher, K. Puschkarsky, and H. Reisinger, "Investigation of threshold voltage stability of SiC MOSFETs", in 2018 IEEE 30th International Symposium on Power Semiconductor Devices and IC's (ISPSD), 40-43, 2018.

Figure 8: (a) Turn-on and (b) turn-off waveforms of a vertical power MOSFET with high-k/metal gate stack at T_j = 125°C and V_G = -10V...+15V under overvoltage and overcurrent conditions of V_D = 800V and I_D = 100 A (4x I_{nom}), respectively (L_σ = 60nH).

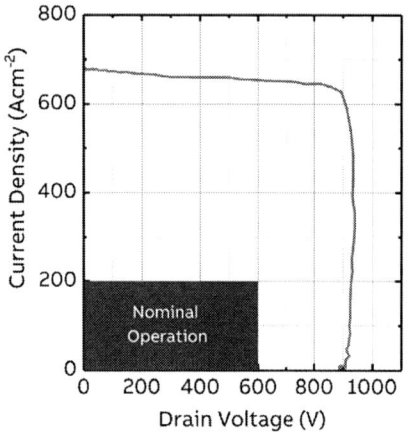

Figure 9: Turn-off reverse bias safe operating area of a vertical power MOSFET with high-k/metal gate stack at 125°C, V_D = 900V, J_D = 675A/cm^2.

Proceedings of the 31st International Symposium on Power Semiconductor Devices & ICs
May 19-23, 2019, Shanghai, China

Feasibility and Limitation of DC/DC Multilevel Converter Power ICs Using Standard CMOS Transistors

Runtao Ning*
Illinois Institute of Technology
Chicago, USA
rning@hawk.iit.edu

*Now with Alpha and Omega
Semiconductor, Portland, OR, USA

Yuanfeng Zhou
Illinois Institute of Technology
Chicago, USA
yzhou103@hawk.iit.edu

Z. John Shen
Illinois Institute of Technology
Chicago, USA
zshen6@iit.edu

Aritra Kundu
Illinois Institute of Technology
Chicago, USA
akundu5@hawk.iit.edu

Abstract— This paper investigates the feasibility and limitation of designing DC/DC converter power ICs using standard CMOS technology. Comparing to custom BCDMOS technologies, this path leads to lower cost and shorter development cycle time, benefiting from the high volume CMOS manufacturing ecosystem. Our study indicates that the flying capacitor multilevel converter (FCMC) topology can facilitate DC/DC conversion using standard CMOS transistors for an input voltage up to 15V for the 90nm/65nm CMOS nodes, and 12V for the 45nm CMOS node. The CMOS FCMC solution offers a much lower switching loss than the common buck converters using LDMOS transistors derived from the same CMOS platform. TCAD and IC circuit simulation are extensively used to explore this design concept for the case of a 7.5V to 1.5V 50MHz converter. The calculated power loss and MOSFET performance FOM are compared between the two solutions. A PCB level FCMC prototype using discrete low voltage MOSFETs is built and tested to validate this concept in lieu of a CMOS chip prototype. The physical limitation of this technical approach is also analyzed with extensive TCAD simulation.

Keywords—BCDMOS, DC/DC Converter, CMOS, Multilevel Converters, power IC

I. INTRODUCTION

DC/DC converter power ICs or Power Supply on Chip (PSoC) are widely used to deliver power to millions of digital ICs. A well-known example is Intel's FIVR (fully integrated voltage regulator) solution which actually integrates 2V-to-1V buck converters into the CPU chip itself [1]. Mixed signal custom BCDMOS technologies using LDMOS or EDMOS as high voltage switching devices are the mainstream choice for most power ICs [2][3]. However, integration of these high voltage devices requires major changes to the cost-effective standard CMOS process, resulting in cost and development cycle time penalty. It would be highly advantageous to facilitate high voltage switching with standard low-voltage CMOS transistors.

Intel's FIVR uses two stacked 1V CMOS transistors to support the 2V input voltage with one transistor biased at a half-rail gate voltage and the other actively switched. However, there is no clear path on how to extend this cascode configuration to more than two CMOS transistors for >2V operation. Recently, integrated 3-level flying capacitor multilevel converters (FCMCs) based on a CMOS

technology were reported [4]-[8]. However, the full benefits and fundamental limitation of the CMOS FCMC concept is not yet comprehensively discussed in the literature, particularly in terms of MOSFET performance and power losses. The following questions remain unanswered: Up to what input voltage level is the FCMC approach feasible for various CMOS technology nodes? What is the performance comparison between this approach and the common BCDMOS technology beside the apparent cost benefit?

This paper intends to study these critical issues with extensive circuit and device TCAD simulation. In addition, a 7.5V to 1.5V, 150mA, 4-level, FCMC PCB prototype using discrete low-voltage MOSFETs is built and characterized to experimentally validate this concept in lieu of a fabricated CMOS chip.

II. FLYING CAPACITOR MULTILEVEL DC/DC CONVERTER: TOPOLOGY AND CONTROL

Fig. 1 shows a 4-level FCMC buck topology and the gate control signals for the three complimentary NMOS pairs that we use for this study. The three complimentary switch pairs turn on and off at a certain duty cycle in sequence to charge and discharge the flying capacitors C1 and C2. C1 and C2 maintain a nominal voltage of 1/3 and 2/3 of the input voltage, respectively, effectively reducing the required

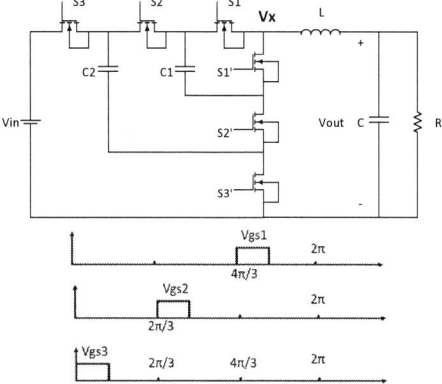

Fig. 1.   4-level Flying Capacitor Multilevel Converter (FCMC) buck topology and PWM gate control signals for the three NMOS pairs.

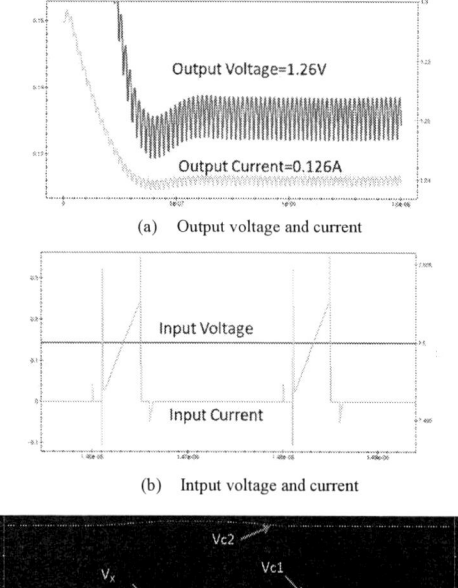

(a) Output voltage and current

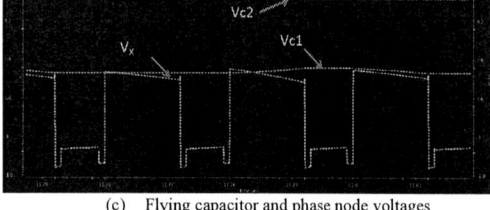

(b) Intput voltage and current

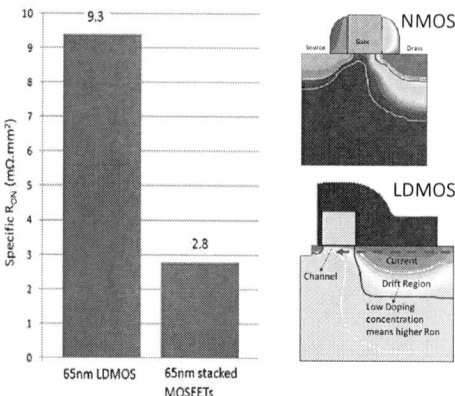

(c) Flying capacitor and phase node voltages

Fig. 2. Switching waveforms of the 4-level FCMC buck converter from Virtuoso circuit simulation.

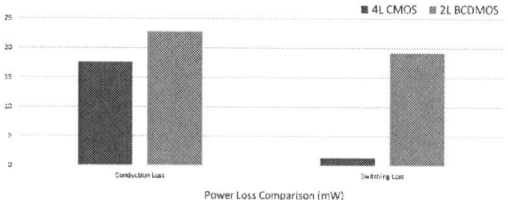

Fig. 3. Specific R_{DSON} comparison for between the LDMOS (BV of 11V) and NMOS (BV of 7.4V, two in stack) transistors based on TCAD simulation.

Fig. 4. Power loss comparison for the 4-level CMOS and 2-level BCDMOS buck converters (in mW) from Virtuoso simulation.

operating voltage of the six MOSFETs to only 1/3 of the input voltage. In a typical case study, we select the input and output voltages to be 7.5V and 1.5V, respectively, and an output current of 150mA. We can use the 2.5V I/O CMOS transistors available in a typical 65nm CMOS technology to implement such a DC/DC converter design. More generally, a FCMC uses N pairs of complementary NMOS switches to handle an input voltage up to N times the NMOS operation voltage (subject to the maximum PN junction breakdown voltage limit, as will be discussed later), and provide N+1 voltage levels at the Vx phase node with an equivalent frequency of N times the switching frequency of the individual NMOS switches. There would be N-1 flying back capacitors with each having a capacitor voltage of Vck=kVin/N. The flying capacitors store energy to control voltage Vx at the phase node.

III. POWER LOSS COMPARISON BETWEEN 4-LEVEL CMOS AND 2-LEVEL LDMOS BUCK CONVERTERS

The 7.5V to 1.5V, 150mA DC/DC converter power IC can be conventionally designed as a 2-level buck converter using BCDMOS technology derived from a baseline 65nm CMOS process. We will compare the total power loss on the active transistors between the conventional 2-level BCDMOS and the proposed 4-level CMOS design options using the Synopsis Virtuoso IC circuit simulation tool. For the 2-level buck converter, both the top and bottom switches are LDMOS transistors with a breakdown voltage of 11V, and a channel width of 1006 µm and 2012 µm, respectively. For the CMOS FCMC design, the top and bottom NMOS

transistors have a breakdown voltage of 7.4V, and a channel width of 450µm and 900 µm, respectively. The two solutions require more or less the same chip area to ensure fair comparison. Fig. 2 shows the Virtuoso simulation waveforms of the output voltage and current, input voltage and current, and flying capacitor and phase node voltages for the 4-level FCMC buck converter. The equivalent phase node PWM frequency is 50MHz, which is the same as the LDMOS switching frequency in the 2-level converter, but 3X the switching frequency of the individual NMOS transistors in the 4-level FCMC converter.

As shown in Fig.3, our TCAD simulation shows that the LDMOS and NMOS transistors (2 in stack) based on the 65nm CMOS baseline process have a specific R_{DSON} of 9.3 and 2.8 mΩmm², respectively. In this regard, the 7.4V NMOS solution shows a 3X advantage over the 11V LDMOS transistor. Fig. 4 shows the comparison of both conduction and switching power losses between the 4-level CMOS and 2-level LDMOS buck converters. The 4L CMOS case exhibits a conduction and switching loss of 17.6mW and 1.4mW respectively while the 2L LDMOS case exhibits 22.8mW and 19.2 mW. The CMOS converter efficiency is 94.3% against 79.2% for the LDMOS converter. This is because 1) the NMOS transistors in the 4L FCMC buck converter only switch at 1/3 of the 50MHz phase node frequency, and therefore introduce a much lower switching loss than the 2L buck converter; and 2) the NMOS transistors have a superior Q_G*R_{DSON} Figure of Merit (FOM) advantage over the LDMOS transistors. They have a lower conduction loss than the LDMOS transistors for the same total chip area.

Fig. 5. PCB-level 4L FCMC 7.5V to 1.5V, 150mA buck converter prototype using 8V discrete power MOSFETs.

TABLE I FCMC Converter Design Parameters

Parameter	Value
Input voltage Vin	7.5V
PWM frequency	500kHz
MOSFET	Si2342DS
Flying capacitors C1 and C2	2.2μF
Inductance L	22μF
Output capacitor C	2.2μF

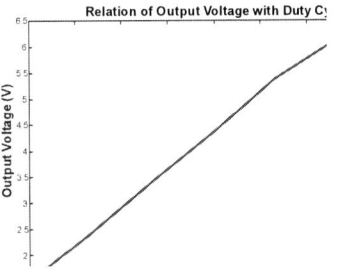

Fig. 6. Measured output voltage as a function of duty cycle at a load resistance of 66Ω.

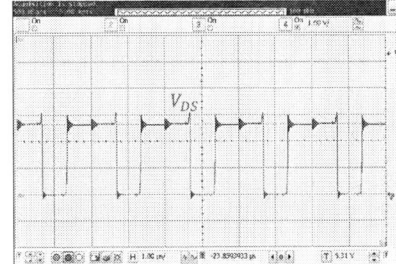

Fig. 7. Measured drain-source voltage switching waveform of NMOS transistor S2 at a duty cycle of 0.33 and a PWM frequency of 500kHz.

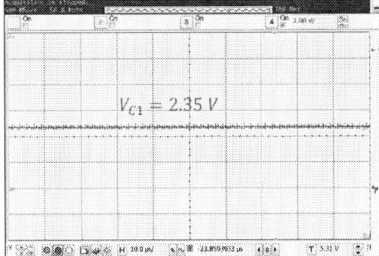

Fig. 8. Measured flying capacitor voltage waveform of C2 at a duty cycle of 0.33 and a PWM frequency of 500kHz.

TABLE II Influence of Flying Capacitances on Voltage Ripples

$C1$ (μF)	$C2$ (μF)	Ripple in V_{C1} (steady state voltage = 2.5 V)	Ripple in V_{C2} (steady state voltage = 5 V)	% Ripple in C1	% Ripple in C2
0.1μF	0.1μF	396.7 mV	403.2 mV	15%	8%
0.57μF	0.57μF	77.4 mV	88.8 mV	3.1%	1.8%
1.04μF	0.57μF	58.10 mV	98 mV	2.3%	1.9%
1.04μF	0.94μF	58.10 mV	64.5 mV	2.3%	1.3%
1.51μF	1.41μF	42 mV	63.0 mV	1.68%	1.26%
2.2μF	2.2μF	40 mV	62.1 mV	1.6%	1.24%
4.4μF	4.4μF	32.3 mV	54.4 mV	1.3%	1.08%

IV. EXPERIMENTAL STUDY

We have designed and built a PCB FCMC prototype using low-voltage discrete NMOS transistors in lieu of IC fabrication to validate the design concept. We use 8V/6A discrete MOSFETs (Si2342DS) to emulate the switching NMOS transistors and a DSP (TMS320f28335) to generate all the gate control signals in this prototype. Fig. 5 and Table 1 show the PCB FCMC prototype and design parameters, respectively. The PWM frequency of the discrete component prototype is limited to 500kHz instead of 50MHz in our simulation study due to noise caused by the PCB and package parasitic impendences. The output voltage of the FCMC can be controlled by varying the duty cycle of the PWM control signal from the DSP, as shown in Fig. 6. Figs. 7 and 8 show the measured switching waveforms of the drain-source voltage of NMOS transistor S2 and the flying capacitor C1 at a duty ratio of 0.33. S2 experiences a full voltage swing of 2.65V, and the voltage on C1 remains around 2.35V with some switching ripples.

It is important to minimize the ripples on the flying capacitors for stable operation of the FCMC. Selection of large C1 and C2 values helps reducing voltage ripples but unfortunately increases the size and the FCMC, and make monolithic integration of the FCMC using on-chip capacitors more difficult. Table II summarizes the measured voltage ripples on C1 and C2 for various C1 and C2 values. When C1 and C2 are both 2.2μF, their voltage ripples are 40mV (1.6% of V_{C1}) and 62.1mV (1.24% of V_{C2}), respectively. Even when C1 and C2 are reduced to 0.57μF, their voltage ripples are only 77.4mV (3.1% of V_{C1}) and 88.8mV (1.8% of V_{C2}), respectively at a PWM frequency of 500kHz. These results show a great promise to realize the FCMC design concept monolithically on a power IC chip using on-chip capacitors (in nF) with a PWM frequency well into a range of hundreds of MHz.

V. VOLTAGE LIMITATION

There is a maximum voltage limit of the CMOS multilevel converter concept, imposed by the P-well/deep N-well (DNW) PN junction D1 instead of the drain-source breakdown of individual NMOS transistors in a triple well CMOS process, as shown in Fig. 9. This is because the a DNW is always tied to the highest positive voltage in the

TABLE III: Typical NMOS Breakdown Voltages in 90nm/65nm/45nm CMOS Technology Nodes

	90nm			65nm			45nm		
	Logic NMOS	I/O NMOS	D1Diode	Logic NMOS	I/O NMOS	D1 Diode	Logic NMOS	I/O NMOS	D1Diode
BV (V)	≥3	7.42	>20	≥2	7.98	>20	≥2	7.62	>15

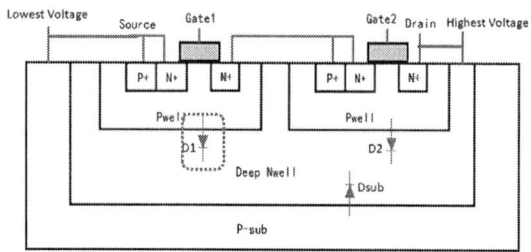

Fig. 9. A triple-well CMOS device structure showing PN junctions that limit the maximum operating voltage of the CMOS FCMC concept.

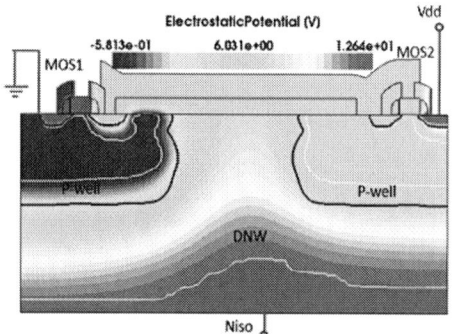

Fig. 10. Potential distribution of two NMOS transistors in a 65nm triple well CMOS process from TCAD simulation.

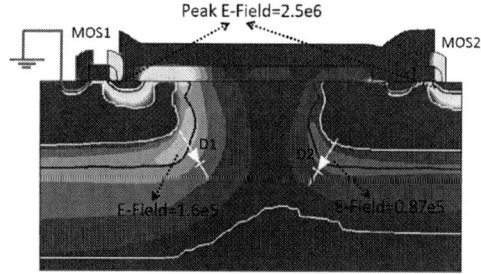

Fig. 11. Electric field distribution of two NMOS transistors in a 65nm triple well CMOS process from TCAD simulation.

system to ensure good isolation. The P-substrate/DNW diode typically has an even higher breakdown voltage than D1. Figs. 10 and 11 show the TCAD simulated electrostatic potential and electric field distributions of two NMOS transistors in a triple-well 65nm CMOS process, indicating the role of D1 (the P-well/DNW diode). Table III summarizes the typical D1 breakdown voltages for the logic and I/O NMOS transistors in common 90nm, 65nm, and 45nm CMOS technology nodes. For the 65nm node, the breakdown voltage of D1 is greater than 20V while the breakdown voltages for the logic and I/O NMOS transistors are specified greater than 2.0V and 7.98V, respectively. It is

concluded that the CMOS FCMC concept can facilitate DC/DC conversion for an input voltage up to 15V for the 90nm and 65nm CMOS nodes, and 12V for the 45nm CMOS node.

VI. SUMMARY

This paper investigates the feasibility and limitation of designing DC/DC converter power ICs using standard CMOS technology. Comparing to custom BCDMOS technologies, this path leads to lower cost and shorter development cycle time, benefiting from the high volume CMOS manufacturing ecosystem. Our study indicates that the flying capacitor multilevel converter (FCMC) topology can facilitate DC/DC conversion using standard CMOS transistors for an input voltage up to 15V for the 90nm/65nm CMOS nodes, and 12V for the 45nm CMOS node. The CMOS FCMC solution offers a much lower switching loss than the common buck converters using LDMOS transistors derived from the same CMOS platform. This concept does need more complex control than two-level buck converters, and additional flying capacitors which may be monolithically integrated on the power IC or externally connected.

ACKNOWLEDGMENT

This work was in part supported by the U.S. National Science Foundation under Grant EECS-1711485.

REFERENCES

[1] Edward A. Burton; Gerhard Schrom; Fabrice Paillet; Jonathan Douglas; William J. Lambert; Kaladhar Radhakrishnan; Michael J. Hill, "FIVR — Fully integrated voltage regulators on 4th generation Intel® Core™ SoCs," APEC2015, pp.432-439.

[2] Tsung-Yi, Huangl et al., "Demonstration of a HV BCD Technology with LV CMOS Process", 2015 IEEE 27th International Symposium on Power Semiconductor Devices & IC's (ISPSD), Hong Kong, China, 2015, pp. 193-196.

[3] H. Fujii, S. Tokumitsu et al., "A 90nm Bulk BiCDMOS Platform Technology with 15-80V LD-MOSFETs for Automotive Applications," 2017 IEEE 29th International Symposium on Power Semiconductor Devices & IC's (ISPSD), Sapporo, Japan, 2017, pp.73-76.

[4] Xun Liu, Philip K. T. Mok, Junmin Jiang and Wing-Hung Ki, "Analysis and Design Considerations of Integrated 3-Level Buck Converters", IEEE Transactions on Circuits and Systems I: Regular Papers, vol. 63 no. 5, 2016.

[5] Xun Liu, Cheng Huang and Philip K.T. Mok, "Dynamic Performance Analysis of 3-Level Integrated Buck Converters" 2015 IEEE International Symposium on Circuits and Systems (ISCAS), Lisbon, Portugal, 2015, pp. 2093-2096

[6] V. Yousefzadeh, E. Alarcon, and D. Maksimovic, "Three-level buck converter for envelope tracking applications," IEEE Trans. Power Electron., vol. 21, no. 2, pp. 549–552, Mar. 2006.

[7] W. Kim, D. Brooks, and G.-Y. Wei, "A fully-integrated 3-level DC-DC converter for nanosecond-scale DVFS," IEEE J. Solid-State Circuits, vol. 47, no. 1, pp. 206–219, Jan. 2012.

[8] G. Villar and E. Alarcon, "Monolithic integration of a 3-level DCM operated low-floating-capacitor buck converter for DC-DC step-down donversion in standard CMOS," in Proc. IEEE Power Electron. Specialists Conf., Rhodes, Greece, Jun. 2008, pp. 4229–4235.

Proceedings of the 31st International Symposium on Power Semiconductor Devices & ICs
May 19-23, 2019, Shanghai, China

Integrated Current Sensing in GaN Power ICs

Stefan Moench*, Richard Reiner*, Patrick Waltereit*,
Rüdiger Quay*, Oliver Ambacher*, Ingmar Kallfass†

*Fraunhofer Institute for Applied Solid State Physics, Freiburg, Germany (stefan.moench@iaf.fraunhofer.de)
†Institute of Robust Power Semiconductor Systems, University of Stuttgart, Germany (ingmar.kallfass@ilh.uni-stuttgart.de)

Abstract—Integrated current sensors at the drain and source of lateral GaN-on-Si power transistors are presented, using the existing resistive metal fingers of large-area comb-structures as shunts. In comb-structures, the finger current flows orthogonal to the channel current, thus the sensor signal is independent from the dynamic channel resistance. At 1 MHz pulsed switching of 75 V in a half-bridge converter, the drain and source sensors are characterized (around 4 mΩ shunt resistance as part of a 100 mΩ transistor) in both high-side and low-side configurations. The measured current sense-ratio temperature coefficient of 0.003/K results mainly from the finger metal and is correctable by integrated linear temperature sensors with a similar coefficient. A readout circuit which proportionally reproduces the continuous external inductor current is implemented by summation of the switched high-side source and low-side drain current sensor signals. Integrated current sensors enable control and protection of GaN HEMTs and ICs, for example in monolithic or co-packaged GaN half-bridges.

Keywords—gallium nitride, current measurement, sensors, power integrated circuits, HEMTs.

I. INTRODUCTION

Background: Control and protection of gallium nitride power integrated circuits (GaN Power ICs) in power converters requires high-bandwidth, low-intrusive, dc-coupled, temperature-compensated and accurate current measurements.

Problem: External current sensors such as hall-sensors, SMD resistors or coaxial shunts are bulky compared to GaN ICs, add circuit parasitics, and are difficult to insert, for example into monolithic or co-packaged half-bridges where the low-side and high-side transistors are already internally connected.

Approach: This work realizes integrated current sensing by measuring the voltage drop across the outermost metalized drain and source fingers of comb-structures in large-area lateral GaN-on-Si power transistors, as illustrated in Fig. 1.

II. GaN-ON-SI POWER IC

Fig. 2 shows the layout of the fabricated GaN Power IC with integrated drain and source current and temperature sensors as part of a large-area 600 V-class GaN-on-Si HEMT ($W_G = 191$ mm, $L_{GD} = 15$ μm) with intrinsic reverse Schottky-diode and gate-driver final stage ($W_{G,PU/PD} = 12/14$ mm, $L_{GD} = 3$ μm). This work extends the GaN-on-Si power technology in [1] by current sensors for source-side and drain-side sensing.

GaN Power IC
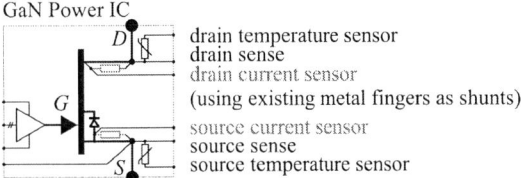

Fig. 1. GaN-on-Si HEMT with drain/source current/temperature sensors and drain/source sense-terminals, intrinsic reverse diode and gate driver final stage. Integrated in a 600 V-class GaN-on-Si Power IC (3×3 mm²).

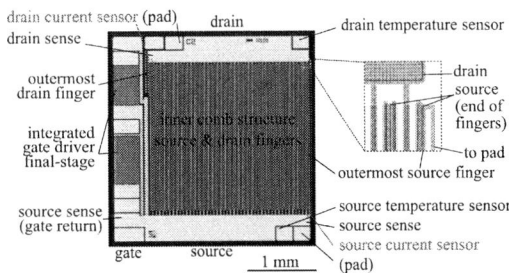

Fig. 2. Layout of fabricated GaN Power IC: Current sensing is realized by sensing the voltage drop across the outermost drain (or source) metal fingers.

A. Integrated Drain/Source Current Sensors

The main power transistor has an area-efficient large-area comb-structure [2]. The source/drain metal fingers of the comb-structure have non-zero series-resistances and contribute each around $< 5\%$ to the on-resistance. This work's current sensing approach is to measure the voltage across the already existing resistive drain/source fingers in comb-structures. No modification of the HEMT is required, except additional sense-traces and pads. The approach enables both drain-side and source-side current sensing, because only passive parts outside the channel are sensed.

B. Analysis

Fig. 3 shows a simplified schematic of the comb-structure. All drain/source fingers, including the outermost, have same trace width and length (as in the fabricated IC). Equations for the voltage drop across a metal finger [2] were simplified for typical large-area geometries, and modified for the outermost fingers: At the connection to the drain/source pads, the

978-1-7281-0582-6/19 $31.00 © 2019 IEEE   111

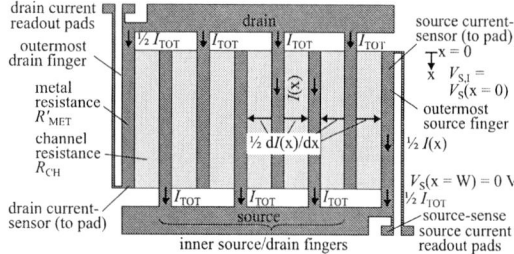

Fig. 3. Layout of comb-structure: The channel current flows orthogonal to the metal finger current, resulting in measurable voltage drop across the fingers proportional to the drain/source current.

outermost fingers carry only around half the current $\frac{1}{2}I_{\text{TOT}}$ compared to inner fingers (I_{TOT}), since no second channel is connected. In Fig. 2 and Fig. 3 the outermost finger on the right is source and on the left drain. Despite a slightly non-uniform current distribution across all $N = 51$ fingers, the analysis still results in a nearly linear dependence of sensed voltage $V_{\text{S,I}}$ from the finger current I_{TOT}. It follows $V_{\text{S,I}} = \frac{1}{4}I_{\text{TOT}}R'_{\text{MET}}W$ (W: width of one finger). In comb-structures, the channel current (sheet resistance R_{CH}) is nearly orthogonal to the finger current, and the channel resistance does not significantly influence the sensor signal. This is advantageous because dynamic on-resistance effects, gate-control, and drain-voltage dependence of the channel are decoupled from the measured signal. The integrated sensor is thus similar to an external shunt. Only temperature T increases the current sense ratio, due to the thermal coefficient α_{MET} of the metal resistance $R'_{\text{MET}}(T)$. Considering all N fingers, the sensors output voltage is proportional to the total drain/source current $V_{\text{S,I}} = R_{\text{SH}}I_{\text{D/S}}$, with a temperature dependent sense-ratio $R_{\text{SH}} = R_{\text{SH}}(T) = R_{\text{SH,0}}(1 + \alpha_{\text{MET}}(T - T_0))$.

C. Integrated Temperature Sensors

Linear temperature sensors [1] with $R_{\text{TEMP,0}} = 38.85\,\Omega$ at $T_0 = 25\,°\text{C}$ were added to the drain and source side referenced to the respective sense pads of the current sensors for temperature-compensation. The temperature sensors are resistive metal traces close and parallel to the outermost source/drain fingers. In an oven, the thermal coefficient of the temperature sensor $\alpha_{\text{TEMP}} = 3.1 \times 10^{-3}\,°\text{C}^{-1}$ was measured. Then at 20 A pulsed switching with external heating of the IC in the steady-state temperature range $25\,°\text{C}...98\,°\text{C}$, the temperature coefficient of the current sensor's sense-ratio $\alpha_{\text{MET}} = 3.0 \times 10^{-3}\,°\text{C}^{-1}$ was derived. α_{MET} and α_{TEMP} agree within $\pm 5\,\%$ which verifies that the current sense-ratio is temperature-dependent similar to the metal fingers, but not significantly influenced by the channel (which has much higher $\alpha = 10.4 \times 10^{-3}\,°\text{C}^{-1}$, calculated from V_{DS} increase).

D. Readout Circuits and Characterization of Sensors

Fig. 4 shows readout circuits used during pulsed measurements in a half-bridge. The low-side (LS) source sensor and

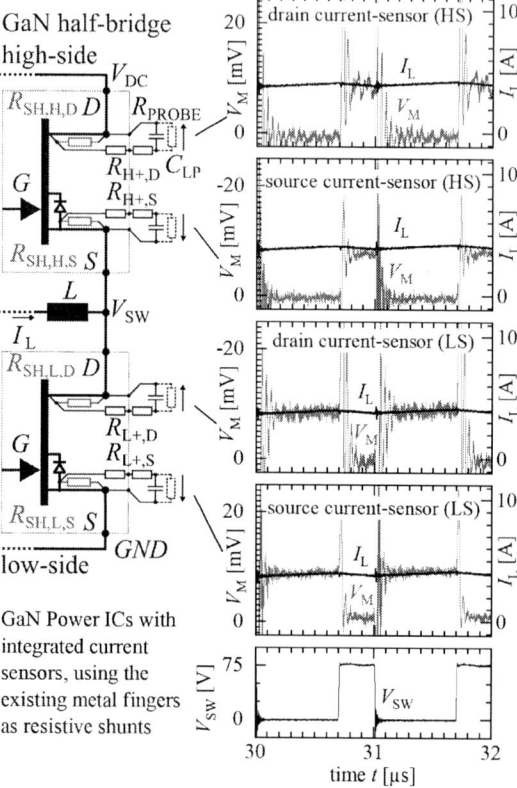

Fig. 4. Integrated drain and source current sensor responses in a GaN half-bridge. The readout circuits use low-pass filters.

the high-side (HS) drain sensor are referenced to a fixed voltage to ground (GND), relaxing the required common-mode transient immunity of connected probes, whereas the LS drain and HS source sensors are referenced to the switch node. 2.5D EM simulations of the comb-structure (including sense wires/pads) showed a high-pass response of the sensor output voltage to drain current (+3 dB at $\approx 100\,\text{MHz}$, resulting from the IC geometry which forms part of the power loop-inductance in a typical half-bridge PCB layout) which is compensated [3] by external RC low-pass filters.

At room-temperature the devices where characterized (four-point measurements): Integrated shunt-resistances $R_{\text{L,SH,S}} = 3.86\,\text{m}\Omega$, $R_{\text{H,SH,S}} = 3.92\,\text{m}\Omega$, $R_{\text{L,SH,D}} = 4.22\,\text{m}\Omega$, $R_{\text{H,SH,D}} = 4.41\,\text{m}\Omega$. The difference between the drain and source sensors is explained by different cross-sections of the vertical metal stack. The difference between the HS and LS devices is from galvanic process variations. The main power transistor has an on-resistance of $\approx 100\,\text{m}\Omega$. The integrated sense-traces also have series resistances: $R_{\text{L+(INT),S}} = 22\,\Omega$,

978-1-7281-0582-6/19 $31.00 © 2019 IEEE

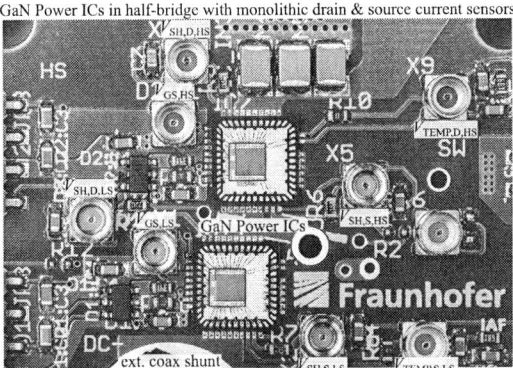

GaN Power ICs in half-bridge with monolithic drain & source current sensors

Fig. 5. Dc-dc half-bridge converter with two 600 V GaN Power ICs, each with integrated drain and source current sensors.

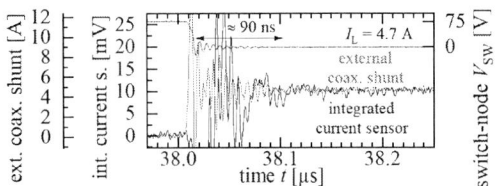

Fig. 6. Comparison: Integrated source current sensor vs. external coaxial shunt.

$R_{\mathrm{L+(INT),D}} = 29\,\Omega$, $R_{\mathrm{H+(INT),S}} = 22\,\Omega$, $R_{\mathrm{H+(INT),D}} = 30\,\Omega$, used as part of the low-pass filter. External filter components are $R_{\mathrm{L/H+(EXT),S/D}} = 22\,\Omega$ and $C_{\mathrm{LP}} = 330\,\mathrm{pF}$. This work used an isolated $50\,\Omega$ input probe (Tektronix TIVM1), which forms a voltage divider with the sense trace resistances and reduces the effective sense-ratio to

$$V_{\mathrm{M}} = I_{\mathrm{D/S}} R_{\mathrm{L/H,SH,D/S}} \frac{1}{1 + \frac{R_{\mathrm{L/H(INT+EXT),D/S}}}{R_{\mathrm{PROBE}}}}.$$

Nevertheless, the measurements showed similar sense-ratios $\approx 2.2\,\mathrm{mV/A}$ for all sensors, since the shunt resistance ratio $\frac{R_{\mathrm{SH,D}}}{R_{\mathrm{SH,S}}} \approx 1.11$ is compensated by the voltage divider ratio $(1 + \frac{R_{\mathrm{(INT+EXT),D}}}{R_{\mathrm{PROBE}}})/(1 + \frac{R_{\mathrm{(INT+EXT),S}}}{R_{\mathrm{PROBE}}}) \approx 1.08$. Instead of the $50\,\Omega$ probing used in this work, a high-impedance probing is advantageous since no external voltage divider is formed.

III. EXPERIMENTAL VERIFICATION

A. DC-DC Converter Demonstrator

Fig. 5 shows the assembled dc-dc converter, using two GaN ICs in 7 mm QFN packages, external gate drivers (UCC27511A, $R_{\mathrm{G,ON/OFF}} = 22/0\,\Omega$, bypassing the integrated gate driver final-stage in the GaN IC), and MMCX connectors to sense the gate, current, and temperature signals. A 100 mΩ coaxial source-current shunt as reference, a load inductor $L = 100\,\mu\mathrm{H}$ and external 10 mA temperature sensor bias-sources were added.

B. Drain and Source Current Measurements

Fig. 4 shows measured responses of the drain and source current sensors at room-temperature, during pulsed switching to avoiding self-heating with $V_{\mathrm{DC}} = 75\,\mathrm{V}$. The low-side sensors follow the inductor current during low-side conduction, and the high-side sensors during high-side conduction. The sense-ratio of $\approx 2.2\,\mathrm{mV/A}$ is reduced by the voltage divider of the probe compared to the shunt resistances as discussed previously. It is also observed that both the drain and source sensors show similar results, despite the fact that

one is referenced to a fixed voltage to ground and the other to the high dV/dt switch node. The settling time of the sensor signal is $< 90\,\mathrm{ns}$, which is also the time in which the switch-node voltage is still oscillating.

Fig. 6 compares the integrated LS source current sensor with the response of an external coaxial shunt. Both sensors show oscillations which decay after around 90 ns. Since the switch-node voltage V_{SW} is also still oscillating during this time, it should be further analyzed in future work what bandwidth can be achieved if V_{SW} is better damped.

C. Sum Current Measurement of Continuous Inductor Current

Control of converters often requires the continuous inductor current I_{L}, not only switched source/drain currents. Fig. 7 shows a readout circuit, which combines the high-side source and low-side drain sensor signals into a continuous measurement signal. The dimensioning

$$\frac{R_{\mathrm{H+(INT+EXT)}}}{R_{\mathrm{L+(INT+EXT)}}} = \frac{R_{\mathrm{H-}}}{R_{\mathrm{L-}}} = \frac{R_{\mathrm{SH,S}}}{R_{\mathrm{SH,D}}}$$

was determined, which equalizes the difference between the drain/source shunt resistances and also removes the external parasitic packaging resistances from the measured signal which is then proportionally to the continuous inductor current $V_{\mathrm{M}} = k I_{\mathrm{L}}$. The measured parasitics were $R_{\mathrm{P,H}} = 4.9\,\mathrm{m\Omega}$, $R_{\mathrm{P,L}} = 10.0\,\mathrm{m\Omega}$, and the added circuit: $R_{\mathrm{L-}} = 10.77\,\Omega$, $R_{\mathrm{H-}} = 10\,\Omega$, $R_{\mathrm{L+(EXT)}} = 34.5\,\Omega$, $R_{\mathrm{H+(EXT)}} = 37\,\Omega$, $C_{\mathrm{LP}} = 1.1\,\mathrm{nF}$. The oscilloscope was additionally bandwidth limited to 20 MHz. Fig. 8 shows the response of the sum current sensor at 250 kHz pulsed switching.

IV. RELATED WORK

A GaN-integrated current sensor using additional sense electrodes between gate/source was demonstrated in [4], based on a modified HEMT structure and partially dependent on the (dynamic) channel resistance. GaN-integrated [5] or externally realized [6] current-mirrors with virtual-grounding require fast (external) amplifiers. Other approaches are hall-sensors [7], flux concentrators [8] magnetic sensors [9] or cascodes [10], all requiring additional area or amplifiers. External current sensor approaches are well known [11], and might be also integrated or added to GaN ICs (AMR [12], TMR [13], magnetic field [14], Rogowski coils [15]–[18]).

978-1-7281-0582-6/19 $31.00 © 2019 IEEE

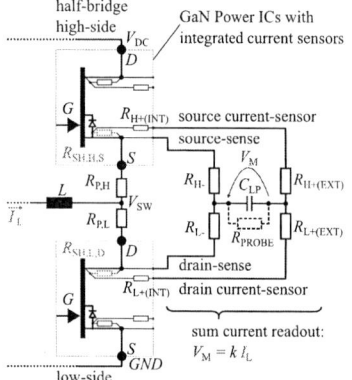

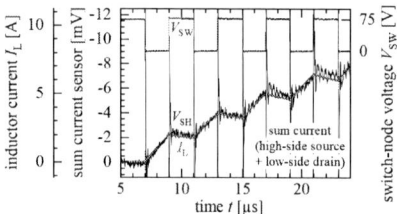

Fig. 7. The continuous inductor current is reproduced by summation of the switched high-side source and low-side drain current sensor signals.

Fig. 8. Integrated sum current sensor response.

V. CONCLUSION

Integrated current sensors enable fully-integrated GaN power converters, and simplify current-control and protection of GaN HEMTs. This work's current sensing approach is shunt-based, measuring already existing passive parts of large-area comb-structures. No modification of the HEMT structure or additional active circuits are required, making this approach a simple solution to realize integrated current sensing in GaN Power ICs.

ACKNOWLEDGMENT

The author thanks Dirk Meder for wire-bonding and assembly of the GaN ICs. The authors thank the staff at Fraunhofer IAF who was involved in the epitaxy, fabrication, and characterization of the devices.

This work was supported by the German Federal Ministry of Education and Research (BMBF) through grant "GaNIAL" (FKZ: 16EMO215K) and "MIIMOSYS" (FKZ: 16ES0670). This work was supported by the German Federal Ministry of Defence (BMVg), Bundeswehr Technical Center for Information Technology and Electronics (WTD 81), and Federal Office of Bundeswehr Equipment, Information Technology and In-Service Support (BAAINBw) within the project Subsys. We kindly acknowledge the support of the Research Fab Microelectronics Germany (FMD) funded by the Federal Ministry of Education and Research (BMBF).

REFERENCES

[1] R. Reiner, P. Waltereit, B. Weiss, S. Moench, M. Wespel, S. Müller, R. Quay, and O. Ambacher, "Monolithically integrated power circuits in high-voltage GaN-on-Si heterojunction technology," *IET Power Electronics*, vol. 11, no. 4, pp. 681–688, 2018.

[2] R. Reiner, F. Benkhelifa, D. Krausse, R. Quay, and O. Ambacher, "Simulation and analysis of low-resistance AlGaN/GaN HFET power switches: Proceedings of the 2011 14th European Conference on Power Electronics and Applications," 2011.

[3] C. M. Johnson and P. R. Palmer, "Current measurement using compensated coaxial shunts," *IEE Proceedings - Science, Measurement and Technology*, vol. 141, no. 6, pp. 471–480, 1994.

[4] R. Sun, Y. C. Liang, Y. Yeo, and C. Zhao, "Au-Free AlGaN/GaN MIS-HEMTs With Embedded Current Sensing Structure for Power Switching Applications," *IEEE Transactions on Electron Devices*, vol. 64, no. 8, pp. 3515–3518, 2017.

[5] J. Roberts, G. Klowak, D. Chen, and A. Mizan, "Drive and protection methods for very high current lateral GaN power transistors: 2015 IEEE Applied Power Electronics Conference and Exposition (APEC)," 2015.

[6] M. Biglarbegian and B. Parkhideh, "Characterization of SenseGaN current-mirroring for power GaN with the virtual grounding in a boost converter: 2017 IEEE Energy Conversion Congress and Exposition (ECCE)," 2017.

[7] T. P. White, S. Shetty, M. E. Ware, H. A. Mantooth, and G. J. Salamo, "AlGaN/GaN Micro-Hall Effect Devices for Simultaneous Current and Temperature Measurements From Line Currents," *IEEE Sensors Journal*, vol. 18, no. 7, pp. 2944–2951, 2018.

[8] N. Poluri and M. M. D. Souza, "An Integrated On-Chip Flux Concentrator for Galvanic Current Sensing," *IEEE Electron Device Letters*, vol. 39, no. 11, pp. 1752–1755, 2018.

[9] P. Igic, N. Jankovic, J. Evans, M. Elwin, S. Batcup, and S. Faramehr, "Dual-Drain GaN Magnetic Sensor Compatible With GaN RF Power Technology," *IEEE Electron Device Letters*, vol. 39, no. 5, pp. 746–748, 2018.

[10] M. Rose, Y. Wen, R. Fernandes, R. van Otten, H. J. Bergveld, and O. Trescases, "A GaN HEMT driver IC with programmable slew rate and monolithic negative gate-drive supply and digital current-mode control: 2015 IEEE 27th International Symposium on Power Semiconductor Devices & IC's (ISPSD)," 2015.

[11] S. Ziegler, R. C. Woodward, H. H. Iu, and L. J. Borle, "Current Sensing Techniques: A Review," *IEEE Sensors Journal*, vol. 9, no. 4, pp. 354–376, 2009.

[12] A. Lauer, S. J. Nibir, M. Biglarbegian, M. Hiller, and B. Parkhideh, "On Integrating Non-Intrusive Current Measurement into GaN Power Modules: 2018 IEEE 6th Workshop on Wide Bandgap Power Devices and Applications (WiPDA)," 2018.

[13] N. Tröster, T. Eisenhardt, M. Zehelein, J. Wölfle, J. Ruthardt, and J. Roth-Stielow, "Improvements of a Coaxial Current Sensor with a Wide Bandwidth Based on the HOKA Principle: 2018 20th European Conference on Power Electronics and Applications (EPE'18 ECCE Europe)," 2018.

[14] J. Wang, M. H. Hedayati, D. Liu, S. Adami, H. C. P. Dymond, J. J. O. Dalton, and B. H. Stark, "Infinity Sensor: Temperature Sensing in GaN Power Devices using Peak di/dt: 2018 IEEE Energy Conversion Congress and Exposition (ECCE)," 2018.

[15] Y. Xue, J. Lu, Z. Wang, L. M. Tolbert, B. J. Blalock, and F. Wang, "A compact planar Rogowski coil current sensor for active current balancing of parallel-connected Silicon Carbide MOSFETs: 2014 IEEE Energy Conversion Congress and Exposition (ECCE)," 2014.

[16] K. Wang, X. Yang, H. Li, L. Wang, and P. Jain, "A High-Bandwidth Integrated Current Measurement for Detecting Switching Current of Fast GaN Devices," *IEEE Transactions on Power Electronics*, vol. 33, no. 7, pp. 6199–6210, 2018.

[17] J. Walter, J. Acuna, and I. Kallfass, "Design and Implementation of an Integrated Current Sensor for a Gallium Nitride Half-Bridge: PCIM Europe 2018; International Exhibition and Conference for Power Electronics, Intelligent Motion, Renewable Energy and Energy Management," 2018.

[18] T. Funk and B. Wicht, "A fully integrated DC to 75 MHz current sensing circuit with on-chip Rogowski coil: 2018 IEEE Custom Integrated Circuits Conference (CICC)," 2018.

Proceedings of the 31st International Symposium on Power Semiconductor Devices & ICs
May 19-23, 2019, Shanghai, China

500°C SiC-based driver IC for SiC power MOSFETs

Saleh Kargarrazi, Hossein Elahipanah, Zikang Tong, Debbie Senesky, and Carl-Mikael Zetterling

Abstract—This paper reports on a SiC BJT-based driver IC for driving SiC power MOSFETs operational from room temperature up to 500°C. The driver features design simplicity, smaller chip size, smaller propagation delay, and relatively high driving currents compared to similar SiC-based drivers. These properties are due to its fewer number of gain buffer stages and relatively higher transconductance of the BJTs as compared to SiC MOSFETs. The high-temperature operation of the driver, as well as the other advantages of the SiC BJT-based driver, suggest it as a viable candidate to be integrated with SiC MOSFETs in a power module.

Index Terms—High temperature electronics, silicon carbide (SiC), driver IC, bipolar junction transistor (BJT).

I. INTRODUCTION

Wide-bandgap (WBG) power devices, namely, silicon carbide (SiC) and gallium nitride (GaN), have gained attentions in power electronics in the last two decades, thanks to their excellent electrical and thermal properties. In spite of the WBG device developments in commercial level, little works have been done in the development of integrated circuits in these technologies. Recently, Navitas Semiconductor [1], provides GaN-based ICs, taking advantage of the lateral GaN structure, with the aim to increase switching speed, power density, power efficiency, and finally reduce the overall power conversion cost. Power devices can benefit from the company of driver circuits in their close proximity in order to minimize the stray parasitic components and thereby reduce the power loss. On the research level, much research has been conducted on the design and integration of driver circuits together with the SiC power devices in a power converter system [2]–[5]. As of today, the commercial drivers used for SiC and GaN power devices are Si based, and are not rated for operation above 150°C. A SiC-based driver for SiC devices will make power electronic systems compatible with high temperature environments.

In earlier studies, before wide-bandgap research starts to gain momentum, silicon-on-insulator (SOI) has been a candidate for designing high-temperature drivers [6], [7]. Owing to the maturity of SOI technology in those studies, several protective features have been also implemented, among which are thermal protection, under-voltage lock out (UVLO), and short circuit protection. However limitation of the operation at temperatures higher than 225°C has encouraged the researchers to consider shifting towards wide-bandgap technologies, especially SiC. In 2014, NMOS SiC drivers have been reported in [3], [4]. In [4], operation of a MOSFET driver up to 400°C, in the presence of cooling system is discussed. Furthermore, [5] is the most recent gate driver microfabricated

in CMOS SiC technology and tested up to 500°C. Owing to the advances in the high-temperature SiC BJT technology [8] development, Schmitt triggers [9], TTL logic [10], operational amplifiers [11], [12], linear voltage regulators [13], [14], SiC lateral power BJTs [15], high-temperature passives [16], active down-conversion mixer [17], and the most recently a high-temperature pulse-width-modulation (PWM) IC [18] have been demonstrated. In this paper we discuss the design, layout specifications, and measurement results of a driver IC in SiC bipolar technology as a high-temperature candidate to drive power MOSFETs. The driver features simple design, and no need for multi-level stages as in [5], nor additional power supplies for overdriving the SiC MOSFET as in [3]. These advantages are essentially the result of the relatively high transconductance of the SiC BJTs, low ON-resistance, and finally the design of the driver, which is explained in the next section. Furthermore, this driver is able to operate successfully up to 500°C without any cooling system, which indicates the possibility of a more reliable operation while operated at lower ambient temperatures.

II. SiC DRIVER IC

A driver circuit interfaces the signal processing/control circuitry and the power device. It has the role of an amplifier for the control signal, and has to maintain sufficient power (voltage and current) to drive the power switch. Similar to Si devices (MOSFETs and IGBTs), SiC MOSFETs can be driven with the widely used push-pull, also known as totem-pole, drivers. SiC MOSFETs are shown to have lower transconductance (g_m) than Si MOSFETs and IGBTs and therefore, turn-ON and OFF occurs at higher drain-source (V_{DS}) voltages. Besides, SiC MOSFETs neither require a constant current in ON state compared to SiC BJTs, nor a complex driver circuitry like that of SiC JFETs. Still, due to their lower transconductance as compared to Si MOSFETs and IGBTs, and similar to SiC JFETs, they demand higher turn-ON voltages (about 20 V) [19]. For example, datasheet of Wolfspeed SiC 900 V MOSFET [20] recommends V_{GS}=18 V for turn-ON. Other manufacturers recommend $V_{GS} = 20$ V or even beyond 20 V [21]–[24].

In this paper, we demonstrate a driver IC in SiC BJT technology that is composed of NPN transistors and resistors. It has to be noted that this driver is the next generation of the driver previously designed and fabricated in SiC BJT technology [2] in order to drive SiC BJTs. As compared to [2], the driver IC of this paper (Fig. 1), is designed for reduced loading effect on the preceding control circuits. It has

978-1-7281-0582-6/19 $31.00 © 2019 IEEE   115

been shown that this driver could be controlled using a PWM designed in the same SiC BJT technology [18]. Furthermore, the driver is designed to provide higher current and voltage levels than [2] in order to drive various power switches. The input stage is designed in Darlington configuration, for higher input impedance and lower required base current. It has to be noted that Darlington configuration in lateral SiC BJTs have been previously shown to be a useful technique specially for achieving high currents [15]. Moreover, the driver output stage is accompanied by a pre-driver stage (Q_4), biased through R_{B1} and R_{B2}, which helps to increase the current and therefore the slew rate of the output pull-up transistor, Q_{P1}, and improves the switching performance of the output transistors (Q_{P1} and Q_{P2}). Particularly, unlike [2], Q_{P1} in this driver fully enters the saturation region, which reduces the total power consumption of the output stage.

The schematic view and microphotograph of the driver is shown in Fig. 1. Using the same orientation of the BJTs helps to alleviate the process variation. The output stage is composed of 10 paralleled BJTs for pull-up and pull-down sides, which enables higher currents. The SiC BJT fabrication process has been discussed in detail in [8], [15]. The driver IC uses the same chip area as of [2], despite its higher number of BJTs.

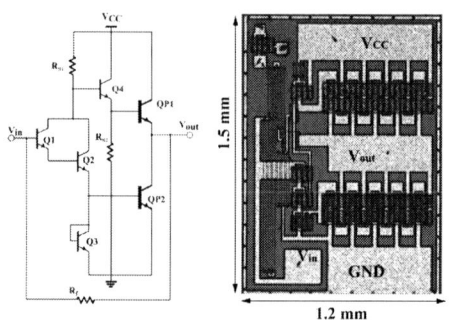

Fig. 1. The schematics of the SiC BJT push-pull (totem-pole) driver IC and the microphotograph of the fabricated chip after 500°C measurement.

III. EXPERIMENTAL RESULTS AND DISCUSSIONS

A. High-temperature measurements

The wafer was placed on a temperature-controlled probe station, where the chuck was exposed at different temperatures from 25°C up to 500°C.

The forward current gain (β) of the single and paralleled BJTs were measured at different temperatures. The peak of the β occurs at the collector currents of 20 mA and 50 mA for the single and paralleled BJTs, respectively. However, the peak of β for the paralleled BJTs has much less variation for the I_C range of 50m A-100 mA as shown in Fig. 2(a), which results in the constancy of the current in the output stage over biasing point variation. Furthermore, the sheet resistance of the collector layer, in which the resistors were realized in, was measured (Fig. 2 (b)). The non-monotonous behavior of

the sheet resistance originates from two opposing phenomena: increasing dopants' ionization degree over temperature and decreasing mobility due to ionized dopants' scattering.

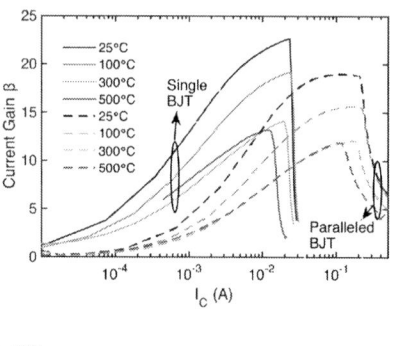

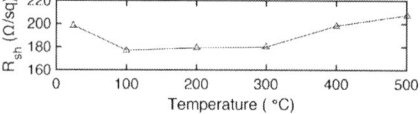

Fig. 2. (Top) Forward current gain of the single BJT and paralleled BJTs in the output stage, (bottom) collector sheet resistance variation over temperature.

The high-temperature tests of the driver IC were performed while the driver IC was placed on the hot-chuck, and connected to a 1.2 kV, 14 A SiC MOSFET (ROHM, SCT2280KE) using a 1-m coaxial cable. A 60 kHz pulse varying between 0 and 12 V was generated using a TMS320F28335 microcontroller, followed by a level-shifter and applied to the input of the driver. The driver IC was supplied with 20 V and the SiC MOSFET was set to switch 250 V (Fig. 3). Despite the slight variation of the voltage levels, the driver IC shows a very robust operation over temperature in the entire temperature range from 25°C up to 500°C (Fig. 4).

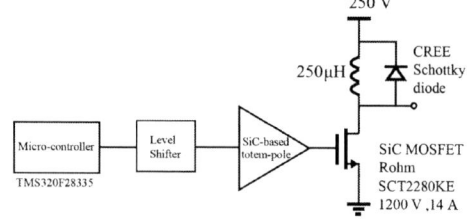

Fig. 3. The measurement setup to drive a 1200 V, 14 A power MOSFET ($SCT2280KE$) using the bipolar SiC driver IC. The input pulse is generated using a TI micro-controller.

B. Driver/switch module

In the second setup, the driver IC was wire-bonded to a SiC CREE MOSFET (CPM2-1200-0025B) on a Spectrum CPG04401 package, which was mounted on a FR4 PCB

978-1-7281-0582-6/19 $31.00 © 2019 IEEE   116

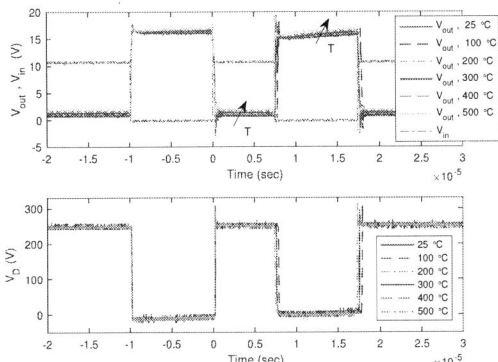

Fig. 4. Measured switching waveform of the bipolar SiC driver IC driving a *Rohm* 1200 V / 14 A SiC power MOSFET *SCT2280KE*, the drain voltage of the power MOSFET (V_D).

(Fig. 5), as an attempt to reduce the stray parasitic inductance. The input pulse was varying between 0 and 14 V and the driver IC was supplied with 25 V. Fig. 6 illustrates the switching operation of the driver at 500 kHz, 1 MHz, and 2 MHz at room temperature, having the drain of the SiC MOSFET floated. Based on the results from Fig. 6, the switching performance of the driver is limited to approximately 2 MHz, as the output voltage becomes triangular, with the rise and fall times dominating the majority of the period. Since the gate driver is developed using SiC BJTs, the intent is not for high-frequency (3-30 MHz) resonant converter applications, where GaN is favored over SiC. Rather, this driver is more suitable for the high power converter space, switching at tens to hundreds of kHz, and where SiC dominates in. It is evident from the 500 kHz and 1 MHz waveforms that the driver functions as appropriate in this frequency regime.

The rise and fall times (t_r, t_f) of 280 ns and 185 ns were recorded for driving CPM2-1200-0025B, with the input capacitance (C_{iss}) of 2.8 nF. The capability of the driver to switch a power device is defined by two factors. The first factor is the available current drive and the second, the equivalent capacitor (C_{eq}) at the output of the driver (or input of the power device). When the power device is sufficiently larger than the pull-up and pull-down transistors of the driver, which is the case of this work (MOSFET: 26 mm², driver's output BJTs: 0.091 mm²), we can assume that $C_{eq} = C_{iss}$. The other factor that contributes to the switching speed is the pull-up and pull-down currents (I_{rise} and I_{fall}) at the output of the driver. I_{rise} and I_{fall} are defined as:

$$I_{rise} = C \cdot \frac{V_{GS}}{t_r} \tag{1}$$

,

$$I_{fall} = C \cdot \frac{V_{GS}}{t_f} \tag{2}$$

where V_{GS} is the voltage at the input of the SiC MOSFET. Since the driver must be capable to drive various sizes of SiC

MOSFETs, these currents have to be high enough to enable high-speed switching. Thanks to the higher transconductance of the SiC BJTs, as compared to SiC NMOS and CMOS devices, relatively high currents could be achieved with simpler design with fewer gain stages, and thus smaller chip size and cost. This current could be calculated from the measured rise and fall times and the C_{iss} of the SiC MOSFET. Also, since there is no need for multiple buffer stages for current amplification, the propagation delays (t_{pLH} and t_{pHL}) are an order of magnitude smaller than the SiC CMOS counterpart [5].

Table. I benchmarks this SiC-based IC against other similar SiC-based driver ICs. I_{rise} and I_{fall} have been calculated taking into account the C_{iss} of the selected SiC MOSFETs in each study, the V_{GS} available for the SiC MOSFETs and the reported t_r and t_f at room temperature. The blank parts in Table. I indicates the data where either were not available or the cases where a fair comparison was not possible. For example, either because of the different test setups [3], lack of information about the model and size of the SiC power device [4], or different nature of selected power device (SiC BJT in case of [2]).

TABLE I
COMPARISON OF THIS WORK WITH RECENT WIDE-BANDGAP DRIVER ICs IMPLEMENTED IN DIFFERENT SiC-BASED TECHNOLOGIES

	SiC NMOS [3]	SiC NMOS [4]	SiC CMOS [5]	SiC BJT [2]	SiC BJT (This work)
Operating Temperature	25	$25 - 420°C$	$25 - 500°C$	$25 - 500°C$	$25 - 500°C$
I_{rise}, I_{fall}	-	-	310 , 495 mA	-	180 , 270 mA
Chip size (Push-pull)	$\sim 6.9\ mm^2$	-	$\sim 9\ mm^2$	$\sim 1.8\ mm^2$	$\sim 1.8\ mm^2$

In future, the pull-up and pull-down devices could be improved for higher current density, by taking advantage of the improved design for the lateral BJTs, similar to the lateral device previously designed and demonstrated in [14], [15].

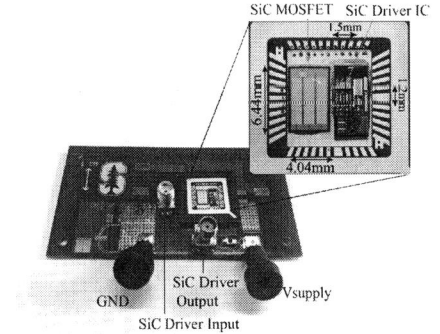

Fig. 5. SiC-based driver and SiC MOSFET low-temperature test circuit board.

IV. CONCLUSIONS

This paper demonstrates a high-temperature driver IC in BJT 4H-SiC technology capable of driving SiC MOSFETs.

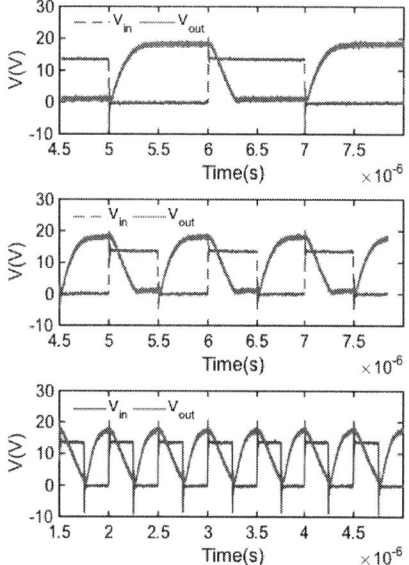

Fig. 6. SiC-based driver, driving a CREE 1.2 kV SiC MOSFET (CPM2-1200-0025B) at 500 kHz, 1 MHz, and 2 MHz at room temperature. The blue and red lines indicate the input and output of the driver, respectively.

The measurement results show a stable and robust measured performance of the driver IC from 25°C up to 500°C. Furthermore, to investigate the driver IC as a part of a power module, the driver IC and a SiC MOSFET were wirebonded and mounted on a test circuit board. The switching performance of the driver is benchmarked against the other SiC-based drivers of power switches, showing the advantages of the this work in terms of simplicity, chip size, and smaller propagation delays.

ACKNOWLEDGMENT

The authors would like to acknowledge the support from Knut and Alice Wallenberg Foundation, and advice from Prof. Jim Plummer and Prof. Juan Rivas-Davila.

REFERENCES

[1] Navitas Semiconductor, https://www.navitassemi.com/.
[2] S. Kargarrazi, L. Lanni, A. Rusu, and C.-M. Zetterling, "A monolithic SiC drive circuit for SiC Power BJTs," in *Proceedings of the International Symposium on Power Semiconductor Devices and ICs*, vol. 2015-June, 2015, pp. 285–288.
[3] N. Ericson, S. Frank, C. Britton, L. Marlino, S. H. Ryu, D. Grider, A. Mantooth, M. Francis, R. Lamichhane, M. Mudholkar, P. Shepherd, M. Glover, J. Valle-Mayorga, T. McNutt, A. Barkley, B. Whitaker, Z. Cole, B. Passmore, and A. Lostetter, "A 4H silicon carbide gate buffer for integrated power systems," *IEEE Transactions on Power Electronics*, vol. 29, no. 2, pp. 539–542, 2014.
[4] R. R. Lamichhane, N. Ericsson, S. Frank, C. Britton, L. Marlino, A. Mantooth, M. Francis, P. Shepherd, M. Glover, S. Perez, T. McNutt, B. Whitaker, and Z. Cole, "A wide bandgap silicon carbide (SiC) gate driver for high-temperature and high-voltage applications," *Proceedings of the International Symposium on Power Semiconductor Devices and ICs*, pp. 414–417, 2014.

[5] M. Barlow, S. Ahmed, A. M. Francis, and A. H. Mantooth, "An Integrated SiC CMOS Gate Driver for Power Module Integration," *IEEE Transactions on Power Electronics*, vol. 8993, no. c, pp. 1–1, 2019. [Online]. Available: https://ieeexplore.ieee.org/document/8644037/
[6] M. A. Huque, S. K. Islam, L. M. Tolbert, and B. J. Blalock, "A 200 C universal gate driver integrated circuit for extreme environment applications," *IEEE Transactions on Power Electronics,*, vol. 27, no. 9, pp. 4153–4162, 2012.
[7] J. Valle-Mayorga, C. P. Gutshall, K. M. Phan, I. Escorcia-Carranza, H. Mantooth, B. Reese, M. Schupbach, and A. Lostetter, "High-temperature silicon-on-insulator gate driver for SiC-FET power modules," *IEEE Transactions on Power Electronics*, vol. 27, no. 11, pp. 4417–4424, 2012.
[8] L. Lanni, R. Ghandi, B. Malm, C.-M. Zetterling, and M. Ostling, "Design and Characterization of High-Temperature ECL-Based Bipolar Integrated Circuits in 4H-SiC," *Electron Devices, IEEE Transactions on*, vol. 59, no. 4, pp. 1076–1083, 2012.
[9] S. Kargarrazi, L. Lanni, and C.-M. Zetterling, "Design and characterization of 500 °C Schmitt trigger in 4H-SiC," in *Materials Science Forum*, vol. 821. Trans Tech Publications, 2015, pp. 897–901.
[10] M. Shakir, S. Hou, B. G. Malm, M. Ostling, and C.-M. Zetterling, "A 600 °C TTL-based 11-stage Ring Oscillator in Bipolar Silicon Carbide Technology," *IEEE Electron Device Letters*, pp. 1–1, 2018. [Online]. Available: https://ieeexplore.ieee.org/document/8429909/
[11] S. Kargarrazi, L. Lanni, and C. M. Zetterling, "A study on positive -feedback configuration of a bipolar SiC high temperature operational amplifier ," *Solid-State Electronics*, vol. 116, pp. 33 – 37, 2016. [Online]. Available: http://www.sciencedirect.com/science/article/pii/S0038110115003470
[12] M. Pourreza and S. Kargarrazi, "A Metaheuristic Approach for an Optimized Design of a Silicon Carbide Operational Amplifier," in *Integral Methods in Science and Engineering, Volume 2: Practical Applications*. Cham: Springer International Publishing, 2017, pp. 211–219.
[13] S. Kargarrazi, L. Lanni, S. Saggini, A. Rusu, and C.-M. Zetterling, "500 °c bipolar SiC linear voltage regulator," *IEEE Transactions on Electron Devices*, vol. 62, no. 6, 2015.
[14] S. Kargarrazi, H. Elahipanah, S. Rodriguez, and C.-M. Zetterling, "500 °c, High Current Linear Voltage Regulator in 4H-SiC BJT Technology," *IEEE Electron Device Letters*, vol. 39, no. 4, 2018.
[15] H. Elahipanah, S. Kargarrazi, A. Salemi, M. Ostling, and C.-M. Zetterling, "500 °c high current 4h-sic lateral bjts for high-Temperature integrated circuits," *IEEE Electron Device Letters*, vol. 38, no. 10, 2017.
[16] J. Colmenares, S. Kargarrazi, H. Elahipanah, H.-P. Nee, and C.-M. Zetterling, "High-temperature passive components for extreme environments," in *WiPDA 2016 - 4th IEEE Workshop on Wide Bandgap Power Devices and Applications*, 2016.
[17] M. W. Hussain, H. Elahipanah, J. E. Zumbro, S. Schroder, S. Rodriguez, B. G. Malm, H. A. Mantooth, and A. Rusu, "A 500 c Active Down-Conversion Mixer in Silicon Carbide Bipolar Technology," *IEEE Electron Device Letters*, vol. 39, no. 6, pp. 855–858, 2018.
[18] S. Kargarrazi, H. Elahipanah, S. Saggini, D. Senesky, and C. Zetterling, "500 °c sic pwm integrated circuit," *IEEE Transactions on Power Electronics*, vol. 34, no. 3, pp. 1997–2001, March 2019.
[19] D. Peftitsis and J. Rabkowski, "Gate and Base Drivers for Silicon Carbide Power Transistors: An Overview," *IEEE Transactions on Power Electronics*, vol. 31, no. 10, pp. 7194–7213, 2016.
[20] Wolfspeed, http://www.wolfspeed.com/.
[21] ROHM Semiconductor, http://www.rohm.com/web/global/.
[22] Littlefuse, https://www.littelfuse.com/.
[23] Global Power, http://gptechgroup.com/.
[24] ST Microelectronics, https://www.st.com/.

978-1-7281-0582-6/19 $31.00 © 2019 IEEE

Proceedings of the 31st International Symposium on Power Semiconductor Devices & ICs
May 19-23, 2019, Shanghai, China

A Total Ionizing Dose Detecting Circuit Based on Off-state Leakage Current of NLDMOS in Power IC

Ping Luo, Rongxun Ling, Xiao Zhou, Pengkai Jiang, Yucao Wu
State key Laboratory of Electronic Thin Films and Integrated Devices
University of Electronic Science and Technology of China
Chengdu, China
Email: pingl@uestc.edu.cn

Abstract—**Due to the demand of aerospace applications, radiation-hard techniques have become critical for power IC. In order to compensate the total ionizing dose (TID) effect on the circuits, detecting it is the precondition. The TID detecting circuit proposed in this paper is able to monitor the TID effect on power IC accurately in the radiated environment. Moreover, it can be integrated with conventional power IC in an ordinary commercial process. Namely, the signals generated by the detecting circuit can be applied to TID compensation for power IC flexibly. The simulation results of the detecting circuit can meet the design requirements accurately from radiation experiment.**

Keywords—*detecting circuit, total ionizing dose, NLDMOS, off-state leakage current*

I. INTRODUCTION

Due to more and more demands of operating in a radiation environment such as aerospace applications and nuclear industry, radiation-hard techniques have become increasingly critical for integrated circuit. The most general types of radiation effects can be classified into total ionizing dose (TID) and single event effects (SEE). Exposure to ionizing particles would lead to significant performance degradation. To make matters worse, this degradation of circuit system is hard to predict because of the randomness of radiation.

Owing to withstand high voltage, the oxide insulator of power IC is pretty thick universally. Because oxide is the most sensitive part to ionizing particles [1], power IC is affected by the TID effects obviously, which would cause system anomalies. In order to compensate the TID effect on the circuits, detecting it is the precondition.

Some discrete devices to measure ionizing dose have been developed before, such as Geiger counter, photodiode, thermos luminescent dosimeter (TLD), radiation sensitive FET (RADFET) [2]-[5]. Nevertheless, few of these discrete devices can be integrated with conventional power IC in an ordinary commercial process. Moreover, the output signals can merely reflect the real-time radiation situation rather than the dose accumulation, which cannot be utilized to trim the TID effect on power IC. Geiger counter can calculate the amount of particles passing through the tube. However, the Geiger-Mueller tube is full of an inert gas such as neon or helium, which makes it impossible to be integrated with IC. Some photodiodes and RADFETs can be applied to detect radiation, yet they can only be fabricated with a specific process and require many auxiliary circuits, such as stable high voltage supply circuit, pulse amplifier, ADC, bandgap and so on. When the crystal of TLD is heated, the intensity of light emitted from the crystal can measure the strength of radiation. This method of application determines that TLD cannot be utilized to IC.

There are two challenges for these discrete TID detecting devices applied to IC. 1) It is difficult to find the criteria that is able to evaluate the TID effect on circuits. Existing radiation detecting technologies applied to detect the real-time radiation situation require complex and large-area signal processing circuits to reflect the accumulation of ionizing dose universally. Their output signals cannot be used to compensate the deviation induced by the TID effect on the circuit directly. 2) Special process is required to produce the detecting device causing that it cannot be flexibly applied to TID compensation of any IC.

In addition to discrete devices, some radiation detecting circuits that can be fabricated in standard CMOS processes have been proposed [6]-[9]. In these papers, some conventional topologies of integrated circuit such as voltage reference, two-stage miller amplifier, CMOS differential amplifier are transformed into radiation detecting circuit. The basic principle of these circuits is that the threshold voltage of MOSFET would shift after radiation which could be adopted as the criterion for evaluating the accumulation of ionizing dose. However, these circuits require complex structures with redundant auxiliary circuits universally. Moreover, complex circuit structures are more sensitive to mismatch which would induced by radiation easily.

There are two difficulties for TID detecting circuits fabricated in commercial process. 1) The detecting circuit will be abnormal or even invalid after being affected by radiation. The more complex the circuit structure is, the more random the effect of radiation is, and the more difficult it is to be used as a criterion for the accumulation of radiation dose. 2) The detecting range of the accumulation of radiation dose should be large enough. The effect of high dose accumulation on the circuit is much more obvious than the low dose accumulation, so the pivotal to the detecting method is how to detect the low dose accumulation on circuit.

A novel circuit is proposed in this paper to detect TID effect on IC and resolve these design challenges mentioned above. In this work, the leakage current induced by TID of NLDMOS is adopted as the criterion to reflect the accumulation of ionizing dose. The physical theoretical analysis of the proposed technique is discussed in Section II.

978-1-7281-0582-6/19 $31.00 © 2019 IEEE
119

Circuit design and experiment are shown in Section III. Implementation and discussion are given in Sections IV and conclusion is illustrated in Section V.

II. PHYSICAL THEORETICAL ANALYSIS

The TID detecting circuit proposed in this paper adopts NLDMOS of BCD process as the basic detecting device. The off-state leakage current of NLDMOS generated by TID effect is applied as the criteria of dose accumulation. In order to illustrate the principle of this circuit thoroughly, some basic physical theoretical analysis are elaborated below.

A. Physical Process of TID Effect on NMOS

As shown in Fig. 1, the TID effect will induce the generation of negative charges under the oxide interface of NMOS [10]. This phenomenon is the basis for adopting NMOS as the detecting device.

Due to energy absorption after ionizing radiation, a large amount of electron-hole pairs are generated at the initial position in oxide. Except for a small portion of the electron-hole pairs recombine rapidly, the remaining free carriers are enforced directional movement. Since a positive voltage is applied to the gate of the NMOS device, the electric field inside the oxide layer is directed to the bulk silicon by the gate. Because the mobility of electrons is much larger than that of holes, they remove away from the oxide through gate fast. Some holes are trapped easily by the oxide defects near the Silica-silicon (SiO$_2$-Si) interface. In addition, some holes form second type of interface ionizing defect: the interface defect by reacting with hydrogen. Because holes are trapped by the defects near the SiO$_2$-Si interface, the electrons are induced at the surface of the bulk silicon to form parasitic channels.

B. Physical Influence of The Thickness of The Oxide

The defects of NMOS include oxide defect and interface defect. Although oxide defect may be found throughout the oxide, most of them are located near the SiO$_2$-Si interface. Most of the interface defects are located at the junction of SiO$_2$-Si and near the SiO$_2$ side. Both of the two defects could capture free holes to form electrons at the surface of the bulk silicon. Namely, the amount of induced electrons is related to the number of defects. The holes captured by defects are named fixed positive oxide charge (N_{ot}). The parameter of N$_{ot}$ in oxide can be expressed as in (1) [11]:

$$\Delta N_{ot} = Kt_{ox} \tag{1}$$

where K is a parameter related to radiation ionizing dose, rate of electron-hole pair generation, rate of hole trapping, etc. And t_{ox} is the thickness of the oxide.

As (1) shows, ΔN_{ot} is proportional to t_{ox}. Namely, the thicker the oxide is, the more obvious the parasitic channel induced is. Therefore, in the BCD process, NLDMOS is the most sensitive device to the total ionizing dose effect.

C. Generation Mechanism of Parasitic Channel of NMOS

Because of the continuous development of Moore's Law, the thickness of the gate oxide is getting thinner and thinner. The electrons induced by ionizing dose under the gate oxide is less and less. Due to the need for device isolation, the thickness of the isolation oxide is much thicker than the gate oxide. Therefore, the isolation oxide (trench oxide) becomes the main cause of the induced negative charge.

As illustrated in Fig. 2, the induced negative charges by radiation form a parasitic channel though which leakage current can flow from drain to source when the state of the device is off. To make matters worse, this parasitic channel cannot be controlled by the gate voltage [12]. The off-state leakage current would be generated under the isolation oxide between the source and the drain. Moreover, this leakage current becomes the dominant contributor to off-state drain-to-source current in NMOS.

III. CIRCUIT DESIGN AND EXPERIMENT

A. TID detector cell

As illustrated in Fig. 3, the basic structure of TID detector cell is a CMOS inverter with constant low state input. The fan-out enhancing part of the detector cell is applied to strengthen the drive capability of the detector.

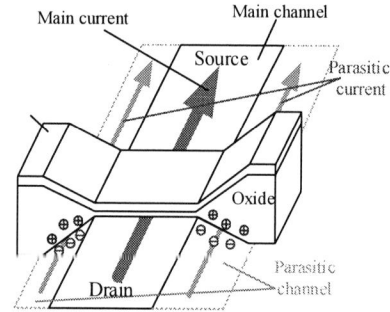

Fig. 2. Generation mechanism of the parasitic channel of NMOS.

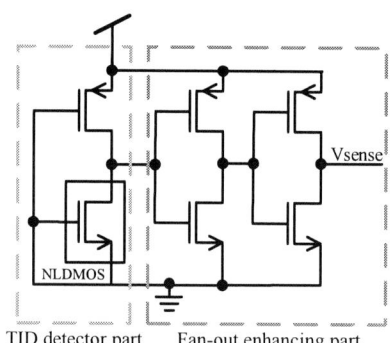

Fig. 3. TID detector cell.

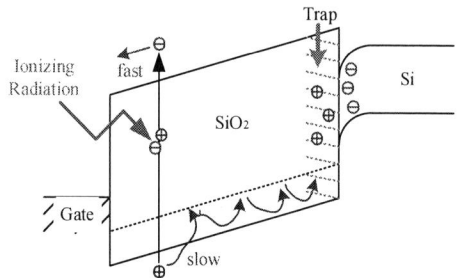

Fig. 1. Physical process of TID effect on NMOS.

In this TID detector part, a low-voltage PMOS is used as the upper transistor with thin gate oxide. And a high-voltage NLDMOS with thick oxide is adopted as the lower transistor. Since the input voltage is low (connected to ground), the output voltage of TID detector cell is high in the no-radiation situation. With the ionizing dose accumulating, the off-state leakage current of the NLDMOS ($I_{leakage(NMOS)}$) gradually increases. When the leakage current is larger than the operating current of PMOS (I_{PMOS}), the output voltage change from high to low. As expressed in (2), the state at which the current of the upper transistor (PMOS) is equal to the lower transistor (NLDMOS) of the cell is named as the Flip Point of the detector.

$$I_{leakage(NMOS)} = I_{PMOS} = k(W/L)(V_{GS}-V_{th})^2 \qquad (2)$$

where k is a parameter related to the process. W is the channel width and L is the channel length of the NLDMOS. V_{GS} and V_{th} is the gate voltage and threshold voltage of the NLDMOS device, respectively. By adjusting the aspect ratio (W/L) of the PMOS in the TID detector part, the Flip Point can be set at different ionizing dose. The radiation ionizing dose can be measured from small to large via a combination of different TID detecting cells.

B. Experiment

As shown in Fig. 4, the circuit has been fabricated in 0.18μm 40V BCD process whose size is 1040×163 μm². And the PCB of test chip is exposed to ⁶⁰Co source with a dose rate of 50 rad/s. This test chip includes low-voltage NMOS (5V) and high-voltage NLDMOS (40V). With the help of radiation experiments, the sensitivity to radiation ionizing doses of this devices can be distinguished. In the process documentation, the essential parameters are: the gate oxide of 5V NMOS and 40V NLDMOS is 13.003nm and 12.42nm, respectively. From the data, the gate oxide thickness of 40V NLDMOS is a bit lower than 5V NMOS. The more critical thickness of the isolating oxide is not available from the documentation.

C. Results of The Experiment

The results of the experiment are illustrated in Fig. 5. The curves shown in Fig. 5(a)(b) reflect transfer characteristics of high-voltage NLDMOS (HM) and low-voltage NMOS (LM), respectively. From these figures, with the radiation ionizing dose accumulating, the curves of the device continues to move upwards in the coordinate axis.

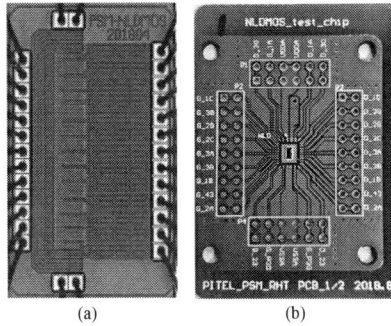

Fig. 4. Experimental integrated circuit and PCB. (a) Micrograph of the NLDMOS test chip. (b) The PCB of NLDMOS test chip for TID experiment.

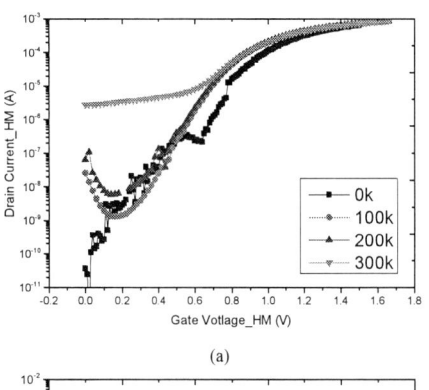

(a)

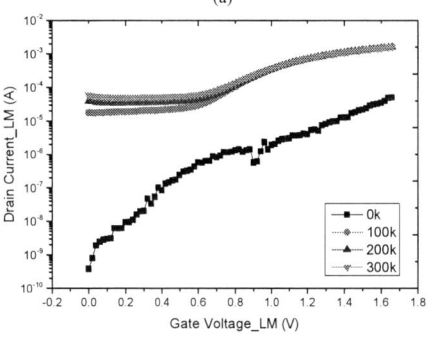

(b)

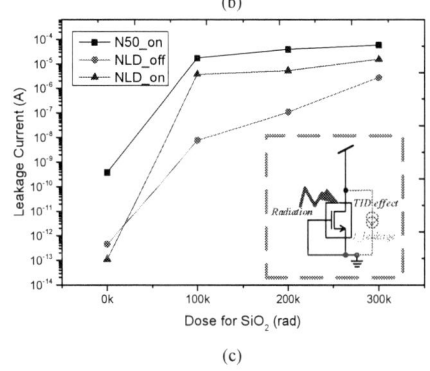

(c)

Fig. 5. Results of the experiment. (a) The current-voltage characteristics of NLDMOS after TID effect. (b) The current-voltage characteristics of 5V NMOS after TID effect. (c)The off-state leakage current of NLDMOS and 5V NMOS induced by TID.

Comparing the curves in Fig. 5(a) and Fig. 5(b), the transfer characteristic of 5V NMOS at different dose have little difference which would cause trouble in detection. The curves of NLDMOS are more distinct at different radiation ionizing dose. Therefore, NLDMOS is a better choice than low-voltage NMOS for applying as a ionizing detecting device. The off-state leakage current at different dose is shown in Fig. 5(c). Namely, the data is measured at the state when the gate electrode of the device is connected to the ground. The curve *N50_on* represents that the gate voltage of the 5V NMOS is high when the device is irradiated. In addition to the 5V NMOS, this figure also shows the off-state leakage current of the NLDMOS after receiving radiation during the on and off states. In Fig. 5(c), comparing these three curves, the curves of *N50_on* and *NLD_on* tend to be flat after 100krad. That is to say, the rising slope of the two curves becomes tiny, which will cause the subsequent signal be difficult to be distinguished after the radiation ionizing dose exceeds 100 krad. Conversely, the *NLD_off* curve varies apparently at different dose points, which means off-state NLDMOS is suitable for radiation detection. From the *NLD_off* curve, the off-state leakage current increases gradually with the accumulation of ionizing dose. In this experiment, the aspect ratio (*W/L*) of the NLDMOS is 90/1.35μm.The off-state current is 7nA, 100nA and 2.78μA at 100krad, 200krad and 300krad (SiO₂) of ionizing dose, respectively. According to the results of the experiment, a parallel current source at both ends of the NLDMOS can be added to simulate the influence of the TID effect on the circuit. The more the radiation total ionizing dose is accumulated, the larger the value of the TID equivalent current source is. By this method, the effect of the ionizing dose on the circuit can be imitated to verify whether the detecting circuit is valid.

IV. IMPLEMENTATION AND DISCUSSION

Different doses can be detected by a combination of basic TID detection cells, as illustrated in Fig.6 and table 1. According to the experiment above, a current source paralleled to NLDMOS can be applied to simulate the ionizing dose environment. As shown in Fig. 6(a), the PMOS of these detector cells have different *W/L*. The larger the *W/L* is, the greater the leakage current is required to reverse the output voltage, which means that more radiation ionizing dose needs to be accumulated. Namely, by adjusting the *W/L* of PMOS in the detector cell, the Flip Point of the cell can be set at different ionizing dose. The output voltage of the detecting circuit need to be transformed to code by an encoder. Table 1 shows the truth table of the detecting circuit. Different codes correspond to different ionizing dose accumulations, and the code can be used as TID trimming code inside circuits. The simulation results of the TID detecting circuit is shown in Fig. 6(b). As the off-state current increases, the output voltage reverse from high to low.

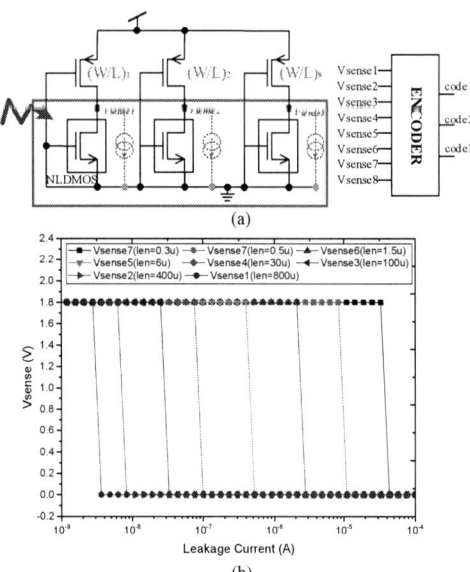

Fig. 6. TID detecting circuit and its simulation result. (a) TID detecting circuit. (b) The simulation results of the TID detecting circuit.

TABLE I. THE TRUTH TABLE OF THE ENCODER

Dose	Vsense Number								Code
	1	2	3	4	5	6	7	8	
0krad	1	1	1	1	1	1	1	1	111
TID1	0	1	1	1	1	1	1	1	110
TID2	0	0	1	1	1	1	1	1	101
TID3	0	0	0	1	1	1	1	1	100
TID4	0	0	0	0	1	1	1	1	011
TID5	0	0	0	0	0	1	1	1	010
TID6	0	0	0	0	0	0	1	1	001
TID7	0	0	0	0	0	0	0	1	000

When the *W/L* of the PMOS in detector cell is 300nm/1.5μm, inversion current of the detecting circuit is 2.8μA, which is the value of the off-state leakage current when the total ionizing dose is 300krad. By adjusting the *W/L* of the PMOS in detector cell, the simulation results can meet the design requirements accurately from the radiation experiment mentioned above.

V. CONCLUSION

A TID detecting circuit based on off-state leakage current of NLDMOS is proposed in this paper. The topology of its basic cell is an inverter. By using combination of these cells with arranged different parameters, this circuit can be applied to detect TID effect on IC in a wide range with high accuracy. Moreover, this circuit can be integrated with conventional power IC in an ordinary commercial process.

REFERENCES

[1] H. J. Barnaby, "Total-Ionizing-Dose effects in modern CMOS technologies," in IEEE Transactions on Nuclear Science, vol. 53, no. 6, pp. 3103-3121, December 2006.

[2] D. R. Corson, and R. R. Wilson, "Particle and quantum counters," Review of Scientific Instruments, vol. 19, no. 4, pp. 207-233, April 1948.

[3] F. S. Goulding, "Transistorized radiation monitors," in IRE Transactions on Nuclear Science, vol. 5, no. 2, pp. 38-43, August 1958.

[4] T. Nakajima, Y. Murayama, T. Matsuzawa, A. Koyana, "Development of a new highly sensitive LiF thermos luminescence dosimeter and its applications," Nuclear Instruments & Methods, vol. 157, no. 1, pp. 155-162, November 1978.

[5] P. Falke, et al., "Cosmic ray dose monitoring using RadFET sensors of the rosetta instruments SESAME and COSIMA," Acta Astronautica, vol. 125, no. C, pp. 22–29, August-September 2016.

[6] E. Carbonetto et al., "CMOS differential and amplified dosimeter with field oxide N-Channel MOSFETs," in IEEE Transactions on Nuclear Science, vol. 61, no. 6, pp. 3466-3471, December 2014.

[7] K. J. Shetler et al., "Total dose measurement circuit design based on a voltage reference topology," in IEEE Transactions on Nuclear Science, vol. 64, no. 1, pp. 559-566, January 2017.

[8] M. Garcia-Inza, S. H. Carbonetto, J. Lipovetzky and A. Faigon, "Radia-tion sensor based on MOSFETs mismatch amplification for radio-therapy applications," in IEEE Transactions on Nuclear Science, vol. 63, no. 3, pp. 1784-1789, June 2016.

[9] G. Salaya, M. Garcia-Inza, S. Carbonetto and A. Faigón, "Design and characterization of a CMOS two-stage miller amplifier for ionizing radiation dosimetry," 2017 Argentine Conference of Micro-Nanoelectronics, Technology and Applications (CAMTA), Buenos Aires, 2017, pp. 32-36.

[10] T. R. Oldham, "Analysis of damage in MOS devices for several radiation environments," in IEEE Transactions on Nuclear Science, vol. 31, no. 6, pp. 1236-1241, December 1984.

[11] D. M. Fleetwood, T. L. Meisenheimer and J. H. Scofield, "1/f noise and radiation effects in MOS devices," in IEEE Transactions on Electron Devices, vol. 41, no. 11, pp. 1953-1964, November 1994.

[12] W. Snoeys, et al., "Layout techniques to enhance the radiation tolerance of standard CMOS technologies demonstrated on a pixel detector readout chip," Nuclear Instruments & Methods in Physics Research A, vol. 439, no.2, pp. 349-360, January 2000.

Proceedings of the 31st International Symposium on Power Semiconductor Devices & ICs
May 19-23, 2019, Shanghai, China

A Novel Cross-Over CMR Transformer Technology for Magnetic Isolation Gate Driver Applications

Yunzhong Luo, Jian Fang, Erli Zhang , Ming Li and Bo Zhang
State Key Laboratory of Electronic Thin Films and Integrated Devices
School of Electronic Science and Engineering, University of Electronic Science and Technology of China
Chengdu 610054, China
Email: fjuestc@uestc.edu.cn

Abstract—A novel Cross-Over CMR transformer technology is proposed. The CMR (common-mode rejection) problem of magnetic isolation signal transmission is solved by the secondary Cross-Over differential coils and ANPTM (alternating narrow pulse transmission mode). Based on CSMC 1um BCD technology platform, the simulation results show that compared with the traditional magnetic isolation, the CODT (Cross-Over differential transformer) technique has a lower mismatch and smaller area. The circuit transmission delay reduces by 30% to 23ns with signal transmission frequency as high as 10MHz. The isolation voltage reaches 5kV while improving CMR (CMR greater than 50kV/μs). The proposed technology can be applied in high voltage floating gate driver for IGBT, CoolMOS, and even GaN HMET devices.

Keywords—magnetic isolation; differential transformer; CMR; process mismatch

I. INTRODUCTION

In power switch fields, the gate driver circuits are used to generate floating gate signals for IGBT, CoolMOS, and even GaN HMET devices, and play a role as isolation barriers. At present, the main isolation techniques include level shift converter isolation, opto-coupler isolation and magnetic isolation. However, the isolation capability of level shift converter is limited, the opto-coupler isolation is difficult to integrate with large in volume, high in power consumption. Magnetic isolation has a higher breakdown voltage, lower power consumption faster speed and higher stability than others. Since the floating signal V_S of the gate driver circuit affects the signal transmission, the CMR capability of the on-chip transformer becomes an unsolved problem. In response to this problem, some solutions have been reported recently. At circuit level, Shunichi Kaeriyama used high-pass filters and voltage threshold to enhance the CMR based on the difference in voltage amplitude and frequency between noise and signal, but this technique has a small CMR and is merely suitable for applications with low amplitude noise [1]. Egidio Ragonese used the conventional differential transformer to enhance the CMR, however, since both the primary and secondary coils are double-spiral inductors, the gate driver circuit is obviously affected by the process mismatch and has larger area than others [2]. Tso-Wei Li reported a fully differential transformer for CMR, but its breakdown voltage is limited and the process is complicated[3]. These reported techniques don't fully solved the problems between CMR, process mismatch, chip speed and area.

In this paper, a novel Cross-Over CMR transformer technology for magnetic isolation gate driver applications is proposed. Fig. 1 shows the schematic diagram of the proposed Cross-Over CMR transformer technology gate driver circuit. The CMR problem of magnetic isolation signal transmission is solved by the novel Cross-Over CMR transformer technology and the pulse transmission technique ANPTM. This technology not only increases CMR, reduces the effects of process mismatch, but also decreases transformer area.

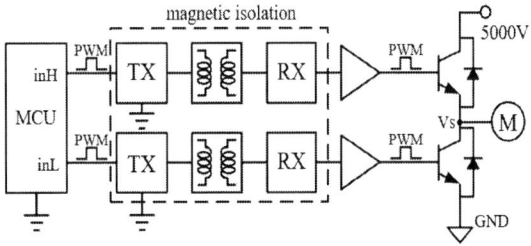

Fig. 1. Schematic diagram of the proposed Cross-Over CMR transformer technique gate driver circuit

II. CROSS-OVER CMR TRANSFORMER DESIGN

The main purpose of CODT is isolating high voltage and obtaining a high CMR. The high voltage is separated by a polyimide layer between the primary and secondary coils, and insulation voltage of the 15μm thick polyimide layer is 5kV[4], the common mode noise is suppressed by the CODT's secondary coils. Fig. 2 shows the structure of the CODT.

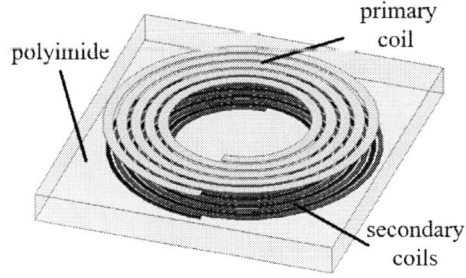

Fig. 2. Structure of the CODT

The CODT is consisted of one primary coil, and two secondary coils which have the same geometry and cross-over layout. The polyimide layer is between the primary coil and secondary coils. Fig. 3 is the schematic diagram of the magnetic isolation coils, which shows all the parameters of the coil. Table I shows structural parameters of the transformer.

978-1-7281-0582-6/19 $31.00 © 2019 IEEE   123

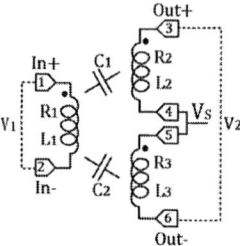

Fig. 3. Schematic diagram of the magnetic isolation coils

TABLE I.   DESIGN PARAMETERS OF THE TRANSFORMER

Turns No. of primary/secondary coils	5/10
Inner radius of primary/secondary coils(μm)	200/200
Width of primary/secondary coils(μm)	15/4
Thickness of primary/secondary coils(μm)	1/1
Spacing of primary/secondary coils(μm)	5/6
Thickness of polyimide(μm)	15

The expression of the transformer output voltage V_2 can be expressed as following equations [4]

$$V_2 \cong k\left(\sqrt{L_1 L_2} + \sqrt{L_1 L_3}\right)\frac{dV_1}{dt} +$$

$$\left[C_1\left(\frac{R_2}{2}\frac{dV_S}{dt} + \frac{L_2}{2}\frac{d^2V_S}{dt^2}\right) - C_2\left(\frac{R_3}{2}\frac{dV_S}{dt} + \frac{L_3}{2}\frac{d^2V_S}{dt^2}\right)\right] \quad (1)$$

Where L_1 and R_1 are the inductance and resistance of the primary coil, respectively. L_2, L_3, R_2, R_3 are the inductance and resistance of the two secondary coils, respectively. C_1 and C_2 are the coupling capacitance between the primary coil and the secondary coils. V_2 is affected by the primary and secondary coils resistances, inductances, the coupling capacitance between the primary and secondary coils, and the floating ground V_S. The first part represents the signal component by magnetic isolation of primary and secondary coils, and the second part represents the common-mode noise components. Since the CODT has two secondary coils, the CMR is higher than that of the magnetic isolation circuit with single secondary coil.

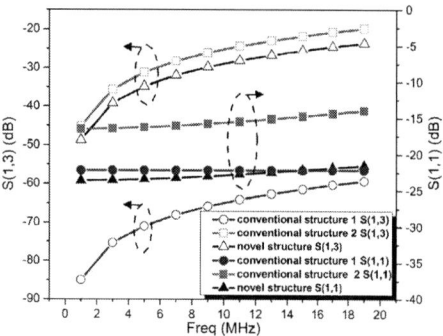

Fig. 4. The variation of S(1,1) and S(1,3) with frequency in the range of 1MHz to 20MHz for three kinds of transformers

Fig. 4 shows the variation of S(1,1) and S(1,3) with frequency in the range of 1MHz to 20MHz for three kinds of

transformers. At 10MHz, the S(1,1) of proposed structure is -23dB, which is close to the -22dB of the conventional structure 1[2], and better than -15dB of the conventional structure 2[5]. At the same time, the S(1,3) of proposed structure is -29dB,which is close to the -25dB of the conventional structure 2, and better than -65dB of the conventional structure 1. It reflects that the S(1,1) and S(1,3) of the CODT are better than conventional structures.

The CODT and conventional structure 2 both have high CMR without considering process mismatch. However, the performance of the circuit is actually affected by manufacturing process. Due to the large area of the transformer, it is difficult to totally avoid the effects of process mismatch. For planar transformers, the performance is mainly affected by the flatness of metal. The metal thickness can be expressed as following equations

$$Z = h + (X + Y) \cdot tan\theta \quad (2)$$

Where Z is the actual metal thickness, h is the design thickness, θ is the process mismatch gradient angle, and (X,Y) is the rectangular coordinate value. The schematic is shown in Fig. 5. The secondary coils process mismatch of the differential transformer will directly result in a reduction in CMR capability.

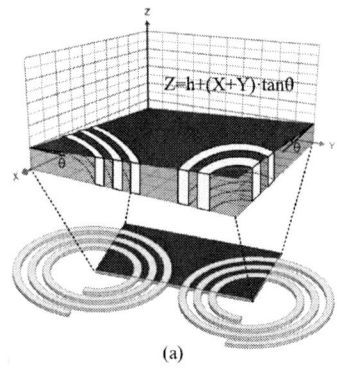

(a)

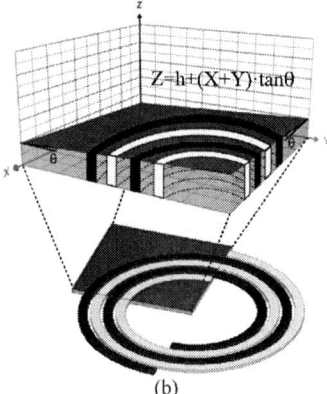

(b)

Fig. 5. Secondary coils thickness process mismatch: (a) conventional structure 2, and (b) novel structure

978-1-7281-0582-6/19 $31.00 © 2019 IEEE   124

TABLE II. TRANSFORMER STRUCTURAL CONTRAST

	conventional structure 1	conventional structure 2	proposed structure
structure	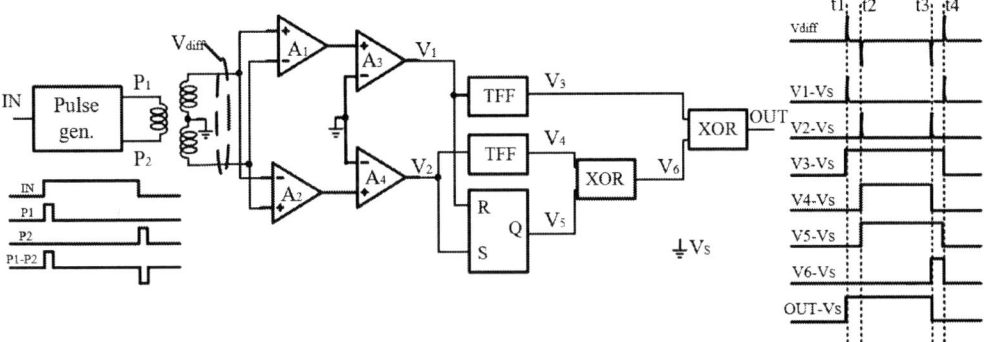		
area ratio	1	2	1
S(1,3) *	-65dB	-25dB	-29dB
S(1,1) *	-22dB	-15dB	-23dB

* S-parameter at 10MHz

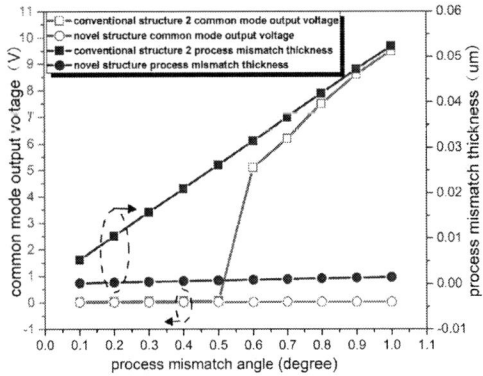

Fig. 6. Schematic diagram and timing diagram of the ANPTM signal transmission circuit

The central spiral of the spiral coil satisfies the polar coordinate equation

$$\rho = w\frac{\alpha}{2\pi} + r \qquad (3)$$

Where ρ is the polar diameter of the spiral, w is the coil pitch, r is the inner diameter of the coil, and α is the polar angle. Since it has been assumed that the metal surface is a plane, the thickness at the center of the coil is equal to the average value of the thickness of the coil in any range of $\Delta\alpha$, so the thickness of the planar spiral transformer is equivalent to the height of the helicoid at the center of the coil. CODT's secondary coil average thickness mismatch value is

$$\triangle Z_c = \frac{2w \cdot tan\theta}{5\pi w + 2\pi r} \qquad (4)$$

The average thickness mismatch value of conventional structure 2

$$\triangle Z_t = \frac{2c \cdot tan\theta}{5\pi w + 2\pi r} \qquad (5)$$

Where c is the spacing between the centers of the secondary coils of structure 2, due to c≫w, $\triangle Z_t \gg \triangle Z_c$. Which means that the process mismatch of CODT is much lower than that of conventional structure 2. Since the process mismatch of the coils thickness directly affects the geometrical dimensions between the secondary coils, which causes an influence on CMR. In order to investigate the ability of the differential transformer to suppress floating ground with process mismatch, the artificially added secondary coil error is simulated and verified at a floating ground of 50kV/µs. Fig. 7 shows the the relationship between the common mode

voltage output value and the secondary coils process mismatch gradient angle. In the range where θ is less than 1 degree, CODT has stronger CMR, when θ is greater than 0.5 degree, the CMR of the conventional structure 2 is greatly reduced. Therefore, CODT has a higher CMR under the same process mismatch condition. Structural contrast of these three transformers are listed in Table II.

Fig. 7. Process mismatch angle and common mode voltage output value

III. CIRCUIT IMPLEMENTATION OF CODT

Four different signal transmission technologies are reported in [5-8], respectively, among which the transmission technique mentioned in [5] has the advantages of small area, low power consumption and short delay. In order to cooperate with CODT, based on the pulse transmission technique in [5], an ANPTM is proposed. A rising edge produces a positive narrow pulse and a falling edge produces a negative narrow pulse. The rising edge of the narrow pulse produces a positive spike after the CODT, corresponding to the rising edge of the output. The falling edge of the negative narrow pulse produces a negative spike after the CODT, corresponding to the falling edge of the output, Meanwhile, the common mode noise was suppressed. Compared with transmission technique in [5], the inductive voltage is increased and the transmission performance is improved by increasing the dV/dt of the transformer input. Fig. 6 shows the schematic diagram and timing diagram of the ANPTM signal transmission circuit.

Fig.8 shows the transmission waveforms under load conditions when input signal is a 10 MHz square wave, the floating ground V_S is 5 kV, and $dV_S/dt = 50$ kV/μs. It can be seen from Fig. 8 that after using ANPTM, the delay of the output signal is shortened from 34ns to 23ns, and the overall delay is reduced by 30%, the transmission mode can meet the requirements of high frequency applications.

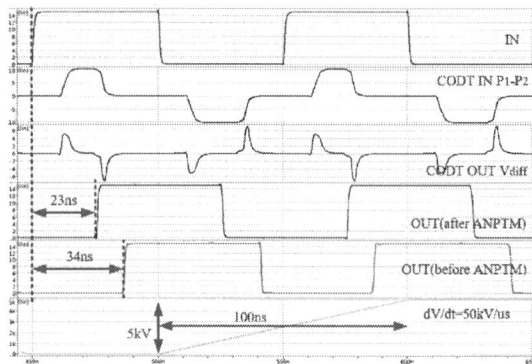

Fig. 8.   Signal transmission waveform

IV. CONCLUSION

In summary, a novel Cross-Over CMR transformer technology is proposed. The CMR problem of magnetic isolation signal transmission is solved by the secondary Cross-Over differential coils and the proposed ANPTM. The breakdown voltage of CODT reaches 5kV and the ability to suppress floating ground is larger than 50kV/μs. In addition, it is less affected by process mismatch and has a smaller area than traditional differential transformers. At the same time, the signal transmission frequency based on ANPTM can be at least 10MHz and the circuit delay is as low as 23ns. The proposed high-voltage magnetic isolation technology based on Cross-Over differential transformer has a higher CMR and faster operational speed, so it can be used in high voltage floating gate driver circuits for IGBT, CoolMOS and even GaN HMET devices.

REFERENCES

[1] Kaeriyama, S. , et al. "A 2.5kV isolation 35kV/us CMR 250Mbps 0.13mA/Mbps digital isolator in standard CMOS with an on-chip small transformer." *Vlsi Circuits* IEEE, 2010.

[2] Ragonese, Egidio , et al. "A Fully Integrated Galvanically Isolated DC-DC Converter With Data Communication." IEEE Transactions on Circuits & Systems I Regular Papers PP.99(2017):1-10.

[3] Li, Tso Wei , and H. Wang . "A millimeter-wave fully differential transformer-based passive reflective-type phase shifter." Custom Integrated Circuits Conference IEEE, 2015.

[4] S. Uchida et al. A face-to-face chip stacking 7kV RMS digital isolator for automotive and industrial motor drive applications[J]. Proc. Int. Symp. Power Semicond. Devices ICs,2014, pp: 442–445.

[5] Kaeriyama, et al. "A 2.5 kV Isolation 35 kV/us CMR 250 Mbps Digital Isolator in Standard CMOS With a Small Transformer Driving Technique." IEEE Journal of Solid-State Circuits 47.2(2012):435-443.

[6] B. Chen, J. Wynne, and R. Lkiger, "High speed digital isolators using microscale on-chip transformers," Elektronik Mag., 2003.

[7] Knoedl, G., Jr, et al. "A monolithic signal isolator [for power convertors]." *IEEE Applied Power Electronics Conference & Exposition* 1989.

[8] Chen, Baoxing . "Isolated half-bridge gate driver with integrated high-side supply." 2008 IEEE Power Electronics Specialists Conference IEEE, 2008.

Proceedings of the 31st International Symposium on Power Semiconductor Devices & ICs
May 19-23, 2019, Shanghai, China

A High-reliability Half-Bridge GaN FET Gate Driver with Advanced Floating Bias Control Techniques

Xin Ming, Xuan Zhang, Zhi-wen Zhang, Li Hu, Yao Qin, Xu-dong Feng, Qi zhou, Zhuo Wang, and Bo Zhang
State key Laboratory of Electronic Thin Films and Integrated Devices
University of Electronic Science and Technology of China
Chengdu, China
Email: mingxin@uestc.edu.cn

Abstract—The advanced floating bias control techniques (FBCT) for GaN devices are proposed in this paper, which aims at solving negative voltage transient issues present at switching node of power stage during dead time. It adopts maximum power-supply tracking to reduce time delay of level shifter and active clamping for bootstrap voltage to avoid overvoltage conditions. This high-reliability half-bridge GaN gate driver has been fabricated in a 0.5μm 80V HV CMOS process and occupies a chip area of 1699×1522μm² where the active area of FBCT with internal bootstrap diode is 1350×620μm². The response time of level shifter is only 1.97ns in the process of turning on high-side power switch. Transmission delay is small and there are no logic errors or latch-up issues in the gate driver due to the negative voltage transient. Bootstrap voltage ripple based on active clamping technique is well controlled over a wide frequency range (166mV@500kHz~797mV@5MHz at C_{boot}=0.1μF). The total quiescent current consumed by FBCT is only 90μA.

Keywords—GaN gate drive; negative voltage transient; floating bias control; maximum power-supply tracking; active clamping.

I. INTRODUCTION

GaN devices achieve a good figure-of-merit (FOM) compared to Si power devices, which can push switching frequency to MHz and break through the bottleneck of power density and efficiency of traditional power supplies. Based on these advantages, it can be widely used in various applications, including miniaturized power supplies, power amplifiers, Lidar and so on. However, the physical particularity of GaN requires customized gate drive circuits and advanced loop control strategies to enhance reliability of GaN switch applications as well as its high-frequency performance.

As can be seen, GaN FETs differ from their silicon counterparts because of their significantly faster switching speeds and consequently have different requirements for gate drive. One important design challenge with half-bridge gate drive is a more negative voltage transient (−2V~−3V) present at switching node of power stage during dead time [1]. This "body diode" voltage drop can be troublesome because it directly affects the floating ground of high-side gate driver and might pull down power supply of the internal circuitry, including high-speed level shifter [2]–[5]. Significant time delay in the top channel can thus be observed when turning on high-side power switch, which limits the minimum-on time

This work was supported by the National Key R&D Program of China under Grant 2017YFB0402800.

and is not suitable for high-frequency PoL applications (i.e. 48-1V PoL DC/DC for CPU, DSP and memory). The other issue is the possibility to develop an overvoltage condition across the bootstrap capacitor, exceeding the maximum allowable gate voltage of E-mode GaN FETs easily (i.e. $V_{GS,max}$=6V). Some active balancing schemes by using ZVS detection to avoid overcharging bootstrap cap have been reported [6]. However, the power supply design of level shifter is still not mentioned right now, which can not fully ensure a high-reliability fast switching operation.

In this work, the novel floating bias control techniques for GaN-based half-bridge gate drive are proposed to solve design challenges mentioned above. Concept and circuit implementations of the proposed technique are discussed in Section II. Experimental results and conclusions are given in Sections III and IV, respectively.

II. DESIGN OF PROPOSED FBCT

Fig. 1 shows detailed circuit analysis of negative voltage effects on a buck converter based on GaN device. During dead time, the low-side eGaN conducts in reverse when gate-to-drain reaches threshold and introduces large negative voltage transient especially at heavy load. It splits voltage potential of V_{SS} and floating ground V_{SW} obviously (phenomenon ①). This behavior will result in two problems in the bootstrap function. For the current-mode level shifter design in ②, the input signal is first transferred to control current by LDMOS (V-I converter) and then goes to the high-side channel to reset output state [3] (I-V converter). Since the current branch from low side to high side channel is referenced to V_{SS}, the diode-connected transistor (blue color) gains low V_{GS} and I_P due to a small V_{BST} during dead time. Because of supply-voltage headroom limitation, the reduced I_P may introduce important time delay or even output-signal loss. ① will also overcharge C_{boot} in the process if no protection circuit is implemented in the charging path. This is due to the fact that V_{BST} is clamped to be close to V_{DD} by HVD1 while V_{SW} is a negative voltage as shown in ③. It may damage the high-side GaN FET when it is turned on and introduces severe reliability problems. Different methods have been reported to resolve overcharging trouble, including active BST balancing technique [6] or zener-diode clamping [7]–[10]. However, the previous ones may sacrifice

978-1-7281-0582-6/19 $31.00 © 2019 IEEE

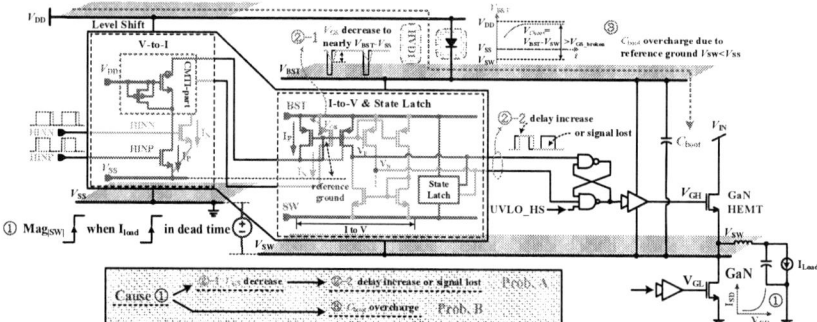

Fig. 1 Circuit analysis of negative voltage effects on a buck converter based on GaN device

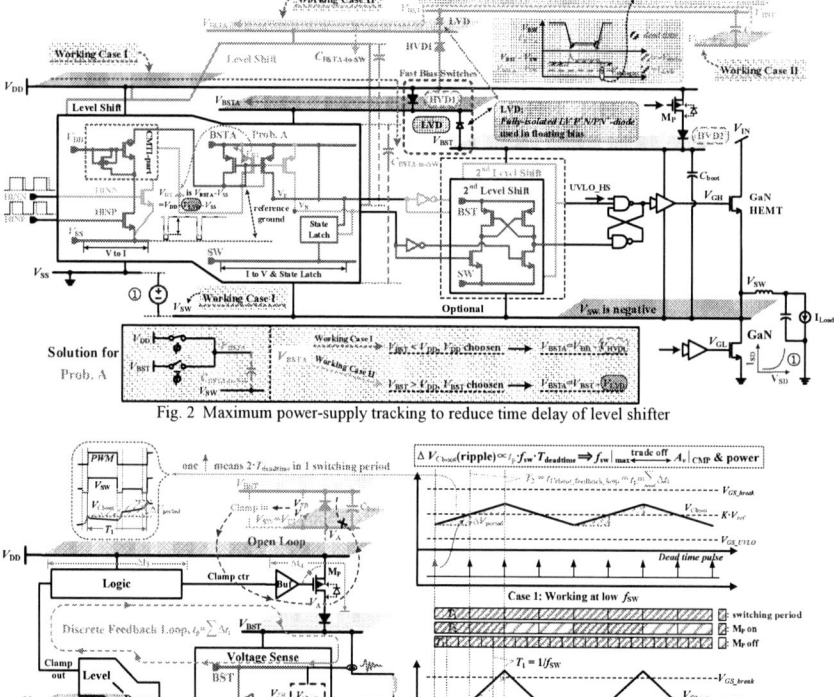

Fig. 2 Maximum power-supply tracking to reduce time delay of level shifter

Fig. 3 Active clamp function for bootstrap voltage

large power loss (zener clamping) or need complex circuit structures (ZVS detection for active balancing). In addition, the power supply design for level shifter based on GaN applications is not mentioned anymore in previous literatures.

The maximum power-supply tracking methodology is shown in Fig. 2 to reduce transmission delay of level shifter. The concept is to split the power supplies of level shifter and high-side gate drive with two different kinds of diodes, where

one is a high-voltage fast recovery diode (HVD) and the other is a fully isolated low-voltage diode (LVD). V_{BSTA} is selected between V_{DD} and V_{BST} with fast bias switches. For example, when V_{SW} is negative during dead time, HVD1 is turned on to supply transient current and V_{BSTA} is close to V_{DD} since $V_{BST} < V_{DD}$. The current branch I_P can then get enough voltage headroom to transfer current information for subsequent digital control. On the other hand, if high-side power switch is turned on, HVD1 is shut down and the bootstrap cap can supply charge to the level shifter through the fully isolated LVD. Therefore, the differential supply voltage (V_{BSTA}–V_{SW}) across level shifter is not changed greatly under different working conditions and small time delay can be ensured. The power-supply headroom for level shifter can then be guaranteed during dead time.

The basic idea to realize active clamping for bootstrap voltage is also shown in Fig. 3. By utilizing a high-precision high-PSR voltage sense circuit to give a feedback control on the power switch M_P in series with bootstrap diode, the bootstrap voltage can be well controlled without overcharging. This method is very simple and consumes small quiescent current and chip area. The detailed working process is analyzed below. Voltage sense circuit is utilized to monitor bootstrap voltage and gives an order to turn on/off the charging path by M_P. Due to time delay (Δt_1-Δt_4) of the discrete negative feedback loop, M_P can not react quickly to the detection and the bootstrap voltage ripple is proportional to f_{sw} ($\Delta V_{Cboot,ripple} \propto t_p \cdot f_{sw} \cdot T_{deadtime}$). V_{boot} may either touch V_{GS_break} of GaN or V_{GS_uvlo} at extremely high f_{sw} and heavy load conditions, which should be carefully designed. The quiescent current is from voltage sense circuit, affecting discrete feedback bandwidth and bootstrap capacitance.

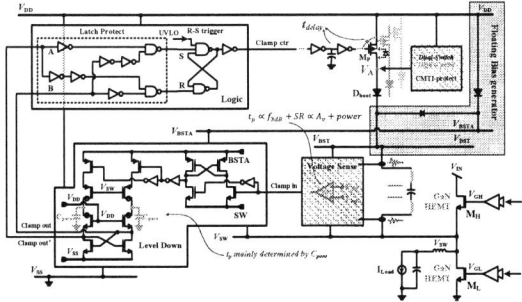

Fig. 4 Circuit realization of bootstrap voltage clamping

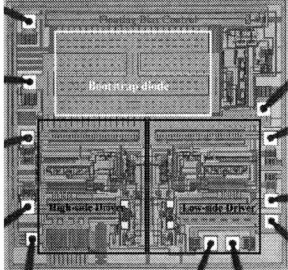

Fig. 5 High precision and high PSR voltage sense circuit

The complete circuit realization of bootstrap voltage clamping is given in Fig. 4. The total time delay includes response time of voltage sense comparator, level down circuit, combinational logic and gate driver to turn on power transistor M_P. Moreover, if overvoltage issue occurs, M_P is shut down and node V_A will enter high-impedance state. When high-side power switch M_H is turned off, the negative dV/dt at V_{SW} can couple to V_A through parasitic capacitance and exceed

breakdown voltage of M_P. Common mode transient immunity (CMTI) protection circuit should thus be implemented to clamp node voltage V_A effectively by providing a large transient current to compensate AC current from D_{boot}.

The proposed high precision and high PSR voltage sense circuit is shown in Fig. 5. First, because the maximum gate voltage clamping margin is limited (<0.5V) and GaN devices can be used in a wide temperature range due to the wide band gap E_g, temperature coefficient of voltage reference V_{ref} should be small in order to control the clamping bootstrap voltage precisely and a floating bandgap reference (BGR) in high-side channel is thus proposed. Second, since the wire bond is connected in series with bootstrap cap, the real power rails V_{BST} and V_{SW} are very noisy and decouple from the bootstrap cap during fast switching operation. Voltage sense circuit should achieve high PSR performance in order not to trigger clamping function falsely. Third, due to the fact that this BGR is a floating architecture, fully isolated BJTs for BGR core circuit are required in the process.

III. EXPERIMENTAL RESULTS AND DISCUSSION

This half-bridge GaN gate drive has been fabricated in 0.5μm 80V HV CMOS process and occupies a chip area of 1699×1522μm², where the active area of FBCT with internal bootstrap diode is 1350×620μm². The size of bootstrap diode is determined by high-frequency charging ability for the external cap. Chip micrograph and floorplan of the gate drive is shown in Fig. 6, where high-side and low-side gate drivers are located symmetrically to reduce mismatch of transmission delay for different channels. The tested E-mode GaN device is EPC2015.

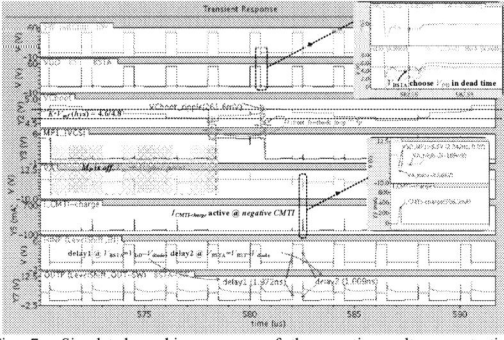

Fig. 6 Chip micrograph of the proposed high-reliability GaN FET gate drive

Fig. 7 Simulated working process of the negative voltage protection (V_{IN}=48V, f_{sw}=500kHz, I_{load}=25A)

The simulated working process of the negative voltage protection is shown in Fig. 7, where the largest load current is applied to emulate the worst condition. During dead time, BSTA is selected by V_{DD} with a diode-voltage loss ($\sim$4.36V) and guarantees small response time of level shifter ($\sim$1.97ns) when turning on high-side power switch. For active clamping circuit, the voltage ripple is larger than hysteresis detection voltage due to feedback bandwidth limitation ($\sim$261.6mV). Moreover, when M_P is shut down and negative dV_{SW}/dt occurs, a large charging current (306mA) from CMTI circuit in Fig. 4 is utilized to clamp the high-impedance node V_A and avoid avalanche breakdown of low-voltage transistor M_P to enhance the reliability.

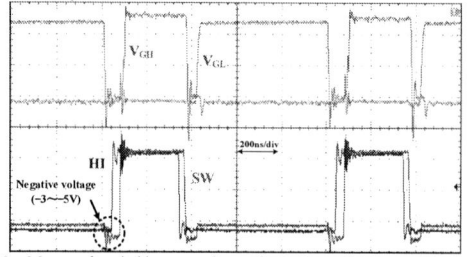

Fig. 8 Measured switching waveforms for PoL applications at V_{IN}=12V, I_{load}=1.5A and f_{sw}=1MHz.

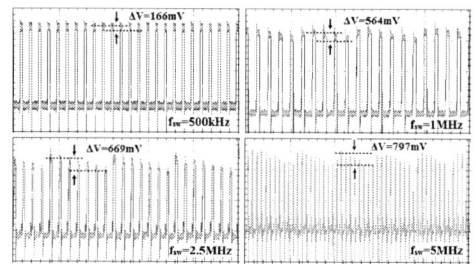

Fig. 9 Bootstrap voltage ripple at different f_{sw} and C_{boot}=0.1μF

The measured switching waveforms for PoL applications at V_{IN}=12V, I_{load}=1.5A and f_{sw}=1MHz is shown in Fig. 8. The gate drive responds well under a large negative voltage spike ($<$−3V), considering both GaN's reverse conduction and parasitic inductances from package and PCB traces. The transmission delay is small and there are no logic errors or latch-up issues in the gate driver due to the negative voltage transient. The active clamping function for bootstrap capacitor is shown in Fig. 9, where the bootstrap voltage ripple at different f_{sw} is given. Because of bandwidth limitation of the discrete feedback loop in Fig. 3, this performance is proportional to f_{sw}, varying from 166mV to 797mV in a wide frequency range, which is set in a safe operating voltage region for E-mode GaNs. Difference between simulation and test results is based on accuracy of HVD charging model.

IV. CONCLUSION

E-mode GaN transistors can conduct in reverse direction during dead time when the drain voltage is higher than the gate voltage by at least one threshold voltage. An advanced floating bias control are thus proposed in this paper to solve issues of large transient negative voltage, including maximum power-supply tracking to improve voltage headroom of level shifter and active clamping strategy to control the charge saturation of bootstrap capacitance. The physicochemical mechanism and relevant circuit analysis are provided in detail. Based on these techniques, good response speed of level shifter and voltage clamping performance at high frequency are easily achieved, while consuming small quiescent current from FBCT. This high-reliability gate driver can thus meet high-frequency high-power-density PoL applications for GaN devices very well.

REFERENCES

[1] Fairchild Semiconductor Products, *Design and Application Guide of Bootstrap Circuit for HV Gate-Drive IC*, Application Note AN-6076, Sep. 2008.

[2] Z. Liu, L. Cong and H. Lee, "Design of on-chip gate drivers with power-efficient high-speed level shifting and dynamic timing control for high-voltage synchronous switching power converters," *IEEE J. Solid-State Circuits*, vol. 50, no. 6, pp. 1463–1477, Jun. 2015.

[3] X. Ming, *et al.*, "A high-voltage half-bridge gate drive circuit for GaN devices with high-speed low-power and high-noise-immunity level shifter," in *Proc. Int. Symp. IEEE Power Semicond. Devices*, Chicago, May 2018, pp. 355–358

[4] Y. Y. Lu, *et al.*, "A 600V high-side gate drive circuit with ultra-low propagation delay for enhancement mode GaN devices," in *Proc. Int. Symp. IEEE Power Semicond. Devices*, Chicago, May 2018, pp. 80–83

[5] D. W. Liu, *et al.*, "Design of 370-ps delay floating-voltage level shifters with 30-V/ns power supply slew tolerance," *IEEE Trans. Circuits Syst. II, Exp. Briefs*, vol. 63, no. 7, pp. 688–692, Jul. 2016.

[6] X. Ke, J. Sankman, M. K. Song, P. Forghani, and D. B. Ma, "A 3-to-40 V 10-to-30 MHz automotive-use GaN driver with active BST balancing and VSW dual-edge dead-time modulation achieving 8.3% efficiency improvement and 3.4 ns constant propagation delay," in *IEEE Int. Solid-State Circuits Conf. (ISSCC) Dig. Tech. Papers*, Feb. 2016, pp. 302–304.

[7] J. Delaine, *et al.*, "High frequency DC-DC converter using GaN device," in *Applied Power Electronics Conference and Exposition*, 2012. APEC '12. pp. 1754•1761

[8] EPC: AN015, "Introducing a family of eGaN FETs for multi-megahertz hard switching applications," 2014, <http://epc-co.com/epc/documents/producttraining/AN105_eGaN_FETs_for_Multi-Megahertz_Applications.pdf>.

[9] M. K. Song, L. Chen, J. Sankman, S. Terry, and D. Ma, "A 20 V 8.4 W 20 MHz four-phase GaN DC-DC converter with fully on-chip dual-SR bootstrapped GaN FET driver achieving 4 ns constant propagation delay and 1 ns switching rise time," in *IEEE Int. Solid-State Circuits Conf. (ISSCC) Dig. Tech. Papers*, Feb. 2015, pp. 302–303.

[10] Xugang Ke, *et al.*, "A 3-to-40V V_{IN} 10-to-50MHz 12W isolated GaN driver with self-excited t_{dead} minimizer achieving 0.2ns/0.3ns t_{dead}, 7.9% minimum duty ratio and 50V/ns CMTI," in *IEEE Int. Solid-State Circuits Conf. (ISSCC) Dig. Tech. Papers*, pp. 386–388, Feb. 2018.

Proceedings of the 31st International Symposium on Power Semiconductor Devices & ICs
May 19-23, 2019, Shanghai, China

A Predictive Gate Driver Suitable for Half-Bridge Applications

Zekun Zhou[*], *IEEE Member*, Yandong Yuan, Yunkun Wang,
Bo Zhang, *IEEE Senior Member*
State key Laboratory of Electronic Thin Films and Integrated Devices
University of Electronic Science and Technology of China,
Chengdu, China
[*]zkzhou@uestc.edu.cn

Yue Shi[+]
College of Communication Engineering,
Chengdu University of Information Technology, Chengdu, China
[+]october@cuit.edu.cn

Abstract—A predictive dead-time (DT) gate driver for half-bridge applications is proposed in this paper. Compared with the conventional DT control methods, the proposed method adopts the DT information of previous cycle to control the state of current cycle, which has distinct merits in avoiding the inherent efficiency loss. Two specifically designed sampling circuits are adopted for the purpose of improving universality and robustness, which enables the proposed gate driver suitable for various applications, e.g. Si, SiC and GaN. The proposed gate driver has been fabricated in a 0.35 μm BiCMOS technology with an active area of 0.836 mm². Simulation and experimental results show precisely controlled dead-time, which is approximately zero, over variable load conditions and input voltages.

Keywords—*gate driver, predictive, universality, approximately zero.*

I. INTRODUCTION

For half-bridge applications, such as a synchronous Buck converter, when the switch power transistor turns off, an active controlled power transistor, as a freewheeling device, is required to flow the load current until the switch power transistor turns on again. However, supposed the two power transistors were simultaneously turned on, a large current passing through both transistors would be induced. This large current would lead to significant power loss or even damage of the converter. Thereby, dead-time (DT) is inevitably needed to be inserted into control signals of the two power transistors in half-bridge applications.

However, a non-optimal dead time will cause additional power loss, which is mainly composed by three parts: (1) cross conduction between high side and low side power MOSFET, (2) forward conduction and (3) reverse recovery at the low-side body diode. Since the dead time highly depends on the load current and the input voltage [1], dead time regulation is required.

Conventional approaches are mainly using a fixed dead-time [2] as for their simplicity. However, this approach results in a quite non-optimal design and large power loss, which are induced by the long-time conduction of parasitic body-diodes for safety.

Considering the above problems of fixed dead-time, other approaches, i.e. adaptive dead-time [3], are used. Unfortunately, these techniques are inaccurate to determine whether switch transistors are off by solely monitoring the switch-node voltage, leading to body-diode conduction. Besides, this efficiency loss associated with system inherent propagation delay in adaptive DT method greatly increases with higher switch frequency and larger loading current, which is development trend in half-bridge applications. This

makes the state-of-the-art power loss relevant to DT non-negligible, especially in GaN and SiC systems.

For the purpose of solving the mentioned problems, a gate driver with predictive dead-time control method is proposed. In Section II, the principle of the proposed gate driver with predictive DT control method is introduced in detail. Then, in section III, the circuit implementation for the proposed gate driver is displayed and discussed. Section IV shows the simulation and experimental results. The conclusion of this paper is given in Section V.

II. PRINCIPLE OF THE PROPOSED GATE DRIVER

Fig. 1 shows the block diagram of the proposed gate driver integrated circuit (IC). The driver chains (DC) in Fig. 1 are used to enhance the drive capability and reduce electro-magnetic interference (EMI). The DT control method is divided into three subintervals for more detailed explanation, i.e., sample, quantization and regulation.

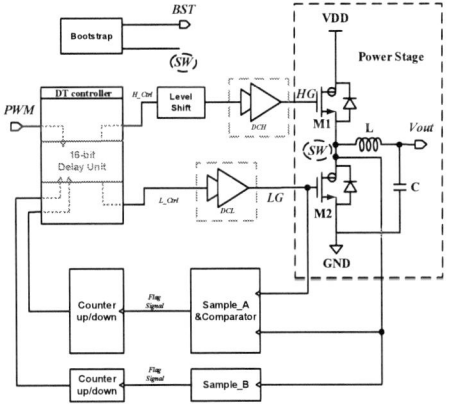

Fig. 1 Block diagram of the proposed gate driver IC

First of all, the information of DT_off (DT of M1 turn-on) and DT_on (DT of M1 turn-off) are detected by Sample_A and Sample_B respectively. The specific detection principle is shown in Fig. 2. As we can see from Fig. 2(a) and Fig. 2(b), Sample_A monitors the falling edge of V_{SW} and rise edge of V_{GS_M2} simultaneously. The comparator then uses the state information obtained from the sample to generate a flag signal. Fig. 2(c) and Fig. 2(d) depict that Sample_B detects the state information of DT_on by monitoring SW node. If the voltage is below a given potential, the sample generates a pulse flag signal, or vice versa.

978-1-7281-0582-6/19 $31.00 © 2019 IEEE   131

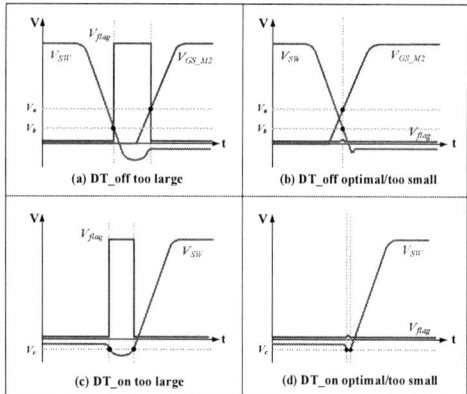

Fig. 2 Illustration graphs of DT detection principle.

The core module DT controller is then controlled by two digital up-down-counters, which increase or decrease DT if the two samples give a pulse flag signal, or a constant low voltage potential.

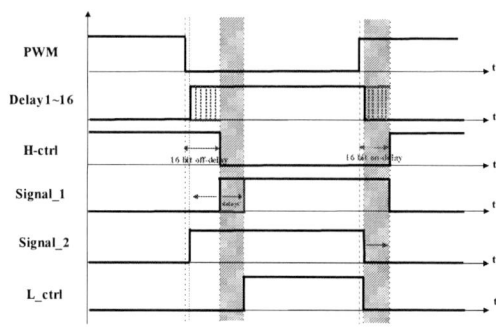

DT_off = (delayC-H_fall+L_rise-SW_fall)-n*per_off_delay；
DT_on = (15*per_on_delay-L_fall+H_rise)-n*per_on_delay。

Fig. 3 Illustration graphs of DT control method.

The dead times are regulated by a 16-bit delay unit with a nominal dead time resolution of 4ns. High side power MOSFET control signal (H-ctrl) is selected as delay-16, which is the longest delay signal. And the low side power MOSFET control signal (L-ctrl) is selected according to the output signal of counter. More specifically, signal_1 and signal_2 in Fig. 3, which are used to generate L-ctrl, are used to regulate DT_off and DT_on, respectively. Furthermore, a constant delay is inserted into signal_1. Thereby, the adjusted DT can be given as shown in Fig.3.

III. CIRCUIT REALIZATION

As mentioned in section II, two DCs are used to enhance the drive capability and reduce EMI. Fig. 4 shows the circuit implementation of the DCs, which have an identical structure. NOR and NAND generate a fixed dead-time [6] for PMOS (PG1, PG2 and PG3) and NMOS (NG1) as for their ampere level peak current. The three PMOS are driven step by step to decrease the risk of dv/dt turn on of M1 and ringing caused by

bonding wire. Moreover, EMI of the power stage is also reduced.

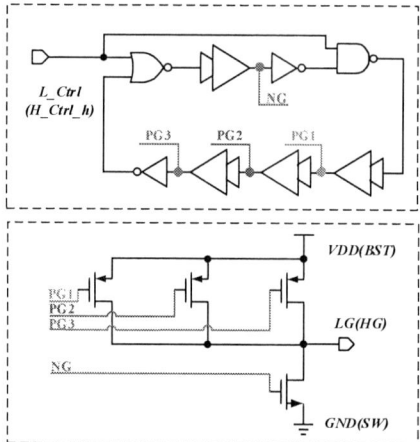

Fig. 4 Circuit implementation of DC

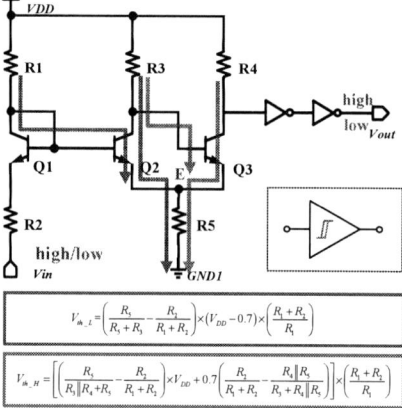

$$V_{in\_L} = \left(\frac{R_5}{R_5+R_3} - \frac{R_2}{R_1+R_2} \right) \times (V_{DD}-0.7) \times \left(\frac{R_1+R_2}{R_2} \right)$$

$$V_{in\_H} = \left[\left(\frac{R_5}{R_5 \| R_4 + R_3} - \frac{R_2}{R_1+R_2} \right) \times V_{DD} + 0.7 \left(\frac{R_2}{R_1+R_2} - \frac{R_4 \| R_5}{R_4 \| R_5} \right) \right] \times \left(\frac{R_1+R_2}{R_2} \right)$$

Fig. 5 Circuit implementation of Sample_A

As shown in Fig. 4 and Fig. 5, two Sample_A are used to monitor V_{SW} and V_{GS_M2} respectively. The flag signal is subsequently generated by a comparator according to the two samples' status. Thereby, the information of DT_off is detected and transformed to a flag signal. In order to prevent the interference of SW ringing, sample is designed with hysteresis.

Unlike the Sample_A, information of DT_on is solely detected by one Sample_B. In Buck convertor, the on-resistance of the power device is affected by the PVT conditions. In order to prevent the negative voltage detection circuit from being erroneously triggered during the conduction of the lower power device, it is necessary to set the negative voltage detection point to be lower than the lowest power device conduction voltage under all PVT conditions. Unfortunately, the design of ensuring the reliability of the circuit seriously affects the detection performance of the circuit. To this end, MNR is added into a conventional source input voltage comparator, as shown in Fig. 6, to ensure the reliability and performance at the same time.

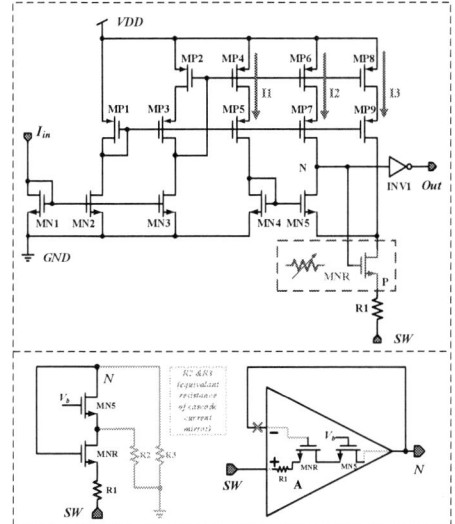

Fig. 6 Circuit implementation of Sample_B

Furthermore, a negative feedback loop is established to prevent the false triggering caused by SW ringing. In Fig. 6, the close-loop gain from point N to point SW can be given as:

$$A_{close} = \frac{A_{OL}}{1+|LG|} = \frac{(g_{mR}ro_R+1)(g_{m5}ro_5+1)}{g_{mR}ro_R(g_{m5}ro_5+1)+1} \quad (1)$$

where the LG can be given as:

$$|LG| = G_{M1}R_{OUT} = g_{mR}ro_R(g_{m5}ro_5+1) \quad (2)$$

When the voltage at node SW is higher than or near the negative voltage detection point, MNR is in the linear region. Meanwhile, the current mirror MP6 and MP7 are also in the linear region. This phenomenon leads to reduction of the loop gain，and the close-loop gain is very large. On the contrary, when the voltage at node SW is lower than the negative voltage detection point, MNR enters the saturation region, and the loop gain increases. The reduction of close-loop gain is more serious when the voltage at node SW is lower than the negative voltage detection point, thereby limiting the amplification effect of the comparator on the ringing signal at the switch node SW. This phenomenon effectively avoids false triggering caused by ringing.

The simulation result of dead-time with respect to the load current and input voltage is shown in Fig. 7. In all cases, dead-time of the proposed gate driver remains optimal.

IV. EXPERIMENTAL RESULTS AND DISCUSSION

The proposed gate driver has been fabricated using a 0.35 μm BiCMOS technology, and the chip area is 0.836 mm².

Chip micrograph, PCB layout and test board of the gate drive are shown in Fig. 9.

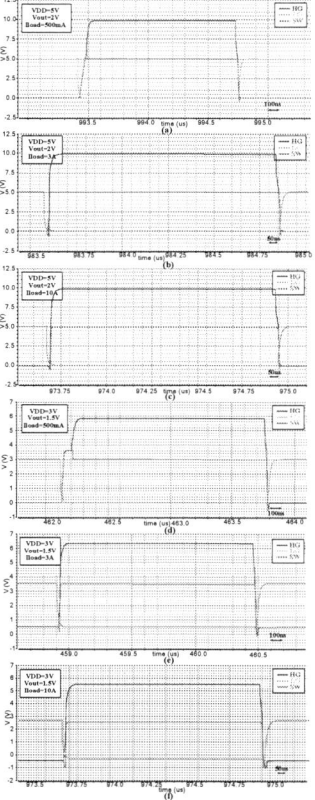

Fig. 7 Simulation results of proposed gate driver.

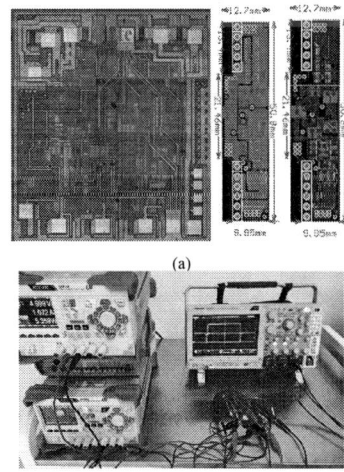

Fig. 8 Chip micrograph, PCB layout and test board of the gate driver.

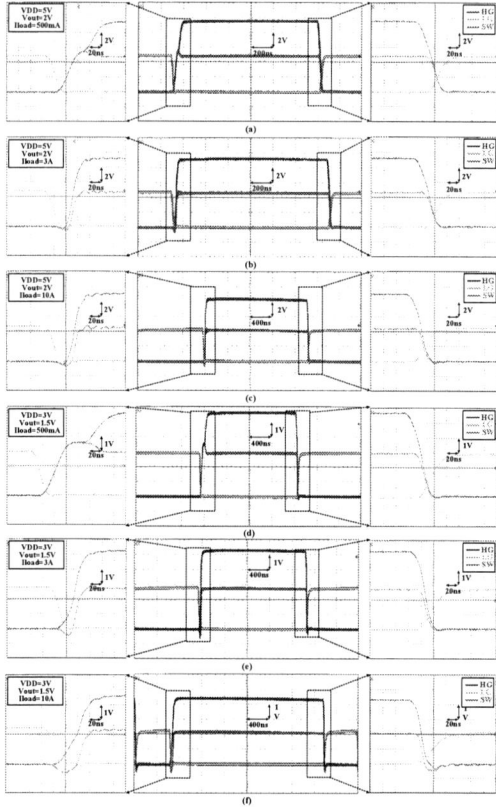

Fig. 9 Measured switching waveform of a Buck converter at f_{sw}=300kHz.

Fig. 9 gives the measured switching waveform of proposed gate driver. The DT is tuned cycle by cycle due to its control method. Initially, the DT is about decades of nanoseconds at both transitions. After dozens of cycles, DT_on and DT_off are both tuned to approximately zero. Fig. 9(a) and Fig. 9(d) show abnormal DT_on due to their FCCM working mode.

V. CONCLUSION

A predictive DT gate driver for half-bridge applications is proposed in this paper. The proposed gate driver IC enables approximately zero DT over variable load status, input voltages and power MOSFET conditions, e.g., type, size or process corner. Two specifically designed simples enable improved universality and robustness, which enables the proposed gate driver suitable for various applications, e.g. Si, SiC and GaN.

ACKNOWLEDGMENT

This work was supported by the National Science Foundation of China under Grant No. 61674025, No. 61306035, by the Fundamental Research Funds for the Central Universities under Grant No. ZYGX2016J056, by the Open Projects of Sichuan Key Laboratory of Meteorological Information and Signal Processing under Grant No. QXXCSYS201504, No. QXXCSYS201603.

REFERENCES

[1] Wittmann J, Barner A, Rosahl T, and Wicht B, "An 18 V Input 10 MHz Buck Converter With 125 ps Mixed-Signal Dead Time Control." IEEE J. Solid-State Circuits, Vol. 51, No. 7, pp. 1705-1715, JUL. 2016.

[2] W. R. Liou, M. L. Yeh, and Y. L. Kuo, "A High Efficiency Dual-Mode Buck Converter IC For Portable Applications", IEEE Trans. Power Electronics, Vol. 23, No. 2, pp. 667-677, MAR. 2008.

[3] T. Y. Man, P. K. T. Mork, and M. Chan, "An Auto-Selectable - Frequency Pulse-Width Modulator for Buck Converters with Improved Light-Load Efficiency", IEEE Int. Solid-State Circuits Conf., Vol. 51, pp. 440-626, 2008.

[4] Lee Hoi, Ryu Seong-Ryong, "An Efficiency-Enhanced DCM Buck Regulator with Improved Switching Timing of Power Transistors", IEEE Trans. Circuits Syst. II, Exp. Briefs, Vol. 57, No. 3, pp. 33-44, MAR. 2010.

[5] M. Zhou, Q. W. Low, S. Liter, "A High Efficiency Synchronous Buck Converter with Adaptive Dead-Time Control", IEEE Int. Symposuim on Itegrated Circuits, DEC. 2016.

[6] S. Lee, S. Jung, H. Jin, and C. Park, "Robust and efficient synchronous buck converter with near-optimal dead-time control", IEEE Int. Solid-State Circuits Conf., pp. 392-394. 2011.

Proceedings of the 31st International Symposium on Power Semiconductor Devices & ICs
May 19-23, 2019, Shanghai, China

SiC MOSFET with Integrated Zener Diode as an Asymmetric Bidirectional Voltage Clamp Between the Gate and Source for Overvoltage Protection

Cheng-Tyng Yen, Fu-Jen Hsu, Kuo-Ting Chu, Chien-Chung Hung, Lurng-Shehng Lee, Chwan-Ying Lee
Hestia Power Inc.
10F-2, 27 Guanxin Rd, Hsinchu, Taiwan, R.O.C.
Email: ct.yen@hestia-power.com

Abstract—**An integrated back-to-back SiC Zener diode is proposed and designed as an asymmetric bidirectional voltage clamp between the gate and source to protect the gate oxide of SiC MOSFET from the overvoltage stress. The leakage current of integrated SiC Zener diode is very low compared to conventional silicon poly Zener diode used for the same purpose and close to that of gate oxide, even at high temperatures. The pulsed current test shows that the overvoltage can be clamped by the integrated Zener and the characteristics of Zener integrated MOSFET did not degrade or change after continuous switching.**

Keywords- SiC MOSFET, Voltage Clamp, Zener, Gate Oxide, Bidirectional, Reliability, Overvoltage Protection

I. INTRODUCTION

SiC MOSFETs have rapidly emerged in many power electronic applications, such as switch mode power supply, fast charging station and DC/DC converter, on board charger (OBC) and main inverter for electric vehicles. The output characteristics of SiC MOSFETs and Si MOSFETs are very different. In Si MOSFETs, the on-resistance ($R_{DS(on)}$) MOSFET saturates after certain gate voltage (V_{GS}), e.g. applying V_{GS} larger than 12V usually gives no extra gains on $R_{DS(on)}$ for almost all commercial Si power MOSFETs. However, the $R_{DS(on)}$ of SiC MOSFET tends to continuously decrease by applying higher V_{GS}, this is why a higher V_{GS} is usually recommended in SiC MOSFET. This practice not only increases the stress of electric field exerted on the gate oxide during the operations of SiC MOSFETs but also reduces the allowable margin for overvoltage. For instance, in SiC MOSFET, the recommended operating V_{GS} is usually 80% of maximum rated V_{GS}, while in Si MOSFET, the recommended operating V_{GS} is usually less than 50% of its maximum rating. In the mean time, the high di/dt and dv/dt of fast switching SiC MOFSET can easily induce gate ringing more significant than Si MOSFET which, if not being taken care, will risk in damaging gate oxide by transient overvoltage. Therefore, a protective measure which can be used to constrain the V_{GS} spikes will be helpful to ensure a more reliable long-term operation of SiC MOSFETs.

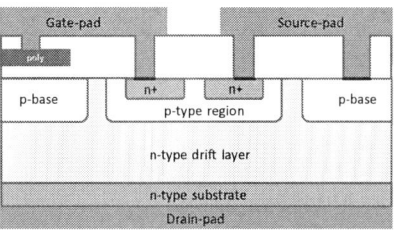

Fig. 1. The cross-section of integrated SiC Zener diode.

II. DEVICE CONCEPT AND STRUCTURE

The conventional approaches to protect the gate oxide include connecting external Zener diodes on the board level or integrating poly Zener diodes [1] in the gate pad area of MOSFET. External Zener diodes may respond slowly due to introduced parasitic inductance and integrated poly Zener diodes usually result in a much higher gate-to-source leakage current compare to plain MOSFET. The poly Zener diodes are also less controllable at elevated temperatures. Those are the reasons Kaguchi et al. of Mitsubishi Electric chose to incorporate an additional lateral SiC PMOS to unidirectionally protect gate oxide from negative overvoltages [2]. In this work, we demonstrate an asymmetric bidirectional voltage clamp between the gate and source by integrating a back-to-back SiC Zener diode, as the cross-section illustrated in Fig.1, to provide both positive and negative overvoltage protections adapted for the asymmetric operating range of V_{GS} in SiC MOSFET. The

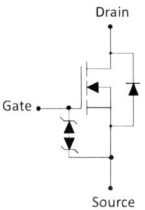

Fig. 2. The inner circuit of SiC MOSFET with integrated Zener diode.

978-1-7281-0582-6/19 $31.00 © 2019 IEEE

integrated Zener diode was formed by a floating p-type region and two separate n+ regions embedded between the gate and the source, where one of the n+ regions was electrically connected to the gate, and the other n+ region was electrically connected to the source through commonly formed Ohmic contacts. The inner circuit of SiC MOSFET with integrated Zener diode is shown in Fig.2. Because this integrated Zener diode was disposed in the space between the gate-pad and the source-pad, it essentially requires no extra area.

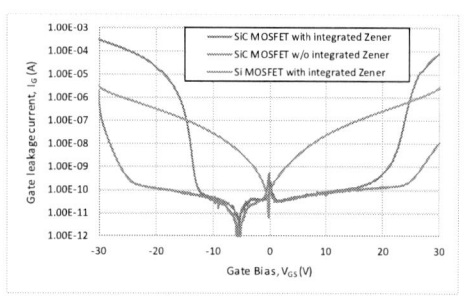

Fig. 3. The gate leakage current vs. gate-source voltage (I_G-V_{GS}) characteristics of SiC MOSFET with integrated Zener at different temperatures.

III. ELECTRIC CHARACTERISTICS

Fig.3 compares the gate leakage current versus gate-source voltage (I_G-V_{GS}) characteristics at 30℃ among SiC MOSFET with integrated Zener diode, without integrated Zener diode and a commercial Si MOSFET with integrated poly Zener diode. The asymmetric breakdowns of integrated back to back SiC

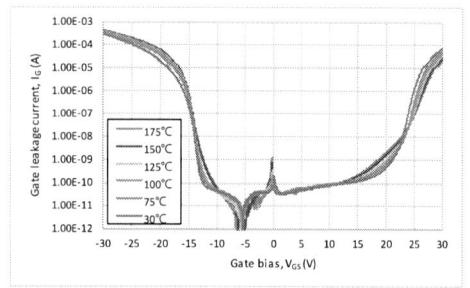

Fig. 4. I_G-V_{GS} characteristics of SiC MOSFET with integrated Zener at different temperatures.

Zener diode are intendedly designed for the asymmetric maximum operational range of V_{GS} in SiC MOSFET datasheet (-10V to +25V in this work). The integrated back-to-back SiC Zener diode showed breakdown around V_{GS}= +21V and V_{GS}= -13.5V if breakdown was defined at I_G=1nA, which happened before the onset of Fowler-Nordheim (FN) tunneling around V_{GS}= +/- 25V. The leakage currents of integrated SiC Zener diode at the recommended operational V_{GS} at 20V and -10V

were about three orders lower than the leakage currents of Si poly Zener at the same gate voltages. The leakage current of integrated Zener diode is almost negligible before the diode breakdown and thus enables SiC MOSFET with integrated

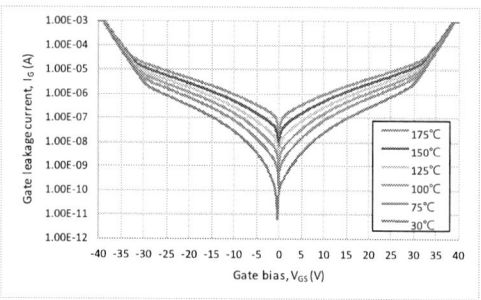

Fig. 4. I_G-V_{GS} characteristics of commercial Si MOSFET with integrated poly Zener.

Zener diode exhibits a very low gate leakage current comparable to the standard SiC MOSFET.

Fig.4 shows that the I_G-V_{GS} characteristics of SiC MOSFET with integrated Zener only changed a little and leakage current stayed very low with increasing temperature. The gate leakage current $I_G@V_{GS}$=20V and $I_G@V_{GS}$=-10V were only 2.15nA and 66pA correspondingly at 175℃.

On the other hand, I_G-V_{GS} of commercial Si MOSFET with integrated poly Zener changed significantly and leakage currents increased rapidly with increasing temperature as shown in Fig.5. The gate leakage current $I_G@V_{GS}$=20V and $I_G@V_{GS}$=-10V increased to 4.27 μA and 0.95 μA at 175℃, which is three orders larger than the integrated Zener diode in this work. A high gate leakage current would create a heavy burden on the gate driver, result in considerable driving loss and limit the highest frequency can be operated.

The very low leakage current of integrated SiC Zener diode at high temperature is because of the wide bandgap and low intrinsic carrier concentration of SiC [3].

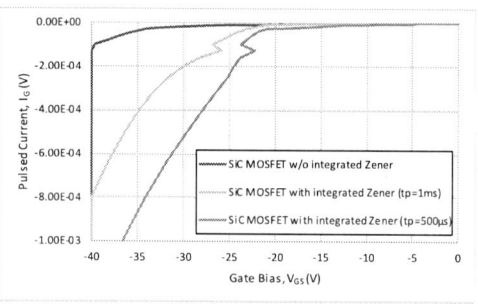

Fig. 6. Transient responses of gate voltage by ramping up negative pulsed currents with 500µs and 1ms durations (tp). The voltage compliance was 40V.

978-1-7281-0582-6/19 $31.00 © 2019 IEEE

Fig.5 and Fig.6 compare the transient responses of gate voltages of SiC MOSFET with and without integrated Zener diode by ramping up positive and negative pulsed currents. Fig.5 and Fig.6 shows that the integrated Zener diode did effective clamp the V_{GS} to a lower value when a certain amount of charge was injected into the gate and the clamping was more effective when the pulse width of injected gate current was shorter. On the other hand, a tiny I_G would be enough to induce a over 40V V_{GS} in standard SiC MOSFET. The asymmetric clamped gate voltages (larger V_{GS} with positive gate current, smaller V_{GS} with negative gate current) consist with the sweeping I_G-V_{GS} curve shown in Fig.3.

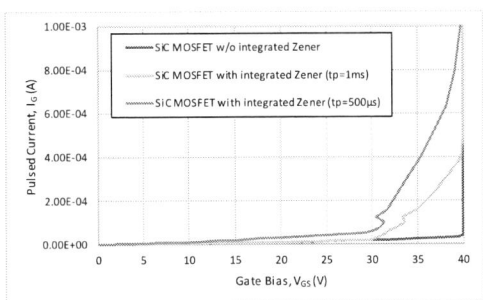

Fig. 5. Transient responses of gate voltage by ramping up positive pulsed currents with 500µs and 1ms durations (tp). The voltage compliance was 40V.

The capability that the integrated SiC Zener diode could successfully clampe the gate voltage by absorbing gate currents would be helpful to reduce the stress induced by the electric field of overvoltage on the gate oxide and improve the long term reliability of SiC MOSFET.

IV. DYAMIC CHARACTERISTICS

Fig.7 shows the input capacitance (Ciss), output capacitance (Coss) and reverse transfer capacitance of SiC MOSFET with integrated Zener diode with respect to V_{DS}. No obvious differences were found suggested that the capacitance contributed by the integrated Zener diode was negligible according to Fig.7. The reverse characteristic of SiC MOSFET with integrated Zener shown in Fig.8 was also normal compared to standard SiC MOSFET.

To investigate if there is any problem during the switching might be caused by the floating p-type region used to form the integrated Zener diode as shown in Fig.1, the resistive load dynamic test of SiC MOSFET with integrated Zener diode was performed. The MOSFET was switched between V_{DS}= 600V and I_D = 3A with V_{GS} = 0V/20V. Fig.9 and Fig.10 show the switching waveforms of the dynamic test. There was no obvious difference as compared to standard SiC MOSFET. No

degradations were found after more than 8.6×10^8 times of switching according to the post-test parametric verification.

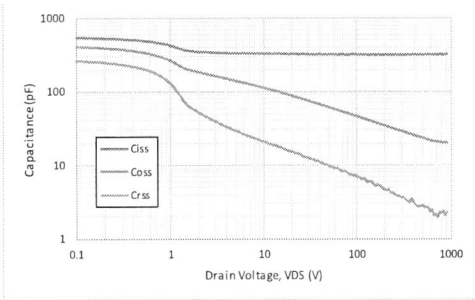

Fig. 7. Capacitance-voltage characteristics of SiC MOSFET with integrated Zener.

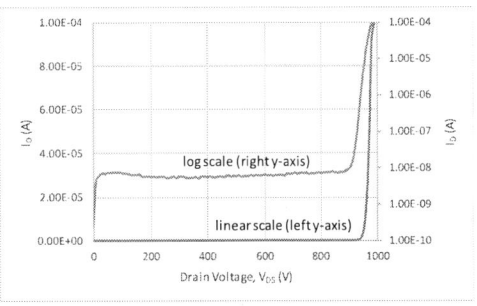

Fig. 8. Zero gate voltage reverse characteristic of Zener integrated MOSFET.

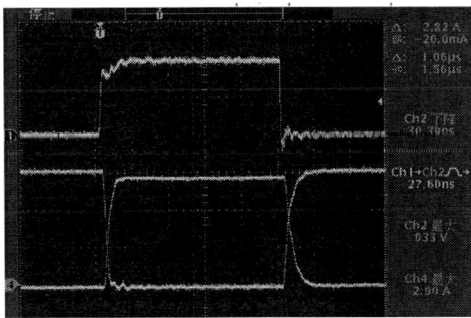

Fig. 9. Resistive load switching waveform of SiC MOSFET with integrated Zener, with V_{GS}=0/20V, V_{DS}=600V, I_D=3A and $R_{G,ext}$=10Ω; where Ch1=V_{GS} where Ch1=V_{GS}, 10V/div; Ch2=V_{DS}, 200V/div, Ch4=I_D, 1A/div; t=200ns/div.

978-1-7281-0582-6/19 $31.00 © 2019 IEEE   137

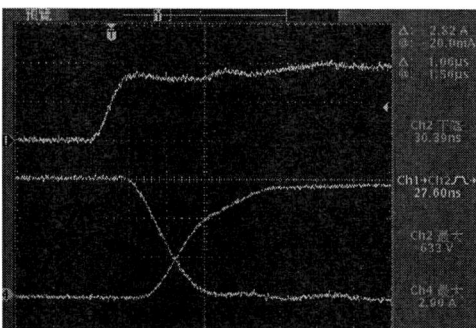

Fig. 10. Zoomed in turn-on waveform of SiC MOSFET with integrated Zener with V_{GS}=0/20V, V_{DS}=600V, I_D=3A and $R_{G,ext}$=10Ω; where Ch1=V_{GS}, 10V/div; Ch2=V_{DS}, 200V/div, Ch4=I_D, 1A/div; t=20ns/div.

compared to silicon poly Zener used in commercial Si MOSFET at both room temperature and high temperatures. A low leakage current can mitigate the loading of gate driver and reduce the energy loss from gate drivers. The pulsed current test showed that the integrated SiC Zener could successfully clamped gate voltage to a value lower than the onset of FN tunneling, which will be helpful to minimize the stress caused by the electric field of overvoltage during the operation of SiC MOSFET and potentially improve the long-term reliability of gate oxide. The concern that a floating p-type region existed in the structure of integrated Zener diode might have an adverse impact was checked by continuous switching test and confirmed that there were neither failures nor degradations was found after more than 8.6×10^8 times of switching. The capability of this integrated SiC Zener diode in absorbing electrostatic discharge (ESD) energies will be studied in the future.

V. CONCLUSION

An integrated back-to-back SiC Zener diode designed as an asymmetric bidirectional voltage clamp between the gate and source to protect the gate oxide of SiC MOSFET from the overvoltage stress was demonstrated in this work. The integrated SiC Zener exhibited very low leakage current

REFERENCES

[1] R. K. Williams, US 6,172,383, Power MOSFET Having Voltage Clamped Gate.

[2] N. Kaguchi, E. Suekawa, M. Ikegami, US 9,627,383, Semiconductor Device.

[3] R. Ishii, H. Tsuchida, K. Nakayama, Y. Sugawara, "20V-400A SiC Zener Diodes with Excellent Temperature Coefficient", Proc. Intl. Symp. Power Sem. & ICs (ISPSD) May 27-30, 2007, Jeju, Korea.

Proceedings of the 31st International Symposium on Power Semiconductor Devices & ICs
May 19-23, 2019, Shanghai, China

Breakthrough in Channel Mobility Limit of Nitrided Gate Insulator for SiC DMOSFET with Novel High-temperature N_2 Annealing

S. Asaba[1], T. Ito[1], S. Fukatsu[1], Y. Nakabayashi[1], T. Shimizu[1], M. Furukawa[2], T. Suzuki[2] and R. Iijima[1]

[1]Corporate R&D Center, Toshiba Corporation, 1, Komukai Toshiba-cho, Saiwai-ku, Kawasaki-shi, 212-8582, Japan
[2]Discrete Semiconductor Div., Toshiba Electronic Devices & Storage Corporation,
300, Ikaruga, Taisi-cho, Ibo-gun, Hyogo 671-1595, Japan
Email: shunsuke.asaba@toshiba.co.jp

Abstract—**We have improved the limited channel mobility of nitridation on silicon carbide Si-face planar MOSFET to 50 cm^2/Vs with the newly developed gate insulator process. Using the novel high-temperature N_2 annealing process, nitrogen whose density is twice that of the nitrogen used in the conventional NO_X process can be introduced to the MOS interface. The developed N_2 annealing is applied to a 1.2 kV class vertical MOSFET and remarkable reduction of the on-resistance has been demonstrated.**

Keywords—SiC, MOSFET, interface, mobility, reliability

I. INTRODUCTION

The SiC-MOSFET is an attractive candidate for high-voltage power devices with low-loss and high switching speed. For a MOS device on SiC, passivation of interface defects of SiO_2 and SiC is required in order to reduce channel resistance. Although venturesome techniques of passivation by P, B and Ba realize high mobility of over 80 cm^2/Vs [1-3], some subjects related to reliability have to be resolved. Nitridation annealing is the standard gate insulator process technique for fabricating SiC-MOSFET products with credible gate reliability. From the viewpoint of their practical use, the conventional thermal annealing process to nitride the interface with NO_X (NO or N_2O) ambient has been the only reasonable method of fabricating MOSFET products [4-8]. Although R&D with a view to reducing the on-resistance is being pursued, the channel mobility is still limited to no more than 30 cm^2/Vs. In order to obtain a MOSFET with lower on-resistance, a breakthrough technique to increase the channel mobility with the passivation based on interface nitridation is desired. We have recently developed a nitridation process with pre-treatment annealing prior to high-temperature N_2 annealing that realizes superior channel mobility of 50 cm^2/Vs and excellent reliability [9-10]. This paper demonstrates that a high-quality nitrided interface is obtained by controlling oxidation and nitridation. In addition, the developed gate insulator process significantly reduces on-resistance of 1.2 kV class vertical power MOSFET because of high mobility attributable to improved interface quality

II. THE DEVELOPED N_2 ANNEALING PROCESS

A. Sequence for Forming Gate Insulator

The gate insulator discussed in this study was formed on Si-face 4H-SiC substrates by means of the process flow shown in Figure 1. As a reference, conventional nitridation annealing was carried out using NO_X ambient following SiO_2 deposition. Instead of the NO_X process, gate insulator was also formed by the developed high-temperature N_2 annealing

process consisting of two steps. The thermal treatment step was carried out prior to the N_2 annealing step, with O_2 ambient at a temperature of less than 900°C, which we call the pre-annealing. In this temperature range, obvious oxide growth tends not to occur on SiC surface. Subsequently, the substrate was annealed in N_2 ambient at the moderately high temperature of 1300°C. Thereafter, the conventional device manufacturing process for MOS capacitor, lateral MOSFET and vertical power MOSFET with planar gate structure was continued, then samples with gate insulator formed by the NO_X annealing and the high-temperature N_2 annealing were fabricated.

B. Effective Nitrogen Incorporation

The developed high-temperature N_2 annealing process is designed to provide superior controllability of both oxidation and nitridation at the interface. Assumed basic reactions in nitridation are illustrated in Figure 2. Oxygen plays an important role in determining the behavior of nitrogen around the interface. Nitrogen incorporation requires the assistance of oxygen. Nitrogen is incorporated in vacant carbon sites that appear after carbon is oxidized to CO. On the other hand, oxygen could be an obstacle because incorporated nitrogen is removed as the oxidation of SiC surface proceeds. The effective density of incorporated nitrogen is determined by the introduction rate and the absorption rate. The developed high-temperature N_2 annealing process probably controls introduction and desorption of nitrogen independently by performing oxidation caused by the pre-annealing and nitridation caused by N_2 annealing separately. Owing to this method, the process window of the nitridation reaction is extended.

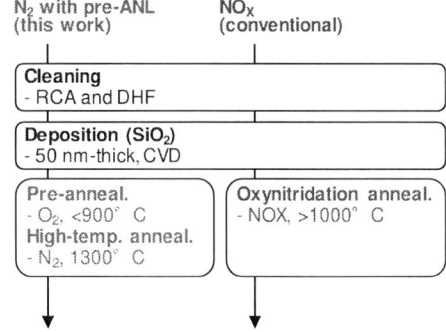

Fig. 1: Process flow to form gate insulator

978-1-7281-0582-6/19 $31.00 © 2019 IEEE

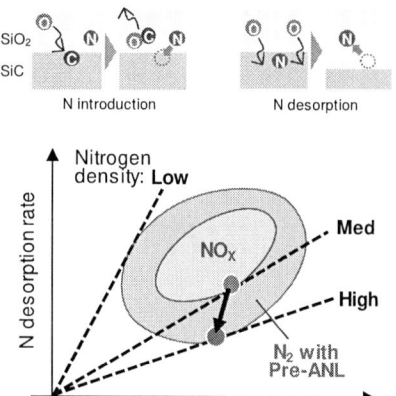

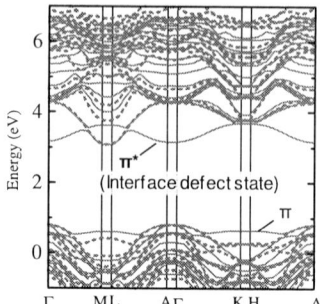

Fig. 4: Energy band diagram of the interface (first-principles calculation). Interface states near the conduction band edge disappear by N incorporation (Dotted red lines: N substitution, solid gray lines: without N)

Fig. 2: Achievable area of interface reaction of nitrogen introduction and absorption for each process. Dots represent optimized condition. Effective density of nitrogen such that substitute carbon sites at the interface are increased by using the high-temperature N_2 annealing.

III. IMPROVED INTERFACE QUALITY

A. *Interface State Reduction by Nitrogen Termination*

Figure 3 shows XPS spectra of N1s measured at the area that is etched sufficiently with dHF. The chemical state of nitrogen is identified as dominating NSi_3 as was also reported in the case of NO annealing [7]. This bonding structure can be explained in terms of nitrogen substitution of carbon sites below the silicon sites of the surface layer of Si-face SiC. When carbon is not replaced by nitrogen, carbon is possibly to form C=C bonding. Substitution of carbon sites by nitrogen leads to suppression of forming C=C defect. Theoretical calculation supports substitution of carbon site by nitrogen is effective to diminish energy state that is crucial to electron scattering (Fig. 4).

We note that the interface structure formed by the developed high-temperature N_2 annealing is qualitatively

similar to that formed by conventional NO_X annealing and has no exotic bonding structure. The confirmed superior reliability of the gate insulator formed by using the high-temperature N_2 annealing, such as credible threshold voltage stability, high breakdown tolerance and practicable long lifetime, are merits for nitridation process technology [10-11].

B. *Mobility Enhancement by High Concentration Nitrogen*

Lateral MOSFETs were fabricated using various process conditions. Mobility characteristics changed in response to the temperature of the pre-annealing, duration of the N_2 annealing and so on. The relation of peak mobilities and interface nitrogen densities of MOSFETs with various process conditions are summarized in Figure 5. We succeeded in controlling interface nitridation by tuning the process conditions. The doubled concentration of nitrogen is incorporated and, as a result, peak mobility is enhanced from 30 to 50 cm^2/Vs, strongly depending on nitrogen density [8]. The structure of the interface formed by the N_2 annealing is considered to be not only qualitatively similar to that formed by NO_X annealing but quantitatively superior owing to higher nitrogen density.

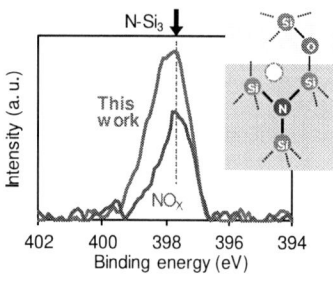

Fig. 3: XPS spectra of N1s measured after sufficient removal of SiO_2 by DHF. Inset figure schematically shows bonding structure of nitrogen.

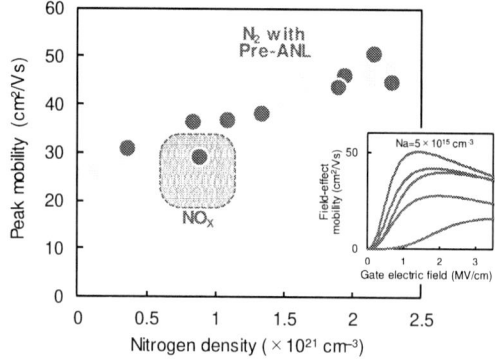

Fig. 5: Dependence of peak mobility of lateral MOSFET on the interface nitrogen density. Inset figure shows mobility characteristics for samples fabricated with some process conditions.

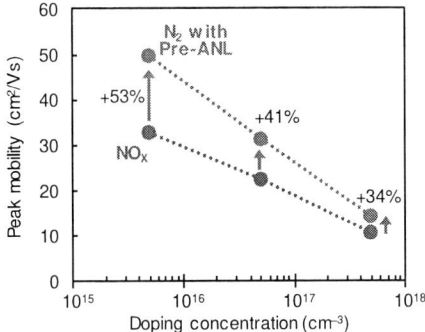

Fig. 6: Peak mobility for several p-base doping concentrations.

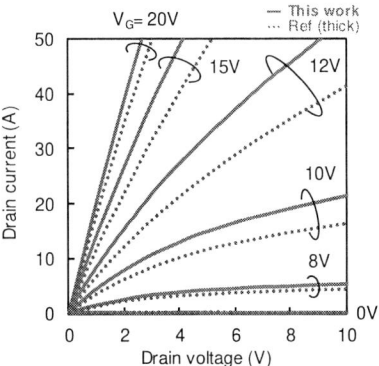

Fig. 8: I_D-V_D characteristics for various V_G. (Solid line: the high-temperature N_2 annealing dotted line: NO_X annealing)

Figure 6 shows peak mobilities of lateral MOSFETs fabricated on the substrates with p-base doping concentration of 5×10^{15}, 5×10^{16} and 5×10^{17} cm^{-3} by the N_2 annealing process with optimized conditions. The mobility enhancement effect was evidently confirmed throughout the doping concentration range. Mobility gain was over 40% at the practical p-base doping concentration for SiC-MOSFET product design. The developed high-temperature N_2 annealing is a promising method for the gate insulator process that reduces channel resistance of SiC-MOSFET.

IV. PERFORMANCE IMPROVEMENT OF SiC-DMOSFET

Finally, the developed N_2 annealing was applied to the gate insulator process of power MOSFETs. The MOSFETs have general planar gate structure and are designed to block drain voltage of 1.2 kV. The cross-sectional TEM observation after the device manufacturing process revealed actual oxide thickness to be 50 nm. Therefore, MOSFETs fabricated by the conventional NO_X annealing with gate oxide thickness thicker and thinner than 50 nm were prepared as reference samples.

Figure 7(a) shows drain current characteristics when high drain voltage is applied. Sharp avalanche currents were measured at similar drain voltage sufficiently higher than 1.2 kV for both MOSFETs. No difference in breakdown characteristics between the two gate insulators was found.

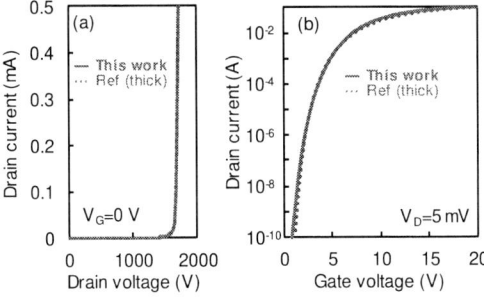

Fig. 7: (a) I_D-V_D characteristic of breakdown. (b) I_D-V_G characteristic. (Solid line: the high-temperature N_2 annealing dotted line: NO_X annealing)

Figure 7(b) shows gate subthreshold characteristics of MOSFETs with similar gate threshold voltage. Abnormal behavior such as drain leakage was not observed. Deterioration of gate properties due to alternation of gate insulator did not occur. I_D-V_D characteristics are compared in Figure 8. Standard characteristics of MOSFETs were measured for both samples.

To clarify the effect of using the developed gate insulator, specific on-resistance and threshold voltage were plotted in figure 9. As mentioned above, oxide thickness of MOSFETs with NO_X annealing is different from that of MOSFETs with the N_2 annealing. Therefore, the reference line for NO_X annealing with oxide thickness of 50 nm, which is equivalent to that of the N_2 annealing, was derived from the characteristic value of reference samples. According to the distribution, on-resistance was remarkably reduced 5-9% by using the gate insulator formed by the N_2 annealing. It is considered that channel resistance improvement due to enhanced channel mobility contributed to reduction of the entire on-resistance of MOSFET. Application of the developed novel gate insulator is demonstrated to be effective for reduction of on-resistance, without oxide thinning essentially leading to some reliability-related drawbacks.

V. CONCLUSION

High-temperature N_2 annealing has been developed to improve the limited channel mobility of gate insulator formed by nitridation for planar-type SiC-MOSFET on Si-face. By controlling the interface reaction of oxidation and nitridation, nitrogen incorporation can be augmented. Although the formed interface structure is qualitatively similar to that obtained by the conventional NO_X process, higher nitrogen density realizes significantly high channel mobility of 50 cm^2/Vs. Reduction of the on-resistance is demonstrated on the 1.2 kV class planar gate vertical power MOSFET. The developed nitridation process to improve the gate insulator performance could boost the advance of the next-generation SiC planar MOSFETs.

978-1-7281-0582-6/19 $31.00 © 2019 IEEE

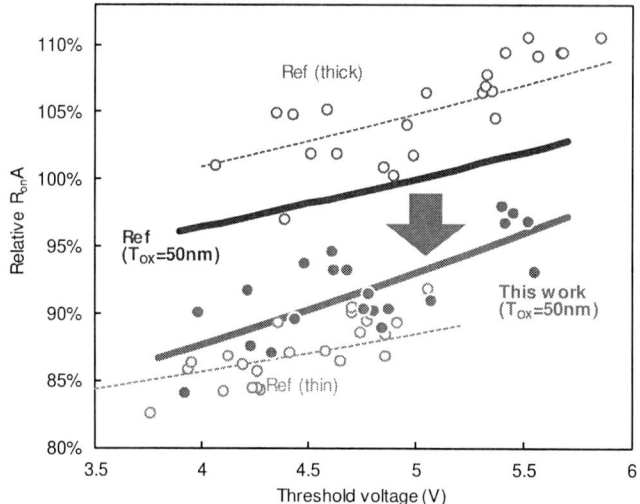

Fig. 9: Relation between specific on-state resistance ($R_{on}A$) and threshold voltage (V_{th}). Calculated line for equivalent thickness is estimated from reference samples with thick and thin oxides.

REFERENCES

[1] D. Okamoto, H. Yano, K. Hirata, T. Hatayama, T. Fuyuki, "Improved Inversion Channel Mobility in 4H-SiC MOSFETs on Si Face Utilizing Phosphorus-Doped Gate Oxide", IEEE Electron Device Lett. vol. 31, pp. 710-712, Jul. 2010

[2] D. Okamoto, M. Sometani, S. Harada, R. Kosugi, Y. Yonezawa, and H. Yano, "Improved Channel Mobility in 4H-SiC MOSFETs by Boron Passivation", IEEE Electron Device Lett. vol. 35, pp. 1176-1178, Dec. 2014

[3] D. J. Lichtenwalner, L. Cheng, S. Dhar, A. Agarwal and J. W. Palmour, "High mobility 4H-SiC (0001) transistors using alkali and alkaline earth interface layers", Appl. Phys. Lett. vol. 105, pp. 182107, Nov. 2014.

[4] G. Y. Chung, C. C. Tin, J. R. Williams, K. McDonald, R. K. Chanana, R. A. Weller, S. T. Pantelides, L. C. Feldman, O.W. Holland, M. K. Das, and J. W. Palmour, "Improved Inversion Channel Mobility for 4H-SiC MOSFETs Following High Temperature Anneals in Nitric Oxide", IEEE Electron Device Lett. vol. 22, pp. 176-178, Apr. 2001

[5] P. Jamet, S. Dimitrijev, and P. Tanner, "Effects of nitridation in gate oxides grown on 4H-SiC" J. Appl. Phys. vol. 90, pp. 5058-5063, Oct. 2001

[6] K. Fujihira, Y. Tarui, K. Ohtsuka, M. Imaizumi and T. Takami," Effects of N₂O Anneal on Channel Mobility of 4H-SiC MOSFET and Gate Oxide Reliability", Mater. Sci. Forum vol. 483-485 pp. 797-600, May 2005

[7] Y. Xu, X. Zhu, H. D. Lee, C. Xu, S. M. Shubeita, A. C. Ahyi, Y. Sharma, J. R. Williams, W. Lu, S. Ceesay, B. R. Tuttle, A. Wan, S. T. Pantelides, T. Gustafsson, E. L. Garfunkel, and L. C. Feldman, "Atomic state and characterization of nitrogen at the SiC/SiO2 interface", J. Appl. Phys. vol. 115, pp. 033502, Jan. 2014

[8] H. Yoshioka, T. Nakamura, and T. Kimoto, "Generation of very fast states by nitridation of the SiO₂/SiC interface", J. Appl. Phys. vol. 112, pp. 024520, Jul. 2012

[9] S. Asaba, T. Shimizu, Y. Nakabayashi, S. Fukatsu, T. Ito and R. Iijima, "Novel Gate Insulator Process by Nitrogen Annealing for Si-Face SiC MOSFET with High-Mobility and High-Reliability", Materials Science Forum vol. 924, pp. 457-460, Jun. 2018

[10] S. Asaba, T. Ito, S. Fukatsu, Y. Nakabayashi, T. Shimizu, M. Furukawa, T. Suzuki and R. Iijima, "Interface Reaction in the High-temperature N₂ Annealing Process for Gate Insulator on SiC with High-Mobility and High-Reliability", ECSCRM 2018, Sep. 2018

Proceedings of the 31st International Symposium on Power Semiconductor Devices & ICs
May 19-23, 2019, Shanghai, China

Trench Field Plate Engineering for High Efficient Edge Termination of 1200 V-class SiC Devices

Yong Liu[1], Wentao Yang[1], Hao Feng[1], Yuichi Onozawa[2], Setsuko Wakimoto[2], Naoto Fujishima[2], Johnny K.O. Sin[1]

[1]Department of Electronic and Computer Engineering, The Hong Kong University of Science and Technology
Clear Water Bay, Kowloon, Hong Kong, China

[2]Fuji Electric Co., Ltd, 4-18-1 Tsukama, Matsumoto, Nagoya 390-0821, Japan
Email: yliueg@connect.ust.hk

Abstract—**Trench field plate engineering for high efficient edge termination of 1200 V-class SiC devices is proposed and experimentally demonstrated. The structure features a deep trench filled with benzocyclobutene (BCB) dielectric, and a U-shaped field plate is incorporated in the deep trench. The deep trench extends through the epitaxial layer into the substrate to eliminate the negative effects of the microtrench on breakdown voltage. The U-shaped field plate modulates the electric field distribution to prevent premature breakdown at the interface between the P-base and trench. The SiC outside the trench is diced away to reduce the peak electric field at the edge of the field plate by spreading the electrostatic potential both horizontally and vertically. The trench field plate edge termination structure is implemented by integrating with a PiN diode. Experimental results show that the structure achieves a breakdown voltage of 1380 V, which is approximately 100% of the ideal planar junction breakdown voltage of the PiN diode. The fabricated edge termination structure has an ultra-short width of 33 μm. Compared with conventional junction termination extension structures used in 1200 V-class SiC devices, the width of the proposed structure is approximately 75% shorter.**

Keywords—*Trench field plate, high efficient, edge termination, deep trench, U-shaped field plate, ideal planar junction breakdown, ultra-short width.*

I. INTRODUCTION

Silicon carbide (SiC) based high voltage devices have shown excellent performance due to the inherent material properties of wide bandgap, high critical electric field, and high thermal conductivity. Edge termination is an indispensable part of the high voltage devices to prevent premature breakdown due to peripheral junction curvature effect. Conventional edge termination structures, including guard ring [1, 2] and junction termination extension (JTE) [3, 4] have been proposed and used for SiC devices. However, the large edge termination width of the guard ring and JTE structures prevents area efficient SiC power devices to be designed, particularly for low current rating devices. Furthermore, JTE structure is very sensitive to dopant activation ratio and surface charge, which requires precise control of the post-implantation annealing process and high quality passivation.

In order to reduce the width of the edge termination structure, a mesa edge termination approach fabricated using a Bosch etching process has been proposed to improve the breakdown voltage by eliminating the microtrench [5]. However, the breakdown voltage is only ~66% of the ideal

This work was supported by Fuji Electric Co., Ltd. under grant FECL16171350.

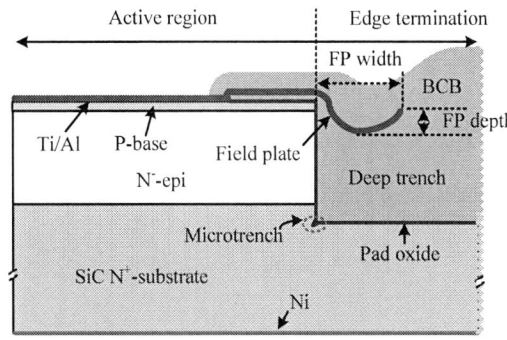

Fig. 1. Schematic cross-section of the proposed trench field plate edge termination (not in scale).

planar junction breakdown voltage. To improve the breakdown voltage, an edge termination structure combined with trench and multiple floating rings (TMFLRs) was proposed [6]. The TMFLRs can achieve 90% of the ideal planar junction breakdown voltage. But, it still needs a large edge termination width. Another edge termination technique called bevel junction termination extension has been proposed to shorten the edge termination width and achieve ~95% of the ideal planar junction breakdown voltage [7]. However, the bevel structure still needs a relatively large edge termination width. Besides, the termination extension part of the structure is in general very sensitive to dopants variation and surface charge. Furthermore, some edge termination structures using multiple and deep trenches have been proposed for silicon devices to reduce the edge width [8, 9, 10, 11]. However, they cannot achieve 100% of the ideal planar junction breakdown voltage. Recently, a sloped field plate edge termination structure has been proposed for silicon devices, which achieves 100% of the ideal planar junction breakdown voltage and with a very short edge termination width [12, 13, 14]. But, this structure cannot be used in SiC devices because of the much higher electric field (>9 MV/cm) in the trench compared to the silicon structure.

In this paper, a trench field plate edge termination structure for 1200 V-class SiC devices is proposed and experimentally demonstrated. The breakdown voltage of the proposed structure is measured to be 1380 V, which is approximately 100% of the ideal planar junction breakdown voltage. Furthermore, the edge termination width has been reduced to 33 μm, which is only ~1/4 of the width of the conventional JTE structure.

978-1-7281-0582-6/19 $31.00 © 2019 IEEE
143

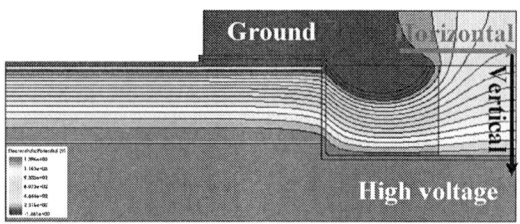

Fig. 2. TCAD simulation results of electrostatic potential distribution of the proposed structure at high reverse bias voltage. The potential at the edge termination area can spread horizontally and vertically due to the removal of the SiC at the right side of the trench.

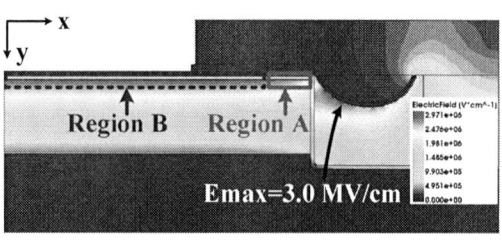

(a)

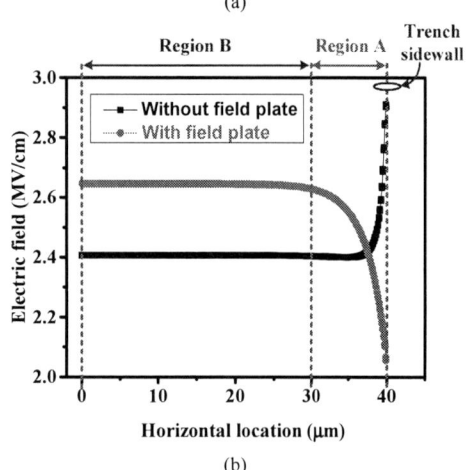

(b)

Fig. 3. TCAD simulation results of (a) electric field distribution of the proposed edge termination structure, and (b) electric field distribution comparisons along P-base/N-epi junction interface of the structure with and without the field plate.

II. DEVICE STRUCTURE, MECHANISMS AND ANALYSIS

A. Device structure and mechanism

Figure 1 shows the schematic cross-section of the proposed trench field plate edge termination structure. There is a deep trench which extends through the epitaxial layer into the substrate. The microtrench, which is usually brought in during the trench etching process, does not affect the reverse blocking voltage since its location is at the trench corner in the substrate below the epi-layer, as shown in Fig. 1. The trench is filled with benzocyclobutene (BCB) dielectric which has the critical electric field of 5.3 MV/cm [15]. A U-shaped field plate is implemented inside the

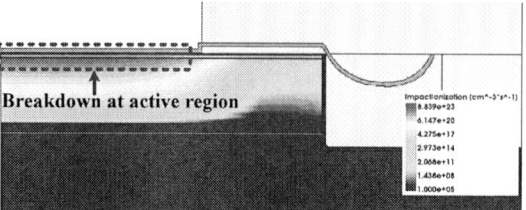

Fig. 4. TCAD simulation result of impact ionization distribution of the proposed edge termination structure.

trench to relieve the electric field crowding near the interface between the P-base and the deep trench. The field plate is designed with a U-shape to better spread the potential lines at the edge of the field plate, which is able to reduce the peak electric field at the edge of the field plate. Furthermore, the SiC at the right side of the trench is removed by dicing to allow the spreading of the potential at the edge area horizontally and vertically, which further reduces the electric field in the trench. In doing so, the avalanche breakdown point is shifted from the edge-termination to the active region. Between the SiC and the trench, there is a pad oxide layer for reducing the interface states.

B. Simulation analysis

The proposed edge termination structure is simulated using the TCAD (Technology Computer-Aided Design) tool Sentaurus Device [16]. In the simulation, the concentration and thickness of the epitaxial layer are set as $1.4E16$ cm^{-3} and 10 μm, respectively. The P-base has a junction depth of 0.6 μm and concentration of $1e19$ cm^{-3}. The width and depth of the trench are 33 μm and 13 μm, respectively. The trench is filled with BCB which has dielectric constant of 2.65. The thickness of pad oxide between the SiC and the BCB is set as 500 Å. The field plate has a depth of 4 μm and width of 18 μm. Figure 2 shows the simulation results of the electrostatic potential distribution of the proposed structure at high reverse bias voltage. The electrostatic potential at the edge area spreads both horizontally and vertically due to the removal of the SiC at the right side of the trench, which improves the uniform distribution of the electrostatic potential. Therefore, the electric field crowding at the edge of the field plate in the trench is relieved. Figure 3 (a) shows the simulation results of the electric field distribution of the proposed edge termination structure. It can be seen that the maximum electric field in the trench is reduced to 3 MV/cm, which is lower than the critical electric field of the BCB. Thus, the BCB in the trench is well protected without dielectric premature breakdown. Figure 3 (b) shows the electric field distribution comparisons along the P-base/N-epi junction interface of the structures with and without the field plate. Due to the modulation of the U-shaped field plate, the electric field in Region A is reduced to be lower than that in Region B, which prevents premature breakdown at Region A. The simulation result of the impact ionization distribution is shown in Fig. 4. The highest impact ionization is located at the active area, which indicates that the breakdown is occurred at the active region (Region B). Thus, the simulation verifies that the proposed structure can achieve ideal planar junction breakdown voltage.

978-1-7281-0582-6/19 $31.00 © 2019 IEEE

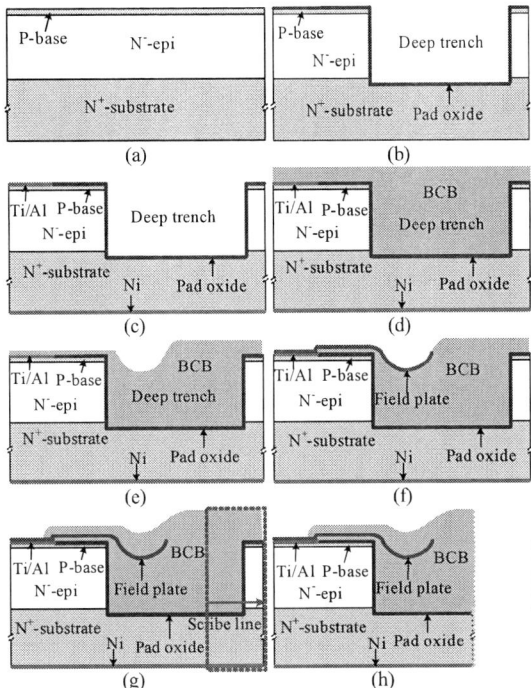

Fig. 5. Major process steps of (a) P-base formation, (b) trench etching and thermal oxide growth, (c) ohmic contact formation, (d) BCB filling, (e) BCB patterning through photolithography, (f) contact region etching, Al field plate sputtering and patterning, (g) second BCB coating and pad opening, and (h) device after dicing. (Not in scale).

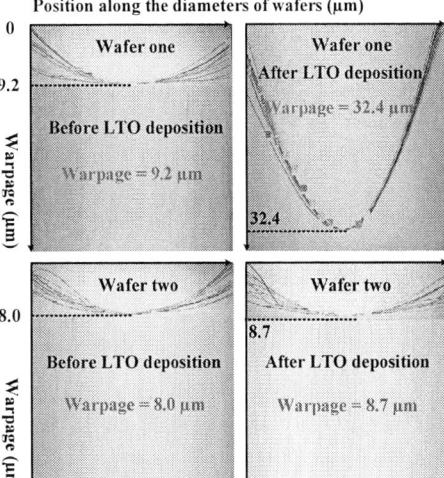

Fig. 6. Warpage measurement of SiC wafers (a) before LTO deposition on wafer one; (b) after single-sided LTO deposition on wafer one, and (c) before LTO deposition on wafer two; (d) after double-sided LTO deposition on wafer two.

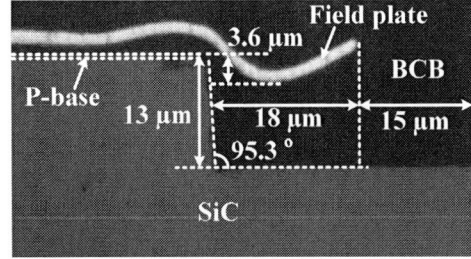

Fig. 7. Optical cross-sectional image of the fabricated devices.

III. FABRICATION PROCESS DESIGN

The proposed trench field plate edge termination structure is implemented with a PiN diode. Figure 5 shows the major fabrication process steps. 4H-SiC wafer with 4° off-axis toward <11-20> is used as the starting material. After standard cleaning, the wafers are carried out with multi-steps Al (aluminum) implantations at 500 °C. Then, post-implantation annealing is carried out at 1700 °C in Ar (argon) for 10 minutes to activate dopants and recover lattice damage, as shown in Fig. 5 (a). To prevent the surface of the SiC wafer becoming rough, a carbon cap layer is sputtered at the surface before high temperature annealing. After that, oxide is deposited as hard mask, and a deep trench (~13 μm) is formed with dry etching. A sacrificial thermal oxide layer is grown and then removed with buffered oxide etch (BOE) to reduce the lattice damage caused by the trench etching process. A thermal oxide is grown again, as shown in Fig. 5 (b). Ti (titanium)/Al metal stack and Ni (nickel) are sputtered at the front-side and backside of the wafers, respectively. The Ti/Al is patterned through a lift-off process. Then, the contact metals are sintered in Ar with RTA (rapid thermal annealing) for two minutes to reduce the contact resistance, as shown in Fig. 5 (c). The trench is filled with BCB dielectric, as shown in Fig. 5 (d). BCB photolithography is done to form the U-shape profile, as shown in Fig. 5 (e). The BCB dielectric is cured at 250 °C in nitrogen ambient for one hour. The contact window is then opened with dry etching. Al is sputtered and patterned as interconnecting metal and field plate, as shown in Fig. 5 (f). The surface is then coated with a second BCB as passivation layer. Then, the pad window is opened with BCB photolithography, as shown in Fig. 5 (g). After finishing all the processes, the device is diced along the scribe line, as shown in Fig. 5 (h).

It is worth mentioning that before the trench etching process, the wafer warpage issue caused by the stress from the oxide hard mask may cause photolithography problems. A simple experiment is designed to solve the problem. One bare wafer is deposited with oxide (2 μm) on a single side, and the other wafer is deposited with oxide (2 μm) on double sides. Figure 6 shows the warpage measurement results of the wafers before and after oxide deposition. It can be seen that when oxide is deposited on a single side of the wafer, the warpage increases 23.2 μm. However, when oxide is deposited on double sides of the wafer, the warpage just increases 0.7 μm. The reason is that the stresses from the oxide on double sides of the wafer cancels out each other. Thus, the warpage issue can be solved by depositing oxide mask on the double sides of the wafer.

After finishing all the processes, the optical cross-sectional image of the fabricated device is checked, as shown in Fig. 7. The trench depth is 13 μm. The trench angle is 95.3°. The field plate length and depth are 18 μm and 3.6 μm, respectively. The total edge termination length is 33 μm, including the field plate length (18 μm) and sidewall passivation length (15 μm). The area of the device is 0.25 mm².

978-1-7281-0582-6/19 $31.00 © 2019 IEEE

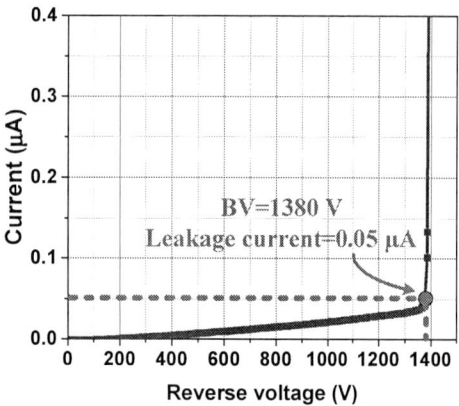

Fig. 8. Measured breakdown characteristics of the fabricated devices.

Table I. MAJOR PARAMETERS OF THE FABRICATED DEVICES.

Device parameters	Value
Trench depth (μm)	13
Trench sidewall angle (°)	95.3
U-shaped field plate depth (μm)	3.6
U-shaped field plate length (μm)	18
Sidewall passivation length (μm)	15
Pad oxide thickness on the trench (Å)	470
P-base total dose (cm^{-2})	5.4e14
P-base junction depth (μm)	0.6
N$^-$-epi layer concentration (cm^{-3})	(1.4±3%)e16
N$^-$-epi layer thickness (μm)	10±0.29
N$^+$ substrate thickness (μm)	350±25
N$^+$ substrate resistivity (Ω·cm)	0.015-0.030
Micropipe density (cm^{-2})	≤1

The major parameters of the fabricated devices are summarized in Table I.

IV. EXPERIMENTAL RESULTS AND DISCUSSION

The fabricated devices are characterized using a Tektronix 370 curve tracer and a probe station. The typical reverse blocking characteristic is shown in Fig. 8. The breakdown voltage is 1380 V at room temperature, which is consistent with the calculated ideal breakdown voltage (1354 V – 1423 V) based on the variations of the N$^-$-epi parameters listed in Table I. It can be seen from Fig. 8 that the average leakage current is approximately 0.05 μA (20 μA/cm^2), which is rather high and may be caused by the contaminations in the fabrication process.

V. CONCLUSION

A field plate edge termination structure for 1200 V-class SiC devices is proposed and experimentally demonstrated. The fabricated SiC diode with the proposed edge termination structure shows a reverse blocking voltage of 1380 V, and achieves approximately 100% of the ideal planar junction breakdown voltage. The edge termination width is 33 μm, which is only ~1/4 of the width of the conventional JTE structure.

ACKNOWLEDGMENT

The authors would like to thank the staff of Fuji Electric Company, Ltd. for their technical advices, the staff of the Nanosystem Fabrication Facility (NFF) of HKUST and the Semiconductor Product Analysis and Design Enhancement Center of HKUST for their help in device fabrication and characterization. Furthermore, we would like to thank CuttingEdge Ions, Ltd. for their excellent implantation service, and Global Power Technology, Ltd. for their good post-implantation annealing service.

REFERENCES

[1] D.C. Sheridan, G. Niu, J.N. Merrett, J.D. Cressler, C. Ellis, and C.C. Tin, "Design and fabrication of planar guard ring termination for high-voltage SiC diodes," Solid-State Electronics, vol. 44, no. 8, pp. 1367-1372, Aug. 2000.

[2] H. Onose, S. Oikawa, T. Yatsuo, and Y. Kobayashi, "Over 2000 V FLR termination technologies for SiC high voltage devices," in Proc. ISPSD, Toulouse, France, pp. 245-248, May, 2000.

[3] R. Perez, D. Tournier, A. Perez-Tomas, P. Godignon, N. Mestres, and J. Millan, "Planar edge termination design and technology considerations for 1.7-kV 4H-SiC PiN diodes," IEEE Trans. Electron Devices, vol. 52, no. 10, pp. 2309-2316, Oct. 2005.

[4] W. Sung, E.V. Brunt, B.J. Baliga, and A.Q. Huang, "A new edge termination technique for high-voltage devices in 4H-SiC–multiple-floating-zone junction termination extension," IEEE Electron Device Lett., vol. 32, no. 7, pp. 880-882, Jul., 2011.

[5] C. Han, Y. Zhang, Q. Song, Y. Zhang, X. Tang, F. Yang, and Y. Niu, "An Improved ICP Etching for Mesa-Terminated 4H-SiC pin Diodes," IEEE Trans. Electron Devices, vol. 62, no. 4, pp. 1223-1229, Apr. 2015.

[6] H. Yuan, Q. Song, X. Tang, L. Yuan, S. Yang, G. Tang, Y. Zhang, and Y. Zhang, "Trench Multiple Floating Limiting Rings Termination for 4H-SiC High-Voltage Devices," IEEE Electron Device Lett., vol. 37, no. 8, pp. 1037-1040, Aug. 2016.

[7] W. Sung, A.Q. Huang and B. J. Baliga, "Bevel Junction Termination Extension-A New Edge Termination Technique for 4H-SiC High-Voltage Devices," IEEE Electron Device Lett., vol. 36, no. 6, pp. 594-596, Jun. 2015.

[8] C. Park, J. Kim, T. Kim, and D. J. Kim, "Deep Trench Terminations Using ICP RIE for Ideal Breakdown Voltages," in Proc. ISPSD, Cambridge, UK, pp. 199-202, Apr. 2003.

[9] L. Theolier, H. Mahfoz-Kotb, K. Isoird, and F. Morancho, "A new junction termination technique: The Deep Trench Termination (DT2)," in Proc. ISPSD, Barcelona, Spain, pp. 176-179, Jul. 2009.

[10] K. Seto, R. Kamibaba, M. Tsukuda, and I. Omura, "Universal Trench Edge Termination Design," in Proc. ISPSD, Bruges, Belgium, pp. 161-164, Jun. 2012.

[11] J.K. Oh, M.W. Ha, M.K. Han, Y.I. Choi, "A New Junction Termination Method Employing Shallow Trenches Filled With Oxide," IEEE Electron Device Lett., vol. 25, no. 1, pp. 16-18, Jan. 2004.

[12] W. Yang, H. Feng, X. Fang, Y. Onozawa, H. Tanaka, and J.K.O. Sin, "A Novel Sloped Field Plate-Enhanced Ultra-Short Edge Termination Structure," IEEE Electron Device Lett., vol. 37, no. 4, pp. 471-473, Apr. 2016.

[13] W. Yang, H. Feng, X. Fang, Y. Onozawa, H. Tanaka, and J.K.O. Sin, "A Novel Edge Termination Structure for Achieving the Ideal Planar Junction Breakdown Voltage," in Proc. ISPSD, Prague, Czech Republic, pp. 287-290, Jun. 2016.

[14] W. Yang, H. Feng, Y. Liu, X. Fang, Y. Onozawa, H. Tanaka, K. Mitsuzuka, and J.K.O. Sin, "A New 1200 V-class Edge Termination Structure with Trench Double Field Plates for High dV/dt Performance," in Proc. ISPSD, Sapporo, Japan, May. pp. 109-112, 2017.

[15] Processing Procedures for CYCLOTENE 4000 Series Photo BCB Resins, Dow Chemical Inc., Midland, MI, USA, Mar. 2009.

[16] Sentaurus Device User Guide, Ver. J-2014.09-SP1, Synopsys, Inc., Mountain View, CA, USA, 2014.

Proceedings of the 31st International Symposium on Power Semiconductor Devices & ICs
May 19-23, 2019, Shanghai, China

Power Cycling Test on 3.3kV SiC MOSFETs and the Effects of Bipolar Degradation on the Temperature Estimation by V_{SD}-Method

Felix Hoffmann*, Victor Soler[1], Andrei Mihaila[+], Nando Kaminski*

*University of Bremen, Institute for Electrical Drives, Power Electronics, and Devices, Bremen, Germany
[1]Centre Nacional de Microelectrònica, CNM-CSIC, Barcelona, Spain
[+]ABB Switzerland Ltd, Corporate Research Center, Baden-Dättwil, Switzerland
felix.hoffmann@uni-bremen.de

Abstract—In this work the impact of bipolar degradation on the temperature estimation during power cycling of high voltage SiC MOSFETs by means of the body-diode voltage drop is investigated. The results indicate that the voltage drop is subject to drift for devices prone to bipolar degradation even at moderate current densities in the regime of common sensing currents for the V_{SD}-method. Hence, the criteria of the sensing current for the V_{SD}-method are not only good sensitivity and linearity while avoiding self-heating but must consider possible drift of the calibration curve.

Keywords—SiC MOSFET, high voltage, reliability, power cycling test, TSEP, junction temperature measurement, bipolar degradation

I. INTRODUCTION

Silicon Carbide (SiC) MOSFETs are gaining market share and devices with up to 1.7kV blocking capability are readily available. Further developments trend to higher voltages, fully exploiting the advantages of SiC [1][2]. A critical aspect of SiC devices is the reliability [3]. Due to its higher stiffness compared to silicon, the power cycling capability is particularly crucial [4][5]. Another concern of SiC devices is bipolar degradation (BD), which causes an increase in voltage drop of the MOSFETs body-diode and can even affect the $R_{DS, on}$ [6][7][8]. For junction temperature (T_{vj}) estimation of SiC MOSFETs during Power Cycling Tests (PCT), the body-diode voltage drop is considered the most suitable temperature

sensitive electrical parameter (TSEP) [9]. However, it has not been investigated regarding possible drift over the course of a PCT due to BD so far. In this work, the impact of BD on the accuracy of the V_{SD}-Method and the influence of the applied source-drain sensing current density (J_{Sense}) are investigated.

II. TEST DEVICES

In order to investigate possible impact of BD on the temperature estimation with the V_{SD}-Method a total of four SiC MOSFETs with a nominal blocking capability of 3.3kV were tested. The devices are packaged in an experimental package consisting of a DBC substrate and a 3D-printed housing with one MOSFET chip per package. The target $R_{DS, on}$ of the chip is 300mΩ. The structure of the MOSFETs under test is shown in Fig. 1. The device features an epilayer of 35μm with an implanted p-well and an active area of 16.26mm². The gate is designed as a planar gate with a boron-doped gate oxide. No specific measures were taken during epi-growth process to reduce stacking fault expansion. Fig. 2 shows an image of a chip under test with a J_{Sense} of 60mA·cm⁻². From the light emission in the junction termination area it is apparent, that even for low current densities bipolar operation occurs at least locally in the drift region close to the p⁺n⁻-junction.

Fig. 1. Schematic structure of the DUT: SiC MOSFET with implanted p-well and an epilayer of 35μm [10].

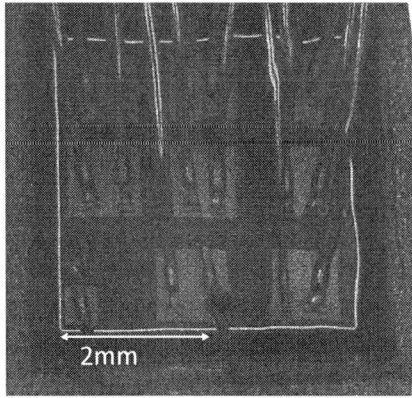

Fig. 2. Microscope image with visible electro-luminescence at the edges (J_{SD} = 60mA·cm⁻²). The image exhibits a distortion caused by refraction in the silicone gel, apparent through the distortion of the chip's contours

978-1-7281-0582-6/19 $31.00 © 2019 IEEE

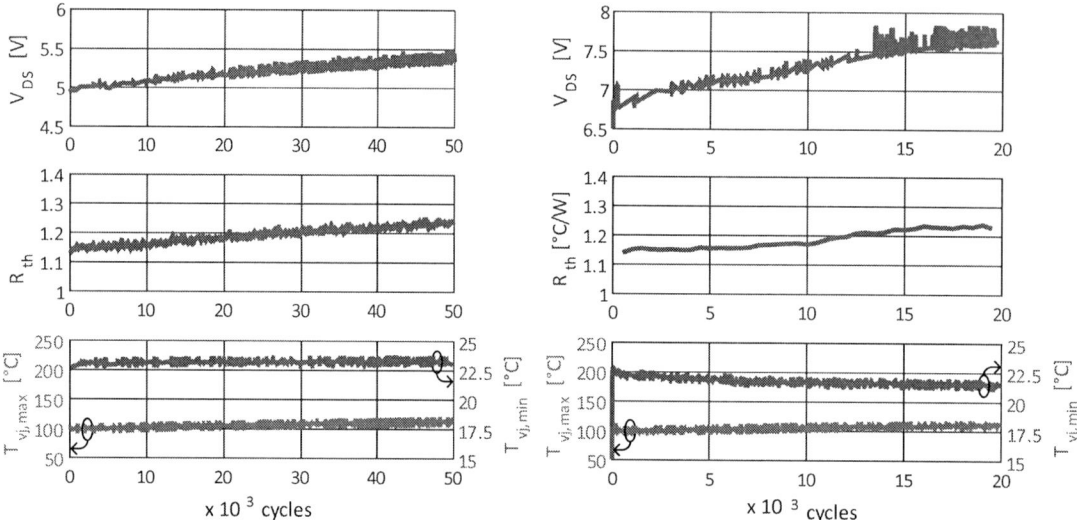

Fig. 3. Monitoring data of the PCT performed on the MOSFETs (forward conduction) over the course of the test on DUT3 (left) and DUT2 (right). DUT2 shows a decrease in $T_{vj, min}$ of ca. 1°C (corresponding to 3.5mV at a sensing current density of 60mA·cm⁻²) already after 20kcycles, whereas the reference DUT3 does not show a decrease even after 50kcycles. Both DUTs show a steady increase in $V_{DS, on}$, R_{th} as well as $T_{vj, max}$ over the course of the PCT

III. TEST EXECUTION AND RESULTS

A PCT was performed with all four 3.3kV SiC MOSFETs. In order to facilitate subsequent tests, the PCT was not conducted until end-of-life but was terminated before significant degradation occurred.

A. Power Cycling Test

The test conditions are specified in Table I. For this test the V_{SD}-Method was utilized to estimate T_{vj} with a sensing current density $J_{Sense} = 60mA·cm^{-2}$ and $V_{GS, off} = -6V$. During the test, the minimum and maximum junction temperature ($T_{vj, min}$ and $T_{vj, max}$), the on-state drain-source voltage (V_{DS}) and the thermal resistance (R_{th}), extracted from the thermal impedance (Z_{th}) monitoring, were logged. The monitored data, shown in Fig. 3, shows that V_{DS}, R_{th} as well as $T_{vj, max}$ are increasing during PCT for all DUTs due to power cycling induced degradation mechanisms. Fig. 3 left shows that for DUT3 (similar for DUT1 and DUT4) $T_{vj, min}$ was stable throughout the whole test. In contrast, Fig. 3 right shows that the $T_{vj, min}$ reading indicates a decrease for DUT2 over the course of the test. Since the power dissipation and the maximum temperature was increasing, the effect cannot be accounted to an actual decrease in $T_{vj, min}$, but to a drift of the TSEP i.e. body-diode voltage

TABLE I. PCT TEST CONDITIONS

	DUT1	DUT2	DUT3	DUT4
On time [s]	3	3	3	3
Off time [s]	3	3	3	3
Load current [A]	9.3	8.0	9.8	7.8
Coolant temperature [°C]	18	18	18	18

drop. For DUT2 the test was terminated after approximately 20kcycles when the decrease in $T_{vj, min}$ reading was exceeding 1°C. For the remaining DUTs, which did not show this behavior, the test was continued until 50kcycles. During this period no decrease in $T_{vj, min}$ could be observed for any of the remaining DUTs.

Due to the NTC behavior of the MOSFET's body diode, a decrease in temperature reading is caused by an increase in voltage drop. For the DUTs used in this test a decrease of 1°C in temperature corresponds to 3.5mV at the applied sensing current density. It is evident that this cannot be caused by an increase in resistivity in both directions since this corresponds to an increase of $R_{DS,on}$ of more than 100%, which cannot be observed in forward characteristic during the test. Hence, this effect predominantly impacts the body-diode characteristic and suggests only local bipolar degradation underneath the p-well in the region of the source contact as possible root cause.

B. Bipolar Degradation Test

In order to investigate the described effect of decrease in $T_{vj, min}$ reading over the course of a PCT, a BD test was conducted on the body-diode of DUT2. As a reference, DUT3, which did not show a decrease in $T_{vj, min}$ reading during PCT, was also tested. The test was performed for 600 minutes with a source-drain current density of 20A·cm⁻². During the test, a gate-source voltage of -6V was applied to avoid subthreshold effects and ensure pin-diode characteristics. The results of this test are shown in Fig. 4. It can be seen that even though both DUTs show an increase of forward voltage drop over time, the effect is much less pronounced for DUT3. DUT2 showed an increase in the body-diode voltage drop of approximately 370mV which is not reversible and did not have a significant

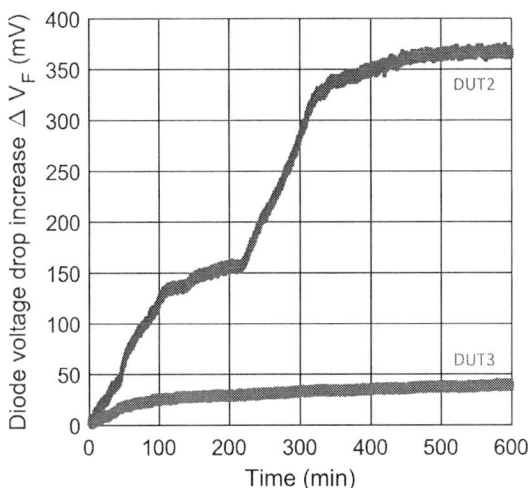

Fig. 4. Results of BD test of the MOSFETs body-diode ($V_{GS, off}$ = -6V) with a load current density of 20A·cm⁻². DUT3 shows just a small increase of less than 40mV, whereas DUT2 shows a considerable increase in V_{SD} over time of around 370mV after 600 min of BD test.

impact on the forward-characteristics of the MOSFET. In contrast, the reference device DUT3, which had not shown a decrease in $T_{vj, min}$ during PCT, only showed an increase of approximately 40mV during BD test. Compared to most recent results of BD free devices this is still a significant degradation but also a remarkable improvement of conventional processing considering that no measures had been applied to avoid stacking faults [11]. Considering the test results of DUT2, which showed a decrease in $T_{vj, min}$ reading, also indicated a more pronounced susceptibility towards BD compared to the other devices.

C. Double Pulse Test

In order to verify that the increase of body-diode voltage drop during the test was caused by BD, a double pulse test was performed. When BD occurs, the areas, in which the stacking faults expand act as recombination centers and therefore, the electron-hole plasma concentration decreases locally in these regions. This leads to a decrease of the total plasma charge of the device and therefore the plasma charge can be used as an indicator for BD. Unfortunately, the plasma charge cannot be measured directly, however it manifests in the reverse recovery charge during commutation and hence, can be estimated during double pulse test. Therefore, it is a suitable test to measure a possible reduction of plasma-concentration of a device after a bipolar degradation test. The reverse recovery characteristics of the body-diodes of all DUTs are shown in Fig. 5. It is evident, that the reverse-recovery peak and the total reverse-recovery charge is lower for DUT2 after BD test compared to the other devices including DUT3, which was equally subject to the bipolar degradation test with the same test conditions. For a current density of 60A·cm⁻², DUT2 shows an 8% lower reverse recovery peak current $I_{rr, max}$ and a 5% lower reverse recovery charge Q_{rr} compared to the other DUTs. The difference in $I_{rr, max}$ is even more pronounced for a higher current density of

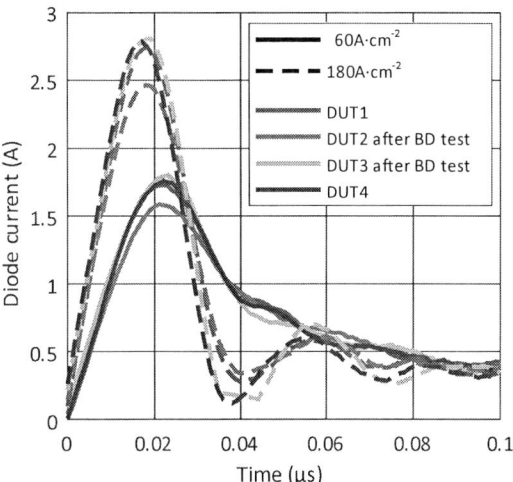

Fig. 5. Reverse recovery of the body-diode during double pulse test at 22°C. ($V_{GS, off}$ = -6V) with a test voltage of 400V. DUT3 did not show BD during BD test and also does not show significantly lower I_{rr} or Q_{rr}.

180A·cm⁻². This is consistent with the results of the BD test and indicates a lower plasma concentration for DUT2 after the bipolar degradation test compared to the other devices which were less prone to BD.

IV. IMPACT ON BODY-DIODE CHARACTERISTICS

After the bipolar degradation test, the TSEP calibration curve was measured again in order to evaluate the effect of bipolar degradation. The initial calibration curve as well as the one obtained after the BD test are shown in Fig. 6. It can be observed, that for very low current densities, the effect of BD on the calibration curve is negligible. For higher current densities, significant change of the calibration curve can be observed. Common sensing current densities when using a diode voltage drop as a TSEP are ranging up to 1000mA·cm⁻² and in some cases even above [9][12]. This means that BD can occur at current densities in the range of typical sensing current densities and Fig. 6 shows that it can have a significant impact on the calibration curve. Also, it is evident that the effect of BD on the calibration curve is highly temperature dependent. For 600mA·cm⁻² the deviation at 20°C is 64mV corresponding to 13°C, whereas for high temperatures the deviation is much smaller. This is supported by the forward characteristic of the body diode for currents in the range of typical sensing current densities, shown in Fig. 7. It can be observed that at 20°C the effect of BD is negligible for smaller current densities whereas it becomes more pronounced when the current density is increasing. For higher temperatures, in this case 160°C, the diode forward characteristics measured before and after the BD test are well aligned, indicating that the impact of BD on the body-diode forward characteristics is decreasing with temperature even at higher current densities. This is consistent with previous publications [13][14], which reported that the effect of BD was reduced at high temperatures whereas on the other hand high temperatures at the same time accelerate the

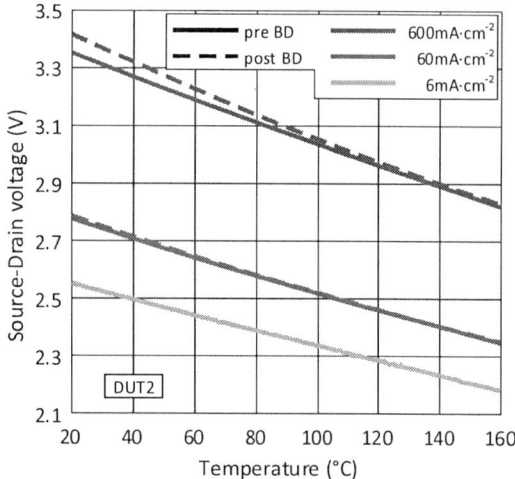

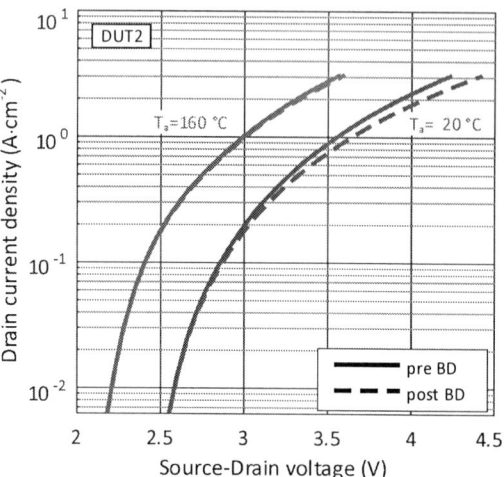

Fig. 6. Calibration curve of DUT2 before and after BD test ($V_{GS,off}$= -6V). For very low J_{Sense}, the effects of BD are negligible. For higher J_{Sense}, significant change of the calibration curve can be observed.

Fig. 7. Forward characteristics of the body-diode of DUT2 before and after BD ($V_{GS,off}$ = -6V).

process of BD. This severely affects the online monitoring of the junction temperature, which will encounter increasing inaccuracy over the course of a PCT and through the strong nonlinearity of the calibration curve shift, the temperature swing. For sufficiently high temperatures the reading of the maximum temperature remains accurate, whereas the minimum temperature is underestimated. This leads to a false reading in temperature swing and therefore to inaccurate lifetime estimation. Moreover, it also distorts the cooling characteristic and therefore significantly affects online-monitoring of the thermal impedance (Z_{th}).

V. CONCLUSION

The results of this work indicate that even though the V_{SD}-Method yields a good T_{vj} reading at the beginning of the test, the TSEP can be subject to drift due to BD even for small J_{Sense}. This is promoted by the high temperatures occurring during PCT [13]. Due to their thicker drift regions, devices with higher blocking voltages are particularly affected [14]. For the DUTs used in this work the drift during PCT was comparatively small, however it can be more pronounced for larger-sized devices [7] or devices more prone to BD and is even worse if the diode is used for heating. The calibration curves before and after BD test reveal, that the drift is nonlinear and depending on the temperature. This is particularly crucial for online-monitoring of the thermal impedance. However, for small sensing current densities the drift is negligible. Hence, when selecting a suitable sensing current for temperature estimation during PCT, not only self-heating, sensitivity and linearity of the calibration curve are important but also avoiding drift to obtain a stable temperature reading. A beneficial aspect is, however, that J_{Sense} can be chosen deliberately high to obtain information about the susceptibility of a device towards BD already during PCT. This might even make dedicated BD tests obsolete.

ACKNOWLEDGMENT

This work was supported by the European FP7 project #604057 "SPEED" (Silicon Carbide Power Technology for Energy Efficient Devices).

REFERENCES

[1] V. Soler *et al.*, "4.5kV SiC MOSFET with boron doped gate dielectric", in *ISPSD* 2016, pp. 283–286.

[2] J. Millán *et al.*, "High-voltage SiC devices: Diodes and MOSFETs", in *CAS* 2015, pp. 11–18.

[3] N. Kaminski, S. Rugen, F. Hoffmann, "Gaining Confidence – A Review of Silicon Carbide's Reliability Status", in *IRPS 2019*

[4] J. Lutz, "Packaging and Reliability of Power Modules", in *CIPS 2014*.

[5] C. Herold et al., "Power cycling capability of Modules with SiC Diodes", in CIPS 2014, pp. 36–41.

[6] J. P. Bergman, et al., "Characterisation and defects in silicon carbide", in Silicon Carbide and Related Materials 2001, vol. 389-3, pp. 9-14.

[7] T. Kimoto *et al.*, "Understanding and reduction of degradation phenomena in SiC power devices", in *IRPS 2017*, pp. 2A-1.1-2A-1.7.

[8] Y. Ebiike *et al.*, "Reliability investigation with accelerated body diode current stress for 3.3 kV 4H-SiC MOSFETs with various buffer epilayer thickness" in *ISPSD* 2018, pp. 447–450.

[9] C. Herold et al., "Power cycling methods for SiC MOSFETs", in *ISPSD 2017*, pp. 367–370.

[10] V. Soler *et al.*, "High-Voltage 4H-SiC Power MOSFETs With Boron-Doped Gate Oxide", in *IEEE Transactions on Industrial Electronics*, vol. 64, no. 11, pp. 8962–8970.

[11] Y. Bu, H. Yoshimoto, K. Konishi, A. Shima, and Y. Shimamoto, "6.5 kV 4H-SiC PiN diodes without bipolar degradation", in *ECSCRM 2016*

[12] J. A. O. Gonzalez, et al., "Electrothermal Considerations for Power Cycling in SiC Technologies", in *CIPS 2016*, pp. 1–6.

[13] P. Brosselard *et al.*, "The effect of the temperature on the Bipolar Degradation of 3.3 kV 4H-SiC PiN diodes", in *ISPSD 2008*, pp. 237–240.

[14] T. Kimoto and J. A. Cooper, Fundamentals of silicon carbide technology: growth, characterization, devices and applications. John Wiley & Sons Singapore Pte. Ltd., 2014

Proceedings of the 31st International Symposium on Power Semiconductor Devices & ICs
May 19-23, 2019, Shanghai, China

175V, > 5.4 MV/cm, 50 mΩ.cm² at 250°C Diamond MOSFET and its reverse conduction

Cédric Masante, Julien Pernot, Juliette Letellier and David Eon
Inst. NEEL
Univ. Grenoble Alpes, CNRS
Grenoble, France
cedric.masante@neel.cnrs.fr

Nicolas Rouger
LAPLACE
Univ. de Toulouse, CNRS
Toulouse, France

Abstract—**A diamond MOSFET has been fabricated and characterized up to 250°C. The fabrication process has been improved in order to significantly reduce the specific on resistance, down to 50 mΩ.cm², and the gate leakage current at high temperature. The maximum electrical field in diamond, at the breakdown value of 175V, is estimated to be higher than 5.4 MV/cm, with a boron doping of 2×10^{17} cm^{-3}.**

Keywords—Diamond, wide band gap, MOSFET, D3MOSFET, deep depletion.

I. INTRODUCTION

The use of wide band gap materials to design high blocking voltage and high current density devices is growing in interest. Amongst them, diamond is characterized by its high breakdown field (estimated between 5 and 15 MV/cm), its high carrier mobility (2000 cm/(V.s) for holes) and its exceptional thermal conductivity (22 W/cm.K at room temperature). These properties are crucial for the development of new power devices, which is expected to be a fast growing market in future years. Some diamond power devices have been previously demonstrated experimentally, such as the 2D hole gas MOSFET [1, 2], JFET [3] and MESFET [4], with blocking voltages in the order of 1kV to 2kV. However, even if their performances are very promising, these devices are not reaching yet the full capability of diamond, only demonstrating critical electric field of around 2MV/cm in diamond.

Complementary to these existing architectures, we proposed the use of the deep depletion concept to design a MOSFET relying on bulk conduction, taking advantage of the full potentialities of diamond. Such concept has been demonstrated in previous studies [5, 6] thanks to the recent efforts to control the diamond/Al$_2$O$_3$ interface [7]. However, improvements had to be done to reach the on state diamond potentialities. The devices fabricated in [5, 6] are exhibiting very large contact resistances. This work presents results obtained with an improved fabrication process.

II. DEEP DEPLETION CONCEPT FOR DIAMOND MOSFET

The working principle of this device is extensively discussed in [8]. It relies on the capability of wide band gap semiconductors to have a very stable depletion regime even for high gate biases, past the inversion threshold (so called deep depletion), and high temperatures. Thanks to such a stable deep depletion regime, it becomes possible to deplete much thicker layers than using low band gap materials without the apparition of an inversion layer. In Fig. 1, we

This work is funded by the French ministry of research and French ANR research agency under grants: ANR-16-CE05-0023, #Diamond-HVDC

a)

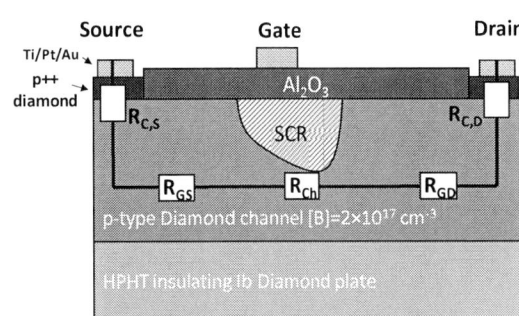

b)

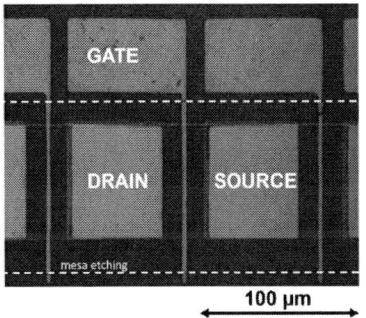

c)

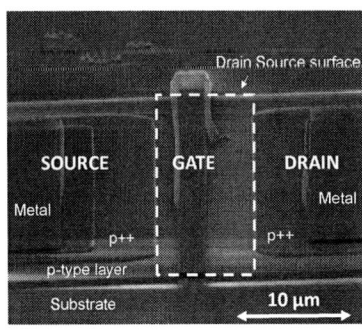

Fig. 1. a) Schematic cross section of a D3MOSFET, with a 430 nm p-type channel layer and 50 nm of gate oxide (zero gate and drain bias). b) Optical top view of a fabricated D3MOSFETs. c) SEM view of the elementary transistor with a tilt angle close to 90°.

978-1-7281-0582-6/19 $31.00 © 2019 IEEE

apply this concept to introduce a lateral, normally-on, deep depletion diamond (D3) MOSFET.

The structure is comprised of a 430nm diamond p-type layer ([Boron]=2×10^{17} cm^{-3}) grown by MPCVD on top of an insulating HPHT Ib 4mmx4mm diamond plate. A selective growth of highly boron doped diamond layer, underneath the drain and source Ti/Pt/Au stack, ensure the good ohmicity of these contacts. It has been verified that the contact resistances $R_{C,D}$ and $R_{C,S}$ are negligible comparatively to the p- layer resistance $R_{DS}ON=R_{GD}+R_{ch}+R_{GS}$ by the characterization of TLM structures on the same chip. Moreover, a mesa etching of the p-type channel layer allows to design electrically insulated MOSFETs on the same chip, in addition to limit the surface area of the gate contact on the channel region. The gate can be safely contacted with the probe tip without damaging the device. An UV ozone surface treatment is then performed to passivate the surface, as diamond surface potential is very sensitive to the surface chemistry. The gate oxide is a 50 nm layer of Al$_2$O$_3$ deposited by ALD at 380°C and followed by an annealing at 500°C in high vacuum during 30min. The gate metallization is a Ti/Pt/Au stack. Several D3MOSFETs of various sizes are fabricated on the same chip, with a constant gate length of 2 μm, and varying gate-source and gate-drain distances between 2 and 10 μm.

III. ON STATE CAPABILITY

Due to the high activation energy of boron acceptors in diamond (380 meV), the resistivity of the p- drift and channel regions is reducing with the temperature down to an optimal temperature, before being increased by the degradation of the hole mobility (Fig. 2.). Therefore, such diamond devices are meant to be operated at high junction temperatures (>150°C). In the case of this study, the optimal operation temperature corresponding to [B]= 2×10^{17} cm^{-3} is

considered to be between 250°C and 300°C, as resistivity is reaching a plateau. The ideal breakdown of the device, assuming avalanche breakdown occurs in diamond, is estimated between 400V (7MV/cm) and 800V (10MV/cm).

The drain-source current characteristic at 250°C is represented in Fig. 3. a), for negative and positive drain biases. As expected, the total on-state resistance $R_{DS}ON$ is higher at room temperature, and reduced from 3300 mΩ.cm² to 50 mΩ.cm² at 250°C. The resistance is normalized with the surface of the active area represented in fig. 1. c). The threshold voltage V_{th} is found to be +30V in the Ist quadrant by fitting the transfer characteristic (fig. 3. b)). The IIIrd quadrant operation is qualitatively exhibiting the expected behaviour of this structure, with a diode-like characteristic

a)

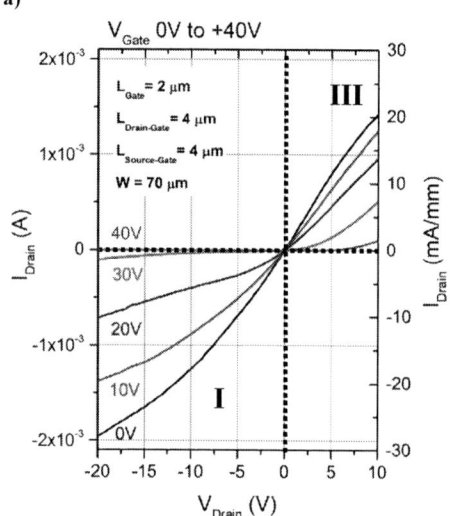

b)

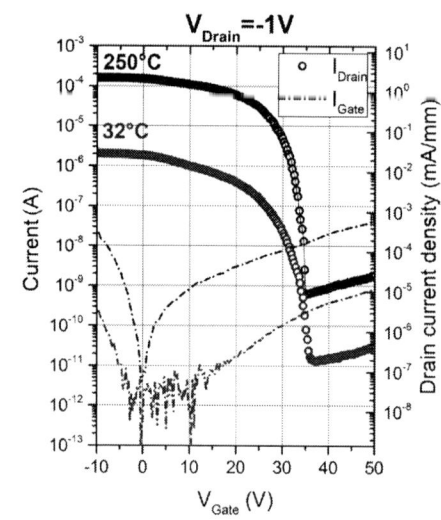

Fig. 3. a) Measured current-voltage characteristic of one of the fabricated D3MOSFETs at 250°C. b) Transfer characteristic in log scale and gate leakage current at room temperature and 250°C.

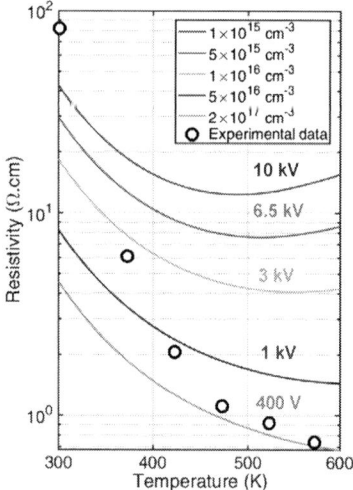

Fig. 2. Resistivity of boron doped diamond as function of temperature for different doping levels with the correspondings estimated maximum breakdown voltage values. Experimental points correspond to the resisitivity extracted using TLM structures on the sample. Difference with the calculated curve at 2.10^{17} cm^{-3} is attributed to compensation.

978-1-7281-0582-6/19 $31.00 © 2019 IEEE

with no pinch-off of the channel. Similarly to GaN Hemts, such a diamond MOSFET has a "body diode"-like behaviour, even when the gate is biased above Vth. In the IIIrd quadrant, biasing the gate below Vth will allow a reduction of the ON state voltage drop under reverse conduction.

At high temperatures, the current density is comparable to what is observed in 2D hole gas diamond power MOSFETs, and can be further improved by shrinking the drain to gate distance down to 1 µm and the source-gate and gate length in the submicrometer range. However, it was not possible in the framework of this study, due to the limiting resolution of the lithography process used.

The Fig. 3. b) presents the transfer characteristic and leakage current of the D3MOSFET in log scale. One can note that even for a temperature as high as 250°C and a gate bias up to 50V, the gate and drain leakage currents remain significantly lower than the on state drain current. As a consequence, a very good I_{on}/I_{off} ratio of around 10^5 is achieved for a wide range of temperatures.

IV. OFF STATE CAPABILITY

The best maximum blocking voltage (Fig. 4) is measured to be -175V both at room temperature and 250°C. It is attributed to the oxide breakdown, which is destructive and cannot be measured again, as already observed in [5, 6]. No correlation with L_{GD} has been observed, since L_{GD} larger than 2µm were fabricated and the maximum SCR extension in diamond toward the drain is estimated to be 1µm at breakdown value. Finite element analyses (Fig. 4. b)) demonstrate that in this lateral design, it exists a strong

electric field crowding effect both in diamond and in the oxide under the edge of the gate contact. This effect is well known to cause premature breakdown, requiring specific architectures to get mitigated such as field plates [9] or floating field rings [10] which are under investigation in diamond devices. Assessments of this maximum electric field at -175V drain bias and +50V gate bias in the ideal case, without any defects taken into consideration, range between 5.4 MV/cm and 6.8 MV/cm in diamond (Fig. 4. c)). The maximum electric field in the oxide is calculated to be 6.3 MV/cm. This value is in good agreement with the limiting breakdown field of the oxide, measured to be between 5.7 MV/cm and 6.3MV/cm in MIMCap structures, fabricated on the same chip. The presence of a large density of traps at the diamond/oxide interface, estimated here in the range of 10^{12}–10^{13} cm^{-2}/eV [11], can also cause the maximum electric field in the oxide and diamond to be different than calculated. Despite the premature breakdown, this critical field value is the highest reported in diamond FETs, and is superior to SiC (~2.8 MV/cm) and GaN (~3.5 MV/cm).

V. DISCUSSION

This lateral design is limited in his conduction capability comparatively to other materials' devices. From the results already demonstrated, performances of a vertical D3MOSFET can be extrapolated and compared to already demonstrated devices (fig. 5). One of the advantages of the vertical design in addition to the much better current density, is to avoid the large electric field crowding effect, which is limiting the off-state blocking capability, even in optimized lateral devices. Moreover, it offers the possibility to use the

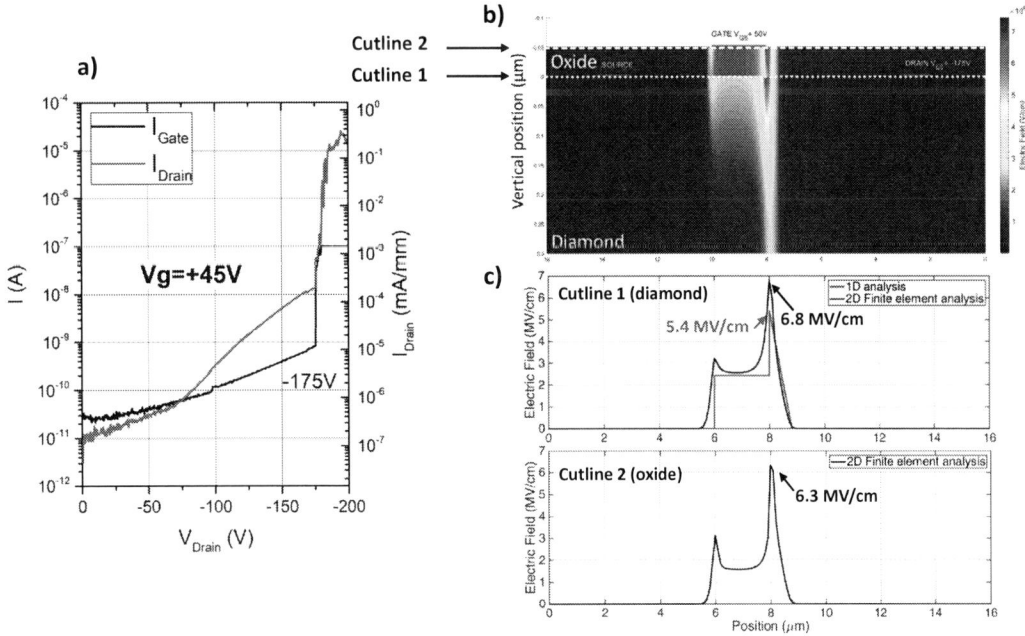

Fig. 4. a) Off state characteristic of the D3MOSFET, the gate leakage current remains small before the hard oxide breakdown occurs. b) Finite element simulation of the electric field distribution in off state with a drain bias of -175V and a gate bias of +50V. Without a specific field plate edge termination, the electric field crowding effect causes a very high electric field peak under the gate, which reduces the maximum voltage breakdown capability. c) Lateral electric field profile in diamond below the gate oxide, extracted from the above simulation (black) and calculated using 1D electrostatic analysis (red). A minimum value of 5.4 MV/cm for the critical field in diamond is extracted.

978-1-7281-0582-6/19 $31.00 © 2019 IEEE

accumulation regime of the MOS stack, to operate the MOSFET in enhancement mode. A vertical design is proposed in fig. 6. The fabrication of such diamond MOSFET is currently a very challenging task, as it requires defect-free epilayers stacks of several micrometers thick and smooth etching techniques, without etch pits. It can be noted that a similar concept has been demonstrated using GaN in a vertical depletion device [12], showing very good results. The same process could be applied with diamond to obtain much better performances than demonstrated here at high temperature operation.

VI. CONCLUSION

A lateral diamond deep depletion MOSFET has been fabricated and characterized. Thanks to an improved fabrication process, a specific $R_{DS}ON$ of 50 mΩ.cm² at 250°C has been observed with a blocking voltage of -175V, associated with a maximum electric field >5.4 MV/cm in the diamond. For such a device, a Boron doping as high as 2×10^{17} cm^{-3} was used for the channel and drift regions, showing great perspectives for future diamond power devices. The 250°C characteristics are exhibiting low drain and gate leakage currents and an Ion/Ioff ratio of 10^5.

REFERENCES

[1] H. Kawarada *et al.*, 'Diamond MOSFETs using 2D hole gas with 1700V breakdown voltage', in *Power Semiconductor Devices and ICs (ISPSD), 2016 28th International Symposium on*, 2016, pp. 483–486.

[2] Y. Kitabayashi *et al.*, 'Normally-Off C–H Diamond MOSFETs With Partial C–O Channel Achieving 2-kV Breakdown Voltage', *IEEE Electron Device Letters*, vol. 38, no. 3, pp. 363–366, Mar. 2017.

[3] T. Iwasaki *et al.*, 'High-Temperature Bipolar-Mode Operation of Normally-Off Diamond JFET', *IEEE Journal of the Electron Devices Society*, vol. 5, no. 1, pp. 95–99, Jan. 2017.

[4] H. Umezawa, T. Matsumoto, and S.-I. Shikata, 'Diamond Metal-Semiconductor Field-Effect Transistor With Breakdown Voltage Over 1.5 kV', *IEEE Electron Device Letters*, vol. 35, no. 11, pp. 1112–1114, Nov. 2014.

[5] T. T. Pham *et al.*, '200V, 4MV/cm lateral diamond MOSFET', in *2017 IEEE International Electron Devices Meeting (IEDM)*, San Francisco, CA, USA, 2017, pp. 25.4.1-25.4.4.

[6] T.-T. Pham, J. Pernot, G. Perez, D. Eon, E. Gheeraert, and N. Rouger, 'Deep-Depletion Mode Boron-Doped Monocrystalline Diamond Metal Oxide Semiconductor Field Effect Transistor', *IEEE Electron Device Letters*, vol. 38, no. 11, pp. 1571–1574, Nov. 2017.

[7] T. T. Pham *et al.*, 'High quality Al₂O₃/(100) oxygen-terminated diamond interface for MOSFETs fabrication', *Applied Physics Letters*, vol. 112, no. 10, p. 102103, Mar. 2018.

[8] T. T. Pham *et al.*, 'Deep depletion concept for diamond MOSFET', *Applied Physics Letters*, vol. 111, no. 17, p. 173503, Oct. 2017.

[9] H. Umezawa, M. Nagase, Y. Kato, and S. Shikata, 'Diamond Vertical Schottky Barrier Diode with Al₂O₃ Field Plate', *Materials Science Forum*, vol. 717–720, pp. 1319–1321, May 2012.

[10] K. Driche, S. Rugen, N. Kaminski, H. Umezawa, H. Okumura, and E. Gheeraert, 'Electric field distribution using floating metal guard rings edge-termination for Schottky diodes', *Diamond and Related Materials*, vol. 82, pp. 160–164, Feb. 2018.

[11] T. T. Pham *et al.*, 'Comprehensive electrical analysis of metal/Al₂O₃/O-terminated diamond capacitance', *Journal of Applied Physics*, vol. 123, no. 16, p. 161523, Apr. 2018.

[12] M. Sun, Y. Zhang, X. Gao, and T. Palacios, 'High-Performance GaN Vertical Fin Power Transistors on Bulk GaN Substrates', *IEEE Electron Device Letters*, vol. 38, no. 4, pp. 509–512, Apr. 2017.

[13] H. Kawarada *et al.*, 'C-H surface diamond field effect transistors for high temperature (400 °C) and high voltage (500 V) operation', *Applied Physics Letters*, vol. 105, no. 1, p. 013510, Jul. 2014.

[14] N. Oi *et al.*, 'Vertical-type two-dimensional hole gas diamond metal oxide semiconductor field-effect transistors', *Scientific Reports*, vol. 8, no. 1, Dec. 2018.

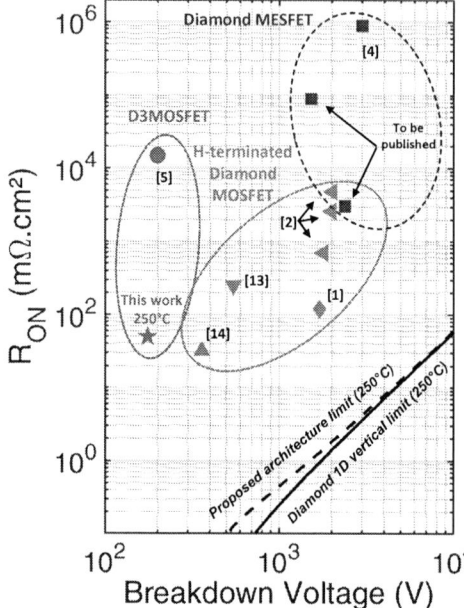

Fig. 5. On state resistance as function of breakdown voltage for various diamond FETs reported in literature at room temperature. The 1D vertical limit at 250°C is represented in full line and the expected performance of the architecture propoesed in fig. 6. in dashed line.

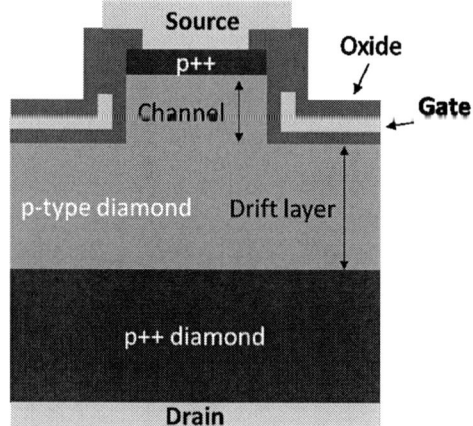

Fig. 6 .Schematic cross section of the proposed vertical architecture of D3MOSFET

Proceedings of the 31st International Symposium on Power Semiconductor Devices & ICs
May 19-23, 2019, Shanghai, China

Practical One-Step Solution of Smoothly Tapered Junction Termination Extension for High Voltage SiC Gate Turn-off Thyristor

Hu Long, Qing Guo, Kuang Sheng*
College of Electrical Engineering, Zhejiang University, Hangzhou, China
*shengk@zju.edu.cn

Abstract—**This work proposes a practical one-step solution to form a smoothly tapered junction termination extension (ST-JTE) for bipolar devices, verified by fabrication of SiC GTOs achieving 8.6 kV. This one-step solution is comprised of direct photolithography and etching with a single graded mask. The mask is improved with three methods for better etched profile. This solution is practical and advanced for (1) greatly reduced fabrication time and cost, (2) processes more controllable and repeatable in mature production, and (3) elimination of implantation damage with the least etching damage around the main junction. Therefore it is promising to be applied in SiC bipolar devices.**

Keywords—SiC, power devices, high voltage, terminations, junction termination extension, JTE, GTO, graded mask

I. INTRODUCTION

The advancing civilization is demanding for growing energy supply, which could hardly be satisfied without a sustainable cycling using renewable energy. As most of the energy conversion depends on electric facilities [1], the efficiency of these facilities is of much importance. Owing to electric facilities progressing into high-voltage and high-power field, a challenge comes that how to keep the energy conversion efficiency at higher rating. In some extremely high-voltage applications like high-voltage direct current (HVDC) transmission, a large number of Si IGBTs connected in series is necessary due to the lack of commercial devices at ultra-high voltage [2]. This situation inevitably brings about increased forward drops and complex control methods, consequently leading to more equipment cost and energy consumption.

Fortunately the emergence of wide-bandgap semiconductors like silicon carbide (SiC) has brought new opportunities in this stalemate. Compared with Si, SiC is endowed by its wide bandgap with a tenfold critical breakdown field [3], making it ideal for high-voltage applications [1, 4]. In addition, with a threefold thermal conductivity than Si, SiC is expected to enable devices operating at high temperatures (>300 °C), which greatly curtails the budget of auxiliary cooling system in high-power applications [1].

In order to take full advantage of SiC in high-voltage and high-power applications, a well-designed termination is essential. There are mainly two categories of terminations commonly used in SiC devices, namely field limiting rings

This work was supported by the National Key Research and Development Program of China (2018YFB0905704) and National Natural Science Foundation of China (No. 51577169)

(FLR) and junction termination extension (JTE). In the high-voltage field, JTEs are adopted mostly because of its higher efficiency in size, especially the ones with smoothly graded dose [5-18]. These kinds of smoothly graded JTE can achieve best electric field distribution theoretically once the optimal doping profile is attained.

No longer practical is it to form a smoothly graded JTE in SiC by impurity diffusion as Si devices typically do, due to the low diffusivity of the favored acceptor aluminum [19]. As a consequence, the attention is attracted by other approaches including (1) implantation with boron instead of aluminum [5-7], (2) multiple cycles of etching/implantation to form separate regions with graded doses [8-12] and (3) graphitization of tapered photoresist as mask for subsequent implantation [5, 7, 13]. Their drawbacks, respectively, are (1) 40% larger activation energy (~280 meV) of boron than aluminum [19] together with problems of deep level (D center) [19, 20] and abnormal diffusion [20, 21], (2) sharply increasing cost and time of fabrication and (3) dependence on graphitization less controllable [7] with extra thermal budget . Therefore, a new solution featuring both reasonable cost and relatively mature process is much awaited.

II. DESIGN OF STRUCTURE AND PROCESSES

This work proposes a practical one-step solution to form a smoothly tapered JTE (ST-JTE) for bipolar devices, which is preferred in high-voltage field. It is verified by fabrication of SJ-JTE in SiC gate turn-off thyristor (GTO) achieving 8.6 kV. This one-step solution, comprised of direct

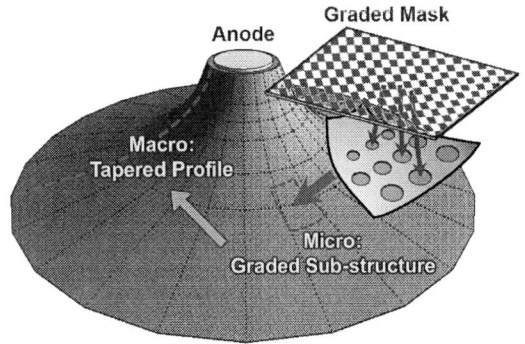

Fig. 1. Schematic diagram of the proposed solution. The graded mask is designed according to the desired profile initially. The graded pattern is then transferred to the device surface as graded sub-structure by photolithography and etching. The etching depths varys with the size of local apertures due to the window-limiting etching, thus leading to a smoothly tapered profile with the aid of lateral etching.

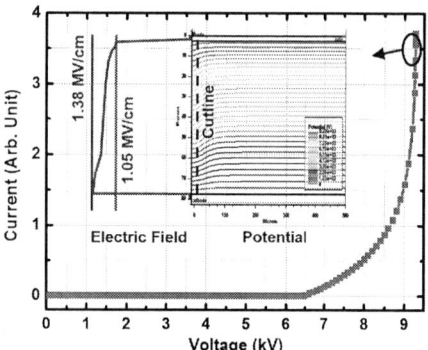

Fig. 2. The ideal open-base breakdown simulated by ATLAS with SELB impact model to be ~9.5 kV. The bending in electric field indicates the bipolar amplification effect in thyristor, resulting premature breakdown earlier than simple P-N junction does.

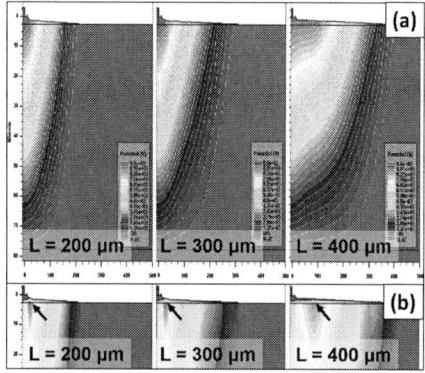

Fig. 3. Distributions of (a) potentials and (b) electric fields near breakdown voltage. Three lengths of ST-JTEs are demonstrated, namely 200, 300 and 400 μm respectively. The electric field peak moves outward with increasing length, as pointed by the black arrows.

photolithography and etching (P/E) with a single graded mask, is free of implantation and able to be applied in most multilayer epitaxial bipolar devices. The key idea is a combination of graded mask and lateral etching (Fig. 1). With the graded mask modulated by the density of tiny apertures, clusters of graded sub-structure are etched out by P/E. Lateral etching benefits the conversion from the graded pattern into a smooth surface, which is a natural ST-JTE finally. The concept of the graded mask is to control local equivalent transmittance, which strongly correlates with the local etching depth, by the local proportion of bright area to the dark one. An analogy might help with the concept of pulse width modulation (PWM) in power electronics, which is to control duty ratio by the proportion of on-state time to off-state one. Based on the reported work with graded mask [14-17, 22, 23], this work extends the concept to multilayer epitaxial structure with an improved mask design. It is improved for three methods to keep the etched surface highly uniform and symmetric after etching, which will be discussed later together with fabrication processes.

The profile of ST-JTE is designed by numerical simulation using finite element method in ATLAS. The avalanche breakdown is simulated with SELB impact model [24], and the ideal open-base breakdown voltage of epitaxy is to be ~9.5 kV (Fig. 2). It is naturally lower than calculated

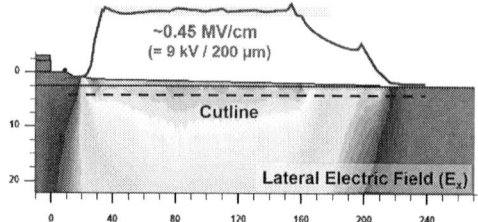

Fig. 4. The lateral electric field is optimized to be rectangular in a long range to keep JTE length minimal initially, while a risk of relatively higher total electric field near the main junction is discussed later, resulting in premature breakdown.

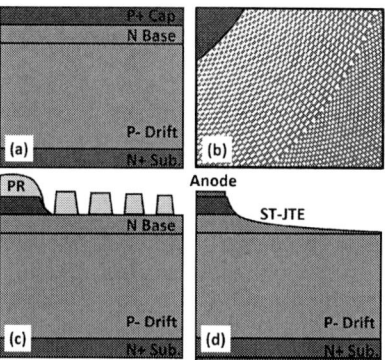

Fig. 5. The fabrication process of ST-JTE in SiC GTO. (a) Structure of the epitaxy for SiC GTO. (b) Graded mask designed according to the final desired profile. (c) Photolithography with this graded mask forming graded windows for etching. (d) Final profile of the fabricated ST-JTE after etching through these graded windows.

simple P-N junction (closed to open-emitter connection) of ~11 kV [25, 26] with the identical doping (74.2 μm, 4.5e14 cm^{-3}), due to the bipolar amplification effect of thyristor [27, 28]. For GTOs with termination, the contours of potential spread further with longer ST-JTE (Fig. 3a), and the electric field peak moves outward accordingly (Fig. 3b). All these results are optimized and present fine-tuned potentials distributions, especially the 200-μm one whose lateral electric field exhibits an ideal rectangular distribution in a long range along surface (Fig. 4). These results are converted into graded mask with improved methods discussed in the next section (Fig. 5b), according to the relationship between the desired JTE height and the actual bright/dark ratio.

III. FABRICATION AND FEATURES

The fabrication processes of ST-JTE in SiC GTO are as follows (Fig. 5). After the active anode region has been completed by conventional steps, a layer of photoresist with graded windows is formed by photolithography with the improved graded mask (Fig. 5c). Then performed is the ICP etching, resulting in a tapered profile as designed (Fig. 5d). No implantation is needed owing to the epitaxial termination design with the mentioned graded mask. The mechanisms here are the window-limiting etching effect and the lateral etching effect. The first effect results in lower etching rate at smaller window, thus achieving a tapered profile. The second effect helps to expand the windows and smooth the surface. Therefore tiny but numerous apertures and isotropic etching condition are favored for a smoothly tapered profile.

978-1-7281-0582-6/19 $31.00 © 2019 IEEE

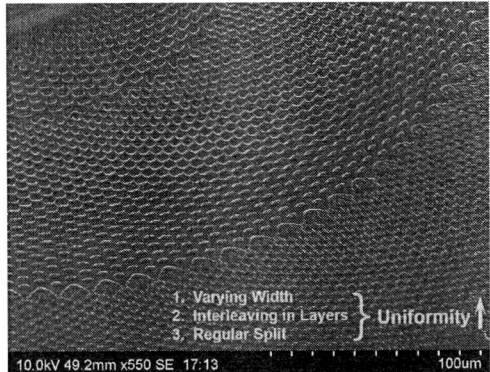

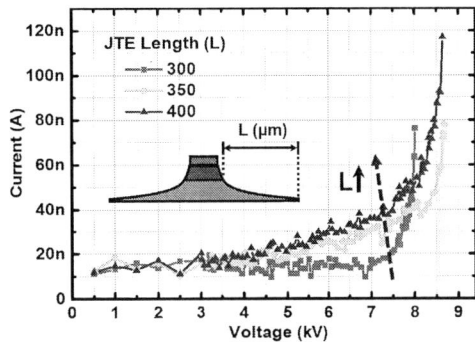

Fig. 8. Breakdown characterization of the fabricated open-base SiC GTOs with ST-JTEs. Three different JTE lengths result in distinguishing breakdown voltage and leakage current. The ones with longer ST-JTE achieve higher breakdown voltage of 8.6 kV.

Fig. 6. Graded sub-structure on the surface of ST-JTE taken by SEM. The pattern is improved by three methods to keep highly uniformity and symmetry as follows. (1) Density of tiny apertures is graded with both sizes and spacings to avoid local sparse region; (2) Adjacent apertures are perfectly interlaced successively to keep distribution well-balanced; (3) Apertures split regularly into smaller but more parts for highly uniformity over a long range.

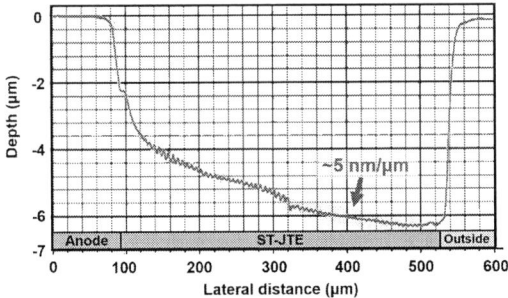

Fig. 7. Surface profile of the fabricated ST-JTE achieving high breakdown voltage, scanned by profiler. The ST-JTE of 400 μm with designed profile is formed by the graded sub-structure limiting the local etching rate. The sub-structure is eventually smoothened due to the lateral etching. Gradual slope of ~5 nm/μm along a distance of 200 μm (and expected to be more) is accessible by the proposed one-step solution.

Surface micro-structure might leave a trace after etching though (Fig. 6), the macro surface is smooth enough according to the results scanned by profiler (Fig. 7). This proposed method achieves a gradual slope over a long distance up to 400 μm by P/E with a single mask, and the slope is accessible to be ~5 nm/μm within 200 μm. It is expected to obtain longer slope simply by a larger mask design, due to the P/E process independent of distance.

Combined with the processes, it is clear to see how the proposed mask design improves the uniformity of the etched surface (Fig. 6). The comparison to early work with graded mask [14-17, 22, 23] is as follows:

(1) In most early work, aperture size is fixed, leading to local sparsity when the gaps between apertures become large, especially in the outer region. This work adopts size-variable aperture, avoiding local sparsity and thus conducive to uniformly etched surface.

(2) In most early work, casual positional relation between apertures of adjacent layers results in occurrence of uneven apertures distributions. This work realizes perfect interleaving layer-by-layer and thus enhance the symmetry of the whole surface.

Fig. 9. Breakdown of the devices with shorter ST-JTE (200 μm). Although an rectangular distribution of lateral electric field is obtained as Fig. 4, the high total electric field near the main junction results in local leakage filament at the foot of anode. This phenomenon could be relieved by lengthened termination or higher base.

(3) An additional innovation improves the uniformity that the apertures are split into smaller ones with half the size but twice the number once single aperture grows large enough at certain layers, allowing the apertures to keep tiny and to alternate frequently over a long distance.

The fabricated devices were characterized under high-voltage condition with platform based on Tektronix DMM4050 and Stanford PS375. The devices were immersed in Fluorinert solution to avoid air sparking. The open-base GTOs with the proposed ST-JTE achieve breakdown voltage of highest 8.6 kV (Fig. 8), which is ~90% of the ideal case. On the one hand, it is expected to find that the device with longer ST-JTE reaches higher breakdown voltage, owing to the lower electrical field along surface as indicated by the simulated results (Fig. 3). On the other hand, slightly increased leakage current accompanies with longer ST-JTE before breakdown, which could be attributed to larger etched surface as the leakage paths.

Regarding devices with shorter termination of 300 μm, the lower breakdown voltage of ~8 kV connotes higher inner electric fields. As for the devices with 200-μm ones, premature failure occurs occasionally due to the high electric fields near the anode as simulated, leading to a burnt filament as exhibited at the foot of anode mesa (Fig. 9). This phenomenon causing reliability problems could be relieved by lengthened termination or higher base dose as a tradeoff.

IV. Summary

Taking the high-voltage GTOs as a demonstration, this work proposes a practical one-step solution for the termination ST-JTE with an improved graded mask, verified by fabrication of SiC GTOs achieving 8.6 kV. The proposed solution enables a smoothly tapered JTE to be formed in a single cycle of direct photolithography and etching with a single mask on multilayer epitaxial devices. Several innovations of JTE patterns on mask are proposed to improve the uniformity of the final etched surface, including varying aperture width, interleaving in layers and regular split. This solution is practical and advanced for (1) greatly reduced fabrication time and cost, (2) processes more controllable and repeatable in mature production, and (3) elimination of implantation damage with the least etching damage around the main junction. Therefore it is promising to be applied in SiC bipolar devices like SiC GTOs, BJTs and IGBTs in high voltage field.

References

[1] A. Q. Huang, "Power Semiconductor Devices for Smart Grid and Renewable Energy Systems," *Proc. IEEE*, vol. 105, no. 11, pp. 2019-2047, 2017.

[2] T. P. Chow, I. Omura, M. Higashiwaki, H. Kawarada, and V. Pala, "Smart Power Devices and ICs Using GaAs and Wide and Extreme Bandgap Semiconductors," *IEEE Trans. Electron Devices*, vol. 64, no. 3, pp. 856-873, 2017.

[3] J. Millán, P. Godignon, X. Perpiñà, A. Pérez-Tomás, and J. Rebollo, "A Survey of Wide Bandgap Power Semiconductor Devices," *IEEE Trans. Power Electronics*, vol. 29, no. 5, pp. 2155-2163, 2014.

[4] T. Kimoto and J. A. Cooper, *Fundamentals of Silicon Carbide Technology: Growth, Characterization, Devices and Applications*. John Wiley & Sons, 2014.

[5] E. A. Imhoff, F. J. Kub, and K. D. Hobart, "Grayscale Junction Termination for High-Voltage SiC Devices," *Mater. Sci. Forum*, vol. 615-617, pp. 691-694, 2009.

[6] A. V. Bolotnikov, P. G. Muzykov, Q. Zhang, A. K. Agarwal, and T. S. Sudarshan, "Junction Termination Extension Implementing Drive-in Diffusion of Boron for High-Voltage SiC Devices," *IEEE Trans. Electron Devices*, vol. 57, no. 8, pp. 1930-1935, 2010.

[7] E. A. Imhoff *et al.*, "High-Performance Smoothly Tapered Junction Termination Extensions for High-Voltage 4H-SiC Devices," *IEEE Trans. Electron Devices*, vol. 58, no. 10, pp. 3395-3400, 2011.

[8] G. Paques, S. Scharnholz, N. Dheilly, D. Planson, and R. W. D. Doneker, "High Voltage 4H SiC Thyristors With a Graded Etched Junction Termination Extension," *IEEE Electron Device Lett.*, vol. 32, no. 10, pp. 1421-1423, 2011.

[9] L. Lin and J. H. Zhao, "Fabrication and Characterization of 4H-SiC 6kV Gate Turn-Off Thyristor," *Mater. Sci. Forum*, vol. 717-720, pp. 1163-1166, 2012.

[10] Q. J. Zhang *et al.*, "12 kV, 1 cm2 SiC GTO Thyristors with Negative Bevel Termination," *Mater. Sci. Forum*, vol. 717-720, pp. 1151-1154, 2012.

[11] A. Salemi, H. Elahipanah, K. Jacobs, C. Zetterling, and M. Östling, "15 kV-Class Implantation-Free 4H-SiC BJTs With Record High Current Gain," *IEEE Electron Device Lett.*, vol. 39, no. 1, pp. 63-66, 2018.

[12] C. Zhou, R. Yue, Y. Wang, J. Zhang, G. Dai, and J. Li, "10-kV 4H-SiC Gate Turn-OFF Thyristors With Space-Modulated Buffer Trench Three-Step JTE," *IEEE Electron Device Lett.*, vol. 39, no. 8, pp. 1199-1202, 2018.

[13] J. N. Merrett, T. Isaacs-Smith, D. C. Sheridan, and J. R. J. o. E. M. Williams, "Fabrication of self-aligned graded junction termination extensions with applications to 4H-SiC P-N diodes," *J. Electron. Mater.*, vol. 31, no. 6, pp. 635-639, 2002.

[14] M. Snook *et al.*, "Single Photolithography/Implantation 120-Zone Junction Termination Extension for High-Voltage SiC Devices," *Mater. Sci. Forum*, vol. 717-720, pp. 977-980, 2012.

[15] A. Bolotnikov *et al.*, "Design of Area-Efficient, Robust and Reliable Junction Termination Extension in SiC Devices," *Mater. Sci. Forum*, vol. 858, pp. 737-740, 2016.

[16] S. D. Arthur *et al.*, "Semiconductor device with junction termination extension," U.S. Patent 9 406 762 B2, Aug., 2016.

[17] V. Veliadis *et al.*, "Process Tolerant Single Photolithography/Implantation 120-Zone Junction Termination Extension," *Mater. Sci. Forum*, vol. 740-742, pp. 855-858, 2013.

[18] W. Sung, A. Q. Huang, B. J. Baliga, I. Ji, H. Ke, and D. C. Hopkins, "The first demonstration of symmetric blocking SiC gate turn-off (GTO) thyristor," in *Proc. Int. Symp. on Power Semiconductor Devices and IC's (ISPSD)*, 2015, pp. 257-260.

[19] T. Troffer *et al.*, "Doping of SiC by Implantation of Boron and Aluminum," *Phys. Stat. Sol. (a)*, vol. 162, no. 1, pp. 277-298, 1997.

[20] N. I. Kuznetsov and A. S. Zubrilov, "Deep centers and electroluminescence in 4H-SiC diodes with a p-type base region," *Mater. Sci. Eng., B*, vol. 29, no. 1, pp. 181-184, 1995.

[21] N. Yuki, K. Tsunenobu, M. Hiroyuki, and P. Gerhard, "Abnormal Out-Diffusion of Epitaxially Doped Boron in 4H-SiC Caused by Implantation and Annealing," *Jpn. J. Appl. Phys.*, vol. 46, no. 8A, p. 5053, 2007.

[22] W. Tantraporn and V. A. K. Temple, "Multiple-zone single-mask junction termination extension—A high-yield near-ideal breakdown voltage technology," *IEEE Trans. Electron Devices*, vol. 34, no. 10, pp. 2200-2210, 1987.

[23] R. Stengl, U. Gosele, C. Fellinger, M. Beyer, and S. Walesch, "Variation of lateral doping as a field terminator for high-voltage power devices," *IEEE Trans. Electron Devices*, vol. 33, no. 3, pp. 426-428, 1986.

[24] S. Selberherr, *Analysis and simulation of semiconductor devices*. Springer Science & Business Media, 2012.

[25] H. Niwa, J. Suda, and T. Kimoto, "Impact Ionization Coefficients in 4H-SiC Toward Ultrahigh-Voltage Power Devices," *IEEE Trans. Electron Devices*, vol. 62, no. 10, pp. 3326-3333, 2015.

[26] A. O. Konstantinov, Q. Wahab, N. Nordell, and U. Lindefelt, "Ionization rates and critical fields in 4H silicon carbide," *Appl. Phys. Lett.*, vol. 71, no. 1, pp. 90-92, 1997.

[27] S. M. Sze and K. K. Ng, *Physics of semiconductor devices*. John wiley & sons, 2006.

[28] B. J. Baliga, *Fundamentals of power semiconductor devices*. Springer Science & Business Media, 2010.

Proceedings of the 31st International Symposium on Power Semiconductor Devices & ICs
May 19-23, 2019, Shanghai, China

Comparison of New Octagonal Cell Topology for 1.2 kV 4H-SiC JBSFETs with Linear and Hexagonal Topologies: Analysis and Experimental Results

Kijeong Han, Aditi Agarwal, and B. Jayant Baliga
Electrical and Computer Engineering
North Carolina State University
Raleigh, NC 27695, USA
khan5@ncsu.edu

Abstract—This paper compares experimentally obtained electrical characteristics of a novel Octagonal (Oct) cell topology for 1.2 kV-rated 4H-SiC JBSFETs with the Linear and Hexagonal (Hex) cell topologies for the first time. The various cell topologies were fabricated using the same process flow at a 6-inch foundry. The third quadrant on-state voltage drop for the JBS diode in the Oct JBSFET was matched with the Linear cell design by using adequate JBS diode area within the cell. Experimental results demonstrate that the Oct JBSFET has 1.7× and 2.2× better HF-FOM [$R_{on}{\times}Q_{gd}$] compared with the Linear and Hex cell JBSFETs, respectively. In addition, the Oct JBSFETs have a much superior [C_{iss}/C_{rss}] ratio to suppress shoot through currents during high frequency switching.

Keywords— *Silicon Carbide, 4H-SiC, JBSFET, JBS Diode, Linear, Hexagonal, Octagonal, C_{iss}, C_{rss}, C_{oss}, Q_{gd}, HF-FOMs*

I. INTRODUCTION

4H-SiC power JBSFETs, a monolithically integrated 4H-SiC MOSFET and JBS diode, have been reported demonstrating third quadrant current flow through a Schottky contact without turning on the body diode [1][2]. It has been demonstrated that the integration of the JBS diode within the SiC power MOSFET is essential to reduce turn-on switching loss at elevated temperatures [3].

It has been established that the 4H-SiC MOSFETs are suitable for high frequency applications due to their superior switching characteristics when compared with Si IGBTs [4]. Lower switching energy loss and higher frequency operation can be achieved by minimizing the reverse transfer capacitance (C_{gd} or C_{rss}) and gate-to-drain charge (Q_{gd}) in the SiC power MOSFETs [5]. In addition, a large [C_{iss}/C_{rss}] ratio is necessary to prevent shoot-through current and false triggering during high frequency operation [6][7]. Consequently, smaller high frequency figures-of-merit (HF-FOMs), defined as [$R_{on}{\times}C_{rss}$] and [$R_{on}{\times}Q_{gd}$], and larger figure-of-merit (FOM) [C_{iss}/C_{rss}], are necessary for the SiC power MOSFETs, especially for the high frequency switching circuits [7].

Different cell topologies on the 4H-SiC planar-gate power MOSFETs have been previously published: Linear [8], Square [9], Hexagonal [10], and Octagonal [11]. Experimental results for the 4H-SiC JBSFETs have been published with the conventional linear cell topology [1][12][13] and the hexagonal cell topology [?].

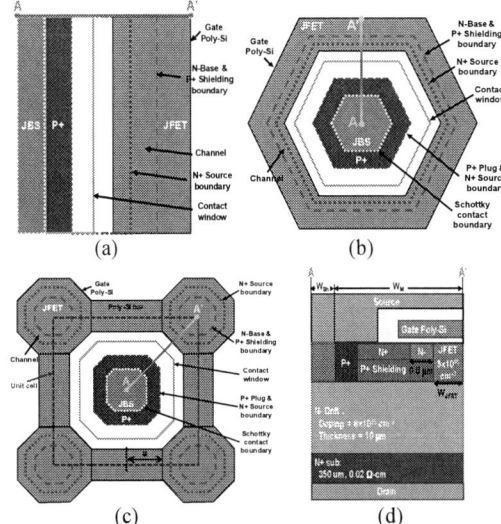

Fig. 1: Cell topologies for the fabricated 1.2 kV 4H-SiC JBSFETs: (a) Linear, (b) Hexagonal, (c) Octagonal, and (d) JBSFET cell cross-section at A-A' for all the cell topologies.

In this paper, a detailed comparison of 1.2 kV rated 4H-SiC JBSFETs with Linear, Hexagonal (Hex), and Octagonal (Oct) cell topologies is provided for the first time. The three different JBSFET cell topologies were simultaneously fabricated at a 6-inch foundry, X-Fab, using the same process with identical edge termination. Two Oct JBSFET structures with different JBS diode density were included for comparison.

II. DEVICE STRUCTURE AND ANALYSIS

The Linear, Hex, and Oct cell topologies for the fabricated accumulation channel (Acc) JBSFETs are shown in Fig. 1(a), (b), (c), respectively with detailed structural and regional information. The cross section for all the fabricated devices along line A-A' is shown in Fig. 1(d). All the structures have the same JFET width (W_{JFET}) of 1.5 μm and channel length of 0.5 μm. In the Oct cell topology, the JFET and channel regions are surrounded by the P$^+$ shielding regions inside the octagonal-shaped poly-Si regions. Poly-Si bars with the same width as the edges of the octagon are used to connect the octagonal poly-Si gate regions as shown in Fig. 1(c). The length of the bar "2a" determines the N$^+$, P$^+$ Ohmic

978-1-7281-0582-6/19 $31.00 © 2019 IEEE   159

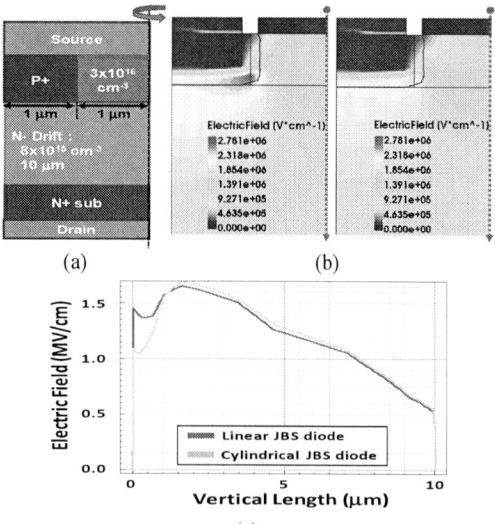

(a)　　　　　　　　(b)

(c)

Fig. 2: (a) TCAD simulation structure for linear and cylindrical (rotated at the right edge) JBS diodes that are monolithically integrated in JBSFETs. (b) Simulation results of electric field distribution at 1.2 kV for linear cell (left) and cylindrical cell JBS diodes (right). (c) Extracted electric field values at 1.2 kV along with the lines marked in red in (b).

contact, and JBS Schottky contact areas as well as the unit cell area in the Oct cell topology. Two bar lengths (a=1.6μm for the Oct design and a=2.8μm for the Oct_D design) were chosen for comparing the electrical performance. The Oct_D design has higher JBS diode density to reduce the third quadrant on-state voltage drop. The JBS diode density is defined as (JBS diode area)/(unit cell area).

A low surface electric field at the Schottky contact is achieved in the JBS diode to reduce the leakage current under the high reverse bias [7]. TCAD numerical simulations were performed to examine the electric field at the Schottky contact in the integrated JBS diodes at high reverse bias (V=1.2 kV). The JBS diodes in the Hex and Oct JBSFETs was modelled as a cylindrical structure by rotating the cross-section on the right-hand-side as indicated in Fig. 2(a). The simulation for the Linear structure was performed without cell rotation. It can be observed from Fig. 2(b) that a much smaller electric field is generated at the corner of the P-N junction in the Oct and Hex cells compared with the Linear cell. More importantly, it can be seen in Fig. 2(c) that a greater reduction of the electric field at the Schottky contact is observed in JBS diodes formed in the Hex and Oct JBSFETs compared with the conventional linear cell topology. This suppresses the large increase in leakage current with increasing reverse bias voltage observed in SiC Schottky diodes [7].

III. FABRICATION TECHNOLOGY AND EXPERIMENTAL RESULTS

Total 11 mask layers were designed for the three different cell topologies using the same design rules for features and alignment tolerances. All the Accumulation-channel (Acc) structures have the same active areas of 0.045cm^2 and same hybrid-JTE edge termination designs. 4H-SiC Si-face (0001) 6-inch wafers (4° tilted toward <11$\bar{2}$0>) with n-type epitaxial layer (8×10^{15} cm^{-3} doped and 10

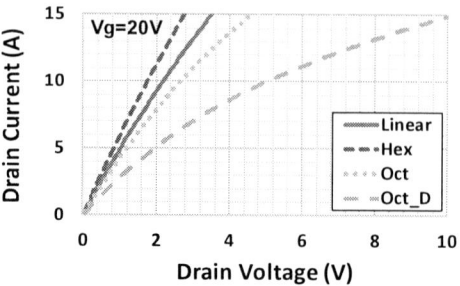

Fig. 3: Typical measured output characteristics at V_g=20V of fabricated 1.2 kV Acc JBSFETs with various cell topologies. Active area is 0.045 cm^2.

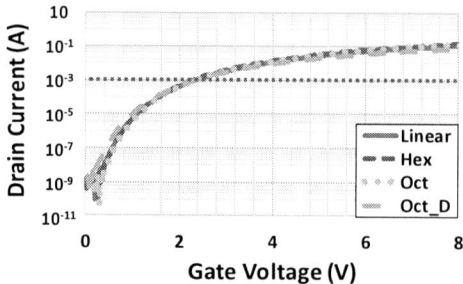

Fig. 4: Measured transfer characteristics for fabricated Acc JBSFETs with various cell topologies. V_{th} was extracted at I_d=1mA.

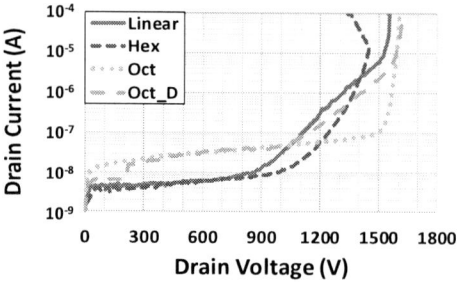

Fig. 5: Typical room temperature blocking characteristics (V_{gs} = 0V) of fabricated JBSFETs. BV is defined at I_d = 100μA.

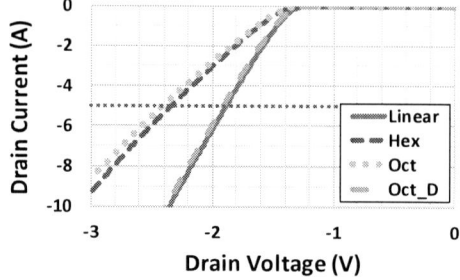

Fig. 6: Measured 3rd quadrant characteristics of fabricated Acc JBSFETs with various cell topologies.

μm thick) on N$^+$ substrates were used as the starting material. All the devices were simultaneously fabricated at a 6-inch commercial foundry, X-Fab, using the same process flow. The n and p-type regions were formed by a series of high temperature implantations of N and Al, followed by an activation annealing process at 1650 °C for 10 min with a carbon cap. A 55nm thick gate oxide (measured with a C-V test element) was grown by dry oxidation at 1175 °C, followed by NO interface annealing. An n-type silicided poly-Si gate was then deposited and patterned. An oxide interlayer dielectric was deposited and patterned, followed by a nickel silicide (NiSi) process with an RTA (900 °C) on the both front and backsides. The Ohmic and Schottky contacts were simultaneously formed with this process [1]. A 4 µm thick Al layer was deposited to serve as the source and gate pads. Nitride and polyimide passivation layers were stacked and patterned on the front side to open the source and gate contact areas. A solderable metal stack was lastly deposited on the backside.

Typical measured output characteristics of the fabricated JBSFETs with various cell topologies are shown in Fig. 3. The on-resistance measured at V_g of 20 V for the Hex cell is 1.28× lower than that for the Linear cell case while those for the Oct and Oct_D cells are 1.25× and 2.46× larger than the Linear cell case. This correlates with their smaller channel and JFET densities provided in Table I. It can be seen from the transfer characteristics in Fig. 4 that all the Acc JBSFETs have the same threshold voltages of 2.32 ± 0.04 V measured at V_d of 0.1 V and I_d of 1 mA.

The measured blocking characteristics of the JBSFETs with identical edge termination are shown in Fig. 5. The breakdown voltage (BV) was extracted at I_d=100µA. The measured BV for the Hex cell design is only 1350V which may be too low for a 1200V rated device. Its BV is much lower than for the linear cell due to localized high electric field concentration at the hexagonal shaped corners [9]. The highest BV of 1620V is observed for the Oct designs due to a reduced electric field in its JFET region. A larger leakage current up to drain voltage of ~1200V is observed in Fig. 5 for the Acc Oct cell designs than those for the Linear and Hex cells. The channel potential barrier is sufficiently large for a channel length of 0.5µm resulting in the low leakage current for the Linear and Hex cell topologies. It is conjectured that the P⁺ shielding and N⁺ source regions were not accurately patterned for the Oct cell topologies during the masking process because the photolithography process was optimized for the Linear cell topology. The Oct and Oct_D JBSFETs may have smaller (<0.5µm) local channel length. It has been demonstrated by numerical simulations that the channel potential barrier gets reduced with smaller channel length [4]. This leads to the higher leakage current observed in the Oct cell designs. Nevertheless, the leakage current for the Oct cell topologies (<0.1µA) is well below the typical leakage current limit of 100 µA in data sheets.

The measured third quadrant characteristics for the different JBSFETs are shown in Fig. 6. The JBS diodes in the Hex and Oct cell structures have higher on-state voltage drops than the Linear cell case. This is because their JBS diode area density of 0.05 is 3× smaller than for the Linear cell. The JBS diode on-state voltage drop in the Oct cell topology can be made equal to that of the Linear cell topology by increasing its area density to 0.167 as shown by the Oct_D cell design.

The measured C_{rss} for all the devices are shown in Fig. 7. The C_{rss} for the Hex cell is larger than that for the Linear cell, while that for the Oct cell is much smaller. This is due to the C_{rss} values being proportional to the JFET area density [7] which is provided in Table 1. The JFET density is defined as

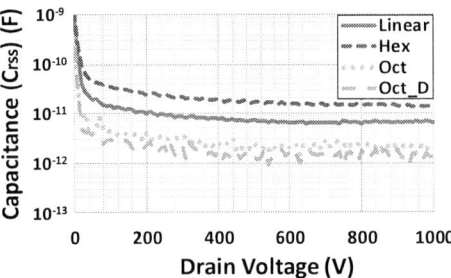

Fig. 7: Typical measured reverse transfer capacitance of fabricated JBSFETs with various cell topologies. The Hex cell has higher C_{rss} values while the Oct cell has lower C_{rss} values than the Linear design.

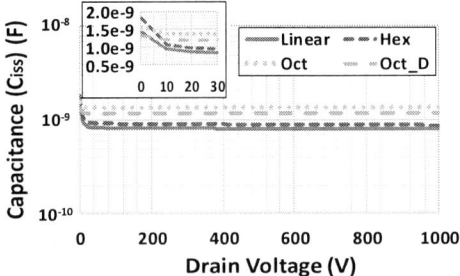

Fig. 8: Measured input capacitance (C_{iss}) of fabricated JBSFETs. The Hexagonal JBSFET show smaller C_{iss} value at V_d=1kV, but the highest value at low V_d as shown in the inset zoom-in graph.

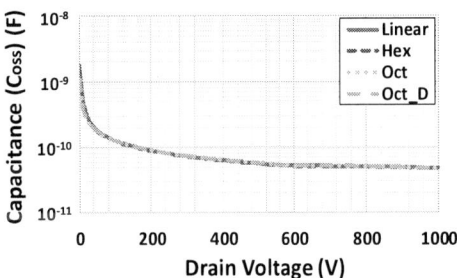

Fig. 9: Typical measured output capacitance (C_{oss}) of fabricated JBSFETs with various cell topologies. All the structures have similar C_{oss} values regardless of the cell topologies.

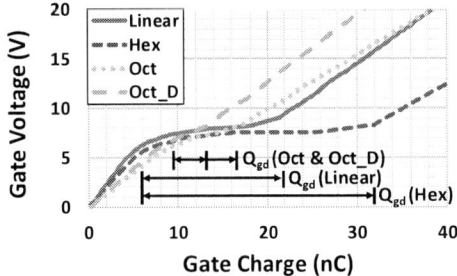

Fig. 10: Measured gate charge at V_d=800V and I_d=10A of fabricated JBSFETs. A large gate plateau (Q_{gd}) is observed in the Hex design. Much smaller gate plateau (Q_{gd}) is observed in the Oct design.

(JFET area)/(unit cell area). The C_{rss} values extracted at V_d=1kV for the Linear, Hex, Oct, and Oct_D are 6.39, 14.11, 2.10, and 1.35 pF, respectively.

The measured C_{iss} for all the devices are shown in Fig. 8.

TABLE I
SUMMARY OF EXPERIMENTAL RESULTS
FOR LINEAR, HEXAGONAL, AND OCTAGONAL JBSFETs

	Linear	Hex	Oct	Oct_D
$W_{A-A'}$ [μm]	6.1	6.6	5.9	7.6
W_{Sh} [μm]	1.0	1.5	1.1	2.8
CH. density [μm^{-1}]	0.164	0.211	0.193	0.116
JFET density	0.246	0.403	0.109	0.065
JBS diode density	0.164	0.052	0.054	0.167
BV [V]	1556	1351	1625	1618
V_{th} [V]	2.34	2.28	2.26	2.36
3rd V_F (@ 5A) [V]	1.89	2.34	2.41	1.91
*$R_{on,sp}$ [mΩ-cm^2]	8.83	6.90	11.00	21.71
$C_{iss,sp}$ (@ 1kV) [nF/cm^2]	17.7	19.7	30.0	26.4
$C_{oss,sp}$ (@ 1kV) [nF/cm^2]	1.08	1.06	1.07	1.09
$C_{rss,sp}$ (@ 1kV) [pF/cm^2]	142	314	47	30
$Q_{gd,sp}$ [nC/cm^2]	356	578	147	84
FOM <C_{iss} / C_{rss}>	125	63	638	880
HF-FOM <$R_{on} \times C_{rss}$> [mΩ-pF]	1254	2167	517	651
HF-FOM <$R_{on} \times Q_{gd}$> [mΩ-nC]	3143	3988	1617	1824

*$R_{on,sp}$ @ V_g=20V, I_d=10A; includes $R_{sub,sp}$ (~0.7 mΩ-cm^2)

The C_{iss} consists of C_{gs} plus C_{rss}. The C_{gs} is determined by the overlap area between the gate and the N-base/N$^+$ source regions [7]. At high drain voltages, the C_{iss} becomes equal to C_{gs} because of negligible C_{rss} values. However, at low drain voltages, the contribution of the C_{rss} becomes significant. The Oct cells have larger C_{iss} values at V_d of 1000V because of the poly-Si bars add to the gate-source overlap area. However, the Hex cell has the greater C_{iss} at V_d of 0V as shown in the inset graph in Fig. 8 due to its larger C_{rss} compared with the Linear and Oct cells.

It can be observed in Fig. 9 that all the cell topologies have the same C_{oss} (=C_{ds}+C_{rss}) of 48.45±0.6 pF at high voltages within measurement accuracy. These results indicate that all the cells have similar C_{ds} and relatively negligible C_{rss} values at V_d of 1000V.

The measured gate charge at V_d=800V and I_d=10A for the various JBSFET cell topologies can be compared in Fig. 10. Much smaller gate charge (Q_{gd}) is observed for the Oct cell JBSFETs compared with the Hex and Linear cell structures due to their significantly smaller JFET area density. The extracted values for $Q_{gd,sp}$ are provided in Table I for all the cell topologies. The $Q_{gd,sp}$ values for the Linear, Hex, Oct, and Oct_D are 356, 578, 147, and 84 nC/cm^2, respectively.

The experimental results are summarized in Table I with structural information. The channel (CH.), JFET, and JBS diode densities for all the cell topologies were calculated as (channel perimeter)/(unit cell area), (JFET area)/(unit cell area), and (JBS diode area)/(unit cell area), respectively. The Hex cell has the lowest $R_{on,sp}$ due to the biggest channel density, but its BV of 1350V may be insufficient for a 1200V rated device.

The Oct cell has excellent HF-FOM [$R_{on} \times Q_{gd}$] by 1.9× and 2.5×, and FOM [C_{iss}/C_{rss}] by 5.1× and 10.1× with superior $C_{rss,sp}$ and $Q_{gd,sp}$ compared with Linear and Hex JBSFETs, respectively. However, larger 3rd quadrant on-state voltage drop is observed for this design than the Linear cell case because of the 3× smaller JBS diode area density.

The Oct_D cell has 2.5× higher $R_{on,sp}$ due to the low channel density, but it has much superior $C_{rss,sp}$ (4.7×) and $Q_{gd,sp}$ (4.2×) than Linear cell JBSFET. As a consequence, the Oct_D JBSFET has superior HF-FOM [$R_{on} \times Q_{gd}$] by 1.7× and 2.2× compared with Linear and Hex JBSFETs, respectively. In addition, its FOM [C_{iss}/C_{rss}] is 7.0× and 14.0× better than the Linear and Hex JBSFETs, respectively.

IV. CONCLUSIONS

A novel 1.2 kV 4H-SiC Octagonal-cell (Oct) JBSFET was successfully fabricated with the Linear and Hexagonal (Hex) cell topologies for the first time using the same process flow at a 6-inch foundry. The on-state voltage drop for the integrated JBS diode in the Oct cell was found to be equal to that of the JBS diode in the Linear cell design when its area density was made the same. It is experimentally demonstrated that the Oct cell JBSFETs show superior HF-FOMs and FOM [C_{iss}/C_{rss}] compared with other cell topologies. This makes it the best choice for high frequency applications where switching losses are dominant.

REFERENCES

[1] W. Sung and B. J. Baliga, "Monolithically Integrated 4H-SiC MOSFET and JBS Diode (JBSFET) Using a Single Ohmic/Schottky Process Scheme," *IEEE Electron Device Letters*, vol. 37, no. 12, 2016.

[2] C.-T. Yen, F.-J. Hsu, C.-C. Hung, C.-Y. Lee, L.-S. Lee, Y.-F. Li, and K.-T. Chu, "Avalanche Ruggedness and Reverse-Bias Reliability of SiC MOSFET with Integrated Junction Barrier Controlled Schottky Rectifier," in *Proc. 30th Int. Symp. Power Semiconductor Devices ICs*, May 2018.

[3] A. Kanale, B. J. Baliga, K. Han, and S. Bhattacharya, "Experimental Study of High-Temperature Switching Performance of 1.2kV SiC JBSFET in Comparison with 1.2kV SiC MOSFET," in *Proc. Int. European Conference on Silicon Carbdie and Related Materials.*, Paper MO.P.MO3, September 2018.

[4] B. J. Baliga, *Gallium Nitride and Silicon Carbide Power Devices.* Hackensack, NJ, USA: World Scientific, 2017, ch. 11, pp. 287-340

[5] J. Wei, M. Zhang, H. Jiang, C.-H. Cheng, and K. J. Chen, "Low ON-Resistance SiC Trench/Planar MOSFET with Reduced OFF-State Oxide Field and Low Gate Charges," *IEEE Electron Device Letters*, vol. 37, no. 11, Nov. 2016.

[6] H. Sheng, Z. Chen, F. Wang, and A. Millner, "Investigation of 1.2 kV SiC MOSFET for high frequency high power applications," in *Proc. 25th Applied Power Elec. Conf. and Exposition (APEC)*, Feb. 2010, pp. 1572-1577.

[7] B. J. Baliga, *Fundamentals of Power Semiconductor Devices 2nd ED.* Gewerbestrasse 11, 6330 Cham, Switzerland: Springer, 2019, ch. 4, pp. 171-206 & ch. 6, pp. 283-440.

[8] J. W. Palmour, L. Cheng, V. Pala, E. V. Brunt, D. J. Lichtenwalner, G.-Y. Wang, J. Richmond, M. O' Loughlin, S. Ryu, S. T. Allen, A. A. Burk, and C. Scozzie, "Silicon Carbide Power MOSFETs: Breakthrough Performance from 900 V up to 15 kV," in *Proc. 26th Int. Symp. Power Semiconductor Devices ICs*, Jun. 2014, pp. 79-82.

[9] P. A. Losee, A. Bolotnikov, L. C. Yu, G. Dunne, D. Esler, J. Erlbaum, B. Rowden, A. Gowda, A. Halverson, R. Ghandi, P. Sandvik, and L. Stevanovic, "SiC MOSFET Design Considerations for Reliable High Voltage Operation," in *IEEE Int. Rel. Phys. Symp.*, Apr. 2017.

[10] R. Singh, D. C. Campbell, J. T Richmond, and J. W. Plamour, "High Channel Density 20 A 4H-SiC ACCUFET with R_{onsp} = 15 mΩ-cm$^{2"}$, *Electronics Letters*, vol. 39, no. 1, pp. 152-154, 2003.

[11] K. Han and B. J. Baliga, "The 1.2 kV 4H-SiC OCTFET: A New Cell Topology with Improved High Frequency Figures-of-Merit," *IEEE Electron Device Letters*, vol. 40, no. 2, Feb. 2018.

[12] C.-T. Yen, C.-C. Hung, H.-T. Hung, L.-S. Lee, C.-Y. Lee, T.-M. Yang, Y.-F. Huang, C.-Y. Cheng, and P.-J. Chuang, "1700V/30A 4H-SiC MOSFET with low cut-in voltage embedded diode and room temperature boron implanted termination," in *Proc. 27th Int. Symp. Power Semiconductor Devices ICs*, May 2015, pp. 265–268.

[13] S. Hino, T. Hatta, K. Sadamatsu, Y. Nagahisa, S. Yamamoto, T. Iwamatsu, Y. Yamamoto, M. Imaizumi, S. Nakata, and S. Yamakawa, "Demonstration of SiC-MOSFET embedding Schottky barrier diode for inactivation of parasitic body diode," in *Proc. 11th Eur. Conf. Silicon Carbide Rel. Mater.*, Sep. 2016, pp. 129–130.

978-1-7281-0582-6/19 $31.00 © 2019 IEEE

Proceedings of the 31st International Symposium on Power Semiconductor Devices & ICs
May 19-23, 2019, Shanghai, China

Differential Variable Base Charge Pumping (Δ-CP) for SiO₂/SiC Interface Characterization

P. Moens[1], A. Constant[1], A. Stockman[1,3], J. Franchi[2] and F. Allerstam[2]

[1]ON Semiconductor, Westerring 15, B-9700 Oudenaarde, Belgium

[2]ON Semiconductor, Isafjordsgatan 32C, SE-16440 Kista, Sweden

[3]Gent University, Technology Park 126, B-9052 Gent, Belgium
peter.moens@onsemi.com

Abstract—We propose an electrical evaluation technique named differential charge pumping (Δ-CP) to extract information on the interface states (D_{it}) at the SiC/SiO₂ interface. This technique allows to obtain information on (1) the physical location of the D_{it} ; (2) the energetic spread of the D_{it} in the bandgap; (3) if the D_{it} are donor or acceptor type. First results indicate contributions from both the "S/D" region and the channel regions to the overall CP signal. Surprisingly a substantial contribution of the "S/D" regions of the transistors to the overall CP signal is observed. In the channel region, both donors and acceptor type states are identified, the majority being acceptor type.

Keywords—SiC ; interface states ; charge pumping ; donor ; acceptor states

I. INTRODUCTION

Due to the large amount of interface states (D_{it}) at the SiO₂/SiC interface, inversion layer channel mobility is low, which impacts the Ron of SiC MOSFETs [1]. For transistors up to 1.2kV, improving Ron is thus to a substantial extent determined by improvements in the channel mobility. Hence it is important to have a measurement technique that allows experimental information on

(1) the physical location of the D_{it}

(2) the energetic spread of the D_{it} over the bandgap

(3) if the D_{it} are donor or acceptor type

To this effect, we extend the variable base Charge Pumping (CP) technique by a differential Charge Pumping (Δ-CP) method on test structures with different L_g. Albeit more difficult to interpret, the variable base CP method [2-4] is preferred over the commonly used variable amplitude CP [5,6], since it contains more information in one single experiment. The Δ-CP method allows to extract the D_{it} profile for both the channel and S/D connect regions separately. From the data, it follows that both regions have a sharply peaked D_{it} distribution close to the conduction band E_c, at $E-E_c=0.11$ eV, but surprisingly the D_{it} in the S/D connect region is higher than in the channel. The D_{it} in the channel region is mainly acceptor type, but depending on the gate oxide process conditions, also donor-like traps can be identified.

II. DIFFERENTIAL CHARGE PUMPING PROCEDURE

Fig.1 shows the test structure used for Δ-CP. A pulse is applied to the gate with amplitude $V_a=10V$, rise and fall times t_{rise}, t_{fall}, and frequency f=10kHz while the base voltage is swept. For selected V_{base} voltages (see Fig. 2), the structure is pulsed from accumulation into inversion. During inversion, electrons are trapped, but are emitted during the accumulation time of the pulse. The charge pumping current I_{cp} is then measured as a hole current at the p+ contact (holes that recombine with the emitted electrons). Measurement is done on structures with different L_g and different W. Due to straggling of the n+ implant and gate alignment limitations, there is a n-type doped region under the gate labeled "S/D".

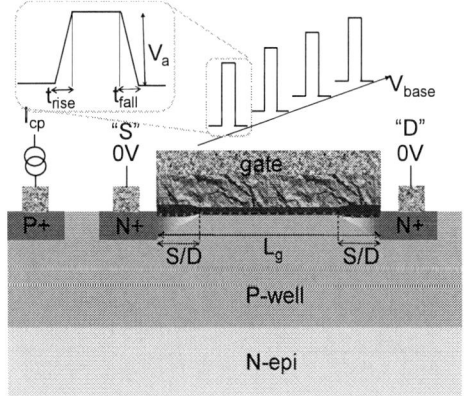

Fig. 1 : Schematic cross-section of the test structure used for variable base CP.

Each region in the transistor will have different gate bias conditions for inversion and accumulation since doping concentration and oxide thickness vary along the SiC/SiO₂ interface. For the pulse conditions applied in this study, inversion and accumulation occurs at a gate voltage V_g for which a carrier concentration of 4.6×10^{12} cm⁻³ minority and majority carriers at the SiO₂/SiC interface, respectively, is reached [2]. These V_g values are denoted by V_e (inversion) and V_h (accumulation).

V_e and V_h can be a priori calculated from simple MOS theory, or derived from a 2D TCAD simulation.

978-1-7281-0582-6/19 $31.00 © 2019 IEEE   163

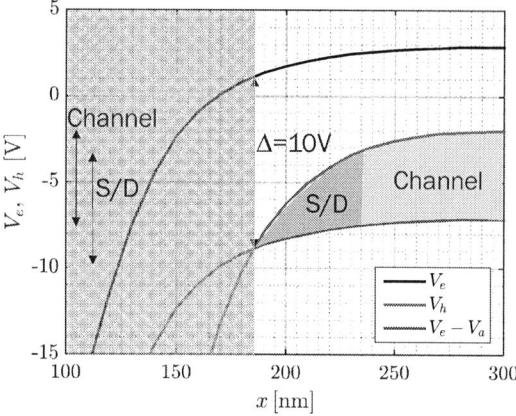

Fig. 2 : V_e and V_h from TCAD, along the SiO$_2$/SiC interface in the vicinity of the "S/D" (see Fig. 1). V_e and V_h are the gate voltages to have a target minority and majority carrier concentration of 4.6×10^{12} cm^{-3} (for our pulse conditions [2]), at the SiO$_2$/SiC, respectively.

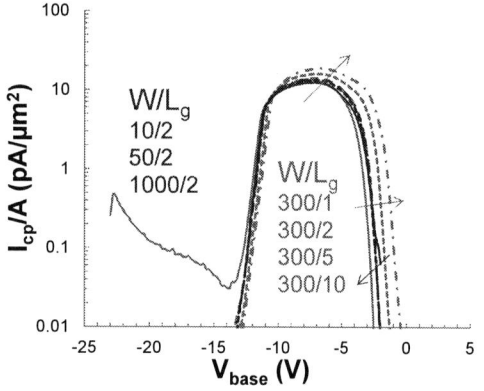

Fig. 3 : Variable base CP, on structures with different W and L_g, normalized to the total channel area (f=10kHz, t_{rise}=100ns, t_{fall}=100ns, V_a=10V).

Fig 2 shows V_e and V_h along the SiO$_2$/SiC interface in the vicinity of the "S/D" (see Fig. 1), as from a 2D TCAD simulation. A CP current I_{cp} will be detected for V_e-V_a<V_{base}<V_h (V_a=10V). Hence, according to Fig. 2, the "S/D" region will have a CP signal between -10V and -3V, whereas the channel CP signal is between -7V and -2V. The gray shaded area is the transistor area that is not pumped, since V_h-V_e>10V. The above is a simplification since it assumes no fixed oxide charge and no charged donors or acceptors (all traps are considered charge neutral).

Comparing the rise and fall edges of the experimental CP signal with Fig. 2, allows to identify the separate contributions from the "S/D" and channel regions, as will be detailed below.

III. EXPERIMENTAL RESULTS

Variable base CP characteristics are measured on transistor structures with different L_g and W, see Fig. 3. The CP current I_{cp} is normalized to the total channel area=W$\times L_g$. In blue are the CP spectra for fixed L_g, but varying W, in red the CP spectra for fixed W but varying

L_g. There is perfect scaling with W (all blue curves are on top of each other), but not with L_g (red curves), hinting to separate contributions along L_g i.e. from the "S/D" and channel regions, since scaling L_g will scale the channel contribution, but not the (fixed) "S/D" contribution. The rising edge of the CP spectra is identical, whereas the falling edge is not. Larger L_g shifts the CP signal to higher base level voltages V_{base}. Also the maximum I_{cp} is not scaling with L_g. This indicates that the CP spectrum consists of (at least) two different contributions along L_g.

In order to identify these contribution, CP is done on structures with different L_g (L_g=2μm and L_g=8μm, W=300μm). By subtracting the CP signals for a fixed pulse condition (f=10kHz, t_{rise}=100ns, t_{fall}=100ns, V_a=10V), the contribution of both the S/D regions (do not change with L_g) and the channel (does change with L_g), can be isolated. The differential CP signal Δ-CP is plotted in Fig. 4(a), which allows to identify the two separate contributions of the channel and "S/D" regions (as shown in Figs 1 and 2). The "S/D" Δ-CP is skewed to lower V_{base} levels in agreement with Fig. 2 (dashed lines represent the "theoretical" locations for V_e-V_a and V_h as from Fig.2, assuming neutral charged states).

To check the validity of our approach, Figs 4(b) and 4(c) show the normalized and absolute CP spectra for L_g=1, 1.5, 2, 3, 5, 8 and 10μm (symbols), respectively. The CP spectrum is "predicted" by summing the contribution from the channel and "S/D" regions, as in Fig. 4(a), from the extraction using the L_g=2 and L_g=8μm data sets. The lines show the model prediction : the blue lines are model fit for the L_g=2 and L_g=8μm datasets, obviously resulting in a perfect fit since these datasets are used for the extraction ; the orange lines are the model predictions for the other L_g CP spectra using the "S/D" and channel Δ-CP components of Fig. 4(a). The channel component is visible as a hump on the falling edge of the CP spectrum, its contribution increasing with increasing L_g. The model fit and predictions (orange lines) agree well, for both rising edge, falling edge, and maximum I_{cp}, supporting the validity of our approach.

Another way to prove that there is a substantial contribution from the "S/D" regions, is by comparing CP characteristics with Source and Drain grounded (standard biasing scheme) with Source open (i.e. only connecting the Drain terminal). This is shown in Fig. 5 where the CP and Δ-CP spectra for different L_g with Source and Drain grounded (green curves, standard measurement), with Source open (red curves) are depicted. Blue curves are the differential signal. The insert shows a zoom-in of the rising edge of the CP signal, clearly showing that until the channel fully opens (at V_{base}=-10V, see Fig, 4a), the red curve is exactly half of the green signal, indicating that only one "S/D" connection is probed. When the channel is open, the Source region I_{cp} is also collected at the Drain terminal. Electrons from the Drain terminal have to drift to the source side of the channel, effectively doubling the distance traveled compared to when both source and drain are grounded. The CP signals are coincident, indicating that there is no geometrical component [7]. In the absence of a geometric component, there is no difference between the spectra for higher V_{base}.

The difference in rising edge of the CP signal for Source and Drain connected, and Source open adds further evidence

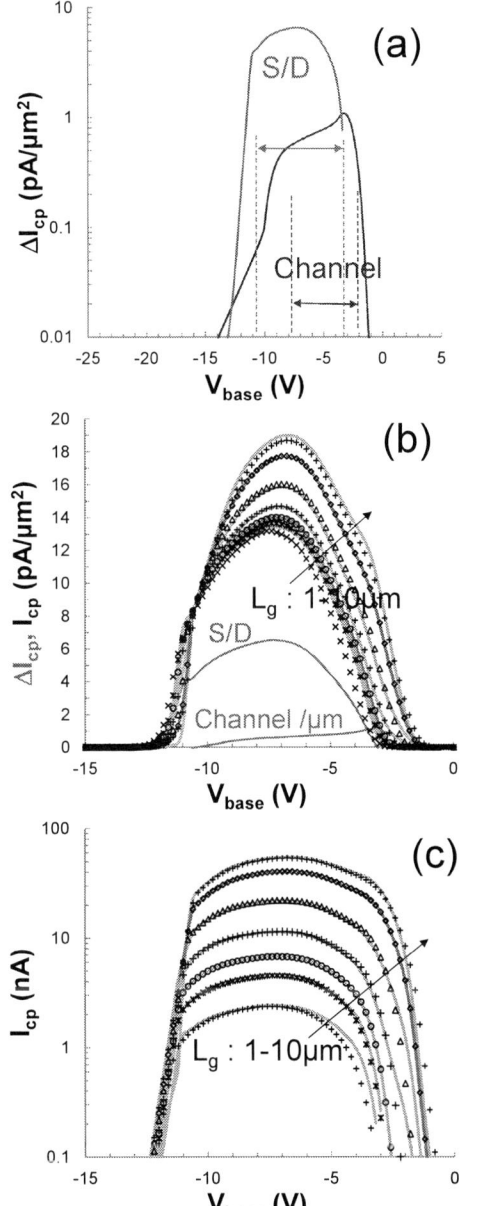

theoretical Δ-CP edges as from Fig. 2, assuming no net positive or negative charge. With decreasing t_{fall}, interface states which are energetically closer to the band edge can be probed, resulting in a higher I_{cp}. The most important observations are :

(1) the D_{it} is significantly higher in the "S/D" region compared to the channel region (~5x-7x)

(2) the channel region shows a clear presence of both donor and acceptor-like traps (explained in the next section)

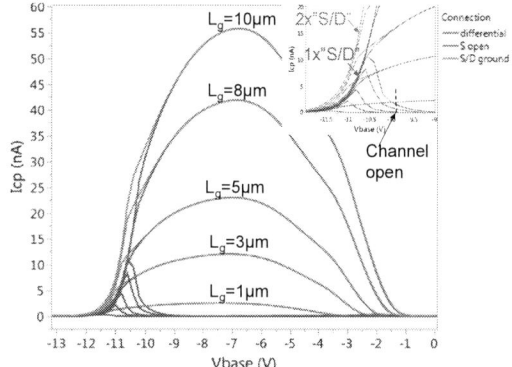

Fig. 5 : CP and Δ-CP spectra for different L_g, comparing S/D grounded (green curves, standard measurement), with S open (hence only connecting the drain—red curves). Blue curves are the differential signal. The insert shows a zoom-in of the rising edge of the CP signal.

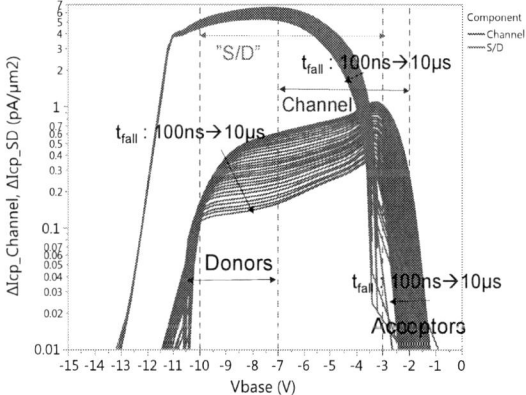

Fig. 6 : Δ-CP spectra for "S/D" and channel component as a function of t_{fall}. Pulse conditions: f=10kHz, t_{rise}=100ns, V_a=10V, t_{fall}=100ns-10μs. CP spectra used for Δ-CP extraction : L_g=2 and L_g=8 μm (See Fig. 4). The dashed lines represent the theoretical Δ-CP edges as from Fig. 2, assuming no net positive or negative charge.

Fig. 4 : (a) Extraction of Channel and "S/D" component, using Δ-CP on devices with W=300μm, L_g=2μm and L_g=8μm. (b) normalized CP spectra for L_g=1, 1.5, 2, 3, 5, 8 and 10μm (symbols), blue line : model fit using the L_g=2 and L_g=8μm dataset. Orange lines : model <u>prediction</u> for the other L_g CP spectra using the S/D and channel Δ-CP components (red lines) ; (c) same data as Fig. 4(b), but in absolute I_{cp}.

to the suggestion that the rising edge is due to the "S/D" region, as already predicted by Figs 1 and 2.

By changing the fall time of the gate pulse t_{fall}, the energy distribution of the D_{it} close to the conduction band edge E_c can be extracted [2,3]. Fig. 6 shows the Δ-CP signals for "S/D" and channel area as a function of t_{fall} separately, as per the procedure explained above. Pulse conditions are f=10kHz, t_{rise}=100ns, V_a=10V, t_{fall}=100ns-10μs. CP spectra used for Δ-CP extraction are L_g=2 and L_g=8 μm (See also Fig. 4). The dashed lines represent the

IV. DONOR OR ACCEPTOR TYPE

The impact of donor and acceptor states on a variable base Δ-CP spectrum for the channel area (n-MOS) is schematically represented in Fig. 7, which shows the variation of the Fermi level E_f during one CP pulse, along with the CP signal for (a) no charged states ; (b) donor states ; (c) acceptor states.

Without any charged states ("neutral"), the CP signal is expected between V_e-V_a and V_h. Donor states are empty

978-1-7281-0582-6/19 $31.00 © 2019 IEEE   165

during the rising edge, representing a net positive charge (blue shaded area in Fig. 7 (b). During the rising time of the pulse, they remain empty since the system is in non-equilibrium and filling of the traps takes time. This net positive charge will be represented by a negative shift of the Δ-CP spectrum to lower V_{base}. The total amount of available donor states (in cm^{-2}) can be estimated from the left-hand voltage shift of the rising edge of the Δ-CP signal. During the falling edge, the majority of the donor traps is filled, hence neutral, and during t_{fall}, they largely remain neutral, reflected in no shift on the falling edge of the Δ-CP spectrum. For acceptor traps, the opposite happens (no shift on the rising edge, but a positive shift on the falling edge). This allows to discriminate between donor and acceptor traps. In the case of a fixed oxide charge, a parallel shift on the total Δ-CP spectrum would be observed.

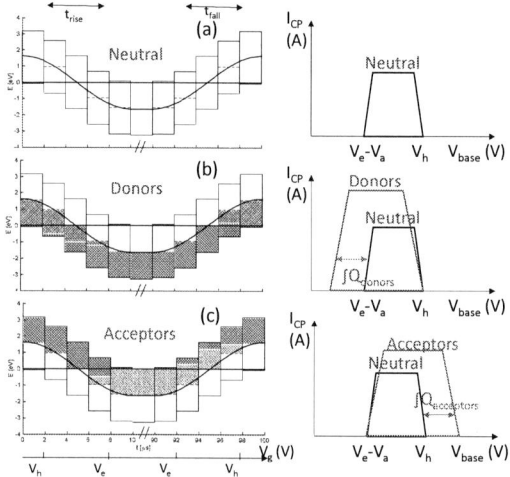

Fig. 7 : Impact of donor and acceptor states on a variable base Δ-CP spectrum for the channel area (n-MOS). Schematic representation of the variation of the E_f during one CP pulse, along with the CP signal : (a) no charged states ; (b) donor states ; (c) acceptor states.

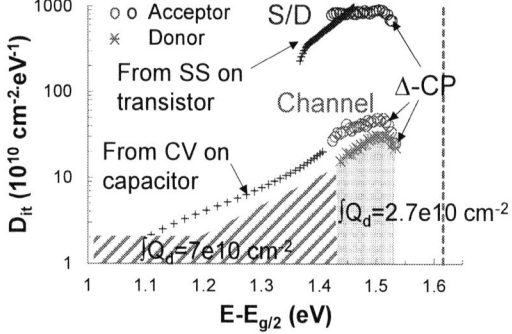

Fig. 8 : Energy distribution of D_{it} for both S/D and channel as from the from Δ-CP of Fig.6, compared to data from subtreshold slope on a transistor, and CV on an n-well capacitor [9,10].

Finally, Fig. 8 shows the extracted D_{it} for both the "S/D" and channel area. From the left-hand shift of the channel Δ-CP signal (Fig. 6), a total ionized donor density of 10^{11} cm^{-2} is calculated. The integrated donor type D_{it} is $\sim 3.10^{10}$ cm^{-2}. The estimated overall donor D_{it} distribution is shown as magenta shaded (extends till E_V). A similar analysis is performed for the channel acceptors, as well as for the "S/D" donors and acceptors, and is also shown in Fig. 8. A sharply peaked distribution at $E-E_c=0.11$ eV is obtained, in good agreement with [8].

V. CONCLUSIONS

In this paper, an electrical evaluation technique named differential charge pumping (Δ-CP) is proposed. This technique is based on the variable base charge pumping method, and allows to extract information on the interface states (D_{it}) at the SiC/SiO$_2$ interface, specifically (1) the physical location of the D_{it} ; (2) the energetic spread of the D_{it} in the bandgap; (3) if the D_{it} are donor or acceptor type. First results indicate contributions from both the "S/D" region and the channel regions to the overall CP signal. Surprisingly a substantial contribution of the "S/D" regions of the transistors to the overall CP signal is observed. In the channel region, both donors and acceptor type states are identified, the majority being acceptor type. A sharply peaked D_{it} distribution at $E-E_c=0.11$ eV is obtained, in good agreement with literature.

REFERENCES

[1] V. Vathulya and M. White,"Characterization and performance comparison of the power DIMOS structure fabricated with a reduced thermal budget in 4H and 6H-SiC, "Solid-State Electronics, vol.**44** (2), pp. 309–315 (2000)

[2] G. Groeseneken, H. E. Maes, N. Beltran and R. F. De Keersmaecker, "A reliable approach to charge-pumping measurements in MOS transistors," in IEEE Transactions on Electron Devices, vol. **31**, no. 1, pp. 42-53 (1984).

[3] G. Groeseneken, "Introduction to Charge Pumping and Its Applications" tutorial at the Semiconductor Interface Specialists Conference (SISC), 10 December, San Diego, CA, USA (2008).

[4] P. Moens and G. Van den bosch, "Characterization of Total Safe Operating Area of Lateral DMOS Transistors (review paper)", IEEE Transactions on Device and Materials Reliability, **6** (3), pp349-357 (2006).

[5] D. Okamoto, H. Yano, T. Hatayama, Y. Uraoka, and T. Fuyuki, "Analysis of Anomalous Charge-Pumping Characteristics on 4H-SiC MOSFETs", IEEE Transactions on Electron Devices, **55** (8) pp. 2013-2020 (2008).;

[6] A. Salinaro, G. Pobegen, T. Aichinger, B. Zippelius, D. Peters, P. Friedrichs and L. Frey, "Charge Pumping Measurements on Differently Passivated Lateral 4H-SiC MOSFETs", IEEE Transactions on Electron Devices, **62**, pp155-163 (2015).

[7] G. Van den bosch, G. Groeseneken and H. Maes, "On the Geometric Component of Charge-Pumping Current in MOSFET's", IEEE Electron Device Letters, **14**, (3), pp107-109 (1993).

[8] J. Berens, G. Pobegen, T. Aichinger, G. Rescher and T. Grasser, "Cryogenic Characterization of NH3 post oxidation annealed 4H-SiC Trench MOSFETs", Proceedings of ESCRM 2018, paper #1097

[9] E. H. Nicollian and J. R. Brews, MOS Physics and Technology (Wiley, New York, 1982) ;

[10] H. Yoshioka, J. Senzaki, A. Shimozato, Y. Tanaka and H. Okumura , "Effects of interface state density on 4H-SiC n-channel field-effect mobility", Appl. Phys. Lett. **104**, 083516 (2014).

Proceedings of the 31st International Symposium on Power Semiconductor Devices & ICs
May 19-23, 2019, Shanghai, China

Experimental and Numerical Investigations of Short-Circuit Failure Mechanisms for State-of-the-Art 1.2kV SiC Trench MOSFETs

Masataka Okawa, Ruito Aiba*, Taiga Kanamori*,
Hiroshi Yano, and Noriyuki Iwamuro
Graduate School of Pure and Applied Sciences
*College of Engineering Science
University of Tsukuba, Tsukuba, Japan
s1820310@s.tsukuba.ac.jp,
iwamuro.noriyuki.fb@u.tsukuba.ac.jp

Shinsuke Harada
National Insutitute of Advanced Industrial Science and
Technology, Tsukuba, Japan
s-harada@aist.go.jp

Abstract— **This paper is focused on the short-circuit capability and analysis of failure mechanism on relatively small DC power supply voltage for state-of-the-art 1.2 kV SiC trench MOSFETs. It is found that the gate-source SiO₂ rapture could be a failure mechanism unique to the SiC MOSFETs. Further, the SiC trench MOSFETs with higher threshold voltage can effectively improve the short-circuit capabilities by reducing the probability of channel normally-on induced by the higher lattice temperature.**

Keywords—SiC MOSFET, Trench MOSFET, Short-Circuit SOA, SiO₂ rapture, Normally-on operation

I. INTRODUCTION

With the characteristics of the wider energy gap and the higher thermal capability, silicon carbide (SiC) has gradually replaced the status of silicon-based devices in many application areas such as the requirements for the low on resistance and high temperature operation capability. In the SiC devices, SiC trench MOSFET has considerable impact on reducing on resistance owing to its small cell pitch and high channel mobility on {11-20} or {1-100} planes; therefore, this device structure is a superior candidate for next power semiconductor devices. However, in the application of power electronics, the robustness, such as short-circuit capability, is an important requirement for stability and reliability of power system. For example, DC input voltage would be frequently changed to the smaller value to control the output AC motor speed [1], so that an investigation of short-circuit failure with the smaller DC input voltage is very important.

In this paper, the authors focused on the short-circuit capability and analyzed its failure mechanisms for state-of-the-art 1.2 kV SiC trench MOSFETs. Especially, the relationship between the gate switch-off voltage and short-circuit withstand capability was investigated by experiments and numerical simulations.

II. EXPERIMANTAL AND NUMERICAL SIMULATION RESULTS

The devices under test (DUTs) are two types of 1.2 kV SiC trench MOSFETs. One is an IE-UMOSFET (see Fig.1) which is SiC trench MOSFET structure having p+ region completely covering trench gate bottom [2]. The other device named "device A" is also SiC trench MOSFET structure having p+ region partially covering trench gate bottom. These p+ regions, connected to the

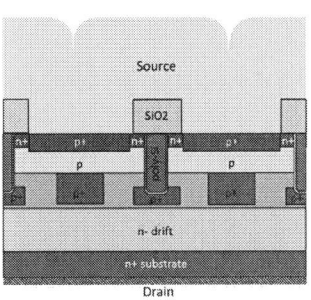

Fig.1 Schematic cross section of an IE-UMOSFET

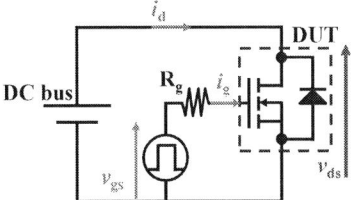

Fig.2 The equivalent circuit of short-circuit test.

source electrodes, prevent from high electric field at the bottom of the trench gate oxide layer at the device forward blocking state. Table 1 shows measured static characteristics of the DUTs. Blocking voltage and specific on resistance ($R_{\text{on sp}}$) were 1576 V and 3.2 mΩcm² in IE-UMOSFET and 1366 V and 3.1 mΩcm² in device A at room temperature, respectively. Also, gate threshold voltages were measured as 3.2 V in IE-UMOSFET and 4.5 V in device A. Figure 2 shows the equivalent circuit of the short-circuit test. The DC bus connected to the DUTs ranges from 400 to 800 V. In a previous research, it was reported that the device failure took place with the gate-source short in the short-circuit test at the lower DC bus of 400V [3]; therefore, the authors focused on the gate current measurement at this time and utilized an AC/DC current probe to measure a trace amount of gate leakage

Table. I Measured results of static characteristics of IE-UMOSFET and Device A.

	IE-UMOSFET	Device A
Die Size	3.0×3.0mm²	3.0×4.0mm²
Breakdown voltage(V, @Id=100μA)	1576	1366
Threshold voltage (V, @R.T)	3.2	4.5
Specific on resistance (mΩcm², @Vgs=20V, R.T.)	3.2	3.1
Thickness of gate oxide (nm)	85	68

978-1-7281-0582-6/19 $31.00 © 2019 IEEE

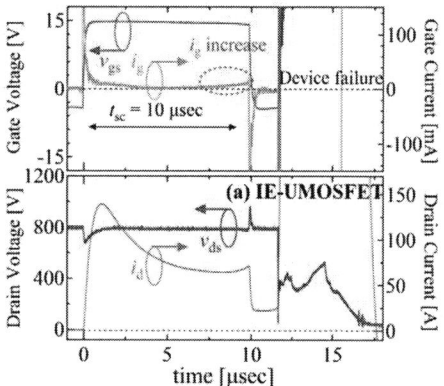
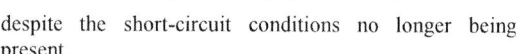

Fig.3 Measured short-circuit waveforms of (a) IE-UMOSFET and (b) Device A. (V_{dd} = 800V, V_{gs} = +15/-4V)

current more precisely. In this paper, all the measurements were carried out in room temperature.

A. Short-circuit failure at high DC bus voltage (800V)

Figure 3 shows the measured short-circuit waveforms of these trench MOSFETs with DC voltage of as high as 800 V, which corresponds to 50.8 % and 58.6 % of the breakdown voltage. It is noted that the gate leakage currents were successfully measured. Both devices failed by thermal runaways because of a relatively large drain tail current just before the gate turn-off [3]. The gate leakage current gradually increased with the short-circuit transient, and it reached about 10 mA just before turning-off the devices. However, in the short-circuit test with the high DC voltage of V_{dd} = 800 V, the device failure occurs due to thermal runaway before the gate-source SiO_2 rapture.

B. Short-circuit failure at low DC bus voltage (400V)

Figure 4 shows the measured short-circuit waveforms of these trench MOSFETs with DC voltage of as small as 400 V, which corresponds to 25.4 % and 29.3 % of the breakdown voltages. It is found that the gate leakage current increased gradually with short-circuit transient and reached a several tens of mA in both devices. In addition, an increased drain current appeared as a tail current after device turn-off. Such a tail current should not be present in unipolar devices. The tail current occurred primarily because the combination of high voltage and increased current generated a severe power dissipation in the device,

despite the short-circuit conditions no longer being present.

Owing to the high lattice temperature and high electric field, the increased gate leakage current which is caused by Fowler-Nordheim tunneling and Poole-Frenkel emission [4, 5] flows resulting in possible damage to the gate SiO_2. Table 2 shows measured resistance between each terminal after short-circuit test. The short-circuit withstand time (t_{sc}) and failure mechanisms are t_{sc} = 32.0 μsec by thermal runaway for IE-UMOSFET and t_{sc} = 15 μsec by SiO_2 rapture for device A. It is clear that the failure modes of two devices are totally different. The authors assumed that this is because of the difference of the gate oxide thickness [6]. Figure 5 shows measured short-circuit waveforms of 1.2 kV silicon trench FS-IGBT for comparison. It is interestingly found that gate leakage current cannot be seen at all during the short-circuit period of 51μsec. This is because of the lower electric field of approximately 2×10^5 V/cm, which is just below the critical field of silicon, and lower lattice temperature in silicon n- drift region.

Recent studies revealed that gate interlayer dielectric breakdown followed by molten aluminum metal diffusion into the cracks was the root reason of failure in SiC planar gate MOSFETs [7, 8]. According to these papers, the gate interlayer dielectric cannot withstand thermal mechanical stress under the high lattice temperature during the short-circuit test. In Ref. [6], it was reported that SiC trench MOSFET with thicker gate oxide layer showed the higher short-circuit capability and can avoid the failure related to the gate-source short before the thermal runaway with the

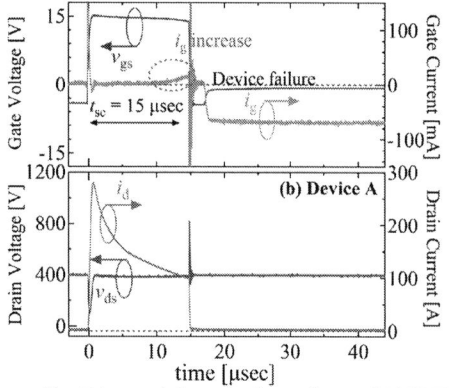
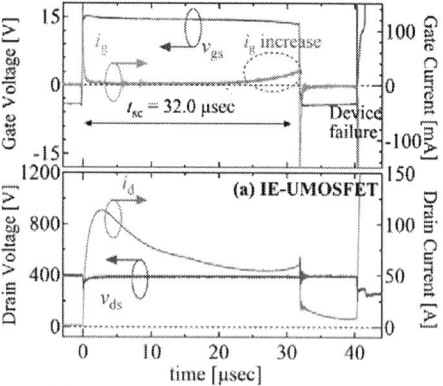

Fig.4 Measured short-circuit waveforms of (a) IE-UMOSFET and (b) Device A. (V_{dd} = 400V, V_{gs} = +15/-4V)

Table.II Measured resistance between each terminal after short-circuit test

Contact	Resistance [Ω]	
	IEU-MOSFET	device A
Gate-source	0.3	9.0
Drain-Source	1.0	∞
Drain-gate	1.0	∞

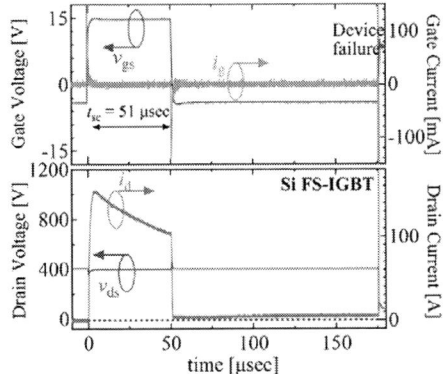

Fig.5 Measured short-circuit waveforms of Si FS-IGBT. (V_{dd} = 400V, V_{gs} = +15/-4V)

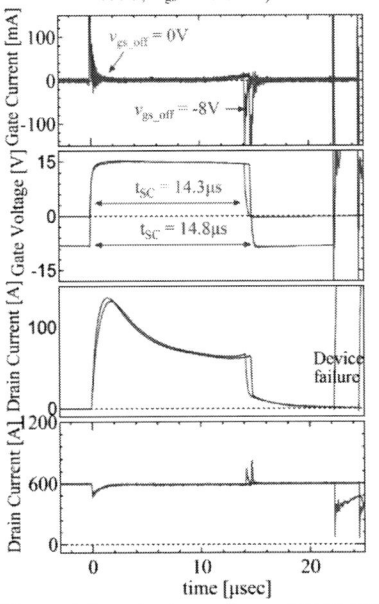

Fig.6 Measured short-circuit waveforms of IE-UMOSFET. (V_{dd} = 600V, V_{gs_off} = 0 and -8V)

DC bus voltage of 400 V. The possibility of these failure mechanisms will be investigated in detail for the SiC trench MOSFETs.

These results represent that the SiO₂ rapture could be the failure mechanism unique to the SiC trench MOSFETs.

C. Depandance of gate switch-off voltage on short-circuit capability

Figure 6 exhibits the measured short-circuit waveforms for the IE-UMOSFET with DC voltage of 600V. It was reported that, in planar gate SiC MOSFETs with gate threshold voltage of 2.0 V, the withstand capability was

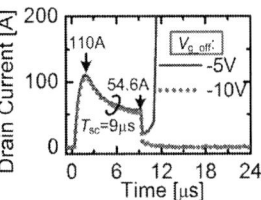

Fig.7 Dependence of short-circuit withstand capability on the gate switch-off voltage (V_{gs_off}) [6]

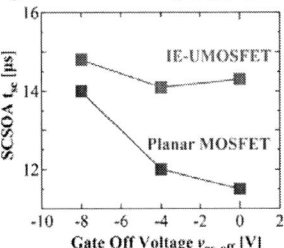

Fig.8 Comparison of dependence of short-circuit capability on V_{gs_off} in IE-UMOSFET and planar MOSFET (Vdd=600V).

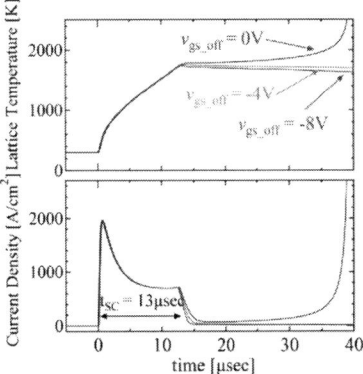

Fig.9 calculated results of the drain current and the maximum lattice temperature of the IE-UMOSFET during short-circuit test. (V_{dd} = 600V, V_{gs_off} = 0, -4, -8V)

dependent on the gate switch-off voltage (V_{gs_off}) and setting the V_{gs_off} of as small as -5 V resulted in deterioration of the short-circuit capability by channel normally-on induced by the higher lattice temperature (see Fig.7) [6]. Figure 8 exhibits the comparison of dependence of measured short-circuit capability on V_{gs_off} in IE-UMOSFET and the planar MOSFET. It is found that the withstand capability of IE-UMOSFET hardly depends on the V_{g_off} of from 0 V to -8V, whereas the planar MOSFET exhibits a strong dependence on V_{gs_off}. This is because the gate threshold voltage of the IE-UMOSFET with MOS channel on {1-100} plane is 3.2 V (see Table I) and is higher than that of the planar MOSFETs by 1.2 V. Figure 9 shows the calculated results of the drain current and the maximum lattice temperature of the IE-UMOSFET as functions of short-circuit transient with the V_{gs_off} of 0, -4, and -8 V. It is found from this figure that the short-circuit capability of IE-UMOSFET can be kept of 13 μsec with no lattice temperature increase after the drain current turned off at the V_{gs_off} of -4 and -8 V. Figure 10 shows the calculated electron current distribution of IE-UMOSFET which correspond to the results shown in Fig.9 during different short-circuit period. It is clear that the electron current keeps flowing through the MOS channel after the device switches off with different V_{gs_off}, and it is found

978-1-7281-0582-6/19 $31.00 © 2019 IEEE

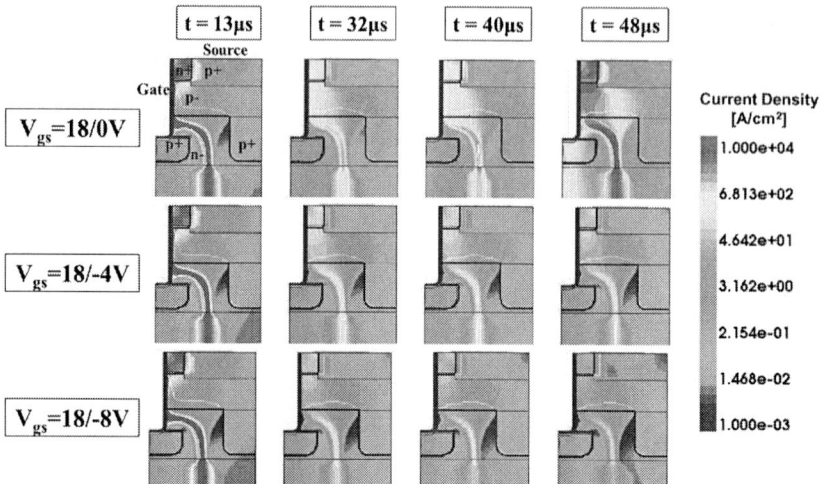

Fig.10 Calculated total current density distribution of the IE-UMOSFET with different short-circuit period. The results correspond to the waveforms shown in Fig.9.

that the current gradually decreases when the V_{gs_off} of -4 and -8 V are applied in the IE-UMOSFET. From these measured and calculated results, it can be concluded that the SiC trench MOSFETs with higher threshold voltage can effectively improve the short-circuit capabilities by reducing the probability of channel normally-on induced by the higher lattice temperature.

M. Namai *et al.* investigated the dependence of short-circuit capability on V_{gs_off} of 1.2kV SiC trench MOSFET experimentally and numerically [9]. Measured breakdown voltage of this device was 1250 V and DC input voltage was set of 800V, which corresponds to as high as 64 %. The paper concluded that larger negative value of V_{g_off} deteriorated the short-circuit capability owing to high drain surge voltage of about 1000V which could cause the avalanche breakdown.

In case of moderate DC input voltage of 600 V which corresponds to 38.1 % of the breakdown voltage, it can be seen a surge voltage of 819 V (see Fig.6) does not affect short-circuit capability even with larger negative value of V_{g_off} (-8 V). This is because the surge voltage is too small to cause the avalanche breakdown. Therefore, it is verified that, in the case of DC voltage of 600 V, the larger V_{g_off} can be effectively improve the short-circuit capability by reducing the possibility of channel normally-on induced by higher lattice temperature.

III. CONCLUSION

In this paper, the investigations of the short-circuit capability and the failure mechanism for the state-of-art SiC trench MOSFET are presented. At this time, a trace amount of gate leakage current was measured in the short-circuit measurement. In the high DC bus voltage of 800 V, the devices were destroyed by thermal runaway before gate-source SiO_2 rapture. However, in the lower DC bus of 400V, state-of-the-art SiC trench MOSFET devices failed via gate-source SiO_2 rapture, which is an unique mechanism to the SiC MOSFETs. Further, the gate switch-off voltage dependences for the SiC MOSFETs are presented. It is verified by experiment and numerical simulation that the SiC trench MOSFETs with higher threshold voltage can effectively improve the short-circuit

capabilities by reducing the probability of channel normally-on induced by the higher lattice temperature.

ACKNOWLEDGMENT

A part of this paper has been implemented under a joint research project of Tsukuba Power Electronics Constellations (TPEC). The authors would like to thank Dr. Y. Kobayashi of Fuji Electric Co., Ltd for discussion.

REFERENCES

[1] J. S. Hsu, C. W. Ayers and C. L. Coomer, *Oak Ridge National Laboratory*, ORNL/TM-2004/137, July, (2004).

[2] S.Harada, K.Kobayashi, K.Kinoshita, N.Ose, T.Kojima, M.Iwaya, H.Shiomi, H.Kitai, S.Kyogoku, K.Ariyoshi, Y.Onishi, and H.Kimura, "1200 V SiC IE-UMOSFET with Low On-Resistance and High Threshold Voltage," *Material Science Forum*, vol.897, pp.497-500, (2017).

[3] M. Namai, J. An, H. Yano, and N. Iwamuro, "Investigation of Short-Circuit Failure Mechanisms of SiC MOSFETs by Varying the DC Bus Voltage, *Japanese Journal of Applied Physics*, vol.57, 074102 1-10, (2018).

[4] M.Sometani, D. Okamoto, S. Harada, H. Ishimori, S. Takasu, T. Hatakeyama, M. Takei, Y. Yonezawa, K. Fukuda, and H. Okumura, "Temperature-dependent analysis of conduction mechanism of leakage current in thermally grown oxide on 4H-SiC," *Journal of Applied Physics*, vol.117, 024505 1-7, (2015).

[5] A.K.Agarwal, S. Seshadri and L. B. Rowland, "Temperature dependence of Fower-Nordheim current in 6H- and 4H-SiC MOS caoacitors," *IEEE Electron Device Lett*, vol.18, no.12, pp.592-594, (1997).

[6] J. An, M. Namai, H. Yano, N. Iwamuro, Y. Kobayashi, and S. Harada, "Methodology for enhanced short-circuit capability of SiC MOSFETs," in *Proceedings of the International Symposium on Power Semiconductor Devices and ICs*, 2018, pp. 391-394.

[7] P. D. Reigosa, F. Iannuzzo, and L. Ceccarelli, "Effect of short-circuit stress on the degradation of the SiO2 dielectric in SiC power MOSFETs," *Microelectronics Reliability*, vol.89-90, pp.577-583, (2018).

[8] X. Jiang, J. Wang, J. Lu, J. Chen, X. Yang, Z. Li, C. Tu, and Z.S. Shen, "Failure mode and mechanism analysis of SiC MOSFET under short-circuit condition," *Microelectronics Reliability*, vol.89-90, pp.593-597, (2018).

[9] M. Namai, J. An, H. Yano and N. Iwamuro, "Experimental and Numerical Demonstration and Optimized Methods for SiC Trench MOSFET Short-Circuit Capability," in *Proceedings of the International Symposium on Power Semiconductor Devices and ICs*, 2017, pp. 363-366

Proceedings of the 31st International Symposium on Power Semiconductor Devices & ICs
May 19-23, 2019, Shanghai, China

Fabrication of a High-Voltage SiC Avalanche Diode with a Superior Voltage Clamp Property

Masayuki Yamamoto*⁺
*Advanced Power Electronics
Research Center
National Institute of Advanced
Industrial Science and Technology
Tsukuba, Japan
⁺Dept. of Electrical and Electronic
Engineering, University of Yamanashi
Kofu, Japan
yamamoto.masayuki@aist.go.jp

Kunio Koseki
Advanced Power Electronics Research
Center
National Institute of Advanced
Industrial Science and Technology
Tsukuba, Japan
kunio.koseki@aist.go.jp

Yasunori Tanaka
Advanced Power Electronics Research
Center
National Institute of Advanced
Industrial Science and Technology
Tsukuba, Japan
yasunori-tanaka@aist.go.jp

Abstract— **The surge voltage management is one of the most important issues in a recent high-voltage and high-frequency power electronics. Aiming to have an outstanding semiconductor surge absorber suited for a high voltage regime, we fabricate the silicon carbide(SiC)-based clamp diodes and evaluate their IV characteristics after the avalanche breakdown. It is shown that the current density can reach 10kA/cm² while the voltage shift is restricted within 50V for the breakdown voltage V_BR=433V.**

Keywords—SiC, avalanche diode, surge absorber, voltage clamp, Joule heating, thermal diffusion, device failure

I. INTRODUCTION

Recently, the switching speed of power devices becomes so fast that the surge voltage during the turn-off operation is hard to ignore. In order to manage this problem, there are considerable efforts on reducing a parasitic inductance in a power module design [1]. An alternative approach is connecting a snubber circuit or a clamp diode parallel to a transistor so as to absorb the surge voltage higher than the clamp voltage.

In an ideal clamp diode, the clamp voltage is fixed no matter how large the current density is. In practice, however, the clamp voltage increases as the current density does. Its slope is theoretically characterized by the specific space charge resistance, $r_{sc} = V_{BR}/qv_sN_d$, where V_{BR} is the breakdown voltage, q the elementary charge, v_s the saturation velocity, and N_d the doping density at the drift layer [2]. If one fabricates a silicon(Si)- and silicon carbide(SiC)-based clamp diode with the same breakdown voltage, the ratio of the specific space charge resistance for the SiC diode to that for the Si one gives

$$\frac{r_{sc}^{SiC}}{r_{sc}^{Si}} = \frac{v_s^{Si} N_d^{Si}(V_{BR})}{v_s^{SiC} N_d^{SiC}(V_{BR})} \sim \frac{1}{430} \quad (1)$$

where we assume $v_s^{SiC} = 2v_s^{Si} = 2 \times 10^7$ cm/s, $N_d^{SiC}(V_{BR}) = \left(\frac{3 \times 10^{15}}{V_{BR}}\right)^{\frac{4}{3}}$, and $N_d^{Si}(V_{BR}) = \left(\frac{5.34 \times 10^{13}}{V_{BR}}\right)^{\frac{4}{3}}$ [3]. This means that the size of a SiC clamp diode can be 430 times smaller than that of a Si one in order to obtain the same performance. From another point of view, SiC clamp diodes can handle the surge voltage in a higher voltage region with

tractable chip sizes while the use of Si clamp diodes is commonly limited up to 300V in the market.

In this work, we fabricate the 433V SiC *pn* diodes and study their IV characteristics after the avalanche breakdown. It is shown that the current density reaches 10kA/cm² while the voltage shift is less than 50V. The specific space charge resistance is given by $r_{sc} = 2.5\text{m}\Omega\text{cm}^2$ for the current density $j_K > 6\text{ kA/cm}^2$. The junction temperature at $j_K = 10\text{kA/cm}^2$ is roughly estimated as 680°C by solving the diffusion equation. The device destruction occurs at $j_K = 10.3\text{kA/cm}^2$. The trace of the local explosion can be found at the edge of the shorted device.

II. METHOD

The schematic cross section of the SiC mesa-type *pn* diode is shown in Fig.1 inset. We use a commercial n^+-type 4H-SiC (c-axis 4° off) substrate and have grown the epitaxial n^--type layer ($6.6 \times 10^{16}\text{cm}^{-3}$, $3.2\mu\text{m}$) by chemical vapor deposition. The *pn* junction is formed by Al⁺ ion implantation with multiple injection energy ($40 - 400\text{keV}$) at 600°C. The junction depth is about $1.0\mu\text{m}$. The substrate is then annealed at 1650°C for 5min in order to activate the dopants. We form the mesa-type termination structure by plasma dry etching. The diameter of the mesa structure is $500\mu\text{m}$ or $600\mu\text{m}$, and the etching depth is $3.8\mu\text{m}$. The surface passivation is performed by the thermal oxidation at 1200°C for 50min. The surface (anode) and backside (cathode) electrodes are formed by the deposition of an Al/Ti and Ni layers, respectively. Finally, the both electrodes are sintered by annealing at 1000 °C for 2min. Figure 1 shows the reverse-biased IV characteristics for one of the diodes we have fabricated, showing the breakdown voltage $V_{BR} = 433\text{V}$. Other dozens of diodes give V_{BR} in the range from 430V to 435V.

In order to study the IV characteristics after the avalanche breakdown, we firstly dice the SiC substrate into 3mm² chips and bonding them on DCB substrates. It is then attached to the avalanche test circuit as shown in Fig. 2. The test circuit consists of a high-voltage power supply (Vdc), capacitor bank (Cb), resistor (Rs), SiC-metal oxide semiconductor field effect transistor (MOSFET) and the clamp diode under test. By transmitting the trigger signal through optical fiber to the gate driving circuit, stored energy in the capacitor bank is applied to the clamp diode. Cathode to anode voltage and cathode current were measured by a high voltage differential probe

978-1-7281-0582-6/19 $31.00 © 2019 IEEE

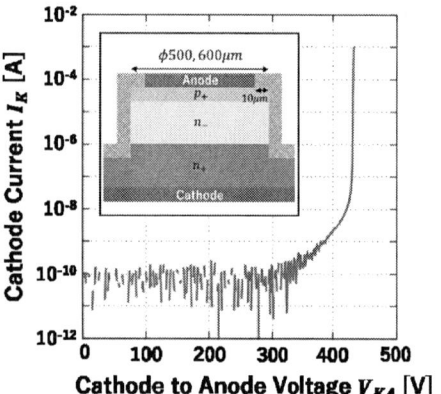

Fig. 1. Reverse-biased IV characteristics. The breakdown voltage is $V_{BR} = 433$V at $I_K = 1$mA. Inset: the cross-sectional structure of the SiC avalanche diode.

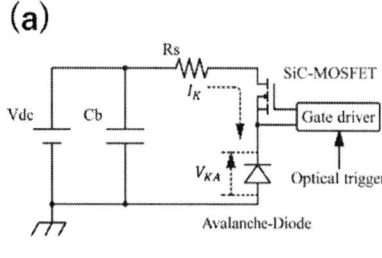

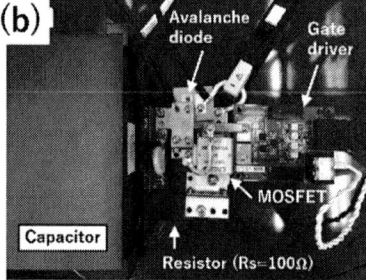

Fig. 2. Circuit for the avalanche test (a) and its actual set up (b).

(Tektronics Inc., THDP0200) and an AC/DC current probe (Tektronics Inc., TCP0030), respectively.

III. RESULTS

Figure 3 shows the cathode to anode voltage V_{KA} (solid lines) and the cathode current I_K (dotted lines) as a function of the time t for different power supply voltages (Vdc). The diameter of diode is 500μm. These waveforms correspond to the current density $j_K = 1$, 5 and 10kA/cm² ($I_K \sim 2$, 10 and 20A) at $t = 400$ns, respectively. The time origin, *i.e.*, $t = 0$ns, is set to be 320ns later than the time at which the optical trigger is applied to the gate driver (see Fig.2 (a)). The V_{KA} increases as I_K does. It also increases as a function of time t during the pulse injection in the case of higher current densities. For example, $V_{KA} = 458$V at $t = 200$ns while $V_{KA} = 491$V at $t = 600$ns for $j_K = 10$kA/cm² (orange solid line in Fig.3). This would be due to the rapid Joule heating, which we will discuss later.

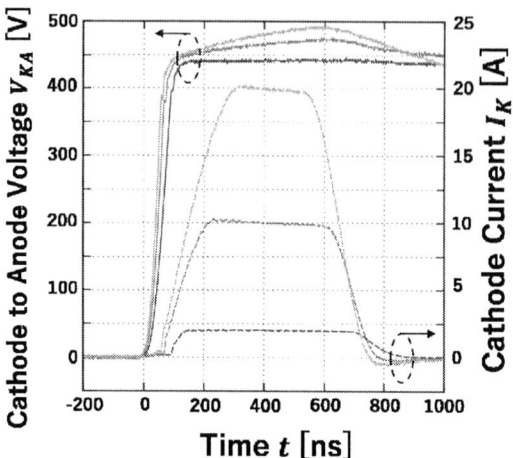

Fig. 3. Waveforms for the cathode to anode voltage V_{KA} (solid lines) and cathode current I_K (dotted lines), corresponding to the current density $j_K = 1$kA/cm² (purple), 5 kA/cm² (green) and 10kA/cm² (orange) at the diode with $\phi500\mu$m. The pulse width is set to be 600ns.

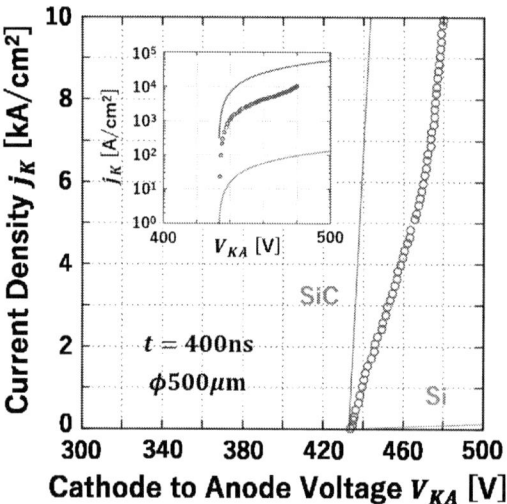

Fig. 4. Current density j_K as a function of the cathode to anode voltage V_{KA} after the avalanche breakdown, evaluated at $t = 400$ns in Fig. 3. Green and light blue lines represent the theoretical limits for SiC and Si diodes, respectively. Inset: the semi-log plot for the same data.

Figure 4 shows the j_K-V_{KA} characteristics (red circles) evaluated at $t = 400$ns for the waveforms as shown in Fig. 3. After the avalanche breakdown, V_{KA} increases from 433 to 480V as j_K does from 0 to 10kA/cm². Interestingly, the slope of j_K-V_{KA} curve becomes sharp after $j_K > 6$kA/cm². One can estimate the specific space charge resistance r_{sc} as 6.0mΩcm² for $j_K < 6$ kA/cm² and 2.5mΩcm² for $j_K > 6$ kA/cm². The green and light blue lines represent the theoretical limits for SiC and Si *pn* diodes, respectively. They are described by [2]

$$j^{SiC(Si)}(V) = q v_s^{SiC(Si)} N_d^{SiC(Si)} \left(1 - \frac{V_{BR}}{V}\right) \qquad (2)$$

978-1-7281-0582-6/19 $31.00 © 2019 IEEE   172

where $v_s^{SiC(Si)}$ and $N_d^{SiC(Si)}$ are given in Eq.(1) and $V_{BR} = 433$V. The specific space charge resistance for SiC is $r_{sc}^{SiC} = 1.0$mΩcm^2 in this theoretical limit. The difference between theoretical and experimental results would attribute to the Joule heating. We note that almost the same j_K-V_{KA} curve has been obtained for the diode whose diameter is 600μm, where $I_K = 25$A corresponds to $j_K = 9$kA/cm^2. This reinforces the fact that the current is not crowding at a certain portion of the diode but is flowing through the whole area.

Now let us consider the junction temperature during the pulse injection. The total dissipation energy at the time t is given by

$$\Delta U(t) = \int_0^t V(\tau) I(\tau) d\tau \qquad (3)$$

For example, $\Delta U(400$ns$) = 2.25$mJ for the waveforms at $j_K = 10$kA/cm^2 (orange lines in Fig. 3). If we assume that all dissipation energy turns into the Joule heat at the drift layer without considering a thermal diffusion, the junction temperature is roughly estimated as

$$T_j^{no\ diffuse} = T_0 + \frac{\Delta U}{\rho C_P^{RT} S d} = 2322°C \qquad (4)$$

where we set the initial temperature $T_0 = 25°$C, the SiC density $\rho = 3.24$ g/cm^3, the SiC specific heat capacity at room temperature $C_P^{RT} = 0.7$J/g/°C, the diode area $S = \pi(0.025)^2$ cm^2, and the drift layer thickness $d = 2.2 \times 10^{-4}$ cm. This estimation seems too high. In fact, $\Delta U(600$ns$) = 4.12$ mJ at $j_K = 10$ kA/cm^2 gives $T_j^{no\ diffuse} = 4241°$C, which is much higher than the SiC melting point 2730°C.

Such an overestimation comes from disregarding a thermal diffusion. In order to include its effect, we numerically calculate the one-dimensional (1D) diffusion equation with a position-dependent thermal diffusivity based on the finite element method

$$\frac{T_i^{l+1} - T_i^l}{\Delta t} = \frac{1}{\Delta x^2} \left(\frac{\left(\kappa_{i+1}^l - \kappa_{i-1}^l\right)\left(T_{i+1}^l - T_{i-1}^l\right)}{4} \right.$$

$$\left. + \kappa_i^l \left(T_{i+1}^l + T_{i-1}^l - 2T_i^l\right) \right) \qquad (5)$$

where T_i^l and κ_i^l denote the temperature and the thermal diffusivity at the dimensionless time l and site i while Δt and Δx represent the time step and the distance between the neighboring sites, respectively [4].

We consider the finite 1D system as shown in Fig. 5(a) with Neumann boundary condition. It consists of three segments, namely, the SiC, Aluminum (Al) and air regions. The total system size is set to be $2N + 1$ sites, where the SiC region is from the site $-N$ to 2, the Al from 3 to 7, and the air from 8 to N, respectively. The site in the drift layer next to the p_+ region is set to be $i = 0$. We set $N = 50$, $\Delta x = 0.55\mu$m and $\Delta t = 0.2$ ns. The temperature-dependent thermal diffusivity in the unit of nm$^2/\mu$s and specific heat capacity in the unit of J/g/C° for n-type 4H-SiC are given by

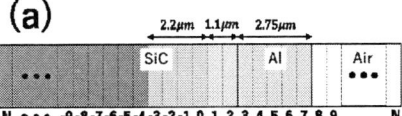

(a)

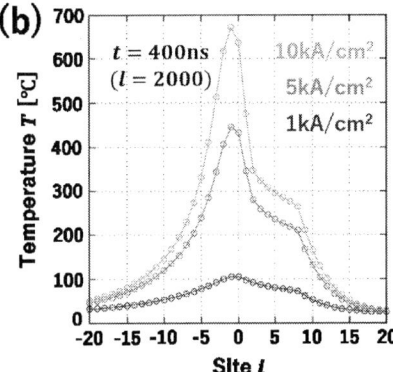

(b)

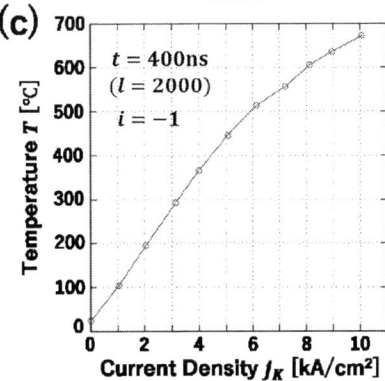

(c)

Fig. 5. (a) One-dimensional model for calculating the thermal diffusion of the Joule heat generated at the SiC drift layer. (b) Temperature T as a function of the site i evaluated at $t = 400$ ns for the power dissipation corresponding to the current density j_K=1kA/cm^2 (purple), 5 kA/cm^2 (green) and 10kA/cm^2 (orange), respectively (see Fig.3). (c) Temperature T as a function of the current density j_K at the site $i = -1$ evaluated at $t = 400$ns.

$$\kappa_{SiC}(T) = 0.01818 + 0.675\exp\left(-\frac{T+273}{150}\right) \quad (6)$$

and

$$C_P(T) = 0.48 + 0.023\exp\left(\frac{T+273}{262}\right) \qquad (7)$$

respectively [5]. For the Al and air regions, the thermal diffusivity is assumed to be temperature-independent, $\kappa_{Al} = 0.097$nm$^2/\mu$s and $\kappa_{air} = 0.022$nm$^2/\mu$s, for simplicity. We also neglect the thermal resistance at material interfaces.

The system incorporates the Joule heat in the drift layer at every time step $\Delta t = 0.2$ns as

$$\left(T_i^l\right)_{new} = \left(T_i^l\right)_{old} + \delta T_i^l \quad \text{for } i = -3, -2, -1, 0 \quad (8)$$

with

$$\delta T_i^l = \frac{\Delta u_i^l}{\rho C_P(T_i^l)} \qquad (9)$$

where Δu_i^l denotes the energy density in the unit of J/cm^3 at the site i and the time $l\Delta t$, and ρ the SiC density (3.24g/cm^3), respectively. We set the energy density as

$$\Delta u_i^l = \frac{7+2i}{16} \cdot \frac{I(l\Delta t)V(l\Delta t)\Delta t}{S\Delta x} \tag{10}$$

where the diode area $S = \pi(0.025)^2$cm^2. This becomes the largest at $i = 0$, where the electric field in the drift layer should be the highest. We set the initial temperature $T_i^{l=0} = 25°C$ for any i.

Figure 5(b) shows the temperatures $T_i^{l=2000}$ evaluated at the time $t = l\Delta t = 400$ ns as a function of the site i for different current densities. For j_K =10kA/cm^2, the temperature reaches 680°C at the site $i = -1$. During this time period, the Joule heat diffuses to the site $i = \pm 20$, corresponding to $\pm 11\mu m$ from the pn junction. The temperature at the site attached to the Al regions ($i = 2$) is about 350°C, implying that the Al layer plays a role for radiating the heat from the SiC drift layer.

Figure 5(c) shows the $T_{i=-1}^{l=2000}$ as a function of j_K. The rising rate of temperature is slow down for $j_K > 6$kA/cm^2. This is because the specific heat capacity $C_P(T)$ increases as the temperature does. This may partly explain the experimental result that the slope of j_K-V_{KA} curve becomes sharp around $j_K = 6$kA/cm^2 (Fig. 4).

Figure 6 shows the voltage and current waveforms corresponding to $j_K = 10.3$ kA/cm^2. It shows the sharp current spike and voltage drop at the time $t = 630$ ns, indicating the device failure. The diode is shorted after this operation. Figure 7(a) shows the top view of the shorted device. One can see the scratches near the edge of the diode. The SEM image (Fig. 7(b)) implies that the local explosion occurred under the edge of the anode contact, located about $20\mu m$ inward from the edge of the mesa structure (see also Fig.1 inset). Figure 7(c) shows the optical microscope image after removing the metallic and SiO$_2$ layers by SPM and BHF washings. One can see not only the hole due to the explosion but also the cracks of SiC, reaching to the mesa edge. We note that the traces of such explosions are also found at other shorted samples. Among them, they are mainly located under the edge of the anode contact. This implies that there may be a certain vulnerability at this location.

IV. SUMMARY

In this work, we fabricate the mesa-structured SiC pn diodes and evaluate their IV characteristics after the avalanche breakdown. It is shown that the current density reaches 10kA/cm^2 while the voltage shift is restricted within 50V for the breakdown voltage 433V. The specific space charge resistances are given by 6.0mΩcm^2 for $j_K < 6$kA/cm^2 and 2.5mΩcm^2 for $j_K > 6$ kA/cm^2, respectively. The junction temperature is roughly estimated as 680 °C for the $j_K = 10$kA/cm^2 pulse by solving the diffusion equation numerically. It can be 2322°C if we neglect the thermal diffusion, indicating that the heat diffusion should be considered even for the short power pulse whose width is less than 1μs. The device destruction occurs at $j_K =10.3$kA/cm^2. The trace of the local explosion can be found at the edge of the shorted device.

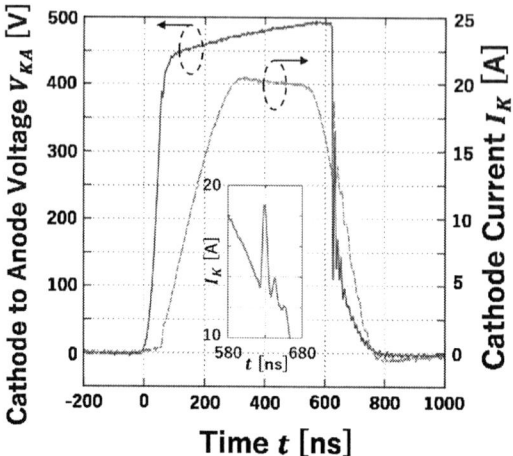

Fig. 6. Waveform for the cathode to anode voltage V_{KA} (solid purple line) and cathode current I_K (dotted green line), corresponding to the current density j_K= 10.3kA/cm^2. The device failure occurs at the time t = 630ns. Inset: the current waveform for 580ns $< t <$ 680ns.

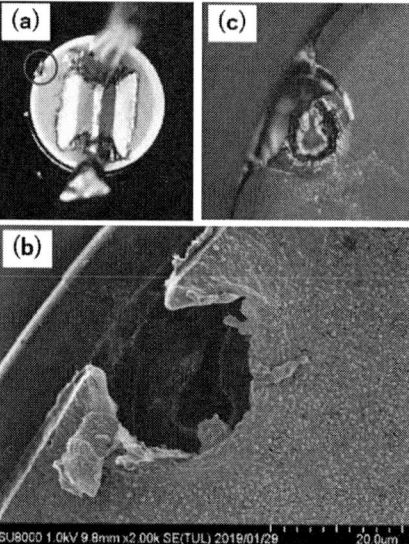

Fig. 7. (a) Top view of the shorted diode. Scratch (enclosed by red circle) can be found at the edge. (b) SEM image focusing on the scratch. (c) Optical microscope image on the same scratch after removing the metallic and SiO$_2$ layers by SPM and BHF washings.

REFERENCES

[1] C. Chen *et al.*, "A review of SiC power module packaging: layout, material system and integration", CPSS Trans. Power Elec. and Appl, vol. 2, pp.170–186, Sep. 2017.

[2] M. Levinshtein *et al.*, Breakdown phenomena in semiconductors and semiconductor devices, World Scientific, 2005, pp. 47–50.

[3] B.J. Baliga, Fundamentals of power semiconductor devices, Springer, 2008, pp.97–98.

[4] J. Lee, "Stability of finite diffenrence schemes on the diffusion equation with discontinuous coefficients", https://math.mit.edu/research/highschool/rsi/documents/2017Lee.pdf, 2017.

[5] R. Wei *et al.*, "Thermal conductivity of 4H-SiC single crystals", J. Appl. Phys., vol. 113, pp.053503(1) − (4), Feb. 2013.

Proceedings of the 31st International Symposium on Power Semiconductor Devices & ICs
May 19-23, 2019, Shanghai, China

Operation of ultra-high voltage (>10kV) SiC IGBTs at elevated temperatures: benefits & constraints

Amit K. Tiwari*, Florin Udrea
Department of Engineering
University of Cambridge
Cambridge, CB2 1PZ, U. K.
*akt40@cam.ac.uk

Neophytos Lophitis
Faculty of Engineering, Environment
and Computing Coventry University,
Coventry, CV1 2JH, U. K.

Marina Antoniou
School of Engineering,
University of Warwick
Coventry, CV4 7AL, U. K.

Abstract—State of the art TCAD simulation models are used to simulate the performance of ultra-high voltage (10-20 kV) SiC IGBTs in the temperature range 300-775 K. We show that unlike Si-based counterparts, ultra-high voltage SiC IGBTs stand to gain from the temperature rise if the limit is not exceeded. We show that whilst an operation at 375 K is highly promising to achieve the most optimum on-state characteristics from SiC IGBTs, no significant degradation in the on-state current and breakdown voltage alongside with negligible rise in leakage current is observed until 550 K. Therefore, ≥10 kV SiC IGBTs are highly promising for Smart Grid and HVDC.

Keywords—SiC IGBT, Ultra-high voltage, Elevated Temperatures, HVDC and Smart Grid

I. INTRODUCTION

Owing to its superlative electrical properties, i.e., a wide bandgap of 3.25 eV, high critical field of ~2×10⁶ V.cm⁻¹, high mobility of 950 cm².V⁻¹.s⁻¹ and saturation velocity of ~2.2×10⁷ cm.s⁻¹, 4H-SiC is of particular interest for high-voltage and high-temperature applications [1]. The 10 times increase in the critical field of 4H-SiC, when compared with that of Si, allows high-voltage devices to be fabricated on a significantly thinner and more-conducting blocking epilayer, which in turn significantly lowers the specific on-resistance [2-6]. In addition, because of the wide bandgap, the intrinsic carrier concentration remains low at high temperatures, being ~10¹⁰ cm⁻³ at 800 K, which is comparable to that of Si at room temperature [7, 8].

However, the advantage of having superior material properties is often overshadowed by the way the device operates. In particular, the strong dependence of both material properties and device operation physics upon temperatures can significantly alter device characteristics, which pose further challenges to designs of gate-drivers and converter topologies. In recent years, SiC IGBTs have attracted considerable attention for ultra-high voltage applications, most notably for low-carbon Smart Grid and HVDC, where elevated temperatures can lead to instability and thermal runaway problems [9-13].

In the case of Si IGBTs, the on-state current rating is chosen to restrict the IGBT temperature below the maximum junction temperature of 175 °C. The maximum junction temperature is less of an issue for SiC, and 300 °C is considered a safe operating temperature for devices, such as Schottky diodes, MOSFETs and JFETs [7, 8], as long as

dedicated high temperature packages are used. However, the case of high temperature operation on ultra-high voltage SiC IGBTs has not been made yet.

A simultaneous high-voltage and high-temperature capability with a higher degree of stability to temperature variations is highly desirable in SiC IGBTs in order to offer a robust alternative to Si-based counterparts in the sector of power transmission and distribution. It is therefore imperative to assess the performance of ultra-high voltage SiC IGBTs at a wide range of temperatures and to identify possible range within which their operation might be compromised.

We have therefore examined the effect of different operating temperatures on the on-state performance of high-voltage (≥10 kV) SiC IGBTs. A very wide range of temperatures (i.e., 300-775 K) is included, and a particular emphasis is placed on key device characteristics, including the on-state current density, breakdown voltage and leakage current.

II. METHODOLOGY

2D-TCAD simulations are performed on ultra-high voltage SiC IGBTs using a simplified test-cell shown in

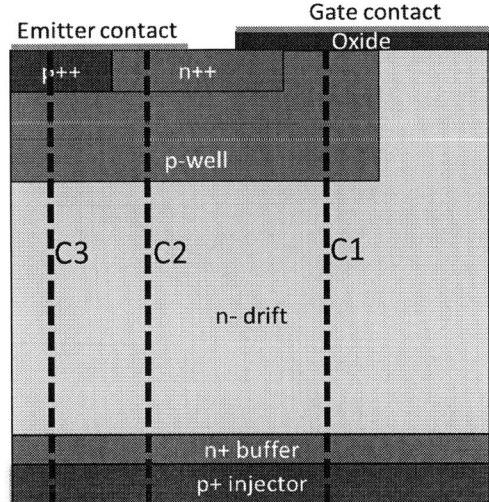

Fig. 1. An n-channel DMOS SiC IGBT test cell schematic.

This work was supported by the EPSRC's Centre for Power Electronics under Underpinning Power Electronics (UPE) 2: Switch Optimization Theme (Ref. EP/R00448X/1).

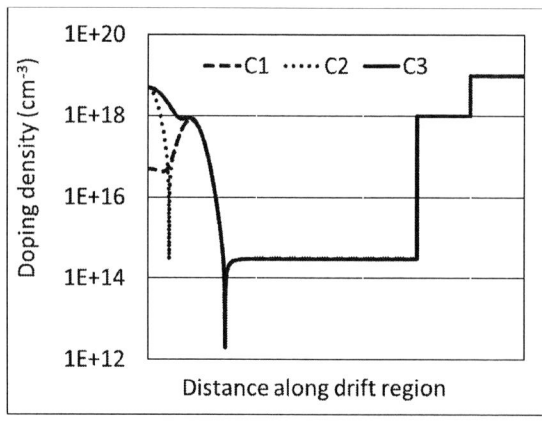

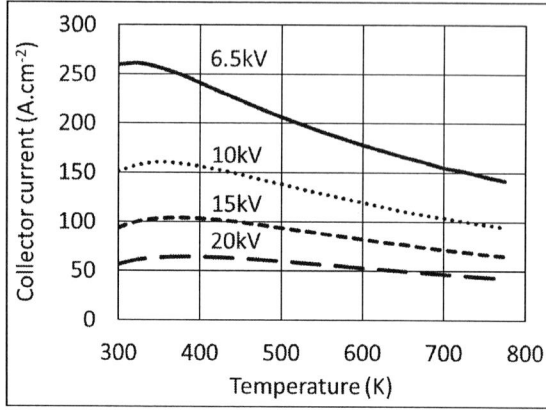

Fig. 2. Doping profiles along different cut-lines in Fig. 1 used to simulate ≥10 kV class SiC IGBTs.

Fig. 4. Collector current density of ≥10 kV SiC IGBTs as a function of device temperature, when V_G=20 V and V_C=5 V

Fig.1. The test cell is an n-channel punch through IGBT design, which consists of doping profiles akin to those in Fig. 2. The drift region of test cell is simulated with the fixed doping concentration of 3×10^{14} cm^{-3} and thicknesses of 50, 100, 150 and 200 μm, to achieve blocking voltages of 6.5, 10, 15, 20 kV, respectively.

To have a high-degree of carrier injection, a p^{++} injector with doping density of 1×10^{19} cm^{-3} is utilized. The n$^+$ buffer doping concentration is set at 1×10^{18} cm^{-3}, adequate to stop the electric field reaching the injector.

A retrograde p-well is utilized to achieve a robust control on threshold voltage and to eliminate the possibility of punch-through. Details are given in our previously published work [14]. The doping density at the surface of the retrograde p-well is 5×10^{16} cm^{-3} and the gate oxide thickness is fixed at 50 nm, yielding a threshold voltage of 7-8 V. The doping profile along the cut line C1 shown in Fig 1 can be seen in Fig 2.

It is worth mentioning here that to achieve adequate conductivity modulation, an electron lifetime of 10 μs is

utilized in simulations. The hole lifetime is considered to be the $1/5^{th}$ of that of electron lifetime. Since low carrier lifetime is often achieved during bulk and epilayer growth of SiC, we have performed additional TCAD simulations with a reduced electron lifetime of 2.5 μs. We noted that the outcome of both set of simulations (i.e., with electron lifetime of 2.5 and 10 μs) is very similar and thus not affecting the conclusion of this work. Furthermore, breakdown and leakage current characteristics are obtained using a background carrier concentration of 1×10^9 cm^{-3}.

Impact ionization as a function of electric field is calculated using the Okuto-Crowell model for 4H-SiC. Doping and temperature dependency of carrier mobility is calculated using the Caughey-Thomas model. The lifetimes of carriers are modeled as a product of a doping-dependent, field-dependent and temperature-dependent factors, as embedded in Scharfetter model utilizing Shockley-Read-Hall (SRH) recombination mechanism.

Simulations consider the dependency of energy bandgap upon temperature and doping. Incomplete ionization of dopants in SiC is also considered. Nitrogen is modelled as a donor trap located 0.0709 eV from the conduction band and Aluminium modelled as an acceptor trap located 0.265 eV from the valance band. Extensive details regarding physics models and parameters used in simulations of 4H-SiC IGBTs can be found in our previous work [15,16].

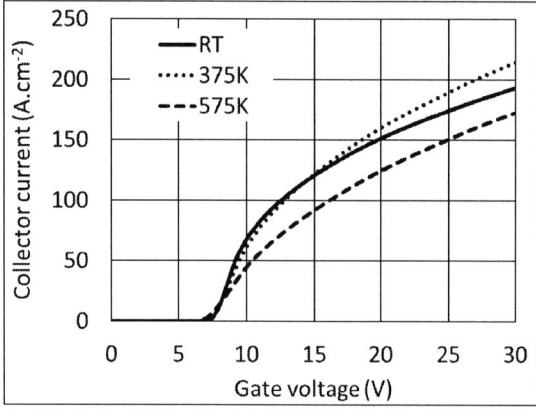

Fig. 3. Transfer characteristics of a 10 kV IGBT at RT, 375 K and 575 K, when collector voltage (V_C) is set at 5 V.

III. RESULTS AND DISCUSSION

Simulated transfer characteristics of a 10 kV SiC IGBT at room temperature (RT), 375 K and 575 K are shown in Fig. 3. We note that similar to Si IGBTs, the gate-threshold voltage (V_T) reduces with increasing temperature. However, for gate voltages higher than 15 V, the collector current density (I_C) is first increased with temperature (i.e., $I_{C,375K}$ > $I_{C,RT}$), followed by a decay with further increase in temperature (i.e., $I_{C,RT}$ > $I_{C,575K}$). For gate voltages in the range V_T–14V, $I_{C,RT}$ is higher than both $I_{C,375K}$ and $I_{C,575K}$. In the case of Si IGBTs, for gate voltages higher than V_T, the collector current consistently reduced with increase in temperature.

978-1-7281-0582-6/19 $31.00 © 2019 IEEE

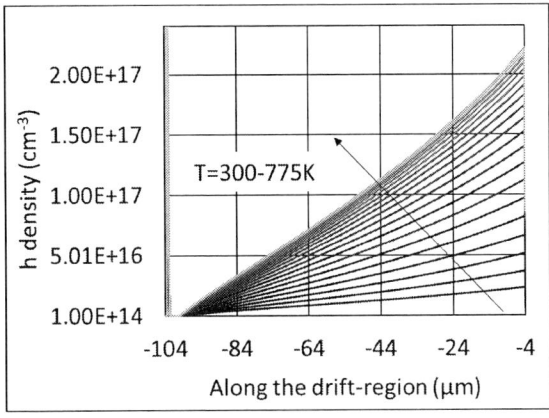 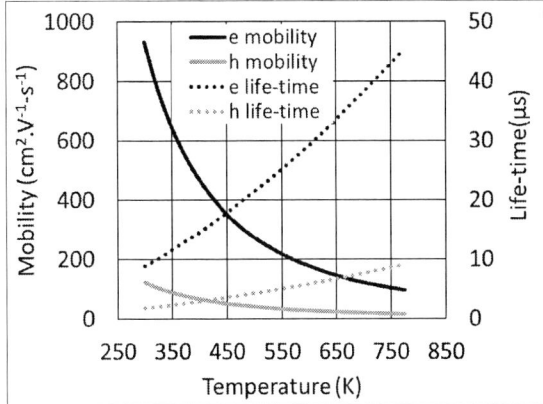

Fig. 5. Hole-injection along the drift region in a 10kV SiC IGBT, as a function of temperature, when V_G=20V and V_C=5V (left). Electron (e) and hole (h) mobilities and lifetimes in a 10kV SiC IGBT as a function of temperature when V_G=20V and V_C=5V (right). Simulation results are obtained using an electron lifetime of 10μs.

The collector current density in the extended temperature range for blocking voltages of 6.5, 10 15 and 20 kV is depicted in Fig. 4. We note that for investigated blocking voltages, consistent with transfer characteristics, a mild increase in temperature is benefitting the on-state current of SiC IGBTs, being the highest at 375 K. For temperatures higher than 375 K, the IGBT current begins to decrease. This constitutes an initial negative temperature coefficient of resistance with temperature up to until 375 K and a positive temperature coefficient for resistance as the temperature becomes higher than that. The latter is an important finding for achieving parallel operation of multiple cells in a chip or multiple chips in a power module. It means that any thermal runaway can be avoided, if the package and the module can allow temperatures higher than 375 K.

The behavior of SiC IGBT output current can be attributed to a complex relationship between temperature-dependent injection efficiency, carrier incomplete ionization, lifetime and mobility. As shown in Fig. 5, the hole injection

significantly increases with temperature. A three order of magnitude difference in the hole density near the p+ injector is evident when the temperature increased from 300 K to 775 K. Carrier lifetimes also increase with temperature, however, carrier mobilities have a significant reduction. Electron and hole mobilities reduced from 950 cm^2.V^{-1}.s^{-1} and 125 cm^2.V^{-1}.s^{-1} to 96 cm^2.V^{-1}.s^{-1} and 16 cm^2.V^{-1}.s^{-1}, respectively, for the temperature range 300-775 K. The change in carrier mobilities is even steeper in the temperature range 300-400 K. Carrier mobilities at 400 K are smaller than half of their respective room temperature values, greatly suppressing the advantage of enhanced injector efficiency and carrier lifetime. In other words, the onus lies with achieving the best trade-off between injector-efficiency, carrier lifetime and mobility to see the advantages of elevated temperatures in SiC IGBTs operation.

Temperature is known to improve the carrier mean-free path, and reduce the avalanche coefficients despite the increased carriers thermal ionization, which in-turn improves

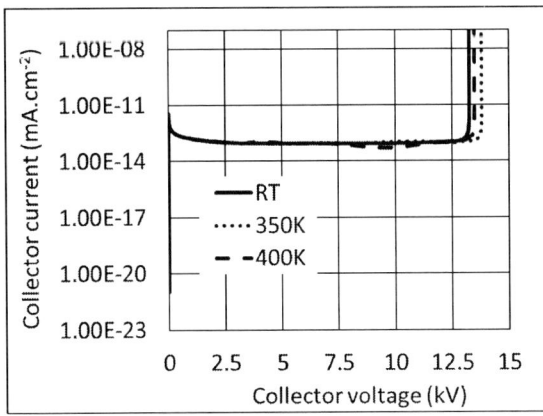 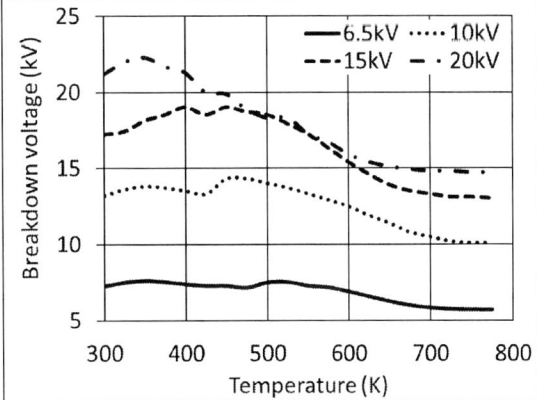

Fig. 6. Breakdown characteristics of a 10 kV SiC IGBT at RT, 325 K and 575 K (left) Breakdown voltage of ≥10 kV SiC IGBTs in extended temperature range (right).

978-1-7281-0582-6/19 $31.00 © 2019 IEEE

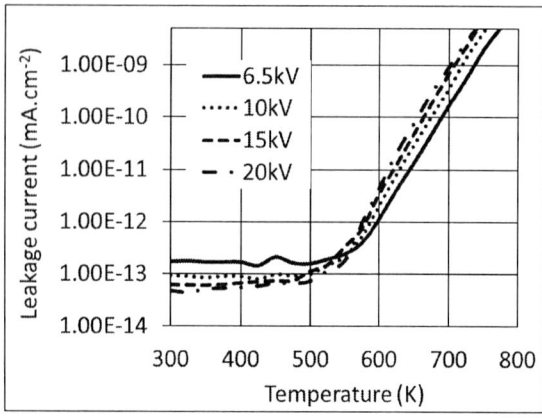

Fig. 7. Leakage current of ≥10 kV SiC IGBTs in the OFF state as a function of temperature.

the blocking capability of a unipolar device. However, significant bipolar effects (i.e., the transistor gain., and the a high-degree of electron-hole pairs concentration near the p-well/drift junction), further augmented by the reduced p-well/drift junction voltage at high-temperatures, can trigger a premature breakdown in IGBTs.

As shown in Fig. 6, the breakdown voltage of a 10 kV SiC IGBT at room temperature is higher than 400 K but lower than 350 K, arising due to the combined effect of prolonged mean-free path and the increased bipolar effect at elevated temperatures.

The variation in breakdown voltage in the extended temperature range is further examined for SiC IGBTs with blocking ratings of 6.5, 10, 15 and 20 kV. Each data set consists of multiple peaks, which can be attributed to the dependence of breakdown voltage on both mean-free path and the bipolar effect as temperature increases. Nevertheless, it is important to note that the breakdown voltage for investigated IGBTs is broadly unaffected by temperature until ~500 K, reducing afterwards.

A qualitative assessment of leakage current characteristics of SiC IGBTs as a function of temperature is also performed. As shown in Fig. 7, the leakage current reduced with the blocking capability of IGBT. A high blocking capability requires a more resistive drift region, i.e., thick or/both low-doped epilayer, which results in reduced ON-state as well as a reduced OFF state current. Nevertheless, it is noteworthy that the leakage current remains low until 550K and begins to rise considerably afterwards.

IV. CONCLUSION

In conclusion, we have extensively studied the performance of ultra-high voltage class (≥10 kV) of SiC IGBTs at elevated temperatures. We show that the ultra-high voltage range of SiC IGBTs is benefiting from elevated-temperatures due to a complex interplay of bipolar effects and SiC material properties. In particular, we show that the SiC IGBT operation in the range 300-475 K is satisfactory

with negligible impact upon the blocking voltage capability and leakage current, whilst 375 K being very promising to achieve the most optimum on-state operation. Therefore, ≥10 kV SiC IGBTs are expected to work reliably and with highly promising performance when used in Smart Grid and HVDC.

REFERENCES

[1] B. J. Baliga, "Silicon carbide power devices" World scientific, 2006.

[2] S. H. Ryu, L. Cheng, S. Dhar, C. Capell, C. Jonas, J. Clayton, M. Donofrio, M. J. O'Loughlin, A. A. Burk, A. K. Agarwal, J. W. Palmour, "Development of 15 kV 4H-SiC IGBTs", Mater. Sci. Forum, vol. 717–720, pp. 1135-1138, May 2012.

[3] E. V. Brunt, L. Cheng, M. J. O'Loughlin, J. Richmond, V. Pala, J. W. Palmour, C. W. Tipton, C. Scozzie, "27 kV, 20 A 4H-SiC n-IGBTs", Mater. Sci. Forum, vol. 821–823, pp. 847-850, June 2015.

[4] A. Kadavelugu, S. Bhattacharya, S. H. Ryu, E. V. Brunt, D. Grider, A. Agarwal, S. Leslie, "Characterization of 15 kV SiC n-IGBT and its application considerations for high power converters", IEEE Energy Conversion Congress and Exposition (ECCE), pp. 2528-2535, October 2013.

[5] S. H. Ryu, C. Capell, C. Jonas, M. O'Loughlin, J. Clayton, K. Lam, E. V. Brunt, Y. Lemma, J. Richmond, D. Grider, S. Allen, J. W. Palmour, "An analysis of forward conduction characteristics of ultrahigh voltage SiC N-IGBTs", Mater. Sci Forum, vol. 858, pp. 945-948, May 2016

[6] K. Fukuda, D. Okamoto, M. Okamoto, T. Deguchi, T. Mizushima, K. Takenaka, H. Fujisawa, S. Harada, Y. Tanaka, Y. Yonezawa, T. Kato, S. Katakami, M. Arai, M. Takei, S. Matsunaga, K. Takao, T. Shinohe, T. Izumi, T. Hayashi, S. Ogata, K. Asano, H. Okumura, and T. Kimoto, "Development of ultrahigh-voltage SiC devices", IEEE Trans. Electron Devices, vol. 62, no. 2, pp. 396-404, February 2015.

[7] P. Neudeck, R. S. Okojie, and L. Y. Chen, "High-temperature electronics-a role for wide bandgap semiconductors?" IEEE proceedings, vol. 90, no. 6, pp. 1065-1076, June 2002.

[8] W. Wondrak, R. Held, E. Niemann, and U. Schmid, "SiC devices for advanced power and high-temperature applications" IEEE Trans. on Industrial Electronics, vol. 48, no. 2, pp. 307-308, April 2001.

[9] N. Iwamuro and T. Laska, "IGBT history, state-of-the-art, and future prospects" IEEE Trans. on Electron Devices, vol. 64, no. 3, pp. 741-752, March 2017.

[10] M. Rahimo, P. Streit, A. Kopta, U. Schlapbach, S. Eicher, and S. Linder, "The status of IGBTs and IGCTs rated over 8kV", In proceedings of PCIM Europe, pp. 1-6, 2003.

[11] J. Wang, A. Q. Huang, W. Sung, Y. LIU, and B. J. Baliga, "Smart grid technologies", IEEE Industrial Electronics Magazine, vol. 3, no. 2, pp. 16-23, June 2009.

[12] J. W. Palmour, J. Q. Zhang, M. K. Das, R. Callanan, A. K. Agarwal, D. E. Grider, "SiC power devices for smart grid systems", In proceedings of IEEE International Power Electronics Conference, pp. 1006-1013, August 2010.

[13] L. Cheng, J. W. Palmour, A. K. Agarwal, S. T. Allen, E. V. Brunt, G. Y. Wang, V. Pala, W. J. Sung, A. Q. Huang, M. J. O'Loughlin, A. A. Burk, D. E. Grider, and C. Scozzie, "Strategic overview of high-voltage SiC power device development aiming at global energy savings" Mater. Sci. Forum, vols. 778-780, pp. 1089-1095, February 2014.

[14] Amit K. Tiwari, M. Antoniou, T. Trajkovic, S. Perkins, N. Lophitis, F. Udrea, "Performance improvement of >10kV SiC IGBTs with retrograde p-well", Mater. Sci Forum (2018) (in press).

[15] S. Perkins, M. Antoniou, Amit K. Tiwari, A. Arvanitopoulos, T. Trajkovic, F. Udrea, N. Lophitis, ">10kV 4H-SiC n-IGBTs for Elevated Temperature Environments", In proceedings of International Seminar on Power Semiconductors (ISPS), August 2018.

[16] N. Lophitis, A. Arvanitopoulos, S. Perkins, M. Antoniou, "TCAD device modelling and simulation of wide bandgap power semiconductors", InTech – Open Access Publisher, pp. 17-44, Feb. 2018.

Proceedings of the 31st International Symposium on Power Semiconductor Devices & ICs
May 19-23, 2019, Shanghai, China

3kV SiC Charge-Balanced Diodes Breaking Unipolar Limit

Reza Ghandi, Alexander Bolotnikov, David Lilienfeld, Stacey Kennerly and Raju Ravisekhar
GE Global Research Center
Niskayuna, NY, USA
Email: ghandi@ge.com

Abstract—**In this work, we report design and fabrication of 3kV SiC Charge-Balanced JBS diodes with differential on-resistance below SiC 1D-unipolar limit. This diode implements a novel drift layer architecture that comprises p-doped buried Charge-Balanced regions inside drift layer. We also report successful high voltage double pulse inductive switching of SiC CB-JBS diode with total estimated recovery loss of 12.5mJ/cm^2 at 150 °C when switched between 1.7kV and 270A/cm^2.**

Keywords—SiC Charge-Balanced diodes, SiC JBS diode

I. INTRODUCTION

Existing silicon carbide (SiC) unipolar solutions for medium voltage and high frequency applications (>3 kV) suffer from high conduction loss at elevated temperatures which make them marginally better than the incumbent silicon IGBTs and diodes [1]–[3] . SiC super-junction (SJ) technology can overcome the limitation of unipolar devices but requires challenging fabrication processes. There are two potential approaches for fabrication of SiC SJs comprising n-doped and p-doped pillars with <10um pitch. The first one is based on formation of deep junctions into SiC by implementing high energy implantation and multi-epitaxial growth [4]. However, this approach requires complex masking solutions for thick drift layers and can also create non-recoverable lattice damage even after high temperature anneal that consequently increases leakage current during reverse operation of the device [5]. The other approach includes formation of trenches inside SiC and filling them using epitaxial growth [6]. Minimizing crystallographic defects at the bottom of the trench during re-fill process and at the same time, achieving uniform dopant distribution of charges inside the re-grown layer requires complex growth condition and therefore make this process challenging.

In this work, we report an alternative solution that implements a novel drift layer architecture for 3kV SiC Charge-Balanced (CB) JBS diodes that outperform the 1-D $R_{on,sp}$ versus breakdown voltage limit through buried p-doped regions inside drift layer. We have characterized the diode from room temperature up to 175 °C and show successful double pulse switching from 1.7kV to 250 A/cm^2.

II. DEVICE FABRICATION

We fabricated Charge-Balanced JBS diodes over 100mm 4°-off 4H-SiC n-type substrate. Drift layer of these diodes consists of three n-type 10um-thick epitaxial layers with doping concentration of 1x10^{16} cm^{-3} that were grown in three consecutive runs followed by top P+ JBS and P-JTE

implantations. We implemented shallow implantation in between growth runs to form two buried P-type Charge-Balance regions inside epitaxial layers (Fig.1). In reverse mode, p-doped CB-regions get depleted and if designed correctly, they can balance n-doped charges adjacent and below the CB-regions inside drift layer and act as an electric field divider.

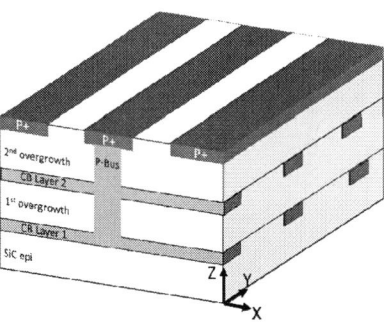

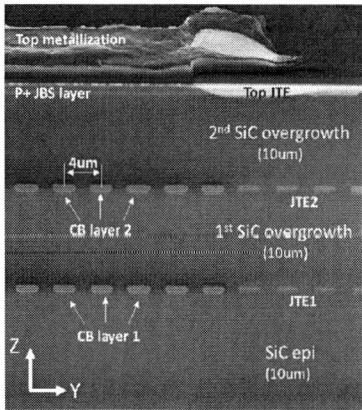

Fig. 1. Perspective view (top) and cross-sectional SEM view (bottom) of 3kV SiC CB JBS diodes.

Consequently, this design allows higher doping concentration of the drift layers for the same breakdown voltage. It has been previously reported that CB diodes with floating CB regions exhibit significant turn-on loss due to slow carrier generation-recombination rate during transition from blocking to conduction [7]. To avoid floating CB regions, intermittent deep vertical P-type pillars (P-Bus) are implemented using high energy implantation [8] to supply holes from P+ JBS anode to the buried CB-regions

978-1-7281-0582-6/19 $31.00 © 2019 IEEE

throughout the active area of the diode. P-Bus pillars need to be designed to minimize electric field alteration and at the same time, provide a conductive path for charges during turn on. We used Al implantation to form CD-regions, P-Bus pillars, P+ JBS regions and JTEs. That was followed by activation anneal, field oxide deposition and frontside and backside metallization. We selected Ni as the Schottky top contact with Al over-layer metallization. The fabrication of diodes was concluded with Nitride and polyimide passivation layers for high voltage surface protection.

III. RESULTS AND DISCUSSION

Forward and blocking characteristics of SiC CB-JBS diodes with turn-on voltage of 1V and breakdown of 3.3kV is plotted in Fig.2. The differential on-resistance of diodes is 4.3 mOhm-cm^2 at room temperature and 9.9 mOhm-cm^2 at 150 °C which is ~40% below 1D-unipolar limit.

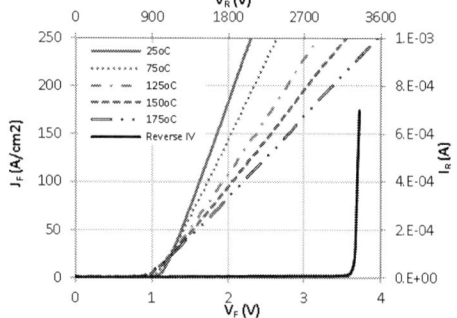

Fig. 2. Forward and reverse characteristics of 3kV SiC CB JBS diodes.

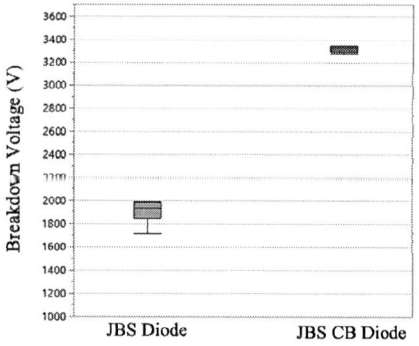

Fig. 3. Comparison of breakdown voltage between SiC CB JBS diode, control JBS diode and co-fabricated on the same wafer.

Fig.3 depicts the effect of charge balanced regions on breakdown voltage of SiC CB diodes in comparison with control JBS diode co-fabricated on the same wafer. Both JBS diodes have a 30um epitaxial drift layer doped to 1×10^{16}cm^{-3} and similar P+ JBS and JTE implantation on the anode side of the diode. The results show that implementing buried CB regions inside the drift layer of the JBS diode improves breakdown voltage from 1.9kV to 3.3kV and suggest p-doped CB-regions are effective in balancing charges and act

as electric field divider during revere operation of the device. Since buried CB layer design is independent of the top portion of the device, we can assume that this form of drift layer modification can be adopted in any type of diodes and transistors allowing such devices to operate at higher reverse biases without compromising the on-resistance.

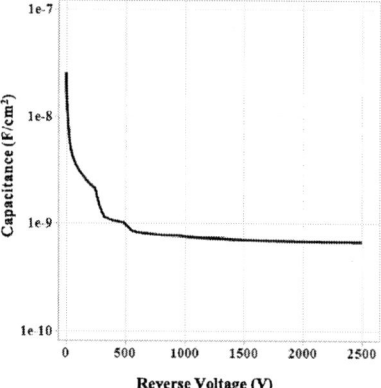

Fig. 4. Capacitance-versus-Voltage (C-V) characteristics of SiC CB-JBS diodes.

The Capacitance-Voltage (C-V) characteristics of CB-JBS diodes have been measured up to 3kV. Fig.4 shows the representative C-V characteristics of the fabricated diodes when normalized to the device active anode area. The presence of two steep drop or "hump" in ccapacitance characteristic of charge-balanced devices can be seen around 250V and 500V that corresponds to depletion of areas around first and second buried p-doped charge balanced layers respectively.

To evaluate dynamic response of CB diodes, double pulse inductive switching test was carried out on fabricated devices. In this test, the active switch is a GE 2.5kV/35A SiC MOSFET and the freewheeling diode is represented by the 3kV CB-JBS diodes (DUT). Fig.5 is Circuit schematic of 1700V double-pulse inductive switching and measured switching waveforms of the SiC CB diode during forward and reverse recovery.

In this circuit, applying positive bias to the gate terminal of the MOSFET (V_G) turns on the active switch, forces the diode into reverse recovery until the diode current returns to 0A and the load current flows through the active switch. The reverse recovery phase of the SiC CB-JBS diodes is virtually indistinguishable compared to conventional SiC Schottky diodes. Beyond the reverse recovery phase, the reverse voltage on the CB-JBS diode (V_R) is 1700V (~50% of rated voltage). After some duration, the active switch is turned off and the load current begins to commute to the SiC CB-JBS diode. With the active switch now in the off state, the forward current (I_F) quickly increases from 0A to full load current of 2A (J_F =250A/cm2). When switched from blocking state to conducting state with a fast ramp, the SiC CB-JBS diodes have an observable forward recovery phase. Fig. 5 shows that the forward voltage across the diode has a large peak then decays back to nominal V_F within approximately 500ns after turn-on. It is believed that this delay is due to the resistance from the p+ JBS anode to the buried CB-regions through the deep P-Bus regions. The estimated switching losses resulted from forward recovery

978-1-7281-0582-6/19 $31.00 © 2019 IEEE

and reverse recovery are 19mJ/cm^2 and 2.5mJ/cm^2 respectively.

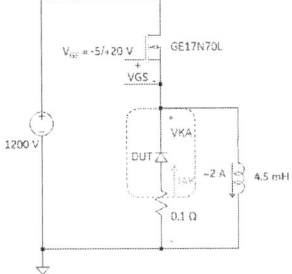

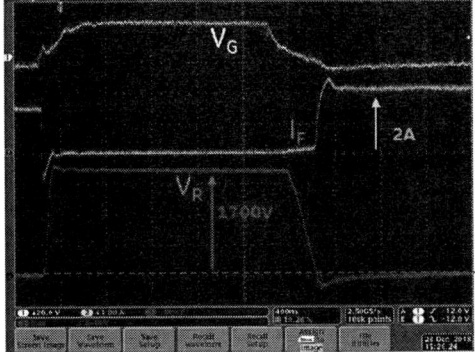

Fig. 5. (top) Circuit schematic of double-pulse inductive switching of SiC CB-JBS diode and (bottom) room temperature switching waveforms of SiC CB-JBS diode between 1700V and 2A (J$_F$=250A/cm2); V$_G$: MOSFET gate voltage, I$_F$: Diode forward current, V$_R$: Diode reverse voltage.

In Fig.6, we show effect of switching voltage on recovery loss for SiC CB-JBS diode. One can observe that forward recovery is linearly dependent on off-state voltage and also dominates total recovery loss across voltage ranges. The data suggests that by increasing switching voltage from 500V to 1700V (3.4X), forward recovery loss and consequently total recovery loss increase by ~7X. Furthermore, we show in Fig.7 that forward recovery loss is also dependent on current density and varying current density from 125A/cm^2 to 250A/cm^2 results in 2X increase in forward recovery loss.

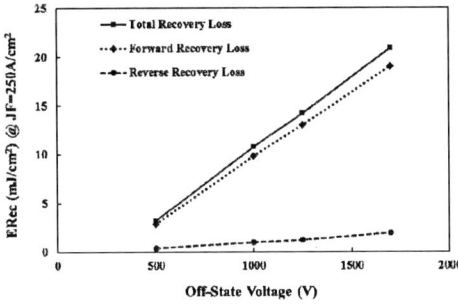

Fig. 6. Switching loss of SiC CB-JBS diode vs. voltage at 250 A/cm^2

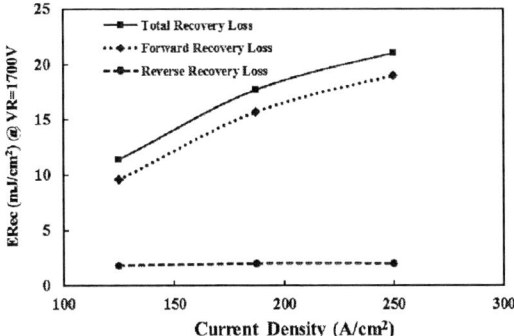

Fig. 7. Switching loss of SiC CB-JBS diode vs. current density at 1700V

Furthermore, we evaluated effect of temperature on forward and reverse recovery losses of CB-JBS diodes and plotted recovery loss versus temperature in Fig.8. We observed that while reverse recovery is independent of temperature, forward recovery loss at room temperature drops by ~40% at 150 °C and reduce total loss from 20mj/cm^2 to 12.5mJ/cm^2.

This behavior agrees with our assumption of P-Bus role in high forward recovery loss. We believe that at high temperatures, P-Bus pillar resistance decreases due to higher ionization rate of P-doped charges. Therefore, at high temperatures and during turn-on, diodes see shorter delay for charge transition from the p+ JBS anode to the buried CB-regions through the deep P-Bus that consequently results in lower forwards recovery loss.

Based on this measurement, the total switching loss observed for nominal conditions of J$_F$=250A/cm^2, 50% duty cycle, T$_j$ = 150 °C, rated dc-link voltage is approximately 12.5mJ/cm^2, which would allow operation of these diodes at frequencies up to f=20kHz (assuming 250W/cm^2 thermal limit).

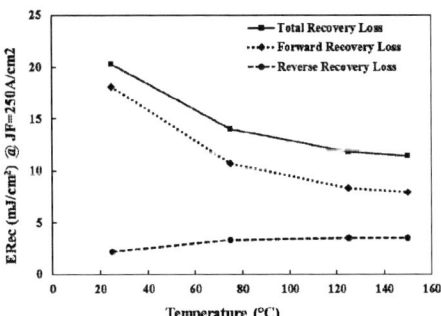

Fig. 8. Switching loss of SiC CB-JBS diode vs. temperature at 1700V and 250 A/cm^2

From switching loss analysis, we can observe that for SiC CB-JBS diodes, forward recovery loss dominates total switching loss across voltage and current density ranges. Since this loss is directly related to P-Bus connecting pillars in CB-JBS diodes, one can assume that increasing P-Bus density (or reducing P-Bus pitch) can provide enough charges to buried CB-regions in shorter time during turn on and therefore enhances switching speed and reduces forward recovery loss. However, this approach consumes active area of the device and therefore increases on-resistance. To independently evaluate effect of P-Bus pitch, we characterized CB diodes with different P-Bus pitches but same P-Bus width. Fig.9 is trade-off between median breakdown voltage, on-resistance and forward recovery loss (when switched between J_F=250A/cm^2 and V_R=1700V) for three different P-Bus pitches. The data suggests that while all designs maintain BV>3kV, one can achieve ~40% improvement in forward recovery loss by implementing smaller P-Bus pitch. However, this approach increases differential on-resistance by 12% and decrease breakdown voltage by 15%.

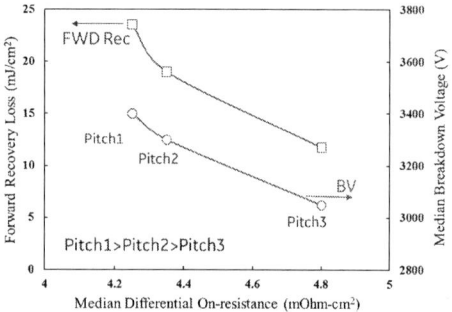

Fig. 9. Tradeoff of measured on-resistance versus switching loss (forward recovery) of 3kV SiC CB-JBS diodes. JF=250A/cm2, VR=1700V

Trade-off curves (BV vs. Ron,sp) of fabricated 3kV diodes are shown in Fig.10 and compared to other SiC reported devices across industry and also SiC super-junction counterparts. We report that SiC CB-JBS didoes surpass unipolar limit and demonstrate breakdown voltage of 3300V with up to ~40% reduction in differential specific on-resistance.

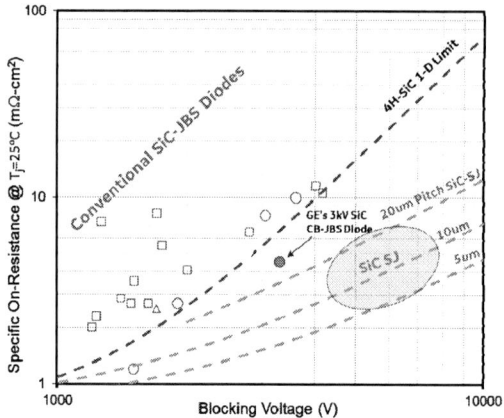

Fig. 10. Differential on-resistance versus breakdown voltage for reported SiC FETs and diodes

IV. CONCLUSION

In this work, we report 3kV SiC Charge-Balanced (CB) JBS diodes with novel drift layer architecture that outperform the 1-D $R_{on,sp}$ versus breakdown voltage limit through a scalable process technology. These diodes implement a novel drift layer architecture that comprises p-doped buried Charge-Balanced layers inside drift layer that are connected to P+ JBS anode though lowly doped P-Bus layer. We have characterized the diode from room temperature up to 175 °C and show successful double pulse switching from 1.7kV to 250 A/cm^2.

ACKNOWLEDGMENT

The information, data, or work presented herein was funded in part by the Advanced Research Projects Agency-Energy (ARPA-E), U.S. Department of Energy, under Award Number DEAR0000674 advised by Program Director Isik Kizilyalli. The views and opinions of authors expressed herein do not necessarily state or reflect those of the United States Government or any agency thereof.

REFERENCES

[1] A. Bolotnikov et al., "Overview of 1.2kV-2.2kV SiC MOSFETs targeted for industrial power conversion applications," *2015 IEEE Appl. Power Electron. Conf. Expo.*, pp. 2445–2452, 2015.

[2] P. A. Losee et al., "High Performance 1.2kV-2.5kV 4H-SiC MOSFETs with Excellent Process Capability and Robustness," *Mater. Sci. Forum*, vol. 858, pp. 876–879, May 2016.

[3] L. Knoll et al., "Dynamic switching and short circuit capability of 6.5kV silicon carbide MOSFETs," in *2018 IEEE 30th International Symposium on Power Semiconductor Devices and ICs (ISPSD)*, 2018, pp. 451–454.

[4] T. Tanaka et al., "First Demonstration of Dynamic Characteristics for SiC Superjunction MOSFET Realized using Multi-epitaxial Growth Method," *2018 IEEE Int. Electron Devices Meet.*, p. 8.2.1-8.2.4, 2019.

[5] T. Kimoto et al., "High-energy (MeV) Al and B ion implantations into 4H-SiC and fabrication of pin diodes," *J. Appl. Phys.*, vol. 91, no. 7, pp. 4242–4248, Apr. 2002.

[6] K. Mochizuki et al., "Strong impact of slight trench direction misalignment from [11-20] on deep trench filling epitaxy for SiC super-junction devices," *Jpn. J. Appl. Phys.*, vol. 56, no. 4S, p. 04CR05, 2017.

[7] A. Bolotnikov, P. Losee, R. Ghandi, S. Kennerly, R. Datta, and X. She, "SiC Charge-Balanced Devices Offering Breakthrough Performance Surpassing the 1-D Ron versus BV Limit," in *European Conference on SiC and Related Material (ECSCRM)*, 2018.

[8] P. Thieberger et al., "Novel high-energy ion implantation facility using a 15 MV Tandem Van de Graaff accelerator," *Nucl. Instruments Methods Phys. Res. Sect. B Beam Interact. with Mater. Atoms*, vol. 442, pp. 36–40, Mar. 2019.

Proceedings of the 31st International Symposium on Power Semiconductor Devices & ICs
May 19-23, 2019, Shanghai, China

Failure Mechanism Analysis of SiC MOSFETs in Unclamped Inductive Switching Conditions

Na Ren, Kang L. Wang
Electrical Engineering Department
University of California Los Angeles
Los Angeles, USA
renna@ucla.edu

Jiupeng Wu, Hongyi Xu, Kuang Sheng
Electrical Engineering Department
Zhejiang University
Hangzhou, China
shengk@zju.edu.cn

Abstract—**In this work, avalanche ruggedness and failure mechanism of SiC MOSFET in single-pulse Unclamped Inductive Switching (UIS) tests are investigated and compared with that of Si IGBT. The experimental results show that the avalanche energy of SiC MOSFET is only 30% that of Si IGBT due to the much smaller chip size (1/7 that of Si IGBT). To improve the avalanche capability of SiC MOSFET, the failure mechanism is analyzed. First, junction temperature in the UIS test is calculated through modeling. The maximum junction temperatures reach 650 K and 490 K in SiC MOSFET and Si IGBT, respectively. Then, BJT latch-up probability is analyzed with an analytical model. It is demonstrated that BJT latch-up can be triggered at the failure temperature (650 K) in SiC MOSFET, whereas it can be eliminated in Si IGBT due to the much deeper P+ body structure. Based on the analyses, the device structure optimization is proposed for SiC MOSFET to prevent the BJT latch-up and enable avalanche capability improvement.**

Keywords—*SiC MOSFET, Si IGBT, avalanche capability, BJT latch-up*

I. INTRODUCTION

Silicon carbide power MOSFETs offer impeccable device features for power electronics applications, such as fast switching speed, low power loss, high power density, and high thermal conductivity [1]. Although the commercial products are technologically mature, the assessment of stability and reliability are essential for the development of this device [2]. Among the reliability issues, the avalanche robustness of SiC MOSFET is critical, and it is sensitive to the chip size design as the maximum energy density of the chip is fixed. While reducing chip size becomes the trend in SiC device for cost reduction purpose, the avalanche robustness will be down rated. The final destruction of the device is always due to high temperature. However, it must be distinguished by which effect the temperature increase was generated. Although BJT latch-up failure mechanism has been reported in some Si devices, in SiC MOSFETs, the device failure mechanism is still unclear. Under UIS test, both thermal generation and parasitic BJT latch-up effects have positive feedback with temperature. And localized high current/energy can also cause high temperature in the device. Whereas the thermal generation effect is only pronounced at intrinsic temperature which is quite high (~600 K for Si and ~1500 K for SiC), BJT latch-up can be triggered at lower temperature which depends on the device design and process parameters. Recently, some paper analyzed that BJT latch-up is less likely to be happened in SiC MOSFET, since the built-in potential barrier of SiC PN junction is higher due to the wide bandgap and the less temperature sensitive resistivity of SiC [1].

To improve the avalanche capability and facilitate future device technology advancements, an in-depth understanding into physics of failure mechanism of SiC MOSFET in UIS test condition is highly required.

II. UNCLAMPED INDUCTIVE SWITCHING TEST

Fig. 1 shows the UIS test bench and circuit diagram. In this circuit, the device under test (DUT) itself is used as the turn-on switch to charge the inductor. With the UIS test circuits, commercial products of 900 V/11.5 A SiC MOSFET and 600 V/16 A Si IGBT are tested. Their chip area are 3 mm^2 (active area is 1.3 mm^2) and 13 mm^2 (active area is 10 mm^2), respectively. The load inductor (L) is ranged from 75 μH to 10 mH to get various avalanche current. DUT devices were tested to their failure at each inductance.

The experimental results of avalanche energy and avalanche current of SiC MOSFET and Si IGBT are compared in Fig. 2. The avalanche energy per area at the same avalanche current is 2.2× higher in SiC MOSFET when compared to that of Si IGBT. As SiC MOSFET has a much smaller chip size (active area and chip area are only 1/7 and 1/4 that of Si IGBT, respectively), the avalanche energy of SiC MOSFET is only 30% that of Si IGBT.

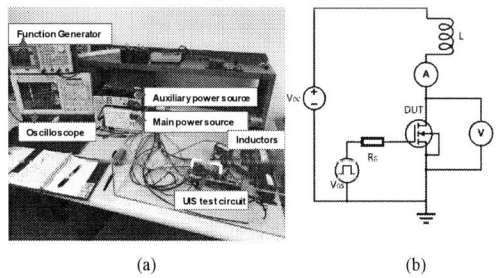

(a)　　　　　　　　　　(b)

Fig. 1 (a) The UIS test bench. (b) A schematic view of the UIS test circuit.

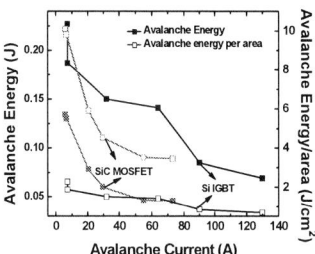

Fig. 2 Comparison of avalanche energy capability between SiC MOSFET and Si IGBT.

978-1-7281-0582-6/19 $31.00 © 2019 IEEE

III. FAILURE MECHANISM ANALYSIS

In this section, the failure mechanisms of SiC MOSFET and Si IGBT in UIS tests will be analyzed and discussed.

A. Junction Temperature Calculation

As electrical mechanisms are all related to temperature, the junction temperature estimation is essential for the avalanche failure mechanism analysis.

As introduced in reference [3], when the reverse bias voltage to a P+-N junction exceeds the breakdown voltage, the junction current starts increasing according to some differential resistance (see Fig. 3(a)). This effect is also observed in SiC devices [4, 5]. Two solid lines in Fig. 3(a) represent I-V characteristics at temperature T_0 and T. ρ_{sc} represents space-charge resistance. The dotted line shows I-V characteristics when junction temperature increases with input power.

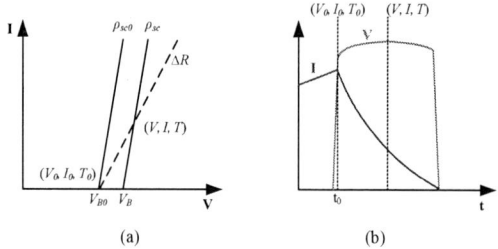

Fig. 3 (a) A schematic diagram of breakdown IV characteristics of P+-N junction. (b) A schematic diagram of UIS test current / voltage waveforms.

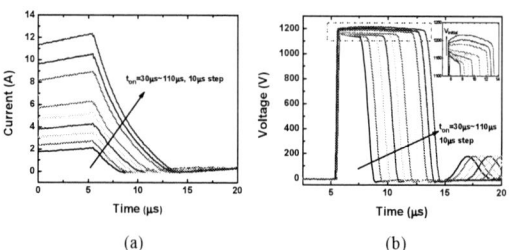

Fig. 4 UIS test current and voltage waveforms of SiC MOSFETs at 1mH inductance.

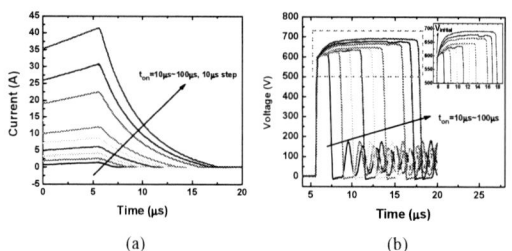

Fig. 5 UIS test current and voltage waveforms of Si IGBT at 1mH inductance.

The applied voltage V at current I can be expressed by

$$V = V_B + I \cdot \rho_{sc} = V_{B0}\{1 + \beta(T - T_0)\} + I \cdot \rho_{sc} \quad (1)$$

where V_{B0} and V_B are the breakdown voltage at T_0 and T, respectively. β is the temperature coefficient of the breakdown voltage. As shown in Fig. 3(b), the initial avalanche voltage (V_0) in UIS tests at different ambient

temperature and different inductor current (I_0) are extracted and used for the V-T calibration.

A series of UIS tests were conducted with an increasing avalanche current for SiC MOSFET and Si IGBT, the current and voltage waveforms of which are shown in Fig. 4 and 5, respectively. The same UIS test procedure was repeated at elevated temperatures (50 °C to 200 °C). For each temperature test, the initial voltage can be extracted from the voltage waveform. The dependence of avalanche voltage on avalanche current and temperature is shown in Fig. 6. The coefficients (V_{B0}, ρ_{sc} and β) can be obtained by the linear fits.

Based on the known coefficients and equation (1), the transient junction temperature during avalanche event can be calculated from the voltage waveforms. The calculation results of SiC MOSFET and Si IGBT are shown in Fig. 7(a) and (b), respectively. It can be concluded that the peak junction temperature before failure reaches 650 K in SiC MOSFET and 490 K in Si IGBT. As this temperature is below the melting point of aluminum metallization and far smaller than the intrinsic temperature of SiC and Si, other mechanism must have been triggered by this temperature and results in the final device destruction.

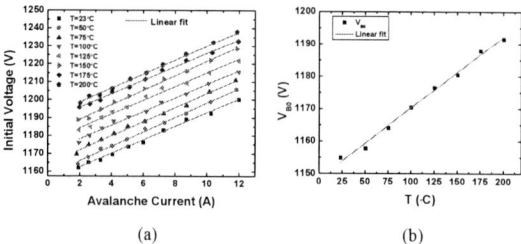

Fig. 6 (a) The extracted initial voltage V_0 from Fig. 4 versus avalanche current I_0 at different ambient temperatures T_0, (b) V_B extracted from (a) has linear relation with temperature. The linear fits from (a) and (b) can give the coefficients ρ_{sc}, V_{B0} and β for the V-T model in Fig. 3.

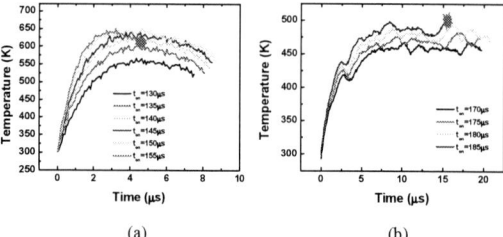

Fig. 7 Temperature calculation results of (a) SiC MOSFET and (b) Si IGBT.

B. BJT Latch-up Analysis

As reported by the literatures [6-11], BJT latch-up is the main failure mechanism in UIS tests for some Si devices. The probability of BJT latch-up failure in SiC MOSFET is analyzed in this work. The cell structure of the SiC MOSFET is shown in Fig. 8(a). It has a shallow P+ body structure. The parasitic BJT portion is depicted in Fig. 8(b). In avalanche mode, the PN junction between P+ body and N-epi are acting as diode blocking the breakdown voltage. From TCAD simulation results of electric field distribution (Fig. 9(a)), due to the radial field component, the electric field inside the device is most intense at the point where the junction bends. This strong electric field causes maximum

978-1-7281-0582-6/19 $31.00 © 2019 IEEE

current flowing near the parasitic BJT [6]. The power dissipation increases temperature, thus increasing base resistance R_B. The avalanche current multiplied by the resistance is voltage drop across the resistor. The voltage drop will increase with temperature. Moreover, the built-in voltage is decreased with temperature. When the voltage drop is sufficient to forward bias the emitter-base PN junction, the BJT can be turned on, which is verified by the TCAD simulation as shown in Fig. 9(b).

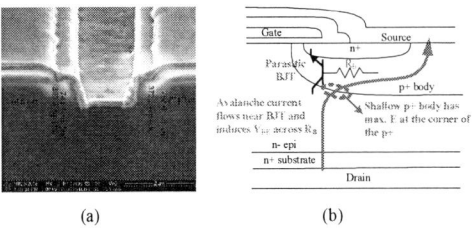

(a)　　　　　　　　(b)

Fig. 8 (a) The cell structure of SiC MOSFET. (b) The schematic view of parasitic BJT portion in MOSFET for the BJT turn-on analysis.

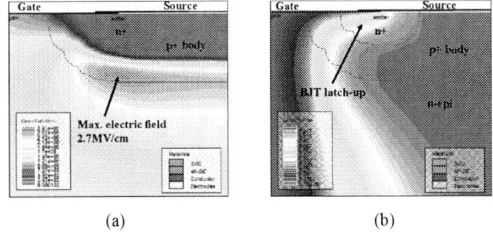

(a)　　　　　　　　(b)

Fig. 9 (a) TCAD simulation results of electric field distribution in SiC MOSFET at avalanche breakdown. (b) TCAD simulation of BJT latch-up.

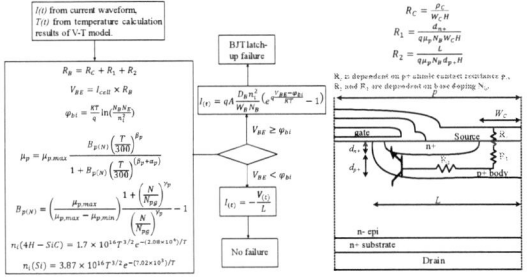

Fig. 10 An analytical model of BJT latch-up

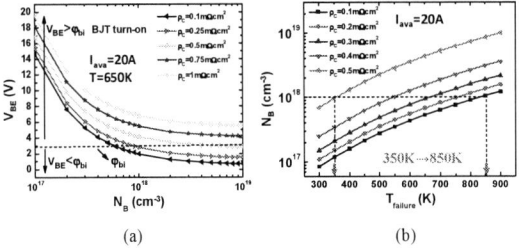

(a)　　　　　　　　(b)

Fig. 11 (a) Modeling results of base-emitter voltage drop at junction temperature of 650K in SiC MOSFET, (b) BJT latch-up failure trigger temperature ($T_{failure}$) for each set of N_B and ρ_c design of SiC MOSFET.

Detailed BJT lath-up analysis is conducted based on an analytical model shown in Fig. 10. The structural parameters used for the modeling are measured from SEM (scanning electron microscope) of the devices (in Fig. 8(a)).

The most important part of the model is the base-to-emitter resistance R_B. As shown in Fig. 10, R_B can be modeled as a series connection of P+ Ohmic contact resistance R_C, resistance between two adjacent N+ regions (R_1) and the resistance under the N+ region (R_2). The P+ Ohmic contact resistance in SiC cannot be ignored since the large energy bandgap of SiC combined with its high electron affinity gives rise to large Schottky barrier height at the metal/P-SiC interface and leads to high P-type Ohmic contact resistance, which is typically at 0.1~1 mΩcm^2 level.

The mobility of holes in the base of the BJT is temperature and doping concentration dependent and can be modeled with (2) [12].

$$\mu_p = \mu_{p,\max} \frac{B_{p(N)}(\frac{T}{300})^{\beta_p}}{1+B_{p(N)}(\frac{T}{300})^{\beta_p+\alpha_p}}$$

$$B_{p(N)} = (\frac{\mu_{p,\max}}{\mu_{p,\max}-\mu_{p,\min}}) \frac{1+(\frac{N}{N_{pg}})^{\gamma_p}}{(\frac{N}{N_{pg}})^{\gamma_p}}-1 \quad (2)$$

where $\mu_{p,\max}$, $\mu_{p,\min}$, N_{pg}, γ_p, α_p and β_p depend on type of the material, and N is the doing concentration.

By multiplying the body resistance in cell current (I_{cell}), the base-emitter voltage (V_{BE}) of the BJT is calculated as

$$V_{BE} = I_{cell} \cdot R_B = \frac{I}{W/p}(\frac{\rho_c}{W_cH}+\frac{d_{n+}}{q\mu_pN_BW_cH}+\frac{L}{q\mu_pN_Bd_{p+}H}) \quad (3)$$

where W and H are the width and height of the active region in SiC MOSFET. p, W_c, d_{n+}, d_{p+}, and L are the structural parameters as shown in Fig. 10. ρ_c and N_B are the Ohmic contact resistance and the base doping concentration, respectively. The influences of ρ_c and N_B on V_{BE} will be analyzed in this work.

On the other hand, the built-in voltage of the parasitic BJT can be calculated using (4) as follows,

$$\varphi_{bi} = \frac{KT}{q}\ln(\frac{N_BN_E}{n_i^2}) \quad (4)$$

where K is the Boltzman constant, N_E is the emitter doping concentration. n_i is the intrinsic carrier concentration, which increases with temperature [13].

When $V_{BE} > \varphi_{bi}$, the base-emitter voltage can forward bias the base-emitter PN junction. The parasitic BJT is turned on and bipolar action takes place. The injected electrons arrive at the drain and could create more electron-hole pairs through avalanche multiplication. The positive feedback between the avalanche breakdown and the parasitic bipolar action results in local breakdown at lower voltage. Therefore, the reduced avalanche voltage at the failing cell will attract more current from neighboring cells and cause the hot-spots and destructive device failure [14].

With equation (3), the base-emitter voltage of the BJT in SiC MOSFET at the UIS failure temperature (650 K) can be calculated. The calculation results are shown in Fig. 11(a). It is shown that, if N_B is $\leq 1 \times 10^{18}$ cm^{-3} and ρ_c is ≥ 0.25 mΩcm^2, $V_{BE} > \varphi_{bi}$ is occurred, thus BJT turn-on is triggered and device is failed.

978-1-7281-0582-6/19 $31.00 © 2019 IEEE

Generally, for a SiC MOSFET with any set of N_B and ρ_c design, the BJT latch-up failure trigger temperature can be predicted by this model. The modeling results are shown in Fig. 11(b). ρ_c is varied from 0.1 mΩcm^2 to 0.5 mΩcm^2, which is the typical level of Al/Ti based P type Ohmic contact system for SiC. As shown in Fig. 11(b), the BJT latch-up failure trigger temperature can be pushed higher with a higher base doping concentration and lower P+ Ohmic contact resistance design. For example, when the base doping concentration N_B is increased from 1×10^{17} to 1×10^{18} cm^{-3}, the BJT lath-up failure trigger temperature can be raised from 350 K to 850 K at a P+ Ohmic contact resistance ρ_c of 0.1 mΩcm^2. On the other hand, when the P+ Ohmic contact resistance is decreased from 0.5 mΩcm^2 to 0.1 mΩcm^2, failure trigger temperature can also be raised from 350 K to 850 K at a base doping concentration N_B of 1×10^{18} cm^{-3}.

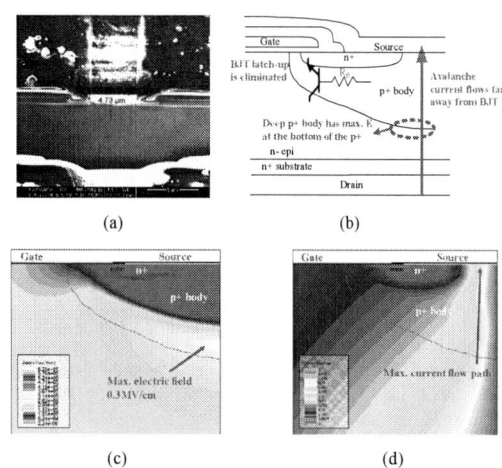

Fig. 12 (a) The cell structure of Si IGBT. (b) The schematic view of parasitic BJT portion in IGBT, (c) TCAD simulation results of electric field distribution at avalanche and (d) current flowlines.

The cell structure of Si IGBT is also examined with SEM, which is shown in Fig. 12(a). Similarly, the BJT portion is depicted in Fig. 12(b). Different from that in SiC, the implantation depth in Si can be much deeper, and the dopants can diffuse into bulk region during activation annealing. It's easier to obtain such a ~5 µm deep P+ body structure. Thus, the most intense electric field is moved to the bottom of the deep P+ region, which is far away from the parasitic BJT. This is verified by the TCAD simulation as shown in Fig. 12(c). Therefore, the BJT portion can be bypassed and the maximum current flows through the deep PN junction (Fig. 12(d)). As BJT latch-up failure is eliminated, the avalanche capability of Si IGBT is dominated by the temperature limit of the device materials (metal, dielectric, bonding wire, Si, etc.).

IV. CONCLUSION

Single pulse UIS tests are conducted on commercial SiC MOSFET and Si IGBT devices. The experimental results show that the avalanche energy of SiC MOSFET is only 30% that of Si IGBT due to the much smaller chip size (1/7 that of Si IGBT).

In order to improve the avalanche capability of SiC MOSFET, the failure mechanism is studied. At first, peak junction temperature in UIS tests are calculated. The maximum junction temperatures reach 650 K and 490 K in SiC MOSFET and Si IGBT, respectively. Then, the probability of BJT latch-up failure is analyzed with an analytical model. The results show that the BJT turn-on can be triggered at the failure temperature 650 K for a shallow P+ body designed SiC MOSFET. For Si IGBT, with a ~5 µm deep P+ body structure, the most intense electric field is diverted to the bottom of the deep P+ region and the maximum current flows through the deep PN junction. As a result, the BJT portion is bypassed and the latch-up failure can be eliminated.

Based on the comparative study of SiC MOSFET and Si IGBT, BJT latch-up failure exists in the state-of-the-art planar gate SiC MOSFET. To improve the avalanche capability, higher base doping concentration, lower P+ Ohmic contact resistance, and/or a deep P+ body structure is required to eliminate the BJT latch-up failure mechanism.

REFERENCES

[1] A. Fayyaz, A. Castellazzi, G. Romano, M. Riccio, A. Irace, J. Urresti and N. Wright, "UIS failure mechanism of SiC power MOSFETs," *2016 IEEE 4th Workshop on Wide Bandgap Power Devices and Applications (WiPDA)*, Fayetteville, AR, 2016, pp. 118-122.

[2] M. Treu, R. Rupp and G. Sölkner, "Reliability of SiC power devices and its influence on their commercialization - review, status, and remaining issues," *2010 IEEE International Reliability Physics Symposium*, Anaheim, CA, 2010, pp. 156-161.

[3] Yuji Okuto, "Junction Temperatures under Breakdown Condition," *Japanese Journal of Applied Physics*, vol. 8, no. 7, pp. 917-922, July, 1969.

[4] R. Rupp, R. Gerlach, A. Kabakow, R. Schorner, Ch. Hecht, R. Elpelt and M. Draghici, "Avalanche behaviour and its temperature dependence of commercial SiC MPS diodes: Influence of design and voltage class," 2014 *IEEE 26th International Symposium on Power Semiconductor Devices & IC's (ISPSD)*, Waikoloa, HI, 2014, pp. 67-70.

[5] Y. Nanen, M. Aketa, H. Asahara and T. Nakamura, "Estimation of junction temperature at failure of SiC DMOSFETs in UIS test," *2016 IEEE International Meeting for Future of Electron Devices*, Kansai (IMFEDK), Kyoto, 2016, pp. 1-2.

[6] *AN 1005, Power MOSFET Avalanche Design Guidelines*, Vishay Siliconix, Santa Clara, CA, USA, 2011.

[7] K. Fischer and K. Shenai, "Electrothermal effects during unclamped inductive switching (UIS) of power MOSFET's," *IEEE Trans. Electron Devices*, vol. 44, no. 5, pp. 874–878, May 1997. DOI: 10.1109/16.568052.

[8] M. S. T. McDonald, A. Murray, and T. Avram, "Power MOSFET avalanche design guidelines," International Rectifier, El Segundo, CA, USA, Tech. Rep. AN-1005, 2011.

[9] AN 601,Unclamped Inductive Switcing Rugged MOSFETs for Rugged Environments, Vishay Siliconix, Santa Clara, CA, USA, 1994.

[10] R. R. Stoltenburg, "Boundary of power-MOSFET, unclamped inductiveswitching (UIS), avalanche-current capability," *in Proc. 4th Annu. IEEE APEC*, Mar. 1989, pp. 359–364. DOI: 10.1109/APEC.1989.36987.

[11] L. Jiang, W. Lixin, L. Shuojin, W. Xuesheng, and H. Zhengsheng, "Avalanche behavior of power MOSFETs under different temperature conditions," *J. Semicond.*, vol. 32, no. 1, pp. 0140011-0140016, 2011.

[12] R. Bonyadi, Olayiwola Alstise, Saeed Jahdi, Ji Hu, Jose Angel Ortiz Gonzalez, Li Ran and Philip A. Mayby, "Compact Electrothermal Reliability Modeling and Experimental Characterization of Bipolar Latchup in SiC and CoolMOS Power MOSFETs," *in IEEE Transactions on Power Electronics*, vol. 30, no. 12, pp. 6978-6992, Dec. 2015. DOI: 10.1109/TPEL.2015.2388512.

[13] B. J. Baliga, *Fundamentals of Power Semiconductor Devices*, 1st ed. New York, NY, USA: Springer-Verlag, 2008, pp. 23-26.

[14] P. R. Bond, "Secondary breakdown in transistors," Retrospective Theses and Dissertations, 1963.

978-1-7281-0582-6/19 $31.00 © 2019 IEEE

Proceedings of the 31st International Symposium on Power Semiconductor Devices & ICs
May 19-23, 2019, Shanghai, China

Investigation of UIS Capability for -600V Class Vertical SiC p-channel MOSFET

Kailun Yao, Hiroshi Yano, and Noriyuki Iwamuro
Graduate School of Pure and Applied Sciences
University of Tsukuba
1-1-1 Tennodai, Tsukuba, 305-8577, Japan
E-mail: s1820360@s.tsukuba.ac.jp

Abstract—**Vertical planar gate 4H-SiC p-ch SiC MOSFET is fabricated as a potential candidate for complementary power applications. Its unclamped inductive switching (UIS) capability is investigated by measurement and TCAD simulation. The p-ch SiC MOSFET shows a low UIS capability rather than n-ch SiC MOSFET because of the unoptimized termination structure. The UIS capability of p-ch MOSFET can be improved by good design of edge region.**

Keywords—*p-channel SiC MOSFET, Unclamped Inductive Switching capability, Edge block capability*

I. Introduction

As one kind of wide band-gap semiconductors, silicon carbide (SiC) has superior physical and electric properties which can lead to excellent device performances such like extremely low on-resistance and fast switching speed. In the past decade, n-ch SiC MOSTETs and SiC Schottky barrier diodes have been successfully commercialized and used in power electronics systems to reduce energy loss. In order to achieve long-term safety operations, the reliability issues like short-circuit and unclamped-inductive-switching (UIS) have also been studied by using the experiment way and numerical simulation [1]-[3]. Various papers have reported that exceed the metal melting temperature is the most likely UIS failure mechanism for n-ch SiC MOSFETs [4]-[6]. However, the UIS capability of p-ch SiC MOSFET still needs to be studied.

Vertical planner gate 4H-SiC p-ch MOSFET was fabricated successfully as a potential candidate for the complementary inverter applications [7]. In this work, the UIS capability of p-ch SiC MOSFET is studied and compared with the n-ch SiC MOSFET counterpart which has the same device structure.

II. Fabrication of Device

The fabrication of SiC p-ch MOSFET is based on the design of SiC n-ch IEMOS which was first presented in 2005 [8]. Figure 1 shows the schematic cross-section of -600 V class SiC p-ch MOSFET. A thickness of 350 μm Si-face p-type 4H-SiC with resistivity of 2 Ω·cm is used for substrate. The drift layer has the thickness of 5 μm and the doping concentration of 1.6×10^{16} cm^{-3}. Highly phosphorous ions with the concentration of 4×10^{18} cm^{-3} are selectively implanted to form the bottom of the n-base region. Then, the n-base region is formed by using the epitaxial method with the doping concentration of 5×10^{15} cm^{-3} and the thickness of 0.5 μm to improve the channel

mobility. The JFET region is formed by selective aluminum ions implantation to reduce its resistance. The gate oxide thickness of 50 nm is grown on the surface and highly doped polycrystalline silicon is deposited and annealed as gate electrodes. At last, metal process is applied to form the source and drain electrodes. The die size of device is 3×3 mm^2. The active area is 5.7mm^2.

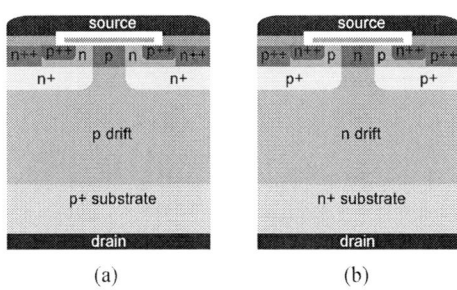

Fig. 1. Cross-section of 4H-SiC IEMOSFET, (a) p-ch, (b) n-ch

The fabrication procedures of tested n-ch MOSFET are the same as p-ch MOSFET. Measured breakdown voltages for n-ch and p-ch MOSFETs are 725V and -633V ($@I_{DS}$=100μA, R.T.), respectively.

III. Experimental and Simulation Results

A. UIS mesurement setting up

The SiC n-ch and p-ch MOSFETs are tested under ±200 V DC voltage with the load inductor L of 1mH. Due to the high on resistance of p-ch MOSFFET, self-heat effect will strongly influence UIS measurement. Figures 2 (a) and (c) give a conventional UIS test circuit and measured waveforms. It is found from this figure that the drain voltage shows dramatically increase during the charging period which could cause some damages to the device, so that a modified test circuit shown in Fig.2 (b) was utilized to solve this problem. In order to avoid heating up devices during charging an inductor, a 1200 V n-ch SiC MOSFET with a breakdown voltage of 1634V ($@I_{DS}$ =100μA, R.T.) is connected in parallel to ramp up the inductor current. Then, because of the higher breakdown voltage of paralleled n-ch MOSFET, only the p-ch MOSFET will work in avalanche mode. To maintain consistency, the test of n-ch MOSFET counterpart also utilizes this setup.

978-1-7281-0582-6/19 $31.00 © 2019 IEEE

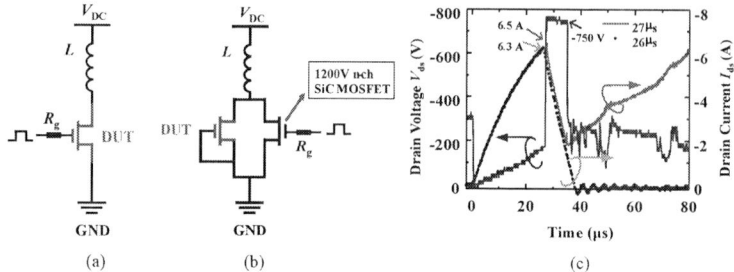

Fig. 2. (a) Conventional UIS test circuit, (b) Modified UIS test circuit, and (c) UIS waveforms measured by conventional test circuit

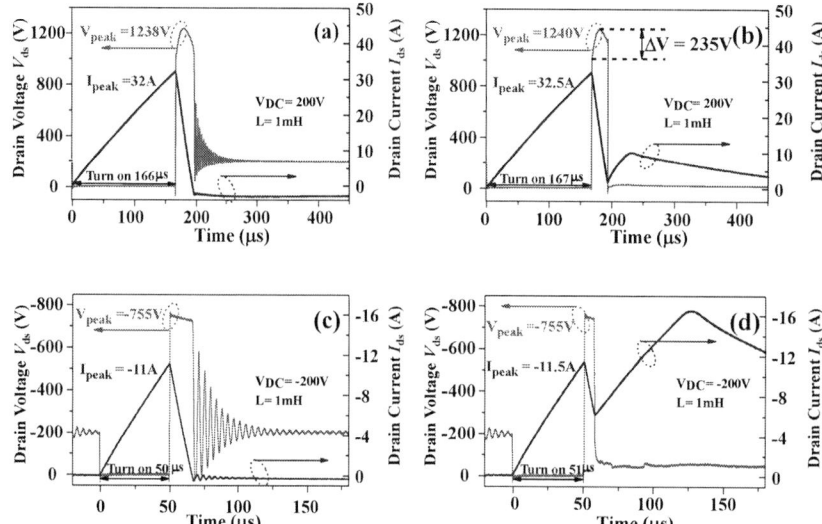

Fig. 3. Measured UIS waveforms, (a) n-ch just before failure, (b) n-ch just failure, (c) p-ch just before failure, (d) p-ch just failure

B. UIS capability of p-ch and n-ch SiC MOSFETs

In the UIS measurement, a drain current is increased step by step via changing turn-on time until the tested devices reach their failure point. Figure 3 shows the UIS waveforms of n-ch MOSFET: (a) just before the failure, (b) just the failure, and p-ch MOSFET: (c) just before the failure (d) just the failure, respectively. The maximum UIS current that n-ch MOSFET can handle is 32.0 A ($J = 561.4$ A/cm^2). However, the maximum tolerable UIS current of p-ch MOSFET is -11.1 A (194.7 A/cm^2) which is much lower than that of n-ch MOSFET.

The characteristics of UIS drain voltage waveforms are also different between n-ch and p-ch MOSFETs. The drain voltage of n-ch MOSFET first reaches 1003 V, followed by showing an increase tendency during avalanche time. Then, the peak value of 1238 V appears at the middle of avalanche period. The increment of drain voltage ΔV is 235 V. The increase of avalanche voltage implies that junction temperature increases by heating up via sudden power dissipation for a very short time. However, the drain voltage of p-ch MOSFET during avalanche time reaches its peak value of -755 V just when the avalanche takes place and then decreases slowly until the current is fully discharged. By which means, the heat-up of p-ch MOSFET during avalanche time is not severe. The calculated UIS energy densities are 8.7 J/cm^2 and 0.83 J/cm^2 for n-ch and p-ch MOSFETs, respectively. Figures 4 (a) and (b) show the breakdown voltage of n-ch and p-ch SiC MOSFET as a function of temperature. The n-ch SiC MOSFET shows an average slope of 0.39 V/°C. On the other hand, the p-ch MOSFET showed similar results that breakdown voltage increases as the temperature rising up; however, the slopes are different in three measured devices. It can be seen that the slope of p-ch MOSFET is lower than that of n-ch MOSFET counterparts.

A method of estimating the maximum junction temperature during UIS has been reported in Ref. [9]. By adopting their method, the maximum junction temperature can be calculated by using the equation (1),

$$T_{\max} = T_{\mathrm{ambient}} + \frac{\Delta V}{0.39\ \mathrm{V/^\circ C}}. \qquad (1)$$

The measurement was carried out at room temperature of 25°C. By using the data of $\Delta V = 235$ V, the maximum junction temperature is calculated to be 627°C, which is closed to the melting point of aluminum 660.3°C [10]. However, it is difficult to estimate the junction temperature of p-ch MOSFET during the UIS measurement because the

978-1-7281-0582-6/19 $31.00 © 2019 IEEE

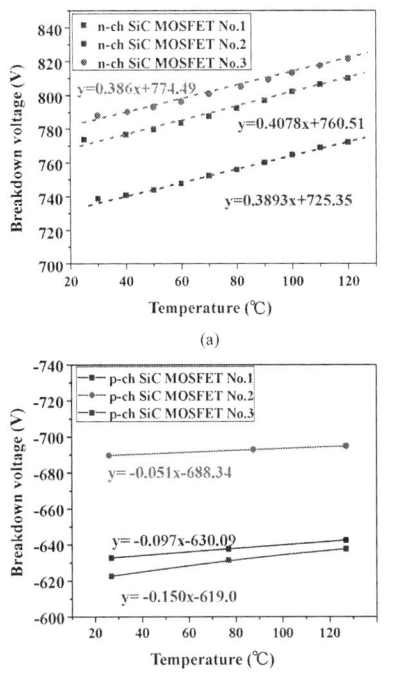

(a)

(b)

Fig. 4. Temperature dependence of breakdown voltage (a) n-ch SiC MOSFET, (b) p-ch SiC MOSFET

drain voltage decreases when avalanche happens as shown in Fig.3

In order to compare the UIS capability between n-ch and p-ch SiC MOSFETs, post-failure devices are decapsulated to observe damage points. Figures 5 (a) and (b) show the failure point of n-ch and p-ch SiC MOSFETs.

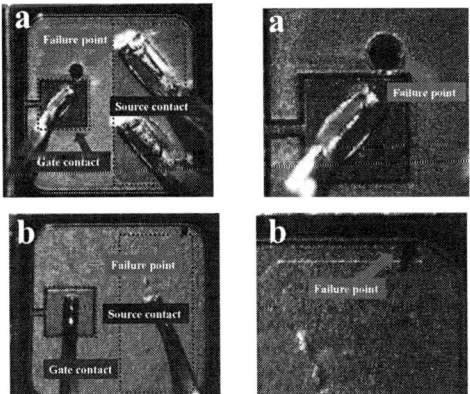

Fig .5. Failure point obvervation of (a) n-ch SiC MOSFET, (b) p-ch SiC MOSFET

For n-ch SiC MOSFET, a damage point is located in the active area of the chip. However, the damage point of p-ch MOSFET is located near the termination region. From the expanded photos, n-ch MOSFET shows a fusion like point, however, the damage point of p-ch MOSFET is smaller and its shape is ambiguous. Thus, authors assumed that the UIS failure of p-ch MOSFET could not be taken place due to the thermal causes in large area but has different failure mechanism.

C. Simulation results

In order to understand the different UIS capability between n-ch and p-ch SiC MOSFETs, a two-dimensional electro-thermal mix-mode device simulation using Sentaurus TCAD was carried out to reveal physical processes inside the MOSFETs during the UIS test [11].

The simulation analyses consist of two parts: the active region and the termination region. Since the damage points of p-ch SiC MOSFETs are mainly located near the termination (21 devices out of 32) as shown in Fig. 5, authors assumed that the termination region of p-ch MOSFET was not in the best design in spite of the fact that they have the same structure and same doping density each other. A possible failure mode is the snap-back characteristic at the edge termination region. Figure 6 shows the V_{DS} - I_{DS} locus during the UIS transient to explain a failure model ((a) failure UIS event, (b) safe UIS event).

First, avalanche breakdown took place in active region. As shown in Fig.6 (a), when a V_{DS} - I_{DS} locus during the UIS transient exceeds the breakdown voltage of the edge region VBE1 (point B), the locus would switch to the breakdown characteristic of the edge region (point C) because the same drain-source voltage was applied to the edge region as well and trace it. Hence, current concentration took place at the edge structure owing to its snap-back characteristics, so that the p-ch SiC MOSFET was in destructive failure with low UIS capability. However, the blocking capability at edge region is much higher than the active region, the device will safely trace the V_{DS} - I_{DS} locus in the active region during the entre UIS event, resulting in turning off the device safely.

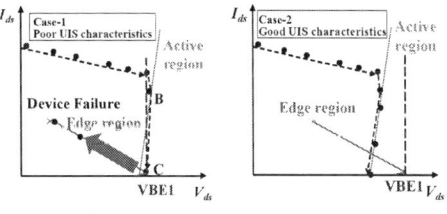

(a)   (b)

Fig. 6. Modeled V_{DS} -I_{DS} locus of p-ch MOSFET during the UIS transient (a) failure locus. (b) safe locus. "VBE1" in this figure means the avalanche breakdown voltage of edge region.

A TCAD simulation was carried out to prove this failure mode. Simulated blocking capability of edge and active regions are shown in Fig. 7. Breakdown voltage of the active region was modeled of -730 V with positive resistance characteristic. Also, the breakdown voltage of edge region was set -770 V as VBE1 with negative resistance characteristic (snapback). In the UIS simulation, this active and edge regions are connected in parallel and calculated at the same time.

978-1-7281-0582-6/19 $31.00 © 2019 IEEE   189

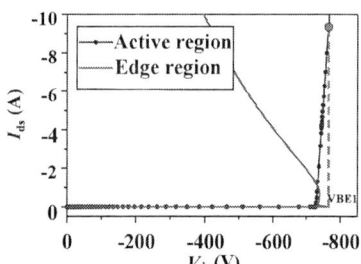

Fig. 7. Simulation results of blocking characteristics at active and edge regions of the SiC p-ch MOSFET. "VBE1" in this figure means the avalanche breakdown voltage of edge region.

Figure 8 (a) shows the calculated UIS waveforms of p-ch MOSFET with the edge region modeled in Fig.7 and Figure 8 (b) shows the waveforms without the edge region. It is found that whereas UIS capability of p-ch MOSFET cell shows much better, the cell with the edge region modeled in Fig. 7 exhibits very poor UIS capability. Also, the simulated UIS waveforms shown in Fig. 8(a) was very similar with the measured results shown in Fig. 3(d). These results suggested that avalanche multiplication has to be distributed uniformly in active region of the device in order to achieve good UIS capability.

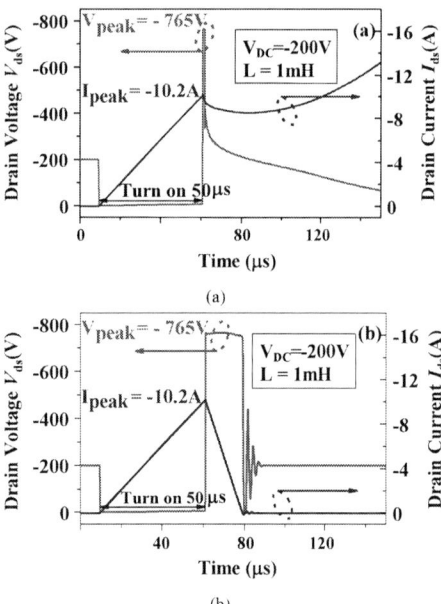

Fig. 8. Simulated UIS waveform: (a) active region with the edge region modeled in Fig. 7, (b) active region without the edge region.

IV. CONCLUSION

The UIS capabilities of SiC n-ch and p-ch MOSFETs are studied by the experiment and TCAD simulation. It is found that the p-ch SiC MOSFET shows a 10 times lower UIS capability (0.83 J/cm^2) than the n-ch SiC MOSFET and the failure points are mainly located near the edge region. This poor UIS capability can be explained by modeling the edge structure with not only a little higher blocking voltage than the active cell region but also the negative resistance characteristics. By using this model, the UIS waveforms of p-ch MOSFET can be successfully calculated and this result was very similar to the measured one.

ACKNOWLEDGMENT

A part of this paper has been implemented under a joint research project of Tsukuba Power Electronic Constellations (TPEC)

REFERENCES

[1] A. Kumar, S. Parashar, J. Baliga and S. Bhattacharya, "Single shot avalanche energy characterization of 10kV, 10A 4H-SiC MOSFETs," in *2018 IEEE Applied Power Electronics Conference and Exposition (APEC)*, March 2018, pp. 2737-2742.

[2] J. An, M.Namai, H.Yano, N.Iwamuro, Y.Kobayashi and S.Harada, "Methodology for enhanced shor-circuit capability of SiC MOSFETs" in *Proc. Int. Symp. Power Semiconductors and ICs*, May 2018, pp. 391-394.

[3] A. Fayyaz, A. Castellazzi, G. Romano et al., "UIS failure mechanism of SiC power MOSFETs," in *Proc. Wide Bandgap Power Devices Appl.*, Nov. 2016 ,pp. 118-122

[4] N.Ren, H.Hu, K.L.Wang, Z.Zuo, R.Li and K.Sheng, "Investigation on Single Pulse Avalanche Falure of 900V SiC MOSFETs" in *Proc. Int. Symp. Power Semiconductors and ICs*, May 2018, pp. 431-434.

[5] A. Kumar, S. Parashar, S.Sabri, E.V.Brunt, S. Bhattacharya andV.Vleliadis, "Ruggedness of 6.5kV, 30A SiC MOSFETs in Extreme Transient Conditions", in *Proc. Int. Symp. Power Semiconductors and ICs*, May 2018, pp. 423-426.

[6] M. D. Kelley, B. N. Pushpakaran and S. B. Bayne, "Single-Pulse Avalanche Mode Robustness of Commercial 1200 V/80 m SiC MOSFETs," in *IEEE Tran. on Power Electronics*, vol. 32, no. 8, pp. 6405-6415, Aug. 2017.

[7] J. An, M. Namai, H. Yano, and N. Iwamuro, "Investigation of robustness capability of −730 V P-channel vertical SiC power MOSFET for complementary inverter applications," *IEEE Trans. Electron Devices*, vol. 64, no. 10, pp. 4219-4225, Oct. 2017.

[8] S. Harada, M. Kato, K. Suzuki, M. Okamoto, T. Yatsuo, K. Fukuda, and K. Arai, "1.8 mΩcm^2 , 10 A Power MOSFET in 4H-SiC", in *IEEE IEDM* Tech., Dig. , 35.1, Dec. 2006.

[9] J. An, M. Namai, D. Okamoto, H. Yano, H. Tadano, and N. Iwamuro, "Investigation of Maximum Junction Temperature for 4H-SiC MOSFET during Unclamped Inductive Switching Test," *Electronics and Communications in Japan*, vol. 101, no. 1, pp. 2431, 2018

[10] National Astronomical Observatory of Japan. Chronological scientific tables 2015, vol.89, pp. 409, Maruzen, Tokyo, 2015 (in Japanese).

[11] *TCAD Sentaurus Device manual*, Synopsys, Inc, Mountain View, CA, USA, 2017

Proceedings of the 31st International Symposium on Power Semiconductor Devices & ICs
May 19-23, 2019, Shanghai, China

Ultra-high speed 7mohm, 650V SiC half-bridge module with robust short circuit capability for EV inverters

Anup Bhalla[1], Melvin Nava[2], Mike Zhu[3], Frank Sudario, Deborah Sumaoang, Peter Alexandrov, Xueqing Li, Peter Losee

United Silicon Carbide, Inc., 7 Deer Park Drive, Monmouth Junction, NJ 08852, USA
[1] abhalla@unitedsic.com, [2] mnava@unitedsic.com, [3] mzhu@unitedsic.com

Abstract— **Ultra-low RdsA SiC JFET technology is used to realize a 7mohm, 650V Stack Cascode switch, rated at >100A, with very fast switching, and 8us rated short circuit capability that is reported here for the first time. These devices meet all the technical requirements for EV inverters, representing the lowest RdsA and highest short circuit rated contemporary switch technology. The devices are assembled in a high performance, low inductance module to demonstrate extremely low switching losses, making the technology suitable for higher frequency DC-DC applications as well.**

Keywords—EV, Inverter, SCWT, SiC, Stack Cascode, Module

I. INTRODUCTION

Growth in battery EV applications is driving much of the latest growth in SiC, and with the projected benefits of incorporating SiC transistors into EV inverters, there is considerable development effort focused on improving SiC MOSFETs and module packaging to address this need. Key requirements for inverter drive switches is low conduction loss, low hard switching losses at moderate frequencies, robust short circuit capability and the possibility of operating at 200C. In this work, we use UnitedSiC's low RdsA 650V JFETs [1], on which are stacked custom designed low voltage Silicon MOSFETs to realize a 3.5x5.5mm stack cascode with a 7mohm measured on-resistance. The SiC JFET Vth is engineered to limit the short circuit current with minimal impact on RdsA, a tradeoff not currently achievable with SiC MOSFETs [3]. To manage the fast switching of these devices, a module is constructed with built-in ceramic

decoupling capacitors and a Cu clip, that together reduce the commutation loop inductance. The compact module achieves excellent thermal and switching performance, low switching losses and 8μs short circuit capability.

II. FABRICATION AND CHARACTERIZATION

The structure of the devices used it the module is shown in Figure 1. The construction begins with a normally-on trench 650V 4H-SiC JFET with a die size of 3.5mm by 5.5mm including the scribe lines, which is thinned to a final thickness of a 100um to provide a net resistance of 6mohm. The active area of the SiC JFET is 15mm². On this is stacked a low voltage Silicon MOSFET, which is designed with an ESD protected gate, a high V_{TH} of 4.7V, a wide V_{GS} operating range, an R_{DSON} =1mohm and 35V breakdown. The MOSFET is provided with built in PN junction clamping for robust repetitive UIS capability. The options for stacking the device include high temperature solder or Ag sintering, and the high temperature solder is used in this case. To facilitate the die stacking, the SiC JFET wafer is prepared with electroless N-Au plating on the 5um topside Aluminum,

Device	UJSC065007
Vgsmax V	+/-25
Vdsmax V	650
TJ(MAX) C	175
ESD KV	4
V_{TH} V (10mA, Vds=5V)	4.7
I_D(rated)A @Tc=100C	100
$R_{DS(ON)}$ 25C	7mohm
$R_{DS(ON)}$ 125C	10.9mohm
$R_{DS(ON)}$ 175C	14mohm
V_{FSD} (100A)	1.4
Ciss pF (400V)	8173
Coss pF (400V)	300
Crss pF (400V)	4
R_G ohm	1
SCWT (400V) Vgs=15V	8us

Table 1. Summary of device characteristics

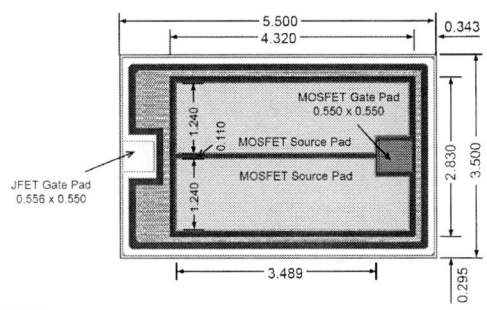

Figure 1: 7mohm, 650V Stack cascode structure. SiC JFET active area is 15mm². The JFET gate pad is bonded to the MOSFET source lead/trace.

978-1-7281-0582-6/19 $31.00 © 2019 IEEE

A. DC and AC characteristics of the switch

Table 1 shows a summary of the basic datasheet parameters of the switch. The MOSFET gate rating of +/25V makes it compatible with a wide range on gate drive options. The preferred driving method is 0V or -5V to +12V. This is because there is no decrease in device $R_{DS(ON)}$ beyond Vgs=10V. The V_{TH} of the LV MOSFET is 4.7V at 25C, and stays above 3V at 200C. This allows low net leakage, and the devices can be operated at 200C when the package technology permits. The gate is ESD protected with back to back Polysilicon diodes. The on resistance of the device is measured to be 7mohm at 25C, and increases to 14mohm at 175C. The measured on-resistance vs current at 25C, 175C is shown in Figure 2, as is the normalized variation with temperature. The conduction losses therefore remain significantly lower than commercial SiC MOSFETs in the 650V-900V class even at 200C. As with other SiC unipolar devices, the absence of a knee voltage creates a significant advantage over IGBTs even in lower frequency applications if low conduction losses are needed below rated current.

From the data table, it is also clear that the VFSD of the diode is just 1.4V at 100A, which is much lower than what is achievable with SiC MOSFETS (3 to 4.5V) and is comparable to SiC Schottky diode, or Si fast recovery diode. In the cascode structure, the body-diode of the low voltage MOSFET, with minimal Q_{RR}, operates in series with the high voltage JFET channel. During diode recovery, the as soon as the low voltage MOSFET diode recovers, the JFET turns-off, and most of the recovery charge in the structure come from the Q_{OSS} of the SiC JFET. Therefore, the Q_{RR} changes very little with temperature even up to 200C, as shown in Figure 3.

The Ciss of the device depends on the low voltage MOSFET, and is 8.17nF, while the Crss of the stack cascode at high voltage is very low, due to the near zero Cds of the SiC trench JFET. Active Miller clamping is generally not

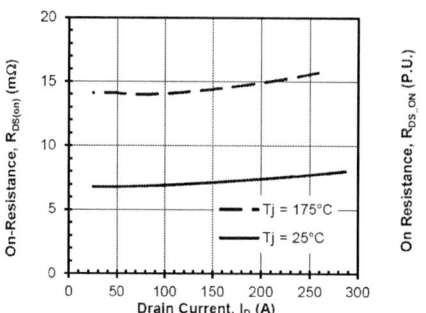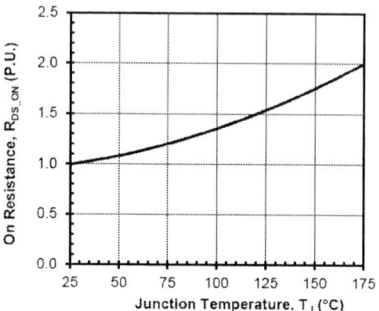

Figure 2: Measured $R_{DS(ON)}$ vs Drain current (V_{GS}=12V) and Junction temperature (V_{GS}=12V, I_D=100A).

necessary for this device.

Figure 4 shows the measured typical short circuit capability of the devices. The devices can be rated at 8us at 400V bus and are seen to fail at 10us. The peak current of 400A is set by the V_{TH} design of the SiC JFET, and is not limited or controlled by the low voltage Si MOSFET in series. Therefore, there is no change in the short circuit

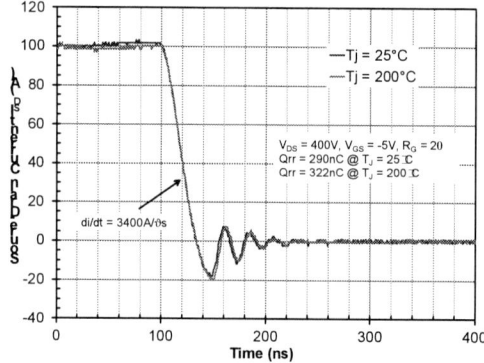

Figure 3: Q_{RR} for UJSC065007 cascode device from 25C to 200C.

current for gate drive voltages from 10-20V, which can be varied to adjust turn-on di/dt without compromising SCWT. During short circuit, the temperature rise is confined to the SiC JFET and the joint material used for the stack as was

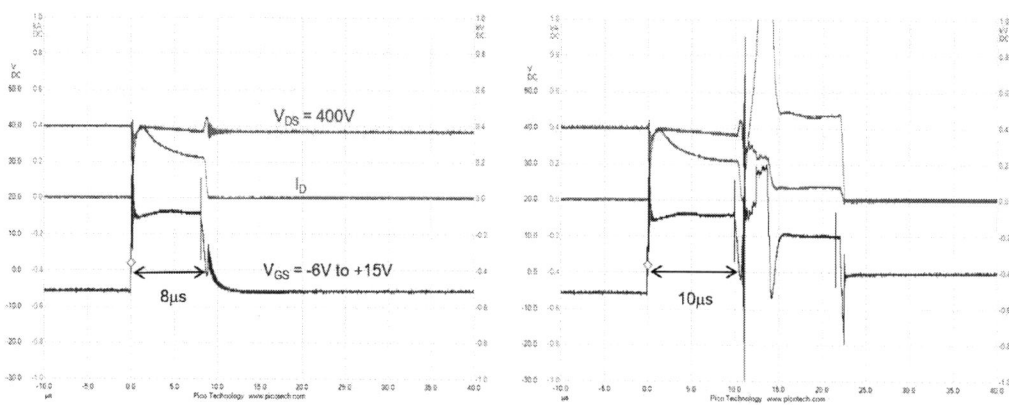

Figure 4: Measured Short circuit waveforms T_{jstart}=25C, 400V Bus. The device passes 8μs and fails 10μs.

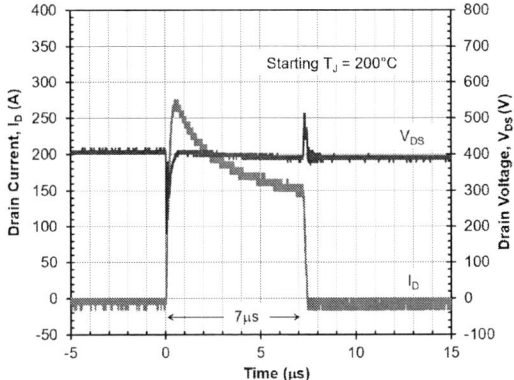

Figure 5: Measured Short circuit waveforms T_{jstart}=200C, 400V Bus. The device passes 7µs and fails 8µs.

shown via TCAD simulations in our previous work[3]. The devices have also been verified to handle 100X short circuit events at the rated condition with no change in any device parameters. They have also been tested for short circuit capability with a starting T_J=200C, 400V bus and are able to pass up to 7us as shown in Figure 5. The peak short circuit current is limited by the SiC JFET, and drops from 400A to 270A when T_{JSTART} is increased from 25C to 200C.

III. MODULE CONSTRUCTION AND SWITCHING

Figure 5 shows a 3D mage of the internal construction of the module. The module is 26x33.5mm and is shown in Figure 6 adjacent to a TO247-4L device for scale. The module uses a standard Alumina DBC material, with higher performance versions with AlN or Si3N4 AMB possible. The module can hold up to 2X 6X6mm chips on the high side and low side. In this work, we show the data for just one chip on the high side and low side.

The module has provision for up to three decoupling capacitors within the module. In the 650V 1HS/1LS data below, we have used 2X22nF, 1000V capacitors. A copper structure is used to route the phase node current and allows for a low loop inductance, which for 1HS/1LS is measured at 4.3nH.

The assembly process begins with die attach of the chips to the DBC and wire bonding, followed by attachments of the gate and source Kelvin pins, leads and de-coupling

ceramic capacitors. Finally, the outer shell is placed and glued into place, Then the encapsulating gel is dispensed and cured to complete the assembly. The module can be mounted to a PCB for lower current application, or connected via bus bars to the power terminals.

A. MODULE SWITCHING RESULTS

The half-bridge module was mounted to a double pulse test board with isolated HS and LS drive using a XXX gate driver. Both the high side and low side switches used an external Rg=1ohm, and Vgs=+12V/-5V. Figure 7 shows the measured Turn-off and Turn-on waveforms at 100A, 400V, 25C. The internal de-coupling capacitor precludes the ability to measured switch currents, but does lead to the low measured loop inductance. The dV/dt at turn-off is about 75V/ns, and even with a low loop inductance shows a voltage overshoot of nearly 200V, indicating a turn-off di/dt of about 40A/ns. The turn-on dV/dt is 25V/ns, and from the initial voltage drop and duration, we estimate a turn-on di/dt of about 12.5A/ns.

The module loop inductance is estimated from the ringing frequency at turn-off (135MHz) and measured Coss values for the stack cascode switch (324pF), which leads to an estimated loop inductance of 4.3nH.

Figure 8 shows the measured turn-on and turn-off V_{DS} waveforms incorporating an RC snubber of 1360pF, 4.7ohm on the driver board to reduce overshoots and damp the power current ringing at turn-off with minimal slowdown of the turn-on and turn-off dV/dt. It is more effective to use a RC snubber in place of the bus de-coupling capacitor within the module, which is left to future improvements.

CONCLUSION

We have demonstrated a high-performance stack cascode SiC device using low RdsA trench SiC JFETs, incorporated into a low inductance half-bridge module that meets all the main targets of a wide bandgap switch technology suitable for EV Traction inverters. The stack cascode offers a uniquely good trade-off between RdsA and short circuit capability, with very low temperature independent switching losses. This allows upto 200A, 1200V half-bridges to be incorporated into this 26x33.5mm module. The measured performance indicates that this technology holds great promise both for EV inverters as well as for DC-DC high

Figure 5: Construction of the half-bridge module. Only 1HS and 1LS chip is populated, but the module is capable of handling 2HS and 2LS chips. Internal decoupling capacitors 2x22nF, 1000V are used. The unique phase node clip routes the commutation loop current to cut the net loop inductance to 4.3nH.

Figure 6: Image of completed half-bridge module which is 26x33.5mm, and able to handle 2X 100A, 1200V devices per switch position. A standard TO247-4L package is shown alongside for reference.

frequency converters.

References

[1] P. Alexandrov et. al., "650V SiC Cascode: A Breakthrough for Wide-Bandgap Switches", ECSCRM 2016, Material Science Forum, Vol 897, pp 673-676

[2] X. Li et. al.,"Investigation of SiC Stack and Discrete Cascodes", PCIM 2014

[3] X. Li et. al., "Short Circuit Capability of SiC Cascode", Material Science Forum, Vol 924, pp. 871-874, 2018

[4] B. Kakarla et. al., "Short circuit ruggedness of New generation 1.2Kv SiC MOSFETs", WiPDA 2018, pp 118-124

[5] R. Green et. al., "Short-Circuit Robustness Testing of SiC MOSFETs." Materials Science Forum, Volume 897, 2017, pp. 525-528.

[6] H. Hatta, et. al. "Suppression of Short-Circuit Current with Embedded Source Resistance in SiC-MOSFET", Material Science Form, Vol 924, pp 727-730

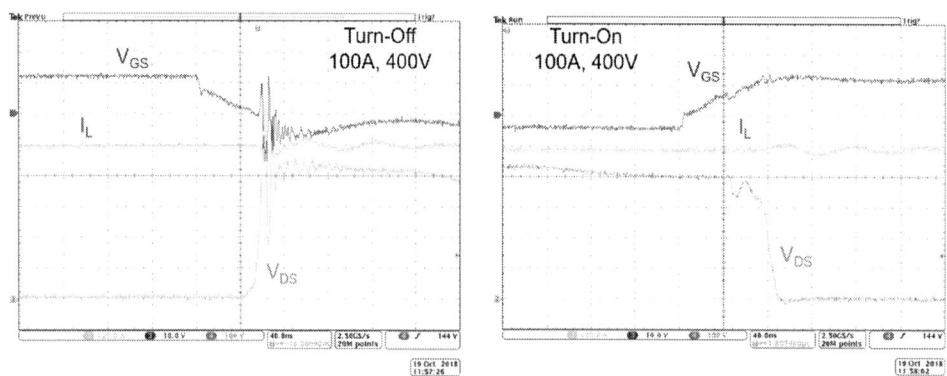

Figure 7: Turn-off and Turn-on double pulse test results of the UJSC065007 in the half-bridge module at 25C, 100A, 400V bus with a gate drive using Rg=1ohm, Vgs=+12/-5V. The Vgs waveform, Vds waveform and Inductor current are monitored, but the loop is not broken to measure switch current. Turn-off dV/dt is about 70V/ns, Turn-on dV/dt is 25V/ns.

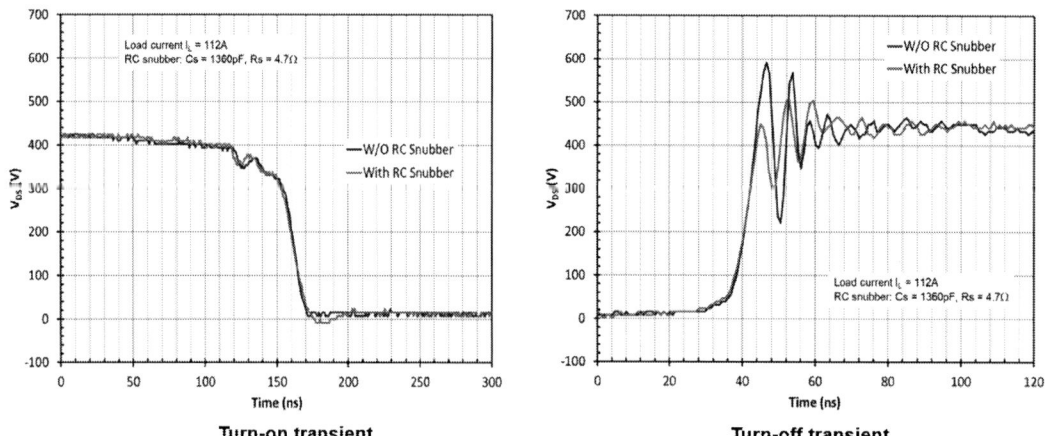

Figure 8: Turn-on and Turn-off double pulse results of the UJSC065007 in the half-bridge module at 25C, 112A, 400V bus with a gate drive using Rg=1ohm, Vgs=+12/-5V. The effect of adding an RC snubber across the switches externally to help reduce the voltage overshoort and damp the ringing faster.

Proceedings of the 31st International Symposium on Power Semiconductor Devices & ICs
May 19-23, 2019, Shanghai, China

Determination of the Transient Threshold Voltage Hysteresis in SiC MOSFETs after Positive and Negative Gate Bias

Christian Unger, Martin Pfost

Chair of Energy Conversion, TU Dortmund

Emil-Figge-Straße 68, Dortmund, Germany - email: christian.unger@tu-dortmund.de

Abstract—**Due to the inherently high trap density at the SiC/SiO$_2$ interface, SiC-MOSFETs are subject to a hysteresis effect depending on the previous bias of the gate. Applying a negative bias to the gate leads to trapping of holes that influence the threshold voltage. We present the application-relevant phenomenon caused by the V_{th}-hysteresis under different operating and measurement conditions. Furthermore, we introduce a measurement technique to determine the transient threshold-voltage shift $\Delta V_{th}(t)$ in different operating points. A simple power-law approach can be used to describe the trap-induced V_{th}-shift. In addition, we show that positive biasing leads to electron trapping which temporarily weakens the conductivity of the channel.**

I. INTRODUCTION

Even though SiC-based MOSFETs have established a firm role in leading-edge power electronic systems, their major weakness remains to be the SiC/SiO$_2$-interface. The inherently high density of interface traps-states is associated with several negative effects. Hence, a lot of effort is put into the reduction of the interface trap density D_{it}. However, even low D_{it} have severe implications because the traps are located right at the interface and are therefore in close proximity to the mobile charges in the channel.

As a consequence, SiC-MOSFETs suffer from a relatively low field-effect mobility due to scattering at the traps. In addition, the trapped charge interacts with the depletion- and inversion-charge under the gate which causes a shift of the threshold voltage V_{th}. The magnitude of the V_{th}-shift depends on the amount of the trapped charge which is determined by the previously applied gate-source voltage V_{GS}. This shift, however, is not permanent but can be reversed depending on the subsequent gate-bias.

The charging and neutralization of trapped charge requires time and can range from µs to several seconds. Several different mechanisms are discussed under the comprehensive term NBTI [1]. In these cases, the V_{th}-shift is accelerated by high temperatures and can become permanent under certain conditions. In contrast to that, the faster trap states observed at room temperature are a fully reversible and can cause relatively high V_{th}-shifts of several volts.

In previous work [2], the effects of the trapped charge on a 2nd generation device were analyzed by measurements of the transient drain current change in saturation. In this work, this approach is complimented by additional measurement techniques to observe the standard switching operation and determine the threshold voltage shift.

II. MECHANISM & SYMPTOMS OF THE V_{th}-HYSTERESIS

The V_{th}-hysteresis is caused by the charging and neutralization of interface states and the influence of the trapped charge on the channel. Fig. 1 shows the device in three consecutive phases of a turn-on sequence: First in accumulation, then, directly after turn-on and finally after steady state is reached.

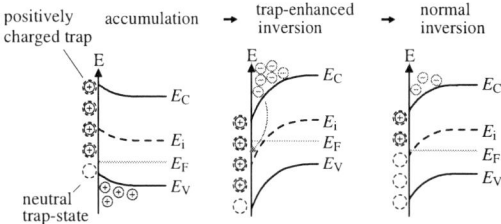

Fig. 1. Energy bands during three states of a turn-on sequence. The positively charged trap states temporarily increase the inversion charge in the channel.

During off-state, trap states are charged by the holes accumulating at the interface. The amount of trapped holes depends on the position of the Fermi level, which is determined by $V_{GS,off}$. Directly after turn-on, the positively charged traps attract additional electrons and thereby enhance the inversion charge in the channel. However, electrons from the channel gradually neutralize the trapped holes. Once the trapped charge is neutralized, the inversion charge normalizes to the steady state value.

The V_{th}-Hysteresis causes several phenomena in dependence of the operating condition. In the following, these symptoms are presented for application- and measurement-relevant cases. The device under investigation is a 3rd generation SiC-MOSFET rated for 1200 V.

A. Drain current in saturation

During characterization measurements, the DUT is often operated in saturation. In this condition, the current flowing through the device is determined by the channel, and is therefore affected by the traps. When the device is turned on with a constant on-state gate-source voltage $V_{GS,on}$, a variation of the drain current I_D in dependence of the off-state gate-source voltage $V_{GS,off}$ can be observed, see Fig. 2.

With increasingly negative $V_{GS,off}$-voltages, an increase of I_D after turn-on is observed. This is caused by the positively charged traps that enhance the inversion charge in the channel

978-1-7281-0582-6/19 $31.00 © 2019 IEEE

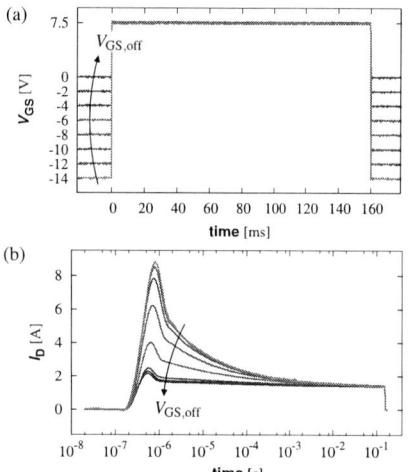

Fig. 2. Transient gate-source voltage V_{GS} in (a) and the corresponding drain current I_D in (b) for $V_{GS,on} = 7.5\,V$ and $V_{GS,off} = -14\,V .. 0\,V$. The device is operated in saturation at $V_{DS} = 15\,V$. Note the time-logarithmic scale in (b). A drain current overshoot followed by a decay with time is observed in dependence of $V_{GS,off}$.

and thereby increase the conductivity of the device. However, with time, the traps are gradually neutralized. In consequence, the inversion charge normalizes and I_D reaches the steady state value. The drain current increase does not depend linearly on the $V_{GS,off}$-value. Only if $V_{GS,off}$ is decreased below $-4\,V$ the I_D-variation becomes significant. Furthermore, for $V_{GS,off} < -10\,V$ the effect saturates, as also reported by [3].

B. V_{GS} during inductive switching

During conventional switching operation, other than in saturation, the current through the device is determined by the load. However, the influence of the traps can still be observed via the gate-source voltage. Fig. 3 shows a comparison of V_{GS} during turn-on of a SiC- and a Si-MOSFET.

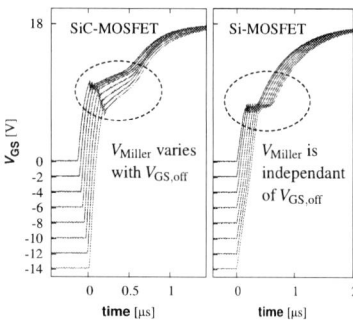

Fig. 3. Comparison of the gate-source voltage during turn-on of a SiC- and a Si-MOSFET at $V_{DS} = 300\,V$ and $I_D = 15\,A$ and a relatively high R_G of $150\,\Omega$.

In case of the SiC-MOSFET the Miller-plateau shifts to lower V_{GS} with increasingly negative $V_{GS,off}$. This is also explained by the trapping phenomenon. The onset of the Miller-plateau V_{Miller} corresponds to a V_{GS}-value at which the conductivity of the device is just high enough to conduct the full load current. Since the traps temporarily increase the conductivity of the channel with increasingly negative $V_{GS,off}$, the Miller-voltage decreases accordingly. For comparison, in Si the Miller-plateau is independent of $V_{GS,off}$, since the conductivity of the device is solely determined by V_{GS}.

III. MEASUREMENT AND EXTRACTION OF ΔV_{th}

The trap-induced variation of the channel conductivity can be represented by a shift of the threshold voltage. In order to measure the temporal progression of ΔV_{th} for different values of $V_{GS,off}$ and $V_{GS,on}$ the setup given in Fig. 4(a) is designed. The device is operated in saturation like in Sec. II-A. However, using a regulator, the current through the device is held constant. As a result, the trap-induced change of the conductivity is compensated and translated into a change of V_{GS} which allows the calculation of ΔV_{th}.

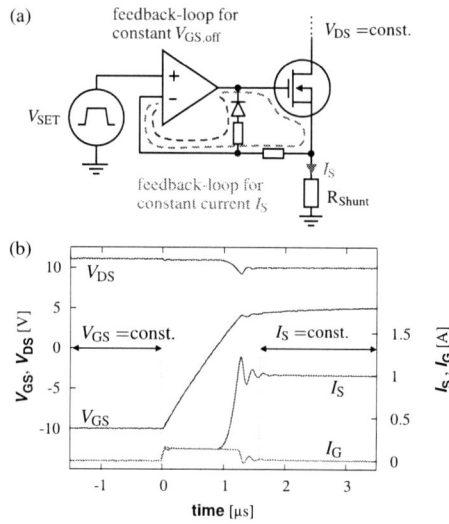

Fig. 4. (a) Circuit to extract the trap-induced threshold voltage shift ΔV_{th} by regulating to a constant source current I_S. (b) Measurement of the settling process for $V_{GS,off} = -10\,V$ and $I_S = 1\,A$ at $V_{DS} = 10\,V$. The time between the constant V_{GS}-phase and the constant I_S-phase is roughly $\Delta t = 1.5\,\mu s$.

In order to maintain, a constant off-state voltage during the biasing-phase and also a constant on-state current during the measurement phase, two feedback loops are required. Since V_{GS} has a different polarity in both phases, a diode can be used to decouple both loops. The neutralization process starts immediately at the end of the biasing-phase. Hence, the transition time Δt between the two phases must be minimized. Fig. 4(b) shows the settling process during an exemplary measurement with $V_{GS,off} = -10\,V$ and $I_S = 1\,A$.

978-1-7281-0582-6/19 $31.00 © 2019 IEEE

A. Extraction of $\Delta V_{\mathrm{th}}(t)$

Performing measurements with the same current of $I_{\mathrm{S}} = 1\,\mathrm{A}$ for various values of $V_{\mathrm{GS,off}}$ yields the array of curves in Fig. 5.

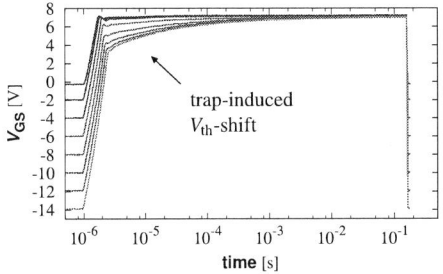

Fig. 5. Gate-source voltages at constant current $I_{\mathrm{S}} = 1\,\mathrm{A}$ for $V_{\mathrm{GS,off}} = -14\,\mathrm{V}..-0.2\,\mathrm{V}$. The trap-induced conductivity increase (in dependence of $V_{\mathrm{GS,off}}$) is now compensated by the regulator and thus translated into a change of V_{GS}.

For increasingly negative values of $V_{\mathrm{GS,off}}$ the increased conductivity is compensated by a lower V_{GS}. As observed before in Fig. 2 for the drain current, the variation in V_{GS} is also low for $-4\,\mathrm{V} < V_{\mathrm{GS,off}}$ and $V_{\mathrm{GS,off}} < -10\,\mathrm{V}$. Especially for $V_{\mathrm{GS,off}} = -0.2\,\mathrm{V}$, the transient behavior is insignificant. Thus, by using the $V_{\mathrm{GS}}(t)$ at $V_{\mathrm{GS,off}} = -0.2\,\mathrm{V}$ as a reference, the threshold voltage shift $\Delta V_{\mathrm{th}}(t)$ is calculated as the difference:

$$\Delta V_{\mathrm{th}}(t)\big|_{V_{\mathrm{GS,off}}} = V_{\mathrm{GS}}(t)\big|_{V_{\mathrm{GS,off}}} - V_{\mathrm{GS}}(t)\big|_{V_{\mathrm{GS,off}}=-0.2\,\mathrm{V}} \quad (1)$$

By performing this calculation for all values of $V_{\mathrm{GS,off}}$, the $\Delta V_{\mathrm{th}}(t)$ curves in Fig. 6 are obtained. The turn-on and turn-off are blanked out.

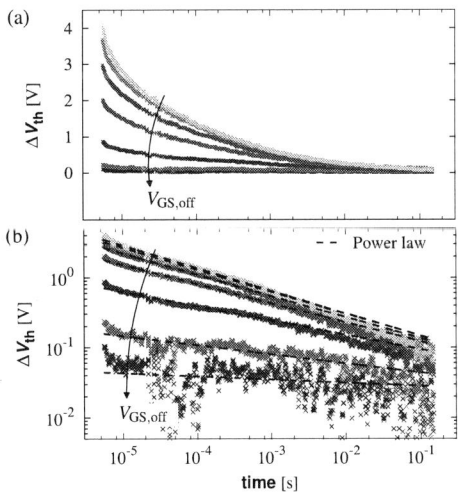

Fig. 6. Temporal progression of the threshold voltage shift for $V_{\mathrm{GS,off}} = -14\,\mathrm{V}..-2\,\mathrm{V}$ at $I_{\mathrm{D}} = 100\,\mathrm{mA}$ and $V_{\mathrm{DS}} = 10\,\mathrm{V}$ on a semi-logarithmic plot in (a) and on a double-logarithmic plot in (b).

In the time-logarithmic plot in Fig. 6(a), ΔV_{th} diminishes exponentially. This is obvious by the linear progression in the double-logarithmic plot in Fig. 6(b). In dependence of $V_{\mathrm{GS,off}}$, a parallel shift of ΔV_{th} and a change of the slope is observed. Consequently, the threshold voltage shift ΔV_{th} can be described by a power law $\Delta V_{\mathrm{th}}(t) = B \cdot t^m$. Although this is non-physical, it is sufficient for a rough approximation. Fig. 7 shows the coefficients of the power law approximation in Fig. 6(b).

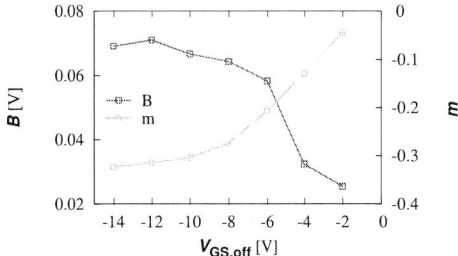

Fig. 7. $V_{\mathrm{GS,off}}$- dependence of the fitting coefficients B and m of a power law approximation $\Delta V_{\mathrm{th}}(t) = B \cdot t^m$.

As expected, only a small variation is observed for $V_{\mathrm{GS,off}} < -10\,\mathrm{V}$ as the saturation of the trapping effect begins. However, ΔV_{th} does not only depend on $V_{\mathrm{GS,off}}$ but also on $V_{\mathrm{GS,on}}$ which corresponds to I_{D}.

B. Drain current dependence of ΔV_{th}

An advantage of this technique is the ability to measure ΔV_{th} with different values of I_{D}. Other approaches often use a single operating point to determine ΔV_{th} with a relatively low measurement current [4]. In Fig. 8, ΔV_{th} is measured for $I_{\mathrm{D}} = \{100\,\mathrm{mA}, 1\,\mathrm{A}, 10\,\mathrm{A}\}$ at $V_{\mathrm{GS,off}} = -14\,\mathrm{V}$.

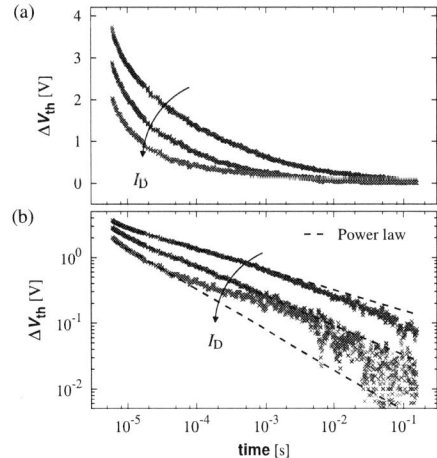

Fig. 8. Temporal progression of the threshold voltage shift between $V_{\mathrm{GS,off}} = -14\,\mathrm{V}$ and $V_{\mathrm{GS,off}} = -0.2\,\mathrm{V}$ for $I_{\mathrm{D}} = \{100\,\mathrm{mA}, 1\,\mathrm{A}, 10\,\mathrm{A}\}$.

With increasing current, the initial value of ΔV_{th} decreases, while at the same time, the decay rate increases, see Fig. 8(a). This translates to a parallel shift and a change in the slope of the traces in the double-logarithmic plot in Fig. 8(b).

The reduction of ΔV_{th} with I_D can be explained by the ratio between the trap- and the inversion charge. With higher I_D and thus higher V_{GS}, the inversion charge increases while the trapped charge remains unchanged. Hence, the impact of the traps reduces as $V_{GS,on}$ increases. The change of the slope corresponds to a faster neutralization of the trapped charge which is a result of the higher electron concentration at the interface. This explains why the influence of the trapping is very small at very high $V_{GS,on}$, e.g. during a short-circuit [2].

IV. POSITIVE BIASING

To better understand the influence of $V_{GS,on}$ on the neutralization of the trapped charge, transient drain current measurements are performed with both negative and positive bias. To suppress the flow of current during the positive-bias phase, an additional high-side switch is utilized. The setup and a sketch of the gate-signals is shown in Fig. 9.

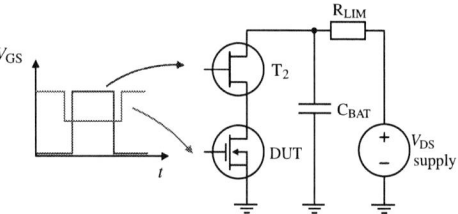

Fig. 9. Modified setup used for suppressing current-flow with positive gate bias.

To ensure a fast turn-on of the high-side, a GaN-HEMT is used for the switch T_2. A symmetrical bias from $-16\,$V to $16\,$V is applied to the DUT. Then, V_{GS} is switched to the on-state voltage of $V_{GS,on} = 6\,$V and, after a short delay Δt of typically $1.5\,\mu$s, the high-side switch is turned on. The gate-source and drain-source voltages at the DUT are given in Fig. 10 while the resulting drain-currents are shown in Fig. 11.

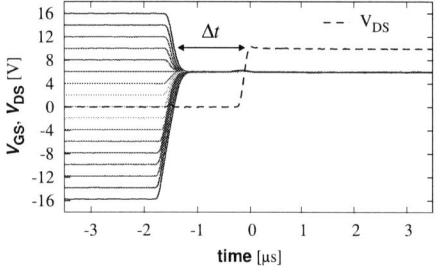

Fig. 10. Gate-source and drain-source voltages of the DUT for positive and negative gate-bias, resulting from the setup given in Fig. 9.

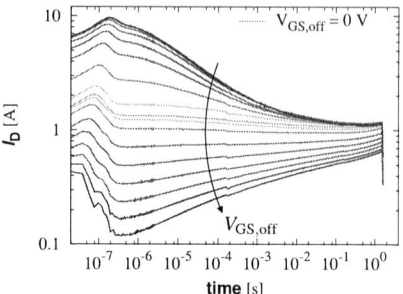

Fig. 11. Transient drain currents in saturation corresponding to the bias conditions given in Fig. 10.

In Fig. 11, a change in I_D is observed for both polarities of the previous bias, after negative gate-biasing I_D increases. Inversely, I_D decreases after positive gate-biasing. Furthermore, the time until convergence differs significantly. For negative $V_{GS,off}$, I_D converges after $\approx 100\,$ms while for positive $V_{GS,off}$ no convergence is observed even after $1.6\,$s.

The reduction of I_D after positive bias can explained by electron trapping. This also concurs with the longer time required for the neutralization. To neutralize the trapped electrons, holes are required which are only available in low concentration in inversion. These results show that the trapping-induced V_{th}-hysteresis can not be explained solely by a neutralization of trapped charge, but must also account for the trapping of inversely polarized charge carriers.

V. CONCLUSIONS

The charge trapping that occurs at the SiC/SiO$_2$ interface affects the conductivity of the channel. Using negative $V_{GS,off}$, this leads to a temporary increase of the pulsed drain current in saturation and to a shift of the Miller-plateau in switching operation. The threshold voltage shift associated with the traps is measured using a constant current approach. An increase with $V_{GS,off}$ and a decrease with I_D is observed. A power law can be used to describe the progression of ΔV_{th} over time. Furthermore, it is shown that positive biasing leads to electron trapping, that causes a decrease of the channel conductivity.

REFERENCES

[1] A. J. Lelis, R. Green, D. B. Habersat, and M. El, "Basic mechanisms of threshold-voltage instability and implications for reliability testing of sic mosfets," *IEEE Transactions on Electron Devices*, vol. 62, no. 2, pp. 316–323, 2015.

[2] C. Unger, and M. Pfost, "Influence of the off-state gate-source voltage on the transient drain current response of SiC MOSFETs," in *Proc. ISPSD*, 2018.

[3] G. Rescher, G. Pobegen, T. Aichinger, and T. Grasser, "On the sub-threshold drain current sweep hysteresis of 4H-SiC nMOSFETs," in *Proc. IEDM*, 2016, pp. 10.8.1–10.8.4.

[4] K. Puschkarsky, T. Grasser, T. Aichinger, W. Gustin, and H. Reisinger, "Understanding and modeling transient threshold voltage instabilities in SiC MOSFETs," in *Proc. IRPS*, 2018, 3B.5–1–3B.5–10.

Proceedings of the 31st International Symposium on Power Semiconductor Devices & ICs
May 19-23, 2019, Shanghai, China

Investigations of p-Shielded SiC Trench IGBT with Considerations on IE Effect, Oxide Protection and Dynamic Degradation

Jin Wei[1], Meng Zhang[2], Huaping Jiang[3], Baikui Li[2], Zheyang Zheng[1], and Kevin J. Chen[1]

[1] Department of Electronic and Computer Engineering, The Hong Kong University of Science and Technology, Hong Kong
[2] Key Laboratory of Optoelectronic Devices and Systems of Ministry of Education and Guangdong Province, College of Physics and Optoelectronic Engineering, Shenzhen University, Shenzhen, China
[3] School of Electrical Engineering, Chongqing University, Chongqing, China
Phone: +852-23588969, Fax: +852-23581485, Email: jweiaf@connect.ust.hk; eekjchen@ust.hk

Abstract—This work investigates the design of gate structure for the SiC trench IGBTs. The conventional trench IGBT structure established in silicon is not suitable for SiC IGBT, since it suffers from high oxide field. The grounded p-shield concept developed for SiC trench MOSFET cannot be readily transferred to SiC trench IGBT, since it counteracts the injection enhancement (IE) effect and nullifies the purpose of a trench gate structure. The floating p-shield structure has been found to cause severe dynamic degradation in SiC trench MOSFETs due to the charge storage effect. However, this dynamic degradation is effectively suppressed in the SiC trench IGBT, since the holes injected from the collector neutralize the stored negative charges in the floating p-shield. In the meanwhile, the floating p-shield is found to effectively suppress the high oxide field while maintains the IE effect. Thus, the floating p-shield solves the key issues and proves a promising solution for SiC trench IGBTs.

Keywords—SiC trench IGBT; floating p-shield; grounded p-shield; gate oxide field; IE effect; charge storage

I. INTRODUCTION

SiC power devices have attracted great interest because of its fundamentally superior material properties [1-4]. During the past two decades, power MOSFETs have been the focus of research in SiC, which covers a voltage range from several hundred volts to ~10 kV. As the SiC MOS channel suffers from a low electron mobility [5, 6], the trench MOSFETs have been widely investigated to reduce the channel resistance [7-9]. However, as the critical breakdown field of SiC is much higher than silicon, the internal oxide field of the SiC trench MOSFET is elevated accordingly, threatening the device reliability. Grounded p-shields have been routinely adopted for the design of SiC trench MOSFET, which provides effective protection for the gate oxide [10-15].

For very high voltage applications (> 10 kV), SiC IGBT is preferred over SiC MOSFET because of the conductivity modulation [16, 17]. In the state-of-the-art silicon IGBTs, trench gate structure is commonly used to fully exploit the conductivity modulation, owing to the injection enhancement

This project is supported by Hong Kong Innovation and Technology Fund under ITS/234/16.

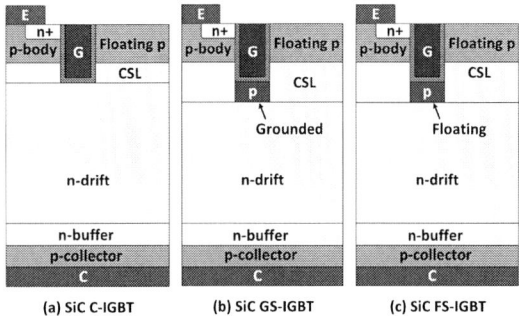

Fig. 1. Schematic structures of (a) conventional SiC IGBT (C-IGBT), (b) SiC grounded shield IGBT (GS-IGBT) and (c) SiC floating shield IGBT (FS-IGBT).

(IE) effect [18, 19]. However, if the conventional trench gate IGBT structure is directly implemented in SiC, the gate oxide of device would be vulnerable as it is subjected to high electric field.

The grounded p-shield structure established for SiC trench MOSFET has been proved effective in reducing the oxide field. However, when adopted in SiC trench IGBTs, the grounded p-shield could serve as a hole extraction path that drains the minority carriers at the top side of the device, resulting in compromised conduction performance [20, 21]. Thus, the grounded p-shield structure could nullify the very purpose of the trench gate structure. The floating p-shield is easy to implement in SiC trench gate devices, and it sacrifices no chip area for contact vias to the buried p-shield. However, the floating p-shield has been found to cause severe dynamic degradation in SiC trench MOSFET, due to the charge storage effect [22-24].

Because of the aforementioned issues, the implementation of SiC trench IGBTs is not established. In this paper, we investigate the SiC trench IGBT structures with numerical TCAD simulations, focusing on the aforementioned issues on IE effect, oxide protection and dynamic degradation. A floating p-shield is found to effectively suppress the high oxide field

978-1-7281-0582-6/19 $31.00 © 2019 IEEE

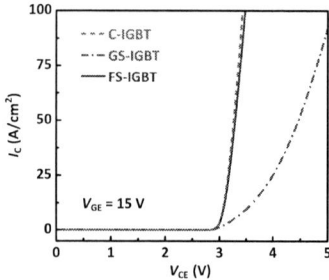

Fig. 2. *I-V* characteristics of the SiC IGBTs. The C-IGBT and the FS-IGBT feature a lower V_{ON} (at I_C = 50 A/cm²) than the GS-IGBT.

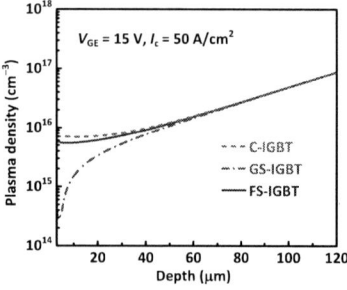

Fig. 3. Plasma distribution along the depth of the drift region. For GS-IGBT, the plasma at the top side is very low, which leads to a high V_{ON}.

while maintains a desirable IE effect. Unlike in a SiC MOSFET, a floating p-shield does not affect the dynamic V_{ON} in the SiC IGBT, since the holes injected from the collector effectively neutralize the stored negative charges in the p-shield. Therefore, a floating p-shield is found to be a promising solution for SiC trench IGBTs.

II. CHARACTERISTICS OF THE SiC IGBTs

Fig. 1 shows the conventional trench IGBT (C-IGBT), the grounded shield IGBT (GS-IGBT), and the floating shield IGBT (FS-IGBT). The doping concentration and thickness of the drift region are 2.5×10^{14} cm⁻³ and 100 μm, respectively. The doping concentration of the carrier storage layer (CSL) is 2×10^{16} cm⁻³. The doping concentration of the p-collector region is 1×10^{19} cm⁻³, unless otherwise specified.

Fig. 2 shows the *I-V* characteristics of the studied SiC IGBTs. All the IGBTs start to conduct when V_{CE} exceed ~3 V, marking the turn-on of the collector side PN junction. For the C-IGBT and FS-IGBT, the current rises sharply. V_{ON} (at I_C = 50 A/cm²) is ~3.3 V for both C-IGBT and FS-IGBT. V_{ON} of the GS-IGBT is ~4.5 V, which is much larger than the other two IGBTs. The large V_{ON} in the GS-IGBT is due to the weakened IE effect at the top of the structure. Fig. 3 presents the plasma distribution along the depth of the drift region in the studied IGBTs. It is clear that the plasma density at the top side of GS-IGBT is very low due to the extraction of holes through the grounded p-shield. Thus, a large portion of the voltage drops at the top portion of the drift region, whose conductivity is much poorer than the bottom portion.

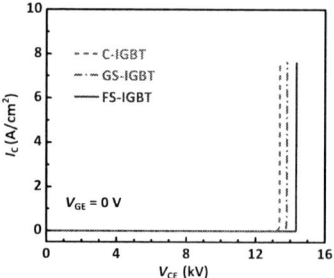

Fig. 4. *BV* characteristics of the studied SiC IGBTs. All devices present a *BV* larger than the designed 12 kV.

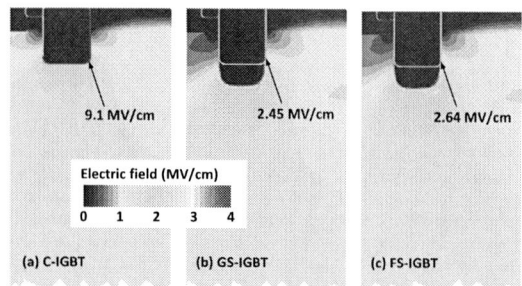

Fig. 5. Off-state electric field distribution in the SiC IGBTs with V_{CE} = 12 kV and V_{GE} = −5 V.

The breakdown characteristics of the studied IGBTs are plotted in Fig. 4. The difference in the gate structures of the studied IGBTs exerts limited influence on the breakdown voltages (*BV*s). All of the studied SiC IGBTs obtain a *BV* larger than 12 kV.

For the SiC trench devices, the oxide field is often the limiting factor for its reliability. Fig. 5 shows the off-state electrical field distribution inside the studied IGBTs with V_{GS} = −5 V and V_{DS} = 12 kV. Due to the lack of shielding structure, the C-IGBT suffers from an extremely large oxide field, since a large portion of the electrical field lines originating from the depleted drift region can penetrate through the gate oxide. The maximum oxide field (E_{ox-m}) in the C-IGBT is as high as 9.1 MV/cm, while a commonly accepted criterion for maximum oxide field in SiC MOS devices is 3~4 MV/cm [25, 26]. Owing to the protection of the grounded p-shield, the GS-IGBT boasts a much lower E_{ox-m} of 2.45 MV/cm. The grounded p-shield and the n-drift region form a reversely biased pn junction. A negatively charged region is created in the p-shield, which terminates a large portion of the electrical field lines from the drift region, thus shielding the gate oxide. In the FS-IGBT, E_{ox-m} is also found to be reduced to a low value of 2.64 MV/cm. In the off-state, the potential of the floating p-shield is boosted up. However, the region between p-body and p-shield forms a parasitic p-channel MOSFET. When the voltage different between gate and p-shield reaches the threshold voltage, the parasitic p-MOSFET is turned on, which clamps the potential of the p-shield. The p-shield becomes virtually grounded. Thus, the floating p-shield can also

978-1-7281-0582-6/19 $31.00 © 2019 IEEE

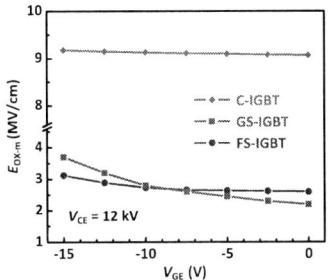

Fig. 6. Maximum oxide field ($E_{\text{ox-m}}$) of the studied SiC IGBTs at $V_{\text{CE}} = 12$ kV.

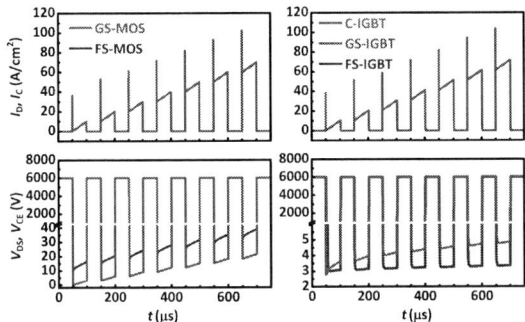

Fig. 7. Switching waveforms of the SiC MOSFETs and IGBTs.

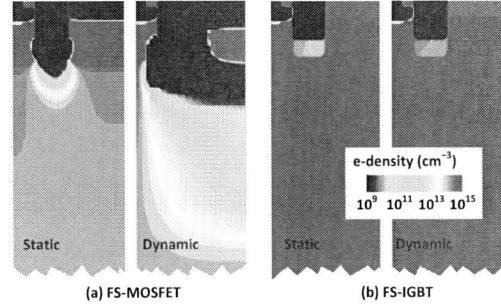

Fig. 8. The electron distribution at $V_{\text{GS}}(V_{\text{GE}}) = 15$ V and $I_{\text{D}}(I_{\text{C}}) = 50$ A/cm^2. The static distribution is obtained without high drain/collector voltage stress. The dynamic distribution is obtained during switching operation after having experienced high drain/collector voltage stress.

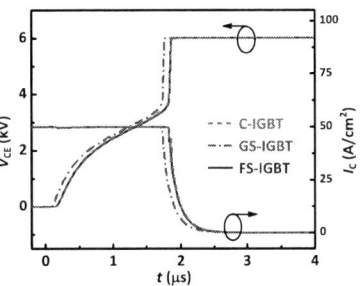

Fig. 9. Turn-off switching waveforms of the studied SiC IGBTs for the same collector dose.

terminate the electrical field lines and protect the gate oxide above it.

In practical applications, a negative gate bias is often used to turn off the power transistors, so as to prevent the possible false turn-on phenomenon. Fig. 6 plots the maximum oxide field in the studied IGBTs when various negative gate biases are applied. The C-IGBT suffers from a nearly constant high oxide field in the whole range of V_{GE} (from -15 to 0 V). For the GS-IGBT and FS-IGBT, $E_{\text{ox-m}}$ is kept to below 4 MV/cm in the whole V_{GE} range. In the GS-IGBT, $E_{\text{ox-m}}$ increases with a more negative V_{GE}, and becomes close to 4 MV/cm with $V_{\text{GE}} = -15$ V. $E_{\text{ox-m}}$ in the FS-IGBT is much less dependent on V_{GE}, since the potential of the floating p-shield is adaptive to the gate voltage.

The switching performances of the IGBTs are studied using the double pulse tester circuit. The supply voltage of the circuit is 6 kV. The gate voltage is switched between $+15$ V and -5 V to set the IGBT to on- and off-state, respectively. Here, for comparison, the SiC grounded shield MOSFET (GS-MOSFET) and SiC floating shield MOSFET (FS-MOSFET) are also studied. The FS-MOSFET and GS-MOSFET have similar static on-resistances. Fig. 7 exhibits the switching waveforms of the MOSFETs and IGBTs. The FS-MOSFET is found to suffer from a severe degradation during the switching operation, as evidenced by the dramatically higher V_{DS} during on-state compared to the GS-MOSFET. The degradation of the dynamic V_{ON} in the FS-MOSFET is caused by the storage of negative charges in the floating p-shield. In the off-state, the

floating p-shield becomes virtually grounded by the parasitic p-MOSFET, as discussed previously. A negatively charged region is formed in the p-shield that blocks the high off-state voltage. When the high V_{DS} is removed, the potential of the floating p-shield becomes negative, and the discharging current for the floating p-shield is blocked by the reversely biased pn junction around the p-shield. Thus, the region around the floating p-shield is depleted, resulting in a narrower conduction path, as depicted in the simulated carrier contours in Fig. 8(a). For the IGBTs, their dynamic V_{ON} agree well with their static values. The FS-IGBT exhibits a low dynamic V_{ON} than GS-IGBT. As shown in Fig. 8(b), the carrier contour of the FS-IGBT keeps unaffected after a high V_{CE} stress. The suppression of the dynamic degradation in the FS-IGBT is attributed to the holes injected from the collector, which neutralizes the stored negative charges in the floating p-shield.

Fig. 9 shows the turn-off transient characteristics of the studied SiC IGBTs. With the same collector doping, the three devices exhibit similar turn-off characteristics. The shallower plasma density at the top side of GS-IGBT slightly accelerates the extraction of minority carriers at the initial stage of the turn-off transient. However, during this phase of the turn-off transient, the voltage of the device is low, so the power loss during this phase is a limited portion of the total turn-off loss (E_{OFF}). Thus, E_{OFF} of the GS-IGBT is similar to the other two IGBTs. The E_{OFF}-V_{ON} trade-off curves of the studied IGBTs are plotted in Fig. 10. The C-IGBT and the FS-IGBT feature much

978-1-7281-0582-6/19 $31.00 © 2019 IEEE

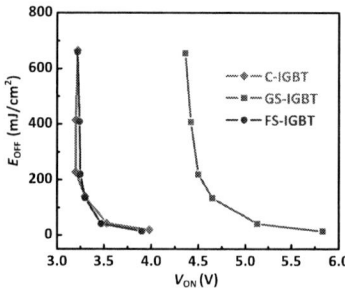

Fig. 10. Trade-off curves for V_{ON} and E_{OFF} of the studied SiC IGBTs. The trade-off is tuned by changing the collector doping dose.

better E_{OFF}-V_{ON} trade-off than the GS-IGBT owing to the strong IE effect.

III. CONCLUSION

The gate structure of SiC trench IGBTs is investigated using numerical TCAD simulations. The conventional trench gate structure established for silicon IGBT is not suitable for SiC IGBT, since it leads to high oxide field. The grounded p-shield concept developed for SiC trench MOSFET cannot be readily transferred to SiC trench IGBT, because it weakens the conductivity modulation and results in higher V_{ON}. A floating p-shield is found to effectively suppress the high oxide field while maintains high conductivity modulation. Unlike in a SiC MOSFET, a floating p-shield does not affect the dynamic V_{ON} in the SiC IGBT, since the holes injected from the collector effectively neutralize the stored negative charges in the p-shield. Furthermore, a floating p-shield is easier to be implemented than a grounded p-shield and is more area-effective. Therefore, a floating p-shield is a promising solution for SiC trench IGBTs.

REFERENCES

[1] B. J. Baliga, *Wide Bandgap Semiconductor Power Devices: Materials, Physics, Design, and Applications*. Elsevier, 2019.

[2] A. Salemi, H. Elahipanah, K. Jacobs, C. Zetterling, and M. Ostling, "15 kV-class implantation-free 4H-SiC BJTs with record high current gain," *IEEE Electron Device Lett.*, vol. 39, no. 1, pp. 63-66, Jan. 2018.

[3] J. Wei, H. Jiang, O. Jiang, and K. J. Chen, "Proposal of a GaN/SiC hybrid field-effect transistor for power switching applications," *IEEE Trans. Electron Devices*, vol. 63, no. 6, pp. 2469-2473, Jun. 2016.

[4] K. Tone, J. H. Zhao, L. Fursin, P. Alexandrov, and M. Weiner, "4H-SiC normally-off vertical junction field-effect transistor with high current density," *IEEE Electron Device Lett.*, vol. 24, no. 7, pp. 463-465, 2003.

[5] G. Y. Chung, C. C. Tin, J. R. Williams, K. McDonald, R. K. Chanana, R. A. Weller, S. T. Pantelides, L. C. Feldman, O. W. Holland, M. K. Das, and J. W. Palmour, "Improved inversion channel mobility for 4H-SiC MOSFETs following high temperature anneals in nitric oxide," *IEEE Electron Device Lett.*, vol. 22, no. 4, pp. 176-178, Apr. 2001.

[6] L. Cheng, A. K. Agarwal, S. Dhar, S. Ryu, and J. W. Palmour, "Static performance of 20 A, 1200 V 4H-SiC power MOSFETs at temperatures of −187 °C to 300 °C," *J. Electron. Mater.*, vol. 41, no. 5, pp. 910-914, May 2012.

[7] A. K. Agarwal, J. B. Casady, L. B. Rowland, W. F. Valek, M. H. White, and C. D. Brandt, "1.1 kV 4H-SiC power UMOSFET's," *IEEE Electron Device Lett.*, vol. 18, no. 12, pp. 586-588, Dec. 1997.

[8] M. Zhang, J. Wei, H. Jiang, K. J. Chen, and C. Cheng, "A new SiC trench MOSFET structure with protruded p-base for low oxide field and enhanced switching performance," *IEEE Trans. Device and Mater. Reliab.*, pp. 2592-2598, Jun. 2017.

[9] H. Jiang, J. Wei, X. Dai, M. Ke, I. Deviny, and P. Mawby, "SiC trench MOSFET with shielded fin-shaped gate to reduce oxide field and switching loss," *IEEE Electron Device Lett.*, vol. 37, no. 10, pp. 1324-1327, Oct. 2016.

[10] J. Tan, J. A. Cooper and M. R. Melloch, "High-voltage accumulation-layer UMOSFET's in 4H-SiC," *IEEE Electron Device Lett.*, vol. 19, no. 12, pp. 487-489, Dec. 1998.

[11] T. Nakamura, Y. Nakano, M. Aketa, R. Nakamura, S. Mitani, H. Sakairi, and Y. Yokotsuji, "High performance SiC trench devices with ultra-low Ron," in *IEDM Tech. Dig.*, Washington, DC, USA, Dec. 2011, pp. 599-601.

[12] J. Wei, M. Zhang, H. Jiang, C. Cheng, and K. J. Chen, "Low ON-resistance SiC trench/planar MOSFET with reduced OFF-state oxide field and low gate charges," *IEEE Electron Device Lett.*, vol. 37, no. 11, pp. 1458-1461, Nov. 2016.

[13] D. Peters, R. Siemieniec, T. Aichinger, and T. Basler, "Performance and ruggedness of 1200V SiC - trench - MOSFET," in *Proc. ISPSD*, Sapporo, Japan, May 2017, pp. 239-242.

[14] S. Harada, Y. Kobayashi, K. Ariyoshi, T. Kojima, J. Senzaki, Y. Tanaka, and H. Okumura, "3.3kV-class 4H-SiC MeV-implanted UMOSFET with reduced gate oxide field," *IEEE Electron Device Lett.*, vol. 37, no. 3, pp. 314-316, Mar. 2016.

[15] M. Zhang, J. Wei, H. Jiang, K. J. Chen, and C. Cheng, "SiC trench MOSFET with self-biased p-shield for low R_{ON} and low OFF-state oxide field," *IET Power Electron.*, vol. 10, no. 10, pp. 1208-1213, Aug. 2017.

[16] Q. Zhang, H. R. Chang, M. Gomez, C. Bui, E. Hanna, J. A. Higgins, T. Isaacs-Smith, and J. R. Williams, "10kV trench gate IGBTs on 4H-SiC," in *Proc. ISPSD*, Santa Barbara, CA, USA, May 2005, pp. 303-306.

[17] X. Wang and J. A. Cooper, "High-voltage n-channel IGBTs on free-standing 4H-SiC epilayers," *IEEE Trans. Electron Devices*, vol. 57, no. 2, pp. 511-515, Feb. 2010.

[18] M. Kitagawa, I. Omura, S. Hasegawa, T. Inoue, and A. Nakagawa, "A 4500 V injection enhanced insulated gate bipolar transistor (IEGT) operating in a mode similar to a thyristor," in *IEDM Tech. Dig.*, Washington, DC, USA, Dec. 1993, pp. 679-682.

[19] M. Sumitomo, J. Asai, H. Sakane, K. Arakawa, Y. Higuchi, and M. Matsui, "Low loss IGBT with partially narrow mesa structure (PNM-IGBT)," in *Proc. ISPSD*, Bruges, Belgium, Jun. 2012, pp. 17-20.

[20] S. Linder, *Power Semiconductors*. Lausanne, Switzerland: EPFL Press, 2006.

[21] J. Wei, M. Zhang, H. Jiang, S. To, S. Kim, J. Kim, and K. J. Chen, "SiC trench IGBT with diode-clamped p-shield for oxide protection and enhanced conductivity modulation," in *Proc. ISPSD*, Chicago, IL, USA, May 2018, pp. 411-414.

[22] Y. Kagawa, N. Fujiwara, K. Sugawara, R. Tanaka, Y. Fukui, Y. Yamamoto, N. Miura, M. Imaizumi, S. Nakata, and S. Yamakawa, "4H-SiC trench MOSFET with bottom oxide protection," *Mat. Sci. Forum*, vol. 778-780, pp. 919-922, Feb. 2014.

[23] J. Wei, M. Zhang, H. Jiang, H. Wang, and K. J. Chen, "Dynamic degradation in SiC trench MOSFET with a floating p-shield revealed with numerical simulations," *IEEE Trans. Electron Devices*, vol. 64, no. 6, pp. 2592-2598, Jun. 2017.

[24] J. Wei, R. Xie, H. Xu, H. Wang, Y. Wang, M. Hua, K. Zhong, G. Tang, J. He, M. Zhang, and K. J. Chen, "Charge storage mechanism of drain induced dynamic threshold voltage shift in *p*-GaN gate HEMTs," *IEEE Electron Device Lett.*, doi: 10.1109/LED.2019.2900154. Early Access.

[25] Y. Sui, T. Tsuji and J. A. Cooper, "On-state characteristics of SiC power UMOSFETs on 115-μm drift layers," *IEEE Electron Device Lett.*, vol. 26, no. 4, pp. 255-257, Apr. 2005.

[26] D. Peters, T. Basler, B. Zippelius, T. Aichinger, W. Bergner, R. Esteve, D. Kueck, and R. Siemieniec, "The new CoolSiC™ trench MOSFET technology for low gate oxide stress and high performance," in *Proc. PCIM*, Nuremberg, Germany, May 2017, pp. 168-174.

Proceedings of the 31st International Symposium on Power Semiconductor Devices & ICs
May 19-23, 2019, Shanghai, China

Design and Experimental Study of 1.2kV 4H-SiC Merged PiN Schottky Diode

Jiupeng Wu, Na Ren*, Kuang Sheng

Electrical Engineering Department
Zhejiang University
Hangzhou, China
renna.zju@gmail.com

Abstract—In this paper, simulation, modeling and experimental studies of 1.2kV/2A 4H-SiC MPS diodes are conducted. First, design considerations for MPS cells and JBS cells are presented. As the pn junction turn-on voltage (V_{turn}) has a significant impact on the surge current capability of MPS diodes, a lower V_{turn} is desirable. Both simulation and modeling results show that there is a trade-off between the forward voltage drop at nominal current (V_F) and the V_{turn} for the MPS cell design. MPS diodes with different designs are fabricated and their experimental results are compared. It is demonstrated that a wider p+ region accompanied with a larger p+ spacing design can get a better trade-off between the V_F and V_{turn}. Moreover, the surge current capability can be improved by 17% when the p+ region width is increased from 8μm to 20μm.

Keywords—4H-SiC, MPS diode, JBS diode, surge capability

I. INTRODUCTION

Silicon Carbide (SiC) Merged PiN Schottky (MPS) diode has already found extensive applications in power systems as it possesses fast switching speed, low forward voltage drop, low reverse leakage current and high surge capability [1, 2]. MPS diodes are characterized by inserting of large p+ stripes into the active region of Junction Barrier Schottky (JBS) diodes [2]. The pn junctions formed by these p+ stripes and adjacent n- stripes will be turned on under large forward voltage bias, offering minority carrier injection to enhance the surge current capability of the diode.

The device layout (shape and arrangement of device cells) and structural parameters (sizes of p+ and n- stripes) have significant influences on the characteristics of MPS diodes [3, 4]. The forward voltage drop at nominal current (V_F) and the pn junction turn-on voltage (V_{turn}) are important specifications for the MPS diode, which determine the current capability in normal operation mode and surge mode, respectively. Low V_{turn} is desirable to achieve high surge current capability. However, there is a trade-off between the V_F and V_{turn} for the designs of the layout, Schottky/p+ ratio, p+ width and p+ spacing. In this paper, design considerations and device performance optimization for 1.2kV/2A SiC MPS diodes will be presented based on one-dimensional stripe layout. The surge current capability of the MPS diodes are evaluated and compared among the different designs.

II. DEVICE SIMULATION

A. JBS Cell Simulation

Static characteristics of the JBS and MPS devices are acquired by TCAD simulation software Silvaco. The simulation structure and definitions of parameters used in

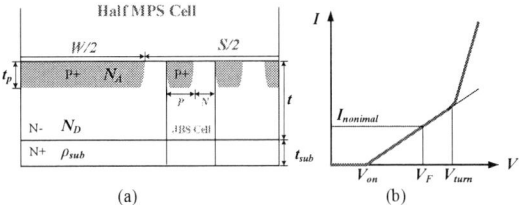

Fig. 1. (a) The schematic view of an MPS diode structure showing half MPS cell and embedded JBS cells, as well as (b) the definitions of forward voltage drop at nominal current V_F and pn junction turn-on voltage V_{turn}.

TABLE I.   STRUCTURAL PARAMETERS USED IN THE SIMULATION

Label	Descriptions	Values
A	Device active area	1 mm²
t	Drift layer thickness	12 μm
t_p	P+ depth	0.5 μm
t_{sub}	Substrate thickness	360 μm
N_D	Drift layer doping concentration	8×10^{15} cm⁻³
N_A	P+ doping concentration	1×10^{18} cm⁻³
ρ_{sub}	Resistivity of the substrate	0.02 Ωcm

simulation are shown in Fig. 1(a) and Fig. 1(b), respectively. The structural parameters used in the simulation are shown in Table I. Four structural parameters are studied, i.e., MPS cell p+ width (W) and p+ spacing (S), narrow p+ width (P) and Schottky width (N).

Each half MPS cell contains half of one wide p+ stripe and one wide n- stripe, while several narrow p+ stripes are inserted into this half wide n- stripe to form embedded JBS cells. The definitions of V_F and V_{turn} are also shown in Fig. 1(b). V_F is the voltage drop at nominal current, and the second knee point of the I-V curve is the PN junction trigger point (V_{turn}).

Forward and reverse characteristics of the JBS cells are simulated and presented in Fig. 2(a) and Fig. 2(b), respectively. It is shown in Fig. 2(a) that the forward voltage drop is decreased with an increasing Schottky width (N). However, the reverse leakage current in Fig. 2(b) is also increased due to a weakened electric field shielding effect. Therefore, there is a trade-off between the forward voltage drop and reverse leakage current for the N design, which is shown in Fig. 2(c). An optimal design of $N=3$μm and $P=3$μm can be concluded for the JBS cell from the figure.

978-1-7281-0582-6/19 $31.00 © 2019 IEEE   203

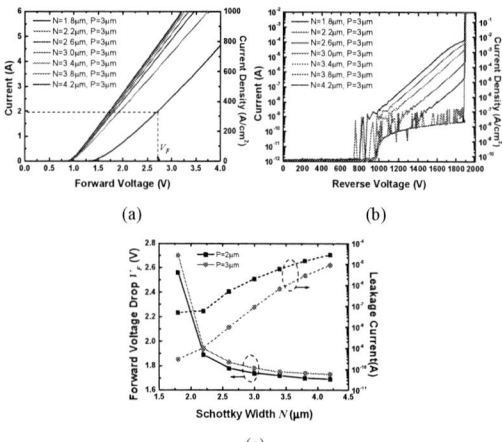

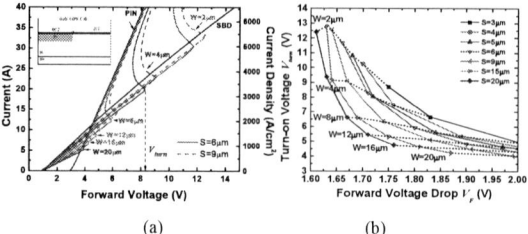

Fig. 3. (a) Simulation results of forward I-V characteristics of MPS diodes with varied W and S designs. (b) The trade-off between forward voltage drop (V_F) at 2A and pn junction turn-on voltage (V_{turn}).

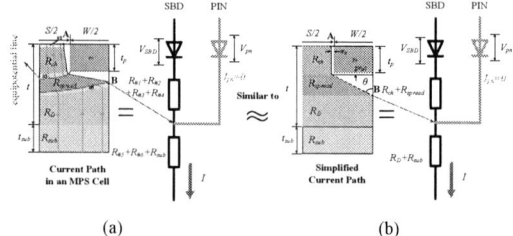

Fig. 4. An analytic model of the MPS cell for calculating the pn junction turn-on voltage.

Fig. 2. Simulation results of (a) forward and (b) reverse I-V characteristics of the JBS cell in MPS diodes with varied Schottky width N designs (narrow p+ width P in JBS cell is fixed at 2µm and 3µm). (c) The trade-off between forward voltage drop and leakage current at reverse voltage bias of 1600V versus Schottky width N.

B. MPS Cell Simulation

Simulation results of forward I-V characteristics of the MPS cell without inserting JBS cells are shown in Fig. 3(a), in which the forward voltage drop V_F is increased up to 15V to include the bipolar mode regime. Two design groups of p+ spacing (S=6µm, S=9µm) are compared in Fig. 3(a). It is shown that the smaller S design will lead to a lower V_F but a higher V_{turn}. At each S design, the p+ width (W) is varied from 2µm to 20µm. With an increasing W design, the V_F is decreased while V_{turn} is increased.

The V_F at 2A and V_{turn} of the MPS diodes with different W and S designs are extracted from Fig. 3(a) and summarized in Fig. 3(b). With a wider p+ stripe design (larger W), the pn junction can be turned on at a lower voltage (V_{turn}), which is desirable for the bipolar mode. However, it has adversely effects on the forward voltage drop V_F in unipolar mode, namely, V_F will be increased accordingly. If the wider W design is accompanied with a larger S design, V_F can be lowered. An optimal design window of W (12–16µm) and S (>15µm) can be concluded from Fig. 3(b).

III. MPS CELL MODELING

In order to reveal the correlation between V_F/V_{turn} and device parameters, an analytical model is established. Schematic diagrams of the model are shown in Fig. 4 [5, 6]. The current path of the MPS diode in unipolar mode and its equivalent circuit diagram are shown in Fig. 4(a). This current path can be divided into different parts by equipotential line, namely channel part (R_{ch}), spreading part (R_{spread}), drift part (R_D), substrate part (R_{sub}), and three transition parts ($R_{\#1}$, $R_{\#3}$, $R_{\#5}$). The current flows from the anode to the cathode and a potential difference between point A and B (V_{AB}) will be established up. The p+ region is treated as an equipotential body, in which way V_{AB} can be regarded as the bias voltage of the pn junction. As long as V_{AB} is larger than V_{turn}, the pn junction turns on. The MPS cell can be modeled as a parallel connection of a Schottky diode and a PiN diode.

Since the shape and position of the equipotential lines are difficult to decide, simplification is used in this study, as shown in Fig. 4(b). Those transition parts ($R_{\#1}$, $R_{\#3}$, $R_{\#5}$) could be ignored and the rest parts can be simply divided by horizontal lines [5, 6]. Thus, the device resistance can be modeled with the series connection of channel resistance (R_{ch}), spreading resistance (R_{spread}), drift layer resistance (R_D) and the substrate resistance (R_{sub}). A spreading angle of θ is assumed for the calculation of R_{spread}. The resistance can be calculated with equations as below,

$$R_{ch} = \rho_{epi} \frac{2t_p + W_D}{z \cdot (S - 2W_D)} \tag{1}$$

$$R_{spread} = \frac{\rho_{epi}}{z \cdot \tan\theta} \ln\left(\frac{S+W}{S - 2W_D}\right) \tag{2}$$

$$R_D = \rho_{epi} \frac{(2t - 2t_p - W_D)\tan\theta - (W + 2W_D)}{z \cdot (S+W)\tan\theta} \tag{3}$$

$$R_{sub} = \rho_{sub} \frac{2t_{sub}}{z \cdot (S+W)} \tag{4}$$

$$R_{total} = R_{ch} + R_{spread} + R_D + R_{sub} \tag{5}$$

where t, t_p, t_{sub}, W_D, ρ_{epi}, ρ_{sub}, z are the thickness of the drift layer, the p+ depth, the thickness of the substrate, the depletion width of the pn junction, the resistivity of the drift and the substrate layer and the width in the third dimension of the MPS cell, respectively. V_F and V_{turn} can be calculated with equations (6) and (7), respectively [6], as below,

$$V_F = I \cdot R_{total} + V_{SBD} \tag{6}$$

$$V_{turn} = \frac{V_{pn} - V_{SBD}}{R_{ch} + R_{spread}} \cdot R_{total} + V_{SBD} \tag{7}$$

978-1-7281-0582-6/19 \$31.00 © 2019 IEEE

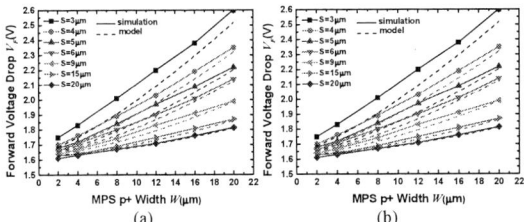

Fig. 5. Comparison of simulation results and modeling results of (a) V_F and (b) V_{turn} for MPS diodes with different W and S designs.

where V_{pn} and V_{SBD} are the voltage drops across the pn and Schottky junction, respectively.

The calculation results of V_F and V_{turn} are plotted in Fig. 5(a) and (b), respectively, where they are compared with the simulation results. The modeling results is deviated from the simulation results at extremely small S design as the 2-dimensional effect produced by the p+ region is pronounced in this condition. For the larger S designs, the modeling results are consistent with the simulation results. Therefore, the simplified model in Fig. 4(b) works well for the prediction of V_F and V_{turn} for the MPS diodes.

IV. DEVICE FABRICATION AND EXPERIMENTAL RESULTS

A. Device Fabrication

Devices are fabricated on 4°-off axis SiC epitaxial wafers purchased from CREE (n-type drift layer doping is $8 \times 10^{15} \text{cm}^{-3}$, drift thickness is 12μm). The p+ stripes in the active region and the floating guarding ring termination are formed by Al ion implantation with a total dose of $2 \times 10^{15} \text{cm}^{-2}$ and maximum energy of 360keV at elevated temperature. Implanted ions are activated at 1550°C in Ar ambient under the protection of carbon. SiO_2/SiN_x layer is used as the passivation layer. Metal layers consisting of Ti/Ni/Pt are used for p-type Ohmic contact and Ti for Schottky contact, while metal layer of Ni is used for back n-type Ohmic contact. MPS diodes with different parameters are fabricated. An image of a fabricated device is shown in Fig. 6, in which JBS cells and the W/S of a MPS cell are marked. The W has five design splits (W=8μm, 12μm, 14μm, 16μm, 20μm) and S has three designs (S=9μm, 15μm, 21μm). Pure PiN diodes and JBS diodes are also fabricated for the references.

Fig. 6. An image of a fabricated device.

B. Static Characterization

Static forward and reverse characteristics of the fabricated JBS, PiN and Schottky diodes are measured and shown in Fig. 7(a) and Fig. 7(b), respectively. In low current regime (<20A), the I-V characteristic of JBS diodes is dominated by the Schottky junction (unipolar mode). On the other hand, when the current is higher than 20A, the bipolar current through the pn junction starts to take a part. In this

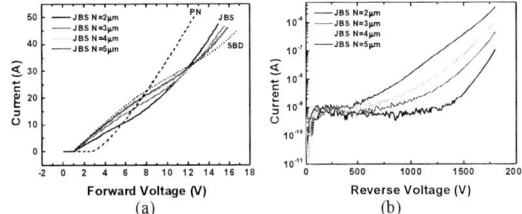

Fig. 7. Experimental results of 2A JBS diodes with different Schottky width N designs. (a) Forward I-V characteristics showing both Schottky conduction regime and bipolar conduction regime, (b) Reverse I-V characteristics showing over 1800V breakdown voltage.

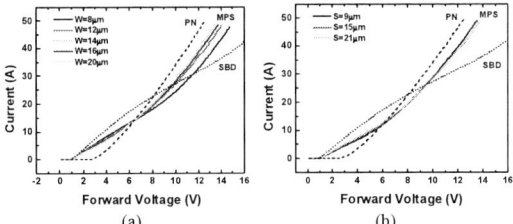

Fig. 8. (a) Experimental results of 2A MPS diodes with different W design, (b) Experimental results of MPS diodes with different S design.

regime, the slope of the I-V curve is increased due to the minority carrier injection.

The forward characteristics of the fabricated MPS diodes with varied W designs and S designs are shown in Fig. 8(a) and Fig. 8(b), respectively. When compared to the JBS diodes (in Fig. 7(a)), the MPS diodes shows lower V_{turn}. Thus, the bipolar conduction capability is enhanced in the MPS diodes. The MPS diodes with larger W have lower forward voltage drop V_F in unipolar mode and higher pn junction turn-on voltage V_{turn}.

On the other hand, for the different S design in Fig. 8(b), the forward I-V curves of the MPS diodes show minimal difference. This is because that JBS cells with fixed Schottky width design are inserted into between the large p+ regions. The channel resistance is unchanged which is predicted by the model in section III.

C. Surge Current Test

After the static I-V characterization of MPS diodes and JBS diodes, the surge current tests are performed. The purpose of the surge current test is to analyze the reliability of SiC MPS diodes under high current stress conditions. A sinusoidal forward current pulse with a pulse width of 10ms is used in this surge test. During the test, peak surge current (I_{surge}) is raised gradually until the device fails. The current and voltage waveforms are measured. The results of the device with W=14μm, S=21μm, P=3μm, N=3μm design are shown in Fig. 9(a). The surge current is increase up to 70A from test 1 to test 5. For the last test (test 5), both the current and voltage waveforms show a collapse at t=5.8ms. This represents the occurrence of device failure. The I-V trajectories for each test is plotted in Fig. 9(b). At a <20A current stress, the device exhibits pure Schottky conduction behavior as the current is almost linearly increased with the voltage. At a current between 20A and 60A, the pn junction is turned on and bipolar effect takes place as the slope of the I-V trajectory is drastically increased. After 60A, the slope of the I-V trajectory is decreased to a very small value.

978-1-7281-0582-6/19 $31.00 © 2019 IEEE

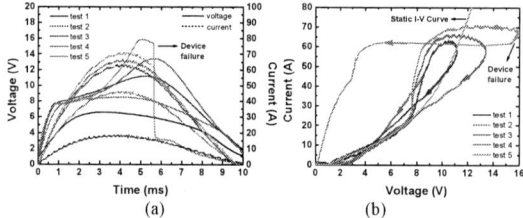

(a)   (b)

Fig. 9. Surge test results of device with design W=14μm, S=21μm, P=3μm, N=3μm. (a) V-t and I-t curves. (b) I-V trajectories. The direction of evolution has been marked by the arrows for each curve.

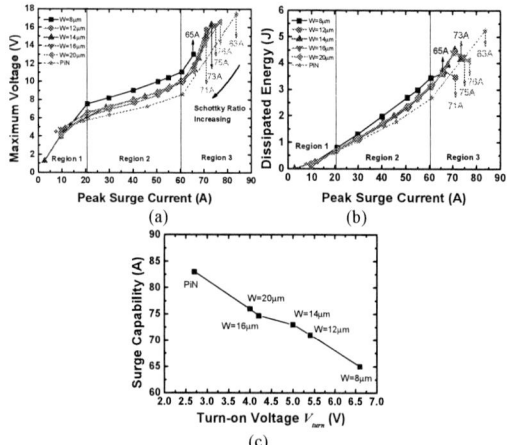

(a)   (b)

(c)

Fig. 10. (a) Maximum voltage and (b) dissipated energy versus peak surge current, as well as (c) maximum surge current capability versus pn junction turn-on voltage V_{turn} for MPS diodes with different W designs.

In this case, the device voltage is increase rapidly with the current. Furthermore, from test 3 to test 5, the maximum voltage that the device reaches is increased by 60% from 10V to 16V. From the surge test results in Fig. 9, the maximum voltage (V_{max}) each device experienced in each test is extracted and summarized in Fig. 10(a). With a larger W design, the V_{max} at a certain surge current can be reduced due to a lower V_{turn}.

Moreover, The V_{max} versus I_{surge} can be divided into three segments (region 1, region 2 and region 3) corresponding to different device electrical behaviors.

When the surge current is lower than 20A, the devices behaves like pure Schottky diodes and resistive voltage is dominated in this current range, therefore there is minor difference of V_{max} among the different MPS diode designs.

When the surge current is between 20A and 60A, the device electrical characteristic is dominated by the pn junction bipolar effect. Due to the lower V_{turn}, the MPS diode with a large W design shows lower V_{max}.

When the surge current is higher than 60A, the device resistance becomes dominant again in this high current regime. The device voltage increases rapidly with the current as the device resistance has positive temperature coefficient and the junction temperature is raised up by the high current/energy.

In addition, the total energy dissipated in the device for each test can be calculated from the current/voltage waveforms, as shown in Fig. 10(b). Accordingly, the plot of energy versus surge current can also be divided into three regions. The calculated energy is lower for the MPS diode with larger W designs due to the lower V_{turn} and V_{max}. Furthermore, the device failures are occurred in region 3 for all the tested devices due to the rapid increase of device voltage in this region.

The relations between maximum surge current capability and pn junction turn-on voltage V_{turn} of the diodes with different W designs are summarized in Fig. 10(c). The PiN diode fails at I_{surge}=83A, while MPS diodes fails at I_{surge}=65–76A. The MPS diode with a larger W design (a higher p+ ratio) shows higher surge current. The surge current capability can be improved by 17% when W is increased from 8μm to 20μm.

V. CONCLUSION

A simulation, analytical modeling and experiments of 1.2kV/2A 4H-SiC MPS diodes are conducted in this work. The influences of the structural parameters, including MPS cell p+ width W, p+ spacing S, narrow p+ width P and Schottky width N, on device performance are studied by simulation. It is demonstrated that there is a trade-off between the forward voltage drop (V_F) at nominal current and the pn junction turn-on voltage (V_{turn}). An analytical model of calculating the pn junction turn-on voltage is established. For a wider W and a lower S design, the V_F is increased whereas the V_{turn} is decreased due to the higher p+ ratio. The diodes are fabricated and characterized. The surge current test is also performed. The results show that the surge current capability of the MPS diode can be improved by 17% when the W design is increased from 8μm to 20μm, due to the 39% lower V_{turn}.

REFERENCES

[1] R. Singh, D. C. Capell, A. R. Hefner, J. Lai, and J. W. Palmour, "High-Power 4H-SiC JBS Rectifiers," IEEE Transactions on Electron Devices, vol. 49, no. 11, pp. 2054–2062, November 2002.

[2] R. Rupp, M. Treu, S. Voss, F. Bjork, and T. Reimann, "2nd Generation SiC Schottky diodes: A new Benchmark in SiC device ruggedness," Proceedings of the International Symposium on Power Semiconductor Devices and ICs (ISPSD), Naples, Italy, pp. 1–4, June 2006.

[3] Y. Huang, and G. Wachutka, "Comparative Study of Contact Topographies of 4.5kV SiC MPS Diodes for Optimizing the Forward Characteristics," International Conference on Simulation of Semiconductor Processes and Devices (SISPAD), Nuremberg, pp. 117-120, September 2016.

[4] V. d'Alessandro et al., "Influence of Layout Geometries on the Behavior of 4H-SiC 600V Merged PiN Schottky (MPS) Rectifiers," Proceedings of the International Symposium on Power Semiconductor Devices and ICs (ISPSD), Naples, Italy, pp. 1–4, June 2006.

[5] Lin Zhu, and T. Paul Chow, "Analytical Modeling of High-Voltage 4H-SiC Junction Barrier Schottky (JBS) Rectifiers," IEEE Transactions on Electron Devices, vol. 55, no. 8, pp. 1857–1863, August 2008.

[6] Na Ren, Jue Wang, and Kuang Sheng, "Design and Experimental Study of 4H-SiC Trenched Junction Barrier Schottky Diodes," IEEE Transactions on Electron Devices, vol. 61, no. 7, pp. 2459–2465, July 2014.

Proceedings of the 31st International Symposium on Power Semiconductor Devices & ICs
May 19-23, 2019, Shanghai, China

Characterization of BTI in SiC MOSFETs Using Third Quadrant Characteristics

Jose Ortiz Gonzalez
School of Engineering
University of Warwick
Coventry, United Kingdom
J.A.Ortiz-Gonzalez@warwick.ac.uk

Olayiwola Alatise
School of Engineering
University of Warwick
Coventry, United Kingdom
O.Alatise@warwick.ac.uk

Philip Mawby
School of Engineering
University of Warwick
Coventry, United Kingdom
P.A.Mawby@warwick.ac.uk

Abstract—Bias Temperature Instability (BTI) is a reliability concern for SiC MOSFETs which can have serious implications in the application if the true extent of the threshold voltage shift is underestimated. In this paper the third quadrant characteristics of SiC MOSFETs are used for characterizing the impact of accelerated gate stresses, evaluating the peak threshold voltage shift and tracking the recovery after stress removal. This method allows the evaluation of the impact of cumulative pulsed stresses of both long and short duration, which can be fundamental for characterizing the dynamics of BTI-induced threshold voltage shift in SiC MOSFETs under repetitive switching at the rated and accelerated gate voltage stresses.

Keywords—Bias Temperature Instability, SiC MOSFET

I. INTRODUCTION

Despite the improvements of the new generation SiC power MOSFETs, Bias Temperature Instability remains a reliability concern hence application engineers using SiC MOSFETs should take threshold voltage (V_{TH}) shift into consideration. BTI is highly relevant to SiC MOSFETs due to a high density of oxide and interface traps at the SiC/SiO$_2$ interface, as well as small band offsets due to the wider bandgap [1-4]. A peculiar characteristic of BTI in SiC is the recovery of V_{TH} after stress removal. This recovery can mislead device engineers causing them to underestimate the true extent of the V_{TH} shift after High Temperature Gate Bias (HTGB) stress. In high current applications where parallel connected SiC MOSFETs can be biased at negative gate voltages for long standby periods, loss of gate synchronization due to non-uniform V_{TH} shift can cause electrothermal destruction from poor current sharing. The lack of recovery time after V_{GS} bias in the application makes the standard reliability tests that allow recovery time unsuitable. It is therefore necessary to devise techniques for evaluating the V_{TH} shift and recovery in real-time. This paper evaluates how a novel method for characterizing V_{TH} shift caused by BTI presented by the authors in [5] can be used for evaluating cumulative stress pulses and capture phenomena that are not apparent during the conventional long stresses for BTI characterization.

II. BIAS TEMPERATURE INSTABILITY OVERVIEW AND EXPERIMENTAL SETUP

BTI in SiC MOSFETs is a topic of interest, given the recent number of publications [1-4]. Depending on the polarity of the stress, the shift of V_{TH} can be positive or

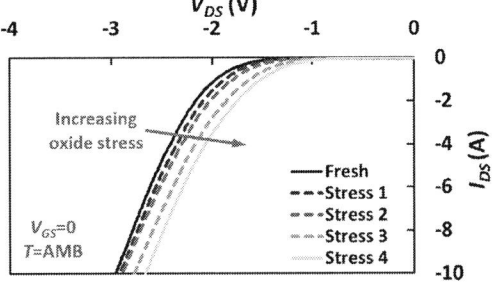

Fig. 1 Impact of NBTI on the 3rd quadrant characteristics of a SiC MOSFET. Cumulative negative HTGB stresses performed at 150 °C, as defined in [5]

negative, giving either Positive Bias Temperature Instability (PBTI) or Negative Bias Temperature Instability (NBTI). Fundamental for evaluating the impact of BTI in SiC MOSFETs is capturing the peak V_{TH} shift after the stress and the subsequent recovery once the stress is removed.

To that end, different methods have been proposed [1, 3], since the erroneous determination of the V_{TH} shift can have negative consequences on the qualification of the devices. Using the methodology presented by the authors in [5], the peak shift and recovery of V_{TH} after the stress removal can be detected. The methodology is based on 3rd quadrant characteristics of SiC MOSFETs. The threshold voltage shift caused by BTI affects the value of the source-drain voltage V_{SD} when $V_{GS} = 0$ V. This is caused by the partial conduction of current through the channel when $V_{GS}=0$ known as the body effect in SiC MOSFETs. Fig. 1 shows the measured 3rd quadrant characteristics of a SiC planar MOSFET at ambient temperature which was subjected to highly accelerated negative HTGB stresses, using a high gate voltages at a temperature of 150 °C as defined in [5]. The 3rd quadrant characteristics were measured at ambient temperature, after 16 hours relaxation at $V_{GS}=0$, so as to characterize only the

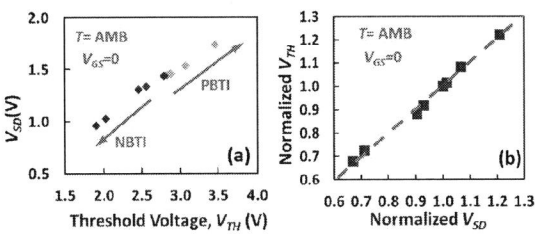

Fig. 2 (a) V_{SD} as a function of threshold voltage V_{TH} during both PBTI and NBTI, (b) Normalized V_{TH} as a function of the normalized V_{SD} (measured at I_{SD}=50 mA) (Stresses defined in [5])

This work was supported by the UK Engineering and Physical Science Research Council (EPSRC) through the grants EP/R004366/1 and EP/L021579/1.

978-1-7281-0582-6/19 $31.00 © 2019 IEEE   207

permanent V_{TH} shift caused by the stress.

Similar to the use of V_{SD} as Temperature Sensitive Electrical Parameter (TSEP), using both positive and negative stresses a relationship between V_{TH} and V_{SD} can been defined. This is shown for the evaluated planar SiC MOSFET in Fig. 2, for a defined temperature (ambient) and a low value sensing current I_{SENSE} of 50 mA (this is the current used to measure V_{SD} during 3rd quadrant operation). As V_{SD} is temperature dependent, it is important that I_{SENSE} does not cause the self-heating of the device hence a suitable cooling method to minimize the impact of temperature is required. The normalized values are used to define a relationship between V_{SD} and V_{TH}, given by (1) [5].

$$V_{TH,norm} = 1.02 \cdot V_{SD,norm} - 1.02 \qquad (1)$$

The test setup used here is similar to the test setup used for determining the junction temperature using V_{SD} as TSEP [6] and it is shown in Fig. 3. It consists of a gate driving circuit used for stressing the gate oxide with a defined V_{GS} stress and a current source which injects the sensing current I_{SENSE} flowing from source to drain. The stress timing signal is generated using a waveform generator model TDS2024C from Tektronix, the transient V_{SD} is measured using a differential probe model TA-043 from Pico Technology and captured using an oscilloscope model TDS5054B from Tektronix. The sensing current I_{SENSE} is measured using a digital multimeter Fluke 175. Depending on the gate voltage used, I_{SENSE} will flow: (a) through the channel only (positive V_{GS} which fully turns ON the device), (b) through the parasitic body diode (sufficient negative V_{GS}) or (c) there is a current divider between the body diode and the MOSFET channel, due to the body effect and depending on V_{TH} (V_{GS}= 0 V). Using the calibration curve obtained in Fig. 2, for a known temperature the threshold voltage shift can be detected by measuring V_{SD}.

Fig 4 shows the application of this technique. In this figure, the V_{GS} stress (17 V) is shown in Fig. 4(a), together with the measured V_{SD} in Fig. 4(b) and the calculated normalized V_{TH} in Fig. 4(c) (equation (1) has been used for $V_{TH,norm}$ calculation). Before the application of V_{GS} (for t < 4 s), the measured V_{SD} is 1.4 V (corresponding to nominal V_{TH}). During V_{GS} stress application (4 s < t < 14 s), V_{SD} falls to the ON-state resistance since the MOSFET is ON. After the V_{GS} stress is removed (t > 14 s), V_{SD} increases to a 6% higher value corresponding to increased V_{TH} from PBTI due to negative charge trapping. Hence, the technique here shows

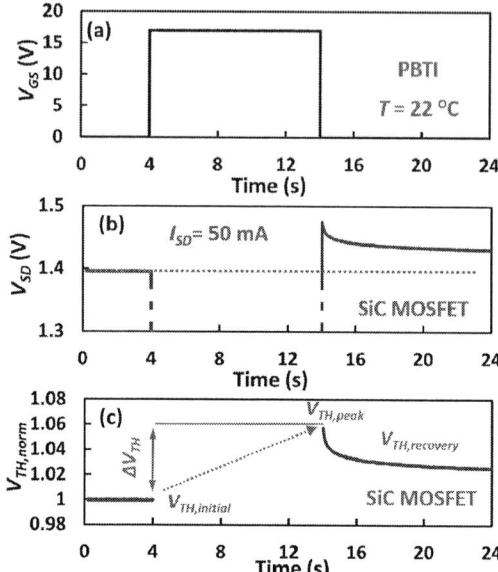

Fig. 4 Use of 3rd quadrant for BTI characterization: (a) Gate voltage stress, (b) 3rd quadrant V_{SD} voltage measured using I_{SD}= 50 mA, (c) Normalized V_{TH} pre and after stress

a 6 % increase in V_{TH} due to a 10 s application of the rated V_{GS}. More details of the measurement technique for NBTI and PBTI are available in [5].

III. IMPACT OF REPETITIVE GATE STRESS PULSES ON BTI

In [5], the method was introduced and evaluated for short stress pulses. However, one of the main benefits of this method is that it enables the investigation of the impact of stress duration on BTI and V_{TH} recovery as well as the impact of repetitive stresses on the dynamics threshold voltage shift. Using the test setup presented in Fig. 3, both the impact of short and long repetitive stress pulses has been evaluated for the selected SiC MOSFET. This can be fundamental for understanding threshold voltage shift during the initial phases of a long stress.

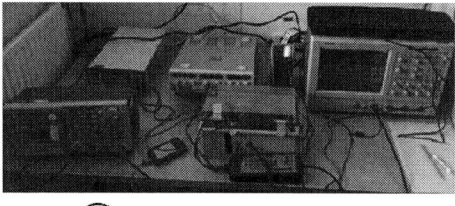

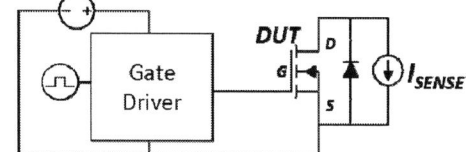

Fig. 3 Experimental setup for characterizing BTI shift using the 3rd quadrant characteristics of a SiC MOSFET

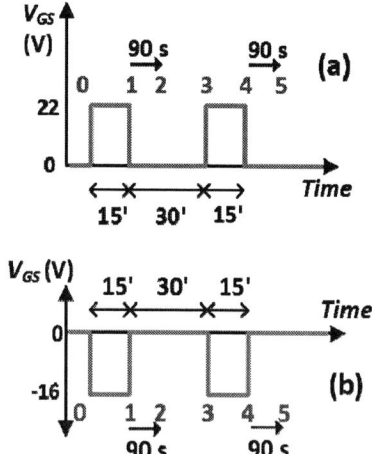

Fig.5 Long duration repetitive stress pulses for evaluation of: (a) PBTI and (b) NBTI

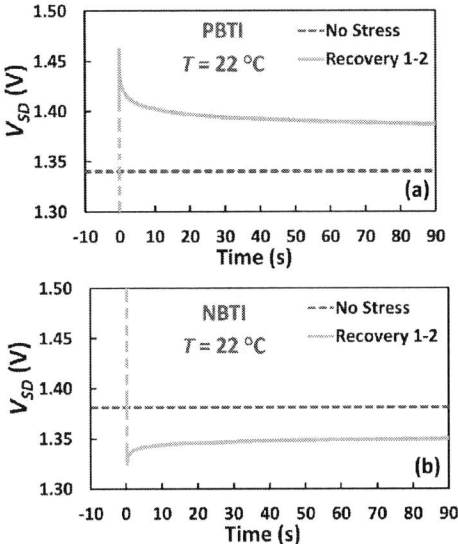

Fig. 6 Measured V_{SD} transient during the recovery transient after the first pulse. V_{GS}=0. (a) PBTI, (b) NBTI

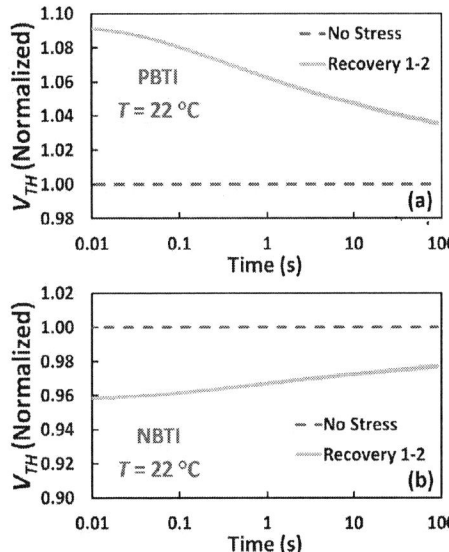

Fig. 7 Recovery of $V_{TH\text{-}NORMALIZED}$ after the stress pulse. (a) PBTI, (b) NBTI

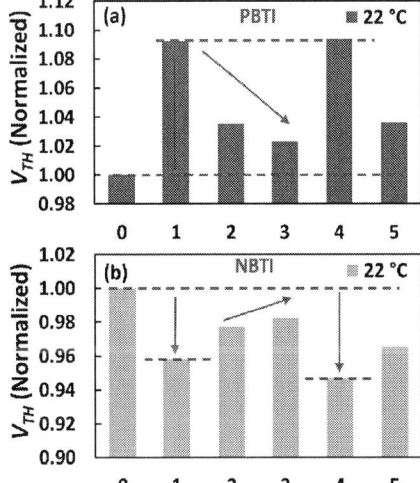

Fig. 8 Normalized V_{TH} shifts after cumulative pulsed stress tests (a) PBTI, (b) NBTI

A. Long duration repetivitve stresses

The impact of longer stress pulses and recovery times (in the range of minutes) was evaluated for both PBTI and NBTI stresses. The stress and characterization sequence is shown in Fig. 5(a) for the evaluation of PBTI and Fig. 5(b) for the evaluation of NBTI. The stress voltages are 22 V and -16 V for the PBTI and NBTI stresses respectively. The points where V_{SD} was measured for tracking the peak shift and recovery of V_{TH} after stress are identified from 0 (unstressed device) to 5 in Fig 5. The recovery transients (1-2) and (4-5) after each stress have been captured during 90 s.

The captured transient after the first pulse (1-2) is shown in Fig. 6(a) for the positive stress and Fig. 6(b) for the negative stress. As defined in the previous section and in more detail in [5], it is clearly observed how the positive stress shifts the measured V_{SD} upwards due to the increased V_{TH} and the negative gate stress shifts the measured V_{SD} downwards due to the reduction of V_{TH}.

The recovery of V_{SD}, thereby recovery of V_{TH}, after the stress after the first pulse is clearly observed in both cases. Using (1) the normalized V_{SD} value can be converted in the normalized V_{TH} value and the recovery of V_{TH} can be evaluated in more detail. This is shown in Fig 7 for both PBTI and NBTI stresses. For the device subjected to the positive stress the normalized increased of V_{TH} is higher, with an increase of around 9 %, than for the device subjected to negative stress, with a reduction of around 4 %.

The recovery is faster for the positive stress than for the negative stress and for both recovery transients there is an initial segment where the measured V_{SD} appears to be stationary. The authors attribute this to the transient response of the power supply during the change of conduction paths after the stress, when the voltage changes from low to high (PBTI) or high to low (NBTI).

The calculated V_{TH} shift for the points defined in Fig. 5 is presented in Fig. 8. For both PBTI and NBTI stresses, it is clearly observed how after 30 minutes recovery at V_{GS}=0, the threshold voltage recovers to a value close to the pre-stress V_{TH}, with the majority of the recovery happening in the first seconds after stress removal, as shown in the transient plots in Fig. 7. As mentioned in [3], this could have serious implications in the qualification of power devices. The impact of the cumulative stress of the second pulse is more apparent for the negative stress. After the first stress pulse, the initial peak shift detected for NBTI evaluation is around -4% whereas for the second pulse, the peak shift is approximately -5.5 %. For the positive stress despite the higher initial positive shift, the cumulative impact of the stresses is less apparent for the evaluated stress voltage. For both stress test, the device was attached to an aluminium block which acted as a heatsink and the impact of self-heating evaluated. In the case of NBTI, the temperature increase during the 15 minutes pulse was 1 °C, whereas in the case of PBTI, the impact of self-heating can be neglected.

978-1-7281-0582-6/19 $31.00 © 2019 IEEE

B. Short duration repetivitve stresses

From the results presented in section III.A it is clearly observed how the shift of V_{TH} is more apparent for the first pulse and how it recovers exponentially after the stress removal. However, there is no information about the transient nature of the shift during the stress period. One of the main benefits of the presented characterization technique is that it will allow its evaluation using cumulative short stress pulses. This has been done for both negative and positive stresses in this paper. The stress time t_{STRESS} selected was 2 s, followed by a recovery time t_{REC} of 2 s at V_{GS}=0. The stresses were performed at the ambient temperature of 22 °C and the number of pulses was 40.

Fig. 9 shows the measured V_{SD} transient for V_{STRESS} voltages of 22 V and -26 V. In both cases, an initial shift of V_{SD} can be observed together with a partial recovery of V_{SD} during the relaxation time of 2 s. As described in the section II, this recovery represents the recovery of V_{TH} after the stress. Comparing Fig. 9(a) and Fig. 9(b), the rate of recovery is apparently higher for the positive stress, as the change of V_{SD} is higher during the recovery phase of the pulsed stress. Comparing both stresses, a continuous reduction of V_{SD} can be observed for the negative gate stress. Normalizing the peak V_{SD} value and the variation of V_{SD} during the recovery, for the PBTI stress a total peak shift of +7.4 % with a recovery of -4 % during the 2 s recovery phase are observed, whereas in the case of NBTI, the peak shift is -8.4 % with a relative recovery of + 1.5 %. In the case of the negative pulsed stresses a continuous reduction of V_{SD} can also be observed.

As BTI is also stress level dependent, another clear benefit of this method is that it allows to evaluate the impact on the V_{TH} shift of highly accelerated stress voltages for short periods of time and its impact on shift and recovery. The same pulsed stress tests of period 4 s was performed using a stress voltage of 35 V, well above the nominal gate voltage of the evaluated planar device and the results are shown in Fig. 10, for a temperature of 22 °C. As the results in Fig. 10 show, during the initial phase of the 35 V pulsed stress, there is an initial V_{SD} reduction indicating a reduction of V_{TH} followed by the expected increase of V_{SD} for PBTI

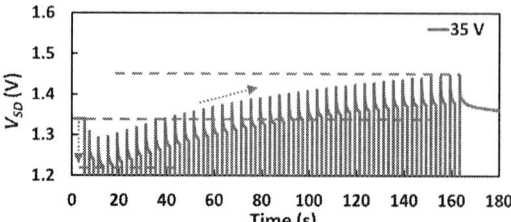

Fig. 10 V_{SD} during PBTI pulsed stress tests. V_{STRESS}=35 V, V_{REC}=0 V, t_{STRESS}=2 s, t_{REC}=2 s. I_{SENSE}=50 mA, T=AMB

during the final stage of the pulsed stress. This phenomenon of dip and rebound was already described in [7] for Si MOSFETs and is caused by the different contribution of the oxide trapped charges (decreasing V_{TH}) and the interface trapped charges (increasing V_{TH}) during the different stages of the stress. This is defined by (2) [7], where N_{ot} is the stress-induced change in the oxide trapped charge, Nit is stress-induced change in the interface trapped charge and C_{OX} is the specific gate oxide capacitance.

$$V_{TH} = V_{TH0} - \frac{qN_{ot}}{C_{OX}} + \frac{qN_{it}}{C_{OX}} \qquad (2)$$

In a traditional long duration stress, this peculiar feature of the highly accelerated stress test would not be captured, hence the benefits of using the 3rd quadrant characteristics for assessing the impact of the BTI in SiC MOSFETs.

IV. CONCLUSION

In this paper it has been shown how the third quadrant characteristics of SiC MOSFETs can be used for evaluating the V_{TH} shift caused by BTI. It is shown how V_{SD} is and effective cursor for detecting the peak shift of V_{TH} and tracking the recovery of V_{TH} after the gate stress is removed. The implementation of this method is similar to the use of V_{SD} as TSEP. Using short duration pulsed stress tests, the phenomenon of dip and recovery of V_{TH} has been captured during initial stages of highly accelerated stress tests, hence demonstrating the importance that this methodology could have for characterizing BTI in SiC MOSFETs.

REFERENCES

[1] T. Aichinger, G. Rescher, and G. Pobegen, "Threshold voltage peculiarities and bias temperature instabilities of SiC MOSFETs," *Microelectronics Reliability*, vol. 80, pp. 68-78 January 2018

[2] A. J. Lelis et Al. "SiC MOSFET threshold-stability issues," *Material Science in Semicond Processing*, vol. 78, pp. 32-37, May 2018

[3] D. B. Habersat, A. J. Lelis, and R. Green, "Measurement considerations for evaluating BTI effects in SiC MOSFETs," *Microelectronics Reliability*, vol. 81,pp. 121-126, February 2018.

[4] K. Puschkarsky, H. Reisinger, T. Aichinger, W. Gustin and T. Grasser, "Understanding BTI in SiC MOSFETs and Its Impact on Circuit Operation," in *IEEE Transactions on Device and Materials Reliability*, vol. 18, no. 2, pp. 144-153, June 2018.

[5] J. A. Ortiz Gonzalez and O. Alatise, "A Novel Non-Intrusive Technique for BTI Characterization in SiC MOSFETs," in*IEEE Transactions on Power Electronics*. Early Access, pp. 1-1, 2018

[6] G. Zeng, H. Cao, W. Chen and J. Lutz, "Difference in Device Temperature Determination Using p-n-Junction Forward Voltage and Gate Threshold Voltage," in IEEE Transactions on Power Electronics, vol. 34, no. 3, pp. 2781-2793, March 2019.

[7] U. Karki and F. Z. Peng, "Effect of Gate-Oxide Degradation on Electrical Parameters of Power MOSFETs," in *IEEE Transactions on Power Electronics*, vol. 33, no. 12, pp. 10764-10773, Dec. 2018.

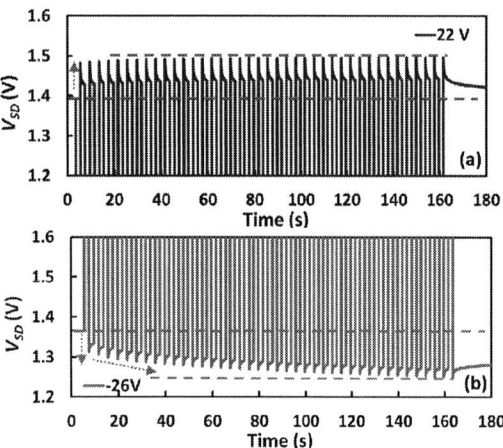

Fig. 9 (a) V_{SD} during PBTI pulsed stress tests. V_{STRESS}=22 V, V_{REC}=0 V, t_{STRESS}=2 s, t_{REC}=2 s. I_{SENSE}=50 mA, T=AMB
(b) V_{SD} during NBTI pulsed stress tests. V_{STRESS}=22 V, V_{REC}=0 V, t_{STRESS}=2 s, t_{REC}=2 s. I_{SENSE}=50 mA, T=AMB

Proceedings of the 31st International Symposium on Power Semiconductor Devices & ICs
May 19-23, 2019, Shanghai, China

Planar 1.2kV SiC MOSFETs with retrograde channel profile for enhanced ruggedness

L. Knoll, A. Mihaila, S. Wirths, Y. Arango, A. Prasmusinto, E. Bianda, L. Kranz, M. Bellini, G. Romano, C. Papadopoulos [*]
ABB Switzerland Ltd., Corporate Research, CH-5405 Baden-Dättwil, Switzerland
[*]ABB Switzerland Ltd., Semiconductors, CH-5600 Lenzburg, Switzerland
email: lars.knoll@ch.abb.com

Abstract— The static and dynamic performance of Silicon Carbide (SiC) MOSFET rated for 1200V applications has been investigated. MOSFETs with a planar design and several different cell pitches have been fabricated. Special attention has been dedicated to the channel design, where a novel retrograde doping profile has been employed. For reference, a more common box profile channel has also been used. Turn-off measurements under high current and over voltage conditions reveal that the MOSFET body diode offers wide safe operating area capability. The MOSFETs feature exceptional ruggedness against long short circuit events, with the retrograde channel designs able to withstand the 10μs industry standard specification.

Keywords—SiC MOSFET, planar cell, retrograde channel, ruggedness

I. INTRODUCTION

Commercially available for several years, SiC power semiconductors are reaching higher volumes as well as gaining more and more market acceptance. In power electronics applications such as electric vehicles (powertrain as well as on- and off-board chargers), photovoltaics and motor drives, SiC MOSFETs have been challenging the dominant position of Si power devices [1, 2].

650V and 1200V-rated SiC MOSFETs are currently available from several vendors. Offered with either planar or trench cell designs, SiC MOSFETs provide competitive static losses, fast dynamic performance and adequate reliability [3, 4, 5]. Regarding the fault handling capability, SiC MOSFETs still fall short of the typical industry standard values (of 10μs) shown by their Si counterparts. This is typically associated with the strong trade-off between conduction losses and short-circuit endurance times [6, 7]. Nevertheless, 1200V-rated SiC MOSFETs typically deliver short circuit withstand pulses in the range of ≤5μs.

This paper presents a 1200V-rated SiC planar MOSFET with a novel retrograde channel design. The doping profiles of the channel and p-well regions have been carefully engineered to provide improved shielding from the high drain potential while maintaining an optimum trade-off between the MOSFET static and dynamic performance. A more standard cell design employing a constant doping profile (BOX) in a short channel has also been used for benchmarking purposes. Both static and dynamic performance of the MOSFET are investigated, with a special focus on the robustness of the body diode as well as the device ruggedness under short circuit operation conditions.

II. EXPERIMENTAL APPROACH

Planar 1.2kV vertical power MOSFETs have been fabricated on 4" 4H-SiC (0001) wafers with cell pitches of 12μm (p12) and 14μm (p14), a constant body diode region of 10μm as well as two channel lengths (80 nm and 250 nm). Here, the doping of the long channel device is reduced drastically at the surface and increased in the deeper regions of the channel (see figure 1), called retrograde channel (RG).

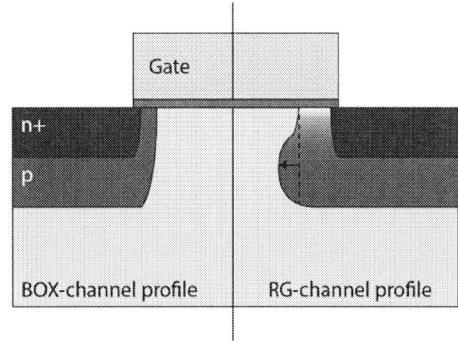

Fig. 1. Schematic x-section of the BOX- and RG-channel profile. The RG channel profile has an extended p-well below the channel region.

To minimize electric fields at the gate oxide in between the cells, i.e. to avoid a JFET region implantation step, we optimize charge and depth distributions in the channel. On the one hand this design allows rugged dynamic operation with a wide SOA and on the other hand extraordinary fault handling capability in case of a short circuit or surge events. In particular, the RG channel profile reduces the saturation current and effectively shields the drain potential at large V_D values.

III. RESULTS AND DISCUSSION

A. Static behaviour

Figure 2 (a) presents the static output of a 1.2kV MOSFET at 125°C. Several pitches and channel lengths are shown (p12-80, p12-250 and p14-250 MOSFETs), at a gate voltage of 15V, respectively. On the same figure, a reference Si SPT+ IGBT curve is also shown. As can be seen, at the nominal current of the Si IGBT (I_{NOM}) of 105A/cm², all three MOSFETs designs provide lower static losses compared to the bipolar Si device. Figure 2 (b) displays the 1.2kV MOSFET body diode static behavior. For positive gate voltage values (15V), the channel is fully turned-on and conducts in the reverse direction until the actual MOSFET

978-1-7281-0582-6/19 $31.00 © 2019 IEEE

body diode kicks in at around 4.5V. When the channel is firmly closed (V_G=-10V), the body diode turns on around -2.5V. The voltage drop of body diode varies slightly for different cell designs, with the p14-250 offering the lowest overall value.

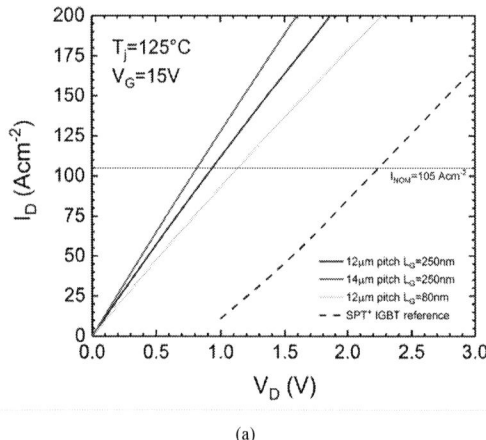

(a)

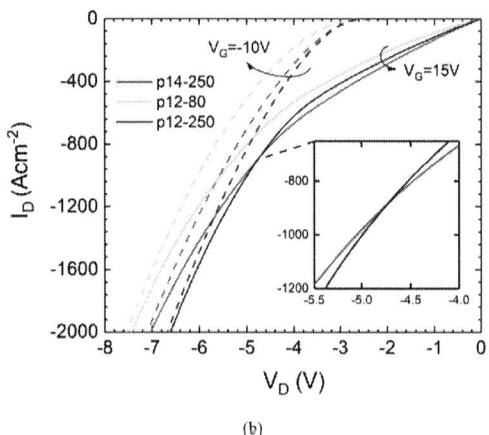

(b)

Fig. 2. (a) MOSFET and (b) body diode output characteristics of a 1.2kV MOSFET with 12μm and 14μm pitch values and channel lengths of 250nm and 80nm.

The variation of the specific on-state resistance with temperature, for various cell pitches, is displayed in figure 3. All designs show a positive temperature coefficient, an essential feature for paralleling devices. The p14-250 device shows the lowest on-state values, followed by p12-250 and p12-80 devices. This behavior can be explained by the stronger JFET contribution in smaller pitch designs, which therefore results in a more pronounced temperature coefficient.

B. Dynamic behaviour

The MOSFET switching behavior has been evaluated in a double pulse test set-up, at 125°C. The typical stray inductance value was Lσ=480nH.

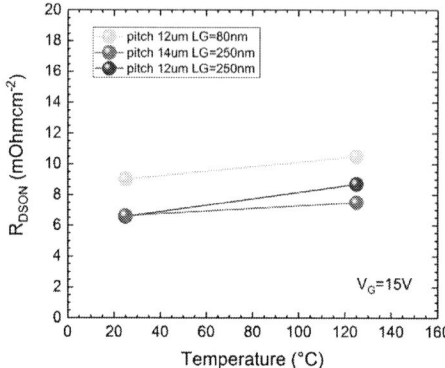

Fig. 3. Specific on-state resistance variation versus temperature for various cell pitches and channel lengths; V_G=15V

The turn-off behavior of the MOSFET body diode is depicted in figure 4. Here, a p12-250 design has been considered. The load current is 50A. Two values have been used for the DC link voltage, namely 800V and 900V, respectively. For the tested current level, the MOSFET body diode offers fast switching performance with very small reverse recovery losses.

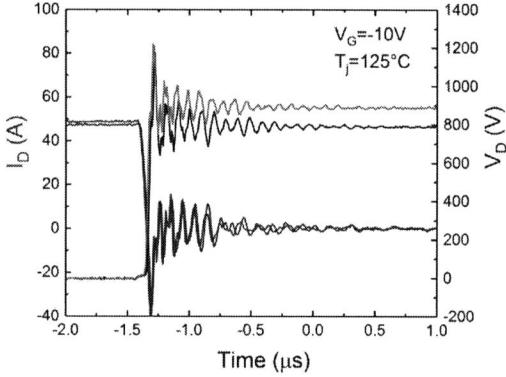

Fig. 4. Body diode turn-off waveforms up to V_D=900V with R_G=22Ohm and 50A demonstrating safe turn-off at 2x I_{NOM} with Lσ=480nH

Figure 5 depicts the turn-on waveforms of p12-250 MOSFETs at 800V and 900V with 50A, hard switched under harsh conditions with a large stray inductance of Ls=480nH. The gate swing was from -10V up to 15V. Notably, the devices can turn-on two times the rated current of 25A (2x I_{NOM}). An overshoot of about 20A is noticeable on the drain current waveform, which is associated with the accumulation of reverse recovery charge coming from the body diode and the charge in the output capacitance of the upper MOSFET used as the free-wheeling diode.

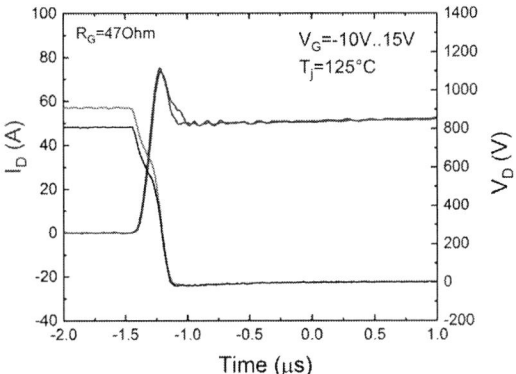

Fig. 5. Turn-on waveforms of p12-250 MOSFETs at 800V and 900V switching 50A, representing 2x I_{NOM} and R_G=47Ohm with $L\sigma$=480nH

The turn-off characteristics of a p12-250 MOSFET for over current and over voltage conditions are indicated in figure 6, at 125°C. The device turns-off twice the load current of 50A, which is twice the nominal value I_{NOM}. Some oscillations are observed on the turn-off waveforms, with the peak drain voltage value exceeding the rated 1200V voltage. However, this is still below the actual device breakdown voltage and, hence, no failures were witnessed during the test.

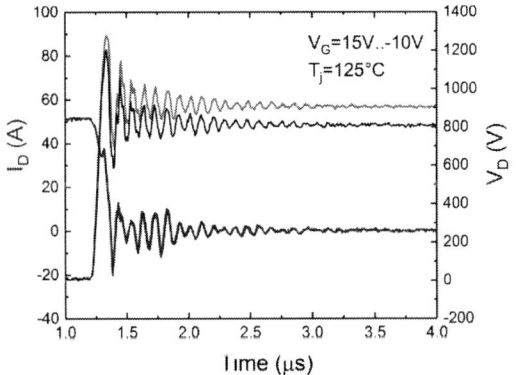

Fig. 6. Turn-off waveforms up to V_D=900V at 50A with R_G=22Ohm demonstrating safe turn-off SOA at 2x I_{NOM}

Owing to their unipolar nature, SiC MOSFETs do not experience dynamic avalanche conditions and, therefore, should exhibit a considerable turn-off safe operating area [7]. The turn-off SOA curves of a p12-250 MOSFET are shown, for several R_G values, in figure 7, at 125°C. The nominal operation area is also schematically indicated on the figure. The gate voltage has been varied from +15V down to -10V. At R_G=100Ohm the maximum DC link voltage could be increased to 1000V. For lower gate resistance values of 47Ohm and 22Ohm, the switching voltage was reduced to 900 V. Remarkably, the device could turn-off more than two times the nominal current while blocking a DC link voltage of 1000V.

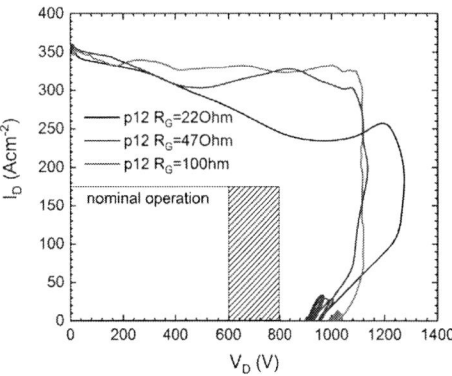

Fig. 7. Turn-off Safe Operating Area of 1.2kV MOSFETs with pitch 12 and L_G=250nm with varying R_G from 22Ohm to 100Ohm; T=125°C

The robustness of the body diode is an important parameter towards increasing the industrial utilization of SiC power MOSFETs. In the turn-off measurements shown in figure 8, the load current of the body diode was increased from 100A up to 190A, while using a DC link voltage of 900V. The auxiliary switch used here was a p12-250 MOSFET. The body diode shows fast commutation behavior with low levels of reverse recovery charge. More importantly, with the channel entirely closed for V_G=-10V, the body diode demonstrates large dynamic overcurrent capability under 190A/900V switching conditions. The turned-off current value represents about 8x I_{NOM}, thus demonstrating the wide SOA capability of the body diode.

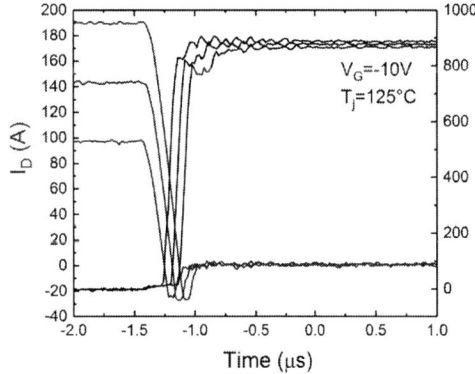

Fig. 8. Body diode turn-off at 900V and R_G=47Ohm with increasing I_D from 100A to 190A representing ~8x I_{NOM}

C. *Short Circuit results*

Figure 9 depicts the short circuit capability of p12-80, p12-250 and p14-250 devices, respectively. The test was performed at 125°C and for V_G=15V. The drain voltage was at the nominal value of 600V. A high saturation current was observed for p14-250 design along with a strong self-heating, limiting the short circuit capability drastically. Both p12 devices have smaller short circuit-peak current levels. The RG channel profile of p12-250 MOSFETs can further reduce the peak current and, therefore, provide more than 10μs short circuit withstand time. A clear trade-off between saturation current and short circuit performance can be observed for devices with the same channel length. Here, it is important to mention that all devices managed to turn-off safely, with no failure experienced after the short circuit pulse.

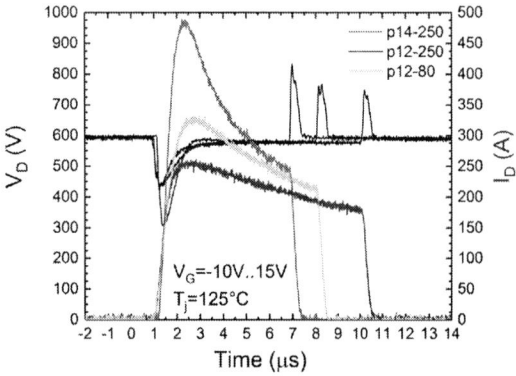

Fig. 9. Short circuit waveforms under nominal conditions with V_D=600V and R_G=100Ohm for p12-250, p12-80 and p14-250 MOSFETs depicting lower peak current with smaller pitches and longer channel length

In figure 10, the short circuit waveforms of a p12-250 MOSFET are shown at 125°C and for V_G=-10V up to +15V. The DC link voltage has been varied from V_D=200V up to

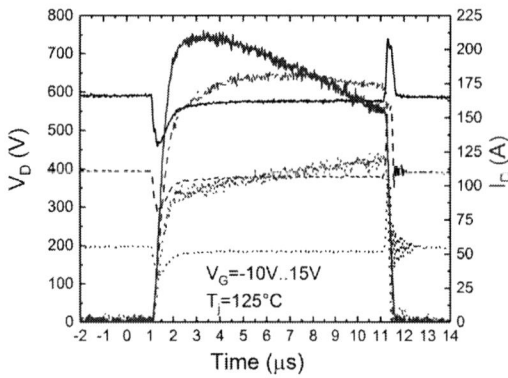

Fig. 10. Short circuit mode of 12um pitch with L_G=250nm under nominal conditions demonstrating short circuit capability up to 10μs.

nominal conditions of V_D=600V.

Under nominal conditions, at the beginning of the pulse, a single chip can conduct more than 200A. The self-heating effect reduces this value by about 25% at the end of the short circuit pulse. The p12-250 device can withstand a short circuit time of more than 10μs while dissipating about 8J/cm² in terms of energy loss. This demonstrates that our 12μm pitch device with a RG channel design can offer a very attractive trade-off in terms of conduction losses versus short circuit capability.

IV. CONCLUSIONS

1200V MOSFETs with cell pitch widths of 12μm and 14μm have been fabricated, respectively. The MOSFET channel has been carefully optimized using a retrograde doping profile. Static and dynamic measurements demonstrate the excellent performance of the RG MOSFETs. The enhanced electrostatic control along with a stronger shielding of the drain potential offer exceptional short circuit and reverse blocking SOA at a moderate increase R_{DSON}. The body diode offered a wide safe operating area performance, turning-off safely 8x I_{NOM}. The RG devices featured excellent ruggedness against long short circuit pulses.

REFERENCES

[1] D. Dujic, "Electric Vehicles Charging – An Ultrafast Overview", PCIM 2018, Nuremberg, June 2018

[2] M. Nawaz, "Moving from Si to SiC from the End User's Perspective", IEEE APEC-2018, San Antonio, Texas, March 2018

[3] D. Peters et al., "Performance and ruggedness of 1200V SiC - Trench - MOSFET", International Symposium on Power Semiconductor Devices and IC's (ISPSD), 2017, p.239

[4] P. Losee et al., "1.2kV class SiC MOSFETs with improved performance over wide operating temperature", International Symposium on Power Semiconductor Devices & IC's (ISPSD), 2014, p. 297

[5] S. Chowdhury et al., "Next generation 1200V, 35mOhmcm2 SiC planar gate MOSFET with excellent HTRB reliability", International Symposium on Power Semiconductor Devices & IC's (ISPSD), 2018, p. 427

[6] J. Yamada et al., "Gaining speed: Mitsubishi electric SiC power modules", Bodo's Power Systems, September 2018, p. 22

[7] M. Rahimo, "Performance Evaluation and Expected Challenges of Silicon Carbide Power MOSFETs and Diodes for High Voltage Applications", *Material Science Forum*, vol. 897, pp. 649-654,201

Proceedings of the 31st International Symposium on Power Semiconductor Devices & ICs
May 19-23, 2019, Shanghai, China

An Experimentally Verified 3.3 kV SiC MOSFET Model Suitable for High-Current Modules Design

A. Borghese, <u>M. Riccio</u>, L. Maresca, G. Breglio and A Irace.
Dept. of Elect. Engineering and Information Technologies
University of Naples Federico II, Naples, Italy
michele.riccio@unina.it

G. Romano, E. Bianda, A. Mihaila,
M. Bellini, L. Knoll, and S. Wirths
ABB Switzerland Ltd.
Corporate Research Center
CH-5405 Baden-Dättwil, Switzerland

Abstract—In this study, an electrothermal compact model for SiC power MOSFETs is experimentally verified for the first time on a 3.3 kV device. Both the parameters determining the static behavior and those controlling the dynamic performances are calibrated over experimental data. The good agreement achieved proves the model scalability form 1.2 kV devices to 3.3 kV ones. Successively, to validate the good convergence properties of the model, an electrothermal simulation of a three-phase inverter consisting of 36 hard switched MOSFETs is performed.

Keywords—electrothermal modeling, power module, power MOSFET, silicon carbide (SiC), SPICE.

I. INTRODUCTION

Silicon carbide (SiC) power MOSFETs are finding widespread adoption in several application areas thanks to many excellent features. 3.3 kV devices have clearly become appealing for several medium voltage applications [1] - [3]. Among all, electric traction is one of the major fields that could possibly much exploit all the advantages brought by such devices [4]. As already demonstrated, 3.3 kV MOSFETs are also able to fulfill the most demanding requirements of such applications, namely short-circuit and surge current [5], [6]. Despite the many indubitable benefits associated with the introduction of wide band gap (WBG) devices, the implementation of systems based on such components also introduces new complications, for instance related to device paralleling [7] or to electromagnetic interference (EMI) [8]. Circuit-level electrothermal (ET) simulations represent a helpful tool for validating the design of power converters and modules [9], [10], thus reducing the need for costly prototypes and improving the time-to-market. To achieve reliable ET simulations, it is essential not only an accurate modelling of the electrical behavior of the device but also a correct description of its dependence on the temperature. After the well-established model presented in [11], several other examples successively emerged in the literature [12], [13]. The models reported in other publications ([14] - [16]) also introduced the effects of the temperature, even though those were not implemented for describing the self-heating phenomenon. Models allowing fully coupled ET simulations [17] - [19], i.e. those where the reciprocal influence of electrical and thermal quantities is considered, are crucial to analyze the current sharing properties of parallel connected devices and thus to assist the design of multichip power modules. In this work, a temperature-dependent model based on the one developed in [20] is experimentally verified, for the first time, on a 3.3 kV device, thus demonstrating the scalability of the model to different voltage ratings. Afterwards, to extensively validate its suitability for

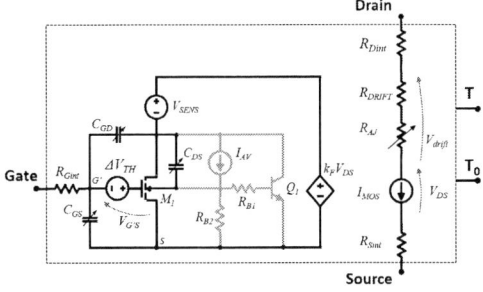

Fig. 1. Developed SPICE subcircuit, with electrical and thermal nodes. Elements in gray model the out-of-SOA operation [20].

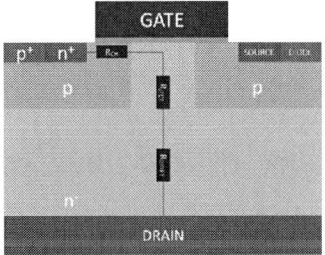

Fig. 2. Cross-section of a MOSFET cell indicating the channel resistance R_{CH}, the JFET resistance R_{JFET}, and the drift resistance R_{DRIFT}.

performing dynamic ET simulations, the model is used to simulate a circuit application employing a complex arrangement of parallel devices.

II. MODEL CALIBRATION

The compact model adopted for this analysis is thoroughly described in [20], therefore, only its main characteristics will be recalled in the following subsections. Experimental data of a 3.3 kV, 4H - SiC planar power MOSFET at different operating temperatures were used as the target of the parameters tuning process.

A. Static behavior validation

The model is implemented as a SPICE netlist and its corresponding subcircuit is depicted in Fig. 1. In addition to the source, drain and gate terminals, it features two more pins (T and T_0) which are used for the connection to an equivalent thermal network, thus enabling the electrothermal feedback. The static behaviour of the model is primarily determined by the non-linear drain resistance, whose value depends on the resistivity associated to three different sections (Fig. 2) of the

978-1-7281-0582-6/19 $31.00 © 2019 IEEE   215

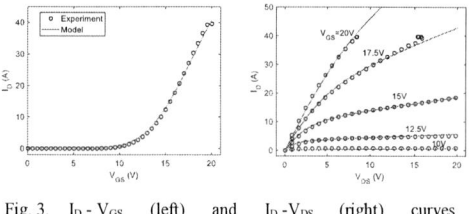

Fig. 3. I_D - V_{GS} (left) and I_D - V_{DS} (right) curves at T=300 K - V_{DS}=10 V. SPICE model: solid line; symbols: experimental data.

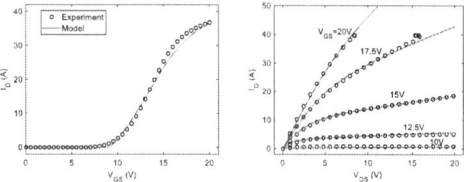

Fig. 4. I_D - V_{GS} (left) and I_D - V_{DS} (right) curves at T=400 K - V_{DS}=10 V. SPICE model: solid line; symbols: experimental data.

device structure: the channel region, the accumulation - JFET region and the drift-layer. The main element describing the channel region is a standard level 1 MOSFET component. As described by (1), the accumulation - JFET and the drift regions are modelled by the series of two resistors, R_{AJ} and R_{EPI}, respectively (where V_{drift} is the voltage drop across R_D).

$$R_D(V_{GS}, V_{drift}, T) = R_{AJ}(V_{GS}, V_{drift}, T) + R_{EPI}(T) \quad (1)$$

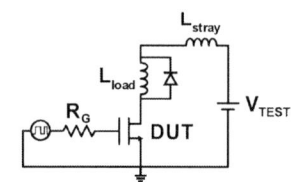

Fig. 5. Schematic of the simulated double pulse test setup. V_{TEST} = 1.8k V; R_G = 4.7 Ω; L_{load} = 12 mH; L_{stray} = 500 nH.

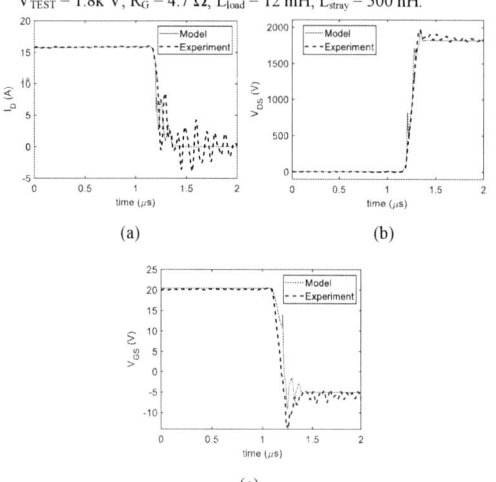

Fig. 6. Inductive turn-off waveforms: a) drain current; b) drain voltage; c) gate voltage. Solid lines: SPICE model; dashed lines: experimental data.

The first step for the DC calibration of the model consists in the estimation of the current factor (K_{P0} [A/V^2]) and threshold voltage (V_{TH0} [V]) at the reference temperature T_0. These were extracted through quadratic extrapolation method (QEM) from the steepest portion of the isothermal transfer characteristic at room temperature (300 K). In addition to these, the channel - length modulation parameter (λ [V^{-1}]) was measured from the output characteristics. To this purpose, the I_D - V_{DS} curve at the lowest V_{GS} bias (i.e. 10 V) was selected in order to guarantee the full operation in saturation mode of the device ($V_{DS} \gg V_{GS}$ - V_{TH}). Subsequently, the so obtained values were set in the level 1 MOSFET component and where used as the starting guess of an iterative optimization procedure implemented using the MATLAB optimization toolbox. Such a routine tunes the values of K_{P0}, V_{TH0}, and of the drain resistance parameters by minimizing the error between the simulated and the experimental I_D - V_{DS} curves at 300 K. A comparison highlighting the excellent agreement between the simulated and experimental DC characteristics at room temperature is shown in Fig. 3. Afterwards, the iterative optimization procedure was repeated to evaluate the temperature coefficients of the static parameters. The comparison reported in Fig. 4 shows that the model can accurately reproduce the DC characteristics also at 400 K, with only a slight deviation in the I_D - V_{GS} curve at high V_{GS}.

B. Dynamic Behavior Validation

An accurate modeling of the MOSFET capacitances is of paramount importance to a correct reproduction of the transient characteristics. The three main parasitic capacitances considered in this model are C_{GS}, C_{GD} and C_{DS}. The only one regarded as constant is C_{GS}, while the others were modelled as non-linear functions of the bias, according to the expressions reported in (2) and (3).

$$C_{DS}(V_{ds}) = \frac{C_{DS0}\left[\frac{\pi}{2} + \arctan\left(-\frac{V_{ds}}{V_{ds}^*}\right)\right]}{\pi/2} + C_{DSMIN} \quad (2)$$

$$C_{GD}(V_{gd}) = (C_{GD0} - C_{GDMIN})\left[1 + \frac{2}{\pi}\arctan\left(-\frac{V_{gd}}{V_{gd}^*}\right)\right] \quad (3)$$

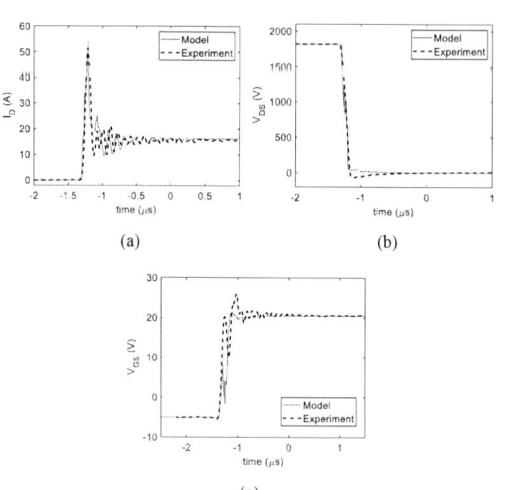

Fig. 7. Inductive turn-on waveforms: a) drain current; b) drain voltage; c) gate voltage. Solid lines: SPICE model; dashed lines: experimental data.

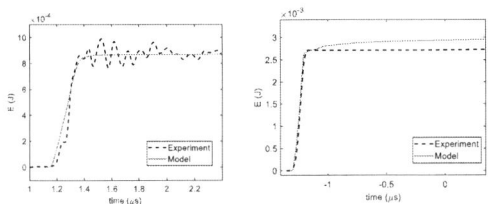

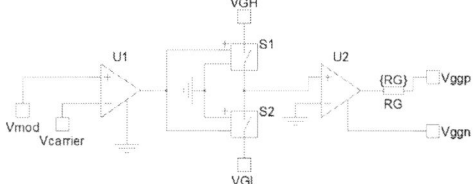

Fig. 10. Circuit schematic of the modelled PWM gate driver.

Fig. 8. Turn-off (left) and turn-on (right) dissipated energy waveforms. Solid line: SPICE model; dashed line: experimental data.

where the fitting parameters V^*_{ds}, V^*_{gd}, C_{DS0}, C_{DSMIN}, C_{GD0} and C_{GDMIN} were calibrated to fit the experimental waveforms. In particular, the model was simulated during double pulse test (DPT) with the following specifications: V_{TEST} = 1.8k V, R_G = 4.7 Ω and L_{load} = 12mH. Fig. 5 reports a schematic circuit of the DPT platform which also includes the inductance L_{stray} = 500 nH to account for the parasitic elements introduced by the cables. The waveforms reported in Fig. 6 and Fig. 7 illustrate that, after the calibration step, both the turn-off and turn-on phases are properly reproduced by the model. A better quantification of the accuracy of the simulated waveforms can be assessed by analyzing the switching dissipated energies of Fig. 8. These results prove that the model is not only able to predict the switching waveforms, but it can also provide a reliable estimation of the switching losses.

III. INVERTER SIMULATION

Subsequently, the model was used to simulate a circuit application containing a large number of MOSFETs. Specifically, a three-phase inverter was considered as a case-study to demonstrate the model suitability to perform such kind of simulations. A schematic representation of the converter is provided in Fig. 9. It consists of three half-bridges (HB) connected in parallel to a DC voltage source of 1800 V, while the midpoints of the HBs are wired to a three-phase load in star configuration. The values of load inductance and resistance are 10 mH and 5 Ω, respectively. The whole inverter contains 36 active devices since the elementary switching cell of each leg is made of six parallel connected MOSFETs. Each model instance was connected to an equivalent thermal network to enable the ET feedback. The inverter was driven by an open-loop sine-triangle pulse-width modulation (PWM) controller depicted in Fig. 10. The carrier signal is a 100 kHz triangular waveform while the modulating one is a 100 Hz sine wave with a modulation index of 0.5. To avoid shoot-through conditions, a dead time of 900 ns was added between the switching edges of opposite transistors of the same leg. The gate voltage swings from -5 V to +20 V and

two different resistors (2.5 Ω and 6 Ω) are connected to the gates of the high-side and low-side devices, respectively.

Despite the circuit complexity related to the high number of hard-switched power devices (36), an operation time of 40 ms (4×10^3 switching events) was simulated in less than 7 hours on a regular PC equipped with an Intel i7-2600 CPU. The waveforms of the output currents over the entire simulation interval are shown in Fig. 11. Fig. 12 reports the individual temperatures of MOSFETs belonging to different sets of parallel devices. After a rise time of approximately 15 ms, the temperatures of both the high-side and low-side MOSFETs reach an average value of 55 °C and oscillate around it with a swing of ±5 °C. These data could successively be used in time-to-failure estimation rules to check whether a given temperature swing ensures or not a desired lifetime.

IV. CONCLUSION

In this paper, a model for SiC power MOSFET, originally developed and calibrated for a 1.2 kV device, has been experimentally verified, for the first time, on a 3.3 kV MOSFET. As a first step, the static behavior of the model has been calibrated over isothermal transfer and output characteristics, both at room and high temperature. In both cases, excellent agreement has been obtained between the simulated and the experimental data. Subsequently, the dynamic section of the model has been tuned and a good match with the experimental waveforms has been achieved as well, thus proving the model scalability from the 1.2 kV rating to the 3.3 kV rating. Finally, to further validate the model flexibility, a three-phase inverter containing a vast arrangement of hard switched MOSFETs has been modelled and simulated. Despite the circuit complexity, the simulation of 40 ms of operation time has been completed in less than 7 hours. These results and performances suggest that the model can be used as an effective tool to assist a thermal-aware design of complex multi-chip power modules or converters. In addition, upon the evaluation of relevant devices parameters or parasitic elements associated to the module structure, the model can be used to evaluate possible current and voltage unbalances determined by fabrication mismatches.

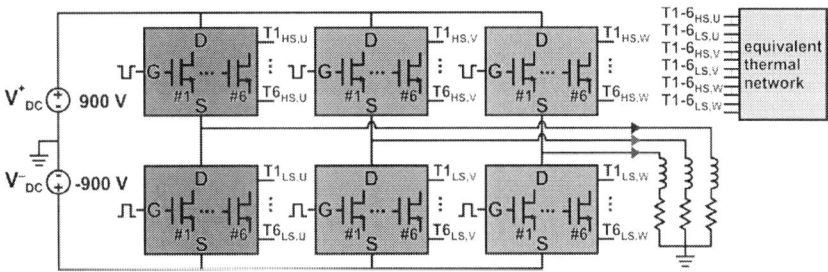

Fig. 9. Circuit schematic of the simulated three-phase inverter.

978-1-7281-0582-6/19 $31.00 © 2019 IEEE

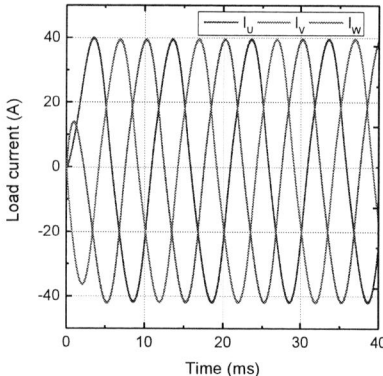

Fig. 11. Load current waveforms.

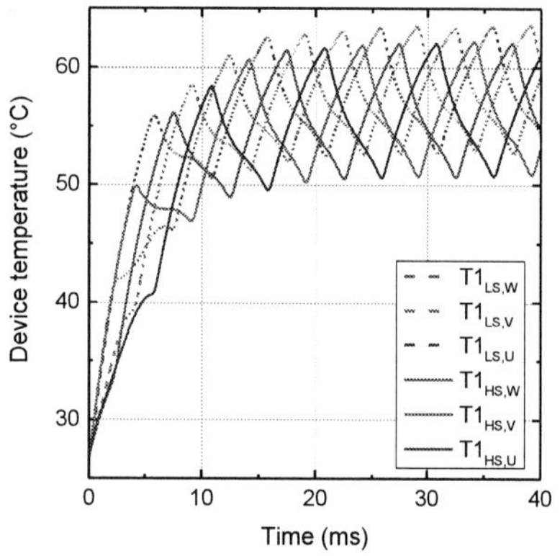

Fig. 12. Individual temperatures of MOSFETs from the six different subsets of parallel devices.

REFERENCES

[1] H. Wen, J. Gong, Y. Han, and J. Lai, "Characterization and Evaluation of 3.3 kV 5 A SiC MOSFET for Solid-State Transformer Applications," in *Proc. Asian Conf. on Energy, Power and Transp. Electrific. (ACEPT)*, 2018.

[2] A. Marzoughi, R. Burgos, and D. Boroyevich, "Characterization and performance evaluation of state-of-the-art 3.3 kV 30 a full-SiC MOSFETs," in *Proc. IEEE Energy Convers. Congr and Expo. (ECCE)*, 2017, pp. 1350-1357.

[3] A. Castellazzi *et al.*, "SiC power MOSFETs performance, robustness and technology maturity," *Microelectron. Reliab.*, vol. 58, pp. 164-176, 2016.

[4] K. Hamada *et al.*, "3.3 kV/1500A power modules for the world's first all-SiC traction inverter," in *Jpn. J. of Appl. Phys.*, vol. 54, no. 4, pp. 04DP07-1-04DP07-4, 2015.

[5] L. Knoll *et al.*, "Robust 3.3kV silicon carbide MOSFETs with surge and short circuit capability," in *Proc. Int. Symp. on Power Semicond. Devices and IC's (ISPSD)*, 2017, pp. 243-246.

[6] G. Romano *et al.*, "A Comprehensive Study of Short-Circuit Ruggedness of Silicon Carbide Power MOSFETs," *IEEE J. Emerg. Sel. Top. Power Electron.*, vol. 4, no. 3, pp. 978–987, Sep. 2016.

[7] A. Borghese *et al.*, "Effect of Parameters Variability on the Performance of SiC MOSFET Modules," *IEEE Int. Conf. on Elect. Syst. for Aircr., Railway, Ship Propulsion and Road Vehicles & Int. Transp Electrification Conf. (ESARS-ITEC)*, 2018.

[8] J. Sakata, M. Taguchi, S. Sasaki, T. Kuroda, and K. Toda, "An EMI-less full-bridge inverter for high speed SiC switching devices," in. Proc. *IEEE Appl. Power Electronics Conf. and Expo. (APEC)*, 2018, pp. 2570-2576.

[9] K. Li, P. Evans, and M. Johnson, "Using multi time-scale electro-thermal simulation approach to evaluate SiC-MOSFET power converter in virtual prototyping design tool," in *Proc. IEEE Workshop on Control and Model. for Power Electronics (COMPEL)*, 2017.

[10] D. Cavaiuolo *et al.*, "A robust electro-thermal IGBT SPICE model: Application to short-circuit protection circuit design," *Microelectron. Reliab.*, vol. 55, no. 9–10, pp. 1971–1975, Aug. 2015.

[11] T. R. McNutt, A. R. Hefner, H. A. Mantooth, D. Berning, and S. Ryu, "Silicon Carbide Power MOSFET Model and Parameter Extraction Sequence," *IEEE Trans. Power Electron.*, vol. 22, no. 2, pp. 353-363, March 2007.

[12] M. Mudholkar *et al.*, "Datasheet Driven Silicon Carbide Power MOSFET Model," *IEEE Trans. Power Electron.*, vol. 29, no. 5, pp. 2220-2228, May 2014.

[13] Z. Duan, T. Fan, X. Wen, and D. Zhang, "Improved SiC Power MOSFET Model Considering Nonlinear Junction Capacitances," *IEEE Trans. Power Electron.*, vol. 33, no. 3, pp. 2509-2517, March 2018.

[14] K. Sun, H. Wu, J. Lu, Y. Xing, and L. Huang, "Improved Modeling of Medium Voltage SiC MOSFET Within Wide Temperature Range," *IEEE Trans. Power Electron.*, vol. 29, no. 5, pp. 2229-2237, May 2014.

[15] M. Shintani *et al.*, "Surface-Potential-Based Silicon Carbide Power MOSFET Model for Circuit Simulation," *IEEE Trans. Power Electron.*, vol. 33, no. 12, pp. 10774-10783, Dec. 2018.

[16] H. Li *et al.*, "A Non-segmented PSpice Model of SiC MOSFET with Temperature-Dependent Parameters," IEEE Trans. Power Electron.,

[17] R. Kraus and A. Castellazzi, "A Physics-Based Compact Model of SiC Power MOSFETs," *IEEE Trans. Power Electron.*, vol. 31, no. 8, pp. 5863-5870, Aug. 2016.

[18] G. Bazzano, D. G. Cavallaro, R. Greco, A. Raffa, and P. P. Veneziano, "A New Analog Behavioral SPICE Macro Model with Thermal and Self-Heating effects for Silicon Carbide Power MOSFETs," in *Proc. PCIM*, 2015.

[19] V. d'Alessandro *et al.*, "SPICE modeling and dynamic electrothermal simulation of SiC power MOSFETs," in *Proc. IEEE Int. Symp. on Power Semicond. Devices & IC's (ISPSD)*, 2014, pp. 285-288.

[20] M. Riccio *et al.*, "A Temperature-Dependent SPICE Model of SiC Power MOSFETs for Within and Out-of-SOA Simulations," *IEEE Trans. Power Electron.*, vol. 33, no. 9, pp. 8020-8029, Sept. 2018.

978-1-7281-0582-6/19 $31.00 © 2019 IEEE

Proceedings of the 31st International Symposium on Power Semiconductor Devices & ICs
May 19-23, 2019, Shanghai, China

Time-Resolved Short Circuit Failure Analysis of SiC MOSFETs

Thomas Ziemann, Alexander Tsibizov, Bhagyalakshmi Kakarla, Lorenz Bort* and Ulrike Grossner

Advanced Power Semiconductor Laboratory
*High Voltage Laboratory
ETH Zurich
Zürich, Switzerland
ziemann@aps.ee.ethz.ch

Abstract—**High-speed optical imaging is used in conjunction with fast electrical measurements to advance the understanding of the development of short circuit failures in silicon carbide power MOSFETs. Special samples are manufactured, which are compatible and comparable to TO-247 packages, but do not have any encapsulation. This allows optical observation of die surface during the test. The information on visible processes on the die allows for a better understanding of the sequence of events leading up to a failure. Imagery of destructive drain-source failures is also obtained, as well as post-failure images of surface and cross-sections. Aluminum metal melting is observed even for very short tests, before electrical indications of damage. The onset and completion of melting are used as information on the temperature of the die surface. Using this data for calibration, a detailed electro-thermal model is then used to simulate the temperature distribution and evolution during the short circuit.**

I. INTRODUCTION

Silicon carbide (SiC) MOSFETs offer fast switching of high voltages with high efficiency, leading to smaller and cheaper converters. In the event of a short circuit, the high power density quickly leads to extreme temperatures. This rapid heating above $1000\,°C$ leads to an often catastrophic failure within less than $10\,\mu s$ at typical operating voltages [1]. It also makes understanding the process difficult, as any measurement needs to be very fast. Simulation is doubly difficult, as neither high temperature material parameters nor the temperatures reached are known. As short circuit behavior is a critical device characteristic in some applications, precise knowledge of the process is desired. This is necessary not only for optimizing device design and improving the robustness of future devices, but also for implementing protection features and diagnosing possible device damage. Electrical measurements of SC testing and simulation of the process have been shown by several authors [2]–[7]. The suspected failure mode is hot-spot formation with local thermal runaway as cause of drain-source failures. There is some uncertainty about the occurring temperatures and the sequence of events, also due to use of low temperature material parameters and simplified models. IR thermography can give an insight into surface temperature distribution, but slow cameras necessitate low V_{DS} and use of equivalent time sampling across many devices [2], [3], [8].

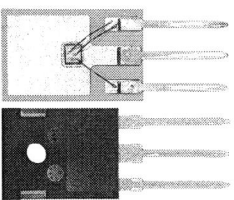

Fig. 1. Finished sample with DCB, die, bond wires and leads, side by side with a commercial TO-247 packaged device housing the same die.

II. APPROACH

During short-circuit, a high-speed camera is used to record the visible processes on the die. This allows imaging the sequence of events during a single test and correlating it with the electrical data for this specific test and device. High frame rates even allow observing highly dynamic tests at $V_{DS} = 800\,V$. After the test, optical and electrical analysis as well as imaging of cross-sections is performed. Simulation is used to further investigate the failure process.

A. Test setup

The short circuit test setup consists of a low inductive capacitor bank, an array of auxiliary switches and a mounting space for the TO-247 DUTs. The gate driver and the measurement connections are placed as close as possible to the DUT. A $400\,MHz$ shunt is used to measure the drain current I_D, $500\,MHz$ voltage probes record V_{DS} and V_{GS}. Measurements are run at $10\,GS/s$ and 10 bit. I_D is recorded with two channels to get optimum resolution both before ($<300\,A$) and during failure ($<3500\,A$). High-speed images are taken at up to 2.1 million frames per second (FPS); lower speeds are used to improve image resolution. Using a macro lens, observation of the whole die or parts of it is possible.

B. Sample manufacturing

Cree 2nd generation $1200\,V$, $80\,m\Omega$ dice are soldered to TO-247 sized DCBs using SAC 305 solder paste. They are then wire-bonded using $300\,\mu m$ aluminium wire, resulting in a $33\,\%$ larger cross-section of the source wires compared to the commercial devices. Finally, the samples are equipped with

978-1-7281-0582-6/19 $31.00 © 2019 IEEE   219

TO-247 leads, see Fig. 1. The construction allows using the same characterization and test setups as for the commercially packaged devices with the same dice. The differences in construction and the lack of encapsulation can be shown not to result in different short circuit behavior, as the heating is confined to the die for the short duration of the test.

C. Sample verification

To verify the samples indeed behave the same as the commercial, TO-247-packaged equivalents, characterization and short-circuit tests are compared. IV-characterization after fabrication shows results equal to the reference devices. In all samples, the breakdown voltage in air is above $1000\,\text{V}$, confirming that encapsulation is not necessary for the tests. Nevertheless, some samples are covered with a clear conformal coating to observe the effect of an organic layer on top of the die. Several short circuit tests show that behavior before and during failure as well as short-circuit withstand time (SCWT) are equivalent for all devices.

III. TESTING

A. Electrical measurement

Short circuit tests of 19 samples plus reference devices are run at $V_{\text{DS}} = 400\,\text{V}$ and $V_{\text{DS}} = 800\,\text{V}$. At $400\,\text{V}$, gate damage is observed starting around $14\,\mu\text{s}$ SC duration, becoming stronger with SC duration. At $18\,\mu\text{s}$ there is significant gate leakage, but delayed drain-source failure has also been observed in Fig. 2a. Up to $20\,\mu\text{s}$ however, the device often does not experience drain-source failure, see Fig. 2b. At $22\,\mu\text{s}$, drain-source breakdown is likely to occur before turn-off and the current quickly rises beyond $1\,\text{kA}$, see Fig. 2c. Packaged devices are usually destroyed at this point. At $800\,\text{V}$, the curves are very similar, but gate damage happens from $3.5\,\mu\text{s}$ and failures occur already at $4.5\,\mu\text{s}$.

B. High-speed imaging

For all non-encapsulated samples, high speed camera footage is obtained, which can be compared to the electrical measurement for each test. All surface events can be seen, which includes the progression of melting of the top metalization and the explosive failure if drain-source breakdown occurs. Fig. 3 and 4 show the same tests as Fig. 2 b and c, respectively. The onset of melting (Fig. 3b) is visible as small change in surface structure, a liquid surface of the source pads (Fig. 3c) can be recognized from their dark appearance, as the light is reflected away from the camera. Further heating again results in a rougher surface (Fig. 3d) and bulging polyimide passivation. The surface then becomes so hot it appears white (Fig. 3e), around the time the power is turned off. After the test, the pads again appear black, as the metal re-solidifies with a smooth surface (Fig. 3f). If a failure occurs, hot metal fragments immediately cover the surface of the die (Fig. 4d) and rapidly expand outwards (Fig. 4e). To see more details of the failure instant, several tests are done without lighting and reduced aperture, but the brightness of the hot metal is

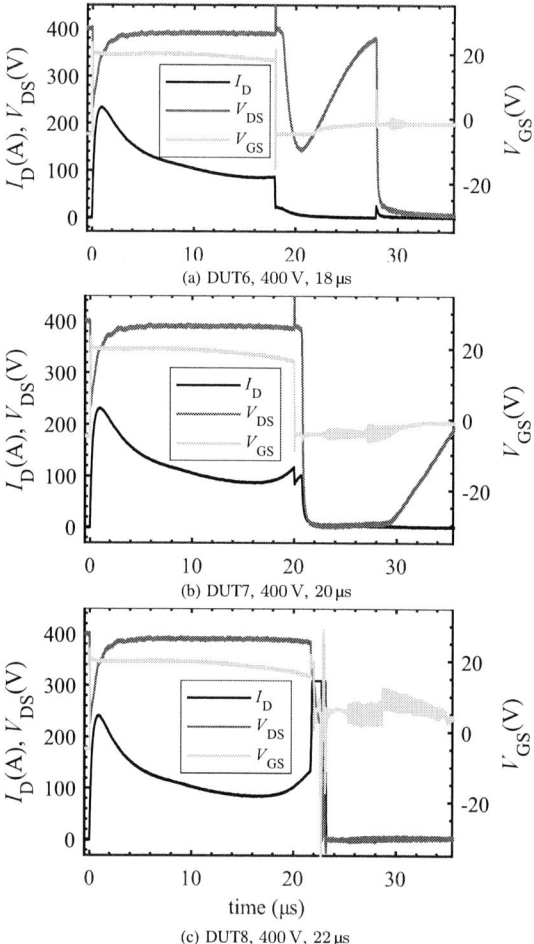

Fig. 2. Electrical measurements of SC tests at $400\,\text{V}$ with progressively longer duration. After a peak of $240\,\text{A}$, the current decreases, then starts to rise again. (a) shows successful turn-off, gate damage and, unusually, delayed failure without current. (b) shows partial turn-off, the auxiliary switch then interrupts the current and V_{DS} rises, indicating intact drain-source isolation but damaged gate. (c) shows gate damage and drain-source failure with high current (clipped at $300\,\text{A}$).

so high that the localized origin of the failure is immediately overexposed.

To accurately capture the melting process and allow conclusions on the surface temperature over time, additional tests are conducted with the maximum possible frame rate, imaging a small section of the source pads. Fig. 5 and 6 show the results for $400\,\text{V}$ and $800\,\text{V}$, respectively. For $400\,\text{V}$, the onset of melting is found at $5.7\,\mu\text{s}$, when the first significant change in brightness is measured on the source pad. After $12.4\,\mu\text{s}$, the surface appears dark, as it is smooth and reflects the light away from the camera, indicating it has completely melted. For $800\,\text{V}$, the same is observed at $2.4\,\mu\text{s}$ and $8.6\,\mu\text{s}$.

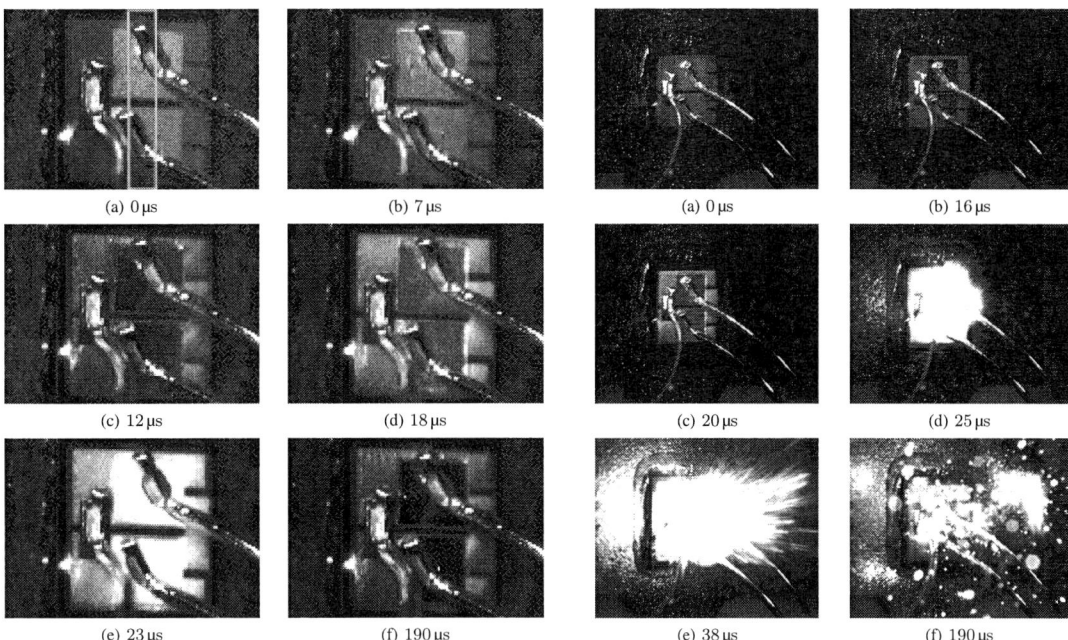

Fig. 3. Images of DUT7 during a test at $400\,\mathrm{V}$ for $20\,\mu\mathrm{s}$, $450\,\mathrm{kFPS}$, corresponds to Fig 2b. (a): beginning of test. Marked area is shown in Fig. 5 and 6. (b): pad surface starts melting. (c): surface is molten. (d): Surface begins to bulge. (e): rough surface, a few small Al fragments are ejected. (f): Ejection has stopped.

Fig. 4. Images of DUT8 during a test at $400\,\mathrm{V}$ for $22\,\mu\mathrm{s}$, $225\,\mathrm{kFPS}$, corresponds to Fig 2c. (a): beginning of test. (b): pad surface is molten. (c): surface is bulging. (d): explosive failure starts. (e): Current is off, Al fragments are ejected. (f): Ejection has stopped, one bond wire is loose.

C. Post mortem analysis

After the tests, the pad surfaces are found to be smooth and highly reflective, as can be seen in Fig. 7. If explosive failure occurred, a localized crater measuring some $100\,\mu\mathrm{m}$ is found (Fig. 7d,e). Typically, the crater forms under or close to a bond wire, detaching it from the pad. Cross-sectioning of such a crater as seen in Fig. 8 shows a depth of $70\,\mu\mathrm{m}$, exposing the highly doped substrate under the $\sim\!17\,\mu\mathrm{m}$ thick active layers.

IV. SIMULATION

A detailed 2D structure of the MOSFET's active cell is generated, including over-layers of dielectrics and $4\,\mu\mathrm{m}$ source aluminum layer, which is necessary for accurate electro-thermal (ET) simulations. 2D ET simulations are done in Sentaurus Device using the "Thermodynamic" model, full anisotropy and temperature dependence of thermal material properties. Aluminum melting ($933\,\mathrm{K}$) is accounted for by increasing its heat capacity at higher temperatures. The simulation results can be compared to measurements and observation of the beginning of melting, see Fig. 9. The simulated melting of aluminium begins at $5.97\,\mu\mathrm{s}$, closely matching the $5.7\,\mu\mathrm{s}$ observed in Fig. 5b. The same is true at $800\,\mathrm{V}$, with $2.48\,\mu\mathrm{s}$ matching $2.4\,\mu\mathrm{s}$ from Fig. 6b. This validates the temperature simulation, which further shows above $1400\,\mathrm{K}$ at the time of drain-source failure. The difference between simulation and measurement

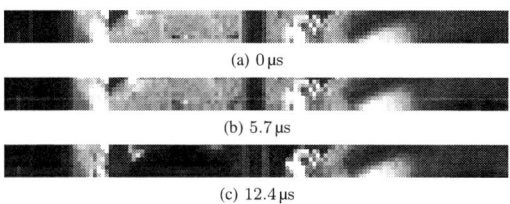

Fig. 5. Images of DUT9 tested at $400\,\mathrm{V}$ for $10\,\mu\mathrm{s}$, $2100\,\mathrm{kFPS}$. Source pads and bond wires shown, see Fig. 3a. Brightness measurement on marked source pad area. (a): beginning of test, brightness $65.2\,\%$. (b): pad surface starts melting, brightness $63.3\,\%$. (c): surface is smooth and reflective, brightness $7.0\,\%$.

close to failure can be explained by deficiencies in the the pn-junction leakage model, as well as individual cells deviating from simulated average behavior.

V. CONCLUSION

High-speed optical images allow determination of surface events affecting the source metal. The combination of optical imaging and electrical measurements, both during and after the test, shows when which damage occurs. At $400\,\mathrm{V}$, the source metal begins melting at $5.7\,\mu\mathrm{s}$, it is completely molten at $12\,\mu\mathrm{s}$ and material is ejected at $23\,\mu\mathrm{s}$. At $800\,\mathrm{V}$, the aluminium melts earlier, starting at $2.4\,\mu\mathrm{s}$. If the short-circuit is maintained, an explosive drain-source failure, accompanied by very high currents, will occur. Permanent drain-source failure can also

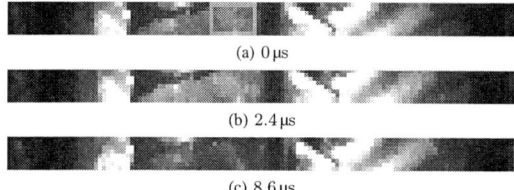

(a) 0 µs

(b) 2.4 µs

(c) 8.6 µs

Fig. 6. Images of DUT14 tested at 800 V for 3.5 µs, 2100 kFPS. Source pads and bond wires shown, see Fig. 3a. Brightness measurement on marked source pad area. (a): beginning of test, 55.9 % brightness. (b): pad surface starts melting, peak brightness at 61.6 %. (c): surface is smooth and reflective, remains this way permanently, brightness 33.9 %.

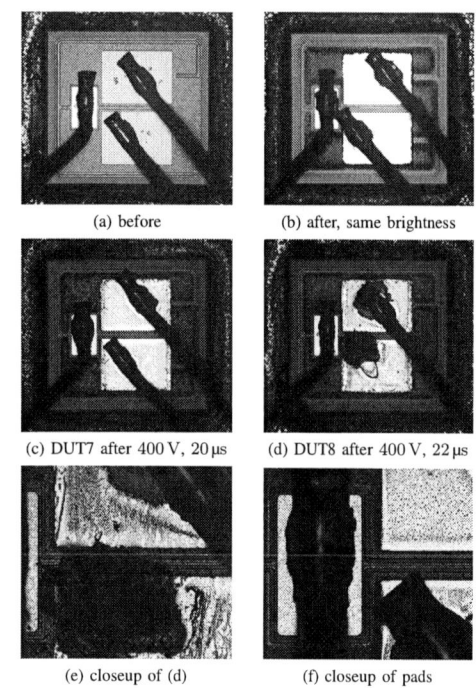

(a) before

(b) after, same brightness

(c) DUT7 after 400 V, 20 µs

(d) DUT8 after 400 V, 22 µs

(e) closeup of (d)

(f) closeup of pads

Fig. 7. (a) shows a device before testing. (b) shows the high reflectivity of the source pads after melting, the images below are taken at lower brightness. (c) and (d) show reflective, uneven metal surface, with (d) also exhibiting an explosion crater. (e) shows more closely the crater on (d), while (f) shows the surface features of source (melted) and gate (not melted) pads.

occur with delay, triggered not by active current and heating but by heat spreading to sensitive regions. This failure creates a localized, highly conductive path between substrate (drain) and top metalization (source). The failure location is visible as bright spot on the die; it cannot be determined if molten metal or plasma in the form of an electrical arc create the conductive path. If catastrophic failure does not occur, the devices sustain gate damage, but retain drain-source isolation. A detailed simulation matches the observed timing of the metal melting and shows the temperature of an average cell at failure reaches 1400 K.

Fig. 8. Cross-section through the crater shown in Fig. 7(e). The die is in the center, with solder and DCB below. The crater is 70 µm deep, allowing current flow from source metal to substrate.

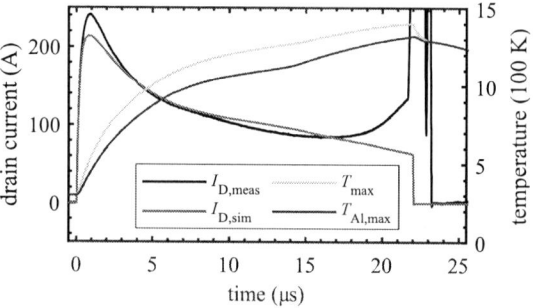

Fig. 9. Simulation of SC in the investigated MOSFET at 400 V. The maximum temperature in the aluminium reaches melting point at the time observed in Fig. 5b. At failure, the temperature of an average cell reaches 1400 K.

REFERENCES

[1] B. Kakarla, T. Ziemann, R. Stark, P. Natzke, and U. Grossner, "Short Circuit Ruggedness of New Generation 1.2 kV SiC MOSFETs," in *2018 IEEE 6th Workshop on Wide Bandgap Power Devices and Applications (WiPDA)*, pp. 118–124.

[2] A. Castellazzi, A. Fayyaz, L. Yang, M. Riccio, and A. Irace, "Short-circuit robustness of SiC Power MOSFETs: Experimental analysis," *Proceedings of the International Symposium on Power Semiconductor Devices and ICs*, pp. 71–74, 2014.

[3] G. Romano, L. Maresca, M. Riccio, V. D'Alessandro, G. Breglio, A. Irace, A. Fayyaz, and A. Castellazzi, "Short-circuit failure mechanism of SiC power MOSFETs," *Proceedings of the International Symposium on Power Semiconductor Devices and ICs*, pp. 345–348, 2015.

[4] S. Mbarek, P. Dherbécourt, O. Latry, and F. Fouquet, "Short-circuit robustness test and in depth microstructural analysis study of SiC MOSFET," *Microelectronics Reliability*, vol. 76-77, pp. 527–531, 2017.

[5] D. Pappis and P. Zacharias, "Failure modes of planar and trench SiC MOSFETs under single and multiple short circuits conditions," *19th European Conference on Power Electronics and Applications*, pp. 1–11, 2017.

[6] B. Kakarla, T. Ziemann, S. Nida, E. Doenni, and U. Grossner, "Planar to Trench: Short Circuit Capability Analysis of 1.2 kV SiC MOSFETs," in *Silicon Carbide and Related Materials 2017*, ser. Materials Science Forum, vol. 924, Trans Tech Publications, pp. 782–785.

[7] C. Tu, J. Chen, Z. J. Shen, X. Yang, J. Lu, X. Jiang, J. Wang, and Z. Li, "Failure modes and mechanism analysis of SiC MOSFET under short-circuit conditions," *Microelectronics Reliability*, vol. 88-90, no. July, pp. 593–597, 2018.

[8] G. Romano, A. Fayyaz, M. Riccio, L. Maresca, G. Breglio, A. Castellazzi, and A. Irace, "A comprehensive study of short-circuit ruggedness of silicon carbide power MOSFETs," *IEEE Journal of Emerging and Selected Topics in Power Electronics*, vol. 4, no. 3, pp. 978–987, 2016.

978-1-7281-0582-6/19 $31.00 © 2019 IEEE

Proceedings of the 31st International Symposium on Power Semiconductor Devices & ICs
May 19-23, 2019, Shanghai, China

Design Considerations for High Voltage SiC Power Devices: An Experimental Investigation into Channel Pinching of 10kV SiC Junction Barrier Schottky (JBS) Diodes

Jusitn Lynch
Colleges of Nanoscale Science and Engineering
State University of New York Polytechnic Institute
Albany, NY 12203, USA
jlynch@sunypoly.edu

Nick Yun
Colleges of Nanoscale Science and Engineering
State University of New York Polytechnic Institute
Albany, NY 12203, USA
nyun@sunypoly.edu

Woongje Sung
Colleges of Nanoscale Science and Engineering
State University of New York Polytechnic Institute
Albany, NY 12203, USA
wsung@sunypoly.edu

Abstract—This paper reports on the design, fabrication and electrical characteristics of 10kV 4H-SiC JBS diodes. Both Ni and Ti JBS diodes were fabricated on 7×10^{14} cm^{-3} doped, 95µm- thick n-type epi on 6-inch 4H-SiC n+ substrates. SEM imaging and SIMS profiling were conducted to investigate the unexpected high knee voltages from Ti JBS diodes. The "extensive" straggle by implanted ions was discovered from this investigation, which was attributed from the low background doping concentration in the drift layer.

Keywords—Junction Barrier Schottky Diode, JBS, 4H-SiC, 10kV, Longitudinal straggle, Lateral straggle, High voltage design considerations, Edge termination

I. INTRODUCTION

The superior electrical and thermal properties of Silicon Carbide (SiC) have long made it the preferred material for the fabrication of high voltage and high temperature electronic devices. The high critical electric field of SiC, a material property common for wide bandgap materials, allows for the voltage supporting drift layers to be designed thin and highly conductive. With low defect density and thick drift layers, SiC devices can be designed to withstand large voltages, on the order of 10kV, while still maintaining low leakage and reasonable chip size. It is for this reason, that 10kV Junction Barrier Schottky (JBS) diodes are being developed to withstand high voltages while maintaining low forward voltage drops.

For SiC power devices, unipolar devices are preferred for applications that call for fast and efficient switching. SiC JBS diodes, unipolar in nature, are able to exploit the advantages of both SiC Schottky Barrier Diodes (SBDs) and P-i-N diodes. The low leakage currents, high breakdown voltages, and high surge current capabilities have made the SiC JBS diode an attractive power device for various high voltage applications. Further development and optimization of JBS diodes in 10kV voltage rated range is a logical next step for future higher voltage applications. Minimizing forward voltage drops and switching losses is needed to capture exciting new markets and for continued improvement of device efficiency.

In this paper we present the electrical characteristics of 10kV SiC JBS diodes. The proposed diodes exhibited impressive blocking characteristics and demonstrated a high

of 88% blocking efficiency due to superior edge termination techniques. SEM imaging and subsequent SIMS profiling of the fabricated diodes showed exceptionally large longitudinal and lateral straggle, nearly 1.0µm in both directions, causing a "channel pinching" of the fabricated JBS diodes. It is widely known that Aluminum (Al) implants show longitudinal straggle, but recently the effect has been shown to increase with multiple energy implants. This effect, termed the "scatter-in channeling effect", was found in moderately doped (~4×10^{15}cm^{-3}) epi layers by *Mochizuki et al* [1]. Lateral straggle has also been previously reported in literature [2, 3], though this paper provides new experimental data and SEM imaging for the increased straggle in a lightly doped drift layer, and its negative effect on the device performance. This paper will provide the following details of the recent findings: Section II, a thorough discussion of device design and fabrication method for the proposed high voltage SiC JBS diodes. Section III, electrical characteristics of the fabricated devices are provided. Additionally, SEM images and relevant data to the channel pinching are presented along with a discussion of design considerations need for the channel pinching. Section IV, conclusion and proposed design considerations.

II. DEVICE DESIGN AND FABRICATION TECHNOLOGY

A. Device Design

Fig. 1 shows the cross-sectional diagrams of the 10kV JBS diodes fabricated on a single 6-inch 4H-SiC wafer. When a positive bias is placed on the anode metal of the diode, the device operates in forward conduction mode. As long as the applied bias is able to overcome the Schottky barrier, electrons flow through the Schottky contact from the n+ Cathode. The JFET width is optimized in a manner that the width is large enough to provide a sufficient "channel" for electron flow during forward conducting operation. On the other hand, a narrow JFET region is preferred to decrease leakage current during the blocking operation, as it decreases the electric field at the Schottky contact. In general, the narrowing of the JFET region causes increased resistance requiring an enhanced doping by using a JFET implantation.

978-1-7281-0582-6/19 $31.00 © 2019 IEEE   223

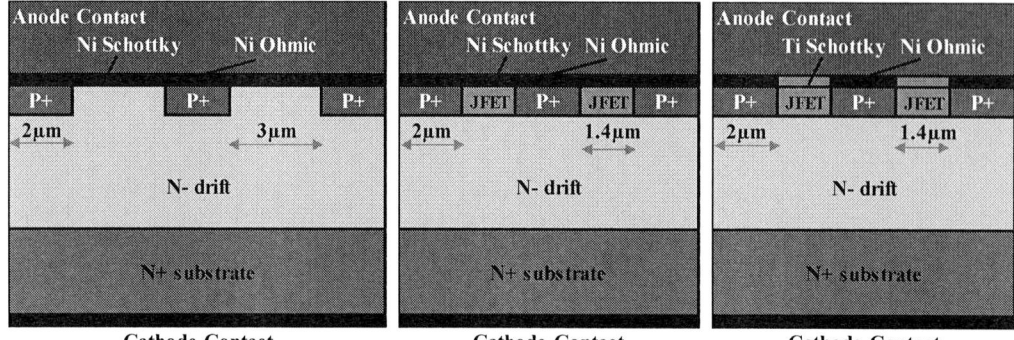

Fig. 1: Schematic diagram of the fabricated JBS diodes. Shown in order of left to right is JBS1, JBS2, and JBS3. Schottky width was designed as 3μm for JBS1 and 1.4μm for JBS2 and JBS3. The width and (expected) depth of the P+ region was 2μm and 0.7μm for each structure. A JFET implantation was only included in JBS2 and JBS3. It is important to note the depth of the JFET implantation was designed at 0.7μm.

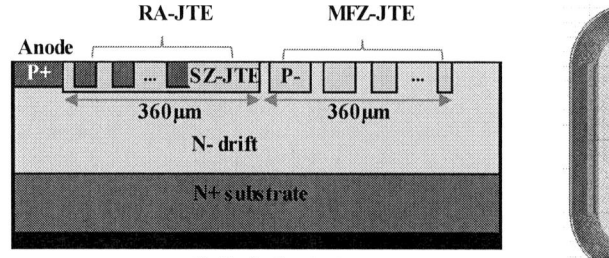

Fig. 2: Schematic diagram of the Hybrid-JTE edge termination used for the JBS didoes. The RA-JTE consists of 18-, 3μm wide P+ rings increasingly spaced in a 360μm P- JTE. The MFZ-JTE consists of 36-zones of decreasing width increasingly spaced to fill 360μm.

Fig. 3: Top view layout of JBS3 with highlighted dimensions of the active area and Hybrid-JTE edge termination structure. The Hybrid-JTE structure spans a total width of 720μm.

It is critical to determine the JFET doping concentration, as too high of a doping can lead to a high electric field at the Schottky contact. JBS2 (Ni Schottky) and JBS3 (Ti Schottky) diodes are designed with a narrow distance between P+ grids (Schottky width or JFET width). In contrast, JBS1 (Ni Schottky) diode did not receive the JFET implants due to the wide JFET width. During blocking operation, a positive cathode bias is supported across the drift layer. The P+ grids provide a shielding effect to the Schottky regions, leading to a decrease in leakage current. Additionally, under a surge operation of the JBS diode, the voltage drop at the P+/drift junction becomes large enough to overcome the built-in potential of the junction, and minority carriers are flooded into the drift allowing increased surge capability. The width of the P+ must be carefully optimized in order to provide both shielding and surge benefits. Device dimensions are included in Fig. 1.

In order to achieve the highest possible breakdown voltages, the devices were designed with a superior Junction Termination Extension (JTE) based edge termination called "Hybrid-JTE" [4]. A cross-sectional schematic of the Hybrid-JTE is provided in Fig. 2. In order to accurately develop devices, it is crucial to use the most efficient and effective edge termination technique. The Hybrid-JTE provides an efficient edge termination for a wide range of JTE implant dose. The edge termination optimization was performed following the techniques described in [4]. Edge termination parameters and device dimensions, including Schottky width, JFET doping concentration, and P+ grid width and depth, were optimized using 2-D device width

and depth, were optimized using 2-D device simulation. The top view schematic layout and a closer look at the novel 10kV edge termination used are presented in Fig. 3.

B. Fabrication Technology

The proposed SiC lateral diodes were fabricated in a 6-inch wafer foundry company, X-FAB, TX, U.S.A. Fig. 4 shows the process flow used for fabrication. A 95μm thick, 7×10^{14} cm^{-3} doped N-type drift layer on a 6-inch, N+ 4H-SiC substrate was used for fabrication. Aluminum ion implants were used for making the P+ region, while Nitrogen ion implants formed the JFET region. Implantation steps were followed by a 1650°C 10-min activation anneal with a carbon cap. Interlayer dielectric was deposited and patterned for ohmic contact and Ni Schottky contacts, followed by Ni deposition and a 900°C RTA for 2-min. Then, a subsequent Schottky opening process (oxide etch) was conducted, which was followed by the top metal (Ti/Al) deposition and patterning process. The Ti/Al stack, which also formed the Ti Schottky contact, was followed by a passivation layer of nitride and polyimide and subsequent opening of the pad area. Finally, the wafer was flipped for the backside Cathode metal deposition

III. RESULTS/DISCUSSION

A. Electrical Charteristics

Fig. 5 shows a typical blocking curve of the fabricated JBS diodes. Near ideal blocking voltages were achieved using the Hybrid-JTE structure. Fig. 6 and Fig.7 display the forward conduction operation of the diodes at room

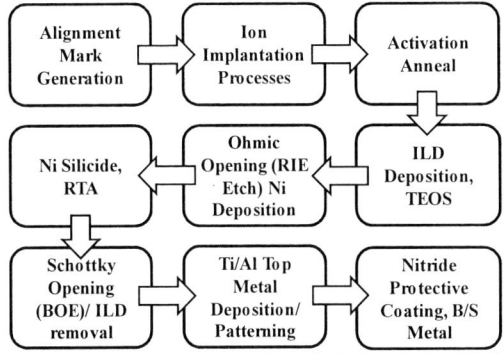

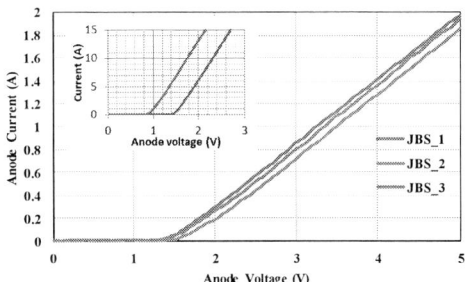

Fig. 4: Process flow for the proposed JBS diodes. For JBS1 and JBS2 a single RTA (900°C) and metal (Ni) was used to simultaneously form Schottky on n- epi and ohmic contact on P+ region. JBS3 utilizes the Ti-based top metal stack as the Schottky on n- epi.

Fig. 6: Forward I-V characteristics of the fabricated JBS diodes at room temperature (25°C). The inset graph shows the forward I-V characteristics of a 1.2kV Ti (green) and Ni (red) JBS diodes.

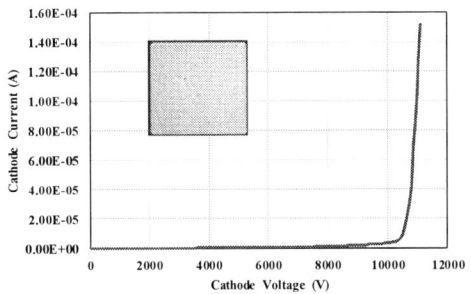

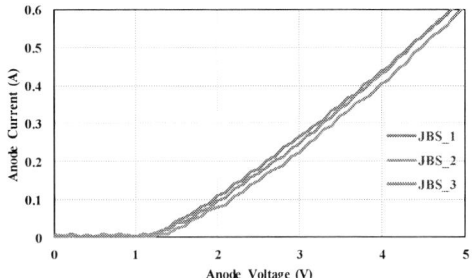

Fig. 5: Typical reverse blocking behaviors of the fabricated JBS diodes. The JBS diode used for this curve is JBS1, which shows breakdown at a blocking voltage of 10800V. An 86% of the ideal breakdown voltage was achieved using Hybrid-JTE.

Fig. 7: Forward I-V characteristics of the fabricated JBS diodes at a high temperature (150°C). Still present is the increased knee voltage for JBS3.

temperature (25°C) and high temperature (150°C), respectively. As observed in the current-voltage (I-V) behavior, JBS3, which utilizes Titanium for the Schottky contact, has an increased knee voltage compared to lower voltage rated devices as shown in the inset of Fig. 6 (from our previous work). The measured knee voltage of the 10kV Ti JBS diode (JBS3), 1.5V, should be comparable to the 1.2kV JBS diode 0.9V. Additionally, an increased knee voltage is seen in JBS2 compared to JBS1, despite the use of the same Schottky metal (Ni). The cause of these unexpectedly high knee voltages is seen in a SEM cross-sectional image of JBS3 in Fig. 8. The SEM image shows the unexpected shape of the P+ implant, which is not present in the 1.2kV device with the same implant schedule. Our fabricated high voltage devices show increased longitudinal and lateral straggle of the P+ grids. As discussed in the introduction, *Mochizuki et al* has shown increased straggle in moderately doped drift layers ($\sim 4 \times 10^{15}$cm^{-3}). From these studies, Al implants experience significant longitudinal straggle in SiC, with the doping concentrations of the straggled implants around 1×10^{15}cm^{-3}. This extreme longitudinal straggle is due to the scatter-in channeling effect [1].

B. Channel Pinching

The implant schedule for the P+ implant contained multiple energy implants with a total does of 5×10^{13}cm^{-2}.

The low energy implants used to make shallow part of the P+ profile in the JBS diode causes amorphization and lattice damage, which is normally cured through the high temperature post-implant anneal process. Before this process is preformed, the high energy implants needed for the deep junctions, a requirement for high voltages are scattered as they enter the substrate and are directed in random directions [1]. This random scattering leads to a channeling effect experienced by a fraction of implanted ions, which causes an increase in longitudinal straggle at relatively low doping concentrations. The high voltage devices experience a longitudinal straggle of 140% of the projected depth of P+ implant compared to the non-existent longitudinal straggle in the low voltage devices. This can be attributed to the drift doping of the low voltage devices being well higher than the 10kV devices. With this high doping, the expected concentration of the straggled implanted dopants is compensated and suppressed. In the high voltage drift layers, the doping concentration is 7×10^{14}cm^{-3}, which is overcompensated by the straggled implants. In addition to the increased longitudinal straggle, a large lateral straggle was also observed. A lateral straggle of 0.2μm can be seen in the portion of the drift region enhanced with JFET doping. This matches up well with lateral straggle seen in implants into drift regions with comparable doping of the JFET doping [3, 4]. A 0.5μm straggle was observed in the region without JFET doping. The images show that when drift doping is low, the straggle of implants increases considerable compared to the moderately doped drift regions.

978-1-7281-0582-6/19 $31.00 © 2019 IEEE

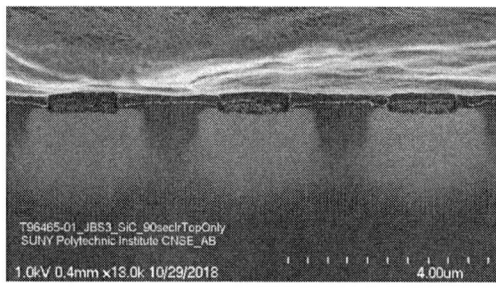 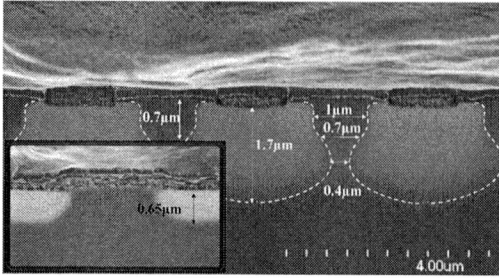

Fig. 8: Cross-sectional SEM images of JBS3. The left image shows the original SEM. The right image has additional white lines to outline the P+ implants and give approximated dimensions. The P+ implants show increased lateral straggle after 0.7μm as well as a larger implant depth than expected. The conduction channel narrows to 0.4μm at a depth below the expected P+ implant depth. The inset in the image on the right is a close up of the same P+ implant in a 1.2kV JBS diode. The implants in the inset show an extreme reduction in both depth and lateral straggle.

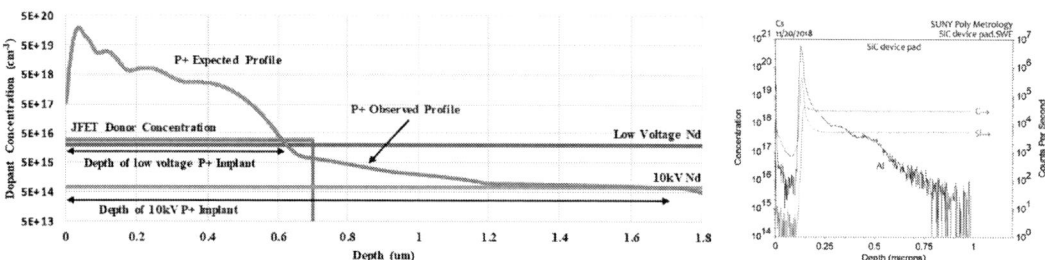

Fig. 9: Diagram of the doping of JBS3 based on the SEM data presented in Fig. 8. The expected P+ profile (orange) was calculated using SRIM software, which does not account for the tail observed, from scatter-in channeling. The measured P+ profile was approximated from the SEM images. Drift (N_d) and JFET concentrations are assumed constant. It is clearly seen the lower donor concentration in the 10kV device allows for the unexpected P+ implant depth not seen in low voltage devices. SIMS pattern is shown on the right, but a concentration below $1 \times 10^{15} \text{cm}^{-3}$ could not be detected.

A schematic of the P+ implant profile is provided in Fig. 9. Due to the drastic difference in drift doping compared to JFET doping, the acceptor concentration in the tail of the P+ implant remains dominant, and expands further vertically, and horizontally (due to lateral straggle). SIMS patterning, shown on the right of Fig. 9, was conducted, but due to a sensitivity limit above $1 \times 10^{15} \text{cm}^{-3}$ and a depth limit of 1μm, conclusive data could not be obtained. This channel pinching effect, not seen in lower blocking voltage rated devices due to the heavier doped drift regions, is thought to cause excess depletion, pinching the JFET channel (underneath the Schottky region). This pinching is believed to be the reasoning for increased knee voltages seen in JBS2 and JBS3, as an additional voltage is required to "un-pinch" the channel. The pinching effect is also thought to increase device resistance, but due to the high resistance of the drift region this increased resistance is diluted

IV. CONCLUSION

10kV SiC JBS diodes were fabricated with unexpectedly high voltage drops compared to lesser rated devices of the same kind. After investigation, it was found that the reason for this high knee voltage was due to the channel pinching of the JBS diodes, due to unexpected longitudinal and lateral straggle. The cause of this unexpected straggle was the low doping concentration of the drift layer, a parameter that is necessary for high rated SiC devices. ***Designers of high voltage power devices must take this effect into consideration*** to ensure the best possible electrical characteristics. Possible solutions to consider are: (1) eliminating the JFET implantation combined with more

relaxed device dimensions (2) increasing JFET implantation depth (3) mid implantation lattice recovery anneal (4) implement an implant schedule that incorporates decreasing implant energies. Even with implantation of the techniques listed above, designers of high voltage devices on lowly doped drift layers need to be wary of the lateral straggle of high energy implants.

ACKNOWLEDGMENT

The information, data, or work presented herein was funded in part by the Office of Energy Efficiency and Renewable Energy (EERE), U.S. Department of Energy, under Award Number DE-EE0006521 with North Carolina State University, PowerAmerica Institute.

REFERENCES

[1] K. Mochizuki, T. Someya, T. Takahama, H. Onose, N. Yokoyama, "Detailed Analysis and Precise Modeling of Multiple-Energy Al Implantations Through SiO2 Layers Into 4H-SiC." IEEE Transactions on Electron Devices, vol. 55, no. 8, Aug. 2008

[2] J. Nishio et al., "Ultralow-Loss SiC Floating Junction Schottky Barrier Diodes (Super-SBDs)", IEEE Trans. On Electron Dev., Vol. 55, No. 8, Aug 2008

[3] K. Mochizuki, N. Kameshiro, H.Onose, N. Yokoyama, "Influence of Lateral Spreading of Implanted Aluminum Ions and Implantation-Induced Defects on Forward Current-Voltage Characteristics of 4H-SiC JBS Diodes", IEEE Trans. On Electron Dev., Vol. 56, No. 5, May 2009

[4] W. Sung, and B.J. Baliga, "A Near Ideal Edge Termination Technique for 4500V 4H-SiC Devices: the Hybrid Junction Termination Extension (Hybrid-JTE)." IEEE Electron Device Letters, Vol. 37, No. 12, December 2016

Proceedings of the 31st International Symposium on Power Semiconductor Devices & ICs
May 19-23, 2019, Shanghai, China

Experimental Investigations of SiC MOSFETs under Short-Circuit Operations

Lei Cao, Zijian Gao, Qing Guo*, Kuang Sheng

College of Electrical Engineering, Zhejiang University, Hangzhou, China

*E-mail: guoqing@zju.edu.cn

Abstract—The short-circuit capability of power MOSFETs is crucial for the fault protection. In this work, short-circuit ruggedness and failure mechanisms of SiC planar and trench MOSFETs are compared and analyzed. Single-event short circuit tests are demonstrated. Experimental results show that thermal effect and gate stress are the two important factors which result in performance degradation and damage of SiC MOSFETs under short circuit operations. High gate stress will lead to hot carrier injection effect as well as tunneling effect under strong perpendicular electric field. The high current density and electric field at Pwell/ Ndrift junction cause a high junction temperature, eventually leading to device failure. A temperature limited single-event failure SOA and drive-voltage limited degradation SOA are demonstrated.

Keywords—silicon carbide (SiC) power MOSFET; short-circuit (SC) capability; failure SOA; degradation SOA

I. INTRODUCTION

As wide bandgap semiconductor device, SiC power MOSFETs deliver superior performance over Si counterparts in high temperature, high voltage and high-speed switching applications[1]. After years of research and study, the development of SiC MOSFETs has rapidly progressed. SiC MOSFETs have been successfully demonstrated for practical use in order to improve power density and efficiency[2],[3]. However, SiC chips cost significantly higher than silicon. Certain robustness and reliability issues remain to be fully quantified to minimize unnecessary and costly over-design practices.

The commercial SiC MOSFETs include planar MOSFETs and trench MOSFETs. Trench MOSFETs offer a better on-state resistance than planar MOSFETs due to the elimination of the JFET region [4]-[6].Their cell pitch can be easily scaled down, leading to a higher cell density. However, the increased electric field in the gate oxide bottom when operating over extended durations leads to long-term reliability issues[7],[8].

In this work, the differences between the short circuit capability of commercially available SiC planar MOSFETs and trench MOSFETs are interpreted. Different phenomena under increasing pulse width of single-event short circuit test are presented. The degradation and failure boundaries of device A under short circuit condition are quantified and analyzed

II. CHARACTERIZATION PLATFORM

A. Test Circuit

The short circuit capabilities of discrete SiC MOSFETs are evaluated on the test bench demonstrated in Fig.1. The short circuit waveforms are measured varying DC bus voltage, gate drive voltage as well as short-circuit pulse width.

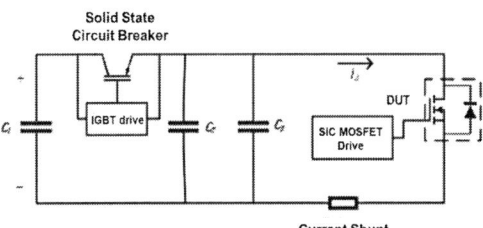

Fig. 1. Schematic circuit for the short circuit test. The short circuit current, drain-source voltage and gate-source voltage are monitored

The capacitor C_1 was firstly charged up by the DC power supply voltage V_{dc}. The SiC MOSFET drive can generate a square wave to go through the device under test (DUT). A coaxial current shunt was used to measure the current flowing through the DUT. After each short circuit test, the leakage current and threshold voltage measurements were carried out to determine DUT degradation during the previous test.

B. Characteristics of DUTs

In this work, the short circuit capabilities of planar and trench SiC MOSFETs with similar ratings are compared. The key parameters of the DUTs are given in Table I. The I-V curves and transfer characteristics curves are shown in Fig. 2(a) and 2(b).

Table I. Key Parameters of the Devices Under Test

Parameters	Device A	Device B	Device C
Corporation	I	II	II
Device Type	Planar	Planar	Trench
Voltage/Current Rating	900V/11.5A	650V/29A	650V/21A
R_{DSON} [a]	250mΩ	200mΩ	195mΩ
Recommended Drive Voltage	15V	18V	18V
Active Region	$0.0090cm^2$	$0.0549cm^2$	$0.0218cm^2$

* The R_{DS_ON} values are extracted from the output curves at the recommended drive voltage.

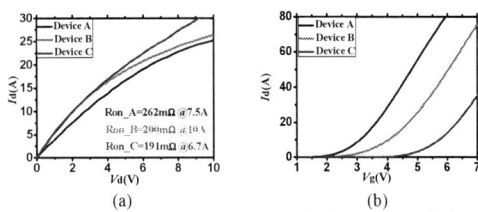

Fig. 2. (a) I-V characteristics and of DUTs under the recommended V_{GS_ON}. (b) Transfer characteristics of DUTs

978-1-7281-0582-6/19 $31.00 © 2019 IEEE   227

III. EXPERIMENTAL RESULTS

A. Differences between DUTs

The short circuit waveforms are measured varying DC bus voltage, gate drive voltage as well as short-circuit pulse width. Through these experiments, it is found that when tested under the same condition, device B (planar MOSFET) is much more robust than device C (trench MOSFET). Higher channel mobility of trench structure and smaller active region area give rise to higher current density. Therefore, the heat generation rate is much higher in trench MOSFET. The junction temperature rises faster and reaches the critical temperature limit of the material in a shorter time. Fig.3 shows the short circuit withstand time of DUTs. The short circuit withstand time of device B and device C increase significantly at a lower gate drive voltage (41% for device B and 44% for device C on average) while the short-circuit withstand time of device A varies little.

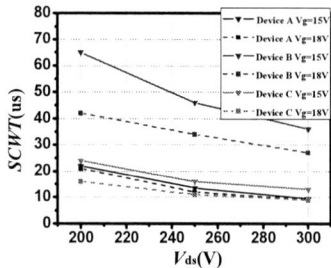

Fig.3 SCWT of three DUTs

Short-circuit currents of different devices during short circuit tests are shown in Fig.4. It can be observed that when operating at a higher gate drive voltage, the short circuit current increases faster to a higher peak current value. The drain current of device A decreases immediately after reaching the peak value at Vg=18V. It can be inferred that the fast increase of junction temperature in device A leads to stronger acoustic scattering and surface roughness scattering, reducing channel mobility of the device immediately [9].

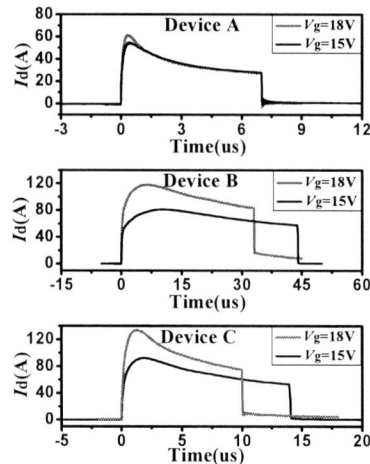

Fig.4 Typical SC waveform of different devices
Test condition: Vdc=300V, Vgs=15/18V

B. Short Circuit Degradation

Device A is tested under Vdc=300V, Vgs=15V condition with increasing SC pulse width (Fig.5). Transfer characteristic curve is measured each time after single pulse short circuit test. The results are shown in Fig. 6(a). It is found that the curve shifts first in negative direction then positive direction.

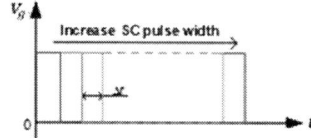

Fig. 5. Single event short circuit test

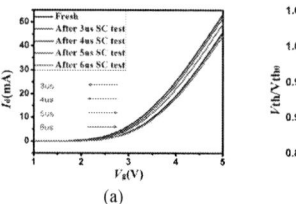
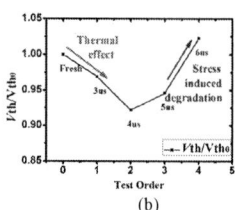

(a)   (b)

Fig. 6. (a)Transfer characteristic of the device before and after SC operation and (b) Vth/Vth0 extracted from the curve (Device A)
Test condition: Vdc=350V, Vgs=15V, Tsc=3us,4us,5us,6us

Junction temperature increases significantly during short circuit operation. The curve shifting in negative direction is attributed to negative temperature coefficient of Vth. The positive shift of transfer characteristics is an indicator of stress induced degradation in gate oxide, which will be illustrated later.

C. Short Circuit Failure Modes

Devices are tested under Vdc=250V, Vgs=15V condition with a starting pulse width of Tsc=10us. We apply successive short circuit pulse with increasing pulse width by the step of 0.5us. As shown in Fig. 7(a), the device still seems to be operational at Tsc=13.5us. However, when we measure the transfer characteristics as well as gate leakage current, it is found that the leakage current has increased from less than 10 pA to above 10 mA, indicating a partial damage of gate isolation. The transfer characteristics curve shifts 0.25V in positive direction. The measurement results are shown in Fig. 7(b). This failure mode exists at different Vdc given an appropriate pulse width. It is used to define failure SOA.

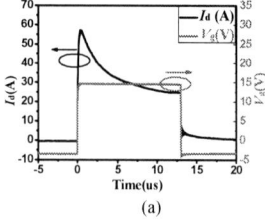
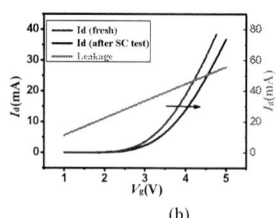

(a)   (b)

Fig. 7. (a) Short circuit current and gate voltage waveform
Test condition: Vdc=250V, Vgs=15V, Tsc=13.5us
(b) Transfer characteristics and leakage current after short circuit test

Two other failure modes can be observed when a longer gate pulse width is applied. Fig.8(a) shows the failure mode when gate junction and source junction are short circuited. The device turns off safely and drain current falls down. However, two microseconds after switch-off, a sudden short-circuit is observed. Gate oxide breakdown and gate interlayer dielectric/ metallization failure [10] are proposed to explain this failure mode. The time between the end of short circuit test and the gate failure may be due to temperature diffusion inside the device, which delays heating inside the gate oxide. Fig.8(b) appears when both gate-source and drain-source junctions are short circuited. Drain current increases several microseconds after switching off

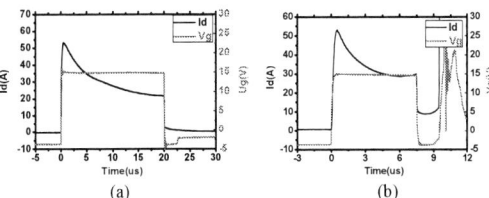

(a)   (b)

Fig. 8. (a) gs shorted failure mode (b) gds shorted failure mode

D. Failure and Degradation SOA

Degradation boundary is defined when the transfer characteristics curve shifts more than 0.15V in positive direction. The failure boundary is defined when an non negligible gate leakage current occurs. The degradation SOA and failure SOA of device A at different test conditions are showed in Fig.9(a) and 9(b). It is found that the device degrades much faster under higher gate stress while the short circuit withstand time doesn't change a lot with gate bias.

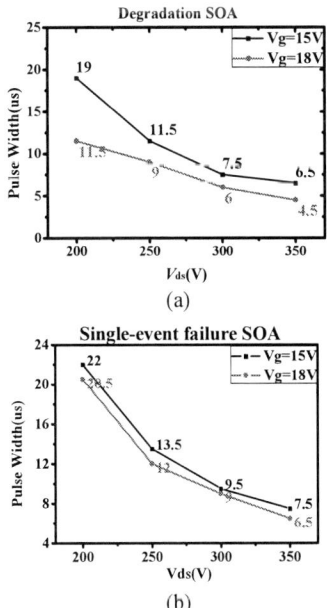

Fig.9. (a) Degradation SOA and (b) Failure SOA of device A

IV. MECHANISM ANALYSIS

A. Impact of The Thermal Effect

Junction temperature increases rapidly during short circuit operation. The single-event failure of device is strongly limited by temperature.

The temperature distribution of the device could be calculated based on the electrothermal model [11]

$$\frac{E(x,t)I(t)}{A} + \frac{\partial}{\partial x}\left(k(T) \cdot \frac{\partial T}{\partial x}\right) = \rho \cdot c(T) \cdot \frac{\partial T}{\partial t}. \quad (1)$$

$\frac{E(x,t)I(t)}{A}$ and $\frac{\partial}{\partial x}\left(k(T) \cdot \frac{\partial T}{\partial x}\right)$ stand for heat generation and dissipation rate during short circuit operation. $\frac{\partial T}{\partial t}$ represents variation in temperature. Similar short circuit current value under Vg=15V and Vg=18V conditions as demonstrated in Fig.10 reflects approximate heat generation rate. As heat generation rate is much higher than heat dissipation rate during the test, it can be inferred that the junction temperature reaches a similar material limit at the end of the short circuit pulse width.

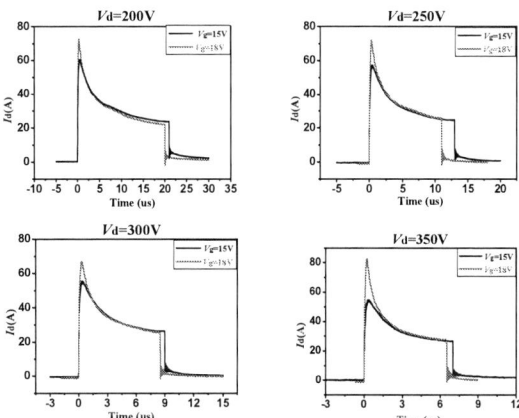

Fig. 10. Last waveform before failure under different DC bus voltage Test condition: Vdc=200/250/300/350V, $V_{gs} = 15V$,

A mathematic model is constructed in Matlab to calculate the temperature distribution over SC operation. It is proved that the maximum junction temperature is similar under Vg=15V and Vg=18V when Vd=300V. The temperature simulation results are provided in Fig.11. It can

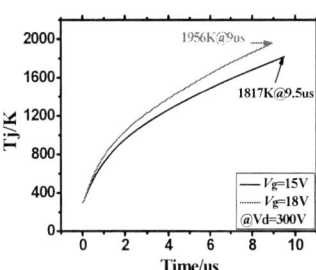

Fig.11. Junction temperature over the last-before-destruction SC operation. Test condition: Vds=300V, Vg=15/18V

be concluded that the single-event short circuit failure is mainly constrained by the temperature limit of SiC material.

B. Impact of The Gate Drive Voltage

Impact ionization at Pwell/ Ndrift junction produces a large amount of electron-hole pairs [12]. When short circuit pulse width continues to increase, carriers gain sufficient energy from high temperature and electric field to cross the interface potential barrier and enter the oxide layer. FN tunneling through the potential barrier also leads to long-term reliability issues. The trapping of electrons within the gate oxide cause shifts in threshold voltage. The transfer characteristics shifts in positive direction. The schematic of these mechanisms are demonstrated in Fig.11. Under higher gate stress, the hot carrier injection effect as well as tunneling effect are more severe. As presented in Fig.9(a), it can be proved in the test results that the device degrades faster when a higher gate stress is applied.

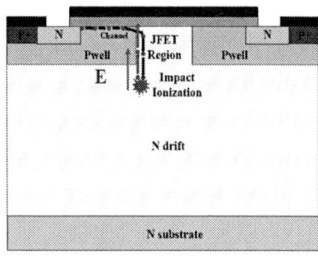

(a)

(b)

Fig. 12. Schematic of (a)impact ionization, (b) hot carrier injection and FN tunneling

V. CONCLUSIONS

The short-circuit capabilities of several state-of-the-art SiC MOSFETs are characterized and investigated in this work. The impacts of high temperature and gate stress on the short circuit performance have been revealed and analyzed. It is found that as we gradually increase the pulse width during single-event short circuit test, a large amount of electron-hole pairs are generated at Pwell/ Ndrift junction.

The transfer characteristics curve shifts in positive direction as a result of hot carrier injection and tunneling effect, indicating the degradation of gate oxide. When we continue to increase the pulse width, a non-negligible gate leakage current can be observed after turning off, representing the failure of device.

The degradation SOA and failure SOA of device A has been provided in this work. It is found that the degradation of SiC MOSFET is strongly influenced by the stress applied on gate oxide, while the final destruction of the device is mainly restricted by the temperature limit of SiC material during short circuit operation.

ACKNOWLEDGMENT

This work is supported by the National Key Research and Development Program of China (No. 2016YFB0100603) and National Natural Science Foundation of China (No. 51577169).

REFERENCES

[1] T. Kimoto and J. A. Cooper, Fundamentals of Silicon Carbide Technology: Growth, Characterization, Devices and Applications: John Wiley & Sons, 2014. (Dec. 2010).

[2] *The World's First Low On-Resistance High-Speed SiC Transistor (DMOSFET)*, Dec. 2010, [online] Available: http://www.rohm.com/web/global/.

[3] *Cree Releases Silicon Carbide Power Devices In Chip Form to Enable More Efficient Power Electronic Modules.*, Dec. 2011, [online] Available: http://www.cree.com/News-and-Events/Cree-News/Press-Releases/2011/December/111208-Sic-Power-Chips.

[4] A. K. Agarwal, J. B. Casady, L. B. Rowland, W. F. Valek, M. H. White, C. D. Brandt, "1.1 kV 4H-SiC power UMOSFETs", IEEE Electron Device Lett., vol. 18, no. 12, pp. 586-588, Dec. 1997.

[5] Q. Zhang, M. Gomez, C. Bui, E. Hanna, "1600V 4H-SiC UMOSFETs with dual buffer layers", Proc. 17th Int. Symp. Power Semiconductor Devices ICs, pp. 211-214, May 2005.

[6] Y. Sui, T. Tsuji, J. A. Cooper, " On-state characteristics of SiC power UMOSFETs on 115-μm drift layers ", IEEE Electron Device Lett., vol. 26, no. 4, pp. 255-257, Apr. 2005.

[7] A. K. Agarwal, R. R. Siergiej, S. Seshadri, M. H. White, P. G. McMullin, A. A. Burk, L. B. Rowland, C. D. Brandt, R. H. Hopkins, "A critical look at the performance advantages and limitations of 4H-Sic power UMOSFET structures", Proc. 8th Int. Symp. Power Semiconductor Devices ICs, pp. 119-122, May 1996.

[8] J. Tan, J. A. Cooper, M. R. Melloch, "High-voltage accumulation-layer UMOSFET's in 4H-SiC", IEEE Electron Device Lett., vol. 19, no. 12, pp. 487-489, Dec. 1998.

[9] Handoko Linewih, Sima Dimitrijev, Kuan Yew Cheong,Channel-carrier mobility parameters for 4H SiC MOSFETs,Microelectronics Reliability,Volume 43, Issue 3,2003,Pages 405-411.

[10] J. Wang, X. Jiang, Z. Li and Z. J. Shen, "Short-Circuit Ruggedness and Failure Mechanisms of Si/SiC Hybrid Switch," in IEEE Transactions on Power Electronics, vol. 34, no. 3, pp. 2771-2780, March 2019.

[11] J. Sun, H. Xu, X. Wu, S. Yang, Q. Guo and K. Sheng, "Short circuit capability and high temperature channel mobility of SiC MOSFETs," 2017 29th International Symposium on Power Semiconductor Devices and IC's (ISPSD), Sapporo, 2017, pp. 399-402.

[12] J. Wei, S. Liu, J. Fang, S. Li, T. Li and W. Sun, "Investigation on degradation mechanism and optimization for SiC power MOSFETs under long-term short-circuit stress," 2018 IEEE 30th International Symposium on Power Semiconductor Devices and ICs (ISPSD), Chicago, IL, 2018, pp. 399-402.

Proceedings of the 31st International Symposium on Power Semiconductor Devices & ICs
May 19-23, 2019, Shanghai, China

Switching Loss Model of SiC MOSFET Promoting High Frequency Applications

Xuan Li[1], Xu Li[1], Liping Yang[1], Alex Q. Huang[2], Pengkun Liu[2], Xiaochuan Deng[1], Bo Zhang[1]

[1]University of Electronic Science and Technology of China, Chengdu, P. R. China.

[2]The University of Texas at Austin, Austin, TX, USA.

Email: andrew_xuanli@foxmail.com, aqhuang@utexas.edu

Abstract—In this work, the influence of parasitic capacitance on the switching loss of SiC MOSFET is investigated. The switching loss models considering C_{gd} and C_{ds} are obtained. Furthermore, the intrinsic minimum switching loss is identified as the energy dissipation resulting from the parasitic capacitance of the SiC MOSFET. Finally, the intrinsic losses of a number of commercial SiC MOSFETs are compared using the proposed model.

Keywords—SiC MOSFET, switching loss model, intrinsic loss, high frequency

I. INTRODUCTION

Primarily due to 10 times larger critical electric field than silicon, the chip size of SiC metal-oxide-semiconductor field-effect transistor (MOSFET) has shrunk significantly when compared with same rating Si MOSFET counterpart. This feature enables faster switching speed and hence SiC's great potentials in high frequency applications[1], [2]. However, if not driven properly with a suitable gate driver, the switching loss of the SiC MOSFET still thermally limits the maximum switching frequency in hard switching applications. Hence how to reduce the switching loss further is a hot topic in power electronics, driving research and

development of new device concept [3], advanced gate driver [4], and low electro-magnetic interference (EMI) packaging concept [5]. However, the theoretical minimum turn-on and turn-off loss of the SiC MOSFET is still not well understood.

In this paper, through analyzing the parasitic capacitance influence on turn-on and turn-off loss, the switching loss model is derived, and intrinsic switching loss is identified and achieved using easily accessible device parameters with reasonable simplifications. Comparison among a number of commercial SiC MOSFET products is carried out using this model to exhibit the correctness of the model.

II. MODELING THE PARASITIC CAPACITANCE

A. Influence of Parasitic Capacitance On The Switching Process

The tested switching performance of a CREE 1200V 80mΩ SiC MOSFET is shown in Fig. 1(a) and (b) using a PCB board based double pulse tester. However, the measured terminal waveforms cannot represent the actual loss during switching process since any discharge of the internal parasitic capacitance can't be measured externally[2].

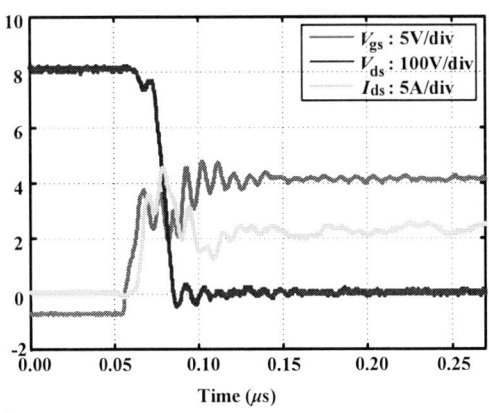 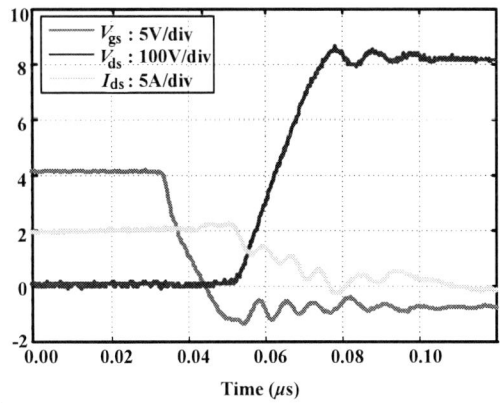

(a)   (b)

Fig. 1. Measured switching waveforms of 1200V 80mΩ CREE SiC MOSFET using a PCB board based test circuit. (a) Turn-on and (b) turn-off process.

National Natural Science Foundation of China (Grant No. 61674026), Natural Science Foundation of Guangdong, China (Grant No. 2017A030313344) and state key laboratory of wide-bandgap semiconductor power electronic devices (Grant No. 2017KF002).

978-1-7281-0582-6/19 $31.00 © 2019 IEEE   231

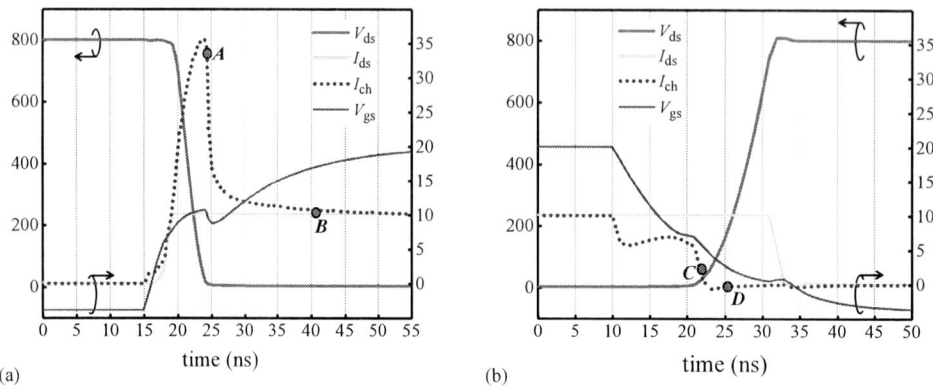

(a)   (b)

Fig. 2. Switching performance of the modeled SiC MOSFET. (a) Turn-on process (b) Turn-off process. The channel current (dash line) is different from typically measured I_{ds} during the switching process. The parameters are set to DC link=800V, load current=10A, R_g=4.6Ω, L_g=0nH, L_s=1nH, L_d=1nH.

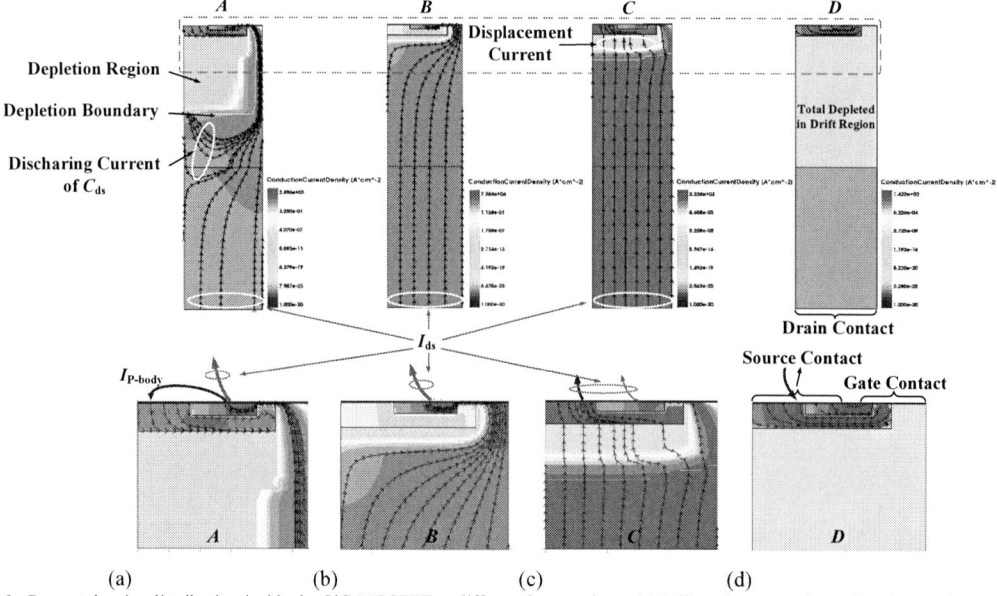

Fig. 3. Current density distribution inside the SiC MOSFET at different key sections. (a) Miller plateau section when the V_{ds} drops to a low conduction voltage drop during turn-on process (Point A of Fig. 2(a)). (b) Total turn-on section (Point B of Fig. 2(a)). (c) Time section when the V_{gs} drops to V_{th} and V_{ds} starts to rise during turn-off process (Point C of Fig. 2(b)). (d) Time section when the channel is in sub-threshold region and the V_{ds} rises to DC voltage during turn-off process (Point D of Fig. 2(b)). $I_{p\text{-body}}$ is the current flowing through body contact.

Therefore, a physical level SiC MOSFET model is developed in Synopsys TCAD carefully to match the fundamental parameters such as BV, R_{on}, G_m, V_{th}, etc. with the tested CREE MOSFET. Fig. 2(a) and (b) show that the simulated terminal voltage and current during switching process are very similar to the measured waveforms except some oscillations caused by the parasitic parameters. In the TCAD model, the carrier distribution in the device can be monitored at each time instance of the switching event. The inner current distributions at A, B, C, and D time points of Fig. 2 are shown in Fig. 3. The simplified switching waveform is shown in Fig. 4. During t_{f_on} section (represented by A), the current distribution is very different

from other time section (represented by B). The energy stored in C_{ds} dissipates in the device. The energy stored in C_{gd} dissipates in total gate loop, in which part of the dissipation generates heat in the internal gate resistance of the SiC MOSFET. During t_{r_off} section, one part of the current from drain contact charges C_{ds} and C_{gd} as displacement current, which does not generate joule heating (represented by point C). When the V_{gs} is lower than the V_{th}, the drift region is almost depleted (represented by D). Based on the Fig. 3, we can also define the current flowing in the channel as I_{ch}. The I_{ch} component is the same as the drain current I_{ds} except during the switching when V_{ds} and V_{gs} are changing as shown by the dashed lines of Fig. 2.

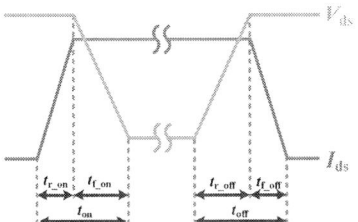

Fig. 4. Simplified switching waveform of SiC MOSFET with an inductive load.

B. Modeling

Based on distinguishing the displacement current and channel current inside the SiC MOSFET, the switching loss during turn-on and -off process can be derived as follows,

$$P_{sw(on)} \approx \int_{t_{on}} I_{ch} V_{ds} f dt = \left[\frac{1}{2} I_{ds} V_{ds} t_{on} + \int_0^{V_{ds}} C_{ds}(V) V \, dV + \delta \int_0^{V_{ds}} C_{gd}(V) V \, dV \right] f \quad (1-1)$$

$$P_{sw(off)} = \int_{t_{off}} I_{ch} V_{ds} f dt = \left[\frac{1}{2} I_{ds} V_{ds} t_{off} - \int_0^{V_{ds}} C_{oss}(V) V \, dV \right] f \quad (1-2)$$

where δ is the portion of the energy stored in the C_{gd} dissipated in the device during t_{f_on}. The first term of (1-1) decreases depending on the C_{gd} charge and the gate driver capability. In the turn-on loss expression, when t_{on} is reduced substantially, this term approaches zero. On the other hand, the C_{ds} and C_{gd} are fixed for a specific device. In other word, the second and third terms cannot be reduced so they form the device's intrinsic or minimum turn-on loss. In the turn-off loss expression, when the gate driver is strong enough, the speed of discharging the charge at the gate terminal can become faster than that of C_{ds} charging by the load current. In other words, the subtraction of the two terms in (1-2) can approximate to zero, and this is therefore the intrinsic or minimum loss during the turn-off.

Therefore, the minimum value of $P_{sw(on)}$ and $P_{sw(off)}$ that can be achieved for a given MOSFET are as follows,

$$\lim_{t_{on} \to 0} P_{sw(on)} \approx E_{oss} f \quad (2-1)$$

$$\lim_{t_{off} \to 0} P_{sw(off)} \to 0 \quad (2-2)$$

The expression indicates that, if an ideal gate driver can be designed so that the turn-on speed and turn-off speed are extremely fast, the energy stored in the C_{oss} still has to be dissipated during the turn-on process and almost all the load current is used to charge C_{oss} losslessly. Therefore, the minimum switching loss of a power MOSFET is E_{oss}.

III. RESULTS AND DISSCUSSIONS

According to the proposed model minimum loss in the power MOSFET, the energy stored in the C_{oss} determines the minimum switching loss, regardless of the load condition. Using this model, a variety of commercial products from CREE, Hestia Power, Littelfuse, ROHM (alphabetic order)

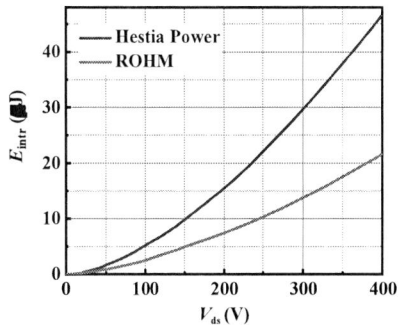

(a)

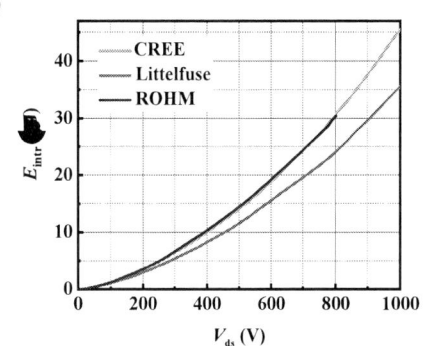

(b)

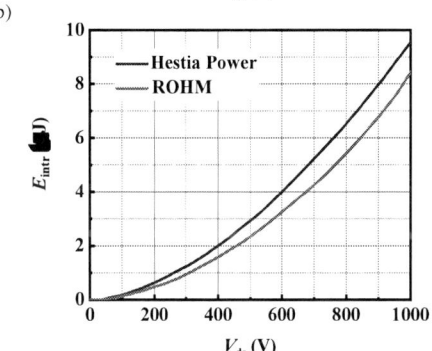

(c)

Fig. 5. Dependency of intrinsic loss on V_{ds} for (a) two 650V 20mΩ SiC MOSFETs (i.e., Hestia power: H1M065F020 and ROHM: SCT3022AL), (b) three 1200V 80mΩ SiC MOSFETs (i.e., CREE: C2M0080120D, Littelfuse: LSIC1MO120E0080, and ROHM: SCT2080KE), and (c) two 1700V 1Ω SiC MOSFET (i.e., Hestia power: H1M170F1K0 and ROHM: SCT2H12NY).

are adopted for the fair comparison of the theoretical minimum switching loss with very similar voltage and current ratings.

For the two 650V 20mΩ SiC MOSFETs, the minimum turn-on loss of the ROHM (SCT3022AL)[6] device is lower than that of Hestia Power (H1M065F020)[7]. The intrinsic loss of ROHM is $22\mu J$ when V_{ds}=400V as shown in Fig. 5(a). For the three 1200V 80mΩ SiC MOSFETs, the intrinsic turn-on loss of the CREE (C2M0080120D)[8] device is almost same with that of ROHM (SCT2080KE)[9], which is

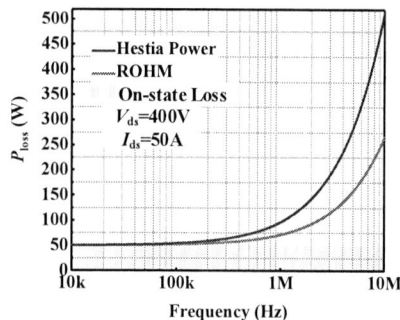

(a)

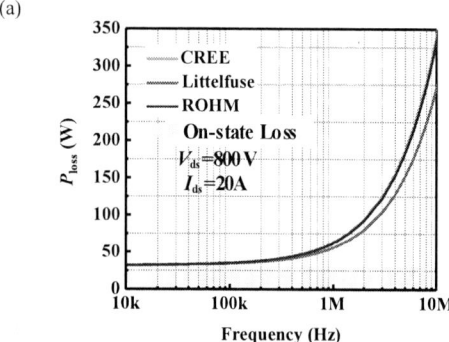

(b)

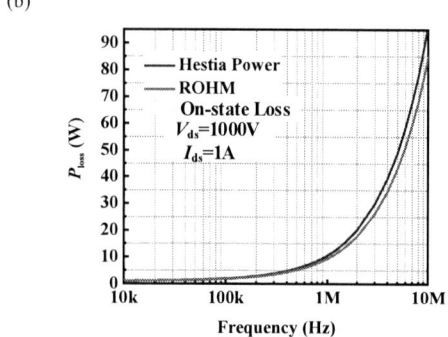

(c)

Fig. 6. Dependency of total intrinsic loss on increasing switching frequency for aforementioned device. The total intrinsic loss including on-state loss with rated R_{on} and minimum switching loss. The yellow region is on-state loss independent of the switching frequency.

slightly higher than that of the Littelfuse (LSIC1MO120E0080)[10] one. The intrinsic loss of Littelfuse is 45μJ when V_{ds}=800V as shown in Fig. 5(b). For the two 1700V 1Ω SiC MOSFETs, the ROHM (SCT2H12NY)[11] is slightly lower than that of the Hestia Power one (H1M170F1K0) [12]. The intrinsic turn-on loss of ROHM is near 8.5μJ when V_{ds}=1000V as shown in Fig 5(c).

Furthermore, the total power losses including the on-state loss and intrinsic switching loss as a function of the switching frequency for the aforementioned devices are shown in Fig. 6(a), (b), and (c). For the 650V 20mΩ SiC

MOSFET, the intrinsic switching loss is equal to the on-state loss at 1MHz for Hestia power and 2MHz for ROHM. It shows the great potential for ultra-high frequency operation. For the 1200V 80mΩ SiC MOSFET, the intrinsic switching loss is equal to on-state loss when 1MHz to 1.3MHz. It still shows a great potential for ultra-high frequency operation, particular at low current condition. For the 1700V 1Ω SiC MOSFET, the intrinsic switching loss is equal with on-state loss when merely around 100kHz. It should be taken into account the switching loss does not include load current generation. This shows less potential to hard-switching high frequency application for the two 1700V device.

IV. CONCLUSION

In conclusion, the influence of parasitic capacitance on SiC MOSFET switching loss is revealed. A switching loss model considering charging or discharging C_{ds} and C_{gd} is obtained. Furthermore, the minimum turn-on and turn-off loss are identified to be E_{oss} and 0 respectively. Using the proposed minimum loss model, the intrinsic losses of several commercial products are compared, showing their potentials for ultra-high frequency high voltage applications.

REFERENCES

[1] T. Kimoto and James A. Cooper, Fundamentals of Silicon Carbide Technology, IEEE Wiley, 2014.

[2] X. Li, J. Jiang, A. Q. Huang, X. Deng, and B. Zhang, "A SiC Power MOSFET Loss Model Suitable for High-Frequency Applications," IEEE Transactions on Industrial Electronics, vol. 64, pp. 8268-8276, Oct. 2017.

[3] K. Han, B. J. Baliga and W. Sung, "Split-Gate 1.2-kV 4H-SiC MOSFET: Analysis and Experimental Validation," IEEE Electron Device Letters, vol. 38, pp. 1437-1440, Oct. 2017.

[4] S. Inamori, J. Furuta and K. Kobayashi, "MHz-switching-speed current-source gate driver for SiC power MOSFETs, " 19th European Conference on Power Electronics and Applications(EPE'17 ECCE Europe), Warsaw, pp. 1-7, Sep, 2017.

[5] L. Zhang, S. Guo, X. Li, Y. Lei, W. Yu, A. Q. Huang, "Integrated SiC MOSFET module with ultra low parasitic inductance for noise free ultra high speed switching," IEEE 3rd Workshop on Wide Bandgap Power Devices and Applications(WiPDA), Blacksburg, pp. 224-229, Nov, 2015.

[6] ROHM, "650V SiC MOSFET" SCT3022AL datasheet, Available: https://www.rohm.com.cn/datasheet/SCT3022AL/sct3022al-e

[7] Hestia power, "650V SiC MOSFET" H1M065F020 datasheet, Available: http://www.hestia-power.com/files/H1M065F020.pdf

[8] CREE, "1200V SiC MOSFET" C2M0080120D datasheet, Available: https://www.wolfspeed.com/media/downloads/157/C2M0080120D.pdf

[9] ROHM, "1200V SiC MOSFET" SCT2080KE datasheet, Available: https://www.rohm.com.cn/datasheet/SCT2080KE/sct2080ke-e

[10] Littelfuse "1200V SiC MOSFET" LSIC1MO120E0080 datasheet, Available: https://www.littelfuse.com/~/media/electronics/datasheets/power_semiconductors/littelfuse_power_semiconductor_silicon_carbide_lsic1mo120e0080_datasheet.pdf.pdf

[11] ROHM, "1700V SiC MOSFET" SCT2H12NY datasheet, Available: https://www.rohm.com.cn/datasheet/SCT2H12NY/sct2h12ny-e

[12] Hestia power, "1700V SiC MOSFET" H1M170F1K0 datasheet, Available: http://www.hestia-power.com/files/H1M

Proceedings of the 31st International Symposium on Power Semiconductor Devices & ICs
May 19-23, 2019, Shanghai, China

New Compact Automotive SiC-Sixpack Converter System with stacked 3D-Gate Driver

S. Buetow, R. Herzer, N. Burani, R. Bittner, M. Kujath

Semikron Elektronik GmbH & Co. KG, Nuremberg, Germany

sven.buetow@semikron.com

Abstract—The paper presents a new developed 1200V/ 60kW$_{peak}$ 3-phase SiC-automotive drive inverter system with a 1200V, 150A SiC-MOSFET-Sixpack module, an integrated gate driver, with all drive and monitoring functions including DC-link voltage, phase current and temperature sensing, and the DC-link itself in an extremely compact housing (600cm³). A novel, low-inductive bus bar arrangement is implemented with a total commutation inductance of less than 5nH, which allows commutations slew rates up to 40kV/µs, extremely low switching losses and a low voltage over-shoot during turn-off.

Keywords—SiC-MOSFET, Sixpack module, automotive converter system, gate driver, monitoring functions, DC-link

I. INTRODUCTION

SiC-MOSFETs allow faster switching with very low dynamic losses and smaller static losses, under low and medium load conditions due to the MOSFET characteristic, than comparable Si-IGBTs. Furthermore, the integrated body diode can be used as freewheeling diode. This and the higher current density of the SiC-MOSFET lead to an outstanding power density, which can be achieved with optimized packages [1, 2]. However, in state of the art motor drive systems the switching speed of inverter has to be limited to approximately 10kV/µs because of the isolation capability of the motor. So the dynamic losses can´t be minimized to the physical device limits.

A next step to increase the power density on system level is

for example the usage of multiphase systems. The use of a higher number of phases reduces the ripple voltage and therefore the losses of the DC-link, which allows smaller DC-links [3]. The increasing power density and complexity of such new power systems are big challenges for the controller and extended driver electronic, because the mandatory clearance and creepage distances remain on high level, and can be a limiting factor for the size reduction of the system. This has to be compensated by an innovative design of the gate driver and high integration level of ICs.

II. SYSTEM OVERVIEW

Fig. 1 presents the schematic of a new developed 1200V/60kW$_{peak}$ 3-phase SiC-automotive drive inverter system with a 1200V, 150A SiC-MOSFET-Sixpack module, a galvanic isolated, integrated gate driver, with all drive and monitoring functions including DC-link voltage, phase current and temperature sensing, and the DC-link itself. All together are implemented in an extreme compact housing (600cm³; see Fig. 2) that has only one plug for all driver functions and sensor signals. Fig. 3 shows the opened DBC layout of the module with six MOSFET switches (3x 50A; 1200V SiC-MOSFETs [4] per switch) back side sintered on DBC, the current sensor shunts and one optional snubber capacitor per half bridge. (The sintered front side contact foil of the chips is not shown in this figure.) The marked green area between low and high side with a width of app. 14mm covers the standardized isolation distance, which is normally

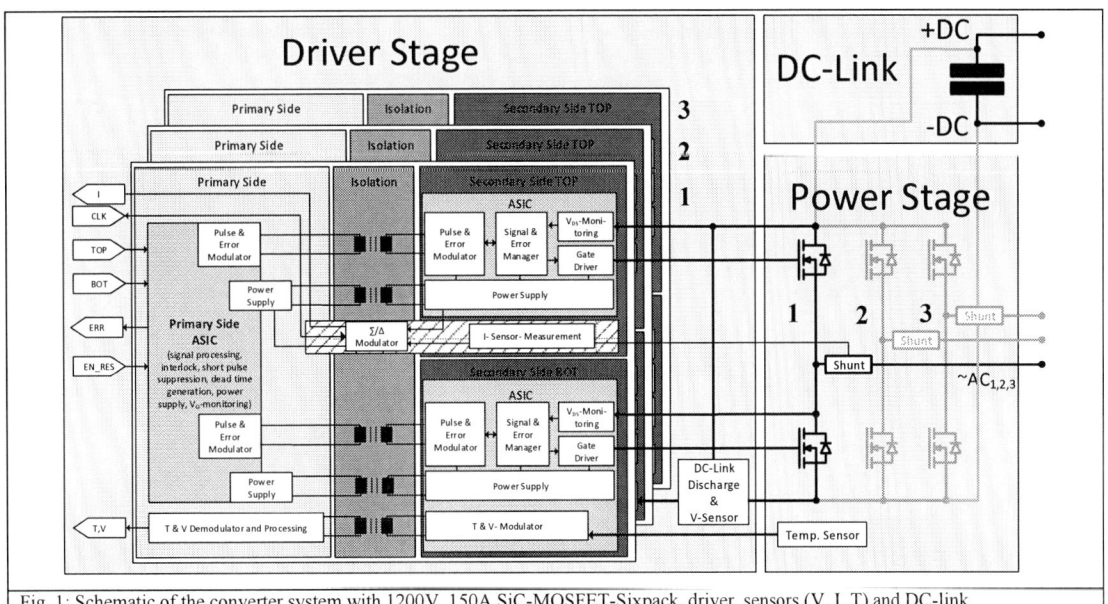

Fig. 1: Schematic of the converter system with 1200V, 150A SiC-MOSFET-Sixpack, driver, sensors (V, I, T) and DC-link

978-1-7281-0582-6/19 $31.00 © 2019 IEEE   235

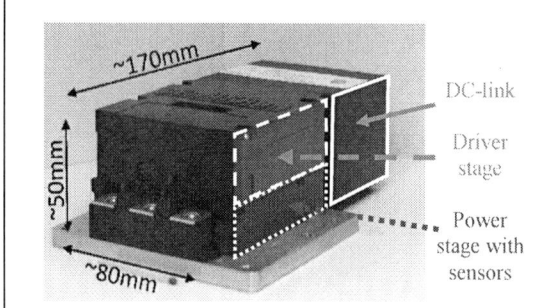

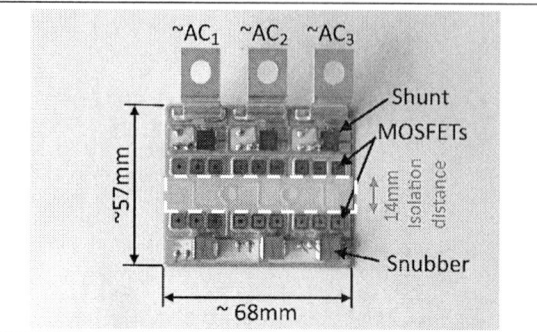

Fig. 2: Photograph of the closed converter system with SiC-MOSFET-Sixpack, driver, sensors and DC-link	Fig. 3: Power hybrid DBC with sintered SiC-MOSFETs and SMD components, without sintered front side contact foil and with marking of the required insulation area (green, dashed)

required for reinforced isolation at PCB-level between primary and secondary side. Inside the module, this area is used to realize an optimized arrangement of the TOP and BOT switch to the shunt-resistor by means of used two-layer contact foil sintered on chip front side. Because of this action, a further reduction of the gate driver size could be achieved and the footprints of the module and the driver could be matched. An innovative, low-inductive bus bar arrangement is introduced for DC-link, which leads to a reduction of total commutation inductance of the system down to only 4.4nH.

III. GATE DRIVER

The gate driver (see block circuit diagram in Fig. 1) provides signal processing, failure management, insulated measurement of analog sensor signals and passive discharge for the DC-link, which is an automotive requirement in case of an accident. Input signal processing includes short pulse suppression, interlock and adjustable dead time generation (0 to 3.5µs in 15 steps). Monitoring management contains under-voltage monitoring of all operation voltages, over-temperature shut down, short circuit protection with soft turn-off for each switch, DC-link voltage measurement and phase current measurement of each phase. Failure, status and digitalized sensor signals are transmitted from the secondary-side driver-ASICs [6] to the primary side controller-ASICs [5] and lead to a perfect control of the whole system. To fit

the gate driver to the extreme small footprint of the module with increased functionality a new concept for the isolation was used. The gate driver was divided in functional groups (low and high sides) and build as a 3D-stack of Printed Circuit Boards (PCB) (see Fig. 4) while the galvanic isolation is realized as an own unit including the signal and power transmission by transformers. The standardized isolation distances for the 1200V class are 3.5mm between the high sides.

A high integration level is necessary in order to realize such an extreme compact Sixpack gate driver with this extended functionality. Therefore application specific driver-ICs (ASIC) in a 0.35µm mixed-signal CMOS technology are developed for the primary [5] and secondary side [6] of the gate driver which, are optimized for low processing times and very fast switching, because of the low tolerances. The high integration level also reduces the required area on PCB level. Fig. 5 shows exemplarily the chip photograph of the secondary side gate driver-IC with its most important sub-circuits. Clearly visible are the large output stages for driving the SiC-MOSFET gates with a peak current of 4.25A for turn-on and -off and an additional stage for soft turn-off in case of a short circuit. An additional external post amplification can be used for applications where higher peak currents are needed. The ASICs also have a thermal pad that improves the cooling performance.

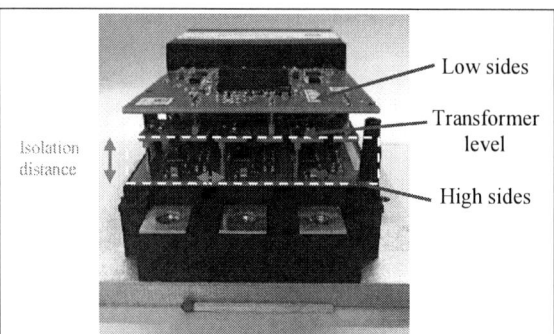

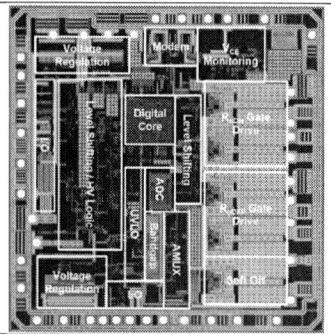

Fig. 4: Photograph of the PCBs of the stacked gate driver placed over the module (transformer unit is in the background of the intermediate board)	Fig. 5: Chip photograph of secondary side driver IC in a 0.35µm mixed signal CMOS technology (app. 3.3 x 3.4 mm²) [5]

IV. MEASUREMENTS

A. Dynamic Characterisation

The achieved system performance is presented with focus on SiC-MOSFETs and the complex gate driver. Fig. 6 shows representatively turn-on and -off characteristic under nominal conditions (V_{DClink}= 600V; I_{load}=150A) at RT. The gate resistors (R_{Gon}= 18.3Ω; R_{Goff}= 15.1Ω) are chosen that the two demanded save operation points (700V; 250A and 850V; 200A) could be switched without exceeding the maximum blocking capability of the dies (1200V) and the maximum slopes (10kV/µs). Thanks to the small parasitic inductance in the commutation path of less than 5nH the MOSFET switching is very soft, the turn-off over-voltages and the reverse recovery peak of the internal freewheeling diode are very low and no oscillations are observed. The behavior under external short circuit conditions is presented in Fig. 7. The short circuit is monitored by V_{DS}-detection of the gate driver. It is turned off after approx. 5µs by the internal short circuit protection of the gate driver. Within this time the current increases to 1250A (8.3 times of the nominal

current), but the voltage over-shoot at turn-off is only 70V.

Without the limitation of the slew rate of 10kV/µs (motor) the MOSFET can be switched much faster. Therefore, other gate resistors (R_{Gon}= 2.1Ω; R_{Goff}= 1.4Ω) were determined which also allow to stay inside the required SOA. The reduction of the gate resistance increases switching speed and leads to a significant reduction of switching losses by 66% (see Fig. 8). With these resistors, high di/dt (15kA/µs) and dv/dt (40kV/µs) are achieved. The losses for several switching conditions (slow, fast and different dead times) are investigated and compared in terms of optimization of the system (see Fig. 8). Beside optimization of the gate resistors also an optimization of the dead times was done. A reduction of switching losses is possible by lowering the dead time between TOP and BOT during switching in a half bridge configuration [7]. Because of the very low-inductive commutation circuit of the system, an extremely fast switching is possible and dead times down to 100ns can be used without cross current over the half bridge. In that case, the switching losses decrease further by 33%. Totally, the switching losses drop down to about 25%.

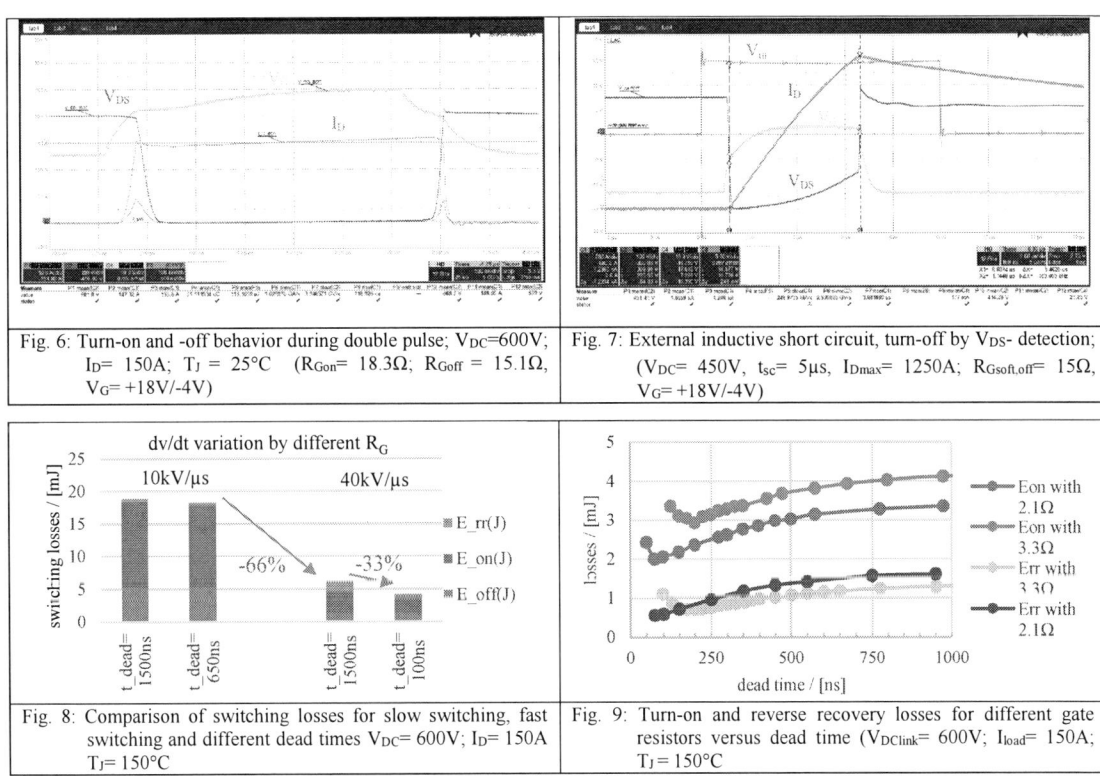

Fig. 6: Turn-on and -off behavior during double pulse; V_{DC}=600V; I_D= 150A; T_J = 25°C (R_{Gon}= 18.3Ω; R_{Goff} = 15.1Ω, V_G= +18V/-4V)

Fig. 7: External inductive short circuit, turn-off by V_{DS}- detection; (V_{DC}= 450V, t_{sc}= 5µs, I_{Dmax}= 1250A; $R_{Gsoft.off}$= 15Ω, V_G= +18V/-4V)

Fig. 8: Comparison of switching losses for slow switching, fast switching and different dead times V_{DC}= 600V; I_D= 150A T_J= 150°C

Fig. 9: Turn-on and reverse recovery losses for different gate resistors versus dead time (V_{DClink}= 600V; I_{load}= 150A; T_J= 150°C

Fig. 9 shows the MOSFET turn-on and diode reverse recovery losses for two small gate resistor combinations versus dead time. While the turn-on losses degrease with lower R_G (faster switching), a slight increase of the reverse recovery losses can be observed. In addition, the reduction of both dynamic losses can be seen with smaller dead times and higher switching speed, respectively. Below about 200ns the losses suddenly increase because of the cross current through the half bridge.

B. Inverter test

Finally, the whole system is tested in a 4Q-inverter test under automotive load conditions. In addition, other tests with varying voltages, currents and dead times are performed, too. The overall losses of the inverter could be reduced with an optimized drive by 33%. Fig. 10 shows the current waveform of one phase (U) that is measured by a Rogowski coil and the three phase voltages are measured by differential probes for the highest tested load conditions of approx. 67kW at a maximum junction temperature of approx. 120°C. According to calculations also 80kW$_{peak}$ for 2.5min are possible with a maximum junction temperature of 150°C

The signals for the currents, the DC-link voltage and temperatures are digitalized (ADC) and transmitted by isolated transformers to the primary side as digital pattern. Fig. 11 shows the reconstruction of the three phase currents within a 4Q-inverter test. During the development the rms-value of the currents and the DC-link voltages are measured and compared with external conventional sensor measurements. The deviations between internal and external measurement are presented in Fig. 12 and Fig. 13. During the signal reconstruction and processing a calibration can be done. This is exemplarily demonstrated for the current in Fig. 12. The measured difference before calibration is already below 1% and after calibration below 0.2%. In addition, tolerance calculations for the DC-link voltage signal are also done. Fig. 13 shows the measured DC-link voltages and voltage tolerance calculations versus the DC-link voltage range. It is clearly to see that the measured values are fine and an additional calibration is not necessary here.

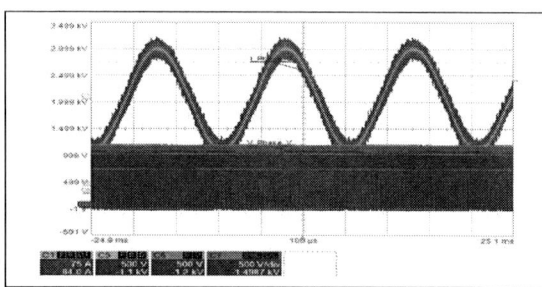

Fig. 10: Oscillogram of 4Q inverter test of the system.
(V$_{DClink}$= 750V; I$_{load}$= 100A$_{rms}$, f$_{sw}$= 20kHz; cosφ= 0.89; P$_{out}$= 67kW; T$_{cooler}$= 65°C; Q$_{cooler}$= 8,8 l/min; t$_{dead}$= 250ns; T$_{Jmax}$≈ 120°C)

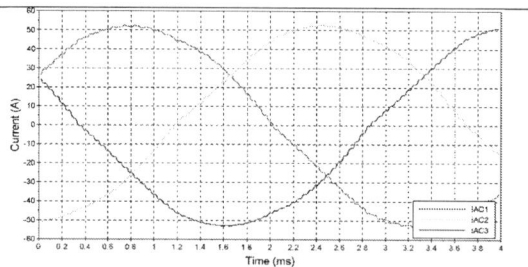

Fig. 11: Reconstructed phase currents, provided from the internal system shunts during inverter test

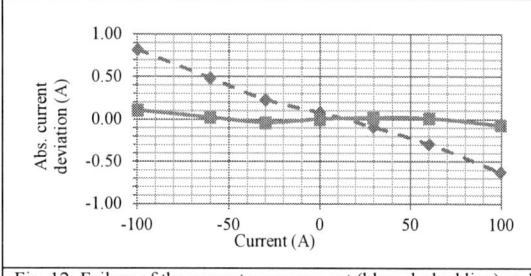
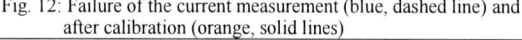

Fig. 12: Failure of the current measurement (blue, dashed line) and after calibration (orange, solid lines)

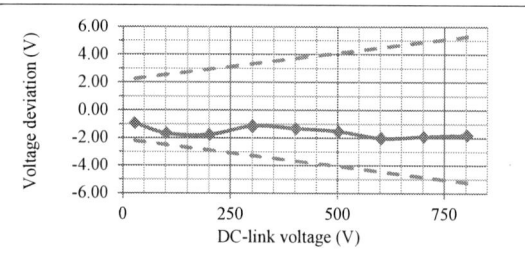

Fig. 13: Failure of the DC-link voltage measurement (solid line) and tolerance calculation (dashed lines)

V. CONCLUSIONS

A 1200V/ 60kW$_{peak}$ 3-phase SiC-automotive drive inverter system with a 1200V, 150A SiC-MOSFET-sintered Sixpack module, an integrated gate driver, with all driver and monitoring functions including DC-link voltage, phase current and temperature sensing, and the DC-link itself in an extreme compact housing (600cm³) is presented. A novel, low-inductive bus bar arrangement is implemented with a total commutation inductance of less than 5nH, which allows commutation slew rates up to 40kV/µs, extreme low switching losses and a low voltage over-shoot during turn-off. The main limits for the power density are the mandatory clearance and creepage distances and the restrictions of the voltage slew rates in the motor drive application. Therefore, a new concept of a stacked gate driver and new gate driver ICs are used, which allows a reduction of the driver footprint. Both footprints, from sintered module and gate driver boards, are matched and optimized. As result, the clean and smooth switching behavior of the system and the benefit of fast switching and optimized gate control is presented. Under this optimized conditions an increased power density of +33% is possible with improved motor concepts.

VI. REFERENCES

[1] C Schmidt, M. Roebiltz (Semikron Elektronik GmbH & Co.KG)"A Performance Comparison of SiC Power Modules with Schottky and Body Diodes"

[2] P. Beckedahl, S. Buetow, A. Maul, M. Roeblitz, M. Spang (Semikron Elektronik GmbH & Co.KG): "400A, 1200V SiC power module with 1nH commutation inductance"; CIPS 2016

[3] S. Piepenbreier (Fraunhofer IISB Erlangen), J. Berlinecke (VW AG), N. Burani, (Semikron Elektronik GmbH& Co. KG), et.al.: "Analysis of a multiphase multi-star PMSM drive system with SiC-based inverter for an automotive application"; PCIM Europe 2018

[4] Data sheet S2307; Rohm; 1200V, 45mΩ SiC-MOSFET

[5] Data sheet primary side controller IC, SKIC2005; Semikron Elektronik GmbH & CoKG

[6] Data sheet secondary side gate driver IC, SKIC1008; Semikron Elektronik GmbH & CoKG

[7] S. Bütow, R. Herzer et.al. (Semikron Elektronik GmbH & Co.KG) "High power, high frequency SiC-MOSFET system with outstanding performance, power density and reliability"; ISPSD2017

VII. ACKNOWLEDGMENT

The work has been performed in the research project HOSKA (High efficient and scalable electronic components for drives and electric vehicles), funded by the German Federal Ministry of Education and Research (BMBF).

Proceedings of the 31st International Symposium on Power Semiconductor Devices & ICs
May 19-23, 2019, Shanghai, China

A Novel All-in-One Digital-Analog Heterogeneous Integrated Intelligent Power Module

Y.T. Lin, K.S. Kao, C.M. Tzeng, H.H. Lin, W.K. Han, J.Y. Chang, T.C. Chang

Electronic and Optoelectronic System Research Laboratories, Industrial Technology Research Institute

Bldg. 14, 195, Sec. 4, Chung-Hsing Road, Chutung, Hsinchu, Taiwan 31040, R. O. C

Email: itriA60042@itri.org.tw

Abstract—**This paper introduces a novel all-in-one digital-analog heterogeneous integrated intelligent power module (IPM) which is composed of Si MOSFETs, Gate Driver and Micro Controller Unit (MCU). This novel all-in-one module realized by heterogeneous integration offers a System in Package (SiP) solution without causing serious Electromagnetic Compatibility (EMC), signal integrity and thermal issues. In addition, we designed the three-phase half-bridge inverter system formed by this IPM to compare with the system consisting of the same circuit and same bare dies of MOSFETs (TO-package), Gate Driver and MCU (SO-package). The performance shows that the inverter system which adopted this intelligent power module is capable of reducing system energy consumption, enhancing system efficiency (2-4%), and successfully shrinking system volume(>33%) at the same time. Besides, this SiP IPM can improve system reliability, electromagnetic compatibility (EMC) performance and pass the international standard of CISPR 11. Keywords—SiP, heterogeneous integration, IPM, EMC, power density.**

Keywords—SiP, heterogeneous integration, IPM, EMC, power density.

I. INTRODUCTION

The semiconductor and the consumer electronics industry are continuously being challenged by demanding consumers nowadays. In order to achieve a dominant position in the market, it's important for manufacturers to offer the high-efficiency, low-cost and small-sized devices. Additionally, it is crucial to meet the rising demand for the multi-functional as well as portable consumer electronics products. However, SoC technology has encountered bottlenecks while trying to integrate the products with both analog and digital functions in the conventional processes. Therefore, the industry has spawned a solution System in Package (SiP) to fulfill the above mentioned goals.

SiP integrates different components through packaging technology to provide a small, versatile solution for the final product. In the SiP field, multi-chips are packaged in a single package and attached to a substrate by wire bonding. Through SiP technology, we could not only improve the products with lower cost and faster speed than integrated System on Chip (SoC) technology, but also reduce production lead times[1-2]. Besides, it has been noted that the system in package (SiP) offers lower switching loss because it has low parasitic inductances [3-5].

However, the integrated circuits for high-speed digital signal and high power trace in the modules would inevitably cause significant electromagnetic interference (EMI) and

radio frequency interference (RFI) problems [6-7] . Therefore, it is necessary for us to put in more effort and resources to promote the assembly process, the material of thermal interface and the layout at package and board levels.

In this paper, we demonstrate a novel intelligent power module (IPM) designed and manufactured by SiP technology. In this intelligent power module, the analog and digital chips are encapsulated together to achieve high density package design with heterogeneous integration technology. Section II presents the architecture, circuit design and assembly process of this novel all-in-one IPM. The system design and performance tests are described in section III. In the final section, we propose some overall conclusions regarding the design as well as the performance of the IPM.

II. APPROACH

A. The architecture and electric circuit of the integrated intelligent power module

This intelligent power module with high density integrated package architecture is composed of six N-Channel Power MOSFETs, one High Voltage 3-Phase Gate Driver and one Microcontroller Unit. Figure 1 shows the schematic diagram of the all-in-one integrated intelligent power module. This IPM was designed with dimensions of 30 mm × 15 mm × 3 mm. And Figure 2 shows the architecture and cross section of the Al substrate.

The electric circuits on the top and bottom layers of Al substrate are manufactured by RDL process. Another point worth mentioning is that, the 2 layers of thermal interface material (TIM) are bonded to both sides of Al, by thermal compression process to form the substrate. Through the two layers of TIM, the substrate could meet the requirements for simultaneous properties of high thermal conductivity and excellent insulation.

Besides, we filled up plug resin around the internal sidewall of PTH to isolate the traces from aluminum. Then we use surface finishing in the final step of the process to obtain the desired surface condition. As for the bonding process, the interconnections of MOSFETs were made by Al wire bonding process. In contrast with MOSFETs, the Gate Driver and Microcontroller chips were bonded by Au wire. Figure 3 shows the top and bottom views of Al substrate with solder mask and metal finishing.

978-1-7281-0582-6/19 $31.00 © 2019 IEEE

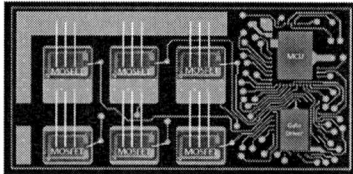

Figure 1. The schematic diagram of the all-in-one integrated intelligent power module

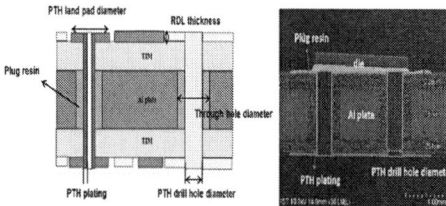

Figure 2. The architecture and cross section of the Al substrate

(a) Top view

(a) Bottom view

Figure 3. (a) The top view and (b) bottom view of Al substrate with solder mask and metal finishing

B. Assembly process

After a series of substrate fabrication process, the assembly process can be divided into 9 steps as shown in Figure 4. First, we prepared the chips and substrates for assembly, then bonded the MOSFET, Gate Driver and MCU by solder printing and adhesive dispensing processes. Next, we use wire binding to connect the MOSFET, Gate Driver and MCU chips with Al and Au wires respectively. After the molding process, we did the solder printing and reflow process. Finally, the molded IPM were divided into individual modules.

(1) Carrier Al substrate (including 2-layers of RDL)

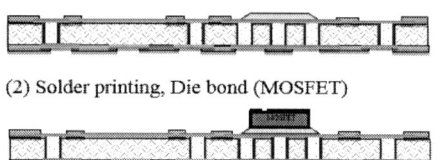

(2) Solder printing, Die bond (MOSFET)

(3) Reflow

(4) Adhesive dispensing, Die bond (Gate driver, MCU)

(5) Al Wire bonding for MOSFET chip

(6) Au Wire bonding for Gate / MCU chip

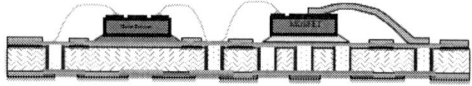

(7) Molding

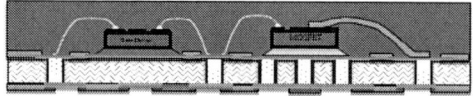

(8) Solder printing & reflow

(9) Singulation

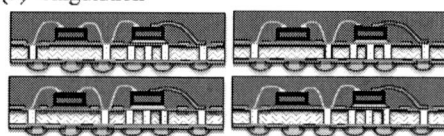

Figure 4. The assembly process of the all-in-one integrated intelligent power module

III. RESULTS AND DISCUSSION

In order to test and verify the performance of this all-in-one integrated intelligent power module, we designed a three-phase half-bridge driving system to analyze the power efficiency, signal integrity and Electromagnetic Compatibility ability. Moreover, we benchmarked against the system which constructed with the same circuit and bare dies of MOSFETs(TO-package), Gate Driver and MCU(SO-package) but utilizing different packaging technology. In this paragraph, we would present the design of the three-phase half-bridge driving system and discuss the experiment results in detail.

A. Three-phase half-bridge driving system

The complete block diagram of the three-phase half-bridge inverter system is shown in Figure 5(a). We designed two specific systems with the same components and electronic schematic, but using two different solutions to form the inverter subsystem. The block diagram of the inverter subsystem is shown in Figure 5(b). For the two specific systems The intelligent power module was adopted to build up our newly designed system while the other one was built by the conventional discrete components with one small Outline (SO) type package MCU, Gate driver and six Transistor Outline (TO) type package MOSFETs, respectively. It is worth to mention that the bare dies inside the SiP IPM are the same as the discrete type, MCU, Gate driver, and MOSFETs at the conventional inverter subsystem. As mentioned previously, the practical systems are shown in Figure 6.

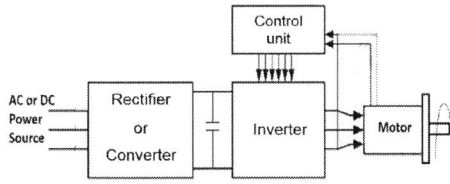

(a) The complete block diagram of the three-phase half-bridge inverter system

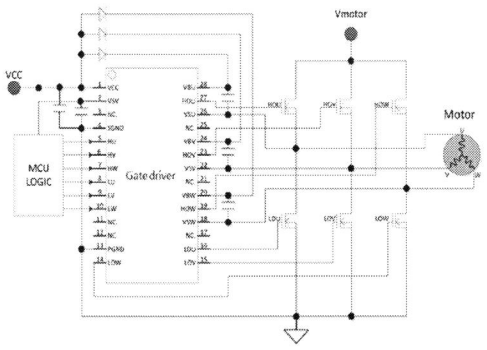

(b) The block diagram of the inverter subsystem

Figure 5. (a) The block diagram of the three-phase half-bridge inverter system (b) The block diagram of the inverter subsystem

B. Experiment results

After conducting a series of tests to verify performance of the two difference systems, we evaluated the system properties described in Table 1. By driving the same brushless DC (BLDC) motor, both systems were tested in ambient conditions of 25 °C with natural convection cooling. And both systems were operated under the conditions of 311V and 70W. After reaching the thermodynamic equilibrium, the experiment results indicated that the maximum temperature of the two systems occur in the MOSFETs devices and IPM respectively. The temperature of the MOSFET devices in the system which adopted conventional discrete devices rose to 68~70 °C, whereas the temperature of the all-in-one IPM rose to 66~70 °C. On the other hand, the measured data of the power meter (Tektronix-PA3000) shows that the system adopted this all-in-one intelligent power module is capable of improving the power efficiency by 2~4 % compared to the conventional discrete system. On the basis of our testing results, this all-in-one IPM

could successfully shrink the system volume (>33%) and increase the power density simultaneously.

C. Electromagnetic compatibility performance

In heterogeneous integration, the analog and digital chips are encapsulated together. Thus, we should consider the issues of Electromagnetic Interference (EMI) more seriously. In order to solve these issues, we optimized the SiP IPM design by improving the wire bonding process, the material of thermal interface and the layout at package and board levels.

Figure 6. The assembly process of the all-in-one integrated intelligent power module

TABLE I. SYSTEM PROPERTIES AND TESTING RESULTS

Inverter subsystem	System testing results		
	Efficiency: $\frac{P_{out}}{P_{in}}$	ΔT: @ 311V ; @ 70W ambient temperature @ 25°C	Size: Total Driving System
Conventional Discrete Devices	90~91%	68~70 °C	125x125 (mm)
All-in-one IPM	93~94%	66~70 °C	95x105 (mm)

Through the optimization of design as mentioned above, the three-phase half-bridge driving system which was comprised by the all-in-one IPM also passed the conducted and radiated emission standards of CISPR 11. CISPR 11 is an international standard commonly referenced in all European EMC standards, which establishes standard measurement methods, measurement equipment, limit lines and interpretation of applicability of limit lines in industrial and medical fields. The peak and average results of conducted emission are shown in Figure 7(a) and Figure 7(b), respectively. Both the quasi-peak and average conducted emission that versus frequency from 0.15 MHz to 30 MHz performance of our EUT were under the limit of standard.

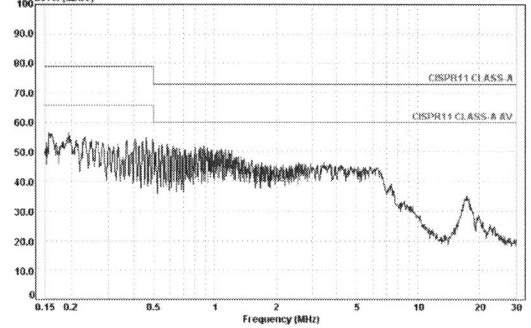

(a) The peak conducted emission versus frequency performance of the EUT.

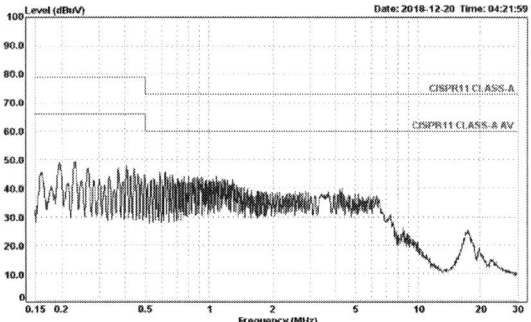

(b) The average conducted emission versus frequency performance of the EUT.

Figure 7. (a) The peak conducted emission versus frequency performance of the EUT (b) The average conducted emission versus frequency performance of the EUT.

The vertical and horizontal results of radiated emission are shown in Figure 8(a) and Figure 8(b). Our EUT results show that both the vertical and horizontal radiated emission performance from 30 MHz to 1GHz could remain below the limit of standard.

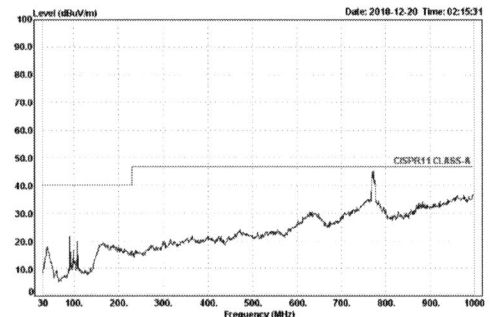

(a) The vertical radiated emission versus frequency performance of the EUT.

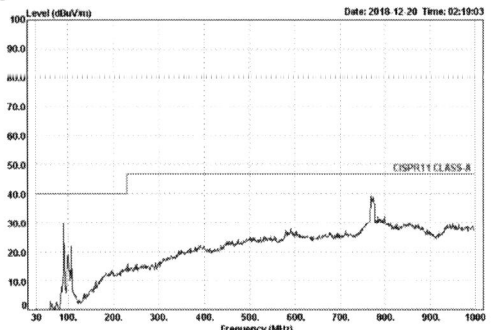

(b) The horizontal radiated emission versus frequency performance of the EUT.

Figure 8. (a) The vertical radiated emission versus frequency performance of the EUT (b) The horizontal radiated emission versus frequency performance of the EUT

IV. CONCLUSIONS

In this paper, we present a novel all-in-one digital-analog heterogeneous integrated intelligent power module (IPM) accomplished by SiP technology. This intelligent power module with high-density integrated package architecture which is composed of six N-Channel Power MOSFETs, one High Voltage 3-Phase Gate Driver and one Microcontroller Unit. Moreover, it is equipped with an exceptional thermal interface material (TIM) and high thermal conductivity aluminum substrate which contribute to high heat dissipation and high insulation characteristic of this intelligent power module. Apart from shrinking the total size, this module could also solve the electromagnetic compatibility, safety, heat dissipation issues. Additionally, we designed two three-phase half-bridge driving system to analyze the power efficiency, signal integrity and Electromagnetic Compatibility ability. As discussed in the previous section, the test results indicate that the driving system which adopted this intelligent power module is capable of reducing system energy consumption, enhancing system efficiency (2-4%), and successfully shrinking system volume(>33%) at the same time. Besides, this SiP IPM can improve system reliability, electromagnetic compatibility (EMC) performance and pass the international standard of CISPR 11.

REFERENCE

[1] H. P. Wei, M. C. Yew, W. K. Yang, K. N. Chiang, "Reliability Analysis of a Package-on-Package Structure using the Novel WLCSP Technology with Fan-Out Capability," EMAP 2006,December 11-14, 2006. Hong Kong.

[2] Yang S., Chiang S.Y., Chou C.Y., Yew M.C., "Reliability analysis of copper interconnections of system-in-packaging," Microsystems,Packaging, Assembly and Circuits Technology Conference. Taipei, pp.52-55, Oct. 2009.

[3] M. Shiraishi, et al., "Low Loss and Small SiP for DC-DC Converters", ISPSD Proceedings, May 2005, pp. 175–178.

[4] Y. Kawaguchi, et al., "Multi Chip Module with Minimum Parasitic Inductance for New Generation Voltage Regulator", ISPSD Proceedings, May 2005, pp. 371–374.

[5] T. Hashimoto, et al., "Advanced Power SiP with Wireless Bonding for Voltage Regulator", ISPSD Proceedings, May 2007, pp. 125–128.

[6] Kuo-Hsien Liao, Alex Chi-Hong Chan, Bradford J. Factor,"Novel EMI Shielding Methodology on SiP Module " in Proc.of 2012 ICEP ,Tokyo, Japan, 17-20 Apr., 2012

[7] J. Miettinen, M. Mantysalo, K. Kaija, and E. O. Ristolainen, BSystem design issues for 3D System-in-Package (SiP),[in Proc. IEEE Int. Electron. Compon. Technol. Conf.,Jun. 2004, pp. 610–615

Proceedings of the 31st International Symposium on Power Semiconductor Devices & ICs
May 19-23, 2019, Shanghai, China

An Integrated Packaging Structure of Press Pack for High Power IGBTs

Erping Deng[1*,2*], Bin Ren[1], Anqi Li[1], Yanhao Wang[1], Zhibin Zhao[1], Yongzhang Huang[1]

[1]State Key Laboratory of Alternate Electrical Power System with Renewable Energy Sources
(North China Electric Power University), Changping District, Beijing 102206, China,
[2]Chemnitz University of Technology, Germany, [1*,2*]dengerpinghit@163.com

Abstract—An Integrated packaging structure of press pack for high power IGBTs is proposed, in view of the shortcomings of Pedestal and Stakpak press pack IGBTs. The simplified model of the three packaging structures were simulated by finite element method. The simulation results reveal the pressure distribution on chip surfaces of the Integrated press pack IGBT is obviously better than the Pedestal one though slightly worse than the Stakpak one. Furthermore, the thermal dissipation ability of the integrated press pack IGBT is the best among the three packaging structures with lowest temperature and uniform temperature distribution. The Integrated packaging structure proposed in this paper meets the expected design requirements with uniform pressure distribution and lower junction temperature.

Keywords—press pack IGBT; Integrated packaging structure; pressure distribution; junction temperature

I. INTRODUCTION

Recently, Press Pack Insulated Gate Bipolar Transistor (PP IGBT) is quite famous for the high power applications as the advantages of short-circuit failure mode, high power density and easy connect in series, especially for the flexible HVDC. Two main packaging styles of PPI are existing in the market and under research [1]. The Pedestal PP IGBT with direct contact has relative low thermal resistance with double side cooling. However, the clamping force balance becomes a big challenge with the current rating increases when the device is heated up [2]. For the Stakpak PP IGBT with spring contact the clamping force is well balanced with separated spring disc but only one side is able to dissipate heat [3]. The internal structure of Pedestal and Stakpak PP IGBTs is shown in Fig 1.

The PP IGBT used in flexible HVDC should have high reliability with uniform pressure distribution and high thermal dissipation efficiency. Thus, the Integrated packaging structure is proposed to avoid the shortcomings of Pedestal and Stakpak structures. And the interested pressure distribution and junction temperature distribution among chips are compared with other two packaging styles. Firstly, the Integrated packaging structure is introduced in more details with the compositions and functions of each part are emphasized. Then, a finite element simulation model of Integrated PP IGBT is established according to [2] and compared with Pedestal and Stakpak structures. The simulation results verify the excellent performance of the proposed Integrated PP IGBT with the well-balanced pressure distribution and low junction temperature.

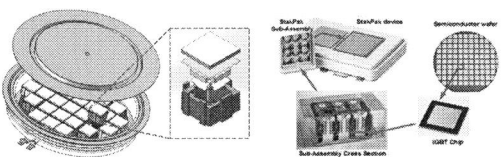

(a) Pedestal PP IGBT   (b) Stakpak PP IGBT
Fig. 1 Internal structure of two main packaging

II. INTEGRATED PACKAGING STRUCTURE

The Integrated PP IGBT proposed in this paper shown in Fig 2 can be divided into seven parts. The structure and function of each part are listed as follows:

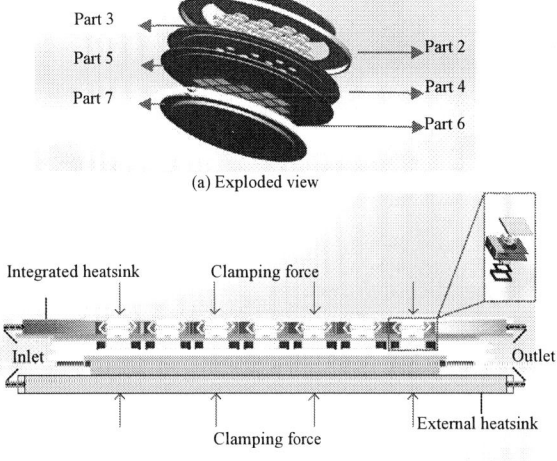

(a) Exploded view

(b) Section view
Fig. 2 Internal structure of the Integrated PP IGBT

1) Part1 is emitter electrode which is directly connected to the external terminal and is subjected to the clamping force.

2) Part2 is the integrated heatsink to dissipate the heat generated by silicon chip through external water cooling.

3) Part3 is separated spring disc. Its main function is to compensate the movement caused by thermal stress to uniform the pressure distribution among chips.

4) Part4 is plastic with high thermal conductivity, which has the following four functions. Firstly, it separates the integrated heatsink from the top-down press pack circuit path (current→part1→part3→part5) to insulate the integrated

978-1-7281-0582-6/19 $31.00 © 2019 IEEE   243

heatsink from the main current loop. Secondly, the plastic assembly is a vertical through structure to ensure that the separated spring disc is directly pressed on the underlying copper electrode, benefiting to pressure difference conduction by copper electrode to make chip pressure being uniform. Thirdly, the height of the overall assembly structure of the plastic with high thermal conductivity is 5 mm lower than the separated spring disc, and the assembly cylinder diameter ratio is 5 mm longer than the longest width of the spring, which can prevent the spring from squeezing and affecting the structure of the plastic during deformation. Finally, the lower substrate of the plastic with high thermal conductivity can effectively transfer the heat of chips to the integrated heatsink, which is beneficial to low chip temperature.

5) Part5 is a thin copper plate used to conduct the current and heat generated by the chips. The cooling water of the integrated heatsink flows directly on this copper surface to cooling down the chips. And the thin copper film is easy to deform under high clamping force to balance the pressure distribution through each separated spring disc.

6) Part6 is the silicon chip submodules within PP IGBT.

7) Part7 is the molybdenum plate with ceramic protection as the electrode of collector side. The molybdenum plate is used to conduct the current, heat and clamping force to chips. And the ceramic is for environment protection and insulation.

III. FINITE ELEMENT MODEL

A. Simplified Packaging Structure Model

The thermal and mechanical behavior of these three types is compared with the electro-thermo-mechanical finite element model [2]. The finite element model and its boundary conditions are shown in Fig 3. For the PP IGBT with Pedestal, Stakpak and Integrated styles, some components that have little influence on the surface pressure distribution and temperature among multi-chips within PP IGBT such as plastic frame, ceramic tube, printed circuit board (PCB), spring pins, sealing ring plate, integrated heatsink and so on can be omitted in the simulation model [4-5]. The silicon chips within PP IGBT are also modeled with active area to generate the heat and terminal area for insulation [6-7].

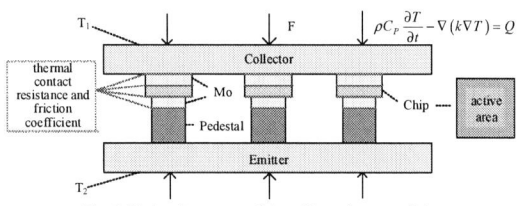

Fig. 3 Finite element model and boundary conditions

B. Boundary Conditions

Through the simplification of the packaging structure, three simulation models of PP IGBTs are established. The boundary conditions of the simulation models include the electrical and thermal contact resistance [8-10], friction coefficient [11-13], power loss and heat convention coefficient [14] and so on. Referring to relevant literatures, the above simulation boundary conditions are set.

IV. SIMULATION RESULTS ANALYSIS

A. Pressure Distribution

The simulation results with different conditions based on the finite element model are presented in Fig 4 and 5. For the *clamping phase* is that the PP IGBT is only clamped by the external clamping system and no load current is conducted. While, the *heating phase* is that the clamped PP IGBT is heated up by the load current to approximate its working condition. It can be seen that the pressure distribution is uniform and the clamping force can be well controlled under the clamping phase of the three proposals. However, the pressure distribution is changed when the PP IGBT is heated up with the heating phase that the StakPak PP IGBT is the best and the proposed Integrated structure is also good in balance. The pressure distribution of the Pedestal structure is totally changed and concentrated in the middle position. The chips at the edge of the packaging have low pressure and even loss of contact.

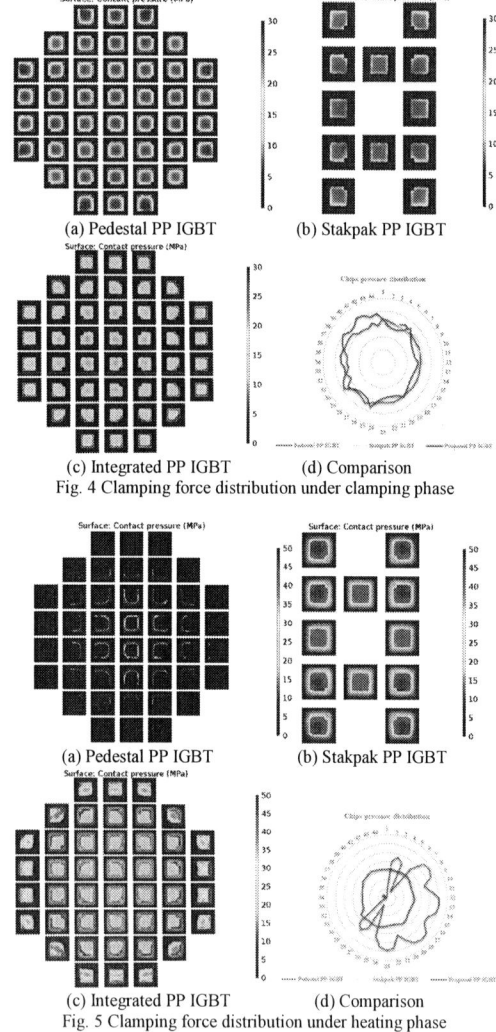

(a) Pedestal PP IGBT   (b) Stakpak PP IGBT

(c) Integrated PP IGBT   (d) Comparison
Fig. 4 Clamping force distribution under clamping phase

(a) Pedestal PP IGBT   (b) Stakpak PP IGBT

(c) Integrated PP IGBT   (d) Comparison
Fig. 5 Clamping force distribution under heating phase

In order to further understand the pressure distribution on the chips in the heating phase of the three types, the chips are numbered in a certain order, and certain extraction paths are selected on the chip, as shown in Fig 6. For Pedestal PP IGBT and Integrated PP IGBT, the chip numbers 1 to 30 are IGBT chips, and the chip numbers 31 to 44 are FRD chips. For Stakpak PP IGBT, chip numbers 1 to 8 are IGBT chips, and chip numbers 9 to 12 are FRD chips. The pressure distribution on the extract paths are shown in Fig 7.

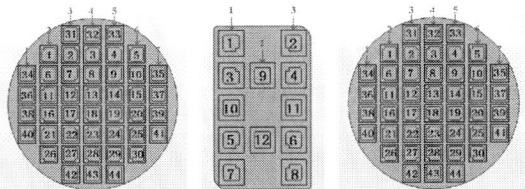

(a) Pedestal PP IGBT (b) Stakpak PP IGBT (c) Integrated PP IGBT
Fig. 6 Number and extract paths of the chips

From the simulation results we can see that the pressure distribution on the chip surface is relative uniform for Stakpak and proposed Integrated structure. However, the pressure is mainly concentrated at the edge contact of the active area of the chips and molybdenum plate in Pedestal and Integrated structure. The maximum pressure on single chip of these three types is 128 MPa, 48 MPa and 45 MPa, respectively. The Integrated structure has uniform pressure distribution among chips and low pressure concentration compared to other two structures. However, we still need to pay more attention to the edge contact of the chip active area.

The average clamping force applied on each silicon chips are extracted through this simulation and shown in Table I. The results show that in heating phase, the average clamping force distribution of Stakpak PP IGBT is the most uniform, followed by proposed Integrated PP IGBT, and then the Pedestal PP IGBT is extremely uneven. The main reason is that the displacement caused by thermal stress is not able to be compensated by the pedestal in the Pedestal structure and this finally leads to the electrodes warpage as shown in Fig 8, similar to [15]. While the independent spring disc for each silicon chip is designed to compensate this displacement in the Stakpak and proposed Integrated structure. The reason why the proposed Integrated structure is not as good as the Stakpak is that the existed copper plate (part5), which is used for the current and heat conducting, constraints the movement of the spring disc at a certain extent.

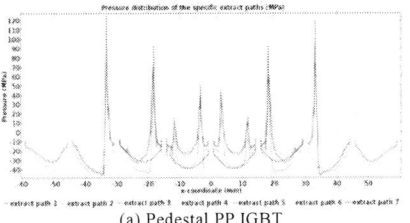

(a) Pedestal PP IGBT

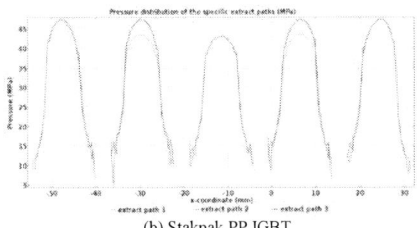

(b) Stakpak PP IGBT

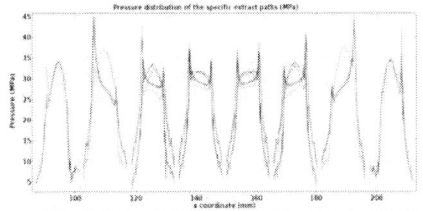

(c) Integrated PP IGBT
Fig. 7 Pressure distribution of the specific extract paths in heating phase

TABLE I.   AVERGAE CLAMPING FORCE COMPARISON IN HEATING PHASE

Packaging styles	Rated (N)	Max. (N)	Error (%)	Min. (N)	Error (%)
Pedestal	1592	3092.2	94.2	7.9419	-99.5
Stakpak	1667	1744.8	4.7	1610.8	-3.4
Integrated	1592	1779.1	11.8	1394.1	-12.4

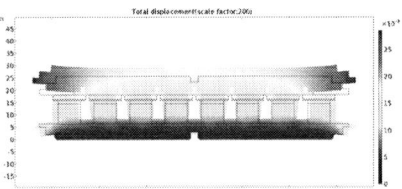

Fig. 8 Emitter electrode edge warping (Scale factor: 200)

B. Junction Temperature Distribution

The junction temperature distribution of three types of PP IGBTs and the calculated maximum junction temperature on each chip surface in the heating phase are shown in Fig 9. The electrical and thermal contact resistance existed in each layer is also included in this simulation.

For the junction temperature distribution among silicon chips within different packaging PP IGBTs we can see that the proposed Integrated and Stakpak packaging have quite uniform temperature distribution among IGBT chips or FRD chips. The temperature difference between the IGBT chips is only about 2.02 K and 1.45 K, respectively as shown in Table II. However, for the temperature distribution within the Pedestal PP IGBT is quite uneven with the maximum of 378 K and the minimum of 357 K. The temperature difference among IGBT chips reaches to 21 K caused by the clamping force imbalance. Furthermore, for the junction temperature of FRD chips the distribution is relative even as they have uniform low clamping force. The junction temperature distribution is strictly related to the clamping force distribution through the electrical and thermal contact resistance. This simulation results also show that the clamping force

978-1-7281-0582-6/19 $31.00 © 2019 IEEE   245

distribution is quite important for PP IGBT and more details about the coupling relationship can be found in [2].

Meanwhile, thanks to the integrated heatsink the maximum junction temperature of the proposed structure is much lower than the Pedestal and Stakpak. The Stakpak structure is only possible to conduct the heat in one side that lead to a higher junction temperature. Thus, from the point view of temperature distribution within PP IGBT, the Pedestal structure is the worst as the clamping force distribution is bad. And from the point view of the maximum junction temperature, the Stakpak is the worst that only one side (collector electrode) is possible to conduct the heat. The proposed packaging structure has relative uniform junction temperature distribution and quite low thermal resistance that lead to low junction temperature.

TABLE II.   JUNCTION TEMPERATURE COMPARISON

Packaging styles	IGBT chip (K)			FRD chip (K)		
	Max.	*Min.*	*Diff.*	*Max.*	*Min.*	*Diff.*
Pedestal	378.38	356.99	21.59	363.38	362.08	1.30
Stakpak	382.92	381.47	1.45	372.52	371.15	1.37
Integrated	361.08	359.06	2.02	350.78	349.78	1.00

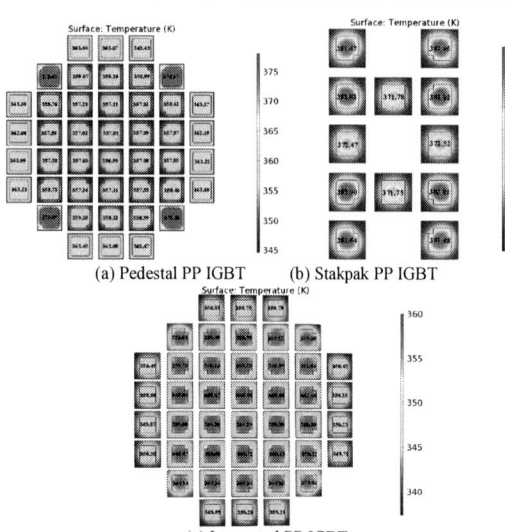

(a) Pedestal PP IGBT   (b) Stakpak PP IGBT

(c) Integrated PP IGBT

Fig. 9 Junction temperature distribution and the maximum junction temperature of chips

V. CONCLUSION

The Integrated packaging structure of press pack for high power IGBTs is proposed in this paper to overcome the disadvantages of imbalanced clamping force of Pedestal and only one side cooling of Stakpak structures. The interested thermal and mechanical behavior of these three types are compared with the electro-thermo-mechanical finite element model. The preliminary conclusions can be draw in below:

1) The pressure distribution among the silicon chips within three packaging structures is uniform under the clamping phase. However the pressure distribution of Pedestal structure changes a lot when the chips are heated up by the load current. The chips located in the edge of the packaging have quite low clamping force and even loss of contact because of the electrode warpage with thermal stress.

Thanks to the separated spring disc, the same structure in Stakpak, the pressure or clamping force distribution of the proposed Integrated structure is acceptable with maximum error of -12.4%.

2) The junction temperature distribution is strongly related to the clamping force distribution with the electrical and thermal contact resistance. The junction temperature distribution within Pedestal PP IGBT is bad though it has a relative low maximum junction temperature. The proposed Integrated packaging has a low junction temperature with the double-side cooling and uniform temperature distribution with balanced clamping force.

In the future works, we will focus on the clamping force and junction temperature measurement within the proposed packaging and power cycling tests to verify the reliability.

REFERENCES

[1] Zhao Z, Deng E, Zhang P, et al. Review of the Difference between the Press Pack IGBT Using for Converter Valve and for DC Breaker [J]. Transactions of China Electrotechnical Society, 2017, 32(19): 125-133.

[2] Deng E, Zhao Z, Lin Z, et al. Influence of Temperature on the Pressure Distribution within Press Pack IGBTs [J]. IEEE Transactions on Power Electronics, 2018, 33(7) :6048-6059.

[3] Eicher S, Rahimo M, Tsyplakov E, et al. 4.5kV press pack IGBT designed for ruggedness and reliability[C]// Industry Applications Conference, 2004. Ias Meeting. Conference Record of the. IEEE, 2004:1534-1539 vol.3.

[4] Deng E, Zhao Z, Xin Q, et al. Analysis on the difference of the characteristic between high power IGBT modules and press pack IGBTs[J]. Microelectronics Reliability, 2017, 78:25-37.

[5] Deng E, Zhao Z, Zhang P, et al. Clamping Force Distribution within Press Pack IGBTs[J]. Transactions of China Electrotechnical Society, 2017,32(06):201-208.

[6] Quirk M, Serda J, Hall P. Semiconductor Manufacturing Technology: United States Edition[J]. Pearson Schweiz Ag, 2000.

[7] Zhang J. Fatigue Life Prediction of Single-chip Submodule of Press-Pack IGBTs Based on Finite Element Method [D]. North China Electric Power University; North China Electric Power University (Beijing), 2018.

[8] Busca C, Teodorescu R, Blaabjerg F, et al. Dynamic thermal modelling and analysis of press-pack IGBTs both at component-level and chip-level [C] // Proceedings of the 39th Annual Conference of the IEEE Industrial Electronics Society, Vienna, Austria, 2013: 677-682.

[9] Deng E, Zhao Z, Zhang P, et al. Influence of the Temperature on the Thermal Contact Resistance Within Press Pack IGBTs[J]. Semiconductor Technology, 2016, 41(12): 906-912.

[10] Poller T, D'Arco S, Hernes M, et al. Influence of the clamping pressure on the electrical, thermal and mechanical behaviour of press-pack IGBTs[J]. Microelectronics Reliability, 2013, 53(9-11):1755-1759.

[11] Lu T, Jin J M. Coupled Electrical–Thermal–Mechanical Simulation for the Reliability Analysis of Large-Scale 3-DInterconnects[J]. IEEE Transactions on Components, Packaging and Manufacturing Technology, 2017, 7(2): 229-237.

[12] Cova P, Nicoletto G, Pirondi A, et al. Power cycling on press-pack IGBTs: measurements and thermomechanical simulation[J]. Microelectronics Reliability, 1999, 39(6): 1165-1170.

[13] Pirondi A, Nicoletto G, Cova P, et al. Thermo-mechanical finite element analysis in press-packed IGBT design[J]. Microelectronics Reliability, 2000, 40(7): 1163-1172.

[14] Shen Y, Deng E, Zhao Z, et al. Influence of Clamping Force on Power Loss of Press-Pack IGBTs for Converter Valve[J]. Semiconductor Technology, 2017, 42(9):687-695.

[15] Tinschert L, Årdal AR, Poller T. Possible failure modes in press-pack IGBTs. Microelectronics Reliability. 2015;55(6):903-911

Proceedings of the 31st International Symposium on Power Semiconductor Devices & ICs
May 19-23, 2019, Shanghai, China

The Development of a 1200V/400A SiC Hybrid Module

Puqi Ning, Tianshu Yuan, Han Cao, Lei Li, Yuhui Kang

Institute of Electrical Engineering
Chinese Academy of Sciences
Beijing, China

University of Chinese Academy of Sciences
Beijing, China

Abstract—**Hybrid switch (HyS) combines low conduction loss of Si IGBT and low switching loss of SiC MOSFET. The cost of HyS is close to pure Si IGBT. The promising high performances of HyS will bring considerable achievement to enhance power density of a converter system. In this paper, a SiC/Si HyS based 1200 V/400 A three phase module was developed, and was tested in a motor drive.**

Keywords—Hybrid switch, silicon carbide

I. INTRODUCTION

In recent years, the silicon carbide (SiC) power semiconductor has emerged as an attractive alternative of silicon (Si) devices [1-3]. Some manufactures have successfully developed SiC MOSFETs, which demonstrated the high blocking voltage, high switching frequency and high temperature abilities. These the advanced properties of SiC MOSFET will lead converters to higher power density, which is a main requirement of automotive application, as shown in Table 1. The successful utilization of SiC devices in Tesla model 3 has demonstrated the trend[4].

TABLE I. CONVERTER DENSITY INCREASING WITH SiC

Organization	Year	Rated power (kW)	Power density
Denso	2014	70	100 kW/L
Mitsubishi	2017	N/A	86 kVA/L
Mitsubishi	2019	N/A	150 kVA/L

For EV cars, the converter requires high speed switching, 60~200 kW power, low on-resistance and reasonable price. These requirements will battle engineers when they design converter with pure SiC devices for low-end EV cars [5].

SiC MOSFETs is 5-8 times more expensive than Si IGBTs with the same power rating. If SiC MOSFET reduce 15% cost every year, it will take 10 year to obtain the same price as SiC IGBT today. However, Si IGBT will reduce the cost and improve performance at the same time.

To overcome the challenges, the combination of Si IGBT and Si MOSFET devices was investigated by compensating disadvantages 20 years ago. In recent years, hybrid switches (HyS) based on Si IGBT and SiC MOSFET were studied [6], as shown in Fig.1. In these paper, the HyS losses, the gate drive pattern, gate drive hardware, HyS module, HyS

---

[1] This work is supported by The National key research and development program of China (2016YFB0100600), the Key Program of Bureau of Frontier Sciences and Education, Chinese Academy of Sciences (QYZDBSSW-JSC044).

converter and costs of hybrid switches have been investigated and verified [6-8].

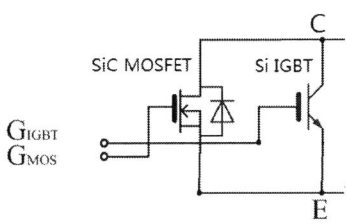

Fig. 1. HyS parallels a SiC MOSET and a Si IGBT

The conduction loss of HyS is close to purse Si IGBT device, and the switching loss of HyS is close to pure SiC MOSFET device [9]. The promising performance of HyS will result in considerable achievement to enhance power density of a converter system. In this paper, a 1200 V/400 A three-phase was developed and tested.

II. POWER MODULE DESIGN

The targeted 1200 V/400 A three phase SiC/Si HyS includes 12 Si IGBTs, 12 Si diodes and 12 SiC MOSFETs. To reduce parasitic parameters and prevent interference from the main power, Kelvin source pin of gates (S1 to S6) were added to legs, as shown in Fig.2.

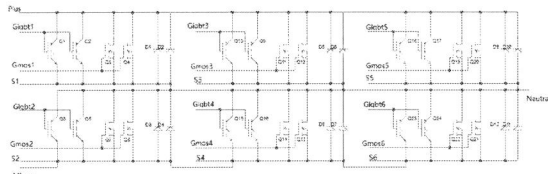

Fig. 2. 1200 V/400 A three phase SiC/Si HyS.

SiC MOSFET dies are selected from available products, as listed in Table 2. The dimension, current rating at 100°C, and the IGBT die matching style are selection points. The current sharing of HyS is set as 1:3. 1200 V/300 A Si IGBT is not that popular and available for regular research organizations. At the same time, regular 1200 V/200 A Si IGBT has a area over 200 mm², which is too large compared with regular SiC MOSFET die. It is difficult to design a compact HyS module. Finally 1200 V/150 A Si IGBT and 50A SiC MOSEFT were selected.

TABLE II. 1200V MOSFET DIE PRODUCT

Vendor	Product	Current
CREE	CPM3-1200-0013A	90 A @100°C
CREE	CPM3-1200-0016A	75 A @100°C
CREE	CPM3-1200-0025B	60 A @100°C
Rohm	S4103	95 A @25°C
Rohm	S4102	72 A @25°C

The selected Si IGBTs are Infineon IRG8CH137K10F, which are 1200 V/150 A devices. The Si diodes are Infineon IRD3CH82DB6, which are 1200 V/150 A devices. The SiC MOSFETs are CREE CPM2-1200-0025B, which are 1200 V/50 A devices.

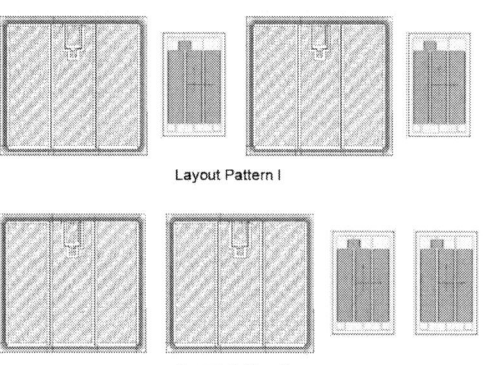

Fig. 3. Basic paralleled die layout.

When paralleling Si IGBT and SiC MOSFET, there are two basic layout patterns, as shown in Fig.3. To better balance current sharing during the transient, pattern I is a better option. However, it will increase the DBC size when routing gate paths. Finally, pattern II was adopted in the module layout. The fabricated prototype module is shown in Fig.4. The size is $140 \times 105.6 \times 18$ mm^3, and it is very close to a 650V/400A HP1 module.

Fig. 4. 1200 V/400 A HyS.

Fig. 5. Dimension comparison.

Fig.5 shows the volume comparison between the 400 A hybrid module and a regular HP1 module.Table 3 shows the comparison among presented 1200 V/400 A hybrid module and product modules with the similar power rating.

TABLE III. 400 A HYBRID MODULE COMPARISON

Modules	Power rating/ device type	Dimension(mm^3), assuming with 8 mm Pinfin	Density
Infineon FS400R07A3E3	650 V/400 A Si	140×110×22	2.3
Infineon FS820R08A6P2B	750 V/820 A Si	155×127×22	4.29
Infineon FS900R08A2P2_B32	750 V/900 A Si	216×100×25	3.75
Fuji M653	750 V/800 A Si	162×116×24	3.99
CREE CAS325M12HM2	1200 V/325 A SiC	110 x 65 x 18	3.03
Presented 400 A module	1200 V/400 A hybrid	140×105.6×18	5.40

Actually, 600~800 V power modules are commonly used for automotive application because DC bus voltage is usually set to 350~450 V for most of pure EV cars. For this DC bus voltage, 600~900 V SiC MOSFET can be selected from table 4. A 650 V/400 A HyS module was also designed.

TABLE IV. 600V/900V MOSFET DIE PRODUCT

Vendor	Product	Voltage	Current @ 150°
CREE	CPM3-0900-0010A	900 V	140 A @100°C
Rohm	S4003	650 V	118 A @25°C
Rohm	S4002	650 V	93 A @25°C

To match CREE CPM3-0900-0010A, Si IGBT was selected as Infineon IGC100T65T8RM, which is a 650 V/200 A devices. Matching Si diodes is Infineon SIDC50D65C8, which is a 650 V/200 A devices. The module size can be reduced to 70%, as shown in Fig.6. This module will be tested in the near future.

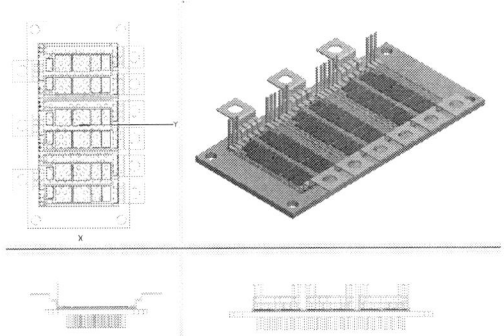

Fig. 6. 650 V/400 A HyS design.

III. GATE DRIVE DESIGN

HyS can keep ZVS mode of Si IGBTs, which is the main advantage. Therefore, the switching pattern shown in Fig.7 should be considered. At the same time, tail current induced by the miller effect should be suppressed by a carefully designed gate drive.

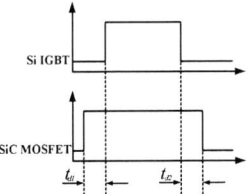

Fig. 7. Switching patterns for HyS.

During the turn-off phase, a miller effect can be clearly found. For a fast switching of SiC MOSFET, high dv/dt will bring electrical interference to Si IGBT drive path. A Miller current appears and the gate voltage of Si IGBT device (V_{GE}) will increase. Once the voltage exceeds the threshold voltage, a fault re-turn-on happens. Si IGBTs will share the current from SiC MOSFET, and the tail current will be observed again. In this paper, commercial chips ACPL-332J with built-in turn-on/turn-off path separators were used. When IGBT is off, a low-impedance path is established inside the gate drive chips and the gate voltage spikes can be reduced, as shown in Fig.8.

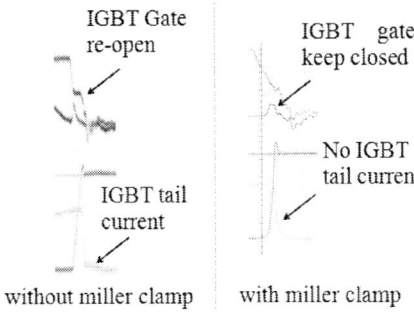

Fig. 8. Miller clamp can limit tail current.

Fig.9 shows the double-pulse-test waveform of presented hybrid module at 1200 V/400 A.

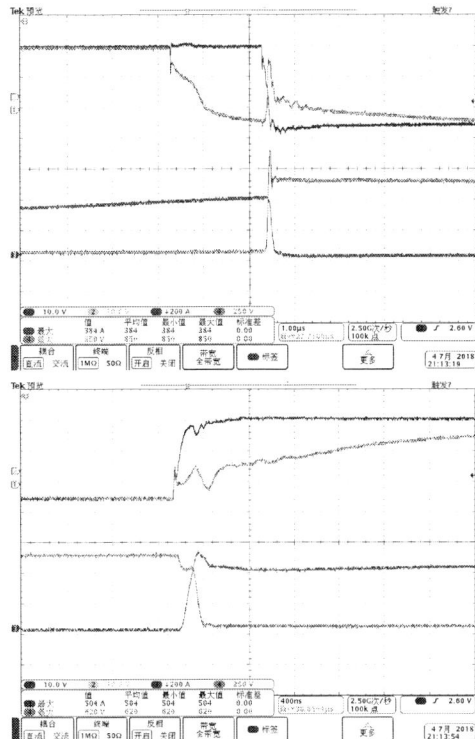

Fig. 9. Double pulse test of 400 A Hybrid module.

During the test, a speed up turn-on was found when SiC MOSET and Si IGBT are driven by a certain time delay. The delay definition is shown in Fig.10.

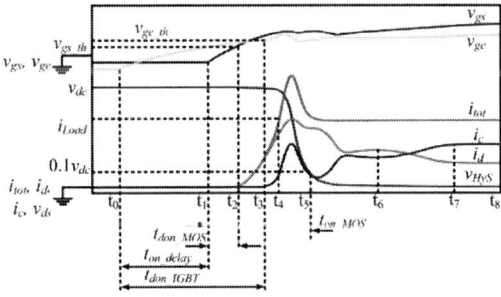

Fig. 10. Delay definition.

In Fig.11, the delay configuration of case 1 is t_{don_IGBT} - t_{on_MOS} < t_{on_delay} ≤ t_{don_IGBT} - t_{don_MOS}. The delay configuration of case 2 is t_{don_IGBT} - t_{don_MOS} < t_{on_delay} ≤ t_{on_IGBT} - t_{don_MOS}. The case 3 is a balanced case of case 1 and case 2, and it obtains the fastest turn-on.

978-1-7281-0582-6/19 $31.00 © 2019 IEEE

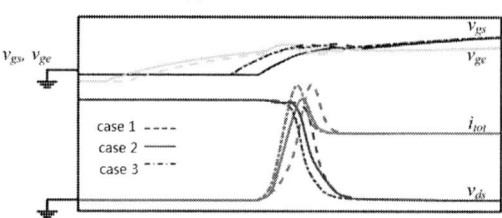

Fig. 11. Turn-on transient.

IV. MOTOR DRIVE

So far, there is not too many HyS based converter designs and related tests. In some literature, demonstrations stopped at double-pulse tests because of the complexity of integrating discrete devices and busbars. The presented hybrid module was used in a motor drive to show the high switching ability and low conduction loss benefit.

The 1200 V/400 A hybrid module was cooled by a cold-plate and controlled by a DSP. The hardware is shown in Fig.12.

Fig. 12. Motor drive hardware.

The motor drive was tested up to 25 kW with 30 kHz switching frequency. Fig.13 depicts the test waveform.

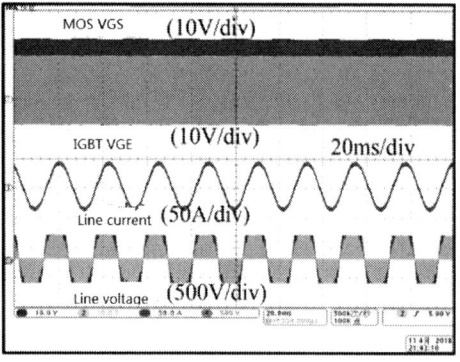

Fig. 13. Motor drive test waveform at 25 kW, 30 k Hz .

To match the presented three module with pinfin, the cold plate is designed as Fig.14. In the next step, single phase 1200 V/400 A HySs will be fabricated and tested, and they have the same DBC layout design as presented three-phase module. To match three separate modules, the water plate is

designed as Fig.9. The thermal performance comparison will be the main task in the next step.

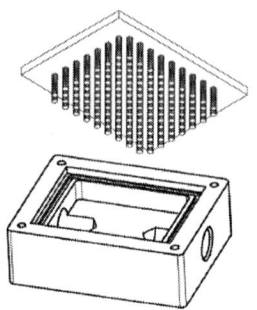

Fig. 14. Cold plate for three-phase module.

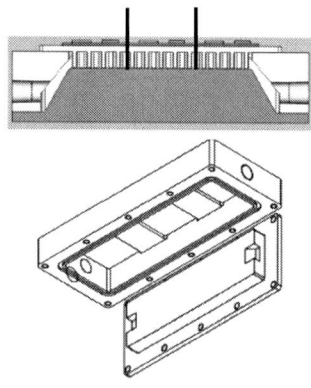

Fig. 15. Cold plate for three single-phase module.

REFERENCES

[1] Rahman, M.F., P. Niknejad, and M.R. Barzegaran. Comparing the performance of Si IGBT and SiC MOSFET switches in modular multilevel converters for medium voltage PMSM speed control. in 2018 IEEE Texas Power and Energy Conference (TPEC). 2018.

[2] Meng, J., et al., A finite difference method modeling for IGBT and diode in PSPICE. Chinese Journal of Electrical Engineering, 2017. 3(3): p. 85-93.

[3] Deshpande, A. and F. Luo. Comprehensive evaluation of a silicon-WBG hybrid switch. in 2016 IEEE Energy Conversion Congress and Exposition (ECCE). 2016.

[4] https://www.mitsubishielectric.com/

[5] Qin, H., et al. Characteristics and Switching Patterns of Si/SiC Hybrid Switch. in PCIM Asia 2017; International Exhibition and Conference for Power Electronics, Intelligent Motion, Renewable Energy and Energy Management. 2017.

[6] Song, X., et al. High voltage Si/SiC hybrid switch: An ideal next step for SiC. in 2015 IEEE 27th International Symposium on Power Semiconductor Devices & IC's (ISPSD). 2015.

[7] Deshpande, A. and F. Luo. Design of a silicon-WBG hybrid switch. in 2015 IEEE 3rd Workshop on Wide Bandgap Power Devices and Applications (WiPDA). 2015.

[8] Lei Li, Puqi Ning, and etc.,"A 1200 V/200 a half-bridge power module based on Si IGBT/SiC MOSFET hybrid switch," IEEE CPSS Transactions on Power Electronics and Applications, Vol.3, issue 4, pp. 292 - 300, 2018.

[9] P. Ning, and etc., "A hybrid Si IGBT and SiC MOSFET module development," IEEE CES Transactions on Electrical Machines and Systems, Vol.1, issue 4, pp. 360-366, 2017.

Proceedings of the 31st International Symposium on Power Semiconductor Devices & ICs
May 19-23, 2019, Shanghai, China

Analog Basis, Low-Cost Inverter Output Current Sensing with Tiny PCB Coil Implemented inside IPM

Battuvshin Bayarkhuu[1], Bat-Otgon Bat-Ochir[2], Kazunori Hasegawa[1],
Masanori Tsukuda[1], Bayasgalan Dugarjav[2], Ichiro Omura[1]

[1]Kyushu Institute of Technology, 2-4 Hibikino, Wakamatsu-ku, Kitakyushu, 808-0196 Japan
[2]National University of Mongolia, Ulaanbaatar, 14201 Mongolia
Email: bayarkhuu.battuvshin201@mail.kyutech.jp

Abstract— **This paper proposes a practical current sensor integration in the intelligent power modules (IPMs) using simple PCB Rogowski coil sensors. The PCB sensors produce signals that proportional to the high frequency switching current from high and low side IGBTs. Then with only general-purpose Op-Amps and photo-couplers based integrator and sample and hold (S/H) circuits reproduce output current of the inverter. Specifically, the "envelop tracking" method has successfully proved on an experimental inverter setup. A significant accomplishment of an improved new analog circuit is the measurement during narrow pulse width around unity modulation index that leads to higher inverter output power.**

Keywords—IPMs, PCB current sensor, Rogowski coil, integration.

I. INTRODUCTION

As an integral part of electric vehicle and energy system, the power semiconductor devices are becoming a trend [1]. It is getting high power density, faster and cheaper. As a result, devices such as SiC and GaN have been actively researched for higher power density and power loss reduction with high switching characteristics [2]. Overall volume and passive components of the system are getting smaller because of properties of those new power semiconductor devices. The properties are including higher voltage withstand, thermal conductivity, operating temperature and current density [3].

The study [4] estimated the power densities of Si and SiC DC-DC converters. It predicted that operating at two times higher temperature, SiC devices based DC-DC converter would have relatively higher efficiency than the Si based counterparts and improve the power density by 50%. [5] also predicted that converters fully utilizing SiC devices will be able to reach power ratings of 1MW that will be 1/50 of their former size by 2025.

IPMs are another representative of the development of the power semiconductor device. IPM is designed to integrate dedicated drive and protection circuitry along with power switching elements (IGBTs and diodes) in an optimized module structure [6]. They provide easy usage in applications and reliability improvement of applied power electronics systems.

Despite the development in both main power semiconductor devices and auxiliary components, magnetic parts of the inverter system remain bulky and costly. The current sensors are a core component in terms of a control system, monitoring, and reliability of the system. Hall-effect-based current sensors and current transformers (CT) are mainly adopted for output current monitoring. However, these current sensors have become a constraint to reduce the volume and overall power density of the systems due to their limited operating range of magnetic core.

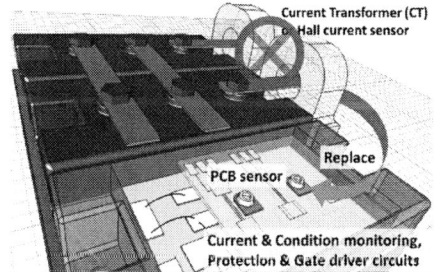

(a) Illustration of the PCB current sensor integration inside IPMs.

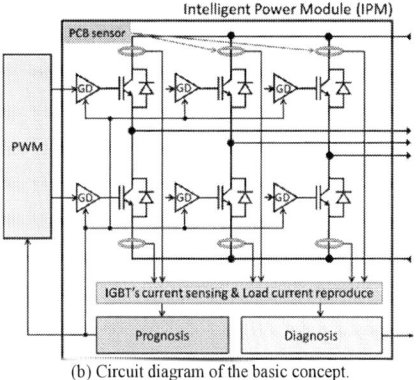

(b) Circuit diagram of the basic concept.

Fig. 2   Basic concept of the current sensor integration inside IPMs.

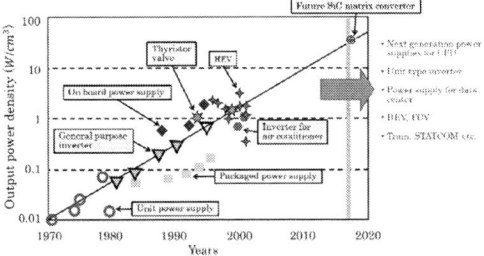

Fig. 1   Power density change in power semiconductor devices [2].

978-1-7281-0582-6/19 $31.00 © 2019 IEEE   251

On the other hand, the PCB "Rogowski coil" current sensor has been developed to reduce the volume and cost of the systems. The PCB sensor has a simple structure with reduced cost. Moreover, the sensor catches up high frequency switching current from the switching device which well suited for the PCB sensor characteristics [7].

Recently, PCB Rogowski -coil approach for current sensing has been demonstrated [8-12]. In those papers, however, FPGAs or digital processors are needed for waveform reproduction which requires a too high cost to integrate into IPMs. The authors' group has proposed a low-cost, analog basis demonstration of "envelop tracking" method [13]. However, the narrow pulse width control signals cause an accuracy problem when the output signal of the integrator circuit does not fall or rise to zero before the next turn-on or turn-off event of the switching device. The method also limited to a low amplitude modulation index.

In this paper, for the first time, tiny PCB coils are embedded inside the IGBT power module package to demonstrate noise immunity of the system and applied new circuit to solve the accuracy problem. Only general-purpose OP-amps and photo-couplers are used in the proposed demonstration circuit to show a practical possibility of integration of current sensing in IPMs.

II. NEW OUTPUT CURRENT MEASUREMENT METHOD

As mentioned before, a simple structured PCB current sensor which based on Rogowski coil shown in Fig.3, measures high frequency switching current of the IGBTs. The sensor fabricated within printed circuit board by a new fishbone pattern and supported by shield layer to eliminate the effect of electromagnetic field inside a power module. Characteristics of the PCB sensor are highly compatible with the high-frequency region such as controller switching signal [7].

The analog approach, named "envelop tracking" method uses two sets of PCB sensors, integrator amplifiers, and S/H circuits. The method detects and tracks the maximum point value of PCB sensor signal at each switching (turn-on or turn-off) time of power semiconductors device. Op amp integrator and S/H circuits are amplifying and sampling the peak points from sensor signals. Separately sampling the current of upper and lower arms of inverter system, two sensors and proposed analog circuits reproduce a envelop measurement signals. Those signals are following an upper edge and a lower edge of the output current, as shown in Fig.4.

In applications of an inverter, mean current and ripple component are the main parameters for controller and monitoring systems. With the method, a calculated average value of the outputs from two analog circuits represents the mean output current. The difference of the two outputs show the ripple amplitude that necessary to monitor motor torque ripple and circuit parameter change.

$$Error[\%] = 100 \; x \; e^{-\frac{t_{on}}{RC}} \qquad (1)$$

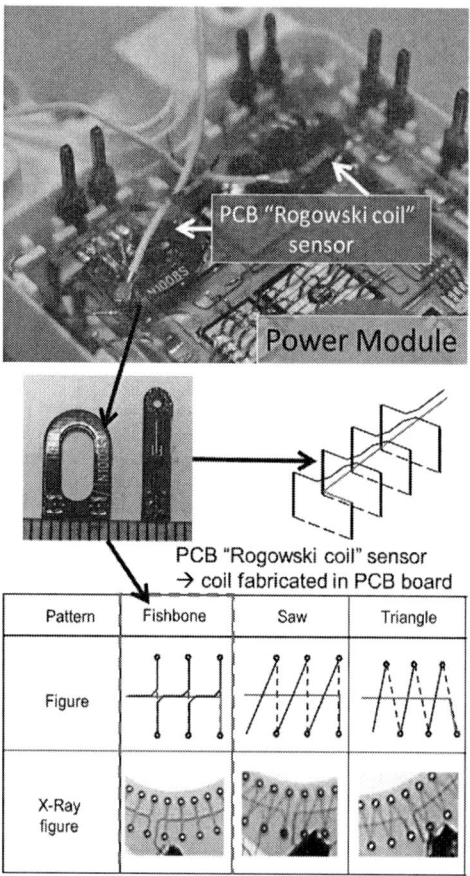

Fig. 3   PCB "Rogowski coil" sensor, placement and structure. [2]

The accuracy error occurs when switching signal pulse (t_{on}) becomes narrower than the integrator time constant (RC), as described in Eq. (1). A photo-coupler has employed as a reset trigger in the integrator circuit which eliminates an unnecessary part of the integrator output signal as shown in Fig.4.

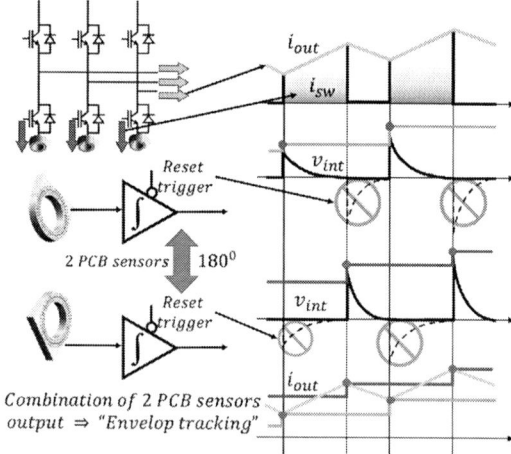

Fig. 4   Analog circuit based current measurement, "Envelop tracking" method. Improvement of the integrator circuit.

III. CURRENT MEASUREMENT

Fig.4 illustrates the main reproduction procedure of the output current that defines the relations between output current i_{out}, switching current i_{SW}, the output signal of the integrator v_{int} and effect of the reset trigger. The output signal generated in the PCB sensor is proportional to the rate of change of current in the IGBT, the output is connected to an integrator circuit to provide an output signal that is proportional to the current.

The integrator can detect rising and falling of the PCB sensor output signals. Since the current reproduction process is using multiple sensors and analog circuits, a photo-coupler cancels out each half of the output signals of the integrator. As described in Eq. (1), the time constant of the integrator circuit is determined by C_2R_4 from Fig 5 (a).

A combination of two PCB sensor outputs still contains the turn-on and turn-off current. Therefore, it is possible to reproduce the output current waveform by means of sampling the turn-on and turn-off current from integrator outputs. Fig.6 shows the measured waveforms of the analog circuit, which the S/H amplifier updates output value just after a rise of the integrator output.

Because of the dead time in switching signal, timing of the sampling trigger must be adjusted properly. So, comparator in the analog circuit determines the correct timing of the sampling. With the small delay time Δt shown in fig. 6, the comparator or switching signal becomes a reference for sampling trigger signals for the S/H. Delay time adjusted to eliminate reverse recovery current effect superimposed in the integrator output signal.

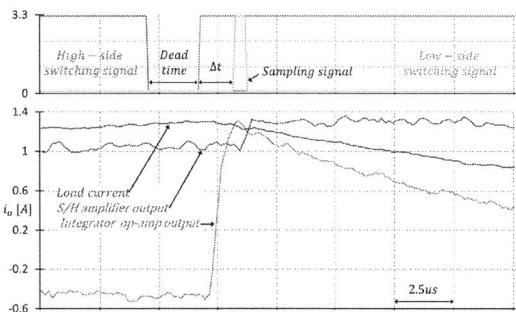

Fig. 6 Measurement waveforms of analog circuit.

IV. EXPERIMENTAL VERIFICATION

An experiment is carried on investigating the effectiveness of the proposed method in practice. The proposed method is implemented on a full bridge inverter setup using an FPGA with sinusoidal PWM control of 50Hz reference and 5kHz carrier signals. The proposed analog circuits reproduce the output current from switching currents of the full-bridge inverter. Output signals of the analog circuit are directly connected to the oscilloscope for monitoring. Fig.7 is the photo of experimental inverter setup where the PCB sensors placed inside the IGBT power module in the bottom right corner of fig.7 (b).

Fig.8 shows the output current measurement without the integrator circuit reset trigger. As discussed before, measurement accuracy error appears when SPWM signal gets narrower around positive or negative peaks. According to Eq. (1), accuracy error value displacement must be around 20% from the actual current value. Furthermore, the measurement value has displaced 19% from the real current value in this experiment.

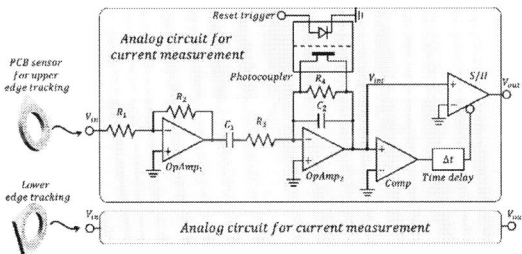

(a) Circuit schematics of the analog circuit for output current reproduction.

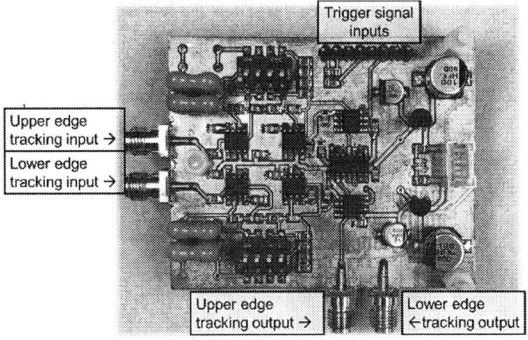

(b) Photo of the proposed analog circuit.

Fig. 5 The proposed analog circuit for current measurement.

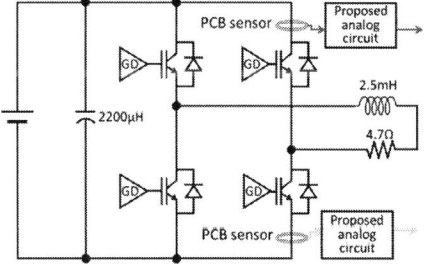

(a) Full-bridge inverter circuit diagram.

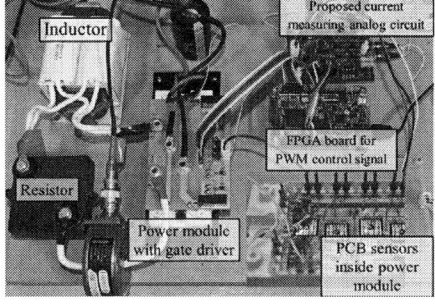

(b) Photo of experimental setup.

Fig. 7 Full-bridge inverter experimental setup.

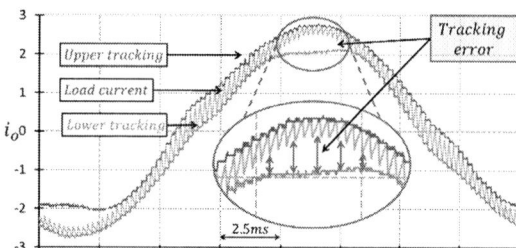

Fig. 8 Output current measurement result without integrator reset. The tracking error appears during narrow pulse width signal.

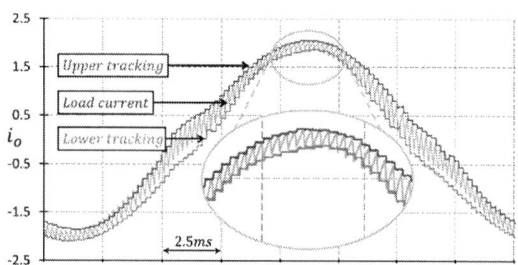

Fig. 9 Output current measurement result of envelop tracking method after integrator reset.

Our final result Fig.9 shows an excellent correspondence between the conventional sensor output and proposed method output. An effect of the photo-coupler in the integrator signal significantly improved the accuracy of "envelop tracking" method for output current measurement.

V. CONCLUSION

In this paper, for the first time, tiny PCB coils are embedded inside the IGBT power module package to demonstrate noise immunity of the system and applied new circuit to solve the accuracy problem. Only general-purpose op-amps and photo-couplers are used in the proposed demonstration circuit to show a practical possibility of integration of current sensing in IPMs. Most importantly, we proved an accurate current measurement with high power output inverter using embedded PCB "Rogowski coil" sensors and low-cost analog circuits.

ACKNOWLEDGEMENTS

The authors gratefully express our thanks to the MJEED project for supporting this study.

REFERENCES

[1] G. Majumdar, T. Oi, T. Terashima, S. Idaka, D. Nakajima, and Y. Goto, "Review of Integration Trends in Power Electronics Systems and Devices," in Proc. Intl. Conf. on Integrated Power Electron. Sys. (CIPS), Mar. 2016.

[2] Yusuke Hayashi, Kazuto Takao, Toshihisa Shimizu, Hiromichi Ohashi, "An Effective Design Method for High Power Density Converters," 13th International Power Electronics and Motion Control Conference, 2008.

[3] Ian Laird, Xibo Yuan, James Scoltock, Andrew J. Forsyth, "Design Optimisation Tool for Maximising the Power Density of 3-Phase DC-AC Converters Using Silicon Carbide (SiC) Devices," in IEEE Trans.Power electronics, Vol. 33, No. 4, pp. 2313-2932, 2018.

[4] K. Shenai, "Silicon carbide power converters for next generation aerospace electronics applications," in National Aerospace and Electronics Conference, 2000. NAECON 2000. Proceedings of the IEEE, pp. 516–523, 2000.

[5] I. Takahashi, "SiC Power Converter Technology in Future," in Electric Machines and Drives Conference, 2003. IEMDC'03. IEEE International, vol. 3, Jun., pp. 1903–1908, 2003.

[6] Gourab Majumdar, Masanori Fukunaga, Toshifumi Ise, "Trends of Intelligent Power Module," Transactions on Electricaland Electronic Engineering IEEJ, pp. 143–153, 2007.

[7] M. Koga, M. Tsukuda, K. Nakashima, and I. Omura, "Application-specific Micro Rogowski Coil for Power Modules -Design Tool, Novel Coil Pattern and Demonstration-," in Proc. IEEE Appl.Power Electron. Conf. (APEC), 2016, pp. 700-703.

[8] N. Langmaack, G. Tareilus, and M. Henke, "Novel Highly Integrated Current Measurement Method for Drive Inverters," in IEEE 1997 Applied Power Electronics Conference, 1997, pp. 455-452.

[9] S. Takahara, K. Hasegawa, M. Tsukuda and I. Omura, "Built in Load Current Measuring Method for IGBT Power Module," in Power technology / Power system technology / Semiconductor power conversion joint research group conference, Mar. 2016, pp. PE-16-039.

[10] S. Tabata, K. Hasegawa, M. Tsukuda, I. Omura, "New Power Module Integrating Output Current Measurement Function," in 29th International Symposium on Power Semiconductor Devices and IC's (ISPSD), Jun. 2017, pp. 267-270.

[11] K. Hasegawa, S. Takahara, S. Tabata, M. Tsukuda and I. Omura, "A New Output Current Measurement Method with Tiny PCB Sensors Capable of Being Embedded in an IGBT Module," in IEEE Trans.Power electronics, vol. 32, no. 3, pp. 1707-1712, Mar. 2017.

[12] Jun Wang, Slavko Mocevic, Yue Xu, Christina DiMarino, Rolando Burgos, Dushan Boroyevich, "A High-Speed Gate Driver with PCB-Embedded Rogowski Switch-Current Sensor for a 10 kV, 240 A, SiC MOSFET Module" in IEEE 3rd Workshop on Wide Bandgap Power Devices and Applications (WiPDA), pp. 5489–5494 Nov. 2015

[13] B. Bat-Ochir, B. Bayarkhuu, K. Hasegawa, M. Tsukuda, B. Dugarjav, I. Omura, "Envelop Tracking Based Embedded Current Measurement for Monitoring of IGBT and Power Converter System," in Microelectronics Rel., Vol. 88–90, p.p 500-504, Sep. 2018

Proceedings of the 31st International Symposium on Power Semiconductor Devices & ICs
May 19-23, 2019, Shanghai, China

Short-Circuit Ruggedness Analysis of SiC JMOS and DMOS

Fu-Jen Hsu, Cheng-Tyng Yen, Chien-Chung Hung, Kuo-Ting Chu, Lurng-Shehng Lee, Chwan-Ying Lee

Hestia Power Incorporated

10F-2, 27 Guanxin Rd, Hsinchu, Taiwan, e-mail: FJ.Hsu@hestia-power.com

Abstract—**To prevent the degradation caused by Bessel plane dislocation and improve the characteristics of body diode in silicon carbide MOSFETs, Schottky barrier diode integrated MOSFET has been developed by many researchers. One of those structures was named JMOS which exhibits much higher chip area efficiency. In this paper, the relationship between short-circuit withstanding time and V_{GS} is presented. After this, a series of comparison between conventional DMOS and JMOS are executed, which covered a single pulse short-circuit event and a repetitive short-circuit test. The result shows that short-circuit withstanding capability of JMOS is still good enough but slightly less than that of DMOS, which means both devices could handle over 3 micro-seconds of short-circuit duration with no failure occurs. With repetitive short-circuit stress in this work, the electrical characteristics have been measured on D.U.T. by monitoring the test waveform after finished 10,000 times of pulses. In this experiment, no significant degradation has been detected after repetitive short-circuit stress. Furthermore, the failure mechanism of JMOS may be different from conventional DMOS and this will be discussed in this work.**

Keywords- short-circuit, SiC MOSFET, ruggedness, body diode

I. INTRODUCTION

Short-circuit ruggedness (also realized as short-circuit withstanding time, SCWT) is a crucial index for power transistors robustness. In order to meet the system application requirement, silicon IGBTs are usually designed to withstand over 10 μsec of short-circuit operation, which leaves enough time for users to turn off the transistor before the tremendous short-circuit current sparking off failure.

Silicon carbide (SiC) FETs have many benefits, including its high thermal conductivity, high breakdown voltage, high current density, and high operating temperature. However, SiC FETs still have a drawback with its poor characteristic of short-channel effect while operating at high current density level. Specifically, SiC FETs will experience higher short-circuit current than conventional silicon transistors since the lower ability of self-current-limited. Therefore, to improve this feature of SiC FETs, researchers verified the short-circuit capability [1], [2] and developed high-speed gate controllers which could turn off the power FETs within 3 μsec [3] to prevent potential dangers in applications, especially in hazard environment and extreme operating conditions.

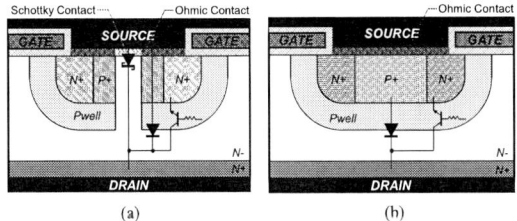

Fig.1: (a) Structure of JMOS and (b) DMOS. The Ohmic contact on n+/p+ regions and the Schottky contact on the drift layer were formed separately.

SiC JMOS [4] in Fig.1 is a solution to mitigate the body diode degradation issue and benefits to improve the power density of the whole system. In precisely, JMOS technology merged a DMOS and JBS in a monolithic device without any additional process and area penalty. In addition, electrical behaviors and avalanche endurance have been presented in previous work [5]. Indeed, to apply JMOS into practical application, other severe operating conditions must also be examined – such as short-circuit capability.

II. EXPERIMENT RESULT

A. Single Pulsed Short-circuit Event

In this work, an SCWT comparison and analysis between SiC DMOS and SiC JMOS are discussed. Fig.2 shows the schematic of single pulsed short-circuit test platform. In detail, an IGBT is placed on the short-circuit current path in between the DC-bus capacitors bank and D.U.T. to disconnect the power source to D.U.T. while a fault signal has been detected.

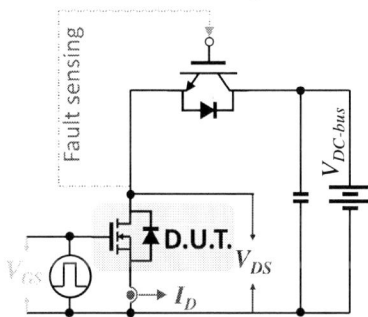

Fig.2: The schematic of the circuitry diagram of the short-circuit test.

To examine the relationship between V_{GS} and SCWT, this work first applies V_{GS} from 10V to 18V on 1700V/45mΩ DMOS, which shares the same process flow and device structure with our 650V and 1200V devices, to verify the trend of short-circuit capability by different gate bias as shown in Fig.3(a). The comparison of the short-circuit peak current (I_{SC}) versus SCWT (t_{sc}) is shown in Fig.3(b) as metrics. From the data shown in Fig.3, it can be discovered that increasing of V_{GS} will reduce SCWT. To be more specific, the short-circuit current drops and the SCWT of the device is improved by reducing V_{GS}.

Nevertheless, the reduction of V_{GS} suppresses efficiency in practical applications, which means users need to find the compromised condition for both characteristics. Considering both the operating efficiency ($R_{DS(on).sp}$) and SCWT capability, this paper chooses V_{GS}=15V as the general test condition to maintain the best balance between system feasibility (i.e. acceptable $R_{DS(on).sp}$) and short-circuit ruggedness. Thus, all DMOS and JMOS in this paper are tested at V_{GS}=15V by single short-circuit event to check the short-circuit capability of each device.

978-1-7281-0582-6/19 $31.00 © 2019 IEEE
255

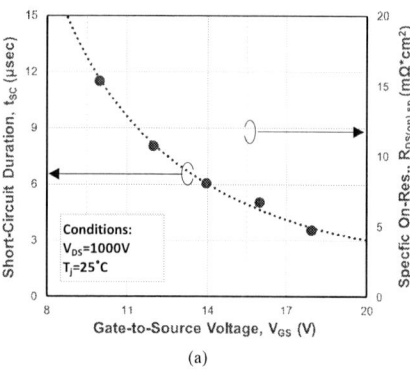

(a)

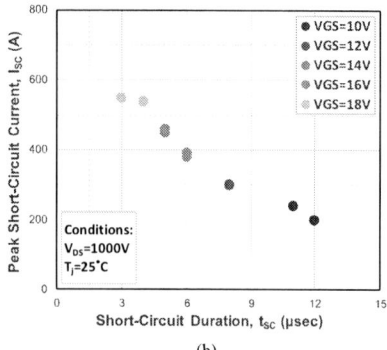

(b)

Fig.3: The (a) t_{SC} and $R_{DS(on),sp}$ vs. V_{GS} (b) I_{SC} vs. t_{SC} of 1700V/45mΩ SiC DMOS. Short-circuit capability be improved by reducing V_{GS}.

B. Repetitive Short-circuit Test

In order to ensure the long-term reliability of SiC DMOS and JMOS with multi-times of short-circuit events, this work also performs a repetitive short-circuit test. The test method and procedure are presented in Fig.4. In specific, a 5 μsec of short-circuit pulse is impacted on D.U.T. in every turn-on event. After this, the D.U.T. was turned-off to force into its blocking mode and sustain the high electric field for one second and subsequently be turned-on and force it into short-circuit mode again. This procedure, which could minimize the effect of switching loss, is operated continuously to 10,000 times of short-circuit impact be done in this test.

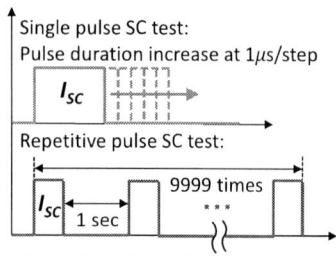

Fig.4: Test method in this work. In order to prevent heat accumulation, a 1sec delay is applied between each pulse.

Fig.5 illustrates I_{SC} and V_{GS} in the repetitive test. The data shown in Fig.5 is recorded in each switching waveforms. In other words, this method could monitor the real-time situation of devices within the test procedure. The variation of I_{SC} and V_{GS} for both types of device is < 2% within 10,000 times of repetitive short-circuit test.

Conditions: Tc=25°C, V$_{DS}$=400V, t_{sc}=5μs, t_int: 1sec

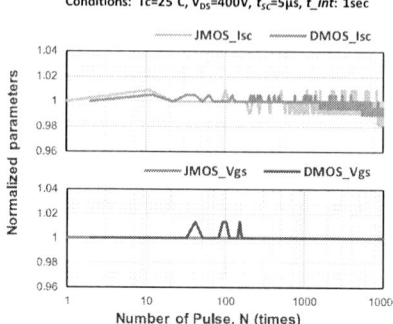

Fig.5: Normalized peak I_{SC} and V_{GS} values in the repetitive test waveform.

III. COMPARISON AND DISCUSSIONS

A. Comparison of DMOS and JMOS

The comparison of experimental results is listed in Fig.6 which includes crucial electrical characteristics before and after 10,000 times of short-circuit stress. The variation of I_{DSS}, $R_{DS(on)}$, and V_{SD} are caused by the non-negative gate bias. With the non-negative gate bias, it can't discharge the electron charge trapping on the interface of gate oxide. Therefore, V_{th} will be recovered after trapped electrons are discharged.

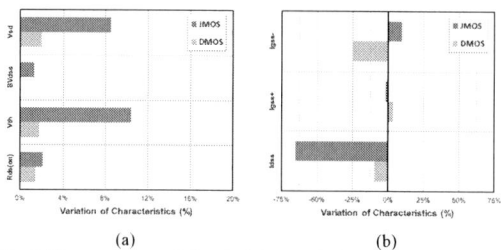

(a)   (b)

Fig.6: The comparison of electrical characteristics variation after 10,000 times of short-circuit events.

Fig.7 illustrated the I_D-V_D curves of JMOS and DMOS. Comparing the I_D-V_D characteristic of DMOS and JMOS, the $R_{DS(on)}$ of JMOS approximates to DMOS in the linear region but higher in the saturation region owing to modification of layout. Theoretically, this behavior suppresses the current and help the device to extend its SCWT due to the smaller operating value of JMOS I_D curve in short-circuit mode at the same V_{DS}. Consequently, this behavior reduces the power consumption on devices. In short, it will produce a lower lattice temperature on the device in the short-circuit event.

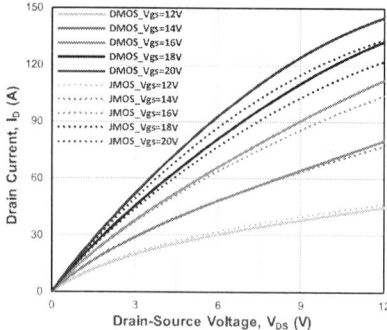

Fig.7: I_D-V_D comparison between 1200V/60mΩ DMOS and JMOS.

978-1-7281-0582-6/19 $31.00 © 2019 IEEE

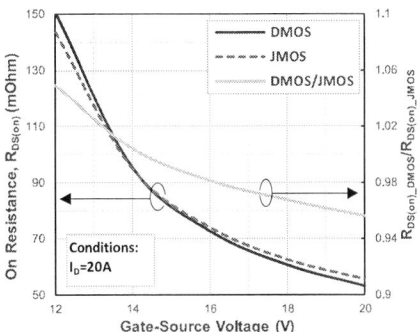

Fig.8: $R_{DS(on)}$ -V_{GS} comparison with I_D=20A between DMOS and JMOS.

In addition, Fig.8 shows the $R_{DS(on)}$ comparison by various V_{GS}, from this chart, the on-resistance of JMOS is lower than DMOS when V_{GS} is smaller than 15V. After V_{GS} exceeding 15V, the $R_{DS(on)}$ of JMOS becomes slightly higher than conventional DMOS, but the difference is quite not significant.

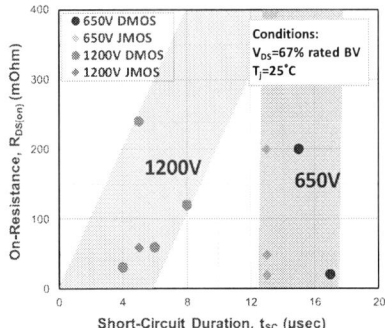

Fig.9: $R_{DS(on)}$ vs. t_{SC} of 650V and 1200V JMOS & DMOS. With the same cell-design, devices' SCWT be suppressed with lower $R_{DS(on)}$. This may due to the inhomogeneity of chips. In specific, since the larger chip contains wider characteristics distribution, there is a higher probability for local heat-spots to trigger localize overheating and penetrate drain and source region. Moreover, the higher voltage rating the device is, the thicker the epilayer the devices have (more defects). Hence, 1200V devices shown a relatively significant trend in $R_{DS(on)}$ vs. t_{SC} compared to 650V devices.

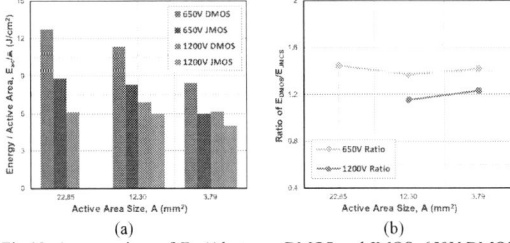

Fig.10: A comparison of E_{SC}/A between DMOS and JMOS. 650V DMOS could handle 40% more energy than JMOS. And 1200V DMOS could handle 15% more energy than JMOS with the same active area.

According to Fig.9 and Fig.10, JMOS could endure almost the same t_{SC} as DMOS in 1200V devices but slightly worse than DMOS in 650V devices. This result shows that even though the $R_{DS(on)}$ of JMOS is higher than DMOS in the saturation region tend to enhance t_{SC}, however, JMOS's energy handling capability still worse than DMOS. This may stem from the weaker thermal tolerance of Schottky contact. To be more specific, even though the typical test waveforms and the fail pattern of failure device are similar between

DMOS and JMOS in Fig.11 and Fig.12, they may have different failure mechanism (see Fig.13 and Fig.14). Fig.14 shows the waveform during D.U.T. fail. DMOS usually follows the failure mode showed in Fig.14(a), in contrast, the majority of JMOS follows the failure mode showed in Fig.14(b). Furthermore, the difference between 650V and 1200V devices may due to the difference of resistance proportion – the effect of the resistance of source area, which we replace some of it to Schottky contact, is more significant in 650V FETs due to the lower drift region resistance.

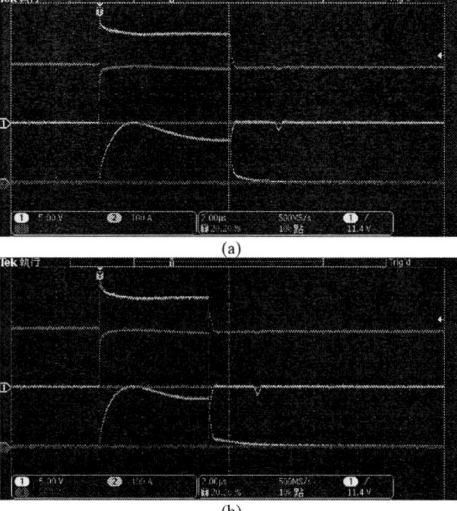

Fig.11: Typical short-circuit waveform of 1200V/60mΩ (a) DMOS (b) JMOS, Ch.1: V_{GS}, Ch.2: I_D, Ch.3: V_{DS}

Fig.12: (a) DMOS (b) JMOS de-cap results after devices fail in single pulse short-circuit test.

B. Comparison of Failure Mechanism

The tail leakage current of conventional DMOS (I_{DMOS_LK}) could be separated by 3 individual terms – the thermal generation current (I_{g_th}), the diffusion current (I_{g_diff}), and the avalanche generation current (I_{g_av}) as (1).

$$I_{DMOS\_LK} = I_{g\_th} + I_{g\_diff} + I_{g\_av} \qquad (1)$$

By prior research [6], since the rating of I_{g_th} exceeds more than 3 orders higher than I_{g_av} and I_{g_diff}, the parameter I_{DMOS_LK} could be represented as (2).

$$I_{DMOS\_LK} \approx I_{g\_th} = \frac{qSn_i}{\tau_g} \sqrt{\frac{2\varepsilon_s}{q}\left(\frac{N_d + N_a}{N_d N_a}\right)V_{dc}} \qquad (2)$$

Where, n_i is intrinsic carrier concentration shown in (3) and τ_g is the Shockley-Read-Hall generation lifetime. In 4H-SiC, τ_g might range from nano-second to micro-second, which depends on the characteristics of epilayers.

$$n_i(T) = \sqrt{N_C N_V} \cdot exp\left(-\frac{E_g}{2kT}\right) = 1.7e16 \times T^{\frac{3}{2}} \times exp\left(-\frac{E_g}{2kT}\right) \quad (3)$$

Because of JMOS shares the same channel structure with DMOS, the tail leakage current of JMOS (I_{JMOS_LK}) could be hypothesized as a DMOS leakage with additional reverse bias JBS equation (I_{Sch_LK}) (4), (5). Although the barrier lowering and the tunneling effect might slightly change the effective I_{DMOS_LK} in JMOS in practical experiments, to simplify the calculation, we ignore the effect of electric field changing and other peripheral effects produced by the leakage current of high-temperature Schottky contact in these equations.

$$I_{JMOS\_LK} = I_{Sch\_LK} + I_{g\_th} + I_{g\_diff} + I_{g\_av} \approx I_{Sch\_LK} + I_{g\_th} \quad (4)$$

$$I_{Sch\_LK} = \left(\frac{p-s}{p}\right) AT^2 \cdot exp\left(-\frac{q\phi_{eff}}{kT}\right) \cdot exp\left(C_T E_{JBS}^2\right) \quad (5)$$

In above equations, p and s are the spacing factor of p+ and Schottky contact, ϕ_{eff} is an effective barrier height including the barrier lowering effect, C_T is a tunneling coefficient, and E_{JBS} is the electric field.

Hence, the relationship of leakage current and temperature could be presented as follows: (6), (7)

$$I_{DMOS\_LK} \propto T^{\frac{3}{2}} \cdot exp\left(-\frac{1}{T}\right) \quad (6)$$

$$I_{JMOS\_LK} \propto \left(c_1 T^2 + c_2 T^{\frac{3}{2}}\right) \cdot exp\left(-\frac{1}{T}\right) \quad (7)$$

Where, c_1 and c_2 is the combination of other constants or parameters which are independent with temperature. The calculation results of these equations present the same trend with the experiment result, as shown in Fig.13.

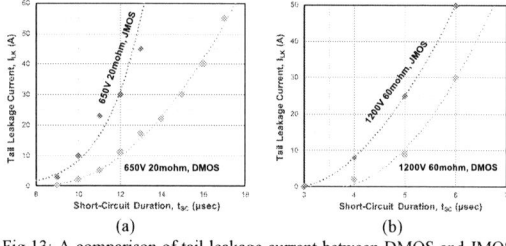

Fig.13: A comparison of tail leakage current between DMOS and JMOS to detect the portent of failure. Apparently, the tail current increasing rate of JMOS is faster than DMOS with a dissimilar trend. (a) 650V/20mΩ, (b) 1200V/60mΩ

According to the discussion, it could be inferred that when JMOS experiences short-circuit event, the heat accumulates in the channel and causes the extremely high temperature in the vicinity of the Schottky contact. This high lattice temperature not only triggers the parasitic BJT to latch-up but also lead to the enormous current leakage from the Schottky contact. Subsequently, more leakage current will generate higher power consumption and causing a higher temperature. As a result, these processes will form an unstoppable positive feedback loop and causing thermal runaway. In other words, the high current nearby the Schottky contact will finally destroy the structure and meltdown corresponding materials. Similarly, in prior research [7], the evidence also indicates the short-circuit failure mechanism of JBSFETs is due to the molten aluminum had penetrated into the drift region.

It's worth to mention that in research [7], it mentions that the SCWT of JBS integrated MOSFET has better performance than typical DMOS. However, that is because the $R_{DS(on).sp}$ of devices in their researching is higher than their DMOS counterparts (8.63 mΩ-cm² vs. 14.98 mΩ-cm² with the same JFET pitch). Once normalizing the SCWT to equivalent

$R_{DS(on).sp}$, the ratio of SCWT of JBS integrated MOSFET and DMOS will be quite similar to the result of this work.

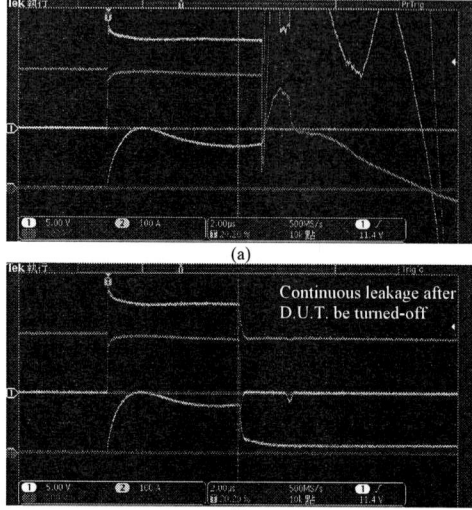

Fig.14: Waveform patterns while D.U.T. fails. (a) 1200V/60mΩ DMOS (b) 1200V/60mΩ JMOS, Pattern (b) shows a significant continuous I_D leakage after the device be turned off. The possible reason is the Schottky contact is already damaged by the high lattice temperature caused by the tremendous power during I_{SC} flow through Schottky contact before negative V_{th} or parasitic BJT be triggered by the high lattice temperature. Ch.1: V_{GS}, Ch.2: I_D, Ch.3: V_{DS}

IV. CONCLUSION

In conclusion, a detailed comparison of short-circuit ruggedness between DMOS and JMOS has been presented in this work. As a result, JMOS has slightly worse short-circuit handling capability and may have different failure mechanisms compared to DMOS. Nevertheless, SCWT of both JMOS and DMOS are longer than 5 μsec, which means they could be used in practical applications without risks in short-circuit capability.

ACKNOWLEDGMENT

The authors would gratefully acknowledge Poworld Elec. co., Ltd. for their support on the experiment instruments.

REFERENCES

[1] A. Castellazzi *et al.*, "Short-circuit robustness of SiC Power MOSFETs: experimental analysis" in *Proc. 26th IEEE International Symposium on Power Semiconductor Devices & IC's (ISPSD)*, pp. 71–74, Jun. 2014.

[2] J. An *et al.*, "Methodology for Enhanced Short-Circuit Capability of SiC MOSFETs" in *Proc. 30th IEEE International Symposium on Power Semiconductor Devices & IC's (ISPSD)*, pp. 391–394, May. 2018.

[3] AN2017-04, "Advanced Gate Drive Options for SiC MOSFETs using EiceDRIVER™" *Infineon Application Note*, Rev.1.1, Jun. 2018.

[4] F. J. Hsu *et al.*, "High Efficiency High Reliability SiC MOSFET with Monolithically Integrated Schottky Rectifier" in *Proc. 29th IEEE International Symposium on Power Semiconductor Devices & IC's (ISPSD)*, pp. 45–48, May. 2017.

[5] C. T. Yen *et al.*, "650V and 1200V SiC MOSFET with Integrated Junction Barrier Controlled Schottky Rectifier" in *Proc. 30th IEEE International Symposium on Power Semiconductor Devices & IC's (ISPSD)*, pp. 56–59, May. 2018.

[6] Z. Wang *et al.*, "Temperature Dependent Short Circuit Capability of Silicon Carbide (SiC) Power MOSFETs" *IEEE Trans. on Power Elec.*, vol. 31, no. 2, pp. 1555–1566, Feb. 2016.

[7] K. Han *et al.*, "New Short Circuit Failure Mechanism for 1.2kV 4H-SiC MOSFETs and JBSFETs" in *Proc. 6th IEEE Workshop on Wide Bandgap Power Devices and Applications (WiPDA)*, pp. 108–113, Nov. 2018.

978-1-7281-0582-6/19 $31.00 © 2019 IEEE

Proceedings of the 31st International Symposium on Power Semiconductor Devices & ICs
May 19-23, 2019, Shanghai, China

Analysis of Degradation Phenomena in Bipolar Degradation Screening Process for SiC-MOSFETs

Takashi Ishigaki, Tatsunori Murata, Koyo Kinoshita,
Takahiro Morikawa, and Tetsuo Oda,
Hitachi Power Semiconductor Device, Ltd.,
Hitachi-shi, Ibaraki, Japan
Email: takashi.ishigaki.ug@hitachi.com

Ryuusei Fujita, Kumiko Konishi, Yuki Mori, and Akio Shima,
Research & Development Group, Hitachi, Ltd.,
Kokubunji-shi, Tokyo, Japan

Abstract—**We have developed a 3.3 kV ultra-high power density SiC power module, which was realized by fulfilment with only SiC-MOSFETs. As a countermeasure for bipolar degradation issues related to body diodes in the MOSFET structure, a high throughput screening process has been introduced. In the screening process, over 10000 chips of 3.3 kV SiC-MOSFET were evaluated. The large amount of data firstly reveals the probability distribution of the degradation phenomena and enables us to study the relationship with defect densities of the state-of-the-art SiC wafers. Furthermore, origins of the degradation were investigated by teardown analysis. The degradation is caused by the expansion or generation of bar-shaped stacking faults. It was identified that their origins are pre-existing bar-shaped stacking faults in a SiC epi-layer, a basal plane dislocation in a SiC substrate, or a closed micropipe.**

Keywords—SiC, MOSFET, bipolar degradation, screening, stacking fault, basal plane dislocation, micropipe

I. Introduction

In recent years, high voltage and large current capacity Silicon Carbide (SiC) power modules have been commercialized and adopted to the railway traction systems instead of conventional Silicon (Si) power modules [1-4]. Especially for railway applications, high power density is strongly required because the equipment shall be deployed within the limited space of the rolling stock, minimizing weight and increasing the opportunity to allocate space for increased passenger numbers. In order to meet these challenges, we have developed a 3.3 kV/800 A SiC power module with a compact next high power density dual ($_n$HPD^2) package [5]. The value of power density is 37.7 kVA/cm^2,

which is 1.8 times as comparison with the conventional value, as shown in Fig. 1. This ultra-high power density was realized by constituting the module with highly integrated SiC- Metal Oxide Semiconductor Field Effect Transistor (MOSFET) chips without using SiC- Schotkky Barrier Diodes (SBDs) to suppress the increase of the chip cost. However, there was a serious concern about this diode (D)-less configuration. The main reason why the D-less design could not be realized so far is due to the bipolar degradation issues peculiar to SiC devices. When a bipolar operation occurs in SiC devices, stacking faults (SFs) may expand from pre-existing basal plane dislocations (BPDs) due to the recombination energy of electrons and holes. The SFs inside a SiC device could cause serious problems such as on-voltage degradation [7]. This bipolar degradation is serious especially in a high voltage class like 3.3 kV. The low doping density of the epi-layer leads to a long hole lifetime, and the thick epi-layer results in large SF areas. Even in SiC-MOSFETs, they have a body diode which is composed of a pn junction inside the own structure. As a countermeasure for this issue, we have introduced a high throughput screening technology. In this paper, the actual reality of the bipolar degradation is firstly clarified by the large amount of data which is obtained in the screening process. In addition, the degradation phenomena in the screening process are investigated in order to identify the origins of the degradation.

II. Bipolar Degradation Screening Technology

Figure 2 shows the schematic illustration of a SiC-MOSFET and the sequence of the bipolar degradation screening process. The degradation quantity of tested chips was evaluated by the on-voltage difference ΔV_{on} which is the

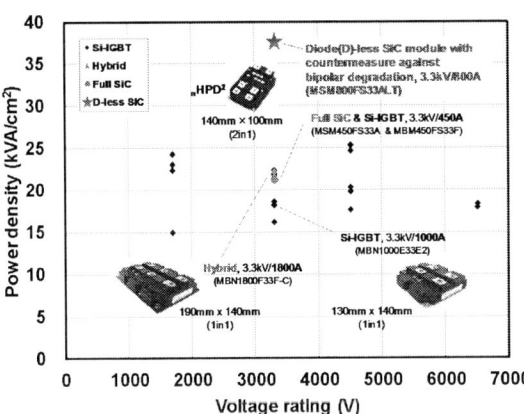

Fig. 1. Power density of Hitachi high power modules [6].

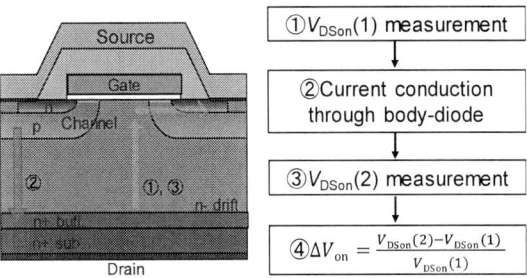

Fig. 2. Sequence of bipolar degradation screening process for SiC-MOSFETs. The degradation quantity was evaluated by ΔV_{on} which is the difference of forward on-voltage before and after the current conduction through the body diode.

978-1-7281-0582-6/19 $31.00 © 2019 IEEE

difference of forward on-voltage before and after the current conduction through the body diode. A carefully determined stress condition based on the accurate modeling of the bipolar degradation [8] and an optimized system design make it possible to process the sequence in high throughput. Figure 3 shows probability distributions of the degraded chips in the test. The chips were made using only high quality the state-of-the-art SiC wafers. And they have the same drift layer (30 μm, 3×10^{15} cm^{-3}) and different buffer layers (1 or 3 μm, 1×10^{18} cm^{-3}). Consequently, more or less 90% of the tested chips show no ΔV_{on} shift. However, it should be noted that there are large ΔV_{on} chips with very low probability. Some chips show more than 100% degradation. As for the difference of the buffer thickness, the thicker buffer layer seems to improve the degradation probability due to the suppression of minority carriers passing through the buffer layer [9]. Furthermore, it is found that some of the degraded chips show lower blocking

voltage than the initial value, as shown in Fig. 4. In this way, these degraded chips are screened out and not mounted into module products by using this screening technology. As a result, reliable SiC power modules with no degradation have been realized.

III. ANALYSIS OF DEGRADATION ORIGINS

The large amount of data obtained in the high throughput screening process firstly makes it possible to analyze origins of the bipolar degradation in wafer-level. Figure 5 shows screening yields per wafer vs. the BPD densities of the epi-layer of the state-of-the-art SiC wafers. The screening yields are calculated as the ratio of the "passed" chips to the "good" chips for each wafer. The "passed" chips passed the screening test. And the "good" chips represent that they passed initial electrical characteristic evaluations. No strong correlation between the yield and the BPD density is observed. This is presumably because the BPD density of epi-layers is already enough low. In addition, almost all the BPD in SiC epi-layers is known to have a Burgers vector of $(1/3)[11\text{-}20]$ or $(1/3)[\text{-}1\text{-}120]$ and becomes to a triangle-shaped SF. Therefore, this kinds of BPDs have small impact on the degradation due to its small area.

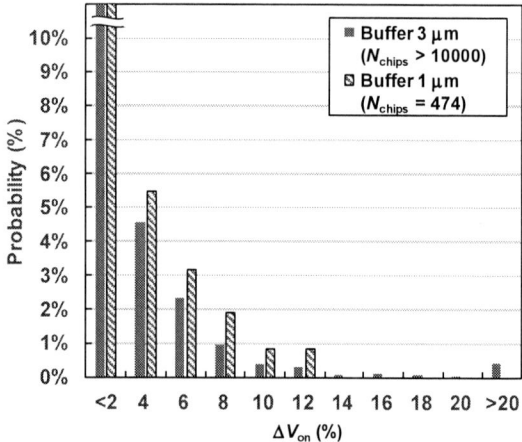

Fig. 3. Probability distributions of the bipolar degradation in 3.3 kV SiC-MOSFET. Numbers of tested chips are over 10000 for 3 μm buffer samples and 474 for 1 μm buffer samples, respectively.

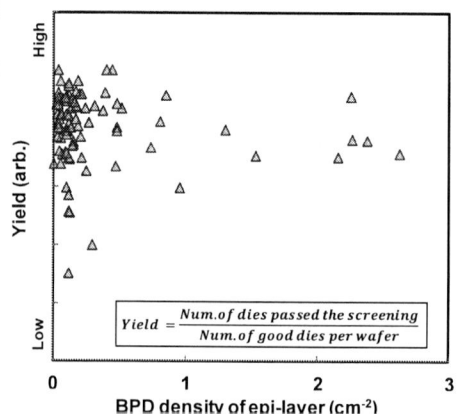

Fig. 5. Yield per wafer in the screening process vs. BPD density of the epi-layer of the state-of-the-art SiC wafers. The yields are calculated as the ratio of the number of dies passed the screening to the initial electrically good dies for each wafer.

Figure 6 shows photoluminescence (PL) images of the degraded chips. The PL images were taken before device fabrication (that is, initial state of the epi-wafer) and after the screening process (followed by removing the device structure). It is found that there are three types in the initial PL images as the origins of the degradation. In the case of type A, pre-existing bar-shaped SFs in the epi-layer clearly expand along the direction through the screening test. Therefore, the expanded areas of the bar-shaped SFs are counted as the degradation amount. In the case of type B, the origins of the bar-shaped SF observed after the degradation are not detected in the initial state. The origins are further investigated later by teardown analysis. A microdot of type C in the initial state seems to be the origin of the broad bar-shaped SF observed in the PL image after the degradation.

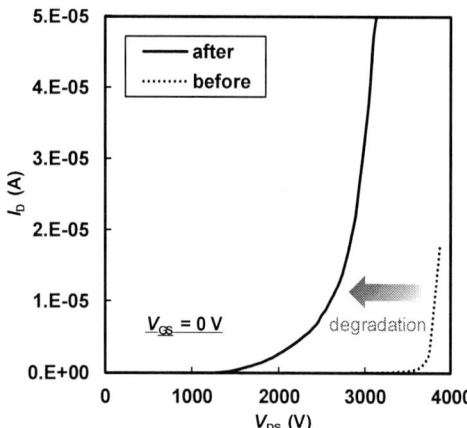

Fig. 4. An example of the degraded blocking voltage characteristic after the bipolar degradation screening process.

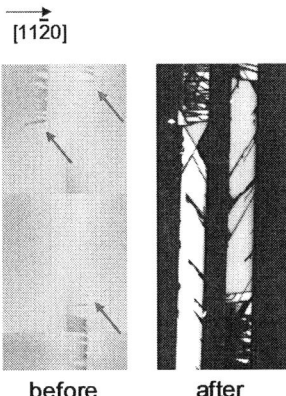

(a) Type A (Expansion of bar-shaped SF)

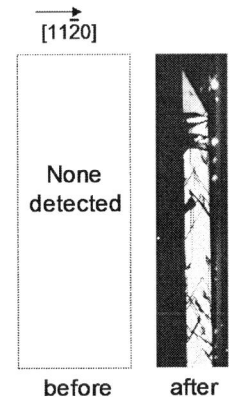

(b) Type B (No detected origin)

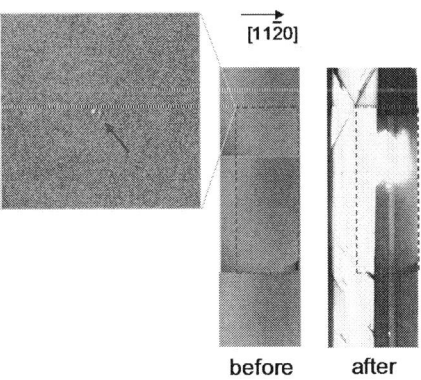

(c) Type C (A microdot-shaped origin)

Fig. 6. Classified PL images of various types of bipolar degradation. The PL images were taken before device fabrication and after the screening process.

Figure 7 (a) shows a cross sectional analysis point of Fig. 6 (b). Figures 7 (b) and (c) show a cross sectional PL image and a cross sectional scanning electron microscope (SEM) image, respectively. As can be seen from Fig. 7 (c), the starting point of the SF is confirmed to be at the buffer/substrate interface. It is known that almost all BPDs in SiC substrates are converted to threading edge dislocations during epitaxial growth and this conversion point becomes an origin of a bar-shaped SF [7]. Therefore, it is presumed that this type of degradation is likely to occur in such thin 1 μm buffer samples.

Figure 8 (a) shows cross sectional analyses points of Fig. 6 (c). Figures 8 (b) and (c) show a cross sectional PL image and a cross sectional SEM image, respectively. We found that there is a closed micropipe in the SiC substrate and a threading dislocation (TD) converted from the micropipe as shown in Fig. 8 (c). This would be the microdot of Fig. 6 (c) which was observed before device fabrication. It is identified that the multiple SFs are generated from the TD as can be seen in Figs. 8 (a) and (c). As for closed micropipes, there is a report that they may cause problems on long term reliability [10]. In this way, this screening technology can also detect this kind of

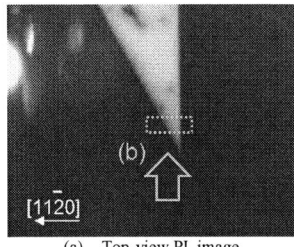

(a) Top-view PL image

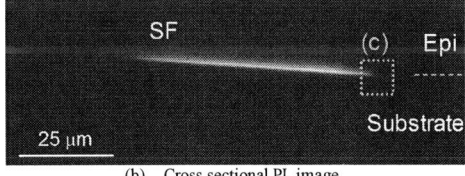

(b) Cross sectional PL image

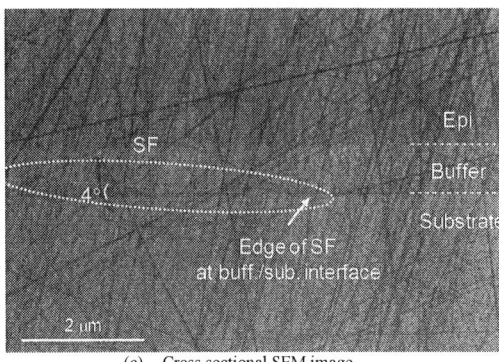

(c) Cross sectional SEM image

Fig. 7. Origin analysis of Fig. 6 (b). (a) Cross sectional analysis point. Cross sectional (b) PL and (c) SEM images.

978-1-7281-0582-6/19 $31.00 © 2019 IEEE

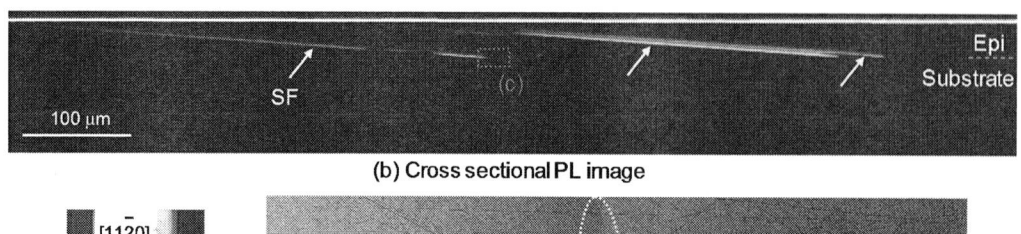

(b) Cross sectional PL image

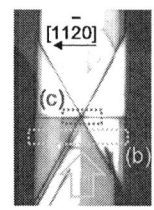

(a) Top-view PL image

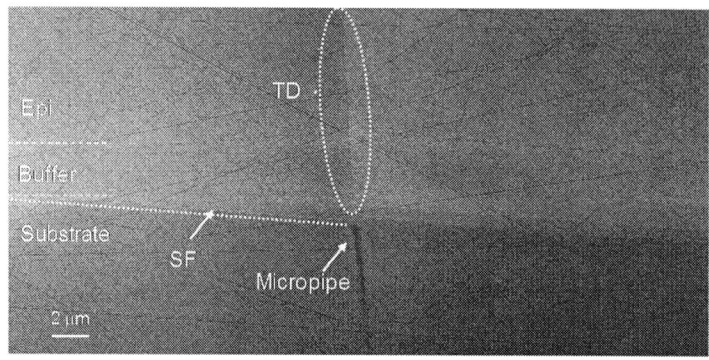

(c) Cross sectional SEM image

Fig. 8. Origin analysis of Fig. 6 (c). (a) Cross sectional analyses points. Cross sectional (b) PL and (c) SEM images.

closed micropipes as a bipolar degradation and prevent the related reliability issues.

IV. CONCLUSIONS

We have introduced a high throughput bipolar degradation screening process to realize a reliable ultra-high power density D-less SiC power module. The large amount of data in the screening process firstly reveals the probability of bipolar degradation. It was found that there is large degradation with very low probability in the state-of-the-art SiC wafers. Furthermore, through in-depth analysis of the degraded chips, the main cause of the degradation was turned out to be not triangle-shaped SFs but bar-shaped SFs. The origins of the bar-shaped SFs were identified and classified to be (a) the expansion of pre-existing bar-shaped SFs, (b) the generation from the BPDs in the buffer layer, and (c) the derivation from TDs related to a closed micropipe.

REFERENCES

[1] T. Ishigaki et al., "3.3 kV/450 A Full-SiC nHPD² (next high power density dual) with smooth switching," PCIM Europe 2017, pp. 33-38, 2017.

[2] T. Negishi et al., "3.3-kV All-SiC Power Module for Traction System Use," PCIM Europe 2017, pp. 51-56, 2017.

[3] K. Terasawa et al., "Development of Lightweight Traction Systems to Reduce Energy Consumption," Hitachi Review Vol. 67, No. 7, pp. 38-43, 2018.

[4] M. Lindahl et al., "Silicon Carbide MOSFET Traction Inverter Operated in the Stockholm Metro System Demonstrating Customer Values," 2018 IEEE Vehicle Power and Propulsion Conference (VPPC), pp. 1-6, 2018.

[5] T. Ishigaki et al., "A 3.3 kV/800 A Ultra-High Power Density SiC Power Module," PCIM Europe 2018, pp. 156-160, 2018.

[6] https://www.hitachi-power-semiconductor-device.co.jp/products/igbt/index.html

[7] T. Kimoto et al., "Understanding and Reduction of Degradation Phenomena in SiC Power Devices," IRPS 2017, pp. 2A-1.1-1.7, 2017.

[8] K. Konishi et al., "Modeling of Stacking Fault Expansion Velocity of Body Diode in 4H-SiC MOSFET,"Material Science Forum, Vol. 897, pp. 214-217, 2017.

[9] Y. Ebiike et al., "Reliability Investigation with Accelerated Body Diode Current Stress for 3.3 kV 4H-SiC MOSFETs with Various Buffer Epilayer Thickness," ISPSD 2018, pp. 447-450, 2018.

[10] R. Rupp et al., "Influence of Overgrown Micropipes in the Active Area of SiC Schottky diodes on long term reliability," Materials Science Forum, Vols. 483-485, pp. 925-928, 2005.

978-1-7281-0582-6/19 $31.00 © 2019 IEEE

Proceedings of the 31st International Symposium on Power Semiconductor Devices & ICs
May 19-23, 2019, Shanghai, China

Repetitive Short Circuit Energy Dependent V_{TH} Instability of 1.2kV SiC Power MOSFETs

Jiahui Sun, Jin Wei, Zheyang Zheng, Yuru Wang, Kevin J. Chen

Dept. of Electronic and Computer Engineering, The Hong Kong University of Science and Technology, Hong Kong, China
Email: jsunaz@connect.ust.hk, eekjchen@ust.hk

Abstract—**Repetitive short circuit energy (E_{SC}) dependent threshold voltage (V_{TH}) instabilities of 1.2kV SiC power MOSFETs are characterized. A bidirectional V_{TH} shift behavior, i.e. negative shift at lower E_{SC} and positive shift at higher E_{SC}, was revealed. The V_{TH} shifts under repetitive SC tests are attributed to the SC pulse phase according to the results of high temperature reverse bias (HTRB) and dynamic high temperature gate bias (HTGB) tests. The underlying mechanisms of the complex V_{TH} shift behavior are explained in a unified framework by taking into account the junction temperature (T_j) increase with higher E_{SC}. TCAD device simulation is used to help analyze the mechanisms.**

Keywords— SiC power MOSFET, repetitive short circuit energy, threshold voltage instability

I. INTRODUCTION

In comparison with conventional silicon (Si) IGBTs, the silicon carbide (SiC) power MOSFET can further boost the efficiency and power density of power electronic systems even at a high temperature environment due to its superior material properties [1]–[4]. The applications of SiC MOSFETs include electric vehicles (EV), hybrid electric vehicles (HEV), railway traction and power grid, where reliability is of utmost importance [5]–[7].

The application-relevant short circuit (SC) event is a critical reliability concern for power transistors as they have to endure high voltage, high current and the resultant high junction temperature (T_j) at the same time. An SC fault can be caused by dielectric breakdown of the load, faulty control, faulty electric wiring, etc. In the long-term operation of SiC power MOSFETs, more than one SC fault may occur. Thus, the influence of repetitive SC events on SiC MOSFETs should be investigated and analyzed.

Several groups have separately analyzed threshold voltage (V_{TH}) shifts of SiC MOSFETs induced by repetitive SC events [8], [9]. The reported V_{TH} shift direction varies with different device samples and SC test conditions, and different mechanisms have been discussed. The negative V_{TH} shift (ΔV_{TH}) has been explained by hot hole injection into silicon dioxide (SiO_2) as a result of impact ionization (I.I.) in the channel region [8]. The positive ΔV_{TH} has been attributed to electron trapping in SiO_2 or at the SiO_2/channel interface resulted from the high I.I. rate and the electric field perpendicular to the channel interface [9]. However, none of these mechanisms involved possible I.I. in the N⁻ drift region and the temperature dependence of the I.I. rate.

In this work, repetitive SC tests are carried out on commercial SiC planar-channel vertical MOSFETs, and the resultant device instability is evaluated by ΔV_{TH} measured

This work is supported by Hong Kong Innovation and Technology Fund under grant ITS/234/16.

after the SC stress. A bi-directional V_{TH} shift behavior is observed and is found to be dependent on the SC energy (E_{SC}). The underlying mechanisms of the complex V_{TH} instability are explained in a unified framework by taking into account the T_j increase with higher E_{SC}. TCAD device simulation is conducted to help understand the mechanisms.

II. V_{TH} INSTABILITY UNDER REPETITIVE SC TESTS

The repetitive SC test platform is mainly composed of an SC test board, a high voltage (HV) power supply, two pulse generators (PG1, PG2) and an HV switch (S1) as shown in Fig. 1(a). The devices under test (DUTs) are 1.2kV, 280mΩ SiC MOSEFTs in TO-247 [10]. Before an SC test, S1 is turned on to charge C_1. As shown in Fig. 1(b), the gate control signal of the DUT (V_{Ctrl1}) is delayed by ~10μs. Thus, the IGBT is turned on before the DUT to charge C_2 and C_3 to the value of V_{dc}. The DUT is then switched on to start an SC test. It operates in the SC mode for a duration of t_p if failure does not occur by the end of the pulse. The IGBT will be turned off several microseconds later than DUT to prevent the burn-out of DUT after failure. To avoid significant T_j increase resulted from the average power in the DUT, the *Period* of repetitive SC tests is set as 1s (Fig. 1(b)). After each set of SC cycles, the DUT will be removed from the test board and connected to B1505A power device analyzer within a time duration less than 1-minute for quasi-static transfer *I-V* characterization.

Figure 2 depicts waveforms of a DUT and a schematic T_j-*Time* curve. Each SC cycle is divided into two phases, Phase I and Phase II. Phase I is an SC phase from 0 to t_p, during which the DUT undergoes high drain current (i_d), drain-source voltage (V_{ds}) and T_j. Phase II is a high

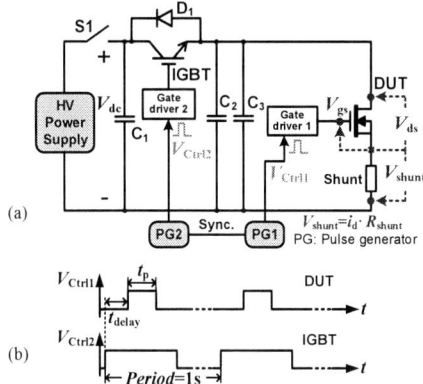

Fig. 1 (a) Schematic diagram of repetitive SC test platform. (b) Timing diagram of gate control signals (V_{Ctrl1} and V_{Ctrl2}) of the DUT and IGBT.

978-1-7281-0582-6/19 $31.00 © 2019 IEEE   263

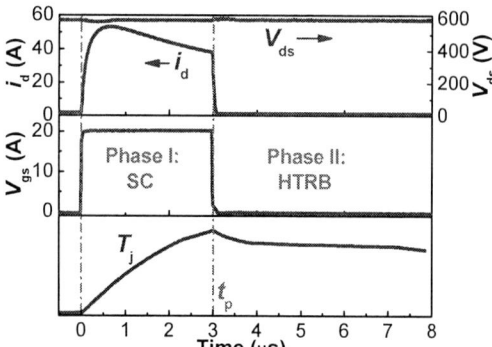

Fig. 2. SC waveforms and schematic T_j of one DUT showing two phases during an SC test.

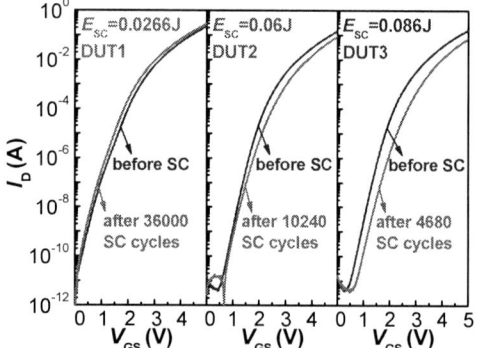

Fig. 3. Transfer I-V curves of DUTs stressed under repetitive SC with V_{gs}=20V/0V, V_{dc}=600V and different E_{SC}. The transfer I-V curves are tested at V_{DS}=10V.

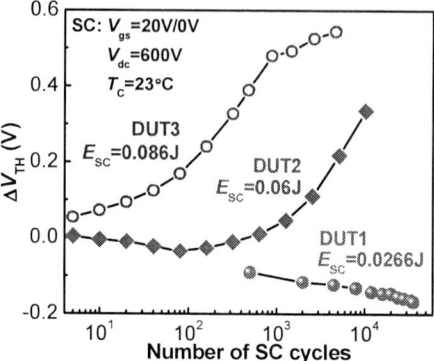

Fig. 4. V_{TH} shifts of DUT1~DUT3 corresponding to Fig. 3. V_{TH} shifts negatively at lower E_{SC} and positively at higher E_{SC}.

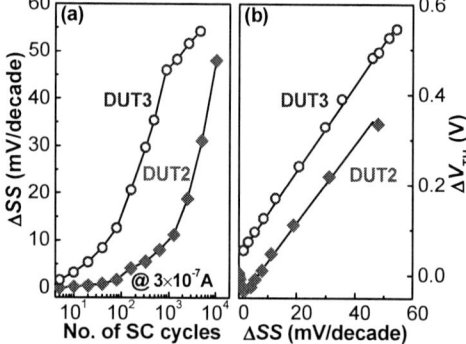

Fig. 5. (a) SS's of DUT2 and DUT3 increase with the accumulation of SC cycles. (b) The correlation between ΔV_{TH} and ΔSS is linear, indicating that positive ΔV_{TH} is mainly caused by oxide interface/border trap generation.

temperature reverse bias (HTRB) phase with high V_{ds} and high T_j from t_p to the end of a *Period*.

Repetitive SC tests are conducted on DUT1~DUT3 at V_{dc}=600V, V_{gs}=20V/0V and T_C=23°C (case temperature). Given different t_p (0.8μs, 2μs and 3μs), E_{SC} is 0.0266J, 0.06J and 0.086J for DUT1, DUT2 and DUT3 respectively. Figure 3 plots the transfer I-V curves of DUT1~DUT3 before and after repetitive SC tests, where evident changes are observed. V_{TH} is defined at I_D=1.25mA with V_{DS}=10V. Different V_{TH} shift behavior is depicted in Fig. 4, changing from a continuous negative shift for DUT1 at lower E_{SC} to a continuous positive shift for DUT3 at higher E_{SC}. As for DUT2 with a medium E_{SC}, V_{TH} decreases slightly within the first 100 SC cycles and then increases from 100 cycles to 10240 cycles. This phenomenon may be caused by the competition of different mechanisms during t_p (2μs).

In order to identify the change in oxide interface/border trap density, subthreshold swing (SS) is extracted from transfer I-V curves at I_D=3×10⁻⁷A. As shown in Fig. 5(a), SS of both DUT2 and DUT3 increases with the accumulation of SC cycles, which indicates generation of new interface/border traps during repetitive SC tests [11]. Moreover, ΔSS has a linear correlation to ΔV_{TH} (Fig. 5(b)), suggesting that the generation of oxide interface/border traps dominates the positive ΔV_{TH} [12].

III. MECHANISM ANALYSIS

Several stress configurations were deployed in the repetitive SC tests. Their effects on V_{TH} should be clarified to reveal the dominant stresses that account for V_{TH} shifts. TCAD device simulation is performed to facilitate the analysis.

A. Auxiliary Reliability Tests

DUT4 is stressed by HTRB for 2500s at V_{DS}=600V, V_{GS}=0V and T_j=150°C. DUT5 is stressed for 32000 cycles by dynamic high temperature gate bias (HTGB) with V_{gs}=20V/0V, V_{ds}=0V, *pulse-width*=0.8μs, *Period*=1ms and T_j=150°C. The biasing scheme of dynamic HTGB is depicted in the inset of Fig. 6. High temperature is applied by connecting the back-side metal of the DUT to a hot plate through silicone grease. The stress on DUT4 and DUT5 is provided by B1505A and B1500A device analyzers respectively. It is interrupted repeatedly to measure transfer I-V curves at V_{DS}=10V, from which V_{TH} is extracted at I_D=1.25mA.

ΔV_{TH}'s of DUT4 and DUT5 are plotted in Fig. 6, and ΔV_{TH} of DUT1 under repetitive SC stress is shown as a reference. No significant ΔV_{TH} is observed in DUT4, indicating that the V_{TH} shifts of DUT1~DUT3 under repetitive SC tests result from stresses in Phase I as defined in Fig. 2. A slight positive ΔV_{TH} of DUT5 is observed,

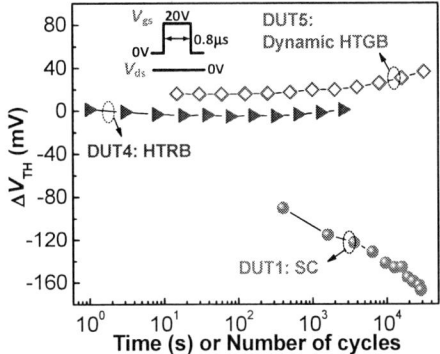

Fig. 6. Comparison of ΔV_{TH} of DUT1, DUT4 and DUT5 under repetitive SC tests, HTRB stress (V_{DS}=600V, V_{GS}=0V and T_j=150°C) and dynamic HTGB stress (V_{gs}=20V/0V, V_{ds}=0V, *pulse-width*=0.8µs, *Period*=1ms and T_j=150°C). For DUT4, the title of the horizontal axis is time. For DUT1 and DUT5, the horizontal axis represents the number of cycles.

suggesting that the negative ΔV_{TH} of DUT1 under repetitive SC tests is not directly related to the large V_{gs} in Phase I.

B. Device Simulation and Mechanism Discussion

In order to understand the physical mechanisms of the V_{TH} shifts in DUT1~DUT3, the SiC planar MOSFET is modeled by using a TCAD simulator. The cross section of the simulated device is shown in Fig. 7. The SC condition at V_{gs}=20V and V_{ds}=600V and the off-state condition at V_{gs}=0V and V_{ds}=600V are simulated for comparison. Figure 8(b) shows that the impact ionization (I.I.) rate is very low in the middle of N⁻ region (Region A) and the P region near the SiC/SiO₂ interface (Region B) due to low current density in the off state. Nevertheless, in the SC condition, both the current density and electric field in Region A and Region B are high as shown in Fig. 8(a), meaning that a large number of carriers are moving in the high electric field. Thus, high I.I. rate appears in Region A and Region B. Holes generated in Region A can be drawn toward the SiO₂ above Region B under the strong electric field (from A to B).

Since the high V_{ds} drops mainly across the lightly-doped N⁻ region, the potential in Region B is lower than the gate potential. Therefore, the electric field points from SiO₂ to channel to pull electrons toward the gate instead of the holes generated from I.I. in Region B as shown in Fig. 9. Thus, negative V_{TH} shift under repetitive SC tests with lower E_{SC} is less likely to result from holes generated in Region B as suggested in [8]. However, holes from I.I. in Region A can be accelerated by the electric field and become energetic enough (hot holes) to overcome the SiC/SiO₂ barrier as shown in Fig. 10. Trapping of holes in SiO₂ would then cause negative ΔV_{TH}.

Impact ionization will be alleviated at higher temperature because of enhanced phonon scattering [13]–[15]. Since T_j increases with higher E_{SC} within several microseconds [16], [17], hot hole injection induced by I.I. will be mitigated at higher E_{SC}. However, more electrons in the channel can obtain adequate energy to overcome the SiC/SiO₂ barrier at higher T_j because the occupancy of states above Fermi level (E_F) is increased. As depicted in Fig. 11, when these electrons drift through SiO₂, new defects (which behave as electron traps) could be generated to capture electrons

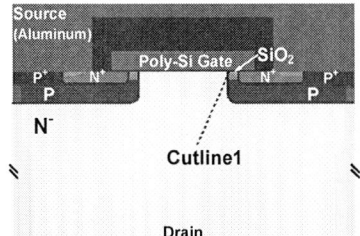

Fig. 7. Cross section of simulated cell structure of SiC MOSFET.

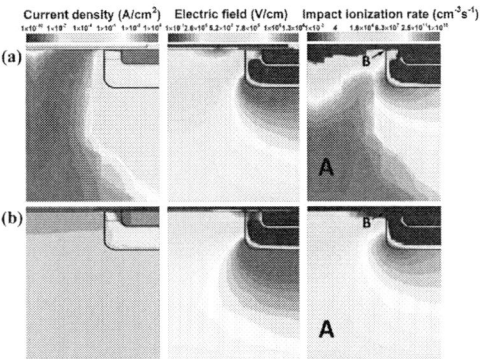

Fig. 8. Simulated current density, electric field and I.I. rate distribution under (a) SC condition: V_{gs}=20V, V_{ds}=600V, T=300K and (b) off-state condition: V_{gs}=0V, V_{ds}=600, T=300K. Temperature increase in the SC condition is not simulated because there are no temperature-dependent models for I.I. coefficients of SiC at temperature higher than 200°C.

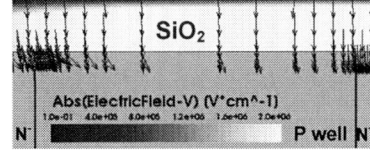

Fig. 9. Electric field in SiO₂ in the SC condition. The electric field points from gate to channel to pull electrons toward the gate instead of the holes generated from I.I. in Region B.

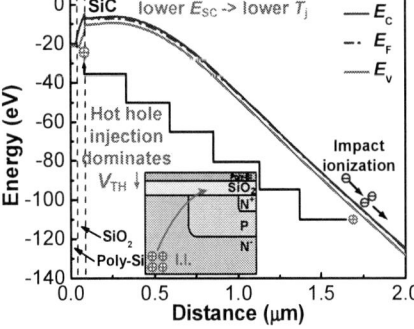

Fig. 10. Band diagram along Cutline1 in Fig. 7 showing hot hole generation and injection to SiO₂.

(negative charge), which lead to the rise of *SS* in Fig. 5(a). Hence, at higher E_{SC}, more negative charges are present in SiO₂ to induce the observed positive ΔV_{TH}. This mechanism is different from what has been proposed in [9] because E_{SC} dependence of ΔV_{TH} is considered in this work.

978-1-7281-0582-6/19 $31.00 © 2019 IEEE

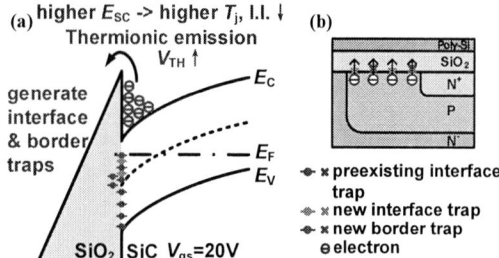

Fig. 11. (a) Schematic band diagram and (b) DUT cross-sectional view illustrating hot electron injection and the resulted electron trap generation at higher E_{SC} compared with that in Fig. 10

Continuous negative V_{TH} shift of DUT1 with a low E_{SC} (t_p=0.8μs) can be explained by hot hole injection under relatively lower T_j during Phase I. The non-monotonic V_{TH} shifts of DUT2 may be induced by the competitive effect of hot hole injection and thermionic electron emission. SC stress during the first 0.8μs on DUT2 is similar to that on DUT1. Thus, hot hole injection is dominant and induces negative V_{TH} shift in the first 0.8μs. Since T_j rises during t_p, thermionic electron emission gradually plays a more important role after 0.8μs. With the accumulation of SC cycles, the effect of hot hole injection on V_{TH} could be stronger in the first few cycles while the influence of thermionic electron emission may be more evident after more SC cycles. Nevertheless, V_{TH} increases monotonically for DUT3 with a higher E_{SC}, which indicates an absolute dominance of thermionic electron emission in the SC phase (Phase I).

IV. CONCLUSION

E_{SC}-dependent V_{TH} instabilities under repetitive SC tests are characterized and two competitive mechanisms (i.e. hot hole injection and thermionic electron emission) during the SC pulse (Phase I) are discussed. At lower E_{SC}, holes generated by impact ionization from the N⁻ drift region are accelerated in the high electric field and injected to SiO_2, leading to negative ΔV_{TH}. At higher E_{SC}, thermally activated electrons in the channel are emitted to SiO_2, which could then induce more electron traps that lead to larger positive ΔV_{TH}.

REFERENCES

[1] T. Kimoto, "Material science and device physics in SiC technology for high-voltage power devices," *Jpn. J. Appl. Phys.*, vol. 54, no. 4, pp. 040103-1–040103-27, Mar. 2015.

[2] J. Wei, M. Zhang, H. Jiang, C. Cheng and K. J. Chen, "Low on-resistance SiC trench/planar MOSFET with reduced off-state oxide field and low gate charges," *IEEE Electron Device Letters*, vol. 37, no. 11, pp. 1458-1461, Nov. 2016.

[3] Z. Chen, Y. Yao, D. Boroyevich, K. D. T. Ngo, P. Mattavelli and K. Rajashekara, "A 1200-V, 60-A SiC MOSFET multichip phase-leg module for high-temperature, high-Frequency applications," *IEEE Trans. Power Electron.*, vol. 29, no. 5, pp. 2307–2320, May 2014.

[4] Y. Shi, L. Wang, R. Xie, Y. Shi and H. Li, "A 60-kW 3-kW/kg five-level T-type SiC PV inverter with 99.2% peak efficiency," *IEEE Trans. Ind. Electron.*, vol. 64, no. 11, pp. 9144–9154, Nov. 2017.

[5] D. Han, J. Noppakunkajorn and B. Sarlioglu, "Comprehensive efficiency, weight, and volume comparison of SiC- and Si-based bidirectional DC–DC converters for hybrid electric vehicles," *IEEE Trans. Vehicular Technology*, vol. 63, no. 7, pp. 3001–3010, Sept. 2014.

[6] J. Fabre, P. Ladoux and M. Piton, "Characterization and implementation of dual-SiC MOSFET modules for future use in traction converters," *IEEE Trans. Power Electron.*, vol. 30, no. 8, pp. 4079–4090, Aug. 2015.

[7] N. He, M. Chen, J. Wu, N. Zhu and D. Xu, "20 kW zero-voltage-switching SiC-MOSFET grid inverter with 300 kHz switching frequency," *IEEE Trans. Power Electron.*, to be published. DOI: 10.1109/TPEL.2018.2866824.

[8] X. Zhou, H. Su, Y. Wang, R. Yue, G. Dai and J. Li, "Investigations on the degradation of 1.2-kV 4H-SiC MOSFETs under repetitive short-circuit tests," *IEEE Trans. Electron Devices*, vol. 63, no. 11, pp. 4346–4351, Nov. 2016.

[9] J. Wei, S. Liu, L. Yang, J. Fang, T. Li, S. Li, and W. Sun, "Comprehensive analysis of electrical parameters degradations for SiC power MOSFETs under repetitive short-circuit stress," *IEEE Trans. Electron Devices*, vol. 65, no. 12, pp. 5440–5447, Dec. 2018.

[10] C2M0280120D, Cree Inc., Durham, NC, USA, 2015. [Online]. Available: http://www.wolfspeed.com

[11] D. K. Schroder, *Semiconductor Material and Device Characterization*, 3rd ed., Hoboken, NJ, USA: Wiley, 2006, pp. 359–360.

[12] C. Hu, S. C. Tam, F.-C. Hsu, P.-K. Ko, T.-Y. Chan and K. W. Terrill, "Hot-electron-induced MOSFET degradation— model, monitor, and improvement," *IEEE J. Solid-State Circuits*, vol. 20, no. 1, pp. 295-305, Feb. 1985.

[13] R. Raghunathan and B. J. Baliga, "Temperature dependence of hole impact ionization coefficients in 4H and 6H-SiC," *Solid-State Electronics*, vol. 43, no. 2, pp. 199–211, Feb. 1999.

[14] H. Niwa, J. Suda and T. Kimoto, "Temperature dependence of impact ionization coefficients in 4H-SiC," *Materials Science Forum*, vol. 778–780, pp. 461–466, Feb. 2014.

[15] C. R. Crowell and S. M. Sze, "Temperature dependence of avalanche multiplication in semiconductors," *Appl. Phys. Lett.*, vol. 9, no. 6, pp. 242–244, Sept. 1966.

[16] J. Sun, H. Xu, X. Wu, S. Yang, Q. Guo and K. Sheng, "Short circuit capability and high temperature channel mobility of SiC MOSFETs," in *Proc. IEEE 29th Int. Symp. Power Semicond. Devices IC's (ISPSD)*, May 2017, pp. 399–402.

[17] Z. Wang, X. Shi, L. M. Tolbert, F. Wang, Z. Liang, D. Costinett and B. J. Blalock, "Temperature-dependent short-circuit capability of silicon carbide power MOSFETs," *IEEE Trans. Power Electron.*, vol. 31, no. 2, pp. 1555–1566, Feb. 2016.

978-1-7281-0582-6/19 $31.00 © 2019 IEEE

Proceedings of the 31st International Symposium on Power Semiconductor Devices & ICs
May 19-23, 2019, Shanghai, China

Short-Circuit Robustness of 4600 V SiC DMOSFETs

Siddarth Sundaresan
GeneSiC Semiconductor
Dulles VA, USA
siddarth.sundaresan@genesicsemi.com

Stoyan Jeliazkov
GeneSiC Semiconductor
Dulles VA, USA
stoyan.jeliazkov@genesicsemi.com

Vamsi Mulpuri
GeneSiC Semiconductor
Dulles VA, USA
vamsi.mulpuri@genesicsemi.com

Ranbir Singh
GeneSiC Semiconductor
Dulles VA, USA
ranbir@ieee.org

Sumit Jadav
GeneSiC Semiconductor
Dulles VA, USA
sumit.jadav@genesicsemi.com

Abstract— **A comprehensive investigation of the operating limits and failure modes for 4600 V/7.78 mm² SiC DMOSFETs under short-circuit conditions is performed. Depending on device design and driving conditions, short-circuit withstand times (t_{SC}) in the range of 3-14 µs are recorded on the DMOSFETs. The impact of various factors such as the device design, gate drive voltage, DC link (drain) voltage, and operating temperature on the t_{SC} is quantified. The evolution of the I-V and C-V characteristics under short-circuit operation is examined, and failure mechanisms are suggested.**

Keywords—Silicon Carbide DMOSFET, Short-Circuit, Reliability, Medium-Voltage

I. INTRODUCTION

This paper is focused on quantifying the short-circuit robustness of 4600 V/7.78 mm² SiC DMOSFETs. The detailed design, fabrication and operational performance characteristics of these MOSFETs were reported earlier [1]. The MOSFETs feature an $R_{DS,ON}$ of 350-550 mΩ, V_{TH} = 2.5 V, maximum DC voltage rating of 4 kV and DC current rating of 4-6 A. For almost any motor control inverter application, short-circuit withstand time of at-least 5 µs is mandatory for fault-detection by the integrated de-saturation protection circuitry associated with commercial gate drivers. This paper examines the stability of the operational characteristics of the DMOSFETs under single-pulse and repetitive short-circuit operating conditions.

II. DEVICE AND STRESS PROCEDURE DESCRIPTION

All devices investigated in this paper were assembled in high-voltage capable packages. The short-circuit measurements were performed on devices with (A) different MOSFET channel lengths resulting in varying drain saturation currents (B) positive gate drive voltages of +16-20 V, (C) negative gate drive voltages of -5 V or -8 V, (D) drain voltage of 1500 V or 2000 V, and (E) baseplate temperature of 25°C or 100°C. The devices were directly turned on to short-circuit at the DC link voltage, thereby simulating a hard-switching fault (HSF) condition. The t_{SC} was ramped in 0.5 µs increments until catastrophic device failure occurred. I-V characteristics were monitored after each short-circuit pulse, until failure. The trade-off between $R_{DS,ON}$ and short-circuit withstand time is quantified. The stability of key electrical transport parameters of the DMOSFETs under short-circuit conditions is investigated.

This work was performed under an US Office of Naval Research SBIR program (Contract# N00014-16-P2060). The support of Capt. Lynn Petersen and Drs. Karl Hobart and Fritz Kub are gratefully acknowledged

III. RESULTS AND DISCUSSION

A. Impact of DC link voltage on short-circuit SOA

The single-pulse short-circuit withstand time is strongly dependent on the DC link voltage applied to the device during the short-circuit pulse. The time evolution of the drain current transients for a MOSFET subjected to short-circuit at DC link voltage of 1.5 kV is shown in Fig. 2. Eventual (catastrophic) device destruction occurred for a 13.5 µs long short-circuit pulse at this DC link voltage. The drain currents reduce after reaching a peak value at ≈ 2 µs due to self-heating during the short-circuit pulse, which degrades the electron mobility in the N- drift layer. For longer short-circuit pulses (>9 µs), three observations can be made from Fig. 2: (1) The peak drain current reached at ≈ 2 µs is reduced (2) The slope of the drain current waveforms (dI/dt) beyond 9 µs is significantly reduced, and (3) noticeable tail currents are observed after device turn-off. The change in dI/dt can be attributed to generation currents in the N-drift layer, as the junction temperature reaches the intrinsic limit (≈ 1050° C) for 4H-SiC. The origin of the tail currents at the longer short-circuit durations are discussed later in this paper.

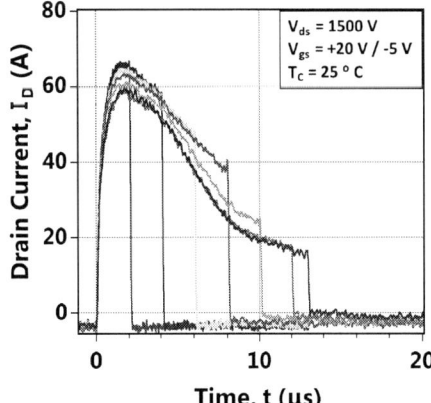

Figure 1: Time evolution of Drain current transients measured during increasing duration short-circuit pulses applied to a DMOSFET at DC link voltage of 1.5 kV, with +20 V/-5 V gate drive voltages, at room-temperature.

The time evolution of the drain current transients for a MOSFET subjected to short-circuit at DC link voltage of 2.0

978-1-7281-0582-6/19 $31.00 © 2019 IEEE

kV is shown in Fig. 3. This device was able to withstand a 6 µs long short-circuit pulse, which is 50% lower than the t_{SC} achieved for the 1.5 kV DC link case. A similar peak drain current (70 A) is observed for this device at ≈ 2 µs, in comparison with the MOSFET subjected to 1.5 kV DC link short-circuit. Drain current tailing is observed for the pre-failure (6 µs) short-circuit pulse. The power waveforms (drain current x drain voltage) from the short-circuit pulses were fed to a R-C thermal network simulating the thermal resistance of the packaged MOSFETs (see ref [2] for more details). Peak junction temperatures in the 1200-1500°C range were simulated during the short-circuit pulses, which is above the intrinsic limit for 4H-SiC. Thus, device failure occurs due to thermal runaway.

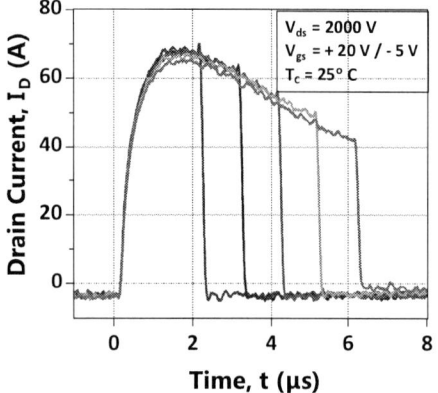

Figure 2: Time evolution of Drain current transients measured during increasing duration short-circuit pulses applied to a DMOSFET at DC link voltage of 1.5 kV, with +20 V/-5 V gate drive voltages, at room-temperature.

B. Impact of MOSFET channel length on short-circuit SOA

The output characteristics of three devices fabricated with different MOSFET channel lengths (L_{CH}) are shown in Fig. 3. The pre-failure short-circuit drain current waveforms are reported in Fig. 4.

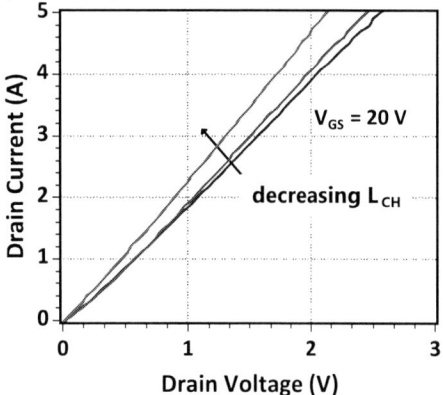

Figure 3: Output characteristics of three 4.6 kV DMOSFETs fabricated with different L_{CH}.

The devices were directly turned on to a short-circuit fault at a DC link of 2000 V. A $V_{GS} = +18$ V/-5 V gate drive scheme was used for driving the MOSFETs. For all devices, the drain current rises to a peak value within ≈ 1-2 µs. The peak drain saturation current reduces from 108 A to 50 A, with increasing L_{CH}. The t_{SC} increases from 4.3 µs for the

shortest L_{CH} device to 8 µs, for the longest L_{CH} device, at a DC link voltage of 2000 V. A corresponding reduction of the peak $I_{D,SAT}$ from 108 A to 50 A is also observed with increasing L_{CH}. The longest L_{CH} MOSFET with t_{SC} of 8 µs incurs an $R_{DS,ON}$ penalty of 20% (I_D=4 A, V_{GS}=20 V), for 2x increase in t_{SC}, when compared with the shortest L_{CH} MOSFET, which achieved a low $t_{SC} = 4.3$ µs.

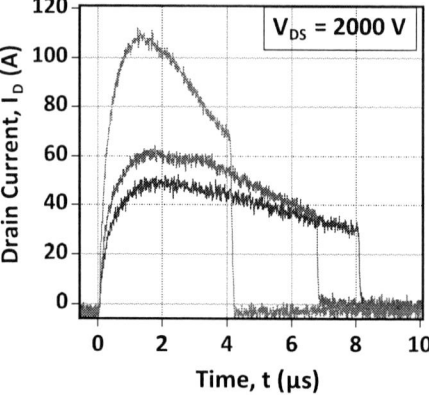

Figure 4: Pre-failure short-circuit waveforms for SiC MOSFETs designed with different channel lengths. The devices were turned on to a short-circuit at a DC link of 2000 V, with a +18 V/-5 V gate drive scheme, at room-temperature.

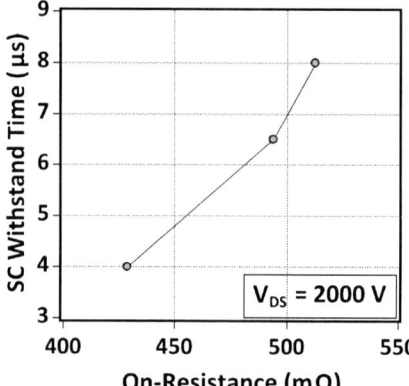

Figure 5: Trade-off plot of MOSFET on-resistance versus short-circuit withstand time from short-circuit measurements performed on devices with different channel lengths.

C. Impact of gate drive voltages on short-circuit SOA

Next, the impact of gate drive voltages on short-circuit withstand time was examined. In a first experiment, three MOSFETs with identical device designs were subjected to HSF at DC link of 2000 V with +20 V/-5 V, +18 V/-5 V and +16 V/-5 V gate drive configurations. For these devices, reducing the gate drive voltage from + 20 V to + 16 V increased the t_{SC} from 6 µs to 9 µs (50% increase), while incurring a 16% deterioration of $R_{DS,ON}$ (see Fig. 14). Similar measurements were performed on devices from different design groups (D1-D4), and the results are summarized in Fig. 6. The different design groups consisted of devices with different channel lengths. For devices within every design group, reducing the gate drive voltage from + 20 V to + 16 V results in increased t_{SC}, while trading off $R_{DS,ON}$ deterioration in the 15-20% range.

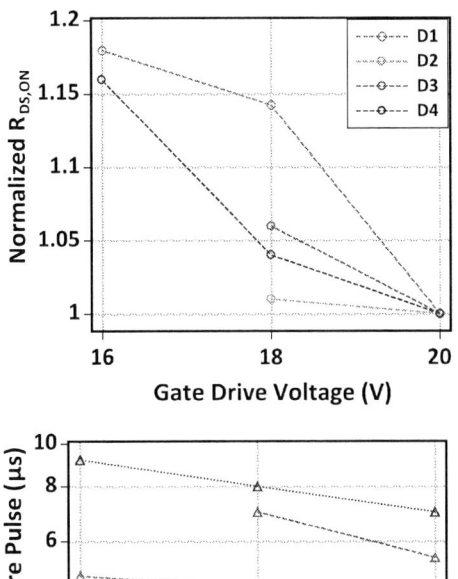

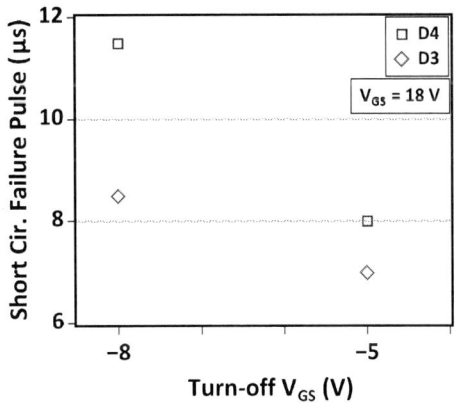

Figure 6 (a) $R_{DS,ON}$ deterioration and (b) Short-circuit withstand time improvement for SiC MOSFETs in different design groups, as the positive gate drive bias magnitude is reduced from +20 V to +16 V, for a DC link voltage of 2.15 kV.

Next, we examined the impact of negative gate drive magnitude on short-circuit withstand time. Lowering the off-state gate drive bias from -5 V to -8 V increased the t_{SC} independent of the device design group, with obviously no on-resistance penalty (Fig. 7).

Figure 7: Short-circuit withstand time of different design groups for different turn-off gate drive conditions. The DC link was 2 kV and the positive gate bias magnitude was +18 V for these tests.

A noticeable reduction of the tail (drain) current at turn-off is observed (Fig. 8), when the turn-off gate drive voltage is reduced from − 5V to -8 V. The MOSFET threshold

voltage is drastically reduced due to the extremely high junction temperatures generated during the SC pulse, which results in current flow through the MOSFET channel even after the gate is "turned off". A lower turn-off gate bias ensures normally-OFF operation even under these extremely high T_J conditions, which manifests as significantly reduced tail currents after turn-off. This behavior is confirmed by electro-thermal simulations (not shown).

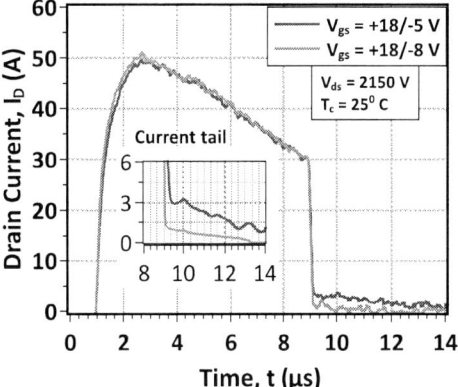

Figure 8: Reduction in tail current magnitude is observed, when the turn-off gate bias is reduced from -5 V to -8 V. The DC link was 2.15 kV and the positive gate bias magnitude was +18 V for these tests.

D. Critical Energy for Short-Circuit Failure

The critical energy for short-circuit failure was determined by integrating the drain current/voltage waveforms from the short-circuit tests. The critical energy for short-circuit failure for the MOSFETs investigated in this work was in the range of 0.5-0.7 J. Further, as shown in Fig. 9, there was no correlation between the critical energy for short-circuit failure and short-circuit withstand time. No impact of device design or drive condition was observed on the critical energy for short-circuit failure, which points to thermal runaway as the major cause of the observed failure for all devices reported in this study.

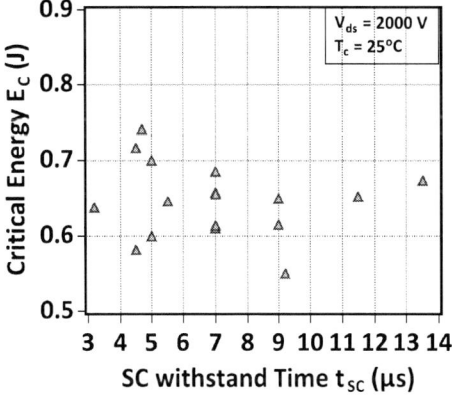

Figure 9: Critical energy for short-circuit failure for multiple devices, designed with different L_{CH}, under a +20 V/-5 V gate drive condition

E. Impact of short-circuit operation on MOSFET device characteristics

The impact of increasing short-circuit operation on the MOSFET I-V characteristics is shown in Fig. 10. A minor increase in the $R_{DS,on}$ can be detected from the output characteristics. The threshold voltage increased, while the gate leakage currents were actually suppressed after application of increasing duration short-circuit pulses. The C-V characteristics of the MOSFETs did not show significant deterioration.

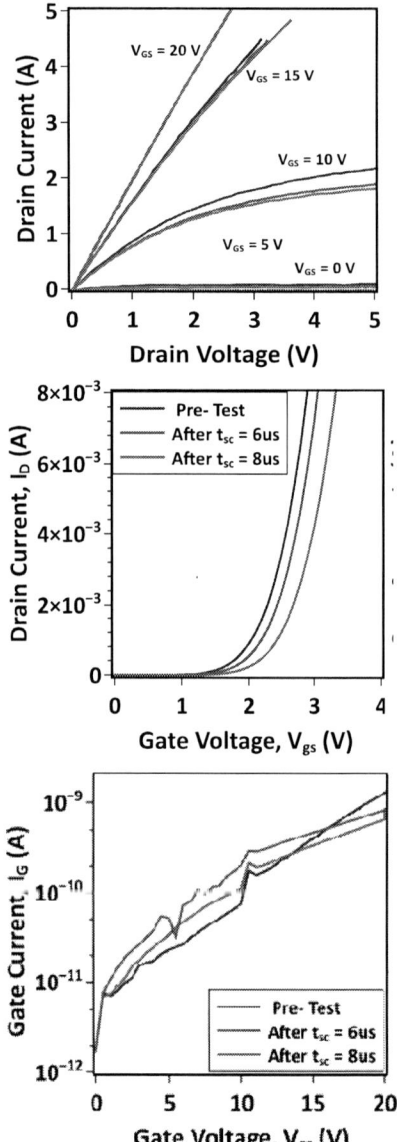

Figure 10: Stability of various MOSFET I-V characteristics are investigated, after increasing duration short-circuit pulses are applied at V_{DS} = 2000 V, for a +18 V/-5 V gate drive configuration.

IV. CONCLUSIONS

A comprehensive evaluation of short-circuit robustness of 4600 V SiC planar DMOSFETs was presented in this paper. The key performance parameter related to short-circuit robustness, i.e. short-circuit withstand time, t_{SC}, was quantified for different operating conditions. Trade-off between short-circuit robustness and device on-resistance (or conduction losses) was quantified. Increasing the MOSFET channel length resulted in 2x increase in short-circuit withstand time form 4 μs to 8 μs for a 20% $R_{DS,ON}$ degradation. Reducing the on-state gate drive voltage from 20 V to 16 V resulted in 15-20% $R_{DS,ON}$ degradation for a $\approx$ 30% increase of short-circuit withstand time within each design group (different L_{CH}). Finally, reduction of the negative gate drive bias from -5 V to -8 V resulted in 25-30% increase of short-circuit withstand time with no $R_{DS,ON}$ degradation. The stability of the key electrical I-V characteristics of the devices were investigated under short-circuit load stress conditions. Possible reasons for the drift of these characteristics was identified and explained.

REFERENCES

[1] S. Sundaresan et al. Mater. Sci. Forum, 924, pp.703-706 (2018)

[2] S. Sundaresan et al. Proceedings of the ECSCRM 2018 conference held in Birmingham, UK (2018).

Proceedings of the 31st International Symposium on Power Semiconductor Devices & ICs
May 19-23, 2019, Shanghai, China

Development of GaN Power IC Platform and All GaN DC-DC Buck Converter IC

Ruize Sun[1, 2, 3*], Y. C. Liang[2, 3], Yee-Chia Yeo[2] and Cezhou Zhao[4], Wanjun Chen[1], Bo Zhang[1]

[1]State Key Laboratory of Electronic Thin Films and Integrated Devices, University of Electronic Science and Technology of China, Chengdu 610054, China
[2]Department of Electrical and Computer Engineering, National University of Singapore, 117580, Singapore.
[3]National University of Singapore (Suzhou) Research Institute, Suzhou 215123, China.
[4]Department of Electrical and Electronic Engineering, Xi'an Jiaotong-Liverpool University, Suzhou 215123, China.
*E-mail: sun.ruize@u.nus.edu

Abstract—In this paper, the GaN power IC platform is developed based on Au-free AlGaN/GaN MIS-HEMTs and a constant-output-ripple all GaN DC-DC buck converter IC is firstly realized with integrated gate driver, pulse width modulation (PWM) feedback control and over-current protection (OCP) circuits. The fabricated all GaN integrated DC-DC buck converter can realize stable 10 V output with constant output ripples when input voltage varies from 15 to 30 V. The output and ripples remain stable with varying load resistance. When subjected to over-current incident, the fabricated power IC can be protected according to desired over-current threshold values. The experiment results have demonstrated the functionality and feasibility of all GaN power ICs in power conversion applications.

Keywords—AlGaN/GaN, Au-free, DC-DC buck converter, integration, MIS-HEMT

I. INTRODUCTION

The development of wide bandgap materials and devices has boosted the advances in power integrated circuits. Especially owing to the superior properties of GaN material and AlGaN/GaN heterostructures, AlGaN/GaN High Electron Mobility Transistors (HEMTs) are widely chosen as the main switches in various forms of power converters. In order to further utilize the advantages of GaN based electronics, the level of integration among the main switches, control circuits and passive components is required to be as high as possible. On-chip controlling, monitoring and protecting circuits can increase functionality and reliability of the fabricated GaN power integrated circuits. Thus power IC platforms are constructed and updated by the advancing GaN based device technology [1].

Previously GaN converter ICs were realized using normally-ON MMIC process [2-3] or normally-OFF HEMT platforms [4-6]. However, the level of integration is limited, the negative normally-ON threshold voltage requires extra matching network, and Schottky gate restricts gate voltage swing.

In this work, the GaN power IC platform is developed based on Au-free AlGaN/GaN MIS-HEMTs and a constant-output-ripple all GaN DC-DC buck converter IC is firstly realized with integrated gate driver, pulse width modulation (PWM) feedback control and over-current protection (OCP) circuits. The fabricated all GaN integrated DC-DC buck converter can realize stable outputs with constant output ripples under various input voltage and load resistance values. When subjected to over-current incident, the fabricated power IC can be protected according to desired over-current threshold values. The experiment results have demonstrated the functionality and feasibility of all GaN power ICs in power conversion applications.

II. GAN POWER IC PLATFORM

In order to facilitate the CMOS-compatible fabrication of all GaN converter ICs for power conversion applications, the Au-free GaN power IC platform is developed and the schematic cross section is shown in Fig. 1. The platform features Au-free metal contacts, normally-OFF gate recess process, MIS gate structure, and embedded current sensing structure.

The AlGaN/GaN epitaxy used in this platform consists of 4.2-μm GaN buffer on Si substrate, 1.6-μm undoped GaN, 20-nm $Al_{0.25}Ga_{0.75}N$ and 1-nm GaN cap as shown in Fig. 1. The electron mobility and sheet carrier concentration are 1627 cm^2/Vs and 0.93×10^{13} cm^{-2} respectively. The fabrication process starts with mesa isolation, then Ti/Al/TiN Au-free ohmic contacts are formed and annealed

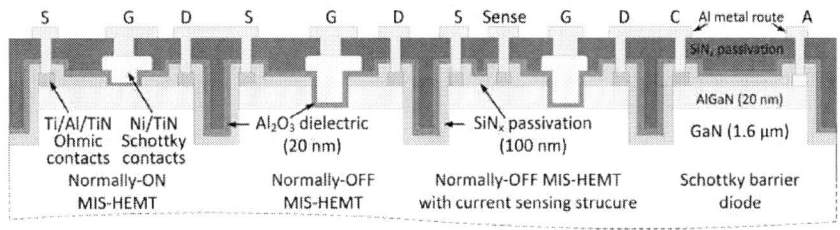

Fig. 1 Schematic cross section of GaN power IC platform based on AlGaN/GaN MIS-HEMTs and Au-free Ti/Al/TiN ohmic contact.

978-1-7281-0582-6/19 $31.00 © 2019 IEEE

at 800 °C for 45 s in N_2 ambient. Then 100-nm SiN_x is deposited as passivation. After gate opening and recess, 20-nm Al_2O_3 gate dielectric is formed by atomic layer deposition. Ni/TiN (100/50 nm) is evaporated as gate metal then Al forms the pads and metal routes.

The developed GaN power IC platform includes low- and high-voltage MIS-HEMT, Schottky barrier diodes and MIS-HEMTs with current sensing structure. The Normally-OFF operation is realized by full recess of AlGaN. Current sensing structure is integrated in normally-OFF MIS-HEMT by inserting floating ohmic electrode between source and gate electrodes [7]. The static and transient performance of MIS-HEMTs and diodes are calibrated in circuit simulation software, facilitating correct design and verification.

III. GaN DC-DC Buck Converter

The all GaN DC-DC converter IC realization approach includes *(a)* calibration of fabricated MIS-HEMTs, diodes and sub-circuits against circuit simulation models to establish the GaN power IC platform; *(b)* design of functional block diagram according to intended modules; *(c)* theoretical verification through numerical analysis on converter operation scheme; *(d)* selection of sub-circuits and layout of detailed circuit diagram; *(e)* fabrication and characterization.

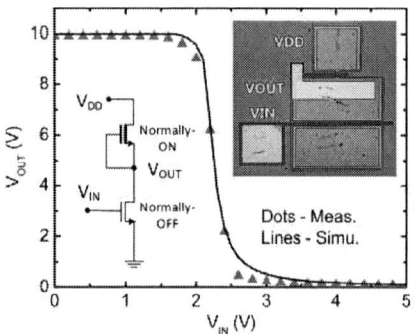

Fig. 2 Calibration verification of example sub-circuit of inverter: measurement and simulation curves. Inset is the microscope image of fabricated inverter with width ratio of W_{OFF}/W_{ON}=200 μm/5 μm.

As a platform verification sub-circuit, the inverter is fabricated using the developed GaN power IC platform and the microscope image is shown in the inset of Fig. 2. The smooth surface of the ohmic contacts is due to the high hardness of TiN cap layer. Meanwhile, the inverter with same configuration is simulated using AlGaN/GaN device models. The good accordance between measured and simulated voltage transferring curves in Fig. 2 demonstrates the successful calibration verification of the fabrication and circuit simulation models, which is the basis of power IC platform.

Fig .3 shows the functional block diagram of proposed all GaN DC-DC buck converter IC, which includes the intended modules of high-side gate driver (including level shifter and buffer), over-current protection circuit (including current sensing MIS-HEMT, over-current comparator and shutdown clamp) and PWM feedback controller (sawtooth generator, PWM comparator, duty cycle switcher and output

comparator). Several voltage reference signals and auxiliary power sources are required to facilitate voltage comparison and high-side bootstrap scheme.

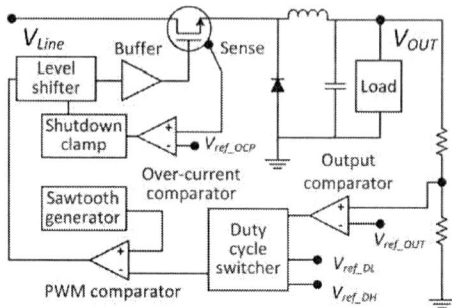

Fig. 3: Functional block diagram of proposed all GaN DC-DC buck converter IC.

In order to avoid premature damage when dealing with sensitive loads, the stable output with constant ripples is preferable [8]. The proposed all GaN DC-DC buck converter IC features stable outputs and constant ripples with the help of the designed feedback scheme. The converter duty cycle will be switched to high duty cycle D_H (70%) when V_{OUT} is lower than desired value, while to low duty cycle D_L (30%) when higher.

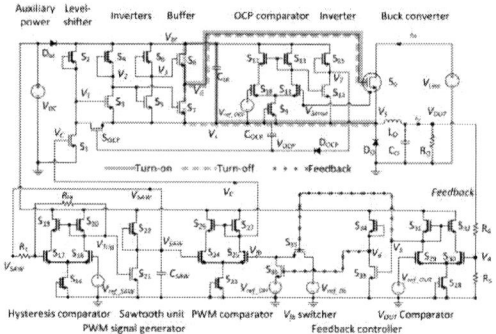

Fig. 4 Circuit diagram of all GaN DC-DC buck converter IC with integrated gate driver, PWM feedback controller and over-current protection (OCP). Line shades are added to explain actions of turn-on, -off, and feedback control.

The design objectives include: (1) output voltage is targeted at 10.0 V and input line voltage range should cover 15 to 30 V; (2) output ripple percentage should be less than 5.0 % for both output voltage and current; (3) converter should be able to maintain stable output with constant ripples against abrupt variations in input line voltage or load conditions; (4) converter should be able to response to over-current incident according to the preset over-current threshold and to be shut down within one duty cycle. After selection of sub-circuits and layout of detailed circuit diagram, the whole circuit diagram is shown in Fig. 4. The diagram includes integrated gate driver, PWM feedback controller [9] and over-current protection (OCP). Line shades are added to explain actions of turn-on, -off, and feedback control. The integrated highside gate driver is

978-1-7281-0582-6/19 $31.00 © 2019 IEEE

realized through NMOS logic and the converter IC is clocked by integrated PWM feedback controller, which eliminates commonly used off-chip FPGA control signals [2, 9]. The over-current protection is integrated owing to embedded current sensing structure in MIS-HEMT main switches.

IV. RESULTS AND DISCUSSION

Based on the developed GaN power IC platform, the converter IC is fabricated and the optical microscope image is shown in Fig. 5. The highside driver, PWM signal generator, feedback controller and over-current protection modules are marked by dashed line box. The off-chip passive components and voltage references are connected by BNC cables (blue arrows). This prototype still needs part-by-part measurement to check its functions. Some of the cable connections can be replaced by integrated Al metal routes in future fabrication batches.

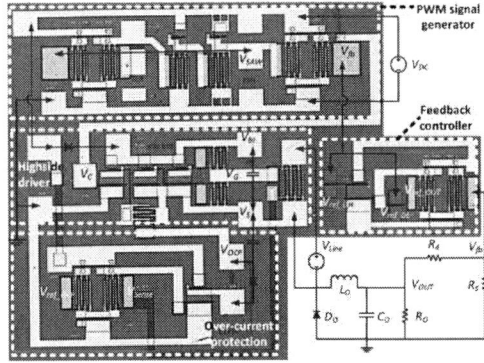

Fig. 5 Optical image of the fabricated all GaN DC-DC buck converter IC prototype with connected external components.

The fabricated all GaN integrated DC-DC buck converter is firstly measured in normal operation with different input line voltages at 10 kHz. The passive components, such as inductor and capacitor, are off-chip due to their large values. The measurement conditions include L_O of 193 mH, C_O of 10 µF, R_{OUT} of 500 Ω. The V_{ref_DH} and V_{ref_DL} are set at 3.5 V and 6.8 V to generate corresponding D_L and D_H of 30.2 % and 70.1 %, which are same as designed values. The V_{Line} is varied from 16 V to 29 V with step of 1 V. Fig. 6(a) shows the typical normal output waveforms with V_{Line} of 24 V to demonstrate the stable output characteristics. Fig. 6(b) shows the average V_{OUT} values and output ripples with input of 16-29 V and the ripples are constantly below 0.4 V (4% of V_{OUT}).

The converter could be subjected to abrupt variations in input line voltage V_{Line}. The designed feedback scheme should maintain stable output with constant ripples under all the variations. Here the V_{Line} is initially set at 20 V, then abruptly switched to values from 16 to 29 V. Fig. 7(a) shows typical stable output and I_L duty cycle response when input abruptly changed from 20 to 16 V. Stable 10 V output with ripples < 4% are obtained with various abrupt changes of input (20 to 16-29 V) as shown in Fig. 7(b). These results demonstrate the successful feedback control.

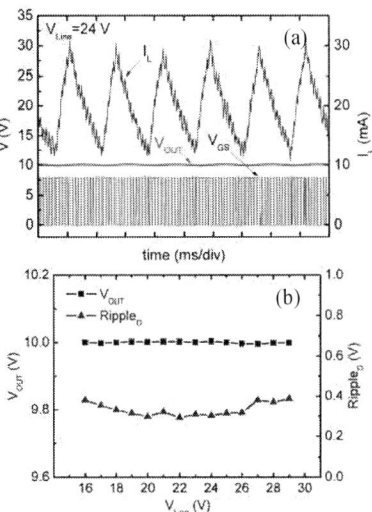

Fig. 6 (a) Stable waveforms of output voltage V_{OUT}, inductor current I_L and gate voltage of power switch V_{GS} with input V_{Line} of 24 V. (b) Average V_{OUT} values and output ripples with input of 16-29 V. Ripples are constantly below 0.4 V (4% of V_{OUT}).

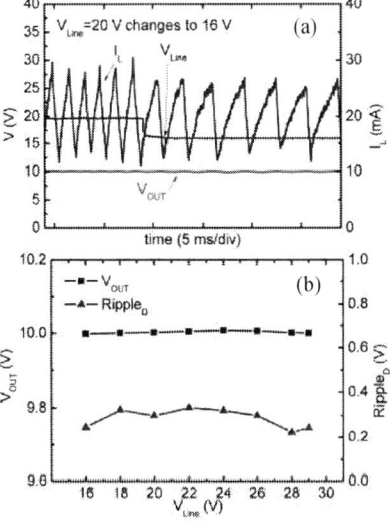

Fig. 7: (a) Transient response waveforms of V_{OUT}, I_L with abrupt variation of input V_{Line} from 20 V to 16 V. (b) Average V_{OUT} and output ripples with abrupt variation of V_{Line} from 20 V to 16, 18, 22, 24, 26, 28 and 29 V (ripples < 4%).

It is essential for DC–DC converters to have stable output against load variations as well. When load R_O is changed from 300 to 500 Ω, curve of I_L in Fig. 8(a) shows that number of periods with D_L and D_H are changed according to feedback scheme, while V_{OUT} remains stable at 10.0 V. Fig. 8(b) shows that load R_O variations covering 300 to 700 Ω range still leave little effects on stable 10 V output and constant ripples below 4%.

When facing over-current incident at t=0, converter IC can be protected according to preset over-current thresholds. Once sensing current of main switch is over a sample

978-1-7281-0582-6/19 $31.00 © 2019 IEEE   273

threshold of 120 mA, OCP shuts down the converter IC within one period as shown in Fig. 9.

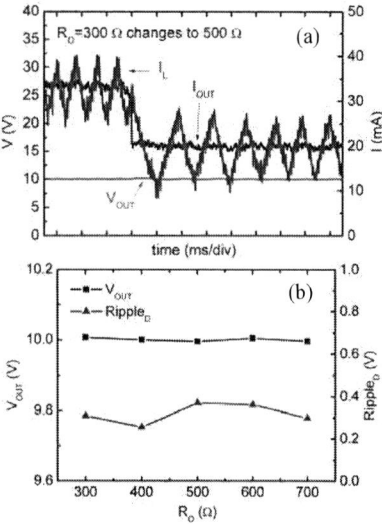

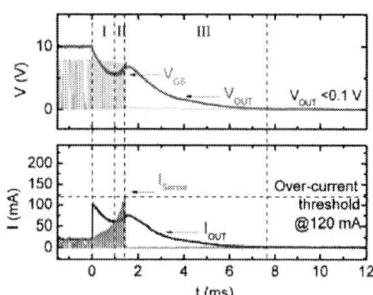

Fig. 8 (a) Transient response waveforms of V_{OUT}, I_L and I_{OUT} with load R_O abrupt variation from 300 to 500 Ω. (b) Average V_{OUT} and output ripples with abrupt variations of load R_O from 500 to 300, 400, 600, 700 Ω (ripples < 4%).

Fig. 9 Transient response waveforms of V_{OUT}, V_{GS}, I_{Sense} and I_{OUT} with over-current incident at t=0 and threshold of 120 mA.

V. CONCLUSION

In this paper, the GaN power IC platform is realized based on normally-OFF AlGaN/GaN MIS-HEMTs and the all GaN integrated DC-DC buck converter has been designed, fabricated and experimentally characterized. The fabricated converter IC realizes stable output at 10 V with constant output ripple and line input voltage ranging from 16 to 29 V.

The converter can maintain stable output against input line voltage variations from 20 V to values of 16~29 V, or load variations from 500 Ω to values of 300~700 Ω. When subjected to over-current incident, the converter can be switched off within one duty cycle period. The proposed GaN power IC platform can be adopted and all GaN DC-DC converters is promising in power conversion applications

ACKNOWLEDGMENT

This research is funded in part by the Sichuan Youth Science and Technology Foundation (No. 2017JQ0020), the Fundamental Research Funds for the Central Universities (No. ZYGX2016Z006), the Major science and technology special projects in Guangdong (No. 2017B010112003) and the National Natural Science Foundation of China (No. 51477108).

REFERENCES

[1] D. Kinzer and S. Oliver, "Monolithic HV GaN Power ICs: Performance and application," *IEEE Power Electronics Magazine*, vol. 3, pp. 14-21, 2016.

[2] Y. Zhang, M. Rodríguez, and D. Maksimović, "Very High Frequency PWM Buck Converters Using Monolithic GaN Half-Bridge Power Stages With Integrated Gate Drivers," *IEEE Transactions on Power Electronics*, vol. 31, pp. 7926-7942, 2016.

[3] B. Weiss, R. Reiner, P. Waltereit, R. Quay, O. Ambacher, A. Sepahvand, et al., "Soft-switching 3 MHz converter based on monolithically integrated half-bridge GaN-chip," *IEEE 4th Workshop on Wide Bandgap Power Devices and Applications (WiPDA)*, 2016, pp. 215-219.

[4] K. J. Chen, "GaN smart power chip technology," *2009 IEEE International Conference of Electron Devices and Solid-State Circuits (EDSSC)*, 2009, pp. 403-407.

[5] C. Tsai, Y. Wang, M. Kwan, P. Chen, F. Yao, S. Liu, et al., "Smart GaN platform: Performance & challenges," *2017 IEEE International Electron Devices Meeting (IEDM)*, 2017, pp. 33.1.1-33.1.4.

[6] S. Ujita, Y. Kinoshita, H. Umeda, T. Morita, S. Tamura, M. Ishida, et al., "A Compact GaN-based DC-DC Converter IC with High-Speed Gate Drivers Enabling High Efficiencies," *IEEE 26th International Symposium on Power Semiconductor Devices & IC's (ISPSD)*, 2014, pp. 51-54.

[7] R. Sun, Y. C. Liang, Y. Yeo and C. Zhao, "Au-Free AlGaN/GaN MIS-HEMTs With Embedded Current Sensing Structure for Power Switching Applications," *IEEE Transactions on Electron Devices*, vol. 64, no. 8, pp. 3515-3518, 2017.

[8] D. G. Lamar, M. Fernandez, M. Arias, M. M. Hernando and J. Sebastian, "Tapped-Inductor Buck HB-LED AC–DC Driver Operating in Boundary Conduction Mode for Replacing Incandescent Bulb Lamps," *IEEE Transactions on Power Electronics*, vol. 27, no. 10, pp. 4329-4337, 2012.

[9] H. Wang, A. M. K. Ho, Q. Jiang and K. J. Chen, "A GaN pulse width modulation integrated circuit," *IEEE 26th Inter3 national Symposium on Power Semiconductor Devices & IC's (ISPSD)*, 2014, pp. 430-43.

Proceedings of the 31st International Symposium on Power Semiconductor Devices & ICs
May 19-23, 2019, Shanghai, China

Integrated High-Speed Over-Current Protection Circuit for GaN Power Transistors

Han Xu, Gaofei Tang, Jin Wei and Kevin J. Chen

Department of Electronic and Computer Engineering, The Hong Kong University of Science and Technology, Hong Kong

Phone: +852-23588530, Fax: +852-23581485, Email: hxuaw@connect.ust.hk eekjchen@ust.hk

Abstract—In this work, a new desaturation-based over-current protection circuit is proposed and monolithically integrated with GaN power HEMTs. Compared to traditional desaturation techniques, this new design features separated sensing branch and blanking time controller. Such a separation allows immediate sensing of over-current (OC) event, while the blanking time can be modified without considering the sensing speed. To mimic real situations in power applications, the circuit was systematically characterized under different operating conditions. It can deliver an accurate OC threshold voltage and fast response without affecting the switching characteristics. These properties will contribute to a more robust and reliable high-speed GaN power systems.

Keywords—Over-current protection, gate driver, integrated circuits, GaN power HMETs

I. INTRODUCTION

GaN power HEMTs are promising to bring higher efficiency and higher power density in power conversion systems, owing to their superior characteristics including high switching speed, low ON-state resistance and high breakdown voltage [1-2]. To improve system-level ruggedness, robust and high-speed over-current protection of the power transistors is of great importance. Compared to Si-based power devices, GaN power devices are much smaller in size for the same R_{DS-ON} and has to endure much higher current density. Thus, they are more vulnerable to over-current events. A report shows that GaN power HEMTs may survive less than 1 μs at high saturation current levels under its rated voltage [3]. Traditional discrete implementations of over-current protection either degrade system performance by adding extra inductance and resistance, or compromise sensing accuracy and response speed for GaN power device. Therefore, it is of great benefit to develop monolithically integrated high-speed over-current protection solutions with fast response time to protect the GaN power transistors.

For Si-based power transistors, several OC protection schemes have been developed [4-5]. The most straightforward method is the resistor-based current sensing that uses a current shunt. However, accurate current shunt with low inductance is costly and will also generate additional power loss. Another approach is based on current sensing of a dummy transistor in close vicinity of the power transistor, which may have limited sensing ratio and accuracy [6]. To address these issues, desaturation techniques have been developed using a sensing diode to monitor the power transistor's drain-to-source voltage V_{DS}

This work is supported by Guangdong Science and Technology Department under Grant 2017B010113002

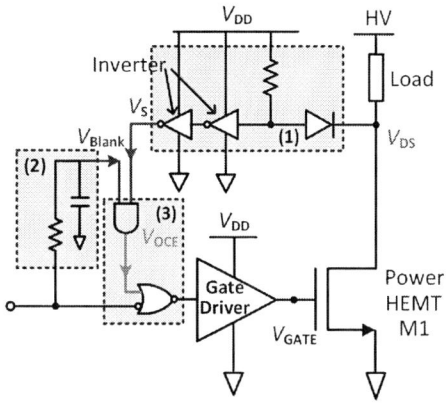

Fig. 1 Schematic of the proposed over-current protection circuit. (1) the sensing branch; (2) blanking controller; (3) driving controller. All components are integrated except the load.

[4], [7]. This OC protection scheme has the benefits of simple configuration and low cost, and requires no extra components in the power loop. However, it is challenging to directly implement this scheme in high-speed GaN power switches due to the relatively long delay time required to prevent the false judgment of a large OFF-state V_{DS}.

In this work, a high-speed over-current protection circuit was realized on an all-GaN platform, on which the protection function block is monolithically integrated with the gate driver and power HEMTs [8-10]. The circuit improves the response speed by separating the sensing branch and blanking time controller. Compared to traditional discrete implementations of OC protection, this integrated protection delivers a fast response time (40 ns) and high sensitivity. The schematic and working principle were illustrated in detail while its functionality was systematically validated under different operating conditions.

II. CIRCUITS DESIGN AND OPERATING PRINCIPLE

The new desaturation-based over-current protection is constructed in a simple but delicate method as shown in Fig. 1. It consists of a sensing branch, a blanking time controller and a driving signal controller. The current information of the power switch is represented by its drain-to-source voltage V_{DS}, which is then further transferred to V_S by the sensing branch. When V_{DS} is below a threshold value (V_{OCT}), V_S = L (low); otherwise V_S = H (high). As illustrated in Fig. 2(a), if V_{DS} is smaller than V_{OCT} in the ON-state, the OC protection circuit generates an enable signal (V_{OCE} = L), so a PWM signal of "H" could be fed to the power transistor

978-1-7281-0582-6/19 $31.00 © 2019 IEEE 275

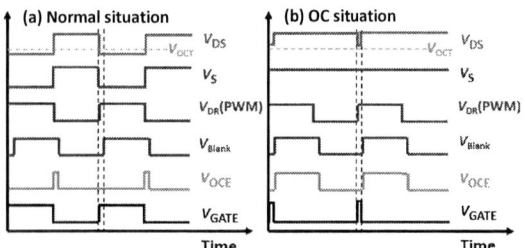

Fig. 2 Illustration of basic waveforms in the over-current protection. (a) Normal situation when ON-state current is below the critical point. (b) OC event with protected Gate signal. Area between dash lines is the designed blanking time.

(M1) to maintain the ON-state. In the OFF-state, V_S is blocked from the gate driver by the PWM signal of "L", so M1 can remain at the OFF-state. For M1 to switch from OFF to ON, the switching transient is completed within the blanking time, during which V_S is still blocked from the driver circuit, so the influence of the high OFF-state V_{DS} is obviated. For an OC event [Fig. 2(b)], V_{DS} will be larger than V_{OCT} once M1 is turned on. V_S (= H) is fed to the gate driver after the blanking time. Then, V_{OCE} = H, and M1 is turned off irrespective of the PWM signal. The current situation is checked every cycle by the pulsed gate signal (V_{GATE}) during the blanking time and an error information can be reported by processing V_S and driving signal V_{DR}.

In traditional desaturation techniques mentioned above, current information is stored as a capacitor voltage while the blanking time is also controlled by this capacitor. To turn on the switch with high V_{DS}, the charging speed must be restricted by a large RC constant to ensure that the capacitor voltage stays below the critical voltage and the protection is not triggered. Therefore, when over-current situation (e.g. a sudden increase of current) occurs, long response time is expected due to the large RC constant. By separating the sensing branch and blanking time controller, as described in this new design, protection can be triggered immediately during ON-state or after the designed blanking time.

III. EXPERIMENT VALIDATION

To verify the functionality of the proposed protection, the circuit was tested in different conditions according to real application situations. Some circuits and chip information were presented in Fig. 3.

A. Basic protection behavior

As described in last section, this protection circuit works by preventing the driving signal from getting into the gate driver and the final result is the regulated V_{GATE}, as shown in Fig. 2. To test the basic protection capability, the protection logic and gate driver were integrated as special function blocks (without power switch). They were tested on a probe station to validate its operating logic.

V_{DD} and GND were connected to DC power supply while a variable voltage can be applied to mimic the drain-source voltage of the power switch. This voltage (V_{DS}) will significantly influence the output of the gate driver (V_{GATE}), as plotted in Fig. 4. When V_{DS} sweeps to about 1V, V_{GATE}

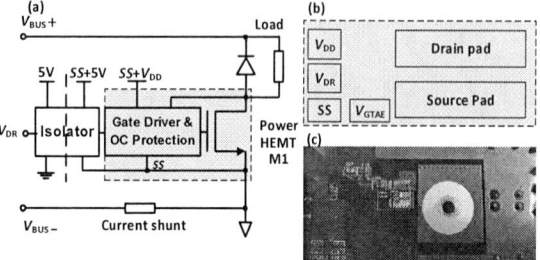

Fig. 3 (a) Schematic circuit of the double pulse tester. (b) Pad information of the integrated circuits. (c) Photo of the PCB with chip on it.

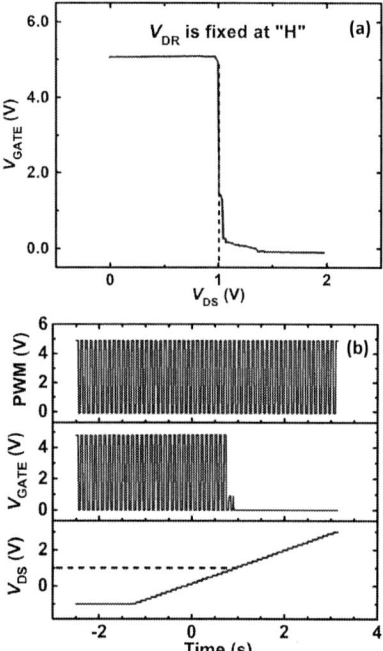

Fig.4 Experiment validation on basic protection capability. (a) Driving signal is fixed at high level. (b) Driving signal is a PWM signal.

will be shut down to zero, even though V_{DR} is fixed at "H" [Fig. 4(a)] or fed with a PWM signal [Fig. 4(b)]. This is the expected protection situation where V_{DS} is an indication of current. A threshold voltage (1V) can be obtained from this experiment and V_{GATE} will be regulated when V_{DS} exceeds this threshold.

B. Protection under switching operation

In real applications, over-current protection must work properly under switching operations. To characterize this property, the new design was tested by a double-pulse test circuits, which refers to previously reported scheme in [8]. The schematic circuit is presented in Fig. 3(a) and the designed IC was mounted on the printed circuit board with silver epoxy and bonding wires. By changing the load to an inductor or power resistor, different conditions in real power applications can be realized.

978-1-7281-0582-6/19 $31.00 © 2019 IEEE

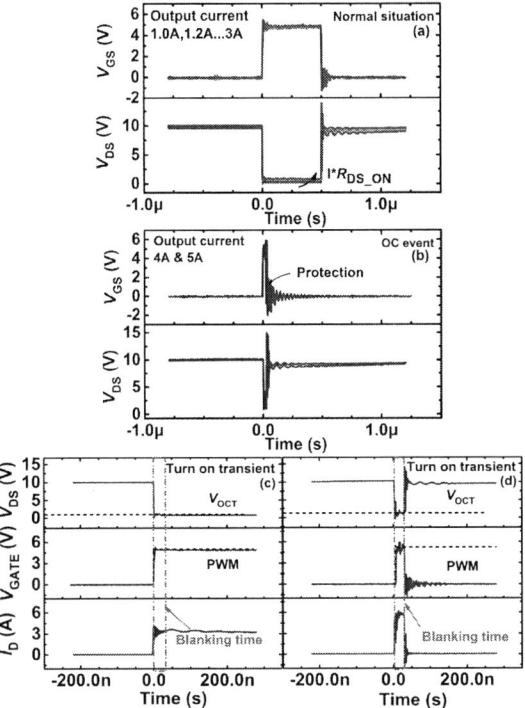

Fig. 5 Resistive switching waveforms. (a) Low current levels and normal situation (b) Higher current level with protection triggered. (c) Zoomed-in turn-on transient for normal switching operation. (d) Zoomed-in turn-on transient for over-current situation.

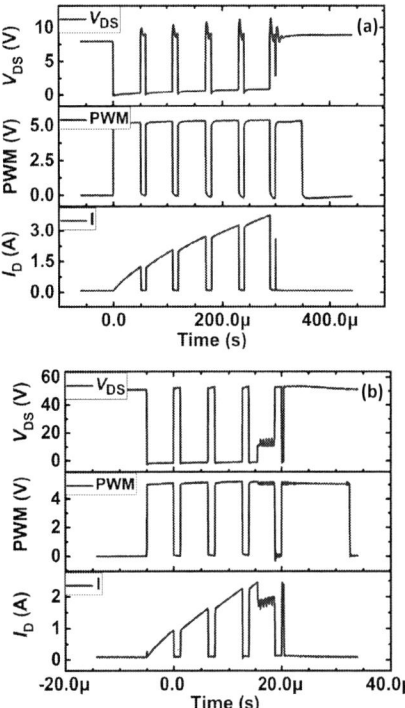

Fig. 6 Inductive switching waveforms (a) Lower voltage with protection at the onset of the turn-on transient. (b) High voltage with the protection in the conduction period.

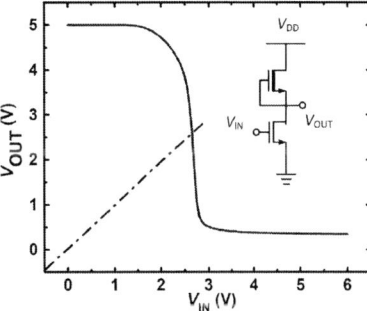

Fig. 7 Static voltage transfer curves of an inverter. The inset is a typical circuits schematic of direct-coupled FET logic.

To address the situation where a sudden or unusual increase of current happens (e.g. short circuits), resistive-load switching was characterized. Different current levels were achieved by changing the value of the load resistor. The bus voltage is fixed at a low level (10V) in order to observe the ON-state drain-source voltage ($V_{DS\text{-}ON}$). Switching waveforms under different current levels are presented in Fig. 5. A significant increase of $V_{DS\text{-}ON}$ can be observed as the current levels rises. When the $V_{DS\text{-}ON}$ exceeds the threshold of 1V(V_{OCT}), the gate voltage of M1 will be pulled to 0V irrespective of the driving signal. Zoomed-in pictures of the turn-on transient are also presented in Fig. 5. Voltage spike induced by ringing may exceed V_{OCT} at the beginning of the turn-on transient, but its influence could be blocked by the designed blanking time and a false judgment will be avoided during normal switching. The blanking time of 30 ns is adopted in this design and the total response time for protection is controlled within 40 ns.

In some cases, over-current happens in a mild way. For instance, when a light load is changed to a heavier load in a buck converter, the feedback system will change the duty cycle of the power switch. If fault occurs in the output stage and the output voltages drops, the feedback loop will continuously increase the duty cycle until the current reaches the limit. To validate the new design in such situations, the load was changed to an inductor. By varying the bus voltage, pulse numbers and pulse duration, the protection can start to function at the onset of a turn-on transient [Fig. 6(a)] or in the duration of the conduction period [Fig. 6(b)].

Different triggered current levels in Fig. 6 can be explained by the different bus voltages and the dynamic R_{ON} in GaN HEMT [11], [12]. When the bus voltage is high, the device is stressed for a long time before the driving signal was fed to the driver. Under such a circumstance, the R_{ON} will increase to 1.1~1.2 of its original value when bus voltage is around 50 V. V_{DS} is the product of R_{ON} and the current I_D and therefore, the current level for two conditions are different.

The middle state in Fig. 6(b) is induced by the inverter in the sensing branch. The inverter is a direct-coupled FET logic and its transfer curve is plotted in Fig. 7 [13]. Although two stages of inverter were used, V_{DS} will have a

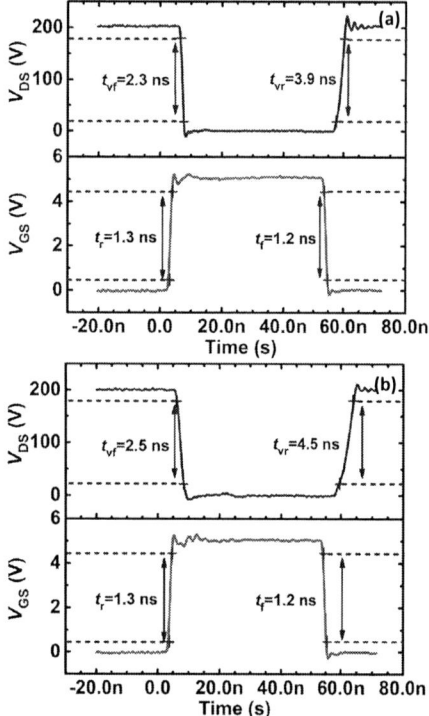

Fig. 8 Combined turn-on and turn-off transient. (a) Circuits with proposed protection. (b) Circuit without protection.

small range of values, where V_S and V_{GATE} are in the middle stage, as shown in Fig. 4. When the continuously increasing V_{DS} reaches the critical point, the power device will move to semi-ON state and the sudden increase of V_{DS} may cause V_{GATE} (in middle state) to swing through C_{GD}. By deploying more stages (e.g. four stages) of inverters with different transfer curves in the sensing branch, V_{DS} will be digitalized completely and the middle state of V_S and V_{GATE} will be eliminated.

C. Influence on switching performance

Another concern related to desaturation techniques is its influence on switching performance. the sensing diode could add additional capacitive load to the main switch, leading to poor switching speed and higher switching loss. However, in this new design, a low capacitive sensing diode (<1% in width compared to M1) is adopted in the OC protection circuit. Switching performance with and without the protection were compared in Fig. 8. They were tested in the same condition and the minor difference is introduced by different PCB traces or chip-to-chip variation. Therefore, it is concluded that implementation of the OC protection circuit cast inappreciable influence on the switching speed of the power transistor.

IV. CONCLUSIONS

In this work, a high-speed over-current protection circuit was realized on an all-GaN platform. Traditional over-current protection circuits either degrade system performance by adding inductance and resistance to the power loop, or compromise sensing accuracy and/or

response speed. The new design features a simple architecture and a quick response, and has been systematically validated at quasi-static conditions as well as switching operations. Superior switching characteristics of the GaN power transistor is maintained, indicating negligible parasitic have been introduced by the implementation of the over-current circuit. Thus, it provides a simple and effective approach to prevent over-current induced failure in GaN power integrated circuits.

REFERENCES

[1] K. J. Chen, O. Häberlen, A. Lidow, C.-L. Tsai, T. Ueda, Y. Uemoto and Y, Wu, "GaN-on-Si power technology: devices and applications," *IEEE Trans. Electron Devices*, vol. 64, no. 3, pp. 779–795, Mar. 2017.

[2] J. Wei et al., "Low on-Resistance normally-off GaN double-channel metal–oxide–semiconductor high-electron-mobility transistor," in *IEEE Electron Device Letters*, vol. 36, no. 12, pp. 1287-1290, Dec. 2015.

[3] C. Abbate, G. Busatto, A. Sanseverino, D. Tedesco, and F. Velardi, "Failure analysis of 650 V enhancement mode GaN HEMT after short circuit tests," *Microelectron. Reliab.*, vol. 88–90, pp. 677–683, Sep. 2018.

[4] R. S. Chokhawala, J. Catt, and L. Kiraly, "A discussion on IGBT short-circuit behavior and fault protection schemes," *IEEE Trans. Ind. Appl.*, vol. 31, no. 2, pp. 256–263, Apr. 1995.

[5] Z. Wang, X. Shi, L. M. Tolbert, F. Wang, and B. J. Blalock, "A di/dt feedback-based active gate driver for smart switching and fast overcurrent protection of IGBT Modules," *IEEE Trans. Power Electron.*, vol. 29, no. 7, pp. 3720–3732, Jul. 2014.

[6] F.-F. Ma, W.-Z. Chen, and J.-C. Wu, "A monolithic current-mode buck converter with advanced control and protection circuits," *IEEE Trans. Power Electron.*, vol. 22, no. 5, pp. 1836–1846, Sep. 2007.

[7] B. Huang, Y. Li, T. Q. Zheng, and Y. Zhang, "Design of overcurrent protection circuit for GaN HEMT," in *2014 IEEE Energy Conversion Congress and Exposition (ECCE)*, Pittsburgh, PA, USA, 2014, pp. 2844–2848.

[8] G. Tang, M.H. Kwan, Z. Zhang, J. He, J. Lei, R. Y. Su, F. W. Yao, Y. M. Lin, J. L. Yu, Tom Tsai, H. C. Tuan, A. Kalnitsky, and K. J. Chen, "High-speed, high-reliability GaN power device with integrated gate driver," in *Proc. 30th Int. Symp. Power Semiconductor Devices IC's (ISPSD)*, Chicago, IL, 2018, pp. 76–79.

[9] S. Ujita, Y. Kinoshita, H. Umeda, T. Morita, S. Tamura, M. Ishida and T. Ueda "A compact GaN-based DC-DC converter IC with high-speed gate drivers enabling high efficiencies," in *Proc. 26th Int. Symp. Power Semiconductor Devices IC's (ISPSD)*, Waikoloa, HI, USA, 2014, pp. 51–54.

[10] D. Kinzer, "GaN power IC technology: past, present, and future," in Proc. 29th Int. Symp. Power Semiconductor Devices IC's (ISPSD), Sapporo, 2017, pp. 19-24.

[11] H. Wang, J. Wei, R. Xie, C. Liu, G. Tang, and K. J. Chen, "Maximizing the performance of 650-V p-GaN gate HEMTs: dynamic R_{ON} characterization and circuit design considerations," *IEEE Trans. Power Electron.*, vol. 32, no. 7, pp. 5539–5549, Jul. 2017.

[12] J. Wei, R. Xie, H. Xu, H. Wang, Y. Wang, M. Hua, K. Zhong, G. Tang, J. He, M. Zhang and K.J. Chen "Charge storage mechanism of drain induced dynamic threshold voltage shift in p-GaN gate HEMTs," *IEEE Electron Device Lett.* in press.

[13] G. Tang, M. H. Kwan, K. Y. Wong, J. Lei, R. Y. Su, F. W. Yao, Y. M. Lin, J. L. Yu, Tom Tsai, H. C. Tuan, A. Kalnitsky, and K. J. Chen, "Digital integrated circuits on an E-Mode GaN power HEMT platform," *IEEE Electron Device Lett.*, vol. 38, no. 9, pp. 1282–1285, Sept. 2017

Proceedings of the 31st International Symposium on Power Semiconductor Devices & ICs
May 19-23, 2019, Shanghai, China

Gate Control Circuit for the LIGBT to Improve the Freewheeling Characteristics in Monolithic IC

Siyuan Yu[1], Jing Zhu[1], Yangyang Lu[1], Weifeng Sun[1,*], Bowei Yang[1], Ding Yan[1], Chuanyi Cheng[1],
Yan Gu[2], Sen Zhang[2], Yunwu Zhang[3]

[1]National ASIC System Engineering Research Center, Southeast University, Nanjing210096, China
[2]Technology development department, CSMC Technologies Corporation, Wuxi, China
[3]Wuxi iDriver Electronic Co., Ltd, Wuxi, China
*Email: *swffrog@seu.edu.cn* Phone: *86-25-83795077* Fax: *86-25-83795077*

Abstract—A gate control circuit for the LIGBT devices in the power stage to improve the freewheeling characteristic is proposed for the first time. Both the measured and simulated results indicate that structure designed have good performance. Compared with conventional circuit, which can achieve the peak reverse recovery current decreased by about 66.7% and the charge shrunk by about 86.6%.

Keywords—*high voltage, gate control circuit, reverse recovery.*

I. INTRODUCTION

High voltage monolithic IC, which integrates the HVIC (High Voltage Integrated Circuit) and the power stage devices in a single die with SOI technology, is widely used in low power motor system instead of the IPM (Intelligent Power Module) because of the size and cost [1]. Since LIGBT has no body diode, another important high voltage device called freewheeling diode (FWD) is required [2], and the FWD's reverse recovery characteristics will severely affect the EMI (Electro-Magnetic Interference) and switching loss of the system [3]. In order to achieve better reverse recovery characteristics, we should reduce the carrier injection efficiency or adopt schottky structure [4-5]. However, the current density of the FWD would be decreased and the cost would be increased because of the additional layer.

Fig.1 shows the block view of the IPM. The IPM can be divided into two components, which are the driver stage and the power stage. The driver stage has two channels: high side channel and low side channel, controlled by the high side and low side input signal respectively. The power stage consists of U,V,W three phases, each phase has low side and high side power switches. Six FWD are placed in parallel.

Fig.2 shows a general double pulse test result when the bus voltage is 100V. The current present a spike pulse at the second pulse rising edge because of the freewheeling current of FWD, meanwhile power loss is caused by the freewheeling current of FWD.

In order to further improve the reverse recovery characteristics without parallel with FWD, a new power stage circuit for the LIGBT is proposed in this paper. By utilizing the proposed power stage circuit, the peak reverse recovery current decreased by about 66.7% and the charge shrunk by about 86.6% compared with the conventional structure.

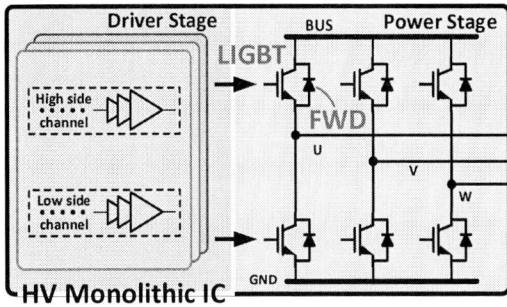

Fig.1 Block view of Intelligent Power Module (IPM) which includes three half-bridge power stages and its drivers. Six FWD are placed in parallel

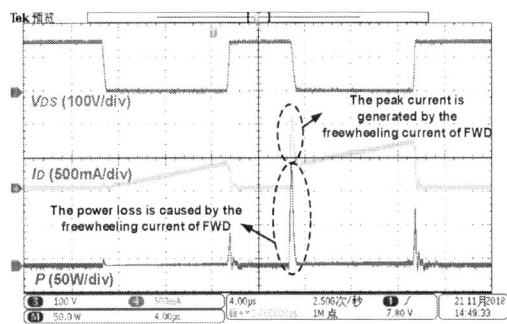

Fig.2 EMI and switching loss of the system caused by the reverse recovery characteristics of the FWD

II. DISCUSSION ON THE PROPOSED POWER STAGE CIRCUIT

A. Operation mechanism of the proposed power stage circuit

The schematic of the proposed power stage circuit is shown in Fig.3 (b). It contains a sensing MOS M4 and a LDMOS M3 instead of the FWD. The comparator connects with its negative terminal to 3.8V, which generated by reference; and the positive terminal samples V_d voltage. In proposed structure, take the high side for example, when the low side LIGBT M2 is turned off, the load current flows through the body diode(D4) of the sensing MOS M4 and the body diode(D3) of the LDMOS M3 first, V_d is -0.7V at this point, then the negative voltage will be raised to the comparator input range by means of the voltage dividing circuit. R2 is twice the resistance of R1.

978-1-7281-0582-6/19 $31.00 © 2019 IEEE   279

The comparator negative terminal voltage can be expressed as:

$$V_- = \frac{R_2}{R_1 + R_2} \cdot (5 - V_d)$$

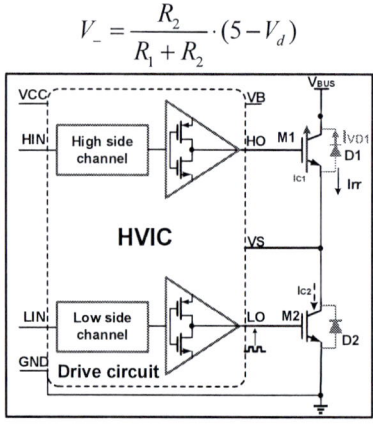

(a)

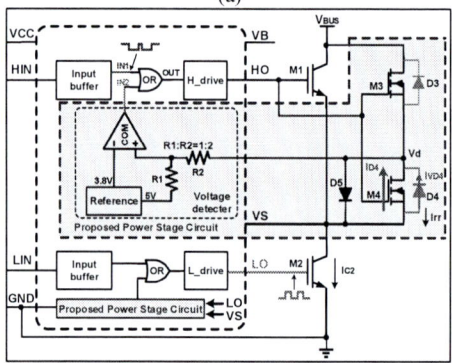

(b)(our patent structure)

Fig.3 The conventional normal circuit (a) and proposed power stage circuit (b).

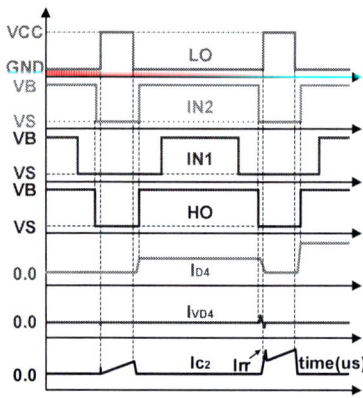

Fig.4 The operating principle of the proposed power stage.

This work was supported by the national natural science foundation of China (61874026, 61804026, 61504025), the national key research and development plan (2017YFB0402904), postdoctoral research funding of Jiangsu (2018K001A), the key research and development plan of Jiangsu (BE2018003-3) and the fundamental research funds for the central universities.

When the conduction voltage drop of D4 is detected by the comparator, a high level IN2 will be outputted, act as input signal of the nor gate together with IN1, then HO will output a high level. Therefore, after a few nanoseconds (the propagation delay of the control circuit) another current path will be triggered because the channel of the LDMOS is turned on by the control circuit, the freewheeling current would bypass the body diode and flow through the channel. Due to the sensing MOS M4 is a low voltage MOSFET, so the reverse recovery characteristics improved a lot. Both the peak reverse recovery current and power losses will decrease.

The basic working principle of the proposed power stage circuit is shown in the Fig. 4 and its working process is illustrated as follow:

(1) When LO is "0", the low side LIGBT M2 is turned off, the load current flows through the body diode of the sensing MOS M4 and the body diode of the LDMOS M3 first. Offset ground VS is "0" and Vd voltage is 0.7V less than VS, the negative voltage will be raised up by means of the voltage dividing circuit which consisted of R1 and R2.

(2) IN2 is going to high voltage once the comparator sensing the conduction voltage drop of D2.

(3) HO is going to high together with IN2, and the conduction channel of the LDMOS M3 and sensing MOS M4 is turned on. Therefore, only a few freewheeling current bypasses the body diode and most flow through the channel, so the I_{rrm} and Q_{rr} will be reduced dramatically.

B. simulation results of the freewheeling current path in LDMOS

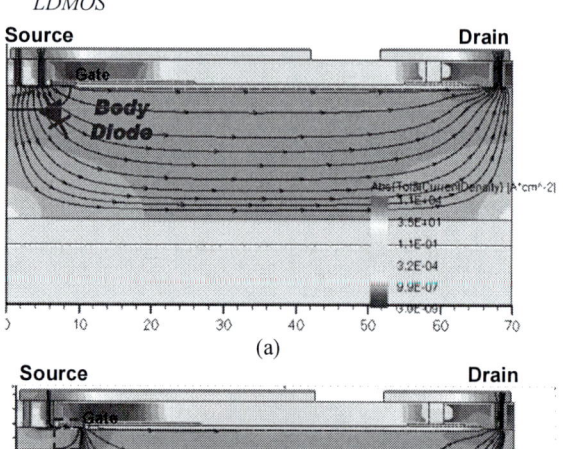

(a)

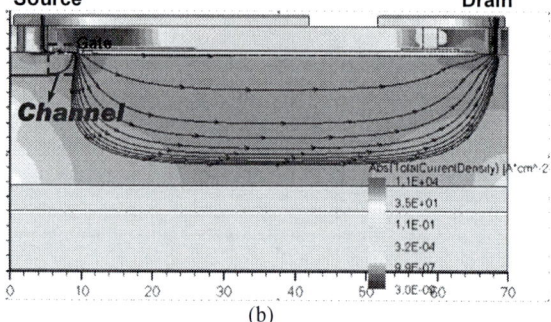

(b)

Fig 5 The simulation results of the freewheeling current path in the LDMOS.

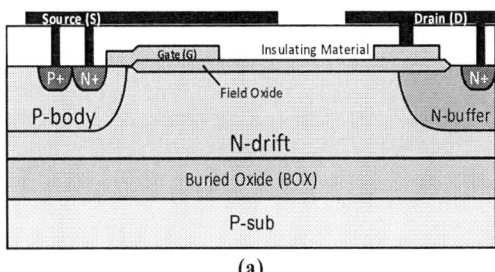

(a)

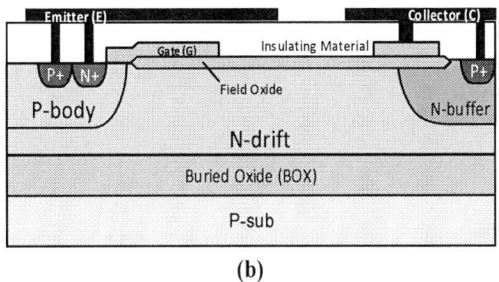

(b)

Fig 6 The cross-section view of the LDMOS(a) and LIGBT(b).

Fig.5 shows the simulation results of the freewheeling current path with TCAD tools. In (a) the freewheeling current flows through the body diode of the LDMOS when the LDMOS is closed, while in (b) it flows through the channel when the LDMOS is opened.

Fig.6 shows the cross-section view of the power devices. The LDMOS structure is the same as LIGBT except that the collector P+ region is replaced by N+ region. Therefore, the breakdown voltage of the LDMOS is almost the same the LIGBT.

III. MEASUREMENT RESULTS

A 600V HVIC adopting the proposed power stage circuit is implemented in the 600V SOI technology and the layout design is shown in Fig.7.

Fig 7 the layout design of the power stage circuit.

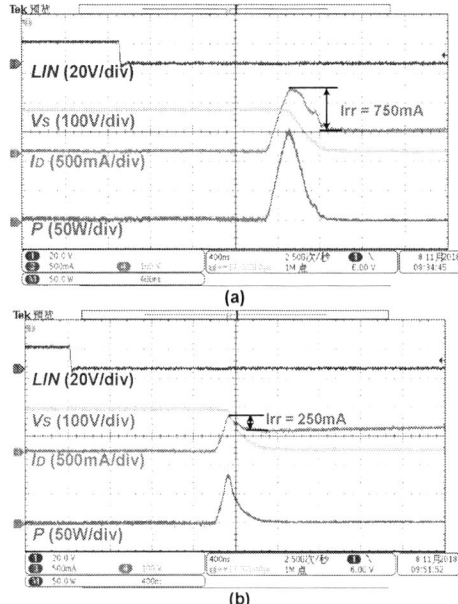

Fig.8 Test results of the conventional power stage (a) and the proposed power stage circuit (b)

The test results of the peak reverse recovery current can be measured in the Fig.8. It shows that the peak reverse recovery current dropped from 750mA to 250mA compared with the conventional normal circuit and the turn-on losses are decreased significantly.

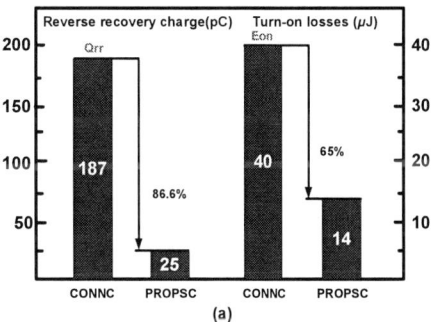

(a)

CONNC: Conventional Normal Circuit
PROPSC: Proposed Power Stage Circuit

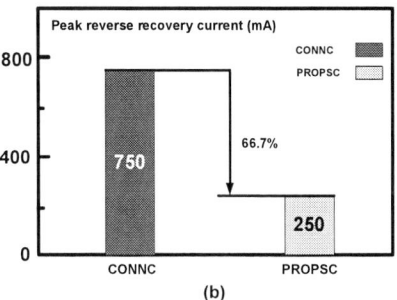

(b)

Fig.9 Proposed power stage circuit reverse recovery characteristic compared with the conventional circuit.

The reverse recovery characteristic is shown in Fig.9 and compared with that of conventional circuit. It can be seen from Fig.9(a) that the reverse recovery charge and the turn-on losses are decreased significantly when utilizing the proposed power stage. Fig.9(b) shows the declines of peak reverse recovery current when utilizing the proposed power stage.

IV. CONCLUSION

In this paper, we have presented a gate control circuit for the LIGBT instead of using FWD parallel with the LIGBT. In this way, the reverse recovery characteristics will improve a lot. The measured results show that the peak reverse recovery current (I_{rrm}) is decreased by about 66.7% and the charge (Q_{rr}) is shrunk by about 86.6%.

REFERENCES

[1]. K. Hara et al., "600V single chip inverter IC with new SOI technology," 2014 IEEE 26th International Symposium on Power Semiconductor Devices & IC's (ISPSD), pp. 418–421, June 2014.

[2]. J. Zhu et al., "Further Study of the U-Shaped Channel SOI-LIGBT With Enhanced Current Density for High-Voltage Monolithic ICs," IEEE Transactions on Electron Devices, vol. 63, no. 3, pp. 1161-1167, 2016

[3]. J.W. Kolar, F.C. Zach, F. Casanellas, "Losses in PWM inverters using IGBTs," IEE Proceedings-Electric Power Applications, 1994.

[4]. Forsythe, E. W., M. A. Abkowitz, and Yongli Gao. "Tuning the carrier injection efficiency for organic light-emitting diodes." The Journal of Physical Chemistry B 104.16 (2000): 3948-3952.

[5]. F. Bjoerk, J. Hancock, M. Treu, R. Rupp, T. Reimann, "2nd generation 600V SiC Schottky diodes use merged pn/Schottky structure for surge overload protection." Twenty-First Annual IEEE Applied Power Electronics Conference and Exposition, 2006. APEC'06.. IEEE, 2006.

Gap in pagination due to unavailable paper.

Pages 283-286

Proceedings of the 31st International Symposium on Power Semiconductor Devices & ICs
May 19-23, 2019, Shanghai, China

Threshold Voltage Instability Mechanisms in p-GaN Gate AlGaN/GaN HEMTs

Arno Stockman[*†], Eleonora Canato[‡], Matteo Meneghini[‡], Gaudenzio Meneghesso[‡], Peter Moens[*] and Benoit Bakeroot[†]

[*]ON Semiconductor, Oudenaarde, Belgium
[†]CMST, imec and Ghent University, Belgium
[‡]Dep. of Information Engineering, University of Padova, Italy
Arno.Stockman@onsemi.com; Arno.Stockman@ugent.be

Abstract—In this study, we propose a technique to evaluate the transient threshold voltage behavior of p-GaN capped AlGaN/GaN high-electron-mobility transistors (HEMTs). The threshold voltage is monitored from 10 μs to 100 s during positive gate bias stress. Technology computer-aided design (TCAD) simulations offer in-depth analysis of the different threshold voltage instability mechanisms: (i) electron trapping at the AlGaN/GaN interface, (ii) hole accumulation and trapping at the p-GaN/AlGaN interface and in the AlGaN barrier, respectively, and (iii) hole depletion of the p-GaN layer.

I. INTRODUCTION

Gallium nitride (GaN) high-electron-mobility transistors (HEMTs) show great promise in high power applications, as a result of their high breakdown field and the existence of a two-dimensional electron gas (2-DEG), yielding a high electron sheet density and mobility [1]. Additionally, they can be used at higher switching frequency compared to traditional Si power devices [2]. Normally-OFF operation is desired in many power applications due to fail-safe operation and for improving the slew rate with direct control of the gate of the AlGaN/GaN device [3]. Local p-type GaN capping of the barrier results in positive threshold voltages. It is shown that ON-state operation induces a threshold voltage shift [4], [5], which could result in a significant operation point drift, degrading the power efficiency. Threshold voltage instabilities in p-GaN gated AlGaN/GaN HEMTs have been reported in [6], [7], [8]. In this study, we present a technique to evaluate the threshold voltage under on-state gate stress. Additionally, the resulting threshold voltage transients are explained using technology computer-aided design (TCAD) simulations.

II. DEVICE AND STRESS PROCEDURE DESCRIPTION

The devices under test in this study are normally-off p-GaN capped AlGaN/GaN HEMTs ($W = 200\,\mu$m), on 6 inch GaN-on-Si wafers. The gate region is capped with a p-type doped GaN layer, on which a Schottky contact is formed. Initially, the enhancement mode device is characterized using a more standardized double pulse measurement setup [5]. A novel testing procedure is established to monitor the threshold voltage of the device during positive gate bias stress, schematically represented in Fig. 1. Before each measurement cycle, the device is kept at zero bias on all terminals for 100 s to promote de-trapping. During the 'stress' phase, a positive gate bias is

applied during a window of 10 μs to 100 s. The stressing phase is interrupted by a very short (10 μs) gate voltage sweep in order to monitor the threshold voltage (at $I_D = 10\,$mA/mm) as a function of time.

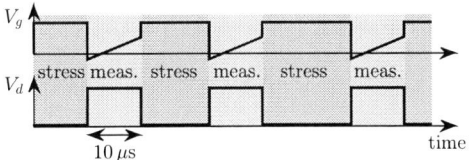

Fig. 1. Schematic representation of the gate and drain voltage during the threshold voltage transient measurement. The 'stress' and 'measurement' phase are shown in red and green, respectively.

III. RESULTS

The initial double pulse characterization is depicted in Fig. 2, using an ON-state stressing phase of $100\,\mu$s and measurement phase of $1\,\mu$s. Pulsed I_DV_G measurements at $V_D = 3\,$V in Fig. 2(a) show a positive threshold voltage shift for gate voltages up to $2\,$V, after which the shift becomes

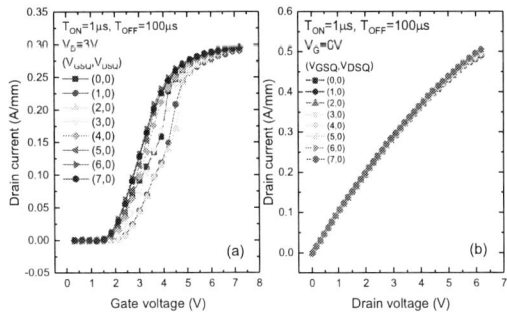

Fig. 2. Double pulse measurements showing the drain current versus (a) gate voltage at $V_D = 3\,$V and (b) drain voltage at $V_G = 6\,$V. No degradation in ON-resistance is observed, indicating that the trapping phenomena are concentrated under the gate region. The stress and measurement pulse last 100 and $1\,\mu$s, respectively.

978-1-7281-0582-6/19 $31.00 © 2019 IEEE

negative and saturates at $V_G = 5\,\mathrm{V}$. The pulsed $\mathrm{I_D V_D}$ characteristic at $V_G = 6\,\mathrm{V}$ is depicted in Fig. 2(b), showing no ON-resistance shift after ON-state stress. From this, we conclude that the trapping phenomena responsible for the threshold voltage shift are located solely in the gate region.

Under the measurement conditions depicted in Fig. 1, the threshold voltage can be monitored as a function of stress time. The experimental threshold voltage is plotted in Fig. 3(a) for gate stress voltages ranging from 1 to 7 V. Note that in order to study the physical mechanisms behind the threshold voltage transients, the former is normalized w.r.t. its value at $10^{-5}\,\mathrm{s}$, represented in Fig. 3(b). Three major threshold voltage variation mechanisms can be distinguished, and we propose the following phenomena at its origin:

(i) electron trapping at the AlGaN/GaN interface, yielding a positive ΔV_{th} shift at low gate bias [4-6];

(ii) hole accumulation at the p-GaN/AlGaN interface and subsequent hole trapping in the AlGaN barrier [7], yielding a negative ΔV_{th} shift which appears at lower stress times with increasing gate bias;

(iii) hole depletion by forward biasing of the p-i-n diode [7], leaving behind a net negatively charged p-GaN layer, resulting in a slow, positive ΔV_{th} shift at high gate bias.

Each mechanism will be investigated in more detail in Section IV. While subject to different stress timings, both the double pulse and the threshold voltage transient measurement show similar threshold voltage behavior, where the latter shifts positive at low gate bias, and becomes negative at higher gate bias.

The temperature dependence of the V_{th} variation transient at $V_G = 2\,\mathrm{V}$ and $5\,\mathrm{V}$ is investigated in Fig. 4(a)-(b), respectively, in which the threshold voltage variation is plotted between $30\,^\circ\mathrm{C}$ to $130\,^\circ\mathrm{C}$. At low gate bias, electron trapping causes a positive threshold voltage variation. Note that due to the

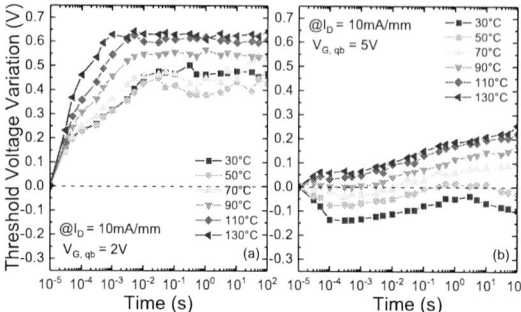

Fig. 4. Experimental threshold voltage variation versus stress time at different temperatures for (a) $V_G = 2\,\mathrm{V}$ and (b) $V_G = 5\,\mathrm{V}$.

presence of the 2-DEG at $V_G = 2\,\mathrm{V}$ the AlGaN/GaN interface states are filled with electrons. On the other hand, with increasing temperature, thermally excited 2-DEG electrons can be trapped in trap states within the AlGaN barrier, resulting in an increase of the total negative charge density in the AlGaN barrier and consequently the threshold voltage variation (see Section IV-A). At higher gate bias, a higher hole current density is injected across the Schottky barrier at the gate contact, resulting in higher rate of hole accumulation and trapping. Consequently, the fast negative slope shifts to time constants below $10^{-5}\,\mathrm{s}$, rendering them invisible on this plot.

IV. DISCUSSION

Fig. 5 depicts the threshold voltage variation extracted from TCAD simulations in the case of a low (Fig. 5(a)) and high (Fig. 5(b)) level of leakage through the p-GaN/AlGaN/GaN barrier. In case (a), similar trends compared to Fig. 3 are seen, i.e. (i) a fast positive slope due to electron trapping at the AlGaN/GaN interface, and (ii) a fast negative slope due to hole trapping in the AlGaN barrier. Both electron and hole trapping saturate when the available states are filled. Only when the p-

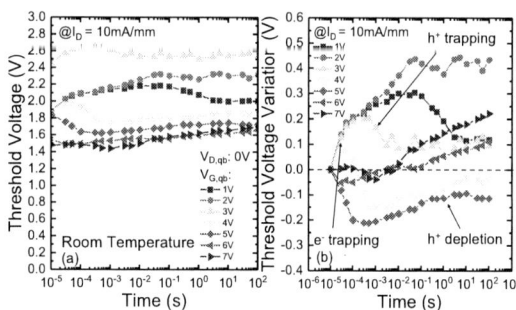

Fig. 3. Experimental (a) threshold voltage and (b) threshold voltage variation versus stress time for gate stress voltages from 1 to 7 V at room temperature. The threshold voltage is extracted at $I_D = 10\,\mathrm{mA/mm}$. Three distinct sets of slopes can be distinguished: (i) a fast positive slope due to electron trapping at the AlGaN/GaN interface, (ii) a fast negative slope due to hole trapping in the AlGaN barrier, and (iii) a slow positive slope due to hole depletion after the p-i-n diode has turned on. All effects show similar slopes, irrespective of gate bias.

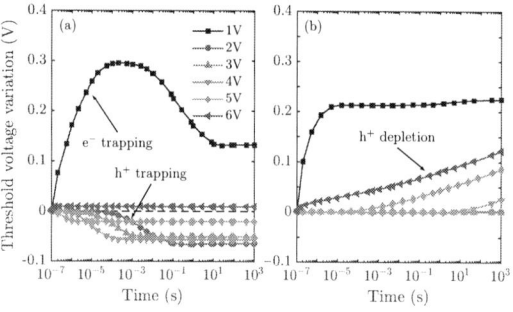

Fig. 5. Threshold voltage variation extracted from TCAD simulations at room temperature in the case of (a) low and (b) high p-i-n diode leakage level. Both cases are simulated with a highly conducting Schottky gate.

i-n leakage level is increased in the TCAD simulations (case b) a slow positive slope due to hole depletion (iii) is observed. In this case, holes recombine with electrons from the 2-DEG and the p-GaN layer is partially depleted of holes, resulting in a net negative charge and hence a positive threshold shift. Saturation of this effect occurs at a much later point in time. In reality, this effect can occur as a band-to-band recombination process within the AlGaN barrier, or recombination as a result of local low-field electron injection over the AlGaN barrier into the p-GaN layer, through dislocation lines or Al clusters in the former [9]. The threshold voltage variation is calculated during positive gate bias stress using the surface potential ψ_s according to [7], so as not to interfere with the stressing phase. The three threshold voltage shift mechanisms are investigated using TCAD simulations and are shown schematically on the energy band diagram of the gate stack in Fig. 9.

A. Electron Trapping

At positive gate stress voltages, electrons within the 2-DEG are injected in acceptor-like interface states at the AlGaN/GaN interface. The capture time constant depends on the density of available electrons, hence a function of the gate voltage. Fig. 6 shows the threshold voltage variation versus stress time at $V_G = 1\,\text{V}$ for three different values of interface traps at the AlGaN/GaN interface. With increasing trap density, more electrons are trapped at the interface during the positive gate stress, yielding a positive threshold voltage variation. This is reflected in the interface electron density shown on the right axis of Fig. 6. As soon as all available trap states are filled with electrons from the 2-DEG, the effect saturates. The apparent shift in time constant with increasing interface trap density is a result of the increased gate capacitance. With increasing temperature, electrons are trapped in the AlGaN barrier, causing a further positive increase of the threshold voltage. At $10^{-1}\,\text{s}$, however, the effect of hole accumulation

(see Section IV-B) can be observed, yielding a negative V_{th} shift. This is accompanied by an increase in the 2-DEG density and subsequent interface trapped electron density.

B. Hole Accumulation/Trapping

Fig. 7(a) shows the simulated V_{th} transients without acceptor traps at the AlGaN/GaN interface. In this case, the electron trapping as discussed in Section IV-A is absent. Fig. 7(b) shows the two-dimensional hole gas (2-DHG) density (located at the p-GaN/AlGaN interface) versus stress time for different gate stress voltages. The time constant in the V_{th} transient corresponds to an accumulation of holes in the 2-DHG, caused by the release of holes trapped in Mg acceptor states and resulting in a slight increase in the depletion region width. This accumulation of holes at the p-GaN/AlGaN interface, both temperature and gate voltage accelerated, yields an initial negative V_{th} shift. Additionally, some of these holes are injected into Mg acceptor states in the AlGaN barrier, which are ionized at equilibrium [10]. The range of hole trapping into Mg trap states, and hence the charge distribution in the

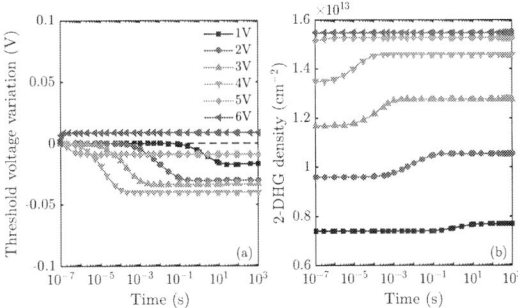

Fig. 7. (a) Simulated threshold voltage variation versus stress time without acceptor traps at the AlGaN/GaN interface. (b) Hole density in the 2-DHG versus stress time for different ON-state stress voltages.

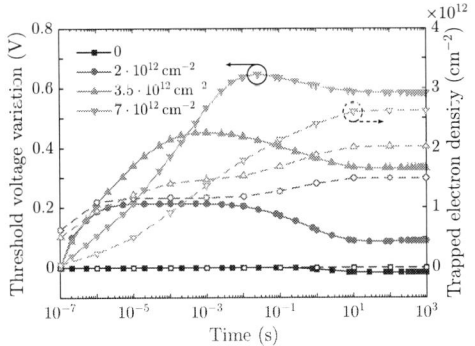

Fig. 6. Left axis: threshold voltage variation versus stress time at $V_G = 1\,\text{V}$ for increasing interface trap density at the AlGaN/GaN interface. These states are uniformly distributed between 0.05 and 0.65 eV below the conduction band. Right axis: interface trapped electron density at the AlGaN/GaN interface versus stress time.

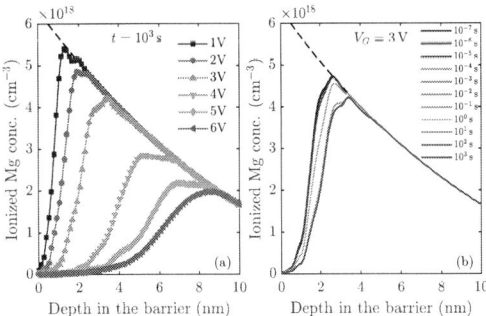

Fig. 8. Simulated ionized Mg charge density versus depth in the barrier, relative to the p-GaN/AlGaN interface, (a) for different gate bias at $t = 10^3$ s, and (b) for different stress time at $V_G = 3\,\text{V}$. The Mg out-diffusion tail is represented in black dashed line.

978-1-7281-0582-6/19 $31.00 © 2019 IEEE

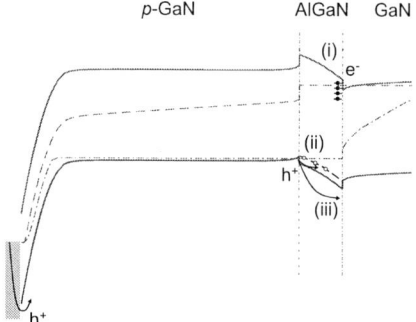

Fig. 9. Energy band diagram of the Schottky metal/p-GaN/AlGaN/GaN gate stack at $V_G = 6\,\text{V}$ with three different threshold voltage variation mechanisms.

barrier is both gate bias and stress time dependent, as can be seen in Fig. 8. With increasing gate bias and stress time, more holes are injected from the Schottky contact through thermionic field emission [11] and consequently over the AlGaN barrier towards the channel. As a result, negatively charged Mg acceptor states in the barrier are neutralized by hole trapping, reducing the total equilibrium charge density and yielding a negative threshold voltage shift, according to [10]. At a fixed gate bias, the trap occupation is time dependent and saturates after a certain stress time.

C. Hole Depletion

At high gate bias, the p-GaN potential is such that the p-i-n diode is forward biased. Holes are emitted from the p-GaN layer into the GaN channel, leaving behind a net negatively charged p-GaN layer. In contrast, electrons injected from the channel into the p-GaN quickly recombine with the high density of available holes in the 2-DHG. In TCAD, this can be simulated by increasing the leakage through the p-i-n diode. The resulting threshold voltage transients are shown in Fig. 5(b). By introducing a trap-assisted-tunneling path in the AlGaN barrier, the 2-DHG accumulation supplied by the Schottky contact or extension of the depletion region is inhibited. As such, the hole accumulation and subsequent trapping discussed in Section IV-B is absent and a positive threshold voltage shift is observed. Note that the slope of the V_{th} transients at high gate bias are constant.

Comparing the TCAD predicted V_{th} transients presented in Fig. 5 to the experimental results in Fig. 3(b), we can comment that in order to obtain all three mechanisms at once, a delicate balance in leakage levels between the Schottky and p-i-n diode is necessary. In reality, this leakage can occur in local 1-D spots, whereas the 2-D simulator assumes an ideal gate stack across the complete width of the device.

V. Conclusion

In this study, we propose a measurement procedure to investigate the threshold voltage during positive gate bias stress in

p-GaN capped HEMTs from $10\,\mu\text{s}$ to $100\,\text{s}$. TCAD simulations provide unique insight in the different mechanisms causing the threshold voltage instability. (i) At all gate positive biases, electron trapping occurs in interface states at the AlGaN/GaN heterojunction interface. (ii) At medium gate voltages, hole injection occurs at the Schottky junction which accumulate at the p-GaN/AlGaN interface, resulting in an enhancement of the 2-DHG. Subsequently, hole trapping occurs in Mg acceptor states located at around $150\,\text{meV}$ above the valence band in the AlGaN barrier. (iii) At high gate bias, the p-GaN potential is high enough to forward bias the p-i-n diode, resulting in injection of holes from the p-GaN layer into the GaN channel. Similarly, electron injection occurs in the opposite direction, and these electrons quickly recombine with the high concentration of holes in the 2-DHG. Both effects result in depletion of holes in the p-GaN layer, yielding a positive threshold voltage shift.

References

[1] N. Ikeda *et al.*, "GaN power transistors on Si substrates for switching applications," *Proc. IEEE*, vol. 98, no. 7, pp. 1151–1161, July 2010.

[2] M. Kuzuhara and H. Tokuda, "Low-loss and high-voltage III-nitride transistors for power switching applications," *IEEE Transactions on Electron Devices*, vol. 62, no. 2, pp. 405–413, February 2015.

[3] K. J. Chen, O. Häberlen, A. Lidow, C. L. Tsai, T. Ueda, Y. Uemoto, and Y. Wu, "GaN-on-Si power technology: devices and applications," *IEEE Transactions on Electron Devices*, vol. 64, no. 3, pp. 779–795, March 2017.

[4] A. Stockman, E. Canato, M. Meneghini, M. Meneghesso, E. Zanoni, P. Moens, and B. Bakeroot, "On the origin of the leakage current of p-gate AlGaN/GaN HEMTs," *IEEE International Reliability Physics Symposium*, pp. 4B.5–1 – 4B.5–4, 2018.

[5] A. Tajalli *et al.*, "Impact of sidewall etching on the dynamic performance of GaN-on-Si E-mode transistors," *Proc. 29th European Symposium on Reliability of electron Devices, Failuse Physics and Analysis*, vol. 88-90, pp. 572–576, 2018.

[6] X. Tang *et al*, "Mechanism of threshold voltage shift in p-GaN gate AlGaN/GaN transistors," *IEEE Electron Devices Letters*, vol. 39, no. 8, pp. 1145–1149, August 2018.

[7] L. Sayadi *et al.*, "Threshold voltage instability in p-GaN gate Al-GaN/GaN HFETs," *IEEE Transactions on Electron Devices*, vol. 65, no. 6, pp. 2454–2460, June 2018.

[8] Y. Shi *et al.*, "Carrier transport mechanisms underlying the bidirectional V_{TH} shift in p-GaN gate HEMTs under forward gate stress," *IEEE Transactions on Electron Devices*, vol. 66, no. 2, pp. 876–882, February 2019.

[9] S. Stoffels *et al.*, "Failure mode for p-GaN gates under forward stress with varying Mg concentration," *Proc. IEEE Internation Reliabilty Phsyics Symposium*, pp. 4B–4, April 2017.

[10] B. Bakeroot, A. Stockman, N. Posthuma, S. Stoffels, and S. Decoutere, "Analytical model for the threshold voltage of p-(Al)GaN high-electron-mobility transistors," *IEEE Transactions On Electron Devices*, vol. 65, pp. 79–86, 2018.

[11] A. Stockman *et al.*, "Gate conduction mechanisms and lifetime modeling of p-gate AlGaN/GaN high-electron-mobility transistors," *IEEE Transactions on Electron Devices*, vol. 65, no. 12, pp. 5365–5372, November 2018.

Proceedings of the 31st International Symposium on Power Semiconductor Devices & ICs
May 19-23, 2019, Shanghai, China

Dynamic Threshold Voltage in p-GaN Gate HEMT

Jin Wei, Han Xu, Ruiliang Xie, Meng Zhang [2], Hanxing Wang, Yuru Wang, Kailun Zhong, Mengyuan Hua,
Jiabei He, and Kevin J. Chen

Department of Electronic and Computer Engineering, The Hong Kong University of Science and Technology, Hong Kong
[2] Key Laboratory of Optoelectronic Devices and Systems of Ministry of Education and Guangdong Province, College of Physics
and Optoelectronic Engineering, Shenzhen University, Shenzhen, China
Phone: +852-23588969, Fax: +852-23581485, Email: jweiaf@connect.ust.hk; eekjchen@ust.hk

Abstract—The p-GaN gate HEMT with a Schottky gate contact is studied in this work. The threshold voltage (V_{th}) of the device is found to have a dynamic nature. When the device experiences a high drain voltage V_{DSQ}, the gate-to-drain capacitance C_{GD} is charged to $Q_{GD}(V_{DSQ})$. The negative part of $Q_{GD}(V_{DSQ})$ is mainly located in the p-GaN layer. When drain voltage drops to a lower value V_{DSM}, the non-equilibrium charges $\Delta Q_{GD} = Q_{GD}(V_{GDQ}) - Q_{GD}(V_{GDM})$ cannot be effectively released, since the discharging current is blocked by the reversely biased metal/p-GaN Schottky junction and p-GaN/2DEG PN junction. An extra V_{GS} is required to counteract the non-equilibrium ΔQ_{GD}, resulting in a change of V_{th}. The change in V_{th} is well predictable by its linear relationship to ΔQ_{GD}. During switching operation, V_{th} is adaptive along the load line. Without considering the dynamic V_{th} effect, there exist large discrepancies in switching transients between the modelled and the experimental results. Incorporation of the dynamic V_{th} in device model result in greatly enhanced accuracy in simulated transient behavior.

Keywords—p-GaN gate HEMT; dynamic threshold voltage; charge storage; device modeling

I. INTRODUCTION

GaN power devices are expected to play an important role in the next-generation power switching applications [1-4]. A critical issue that had hindered the GaN power devices from being accepted by the market was their inherent normally-on operation mode, while normally-off operation is virtually a must-have feature for a successful power device apart from some niche applications [5-8]. The p-GaN gate structure has been widely investigated to realize normally-off operation in GaN devices. At present, normally-off p-GaN gate HEMTs are commercially available [9-11]. A representative p-GaN gate HEMT features a Schottky contact between the gate metal and p-GaN layer; thereby, reduced gate leakage current and enlarged gate swing are obtained [12, 13].

The gate stress induced threshold voltage (V_{th}) instabilities have been widely reported for p-GaN gate HEMTs with Schottky gate contact, and has been explained by hole deficiency or electron trapping [14-17]. The high drain voltage stress has been found to cast a more significant impact on the V_{th} shift in the p-GaN gate HEMTs, as suggested to be caused by hole emission from the p-GaN layer [17]. However, this phenomenon is less reported in literature, and a quantitative

This work is supported by Hong Kong Innovation Technology Fund under ITS/412/17FP.

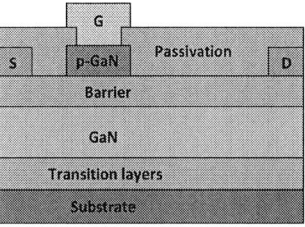

Fig. 1. Schematic structure of a p-GaN gate HEMT. The device in this study features a Schottky gate contact to p-GaN layer unless otherwise specified.

explanation is lacking.

In this work, the p-GaN gate HEMT featuring a Schottky gate contact is found to have an intrinsically *dynamic* threshold voltage (V_{th}), with the V_{th} being a function of the pre-experienced or quiescent drain voltage stress (V_{DSQ}) and the drain voltage where the V_{th} is measured (V_{DSM}). The principle of the dynamic V_{th} is explained with a charge storage model. When the device has experienced a high drain voltage stress (V_{DSQ}), the gate-to-drain capacitance C_{GD} is charged to $Q_{GD}(V_{DSQ})$. As drain voltage drops to V_{DSM} where V_{th} is measured, the non-equilibrium charges $\Delta Q_{GD} = Q_{GD}(V_{GDQ}) - Q_{GD}(V_{GDM})$ in the p-GaN layer cannot be effectively removed, since the reversely biased gate metal/p-GaN Schottky junction and p-GaN/2DEG PN junction block the discharging current. To render the device to on-state, additional gate voltage is required, resulting in a change in the V_{th}. As revealed from experiment, the dynamic V_{th} is linearly correlated to the stored ΔQ_{GD}. Unlike the V_{th} instabilities induced by deep-level trapping [18, 19], the dynamic V_{th} in the p-GaN gate HEMT is an intrinsic property of the device with a *floating* p-GaN layer, and it is well predictable by its dependence on the experienced drain voltage stress and the present drain voltage.

The dynamic V_{th} of the p-GaN gate HEMT indicates that V_{th} is adaptive during a switching transient. This work further demonstrates the influences of the dynamic V_{th} upon the switching transients in a phase-leg circuit. Ignorance of the dynamic V_{th} effect during circuit design would result in inaccurate prediction of the switching characteristics, while incorporation of the dynamic V_{th} effect cures the inaccuracy.

II. CHARACTERISTICS OF THE P-GAN GATE HEMT

Fig. 1 presents a schematic cross-sectional structure of a p-GaN gate HEMT with a Schottky gate contact. The device

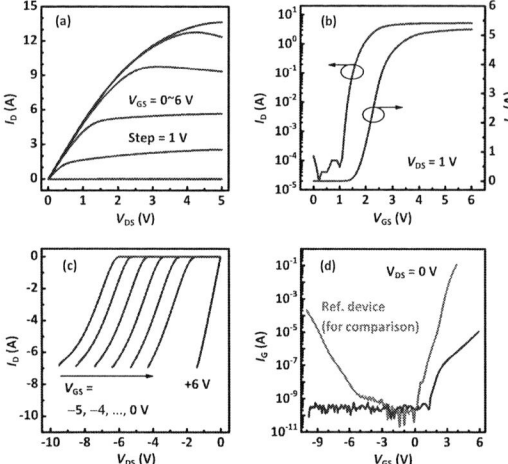

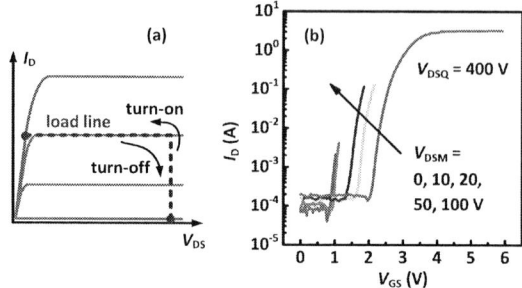

Fig. 4. (a) Illustration of a typical inductive switching load line. (b) Pulse *I-V* characteristics with various V_{DSM}. Maximum measurable currents are limited by equipment.

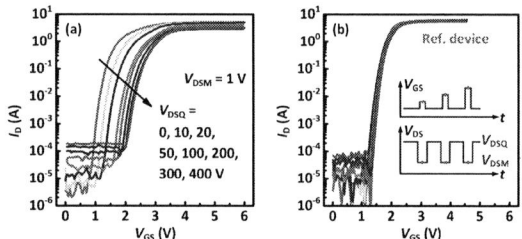

Fig. 2. *I-V* characteristics of the *p*-GaN gate HEMT. (a) Output curves. (b) Transfer curves. (c) Reverse conduction curves. (d) I_{G}-V_{G} curve. I_{G} of the studied *p*-GaN gate HEMT is much smaller than a reference *p*-GaN gate HEMT that features an Ohmic gate contact to *p*-GaN layer.

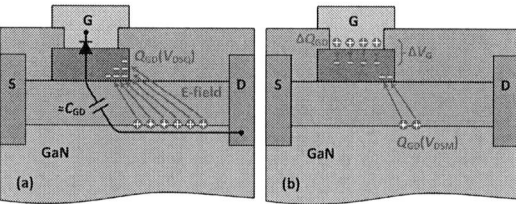

Fig. 5. Mechanism of the dynamic V_{th}. (a) At V_{DSQ} stress, C_{GD} is charged to $Q_{\text{GD}}(V_{\text{DSQ}})$. (b) When V_{DS} drops to V_{DSM} from V_{DSQ}, $\Delta Q_{\text{GD}} = Q_{\text{GD}}(V_{\text{DSQ}}) - Q_{\text{GD}}(V_{\text{DSM}})$ is expected to be discharged, but blocked by the Schottky contact. ΔV_{G} is required to counteract this ΔQ_{GD}, resulting in V_{th} shift.

$V_{\text{DSM}} = 1$ V. As V_{DSQ} increases, V_{th} shifts more positively, and the dependence of V_{th} upon V_{DSQ} becomes weakened for $V_{\text{DSQ}} > 50$ V. As a comparison, the reference *p*-GaN gate HEMT with an Ohmic gate contact exhibits much weaker V_{th} dependence upon V_{DSQ}, as presented in Fig. 3(b).

During a switching transient, V_{DS} of a power transistor sweeps between the high drain voltage in OFF-state bias and the low drain voltage in ON-state, as illustrated in Fig. 4(a). Therefore, V_{th} at different drain voltage is of importance to the switching characteristics. To study this effect, V_{th} of the *p*-GaN gate HEMT is measured at different drain voltage (V_{DSM}) with a fixed $V_{\text{DSQ}} = 400$ V, as plotted in Fig. 4(b). The V_{th} shift is found to be largest when measured at a small V_{DSM} (e.g. 1 V) and becomes less pronounced when $V_{\text{DSM}} > 50$ V.

III. MECHNISM AN AND INFLUENCE OF DYNAMIC V_{TH}

As revealed above, the dynamic V_{th} is dependent on both the experienced high V_{DSQ} stress and the momentary V_{DSM} where V_{th} is measured. The phenomenon is caused by the floating nature of the *p*-GaN layer. Dynamic instabilities caused by floating *p*-regions have also been studied in SiC devices [23, 24]. In the *p*-GaN gate HEMT, when a high V_{DSQ} is applied, C_{GD} of the device is charged to $Q_{\text{GD}}(V_{\text{DSQ}})$, as illustrated in Fig. 5(a). The negative part of $Q_{\text{GD}}(V_{\text{DSQ}})$ is mainly located in the *p*-GaN layer, as the charging current flows across the forwardly biased gate metal/*p*-GaN Schottky junction. The positive part of $Q_{\text{GD}}(V_{\text{DSQ}})$ is located in the depleted access region [25-27]. As illustrated in Fig. 5(b), with drain voltage dropping to V_{DSM}, C_{GD} is expected to be discharged to $Q_{\text{GD}}(V_{\text{DSM}})$. However, the discharging current is

Fig. 3. Pulse transfer *I-V* characteristics. (a) The studied *p*-GaN gate HEMT with a Schottky gate contact. (b) The reference *p*-GaN gate HEMT with an Ohmic gate contact.

studied in this work is a commercially available 650-V/200-mΩ *p*-GaN gate HEMT. The output and transfer characteristics of the *p*-GaN gate HEMT are plotted in Fig. 2(a) and Fig. 2(b). The reverse conduction characteristics of the device are plotted in Fig. 2(c). Due to the lack of body diode, the reverse conduction of the *p*-GaN gate HEMT is through the gated channel [20, 21]; thus, the reverse conduction characteristics are heavily affected by the dynamic threshold voltage studied in this work. Fig. 2(d) shows the gate leakage current of the studied *p*-GaN gate HEMT. The Schottky gate contact helps reduce the forward gate leakage current and allows a large forward gate bias. As a comparison, the gate current of a reference *p*-GaN gate HEMT with an Ohmic gate contact is also plotted, which is much larger than that with Schottky gate contact [22].

A pulse *I-V* characterization is adopted to study the dynamic V_{th} of the *p*-GaN gate HEMT. The inset in Fig. 3(b) illustrates the dynamic biasing waveforms in the pulse *I-V* characterization. By setting the quiescent drain bias (V_{DSQ}) to different levels, the impact of drain stress upon V_{th} is studied, as plotted in Fig. 3(a). The measurement is performed with

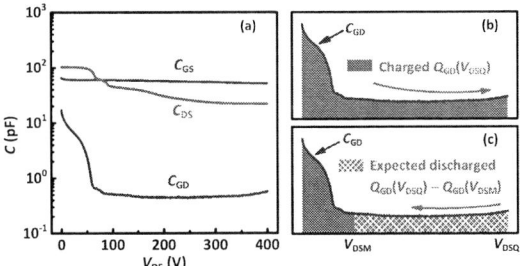

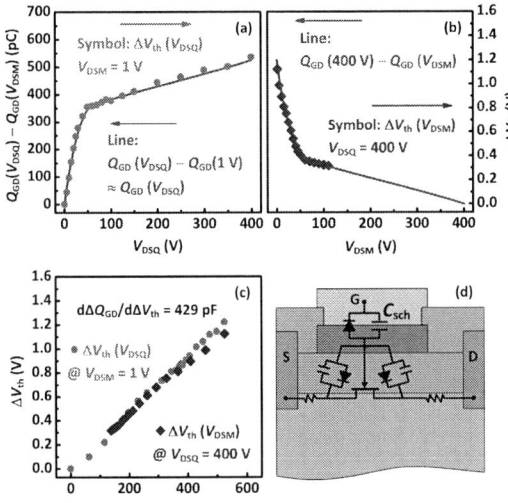

Fig. 6. (a) C-V characteristics of the p-GaN gate HEMT. (b) Illustration of $Q_{GD}(V_{DSQ})$, i.e. by integrating C_{GD} from 0 V to V_{DSQ}. (c) Illustration of $\Delta Q_{GD} = Q_{GD}(V_{DSQ}) - Q_{GD}(V_{DSM})$, i.e. by integrating C_{GD} from V_{DSM} to V_{DSQ}.

Fig. 7. (a) $\Delta Q_{GD} = Q_{GD}(V_{DSQ}) - Q_{GD}(V_{DSM})$ and ΔV_{th} as functions of V_{DSQ} with a fixed $V_{DSM} = 1$ V. (b) ΔQ_{GD} and ΔV_{th} shift as functions of V_{DSM} with a fixed $V_{DSQ} = 400$ V. (c) Correlation of ΔV_{th} and ΔQ_{GD}. (d) Illustration of the equivalent circuit of the p-GaN gate HMET.

blocked by the reversely biased metal/p-GaN Schottky junction and the reversely biased p-GaN/2DEG PN junction. Thus, an additional V_{GS} is required to counteract the non-equilibrium charges $\Delta Q_{GD} = Q_{GD}(V_{GDQ}) - Q_{GD}(V_{GDM})$, as manifested by a V_{th} shift.

To confirm the mechanism of the dynamic V_{th}, C-V characteristics of the p-GaN gate HEMT are measured, as plotted in Fig. 6(a). When the drain voltage sweeps from zero to V_{DSQ}, C_{GD} is charged accordingly, and $Q_{GD}(V_{DSQ})$ could be obtained by integrating C_{GD} from 0 to V_{DSQ}, as illustrated in Fig. 6(b). When the drain voltage drops to V_{DSM}, the non-equilibrium charge $\Delta Q_{GD} = Q_{GD}(V_{GDQ}) - Q_{GD}(V_{GDM})$ can be obtained by integrating C_{GD} from V_{DSM} to V_{DSQ}, as illustrated in Fig. 6(c). The $Q_{GD}(V_{DSQ})$-V_{DSQ} curve is plotted in Fig. 7(a), as is the ΔV_{th}-V_{DSQ} relationship. Here, ΔV_{th} is defined as the difference between dynamic V_{th} and static V_{th} measured at the same V_{DSM}. V_{DSM} is fixed at 1 V. From the results, the V_{DSQ} induced $Q_{GD}(V_{DSQ})$ dovetails the V_{DSQ}-induced ΔV_{th}. In

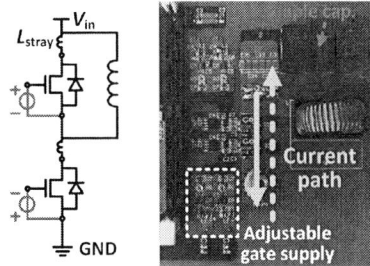

Fig. 8. A phase-leg circuit for testing the switching characteristics of the p-GaN gate HEMT.

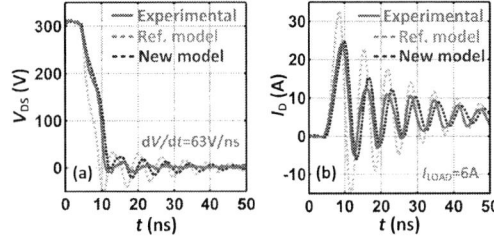

Fig. 9. (a) V_{DS} waveform during turn-on transient. (b) I_D waveform during turn-on transient.

Fig. 7(b), $Q_{GD}(400) - Q_{GD}(V_{DSM})$ and ΔV_{th} are plotted as functions of V_{DSM}. In this case, V_{DSQ} is fixed at 400 V. The change of dynamic V_{th} agrees with $Q_{GD}(400) - Q_{GD}(V_{DSM})$. By summarizing the data from Fig. 7(a) and Fig. 7(b), the relationship between ΔV_{th} and $\Delta Q_{GD} = Q_{GD}(V_{GDQ}) - Q_{GD}(V_{GDM})$ is plotted in Fig. 7(c). A linear correlation is obtained, with $d\Delta Q_{GD}/d\Delta V_{th} = 429$ pF. According to the mechanism explained in Fig. 5, ΔV_{th} is the additional gate voltage required to counteract ΔQ_{GD}. Thus, $d\Delta Q_{GD}/d\Delta V_{th}$ represents the Schottky capacitance (C_{sch}) at $V_{GS} = \sim V_{th}$, as illustrated in the equivalent circuit in Fig. 7(d), under the assumption that the direct capacitance coupling between gate metal and drain is negligible.

The dynamic V_{th} in the p-GaN gate HEMT indicates that the device features an adaptive V_{th} during the switching process as C_{GD} is charged/discharged, which is expected to cast significant influence upon switching transient waveforms. A phase-leg circuit is built to measure the switching transient of the p-GaN gate HEMT, as shown in Fig. 8. The switching waveforms are simulated using a behavioral model [28]. Fig. 9(a) and Fig. 9(b) present the voltage and current waveforms of the p-GaN gate HEMT during turn-on waveforms. Using the reference model without considering the dynamic V_{th} effect, the simulated switching transients exhibit an appreciable discrepancy to the experimental waveforms. By adopting a new model with the dynamic V_{th} effect incorporated, a much improved accuracy is obtained.

IV. CONCLUSION

The V_{th} of p-GaN gate HEMT with a Schottky gate contact is found to be dynamic in nature. The dynamic V_{th} is correlated to the experienced high drain voltage V_{DSQ} and the momentary drain bias V_{DSM} where V_{th} is defined. A closer look of the internal dynamics of the p-GaN gate reveals that the non-equilibrium charges $\Delta Q_{GD} = Q_{GD}(V_{GDQ}) - Q_{GD}(V_{GDM})$ is the cause for the dynamic V_{th}, as evidenced by their linear dependence. Thereby, the dynamic V_{th} is well predictable. Ignorance of the dynamic V_{th} effect in the device model leads to discrepancies between the simulated switching waveforms and the experimental ones, while incorporation of this effect cures this discrepancy. Thus, the dynamic threshold voltage is an inherent device characteristic for the p-GaN gate HEMT with Schottky gate contact, and should be considered in the circuit design.

REFERENCES

[1] K. J. Chen, O. Häberlen, A. Lidow, C. L. Tsai, T. Ueda, Y. Uemoto, and Y. Wu, "GaN-on-Si power technology: devices and applications," *IEEE Trans. Electron Devices,* vol. 64, no. 3, pp. 779-795, Mar. 2017.

[2] Y. Wu, M. Jacob-Mitos, M. L. Moore, and S. Heikman, "A 97.8% efficient GaN HEMT boost converter with 300-W output power at 1 MHz," *IEEE Electron Device Lett.,* vol. 29, no. 8, pp. 824-826, Aug. 2008.

[3] J. Wei, H. Jiang, Q. Jiang, and K. J. Chen, "Proposal of a GaN/SiC hybrid field-effect transistor for power switching applications," *IEEE Trans. Electron Devices,* vol. 63, no. 6, pp. 2469-2473, Jun. 2016.

[4] R. Rupp, T. Laska, O. Häberlen, and M. Treu, "Application specific trade-offs for WBG SiC, GaN and high end Si power switch technologies," in *IEDM Tech. Dig.,* San Francisco, CA, USA, Dec. 2014, pp. 28-31.

[5] X. Hu, G. Simin, J. Yang, M. A. Khan, R. Gaska, and M. S. Shur, "Enhancement mode AlGaN/GaN HFET with selectively grown pn junction gate," *Electronics Lett.,* vol. 36, no. 8, pp. 753-754, Apr. 2000.

[6] Y. Cai, Y. Zhou, K. J. Chen, and K. M. Lau, "High-performance enhancement-mode AlGaN/GaN HEMTs using fluoride-based plasma treatment," *IEEE Electron Device Lett.,* vol. 26, no. 7, pp. 435-437, Jul. 2005.

[7] J. Wei, S. Liu, B. Li, X. Tang, Y. Lu, C. Liu, M. Hua, Z. Zhang, G. Tang, and K. J. Chen, "Low on-resistance normally-off GaN double-channel metal-oxide-semiconductor high-electron-mobility transistor," *IEEE Electron Device Lett.,* vol. 36, no. 12, pp. 1287-1290, Dec. 2015.

[8] T. Kikkawa, *et al.,* "600 V JEDEC-qualified highly reliable GaN HEMTs on Si substrates," in *IEDM Tech. Dig.,* Dec. 2014, pp. 40-43.

[9] I. Hwang, H. Choi, J. Lee, H. S. Choi, J. Kim, J. Ha, C. Um, S. Hwang, J. Oh, J. Kim, J. K. Shin, Y. Park, U. Chung, I. Yoo, and K. Kim, "1.6kV, 2.9 mΩ cm² normally-off p-GaN HEMT device," in *Proc. ISPSD,* Bruges, Belgium, Jun. 2012, pp. 41-44.

[10] Y. Uemoto, M. Hikita, H. Ueno, H. Matsuo, H. Ishida, M. Yanagihara, T. Ueda, T. Tanaka, and D. Ueda, "A normally-off AlGaN/GaN transistor with R$_{on}$A=2.6mΩcm² and BV$_{ds}$=640V using conductivity modulation," in *IEDM Tech. Dig.,* San Francisco, Dec. 2006, pp. 907-910.

[11] K. Y. R. Wong, *et al.,* "A next generation CMOS-compatible GaN-on-Si transistors for high efficiency energy systems," in *IEDM Tech. Dig.,* Washington, DC, USA, Dec. 2015, pp. 229-232.

[12] G. Lukens, H. Hahn, H. Kalisch, and A. Vescan, "Self-aligned process for selectively etched p-GaN-gated AlGaN/GaN-on-Si HFETs," *IEEE Trans. Electron Devices,* vol. 65, no. 9, pp. 3732-3738, Sep. 2018.

[13] T. Wu, D. Marcon, S. You, N. Posthuma, B. Bakeroot, S. Stoffels, M. Van Hove, and G. Groeseneken, "Forward bias gate breakdown mechanism in enhancement-mode p-GaN gate AlGaN/GaN high-electron mobility transistors," *IEEE Electron Device Lett.,* vol. 36, no. 10, pp. 1001-1003, Oct. 2015.

[14] L. Sayadi, G. Iannaccone, S. Sicre, O. Haberlen, and G. Curatola, "Threshold voltage instability in p-GaN gate AlGaN/GaN HFETs," *IEEE Trans. Electron Devices,* vol. 65, no. 6, pp. 2454-2460, Jun. 2018.

[15] X. Tang, B. Li, H. A. Moghadam, P. Tanner, J. Han, and S. Dimitrijev, "Mechanism of threshold voltage shift in p-GaN gate AlGaN/GaN transistors," *IEEE Electron Device Lett.,* vol. 39, no. 8, pp. 1145-1148, Aug. 2018.

[16] J. He, G. Tang and K. J. Chen, "V_{TH} instability of p-GaN gate HEMTs under static and dynamic gate stress," *IEEE Electron Device Lett.,* vol. 39, no. 10, pp. 1576-1579, Oct. 2018.

[17] H. Wang, J. Wei, R. Xie, C. Liu, G. Tang, and K. J. Chen, "Maximizing the performance of 650-V p-GaN gate HEMTs: dynamic R_{ON} characterization and circuit design considerations," *IEEE Trans. Power Electron.,* vol. 32, no. 7, pp. 5539-5549, Jul. 2017.

[18] S. Yang, Z. Tang, K. Wong, Y. Lin, Y. Lu, S. Huang, and K. J. Chen, "Mapping of interface traps in high-performance Al$_2$O$_3$/AlGaN/GaN MIS-heterostructures using frequency- and temperature-dependent C-V techniques," in *IEDM Tech. Dig.,* Washington, DC, USA, Dec. 2013, pp. 152-155.

[19] S. Huang, S. Yang, J. Roberts, and K. J. Chen, "Threshold voltage instability in Al$_2$O$_3$/GaN/AlGaN/GaN metal-insulator-semiconductor high-electron mobility transistors," *Jpn. J. Appl. Phys.,* vol. 50, no. 11R, p. 110202, Nov. 2011.

[20] M. Zhang, J. Wei, X. Zhou, H. Jiang, B. Li, and K. J. Chen, "Simulation study of a power MOSFET with built-in channel diode for enhanced reverse recovery performance," *IEEE Electron Device Lett.,* vol. 40, no. 1, pp. 79-82, Jan. 2019.

[21] A. Lidow, J. Strydom, M. de Rooij, and D. Reusch, *GaN Transistors for Efficient Power Conversion,* 2nd ed. New York, NY, USA: Wiley, 2014.

[22] H. Okita, M. Hikita, A. Nishio, T. Sato, K. Matsunaga, H. Matsuo, M. Tsuda, M. Mannoh, S. Kaneko, M. Kuroda, M. Yanagihara, A. Ikoshi, T. Morita, K. Tanaka, and Y. Uemoto, "Through recess and regrowth gate technology for realizing process stability of GaN-based gate injection transistors," *IEEE Trans. Electron Devices,* vol. 64, no. 3, pp. 1026-1031, Mar. 2017.

[23] J. Wei, M. Zhang, H. Jiang, H. Wang, and K. J. Chen, "Dynamic degradation in SiC trench MOSFET with a floating p-shield revealed with numerical simulations," *IEEE Trans. Electron Devices,* vol. 64, no. 6, pp. 2592-2598, Jun. 2017.

[24] Y. Kagawa, N. Fujiwara, K. Sugawara, R. Tanaka, Y. Fukui, Y. Yamamoto, N. Miura, M. Imaizumi, S. Nakata, and S. Yamakawa, "4H-SiC trench MOSFET with bottom oxide protection," *Mat. Sci. Forum,* vol. 778-780, pp. 919-922, Feb. 2014.

[25] J. Si, J. Wei, W. Chen, and B. Zhang, "Electric field distribution around drain-side gate edge in AlGaN/GaN HEMTs: analytical approach," *IEEE Trans. Electron Devices,* vol. 60, no. 10, pp. 3223-3229, Oct. 2013.

[26] J. Wei, M. Zhang, B. Li, X. Tang, and K. J. Chen, "An analytical investigation on the charge distribution and gate control in the normally-off GaN double-channel MOS-HEMT," *IEEE Trans. Electron Devices,* vol. 65, no. 7, pp. 2757-2764, Jul. 2018.

[27] N. Q. Zhang, B. Moran, S. P. DenBaars, U. K. Mishra, X. W. Wang, and T. P. Ma, "Effects of surface traps on breakdown voltage and switching speed of GaN power switching HEMTs," in *IEDM Tech. Dig.,* Washington, DC, USA, Dec. 2001, pp. 589-592.

[28] R. Xie, X. Yang, G. Xu, J. Wei, Y. Wang, H. Wang, M. Tian, F. Zhang, W. Chen, L. Wang, and K. J. Chen, "Switching transient analysis for normally-off GaN transistors with p-GaN gate in a phase-leg circuit," *IEEE Trans. Power Electron.,* vol. 34, no. 4, pp. 3711-3728, Apr. 2019.

978-1-7281-0582-6/19 $31.00 © 2019 IEEE

Proceedings of the 31st International Symposium on Power Semiconductor Devices & ICs
May 19-23, 2019, Shanghai, China

Temperature-Dependent Gate Degradation of p-GaN Gate HEMTs under Static and Dynamic Positive Gate Stress

Jiabei He, Jin Wei, Song Yang, Mengyuan Hua, Kaikun Zhong, and Kevin J. Chen

Department of Electronic and Computer Engineering, The Hong Kong University of Science and Technology

Hong Kong SAR, China

Email: jhear@connect.ust.hk, eekjchen@ust.hk

Abstract—This paper experimentally investigates the time-dependent gate degradation of Schottky-type p-GaN gate transistors subjected to positive gate voltage stress. By means of combined static/dynamic gate stress and temperature-dependent analysis, the dependence of time-to-breakdown (t_{BD}) on stress mode and temperature are unveiled. It is demonstrated that t_{BD} is Weibull distributed and the mean-time-to-failure (MTTF) is comparable under static and dynamic stress conditions. Both the gate breakdown voltage and MTTF exhibit positive temperature dependence. The maximum applicable gate voltage for a 10-year lifetime is extrapolated at different stress conditions. Moreover, the mechanism of the gate degradation is discussed by comparing the devices' performance before and after the progressive breakdown. It is revealed that electrons accelerated in the depletion region of the p-GaN layer under large forward gate bias would gain enough energy and induce defects near the metal/p-GaN interface, resulting in the time-dependent gate degradation.

Keywords—Dynamic stress, Gate degradation, p-GaN gate, Schottky-type contact, temperature dependent, TDDB, Weibull distribution

I. INTRODUCTION

P-type GaN gate high-electron-mobility transistors (HEMTs) currently dominate the commercial normally-OFF GaN-based power devices market and have shown remarkable efficiency and power density in high-frequency power switching applications [1], [2]. Two approaches are adopted in commercial products to achieve the p-GaN gate HEMTs depending on the gate metal/p-GaN contact scheme. One is ohmic-type contact to the p-GaN layer. [3]. This type of p-GaN gate HEMTs have demonstrated impressive static and dynamic performance[4], and reliability [5]. The other approach is to employ a Schottky gate contact for the virtues of reduced gate leakage current and enlarged gate swing [6]. There are extensive reports on the stability and reliability of Schottky-type p-GaN gate HEMTs. In particular, the floating nature of the p-GaN layer which is sandwiched between the Schottky junction and the heterojunction would cause threshold voltage (V_{TH}) instability under both drain [7], [8] and gate stress [9], [10]. The dynamic V_{TH} shift has been ascribed to. charges storage/emission in the electrically floating p-GaN layer. In addition, the metal/p-GaN Schottky junction under a positive (forward) gate bias is reversely biased. The electric field in the depletion region formed at the metal/p-GaN interface could be

This project is funded by Hong Kong Innovation and Technology Fund under grant ITS/412/17FP.

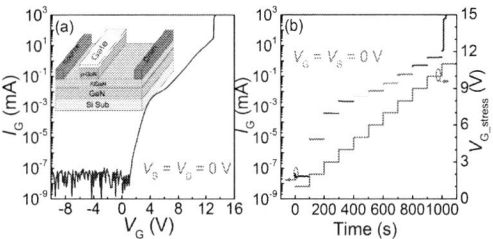

Fig. 1 (a) Quasi-static gate forward and reverse I_G-V_G characteristics at 25˚C. The inset is the schematic cross-sectional view of p-GaN gate AlGaN/GaN HEMT. (b) Gate voltage step-stress measurement: grey staircase line represents gate stress voltage and the colorful lines represent gate leakage.

significantly high, which in turn lead to gate stack breakdown. In [11] the gate breakdown has been studied and explained by avalanche multiplication in the space charge region near the Schottky metal/p-GaN interface. Static forward gate stress measurements have been the subject of wide research efforts. Some works have shown that the p-GaN gate exhibits a time-dependent dielectric breakdown (TDDB) like degradation, exhibiting Weibull [12] or lognormal [13] distribution. The breakdown event was ascribed to the creation of a percolation path in the depletion region of the p-GaN layer. Apart from the gate stress under static conditions, studies on the gate degradation under dynamic gate stress are necessary, which is of particular relevance to lifetime prediction in practical power switching applications.

In this work, time-dependent gate degradation under both static and dynamic stress in Schottky-type p-GaN gate HEMTs is studied. Moreover, as power transistors are typically operated at elevated temperatures, temperature-dependent gate stress measurements are also conducted. For reliable operation of the GaN switching devices, the maximum allowed forward gate voltage is statistically estimated under different stress conditions. By applying gate forward bias stress at different gate voltage, frequency, and temperature, the breakdown events recorded as a function of time are analyzed. Finally, the physical mechanisms of gate degradation are further discussed.

II. MEASUREMENT AND ANALYSIS

The p-GaN gate devices studied in this work feature a metal/p-GaN/AlGaN/GaN gate stack with a Schottky-type gate contact scheme, as schematically shown in the inset of Fig. 1(a). The equivalent circuit of the gate consists of two back to back

978-1-7281-0582-6/19 $31.00 © 2019 IEEE 295

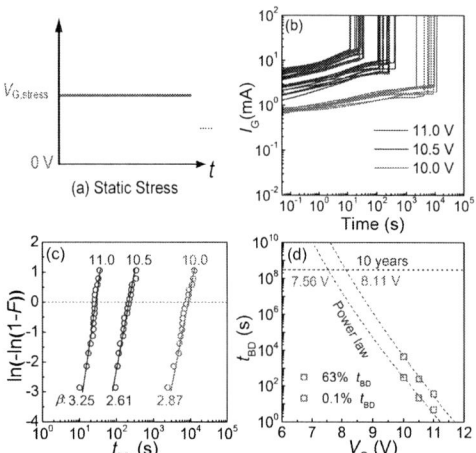

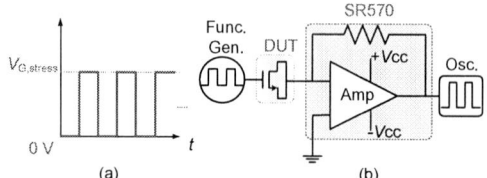

Fig. 3 Dynamic gate stress measurements: (a) test waveform; (b) test setup.

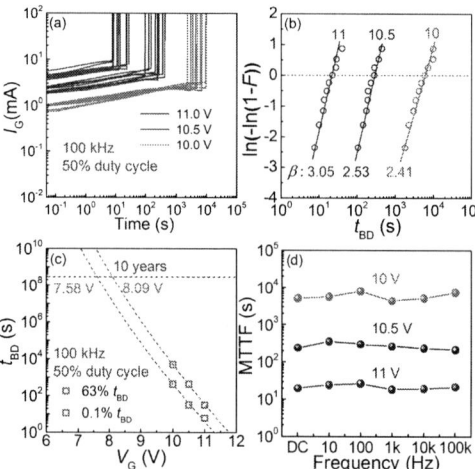

Fig. 2 Static gate stress measurements: (a) Test waveform; (b) Time-to-breakdown (t_{BD}) plot with different V_{G_stress} (10.0 V, 10.5 V and 11 V) at 25 ˚C. (c) Weibull plot of t_{BD} distribution. (d) Maximum applicable V_G by choosing 63 % and 0.1% failure rate for a 10-year lifetime of 8.11 V and 7.56 V, respectively.

diodes (i.e., a metal/p-GaN Schottky junction and a p-GaN/AlGaN/GaN p-i-n heterojunction, respectively). Fig. 1(a) plots the quasi-static gate leakage (I_G-V_G) characteristic at room temperature ($V_S = V_D =0$ V), exhibiting rectifying behavior with low reverse I_G and exponential I_G dependence on the forward V_G. The gate leakage current is still more than five orders of magnitude lower than the ON-state I_D (at $V_{DS} = 1$ V, $V_{GS} = 6$ V) attributed to the reverse-biased Schottky junction. The measured forward gate breakdown voltage is about 12.5 V. The robustness of p-GaN gate against positive gate bias could be pre-evaluated by a step-stress measurement. As shown in Fig. 1(b), gate stress voltage (V_{G_stress}) is increased in the step of 1 V every 100 s. The staircase line represents V_{G_stress} and the colored lines show the gate current. Progressive-steps sudden I_G increase is observed when V_{G_stress} is up to 11 V. To make sure the same degradation mechanism is analyzed, time-to-breakdown (t_{BD}) is defined at the critical point when the 1st-step abrupt increase of I_G occurs in the following discussion.

A. Static Gate Stress

The static stress measurements are carried out by applying a constant dc positive gate voltage (V_{G_stress}) as waveform shown in Fig. 2(a). I_G of the devices is continuously monitored during the stress. Statistical tests have been performed to increase the accuracy of our analysis. Namely, at least ten devices have been characterized by a single stress condition. Fig. 2(b) shows I_G transients of three groups of devices stressed at different V_{G_stress} (i.e., 10 V, 10.5 V, and 11 V) at 25 ˚C. As can be noticed, in the initial stress phase, gate leakage is slightly increasing. For longer stress time, a sudden I_G increase occurs. As widely reported for gate dielectric reliability investigations in MOS/MIS-gated devices and also adopted for GaN-based devices, t_{BD} follows Weibull distribution with the shape

parameter β between 2.6 to 3.3 [Fig. 2(c)], suggesting a single dominant degradation mechanism. Based on the power law fitting, a maximum gate voltage of 8.11 V with a 63% failure rate by setting a 10-years' lifetime is predicted at 25 ˚C, as shown in Fig. 2(d).

B. Dynamic Gate Stress

The dynamic positive gate stress tests are conducted by applying consecutive square-wave pulses to the gate terminal as the waveform depicted in Fig.3(a). Within one cycle, a gate stress voltage and a 0-V voltage are switched, and pulse frequency varies from 10 Hz to 100 kHz. The duty cycle is fixed at 50%, so the effective stress time is always one half of the total stress time. As the setup illustrated in Fig. 3(b), a function generator delivers the square-wave pulses to the gate. Although I_G is not directly measured, gate breakdown detection can be performed on I_{S+D}, which is amplified by a low noise current preamplifier (SR570) and recorded by an oscilloscope.

Figure 4(a) shows the time-resolved gate degradation behavior under dynamic gate stress at the frequency of 100 kHz. Besides, t_{BD} is also Weibull-distributed [Fig. 4(b)] and a maximum gate voltage of 8.09 V is extracted for a 10-year

Fig. 4 Dynamic measurements: (a) t_{BD} characteristics at different forward gate stress of 10 V, 10.5 V and 11 V at room temperature with 100 kHz frequency; (b) Weibull plot of t_{BD} distribution with 100 kHz frequency and 50% duty cycle; (c) Lifetime prediction with 100 kHz frequency: maximum applicable V_G by choosing 63 % and 0.1% failure rate for a 10-year lifetime of 8.09 V and 7.58 V, respectively; (d) MTTF at different frequencies.

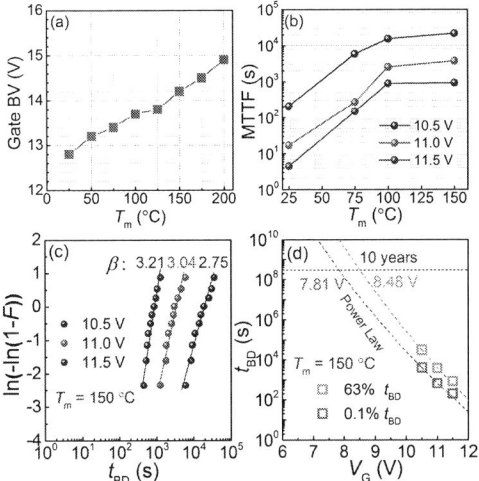

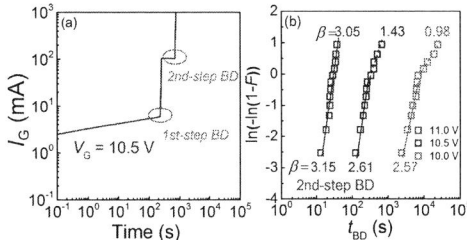

Fig. 6 (a) Time-to-breakdown behavior at forward gate stress voltage of 10.5 V. (b) Weibull plot of 2nd-step t_{BD} distribution at temperature of 25 ˚C.

Fig. 5 (a) Temperature-dependent gate breakdown voltage. (b) Temperature-dependent MTTF at different V_{G_stress} of 10.5 V, 11 V, and 11.5 V. (c) Weibull plot of t_{BD} distribution at temperature of 150 ˚C. (d) Maximum applicable V_G by choosing 63 % and 0.1% failure rate for a 10-year lifetime of 8.48 V and 7.81 V, respectively.

lifetime choosing a 63% failure rate [Fig. 4(c)], which is comparable with the static situation.

The mean-time-to-failure (MTTF) defined at t_{BD} where 63% failure rate is considered exhibits no noticeable differences by varying frequencies shown in Fig. 4(d). Here the uncertainty regarding the choice of fitting law in lifetime extrapolation is minimal because the same power law is adopted under different gate stress conditions. The similar MTTF under static and dynamic gate stress differs from that in MOS-gated devices, where the gate dielectric degradation depends on the stress mode. During stress, traps are generated in the dielectric or the interface. Trapped charges would alter the local electric field. The time-to-breakdown in dielectric depends on both charge trapping and de-trapping rate, and thus, depends on stress frequency [14], [15]. Our results indicate that the p-GaN gate degradation is determined by the effective stress time and the modification of the electric field due to charges trapping/de-trapping is negligible.

C. Temperature-Dependent Gate Stress

A temperature (T_m)-dependent measurements of gate breakdown voltage and t_{BD} are performed to reveal the practical application situation and understand the degradation mechanism. I_G-V_G dc characteristics are measured from 25 ˚C to 200 ˚C in a step of 25 ˚C. Before the gate breakdown occurs, I_G increases as T_m rises. The positive temperature dependence of the gate breakdown voltage is observed and summarized in Fig. 5(a), which is also reported in ref. [11]. Furthermore, time-dependent gate degradation tests are also conducted at elevated T_m [Fig. 5(b)]. MTTF shows a positive T_m dependence, i.e., MTTF increases ~2 orders from 25 ˚C to 150˚C. t_{BD} also exhibits a Weibull distribution at a high temperature of 150 ˚C

[Fig. 5(c)]. By considering a lifetime of 10 years at a percentile of 63%, the maximum estimated applicable gate voltage is 8.46 V at 150 ˚C, which is higher than that at 25 ˚C. [Fig. 5(d)].

D. Gate Degradation Mechanism

The weak dependence on stress mode and positive dependence on temperature of MTTF observed in Schottky-type p-GaN gate demonstrates a unique gate degradation mechanism. As can be seen in Fig. 6(a), the device shows a progressively two-step abrupt I_G increase. To further investigate the mechanism of the gate degradation, the time to 2nd-step breakdown is statically characterized under different V_{G_stress}. Weibull distribution with two shape factors β is observed as plotted in Fig. 6(b), indicating two different failure mechanisms.

Devices' *I-V* performance and V_{TH} stability are then characterized after the progressive-step gate breakdown as demonstrated in Fig. 7. After 1st-step gate breakdown, I_G increases about two orders of magnitude at $V_G > 3$ V [Fig. 7(a)]. Nevertheless, the device can still maintain a low reverse I_G, suggesting that the *p-i-n* heterojunction remains functional. "Gate control" of the channel is also preserved. The degraded Schottky junction indicates an ohmic-like gate behavior (i.e., higher g_m and larger I_D density) [Fig. 7(b)-(c)].

To further consolidate our results and explanation, gate bias stress induced V_{TH} shift (ΔV_{TH}) has been measured under static gate stress as described in ref [10]. V_{TH} shows negative shifts with V_{G_stress} larger than 4 V after 1st-step gate breakdown, while ΔV_{TH} becomes negative at a higher V_{G_stress} (i.e., > 5 V) in fresh devices [Fig. 7(d)]. The negative V_{TH} shifts stem from net additional positive charges in the gate stack (e.g., hole accumulation at p-GaN/AlGaN interface) or the buffer layer near the channel. These indicate that holes injection from gate metal into the p-GaN layer is enhanced after the 1st-step gate breakdown. Eventually, after the final (2nd-step) gate breakdown, the gate leakage exhibits a significant increase in both forward and reverse directions, and gate modulation/control of the channel is lost after the breakdown.

Two mechanisms could contribute to the degradation of the Schottky-type p-GaN gate subjected to forward gate stress. The initial time-dependent gate degradation could be linked to the formation of defect levels in the depleted p-GaN layer close to the metal/p-GaN interface. Fig. 8(a) illustrates the schematic energy band diagram of the gate stack of the p-GaN gate HEMT under positive gate stress. The metal/p-GaN Schottky junction is reverse-biased with an extended depletion region within the

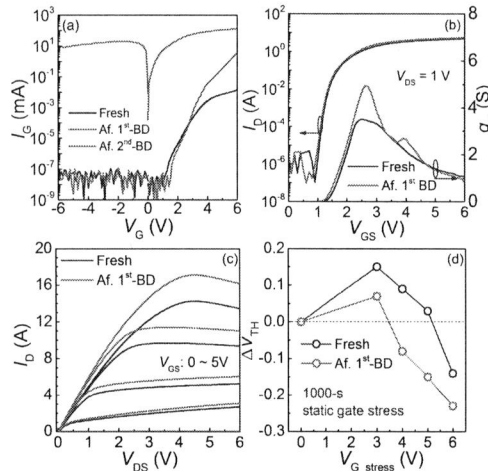

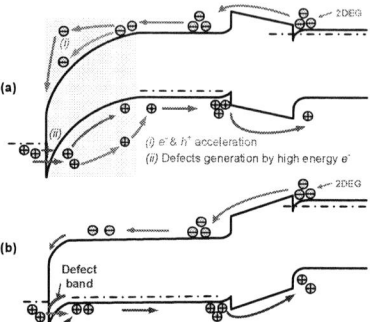

Fig. 7 Device *I-V* performances before and after gate breakdown: (a) Gate I_G-V_G characteristics; (b) transfer, and (c) output characteristics. (d) V_{TH} shift as a function of gate stress voltage in static gate stress measurement before and after 1st-step gate breakdown.

Fig. 8: Schematic band diagrams of the gate stack under high V_{G_stress}: (a) before gate breakdown; (b) enough defects are generated and convert the gate from Schottky to ohmic-like, leading to the 1st-step gate breakdown.

p-GaN layer, while the *p-i-n* heterojunction is forward biased. Electrons in the 2-DEG channel would spill over the AlGaN barrier and are injected into the *p*-GaN layer. Meanwhile, holes injection from the gate electrode to *p*-GaN is also significant under a large V_{G_stress}. Carriers (electrons and holes) could be accelerated and become highly energetic in the depleted *p*-GaN region where a high electric field is presented. Therefore, the high energy electrons would bombard the metal/*p*-GaN interface or the *p*-GaN layer near the interface to induce defect levels. With sufficient accumulated stress time, an adequate concentration of defects is available to convert the gate contact from Schottky-type to Ohmic-like type, leading to the 1st-step gate breakdown as shown in Fig. 8(b). Such a process shall exhibit positive temperature coefficients, due to stronger lattice scattering at a higher temperature. Afterward, most of the gate voltage drops across the AlGaN barrier. Subsequent high gate leakage density can lead to the ultimate 2nd-step gate breakdown caused by defect generation and the resultant degradation of the AlGaN barrier.

III. CONCLUSIONS

In conclusion, gate degradation induced by positive gate stress of Schottky-type *p*-GaN gate HEMTs has been investigated under different operation conditions (i.e., static, dynamic, and elevated temperatures). The lifetime predicted under dynamic gate stress is comparable with the static stress conditions. A positive temperature-dependence of the gate breakdown voltage and MTTF has been observed. In addition, gate degradation mechanism is revealed and ascribed to the creation of defect levels by high energy electrons accelerated in the depleted *p*-GaN layer. The dynamic stress and temperature dependent lifetime extraction in this work would provide a more application-relevant reliability projection of *p*-GaN transistors.

REFERENCES

[1] K. J. Chen *et al.*, "GaN-on-Si Power Technology: Devices and Applications," *IEEE Trans. Electron Devices*, vol. 64, no. 3, pp. 779–795, Mar. 2017.

[2] C.-L. Tsai *et al.*, "Smart GaN Platform: Performance & Challenges," in *Proc. IEDM*, pp. 737–740. Dec. 2017.

[3] Y. Uemoto *et al.*, "Gate Injection Transistor (GIT)—A Normally-Off AlGaN/GaN Power Transistor Using Conductivity Modulation," *IEEE Trans. Electron Devices.*, vol. 54, no. 12, pp. 3393–3399, Dec. 2007.

[4] S. Kaneko *et al.*, "Current-collapse-free operations up to 850 V by GaN-GIT utilizing hole injection from drain," in *Proc. IEEE 27th ISPSD*, May 2015, pp. 41–44.

[5] K. Tanaka *et al.*, "Reliability of hybrid-drain-embedded gate injection transistor," in *Proc. IEEE IRPS*, Apr. 2017, pp. 4B-2.1–4B-2.10.

[6] I. Hwang *et al.*, "*p*-GaN gate HEMTs with tungsten gate metal for high threshold voltage and low gate current," *IEEE Electron Dev. Lett.*, vol. 34, no. 2, pp. 202–204, Feb. 2013.

[7] H. Wang *et al.*, "Maximizing the performance of 650-V *p*-GaN gate HEMTs: Dynamic R_{ON} degradation and circuit design considerations," *IEEE Trans. Power Electron.*, vol. 32, no. 7, pp. 5539–5549, July 2017.

[8] J. Wei *et al.*, "Charge storage mechanism of drain induced dynamic threshold voltage shift in *p*-GaN gate HEMTs," *IEEE Electron Dev. Lett.*, early access, doi: 10.1109/LED.2019.2900154.

[9] L. Sayadi, *et al.*, "Threshold voltage instability in *p*-GaN gate AlGaN/GaN HFETs", *IEEE Trans. Electron Device*, vol. 65, no. 6, pp. 2454-2460, Jun. 2018.

[10] J. He *et al.*, "V_{TH} Instability of p-GaN Gate HEMTs under Static and Dynamic Gate Stress," *IEEE Electron Dev. Lett.*, vol. 39, no. 10, pp. 1576-1579, Aug. 2018.

[11] T.-L. Wu *et al.*, "Forward Bias Gate Breakdown Mechanism in Enhancement-mode *p*-GaN Gate AlGaN/GaN High-Electron-Mobility Transistors," IEEE Electron Dev. Lett., vol. 36, no. 10, pp. 1001–1003, Oct. 2015.

[12] A N. Tallarico *et al.*, "Investigation of the Gate Breakdown in Forward-Biased GaN-Based Power HEMTs," *IEEE Electron Device Lett.*, vol. 38, no. 1, pp. 99–102, Jan. 2017.

[13] A. Stockman *et al.*, "Gate Conduction Mechanisms and Lifetime Modeling of p-Gate AlGaN/GaN High-Electron-Mobility Transistors," *IEEE Trans. on Electron Dev.*, pp. 1–8, 2018.

[14] C. L. Chen *et al.*, "The physical mechanism investigation of AC TDDB behavior in advanced gate stack," in *Proc. IRPS*, 2014, pp. 5B.5.1-5B.5.5.

[15] M. N. Chang *et al.*, "A fundamental AC TDDB study of BEOL ELK in advanced technology," in *2016 IEDM*, 2016, pp. 31.7.1-31.7.4.

Proceedings of the 31st International Symposium on Power Semiconductor Devices & ICs
May 19-23, 2019, Shanghai, China

New Circuit Topology for System-Level Reliability of GaN

Ming-Cheng Lin, Wen-Che Chang, Haw-Yun Wu, Gabriel Petrus Lansbergen, Man-Ho Kwan, Jiun-Lei Yu, Cheng-Pao Wu, Chun-Lin Tsai, Hsiao-Chin Tuan, and Alex Kalnitsky
Analog Power & Specialty Technology Division
Taiwan Semiconductor Manufacturing Company
Hsin-Chu, Taiwan
Email: mclint@tsmc.com

Abstract—To accelerate GaN adoption, beyond-JEDEC system-level reliability should be done to prove the robustness of GaN in applications. In this paper, a new hard switching test vehicle (half-bridge RC load) was proposed & demonstrated to achieve system-like stress, flexibility of acceleration test, low system power consumption with large sample size, easy setup & control which can meet system-level reliability requirement.

Keywords—GaN, MTTF, System Lifetime

I. INTRODUCTION

Due to its superior material properties, high carrier mobility & high band-gap, GaN devices have 5X~10X better FOM than Si which shows their high potential for replacing their Si counterparts in high voltage & high power applications. Commercial availability of GaN devices rated at 600V or 650V is growing from suppliers like GaN Systems, Panasonic, Transphorm, TI, Infineon…making GaN device status change from being nice-to-have to must-have for next generation power conversion. Although GaN devices have passed Si based JEDEC reliability standards, it will be still taking time to build confidence from system field tests. To accelerate GaN adoption, beyond-JEDEC system-level reliability should be done to prove the robustness of GaN in applications. Most works published on GaN system-level reliability focused on hard switching operation due to the more stressful condition it applies to the active device. However none of these works provided a methodology to predict the lifetime of GaN device in a wide

array of separate applications. To achieve that, we believe a methodology should provide following capabilities,

1. System-like stress condition for HTOL,

2. Flexibility of acceleration test (Temperature, Voltage, Current, Frequency & Duty),

3. Low system power consumption for large sample size,

4. Easy to setup & control.

Here, a new hard switching test vehicle (half-bridge RC load circuit) with innovated Gate pattern control was proposed to meet the above requirements.

II. TOPOLOGY & METHODOLOGY

Schematic benchmarks with other state-of-the-art test vehicles are shown in Fig. 1. We note Panasonic's use of a RL load[1] reduces on-duty (3% duty makes non-representative for real system) and limits frequency in order to reduce system power to 100W. Using only L load[2] of TI reduces both on-duty & frequency resulting in lower power consumption. On the other hand, Transphorm uses 400W Boost[3] for limited samples.

Table 1. summarized advantages from HB RC load test vehicle. Lower power consumption helps to achieve large sample size. Acceleration flexibility helps to identify independent aging effect of corresponding factors & helps to predict the lifetime of separate applications.

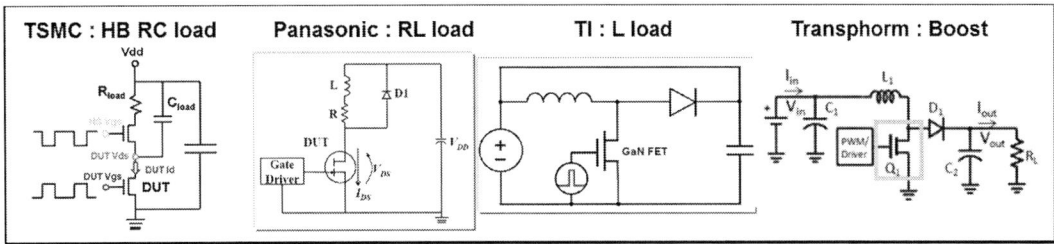

Fig. 1. Test vehicle schematic benchmark.

978-1-7281-0582-6/19 $31.00 © 2019 IEEE

TABLE I. CAPABILITY & FLEXIBILITY COMPARISON OF TEST VEHICLES

Hard Switching Test Vehicle		TSMC	Panasonic	TI	Transphorm
Setup		Half-Bridge RC load	LR load	L load	Boost Convertor
Power consumption		Low (<10W)	High (>100W)	Medium (>10W)	High (400W)
Multi-DUT (sample size)		V	limited	V	limited
Acceleration Flexibility	Temperature	V	V	V	V
	Voltage	V	V	V	V
	Current	V	V	V	V
	Frequency	V	limited	limited	limited
	Duty	V	limited	limited	limited

III. APPROACH

To demonstrate new proposed method can provide system-like stress condition, PFC-like switching stress was selected to show varying on-duty of PFC with responding AC sine wave input in Fig. 2.

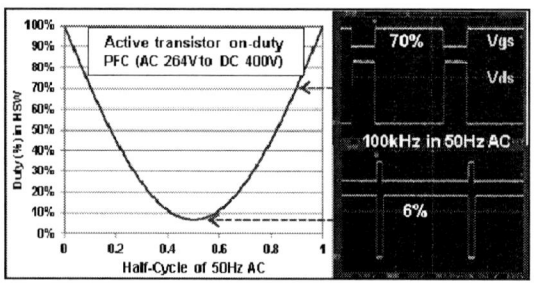

Fig. 2. Half-bridge RC load can perform PFC-like varying on-duty

To demonstrate voltage & current acceleration flexibility, Vdd & Cload were changed in Fig. 3. to perform similar harshness & loci as inductive load switching from RC load.

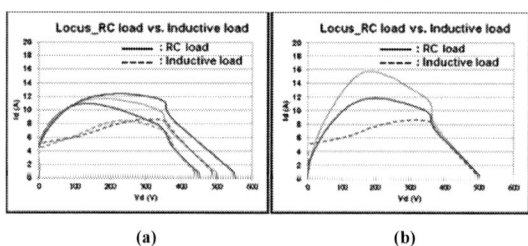

Fig. 3. Similar harshness & loci from RC load compared with L load under Voltage acceleration in (a) & Current acceleration in (b).

Frequency & duty flexibilities were showed in Fig. 4. with linear system power dependence to frequency but power independence to duty.

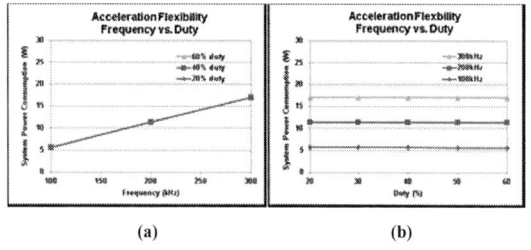

Fig. 4. Frequency acceleration in (a) & On-duty acceleration in (b) with low system power consumption.

In typical HTOL system, thermal feedback & control are needed to keep stable device temperature since device Ron shift will induce conduction loss & temperature shift. In HB RC load, stable Tc (ΔTc < ± 1C) could be achieved with drifting dynamic Ron but without Tc feedback & control for the device under acceleration stress as showed in Fig. 5.

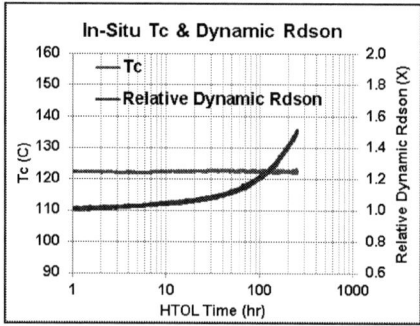

Fig. 5. Tc keeps constant with device Ron shift but without Tc feedback control

With help of easy setup, control and low system power consumption, HB RC loads can be setup in parallel with large sample size. 10 DUTs were stressed & monitored at the same time, showed in Fig. 6.

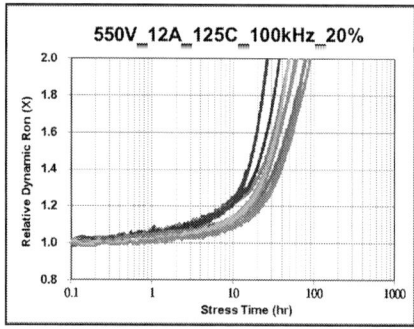

Fig. 6. In-situ dynamic Ron monitoring for 10 DUT

The harshness of the hard switching is tuned by the switching locus of the turn-on transient. Fig. 7. demonstrates that an RC load can created similar stressful loci as an inductive load since the wheeling diode of an inductive load behaves like a voltage dependent capacitor during the turn-on transient of the active transistor. This indicates harshness can be controlled by Cload*dV/dt. An RC load generates

978-1-7281-0582-6/19 $31.00 © 2019 IEEE   300

stressful turn-on transient but suffers from duty dependent system power, just like RL load & L load switching.

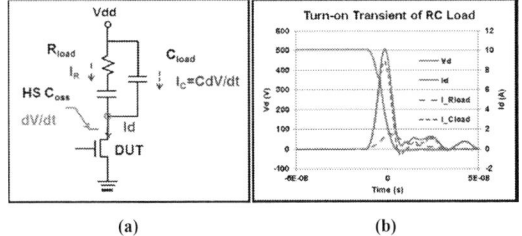

(a)   (b)

Fig. 7. RC load schematic in (a) can create stressful locus in (b) by controlling Cload*dV/dt

Without sacrificing harshness of stress, a half bridge RC load test vehicle was proposed to remove Rload power dissipated during on-duty of DUT (Fig. 8). Two benefits from this approach help it to be a multi-DUT test platform. First, system power was reduced below 10W & independent to on-duty which helps to simplify setup when simulating stress condition of 1kW system operation. Second, the half-bridge schematic removes conduction loss of DUT which makes thermal control easier; DUT Tc feedback is not needed for extra self-heating from Ron shift.

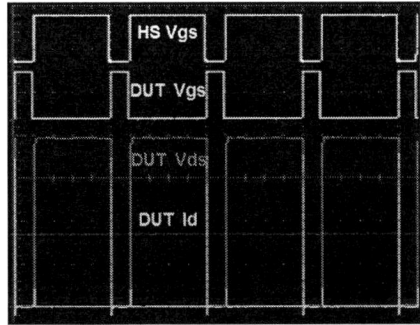

Fig. 8. Half-bridge RC load removed Rload power dissipation during on-duty of DUT

One issue with this conduction loss free schematic is that in-situ dynamic Ron extraction for MTTF is not available without conduction current. To solve this, the anti-state of the high side & low side transistors was broken in 1/1000 pulses with patterned Gate control, as showed in Fig. 9, to implant conduction current for extracting in-situ dynamic Ron of the DUT.

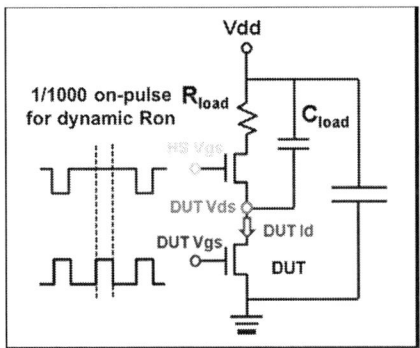

Fig. 9. Anti-state of high side & low side FETs was broken for 1/1000 pulses

Result & Discussion

To demonstrate the capability of this methodology, temperature, current & voltage accelerations were tested under f=100kHz & on-duty=20% for 10 DUTs of TSMC 650V E-HEMT (130mOhm). The failure criteria was set at 50% dynamic Ron shift. Three temperatures, 100C, 125C & 150C, were applied to extract the activation energy, Ea, under conditions of Vdd=550V, I_{pk}=12A. Results were plotted in a Weibull graph. Ea was fitted as 0.39eV from temperature dependent MTTF as showed in Fig.10.

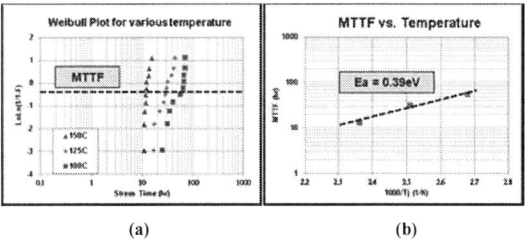

(a)   (b)

Fig. 10. Temperature acceleration was tested in (a) to extract Ea (0.39eV)in (b) for Tj=100C, 125C & 150C under conditions of Vdd=550V, Ipk=12A, f=100kHz & on-duty=20%

Current acceleration factor A_{Ipk} was extracted as 0.068A^{-1} in Fig. 11. from stress under I_{pk}=12A, 16A & 20A at Tj=125C & Vdd=550V.

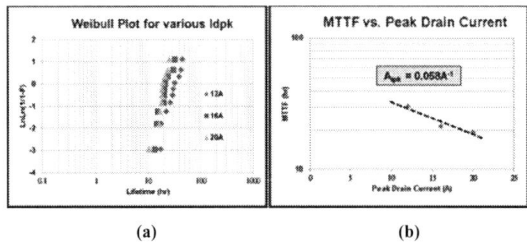

(a)   (b)

Fig. 11. Current acceleration was tested in (a) to extract acceleration factor AIpk (0.068A^{-1}) in (b) for Ipk=12A, 16A & 20A under conditions of Tj=125C, Vdd=550, f=100kHz & on-duty=20%

Voltage acceleration factor A_{Vdd} was extracted as 0.035V^{-1} in Fig 12. from stress under Vdd=525V, 550V & 575V at Tj=125C & Ipk=12A.

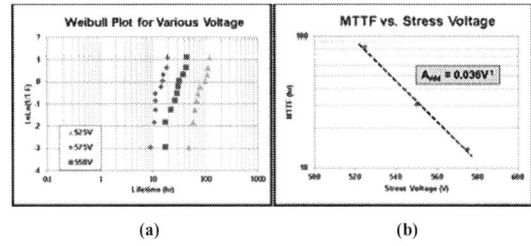

(a)   (b)

Fig. 12. Voltage acceleration was tested in (a) to extract acceleration factor AVdd (0.036V^{-1}) in (b) for Vdd=525V, 550V & 575V under conditions of Tj=125C, Ipk=12A, f=100kHz & on-duty=20%

IV. CONCLUSION

Thus, our Half-bridge RC load test vehicle can extract 5 independent acceleration factors with low power consumption, allowing for significantly improved lifetime prediction & understanding of failure mechanism for separated applications.

REFERENCES

[1] Ayanori Ikoshi *et. al*, "Lifetime Evaluation for Hybrid-Drain-embedded Gate Injection Transistor (HD-GIT) under Practical Switching Operations", IEEE International Reliability Physics Symposium (IRPS), 4E.2, 2018

[2] Sandeep R. Bahl *et. al*, "Product-level Reliability of GaN Devices", IEEE International Reliability Physics Symposium (IRPS), invited, 2016.

[3] P. Parikh *et. al*, "650 Volt GaN Commercialization Reached Automotive Standards", ECS Transactions, 80 (7), p. 17-28, 2017.

Proceedings of the 31st International Symposium on Power Semiconductor Devices & ICs
May 19-23, 2019, Shanghai, China

100 A Vertical GaN Trench MOSFETs with a Current Distribution Layer

Tohru Oka, Tsutomu Ina, Yukihisa Ueno, and Junya Nishii,

Research and Development Headquarters, TOYODA GOSEI Co., Ltd.
Ama, Aichi 490-1207, Japan
toru.oka@toyoda-gosei.co.jp

Abstract—**This paper reports on vertical GaN-based trench MOSFETs operating at 100 A for the first time. A current distribution layer (CDL) in a drift layer is employed for the high current operation. An effective insert position of the CDL is designed and, thereby, the current density of the MOSFET with the CDL is increased about 1.17 times higher than that of the MOSFET without the CDL. Large MOSFET chips with a drain current of up to 100 A are fabricated and their switching characteristics are demonstrated.**

Keywords—*Gallium nitride, vertical transistors, MOSFETs, high current, switching, current distribution layer*

I. INTRODUCTION

Gallium Nitride (GaN) draws attention as a material of power devices replacing Silicon. As GaN power devices with low specific on-resistance, lateral field-effect transistors (FETs) based on AlGaN/GaN are attractive due to high-density and high-mobility two-dimensional electron gas (2DEG) at the AlGaN/GaN interface. However, to achieve high breakdown voltages in the lateral GaN devices, considerable spacing between a gate terminal and a drain terminal is required [1], which increase chip size and thus cost for a required amperage rating. By contrast, the gate-drain spacing of vertical FETs can be designed without increasing the chip size. This is because the gate-drain spacing of the vertical FETs is determined by the thickness of the drift layer. For this reason, the vertical GaN FETs prevalent in Si and SiC power devices are suitable as a structure for realizing high current density and high breakdown voltage operations. Thanks to the availability of relatively high-quality free-standing bulk GaN substrates [2]−[4], vertical GaN FETs with high breakdown voltage of over 600 V have been reported thus far [5]−[13]. However, there are a few reports of vertical GaN FETs with high current operations exceeding 10 A [9], [11]. In this paper, we demonstrate vertical GaN trench MOSFETs with 100 A operation for the first time. In order to decrease a resistance of drift layer and, thus, to increase a current density of the MOSFETs, we employed a current distribution layer (CDL) and designed its insert position in the drift layer. In addition, to demonstrate high current operations, we worked to fabricate large MOSFET chips with a size of 3 mm × 3 mm.

II. DEVICE DESIGN

A. Design of CDL

The CDL is used as a method for lowering a resistance of a drift layer of vertical trench MOSFETs [14], [15]. A typical

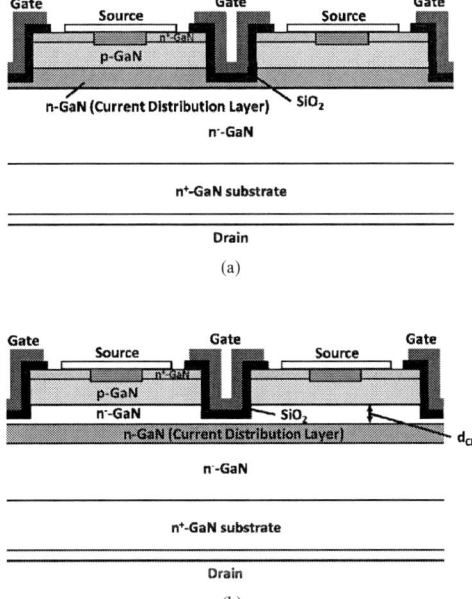

Fig.1. Schematic cross sections of trench MOSFETs with (a) CDL connected to p-GaN and (b) CDL arranged apart from p-GaN.

schematic cross section of a vertical trench MOSFET with the CDL is shown in Fig. 1 (a). The CDL has a higher doping concentration than that of the drift layer, and is generally arranged connected to the p-channel layer, as shown Fig. 1 (a) [14], [15]. The CDL spreads the current flowing in the drift layer by extending electrons from the MOS gate channel to in-plane direction. This increases the current density, and thereby, decreases the drift resistance. Here we are concerned about two matters for this structure. One is the degradation of the breakdown voltage. This is because the electric field at the bottom of the gate trench may increase due to the high doping concentration of the CDL. The other concern is the weakening of the effect of the current distribution. When the CDL is connected to the p-channel layer, effective carriers in a part of the CDL may decrease due to the depletion from the p-channel layer, and this leads to the reduction in the effect of the CDL. These concerns motivated us to design the CDL arranged apart from the p-channel layer, as shown in Fig. 1 (b).

978-1-7281-0582-6/19 $31.00 © 2019 IEEE   303

An effective insert position of the CDL was designed using a device simulator. The layer structures of the simulated trench MOSFETs consisted of total 10-μm-thick n⁻-GaN drift layer that included 0.5-μm-thick CDL, 0.7-μm-thick p-GaN channel layer, 0.2-μm-thick n⁺-GaN source contact layer. The donor concentrations of the drift layer and the CDL were 6×10^{15} cm⁻³ and 2×10^{16} cm⁻³, respectively. An acceptor concentration of p-GaN was 2×10^{18} cm⁻³, and a donor concentration of n⁺-GaN was 6×10^{18} cm⁻³. The distance of the CDL from the p-GaN (d_{CD}) was varied from 0 μm to 7.5 μm. The gate insulator was an 80-nm-thick SiO₂. For comparison, the characteristics of the trench MOSFETs without the CDL were also simulated, where the thickness of the drift layer was 10 μm and the donor concentration was varied from 6×10^{15} cm⁻³ to 8×10^{15} cm⁻³. The on-state drain current density and the off-state electric field were monitored and compared. We monitored the drain current densities at a gate voltage (V_G) of + 20 V from the threshold voltage (V_{th}) and a drain-source voltage (V_{DS}) of 0.5 V. We monitored the maximum electric field intensity in SiO₂ at a bottom corner of the gate trench at V_{DS} of 600 V.

B. Simulated results

Figure 2 shows the cross sectional views of simulated on-state drain current density of the MOSFETs with and without the CDL. The MOSFETs with CDL exhibited higher and more widely-spread current density than that without CDL. As shown in Fig. 2 (b), most part of the CDL connected to p-GaN was depleted, which will become the cause of the reduction in the current distribution effect. In contrast, we found that the CDL apart from the p-GaN spread the drain current more efficiently, as shown in Fig. 2 (c). In order to effectively increase the drain current density, it is important to design the CDL position deeper than the depletion width, the value of which is estimated to be 0.3-0.7 μm according to the concentrations of the p-GaN channel layer and the n⁻-GaN drift layer.

Figure 3 shows the relations between the on-state drain current density and the off-state maximum electric field intensity for the MOSFETs with and without the CDL. Both the drain current density and the electric field intensity for the MOSFETs without the CDL monotonically increase as the donor concentration in the drift layer (N_D) increases. For the MOSFETs with the CDL except for the result at d_{CD} of 0 μm, both the drain current density and the electric field intensity also increase but its variation is different to that for MOSFETs without the CDL. The electric field intensity tends to increase according to d_{CD}, signifying that shallower position of the CDL induces higher electric field. For the MOSFETs with the CDL at d_{CD} of 7.5 μm and 5 μm, the drain current densities are the same as that of the MOSFET without CDL with the same donor concentration of 6×10^{15} cm⁻³. This suggests that the CDL far from the p-channel layer vanishingly influences the current distribution. The drain current density at d_{CD} of 2 μm or less tends to increase rapidly, indicating that the CDL greatly affects the current distribution. Note that the increment of the drain current densities is much larger than that of the electric field intensities. Consequently, compared at the same electric field intensity, the drain current densities of MOSFETs with the CDL at d_{CD} of 0.5 μm and 0 μm become higher than that without the CDL. The result verifies that the CDL has effective positions for increasing drain current density

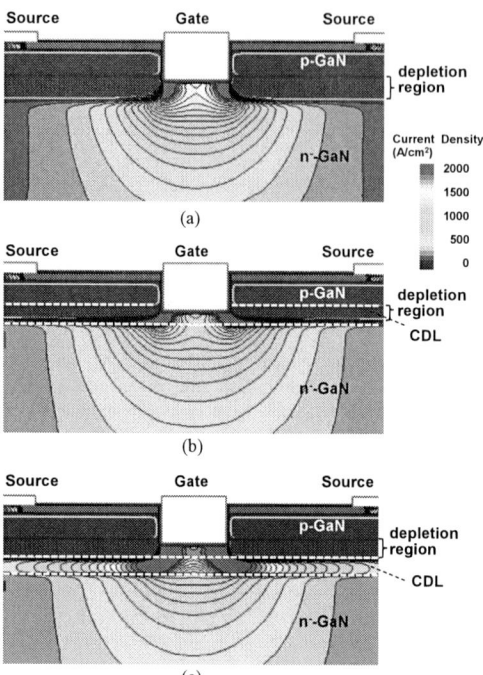

Fig.2. Simulated iso-current-density contours of trench MOSFETs (a) without CDL, (b) with CDL connected to p-GaN, and (c) with CDL at d_{CD} = 0.5 μm. The dashed lines in Fig. 2 (b) and (c) indicte the upper and bottom surfaces of the CDL.

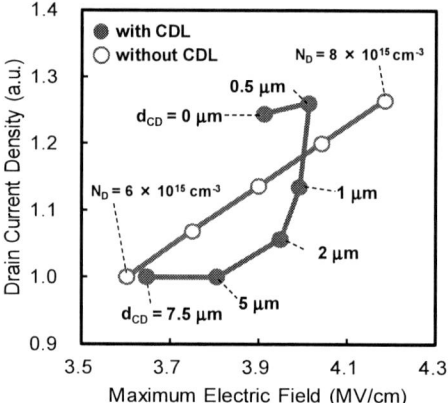

Fig.3. Simulated drain current density vs. maximum electric field intensity of MOSFETs (filled circle: MOSFETs with CDL, open circle: MOSFETs without CDL). The drain current densities were monitored at a gate voltage (V_G) of + 20 V from the threshold voltage (V_{th}) and a drain-source voltage (V_{DS}) of 0.5 V. The maximum electric field intensity in SiO₂ was monitored at a bottom corner of the gate trench at V_{DS} of 600 V.

without sacrificing a breakdown voltage. It should be also noted that the drain current density and the maximum electric field intensity at d_{CD} of 0 μm become lower than those at d_{CD} of 0.5 μm; namely, one concern regarding the degradation of the breakdown voltage does not matter, and the other concern regarding the reduction in the effect of the current distribution is correct.

III. DEVICE FABRICATION AND EXPERIMENTAL RESULTS

A. Investigation of effect of CDL

For the confirmation of the efficacy of the designed CDL, vertical GaN trench MOSFETs with and without the CDL were fabricated with consideration for the simulated results. The epitaxial layers grown on an n⁺-GaN substrate consisted of a 0.2-μm-thick n⁺-GaN source contact layer, a 0.7-μm-thick p-GaN channel layer, and a total 10-μm-thick n⁻-GaN drift layer including a 0.5-μm-thick CDL. The CDL was inserted 0.5 μm beneath the p-GaN layer (i.e., $d_{CD} = 0.5$ μm). The donor concentrations of the drift layer and the CDL were 6×10^{15} cm⁻³ and 2×10^{16} cm⁻³, respectively. In contrast, a donor concentration of the drift layer for the MOSFET without the CDL was 7.5×10^{15} cm⁻³, which was designed so that the simulated maximum electric field intensity at the bottom corner of the gate trench becomes almost the same value for the both MOSFETs. Other device designs and process conditions can be found in our previous reports [8], [9]. For this verification, 0.2 mm × 0.2 mm multi-cell MOSFETs were utilized.

Figure 4 shows the off-state I-V characteristics of the fabricated MOSFETs with and without CDL measured at a V_G of -10 V. A breakdown voltage of the MOSFETs with the CDL was 730 V, and the value was the same as that of the MOSFET without the CDL. As shown in Fig. 4, a slightly high leakage current was observed for the MOSFET without the CDL. The cause is currently under investigation. Since the leakage current did not varied with V_G, we speculate that the leakage current does not result from the difference of the layer designs. Figure 5 shows the transfer I-V characteristics (I_D-V_{GS}) of the MOSFETs measured at a V_{DS} of 0.5 V. At V_G of + 20 V from V_{th}, the drain current density of the MOSFET with the CDL was about 1.17 times higher than that of the MOSFET without the CDL. These results indicate that our design of the CDL apart from the p-channel layer for trench MOSFETs effectively increase a drain current density without sacrificing the breakdown voltage.

B. Demonstration of large chip characteristics

To demonstrate high current operations, large MOSFET chips with the CDL were also fabricated. Figure 6 shows a chip micrograph of the fabricated multi-cell vertical GaN trench MOSFET for the high current operation. The chip dimension was 3 mm × 3 mm.

The output I-V characteristics (I_D-V_{DS}) of the fabricated large MOSFET chip is shown in Fig. 7. The drain current achieved 50 A at V_{DS} of 1.0 V and V_G of 25 V, and reached 100 A at V_{DS} of 2.2 V and V_G of 25 V. As far as we know, this is the first demonstration for vertical GaN-based transistors on GaN substrates with high current operation of up to 100 A.

Switching performance in turn-on and turn-off characteristics was investigated for the fabricated multi-cell trench MOSFETs with a resistive load. Figure 8 shows the measured switching waveforms of turn-on and turn-off operations at a supply voltage V_{DD} of 300 V and a drain current I_D of 30 A. A turn-on delay time ($t_{d(on)}$) and a rise time (t_r) evaluated using the waveform shown in Fig. 8 (a)

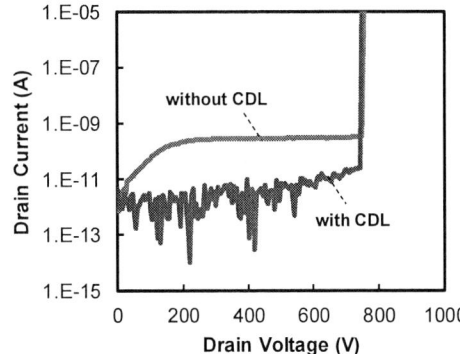

Fig.4. Off-state I-V characteristics measured at V_G of -10 V for the fabricated trench MOSFETs with and without CLD.

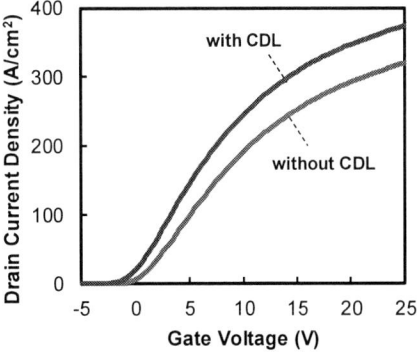

Fig.5. Transfer I-V characteristics (I_D-V_{GS}) measured at V_{DS} of 0.5V for the fabricated trench MOSFETs with and without CLD.

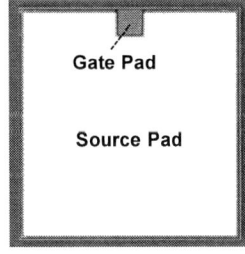

Fig.6. Chip micrograph of a fabricated multi-cell vertical GaN trench MOSFET with CDL. The chip dimension is 3 mm × 3 mm.

were 8 ns and 42 ns, respectively, and a turn-off delay time ($t_{d(off)}$) and a fall time (t_f) evaluated using the waveform shown in Fig. 8 (b) were 49 ns and 18 ns, respectively. Switching loss energy during turn-on and turn-off operations were estimated to be $E_{on} = 84$ μJ and $E_{off} = 30$ μJ. These results suggest that the fabricated vertical GaN trench MOSFETs with the CDL are effective for use in high power and high-speed switching applications.

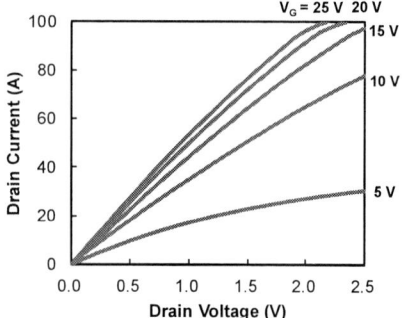

Fig.7. Output I-V characteristics (I_D-V_{DS}) of the fabricated vertical GaN MOSFET chip with CDL.

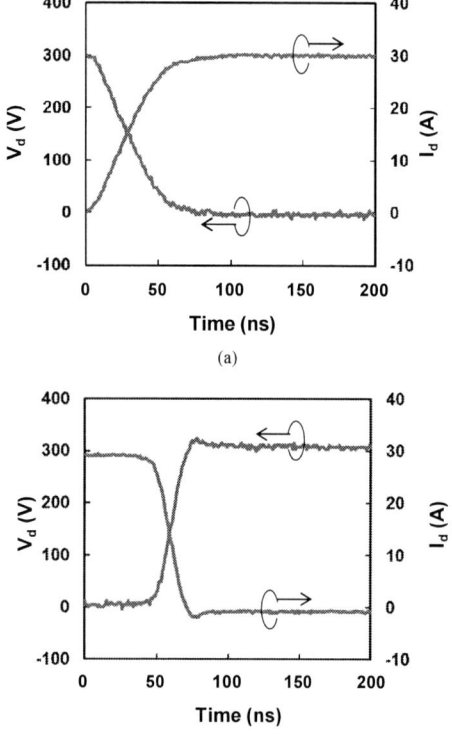

Fig. 8. Measured switching characteristics of the fabricated multi-cell vertical GaN MOSFET chip with resistive load: (a) turn-on waveform, and (b) turn-off waveform.

IV. CONCLUSION

In summary, we have demonstrated vertical GaN-based trench MOSFETs on GaN substrates with an operating current of up to 100 A for the first time. Using a device simulator, we designed and found an effective insert position of a CDL in the drift layer. The CDL should be arranged apart from the p-GaN channel layer so as to avoid the depletion of the CDL caused by the p-GaN channel, leading to the effective current spread in the drift layer without

sacrificing the breakdown voltage. A drain current density of the designed MOSFET with the CDL was about 1.17 times higher than that of the MOSFET without the CDL without degrading the breakdown voltage. A fabricated MOSFET chip with the size of 3 mm × 3 mm operated at a drain current of up to 100 A at V_{DS} of 2.2 V and V_G of 25 V with fast switching characteristics. These results reveal a potential of vertical GaN trench MOSFETs with well-designed CDL for high power and high speed switching applications.

ACKNOWLEDGMENT

The authors would like to gratefully thank process and epitaxial growth staff for their contribution to the device fabrication.

REFERENCES

[1] T. Egawa, "Heteroepitaxial growth and power electronics using AlGaN/GaN HEMT on Si", IEDM Tech. Dig., pp. 613-616, 2012.

[2] K. Motoki, T. Okahisa, R. Hirota, S. Nakahata, K. Uematasu, and N. Matsumoto, "Dislocation reduction in GaN crystal by advanced-DEEP", J. Cryst. Growth, 305, pp. 377-383, 2007.

[3] H. Fujikura, T. Konno, T. Yoshida, and F. Horikiri, "Hydride-vapor-phase epitaxial growth of highly pure GaN layers with smooth as-grown surfaces on freestanding GaN substrates", J. Appl. Phys., vol. 56, 085503, 2017.

[4] J. Wang, G. Ren, Y. Xu, D. Cai, M. Wang, Y. Zhang, X. Hu, and K. Xu, "Low-dislocation density and 6 inch GaN substrates grown by hydride vapor phase epitaxy", Abstr. 12th Int. Conf. Nitride Semiconductors, A.1.2, 2017.

[5] M. Okada, Y. Saitoh, M. Yokoyama, K. Nakata, S. Yaegassi, K. Katayama, M. Ueno, M. Kiyama, T. Katsuyama, and T. Nakamura, "Novel Vertical Heterojunction Field-Effect Transistors with Re-grown AlGaN/GaN Two-Dimensional Electron Gas Channels on GaN Substrates," Appl. Phys. Express, vol. 3, 054201, 2010.

[6] H. Nie, Q. Diduck, B. Alvarez, A. P. Edwards, B. M. Kayes, M. Zhang, G. Ye, T. Prunty, D. Bour, and I. C. Kizilyalli, "1.5-kV and 2.2-mΩ-cm vertical GaN transistors on bulk-GaN substrates", IEEE Electron Device Lett., vol. 35. pp. 939-941, 2014.

[7] T. Oka, Y. Ueno, T. Ina, and K. Hasegawa, "Vertical GaN-based trench metal oxide semiconductor field-effect transistors on a free-standing GaN substrate with blocking voltage of 1.6 kV," Appl. Phys. Express, vol. 7, 021022, 2014.

[8] T. Oka, T. Ina, Y. Ueno, and J. Nishii, "1.8 mΩ•cm² Vertical GaN-based trench metal oxide semiconductor field-effect transistors on a free-standing GaN substrate for 1.2 kV-class operation," Appl. Phys. Express, vol. 8, 054101, 2015.

[9] T. Oka, T. Ina, Y. Ueno, and J. Nishii, "Over 10 A operation with switching characteristics of 1.2 kV-class vertical GaN trench MOSFETs on a bulk GaN substrate ", Proc. 28th Int. Symp. Power Semiconductor Devices and ICs, pp. 459-462, 2016.

[10] R. Li, Y Cao, M. Chen, and R. Chu, " 600 V/1.7 W normally-off GaN vertical trench metal–oxide–semiconductor field-effect transistor", IEEE Electron Device Lett., vol. 37, pp. 1466-1469, 2016.

[11] D. Shibata, R. Kajitani, M. Ogawa, K. Tanaka, S. Tamura, T. Hatsuda, M. Ishida, and T. Ueda, " 1.7 kV / 1.0 mΩcm² normally-off vertical GaN transistor on GaN substrate with regrown p-GaN/AlGaN/GaN semipolar gate structure", IEDM Tech. Dig., pp. 248-251, 2016.

[12] Y. Zhang, M. Sun, D Piedra, J. Hu, Z. Liu, Y. Lin, X. Gao, K. Shepard, and T. Palacios, "1200 V vertical fin power field-effect transistors ", IEDM Tech. Dig., pp. 215-218, 2017.

[13] D. Ji, C. Gupta, S. H. Chan, A. Agarwal, W. Li, S. Keller, U. K. Mishra, and S. Chowdhury, "Demonstrating >1.4 kV OG-FET performance with a novel double field-plated geometry and the successful scaling of large-area devices", IEDM Tech. Dig., pp. 223-226, 2017.

[14] K. Yamamoto and E. Okuno, U.S. Patent, 7994513, 2011.

[15] H. Takeda, U.S. Patent, 8796763, 2014.

Proceedings of the 31st International Symposium on Power Semiconductor Devices & ICs
May 19-23, 2019, Shanghai, China

High Accurate IGBT/IEGT Compact Modeling for Prediction of Power Efficiency and EMI Noise

Takeshi Mizoguchi, Yoko Sakiyama, Naoto Tsukamoto and Wataru Saito

Toshiba Electronic Devices & Storage Corporation, Saiwai-Ku, Kawasaki 212-8520, Japan

Email: takeshi3.mizoguchi@glb.toshiba.co.jp

Abstract—This paper presents a newly developed compact model of IGBT/IEGTs for prediction of power-loss and Electro-Magnetic-Interference (EMI) noise accurately. The proposed model focuses on the capacitance changes between each terminal during the switching operation and has two specific features, (1) the gate-emitter capacitance C_{ge} formed by non-linear functions which consider the negative capacitance for reproducing the turn-on dI/dt and (2) sub-circuits with ideal-diode and CR connected to the gate-collector and the collector-emitter for reproducing the turn-off dV/dt and the tail current. Compared to the conventional model, it was concluded that the proposed model is able to reproduce the measured turn-on and turn-off switching waveform accurately with high convergence.

Keywords—IGBT, IEGT, compact model, negative capacitance, tail current.

I. INTRODUCTION

Power semiconductor devices are the key components of various inverter and converter circuits for high power applications, such as voltage conversion and motor control. Recently, a model-based-design (MBD) is focused on automotive and power electronics systems. In order to realize the MBD process successfully, accurate circuit simulation is essential to predict the power efficiency and the EMI noise. A compact model that accurately describes the device characteristics is a prerequisite in order to predict the circuit performance [1].

For high power applications, the insulated gate bipolar transistor (IGBT) is one of the most important power semiconductor device which integrate MOS-gate control with bipolar conductivity modulation to achieve high input impedance and low on-resistance. The bipolar action originates the complex switching performance, although the conduction loss can be reduced. For an example, the tail current is shown during the turn-off switching because of the remove of the stored carriers in the drift region. In the previous works, several IGBT compact models have been proposed [2]–[4]. However, the switching behaviors have not been reproduced sufficiently, especially, the turn-on behavior due to the negative capacitance. In addition, the convergence in a simulation is also problematic for the MBD process. This paper reports the newly developed compact model for IGBT/IEGTs that realizes the accurate prediction of not only the power-loss but also the EMI noise (dI/dt and dV/dt) with high convergence. The evaluation of the developed model is done by verifying the power-loss, dI/dt and dV/dt of the inductive load switching circuit.

II. MODEL DESCRIPTIONS

A. Conventional model

The conventional IGBT model consists of a MOSFET part and a bipolar part as shown in Fig. 1 [5]. For the MOSFET part, a standard MOSFET model is used, for the bipolar part, however, a standard BJT model is not suitable, since it cannot correctly reproduce the switching characteristics due to high-level carrier injection and the non-quasi-static effects. Therefore the conventional model employs a special equivalent sub-circuit as shown in Fig. 1(b) for the tail current behavior. The additional tail current source G_t represents by the collector current I(V_{sen}) and the current source G_{RQB} as follows:

$$G_t = G_{RQB} - I(V_{sen}) = \{(C_{QB} \cdot V_q)/\text{Taub}\} - I(V_{sen}) \quad (1)$$

where the capacitor C_{QB} is the charge storage element, and "Taub" is the model parameter of the time-constant for the tail current.

B. Proposed model

Fig. 2 shows the proposed model structure of IGBT/IEGT devices in this work. The proposed model also consists of

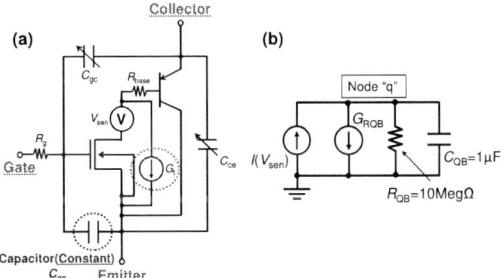

Fig. 1. (a) Conventional model structure of IGBT devices and (b) sub-circuit for the tail current calculation with the carrier lifetime.

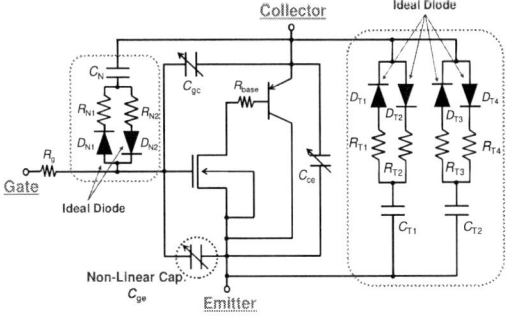

Fig. 2. Proposed model structure of IGBT/IEGT devices. The proposed model consists of a MOSFET part, a bipolar part, C_{ge} formed by non-linear functions and the sub-circuits with ideal-diode connected to the gate-collector and the collector-emitter.

978-1-7281-0582-6/19 $31.00 © 2019 IEEE

a MOSFET part and a bipolar part. In order to the fair comparison of the conventional and the proposed models, model parameters of a MOSFET and a bipolar parts without the sub-circuits are shared each other in this work. The proposed model focuses on the capacitance changes between each terminal during the switching operation and the tail current during the turn-off switching.

The proposed model has two specific features. One is the C_{ge} formed by non-linear functions which consider the negative capacitance [6] for reproducing the turn-on dI/dt. Fig. 3 shows the comparison of the C_{ge} characteristics between the conventional and the proposed models. V_{ce} and V_{ge} dependencies of the C_{ge} by the conventional model are small. On the other hand, the proposed model considers the negative capacitance effect caused by the hole accumulation in the floating p-region of IEGTs. Consequently, the proposed model corresponds the C_{ge} change during the switching and can adjust the turn-on dI/dt.

Another feature is sub-circuits with ideal-diode and CR connected to the gate-collector and the collector-emitter for reproducing the turn-off dV/dt and the tail current. The sub-circuit connected to gate-collector represents the effective gate-collector capacitance C_{gc} during the turn-off switching and adjusts the turn-off dV/dt. Moreover two parallel sub-circuits connected to the collector-emitter represents the tail current due to these different time-constant values of the sub-circuits.

Furthermore, the proposed model has not current source which plays the tail current at the turn-off, and so the simulation convergence is superior compared to the conventional model. The simulation time by the proposed model for the inductive load switching is about 80 times shorter than that by the conventional model.

III. CIRCUIT SIMULATION RESULTS AND DISCUSSIONS

The model parameters were extracted from 4.5kV/1500A IEGT (Toshiba ST1500GXH24) characteristics. The model accuracy was evaluated by the inductive load switching characteristics (Input voltage V_{in}=15V, Supply voltage V_{cc}=2800V, Load inductor L_{load}=100μH and Ambient temperature T_a=125°C) considering with parasitic inductance in the measurement circuit. The free-wheel-diode (FWD) model [7] were chosen to validate the proposed model.

Fig. 4 shows the results for reproduction of measured static I-V and C-V characteristics after parameter extraction using both the conventional and the proposed models. All measured device characteristics are well reproduced.

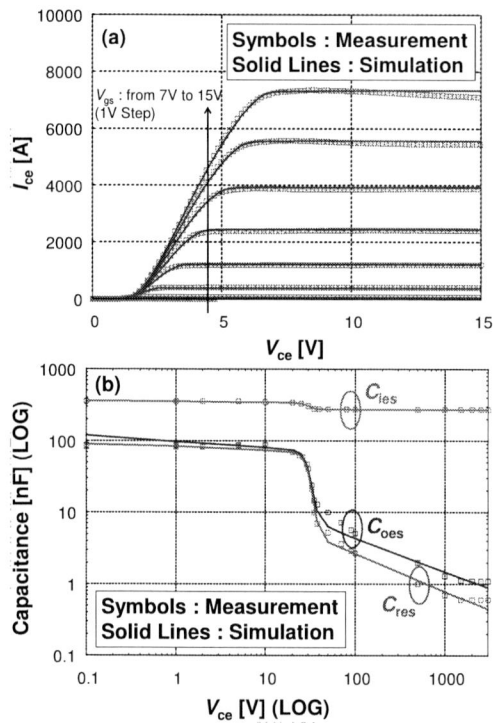

Fig. 4. (a) I_{ce}-V_{ce} characteristics at room temperature from V_{ge}=7V to V_{ge}=15V. Symbols are measurements and lines are simulation results. And (b) Comparison of the modeled $C_{ies}/C_{oes}/C_{res}$-V_{ce} with measurements at room temperature at V_{ge}=0V. Symbols are measurements and lines are simulation results.

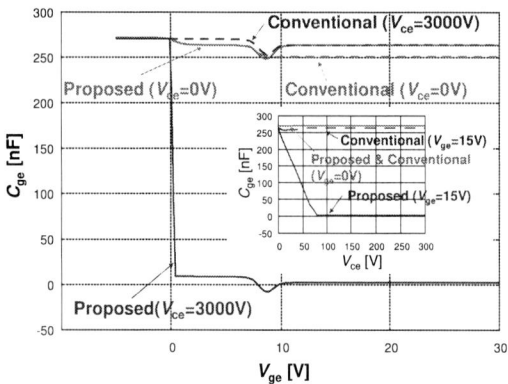

Fig. 3. Comparison of modeled C_{ge}-V_{ge} at V_{ce}=0V, 3000V and C_{ge}-V_{ce} characteristics at V_{ge}=0V, 15V with conventional model and proposed model. Dashed and solid lines are simulation with the conventional model and simulation with the proposed model, respectively.

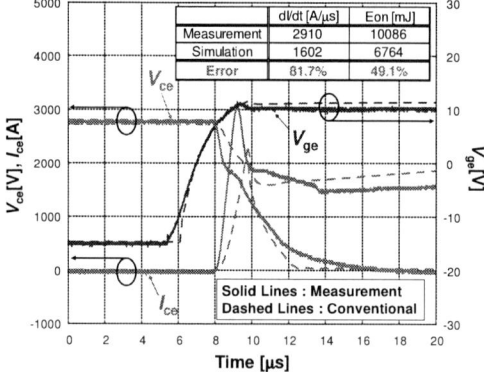

Fig. 5. Comparison of the measured and the simulated turn-on switching waveforms with the conventional model. Gate resistance R_g=7.5Ω and collector current I_c=1500A are applied.

978-1-7281-0582-6/19 $31.00 © 2019 IEEE

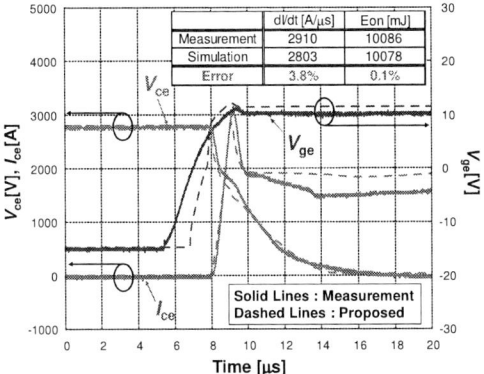

Fig. 6. Comparison of the measured and the simulated turn-on switching waveforms with the proposed model. Gate resistance $R_g=7.5\Omega$ and collector current $I_c=1500A$ are applied.

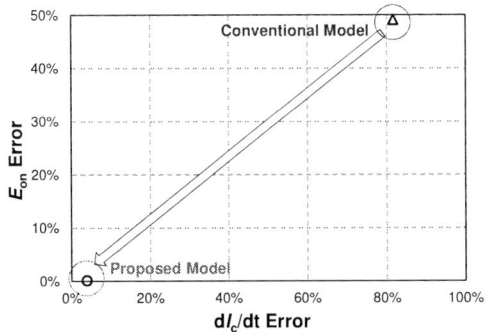

Fig. 7. Improvement of dI_c/dt and E_{on} error rate from the conventional model to the proposed models.

A. Turn-on switching characteristics

Fig. 5 shows a comparison of the measured and the simulated turn-on switching waveforms by the conventional model. Since the conventional model does not consider the negative capacitance caused by the hole accumulation around the gate and the floating p-region of IEGTs, the simulated dI/dt value is much smaller than the measured one and the turn-on loss E_{on} is underestimated from the measured value. In contrast, the proposed model includes the negative capacitance function in the C_{ge} model as shown in Fig. 3. Therefore it is possible to reproduce the measured turn-on switching waveform as shown in Fig. 6. Fig. 7 shows a comparison of error rate of the dI/dt and the E_{on} with the conventional model and the proposed model. As a result, the error rates of the dI/dt and the E_{on} are less than 4%, which is less than 1/20 of the error rate at the conventional model.

B. Turn-off switching characteristics

Fig. 8 shows a comparison of the measured and the simulated turn-off switching waveforms with the conventional model, where the collector current $I_c=1500A$ is applied. The simulated dV/dt value by the conventional model is much higher than the measured one. Because increasing the effective capacitance due to the stored carriers in the drift region

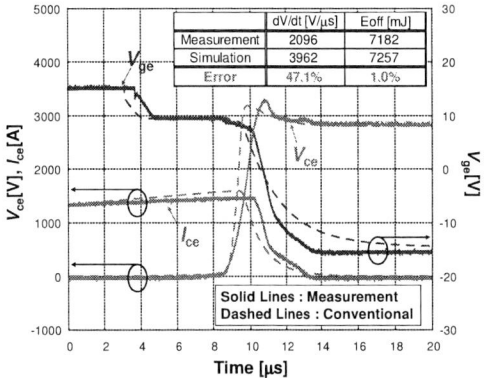

Fig. 8. Comparison of the measured and the simulated turn-off switching waveforms with the conventional model. Gate resistance $R_g=7.5\Omega$ and collector current $I_c=1500A$ are applied.

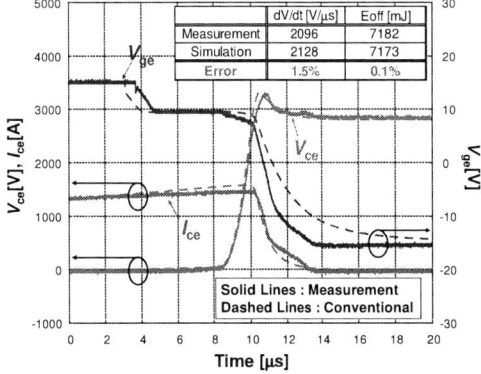

Fig. 9. Comparison of the measured and the simulated turn-off switching waveforms with the conventional model. Gate resistance $R_g=7.5\Omega$ and collector current $I_c=1500A$ are applied. Reasonable values of the C_{gc} and the carrier lifetime parameter value have been considered.

by a bipolar action is not considered in the conventional model. Even in the conventional model, the turn-off switching waveforms can be reproduced by the increase of the C_{gc} value more than the measured static C_{gc} characteristic and the adjustment of the model parameter "Taub" for the tail current. Fig. 9 shows an adjustment result of the turn-off switching waveform by the conventional model, where the collector current $I_c=1500A$ is applied. However, the accuracy is degraded at different current conditions. At the high collector current of $I_c=2600A$, the E_{off} error rate is increased to about 10% due to gaps of the tail current and the dV/dt after the V_{ce} overshoot as shown in Fig. 10.

In contrast, in the proposed model, the dV/dt has been fitted by the sub-circuit between the gate and the collector, and the tail current has been reproduced by two paralleled sub-circuits between the collector and the emitter. Even at the high collector current condition, the proposed model achieves very low error rate of less than 4% for both the dV/dt and the E_{off} as shown in Figs. 11 and 12. This is because that the dV/dt after the V_{ce} overshoot depend on the tail current waveform. Fig. 13 shows the simulated all current components at the turn-

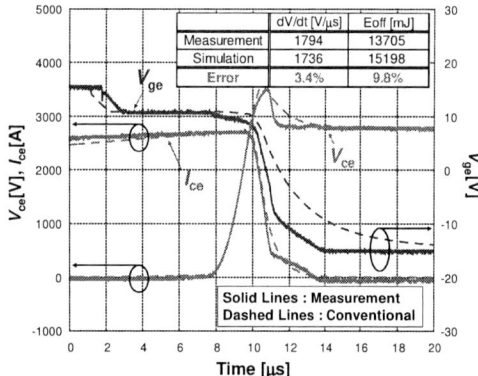

Fig. 10. Comparison of the measured and the simulated turn-off switching waveforms with the conventional model. Gate resistance R_{g}=7.5Ω and collector current I_{c}=2600A are applied.

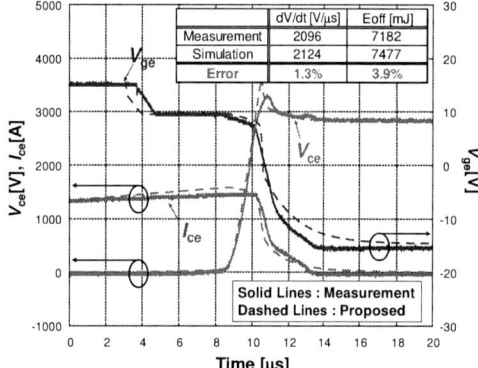

Fig. 11. Comparison of the measured and the simulated turn-off switching waveforms with the proposed model. Gate resistance R_{g}=7.5Ω and collector current I_{c}=1500A are applied.

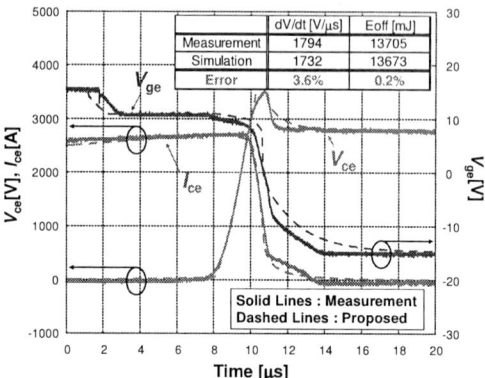

Fig. 12. Comparison of the measured and the simulated turn-off switching waveforms with the proposed model. Gate resistance R_{g}=7.5Ω and collector current I_{c}=2600A are applied.

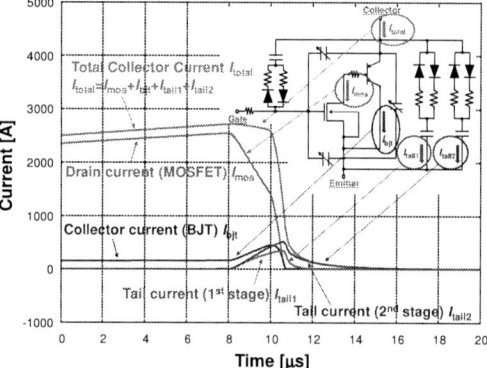

Fig. 13. The simulated all current components at the turn-off switching in the proposed model. Gate resistance R_{g}=7.5Ω and collector current I_{c}=2600A are applied.

off switching waveform in the proposed model, where the high collector current of 2600A is applied. The two paralleled sub-circuits between the collector and the emitter have different time-constant values and reproduce the tail current waveform accurately. Therefore the low error rate can be obtained by the proposed model even for different current conditions.

IV. CONCLUSION

We have developed the newly compact model of IGBT/IEGTs which includes the negative capacitance effect and sub-circuits with ideal-diodes and CR for high accurate prediction of not only the power-loss but also EMI noise. The proposed model achieved to reproduce the switching characteristics in the inductive load. The error rate of the switching loss and $\mathrm{d}I/\mathrm{d}t$ for turn-on switching is 20 times smaller than the conventional model. The error rate of the turn-off switching characteristics was also reduced to less than 4% even for the various current condition. Furthermore, the simulation time of the proposed model is about 80 times faster than that of the conventional model. From these results, the proposed model is useful for high accurate prediction of power electronics circuit performances and MBDs.

REFERENCES

[1] T. Mizoguchi. T. Naka, Y. Tanimoto, Y. Okada, W. Saito, M. Miura-Mattausch, and H. J. Mattausch, "Analysis of GaN-HEMT Switching Characteristics for High-Power Applications", Proc. the 28th ISPSD, pp. 267–270, 2016.

[2] Z. Shen and T. P. Chow, "An Analytical IGBT Model for Power Circuit Simulation", Proc. the 3rd ISPSD, pp. 79–82, 1991.

[3] R. Azar, F. Udrea, M. De Silva, G. Amaratunga, W. T. Ng, F. Dawson, W. Findlay, P. Waind, "Advanced SPICE Modeling of Large Power IGBT Modules", 37th IAS Annual Meeting, vol. 4, pp. 2433–2436, 2002.

[4] D. Navarro, T. Sano and Y. Furui, "A Sequential Model Parameter Extraction Technique for Physcal-Based IGBT Compact Models", IEEE Trans. Electron Devices, vol. 60, no. 2, pp. 580–586, 2013.

[5] R. Kraus, P. Türkes, J. Sigg, "Physic-Based Models of Power Semiconductor Devices for the Circuit Simulator SPICE", 29th Annual Power Electronics Specialists Conference, pp.1726–1731, 1998.

[6] M. Yamaguchi, I. Omura, S. Urano, S. Umekawa, M. Tanaka, T. Okuno, T. Tsunoda and T. Ogura, "IEGT Design Criterion for Reducing EMI Noise", Proc. the 16th ISPSD, pp. 115–118, 2004.

[7] A. Dastfan, "A New Macro-Model for Power Diodes Reverse Recovery", Proc. the 7th WSEAS International Conference on Power Systems, pp. 15–17, 2007.

Proceedings of the 31st International Symposium on Power Semiconductor Devices & ICs
May 19-23, 2019, Shanghai, China

Impact of three-dimensional current flow on accurate TCAD simulation for trench-gate IGBTs

Masahiro Watanabe[1], Naoyuki Shigyo[1], Takuya Hoshii[1], Kazuyoshi Furukawa[1], Kuniyuki Kakushima[1],
Katsumi Satoh[2], Tomoko Matsudai[3], Takuya Saraya[4], Toshihiro Takakura[4], Kazuo Itou[4], Munetoshi Fukui[4],
Shinichi Suzuki[4], Kiyoshi Takeuchi[4], Iriya Muneta[1], Hitoshi Wakabayashi[1], Akira Nakajima[5],
Shin-ichi Nishizawa[6], Kazuo Tsutsui[1], Toshiro Hiramoto[4], Hiromichi Ohashi[1], and Hiroshi Iwai[1]
E-mail: watanabe@ee.e.titech.ac.jp
[1]Tokyo Institute of Technology, Yokohama, Japan, [2]Mitsubishi Electric Corp., Fukuoka, Japan,
[3]Toshiba Electronic Devices & Storage Corp., Tokyo, Japan, [4]The University of Tokyo, Tokyo, Japan,
[5]Nat. Inst. Advanced Industrial Science and Technology, Tsukuba, Japan,[6]Kyushu University, Kasuga, Japan.

Abstract— TCAD simulation has been recognized as a powerful design tool for insulated gate bipolar transistors (IGBTs). In this work, excellent agreement between 3D TCAD simulations and experimental current-voltage characteristics were obtained in the up to 1000 A/cm^2 region for IGBTs with scaled trench-gates. The results of 2D and 3D simulations are compared to discuss the difference in current-voltage characteristics and their physical origins. A method to evaluate the saturation current (J_{Csat}) using a 2D simulation is also presented with an appropriate correction.

Keywords—IGBT, trench-gate, TCAD simulation, three-dimension current flow, scaling

I. INTRODUCTION

High-performance operation of insulated gate bipolar transistors (IGBTs) with scaled trench-gate [1] has been experimentally demonstrated [2]-[4]. Appropriate utilization of 2D and 3D technology computer-aided design (TCAD) simulation is important to optimize the performance of IGBTs. In previous studies [3], [5], good agreement was found between simulations and experiments for trench-gate IGBTs in the *low*-current region, close to the collector emitter saturation voltage (V_{CEsat}). A few works [6] were reported for comparisons of J_C-V_{CE} characteristics in the *high*-current region, however, the difference between 2D- and 3D-simulations was not clarified.

In the present study, 2D- and 3D-TCAD simulations of experimentally obtained J_C-V_{CE} characteristics [2] were performed with structural design parameters that were experimentally measured. The 2D- and 3D-simulation results were compared in the current range of 10 – 1000 A/cm^2 to discuss the origin of the difference between J_C-V_{CE} characteristics by 2D- and 3D-simulation. A method to evaluate saturation current (J_{Csat}) using 2D-simulation is also presented with an appropriate correction.

II. DEVICE STRUCTURE AND ANALYSIS

A. Device Structure

Fig. 1 shows a schematic diagram of the structure of the trench-gate IGBT analyzed in this work. Important structural design parameters are shown in Table I for scaled IGBTs [2]

This work is based on results obtained from a project commissioned by New Energy and Industrial Technology Development Organization (NEDO).

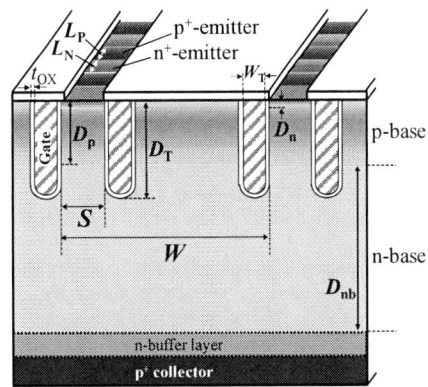

Fig.1: Schematic diagram of trench-gate IGBT structure [2].

with k = 1 and 3.

TABLE I. DEVICE PARAMETERS

Parameters	k=1	k=3
Device width W [µm]	16	16
Mesa width S [µm]	3	1
Trench depth D_T [µm]	6	2
Trench width W_T [µm]	1	0.33
p-base depth D_p [µm]	3	1.4
n$^+$emitter depth D_n [µm]	0.4	0.13
Oxide thickness t_{ox} [nm]	100	33
n-base depth D_{nb} [µm]	120	120
Emitter length L_N=L_P [µm]	4.5	1.5
Peak concentration of p-base [cm^{-3}]	1.0×10^{17}	3.8×10^{17}
n-base concentration [cm^{-3}]	8.5×10^{13}	8.5×10^{13}
n-buffer concentration [cm^{-3}]	9.0×10^{15}	9.0×10^{15}
p$^+$collector concentration [cm^{-3}]	3.7×10^{18}	3.7×10^{18}

The impurity concentration for n$^+$- and p$^+$- emitters was assumed to be 10^{20} cm^{-3}. The peak impurity concentration of the p-base region P_b was set to the measurement results by secondary ion mass spectroscopy (SIMS) depth profile measurements. The carrier life time was assumed to be 10 µs.

978-1-7281-0582-6/19 $31.00 © 2019 IEEE

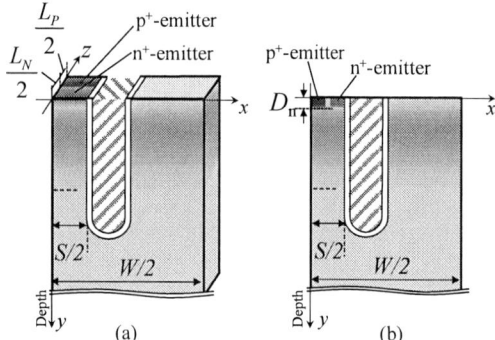

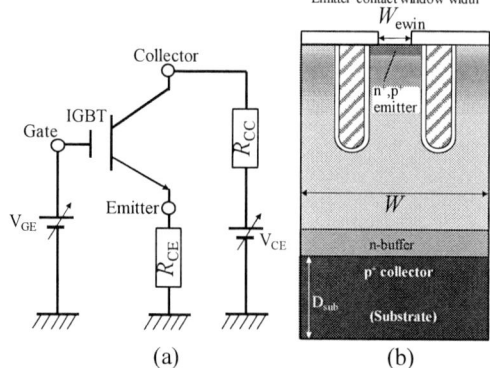

Fig.2: Half-cell models for (a) 3D- and (b) 2D-TCAD device simulations of the trench-gate IGBT.

Fig. 3: (a) The circuit model for mixed mode simulation used in this study. R_{CC} and R_{CE} indicate serial parasitic resistance.
(b) Structural geometry around emitter and collector contacts and substrate.

Figs. 2(a) and (b) show half-cell models for the 3D- and 2D-simulations, respectively. The most notable difference between the two models is the structure of the n$^+$- and p$^+$-emitters at the top of the emitter mesa. In the 2D-simulation, the p$^+$- and n$^+$-emitters were assumed to extend in the direction normal to the page (along the z-axis). In contrast, for the 3D-model, an n$^+$-emitter with a length of L_N and a p$^+$-emitter with a length of L_P were placed alternately where L_N+L_P indicates the pitch of the n$^+$- and p$^+$-emitter structure. A comparison of the 2D- and 3D-simulation results reveals the mechanism to determine the IGBT current by focusing on the current density distribution in the p-base region with the n$^+$- and p$^+$-emitters, which is an essential principle in the 3D scaling of IGBTs.

B. Analysis method

In this study, simulations were conducted using Synopsys TCAD Sentaurus™ M-2016.12-SP2. For physical models,

the bandgap narrowing proposed by Klaassen et al. [7] was modified to accord with the intrinsic carrier concentration model presented by Sproul and Green [8]. Klaassen's mobility model [9] was also adopted to consider the electron-hole scattering, because IGBTs are operated under high-level injection.

C. Parasitic resistances for mixed mode simulation

Fig. 3(a) shows the circuit model for the mixed-mode simulation used in this study. R_{CC} and R_{CE} indicate parasitic serial resistance related to collector and emitter electrodes, respectively. R_{CC} was evaluated to be $1.6 \times 10^{-4}\,\Omega\,cm^2$, which represents the resistance of the p$^+$-substrate with a thickness of 525 µm, and which was numerically extracted by 2D-simulation, while the current density was calculated from the cross sectional area of the n-base region. For simulation of the IGBT device, the p$^+$-collector thickness D_{sub} was

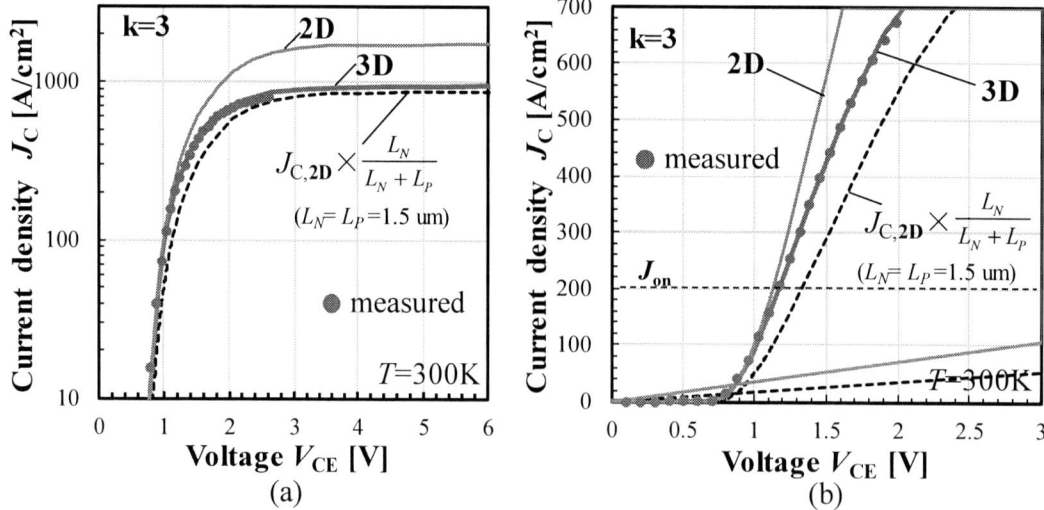

Fig.4: J_C-V_{CE} characteristics calculated by 2D- and 3D-TCAD simulations and corresponding experimentally measured results for an IGBT with k = 3 in (a) logarithmic and (b) linear format.

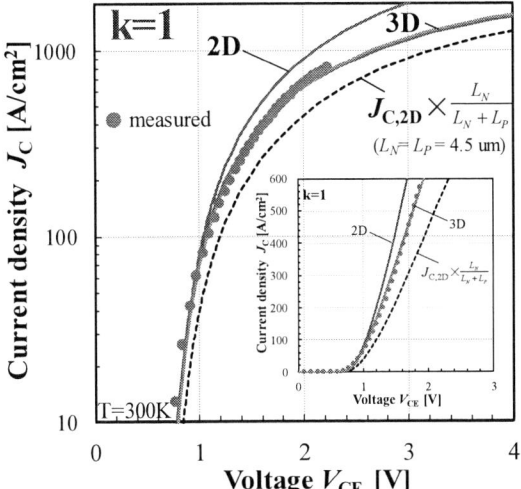

Fig.5: J_C-V_{CE} characteristics obtained from 2D- and 3D-TCAD simulations, and corresponding experimental results for IGBT with k = 1.

assumed to be 10 μm and the remaining thickness of 515 μm was considered as a serial parasitic resistance of R_{CC}. R_{CE} is mainly attributed to the contact resistance of the emitter electrode. The emitter contact resistivity was evaluated by measurement of the test element group (TEG) structure for a transmission line model (TLM) fabricated on the same wafer of the measured IGBT [2]. Thus, the contact resistivity ρ_{CE} was evaluated to be 4.7×10^{-6} Ωcm^2, therefore, R_{CE} was estimated by $\rho_{CE} \times W/W_{ewin} \sim 4.7 \times 10^{-4}$ Ωcm^2, where the device width W = 16 μm, the emitter contact window width $W_{ewin} \sim 0.16$ μm, which was confirmed by transmission electron microscopy (TEM) cross sectional image of k=3 IGBT structures. In the same way, for an IGBT with k=1, $R_{CE} \sim 1.8 \times 10^{-3}$ Ωcm^2 was estimated and used for the mixed-mode simulation.

III. RESULTS AND DISCUSSION

Figs. 4(a) and (b) show the J_C-V_{CE} characteristics obtained by 2D- and 3D-simulations, and the corresponding experimental results [2] for k = 3 ($L_N + L_P$ = 3 μm) on logarithmic and linear scales, respectively. The 3D-simulation results show excellent agreement with the experimental data for a wide range of J_C up to 1000 A/cm^2. However, the 2D-simulation reproduces the experimental data only for J_C < 100 A/cm^2. In the high current region, both J_{Csat} values obtained by both the 3D-simulation ($J_{Csat,3D}$) and those experimentally measured appear to be well fit by the expression $J_{Csat,2D} \times L_N/(L_N+L_P)$, where $J_{Csat,2D}$ is the J_{Csat} value obtained by the 2D-simulation.

Fig.5 shows similar results obtained for k = 1 ($L_N + L_P$ = 9 μm). In the low current region for J_C < 100 A/cm^2, the current is almost determined by the current supply limit of pn junction of bottom p$^+$-collector and n-buffer layer. On the other hand, for high currents greater than 1000 A/cm^2, the current is mainly restricted by the electron supply limit of the top metal oxide semiconductor (MOS) channel; therefore, the current determination mechanism gradually switches in the region between 100 A/cm^2 and 1000 A/cm^2, which can be well understood by the distribution of electron current flow in the emitter mesa, especially near the MOS channel.

Fig. 6 shows the distribution of the electron current density in the vertical direction of the a cross-section close to the channel surface. When J_C is less than the on-state current density (J_{on} = 200 A/cm^2), the injected electrons are distributed uniformly as shown in Fig. 6(a) and electrons accumulate slightly at a depth of around 1 μm in the p-base region. In this situation, the current is almost determined by the carrier supply limit of the bottom p$^+$-collector instead of the top n$^+$-emitter. Therefore, the p-base region where electrons are uniformly distributed acts as a passive electron emitter. This situation is almost the same in the 2D-simulation, so that $J_{C, 3D}$ is similar to $J_{C, 2D}$ in the low-current region although the emitter structure is significantly different between the 2D- and 3D-simulations. On the other hand, saturation current by 3D-simulation $J_{Csat,3D}$ in the high-current region is restricted approximately to $J_{Csat,2D} \times L_N/(L_N+L_P)$, as shown in Figs. 4(a) and 5, because

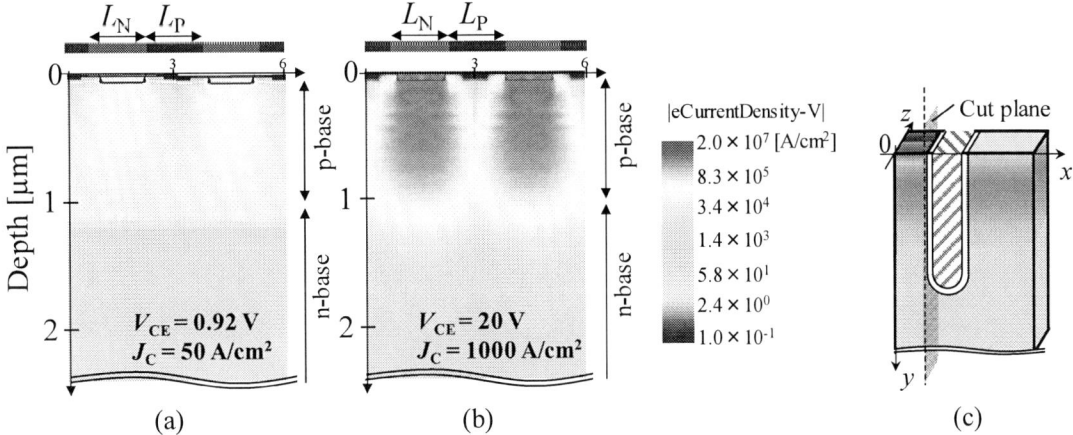

Fig.6: 3D-simulation results for distribution of electron current density in vertical direction in cross-section close to channel surface for IGBT with k = 3 at V_G = 5 V; (a) V_{CE} = 0.92 V and J_C = 50 A/cm^2, (b) V_{CE} = 20 V and J_C = J_{SC} =1000 A/cm^2, (c) position of cross-section.

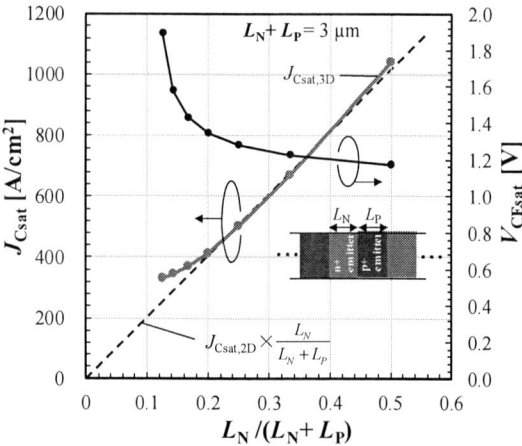

Fig. 7: Dependence of J_{Csat} and V_{CEsat} on $L_N/(L_N+L_P)$ calculated by 3D-simulation for $L_N+L_P= 3\mu m$.

electrons are injected only from the n$^+$-emitter, as shown in Fig. 6(b). When J_C approaches J_{Csat} as shown in Fig. 6(b), a beam-shaped electron current is injected only from the n$^+$-emitter. The electrons gradually disperse as they move through the p-base, because the electron velocity decreases close to the junction of the p-base and n-base regions. In this situation, the current is determined by the electron supply limit of the top MOS channel, i.e., the n$^+$-emitter. Therefore, the ratio of the length of n$^+$-emitter to the pitch (= $L_N/(L_N+L_P)$) directly restricts J_{Csat}.

Fig. 7 shows the dependence of J_{Csat} on $L_N/(L_N+L_P)$ determined from the 3D-simulation. It can be seen that J_{Csat} was decreased and V_{CEsat} was increased as $L_N/(L_N+L_P)$ decreased. A trade-off relation [10],[11] between the short-circuit capacity related to J_{Csat} and V_{CEsat} was observed. As $L_N/(L_N+L_P)$ decreased to around 0.3, a reduction of the saturation current can be expected with suppression of the increase in V_{CEsat}, which could contribute to an improvement of the short-circuit reliability.

The broken line in Fig. 7 shows a plot of $J_{Csat,2D}\times L_N/(L_N+L_P)$, which approximately reproduced $J_{Csat,3D}$. Thus, a 2D-simulation can be employed to roughly evaluate J_{Csat} by the use of an appropriate correction factor. For a deviation of J_{Csat} in the 2D- and 3D-simulations in the range of $L_N/(L_N+L_P) < 0.2$, a more detailed analysis is required. To achieve an accurate TCAD simulation of trench-gate IGBTs especially above the on-state current, the 3D current flow and carrier distribution in the p-base region should be taken into account.

IV. CONCLUSION

Excellent agreement of the current-voltage characteristics for scaled trench-gate IGBTs was obtained between the 3D-TCAD simulations and experimentally measured results up to 1000 A/cm^2. Structural parameters such as the peak impurity concentration and the electrode window width were measured and confirmed for the measured IGBTs. The application of 2D-simulations is limited to only the low-current region for trench-gate IGBTs because of the bias-dependent three-dimensional current flow and carrier distribution. A method to evaluate the saturation current (J_{Csat}) using 2D-simulation was also revealed with an appropriate correction.

ACKNOWLEDGMENT

The authors would like to thank W. Saito of Toshiba Electronic Devices & Storage Corp., I. Omura of Kyushu Institute of Technology, and M. Tsukuda of Green Electronics Research Institute for fruitful discussions.

This work is based on results obtained from a project commissioned by the New Energy and Industrial Technology Development Organization (NEDO).

REFERENCES

[1] M. Tanaka and I. Omura, "Scaling rule for very shallow trench IGBT toward CMOS process compatibility," *Proc. ISPSD 2016*, pp.177-179.

[2] K. Kakushima et al., "Experimental verification of a 3D scaling principle for low Vce(sat) IGBT," *Tech. Dig. IEDM* 2016, pp. 10.6.1-10.6.4.

[3] T. Hoshii *et al.*, "Verification of the injection enhancement effect in IGBTs by measuring the electron and hole currents separately," *Proc. ESSDERC 2018*, pp.26-29.

[4] T. Saraya et al., "Demonstration of 1200V scaled IGBTs driven by 5V gate voltage with superiorly low switching loss,", *Tech. Dig. IEDM 2018*, pp. 8.4.1-8.4.4.

[5] P. Luo et al., "Numerical analysis of 3-dimensional scaling rules on a 1.2-kV trench clustered IGBT," *IEEE Tran. Electron Devices*, vol. 65, no. 4, pp.1440-1446, 2018

[6] H. Feng et al., "A 1200 V-class fin p-body IGBT with ultra-narrow mesas for low conduction loss," *Proc., ISPSD 2016*, p.203-205.

[7] D. B. M. Klaassen, J. W. Slotboom and H. C. de Graaf, "Unified apparent bandgap narrowing in n and p-type silicon," *Solid-St. Electron.*, vol. 35, pp. 125-129, 1992.

[8] A. B. Sproul and M. A. Green, "Improved value for the silicon intrinsic carrier concentration from 275 to 375 K," *J. Appl. Phys.*, vol. 70, pp. 846-854, 1991.

[9] D. B. M. Klaassen, "Unified mobility model for device simulation," *Solid-St. Electron.*, vol. 35, pp. 953-959, 1992.

[10] R. S. Chokhawala, J. Catt and L. Kiraly, "A discussion on IGBT short-circuit behavior and fault protection schemes," *IEEE Trans. Ind. Appl.*, vol. 31, pp.256-263, 1995.

[11] K. Satoh et al., "New chip design technology for next generation power module," *Proc. PCIM 2008*, p.673.

978-1-7281-0582-6/19 $31.00 © 2019 IEEE

Proceedings of the 31st International Symposium on Power Semiconductor Devices & ICs
May 19-23, 2019, Shanghai, China

Self-sustained Oscillation of Superjunction MOSFET Intrinsic Diode During Reverse Recovery Transient

Peng Xue, Luca Maresca, Michele Riccio, Giovanni Breglio and Andrea Irace

Dept. of Electrical Engineering and Information Technologies, University of Naples Federico II, Naples, Italy

demosupen@gmail.com

Abstract—**In this paper, the self-sustained oscillation occurs on the reverse recovery transient of the superjunction MOS-FET intrinsic diode is studied. Based on the double-pulse test, the characteristics of the self-sustained oscillation is identified. Utilizing the Senturus TCAD simulation, the superjunction MOSFET intrinsic diode's reverse recovery oscillation behavior is reproduced. By analyzing the oscillation waveforms, the positive feedback mechanism which excites the oscillation is revealed at the end of the paper.**

Index Terms—**SuperJunction MOSFET, self-sustained oscillation, MOSFET intrinsic diode, reverse recovery**

I. INTRODUCTION

It is well known that the superjunction MOSFET intrinsic diode has very poor reverse recovery performance [1], [2]. With high peak reverse current, the snappy reverse recovery of superjunction MOSFET intrinsic diode gives rise to voltage and current oscillation [3], [4]. Under certain test conditions, the reverse recovery oscillation can maintain self-sustaining [5]. The self-sustained oscillation oscillation causes severe electromagnetic interface (EMI) problems and may completely disrupt the inverter operation in the worst case.

In this paper, the characteristics of the self-sustained oscillation is reported. Based on the TCAD mixed-mode simulation, the positive feedback mechanism which excites the self-sustained oscillation is revealed.

II. EXPERIMENTAL MEASUREMENT

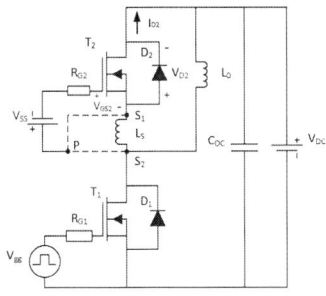

Fig. 1. The test circuit for the double pulse test.

The double pulse switching test is performed on the 650V/47A superjunction MOSFET to obtain the reverse recovery characteristics of the intrinsic diode. The schematic circuit for the double pulse test is shown in Fig. 1. In the test circuit, the inductor $L_0 = 350\mu H$ is utilized to maintain a nearly constant current throughout the switching cycle. R_{G1} and R_{G2} are the gate resistances. V_{DC} is the 1.2KV high voltage supply, which is connected to a capacitor bank C_{DC}. V_{ss} is the gate biased voltage. $-V_{ss}$ is equivalent to the off-state gate driving voltage of the high-side transistor T_2. By connecting the gate to its source, the body diode D_2 of the transistor T_2 is utilized as the freewheeling diode. In the initial attempts, the node P is connected to node S_2 (the front of the source pin). In this scenario, the stray inductance L_S of the device's source pin is utilized as the common source inductance in the test circuit.

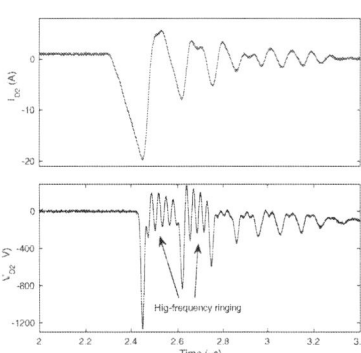

Fig. 2. The reverse recovery waveforms of superjunction MOSFET intrinsic diode with $V_{DC} = 100V$, $V_{ss} = 15V$, $R_{G1} = 10\Omega$ and $R_{G2} = 1\Omega$.

Fig. 2 presents the reverse recovery waveforms of superjunction MOSFET with with $V_{DC} = 100V$, $V_{ss} = 15V$, $R_{G1} = 10\Omega$ and $R_{G2} = 1\Omega$. The gate bias voltage V_{ss} is utilized to make sure the gate of the transistor T_2 is turned off. In the reverse recovery waveforms, the extremely high dV_{D2}/dt give rise to high-frequency voltage ringing when the diode voltage snap towards zero. However, except for the high-frequency voltage ringing, the oscillation is still primarily a nature damped ring-down.

978-1-7281-0582-6/19 $31.00 © 2019 IEEE   315

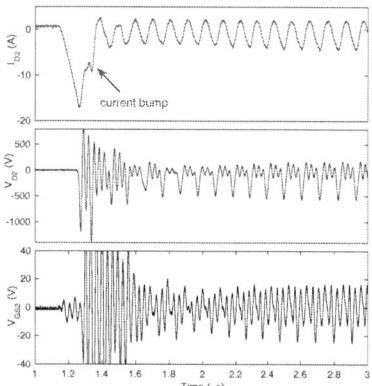

Fig. 3. The reverse recovery waveforms of superjunction MOSFET intrinsic diode with $V_{DC} = 100V$, $V_{ss} = 0V$, $R_{G1} = 10\Omega$ and $R_{G2} = 1\Omega$.

The double pulse test is also performed twice with the gate bias voltage V_{ss} removed (short circuited), whereas the values of V_{DC} and R_{G1} remain unchanged. In the first test, the gate resistance R_{G2} of the high-side transistor T_2 is still set to 1Ω. The obtained reverse recovery waveforms is shown in Fig. 3. In the second test, the gate resistance R_{G2} is removed. So the only resistance left in the gate loop is the internal gate resistance R_{Gi}, which is about 620 mΩ according to the data sheet. Fig. 4 presents the reverse recovery waveforms. Notice that the self-sustained oscillation with self-amplification effect occurs in both of the test waveforms. After the initial oscillation induced by the snap reverse recovery, the voltage and current oscillations do not damp down as present in Fig. 2. Instead, their oscillation amplitudes quickly grow and become self-sustaining in the end.

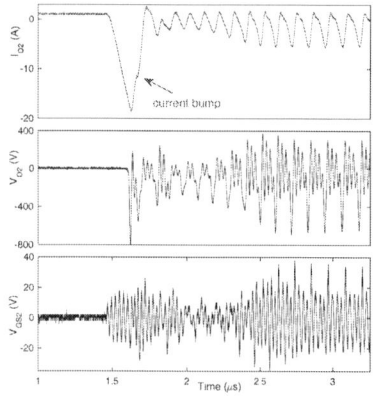

Fig. 4. The reverse recovery waveforms of superjunction MOSFET intrinsic diode with $V_{DC} = 100V$, $V_{ss} = 0V$, $R_{G1} = 10\Omega$ and $R_{G2} = 0\Omega$ (removed).

Another interesting phenomenon appears in both of the reverse recovery waveforms is the anomalous current bump of diode current I_{D2}, as shown in Fig. 3 and Fig. 4. The diode reverse current increase unexpectedly after it snaps off from the peak reverse value towards zero. This implies that the transistor T_2 is turned on during the reverse recovery transient and the MOS channel provide an additional current path for the current conduction. In order to validate this assumption, the gate voltage is measured in the test. As shown in Fig. 3 and Fig. 4, during the oscillatory transient, the gate-source voltage V_{GS2} rings above the MOSFET threshold voltage, which causes the unexpected turn-on of the MOSFET T_2. Moreover, the gate voltage oscillation also magnifies in accordance with the amplification of diode voltage and current oscillation. This implies that the gate voltage oscillation is related to the diode voltage and current oscillation.

In the last attempt, the node P is connected to node S_1 (the end of the source pin) to eliminate the common source inductance L_S. Other test setups are the same as that presented in Fig. 4. As presented in Fig. 5, the oscillation is suppressed to a ring-down which is very similar to the test result presented in Figure 2. The self-sustained oscillation totally vanishes. This demonstrates that the self-sustained oscillation also correlates with the common source inductance L_S.

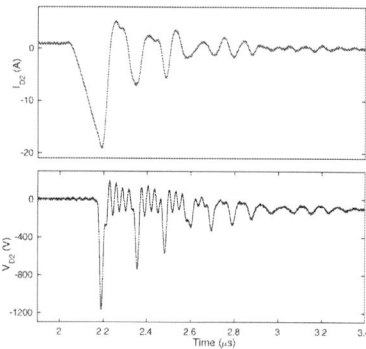

Fig. 5. The reverse recovery waveforms of I_{D2} and V_{D2} with $L_S = 0nH$.

III. TCAD SIMULATION

A. Simulation setup

In order to investigate the physical operation of superjunction MOSFET inartistic diode during the reverse recovery oscillation, the mixed-mode numerical simulation is performed utilizing Sentaurus TCAD simulator. In the simulation, the clamped inductive switching test circuit given in Fig. 1 is modified to include all the parasitic elements, as shown in Fig. 6. In the circuit, R_{G1} and R_{G2} are the gate resistances. R_C and L_C are the stray resistance and inductance in the power loop. L_G is the gate loop stray inductance. L_S is the common source stray inductance. In the simulation, the transistor T_1, T_2 and their inartistic diode are implemented by the numerical device model, whereas the other circuit elements are implemented by the spice model. All the circuit parameters utilized in the simulation are summarized in Table I.

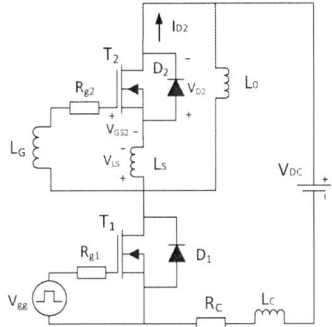

Fig. 6. The clamped inductive load test circuit for TCAD simulation with parasitic elements.

TABLE I
THE CIRCUIT PARAMETERS

Parameter	value	Parameter	value
R_C	0.2 Ω	L_C	200 nH
L_S	7 nH	L_G	14 nH
R_{G1}	10 Ω	R_{G2}	0.4 Ω
V_{DC}	100 V	L_0	350 μ H

B. Analysis of the reverse recovery Oscillations

With the setup presented in the previous subsection, the double pulse test is simulated. The simulated reverse recovery waveforms are presented in Fig. 7. After the diode reverse recovery process, significant self-amplified oscillation phenomenon is observed on the waveforms of V_{D2}, I_{D2} and V_{GS2}. Similar to the experimental results presented in Fig. 4, the divergent oscillation quickly attenuates to constant-amplitude oscillation. Since no resonant noise is utilized in the simulation to excite the oscillation, the self-amplification phenomenon is induced by the interaction between the device and circuit components.

To identify the mechanism of the self-sustained oscillation, it is of interest to find out the phase shift relationship between the related electrical parameters. Fig. 8 shows the waveforms of the voltage drop V_{LS} across the parasitic inductance L_s, gate current I_{G2}, gate-source voltage V_{GS2} and diode current I_{D2}. As shown in Fig. 8, high amplitude voltage V_{LS} oscillation occurs during the oscillatory transient. The V_{LS} oscillation drives the gate voltage V_{GS2} to resonate. When the gate voltage surpasses the MOSFET threshold voltage, the MOSFET is activated and starts to conduct current. When the MOSFET is turned off, the intrinsic diode begins to reverse recover. Therefore, the transistor T_2 operates under two type of conditions: (A). MOSFET conduction phase. (B). Intrinsic

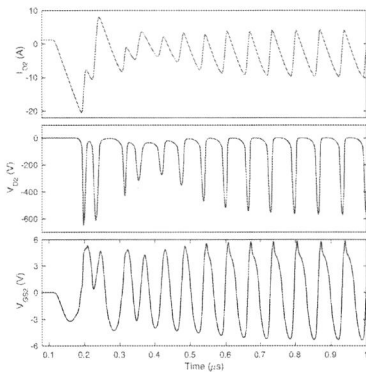

Fig. 7. Simulated waveforms of V_{D2}, I_{D2} and V_{GS2}.

diode reverse recovery phase, as shown in Fig. 8. Both of the phases are illustrated as follows:

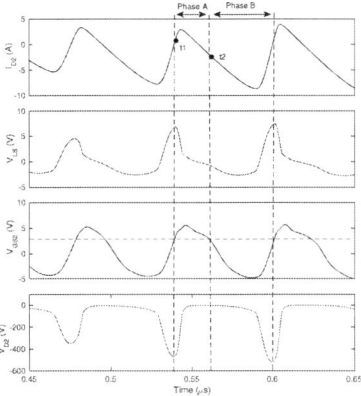

Fig. 8. The simulated waveforms of I_{D2}, V_{LS}, I_{G2} and V_{GS2}.

A). MOSFET conduction phase: When V_{GS2} takes higher value than the MOSFET threshold voltage at time point t_1, this phase starts. At this time, the MOS conductive channel is open, which enables the electron current injection in the depleted drift region. The diode current slope dI_{D2}/dt thereby starts to decrease and eventually changes its polarity from positive to negative. As a result, V_{LS} starts to decrease and becomes a negative value at the end of this phase. Driven by the V_{LS}, the gate voltage V_{GS2} start to decrease. When the V_{GS2} drops down below the threshold voltage, the MOSFET is turned off, which marks the end of this phase.

B). Intrinsic diode reverse recovery phase: When the MOSFET is turned off, the body diode is activated to support reverse voltage. In the beginning, the diode current slope dI_{D2}/dt still has a negative value. When diode current begins to snap off from its peak value toward zero, the current slope

978-1-7281-0582-6/19 $31.00 © 2019 IEEE   317

dI_{D2}/dt changes its polarity again from negative to positive. Accordingly, V_{LS} transfers from a negative value to a positive value. Driven by the positive V_{LS}, V_{GS2} starts to increase. When V_{GS2} surpass the threshold voltage, the MOSFET is activated again. After that, the MOSFET conduction phase and body diode recovery phase will be repeated and the self-sustained oscillation is generated.

IV. THE POSITIVE FEEDBACK MECHANISM

The oscillation process depicted in the previous subsection suggests an underlying positive feedback mechanism. The voltage V_{GS} determines the carrier injection level in the drift region when the MOSFET is turned on. This will influence the peak reverse current I_{RRM}, and eventually change the current slope S_B when the diode starts to snap off. On the other hand, excited by the current slope S_B, the positive voltage V_{LS} on the common source inductance can also drive the gate-source voltage V_{GS}. As shown in Fig. 9, suppose the gate-source voltage V_{GS} has a slight value jump, the MOS channel conduction time T_C thereby increases. During the MOSFET conduction phase, with larger V_{GS} and longer T_C, a higher carrier injection levels in the drift region can be obtained, which can support higher peak reverse current I_{RRM}. Noting that the current slope S_B can be expressed [6]:

$$S_B = I_{RRM}^2/2Q_R \qquad (1)$$

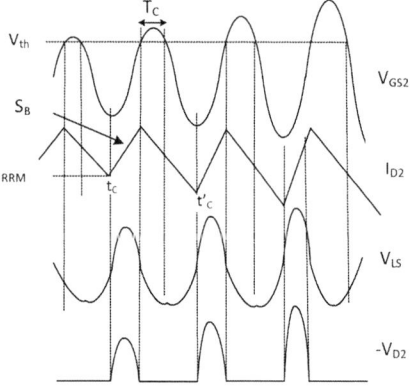

Fig. 9. The positive feedback process between the gate loop and power loop.

When the intrinsic diode starts to snap off at t_C or t'_C (t_C and t'_C are the time when the diode current I_D achieves to its peak reverse value, as shown in Fig. 9, the drift region depletes both vertically and horizontally [7]. Due to the two-dimensional depletion behavior, the injected carriers in the drift region are almost completely removed. Thus, despite the enhanced carrier injection during the MOSFET conduction process, the residential charge Q_R at t'_C isn't significantly enhanced compared to that at t_C. Therefore, it can be concluded from (1) that dI_{RR}/dt increase quadratically with the growth

of I_{RRM}. This magnifies the positive oscillation amplitude of the voltage V_{LS}, which drives the V_{GS} to a higher peak value. After that, the same process will be repeated and the positive feedback process is obtained. Noting that the reverse diode voltage is $-V_{D2} = V_{DC} + L_cS_B$. The reverse diode voltage $-V_{D2}$ also rises in accordance with the continued increasing of S_B. As a result, the self-amplification phenomenons observed in the experiment and simulation are generated.

V. CONCLUSION AND DISCUSSION

In this paper, the reverse recovery oscillation of the superjunction MOSFET intrinsic diode is comprehensively investigated. In the double-pulse test, it is identified that the self-sustained voltage and current oscillations can be excited during the reverse recovery transient of the superjunction MOSFET intrinsic diode. Based on the TCAD simulation, the mechanism of the self-sustained oscillation is revealed.

REFERENCES

[1] X. Cheng, X. M. Liu, J. K. O. Sin, and B. W. Kang, "Improving the CoolMOS body-diode switching performance with integrated schottky contacts," in *IEEE International Symposium on Power Semiconductor Devices and Ics, 2003. Proceedings. Ispsd*, 2003, pp. 304–307.

[2] W. Saito, I. Omura, S. Aida, S. Koduki, M. Izumisawa, and T. Ogura, "Semisuperjunction MOSFETs: new design concept for lower on-resistance and softer reverse-recovery body diode," *IEEE Transactions on Electron Devices*, vol. 50, no. 8, pp. 1801–1806, 2003.

[3] W. C. S. Kim, "650V fast recovery SuperFET II MOSFET for high system efficiency and reliability in resonant topologies," Fairchild application notes, AN-5235, 2015.

[4] W. C. D. Son, "New generation super-junction MOSFETs, SuperFET II and SuperFET II easy drive MOSFETs for high efficiency and lower switching noise," Fairchild application notes, AN-5232, 2013.

[5] P. Xue, L. Maresca, M. Riccio, G. Breglio, and A. Irace, "Investigation on the self-sustained oscillation of superjunction MOSFET intrinsic diode," *IEEE Transactions on Electron Devices*, vol. 66, no. 1, pp. 605–612, 2019.

[6] B. J. Baliga, *Fundamentals of Power Semiconductor Devices*. New York: Springer, 2008.

[7] R. Ng, F. Udrea, K. Sheng, and G. A. J. Amaratunga, "A study of the CoolMOS integral diode: Analysis and optimisation," in *Proceedings International Semiconductor Conference*, 2001, pp. 461–464 vol.2.

Proceedings of the 31st International Symposium on Power Semiconductor Devices & ICs
May 19-23, 2019, Shanghai, China

Static and Dynamic Figures of Merits (FOM) for Superjunction MOSFETs

H. Kang, and F. Udrea, *Member, IEEE*
Electrical Engineering
University of Cambridge
Cambridge, U.K.
email: hk428@cam.ac.uk, fu@eng.cam.ac.uk

Abstract— The introduction of the superjunction has challenged the well-known limit of silicon. In spite of its widespread use, and despite its correct application in classical power MOSFETs, Baliga's figure of merit no longer can be used to give an accurate picture of the material potential in power electronics. Fujihira has updated both the static and dynamic figures of merit, but, as the cell pitch of the SJ scale down, these figure of merits need to include the parasitic JFET effect and an analytical formulation of the gate-drain capacitance and its intricate dependence on the drain voltage. This work presents for the first-time new static and dynamic figure of merits which take into account the spectacular advances in the SJ technology, where the cell dimensions are approaching the sub-micrometer level.

Index Terms—Superjunction, Figure of merit, Switching loss.

I. INTRODUCTION

As the silicon superjunction (SJ) metal-oxide-semiconductor field effect transistors (MOSFETs) have been dominating most of the market for switched mode power supplies (SMPS) in the range of 600 ~ 900V [1], the switching loss of the SJ power MOSFETs has been one of the main concerns for the efficiency of power converters [2]–[4]. To assess the performance of the power MOSFETs including the switching loss (hard switching), two dynamic figure of merits (*FOM*) have been widely utilized [5]–[7].

$$FOM\left(GD\right) = R_{sp} \cdot Q_{GD.sp}, \tag{1}$$

$$FOM\left(DS\right) = R_{sp} \cdot Q_{Oss.sp}. \tag{2}$$

where R_{sp}, $Q_{GD.sp}$, and $Q_{Oss.sp}$ are the specific resistance, specific switching charge for charging the gate-drain capacitance, C_{GD}, and specific switching charge for charging the output capacitance, C_{Oss}, respectively. In MOSFET's hard switching, the term $R_{sp} \cdot Q_{GD.sp}$ encompasses both the switching losses (turn-on and turn-off) and the on-state conduction loss as derived by Huang [8]. Another important *FOM* for measuring the superjunction device performance is $R_{sp} \cdot Q_{Oss}$. These two FOMs have played a crucial role in assessing the performance of different power MOSFET technologies.

Superjunction MOSFET are based on deep n-pillar and p-pillar arrays placed in the drift region. The junction between the pillars presents a very high drain-source capacitance, C_{DS} [9] and, therefore, the turn-off switching loss (off-loss) of the SJ MOSFET should be mostly attributed to the switching charge associated with C_{DS}. The turn off switching loss is more important than turn-on loss for some applications such as data servers, telecom, [10] and electric vehicles [11], [12], where zero-voltage-switching (ZVS) assisted resonant converters are widely used. Typical waveforms for hard and soft (ZVS) switching converter are shown in Fig. 1.

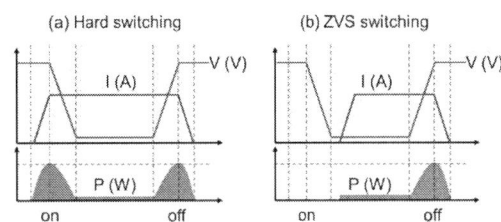

Fig. 1 Schematic waveforms of (a) hard switching and (b) zero voltage switching.

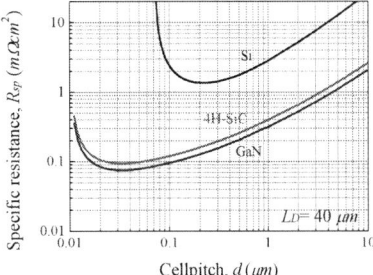

Fig. 2. R_{sp} vs cell pitch at 40 μm pillar length given by equation (3) for Si, 4H-SiC, and GaN vertical superjunction MOSFETs.

To establish the figure of merits given by equations (1) and (2) for superjunctions, the R_{sp} and the capacitance values should be derived as a function of material parameters. The R_{sp}, which includes the parasitic JFET present within the superjunction pillars, has the following expression, earlier derived by us in [12]:

$$R_{sp.real} = \frac{2L}{q\mu_n N_D} \cdot \left(\frac{d}{d - W_{bi}}\right). \tag{3}$$

where, L_D, q, μ_n, and N_D are the length of the pillar, unit charge, electron mobility, and doping concentration in the pillar, respectively. d and W_{bi} are the cell pitch and built-in depletion width towards the n-pillar, respectively. Fig. 2 shows the R_{sp} variation with the cell pitch. The minimum superjunction R_{sp} occurs at a certain cell pitch for a given material and a given pillar length (which determines the breakdown voltage). A detailed calculation procedure is described in reference [13]. The analytical expression of R_{sp} given by equation (3) is not too different from the ideal R_{sp} quoted in commercially available devices (with cell pitches

978-1-7281-0582-6/19 $31.00 © 2019 IEEE

$> 1 \, \mu m$).

$$R_{sp.ideal} = \frac{2L_D}{q\mu_n N_D} . \tag{4}$$

However, owing to the nonlinearity of the superjunction capacitances (C_{GD} and C_{DS}), there have been no specific dynamic *FOM*s for superjunctions. In this study, by suggesting inner capacitance circuit models in accordance with the depletion behavior of a superjunction structure, we establish analytic models for the $Q_{GD.sp}$ and $Q_{DS.sp}$ in a SJ and define the dynamic *FOM*s.

II. C_{GD} CAPACITANCE MODEL

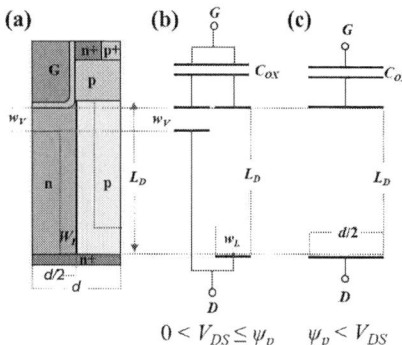

Fig. 3. Internal circuit configuration of Gate-to-Drain capacitance, C_{GD}, with respect to V_{DS} in a superjunction MOSFET.

The cell pitch for the SJ in Fig. 3(a) is d, and the width of the n-pillar and the p-pillar are the same as $d/2$. The lateral oxide above the n-pillar covers the entire area of the n-pillar. The non-linear behavior of the C_{GD} for the SJ should be attributed to the pillar's depletion process. When the V_{DS} is less than the pinch-off potential, ψ_p (the voltage when the pillars are fully depleted along the lateral direction), the C_{GD} is divided into two components: (1) gate oxide – accumulation layer capacitor connected in series, (2) gate oxide - n-pillar depletion capacitor connected in series. The capacitors (1) and (2) are connected in parallel. The oxide - accumulation layer capacitance, C_{ox-ac}, can be written as

$$C_{ox-ac} = \frac{\frac{\varepsilon_{ox}}{t_{ox}}\frac{\varepsilon_S}{w_V}\left(\frac{d}{2}-w_L\right)^2 Z^2}{\frac{\varepsilon_{ox}}{t_{ox}}\left(\frac{d}{2}-w_L\right)Z + \frac{\varepsilon_S}{w_V}\left(\frac{d}{2}-w_L\right)Z} . \tag{5}$$

where, ε_{ox}, ε_S, and t_{ox} are the permittivity of the oxide, the permittivity of the semiconductor, and the oxide thickness (here 100 nm), respectively. w_V, w_L, and Z are the depletion width towards the accumulation layer, the depletion width towards the n-pillar, and width of the cell (third dimension into the paper). In the same way, the oxide – n-pillar capacitance, C_{ox-pi}, can be written as

$$C_{ox-pi} = \frac{\frac{\varepsilon_{ox}}{t_{ox}}\frac{\varepsilon_S}{L_D}w_L{}^2 Z^2}{\frac{\varepsilon_{ox}}{t_{ox}}w_L Z + \frac{\varepsilon_S}{L_D}w_L Z} \approx \frac{\varepsilon_S}{L_D}w_L Z . \tag{6}$$

Where L_D is the length of the pillar. Since the length of the pillar is long enough, the denominator given by equation (6) can be equalized as the oxide capacitance, $\varepsilon_{ox}w_L Z/t_{ox}$, and the

equation becomes $\varepsilon_s w_L Z/L_D$. The depletion widths given by equations (5) and (6) have the following relationships:

$$w_V = \sqrt{\frac{2\varepsilon_S V_{DS}}{qN_D}} , \tag{7-1}$$

$$w_L = \sqrt{\frac{\varepsilon_S V_{DS}}{qN_D}} . \tag{7-2}$$

Where q and N_D are unit charge and the doping concentration in the n-pillar and the p-pillar. From equation (7-2), the pinch-off potential (when the depletion width is $d/2$) for the pillars can be written as:

$$\psi_p = \frac{qN_D}{4\varepsilon_s}d^2 . \tag{8}$$

After the pinch-off potential, the C_{GD} circuit becomes Fig. 1(c) and the capacitance is

$$C_{GD} \approx \frac{\varepsilon_S}{L_D}\frac{dZ}{2} \quad (\psi_p < V_{DS}). \tag{9}$$

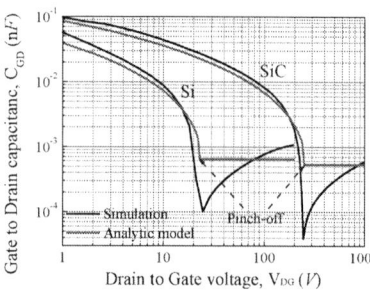

Fig. 4. The analytical model and the simulation data of Gate-to-Dwwrain capacitance. The cell pitch, d= 5 μm, the depth of pillar Z= 1.0 μm, the length of the pillar, L_D = 40 μm, the number of cells=2×10^5, the doping concentration in the pillars= 2.31×10^{15} cm^{-3} for Si, and 2.09×10^{16} cm^{-3} for SiC.

Fig. 4 shows the analytical model and the simulation results for both Si and SiC superjunction MOSFETs. The concentration of the pillars was one-third of the theoretical maximum value for a practical approach. [14]. The analytical model shows very similar drain to gate capacitance values and shapes with respect to the V_{DG}. As it will be discussed later, the high value of the pinch-off potential in the SiC device owing to the high doping pillar doping concentration increases the switching loss because the larger area below the capacitance curve presents a higher energy loss. The drain to gate capacitance of a superjunction device must have a valley where the value decreases rapidly with respect to the applied voltage. This region corresponds to the situation where the pillars of a superjunction become fully depleted laterally (pinch-off potential). This is associated with the transition point from Fig. 3(b) to (c).

III. C_{OSS} CAPACITANCE MODEL

As mentioned above, owing to the long enough length (depth) of the pillars, the drain to source capacitance can be a pproximated as the output capacitance, $C_{Oss}=C_{DS}$. By using t he inner circuit model provided in Fig. 5(b) and (c), the outp ut capacitance before and after the pinch off potential, ψ_p, ca n be written as

$$C_{Oss} = \frac{\varepsilon_S}{2w_L}\left(L_D - 2w_V\right)Z + 2\frac{\varepsilon_S}{\left(\dfrac{d}{2} + w_L\right)}w_V Z$$

$$(\psi_p > V_{DS}), \quad (10)$$

$$C_{Oss} = \frac{\varepsilon_S}{L_D}dZ \quad (\psi_p < V_{DS}). \tag{11}$$

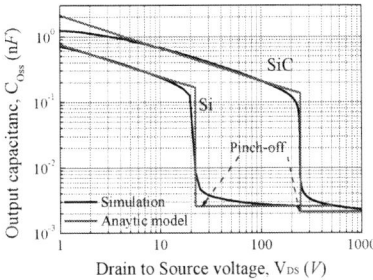

(a)　　　**(b)**　　　**(c)**

$$0 < V_{DS} \le \psi_p \quad\quad \psi_p < V_{DS}$$

Fig. 5. Internal circuit configuration of Drain-to-Source (Output) capacitance, $C_{Oss} = C_{DS}$, with respect to V_{DS} in a superjunction MOSFET.

Fig. 6 shows the simulation data and the analytic models for both Si and SiC devices.

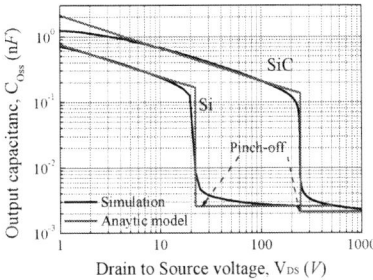

Fig. 6. The analytical model and the simulation data of output capacitance. The detailed material and geometrical values are the same as Fig. 4.

IV. FIGURE OF MERIT, *FOM(GD)*

The total switching charge, $Q_{GD.sp}$ for charging the C_{GD} during the off-state is the area below the $C_{GD}(V_{DS})$ curve as shown in Fig. 4:

$$Q_{GD.sp} = \left(\int_0^{\psi_p} C_{GD}dV_{DS} + \int_{\psi_p}^{V_{DS}} C_{GD}dV_{DS}\right)/dZ. \tag{12}$$

The integral of equation (12) leads to the following results:

$$Q_{GD.sp} \approx qN_D d\frac{\sqrt{2}}{8}\frac{d\varepsilon_{ox}}{t_{ox}\varepsilon_S} + \frac{\varepsilon_S}{2L_D}V_{DS}. \tag{13}$$

By inserting the pinch-off potential, ψ_p, into V_{DS} given by equation (8) into equation (13), $Q_{GD.sp}$ becomes

$$Q_{GD.sp} = \left(\frac{\sqrt{2}}{2}\frac{\varepsilon_{ox}}{t_{ox}} + \frac{\varepsilon_S}{2L_D}\right)\psi_p \approx \frac{\sqrt{2}}{8}\frac{\varepsilon_{ox}}{t_{ox}}\frac{qN_D}{\varepsilon_s}d^2. \tag{14}$$

As shown in Fig. 4, due to the relatively large area below the C_{GD} curves until ψ_p, most of the switching loss comes from the range of $0 < V_{DS} < \psi_p$. Equation (14) also indicates that the switching loss by C_{GD} for the SiC superjunction devices will be much higher than that of the Si device owing to the higher pinch-off potential (or higher doping concentration in the pillars)

As mentioned above, *FOM(GD)* can be obtained by multiplying $C_{GD.sp}$ with the superjunction ideal specific resistance, R_{sp}:

$$FOM\left(GD\right) = \frac{\sqrt{2}}{4}\frac{\varepsilon_{ox}}{t_{ox}}\frac{L_D d^2}{\varepsilon_S\mu_n}. \tag{15}$$

Equation (15) indicates that the power loss from C_{GD} changes little by changing the materials. i.e. the benefit from the lowered on-state resistance of a SiC superjunction is canceled out by the increased switching loss. In our previous study, the breakdown voltage, V_B, and the critical electric field, E_C, had the following relationship [14]:

$$L_D = \frac{\sqrt{2}V_B}{E_C}. \tag{16}$$

By inserting equation (16) into equation (15), the *FOM(GD)* becomes

$$FOM\left(GD\right) = \frac{1}{2}\frac{\varepsilon_{ox}}{t_{ox}}\frac{V_B}{\mu_n\varepsilon_S E_C}d^2. \tag{17}$$

Equation (17) contains the inverse of the product of two sub-figure of merits, which we term (i) the technology dependent figure of merit at a given material, $FOM(GD)_T$,

$$FOM\left(GD\right)_T = \frac{t_{ox}}{\varepsilon_{ox}d^2}, \tag{18-1}$$

and (ii) the material dependent Figure of merit at a given technology, $FOM(GD)_M$,

$$FOM\left(GD\right)_M = \mu_n\varepsilon_S E_C. \tag{18-2}$$

The higher these sub-*FOMs*, the better the dynamic performance of the SJ device. One can use the $FOM(GD)_M$ to assess different materials, and $FOM(GD)_T$ to assess the impact given by the geometrical/technology capability of the technology.

V. FIGURE OF MERIT, *FOM(DS)*

By using the same way shown in equation (12), $Q_{Oss.sp}$ can be obtained by integrating $C_{Oss.sp}$:

$$Q_{Oss.sp} = \left(\int_0^{\psi_p} C_{Oss}dV_{DS} + \int_{\psi_p}^{V_{DS}} C_{Oss}dV_{DS}\right)/dZ. \tag{19}$$

Owing to the small value after the pinch-off potential, the second term given by equation (19) was ignored:

$$Q_{Oss.sp} = \frac{qN_D}{2}L_D - \frac{\sqrt{2}qN_D}{20}d \approx \frac{qN_D}{2}L_D. \tag{20}$$

By multiplying the ideal specific resistance, the final switching figure of merit becomes

$$FOM\left(DS\right) = \frac{qN_D}{2}L_D \cdot \frac{2L_D}{qN_D\mu_n} = \frac{L_D^2}{\mu_n}. \tag{21}$$

The figure of merit can be expressed as material forms by inserting equation (16) into (21):

$$FOM(DS) = 2\frac{V_B{}^2}{\mu_n E_C{}^2}. \tag{22}$$

From equation (22), at a given technology (given voltage rating, V_B) the material dependent figure of merit, $FOM(DS)_M$ can be written as

$$FOM(DS)_M = \mu_n E_C{}^2 \tag{23}$$

In the previous study reported by Fujihira and Miyasaka, a similar figure of merit $R_{sp}Q_{Oss.sp}$ for a superjunction MOSFET can be found [15]:

$$FOM = \frac{81}{64}\frac{V_B{}^2}{\mu_n E_C{}^2} \text{ (Fujihira and Miyasaka)}. \tag{24}$$

The exact model employed for deriving the FOM in [14] is unclear. In spite of the fact that the coefficients given by equation (22) resulted from this work and (24) from reference [14], the FOM derived here and in [14] are consistent. Furthermore, the analytical results given by equation (22) show a good agreement with simulation data for both Si and SiC.

VI. CONCLUSION

The suggested inner capacitance circuit and the analytical models for switching of superjunction MOSFETs showed good agreement with the simulation results for both Si and SiC devices. Owing to the high concentration of a SiC superjunction MOSFET, the pinch-off potential was 9 times higher than that of silicon. This high pinch-off potential contributed to a higher switching loss. The benefit from the decreased specific resistance of a SiC device was offset to some extent by the increased switching loss. Nevertheless, SiC and vertical GaN still offer superior FOMs when compared to silicon. In this work we derived new dynamic FOMs based on advanced calculations of the specific on-state resistance of the superjunction and analytical of the non-linear parasitic capacitances in the drift region.

ACKNOWLEDGMENT

This research is supported by On Semiconductor Corporation as a part of a future power electronics technology. The authors are particularly grateful to Thomas Neyer for his valuable advice on this study and his general support of this research.

REFERENCES

[1] F. Udrea, G. Deboy, S. Member, and T. Fujihira, "Superjunction Power Devices, History, Development, and Future Prospects," *IEEE Trans. Electron Devices*, vol. 64, no. 3, pp. 713–727, Jan. 2017.

[2] D. W. Hart, *Power Electronics*. Mc Graw-Hill, 2010.

[3] J. B. Fedison, M. Fornage, M. J. Harrison, and D. R. Zimmanck, "Coss related energy loss in power MOSFETs used in zero-voltage-switched applications," *2014 IEEE Appl. Power Electron. Conf. Expo. - APEC 2014*, pp. 150–156, Mar. 2014.

[4] J. Roig and F. Bauwens, "Origin of Anomalous Coss Hysteresis in Resonant Converters With Superjunction FETs," *IEEE Trans. Electron Devices*, vol. 62, no. 9, pp. 3092–3094, Sep. 2015.

[5] B. J. Baliga, "Power Semiconductor Device Figure of Merit for High-Frequency Applications," *IEEE Electron Device Lett.*, vol. 10, no. 10, pp. 455–457, Oct. 1989.

[6] I.-J. Kim, S. Matsumoto, T. Sakai, and T. Yachi, "New power device figure of merit for high-frequency applications," in *Proceedings of International Symposium on Power Semiconductor Devices and IC's: ISPSD '95*, 1995, pp. 309–314.

[7] A. Hopkins, N. McNeill, P. Anthony, and P. Mellor, "Figure of merit for selecting super-junction MOSFETs in high efficiency voltage source converters," in *2015 IEEE Energy Conversion Congress and Exposition (ECCE)*, 2015, pp. 3788–3793.

[8] A. Q. Huang and S. Member, "New Unipolar Switching Power Device Figures of Merit," *IEEE Electron Device Lett.*, vol. 25, no. 5, pp. 298–301, Jun. 2004.

[9] Infineon Technologies, "CoolMOS™ C7 : Mastering the Art of Quickness," Apr. 2013.

[10] H.-J. Chiu, L.-W. Lin, P.-L. Pan, and M.-H. Tseng, "A novel rapid charger for lead-acid batteries with energy recovery," *IEEE Trans. Power Electron.*, vol. 21, no. 3, pp. 640–647, May 2006.

[11] M. Yilmaz and P. T. Krein, "Review of Battery Charger Topologies, Charging Power Levels, and Infrastructure for Plug-In Electric and Hybrid Vehicles," *IEEE Trans. Power Electron.*, vol. 28, no. 5, pp. 2151–2169, Aug. 2013.

[12] N. Shafiei, M. Ordonez, M. Craciun, C. Botting, and M. Edington, "Burst Mode Elimination in High-Power-Resonant Battery Charger for Electric Vehicles," *IEEE Trans. Power Electron.*, vol. 31, no. 2, pp. 1173–1188, Apr. 2016.

[13] H. Kang and F. Udrea, "True Material Limit of Power Devices -Applied to 2-D Superjunction MOSFET," *IEEE Trans. Electron Devices*, vol. 65, no. 4, pp. 1432–1439, Mar. 2018.

[14] H. Kang and F. Udrea, "Material Limit of Power Devices--Applied to Asymmetric 2-D Superjunction MOSFET," *IEEE Trans. Electron Devices*, vol. 68, no. 8, pp. 1–7, May 2018.

[15] T. Fujihira and Y. Miyasaka, "Simulated superior performances of semiconductor superjunction devices," *Proc. 10th Int. Symp. Power Semicond. Devices ICs.*, no. V, pp. 423–426, Jul. 1998.

Proceedings of the 31st International Symposium on Power Semiconductor Devices & ICs
May 19-23, 2019, Shanghai, China

New Locos Trench Oxide IGBT Enables 25% Higher Current Density in 4.5kV/1500A Module

[1]L. Ngwendson, [1]I. Deviny, [1]C. Zhu, [1]I. Saddiqui, [1]J. Hutchings, [1]C. Kong, [1]Y. Wang [2]H. Luo

[1]Research and Development Centre, Dynex Semiconductor Ltd., Lincoln, UK

[2] Zhuzhou CRRC Times Electric Co., Ltd., Zhuzhou, China

Email: Luther.Ngwendson@dynexsemi.com

Abstract—**In this paper we present first results of a 4.5kV/1500A IGBT module using new LOCOS (Local Oxidation of Silicon) Trench Oxide IGBTs (LTO–IGBT). It is shown that this new module with increased current density compared to the previous DMOS generation, also show very robust dynamic and competitive performance at $T_J = 150^0C$.**

Keywords: IGBT, LOCOS, Trench IGBT

I. INTRODUCTION

The recent trend of increasing current density and junction temperature in high voltage IGBT modules (3.3kV to 6.5kV) is to satisfy the demands of modern traction and other high power systems design. The key features in these systems include high power density, high efficiency, high reliability and electrical/environmental robustness. The trench gate IGBT technology is very attractive at high voltages compared to DMOS, because it offers much reduced Vce(sat) at increasing current densities as silicon thickness increases. However, It is important to design High Voltage trench gate IGBTs with inbuilt robustness.

High voltage trench gate IGBTs with conventional uniform thin oxide can suffer from two well-known which are a) High switching energy (Eon+Eoff) loss due to high Cgc and Cge and b) high electric field at the trench bottom regions which affects the reliability of the gate oxide. Also, the addition, of a carrier stop or nwell-layer (to reduce Vceon) can have the adverse effect of further increasing electric field in this region. High electric fields will cause accelerated degradation of oxide in the trench bottom areas which impacts long term reliability [1,2].

It was been demonstrated in [1] that IGBTs with uniformly thin gate oxide can suffer from dI/dt changes after thousands of switching events. The phenomenon is due to injection of hot hot carriers into oxide at high voltages. Also high electric field will increase the tunnelling current into gate oxide and affect long term reliability performance.

One of the ways to improve long term reliability of Trench gate IGBTs is to increase gate oxide thickness especially in the bottom region of the trench using LOCOS (Local Oxidation of Silicon) process[3]. In this paper we show the electrical performance of the new 4.5kV/1500A module with LOCOS Trench oxide technology IGBTs in the 190mm x 140mm module footprint, designed for long term reliability. It is shown that very robust performance can be achieved using this technology.

II. LTO-IGBT SIMULATION AND DESIGN

The schematic of the LTO-IGBT device is shown in fig. 1a. The device has an nwell or carrier stop layer to achieve improved Vceon/Eoff trade-off. Extensive TCAD simulations have been performed to investigate the benefits of the configuration of thick oxide in the trench bottom region.

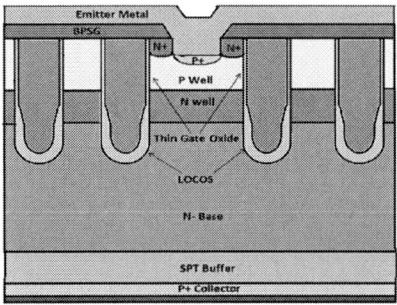

Fig1: LOCOS Trench Oxide IGBT active cell design. Thickness of thick oxide or LOCOS regions is <5000nA.

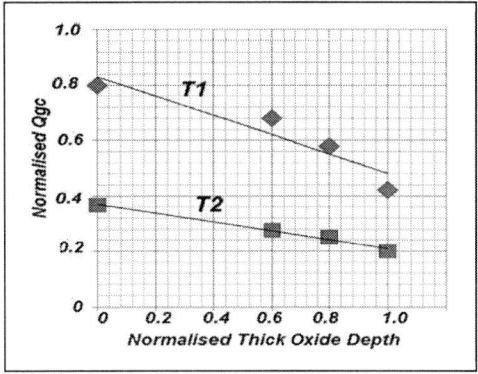

Fig.2: Influence of the depth of thick oxide portion for two different total trench depths (T1>T2) on Qgc Vce=3600V, I=62.5A per chip, Tj=25[0]C

Fig.2 and Fig.3 respectivley show simulated influence on Qgc(miller charge) and Qg(gate charge) for different thick oxide portion depths

978-1-7281-0582-6/19 $31.00 © 2019 IEEE

relative to total trench depth (T), for T1>T2. It can be seen that Qgc (miller charge) reduction with increasing thick oxide portion is more aggressively for the deep trench T1. Fig.4 where it is shown that Qgc or miller chage reduction of 30% has been achieved. This also coresponds to 30% reduction in Cgc.

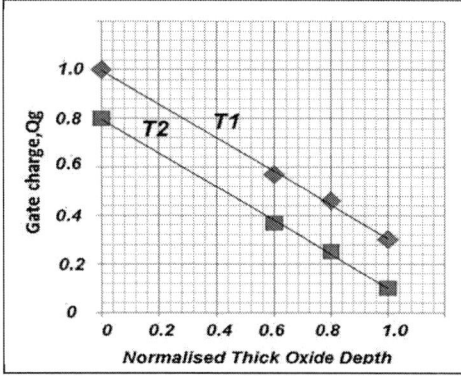

Fig.3: The influence of the depth of the thick oxide portion for two different total trench depths (T1>T2) on Qg. Vce=3600, I=42A per chip, Tj=25⁰C

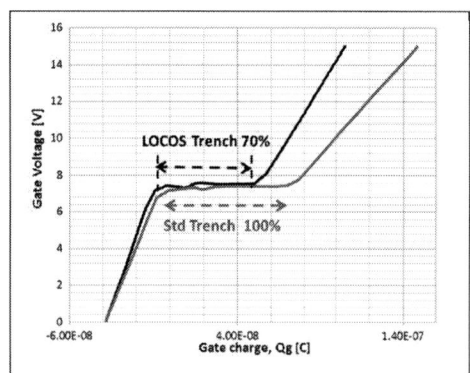

Fig.4: Simulated gate charge (Q_G). Miller charge reduction with the LTO-IGBT technology is ~30%.

The IGBT chip dimension is similar to the previous DMOS generation of 13.5mm x 13.5mm. The 25% increased current density in LTO-IGBT is achieved by having a higher MOS channel packing density in the same device active area thanks to the trench gate technology.

Therefore the new LTO chips enable 25% higher current density in the same 190mm x 140 mm module footprint as in Dynex's previous generation DMOS IGBT (Fig.5). To achieve LTO-IGBT in silicon, a low wafer bow nitride process has been developed . This takes into consideration, nitride thickness and stoichiometry as well as LOCOS temperature to achieve a manufacturabe process.

Fig.5: The 4.5kV/1500A LOCOS Trench Oxide IGBT module in the 190x140mm footprint (24 IGBT + 12 FRD in parallel)

III. EXPERIMENTAL RESULTS

All experimental testing has been done in a module shown in Fig.5 where 24xLTO-IGBTs and 12xFRDs in connected parallel. Fig.6 shows the blocking voltage performance of the 4.5kV/1500A module at 25⁰C. Fig.7 shows that due to optimized top cell design using enhanced Nwell, low Vce(sat) at 1500A (62.5A/chip) of 2.1V, 2.7V and 2.9V at 25⁰C, 125⁰C and 150⁰C respectively, giving a low Vce(sat) temperature coefficient <1.0V from 25⁰C to 150⁰C. Fig8a and 8b show typical turn-on and turn-off waveforms of the module at nominal line voltage (Vce=2800V) at 150⁰C. They show that full switch on and off times of less than 10µs is achieved due to reduced Cge and Cgc hence lower gate charge. Also the module can withstand turn-on dI/dt of 5000A/µs which indicates the robustness of the FRD used.

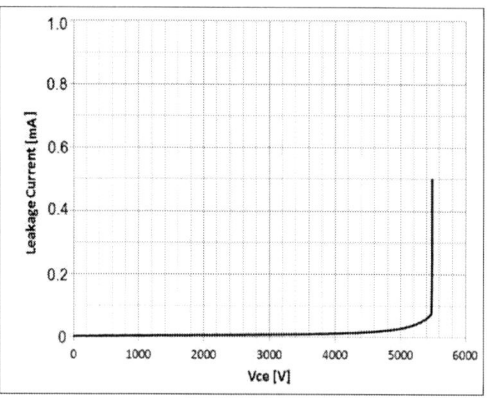

Fig.6: Measured at the module level blocking Voltage performance at T=25⁰C

Fig. 9 and 10 show respectively module SCSOA and RBSOA waveforms at 150⁰C and Vline=3500V (77.8% of BV). It can be seen that 3xInom (or 4500A) can be successfully turned off

978-1-7281-0582-6/19 $31.00 © 2019 IEEE   324

and short circuit withstand time of more than 10µs is achieved.

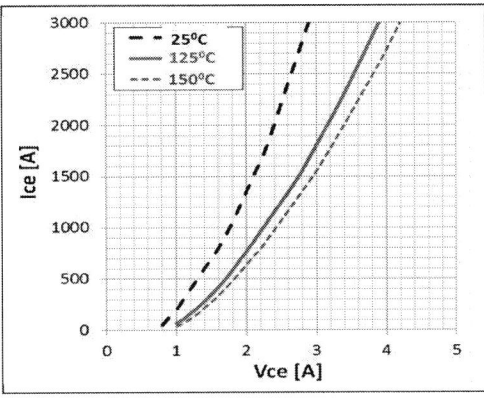

(b)

Fig.7: Measured at the module level Vce-Ice curves at T=25^{0}C, 125^{0}C and I50^{0}C at Vg=15V.

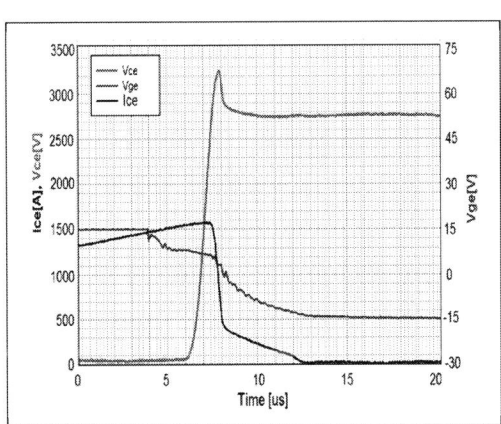

(a)

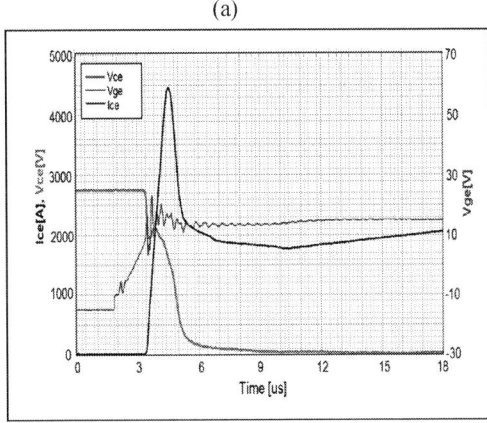

(b)

Fig.8: Measured nominal switching conditions module waveforms **(a)** Turn-off and **(b)** Turn-on. V_{ce}=2800V, T=150°C, V_{GE}=+/-15V, Rgon/off =1.0Ω/3.9Ω. Turn-on dI/dt=5100A/µs

Table 1 summarizes performance of the new LTO IGBT module, in comparison with the previous DMOS generation. In addition to lower Vce(sat) at higher current density, increased T$_{jmax}$ to 150^{0}C, improved Vce(sat)/Eoff trade off and lower Q$_G$ are achieved.

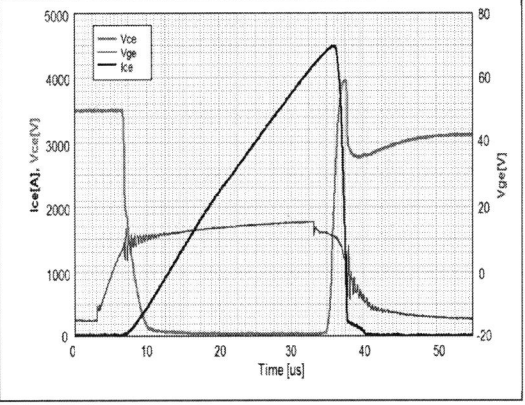

Fig.9: Measured RBSOA waveforms- successful 4500A (3xInom) turn-off. V_{cc}=3500V, T=150°C, V_{GE}=+/-15V, Rgon/off =1.0Ω/3.9Ω

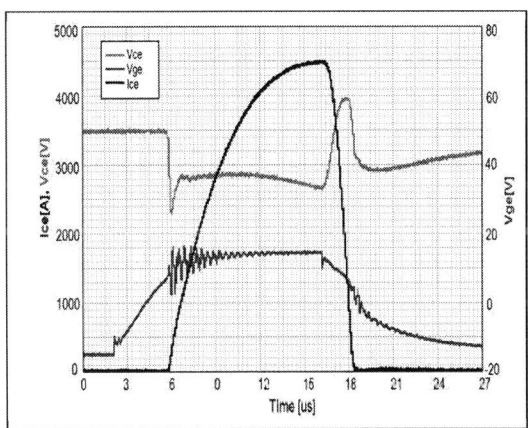

Fig.10: SCSOA waveforms showing more than 10µs short-circuit withstand capability. V_{cc}=3500V, T=150°C, V_{GE}=+/-15V, Rgon/off =1.0Ω/3.9Ω

Table1: Comparing the performance of previous and new generation modules

Parameters	Dynex DMOS IGBT Module	*New* Dynex LTO IGBT Module
IGBT Chip Dimension	13.5mm x 13.5mm	13.5mm x 13.5mm
IGBT Chip Current [A]	50	62.5
Module Current [A]	1200	1500
Tjmax [^{0}C]	125	150
Vce (sat) [V]	3.5	3.0
Eoff [mJ]	4.65	5.5
Eon [mJ]	6.45	6.5

978-1-7281-0582-6/19 $31.00 © 2019 IEEE

IV. CONCLUSION

In this paper we have shown simulation and experimental results of new LOCOS Trench Oxide (LTO) IGBT chips in 1500A/4.5kV modules at $T_J=150^0$C. The results show that the LTO technology can be used to increase current density by 25% without proportiional increase in Eon and Eoff. This has been made possible be reduced capacitances and gate charge. A significant benefit of the LTO-IGBT design is that thick oxide in the trench bottom regions which addresses long term reliabilty concerns ehich can be caused by high electric field within the gate oxide and high voltages.

V. REFERENCES

[1] T. Laska, et.al, "Long Term Stability and Drift Phenomena of different Trench IGBT Structures under Repetitive Switching Tests", Proc ISPSD'07 pp. 1-4, 2007.

[2] C. Sandow, et al., "IGBT with superior long-term switching behavior by asymmetric trench oxide", ISPSD'18

3] K. Miyagi, et. al., "Floating Island and Thick Bottom Oxide Trench Gate MOSFET (FITMOS) Ultra-Low On-Resistance Power MOSFET for Automotive Applications", Power Conversion Conference, 2007

Proceedings of the 31st International Symposium on Power Semiconductor Devices & ICs
May 19-23, 2019, Shanghai, China

Broadband TCAD mixed-mode simulation framework for predictive modeling of fast dynamic switching events

Dan Popescu
IFAG PMM ACDC RDA TI CC
Infineon Technologies AG
85579 Neubiberg, Germany
danhoria.popescu@infineon.com

Maximilian Treiber
IFAG PMM ACDC RDA TI CC
Infineon Technologies AG
85579 Neubiberg, Germany
maximilian.treiber@infineon.com

Abstract—**In this paper, we develop a simulation framework for analyzing and optimizing the large signal, dynamic switching behavior of power devices. This framework enables us to use TCAD models of our active devices, together with a broadband description of the application board. All elements are coupled through a mixed-mode SPICE instance. Simulated results, obtained with this new method for two application boards, are compared to measurement data and their agreement is analyzed.**

Keywords—*TCAD, power switch, EM simulation, mixed-mode simulation, broad band SPICE*

I. Introduction

In the field of power electronics, silicon dies have to shrink aggressively in each generation of new devices to maintain a competitive cost position. Each new shrink helps reducing the switching losses of new power MOSFETS, but, on the other hand, the faster switching behavior usually gives rise to unwanted "ringing" phenomena, which degrade the so-called ease-of-use of the device. However, assessing the ringing behavior or even predicting it for new device generations is a cumbersome task. Thus, an accurate virtual model of the switching environment including the active devices would be very helpful for benchmarking and optimization purposes. Pure TCAD oriented device development fails to take into account any realistic parasitic environment of the semiconductor die. The standard industry flow for such cases relies on electromagnetic (EM) simulation of the passive environment (S-parameters). Active elements, like diodes or transistors can be included e.g. as linearized, small signal models. Modern circuit simulators can partially solve the problem of the chip-package-board co-design, by using convolution techniques and allowing S-parameter descriptions of passives during large signal, transient SPICE simulations [1]. Using this path, depicted in Fig.1 as the SPICE flow, interesting advances and new results have been reported recently [2]-[3].

To our knowledge, however, a method that combines a broadband description of PCBs and packages together with a physical TCAD model of switches and diodes has not been employed before. The advantages of such an approach – see Fig. 2 – are promising:

- Compact models usually use the (depletion) capacitances extracted from small signal measurements. This assumption is usually not valid

during a dynamic switching event with varying gate voltage and flowing charge carriers.

- Having all the device physics inside a TCAD model, one can be sure that the model is also valid for high frequencies; this is usually not the case for compact models where extra care is necessary to accurately describe the high frequency behavior.

- Physical parameters, such as the doping concentration, oxide thickness, etc. can be varied systematically in order to quantify their effect on the device capacitances and e.g. the switching efficiency and ringing behavior.

This paper is organized as follows: section II presents the simulation models and tools used in the flow. In section III, we discuss the two application boards that have been analyzed in this work. The obtained results are compared with measured waveforms and values. Furthermore, a discussion of our results follows in section IV. Finally, in the last section, the conclusions are drawn and an outlook on possible future improvements is given.

II. Simulation Framework Description

A. Electromagnetic Simulation

The starting point for all our simulations is an accurate 3D model of the application board. Layout data is used for the PCB modeling and a comprehensive library of packages, sockets, etc. helps us to take into account e.g. bond wire inductances.

The Ansys Electronic Desktop framework is used for this task. Since we are interested in an accurate parasitic extraction at DC as well as for the RF domain, two different solvers are used and the obtained results are combined. For DC we use Q3D [4] as the reference simulator; the quasi-static approximation employed is perfectly valid for DC as well as low operating frequencies and the accuracy vs. speed trade-off is very good. Subsequently, the same geometry is imported in HFSS [5], a full wave field solver that takes into account the two-way coupling between electric and magnetic fields, typical for EM wave propagation. A stop frequency of 1 GHz has been used for all results presented in this paper.

Both the Q3D and HFSS simulations output Touchstone files containing the S parameters (DC to 1 GHz) of the analyzed network. It is beyond the scope of this paper to give

978-1-7281-0582-6/19 $31.00 © 2019 IEEE

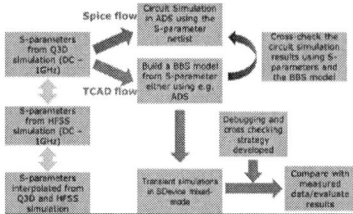

Fig. 1. The simulation flow; we concentrate mainly on the TCAD flow.

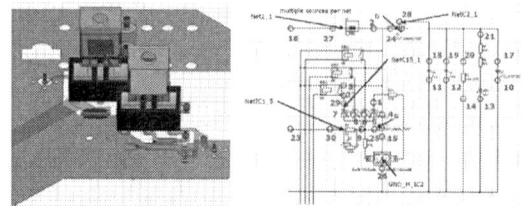

Fig. 3. 3D HFSS model and schematic of our "ringing board".

a detailed description on network theory and scattering parameters. The interested reader is referred to [6] to gain more insights on this topic. In a final step, using some in-house developed MATLAB scripting, the two Touchstone files are combined. In this way, we can ensure best accuracy of our parasitic passive elements for the low frequency as well as the high frequency domain.

B. Broadband Spice Model

Once an accurate frequency description of the passive elements is available, the next step is to prepare this network description also for mixed-mode SPICE – TCAD simulations. While it is true that some SPICE simulators available today on the market – like Keysight ADS [7] – can deal with component frequency description during transient simulations, such features usually require more complex convolution techniques, prone to convergence and performance problems.

Our TCAD mixed-mode approach however cannot deal with S parameter descriptions during transient large signal simulations, thus a conversion to a standard SPICE netlist is needed. One can use e.g. the broad-band SPICE (BBS) model generator included in the ADS simulation environment. However, special care has to be taken, in order to have an accurate, passive and causal BBS netlist. The best trade-off between model order and accuracy has to be found, since a high model order translates to a large number of elements in the netlist that will decrease simulation speed and worsen convergence; on the other hand, a physically accurate behavior over the whole frequency range is, of course, desired.

C. TCAD Modeling

The last software tool used in our simulation framework is the Synopsys TCAD package [8]. Its purpose is to define our active devices and couple them to the passive BBS model of the application board.

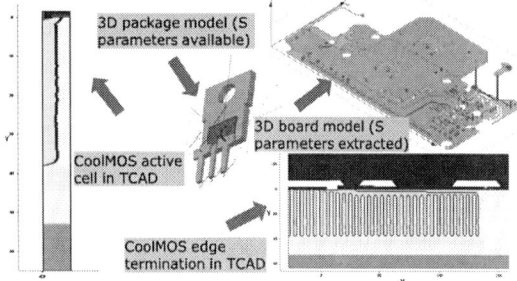

Fig. 2. 3D passive and 2D active structures included in our TCAD flow.

The SDevice tool, a part of the TCAD package, is a physical simulator that uses a finite volumes method to find a self-consistent, coupled solution of the Poisson equation, together with the electron and hole continuity equations. It can handle different device structures such as 2D cuts or even small 3D cells. A lot of physical models are available in the TCAD package, ranging from mobility models for charge carriers to mechanical stress of the dies. This allows us to simulate and explain complex phenomena observed in the measurements.

We use 2D device cuts, along symmetry axes, for all the active devices we want to include in the simulation. Typically, these are silicon power MOSFETs and SiC diodes. Even complex edge termination structures can be included to ensure an accurate description of dynamic events.

As already mentioned in the previous section, a physical description of the active devices gives additional advantages besides the increased accuracy: it provides an efficient feedback tool for front-end device development. The impact of various process parameters, like implant concentration and doping profile shapes, on relevant application parameters can be assessed quickly.

III. APPLICATION BOARD MODELS

A. Ease-of-use

One important property of next generation power MOSFETS is their so-called ease-of-use. A high ease-of-use generally means that the device can be used in various layouts and applications (almost) without special care from the design point of view. Such devices are preferred by customers who follow a plug and play philosophy. On device level, obtaining this feature requires careful engineering of the doping profiles and thus capacitance ratios or the integration of additional devices such as resistors. In order to test the ease-of-use, a board has been designed, fabricated and modeled. The 3D HFSS model and the corresponding schematic is shown in Fig. 3. Several ports are defined in the model, typically where the lumped SMD elements are soldered. Thus, the values of e.g. lumped capacitors and inductors can be varied in the simulation without the need of

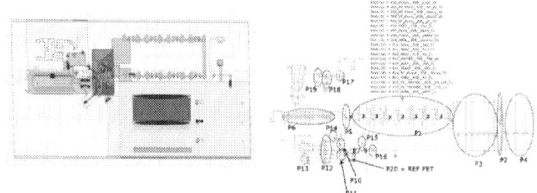

Fig. 4. 3D HFSS model and schematic of our "efficiency board".

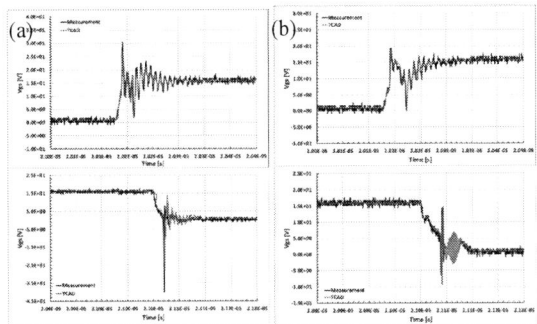
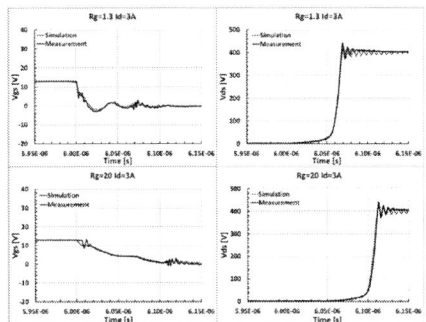

Fig. 5. Turn-on and turn-off V_{gs} transients for a C7 650V 190 mΩ with $C_{gd,ext}$= 5 pF @ ID=33 A and $R_{g,ext}$=0.5 Ω (a) or $R_{g,ext}$=10 Ω (b).

Fig. 6. Turn-off V_{gs} and V_{ds} transients of the "efficiency board" for a C7 650V 190 mΩ with $R_{g,ext}$=1.3 Ω or $R_{g,ext}$=20 Ω.

performing a new EM extract of the passive network.

Our test board is designed to perform inductive switching, typically common for hard switching PFC topologies. A large coil acts as a current source and the power MOSFET acts like a switch, either carrying the load current or directing it through the SiC diode. A "bad layout" is chosen, with large stray inductances and generally large loops on the PCB for the power as well as the control loop. The reasoning behind this is to allow us to test a worst case scenario of the passive environment. Additional relays and lumped components on the board allow us to test various devices and several switching conditions. Mainly the gate resistor and the external gate-drain capacitance have been varied in our measurements. As stated earlier, a BBS model has been generated for this application board and pulsed switching simulations in TCAD, for several currents, have been performed. These results will be compared to experimental data in the next section.

B. Efficiency

The second application board we analyze is designed to accurately measure switching losses under well-defined switching conditions. Our proposed design is shown in Fig. 4, together with the corresponding schematic. Again, some ports for lumped SMD components are included in the 3D EM extract, allowing us to adapt the simulations to various external conditions. Short signal and power loops have been designed in order to keep parasitic influences as low as possible. Additionally, a fast gate driver together with compact SMD packages have been chosen for the passive components as well as for the power switch and the SiC diode.

IV. RESULTS AND DISCUSSION

We will first discuss the results obtained for our ease-of-use test board. Turn-on and turn-off transients are simulated for three CoolMOS® technologies, namely the C7 600V, C7 650V and the P7 600V. All devices are 180 mΩ power switches. The switching conditions are as follows: external gate resistors of 0.5 and 10 Ω together with an external gate-drain capacitor of either 5 pF or 7 pF. The gate voltage is switched on and off several times with an ever increasing load current forced by a large coil. Typically, the current reaches 30A before the measurement is aborted. Only selected results will be shown in the following.

In Fig. 5 one can observe a comparison of the V_{gs} turn-on and turn-off waveforms for a C7 650V device for the two R_g values mentioned above. There is a good agreement between the measured and simulated waveforms, suggesting that our simulation framework is accurate to capture oscillations at $f_{osc} \sim 200$ MHz. Using a compact model (not shown here), such an agreement would be virtually impossible.

TABLE I. MEASURED VS. SIMULATED RINGING AMPLITUDES FOR SEVERAL COOLMOS® TECHNOLOGIES, MEASURED ON OUR "RINGING" BOARD

Technology Condition		C7 600V 180 mOhm		P7 600V 180 mOhm		C7 650V 190 mOhm	
$C_{gd,ext}$ =5pF	$R_{g,ext}$= 0.5 Ω	$V_{gs,min}$ =-62V $V_{gs,max}$ =36V	@ I_D= 26A	$V_{gs,min}$= -11V $V_{gs,max}$= 24V	@ I_D= 27A	$V_{gs,min}$= -40V $V_{gs,max}$= 30V	@ I_D= 33A
	$R_{g,ext}$= 0.5 Ω	$V_{gs,min}$ =-59V $V_{gs,max}$ =60V	@ I_D= 52A	$V_{gs,min}$= -12V $V_{gs,max}$= 22V	@ I_D= 29A	$V_{gs,min}$= -31V $V_{gs,max}$= 26V	@ I_D= 34A
$C_{gd,ext}$ =5pF	$R_{g,ext}$= 10 Ω	$V_{gs,min}$ =-24V $V_{gs,max}$ =25V	@ I_D= 27A	$V_{gs,min}$= -4V $V_{gs,max}$= 19V	@ I_D= 27A	$V_{gs,min}$= -7V $V_{gs,max}$= 20V	@ I_D= 33A
	$R_{g,ext}$= 10 Ω	$V_{gs,min}$ =-39V $V_{gs,max}$ =42V	@ I_D= 46A	$V_{gs,min}$= -4V $V_{gs,max}$= 16V	@ I_D= 26A	$V_{gs,min}$= -10V $V_{gs,max}$= 18V	@ I_D= 34A

*Grey fields denote simulated values while white fields denote measured ones.

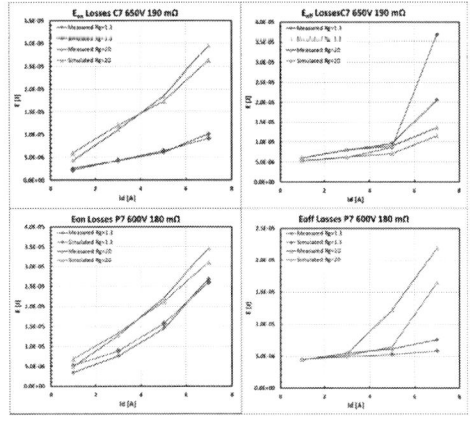

Fig. 7. Measured vs simulated switching losses on the "efficiency board", for two technologies and with $R_{g,ext}$=1.3 Ω or $R_{g,ext}$=20 Ω.

Concluding the „ringing" topic, the maximum and minimum V_{gs} ringing amplitudes at some specific switching condition, for the three mentioned technologies, are summarized in Table I. Again, the simulated values are very close to the actually measured ones. Larger discrepancies, as e.g. for the C7 600V for a large external gate resistor can be explained by our simplified treatment of the device layout. A silicon chip, having a complex 3D geometry, will exhibit signal propagation phenomena due to distributed RC networks. These effects can be neglected in most practical cases, simply due to the small geometrical size of the chip compared to the wavelength of a typical gate signal. However, there are cases, as can be seen in Table I, where a full 3D simulation approach could lead to even higher accuracies.

Addressing the efficiency topic, simulations are also performed using our second application board. We use a double pulse analysis in order to calculate the switching losses at some desired load currents. Switching conditions are chosen similar to the "ringing" board and, again, measured waveforms are used in the comparison.

In Fig. 6, some selected V_{gs} and V_{ds} transients are shown for two external gate resistors as an additional verification of the quality of our model. Our main focus, however, lies on the values of the switching energies. These energies, for a C7 650V and P7 600V, are calculated in Fig. 7. The agreement with measured values for the turn-on energies is very good. For turn-off and higher load currents one can observe a certain discrepancy. Again, we attribute this to local effects due to distributed RC networks. It is outside of the scope of this paper to further discuss this topic, but these "3D topics" pose an interesting challenge for future works.

V. CONCLUSION

We have presented a complex simulation framework that is capable of reproducing large signal, dynamic switching events. This simulation framework is applied to two manufactured application boards, optimized to test the ease-of-use and the efficiency of various power MOSFETS.

Measurements have been performed for three CoolMOS® technologies under several conditions. The good agreement between measured and simulated values validates our framework. Being able to reproduce very fast switching events, using a physical simulator (TCAD), can help us gain useful insights into the device physics. Furthermore, such a "virtual prototyping" approach can pave the way for front-end device optimization, shortening the learning loops needed to generate optimized products.

REFERENCES

[1] J. E. Bracken, Din-Kow Sun and Z. J. Cendes, "S-domain methods for simultaneous time and frequency characterization of electromagnetic devices," in IEEE Transactions on Microwave Theory and Techniques, vol. 46, no. 9, pp. 1277-1290, Sept. 1998.

[2] B. Wei and S. G. Pytel, "New integrated workflow for EMI simulation," 2015 Asia-Pacific Symposium on Electromagnetic Compatibility (APEMC), Taipei, 2015, pp. 162-165.

[3] R. Robutel et al., "Design and Implementation of Integrated Common Mode Capacitors for SiC-JFET Inverters," in IEEE Transactions on Power Electronics, vol. 29, no. 7, pp. 3625-3636, July 2014.

[4] "ANSYS Q3D Extractor: High-Performance Parasitic Extraction," Ansys.com, 2019. [Online]. Available: https://www.ansys.com/products/electronics/ansys-q3d-extractor [Accessed 11 Mar. 2019].

[5] "ANSYS HFSS: High Frequency Electromagnetic Field Simulation Software," Ansys.com, 2019. [Online]. Available: https://www.ansys.com/products/electronics/ansys-hfss [Accessed 11 Mar. 2019].

[6] D. Pozar, Microwave Engineering, 3rd ed., Hoboken, NJ: John Wiley, 2012.

[7] "Advanced Design System (ADS)," Keysight.com, 2019. [Online]. Available: https://www.keysight.com/en/pc-1297113/advanced-design-system-ads?cc=US&lc=eng. [Accessed: 11- Mar- 2019].

[8] M. Tsiklauri, "P370:Electrical Characterization of Printed Circuit Board and Related Interconnects at Frequencies up to 50 GHz," 2017 IEEE International Symposium on Electromagnetic Compatibility & Signal/Power Integrity (EMCSI), Washington, DC, 2017, pp. 1-15.

[9] "TCAD", Synopsys.com, 2019. [Online]. Available: https://www.synopsys.com/silicon/tcad.html. [Accessed: 11- Mar- 2019].

978-1-7281-0582-6/19 $31.00 © 2019 IEEE

Proceedings of the 31st International Symposium on Power Semiconductor Devices & ICs
May 19-23, 2019, Shanghai, China

A Novel CSTBT with Hole Barrier for High dV/dt Controllability and Low EMI Noise

Xiaorui Xu, Wanjun Chen*, Chao Liu, Yuan Wang, Nan Chen, Fangzhou Wang, Qi Shi, Kenan Zhang, Qi Zhou, Zhaoji Li, Bo Zhang
State Key Laboratory of Electronic Thin Films and Integrated Devices, University of Electronic Science and Technology of
China, Chengdu 610054, China
(E-mail: wjchen@uestc.edu.cn)

Abstract—In this work, a novel CSTBT with hole barrier (HB-CSTBT) is proposed for power switching applications. Different from the conventional CSTBT, the HB-CSTBT features the buried P/N junctions under the CS-layer. During the turn-on transient, the buried P/N junctions are depleted laterally to set hole barrier near the gate bottom, suppressing the accumulation of hole carriers in this region. As a result, the gate self-charging effect caused by reverse displacement current is suppressed, contributing to high dV/dt controllability and low electromagnetic interference (EMI) noise. It is found from simulation results that the HB-CSTBT reduces the maximum reverse-recovery dV_{KA}/dt by 66% in comparison with that of the CSTBT, which is of great merit in the EMI noise suppression. Moreover, the HB-CSTBT also overcomes the blocking capability degradation brought by the CS-layer without compromising the other device characteristics.

Keywords—Carrier Stored Trench Bipolar Transistor; Electromagnetic interference (EMI) noise; dV/dt controllability; Blocking capability

I. INTRODUCTION

Insulated Gate Bipolar Transistors (IGBTs) used in power switching applications are required to provide more optimized overall performances [1]. With the combination of the Field Stop technology and the Carrier Store (CS) effect, Carrier Stored Trench Bipolar Transistor (CSTBT) achieves great improvement on the trade-off relationship between forward voltage drop (V_{on}) and turn-off energy loss (E_{off}) [2-4]. As a result, the CSTBT becomes one of the most competitive candidate in power switching applications. However, compared with the conventional trench IGBT (CIGBT), the introduction of the CS-layer in the CSTBT increases the curvature of the potential contours near the gate bottom during the turn-on transient, leading to a larger hole well (HW) under the gate oxide. As a result, more holes tend to accumulate under the gate oxide, which forms a huge displacement current to charge the gate capacitance [5-6]. Consequently, the dV/dt controllability of the CSTBT is poorer than that of the C-IGBT, which restrains the CSTBT from being used in high-frequency fields. Moreover, for the CSTBT, the higher CS-layer doping degrades the blocking capability of the device, which limits the potential in reducing energy loss [7-8]. In order to overcome the limitations brought by the CS-layer, a novel CSTBT with hole barrier (HB) is proposed (HB-CSTBT).

This work was supported in part by the National Natural Science Foundation of China (No. 51877030), the Sichuan Youth Science and Technology Foundation (No. 2017JQ0020), the Major Science and Technology Special Projects in Guangdong under Grant (No. 2017B010112003) and the Fundamental Research Funds for the Central Universities (No. ZYGX2016Z006).

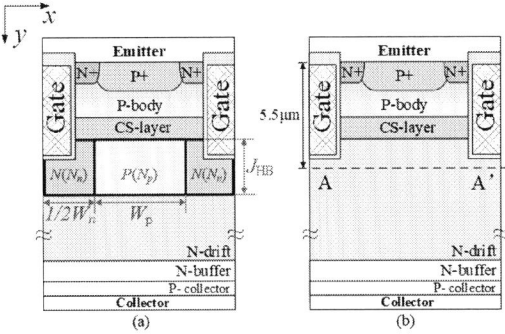

Fig. 1. (a) Schematic section views of (a) the proposed HB-CSTBT and (b) the conventional CSTBT. The cutline AA' lies under the gate oxide.

II. DEVICE STRUCTURE

The schematic cross-section of the proposed HB-CSTBT is shown in Fig. 1(a). Compared with the CSTBT shown in Fig. 1(b), the HB-CSTBT features the buried lateral P/N junctions underneath the CS-layer. Note that the buried lateral P/N junctions can be formed by the ion implantation before the epitaxial growth for the CS-layer.

During the initial stage of the turn-on transient, the collector-emitter voltage (V_{CE}) remains almost constant (equal to the bus voltage). Thus, the gate-emitter voltage (V_{GE}) is lower than the potential of the drift region. The potential difference forms electric field vectors that end to the gate electrode, generating the lateral electric field (LE-Field) and HW under the gate oxide. As a result, hole carriers that are injected from the collector side tend to accumulate underneath the gate oxide, raising the potential of these regions (V_{acc}). This leads to a huge reverse displacement current (I_{dis}), which charges up the gate capacitance and degrades the controllability of R_G over the turn-on dV/dt. A simple expression of the I_{dis} is given as [1, 5, 9]:

$$I_{dis} = C_{GC\_ox} \frac{d(V_{acc} - V_{GE})}{dt} \quad (1)$$

where C_{GC_ox} is the gate oxide component of C_{GC}.

As shown in Fig. 2, the HB-CSTBT forms the reversed LE-Field vectors in addition to the original one utilizing the lateral depleting of the buried P/N junctions, setting the hole barrier (HB) at these regions instead of HW. Consequently, hole carriers are extracted out of the device directly (see in Fig. 3(a)), reducing dV_{acc}/dt drastically. As a result, the HB-CSTBT delivers high controllability of R_G over the turn-on dV/dt and

978-1-7281-0582-6/19 $31.00 © 2019 IEEE   331

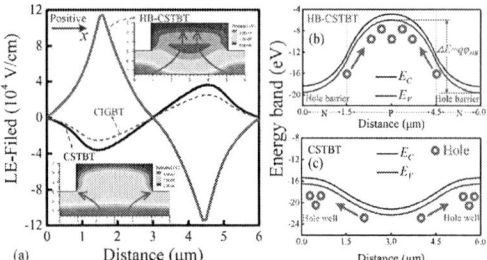

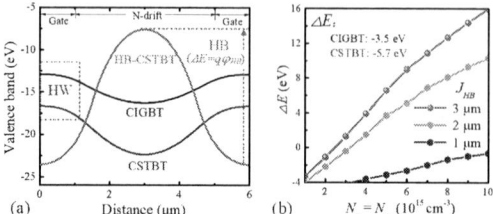

Fig. 2. (a) The lateral electric field (LE-Field) versus distance along cutline AA' shown in Fig. 1. The positive direction is show in the Figure. Inset: the potential distribution near the emitter surface at $V_{CE} = 600$ V. The rad arrows represent the hole current movement. The energy band along cutline AA' of (b) the HB-CSTBT and (c) CSTBT.

Fig. 4. (4) The comparison of valence band versus distance along cutline AA' shown in Fig. 1 and (b) the influence of N_n, N_p ($N_n = N_p$) and J_{HB} on ΔE. The negative and positive ΔE represent the hole well and hole barrier, respectively.

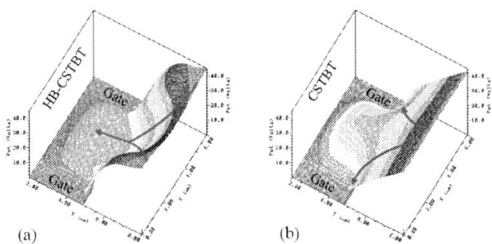

Fig. 3. The 3D view of the potential in volts near the emitter surface of: (a) HB-CSTBT and (b) CSTBT at $V_{CE} = 600$ V. The red arrows represent the movement of hole current.

reduces the electromagnetic interference (EMI) noise according to Eq. (1). However, the CSTBT increases LE-Filed due to the existence of the CS-layer, which leads to large HW under the gate oxide, and thus a poor turn-on dV/dt controllability and high EMI noise.

III. RESULTS AND DISCUSSION

In order to verify the operations and the advantages of the HB-CSTBT, numerical simulation using MEDICI is performed [10]. The major optimized parameters of the HB-CSTBT and the CSTBT are summarized in Table I.

TABLE I. MAJOR STRUCTURAL PARAMETERS

Structure Parameters	Symbol	HB-CSTBT	CSTBT
Cell pitch (μm)	W_C	6	6
Mesa width (μm)	W_M	4	4
Trench depth (μm)	D_T	5	5
Oxide thickness (nm)	T_O	100	100
Wafer thickness (μm)	T_W	120	120
P-body doping (cm⁻³)	N_{PB}	2×10^{17}	2×10^{17}
CS-layer doping (cm⁻³)	N_{CS}	1×10^{16}	1×10^{16}
Buried layer doping (cm⁻³)	N_p & N_n	1×10^{16}	-
Buried layer width (μm)	W_p & W_n	3	-
Buried junction depth (μm)	J_{HB}	3	-
N-drift doping (cm⁻³)	N_{DR}	5×10^{13}	5×10^{13}
N-buffer doping (cm⁻³)	N_{BU}	5×10^{16}	5×10^{16}
P⁺ collector doping (cm⁻³)	N_{PC}	5×10^{17}	5×10^{17}

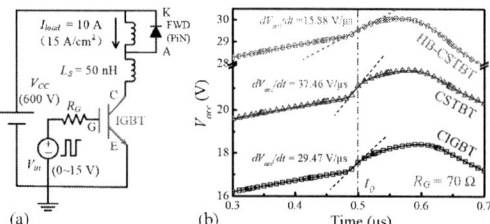

(a)   (b)

Fig. 5. (a) Circuit schematic for transient analysis. (b) V_{acc} (the potential of the point shown in Fig. 6) curves during the turn-on transient.

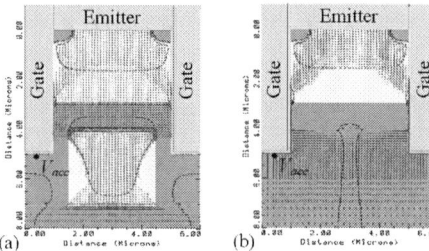

Fig. 6. Simulation results for difference of holes action between (a) HB-CSTBT and (b) CSTBT at t_0. The t_0 is shown in the Fig. 5(b). The red arrows represent the flowing paths of hole current and the black dashed lines represent the depletion edges.

Fig. 4(a) shows the lateral valence band distributions of IGBTs. It can be seen that there exists HB or HW under the gate oxide due to the LE-Field. The height of the HB (HW) is defined as: $\Delta E = q\varphi_{HB}$, where φ_{HB} is the maximum lateral potential difference between the gate bottom and the drift region. As the analysis above, both CIGBT and CSTBT features HW under the gate oxide, ΔE of which are -3.5 eV and -5.7 eV, respectively. Due to the reversed compensatory LE-field brought by the buried P/N junctions, the HB-CSTBT sets HB under the gate oxide instead of the HW. Fig. 4(b) shows that the larger junction doping (N_p, N_n) and junction depth (J_{HB}) induce higher HB.

A hard-switching circuit with a PiN diode serving as the free-wheeling diode (FWD), as shown in Fig. 5(a), is used for transient simulations. Fig. 5(b) shows V_{acc} curves during the turn-on transient. The maximum dV_{acc}/dt of the HB-CSTBT is reduced by 58% in comparison with the CSTBT (reduced from 37.46 V/μs to 15.88 V/μs). Such significant reduction is because

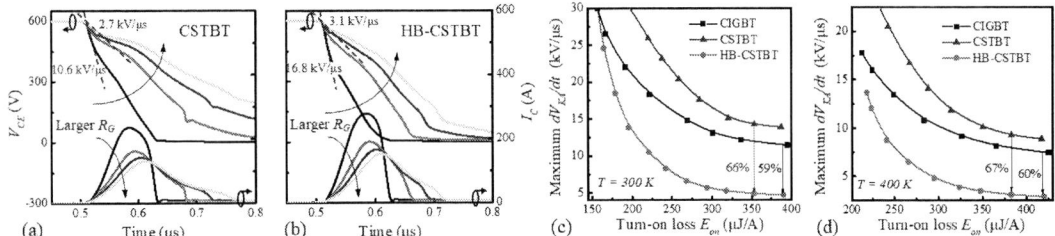

Fig. 7. Turn-on curves at the small current (1/10 rated current) of (a) the CSTBT and (b) the HB-CSTBT. Trade-off relationships between the maximum reverse recovery dV_{KA}/dt of FWD (at 1/10 rated current) and E_{on} of IGBT (at rated current) at (c) T = 300 K and (d) T = 400 K.

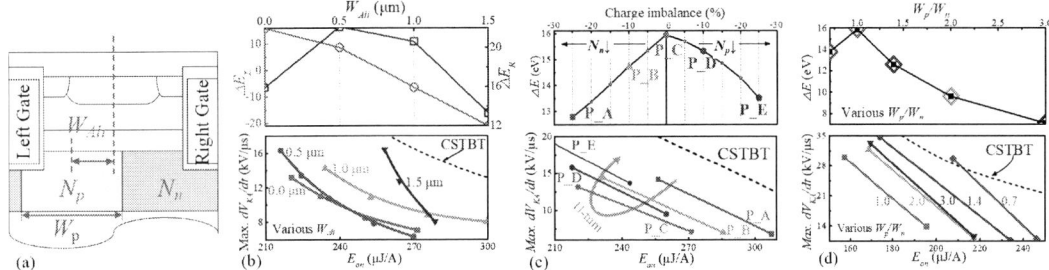

Fig. 8 (a) The key parameters of the HB-CSTBT: W_{Ali} is the lateral distance between the symmetry axes of the IGBT cell and buried P junction. N_p and N_n are the doping concentration of the buried P and buried N junctions, respectively. W_p and W_n are the width of the buried P and buried N junctions, respectively. (b) The influence of W_{Ali} on the HB under the left gate (ΔE_L) and right gate (ΔE_R) and the dV_{KA}/dt-E_{on} trade-off characteristics. (c) The influence of charge imbalance of the buried P/N junctions on the HB and the dV_{KA}/dt-E_{on} trade-off characteristics. (d) The influence of the W_p/W_n on the the HB and the dV_{KA}/dt-E_{on} trade-off characteristics.

the hole accumulation under the gate oxide is suppressed by HB (see Fig. 6).

Fig. 7 (a) and (b) exhibit turn-on I-V curves of the CSTBT and the HB-CSTBT under a small current (with R_G of 10 Ω, 30 Ω, 50 Ω and 70 Ω) [6]. Note that the turn-on starting time has been adjusted for an explicit comparison. It is clear that the maximum $|dV_{CE}/dt|$ of the HB-CSTBT is more sensitive to the increase of R_G in comparison with the CSTBT. This means that the HB-CSTBT shows excellent controllability of R_G over the turn-on dV/dt. Fig. 7 (c) and (d) show the trade-off relationships between the maximum reverse-recovery dV_{KA}/dt of the FWD (at 1/10 rated current) and the turn on loss (E_{on}) of the IGBT (at rated current). Due to the induced electromotive force of the stray inductance (L_S), the highest dV/dt noise is mainly observed from the FWD at the opposite arm [5]. Clearly, the trade-off relationship of the HB-CSTBT is better than that of the CIGBT and the CSTBT. When compared with the CSTBT, the maximum reverse-recovery dV_{KA}/dt of the HB-CSTBT could be reduced by 66% at T = 300 K. Such superiority can also be achieved even at T = 400 K, which is of great merit in suppressing the EMI noise.

The influences of the key parameters on the ΔE and maximum dV_{KA}/dt-E_{on} trade-off characteristics are shown in the Fig. 8. In Fig. 8(b), the HB-CSTBT delivers an asymmetric energy band distribution when $W_{Ali} \neq 0$. Note that W_{Ali} represents the alignment precision between gate trench and buried P/N junctions. With the increase of W_{Ali}, the hole barrier under the left gate oxide (ΔE_L) decreases significantly, thus degrading the turn-on performance. The simulations of the HB-CSTBT above

are all based on the hypothesis that charges in the buried P/N junctions are balanced ($N_p \times W_p = N_n \times W_n$). However, perfect charge balance is difficult to achieve in a practical manufacturing. Consequently, the research about the effect of

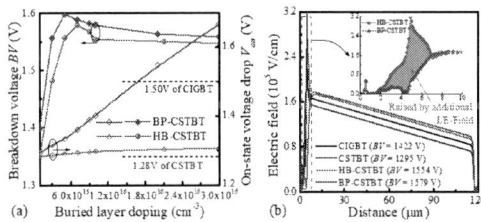

Fig. 9. (a) BV and V_{on} trade-off performances of IGBTs. (b) Vertical electric field distributions of IGBTs at the breakdown state.

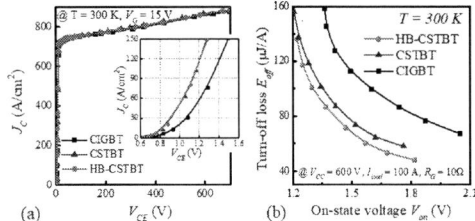

Fig. 10. (a) Saturation characteristics and (b) the trade-off performances between V_{on} and E_{off} of IGBTs. Inset: zoomed-in on-state I-V curves.

charge imbalance on the characteristics of the HB-CSTBT is meaningful and crucial. Fig. 8(c) and (d) show the ΔE and the turn-on trade-off relationships under the charge imbalance situations. It can be seen that the highest ΔE can be obtained when the charge is balanced ($N_p = N_n$ and $W_p = W_n$), which results in the most optimized dV_{KA}/dt-E_{on} trade-off characteristic. Once the charge balance is broken, the dV_{KA}/dt-E_{on} trade-off characteristic will be degraded. However, all dV_{KA}/dt-E_{on} trade-off relationships, as shown in Fig. 8, are better than that of the CSTBT, which means that the buried P/N junctions shows a good fabrication-error tolerance.

Another advantage of the HB-IGBT is that it overcomes the breakdown voltage (BV) degradation brought by the CS-layer, as shown in the Fig. 9. BV is increased from 1422 V of the CIGBT to 1544 V of the HB-CSTBT. This is because, for the HB-CSTBT, the lateral depletion of the buried P/N junction forms potential barrier to optimize the electric field crowding near the gate bottom. However, for the CSTBT, the BV is degraded to 1295 V because of the CS-layer [7]. Besides, the saturation current and V_{on} of the HB-CSTBT is almost identical to that of the CSTBT, which indicates that the HB-CSTBT still retains a better V_{on}-E_{off} trade-off characteristics compared with the CIGBT (see in Fig. 10). Consequently, the HB-CSTBT breaks through the BV limitation brought by the CS-layer while still featuring significant carrier store effect. On the other hand, it can be also seen in Fig. 9 that the BV and V_{on} are less sensitive to the change of the buried layer doping than that of the BP-CSTBT [10]. Such difference is because the larger additional LE-Field can be achieved in the HB-CSTBT (see in Fig. 10(b)).

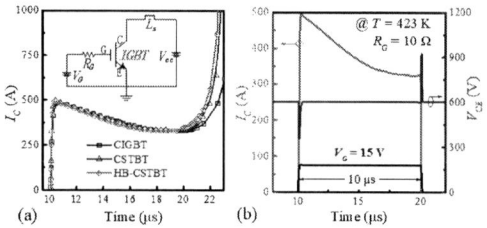

Fig. 11. Simulated (a) short circuit failures of IGBTs and (b) short circuit ruggedness of the HB-CSTBT. A circuit with stray inductance (L_S) of 50 nH is used to study the short-circuit behavior of IGBTs. Considering the self-heating effect, an electro-thermal switching simulation with the initial temperature of 423 K and the thermal resistance of 0.15 K/W is used.

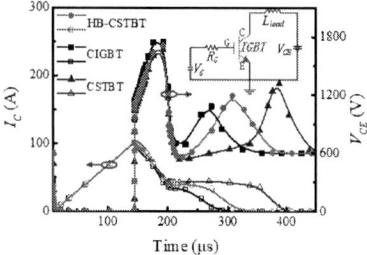

Fig. 12. Current and voltage waveforms during the unclamped inductive switching (UIS). A circuit with inductive load (L_{load}) of 0.8 mH is used to study the dynamic blocking capability of IGBTs. Considering the self-heating effect, an electro-thermal switching simulation with the initial temperature of 300 K and the thermal resistance of 0.15 K/W is used.

The thermal stability is one of the most important indicator to evaluate the reliability of the device. Consequently, both the short-circuit ruggedness and the UIS performance are simulated in Fig. 11 and Fig. 12, respectively. Fig. 11 shows that the HB-CSTBT can be turned off successfully under a 10 μs short circuit pulse, which meets the standard requirement of IGBT products [11]. It can be concluded in Fig. 12 that, although the potential distribution would be altered during the switching stage, the dynamic blocking capability of the HB-CSTBT does not degrade when compared with the CSTBT.

IV. CONCLUSION

A novel HB-CSTBT is proposed and demonstrated by numerical simulations. The HB-CSTBT features hole barrier to suppress the hole accumulation under the gate oxide during the turn-on transient, contributing to high dV/dt controllability and low EMI noise. Simulation results show that, when compared with the CSTBT, a 66% lower reverse-recovery dV_{KA}/dt can be achieved in the HB-CSTBT. Besides, the HB-CSTBT also breaks through the BV degradation brought by CS-layer, thus featuring a high BV while still maintaining low energy and high reliability.

REFERENCES

[1] X. Xu, et al., "A Novel IGBT With Self-Regulated Potential for Extreme Low EMI Noise", *IEEE electron device letters*, vol. 40, no. 1, pp.71-74, January 2019.

[2] H. Takahashi, et al., "Carrier Stored Trench-Gate Bipolar Transistor (CSTBT)-A Novel Power Device for High Voltage Application-" *Proc. ISPSD*, Maui, USA, pp: 349-352, May 1996.

[3] R. Kamibaba, et al., "Next generation 650V CSTBT™ with improved SOA fabricated by an advanced thin wafer technology" *Proc. ISPSD*, Hong Kong, China, pp: 29-32, May 2015.

[4] K. Konishi, et al., "Experimental demonstration of the active trench layout tuned 1200V CSTBT™ for lower dV/dt surge and turn-on switching loss" *Proc. ISPSD*, Prague, Czech Republic, pp: 363-366, June 2016.

[5] H. Feng, et al., "Transient Turn- ON Characteristics of the Fin p-Body IGBT" *IEEE transactions on electron devices*, vol. 62, no. 8, pp: 2555-2561, August 2015.

[6] S. Momota, et al., "Analysis on the low current turn-on behavior of IGBT module " *Proc. ISPSD*, Toulouse, France, pp: 359-362, May 2000.

[7] P. Li, et al., "A Novel Diode-Clamped CSTBT with Ultra-low On-state Voltage and Saturation Current," *Proc. ISPSD*, Prague, Czech Republic, pp. 307-310, June 2016.

[8] J. Zhang, et al., "A novel high-voltage light punch-through carrier stored trench bipolar transistor with buried p-layer" *Chin. Phys. B*, vol. 21, pp: 0685041-0685046, June 2012.

[9] I. Omura, et al., "Oscillation effects in IGBT's related to negative capacitance phenomena," *IEEE Trans. Electron Devices*, vol. 46, no. 1, pp. 237–244, January 1999.

[10] Taurus Medici, Medici User Guide, Version D-2010.03, Synopsys Inc. 2010

[11] M. Trivedi, K. Shenai, "Investigation of the Short-Circuit Performance of an IGBT," *IEEE Trans. Electron Devices*, vol. 45, no.1, pp. 313-320. January 1998.

Proceedings of the 31st International Symposium on Power Semiconductor Devices & ICs
May 19-23, 2019, Shanghai, China

Influence of External Gate Resistance on UIS Capability in Superjunction MOSFET

Masaaki Honda, Mizue Yamaji, Daisuke Arai, Noriaki Suzuki, Takeshi Asada, Wataru Hirasawa, Takeshi Yamaguchi and Yuji Watanabe
Electronic Device Div., Shindengen Electric Mfg. Co. Ltd.
10-13, Minami-cho, Hanno-city, Saitama, Japan
Email: masaaki_honda@shindengen.co.jp

Abstract— We investigated the UIS capability for superjunction MOSFETs with respect to the external gate resistance (R_g) and the charge imbalance (CIB) by experiments as well as device simulations. Measured UIS capability depends not only on the CIB but also on the R_g. Higher UIS capability was exhibited for small R_g than for large R_g as deviating the CIB to $Q_n<Q_p$ while the difference by R_g was small for $Q_n=Q_p$. The device simulations indicated that the device destruction by UIS is assumed to be related to the re-conduction of channel current (I_{ch}). When the R_g is small, I_{ch} turns off once while V_{ds} has not increased enough to generate impact ionization. This transient behavior is different from the one for large R_g. After that, I_{ch} re-conduction occurs due to the high electric field beneath the MOS gate for $Q_n=Q_p$ but suppressed for $Q_n<Q_p$. The graded doping for N-column which reduces phosphorus concentration near the surface helps to suppress the re-conduction and improves the UIS capability for small R_g condition.

Keywords—superjunction MOSFET, UIS, external gate resistance

I. INTRODUCTION

Robustness of power semiconductor device is crucially important under harsh electrical stress condition such as abrupt voltage surge to protect the whole circuit from the destruction. While superjunction (SJ-) MOSFETs can drastically improve conventional trade-off relationship between specific on resistance and breakdown voltage [1-3], it is difficult to achieve fully high avalanche capability due to large current density compared with conventional MOSFETs. Furthermore, it is known that the unclamped inductive switching (UIS) capability of SJ-MOSFETs largely depends on the charge imbalance (CIB) between P-column and N-column in SJ structure [3]. It also has been reported that UIS capability dependence on the CIB is improved by utilizing the deep trench contact in low voltage SJ-MOSFETs [4]. It is necessary to consider the design which can ensure fully high UIS capability for commercialization of SJ-MOSFETs.

In this work, we investigated the UIS capability in 600 V-class SJ-MOSFETs with respect to the external gate resistance (R_g) and the CIB by experiments and device simulations as well. The CIB was deviated from $Q_n=Q_p$ to $Q_n<Q_p$ since UIS capability becomes relatively higher than $Q_n>Q_p$ [3]. The experiments showed that the UIS capability has a dependence on R_g which is enhanced for $Q_n<Q_p$ devices than for $Q_n=Q_p$ ones. The device simulations indicated that the total amount of the re-conduction of the channel current (I_{ch}) is related to the impact ionization beneath the MOS gate structure which leads to the destruction of the device. Re-conduction phenomena of I_{ch} is already reported previously [5], which is caused by high

electric field beneath the gate structure. We also demonstrated that the graded doping profile along the N-column depth with lower phosphorus concentration near the surface helps to suppress the I_{ch} re-conduction and improves the UIS ruggedness.

II. UIS MEASUREMENTS

Measured dependences of UIS capability on CIB and R_g for 600V/20A SJ-MOSFETs are shown in Fig. 1. It exhibited higher UIS capability for small R_g than for large R_g as deviating the CIB to $Q_n<Q_p$, while the difference by R_g was small for $Q_n=Q_p$. Figures 2 (a) to (c) show measured UIS waveforms. In Fig. 2 (a), the slope of drain voltage (V_{ds}) is rounded in the latter half stage of the V_{ds} rise. However, in Fig. 2 (b) V_{ds} rises straightly until V_{ds} reaches the sustain voltage (V_{sus}). Destruction occurred just before V_{ds} reaches V_{sus} for both large and small R_g [e.g. Fig. 2 (c)].

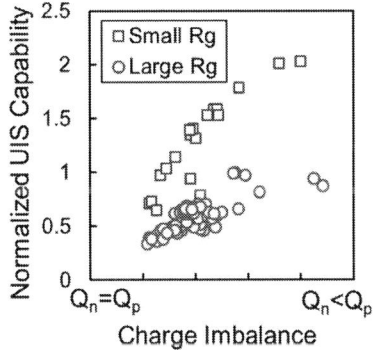

Fig. 1 Experimentally measured dependence of UIS capability on CIB and R_g.

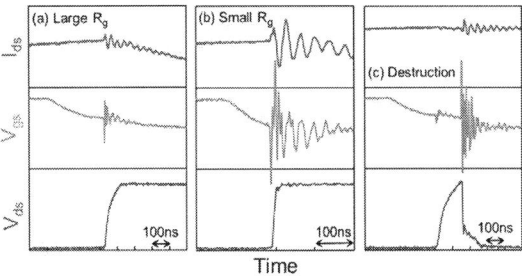

Fig. 2 Typical UIS waveforms in measurement. (a) the waveform for large R_g, (b) the waveform for small R_g, (c) destructive waveform for large R_g.

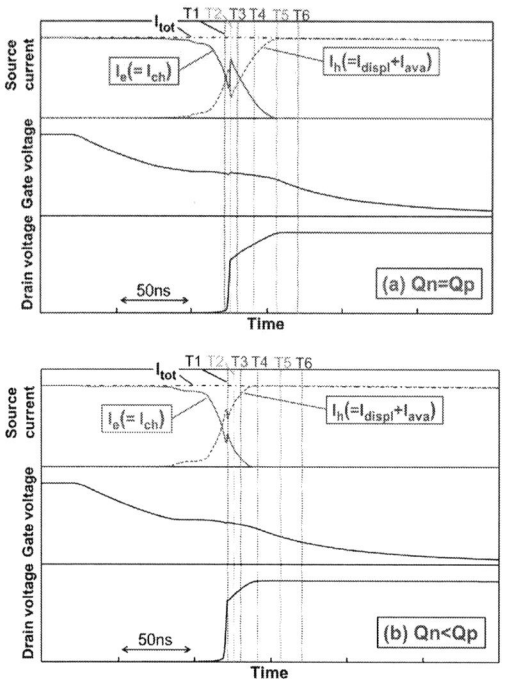

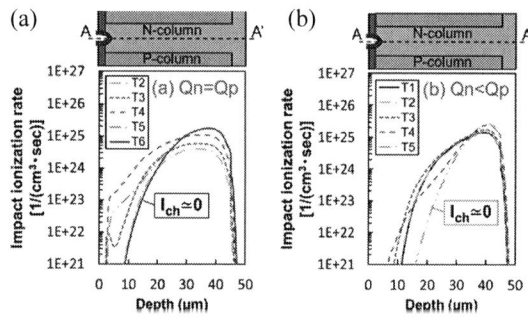

Fig. 3 Simulated UIS waveforms for large R_g condition. (a) $Q_n=Q_p$, (b) $Q_n<Q_p$.

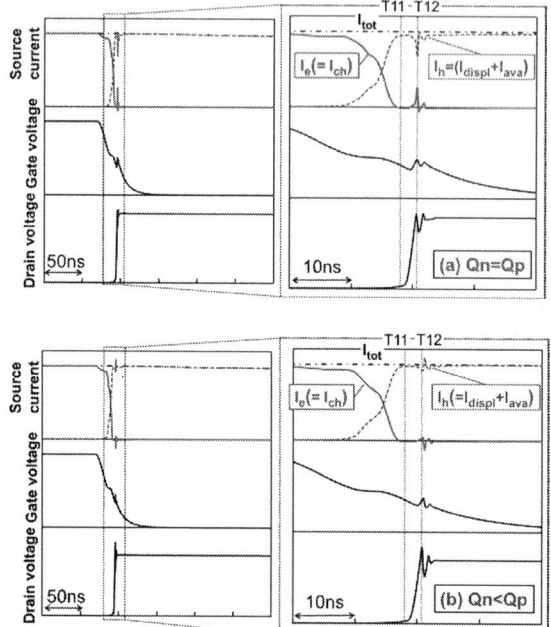

Fig. 5 Simulated UIS waveforms for small R_g condition. (a) $Q_n=Q_p$, (b) $Q_n<Q_p$.

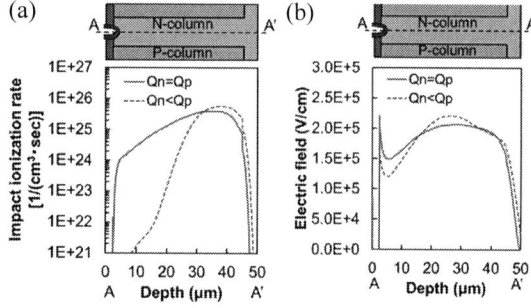

Fig. 4 Impact ionization rate at the moment T1-T6 indicated in Fig. 3. (a) $Q_n=Q_p$, (b) $Q_n<Q_p$.

Fig. 6 (a) Impact ionization rate and (b) electric field at the moment T12 indicated in Fig. 5.

III. DEVICE SIMULATION ANALISIS

We demonstrated the UIS simulations for the conditions of large R_g and small R_g with varied CIB ($Q_n=Q_p$ and $Q_n<Q_p$) to reproduce the experimental results as described in above section. The internal states of the silicon devices were investigated at the moments of just before to after the V_{ds} reaches V_{sus} in order to clarify the mechanism of UIS capability dependence on R_g and CIB found by the experiments.

A. Large R_g condition

Figures 3 (a) and (b) show simulated UIS waveforms for large R_g with the CIB as $Q_n=Q_p$ and $Q_n<Q_p$. We decomposed the total current (black-dotted and dashed line) into electron (red-solid line) and hole (blue-dashed line) current elements. In the drain electrode, the total current (I_{tot}) is equal to the electron current (I_e) which is the summation of the channel current (I_{ch}), the displacement current (I_{displ}), and the avalanche current (I_{ava}). On the other hand, in the source electrode I_{tot} can be divided into I_e which is consisted of only I_{ch} and I_h which is consisted of I_{displ} and I_{ava}. In turn-off transient of the UIS, when decreasing gate voltage (V_{gs}) reaches the beginning of miller period, I_{ch} also starts to decrease. Due to the nature of inductive load, I_{displ} flows to keep I_{tot} constant by extending the depletion region (I_h is dominated by I_{displ}). Accompanying to depletion region extension of the SJ structure, V_{ds} starts to increase. During the further increase of V_{ds}, main component of I_h gradually shifts from I_{displ} to I_{ava}. I_{ch} turns off almost simultaneously V_{ds} reaches V_{sus} and finally I_{tot} is dominated by I_{ava}.

By comparing Fig. 2 (a) and Fig. 3, the rounded V_{ds} rise which was experimentally observed was successfully replicated by the device simulation. Although V_{ds} starts to increase considerably at the end of the miller period, the

978-1-7281-0582-6/19 $31.00 © 2019 IEEE   336

slope of V_{ds} becomes lower in the latter half stage of V_{ds} rise. This dV_{ds}/dt lowering is assumed to be caused by the re-conduction of I_{ch} which is once decreased and then increase temporarily [5, 6]. Ref. [5] describes that I_{ch} is re-conducted since V_{gs} is pulled up due to high electric field beneath the gate structure in the turn-off transient. In the experimental UIS test, actual devices are destructed just before V_{ds} reaches V_{sus}, meanwhile the simulations show that the re-conducted I_{ch} is still flowing at this moment. Therefore, the device destruction is assumed to be related to the re-conduction of I_{ch}.

Figures 4 (a) and (b) show the impact ionization rate in the N-column for each T1 to T6 indicated in Figs. 3 (a) and (b) with $Q_n=Q_p$ and $Q_n<Q_p$, respectively. They indicate that more impact ionizations beneath the MOS gate structure are caused by the remaining I_{ch} which is larger for $Q_n=Q_p$ than for $Q_n<Q_p$. Furthermore, the electric field beneath the gate structure is mitigated as deviating the CIB to $Q_n<Q_p$ which results in suppression of impact ionizations near the surface [3, 5]. From the experimental results and the device simulation analysis, it is considered that more impact ionizations beneath the MOS gate lead to device destruction which results in less UIS capability.

B. Small R_g condition

Figures 5 (a) and (b) show simulated waveforms for small R_g with the CIB as $Q_n=Q_p$ and $Q_n<Q_p$, respectively. I_{ch} disappears once before V_{ds} increases enough to induce impact ionization at the moment T11. After T11, I_{ch} re-conduction spike appears for $Q_n=Q_p$ but suppressed for $Q_n<Q_p$ (T12) upon the increase of V_{ds}. It is different from the case for large R_g, there is no time period that I_{ch} temporarily disappears because V_{ds} reaches V_{sus} just when I_{ch} is perfectly shutdown.

Impact ionization rates and electric fields at the moment T12 for $Q_n=Q_p$ and $Q_n<Q_p$ are shown in Figs. 6 (a) and (b). At T12, more impact ionizations near the MOS gate occur for $Q_n=Q_p$ than for $Q_n<Q_p$ as shown in Fig. 6 (a). For $Q_n=Q_p$, the situation that more impact ionizations occur near the surface will be the same as the condition for large R_g although the I_{ch} re-conduction appears instantaneously. However, for $Q_n<Q_p$ impact ionizations are mitigated near the surface because of suppression of I_{ch} re-conduction. Figure 6 (b) indicates that the total amount of re-conduction of I_{ch} is related to the electric field beneath the gate structure. From these reasons, it is assumed that the UIS capability has a dependence on R_g which is enhanced for $Q_n<Q_p$ devices than for $Q_n=Q_p$ ones.

IV. PROPOSED STRUCTURE FOR $Q_N=Q_P$

In this section, we propose the device structure which can improve the UIS capability for small R_g with $Q_n=Q_p$. From experimental results and device simulation analysis, improvement of the UIS capability is expected by suppressing the re-conduction of I_{ch} for small R_g. We propose the modulation of N-column doping profile. Graded doping profile along the N-column depth with lower phosphorus concentration near the surface as shown in Fig. 7 was considered. Figure 8 shows simulated UIS waveforms for both graded and flat N-column profiles with small R_g as $Q_n=Q_p$. For graded profile, the re-conduction of I_{ch} is suppressed when V_{ds} reaches V_{sus} in contrast to the waveform for flat profile. Figure 9 shows the electric field distributions in the devices with graded profile and with flat

profile at the moment T12' and T12, respectively indicated in Fig. 8. The re-conduction of I_{ch} in the graded profile device is suppressed in comparison with flat profile because of the relaxation of the electric field beneath the MOS gate. Furthermore, capacitance-V_{ds} waveforms for graded and flat profile is shown in Figs. 10 (a) and (b), respectively. We can separate the pinch-off V_{ds} bias for C_{gd} and C_{ds} by adopting graded profile. For graded profile, V_{gs} decreases more before V_{ds} reaching V_{sus} by which C_{gd} finishes to discharge before complete discharge of C_{ds}. As shown in Fig. 11, the V_{gs} difference between miller plateau voltage and valley voltage just before pull up becomes larger by utilizing the graded profile. Therefore, re-conduction of I_{ch} is suppressed by both the relaxation of electric field and the difference of the pinch-off timing between C_{gd} and C_{ds}.

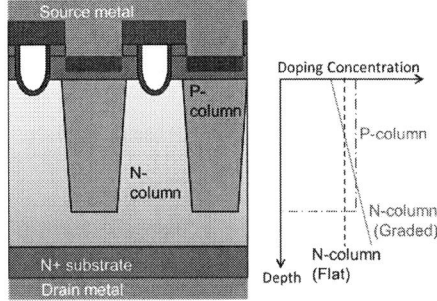

Fig. 7 Doping profiles for graded (solid line) and flat (dashed line) doping in the N-column for SJ structure.

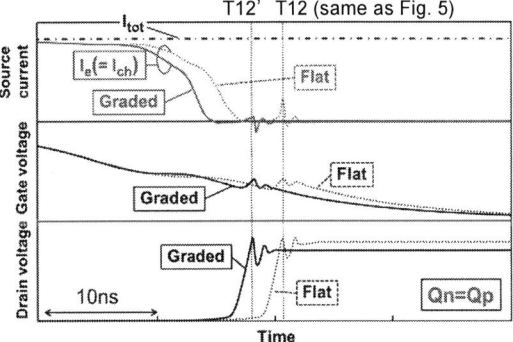

Fig. 8 Simulated UIS waveforms of graded (solid line) and flat (dashed line) doping in the N-column for small R_g condition.

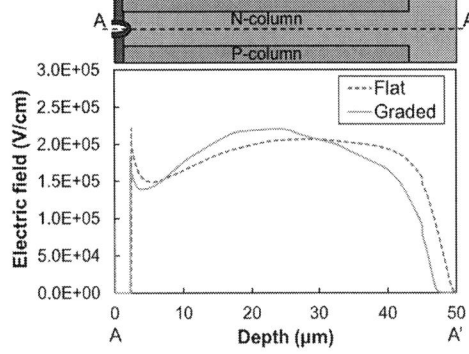

Fig. 9 Simulated electric fields at the moment T12' and T12 indicated in Fig. 8 for graded and flat profile respectively.

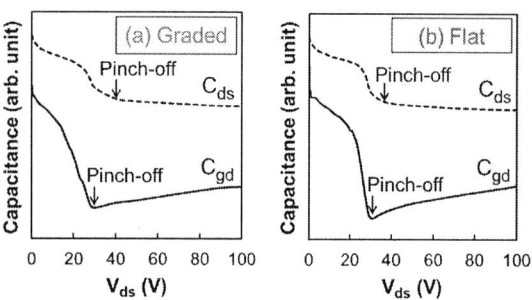

Fig. 10 Simulated capacitance-V_{ds} dependency for both (a) graded and (b) flat profile with $Q_n=Q_p$.

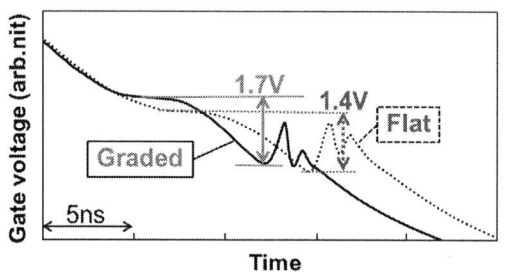

Fig. 11 Magnified simulated gate voltage waveforms for both graded and flat profile shown in Fig. 8.

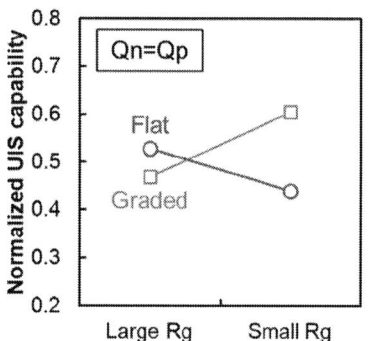

Fig. 12 Measured UIS capability for flat-epi profile and proposed graded profile.

We prepared two types of SJ-MOSFET with the graded and the flat N-column profiles from same fabrication lot except for the N-type epitaxial growth process. The fabricated devices have the same CIB as $Q_n=Q_p$. As shown in Fig. 12, the UIS capability with small R_g was improved by altered N-epi profiles. However, improvement of UIS capability is smaller than we expected, further studies are necessary.

V. CONCLUSION

We investigated the UIS capability for SJ-MOSFETs with respect to the external R_g and the CIB by experiments as well as device simulations. Measured UIS capability depends not only on the CIB but also on the R_g. The device simulations indicated that the device destruction by UIS is assumed to be related to the re-conduction of I_{ch}. The re-conduction causes more impact ionization near the surface which leads to the destruction of the device. For large R_g, I_{ch} remains longer and it turns off almost simultaneously V_{ds} reaches V_{sus}. On the other hand, for small R_g, I_{ch} turns off once while V_{ds} has not increased enough to generate ionization carrier. After that, I_{ch} re-conduction occurs due to the high electric field beneath the MOS gate structure for $Q_n=Q_p$ but suppressed for $Q_n<Q_p$. UIS capability is much larger for small R_g than for large R_g as deviating the CIB to $Q_n<Q_p$. Reducing N-column phosphorus concentration near the surface helps to suppress the re-conduction and improves the UIS capability for small R_g condition.

ACKNOWLEDGMENT

The authors would like to thank Ms. Makiko Noma and Mr. Hiromi Itoh for numerous valuable discussions on this works.

REFERENCES

[1] T. Fujihira, "Theory of Semiconductor Superjunction Devices," Jpn. J. Appl. Phys., vol. 36, pp. 6254-6262, 1997

[2] G. Deboy, M. Marz, J.-P. Stengl, H. Strack, J. Tihanyi and H. Weber, "A new generation of high voltage MOSFETs breaks the limit of silicon," in Tech. Digests of IEDM'98, pp.683-685, 1998

[3] W. Saito, I. Omura, S. Aida, S. Koduki, M. Izumisawa, H. Yoshioka and T. Ogura, "A 20 mΩcm² 600 V-class Superjunction MOSFET," in Proceedings of ISPSD'04, pp. 459-462, 2004.

[4] Y. Kawashima, H. Inomata, K. Murakawa and Y. Miura, "Narrow-Pitch N-Channel Superjunction UMOSFET for 40-60 V Automotive Application," in Proceedings of ISPSD'10, pp. 329-332, 2010.

[5] D. Arai, S. Hisada, M. Yamaji and S. Kunori, "Dependence of Switching Waveform on Charge Imbalance in Superjunction MOSFET Used in Inductive Load Circuit," in Proceedings of ISPSD'17, pp. 487-490, 2017.

[6] H. Yamashita, H. Ura, S. Ono, M. Nashiki, K. Mii, W. Saito, J. Onodera and Y. Hokomoto, "Supression of Switching Loss Dependence on Charge Imbalance of Superjunction MOSFET," in proceedings of ISPSD'15, pp.405-408, 2015

Proceedings of the 31st International Symposium on Power Semiconductor Devices & ICs
May 19-23, 2019, Shanghai, China

Self-Turn-on-Free 5V Gate Driving for 1200V Scaled IGBT

Masanori Tsukuda[*†], Masaki Sudo[†], Kazunori Hasegawa[†], Seiya Abe[†], Takuya Saraya[‡], Toshihiko Takakura[‡],
Munetoshi Fukui[‡], Kazuo Itou[‡], Shinichi Suzuki[‡], Kiyoshi Takeuchi[‡], Tamotsu Ninomiya[*], Toshiro Hiramoto[‡], Ichiro Omura[†]

Email: tsukuda@life.kyutech.ac.jp

[*]Green Electronics Research Institute, Kitakyushu, Fukuoka, Japan
[†]Kyushu Institute of Technology, Kitakyushu, Fukuoka, Japan
[‡]The University of Tokyo, Tokyo, Japan

Abstract—**Negative biasing of the gate voltage in a scaled insulated gate bipolar transistor (IGBT) during the off-state was modeled and found to be effective against self-turn-on failures. The required self-turn-on-free criteria were verified experimentally.**

Keywords—scaled IGBT, self-turn-on, gate shielding layer

I. INTRODUCTION

This paper reports the modeling and demonstration of a self-turn-on-free 5V gate drive (V_{GE_th} = 1.7 V) for a 1200V scaled insulated gate bipolar transistor (IGBT), in which the MOS gate shielding layer prevents self-turn-on failures [1–3]. Since 2011, new IGBTs with a scaling (miniaturizing) factor of k = 3 have been proposed and fabricated [4–6]. The scaled IGBT features a high conduction current capability and a low gate drive voltage of 5V. A low gate drive voltage dramatically reduces the gate drive power in proportion to the square of the voltage swing; that is, by a factor of k^2. However, it has the disadvantage of self-turn-on failures due to switching noise, especially when driven at high dV_{CE}/dt [7, 8], because of the low gate threshold voltage. In this paper, we first analyze the gate shielding mechanism and

propose gate voltage model for the planer gate IGBT. Then we demonstrate a shielding mechanism and present the criteria for a scaled trench gate IGBT that is free from self-turn-on failures.

II. ANALYSIS OF THE MOS GATE SHIELDING MECHANISM

Off-state negative gate biasing has been employed for IGBT gate drives because it was believed that the negative voltage adds a margin to the gate threshold voltage, and thus

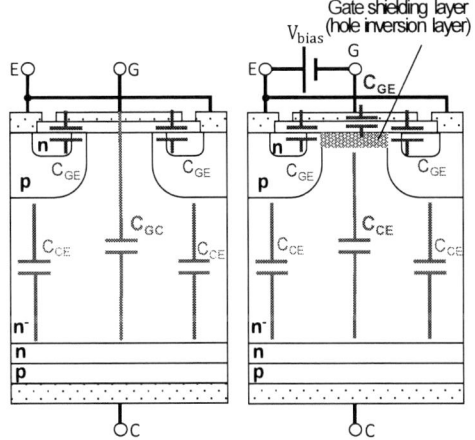

Fig. 2. Schematic view of parasitic capacitance in a planar IGBT with/without a gate shielding layer.

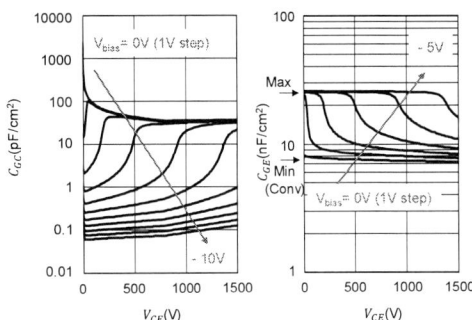

Fig. 3. TCAD simulation of C_{GC} and C_{GE} in a planar IGBT.

Fig. 1. TCAD simulation of a planar gate IGBT with a negative bias voltage and the electrostatic potential distribution in the device.

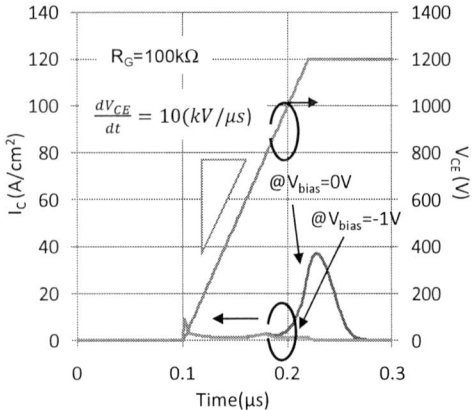

Fig. 4. TCAD simulation of the collector voltage and current waveforms during the off-state under high dV_{CE}/dt conditions for a planar gate IGBT.

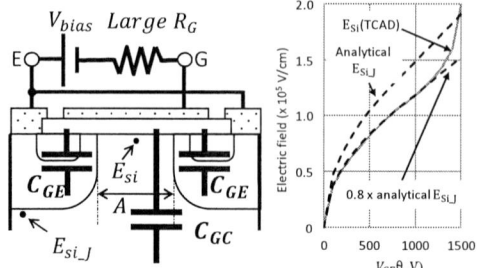

Fig. 5. Schematic of V_{GE} model of Eq. (2) and the electric field as a function of V_{CE} ($E_{si} = \alpha E_{si\_J}$, $\alpha = 0.8$).

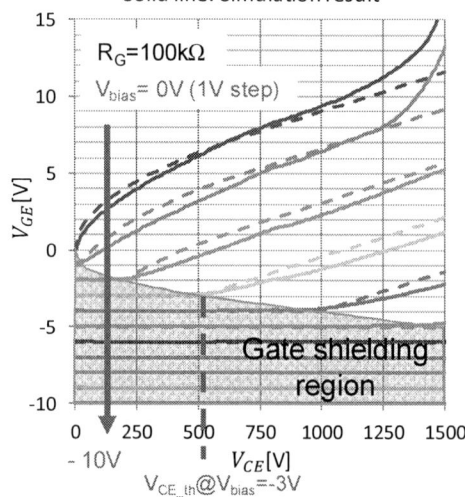

Fig. 6. Off-state V_{GE} calculated using the proposed model compared with the TCAD simulation results.

reduces the self turn-on risk. Recently, it has been reported that a negative gate bias induces a shielding (hole inversion) layer beneath the MOS gate, so that the electric field when driven at high dV_{CE}/dt is completely terminated to the source/emitter voltage, and thus the MOS device is strongly protected from self-turn-on failures. However, the physics-based mechanism has not been explained or modeled.

A gate shielding layer appears with a negative gate bias at a low V_{CE}. This layer is removed at a high V_{CE} with the increase in the electric field in the N-base (N-drift) surface E_{Si}, and this was confirmed TCAD simulation (Fig. 1). The layer appears when the initial($V_{CE}=0V$) shielding layer charge Q_{shield} is larger than the charge from V_{CE}. The shielding layer changes C_{GC}, C_{GE} and C_{CE}. C_{GC} changes in a part of C_{CE} and C_{GE} appears between the MOS gate and the N-base (N-drift) surface. As a result, C_{GC} becomes very small and C_{CE} and C_{GE} become large (Figs. 2 and 3). We used TCAD to simulate the waveforms at a very high dV_{CE}/dt for an IGBT driven at 10kV/us in the off-state (Fig. 4). Applying a negative gate bias of –1V dramatically reduced the self-turn-on current.

III. GATE VOLTAGE MODEL IN THE OFF-STATE

Gate voltage was modeled based on the shielding layer mechanism. When a shielding layer exists, V_{GE} can be expressed by the following equation, because C_{GE} is completely shielded by the shielding layer.

$$V_{GE}(Q_{shield} - \epsilon_{si}E_{si} > 0) = V_{bias} \qquad (1)$$

where $Q_{shield} = \epsilon_{ox}E_{ox} = \epsilon_{ox}\dfrac{V_{bias}}{t_{ox}}$

When no shielding layer exists (Fig. 5), V_{GE} is expressed by the following equation, because C_{GC} and C_{GE} are connected in series between the collector and emitter terminals via the gate terminal.

$$\int_{V_{CE\_th}}^{V_{CE}} C_{GC}dV_{CE} = \int_{V_{bias}}^{V_{GE}} C_{GE}dV_{GE}$$

$$A(\epsilon_{si}E_{si} - Q_{shield}) = (V_{GE} - V_{bias})C_{GE}$$

$$V_{GE}(Q_{shield} - \epsilon_{si}E_{si} \le 0) = V_{bias} + \frac{A(\epsilon_{si}E_{si}-Q_{shield})}{C_{GE}} \qquad (2)$$

where $E_{si} = \alpha E_{si\_J} = \alpha\sqrt{\dfrac{2qN_BV_{CE}}{\epsilon_{si}}}$ or $\alpha\left(\dfrac{V_{CE}}{W_B} + \dfrac{qN_BW_B}{2\epsilon_{si}}\right)$

The V_{GE} calculated from the analytical model was compared with the V_{GE} from the TCAD simulation. The values for V_{GE} were in close agreement, confirming the accuracy of the analytical model (Fig. 6).

IV. SELF-TURN-ON-FREE CRITERIA FOR THE SCALED IGBT

The self-turn-on-free criteria for a scaled 1200V IGBT were confirmed experimentally. An IGBT with a gate drive voltage of 5V (k = 3) was used as the DUT because we assumed that a low V_{GE_th} of 1.7 V would facilitate self-turn-on failures (Fig. 7). Measurement of C_{GC} and C_{GE} confirmed the appearance and disappearance of the gate shielding layer (Fig. 8). C_{GC} decreased on applying a negative bias voltage

978-1-7281-0582-6/19 $31.00 © 2019 IEEE

with a low V_{CE}. Conversely, C_{GE} increased on applying a negative bias voltage. The characteristics agreed with those from the TCAD simulation analysis (Fig. 3). The value of

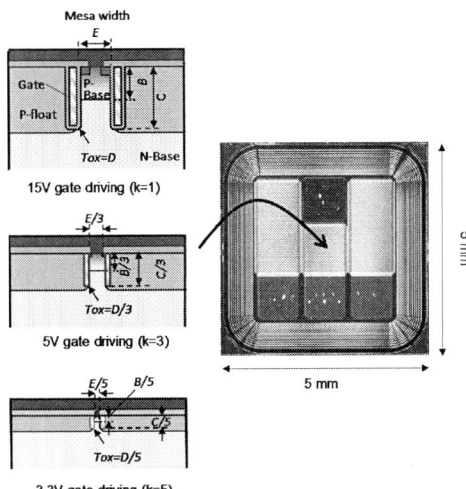

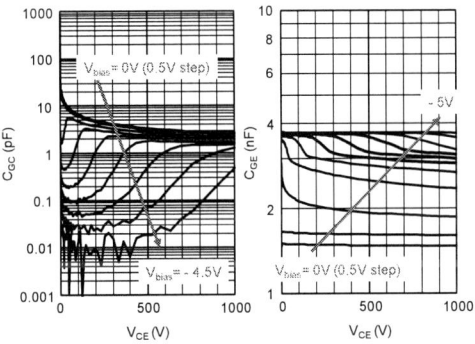

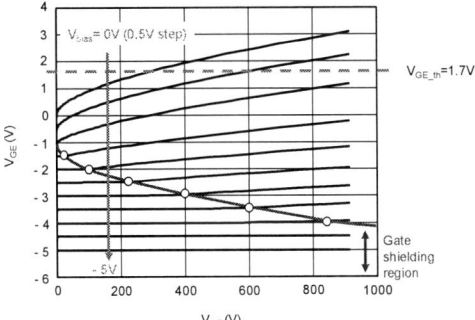

Fig. 7. Scaled IGBT and the fabricated 5V gate driving trench gate IGBT chip.

Fig. 8. Measured C_{GC} and C_{GE} for the 5V gate trench gate IGBT chip (C_{GC} was calibrated assuming that C_{GC} was 0F at $V_{bias} = -5V$).

Fig. 9. Expected V_{GE} for the 5V gate driving IGBT chip from the measured C_{GC} and C_{GE} (Fig. 8)

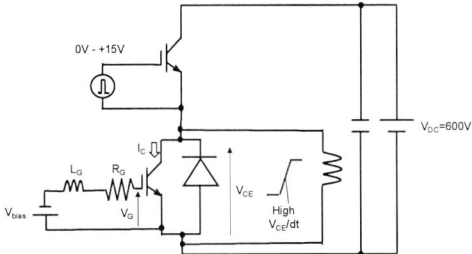

Fig. 10. Test circuit for self-turn-on-free driving (Inductive load and single pulse switching).

V_{GE} was obtained from C_{GC} and C_{GE} using the following equation.

$$V_{GE} = V_{bias} + \frac{\int_0^{V_{CE}} C_{GC} dV_{CE}}{C_{GE}} \qquad (3)$$

The results of the TCAD simulation and the model generally agreed well (Figs. 6 and 9). The required V_{bias} of $-$1V was expected from Eq. 3.

We confirmed experimentally that the scaled IGBT operated free from self-turn-on-failures in a single pulse test. The gate resistance of the switching IGBT was 0Ω for a high dV_{CE}/dt of 25kV/μs (Fig. 10). Even under high impedance gate circuit conditions ($L_G = 400$nH and R_G up to 1kΩ), no self-turn-on failures occurred when applying a negative gate bias of -2V or below (Fig. 11). The difference from the expected value of V_{bias} is assumed to be due to the short expansion of the depression layer with the high dV_{CE}/dt.

We also tried self-turn-on-free driving with no V_{bias} to simplify the gate drive circuit. With L_G as low as 10nH, self-turn-on-free operation was established without a negative gate bias (Fig. 12). Thanks to the low impedance of the gate circuit, the charge on C_{GC} quickly discharged to the emitter terminal.

V. CONCLUSION

We found by modeling and by experiment that a negatively biased gate voltage was effective against self-turn-on failures. Simple yet accurate model equations were proposed and the results showed good agreement with the TCAD simulation. The model can be easily implemented in a SPICE device model (see Appendix), and the parameters can be easily extracted from measurement and the N-base structure. The self-turn-on-free criteria we obtained by experiment demonstrated that completely self-turn-on-free operation is possible with a negative V_{bias} even for a 1200V scaled IGBT with the gate driven at 5V. The practicality of a gate drive voltage at a CMOS logic level of 5 V shows the possibility of new functionality for gate drives using digital IoT/AI technology [9].

ACKNOWLEDGMENTS

This paper is based on the results obtained from a project commissioned by the New Energy and Industrial Technology Development Organization (NEDO) (P10022).

978-1-7281-0582-6/19 $31.00 © 2019 IEEE

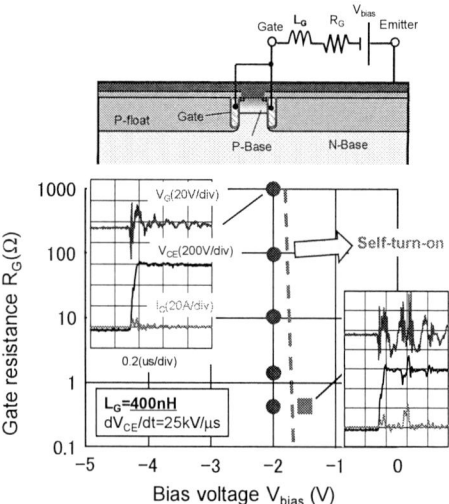

Fig. 11. Absolute criterion for self-turn-on-free driving for 1200V scaled trench gate IGBT and the waveforms.

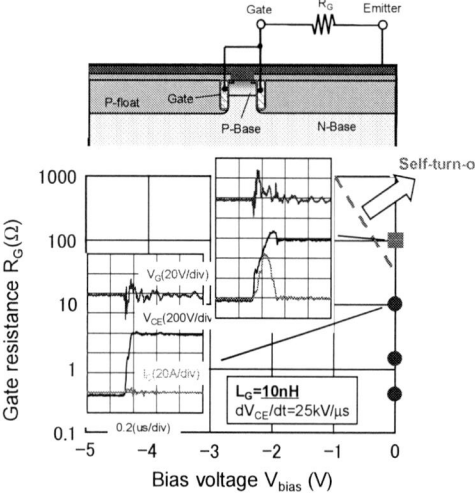

Fig. 12. Criteria for a low gate impedance for self-turn-on-free driving of a 1200V trench gate scaled IGBT and the waveforms.

REFERENCES

[1] T. Nishiwaki, T. Hara, K. Kaganoi, M. Yokota, Y. Hokomoto, Y. Kawaguchi, " Design Criteria for Shoot-Through Elimination in Trench Field Plate Power MOSFET," Proc. of ISPSD'14, pp.382–385, 2014.

[2] K. Murata, K. Harada, "Analysis of a Self Turn-on Phenomenon on the Syncronus Rectifier in a DC-DC Converter," Proc. of INTELEC, pp. 199–204, 2003.

[3] B. Yang, S. Xu, J. Korec, J. Shen, "Design Considerations on Low Voltage Synchronous Power MOSFETs with Monolithically Integrated Gate Voltage Pull-down Circuitry," Proc. of ISPSD'12, pp. 121-124, 2012.

[4] M. Tanaka, I. Omura, "Structure oriented compact model for advanced trench IGBTs without fitting parameters for extreme condition: Part I," Microelectronics Reliability 51, pp. 1933-1937, 2011.

[5] K. Kakushima,T. Hoshii, K. Tsutsui, A. Nakajima, S. Nishizawa, H. Wakabayashi, I. Muneta, K. Sato, T. Matsudai, W. Saito, T. Saraya, K. Itou, M. Fukui, S. Suzuki, M. Kobayashi, T. Takakura, T. Hiramoto, A. Ogura, Y. Numasawa, I. Omura, H. Ohashi, H. Iwai, "Experimental verification of a 3D scaling principle for low Vce(sat) IGBT," Proc. of IEDM, pp. 10.6.1-10.6.4, 2016.

[6] T. Saraya, K. Itou, T. Takakura, M. Fukui, S. Suzuki, K. Takeuchi, M. Tsukuda, Y. Numasawa, K. Satoh, T. Matsudai, W. Saito, K. Kakushima, T. Hoshii, K. Furukawa, M. Watanabe, N. Shigyo, K. Tsutsui, H. Iwai, A. Ogura, S. Nishizawa, I. Omura, H. Ohashi, T. Hiramoto, "Demonstration of 1200V Scaled IGBTs Driven by 5V Gate Voltage with Superiorly Low Switching Loss," Proc. of IEDM, 2018.

[7] S. Abe, K. Hasegawa, M. Tsukuda, I. Omura, T. Ninomiya, "Modelling of the shoot-through phenomenon introduced by the next generation IGBT in inverter applications," Microelectronics Reliability 76-77, pp. 465-469, 2017.

[8] M. Tsukuda, S. Abe, K. Hasegawa, T. Ninomiya, I. Omura, "Bias voltage criteria of gate shielding effect for protecting IGBTs from shoot-through phenomena," Microelectronics Reliability, Vol. 88-90, pp. 482-485, 2018.

[9] K. Miyazaki, S. Abe, M. Tsukuda, I. Omura, K. Wada, M. Takamiya, T. Sakurai, "General-purpose clocked gate driver (CGD) IC with programmable 63-level drivability to reduce Ic overshoot and switching loss of various power transistors," Proc. of APEC, pp. 2350-2357, 2016.

APPENDIX

We propose a configuration for the capacitances in the device for a SPICE device model (Fig. 13). The appearance and disappearance of the shielding layer is at the border of the capacitance value. The required V_{bias} is expressed by the following equation.

$$Required\ V_{bias} = \frac{\epsilon_{si} E_{si} t_{ox}}{\epsilon_{ox}} = \frac{\epsilon_{si} E_{si} A}{C_{GE\_max} - C_{GE\_conv}} \quad (4)$$

This SPICE model is applicable to all MOS gate devices.

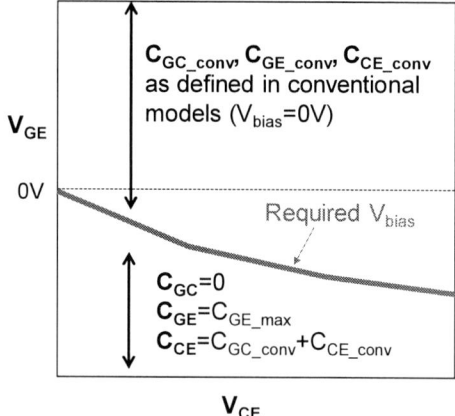

Fig. 13. Capacitance parameters for the SPICE device model from the gate shielding layer for MOS gate devices.

Proceedings of the 31st International Symposium on Power Semiconductor Devices & ICs
May 19-23, 2019, Shanghai, China

Insulated Gate Bipolar Transistors based on Pure Boron Collectors

Ahmed Elsayed, Jan Frederik Dick and Joerg Schulze

Institute of Semiconductor Engineering

University of Stuttgart

Stuttgart 70569, Germany

Email: ahmed.elsayed@iht.uni-stuttgart.de

Abstract—**Continuous efforts are invested to improve mid- and high-voltage devices to improve on-state resistances, switching performance and overall power losses. However, given the current maturity of Silicon technologies, significant improvements are difficult to achieve. In this work, we attempt to improve on-state resistances of Insulated Gate Bipolar Transistors through utilizing ultra-thin pure Boron layers for collector junctions as well as introducing modified channel doping schemes. We also include DC characteristics, comparisons and analysis of the achieved results.**

Keywords—IGBT, Boron, ultra-thin, planar-doping.

I. Introduction

Since their introduction in the 1980s, Insulated Gate Bipolar Transistors (IGBTs) have established themselves strongly for mid- and high-voltage power applications. Due to their improved performance as compared to their predecessors, namely Bipolar Junction Transistors (BJTs) and Metal-Oxide-Semiconductor Field Effect Transistors for power applications (Power-MOSFETs), IGBTs are utilized in many areas of power electronics [1][2].

The IGBT is a minority carrier device displaying typical MOS high input impedance and large current driving abilities typical of BJTs. The conduction mechanism behind IGBT operation allows for large currents driven with minimized on-resistance in comparison to Power-MOSFETs of similar dimensions.

Many efforts have been made to improve device characteristics and performance to optimize conduction and switching losses for the best operation achievable. Currently, Silicon (Si) based device characteristics have reached their theoretical potential due to the current maturity of Si production technologies. Thus, meaning further improvements are made possible through structural modifications and introduction of new material systems [3].

When considering IGBTs, like other vertical Power-MOSFETs, the on-resistance is a sum of the individual device section resistances. Equation (1) shows some of the main contributions within a typical vertical Power-MOSFET; where R_{On}, R_{Sub}, R_{Ch}, and R_{dz} represent the total device on-resistance, substrate resistance (drain for Power-MOSFETs and collector in case of IGBTs), channel resistance and drift zone resistance respectively.

$$R_{On} = R_{Sub} + R_{Ch} + R_{dz} + ... \qquad (1)$$

In case of Power-MOSFETs for high-power applications, R_{dz} dominates during operation due to the necessary material thickness. However, when scaling down devices, that no longer holds true and other contributions should also be considered.

Typical methods of IGBT production start with Si p+ substrates for the injection layer (typically hundreds of microns thick) and subsequent epitaxial growth of following layers [1]. This imposes a higher device resistance and overall higher operational loss. Some techniques further include thinning of this layer (down to tens of microns thick) by grinding once the epitaxial layers achieve sufficient rigidity. However, this adds more cost and complexity to the production process. Furthermore, there is still potential for improvement.

In this work, we attempt to improve IGBT characteristics through employing nanometer thick pure Boron (B) layers to replace traditional thick p-doped Si layers. We further attempt to improve on IGBT characteristics by investigating the effect of planar channel doping schemes as compared to traditional homogeneously doped channels.

II. Ultra-Thin Boron Layers

For decades, B has been utilized in the semiconductor industry due to its adequacy as a p-type doping species and attractive physical properties of its compounds [4]. Recently, driven by the demand for ultra-thin Si-based photodetectors, pure B layers have been utilized for deep Ultra-Violet (UV) photodiodes [5]. This was realized through pure B deposition using Chemical Vapor Deposition (CVD) onto Si surfaces leading to sub 10 nm junctions with high idealities [5].

The mechanism behind being that some B atoms within the first monolayer at the Si-interface will form acceptor states. These states will then fill with electrons forming a monolayer of negative charge. Thus, forming an inversion layer within the underlying Si with a build-up of holes [6].

For the purpose of this work, pure B layers are utilized for IGBT injection layers and are realized using Molecular Beam Epitaxy (MBE). The pure B layers were deposited at a temperature of 650 °C. However, we have further realized diodes fabricated with pure B layers deposited at a much wider range of temperatures. Such devices can be seen in Fig. 1 and Fig. 2. The utilized MBE system as well as the fabrication procedure are described briefly in [7].

However, in comparison to CVD deposited B layers, MBE offers extreme deposition temperature flexibility with substrate deposition temperatures down to almost room temperature, resulting in fully electrically functional ultra-thin B junctions. Fig. 2 shows the IV-characteristics of a series of ultra-thin B p^+-i-n^+ junctions realized with temperatures

978-1-7281-0582-6/19 $31.00 © 2019 IEEE

ranging 700 °C down to 50 °C. Furthermore, Atomic Force Microscopy (AFM) performed on the deposited B layers showed extreme layer uniformity across the full range with average roughness values, Ra = 275 pm.

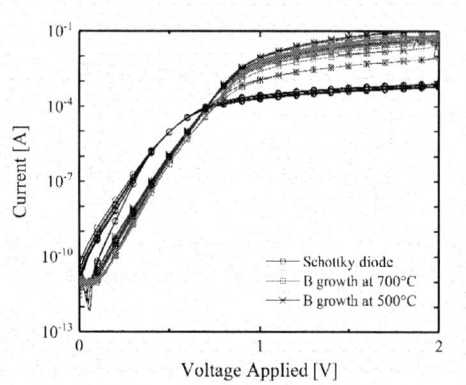

Fig. 1. IV-characteristics of p⁺-i-n⁺ diodes based on ultra-thin B layers deposited at 500 °C and 700 °C into oxide window structures compared to Schottky diodes [7].

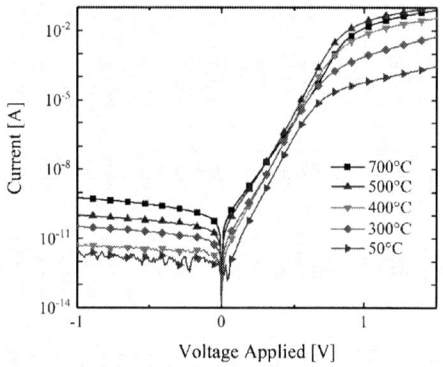

Fig. 2. IV-characteristics of p⁺-i-n⁺ mesa structure diodes based on ultra-thin B layers deposited at temperatures ranging 700 °C down to 50 °C.

III. PLANAR-DOPED CHANNEL IGBTS

Also driven by the need for reduced power losses, it has been previously theoretically and practically proven that local/planar confinement of channel doping in vertical concept MOSFETs can significantly improve device on-resistance. This has been investigated for sub 100 nm devices and smart power application MOSFETs with breakdown voltages ranging 12 V - 40 V corresponding to drift zone thickness of 3 µm [8]. However, to the best of our knowledge, such concept was never investigated for IGBTs. An illustration of homogeneous vs. planar-doped IGBTs can be seen in Fig. 3.

For the homogeneously doped channel IGBT, similar to vertical Power-MOSFETs, the electron current across the channel is confined to the gate oxide interface due to inversion of the immediately underlying layer through gate bias. Subsequently, electrons experience significant surface

scattering due to imperfections along this interface leading to reduced mobility.

Furthermore, due to the spread of the doping atoms across the entire length of the channel, electrons also experience impurity (doping) related scattering, leading to a further reduced mobility. Conversely, for the case of planar-doped channel FETs, confining the doping to a localized region within the channel means the majority of the channel will remain intrinsic. Consequently, electron flow across the channel will further spread and is no longer constricted to the gate-oxide interface. Additionally, experiencing less impurity related scattering hence improving the on-state current.

In order to validate our design and establish a frame of reference, we fabricated vertical Power-MOSFETs with 300 µm thick drift zones in order to evaluate potential improvement of on-state current. Shown in Fig. 4 are the output characteristics of a homogeneous and planar-doped Power-MOSFET with the same geometry.

These MOSFETs were fabricated with 500 nm long channels. For the homogeneously doped devices, the channel was epitaxial grown with a B doping of $5 \cdot 10^{18}$ cm⁻³. For the sake of comparison, the planar-doped devices were designed to have the same amount of dopant atoms, however, only confined within 50 nm in the channel region. Meaning, the planar-doped region had a concentration of $5 \cdot 10^{19}$ cm⁻³.

Although the planar-doped devices displayed higher on-state current, the threshold voltage, V_T and sub-threshold swing, SS were adversely affected due to the higher peak doping within the channel. These values are compared in Table. 1.

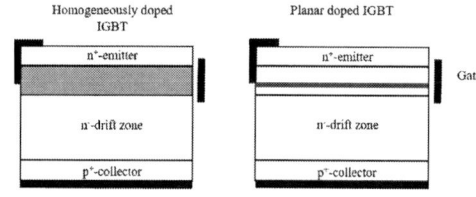

Fig. 3. Illustration of a homogenuos channel IGBT (left) as compared to planar-doped channel IGBT (δ-channel) (right).

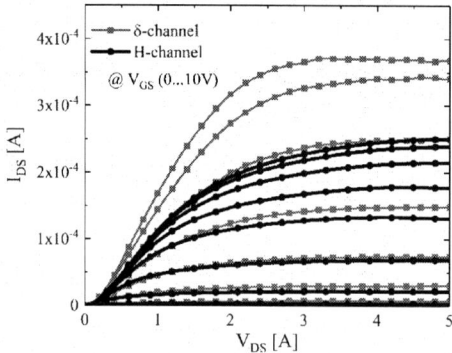

Fig. 4. Output characteristics of homogeneuosly vs. planar-doped channel Power-MOSFETs based on similar geometry devices.

	R_{On}	V_T	SS
Homogeneously doped N-channel Power-MOSFET	$5.43\ k\Omega$	$3.5V$	$370\ mV/dec$
Planar-doped N-channel Power-MOSFET	$8.4\ k\Omega$	$4.5V$	$500\ mV/dec$

TABLE I. HOMOGENEOUSLY VS. PLANAR-DOPED N-CHANNEL POWER-MOSFET

IV. IGBT FABRICATION

For the purposes of this work, commercially acquired Si (100) double-polished substrates were utilized as a base for the subsequent epitaxial growth of the active layers. These substrates had a n-type doping of $1.5 \cdot 10^{15}$ cm^{-3} and were thinned to a thickness of 300 µm. They served as the drift zone for the IGBTs and Power-MOSFETs realized during this work.

Four different IGBT structures were fabricated and compared. These are outlined below:

- Homogeneous channel IGBTs with a 200 nm epitaxial grown p$^+$-Si collector.

- Planar-doped channel IGBTs with a 200 nm epitaxial grown p$^+$-Si collector.

- Homogeneous channel IGBTs with a ~ 4 nm B layer deposited at 650 °C.

- Planar-doped channel IGBTs with a ~ 4 nm B layer deposited at 650 °C.

All four IGBT structures were fabricated with 500 nm epitaxial grown channels similar to those employed for the Power-MOSFETs in the previous section. This was followed by 200 nm of n$^+$-Si for the emitter top contact.

Following growth, substrates were structured through lithography and etched to form mesa structures. Silicon dioxide (SiO$_2$) gate oxides were then deposited through Plasma-Enhanced CVD using a Tetraethyl Orthosilicate (TEOS) precursor, gate structures were then formed, and a passivation oxide layer was deposited. Deep contacts were then etched into the mesa structures to contact both channel and emitter layers. An illustration of a vertical IGBT can be seen in Fig. 5. Special attention was necessary to avoid over etching the deep contact through the planar-doped region to avoid shorting the channel.

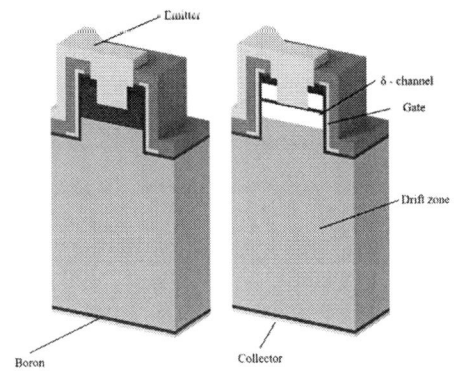

Fig. 5. Illustration of homogeneous channel IGBT (left) as compared to planar-doped channel IGBT with localized doping (right).

V. DC CHARACTERIZATION

Following the fabrication of the different IGBT structures, they were characterized using our DC measurement setup. All output characteristics where normalized to transistor gated length. Shown in Fig. 6, are the output characteristics for a homogeneous channel IGBT plotted against a planar-doped channel IGBT, both based on p$^+$-Si collectors. It was observed that including the planar doping of the channel into our design of a traditional IGBT while keeping other design parameters constant, led to an enhancement of on-state current. A further comparison of the different IGBT structures can be seen in Table. 2.

In addition, shown in Fig. 7 are the output characteristics of a planar-doped IGBT based on pure B collector plotted against a planar-doped channel IGBT based on a p$^+$-Si collector. Fig. 8 shows the output characteristics of IGBTs based on pure B collectors with both a homogeneous and planar-doped channel. Lastly, shown in Fig. 9 is a comparison of a traditional IGBT with a homogeneous channel and p$^+$-Si collector vs. a planar-doped channel IGBT with a p$^+$-Si collector vs. a planar-doped IGBT with pure B collector.

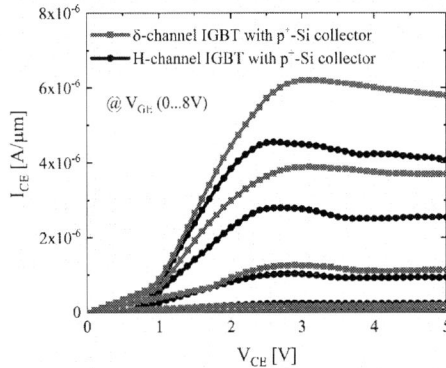

Fig. 6. Output characteristics of planar doped channel IGBTs (in red) vs. homogeneous doped IGBTs (in black) based on p$^+$-Si collectors at gate biases ranging 0 V to 8 V with 1 V steps.

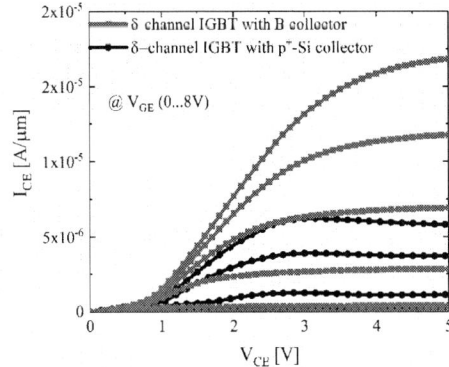

Fig. 7. Output characteristics of planar doped channel IGBTs based on pure B collectors (in red) vs. planar doped IGBTs based on p$^+$-Si collectors (in black) at gate biases ranging 0 V to 8 V with 1 V steps.

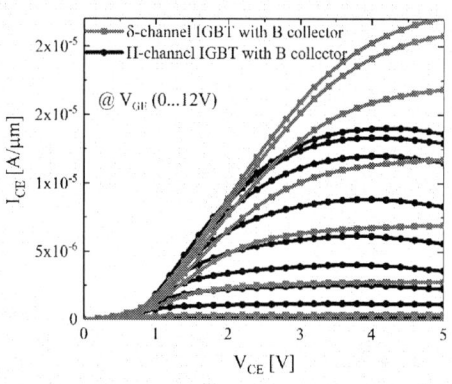

Fig. 8. Output characteristics of planar doped channel IGBTs based on pure B collectors (in red) vs. homogeneous IGBTs based on pure B collectors (in black) at gate biases ranging 0 V to 12 V with 1 V steps.

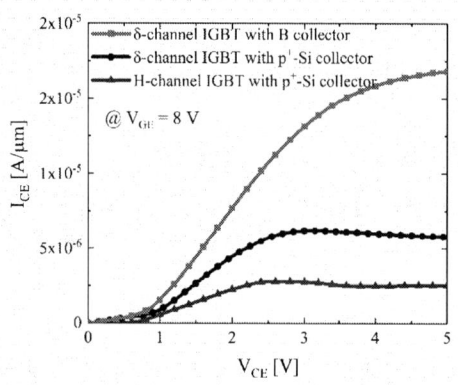

Fig. 9. Output characteristics of planar doped channel IGBTs based on pure B collectors (in red), planar doped channel IGBTs based on p⁺-Si collectors (in black) and homogeneous channel IGBTs based on p⁺-Si collectors (in blue) at a gate bias of 8 V.

TABLE II.   COMPARISON OF R_{On} OF DIFFERENT IGBTS

	R_{On}
Homogeneous channel with p⁺-Si collector	100.74 Ω
Planar-doped channel with p⁺-Si collector	46.50 Ω
Homogeneous channel with pure B collector	29.56 Ω
Planar-doped channel with pure B collector	24.86 Ω

As seen in Fig. 6 and Fig. 9 and similar to results observed for N-channel Power-MOSFETs, through channel doping confinement we achieved a significant enhancement of on-state currents of our IGBTs with a reduction of on-state resistance by approximately half.

Furthermore, as seen in Fig. 9, introduction of the pure B collector further improves the on-state current by approximately one order of magnitude and reduces the on-state resistance by almost 75 % when compared to traditional homogeneous channel IGBTs with p⁺-Si collectors.

VI. CONCLUSION

We have shown that implementing planar-doping schemes for IGBT channels as well as utilizing pure B layers for collector junctions can offer significant performance enhancement in terms of on-state resistances. Coupled with the extreme processing flexibility shown of the B layers in terms of deposition temperatures, this method offers a large potential for reduction of fabrication costs and complexity without compromising performance.

ACKNOWLEDGMENT

This research work was funded by the following research grants: 1) Bosch-MWK RBZ project #6 on "Si-SiC-GaN Benchmarking, Entwurfs- und Bewertungsmethodik, Prototyping, Systemvalidierung"and 2) DAAD PPP Croatia 2017 project on "Characterization of ultra-thin BoSi photodetectors" (ID 57335170).

REFERENCES

[1] V. Khanna, Insulated Gate Bipolar Transistor (IGBT). Hoboken: Wiley, 2004.

[2] B. Baliga, Fundamentals of power semiconductor devices. New York: Springer Science + Business Media,2008.

[3] A. Nakagawa, "Theoretical Investigation of Si Limit Characteristics of IGBT," 2006 IEEE International Symposium on Power Semiconductor Devices and IC's, Naples, pp. 1-4.

[4] O. A. Golikova (1979), Boron and Boron-based semiconductors. phys. stat. sol. (a), 51: 11–40.

[5] F. Sarubbi, L. K. Nanver, T. L. M. Scholtes and S. N. Nihtianov, "Extremely Ultra -Shallow p+-n Boron-Deposited Silicon Diodes Applied to DUV Photodiodes," 2008 Device Research Conference, Santa Barbara, CA, 2008, pp. 143-144.

[6] L. Qi and L. K. Nanver, "Conductance Along the Interface Formed by 400 °C Pure Boron Deposition on Silicon," in IEEE Electron Device Letters, vol. 36, no. 2, pp. 102-104, Feb. 2015.

[7] A. Elsayed and J. Schulze, "Characterization of thin Boron layers grown on Silicon utilizing molecular beam epitaxy for ultra-shallow pn-junctions," 2018 41st International Convention on Information and Communication Technology, Electronics and Microelectronics (MIPRO), Opatija, 2018, pp. 0007-0011.

[8] C. Fink, J. Schulze, I. Eisele, W. Hansch, W. Werner and W. Kanert, "Reducing of ROn in Vertical Power-MOSFETs due to Local Channel Doping" 2001 Jpn. J. Appl. Phys. 40 2637.

Proceedings of the 31st International Symposium on Power Semiconductor Devices & ICs
May 19-23, 2019, Shanghai, China

4.5kV Insulated Gate Triggered Thyristor (IGTT) with High *di/dt* Characteristics for Pulse Power Applications

Chao Liu, Wanjun Chen, Yijun Shi, Bin Qiao, Qian Jiang, Yun Xia, Qijun Zhou, Xiaorui Xu, Qi Zhou, Zhaoji Li, Bo Zhang

State Key Laboratory of Electronic Thin Films and Integrated Devices, University of Electronic Science and Technology of China, Chengdu610054, China

(e-mail: wjchen@uestc.edu.cn)

Abstract—**In this work, we propose a high-voltage (HV) insulated gate triggered thyristor (IGTT) as solid-state closing switch (SSCS) for pulse power applications. The mechanism of proposed HV IGTT is analyzed by TCAD simulation. Special design with consideration of the transient 2-D effect are carried out for enhancing conductivity modulation and achieving high *di/dt* characteristics. With experimental measurements, the characterizations of the fabricated HV IGTT are presented. The experimental results show that the fabricated HV IGTT features blocking voltage over 4500V and performs a *di/dt* up to 239 kA/cm²/μs with peak current over 5.5kA.**

Keywords—Pulse power, insulated gate thyristor, di/dt

I. INTRODUCTION

Pulse power systems, such as particle acceleration and electromechanical launcher, are undergoing the trend of miniaturization and solidification, which leads to the solid-state closing switches (SSCSs) as the replacement of the traditional gas/spark-type switches [1]-[5]. In such applications, the SSCSs should feature properties of high DC voltage (usually >4kV) and high pulse performances (peak current and *di/dt*).

In the mainstream power semiconductor devices, the power MOSFETs operating at unipolar mode have large conduction resistance, preventing it from being applied in high-voltage fields. The insulated gate bipolar transistors (IGBTs) can be evolved for high voltage applications due to the conductivity modulation effect. But the degree of conductivity modulation is limited by the finite MOS driving current [6]-[9]. Thyristors and super GTOs have shown high surge current property [10]-[12]. Nevertheless, they always need large active area and strong gate

This work was supported in part by the National Natural Science Foundation of China (No. 51877030), the Sichuan Youth Science and Technology Foundation (No. 2017JQ0020), the Major Science and Technology Special Projects in Guangdong under Grant (No. 2017B010112003) and the Fundamental Research Funds for the Central Universities (No. ZYGX2016Z006).

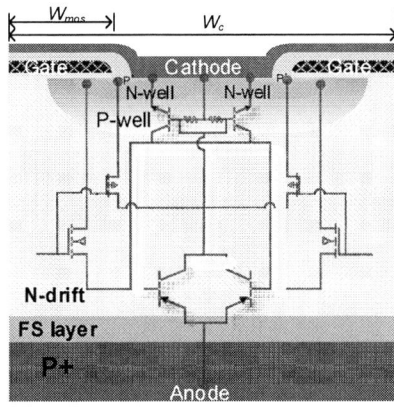

Fig. 1: The schematic cross-section of the proposed HV IGTT with high-*di/dt* characteristics. The equivalent circuit of HV IGTT is drawn in the structure.

pulse for realizing high *di/dt* capability [12], [13]. Silicon-Carbide (SiC) thyristors have potential to realize high pulse performances and simultaneously small chip size, due to the thin voltage sustaining layer brought by the large band gap. Unfortunately, there are many problems to be solved for their practical applications. Till now, Si-based devices, enable mature technology may be pursued as the proper SSCS in high voltage field (1200-6500V). In our previous work [3], [6], [10], a high-*di/dt* MOS-controlled thyristor with BV of 1600V has been presented. However, their blocking voltage is too low in higher voltage fields.

In this work, a high voltage (HV) insulated gate triggered thyristor (IGTT) with high-*di/dt* characteristics for SSCS applications is presented. The mechanism of the HV IGTT is analyzed. Moreover, the design criterion with consideration of the transient 2-D effect is studied by using TCAD simulation.

978-1-7281-0582-6/19 $31.00 © 2019 IEEE   347

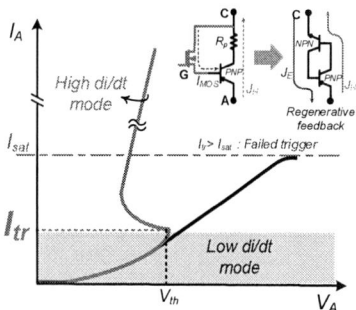

Fig. 2: The operating mechanism of proposed HV IGTT with high-*di/dt* characteristics. I_{MOS}, J_A, J_E is the MOS current, the hole current and the electron current, respectively

With experiments, the pulse performances of the HV IGTT are demonstrated.

II. DEVICE STRUCTURE AND DESIGN

The schematic cross-section of the proposed high voltage IGTT with high-*di/dt* characteristics is shown in Fig. 1. The proposed HV IGTT is characterized by alternating MOS and PNPN thyristor sections. Cathode shunts are implemented so that the HV IGTT can realize high blocking voltage at gate ground and improved *dV/dt* immunity [10]-[11]. Driving the proposed HV IGTT is similar to driving power MOSFET. When the gate voltage (V_g) exceeds the MOS threshold voltage (V_{th}), the MOS channel is opened and provides a current (I_{MOS}). This MOS current works as the base current of the PNP transistor and promotes a higher hole current (J_H). Once this J_H exceeds the trigger current of thyristor (I_{tr}), the upper NPN transistor is turned on and the regenerative feedback mechanism of thyristor is triggered, as demonstrated in Fig. 2. Consequently, almost

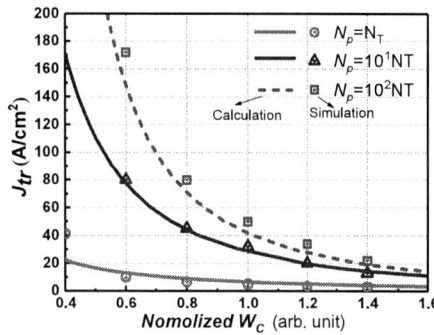

Fig. 3: The simulated and calculated trigger current (I_{tr}) versus cell pitch and P-well doping of proposed HV IGTT with high-*di/dt* characteristics.

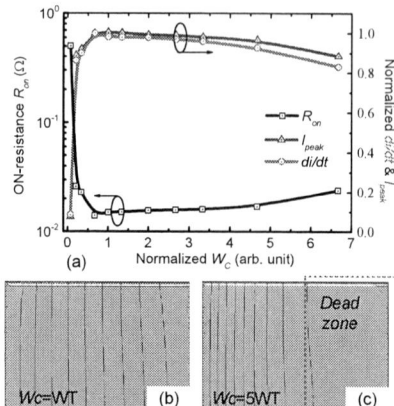

Fig. 4: The pulse performance versus cell pitch. (a) The transient R_{on}, peak current (I_{peak}) and *di/dt* versus cell pitch. The current distributions of (b) W_T device and (c) $5W_T$ device during short pulse are extracted at a high pulse current of 1000A. The cell pitches are normalized by a typical value (W_T).

infinite electron current (J_E) is supplied for strong conductivity modulation. And the HV IGTT can achieve high *di/dt* characteristics due to that ultralow ON-resistance is realized.

Obviously, the I_{tr} should be low so that the HV IGTT can enter the high *di/dt* mode as fast as possible. According to our previous work [6], the set-up of the regenerative feedback action in the PNPN thyristor requires the lateral voltage in the P-well exceeding the built-in voltage (V_{bi}) of P-well/N-well junction. Theoretically, V_P is closely related to the MOS current (I_{MOS}), cell pitch (W_C) and P-well sheet resistance ($\rho_\square$). Thus, the I_{tr} can be given by

$$J_{tr} = \frac{V_{bi}}{2 I_{MOS} \alpha_p \rho_\square \left(\frac{1}{4} W_C^2 - W_{MOS}^2 \right)} \quad (1)$$

where α_P is the current gain of the wide base P-N-P transistor. The P-well sheet resistance can be calculated by [14]

$$\rho_\square = \frac{1}{q \mu_p N_P \int \exp[-(\frac{y}{L_{cha}})^2] dy} \quad (2)$$

where N_P, μ_p and L_{cha} are the doping concentration of P-well, the mobility and the character length, respectively.

According to Eq. (1), a wide cell pitch as well as high P-well sheet resistance contribute to a low I_{tr}. As shown in the simulated and calculated results in Fig. 3, with the increase of W_C, the I_{tr} decreases. And with the decrease of the N_P, the $\rho_\square$ increases and

978-1-7281-0582-6/19 $31.00 © 2019 IEEE   348

the I_{tr} decreases. It should be pointed out that the N_P cannot be too low for maintaining a high blocking voltage. Thus, lengthening the W_C is usually used to lower the I_{tr}. Wide cell device with lower I_{tr} features higher di/dt, as simulated in Fig. 4 (a). However, the devices with large W_C would suffer from a degradation of di/dt, which results from that the transient 2-D effect brings about large area of dead zone (where no regenerative feedback action occurs) during the short pulse and reduces the effective turn-on area. As shown in Fig. 4 (b) and (c), large area of dead zone exists when the cell pitch is too wide, for example $5W_T$. According to the simulated results, over 20% degradation of pulse current can be brought about by the transient 2-D effect. Thus, the cell pitch should be carefully designed for low I_{tr} and simultaneously avoiding 2-D effect.

III. EXPERIMENTAL RESULTS AND DISCUSSION

Using modern IC fabrication process, the proposed HV IGTT is fabricated. The blocking voltage of fabricated devices is rate to 4500V, as shown in Fig. 5(a). No walk-back phenomenon is detected in the measurement, indicating the fabricated HV IGTT has stability under high voltage. With careful design of W_C and N_p, the fabricated HV IGTT features a relatively low I_{tr} ($\approx$3A), as shown in Fig. 5(b). Low I_{tr} helps the proposed HV IGTT fast enter the regenerative feedback action and obtain a low on-state voltage V_{on} (5V@50A).

The pulse performances of the fabricated devices are tested in the RLC circuit, as shown in Fig. 6. A commercial power MOSFET driver (TC4427) is used to drive the device under tests (DUTs). The pulse current and voltage waveforms of the IGTT are shown in Fig. 7. As shown, the IGTT immediately gets a peak current of 5510A with a main pulse width of 0.3μs, resulting in a di/dt of 50kA/μs (239kA/cm²/μs). The high di/dt

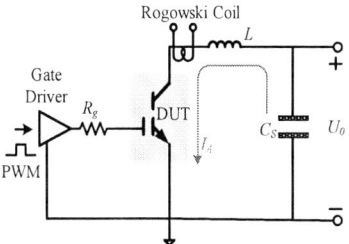

Fig. 6: The schematic diagram of RLC pulse ring down test circuit. The device under test (DUT) is the fabricated HV IGTT. The pulse current is measured by a rogowski coil and the device voltage can be detected by using high-voltage probe.

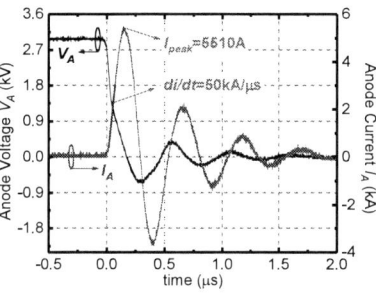

Fig. 7: The typical pulse current and voltage of proposed IGTT with high-di/dt characteristics. In this measurement, the supplying voltage is 3000V and the storage capacitor is 0.2μF. The circuit inductance (L) is estimated to be 45nH.

characteristics comes from the fast entrance of regenerative feedback action due to the low trigger current. When working at thyristor mode, the HV IGTT features low ON-resistance (R_{on}) brought by the strong conductivity modulation. According to [15], when the R_{on} is low to near 0Ω, the di/dt and I_{peak} can be estimated by $I_{peak} \approx U_0 \times \sqrt{C/L}$ and $di/dt \approx U_0/L$, respectively. It can be seen that if the $R_{on}=0$, the di/dt value is only determined by the U_0 and L. Fig. 8 demonstrates the pulse current measured at different capacitance. As shown, the HV IGTT exhibit same di/dt at condition of both 0.2μF and 0.68 μF, experimentally indicating the R_{on} of IGTT is near 0. For further evaluating the on-resistance of IGTT, the measured results of di/dt and I_{peak} are compared with the calculated results at low R_{on}. As demonstrated in Fig. 9, the di/dt and I_{peak} increase proportionately with increasing U_0 and the experimental results are consistent with the analytical results indicating the special design carried out above do be good for ultralow R_{on} and high di/dt characteristics.

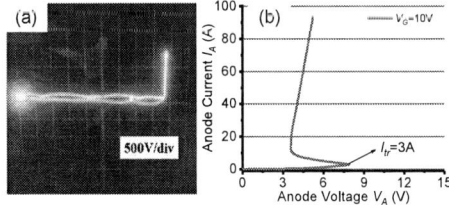

Fig. 5: The static characteristics of fabricated HV IGTT: (a) forward blocking characteristics; (b) the forward conducting characteristics s. The added gate voltage is 0V in (a) and 10V in (b). No walk-back phenomenon is observed in the BV test.

978-1-7281-0582-6/19 $31.00 © 2019 IEEE

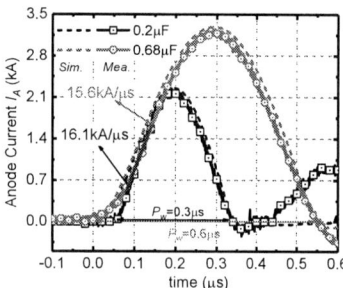

Fig. 8: The tested and simulated pulse current waveforms with varying pulse width (P_w). In this case, different capacitances (0.2μF and 0.68 μF) are used for obtaining different pulse width. The dotted lines are simulated results and the solid lines are the experimentally measured results.

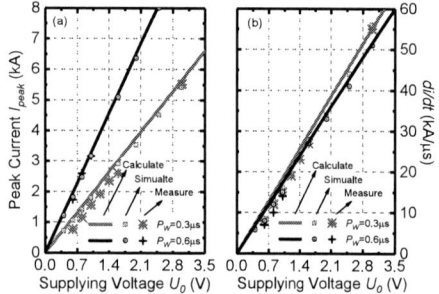

Fig. 9: The comparison of calculated, simulated and measured results of I_{peak} and di/dt. In the calculation, the device ON-resistance is set to be 0.05Ω, the circuit inductance is 45nH and the capacitance is different.

IV. CONCLUSION

In this paper, a high voltage IGTT with superior pulse performances are proposed and fabricated. The mechanism of proposed HV IGTT is analyzed by using TCAD simulation. The trigger current of the inherent thyristor should be as low as possible. With consideration of the transient 2-D effect, the main structure parameters of proposed HV IGTT are designed and optimized. Been measured, the proposed HV IGTT has a breakdown voltage of more than 4500 V at V_g=0 V. The peak current and di/dt are 5.5 kA and 239 kA/cm^2/μs for HV IGTT. The good pulse performances indicate that the proposed HV IGTT is a promising candidate as solid-state closing switch (SSCS) for pulse power applications.

REFERENCES

[1] S. C. Glidden and H. D. Sanders, "Solid State Spark Gap Replacement Switches," in *PPC*, Monterey, CA, 2005, pp. 923-926. DOI: 10.1109/PPC.2005.300444

[2] D. D. Wang, J. Qiu and K. f. Liu, "All solid-state pulsed power generator with semiconductor and magnetic compression switches," in *PPC*, Washington, DC, 2009, pp. 1233 - 1238. DOI: 10.1109/PPC.2009.5386270

[3] Wanjun Chen, , Chao Liu, Xuefeng Tang, Lunfei Lou, Wu Cheng, Qi Zhou, Zhaoji Li, and Bo Zhang, "High Peak Current MOS Gate-Triggered Thyristor with Fast Turn-on Characteristics for Solid-State Closing Switch Applications," *IEEE* Electron Device Letters, vol. 37, no. 2, pp. 205-208, Jan. 2001. DOI: 10.1109/LED.2015.2511182

[4] T. F. Podlesak, F. M. Simon and S. Schneider, "Single shot and burst repetitive operation of thyristors for electric launch applications" *IEEE Trans. on Magnetics*, vol. 37, no. 1, pp. 385 - 388, Jan. 2001. DOI: 10.1109/20.911860

[5] Haiyang Wang, Xiaoping He, Weiqing Chen, Binjie Xue, Aici Qiu, "A High-Current High-*di/dt* Pulse Generator Based on Reverse Switching Dynistors," IEEE Trans. on Plasma Sci., vol. 37, no. 2, pp. 356-358, Feb. 2009. DOI: 10.1109/TPS.2009.2012553

[6] Wanjun Chen, Chao Liu, Yijun Shi, Yawei Liu, Hong Tao, Chengfang Liu, Qi Zhou, Zhaoji Li, and Bo Zhang, "Design and Characterization of High *di/dt* CS-MCT for Pulse Power Applications," IEEE Transactions on Electron Devices, vol. 64, no. 10, pp. 4206-4212, Oct. 2017. DOI: 10.1109/LED.2015.2511182

[7] James A. VanGordon, Scott D. Kovaleski, "CHARACTERIZATION OF POWER IGBTS UNDER PULSED POWER CONDITINOS," in *PPC*, Washington, DC, 2009, pp.280-282. DOI: 10.1109/PPC.2009.5386250

[8] Wanjun Chen, Jinhan Zhang, Bo Zhang, Huaping Jiang, Zhaoji Li, "The SuperJunction MOS-controlled thyristor (SJ-MCT) with low power loss for high-power switching applications," in *ICSICT*, Xi'an, 2012, pp.1-3, DOI: 10.1109/ICSICT.2012.6466725

[9] W. J. Chen, R. Z. Sun, C. F. Peng, B. Zhang, "High dV/dt immunity MOS controlled thyristor using double variable lateral doping technique for capacitor discharge applications," *CHINESE PHYSICS B*, vol. 23, no. 7, Jul. 2014.DOI: 10.1088/1674-1056/23/7/077307

[10] Wanjun Chen, Chao Liu, Xuefeng Tang, Lunfei Lou, Wu Cheng, Hongquan Liu, Qi Zhou, Zhaoji Li and Bo Zhang, "Experimentally Demonstrate a Cathode Short MOS-Controlled Thyristor for Single or Repetitive Pulse Applications," in ISPSD, Prague, Czech Republic,2016, pp. 311-314. DOI: 10.1109/ISPSD.2016.7520840

[11] Tyler Flack, Cameron Hettler, and Stephen Bayne, "Characterization of an n-Type 4-kV GTO for Pulsed Power Applications," IEEE Transactions on Plasma Science, vol. 44, no. 10, pp. 1947-1955, Oct. 2016

[12] S. Ikeda and T. Araki, "The *di/dt* capability of thyristors," *Proceedings of the IEEE*, vol. 55, no. 8, pp. 1180-1196, Aug. 1967. DOI: 10.1109/PROC.1967.5830

[13] Chao Liu, Wanjun Chen, Yijun Shi, Hong Tao, Qijun Zhou, Huiling Zuo, Bin Qiao, Yun Xia, Ziyan Xiao, Wuhao Gao, Nan Chen, Xiaorui Xu, Qi Zhou, Zhaoji Li, Bo Zhang, "A Novel Insulated Gate Triggered Thyristor with Schottky Barrier for Improved Repetitive Pulse Life and High-*di/dt* Characteristics," IEEE Transactions on Electron Devices, vol. 66, no. 2, pp. 1018-1025, Feb. 2019. DOI: 10.1109/TED.2018.2887137

[14] B. J. Baliga, *Fundamentals of power semiconductor devices*, Springer Publication, 2003, pp.508-657.

[15] W. J. Chen, R. Z. Sun, K. Xiao, H. Z. Zhu, C. F. Peng, Z. Y. Ruan, J. X. Ruan, B. Zhang, Z. J. Li, "A behavioral model for MCT surge current analysis in pulse discharge," *Solid-State Electronics*, vol. 99, pp. 31-37, Sep. 2014.DOI:10.1016/j.sse.2014.04.044

Proceedings of the 31st International Symposium on Power Semiconductor Devices & ICs
May 19-23, 2019, Shanghai, China

Silicon RC-Snubber for 900 V Applications Utilising non-Stoichiometric Silicon Nitride

N. Boettcher, T. Heckel and T. Erlbacher
norman.boettcher@iisb.fraunhofer.de
Fraunhofer Institute for Integrated Systems and Device Technology IISB
Schottkystraße 10, 91058 Erlangen, Germany

K. Pelaic
Friedrich-Alexander University
Chair of Electron Devices
Cauerstraße 6, 91058 Erlangen, Germany

Abstract—This paper presents a novel approach for the realisation of silicon RC-snubbers suitable for 900 V applications. The fabrication of dielectric layers with a thickness feasible for this voltage class poses challenges in terms of mechanical stress management, which have not been overcome by other approaches so far. The presented technology focuses on stress reduction during fabrication by utilisation of non-stoichiometric silicon nitride. A wide case study is performed in order to characterise dielectric layer stacks including non-stoichiometric silicon nitride with respect to manufacturability and electric properties. By keeping the amount of non-stoichiometric silicon nitride in dielectric layer stack small, silicon RC-snubbers exhibiting dielectric strength of more than 1500 V and leakage current of less than 100 nA at 900 V can be fabricated. The feasibility for 900 V switching application is demonstrated in a double-pulse experiment.

I. INTRODUCTION

Reducing switching time is a key measure of minimising switching losses in power modules. As a consequence, steep current slopes generate high voltage peaks during turn-off due to the parasitic inductance of the DC-link. Apart from possible catastrophic device breakdowns, these voltage peaks excite parasitic LC circuits. The emerging oscillations lead to electromagnetic emission, which potentially can critically interfere with adjacent components and systems. If properly designed and placed close to the switching cell, RC-snubbers pose a potent and elegant solution to overcome these challenges [1]. Silicon RC-snubbers (SiRC) in particular can be integrated directly into the power module using conventional packaging technologies and the existing heat management [2]. Furthermore, in recent investigations, the improvement of 200 V and 400 V power modules by integration of SiRCs is demonstrated in terms of efficiency and oscillation behaviour [3], [4]. For 900 V applications, however, no successfully manufactured concept of SiRCs (or silicon capacitors) exhibiting sufficient dielectric strength and reasonable capacitance value has been published so far.

In this work, a novel generation of our 600 V trench SiRC technology [5], which is suitable for 900 V applications, utilising a low stress interlayer in the dielectric layer stack is introduced. In chapter II the SiRC technology along with its challenges is introduced. Chapter III presents the design of experiment and characterisations in terms of manufacturability and electrical properties and feasibility.

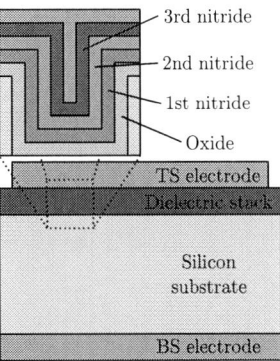

Fig. 1: Schematic cross section (not in scale) of the SiRC. The zoom highlights the dielectric layer stack consisting of a silicon dioxide layer and three consecutive silicon nitride layers, as well as the surface magnification.

II. SILICON RC-SNUBBER TECHNOLOGY

In order to upscale the dielectric strength of silicon capacitors, generally the dielectric layer thickness is increased. However, this results in a lower capacitance per active area value. Therefore, it is common practice to increase the area of the substrate surface with the aid of trench structures [6]. However, increasing both dielectric layer thickness and surface magnification leads to more mechanical stress [7], which in turn bends the semiconductor substrate. In case of SiRC fabrication, film stress eventually results in semiconductor substrate cracking or delamination of the dielectric layer during fabrication. Additionally, the curvature itself prevents process steps requiring vacuum fixation, as for instance polyimide spinning, from execution. Therefore, stress compensation methods as in [8] and [9] have been introduced. In contrast, the presented approach focuses on stress reduction during fabrication instead of stress compensation.

Fig. 1 shows a schematic cross section of the novel 900 V SiRC technology. The structure consists of a silicon substrate, a top side (TS) electrode, a bottom side (BS) electrode and the dielectric layer stack between top side electrode and substrate. The refinement of the novel SiRC-technology lies in the

978-1-7281-0582-6/19 $31.00 © 2019 IEEE

TABLE I: Realised device parameter distribution and corresponding electric properties.

Fabricated samples		Calculated electric properties			Measured electric properties		
Variant	Dielectric layer stack	$C_{Sn,calc}$ in nF	$R_{Sn,calc}$ in Ω	$V_{bd,calc}$ in V	$C_{Sn,Inv}$ in nF	$C_{Sn,Acc}$ in nF	R_{Sn} in Ω
A10	SiO_2–Si_3N_4–Si_3N_4–Si_3N_4	1.11	4.3	2069	Semiconductor substrate cracked.		
B10	SiO_2–Si_3N_4–Si_3N_4–Si_xN_y	1.01	4.3	1723	0.95	1.08	4.0
B30	SiO_2–Si_3N_4–Si_3N_4–Si_xN_y	Fabricated only for process evaluation.					
C10	SiO_2–Si_3N_4–Si_xN_y–Si_3N_4	1.03	4.3	1734	0.90	1.02	4.0
C20	SiO_2–Si_3N_4–Si_xN_y–Si_3N_4	2.09	4.3	1716	1.77	2.01	4.0
C30	SiO_2–Si_3N_4–Si_xN_y–Si_3N_4	Fabricated only for process evaluation.					
D10	SiO_2–Si_3N_4–Si_xN_y–Si_xN_y	1.03	4.3	1393	0.88	1.03	4.0
E10	SiO_2–Si_xN_y–Si_3N_4–Si_3N_4	1.05	4.3	1749	Dielectric layer delaminated.		
E20	SiO_2–Si_xN_y–Si_3N_4–Si_3N_4	2.04	4.3	1729	Dielectric layer delaminated.		
F10	SiO_2–Si_xN_y–Si_3N_4–Si_xN_y	1.04	4.3	1418	0.89	1.02	4.0
G10	SiO_2–Si_xN_y–Si_xN_y–Si_3N_4	1.07	4.3	1402	0.94	1.05	4.0
G20	SiO_2–Si_xN_y–Si_xN_y–Si_3N_4	2.07	4.3	1345	1.67	1.87	4.0
G30	SiO_2–Si_xN_y–Si_xN_y–Si_3N_4	Fabricated only for process evaluation.					
H10	SiO_2–Si_xN_y–Si_xN_y–Si_xN_y	1.02	4.3	1063	0.92	1.04	4.0

composition of the dielectric layer stack. For the reduction of mechanical stress in silicon nitride layers, the utilisation of LPCVD non-stoichiometric silicon nitride (Si_xN_y) is known [10]. As shown in Fig. 1, the dielectric layer stack is composed of a SiO_2 interface layer and three consecutive silicon nitride layers.

III. EXPERIMENTAL METHODS

A. Design of Experiment

The presented design of experiment focuses on the evaluation of the manufacturability and the electric properties of SiRCs utilising a single variation of silicon-rich non-stoichiometric silicon nitride in the dielectric layer stack. The investigated process splits and corresponding electric properties can be obtained from Tab. I. Merely, the composition of the dielectric layer stack and the trench depth are varied. Hence, every realistic combination of Si_3N_4 and Si_xN_y in the three-layer nitride stack is realised (A to H). Additionally, selected combinations are fabricated in a more aggressive design, exhibiting deeper trench structures (10 μm, 20 μm and 30 μm). Each variant consists of two identically processed wafers, whose characteristics can be assumed equivalent.

B. Characterisation of Manufacturability

In Fig. 2, the relative height from wafer edge to wafer center of variant C10 after significant process steps is depicted. For both, thermal oxidation and LPCVD, the whole substrate is covered approximately homogeneously by the respective layer. However, since the top side surface is magnified, stress inflicted on the top side is dominating the wafer bow.

Therefore, the wafer curvature is convex after oxidation due to intrinsic compression stress of SiO_2. Moreover, in contrast to [5], the wafer generally bends convexly during all dielectric layer depositions, which implies additional net compression caused by the silicon nitride layers. However, after etching the dielectric layer stack on the wafer back side, the curvature is concave indicating net tensile stress in the dielectric layer stack. As emphasised by the dotted fit curve, the absolute wafer bow maximum is reduced to below 150 μm at the most critical process stage.

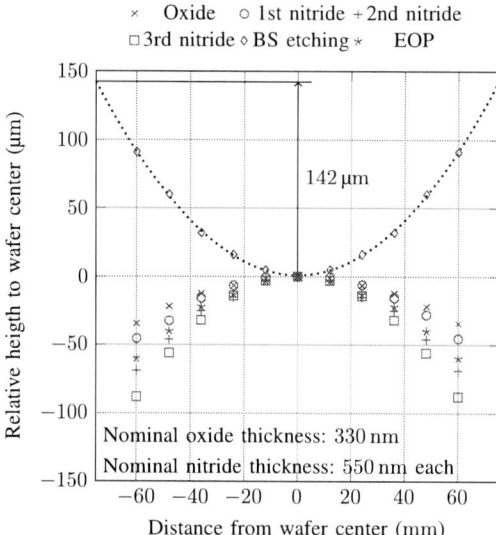

Fig. 2: Measured wafer bow of variant C10 after oxidation, each silicon nitride deposition, back side (BS) etching and at the end of process (EOP).

In Fig. 3, the curvature measurement results of selected splits are shown. Please note, that the wafers of C30 and G30 served only as process evaluation for dielectric deposition and have not been fabricated any further. Comparing C10, C20 and C30 or G10, G20 and G30, respectively, a correlation of surface magnification and wafer bow is clearly distinct. Additionally, the characteristic of Si_3N_4 continuously increasing wafer bow can be observed, whereas the direct influence of Si_xN_y, is apparently dependent on the order of the silicon nitride depositions. Both substrates of A10 cracked after the 3rd Si_3N_4 deposition due to mechanical stress, although the measured curvature is comparably small. Furthermore, the silicon nitride layer stacks of E10 and E20 consistently delaminated during the 3rd silicon nitride deposition.

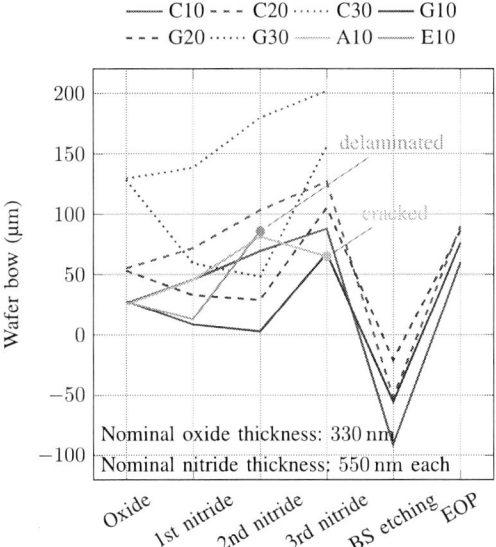

Fig. 3: Measured wafer bow of selected splits after oxidation, each silicon nitride deposition, back side (BS) etching and at the end of process (EOP).

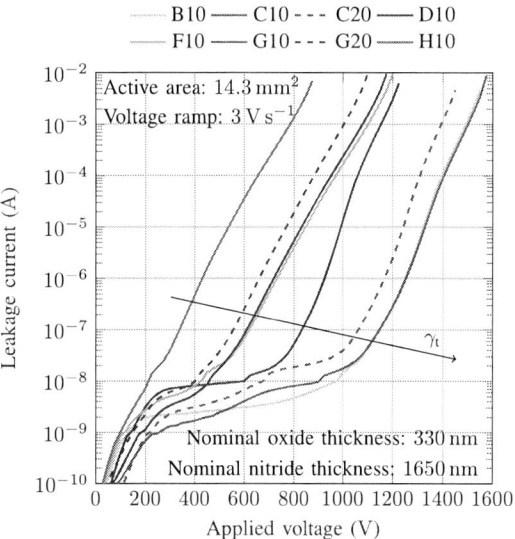

Fig. 4: Measured current-voltage characteristics in accumulation of the novel SiRCs.

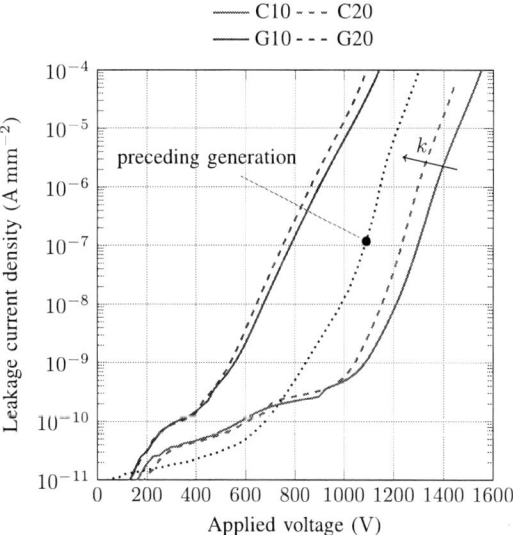

Fig. 5: Measured current density-voltage characteristics in accumulation of the novel and the preceding technology [5].

C. Characterisation of Electric Properties

Since the SiRC exhibits an insulator-semiconductor interface, its capacitance value is voltage-dependent. Therefore, impedance measurements at $100\,\mathrm{kHz}$ in a DC bias range from $-40\,\mathrm{V}$ to $40\,\mathrm{V}$ are carried out. The capacitance $C_{\mathrm{Sn,Inv}}$ and $C_{\mathrm{Sn,Acc}}$ are obtained in deep inversion and accumulation, respectively. The results are listed in Tab. I and are in good accordance with the prediction gained from the simulation models.

In Fig. 4, the leakage current measurement results in accumulation are shown. Leakage current determines the device lifetime and, hence, is a key figure of SiRC technologies. All devices under test (DUT) exhibit the characteristic exponential current increase of silicon capacitors after a certain knee voltage is reached. Therefore, in respect to a desired voltage class, leakage current is the dominating design target. Since $\mathrm{Si}_x\mathrm{N}_y$ is non-stoichiometric, it exhibits a higher defect density than $\mathrm{Si}_3\mathrm{N}_4$ and, consequently, higher leakage current. Hence, the leakage current is decreasing with increasing thickness ratio of $\mathrm{Si}_3\mathrm{N}_4$ and $\mathrm{Si}_x\mathrm{N}_y$ in the dielectric layer stack ($\gamma_\mathrm{t} = t_{\mathrm{Si}_3\mathrm{N}_4} : t_{\mathrm{Si}_x\mathrm{N}_y}$). Please note, due to a current limit of $10\,\mathrm{mA}$, the calculated breakdown voltage values (see Tab. I) are not reached. Thus, no dielectric breakdown is observed.

Expectedly, the absolute leakage current increases with the surface magnification. Moreover, this also applies for the leakage current density above knee voltage, as can be seen in Fig 5. Additionally, the leakage current density of our preceding SiRC technology is shown [5]. Here, the magnified surface area

is used for evaluation. As shown, in terms of leakage current density at voltages below $700\,\mathrm{V}$, the refined SiRC technology is inferior to the preceding generation, due to the presence of $\mathrm{Si}_x\mathrm{N}_y$ in the dielectric stack. However, the leakage current density at $900\,\mathrm{V}$ is reduced by approximately one order of magnitude.

978-1-7281-0582-6/19 $31.00 © 2019 IEEE

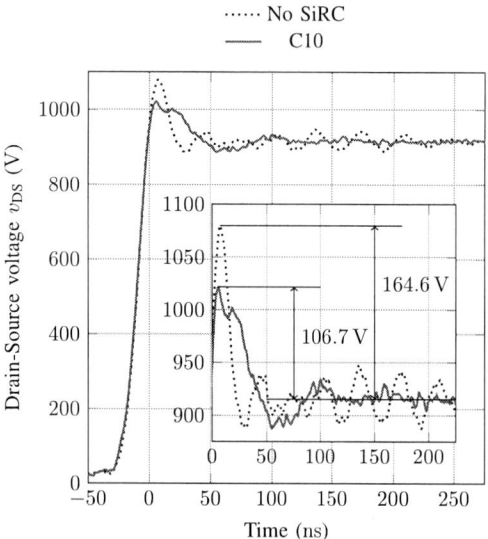

Fig. 6: Measured turn-off behaviour of the double-pulse setup with and without applied SiRC.

In order to validate the feasibility for switching applications with 900 V DC-link voltage, a double-pulse setup is employed. The SiRC is applied in parallel to the DC-link capacitor, directly beside the commutation cell. Thus, the inductance in the attenuation path is dominated by the parasitic SiRC inductance, which is primarily defined by the bond loops. In the experiments, it is determined to $4\,\text{nF}$. Two turn-off experiments are performed at $V_{\text{DC}} = 900\,\text{V}$ and $i_{\text{L}} = 45\,\text{A}$: one without applied SiRC and one with an SiRC of variant C10. As can be obtained from Fig. 6, a voltage slope of $\frac{\partial V_{\text{DS}}}{\partial t} = 44\,\text{V\,ns}^{-1}$ is achieved, which is unaffected by the SiRC. As emphasised in the zoom, the voltage peak is reduced to less than 65 %. Moreover, the oscillation time is decreased drastically.

IV. CONCLUSION

This work presents a wide case study about silicon capacitors utilising a low stress silicon nitride interlayer in the dielectric layer stack. As a result, a concept for a silicon RC-snubber feasible for high power switching applications up to 900 V DC-link voltage emerged. The improvement of manufacturability in terms of mechanical stress is excelling known approaches. Therefore, silicon RC-snubbers with higher dielectric layer thickness and, hence, dielectric strength of more than $1500\,\text{V}$ and leakage current of less than $100\,\text{nA}$ at $900\,\text{V}$ can be fabricated. Since the SiRC is easily integrable into power modules, its parasitic inductance value is neglectable and, therefore, is perfectly suitable for voltage peak reduction and electromagnetic filtering. Subsequently, power modules can be operated more efficiently and output filters can be reduced, if not avoided.

ACKNOWLEDGMENT

The authors would like to thank Ms. Gudrun Rattmann, Mr. Florian Krach, Mr. Achim Endruschat and the π-Fab personnel for fruitful discussions and supporting the fabrication and characterisation of the novel SiRCs. This project is sponsored by the "Federal Ministry of Education and Research" (BMBF) under grant 16EMO0211K "KOOPERATION".

REFERENCES

[1] N. Boettcher et al. "A novel approach for optimal design of monolithic integrated RC snubbers". In: International Symposium and Exhibtion on Electromagnetic Compatability EMC Europe, Anger, France, 2017.

[2] S. Matlok et al. "Switching SiC devices faster and more efficient using a DBC mounted terminal decoupling SiRC element". In: 11th European Conference on Silicon Carbide and Related Materials ECSCRM, Halkidiki, Greece, 2016.

[3] T. Erlbacher et al. "Improving module performance and reliability in power electronic applications by monolithic integration of RC-snubbers". In: 24th International Symposium on Power Semiconductor Devices and ICs ISPSD, Bruges, Belgium, 2012.

[4] F. Krach et al. "Innovative monolithic RC-snubber for fast switching power modules". In: 9th International Conference on Integrated Power Electronics Systems CIPS, Nuremberg, Germany, 2016.

[5] F. Krach et al. "Silicon integrated RC snubber for application up to 900V with reduced mechanical stress and high manufacturability". In: 74th Annual Device Research Conference DRC, Newark, Delaware, USA, 2016.

[6] J. vom Dorp. "Monolithisches RC-Element für leistungselektronische Anwendungen". Ph.D. thesis. University Erlangen-Nuremberg, 2011.

[7] A. Wright et al. "Simulating wafer bow for integrated capacitors using a multiscale approach". In: The 9th Eurosim Congress on Modelling and Simulation Eurosim, Oulu, Finland, 2016.

[8] S. Banzhaf et al. "Stress reduction in high voltage MIS capacitor fabrication". In: 19th International Symposium POWER ELECTRONICS Ee, Novi Sad, Serbia, 2017.

[9] F. Krach et al. "Silicon nitride, a high potential dielectric for 600 V integrated RC-snubber applications". In: *Journal of Vacuum Science and Technology B* 33 (2015), pp. 01A1112.

[10] J. Gardeniers, H. Tilmans, and C. Visser. "LPCVD silicon-rich silicon nitride films for applications in micromechanics, studied with statistical experimental design". In: *Journal of Vacuum Science and Technology A* 14(5) (1996), pp. 2879–2892.

Proceedings of the 31st International Symposium on Power Semiconductor Devices & ICs
May 19-23, 2019, Shanghai, China

Condition Monitoring of High Voltage IGBT Devices Based on Controllable RF Oscillations

Miaosong Gu*, Xiang Cui*, Xinling Tang†, Cheng Peng*, Xuebao Li*, Rui Jin†,Zhibin Zhao*
Email: gumiaosong@ncepu.edu.cn, x.cui@ncepu.edu.cn
*State Key Laboratory of Alternate Electrical Power System with Renewable Energy Sources
North China Electric Power University, Changping District, Beijing, China
†Global Energy Interconnection Research Institute, Changping District, Beijing, China

Abstract—Plasma-Extraction Transit-Time Oscillation has a very significant dependence on voltage, current and temperature. This paper presents a condition monitoring method for high-voltage IGBT based on some characteristic parameters of the PETT oscillation. In order to overcome the negative impact that PETT may have on electromagnetic compatibility, a high-voltage IGBT with flexible and controllable PETT oscillation characteristics is developed.

Keywords—Condition monitoring, PETT, Press Pack IGBT, TSEP

I. INTRODUCTION

Condition monitoring of power semiconductor devices is of great significance for improving the reliability of power electronic converters[1-3]. Although several condition monitoring methods have been proposed so far, each technique has merits and disadvantages in terms of accuracy and complexity. The methods that make the industry widely accepted and applied have not yet been obtained[1], especially for ultra-high voltage power electronics equipment in power systems, such as HVDC converter valves and HVDC circuit breakers.

Plasma-Extraction Transit-Time (PETT) Oscillation is a radio frequency (RF) phenomenon that occurs in bipolar devices at turn-off during the tail current interval[4-6]. Some scholars have conducted qualitative and quantitative research on the characteristics, effects and inhibition of PETT oscillations[6-8].

The negative impact of PETT is mainly the degradation of EMC performance and the error signals it may bring in power electronic converters[6]. It is necessary to limit the PETT，but this does not mean that the PETT must be eliminated; it is enough to suppress it within limits allowed by EMC. There are many methods to improve the performance of EMC. Removing the source of electromagnetic radiation is not the only option, such as adding shielding[6].

It is interesting to note that the PETT oscillation has a very significant dependence on voltage, current and temperature. Why not use some parameters of the PETT oscillation as a temperature sensitive electrical parameter TSEP or other means of condition monitoring? The core work of this paper is to develop a high-voltage IGBT with flexible and controllable PETT oscillation characteristics and achieve condition monitoring using PETT.

Supported by National Natural Science Foundation of China - State Grid Corporation Joint Fund for Smart Grid under Grant U1766219.

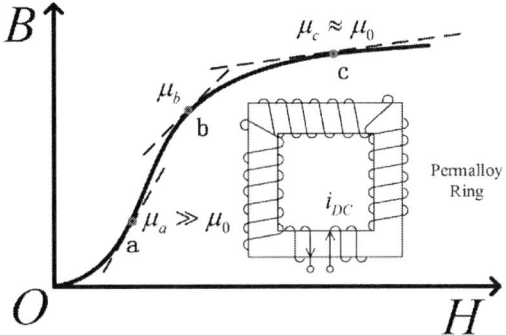

Fig. 1: Adjusting the magnetic permeability of the permalloy ring by DC bias current.

Fig. 2: A prototype of 3.3kV 200A Press-Pack IGBT device that can flexibly regulate PETT RF oscillation.

II. FLEXIBLE CONTROL OF PETT

Considering the adverse effects of this kind of RF oscillation, the premise of condition monitoring by PETT is to achieve flexible control of this oscillation, which needs to limit its intensity and controls the time it occurs. Because the acquisition of the PETT oscillating signal is not required every moment; it is sufficient to let the PETT oscillation appear during the signal sampling period.

The most available method to control the PETT is to regulate the parasitic inductance of the oscillator circuit. Based on magnetic saturation, this paper designs a permalloy ring with DC bias current to flexibly regulate the magnetic

978-1-7281-0582-6/19 $31.00 © 2019 IEEE   355

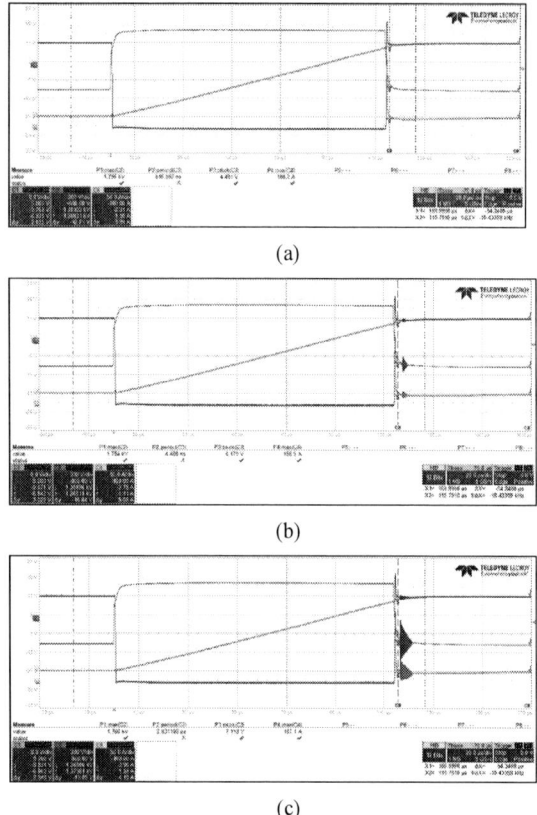

(a)

(b)

(c)

Fig. 3: The 3.3kV 200A Press-Pack IGBT single pulse test results under different magnetic configuration. Vdc=1400V Icmax=200A

(a) complete oscillation suppression; (b) Incomplete oscillation suppression; (c) oscillation without inhibit. (C1:Vge,C2:Vce,C4:Ic).

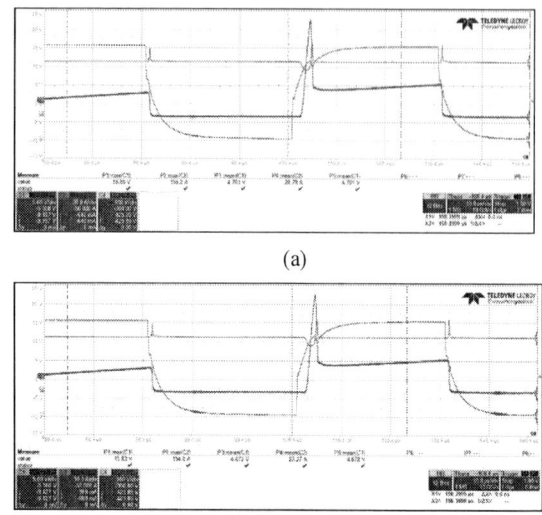

(a)

(b)

Fig. 4: An improved design limiting the PETT extremely weak on the magnetic saturation state
(a) the excitation current is zero; (b) the excitation current reaches saturation current (C1:Vge,C2:Ic,C4:Vdc-link).

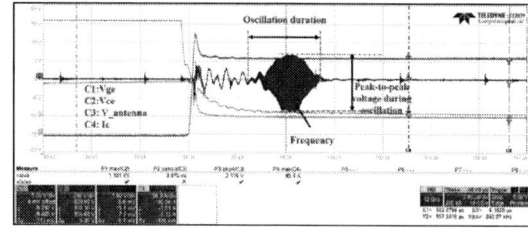

Fig. 5: Weak PETT signal measurement using a near field antenna and definition of some parameters.

permeability. As shown in Figure 1, when the DC bias current is set at point **A**, the magnetic permeability is the largest, which brings the maximum parasitic inductance to the oscillating circuit, thereby achieving complete suppression of the PETT. If the DC bias point is set at point **C**, then the magnetic permeability is the smallest, and the PETT oscillation may be released.

In order to avoid the excessive increase of the parasitic inductance of the current conducting path, the magnetic ring is thinned and a permalloy plate was placed under the PCB to suppress the PETT oscillation effectively.

After that, combined with the 3.3kV 50A NPT-IGBT sub-modules designed by the Global Energy Interconnection Research Institute, a prototype of the Press-Pack high-voltage IGBT device that can flexibly regulate the PETT was developed, as shown in Fig. 2.

With these permalloy loops and plate, the parasitic inductance parameters in the package can be flexibly controlled, achieving complete or incomplete suppression of the PETT, as shown in Fig. 3.

Fig. 3 clearly shows the effective control of this magnetic design method for the PETT oscillation. But in fact, the PETT in magnetic saturation mode does not need to be as dramatic as (c), which leads to the risk of IGBT false triggering. Therefore, the magnetic configuration is optimized subsequently, and Figure 4 shows a final improved design.

When the excitation current is not applied, the PETT is wholly suppressed. After the magnetic ring enters the saturation state, it only makes the PETT extremely weak, so as to meet the requirements of EMC and reduce the cost of electromagnetic shielding.

Limiting the PETT to such a weak level that it is difficult to find oscillations on the Vge, does this not increase the difficulty of measurement and data analysis? This is not necessary to worry, because for RF oscillations, more convenient and accurate measurement can be achieved by the antenna compared to the voltage probe. As shown in Fig. 5, although the PETT in the Vge has been difficult to distinguish by the naked eye. But on the near-field antenna, the signal is still apparent.

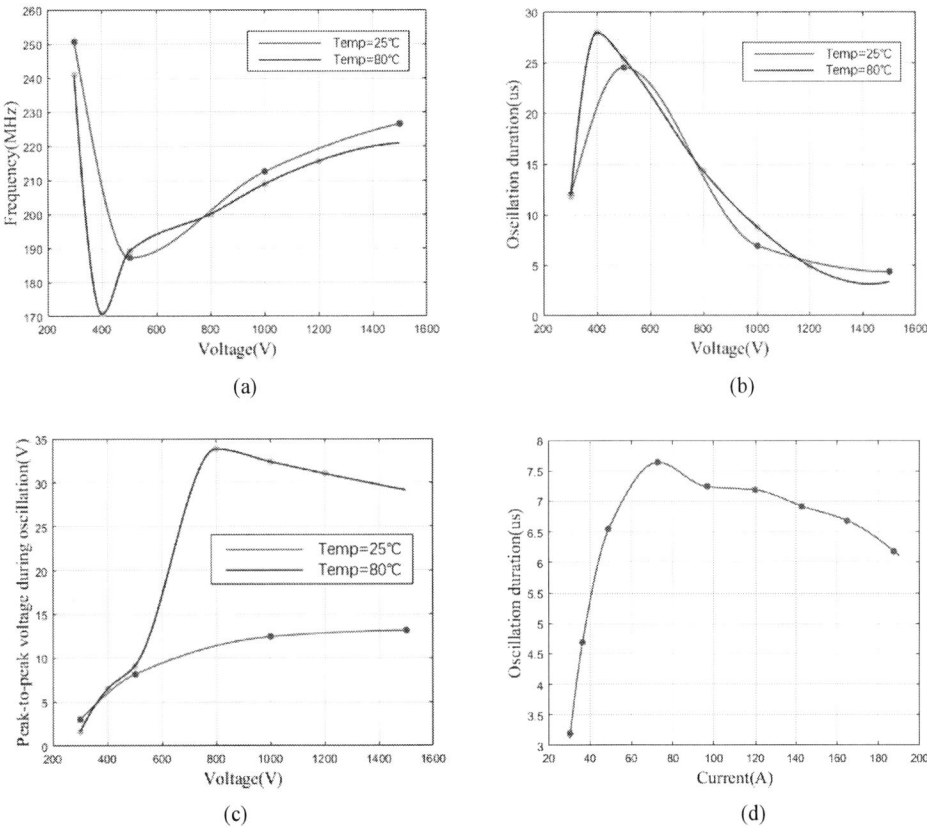

Fig. 6: Dependence between some characteristic parameters of PETT oscillations and voltage, current and temperature, (a) Voltage Vce versus PETT oscillation frequency. (Ic=200A) (b) Voltage Vce versus PETT oscillation duration. (Ic=200A) (c) Voltage Vce versus PETT oscillation peak-to peak voltage. (Ic=200A) (d) Dependence between current Ic and PETT oscillation duration. (Vce=500V).

III. CONDITION MONITORING USING PETT

After the PETT can be effectively controlled, the potential of this phenomenon in the condition monitoring of semiconductor devices can be discussed.

The core work of implementing IGBT condition monitoring based on the PETT oscillation characteristic is to obtain the relationship between the characteristic index of the PETT oscillation waveform and the physical quantity to be monitored such as junction temperature, voltage or current. Once these relationships are determined, then Condition monitoring of IGBTs can be accomplished using a primary data set and signal identification.

In the fig.5, some fundamental characteristic quantities related to the PETT oscillation are defined. In previous studies, only the conditions under which the PETT oscillation occurred and the frequency of oscillations were concerned, but in reality, the PETT oscillation waveform implies more abundant information. This article only selects several intuitive time domain metrics. It must be pointed out that in the data analysis, the oscillation frequency of the PETT also slightly changed with time in one tailing process.

Therefore, the PETT oscillation frequency is taken as the average value during an oscillation.

After a large number of switching tests on this IGBT device under different temperature, voltage, and current conditions, the dependence between the index of the PETT oscillation waveform and the experiment condition is statistically analyzed, as shown in Fig. 6.

The dependence between the PETT oscillation and the voltage, current, and temperatures is a bit complicated. It is challenging to explain these rules with a simple analytical model. But there are some experimental rules that are worth noting:

Fig. 6(a) and Fig. 6(b) show a significant correlation between the frequency of the oscillation and the oscillation duration and voltage. In the voltage range of 800V-1500V, there is a good linear relationship between the oscillation frequency and the voltage. In practical device applications, the amplitude of voltage fluctuations is limited, so it is very probable to estimate the device voltage at this time using the PETT oscillation frequency and duration.

Fig. 6(c) shows the relationship between the peak-to-peak of the PETT oscillation Vge(p-p) and the voltage Vce under different temperature conditions. The temperature has a significant influence on the oscillation amplitude, and the monitoring of the junction temperature of the IGBT could be realized based on this. It is worth noting that when T=80°C Vdc=1500V, no oscillating signal is collected. PETT oscillations will disappear under high-temperature conditions, which can be applied in the high-temperature protection of devices.

Fig. 6(d) shows the PETT oscillation duration for different turn-off current under the same voltage conditions. The dependence between the two is evident, but there is also a non-monotonicity that is unwilling to see. It is possible to estimate the turn-off current of the device by using the duration of the PETT oscillation, but considering the broad range of current variation in practical applications, in order to overcome the effects of non-monotonicity on monitoring, more complex and advanced pattern identification techniques need to be employed, which makes this current monitoring technology limited .

In general, the PETT oscillation waveform contains a lot of device status information, which is worthy of attention. This kind of signal can be the non-contact measurement by an antenna, which is very advantageous for the condition monitoring of power electronic equipment above 100kV in the power system, such as HVDC converter valve and HVDC breaker.

In the previous section, the release of the PETT by the excitation current has been achieved, which makes it possible to detect multiple IGBT devices using only one antenna depending on the order of excitation, significantly reducing the cost of monitoring. However, there are many factors that affect the PETT oscillation, such as the drive circuit, the age of the device, and so on. In addition, the PETT oscillation does not exist in all voltage ranges. The IGBT designed in this paper only oscillates at 1500V. These all restrict the development of this technology to the industry.

IV. CONCLUSIONS

This paper presents a condition monitoring method for high-voltage IGBT based on some characteristic parameters of the PETT oscillation. In order to overcome the negative impact that PETT may have on electromagnetic compatibility, a high-voltage IGBT with flexible and controllable PETT oscillation characteristics is developed. After that, offline experimental research was carried out, and the relationship between the oscillation characteristic index and the device state was analyzed, containing voltage, turn-off current, and junction temperature. The advantages and limitations of this method were discussed.

The advantages of this method:

(A)The PETT can be suppressed to a weak level by regulating the magnetic configuration to meet EMC requirements. (B) PETT oscillation wave contains a lot of device status information, such as junction temperature. (C)

Non-contact measurements can be achieved using an antenna. (D) This method can monitor multiple IGBT devices using only one antenna depending on the order of excitation. (E) The method based on magnetic permeability regulation is convenient to implement and low in cost.

The limitations of this method:

(A) There is non-monotonic dependence and the complexity of signal identification;(B) There are many factors affecting the waveform of the PETT oscillation, and they affect each other;(C) Lack of a precise and practical PETT simulation model;(D) Magnetic modeling and regulation under radio frequency conditions are difficult.

ACKNOWLEDGMENT

This work was supported by National Natural Science Foundation of China - State Grid Corporation Joint Fund for Smart Grid under Grant U1766219 and the Fundamental Research Funds for the Central Universities.

The authors would like to thank Pengyu Fu for his help during the measurement and also to Jinyuan Li for device customization.

REFERENCES

[1] R. Moeini, P. Tricoli, H. Hemida and C. Baniotopoulos, "Increasing the reliability of wind turbines using condition monitoring of semiconductor devices: a review," in IET Renewable Power Generation, vol. 12, no. 2, pp. 182-189, 5 2 2018.

[2] Z. Xu, F. Xu and F. Wang, "Junction Temperature Measurement of IGBTs Using Short-Circuit Current as a Temperature-Sensitive Electrical Parameter for Converter Prototype Evaluation," in IEEE Transactions on Industrial Electronics, vol. 62, no. 6, pp. 3419-3429, June 2015.

[3] U. Choi, F. Blaabjerg, S. Jørgensen, S. Munk-Nielsen and B. Rannestad, "Reliability Improvement of Power Converters by Means of Condition Monitoring of IGBT Modules," in IEEE Transactions on Power Electronics, vol. 32, no. 10, pp. 7990-7997, Oct. 2017.

[4] Y. Takahashi, K. Yoshikawa, M. Soutome, T. Fujii, H. Kirihata and Y. Seki, "2.5 kV-1000 A power pack IGBT (high power flat-packaged NPT type RC-IGBT)," in IEEE Transactions on Electron Devices, vol. 46, no. 1, pp. 245-250, Jan. 1999.

[5] Gutsmann, B., P. Mourick and D. Silber, Plasma extraction transit time oscillations in bipolar power devices. Solid State Electronics, 2002. 46(1): p. 133-138.

[6] R. Siemieniec, P. Mourick, M. Netzel and J. Lutz, "The plasma extraction transit-time oscillation in bipolar power Devices-Mechanism,EMC effects, and prevention," in IEEE Transactions on Electron Devices, vol. 53, no. 2, pp. 369-379, Feb. 2006.

[7] Xinling T , Xiang C , Zhibin Z , et al. Mechanism and Characteristics of Plasma Extraction Transit Time Oscillation of Paralleled IGBT Chips[J]. Transactions of China Electrotechnical Society, 2018.

[8] Saito K , Wada T , Sasajima Y . General Expression for Plasma Extraction Transit-Time Oscillations From Silicon-Bipolar Power Semiconductor Devices[J]. IEEE Transactions on Electron Devices, 2018, PP(99):1-7.

Proceedings of the 31st International Symposium on Power Semiconductor Devices & ICs
May 19-23, 2019, Shanghai, China

An Injection Enhanced LIGBT on Thin SOI Layer with Low ON-state Voltage

Gaoqiang Deng, Xiaorong Luo*, Diao Fan, Tao Sun, Bo Zhang

University of Electronic Science and Technology of China

Chengdu, China

Email: xrluo@uestc.edu.cn   gaoqiang_deng@163.com

Abstract—A Lateral Injection Enhanced Insulated Gate Bipolar Transistor (LIEGT) on thin SOI layer is proposed and investigated by simulations. The LIEGT features a recessed trench at cathode side of the drift region formed by LOCOS process. The LIEGT shows very low loss in the ON-state because the trench suppresses holes being extracted and enhances the electron injection. The saturation current of the LIEGT is 1.4 times that of the conventional LIGBT and the ON-state voltage (V_{ON}) is reduced by 24% at current density of 200A/cm². The LIEGT exhibits improved trade-off between V_{ON} and turn-off loss (E_{OFF}). For the same E_{OFF}, the V_{ON} is decreased by 20% compared with that of the conventional LIGBT. The fabrication of the proposed LIEGT is compatible with the CMOS process.

Index Terms—LIGBT, Injection Enhancement, SOI, forward voltage, turn-off loss, LOCOS.

I. INTRODUCTION

Thin layer Silicon-on-Insulator (SOI) offers low parasitic effect and good process compatibility [1][2][6] and thus is attractive to Power Integrated Circuits. Compared with power MOSFETs, high current capability and low conduction loss make LIGBT an ideal choice for power conversion circuits [3] [4] [5]. However, LIGBTs on thin layer SOI are subject to the degraded ON-state characteristic owing to the greatly narrowed current conduction path.

The Injection Enhancement technology is a typical method to improve the trade-off relationship between ON-state voltage (V_{ON}) and turn-off loss (E_{OFF}) of lateral and vertical IGBTs [8]-[14]. The narrow mesa structures [9] [13] [14] have been experimentally demonstrated ultra-low V_{ON} yet slightly higher E_{OFF} than that of the conventional devices due to the significant increase in carrier density at the Cathode side of the drift region in the ON-state. Nevertheless, the formation of narrow mesa requires high photolithography precision and the metal-semiconductor contact area is limited. IGBTs with partially narrow mesa [9] alleviate these problems but suffer from complicated processes. In [11], a highly-doped N carrier storage layer is introduced within the mesa to further enhance the plasma concentration at the Cathode side.

In this paper, we propose a Lateral Injection Enhanced Insulated Gate Bipolar Transistor (LIEGT) by using LOCOS process on thin SOI. The device mechanism and fabrication processes are simulated and investigated using

This work was supported in part by the 13th Five-year Plan for Microelectronics Advanced Research Program under project 31513030201-2, and by the National Defense Basic Scientific Research JCKY2016210B008.

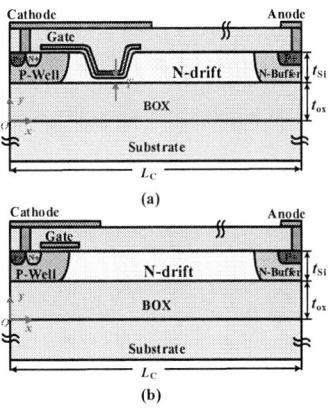

Fig. 1 Schematic cross section of (a) the proposed LIEGT and (b) conventional LIGBT.

Tsuprem4&Medici. The proposed LIEGT exhibits superior ON-state performance and reduces the total energy loss considerably. Also, a compact analytical model for the drift region voltage-drop of the LIEGT is presented.

II. STRUCTURE AND MECHANISM

The schematic cross section of the proposed LIEGT is shown in Fig.1 (a). It is characterized by a trench formed by Local oxidation of silicon (LOCOS) in the drift region near the gate. The poly gate covers the bottom and sidewalls of the trench. The extended source field plate is employed to weaken the peak electric field induced by the extended gate in the blocking state and maintain high BV above 600V. T represents the thinnest silicon thickness under the trench. By adjusting the duration time and temperature of oxidation, the T value can be accurately controlled. Fig.1 (b) shows the sectional view of the conventional LIGBT. The dimension and process parameters used in simulations for the two structures are as follows unless otherwise specified: t_{si}=1µm, t_{ox}=4µm, gate oxide thickness t_{gox}=100nm, L_C=63µm (LIEGT) and 55.5µm (Conv.), P-well dose D_{pw}=2×10¹³cm⁻², N-buffer dose D_{bf}=8×10¹²cm⁻², P-anode dose D_{pa}=8×10¹³cm⁻², carrier life time τ=1µs. Models used in simulations are as follows: CONMOB, FLDMOB, SRFMOB, SRH, AUGER, BGN and IMPACT.I.

Fig.2 (a) shows the ON-state characteristics of the proposed LIEGTs and the conventional LIGBT. The saturation current of the LIEGT is 1.4 times that of the conventional LIGBT. The inset shows that the V_{ON} of the LIEGT with T=90nm decreases

978-1-7281-0582-6/19 $31.00 © 2019 IEEE

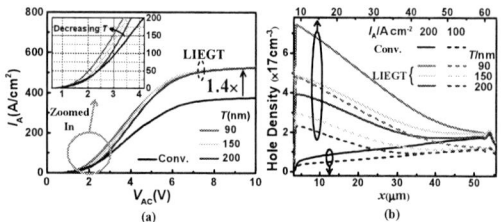

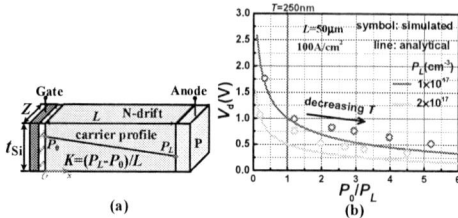

Fig.2 (a) ON-state characteristics of the proposed LIEGTs and the conventional LIGBT (b) Hole density distribution along y=4.5μm in the drift region for LIEGTs and the Conv. LIGBT at I_A=100&200A/cm² (V_{GC}=15V).

Fig.4 (a) Equivalent structure of the LIEGT for modeling (b) Analytical and simulated results for drift region voltage-drop (V_d) under different plasma concentration distributions.

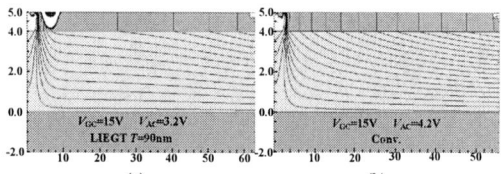

Fig.3 Comparison of potential line distribution in the ON-state at I_A=200A/cm² for (a) LIEGT (T=90nm) and (b) Conv. LIGBT (0.2V/contour)

where k and t are the Boltzmann constant and temperature, respectively. Assuming that the carrier profile in the drift region exhibits linear distribution [8] [15] [16] and there is $p(x)=Kx+P_0$, where $K=(P_L-P_0)/L$. Consequently, the voltage-drop across the drift region can be achieved

$$V_d = \begin{cases} \int_0^L Edx = \dfrac{I_A/Zt_{Si} + k_0t(\mu_n - \mu_p)K}{q(\mu_n + \mu_p)K}\ln\dfrac{P_L}{P_0} & (K \neq 0) \\ \int_0^L \dfrac{I_A}{qp_0(\mu_n + \mu_p)Zt_{Si}}dx = \dfrac{LI_A}{qp_0(\mu_n + \mu_p)Zt_{Si}} & (K = 0) \end{cases} \quad (4)$$

to 3.2V from 4.2V of the conventional structure at current density of 200A/cm². As T decreases, the V_{ON} of the LIEGT reduces because the electron injection is further enhanced. The trench hinders holes from being extracted by cathode and the hole density at the cathode-side drift region is raised up significantly for the LIEGTs, as illustrated in Fig.2 (b). The electron injection is enhanced correspondingly to maintain the electrical neutrality. The device with smaller T exhibits higher cathode-side carrier density. For T =90nm, the cathode-side carrier density of the LIEGT is one order of magnitude higher than that of the Conv. LIGBT. Consequently, the voltage-drop across the drift region is greatly reduced. Fig.3 shows the potential line distribution in the ON-state at I_A=200A/cm² and V_{GC}=15V (0.2V/contour). The LIEGT shows fewer equipotential lines and thus lower voltage-drop than those of the Conv. LIGBT. Moreover, the equipotential line is especially sparse near the cathode side owing to high carrier concentration.

To approximate the voltage-drop across the drift region (V_d) of the LIEGT, an analytical model is developed. Fig.4 (a) gives an equivalent structure of the LIEGT for modelling. Z and L represent the channel width and drift region length respectively. P_L and P_0 represent the anode-side and cathode-side carrier density at 100A/cm², respectively.

The electron and hole currents flowing through the drift region are given as

$$J_p = qp\mu_p E - qD_p(-K) \quad (1)$$
$$J_n = qn\mu_n E + qD_n(-K) \quad (2)$$

K is the concentration gradient. The total current is written as $J=J_n+J_p$. Under the high-level injection, there is $n\approx p$. From (1), (2) and the Einstein relation, the electric field can be given as:

$$E = \frac{J + k_0t(\mu_n - \mu_p)K}{qp(\mu_n + \mu_p)} \quad (3)$$

Fig.4 (b) shows the analytical and simulation results of V_d. Both the analytical and simulation results show that a higher P_0/ P_L contributes to lower V_d. P_L is generally determined by the dose of P-Anode (D_{pa}) and P_0 is dependent on T. As T is smaller than 250nm, P_0 gets higher than P_L according to simulation. The simulation results of V_d are higher than the analytical values because the carrier profile in the drift region does not strictly comply with a linear distribution.

Fig.5 (a) shows the turn-off waveforms for the LIEGT and the conventional LIGBT under clamped inductive load. The inset gives the circuit diagram used for the switching simulations. The gate voltage is switched from 15V to 0V at t=30μs. The LIEGT exhibits longer turn-off time than the conventional LIGBT due to the higher density of the stored carriers. There is a delay during turn-off of the LIEGT before V_{AC} rises for following two reasons. First, the extended gate contributes to larger Miller capacitance (C_{GA}). Second, more carriers at the cathode side have to be extracted before the drift region begins to be depleted as shown in Fig.5 (b). For the LIEGT, the turn-off process is divided into three stages: the cathode-side charge extracting stage (30μs <t< t_3), the voltage rising stage (t_3 <t< t_5) and the current decreasing stage (t> t_5). The drift region begins to be depleted and I_A falls at t_3 and t_5, respectively. Fig.5 (c) compares the power dissipation of two structures. The energy loss during t< t_3 (E_{ce}) is small and contributes little to the total loss. The energy losses for the next two stages (E_{vr} and E_{cd}) are almost the same as those of the Conv. LIGBT.

Fig.6 gives the trade-off relationships between V_{ON} and E_{OFF} for two structures. For T=90nm, the electron injection is extremely enhanced and the V_{ON} decreases greatly but the E_{OFF} is higher than that of the Conv. LIGBT. By increasing T, E_{OFF} of the LIEGT is reduced while the V_{ON} maintains advantage over that of the Conv. LIGBT. The LIEGT with T=200nm shows the best trade-off. For the same E_{OFF}, the V_{ON} of the LIEGT is decreased by 20%. For the same V_{ON}, the E_{OFF} of the

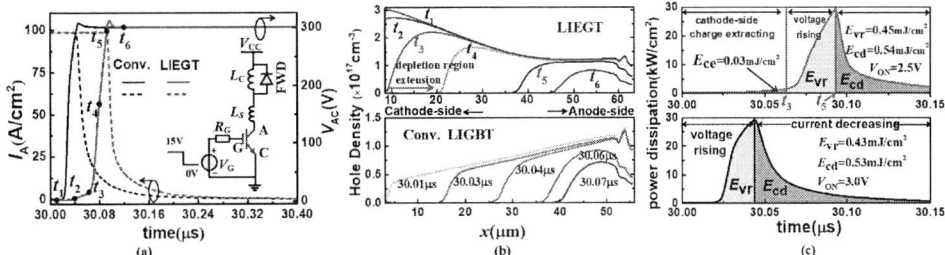

Fig.5 (a) Turn-OFF waveforms for the LIEGT with T=150nm and the Conv. LIGBT under clamped inductive load. L_C=10µH, L_S=1nH, R_G=5Ω, V_{CC}=300V. (b) Transient carrier profile in the drift region for the two structures during turn-OFF. (c) Power dissipation against time for both structures during turn-OFF.

LIEGT is decreased by 38%. For the LIEGT, D_{pa} is smaller than that of the Conv. LIGBT to achieve the same performance, which helps reduce the leakage current and increase latch-up immunity. As D_{pa} is reduced to $3 \times 10^{13} \text{cm}^{-2}$, E_{OFF} for both structures remains at low level. However, the V_{ON} for the Conv. LIGBT gets unacceptably high (11.7V) compared with that of the LIEGT (4.3V).

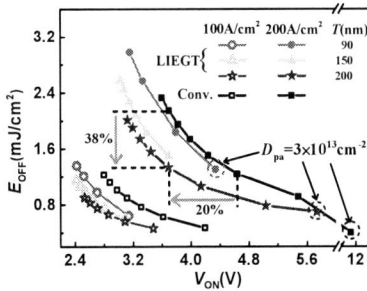

Fig. 6 E_{OFF} versus V_{ON} for LIEGTs and conventional LIGBT at 100A/cm^2 and 200 A/cm^2.

Fig.7 shows the key steps to fabricate the LIEGT. The trench is firstly formed by LOCOS. Phosphorus implantation via one mask followed by 850min anneal at 1200℃ is implemented to realize linear doping profile for drift region to achieve high BV [7] [17]. The subsequent steps are compatible with the CMOS process.

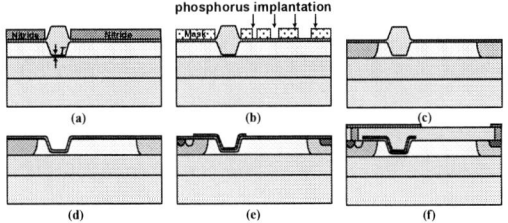

Fig. 7 Key fabrication process steps for the proposed LIEGT. (a) LOCOS. (b). Implantation for linear doping profile. (c) Implantation for P-Well and N-Buffer. (d) Oxide removal and growing the gate oxide. (e) Poly deposition and patterned, followed by implantation for N$^+$, P$^+$ regions. (f) Contact formation and metallization.

III. CONCLUSION

A novel injection enhanced LIGBT on thin SOI compatible with CMOS process is proposed and investigated by simulations. The cathode-side carrier density of the LIEGT is one order of magnitude higher than that of the conventional LIGBT due to the extreme injection-enhancement effect. The proposed LIEGT is demonstrated lower V_{ON} and higher saturation current than those of the conventional LIGBT. For the same E_{OFF}, the V_{ON} of the LIEGT is decreased by 20%. A compact analytical model is also developed to calculate the voltage-drop of the drift region for LIEGT.

REFERENCES

[1] J. Petruzzello, T. Letavic, H. van Zwol, M. Simpson and S. Mukherjee, "A thin-layer high-voltage silicon-on-insulator hybrid LDMOS LIGBT device," in Proc. ISPSD, Jun. 2002, pp.117-120. DOI: 10.1109/ISPSD.2002.1016185

[2] T. Letavic, J. Petruzzellol, J. Claes, P. Eggenkamp, E. Janssen, A. van der Wal, "650V SOI LIGBT for Switch-Mode Power Supply Application," Jun. 2006, pp.1-4. DOI: 10.1109/ISPSD.2006.1666145

[3] F. Udrea, T. Trajkovic and G. A. J. Amaratunga, "Membrane High Voltage Devices - A Milestone Concept in Power ICs," in Proc. IEDM, May 2004, pp. 451–454. DOI: 10.1109/IEDM.2004.1419185

[4] T. Letavic, M. Simpson, E. Arnold, E. Peters, R. Aquino, J. Curcio, S. Herko, S. Mukherjee, "600V Power Conversion System-on-a-Chip based on Thin Layer Silicon-on-Insulator," in Proc. ISPSD, May 1999 pp. 325–328. DOI: 10.1109/ISPSD.1999.764126

[5] T. Letavic, J. Petruzzello, M. Simpson, J. Curcio, S. Mukherjee, "Lateral Smart-Discrete Process and Devices based on Thin-Layer Silicon-on-Insulator," in Proc. ISPSD, Jun. 2001, pp. 407–410. DOI: 10.1109/ISPSD.2001.934640

[6] Ming Qiao, Bo Zhang, Zhiqiang Xiao, Jian Fang, Zhaoji Li, "High-Voltage Technology Based on Thin Layer SOI for Driving Plasma Display Panels," in Proc. ISPSD, May 2008, pp. 52–55. DOI: 10.1109/ISPSD.2008.4538895

[7] Jing Zhu, Weifeng Sun, Qinsong Qian, Lu Cao, Nailong He, Sen Zhang, "700V Thin SOI-LIGBT with High Current Capability," in Proc. ISPSD, May 2013, pp. 119–122. DOI: 10.1109/ISPSD.2013.6694443

[8] Akio Nakagawa, "Theoretical Investigation of Silicon Limit Characteristics of IGBT," in Proc. ISPSD, Jun. 2006, pp. 5–8. DOI: 10.1109/ISPSD.2006.1666057

[9] Masakiyo Sumitomo, Junichi Asai, Hiroki Sakane, Kazuki Arakawa, Yasushi Higuchi and Masaki Matsui, "Low loss IGBT with Partially Narrow Mesa Structure (PNM-IGBT)," in Proc. ISPSD, Jun. 2012, pp. 17–20. DOI: 10.1109/ISPSD.2012.6229012

[10] Manabu Takei, Shinji Fujikake, Haruo Nakazawa, Tatsuya Naito, Tomoyuki Kawashima, Kazuo Shimoyama, Hitoshi Kuribayashi, "DB (Dielectric Barrier) IGBT with Extreme Injection Enhancement," in Proc. ISPSD, Jun. 2010, pp. 383–386.

[11] M. Antoniou, F. Udrea, F. Bauer, A. Mihaila, I. Nistor, "Point Injection in Trench Insulated Gate Bipolar Transistor for ultralow losses," in Proc. ISPSD, Jun. 2012, pp. 21–24. DOI: 10.1109/ISPSD.2012.6229013

[12] M. Antoniou, N. Lophitis, F. Udrea, F. Bauer, I. Nistor, M. Bellini, M. Rahimo, "Experimental demonstration of the p-ring FS+ Trench IGBT concept: A new design for minimizing the conduction losses," in Proc. ISPSD, May 2015, pp. 21–24. DOI: 10.1109/ISPSD.2015.7123379

[13] Tomoko Matsudai, Mitsuhiko Kitagawa and Akis Nakagawa, "A Trench-Gate Injection Enhanced Lateral IEGT on SOI," in Proc. ISPSD, May 1995, pp. 141–145. DOI: 10.1109/ISPSD.1995.515024

[14] Norio Yasuhara, Hideyuki Funaki, Tomoko Matsudai and Akio Nakagawa, "Experimental Verification of Large Current Capability of Lateral IEGTs on SOI," in Proc. ISPSD, May 1996, pp. 97–100. DOI: 10.1109/ISPSD.1996.509457

[15] K. Sheng, F. Udrea, G. A. J. Amaratunga, "Optimum carrier distribution of the IGBT," Solid-State Electronics, vol.44, issue 9, Sep. 2000, pp.1573–1583.

[16] Hengyu Wang, Ming Su, and Kuang Sheng "Theoretical Performance Limit of the IGBT," IEEE Transactions on Electron Devices, vol. 64, no.10, Oct. 2017, pp.4184-4192. DOI: 10.1109/TED.2017.2737021

[17] Shengdong Zhang, Johnny K. O. Sin, Tommy M. L. Lai, and Ping K. Ko. "Numerical Modeling of Linear Doping Profiles for High-Voltage Thin-Film SOI Devices," IEEE Transactions on Electron Devices, vol. 46, no.5, May 1999, pp.1036-1040. DOI: 10.1109/16.760414

978-1-7281-0582-6/19 $31.00 © 2019 IEEE

Proceedings of the 31st International Symposium on Power Semiconductor Devices & ICs
May 19-23, 2019, Shanghai, China

Design Method and Mechanism Study of LDMOS to Conquer Stress Induced Degradation of Leakage Current and HTRB Reliability

Kanako Komatsu, Tomoko Kinoshita, Saori Shioda, Toshihiro
Sakamoto, Koji Kimura, Koji Yonemura
and Fumitomo Matsuoka
Toshiba Electronic Devices & Storage Corporation
580-1 Horikawa-cho, Saiwai-ku, Kawasaki-city, Kanagawa
212-8520, Japan
E-mail: kanako.komatsu@toshiba.co.jp

Keita Takahashi, Akihiro Urata, Shoichi Sakaguchi
and Takahito Nagamatsu

Japan Semiconductor Corporation Oita Operations
3500 Oaza Matsuoka, Oita-city, Oita
870-0197, Japan
E-mail: keita2.takahashi@toshiba.co.jp

Abstract—**8V NchLDMOS, which has an enough tolerance against both a leakage current (Idss) increase and high temperature reverse bias (HTRB) degradation is proposed. Dislocation growth, which results in Idss increase and has a negative influence on HTRB reliability, occurs because of a high-dose ion implantation and/or mechanical stress due to shallow trench isolation (STI). The studied 8V NchLDMOS was developed by taking into account the mechanism understanding to prevent the dislocation growth affecting the device characteristics, and as a result, achieved a competitive low resistance (RonA=1.67mΩ.mm²).**

Keywords—LDMOS, dislocation growth, leakage current, high temperature reverse bias

I. INTRODUCTION

LDMOS is widely used for Mixed-signal ICs. In case of using for output devices, LDMOS's size, which has a considerable impact on the total chip size, is an important factor. Therefore, low on-resistance (Ron) is an indispensable characteristics of LDMOS [1-5]. At the same time, keeping the LDMOS's Idss low is also significant in order to realize a low stand-by power. Moreover, Idss increase during the product operation leads to circuit operation failure and IC's lifetime degradation. A number of studies have been done to design a low-Idss device [6] with obtaining a high yield [7] and also to suppress Idss increase from the reliability viewpoint [8-9]. In this study, we focus on the relation between device characteristics and the impact of the mechanical stress due to STI, and propose the device design which is tolerate against both Idss increase and high HTRB degradation with keeping a low Ron.

II. DEVICE STRUCTURES AND EXPERIMENT

The device was fabricated using 0.13 μm CMOS-DMOS technology [8]. Plane and cross-sectional schematic views of conventional 8V NchLDMOS are shown in Fig.1. In the evaluated device, the channel length is 0.3 μm and the gate oxide thickness is 12.5 nm. The off-state drain-source breakdown voltage (BVdss) is 10 V, which can ensure 8 V operation considering BVdss variance and temperature dependence. The source and the backgate electrodes were shorted by layout design. STI is located only at the outer region of the device. In this study, two kinds of STI filling materials were studied, since STI filling material remains in

the device structure, that it has a great influence on the device characteristics, yield, and reliability. Material-A is deposited by a method of a high density plasma assisted chemical vapor deposition (CVD) processing and material-B is deposited by an atmospheric pressure CVD processing.

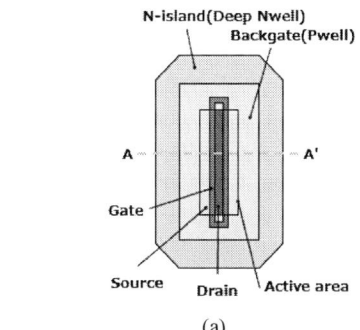

(a)

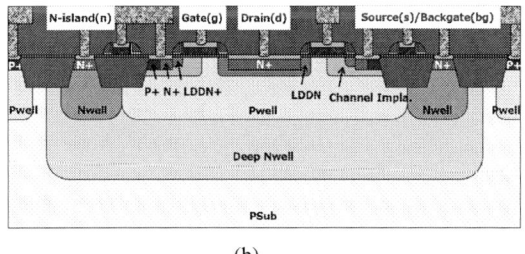

(b)

Fig. 1: (a) Plane and (b) cross-sectional views (A-A') of a conventional 8V NchLDMOS structure.

III. RESULTS AND DISCUSSIONS

A. Initial leakage current

Id-Vd curves of the conventional structure which applied material-A as a STI filling material (Fig. 2(a)) show that some of the chips have Idss which excesses 3pA/cell below Vds=8V. Fig. 2(b) is a TEM image of the fail chip. The dislocation was observed under the contact hole near the STI. A high-dose ion implantation forms an amorphous Si layer. The annealing follows to activate the amorphous Si layer, and during a recrystallization, the Si is activated under the

978-1-7281-0582-6/19 $31.00 © 2019 IEEE

high stress due to the STI filling material, which has a different thermal expansion coefficient compared to Si. While the recrystallization carries on, the value of STI induced stress which affects the Si might excess a critical stress, which results in the dislocation formation. We assume that the origin of the dislocation might be the high-dose ion implantation damage combined with STI stress, and thus, we investigated an influence of high-dose ion implantation damage on the Idss failure rate and the Ron (Fig. 3). As the phosphorus dose decreases, the defect generation decreases. However, decreasing the phosphorus dose results in the Ron increase simultaneously, thus this approach is not preferable for LDMOS performance. Although applying a thermal process for a long time to recover the Si damage caused by high-dose ion-implantation is not suitable for a shrunk process because of a design restriction. Meanwhile, we also studied two kinds of STI filling material having different stress against Si, material-A shows higher stress during annealing than material-B, and stress value are dependent on their film density. In the case of applying the material-B, no Idss fail chip was observed even at 4.6×10^{15} cm^{-2} dose condition (Fig. 4), although some fail chips were observed at the same dose condition in the case of material-A (Fig. 2(a)).

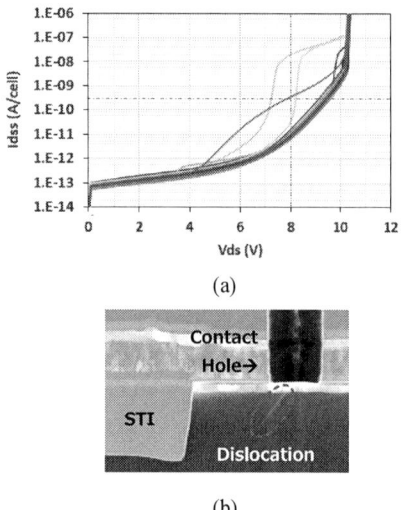

(a)

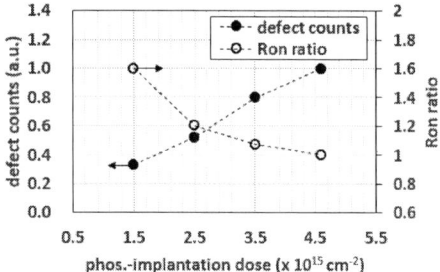

(b)

Fig. 2: (a) Idss-Vds curves (Vgs=0V, T=27°C) of conventional structure with STI filling material-A at 4.6×10^{15} cm^{-2} dose condition (b) cross-sectional TEM image of the fail chip.

Fig. 3: Dependence of the *Idss failure rate and the Ron of conventional structure with STI filling material-A on Phos. dose. *3200cell array pattern of 8V NchLDMOS

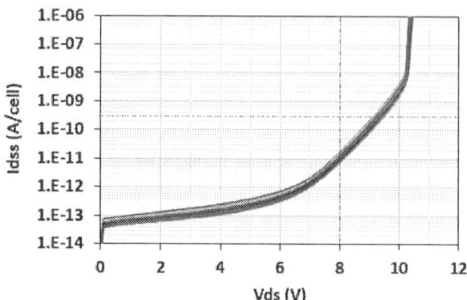

Fig. 4: Idss-Vds curves (Vgs=0V, T=27°C) of conventional structure with STI filling material-B at 4.6×10^{15} cm^{-2} dose condition.

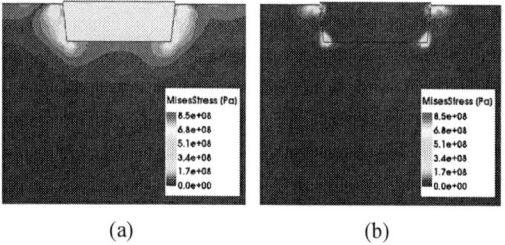

(a)   (b)

Fig. 5: Simulation results of Mises stress applying (a) material-A and (b) material-B as a STI filling material. Applied stress is -300MPa and 0MPa, respectively.

To evaluate the stress difference between material-A and B, a process simulation was carried out based on the estimated stress value, -300MPa in material-A and 0Mpa in material-B, respectively. These values were calculated from the warp evaluation on bare wafers [9]. Fig. 5 shows a calculated stress distribution during the densification under the high temperature annealing which has a strong relation to the dislocation growth. In the case of material-A, higher stress value is observed at the Si substrate near a STI, and it is suppressed in material-B case. This high stress prevents the recrystallization of the Si, thus leaves the dislocation. The simulated results well describe the Idss-Vds characteristics difference between material-A and B (Fig. 2(a) and Fig. 4).

B. High temperature reverse bias (HTRB)

We assume that this Idss failure rate difference occurs from dislocation growth rate related to Si stress from STI filling materials. HTRB test was also carried out to evaluate the influence of STI stress. The Idss continuously increased under the HTRB stress for both material-A and B (Fig. 6). The results indicate that reducing the stress from the STI filling material is not enough to ensure the reliability. Further approach is needed to prevent the Idss increase under the HTRB stress. The HTRB degradation could be caused by the trapped charge at STI/Si-interface [10], in that case, the trapped charge is de-trapped by applying wafer baking after HTRB test, which leads to Idss increase suppression. However, the result of wafer baking showed Idss increase (Fig. 7). The result is unreasonable with the assumption, and thus we propose another mechanism to explain this phenomenon.

978-1-7281-0582-6/19 $31.00 © 2019 IEEE   364

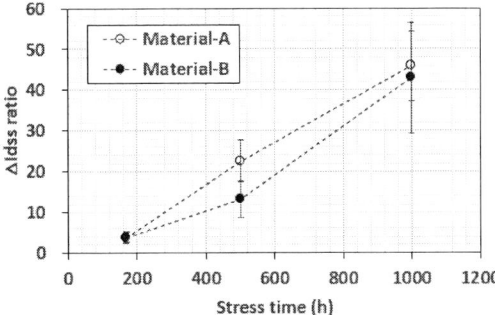

Fig. 6: Measured Idss/Idss0 (Vds=8V, Vgs=0V, T=27°C, 8 points) under the HTRB stress (Vds=8V, Vgs=0V, T=150°C) of conventional structure with STI filling material-A and B.

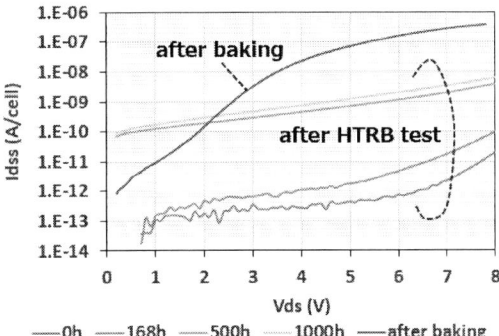

Fig. 7: Measured Idss-Vds (Vgs=0V, T=27°C) of conventional structure with STI filling material-A under the HTRB stress (Vds=8V, Vgs=0V, T=150°C) and after wafer baking for 24h at 250°C.

IV. MECHANISM OF IDSS INCREASE UNDER THE HTRB STRESS

Fig. 8 shows the explanation of the mechanism of Idss increase under the HTRB stress. Idss increase after HTRB is considered to relate to both H+ termination to a dislocation and de-termination from the dislocation. Firstly, the dislocation due to high-dose implantation near the STI/Si edge grows. Secondly, the dislocation is treated by hydrogen (H+) termination to the dangling bond under the N_2/H_2 anneal. This is because the Idss at 0s show low Idss (Fig.7). In spite of this H+ termination during N_2/H_2 anneal, H+ de-terminates from the dangling bond under the HTRB stress. As a result, the leakage current after the HTRB stress increases. From this assumption, the cause of the Idss increase under the HTRB stress is as same as the Idss failure rate increase due to a dislocation growth. In order to prevent the dislocation growth, the first approach is to select a preferable STI filling material, which to suppress a stress inside the Si less than its critical stress. However, just modifying the STI filling material was not enough to prevent HTRB degradation (Fig.6). Additional approach is needed, that is to suppress the value of the critical stress of the specific area inside the Si near the STI, where the high

stress exists. In particular, the high-dose implanted area should be kept away from the highly stressed area close to the STI by modifying the LDMOS layout design. Fig. 9 shows plane and cross-sectional schematic views of (a) conventional and (b) GC-covered structure, which STI/Si-interface is covered by gate poly Si. Modified structure (Fig. 9(b) and (d)) keeps high-dose ion implantation area away from STI/Si-interface. The HTRB result of GC-covered structure with STI filling material-B is shown in Fig. 10. In comparison to the HTRB result of the conventional structure (Fig. 6), in spite of the stress condition is severe for the GC-covered structure, which the HTRB test of conventional structure took place under 150°C/1000h, while new structure was 175 °C/2000h, the ratio of Idss compared to initial Idss (Idss/Idss0) was 45 and <1. These results demonstrate that the solution came out from the HTRB degradation mechanism (Fig.8), which to keep the high-dose implanted area away from the STI was meaningful to suppress Idss increase under the HTRB stress. In conclusion, it is necessary to take into account the high dose ion implantation damage as well as STI's mechanical stress due to STI filling material's density at the same time. As a result, competitive low resistance (RonA =1.67mΩ.mm²) LDMOS has been successfully fabricated (Fig. 11).

Mechanism of leakage current increase under HTRB stress

(1) Growth of the dislocation due to high-dose implantation near the STI/Si edge
(2) Treatment of the dislocation by hydrogen (H+) termination to the dangling bonds under the N_2/H_2 anneal
(3) H+ de-termination from dangling bonds under the HTRB
(4) Increase of the leakage current after HTRB stress

*Wafer baking after HTRB accelerates the H+ de-termination and increases the leakage current

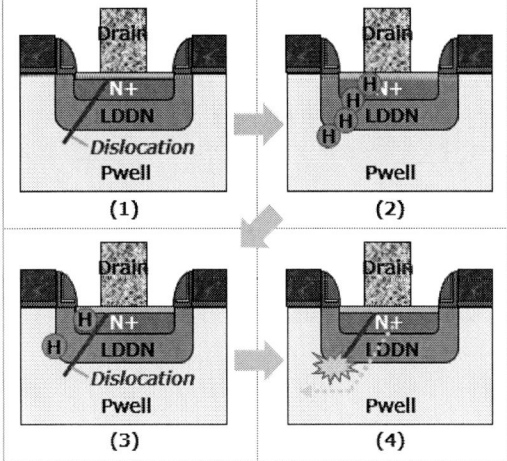

Fig. 8: Explanation of the mechanism of leakage current increase under the HTRB stress.

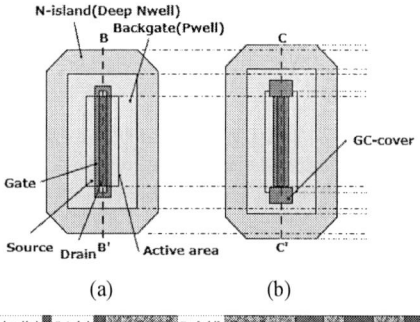

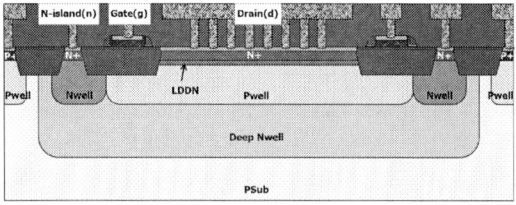

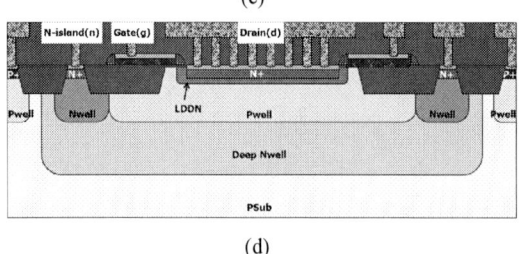

Fig. 9: Plane and cross-sectional views of (a) and (c) conventional and (b) and (d) GC -covered structure.

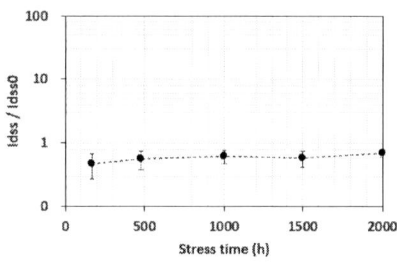

Fig. 10: Measured Idss/Idss0 (Vds=8V, Vgs=0V, T=27°C, 8 points) under the HTRB stress (Vds=8V, Vgs=0V, T=175°C) of GC-covered structure with STI filling material-B.

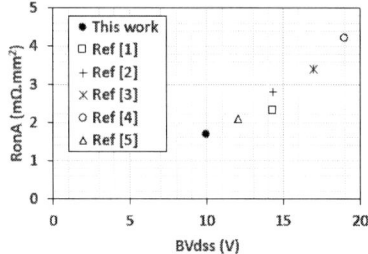

Fig. 11: Comparison of the RonA vs. BVdss for NchLDMOS.

V. CONCLUSIONS

8V NchLDMOS, which has a tolerance against a leakage current (Idss) increase and high temperature reverse bias (HTRB) degradation is proposed. Dislocation growth, which results in Idss increase and has a negative influence on HTRB reliability, occurs because of a high-dose ion implantation and/or mechanical stress due to shallow trench isolation (STI). The studied 8V NchLDMOS was developed by taking into account the mechanism understanding, which to prevent the dislocation growth affecting the device characteristics. As a result, both enough tolerance against HTRB stress and low leakage current with a competitive low resistance (RonA=1.67mΩ.mm2) has been achieved.

ACKNOWLEDGMENT

The authors would like to thank Mr. Kenya Kobayashi and Mr. Shinzo Tsuboi for their continuous encouragement and valuable support.

REFERENCES

[1] H. L. Chou, P. C. Su1, J. C. W. Ng, P. L. Wang, H. T. Lu, C. J. Lee1, W. J. Syue, S. Y. Yang, Y. C. Tseng, C. C. Cheng, C. W. Yao, R. S. Liou, Y. C. Jong, J. L. Tsai, Jun Cai, H. C. Tuan, Chih-Fang Huang, Jeng Gong, "0.18 μm BCD Technology Platform with Best-in-Class 6 V to 70 V Power MOSFETs", Proceeding of the 24th ISPSD, pp. 401–404, Jun. 2012.

[2] R. Roggero, G. Croce, P. Gattari, E. Castellana, A. Molfese, G. Marchesi, L. Atzeni, C. Buran, A. Paleari, G. Ballarin, S. Manzini, F. Alagi and G. Pizzo, "An Advanced 0.16 μm Technology Platform with State of the Art Power Devices", Proceeding of the 25th ISPSD, pp. 361–364, May 2013.

[3] F. Jin, D. Liu, J. Xing, X. Yang, J. Yang, W. Qian, W. Yue, P. Wang, M. Qiao and B. Zhang, "Best-in-Class LDMOS with Ultra-Shallow Trench Isolation and P-Buried Layer from 18V to 40V in 0.18μm BCD Technology", Proceeding of the 29th ISPSD, pp. 295–298, May 2017.

[4] L. Wei, C. Chao, U. Singh, R. Jain, L.L. Goh and P.R. Verma, "A Novel Contact Field Plate Application in Drain-Extended-MOSFET Transistors", Proceeding of the 29th ISPSD, pp. 335–337, May 2017.

[5] L. Wei, U. Singh, J. M. Koo, J. Huihua, "A Novel High Performance Medium-Voltage DEnMOS in 40nm CMOS Technology", Proceeding of the 30th ISPSD, pp. 292–294, May 2018.

[6] C. Pacha, B. Martin, K. Arnim, R. Brenderlow, D. Schmitt-Landsiedel, P. Seegebrecht, J. Berthold, and R. Thewes, "Impact of STI-induced stress, inverse narrow width effect, and statistical V/sub TH/ variations on leakage currents in 120 nm CMOS", Proceedings of the 30th ESSCIRC, pp. 397–400, Sept. 2004.

[7] F. W. Saris, J. S. Custer, R. J. Schreutelkamp, R. J. Liefting, R. Wijburg, and H. Wallinga, "Avoiding dislocations in ion-implanted silicon" in Microelectronic Engineering, vol. 19, 1992, pp. 357–362.

[8] K. Takahashi, K. Komatsu, T. Sakamoto, K. Kimura, and F. Matsuoka, "Hot-Carrier Induced Off-State Leakage Current Increase of LDMOS and Approach to Overcome the Phenomenon", Proceeding of the 30th ISPSD, pp. 303–306, May 2018.

[9] M. H. Park, S. H. Hong, S. J. Hong, T. Park, S. Song, J. H. Park, H. S. Kim, Y. G. Shin, H. K. Kang, and M. Y. Lee, "Stress Minimization in Deep Sub-Micron Full CMOS Devices by Using an Optimized Combination of the Trench Filling CVD Oxides", IEDM Technical Digest, 1997, pp. 669-672.

[10] C. Lin, Y. Jin, C. H. Jan, C. W. Hu, K. Chang, and H. Kao, "Novel Current Re-Distribution Structure for Improved and Easy-to-Manufacturing 24V LDMOS", Proceeding of the 30th ISPSD, pp. 295–298, May 2018.

978-1-7281-0582-6/19 $31.00 © 2019 IEEE

Proceedings of the 31st International Symposium on Power Semiconductor Devices & ICs
May 19-23, 2019, Shanghai, China

Low On-state Voltage and Latch-up Immunity Thin SOI LIGBT with Multi-Segmented Trench Gates

ChaoYang, Xiaorong Luo*, Tao Sun, Dongfa Ouyang, Anbang Zhang, Zhaoji Li and Bo Zhang
State Key Laboratory of Electronic Thin Films and Integrated Devices
University of Electronic Science and Technology of China
Chengdu, China
Email: xrluo@uestc.edu.cn

Abstract—A novel thin SOI lateral insulated gate bipolar transistor (LIGBT) with low on-state voltage drop (V_{on}) and latch-up immunity is proposed. The device features the multi-segmented trench gates in the z-direction, named MST LIGBT. The multi-segmented gates form multi conduction channels along the sidewalls, which increases the channel density so as to reduce V_{on} and increase saturation current. Furthermore, the multi-segmented gates hinder the holes from being extracted by the cathode in the x- and z- direction, which attracts a large number of electrons injected from the cathode, and therefore the conduction modulation effect is enhanced. The V_{on} of MST LIGBT reduces by 34.8 % and 27.6 % in comparison with those of the conventional (Con). LIGBT and the tridimensional channel (TC) LIGBT. In addition, the short circuit time of MST LIGBT increases to 1.92 µs in comparison with 1.73 µs of the Con. LIGBT and 1.81 µs of the TC LIGBT, owing to the smallest shorted cathode resistance (R_{SC}).

Keywords— silicon on insulator (SOI), low on-state voltage, high current density, latch-up immunity.

I. INTRODUCTION

The silicon on insulator Introduction (SOI) lateral insulated gate bipolar transistor (LIGBT) is an attractive candidate for high voltage integrated circuits (HVIC) because of dielectric isolation and low leakage current1. Moreover, the SOI LIGBT is considered as the suitable device for the three-phase monolithic inverter ICs [1]-[4]. Owing to the requirements of the power consumption and the chip size, the SOI LIGBT with ultra-low on-state voltage drop and high current capability is necessary. The thin SOI can significantly simplify the isolation process [5]. In addition, the thin SOI device exhibits lower switching loss than that of the thicker one [6]. However, the thin SOI LIGBT behaves low current capability and high on-state voltage (V_{on}) owing to both the narrow conduction path and the carrier recombination on the top and bottom interfaces of the SOI layer [5]. Several structures have been proposed to improve current capability [7]-[12]. The multi-channels technique is used in TC LIGBT on thin SOI layer to achieve high current capability and good latch-up immunity by setting numerous separated P-body cells located in the emitter region while the V_{on} is not low enough to achieve low conduction consumption [11]. The carrier stored effect is used in several structures to achieve ultra-low V_{on} by enhancing the conductivity modulation effect [13]-[15]. Injection enhanced technology is another effective method to reduce V_{on}. Nevertheless, most of studies focus on the vertical IGBTs [16]-[20].

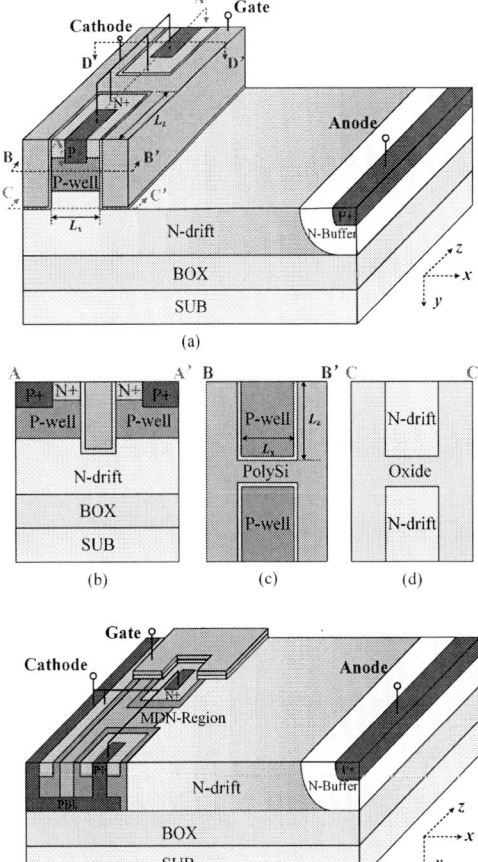

Fig. 1 (a) 3-D schematic of proposed MST LIGBT. Cross section of MST LIGBT along the cut lines: (b) *A-A'*, (c) *B-B'* (d) *C-C'*. (e) 3-D schematic of TC LIGBT[11].

In this paper, we propose a novel thin SOI LIGBT based on both the multi-channels technique and injection enhanced technology. The device dramatically reduces V_{on} and enhances current capability by employing the multi-segmented trench (MST) gates in the z-direction. The proposed structure also exhibits excellent switching performance and latch-up immunity.

978-1-7281-0582-6/19 $31.00 © 2019 IEEE

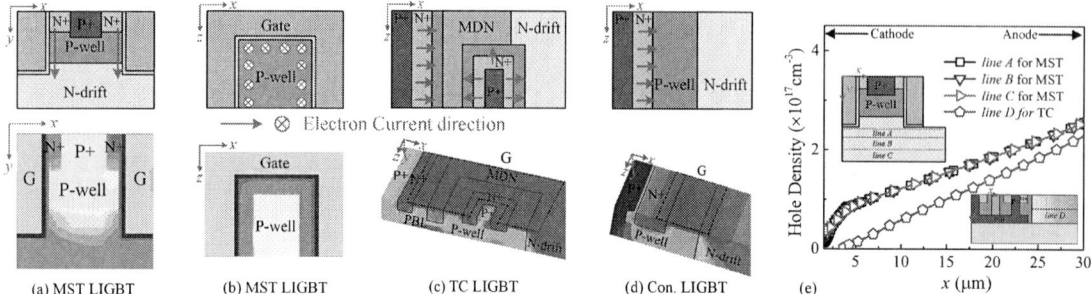

Fig. 2. On-state current path: (a-b)MST LIGBT, (c)TC LIGBT and (d)Con. LIGBT. (e) Hole density distribution at J=100A/cm^2 along the cutlineA,B,C and D.

TABLE I
DEVICE PARAMETERS SPECIFICATION

Parameters	MST	TC	Con.
SOI layer thickness, t_s	1.5 μm	1.5 μm	1.5 μm
Buried oxide thickness, t_{box}	3 μm	3 μm	3 μm
N-drift length, L_d	35 μm	30 μm	30 μm
P-well doping, N_{pwell}	1×10^{17} cm^{-3}	1×10^{17} cm^{-3}	1×10^{17} cm^{-3}
N-buffer doping, N_{buffer}	2×10^{17} cm^{-3}	2×10^{17} cm^{-3}	2×10^{17} cm^{-3}
Gate oxide thickness, t_{ox}	100 nm	100 nm	100 nm
Trench length along x direction, L_x	1.5 μm	---	---
Trench length along z-direction, L_z	1.5 μm	---	---

II. STRUCTURE AND MECHANISM

The schematic views and the cross-sections of the proposed MST LIGBT are shown in Fig. 1(a-d). The optimized parameters are shown in Table I. The proposed LIGBT features the multi-segmented trench gates in the z-direction. This study is carried out with TCAD Sentaurus Device simulation [21]. Philips unified mobility, Auger recombination, SRH recombination, High Field Saturation mobility, and Thermodynamic physical models are included. In Fig. 2 (a-d), compared with the on-state electron current path of TC LIGBT and Con. LIGBT, the multi conduction channels of MST LIGBT are formed along the sidewalls because of the multi-segmented trench gates, which increases the channel density significantly so as to reduce V_{on} and increase saturation current. Furthermore, the multi-segmented trench gates hinders holes from being extracted by the cathode in the x- and z- direction, which attracts a large number of electrons injected from the cathode, and thus enhances the conduction modulation effect, as shown in Fig.2 (e). In conclusion, the MST LIGBT exhibits a low V_{on} and a high current capability.

III. RESULTS AND DISCUSSION

Benefiting from the mentioned multi-channels technique and injection enhanced technology, a lower V_{on} and a higher current density are achieved in MST LIGBT. Fig. 3 shows the I-V characteristics of the three devices. The V_{on} are 1.60 V, 1.79 V and 2.12 V for the MST, TC and Con. LIGBTs, respectively (@ J = 100 A/cm^2 and 300 K). Fig. 4 shows the forward conduction characteristics of the three devices. the MST LIGBT delivers a current density 132 % and 17.3 % larger than those of the Con. LIGBT and the TC LIGBT at V_g = 10 V, respectively. Fig. 5 is the blocking characteristics

of the three structures. The linearly graded dopant profile is employed to achieve a high BV in the three LIGBTs. The BV of the MST, TC and Con. LIGBTs, with the same linearly graded dopant profile of the drift region, is 343 V, 340 V and 347 V, respectively [22].

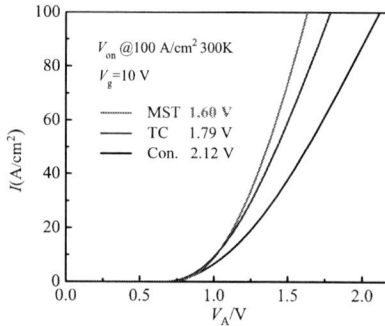

Fig. 3. I-V characteristics of the MST, TC, and Con. LIGBTs.

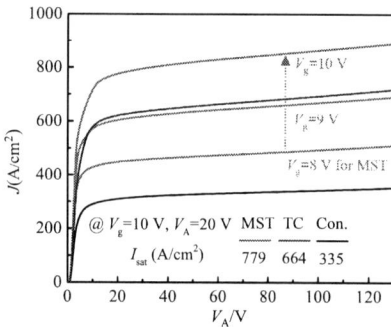

Fig. 4. Forward conduction characteristics of the three LIGBTs.

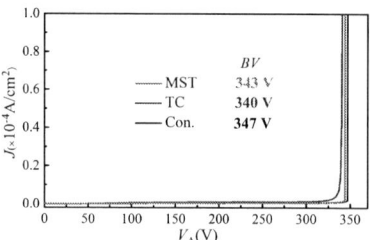

Fig. 5. Blocking characteristics of the three LIGBTs.

978-1-7281-0582-6/19 $31.00 © 2019 IEEE

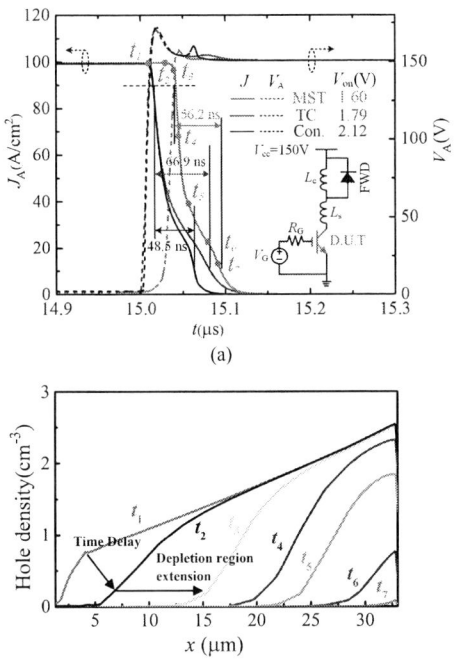

(a)

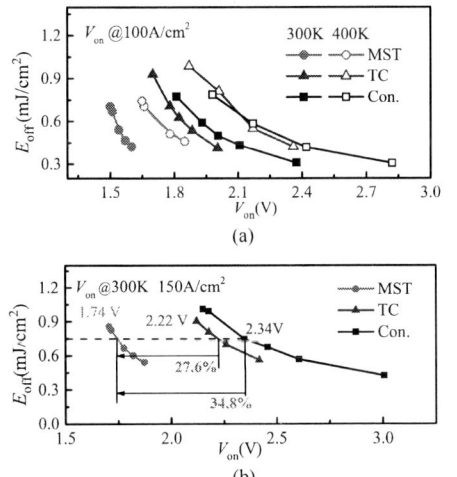

(b)

Fig. 6. (a) Inductive turn-off waves at T = 300 K for MST, TC, and Con. LIGBTs. The bus voltage of 150V, R_g of 10 Ω, load inductance of 50 μH, stray inductance of 10 nH, and carrier lifetime of 1 μs. (b) Hole density in the drift region for MST LIGBT during turn-off.

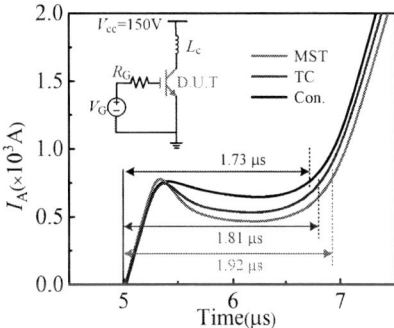

(a)

(b)

Fig. 7 (a) E_{off} versus V_{on} for different LIGBTs at J=100A/cm² (300K and 400K) (b) E_{off} versus V_{on} for different LIGBTs at J=150 A/cm².

Fig. 6 (a) illuminates the dynamic characteristics of the three LIGBTs. The three LIGBTs are simulated at the same P+ anode doping concentration. Though a turn-off delay time appears in MST LIGBT because of the larger gate capacitance, there has little influence on turnoff energy loss (E_{off}). Extra carriers at the cathode is extracted before the drift region begins to be depleted (V_A→0) during the delay time, as shown in Fig. 6 (b). In Fig. 7, the tradeoff performance between E_{off} and V_{on} of the MST LIGBT is superior to that of the TC and Con. LIGBTs. In addition,

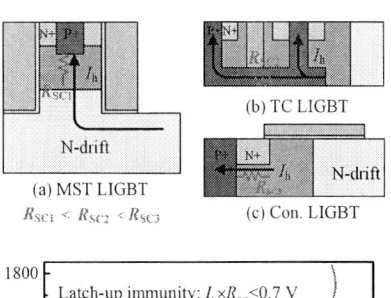

(a) MST LIGBT
$R_{SC1} < R_{SC2} < R_{SC3}$

(b) TC LIGBT

(c) Con. LIGBT

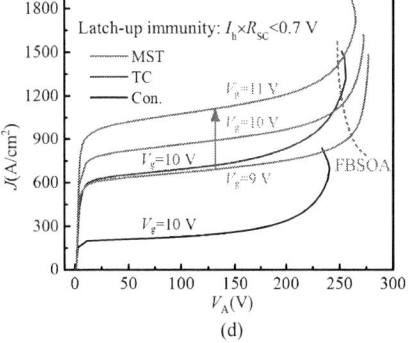

(d)

Fig. 8 Shorted cathode resistance: (a) MST LIGBT, (b) TC LIGBT and (c) Con. LIGBT. (d) Comparison of the static latch-up characteristics for the MST, TC, and Con. LIGBTs.

Fig. 9 Short circuit waves for the three IGBTs. In simulation, the junction temperature is 400K and the thermal resistor is 0.125K/W.

both the E_{off} and V_{on} deteriorate with the increase in the temperature. In Fig. 7 (b), the V_{on} of MST reduces by 34.8 % and 27.6 % in comparison with those of the Con. LIGBT and TC LIGBT at J=150 A/cm² and T=300 K.

Fig. 8 shows the latch-up immunity for the three LIGBTs. The smallest shorted cathode resistance (R_{SC}) for the hole current is achieved in the MST LIBT because of the short vertical current path and extend P+ region, as shown in Fig. 8 (a-c). The increased R_{SC} goes against the forward bias of the parasitic NPN and thus the MST LIGBT exhibits the latch up immunity, as shown in Fig. 8 (d). Fig. 9 shows the short circuit capability for the proposed structure. The short-circuit time of MST LIGBTs increases to 1.92 μs in comparison with 1.73 μs of the Con. LIGBT and 1.81 μs of the TC LIGBT.

IV. CONCLUSION

The proposed MST LIGBT achieves the good latch-up immunity owing to the smallest shorted cathode resistance.

The short circuit time of MST LIGBT increases to 1.92 μs in comparison with 1.73 μs of the Con. LIGBT and 1.81 μs of the TC LIGBT. At the same E_{off}, the proposed MST LIGBT reduces the on-state voltage by 27.6 % and 34.8 % compared with those of the TC and the Con. LIGBTs. In addition, The TGCS LIGBT exhibits a density of over 779 A/cm^2, which has an increase of 17.3 % and 132 % compared with those of the TC and the Con. LIGBTs.

ACKNOWLEDGMENT

This work was supported in part by the 13th Five-year Plan for Microelectronics Advanced Research Program under project 31513030201-2, in part by the National Defense Basic Scientific Research JCKY2016210B008, in part by the National Natural Science Foundation of China under Grant 51677021, and in part by the National Natural Science Foundation of China under Grant 61874149.

REFERENCES

[1] D. Disney, T. Letavic, T. Trajkovic, T. Terashima, and A. Nakagawa, "High-voltage integrated circuits: History, state of the art, and future prospects," *IEEE Trans. Electron Devices*, vol. 64, no. 3, pp. 659–673

[2] A. Nakagawa, H. Funaki, Y. Yamaguchi, and F. Suzuki, "Improvement in lateral IGBT design for 500 V 3 A one chip inverter ICs," in *Proc. ISPSD*, 1999, pp. 321–324

[3] K Hara K, S Wada, J Sakano, *et al.* "600V single chip inverter IC with new SOI technology, " in *Proc. ISPSD*, 2014, pp. 418-421

[4] K. Sakurai, D. Maeda, and H. Hasegawa, "Three-input type single-chip inverter IC including a function to generate six signals and dead time," in *Proc. ISPSD*, 2008, pp. 323–326

[5] Y.-K. Leung, A. K. Paul, J. D. Plummer, and S. S. Wong, "Lateral IGBT in thin SOI for high voltage, high speed power IC," *IEEE Trans. Electron Devices*, vol. 45, no. 10, pp. 2251–2254

[6] Y. K. Leung *et al.* "High voltage, high speed lateral IGBT in thin SOI for power IC," in *Proc. ISPSD*, 1996, pp. 132–133

[7] H. Funaki, T. Matsudai, A. Nakagawa, N. Yasuhara, and Y. Yamaguchi, "Multi-channel SOI lateral IGBTs with large SOA," in *Proc. ISPSD*, 1997, pp. 33–36

[8] T. Shigeki, A. Akio, Y. Youichi, S. Satoshi, and T. Norihito, "Carrierstorage effect and extraction-enhanced lateral IGBT (E2LIGBT): A super-high speed and low on-state voltage LIGBT superior to LDMOSFET," in *Proc. ISPSD*, 2012, pp. 393–396

[9] J. Zhu, W. Sun, Q. Qian, L. Cao, N. He, and S. Zhang, "700V thin SOI-LIGBT with high current capability," in *Proc. ISPSD*, 2013, pp. 119–122

[10] W. Sun *et al.*, "Investigation on Self-Adjust Conductivity Modulation SOI-LIGBT Structure (SCM-LIGBT) for Monolithic High-Voltage IC," *IEEE Trans. Electron Devices*, vol. 64, no .9, pp 3762-3767

[11] J. Zhu *et al.*, "TC-LIGBTs on the thin SoI layer for the high voltage monolithic ICs with high current density and latch-up immunity," *IEEE Trans. Electron Devices*, vol. 64, no. 11, pp. 3814–3820

[12] J. Zhu *et al.* "Further study of the U-shaped channel SOI-LIGBT with enhanced current density for high-voltage monolithic ICs," *IEEE Trans. Electron Devices*, vol. 63, no. 3, pp. 1161–1167

[13] H Takahashi , H Haruguchi, H Hagino, *et al.*, "Carrier stored trench-gate bipolar transistor (CSTBT)-a novel power device for high voltage application," in *Proc. ISPSD*, 1996 , pp. 349-352

[14] S Honda, Y Haraguchi , A Narazaki, *et al.* "Next generation 600V CSTBT™ with an advanced fine pattern and a thin wafer process technologies," in *Proc. ISPSD*, 2012 , pp. 149-152

[15] T Sun T, X Luo, J Wei J, *et al.* "A Carrier Stored SOI LIGBT With Ultralow ON-State Voltage and High Current Capability," *IEEE Trans. Electron Devices*, vol. 65, no 8, pp. 3365-3370

[16] M Kitagawa, I Omura, S Hasegawa, *et al.* "A 4500 V injection enhanced insulated gate bipolar transistor (IEGT) operating in a mode similar to a thyristor" in *Proc. IEDM*, 1993, pp. 679-682

[17] Y Miyaoku, K Matsuura, A Saito, *et al.* "Compact modeling and analysis of the partially-narrow-mesa IGBT featuring low on-resistance and low switching loss," in *Proc. ISPSD*, 2015 pp. 101-104

[18] M Tanaka, A Nakagawa, "Conductivity modulation in the channel inversion layer of very narrow mesa IGBT," in *Proc. ISPSD*, 2017 , pp. 61-64

[19] M Tanaka, A Nakagawa, "Novel 3D narrow mesa IGBT suppressing CIBL," in *Proc. ISPSD*, 2018 , pp. 124-127

[20] J Wei J, X Luo, G Deng, *et al.* "Ultrafast and Low-Turn-OFF Loss Lateral IEGT With a MOS-Controlled Shorted Anode," *IEEE Transactions on Electron Devices*, vol. 66, no. 1, pp. 533-538

[21] *TCAD Sentaurus Device Manual*, Synopsys, Inc., Mountain View, CA,USA, 2013.

[22] S. Zhang, J. K. O. Sin, T. M. L. Lai, and P. K. Ko, "Numerical modeling of linear doping profiles for high-voltage thin-film SOI devices," *IEEE Trans. Electron Devices*, vol. 46, no. 9, pp. 1036–1041

Proceedings of the 31st International Symposium on Power Semiconductor Devices & ICs
May 19-23, 2019, Shanghai, China

Full-chip simulation analysis of power MOSFET's during unclamped inductive switching with physics-base device models

Tsuyoshi Kachi
Power Device Technology Dept.
Renesas Electronics Corp.
Saijo, Ehime, Japan
tsuyoshi.kachi.yh@renesas.com

Katsumi Eikyu
Advanced Device Technology Dept.
Renesas Electronics Corp.
Hitachinaka, Ibaraki, Japan
katsumi.eikyu.ud@renesas.com

Takashi Saito
Design Platform Technology Dept. 2
Renesas Electronics Corp.
Takasaki, Gunma, Japan
takashi.saito.wj@renesas.com

Abstract— **To clarify current and temperature distribution inside single power MOSFET chip, we have developed technology-based computer-aided-design (TCAD)-based, full-chip transient-analysis framework. In conventional design, a single power device chip is assumed as a "lumped element" i.e. a spatial distribution inside each device is neglected. In some cases, the assumption is inadequate. So, we have constructed a simulation framework that takes into account the RC distributions inside a chip, and analyzed the internal state of power MOSFET chips during unclamped inductive switching (UIS).**

Keywords— *MOSFET, unclamped inductive switching, electro thermal simulation*

I. INTRODUCTION

In conventional design, a single power device chip is assumed as a "lumped element" i.e. a spatial distribution inside each device is neglected. In some cases, the assumption is inadequate, especially in high-frequency switching and high-power handling systems which tend to be realized by large-size power chips. So, a simulation framework that takes into account the RC distributions inside a chip is needed.

To clarify current and temperature distribution inside single power MOSFET chip, and their dependency on circuit conditions and chip layouts, we have developed technology-based computer-aided-design TCAD-based, full-chip transient-analysis framework, and analyzed the internal state of power MOSFET chips during UIS.

II. APPROACH

Two approaches are considered. One is compact model based and the other is TCAD based. A combination of RC parasitic network and compact device models are assumed as suitable for such an analysis since it is possible to calculate spatially high-granularity models with relatively few H/W resources [2] [3], however there are various restrictions in compact models; reproducibility of nonlinear body-diode capacitance characteristics, temperature dependency of electrical characteristics, and so on. In contrast, TCAD models are based on physics and has no limitation on reproducibility [4]. We have constructed chip model with TCAD device model including gate and source spreading resistances (Fig. 1).

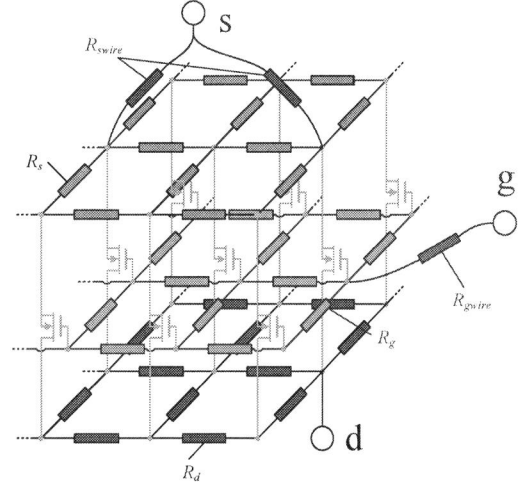

Fig. 1. Schematic diagram of the full-chip model. The model includes the spreading resistance of source metal (R_s), drain metal (R_d), gate poly-Si (R_g), source wires (R_{swire}) and gate wires (R_{gwire}). The MOSFETs are modeled by TCAD

Transient analysis of UIS are performed using device-circuit direct coupled (mixed-mode) simulation. Both self-heat generation and heat transportation outside a chip are calculated in self-consisted manner [5]. Since all parasitic parameters other than spreading resistances of electrode are included in the TCAD model, the number of extracted parasitic parameters are reduced significantly.

III. RESULTS AND DISCUSSION

Calculated MOSFET characteristics show good agreement with measurement results (Fig. 2). In the transient analysis of UIS, as far as the calculation results of current/voltage at external terminal, there are no significant differences between this distribution model and single device model (Fig. 3).

978-1-7281-0582-6/19 $31.00 © 2019 IEEE

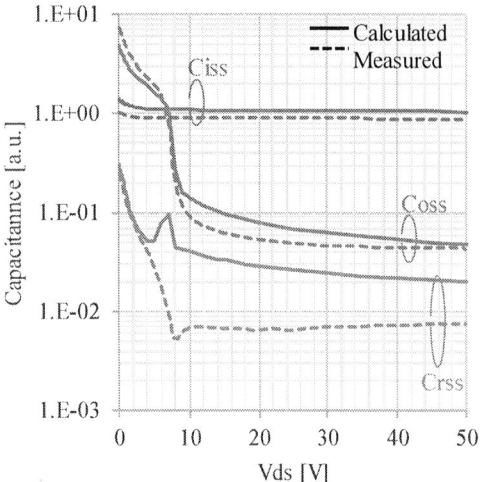

Fig. 2. Calculated Capacitance-V_{ds} characteristics of the full-chip model in comparison with measured data. An abrupt change in output capacitance (C_{oss}) is well reproduced. Feedback capacitance (C_{rss}) dose not match the measured value, but enough small to affect switching behavior.

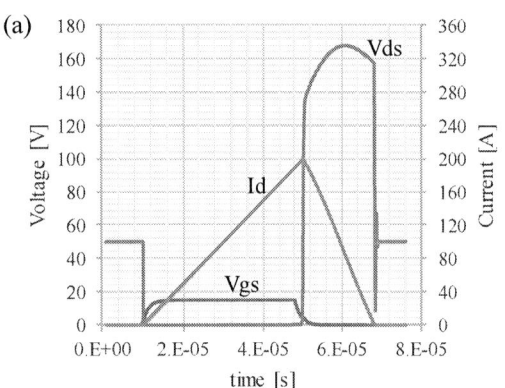

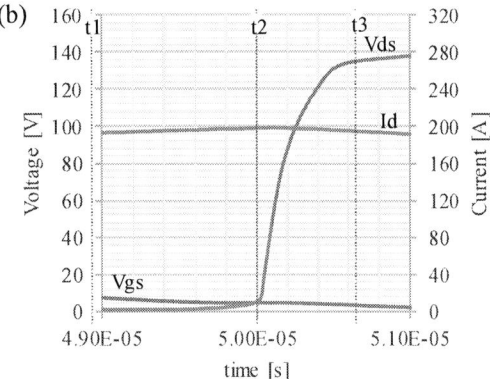

Fig. 3. Calculated UIS waveforms of the of the full-chip model. Voltage & current of outer terminals (indicated by suffixes "g", "s" and "d") are shown. (a) shows an entire period of waveforms from turn-on to turn-off and (b) shows an enlarged view around the turn-off point. The maximum avalanche current (I_{ap}) is set to 200A.

However, it becomes clear that a significantly inhomogeneous current distribution is generated inside a single chip during turn-off period (Fig. 4) by the full-chip model. In the outer periphery of the chip, there are active regions where no current flows throughout the switching period. Setting switching speed (dV_{ds}/dt) higher with smaller external gate resistor causes significant current concentration (Fig. 5).

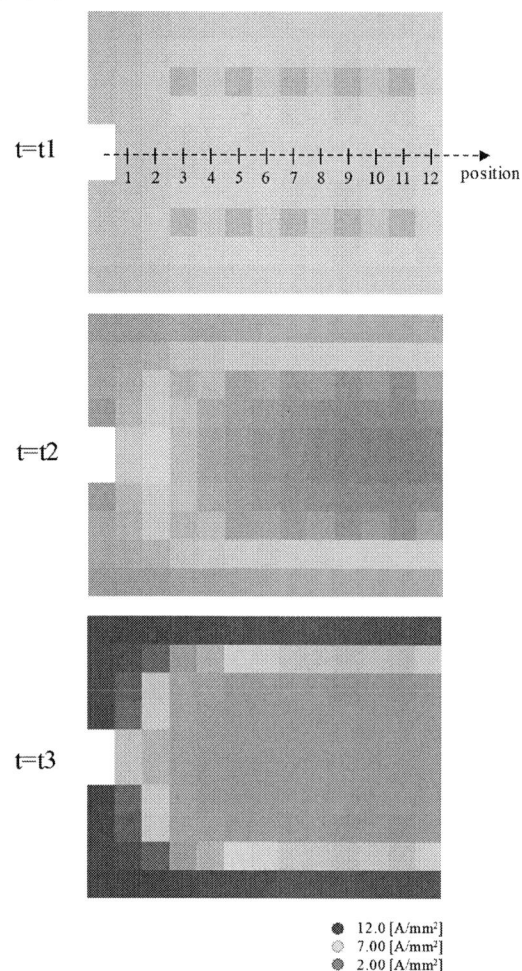

Fig. 4. Drain current distributions inside the chip at the time of corresponding to $t1$, $t2$ and $t3$ in Fig. 3(b) respectively. Current distribution at $t=t1$ reflects the locations of source wire bonds. At $t=t2$ (maximum current point), current concentrates near the center of the chip. As V_{ds} rises, current tends to flow outer periphery of the chip.

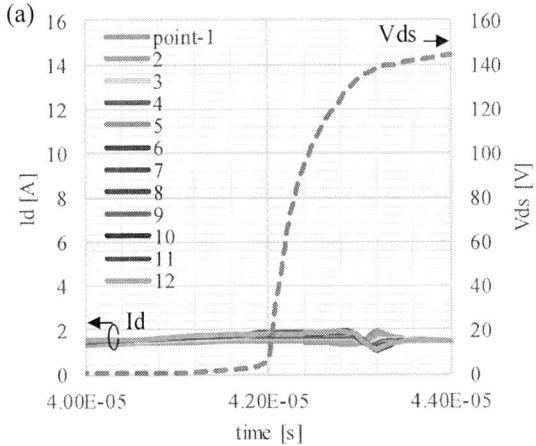

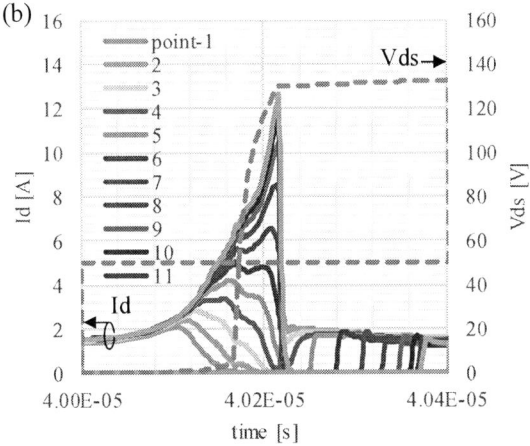

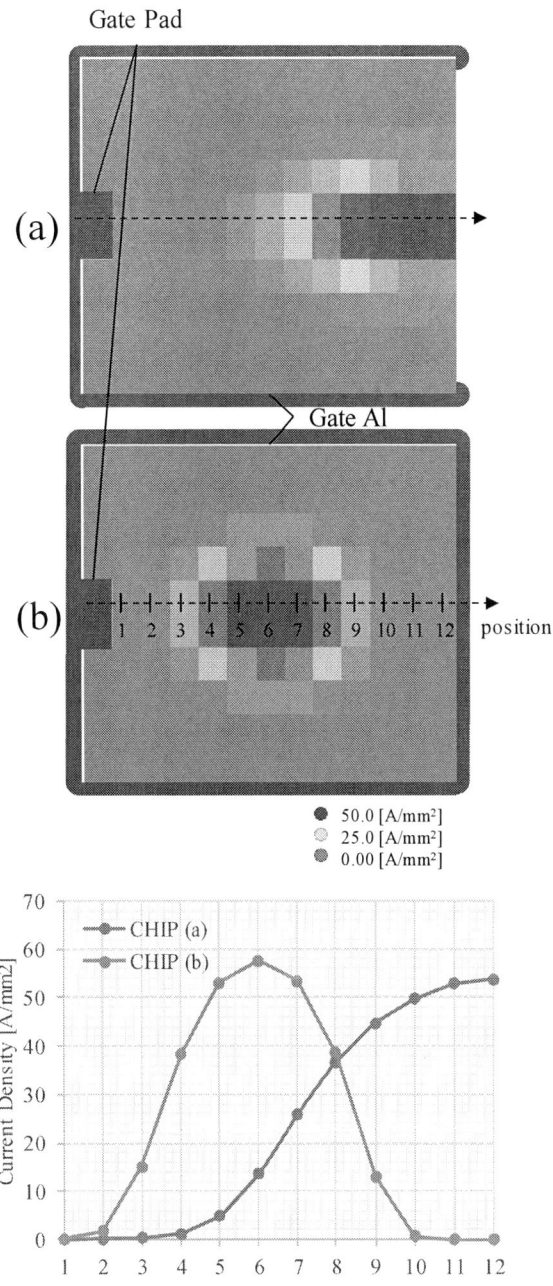

Fig. 5. Drain current (I_d) of each point (shown in Fig. 4) inside the chip and drain voltage (V_{ds}) waveforms during turn-off. (a) is a 200-Ω external gate resistor case and (b) is a 4-Ω one. In the condition (b), the maximum I_d in the chip reaches approximately 6 times of the average current, and no current flows in the chip outer area at all. I_d rise in several divided cell is begun before the V_{ds} rises, not due to the dV_{ds}/dt.

Then, for further investigation on the intrinsic gate resistance network impact, we have calculated another chip which has 1Ω smaller equivalent lump resistance ($R_{g.int}$), by changing gate Aluminum layout. Our distributed full chip model has shown $R_{g.int}$ has not directly correlated with maximum current density (Fig. 6). These results indicate that the networks of intrinsic gate resistance are essential to predict maximum current density inside the chip and equivalent lumped element cannot represent accurate switching behavior.

Temperature distribution inside the chip is also calculated (Fig. 7). The breakdown by thermal instability due to the current concentration on the hot spot is captured clearly.

Fig. 6. I_d distribution maps of two different gate Aluminum layouts. The layout (a) has an intrinsic gate resistance (equivalent series gate resistance, $R_{g.int}$) of 3.7Ω, and the layout (b) has that of 2.7Ω. Small $R_{g.int}$ dose not reduce maximum current inside the chips during turn-off

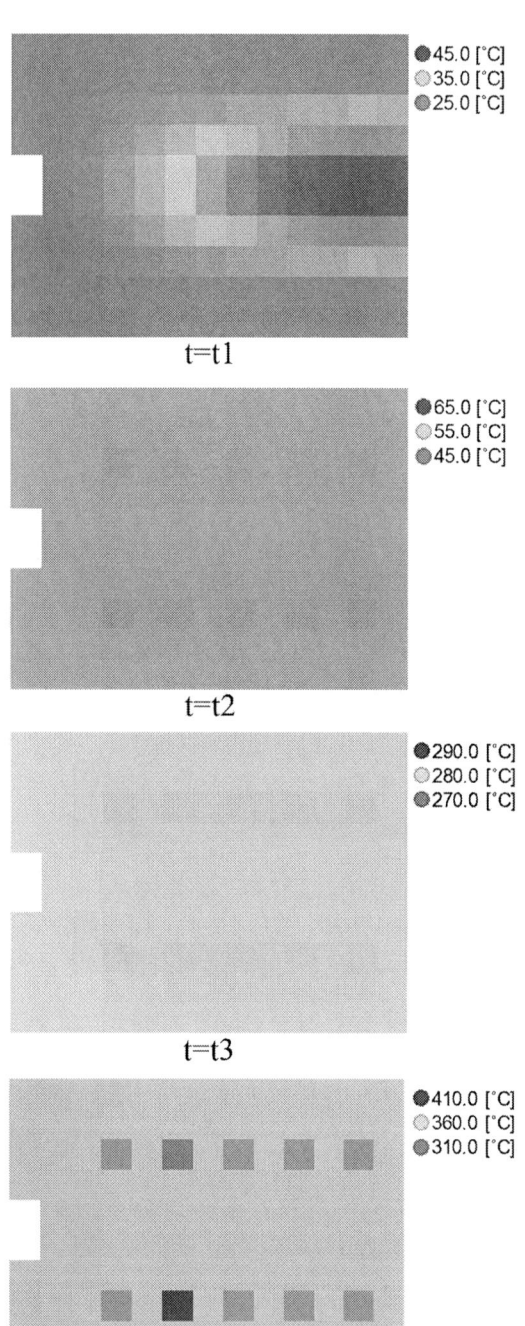

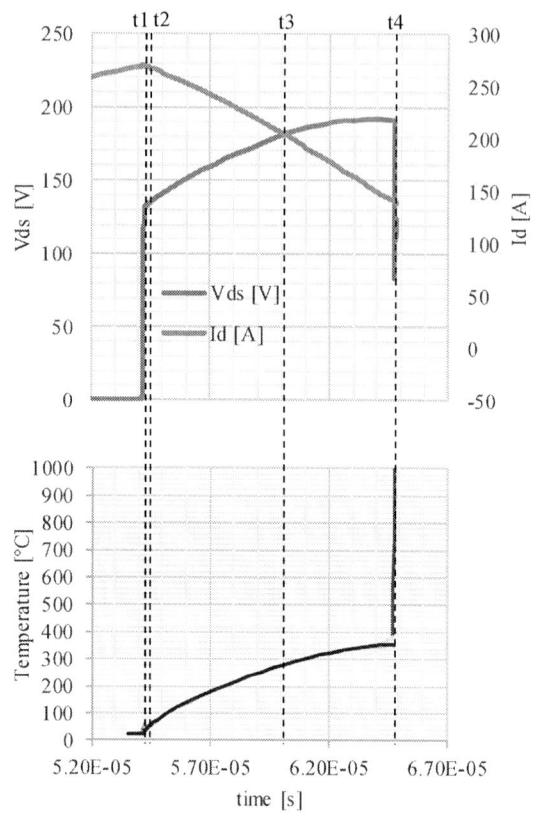

Fig. 7. Temperature distribution maps from turn-off to thermal destruction (left). $t1~t4$ is corresponding to time $t1~t4$ in waveforms of V_{ds}, I_d and maximum lattice temperature inside the divided cell (right). At the destruction time ($t4$), current concentrates at a single bonding point and rapid temperature increasing occurs simultaneously.

IV. CONCLUTION

As shown above, the proposed TCAD mixed-mode based full chip simulation can be a powerful tool for the DTCO (Design Technology Co-Optimization) in power device.

REFERENCES

[1] "Sentaurus Device User Guide Version O-2018.06," Synopsys Inc., June 2018.

[2] A. Chvála et al., IEEE Transactions on Electron Devices, Volume: 61 , Issue: 4, 2014.

[3] T. Biondi et al., IEEE Transactions on Power Electronics, Volume. 22, No. 5, 2007.

[4] T. Sarkar et al., in 2018 IEEE 30th International Symposium on Power Semiconductor Devices and ICs.

[5] W. Schoenmaker et al., in 19th International Workshop on Thermal Investigations of ICs and Systems, 2013.

Proceedings of the 31st International Symposium on Power Semiconductor Devices & ICs
May 19-23, 2019, Shanghai, China

Experimental Study on the Effect of Recessed Gates in Drain STI Regions of nLDMOSFETs

Takahiro Mori, Hirokazu Sayama, Takashi Ipposhi and Koji Iizuka

Front End Production Technology Division, Renesas Electronics Corporation
751 Horiguchi, Hitachinaka, Ibaraki 312-8504, Japan
E-mail: takahiro.mori.cj@renesas.com

Abstract— The introduction of a recessed gate into a shallow trench isolation (STI) section used as an insulating film in a drift region of an nLDMOSFET is a well-known means of improving the trade-off between the on-resistance (R_{sp}) and the off-state breakdown voltage (BV_{off}), as well as reducing hot-carrier injection (HCI) degradation in the maximum substrate current (I_{submax}) condition. Moreover, a drain multi-finger STI nLDMOSFET structure in which an STI section and a silicon surface section in the drift region are repeated in the channel width direction can improve the R_{sp}-BV_{off} trade-off. However, the effect of combining these structures on a silicon substrate has not been sufficiently investigated. In the present study, the effect of recessed gates on R_{sp}, BV_{off}, and HCI degradation in two types of nLDMOSFET are investigated.

I. INTRODUCTION

LDMOSFETs are widely used in power IC chips for automotive and industrial applications. Shallow trench isolation (STI) sections are increasingly being used as insulating films in the drift regions of nLDMOSFETs because of their high compatibility with CMOS logic processes and advances in CMOS process nodes. Recently, the long-term reliability of LDMOSFETs, especially in terms of hot-carrier injection (HCI) degradation, has become as important as the trade-off between on-resistance (R_{sp}) and the off-state breakdown voltage (BV_{off}) [1-5].

One approach for suppressing HCI degradation in the maximum substrate current (I_{submax}) condition in an nLDMOSFET with an STI insulating film is to introduce a recessed gate or a shielding contact into the STI insulating film [1-3]. This approach can also improve the R_{sp}-BV_{off} trade-off. Moreover, a drain multi-finger STI nLDMOSFET structure with an STI section and a silicon surface section repeated in a drift region along the channel width direction can improve the R_{sp}-BV_{off} trade-off [6-8].

Although the effect of a recessed gate on a normal nLDMOSFET has been elucidated in previous studies, that on a drain multi-finger STI nLDMOSFET is not clear. Therefore, the effect of recessed gates on two types of nLDMOSFET is experimentally investigated in the present study.

II. DEVICES AND SIMULATIONS

A. Device Structures

Figures 1 and 2 show cross-sections of the nLDMOSFET structures investigated in this work. Figures 1(a) and 1(b) respectively show an nLDMOSFET with a conventional structure and one with a recessed gate in the STI section of the drain drift region. Figures 2(b) and 2(c) respectively show an nLDMOSFET with a conventional drain multi-finger STI structure and one with a drain multi-finger STI structure that contains a recessed gate. The recessed gate is formed in a drain STI region near the channel side using a dry etching processing step followed by a polysilicon deposition step and a dry etching step [1].

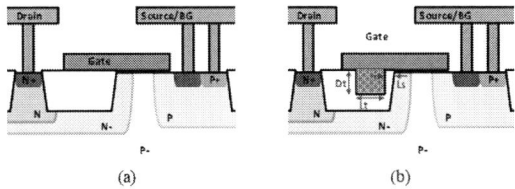

Fig. 1. Cross-sections of (a) conventional normal nLDMOSFET and (b) normal nLDMOSFET with a recessed gate.

20-V nLDMOSFETs with a gate voltage (V_g) rating of 3.3 V are used for investigating the electrical characteristics and HCI degradation. The most important design parameters for nLDMOSFETs (Figs. 1(b) and 2(c)) with recessed gates are the distance from the channel-side STI edge to the recessed gate (Ls), the recessed gate width (Lt) and the recessed gate depth from the Si surface (Dt). Dt is set to half the depth of the STI from the Si surface. The most important design parameters for drain multi-finger STI nLDMOSFETs (Figs. 2(b) and 2(c)) are the width of the active areas between the STI in the drift region (Ws) and the overlap of the gate polysilicon with the active areas in the drift region (Gd). Other parameters, including the drift concentration, are the same for all structures shown in Figs. 1 and 2.

978-1-7281-0582-6/19 $31.00 © 2019 IEEE   375

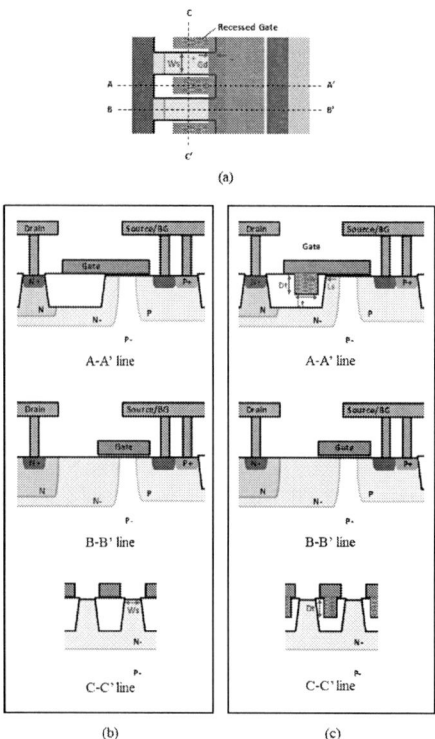

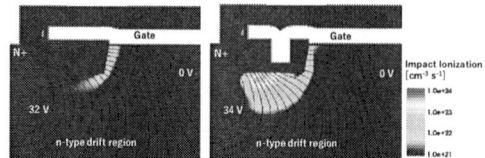

Fig. 3. TCAD simulation results for impact ionization (color contours) and electrostatic potential (line contours; 2-V steps) values for normal 20-V nLDMOSFETs in Figs. 1(a) and 1(b) biased at $V_g = 0$ V and $V_d = 32$ or 34 V.

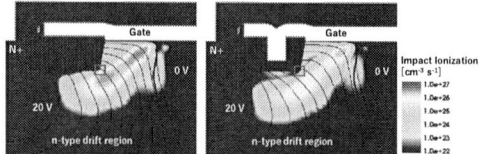

Fig. 4. TCAD simulation results for impact ionization (color contours) and electrostatic potential (line contours; 2-V steps) values for normal 20-V nLDMOSFETs in Figs. 1(a) and 1(b) biased at $V_g = 1/2$ V_{gmax} and $V_d = 20$ V.

Fig. 2. (a) Top view of drain multi-finger STI nLDMOSFET and cross sections of (b) nLDMOSFET with a conventional drain multi-finger STI structure and (c) nLDMOSFET with a drain multi-finger STI structure that contains a recessed gate.

B. TCAD study

Technology computer-aided design (TCAD) simulations were carried out to investigate the effect of recessed gates on 20-V nLDMOSFETs in terms of the BV_{off} state and the HCI degradation state. The impact ionization and electrostatic potential results for normal nLDMOSFETs in the BV_{off} state (at a V_g of 0 V and a drain voltage (V_d) of 32 V in Fig. 1(a) and 34 V in Fig. 1(b)) and the HCI degradation state (at I_{submax}, a V_g equal to approximately half of V_{gmax}, and a V_d of 20 V) are shown in Figs. 3 and 4, respectively. The values of Ls and Lt were both set to 0.2 μm for the structure in Fig. 1(b). BV_{off} for the structure in Fig. 1(b) is improved because the worst impact ionization point shifts from the channel-side STI edge to the drain-side recessed gate edge when a recessed gate is incorporated, as shown in Fig. 3, and thus the electric field at the channel side STI edge is relaxed. Moreover, the impact ionization near the channel-side STI edge is suppressed by the recessed gate in the HCI state, as shown in Fig. 4. Generally, HCI degradation occurs near the channel side STI edge. These TCAD results show that the recessed gate improves BV_{off} and HCI degradation suppression in normal nLDMOSFETs, which is consistent with previous studies [1-3].

TCAD simulations for 20-V drain multi-finger STI nLDMOSFETs (Fig. 2) in the BV_{off} state (at a V_g of 0 V and a V_d of 24 V) and the HCI degradation state (at I_{submax}, a V_g equal to approximately half of V_{gmax}, and a V_d of 20 V) are shown in Figs. 5 and 6, respectively. The values for Ws and Gd were set to 0.16 and 0.12 μm, respectively. The values for Ls and Lt were both set to 0.2 μm for the structure in Fig. 2(c). The worst impact ionization area in the BV_{off} state shifts from the channel-side STI edge to the drain-side when a recessed gate is incorporated (Fig. 5). However, the value of BV_{off} is almost the same for the two structures. The impact ionization near the channel-side STI edge is suppressed by the recessed gate in the HCI state, as shown in Fig. 6. If the HCI degradation for the drain multi-finger STI nLDMOSFET is dominated by the impact ionization near the channel-side STI edge, as in normal nLDMOSFETs, it can be suppressed by incorporating a recessed gate. Section III presents experimental results for the nLDMOSFETs in Figs. 1 and 2.

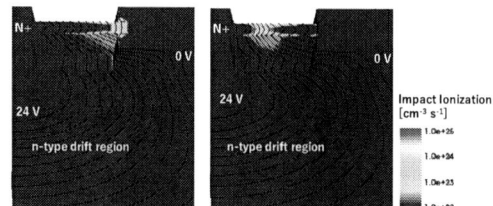

Fig. 5. TCAD simulation results for impact ionization (color contours) and electrostatic potential (line contours; 2-V steps) values for drain multi-finger STI 20-V nLDMOSFETs in Figs. 2(b) and (c) biased at $V_g = 0$ V and $V_d = 24$ V. Only the Si region is shown in these 3D simulation results.

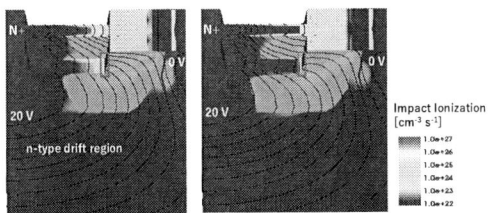

Fig. 6. TCAD simulation results for impact ionization (color contours) and electrostatic potential (line contours; 2-V steps) values for drain multi-finger STI 20-V nLDMOSFETs in Figs. 2(b) and (c) biased at $V_g = 1/2\ V_{gmax}$ and $V_d = 20$ V. Only the Si region is shown in these 3D simulation results.

III. MEASUREMENT RESULTS AND DISCUSSION

20-V nLDMOSFETs with recessed gates and a gate voltage rating of 3.3 V were fabricated using our 90 nm bulk BiCDMOS process [9]. Their DC characteristics and HCI degradation were evaluated.

A. DC characteristics

The changes in BV_{off} and R_{sp} for drain multi-finger STI nLDMOSFETs with the structures in Fig. 2 are plotted against Ws and Gd in Figs. 7 and 8, respectively. The result for nLDMOSFETs with the structures in Fig. 1 is shown for reference. It is apparent that Ws had a small influence on BV_{off} within the measurement range (Fig. 7(a)). Moreover, R_{sp} decreased with increasing Ws because the ratio of the N-type drift silicon region increased (Fig. 7(b)).

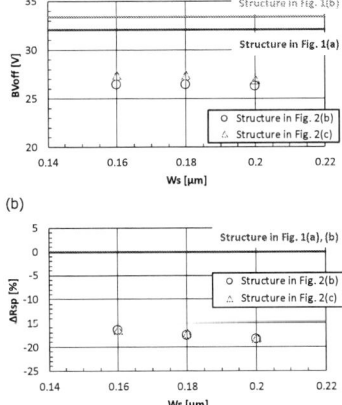

Fig. 7. Experimental (a) BV_{off} and (b) R_{sp} shifts as functions of Ws in drain multi-finger STI 20-V nLDMOSFETs measured at room temperature. The value of Gd was set to -0.12 μm.

When Gd becomes approximately 0 or more, BV_{off} decreases because the electric field near the channel side becomes higher (Fig. 8(a)). Furthermore, R_{sp} decreases with increasing Gd because the gate polysilicon in the drain drift silicon region forms an accumulation layer on the surface (Fig.

8(b)). However, there is only a slight difference in the parameter dependence between the structures with and without recessed gates. Thus, the recessed gate hardly affects the DC characteristics of nLDMOSFETs. A possible reason why the BV_{off} dependence is the same for the two structures is that the drain STI width is too small to relax the surface electric field despite the recessed gate shifting the worst impact ionization point (see Fig. 5). Moreover, a possible reason why R_{sp} is the same for the two structures is that the STI beside the recessed gate (about 0.1 μm) is too thick to form an accumulation layer on the silicon surface on the drain STI side.

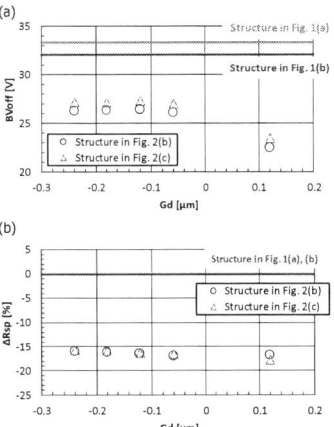

Fig. 8. Experimental (a) BV_{off} and (b) R_{sp} shifts as functions of Gd in drain multi-finger STI 20-V nLDMOSFETs measured at room temperature. The value of Ws was set to 0.16 μm.

The R_{sp}-BV_{off} trade-off for the structures in Figs. 1 and 2 is shown in Fig. 9. The BV_{off} value for the structure in Fig. 1(b) (Ls = Lt = 0.2 μm) is 1.5 V higher than that for the structure in Fig. 1(a), and R_{sp} is the same for the two structures. The improvement of BV_{off} is consistent with the TCAD simulation results shown in Fig. 3. The BV_{off} value for the structure in Fig. 2(c) (Ws = 0.16 μm, Gd = -0.12 μm) is 5 V lower and R_{sp} is 16.4% smaller compared to those for the structure in Fig. 1(a).

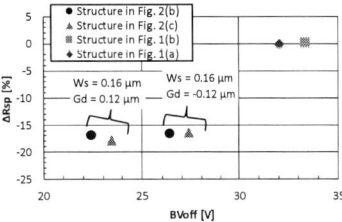

Fig. 9. Experimental R_{sp}-BV_{off} trade-off for 20-V nLDMOSFET structures in Figs. 1 and 2.

978-1-7281-0582-6/19 $31.00 © 2019 IEEE

B. HCI characteristics

HCI stress tests were performed at $T_a = -40$ °C, $V_d = 20$ V and I_{submax} because it is well known that lower temperatures accelerate HCI degradation. In Figs. 10 and 11, ΔI_d the maximum drain current (I_d) decrease in the monitored conditions ($V_d = 0$-8 V, $V_g = 3.3$ V) is plotted against stress time. HCI degradation for all 20-V nLDMOSFET structures is compared in Fig. 10. The values of Ls and Lt for the structures in Fig. 1(b) were both set to 0.2 μm. The values of Ws and Gd for the structures in Fig. 2 were set to 0.16 and -0.12 μm, respectively. As predicted by the TCAD simulations (Fig. 4), HCI degradation in the normal structure with a recessed gate (Fig. 1(b)) is suppressed compared with that in the conventional normal structure (Fig. 1(a)). The ΔI_d values after 1×10^4 s were 11% and 6% for the structures in Figs. 1(a) and (b), respectively. The HCI degradation in the drain multi-finger STI nLDMOSFETs (Fig. 2) is also suppressed compared with that in the conventional normal structure (Fig. 1(a)). However, there is no benefit to incorporating a recessed gate into the drain multi-finger STI nLDMOSFET. The ΔI_d values for these two structures after 1×10^4 s were both 8%.

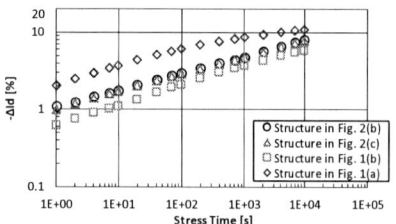

Fig. 10. Experimental ΔI_d values as a function of stress time at $V_d = 20$ V (I_{submax} condition) and $T_a = -40$ °C. The measurements were performed at a monitor gate bias of 3.3 V, with the drain voltage swept from 0 to 8 V.

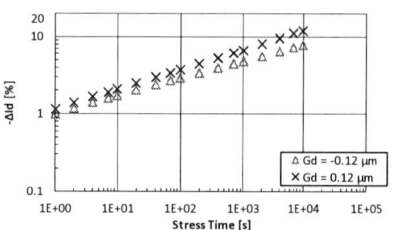

Fig. 11. Experimental ΔI_d values as a function of stress time at $V_d = 20$ V (I_{submax} condition) at $T_a = -40$ °C, with the same monitor conditions as those in Fig. 10.

HCI degradation was compared between Gd = -0.12 and 0.12 μm in the drain multi-finger STI nLDMOSFETs with recessed gates (Fig. 11). The HCI degradation at Gd = 0.12 μm is worse than that at Gd = -0.12 μm. The results in Figs. 10 and 11 indicate that the silicon surface near the channel is the main degradation area, not the channel-side STI edge, as in the normal structures in Fig. 1. It is presumed that the recessed gate does not affect the electric field at the silicon surface.

IV. CONCLUSION

The effects of recessed gates on nLDMOSFETs with an STI section in drain regions were investigated using TCAD simulations and experiments. An improvement in BV_{off} and a suppression of HCI degradation by the recessed gate were found for a normal nLDMOSFET structure because the recessed gate relaxes the electric field near the channel-side STI edge. No improvement by the recessed gate was found for a drain multi-finger STI nLDMOSFET structure because the transistor dimensions, including the STI width and the thickness of the STI beside the recessed gate, are not designed to optimize the DC characteristics and the recessed gate does not relax the electric field at the silicon surface near the channel. The results show that incorporating a recessed gate is beneficial for a normal STI nLDMOSFET but not for a drain multi-finger STI nLDMOSFET.

ACKNOWLEDGMENTS

The authors would like to thank H. Sugiura, Y. Iwanaga, A. Komuro, K. Nitta and S. Tokumitsu for BiCDMOS process integration.

REFERENCES

[1] H. Fujii, T. Mori and T. Ipposhi, "A Recessed Gate LDMOSFET for Alleviating HCI Effects," Proc. of ISPSD16, pp.167-170.

[2] H. Liu, Z. Jhou et al., "A Novel High-Voltage LDMOS with Shielding-Contact Structure for HCI SOA Enhancement," Proc. of ISPSD17, pp.311-314.

[3] T. Mori, H. Fujii, S. Kubo and T. Ipposhi, "Investigation into HCI Improvement by a Split-Recessed-Gate Structure in an STI-based nLDMOSFET," Proc. of ISPSD17, pp.459-462.

[4] T. Mori, S. Kubo and T. Ipposhi, "A Novel Divided STI-based nLDMOSFET for Suppressing HCI Degradation under High Gate Bias Stress," Proc. of ISPSD18, pp.299-302.

[5] K. Takahashi, K. Komatsu et al., "Hot-carrier Induced Off-state Leakage Current Increase of LDMOS and Approach to Overcome the Phenomenon," Proc. of ISPSD18, pp.303-306.

[6] S. Poli, S. Reggiani et al., "Numerical investigation of the total SOA of trench field-plate LDMOS devices," Proc. of SISPAD10, pp.111-114.

[7] J. Jang, K. Cho et al., "Interdigitated LDMOS," Proc. of ISPSD13, pp. 245-248.

[8] R. Ye, S. Liu et al., "ESD Failure Analysis and Robustness Improvement for Multi-STI-Finger LDMOS Used as Output Device," Proc. of ISPSD18, pp.339-342.

[9] H. Fujii, S. Tokumitsu et al., "A 90nm Bulk BiCDMOS Platform Technology with 15-80V LD-MOSFETs for Automotive Applications," Proc. of ISPSD17, pp.73-76.

Proceedings of the 31st International Symposium on Power Semiconductor Devices & ICs
May 19-23, 2019, Shanghai, China

Fast-Switching Lateral IGBT with Trench/Planar Gate and Integrated Schottky Barrier Diode (SBD)

Licheng Sun, Baoxing Duan*, Yandong Wang and Yintang Yang

Key Laboratory of the Ministry of Education for Wide Band-Gap Semiconductor Materials and Devices, School
of Microelectronics, Xidian University,
Xi'an 710071, China
email: bxduan@163.com

Abstract—A novel fast-switching lateral IGBT with trench/planar gate and integrated Schottky barrier diode (SBD) is proposed and studied in this paper by TCAD simulation. The proposed LIGBT consists of the trench/planar gate (TP) at the cathode and an integrated SBD at the anode to reduce turn-off time and maintain a low forward voltage drop. The integrated SBD provides an extra electron extraction path and the additional trench gate enhances the injection of the N^+-cathode. The insulated oxide pillar between the N-buffer and integrated SBD further reduces the snapback voltage. The simulation results show that the turn-off time of the conventional LIGBT is 52.4% larger than that of the proposed LIGBT. Moreover, the latch current density of the proposed LIGBT is increased by nearly 200% compared to that of the conventional LIGBT which means that the proposed LIGBT can improve the latch immunity.

Keywords—Lateral IGBT, fast-switching, integrated Schottky barrier diode (SBD), trench/planar gate

I. INTRODUCTION

The insulated-gate bipolar transistor (IGBT) concept was discovered and developed in the early 1980s to provide an improved alternate power device to bipolar power transistors, and the lateral IGBT has been widely used in power ICs, due to its characteristics of low on-state voltage drop under the condition of high current and input impedance [1-4].

In order to reduce power consumption and the turn-off losses of the IC, LIGBT should maintain a low forward voltage drop while minimizing switching time. Several different structures have been reported to make a trade-off between on-state voltage drop and turn-off time [5-11]. In order to improve turn-off speed of LIGBT, one principle is providing an effective path for electron extraction. Although some of the solutions reported in existing references can achieve faster switching speed, the cost is often higher forward voltage drop or larger snapback voltage. In this paper, based on extra extraction principle, the innovative lateral IGBT, as shown in Fig.1, with trench/planar gate and integrated Schottky barrier diode is proposed for the first time.

The proposed structure with trench/planar gate and integrated SBD in this paper can be applied to high-voltage monolithic IC applications for its fast-switching and high latch-up immunity, compared to the conventional LIGBT,

This work was supported by National Basic Research Program of China (Grant No. 2015CB351906), Science Foundation for Distinguished Young Scholars of Shaanxi Province (Grant No. 2018JC-017).

which shows that great improvements in the tradeoff between forward voltage drop and turn-off time can be achieved.

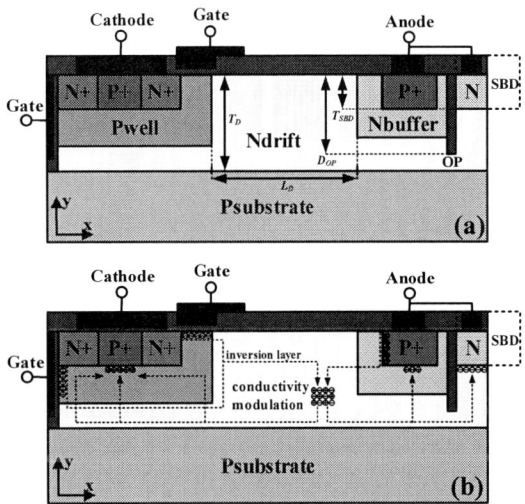

Fig. 1. The schematic cross-section and flowing paths of carriers for TP-SBD LIGBT.

II. DVICE STRUCTURE

The proposed LIGBT consists of the trench/planar gate (TP) at the cathode and an integrated Schottky barrier diode (SBD) at the anode. At the cathode region, the TP structure can enhance the conductivity modulation of the drift region in the on-state, further improving the tradeoff between on-state voltage drop (V_{on}) and turn-off time (T_{off}). At the anode region, the integrated SBD structure can provide one effective path for electron extraction, and the insulated oxide pillar between the N-buffer and SBD can make the snapback voltage (V_{SB}) of the forward I-V characteristics easy to reduce to a low value.

The key process steps presented here for TP-SBD LIGBT can be implemented by feasible and mature trench BCD technology. (a) N-buffer implantation. (b) P^+-anode and P-well implantation. (c) Implantations for the cathode and anode region. (d) Trenches etching and oxidation (Formation

978-1-7281-0582-6/19 $31.00 © 2019 IEEE   379

of trench gate and insulated oxide pillar). (e) Formation of ohmic contacts and Schottky contact.

Table I. KEY PARAMETERS IN DEVICE SIMULATIONS

Symbol	Description	Con	P-SBD	TP-SBD
BV	Breakdown voltage (V)	227	227	231
N_D	N-drift doping (cm^{-3})	1.5×10^{15}	1.5×10^{15}	1.5×10^{15}
N_{SBD}	N-region doping in integrated SBD (cm^{-3})	/	4×10^{17}	4×10^{17}
N_{BUF}	N-buffer doping (cm^{-3})	5×10^{17}	5×10^{17}	5×10^{17}
N_{SUB}	P-substrate doping (cm^{-3})	1×10^{14}	1×10^{14}	1×10^{14}
D_{OP}	Insulated oxide pillar depth (μm)	/	3.5	3.5
T_{SBD}	N-region thickness in integrated SBD (μm)	/	1.0	1.0

Fig.1(a) shows the schematic cross-section of TP-SBD LIGBT. Compared with the conventional shorted anode structure [5-7], during the turn-on, the integrated SBD also have a small turn-on voltage, which will be helpful for avoiding obtaining a large V_{SB}. Meanwhile, the integrated SBD also can provide an extra path for electron extraction which is the same as the conventional shorted anode structure. The approach of TP has already been explored for different references [12-14]. The trench gate can reduce layout area and remove the JFET effect existing in the pinch-off region. It is clear that the TP can increase an extra path for electron injection, compared to the conventional trench or planar gate structure, which will be helpful for improving the injection efficiency of N$^+$-cathode.

Fig.1(b) shows the schematic flowing paths of electrons and holes for TP-SBD LIGBT. When the device is turned on, at the cathode region, the phenomenon of hole current concentration is alleviated by the introduction of TP, which can improve the latch-up immunity of TP-SBD LIGBT. When the device is turned off, at the anode region, the integrated SBD is helpful for extracting electrons out of TP-SBD LIGBT, which can reduce the T_{off} increased by the recombination of electron-hole pairs.

III. RESULTS AND DISCUSSION

To study the characteristics of the novel LIGBT structure, two-dimensional numerical simulation was performed by ISE TCAD [15] and key parameters in device simulations are listed in Table I. The simulated I-V characteristics of different devices are shown in Fig.2(a). Due to the existence of integrated SBD, just like the shorted anode structure, the negative differential resistance (NDR) region is also introduced into the I-V characteristics of P-SBD LIGBT and TP-SBD LIGBT. Thus, the insulated oxide pillar is used to further reduce the V_{SB}. As shown in Fig.2(b), without integrated SBD and insulated oxide pillar, V_{SB} will get a considerably large value. Moreover, the TP structure brings a lower V_{on}, compared with conventional LIGBT and P-SBD LIGBT, while it is clear that the V_{SB} of TP-SBD LIGBT is maintained at a lower value. When the drift region length remains the same, the breakdown voltage (BV) of the three devices is almost the same as 230V, as shown in Fig.2(c). The introduction of TP structure and integrated SBD can improve the performance of forward conduction and switching characteristics of the device without bringing great effect on breakdown characteristics.

Fig.3 analyzes the influences of D_{OP} value and T_{SBD} value on V_{SB} for TP-SBD LIGBT. The electron path at the anode is rerouted by the insulated oxide pillar. The electrons from the N$^+$-cathode can only flow toward the integrated SBD alongside the insulated oxide pillar. Thus, the voltage drop across the integrated SBD and cathode end can easily be adjusted. The increased D_{OP} can bring a decreased V_{SB} as shown in Fig.3(a), because the increased D_{OP} makes the electron path length longer. Similarly, the decreased T_{SBD} also increases the length of the electron path. As T_{SBD} decreases, the curve in Fig.3(b) shows a decreasing trend. In order to suppress V_{SB} to a lower value, a larger D_{OP} value and a smaller T_{SBD} value should be applied in TP-SBD LIGBT.

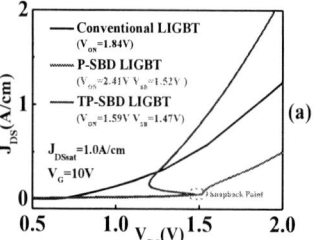

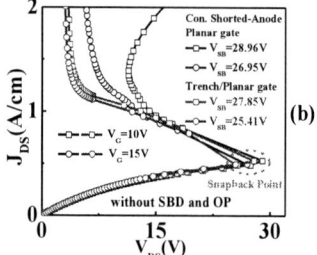

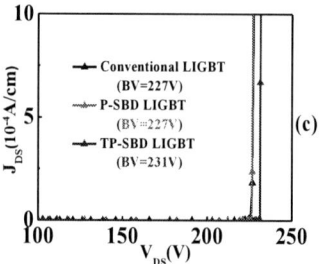

Fig. 2. (a) Simulated forward I-V characteristics and (c) simulated breakdown characteristics for conventional LIGBT, P-SBD LIGBT, and TP-SBD LIGBT. (b) Simulated forward I-V characteristics for P-SBD LIGBT and TP-SBD LIGBT without the insulated oxide pillar.

Fig.4 shows the dynamic characteristics of different LIGBTs. All devices are compared under the same drift region length, that is, the BVs of the three devices are roughly equivalent. The turn-off waveforms of three LIGBT devices with a resistive load are shown in Fig.4(b). It is remarkable that TP-SBD LIGBT achieves the best tradeoff between V_{on} and T_{off}. Compared to conventional LIGBT, TP-SBD LIGBT can hold a lower V_{on} of 1.59V while gain a lower T_{off} of 131.9ns. Although P-SBD LIGBT achieves the lowest T_{off} of 124.9ns, while brings the highest V_{on} of 2.41V. Different from other LIGBTs, TP-SBD LIGBT not only enhances the conductivity modulation of the drift region

978-1-7281-0582-6/19 $31.00 © 2019 IEEE

during turn-on, but also accelerates the extraction of electrons during turn-off.

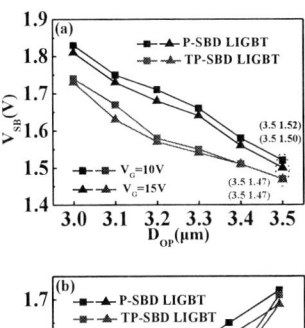

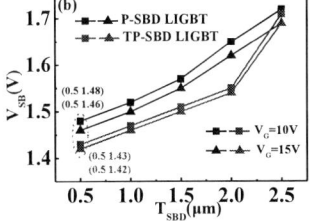

Fig. 3. (a) Simulated dependence of V_{SB} on D_{OP} for P-SBD LIGBT and TP-SBD LIGBT. (b) Simulated dependence of V_{SB} on T_{SBD} for P-SBD LIGBT and TP-SBD LIGBT. L_D=16μm, T_D=4μm.

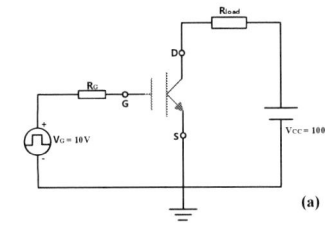

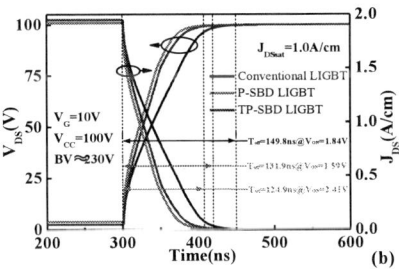

Fig. 4. (a) The switching circuit with a resistive load for LIGBT. (b) Turn-off characteristics for conventional LIGBT, P-SBD LIGBT, and TP-SBD LIGBT.

Fig.5(a) shows that the TP-SBD LIGBT features the best tradeoff performance between V_{on} and T_{off} among all three LIGBTs. Under the same V_{on}, the T_{off} value of conventional LIGBT is 52.4% larger than that of TP-SBD LIGBT at the forward current density (J_{DSsat}) of 1.0A/cm. Meanwhile, compared with TP-SBD LIGBT, P-SBD LIGBT gains a 38.9% increase in V_{on} at J_{DSsat} = 1.0A/cm with almost the same T_{off}. Fig.5(b) shows the simulated static latch-up I–V characteristics for three devices. Due to the hole current path at the cathode is optimized as shown in Fig.1(b), the current concentration phenomenon at the time of large current is avoided. It shows that the introduction of TP brings almost the same latch-up voltage (V_{LU}), compared with conventional

LIGBT, while gains a much larger latch-up current density (J_{LU}).

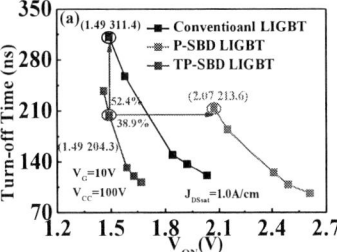

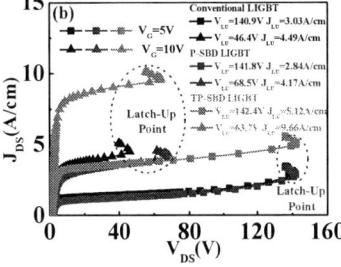

Fig. 5. (a) Tradeoff curves between V_{on} and turn-off time for conventional LIGBT, P-SBD LIGBT, and TP-SBD LIGBT. (b) Simulated static latch-up I–V characteristics for conventional LIGBT, P-SBD LIGBT, and TP-SBD LIGBT.

IV. CONCLUSION

The novel LIGBT with trench/planar gate and integrated SBD is proposed in this paper. The presence of trench/planar gate at the cathode is helpful for obtaining lower forward voltage and higher latch-up immunity. During turn-off, the integrated SBD structure provides an extra electron extraction path while suppressing snapback voltage. The proposed structure meets the requirements for fast switching speed while maintaining low on-state losses and high latch-up immunity.

REFERENCES

[1] B. J. Baliga, "Fundamentals of Power Semiconductor Devices", Springer-Science, New York, 2008.

[2] Wilson W.T. Chan, Jonhnny K.O. Sin, and S. Simon Wong, "An effective Cross-Talk Isolation Structure for Power IC Applications," Proceedings of International Electron Devices Meeting, pp. 971-947, Dec. 1995, doi: 10.1109/IEDM.1995.499378.

[3] Sakurai. N, M. Mori and T. Yatsuo, "High speed, high current capacity LIGBT and diode for output stage of high voltage monolithic three-phase inverter IC," Proceedings of the 2nd International Symposium on Power Semiconductor Devices and Ics. ISPSD '90, pp. 66-71, April. 1990, doi: 10.1109/ISPSD.1990.991060.

[4] Gough P A, Simpson M R, Rumenik V, "Fast switching lateral insulated gate transistor". 1986 International Electron Devices Meeting, pp. 218-221, Dec. 1986, doi: 10.1109/IEDM.1986.191153.

[5] Simpson M R. "Analysis of negative differential resistance in the I–V characteristics of shorted-anode LIGBT". IEEE Transactions on Electron Devices, vol. 38, no. 7, pp. 1633-1640, Jul. 1991, doi: 10.1109/16.85160.

[6] Zhu, Jing, et al. "Electrical Characteristic Study of an SOI-LIGBT With Segmented Trenches in the Anode Region". IEEE Transactions on Electron Devices, vol. 63, no. 5, pp. 2003-2008, April. 2016, doi: 10.1109/TED.2016.2545681.

[7] Juhyun Oh, Dae Hwan Chun, Reum Oh, Hyun Soo Kim, "A snap-back suppressed shorted-anode lateral trench insulated gate bipolar transistor (LTIGBT) with insulated trench collector," 2011 IEEE International Symposium on Industrial Electronics, pp.1367-1370, June. 2011, doi: 10.1109/ISIE.2011.5984358.

[8] Qin Z, Sankara Narayanan E M, "npn controlled lateral insulated gate bipolar transistor," Electronics Letters, vol. 31, no. 23, pp. 2045-2047, Nov. 1995, doi: 10.1049/el:19951344.

[9] Nakagawa A, Yamaguchi Y, Watanabe K. "Two types of 500 V double gate lateral n-ch bipolar-mode MOSFETs in dielectrically isolated P and N silicon islands". Technical Digest, International Electron Devices Meeting, pp.817-820, Dec. 1988, doi: 10.1109/IEDM.1988.32936.

[10] Fang J, Li Z, Li H. "High speed LIGBT with localized lifetime control by using high dose and low energy Helium implantation". 2001 6th International Conference on Solid-State and Integrated Circuit Technology. Proceedings (Cat. No.01EX443), pp.166-169, Oct. 2001, doi: 10.1109/ICSICT.2001.981448.

[11] Chen Wensuo, Xie Gang, Zhang Bo, Li Zehong, Li Zhaoji, "Novel lateral IGBT with n-region controlled anode on SOI substrate".

Chinese Journal of Semiconductors, pp.47-50, 2009, doi: 10.1088/1674-4926/30/11/114005.

[12] Fu Qiang, Zhang Bo, Luo Xiaorong, and Li Zhaoji, "A dual-gate and dielectric-inserted lateral trench insulated gate bipolar transistor on a silicon-on-insulator substrate," Chin. Phys. B, vol. 22, no. 7, pp. 473-477, July. 2013, doi: 10.1088/1674-1056/22/7/07730.

[13] Liu, Meihua, F. X. C. Jiang, and X. Lin, "Dual trench gates SOI LIGBT with low conduction loss," 2014 12th IEEE International Conference on Solid-State and Integrated Circuit Technology (ICSICT), Oct. 2014, doi: 10.1109/ICSICT.2014.7021687.

[14] Yi-Tao He, Ming Qiao, Zhuo Wang, Bo Zhang, Zhao-Ji Li, "A low turnoff loss SOI LIGBT with p-buried layer and double gates," 2016 13th IEEE International Conference on Solid-State and Integrated Circuit Technology (ICSICT), pp. 1092 - 1094, Oct. 2016, doi: 10.1109/ICSICT.2016.7998660.

[15] Integrated Systems Engineering, Zurich, Switzerland, DESSIS, ISE TCAD Release 10.0 User's Manual, 2004.

Proceedings of the 31st International Symposium on Power Semiconductor Devices & ICs
May 19-23, 2019, Shanghai, China

Diode Reverse Recovery Characteristics of a Shielded-Gate Trench Power MOSFET

Zia Hossain, Raghuram Mullapudi*, Harshad Surdi**

ON Semiconductor, 5005 East McDowell Road, Phoenix, AZ 85008
E-Mail: Zia.Hossain@onsemi.com, Tel: 602-244-4612, Fax: 602-244-6716
*Rochester Institute of Technology, Rochester, NY
**Arizona State University, Tempe, AZ

Abstract – A better understanding of the diode reverse-recovery behavior of power MOSFET is critical when designing a power electronics systems with faster switching speed, minimum switching losses, and higher system efficiency. In order to study the switching transient times, one needs to focus on both switching device architecture and circuit parasitic inductances and capacitances. The goal of this paper is to study the MOSFET device structure and circuit elements, and their impacts on reverse recovery characteristics when MOSFET is used as a Synchronous Rectifier (SR). This paper will present the role of the drain-to-shield capacitance of the shielded-gate MOSFET on the reverse recovery current and voltage transients. Mixed-mode TCAD simulation is performed to study the intricate reverse-recovery behavior observed in the shielded-gate trench MOSFETs.

I. Introduction

Shielded-gate trench MOSFETs are well known in the industry for its lower $R_{DS(ON)} \times$ Area and lower $R_{DS(ON)} \times Q_{GD}$ figure of merits (FoMs), and are also widely used in the low to medium voltage power applications ranging from 25 V to 200 V. The typical cross sections of such FETs are shown in Fig. 1, one with shallower trench depth and NEPI-1 structure, and the other with deeper trench depth and NEPI-2 structure. Both these structures are optimized to achieve a targeted breakdown voltage of 60V. In power device applications, transient switching times and its losses are critical factors to be concerned about when designing a power electronics systems. Transient switching times, both turn on and turn off transients limit the operating frequency of the circuit and cause switching losses, affecting both switching power device and circuit efficiency. It is important to achieve faster switching with minimal concomitant losses when power devices are in operation. Hence, a solid understanding of diode reverse recovery characteristics is imperative for design engineers to avoid unintended transients and their losses. It is worth mentioning that diode reverse-recovery characteristics are largely dependent on both device and circuit parameters, such as gate resistance, gate capacitances, and parasitic inductances. The reverse recovery test circuit used for both simulation and experiment is shown in Fig. 2. The test circuit is built into the mixed-mode TCAD simulation environment

using Synopsys Sentaurus tool [1]. The test is carried out with a two-pulse signal applied on the gate terminal of the control FET (bottom FET) as a switching device, and the top FET as a synchronous rectifier, by shorting the gate terminal to the source terminal and utilizing the pbody-to-drain (*p-n*) diode of the FET as a DUT to study the reverse recovery effect.

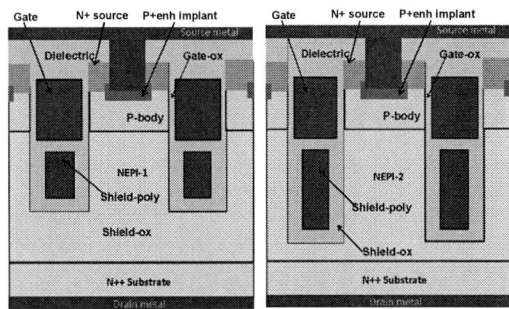

Fig.1. Typical cross-sections of a shielded-gate trench MOSFET (left - shallow trench depth structure, right – deep trench depth structure).

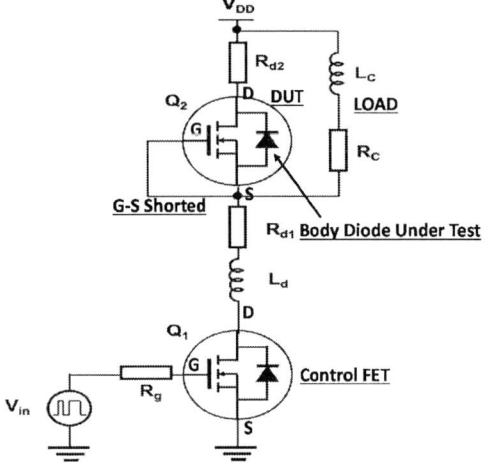

Fig.2. Test circuit used for both TCAD and experiment, showing control FET (bottom) and G-S shorted FET (top) under test (DUT).

978-1-7281-0582-6/19 $31.00 © 2019 IEEE

At the instant of 1st pulse start, the control FET is turned on to develop the load current through the load inductor, L_c ($L_c \gg L_d$), and as soon as the desired load current (I_F) is reached, the control FET is turned off to commutate the current from the control FET to the pbody-to-drain (p-n) diode of the top FET as forward current (I_F). At the instant of 2nd pulse start, the control FET is turned on again to commutate the diode current gradually to the control FET current when the full supply voltage, V_{DD} is supported by the parasitic inductance, L_d. As a result, the current through the control FET rises while the current through the diode falls at the same rate of $(di/d)_{Forward} = V_{DD}/L_d$. When the diode current falls or crosses the zero current line, the diode turns off and begins to block the supply voltage. However the diode current keeps on rising in the negative direction and reaches to the peak reverse-recovery current, I_{RM}, when the voltage across the diode reaches to V_{DD}. The reversal of the current direction is due to the sweep-out of the plasma from the epi region, established when the diode was in conducting state. The diode finally turns off when the diode current finally drops to zero at the rate of $(di/dt)_{Reverse}$, while blocking the full supply voltage plus voltage overshoot. The reverse-recovery process of the p-i-n diode is described in detail in [2].

II. Shield-to-Drain Capacitance and its Significance on Reverse Recovery

During the period of reverse recovery action, current through all four terminals of the trench FET (gate, p-body, shield, drain electrodes) is plotted in Fig. 3 to illustrate the reverse-recovery current waveform. It is observed that gate current is very minimal compared to p-body junction current and shield current. Shield current is the displacement current between the drain electrode and the shield electrode, in which the shield electrode has the same potential as source electrode of the FET. The shield current seems to be the most dominant current in the total reverse recovery current. Shield terminal or shield current plays a major role in the total reverse recovery waveform in terms of both peak reverse recovery current (I_{RM}) and shape or softness, "S" factor of the waveform ($S = t_b/t_a$). As the value of "S" increases, the slope $(di/dt)_{Reverse}$ also transitions from snappier to softer recovery of the diode. Figs. 4 and 5 show the reverse-recovery current components for shallow and deep trench depth structures, respectively, for a given $(di/dt)_{Forward}$ of 2000 $A/\mu s$. Due to the higher drain-to-shield capacitance of the deep trench structure, the total reverse-recovery current is basically the shield current as shown in Fig. 5. In contrast, in the shallow trench depth structure, the total reverse-recovery current is comprised of both shield and p-n (pbody-to-drain) junction currents. Capacitances of the drain-to-shield and pbody-to-drain junction are plotted with the function of drain voltage for both deeper and shallower trench depths in Fig. 6 for further investigation. Shield-to-drain capacitance is significantly higher than the pbody-to-drain junction capacitance, and hence the shield current has the dominant effect on the total reverse-recovery current waveform, as compared to the p-n junction current. Deeper trench has a higher and more uniform capacitance profile compared to the

shallower trench with respect to the drain voltage variation, contributing to softer reverse recovery and reduced voltage overshoot.

Fig. 7 shows the experimental reverse recovery current and voltage waveforms for a di/dt of 1000 $A/\mu s$.

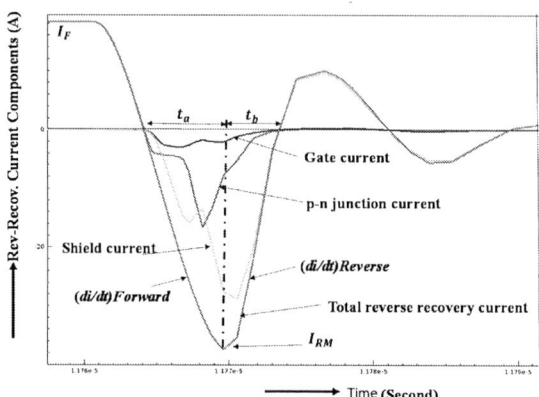

Fig. 3. Reverse-recovery current components through all terminals (source/body, shield, gate, drain electrodes).

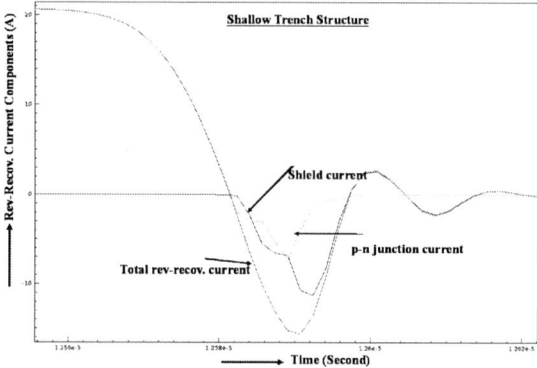

Fig. 4. Reverse-recovery current components for shallow trench depth structure.

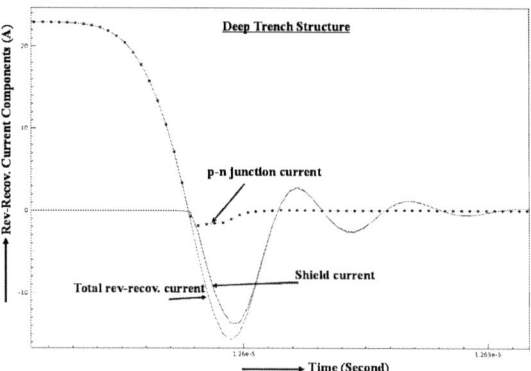

Fig. 5. Reverse-recovery current components for deep trench depth structure.

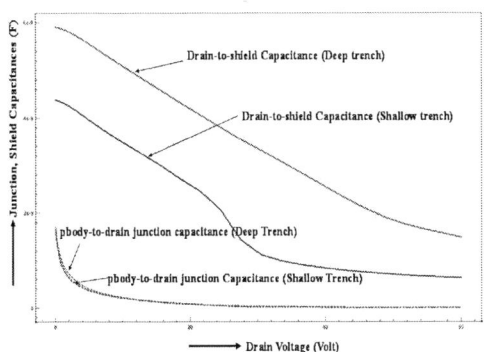

Fig. 6. Variation of drain-to-shield and pbody-to-drain capacitances with drain voltage sweep for deep and shallow trench structures.

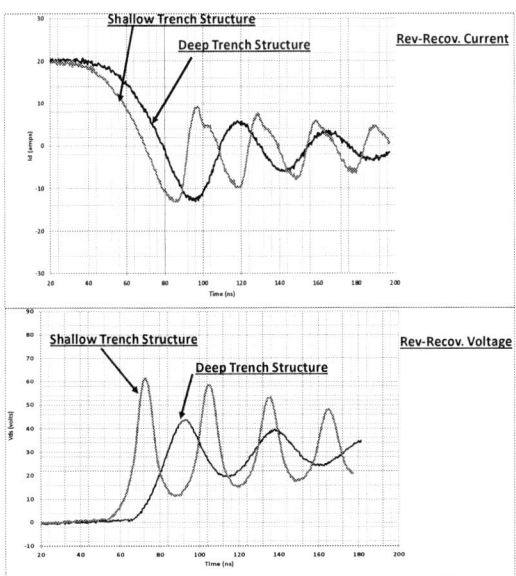

Fig. 7. Experimental reverse-recovery current and voltage waveforms for a di/dt of 1000 $A/\mu s$.

The experimental data has the same trend as TCAD simulation data, validating the idea that deeper trench with higher shield-to-drain capacitance has a softer recovery, resulting in slower voltage rise and reduced voltage overshoot. The forward di/dt slope can also be controlled by adjusting the gate signal rise time or the gate resistance of the control FET. Forward di/dt slope has an important role in the reverse-recovery waveform in terms of peak reverse current, I_{RM}, reverse di/dt, voltage overshoot, V_{RM}. TCAD simulation is performed to see the impacts of forward di/dt on the reverse-recovery characteristics, as shown in Table 1 for both shallow and deep trench structures. It is observed that with decreasing rise time, control FET is turning on and off faster, resulting in faster forward and reverse di/dt, higher peak

reverse current, I_{RM}, and higher voltage overshoot, V_{RM}. These changes, such as reverse di/dt and voltage overshoot become worse for shallower trench structure, as compared to the deeper trench structure.

Rise time (μs)	Forward Current I_F (A)	Forward di/dt (F) (A/μs)	Reverse di/dt (R) (A/us)	Rev. Peak Current I_{RM} (A)	Rev. Peak Voltage V_{RM} (V)
23.4	17.7	7200	25000	-37	-111
46.8	18.5	7000	20000	-36	-108
93.6	19.3	5000	16000	-32	-86
187.2	20.4	3700	8800	-24	-55
370	20.5	2200	3300	-15	-32

Rise time (μs)	Forward Current I_F (A)	Forward di/dt (F) (A/μs)	Reverse di/dt (R) (A/us)	Rev. Peak Current I_{RM} (A)	Rev. Peak Voltage V_{RM} (V)
23.4	18.5	8100	15400	-41	-85
46.8	18.7	6300	14000	-39	-83
93.6	19.3	5350	10000	-33	-69
187.2	20.5	3900	5500	-24	-47
370	22.9	2460	2000	-16	-30

Table I. Impact of forward di/dt variation on the reverse-recovery current and voltage characteristics (Top- Shallow trench depth structure; Bottom- Deep trench depth structure).

III. Conclusion

It is evident from both experimental and TCAD simulation data that the shield capacitance plays a major role in reverse-recovery characteristics when shielded-gate trench MOSFET is used as a synchronous FET. The reverse recovery time or the shape of the waveform can be altered by changing the epi doping profile and the shield structure of the MOSFET, essentially altering the shield capacitance profile of the MOSFET. Nonetheless, a trade-off between lowest R_{dson} and optimum shield capacitance profile exists while optimizing the reverse-recovery characteristics. Moreover, the reverse-recovery parameters, such as peak reverse-recovery current, reverse di/dt, and voltage overshoot can also be controlled by adjusting the forward di/dt. Faster switching time or decreasing rise time reduces reverse-recovery time and loss, however at the expense of snappier diode reverse recovery and increased voltage overshoot. Design engineers have to make a proper choice of selecting epi doping and trench cell structure when designing a MOSFET to meet the device and circuit tradeoffs.

Acknowledgment

The authors would like to thank Roman Gurevich (ON Semiconductor) for the reverse-recovery test setup and the measurement data.

References

[1] Sentaurus TCAD Process/Device Simulation Tools – SPROCESS, SDEVICE.
[2] Stephen Linder, *Power Semiconductors*, EPFL Press, 2006, ISBN 2-940222-09-6.

(This page is intentionally left blank.)

Proceedings of the 31st International Symposium on Power Semiconductor Devices & ICs
May 19-23, 2019, Shanghai, China

Channel-off Avalanche Instability in SOI Lateral IGBT at Low Temperature: Mechanism and Optimization Schemes

Jie Ma[1], Long Zhang[1], Jing Zhu[1], Shilin Cao[1], Ankang Li[1], Yanqin Zou[1], Weifeng Sun[1]*,
Yan Gu[2], Sen Zhang[2], Yunwu Zhang[3] and Zhuo Yang[4]

[1]National ASIC System Engineering Research Center, School of Electronic Science & Engineering,
Southeast University, Nanjing, China. *Email: swffrog@seu.edu.cn.
[2]CSMC Technologies Corporation, Wuxi, China.
[3]Wuxi i-driver Electronics Co., Ltd., Wuxi, China.
[4]Wuxi NCE POWER Co., Ltd., Wuxi, China.

Abstract—In this paper, the off-state avalanche stability is investigated and a time-dependent avalanche instability is observed at -40°C in SOI lateral IGBT. Combining results from measurements with those from TCAD simulations, the mechanism for avalanche instability is revealed. Two optimization schemes are proposed to suppress or eliminate the unstable avalanche phenomenon. The unstable avalanche can be suppressed by employing a thick buried oxide layer and V_{CE} shift can be reduced from 54V to 11V. Another scheme is that replacing P-type substrate by N-type substrate, and then V_{CE} shift can be completely eliminated.

Keywords—SOI, IGBT, off-state, avalanche stability, low temperature.

I. INTRODUCTION

Silicon-on-insulator lateral insulated gate bipolar transistors (SOI-LIGBTs) used as power stage devices in monolithic inverter ICs [1-3] have drawn much attention for their multiple merits: high current density, easy gate control and excellent isolation properties. In recent years, many efforts have been devoted to promoting performances of SOI-LIGBTs at room temperature and high temperature (125°C or 150°C) while few literatures focus on its electrical characteristics at low temperature (e.g. -40°C). Inductive load switching is one of the most common operating conditions for IGBTs (vertical or lateral). During the clamped or unclamped inductive turn-off, the IGBT endures an avalanche impact process (usually called dynamic avalanche) to conduct the load current at the time instant when gate-emitter voltage falls below the threshold voltage. The unstable avalanche may eventually induce current filaments or waveforms oscillations [4-5], which strongly affects the ruggedness of devices.

This paper investigated the off-state avalanche stability at low temperature (-40°C) in SOI-LIGBT. The mechanism and origin for collector-emitter voltage (V_{CE}) shift during dynamic avalanche are discussed combining results from measurements with those from TCAD simulations.

II. DEVICE STRUCTURE AND MEASUREMENT SETTING

Fig. 1 shows the schematic view of SOI-LIGBT and photograph of the fabricated device. P-sub and N-drift doping concentration are $1.36\times10^{15}cm^{-3}$ and $8.3\times10^{14}cm^{-3}$, respectively. Filed plates are adopted to alleviate electric

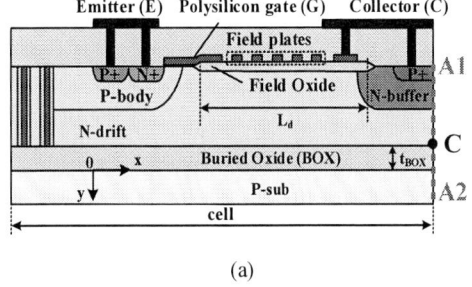

(a)

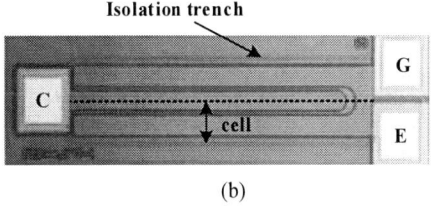

(b)

Fig. 1. (a) Schematic view of the SOI-LIGBT cell. (b) Photograph of the fabricated device.

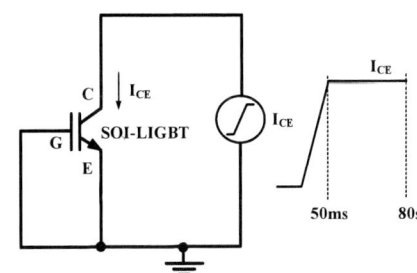

Fig. 2. Circuit used for avalanche stability measurements.

field crowding on the silicon surface. Two deep-oxide trenches are arranged at the emitter side to isolate the SOI-LIGBT from other devices. As shown in the Fig. 2, a convenient method is adopted to evaluate the avalanche stability. Collector-emitter current (I_{CE}) of 1μA is imposed to collector electrode. I_{CE} increases linearly and reaches 1μA in 50ms. The samples are measured at ambient temperatures of -40°C, 25°C and 125°C. At each temperature condition, fresh

978-1-7281-0582-6/19 $31.00 © 2019 IEEE   387

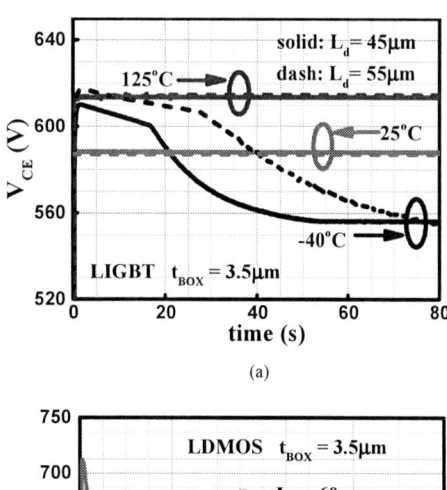

(a)

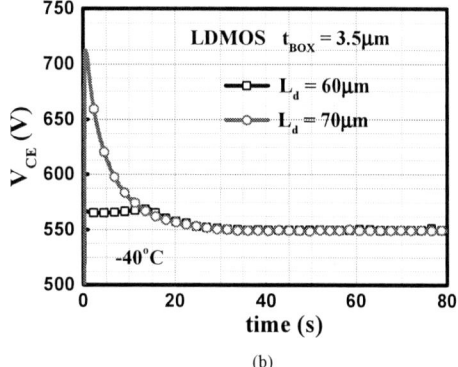

(b)

Fig. 3. Measured dependence of V_{CE} on time of (b) SOI-LIGBT and (c) SOI-LDMOS.

samples are used. The devices are forced into channel-off avalanche condition with gate and emitter grounded. Note that the original SOI-LIGBT with buried oxide layer thickness (t_{BOX}) of 3.5μm and N-drift length (L_d) of 45μm is designed to break down at collector-side bottom (upper surface of the BOX layer) to avoid effects induced by surface breakdown [6]. The collector-emitter voltage (V_{CE}) is detected and recorded in 80s.

III. RESULTS AND DISCUSSION

Fig. 3 (a) and (b) show the measured dependence of V_{CE} on time. It is found that time-dependent V_{CE} shift phenomenon occurs only at -40 °C while the devices exhibit stable V_{CE}s at 25°C and 125°C. At -40 °C, V_{CE} shifts down over time and eventually can be stabilized. Devices with different L_d exhibit different initial V_{CE} but a same stable value. The initial V_{CE} can be increased with the increase of L_d, indicating that the initial V_{CE} closely depends on the lateral breakdown voltage (BVL). This time-dependent phenomenon also can be observed in SOI-LDMOS. At -40 °C, V_{CE} shifts down over time and eventually can be stabilized.

Fig. 4 (a) defines the time instants on the V_{CE} curve. t1 is the time instant when V_{CE} rises up to its initial value at -40ºC and t2 is the time instant when V_{CE} begins to shift down. t4 is

This work was supported by the national natural science foundation of China (61874026, 61804026, 61504025), the national key research and development plan (2017YFB0402904), postdoctoral research funding of Jiangsu (2018K001A), the key research and development plan of Jiangsu (BE2018003-3) and the fundamental research funds for the central universities.

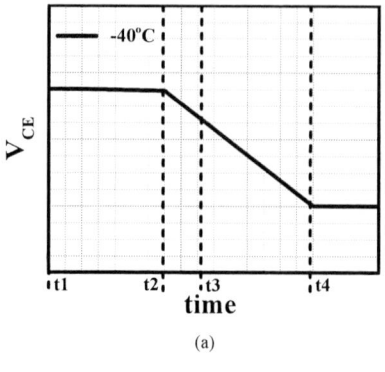

(a)

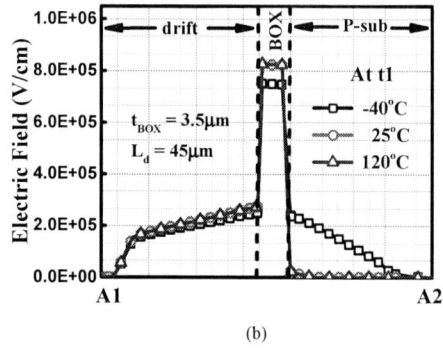

(b)

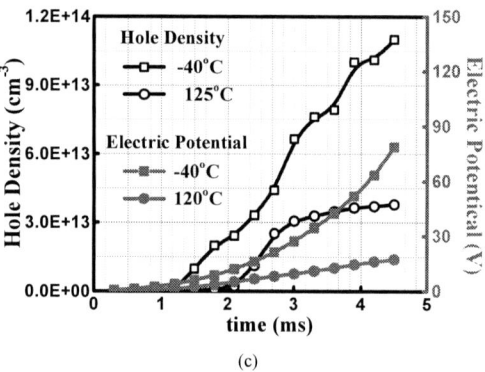

(c)

Fig. 4. (a) Definition of time instants during V_{CE} shifting down. (b) Electric filed profiles along the line A1-A2 at t1 in SOI-LIGBT with t_{BOX} = 3.5μm and L_d = 45μm. (c) Dependences of hole density and electric potential on time at point C in SOI-LIGBT with t_{BOX} = 3.5μm and L_d = 45μm.

the time instant when V_{CE} begins to be stabilized and t3 is a time instant between t2 and t4. At 25°C and 125°C, V_{CE} is always stable at t1, t2, t3 and t4. Fig. 4 (b) gives the vertical electric field (E-field) profile along the line A1-A2 (illustrated in Fig. 1). The E-field beneath the BOX indicates a depletion expansion in P-sub at -40°C. Fig. 4 (c) compares the hole density and electric potential at point C (illustrated in Fig. 1) which locates at the upper surface of BOX layer of -40°C and 125°C conditions. The accumulation speed of holes at -40°C is much higher than that at 125°C. The fast hole accumulation leads to a rapid elevating of electric potential on the upper surface of BOX and thereby induces the depletion of P-sub.

978-1-7281-0582-6/19 $31.00 © 2019 IEEE

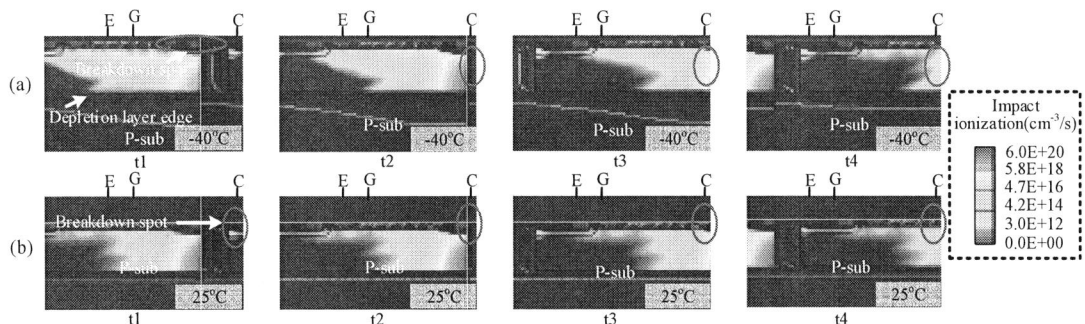

Fig. 5. Impact ionization rate and depletion layer distributions in SOI-LIGBT with t_{BOX} = 3.5μm and L_d = 45μm at (a) -40°C and (b) 125°C.

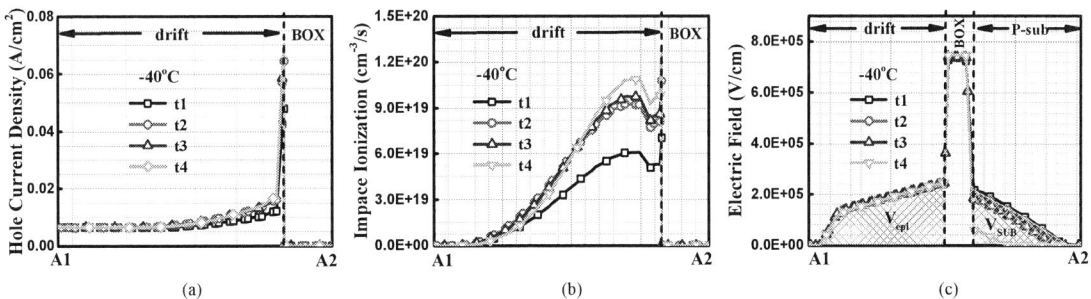

Fig. 6. (a) Hole current density, (b) impact ionization rate and (c) electric field profiles along the line A1-A2 in SOI-LIGBT with t_{BOX} =3.5μm and L_d = 45μm.

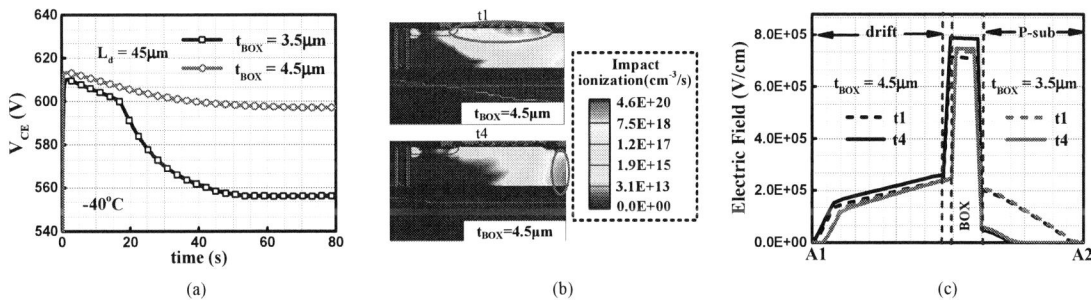

Fig. 7. Optimization scheme I. (a) Measured dependence of V_{CE} on time of SOI-LIGBTs with t_{BOX} =3.5μm and t_{BOX} =4.5μm. (b) Impact ionization rate and depletion layer distributions in SOI-LIGBT with t_{BOX} = 4.5μm at -40°C. (c) Electric field profile along the line A1-A2 in SOI-LIGBTs with t_{BOX} =3.5μm and t_{BOX} =4.5μm at -40°C.

Fig. 5 compares the impact ionization rate and depletion layer distributions of -40°C and 125°C conditions. White lines are depletion layer edge. At -40 °C, the depletion layer has expanded into P-sub when V_{CE} rises up to its initial value at t1. During the period when V_{CE} shifts down (t2-t4), the maximum impact ionization spot (breakdown point) transfers from silicon surface to bottom companying with the shrink of P-sub depletion. At 25°C, no P-sub depletion is observed and the breakdown spot is stabilized at bottom. Vertical breakdown voltage (BVV) can be increased through the expansion of P-sub depletion, leading to an initial surface breakdown (BVV>BVL). V_{CE} of the devices with a fixed t_{BOX} shift to a same stable value (see Fig. 3), indicating that the stabilized value of V_{CE} closely depends on BVV. The evolution of hole density, impact ionization and E-field along the A1-A2 during the V_{CE} shift in Fig. 6 is consistent with

the transfer of maximum impact ionization spot and the depletion expansion-shrink behavior in Fig. 5.

Two optimization schemes are provided in Fig. 7 and Fig.8. As Fig. 7 (a) shows, the V_{CE} shift can be suppressed and the stabilized value of V_{CE} is increased when BOX of 4.5μm is adopted. Unless an exact BVL=BVV is satisfied, there still exist a breakdown spot transfer and a P-sub depletion behavior (see Fig. 7 (b) and (c)). As Fig. 8 shows, another scheme that replacing P-type substrate by N-type substrate can achieve a very stable V_{CE}. It means that the depletion between substrate and N-drift is fully eliminated. Table I summarizes the avalanche instability of SOI-LIGBTs with different dimensions and substrate types. V_{CE} shift can be completely eliminated (ΔV_{CE}=0V) by employing N-type substrate.

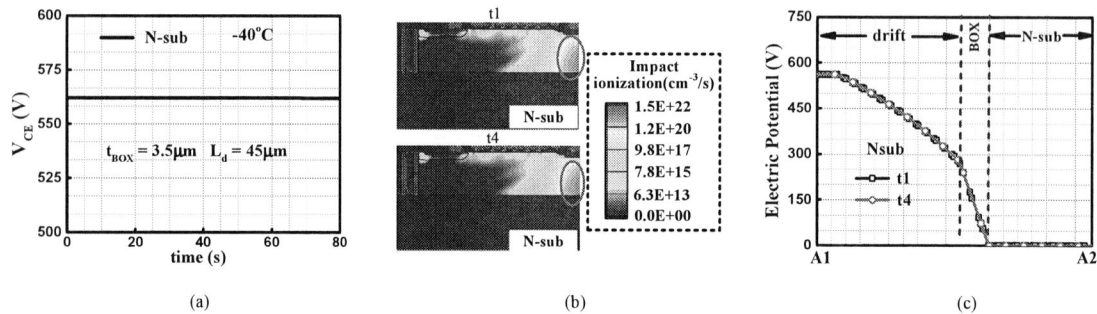

Fig. 8. Optimization scheme II. (a) Simulated dependence of V_{CE} on time of SOI-LIGBT with N-sub. (b) Impact ionization rate and depletion layer distribution at -40°C. (c) Electric potential profile along the line A1-A2.

TABLE I.   SUMMARY OF AVALANCHE INSTABILITY OF SOI-IGBTs

L_d (μm)	t_{BOX} (μm)	Sub type	Sub doping (cm⁻³)	Initial breakdown spot	Stabilized breakdown spot	Initial V_{CE} (V)	Stabilized V_{CE} (V)	ΔV_{CE}(V)
55	3.5	P	1.36×10^{15}	Surface	Bottom	615	557	58
60	3.5	P	1.36×10^{15}	Bottom	Bottom	633	557	76
70	3.5	P	1.36×10^{15}	Bottom	Bottom	659	557	102
45	3.5	P	1.36×10^{15}	Surface	Bottom	610	556	54
45	4.5	P	1.36×10^{15}	Surface	Bottom	613	602	11
45	3.5	N	1.36×10^{15}	Bottom	Bottom	562	562	0

IV. CONCLUSION

The off-state avalanche stability of SOI-LIGBT at low temperature is investigated in this paper. It is found that V_{CE} shift is related to the depletion layer expansion in P-sub. The initial depletion layer expansion is originated from the fast accumulation upon the BOX upper surface. This paper proposed two optimization schemes to suppress or eliminate the unstable avalanche phenomena. The V_{CE} shift can be suppressed when BOX with thickness of 4.5μm is adopted. Another scheme using N-type sub can achieve a very stable V_{CE}. The depletion between sub and N-drift can be fully eliminated.

REFERENCE

[1] K. Hara, S. Wada, J. Sakano, et al, "600V Single Chip Inverter IC with New SOI Technology," in Proc. 26th ISPSD, June 2014, pp. 418-421.

[2] Nakagawa A, Funaki H, Yamaguchi Y, et al. "Improvement in lateral IGBT design for 500 V 3 A one chip inverter ICs," in Proc. 19th ISPSD, May 1999, pp. 321-324.

[3] Long Zhang, Jing Zhu, Jie Ma, et al. "500-V Silicon-On-Insulator Lateral IGBT With W-Shaped n-Typed Buffer and Composite p-Typed Collectors," IEEE Trans. on Electron Devices, vol. 66, pp. 1430-1434, Jan. 2019.

[4] X. Perpina, I. Cortes, J. Urresti-Ibanez, et al, "Layout Role in Failure Physics of IGBTs Under Overloading Clamped Inductive Turnoff," IEEE Trans. Electron Device, vol. 60, pp. 598-605, Feb. 2013.

[5] Yuxiang Chen, Wuhua Li, Francesco Iannuzzo, et al, "Investigation and Classification of Short-Circuit Failure Modes Based on Three-Dimensional Safe Operating Area for High-Power IGBT Modules," IEEE Transactions on Power Electronics, vol. 33, pp. 1075-1082, Feb. 2018.

[6] D. Brisbin, A. Strachan, and P. Chaparala, "PMOS Drain Breakdown Voltage Walk-in: A New Failure Mode in High Power BiCMOS Applications," in Proc.42th IRPS, April 2004, pp. 265-268.

Proceedings of the 31st International Symposium on Power Semiconductor Devices & ICs
May 19-23, 2019, Shanghai, China

The Lowest On-Resistance and Robust 130nm BCDMOS Technology implementation utilizing HFP and DPN for mobile PMIC applications

Daehoon Kim, Kuemju Lee, Jaeeuk Kim, Junghun Choi, Jaehee Lee and Inwook Cho

Technology Development office1, SK Hynix system ic Inc
Cheong-ju, Republic of Korea
daehoon.kim@sk.com

Abstract—**BCD technology has been the workhorse of several critical products in the mobile market. Aggressive design rule and architectural modifications are being exploited to achieve significant improvements in the high voltage device performance such as low Ron.sp and high BVDSS which are crucial for improved switching efficiency and product robustness. In order to meet the requirements, we applied High temperature oxide field plate (HFP) structure with heavily doped poly and Double RESURF with P-Burid Layer (PBL) for the performance. And we used the self-aligned P-Body process and the decoupled plasma nitridation (DPN) gate oxide process for the robust reliability. From these evaluations, for the critical 12V N-LDMOS device, we have been able to achieve an ON-resistance of 3.0 mΩ-mm2 with BVDSS of 21.5V which is considered world-class. The new features make our 130nm BCD technology a powerful platform for future mobile PMIC.**

Keywords—PMIC, LDMOS, Field plate, RESURF, DPN

I. INTRODUCTION

In modern society, various mobile products are produced. In mobile market, 130nm BCD Technology has high potential and many needs. Because this technology can get smaller size chips than the existing 180nm BCD. There has been a significant push in the recent years to integrate high performance power devices using 130nm platform to shrink the die-size while providing high-density memory capability needed for applications such as USB Type-C. Also high voltage device performance, such as low Ron.sp and high BVdss, is still important for high switching efficiency in this product. And robust characteristics and reliability are significant for having stable distribution. This paper describes a high-performance, yet low-cost 130nm BCD technology developed at SK Hynix that uses a highly optimized field plate (FP) structure in conjunction with heavily doped poly and Double RESURF with P-buried layer. In addition, best-in-class performance for the high-voltage devices are achieved by incorporating self-aligned P-body implants and decoupled plasma nitride (DPN) gate oxide for robust reliability.

II. EXPERIMENTAL

Current STI field plate structured LDMOS could obtain a satisfactory BVdss value as reducing lateral E-field by shallow trench isolation, however there was a flaw that Ron.sp increased as much as current length become longer as current path was flowing along the boundary of STI compared to device length in Ron.sp. Horizontal (pitch) and Vertical scaling (diffusion) of the high voltage devices are required to achieve world-class performance for Ron.sp and BVdss. Horizontal scaling was achieved by replacing shallow trench isolation (STI) in the drift region with a field plate structure, which allowed for aggressive reduction of the spacing between source and drain while maintaining sufficient BVDSS(1).

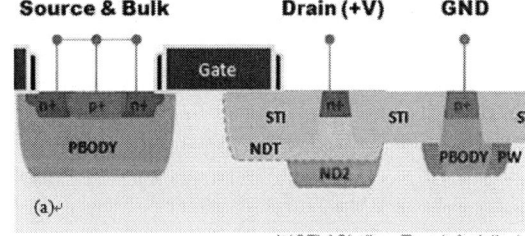

※ STI (Shallow Trench Isolation)

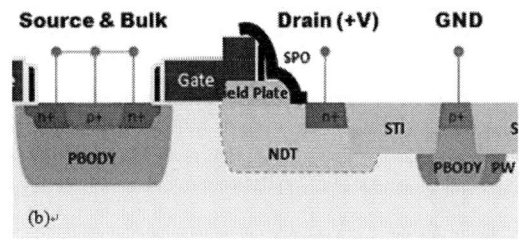

※ HTO (High Temperature Oxide)

Fig. 1 Schematic cross section of (a) STI field plate LDNMOS (b) HTO field plate LDNMOS

978-1-7281-0582-6/19 $31.00 © 2019 IEEE

Vertical scaling consists of HFP structure, Double RESURF P-buried layer(2), self-aligned P-body implants, optimized drift implants and minimal heat cycle. Field plate structure consisted of high temperature oxide with heavily doped polysilicon gate to maximize the depletion in the drift region under off-conditions and hence achieve high BVDSS. To further improve the electric field distribution under off conditions, a double RESURF using PBL was implemented. Integration of both the PBL and HFP allowed for significant increase in the drift doping while being able to sustain electric fields to achieve high BVDSS. This approach resulted in world-class on-resistance and BVDSS performance for the high voltage devices in our 130nm technology(3)

relieve electric field by optimizing poly plate and HFP width to be well matched with drift concentration.

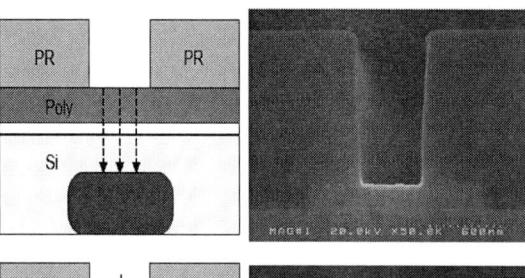

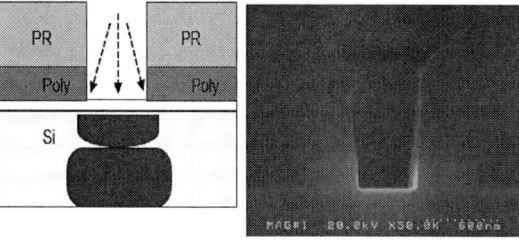

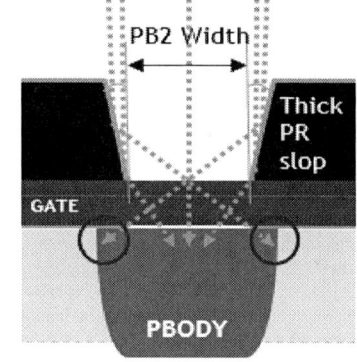

Fig. 3 : The sequential vertical structure of PB2 module process and imagine of PB2 well proximity effect

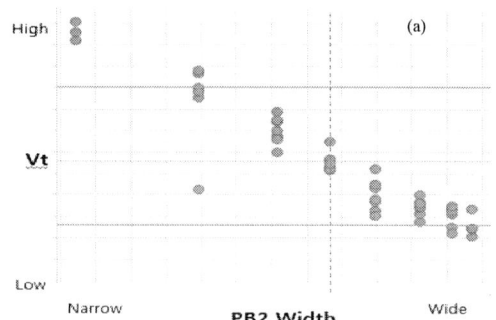

Fig. 2 Simulated depletion status, net doping, impact ionization of 24V LDNMOS (a) PBL IMP high energy and (b) low energy

To address both manufacturability and HV device robustness, we went to get the robust process for HV device. So we used the self-align P-Body process that helps a stable trend by eliminating potential of miss align. And the deeper P-Body junction depth is, the better the device SOA is. Therefore we used high energy implant and height photo resist. But when P-Body width size was less than some value, LDNMOS devices had big variation of Ron.sp and Vtext parameters because of well proximity effect(4). Additionally when the drift concentration is increased, DC SOA gets worse due to electric field crowding while Ron.sp gets lower. So we did to optimize the P-Body width for removing well proximity effect and

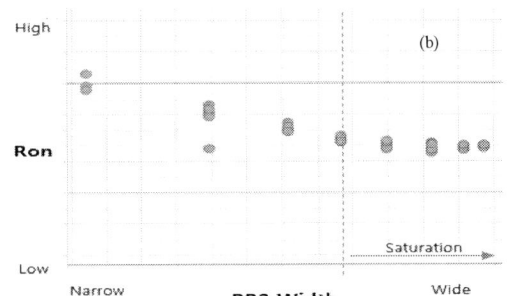

Fig. 4 : LDNMOS Characteristics depending on PB2 Width (a) Vtext Trend (b) Ron Trend

As a part of overall technology shrink, we tested two different gate oxide processes, thermal nitrided (NO) oxide and DPN oxide. Pure thermal oxide was included as a control group. Interface charge density was significantly lower for DPN oxide compared to that of the NO oxide(5). Wafer-level reliability (WLR) showed best results for the DPN oxide, consistent with interface trap density measurements(6). In addition, for the same BVdss class of the high-voltage devices, DPN oxide group has lower Ron.sp compared to pure thermal oxide due to thinner electrical oxide caused by nitridation. Therefore we applied DPN gate oxide process for high performance and robust reliability.

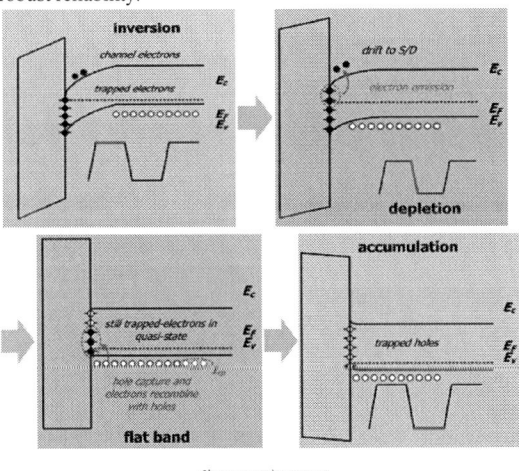

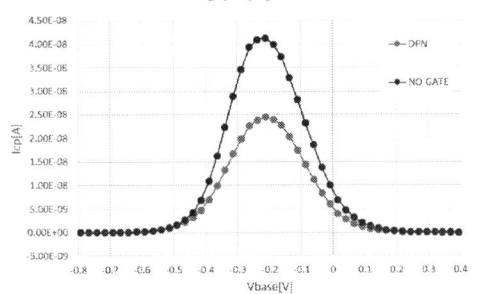

Fig. 5 : Model of charge pumping current and measure data of DPN and NO gate oxide processes

Finally, as a result of measuring Transmission Line Pulse(TLP) curve of relevant element, we could see that safe operation area was around 120% bigger compared to operation voltage.

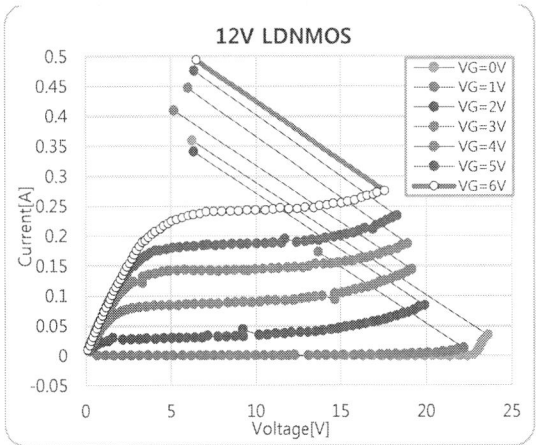

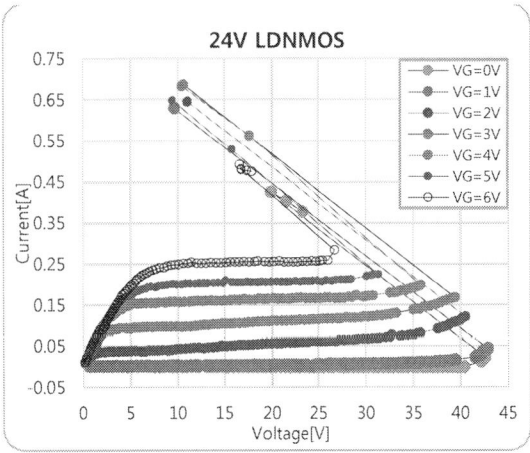

Fig. 6 Measured TLP characteristics of optimized 12V & 24V LDNMOS in this work

Comparing the elements chosen above with the conventional platform with STI structure, we can see that Ron.sp of this work has been decreased by around 20~40%. The graph below compares the tendency of Ron.sp and BVdss of the existing STI structure and new HFP structure & the combination structure of STI and HFP.

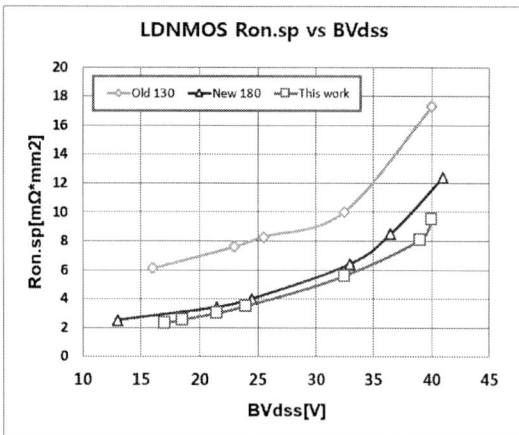

Fig.7 The trade-off between BV and Ron.sp of LDNMOS

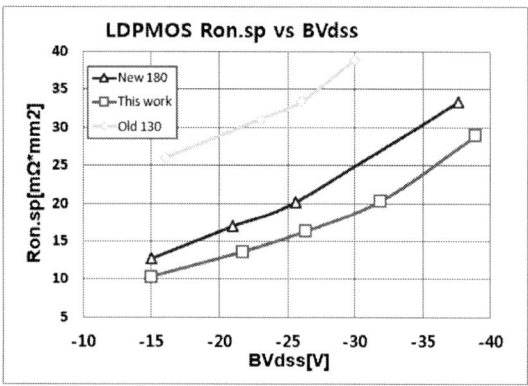

Fig.8 The trade-off between BV and Ron.sp of LDPMOS

III. RESULT

From these evaluations, we made sure that there was an significant improvement in performance compared to the existing 180nm tech of our company in this tech platform. For example, for the critical 12V N-LDMOS device, we have been able to achieve an ON-resistance of 3.0 mΩ-mm2 with BVDSS of 21.5V which is considered world-class. And our new 130nm BCD process in 1.5/5V CMOS process platform can also provide analog components such as DEMOS, BJT, MIM capacitor, Zener diode, power diode and high resistor. Additionally, optional high density EEPROM up to 64K-Byte are enabled with just two additional masks. All the new features make our 130nm BCD technology a powerful platform for future mobile PMIC.

REFERENCES

[1] Feng Jin, "Best-in-class LDMOS with ultra-shallow trench isolation and p-buried ...", ISPSD, p. 205-298, 2017

[2] K.S.Ko et al., "HB1340, Advanced 0.13um BCDMOS technology of cpmplimentary LDMOS including fully isolated transistors" ISPSD2013, pp.159-162.

[3] B.Jayant Baliga, "Fundamentals of Power Semiconductor Devices" chapter3 Breakdown Voltage

[4] Yi-Ming Sheu, "Modeling the Well-Edge Proximity Effect in ..." IEEE Trans.Electron Devices, Nov, 2006

[5] J.B. Yang, "Modeling and Characterization on Negative Bias Temperature Instability in p-channel MOSFETs, ECS Transaction, p283~299, 2007

[6] G. Groeseneken, H. Maes, N. Beltran, and R. F. De Keersmaecker, ``Reliable Approach to Charge-Pumping Measurements in MOS Transistors," IEEE Trans.Electron Devices, vol. 31, no. 1, pp. 42-53, 1984.

Proceedings of the 31st International Symposium on Power Semiconductor Devices & ICs
May 19-23, 2019, Shanghai, China

Experimental Study on the Electrical Properties of Lateral IGBT Under the Mechanical Strain

Wangran Wu, Yaohui Wang, Long Zhang, Guangan
Yang, Siyang Liu, Jing Zhu, Weifeng Sun *
National ASIC System Engineering Research Center
Southeast University, Nanjing, China
*E-mail: swffrog@seu.edu.cn

Abstract— In this paper, we present the experimental study on the electrical properties of lateral IGBT (LIGBT) under the mechanical strain. The on-state characteristics and turn-off characteristics of LIGBT under uniaxial strains parallel to the channel direction are studied extensively. The uniaxial tensile strain will increase the collector current (I_C) and the uniaxial compressive strain will decrease I_C, while the breakdown voltage (V_{BD}) does not degrade. Both the gate voltage (V_g) and collector voltage (V_{CE}) play important roles on the strain effects. The piezoresistance (π-) coefficients of LIGBT are evaluated. The π-coefficient of LIGBT is more than three times of nMOSFET and nDMOS. The uniaxial tensile strain will decrease the on-state voltage drop (V_{on} @ 100 mA) while increase the turn-off time (t_{off}) and turn-off loss (E_{off}). The uniaxial compressive strain is the opposite.

Keywords— *mechanical strain, LIGBT, performance enhancement, piezoresistance coefficient*

I. INTRODUCTION

The lateral IGBT (LIGBT) is the core device in the high power density Si-based integrated circuits as the switching device [1, 2]. The device structure has been continuously optimized in order to increase the current density without sacrificing the turn-off loss and latch-up immunity [3-5]. The strained Si technology will provide a novel way to improve the LIGBT's performance by increasing the carriers' mobility. Process-induced strain was first introduced into Si MOSFETs by Intel in 2002 and began to play a major role in the mainstream VLSI semiconductor industry [6-8]. Now, strain is induced in various ways to enhance the device performance in almost every modern semiconductor workshop [9-11]. However, the study and application of strained Si technology in power devices have experienced much slower development. It has been proved that the strained power MOSFETs could break through the "Si-limit" and the mechanisms of the strained lateral DMOS (LDMOS) have been studied [12-14]. The mechanisms of the strained IGBT must be complex because the IGBT is the bipolar device and the strain affects electrons and holes quite differently. The LIGBT can be fabricated through the bipolar-CMOS-DMOS (BCD) process, which is highly compatible with the strained Si techniques in CMOS process. Thus, the investigation on electrical properties of LIGBT under the mechanical strain is of great importance from the perspective of application.

In this paper, the mechanical strain is introduced into the LIGBT to avoid the unexpected variance. The on-sate and turn-off characteristics of the LIGBT under uniaxial tensile and compressive strains along the channel direction are examined thoroughly. Both the strain effects of the uniaxial tensile strain and uniaxial compressive strain are studied.

II. DEVICE STRUCTURE AND EXPERIMENTS

The uniaxial mechanical strain can be applied using the wafer bending system (Fig. 1(a) and 1(b)). The relative displacements of the top and bottom plates are measured to calculate the amount of strains. The electrical properties of LIGBT are measured under the uniaxial strain (Fig. 1(c)). The LIGBT used in this study is fabricated on the (001) SOI substrate with a <110> channel direction through the standard BCD process. The schematic of the LIGBT is shown in Fig. 2. The device has a break down voltage of 580 V. The values of the key parameters are shown in Fig. 2. The thicknesses of the buried oxide (t_{BOX}) and Si layer (t_{Si}) of the SOI substrate are 3.5 μm and 18 μm. The thickness of the gate oxide (t_{GOX}) is 0.05 μm. The lengths of drift region (L_{drift}), gate region (L_g), and N-buffer ($L_{nbuffer}$) are 53 μm, 3.2 μm and 8 μm. The uniaxial tensile strains are applied parallel to the channel direction in this study.

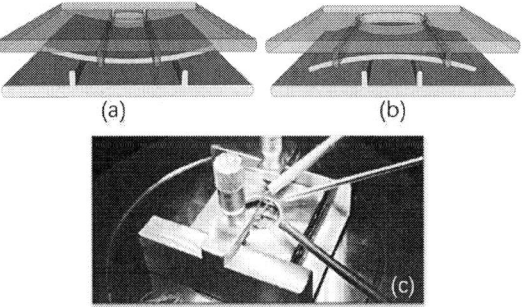

Fig. 1: Illustrations of wafer bending system to apply (a) uniaxial tensile stress, (b) uniaxial compressive stress and (c) the setup of the measurement system. The strain with a positive number represents the uniaxial tensile strain and the strain with a negative number represents the uniaxial compressive strain.

978-1-7281-0582-6/19 $31.00 © 2019 IEEE

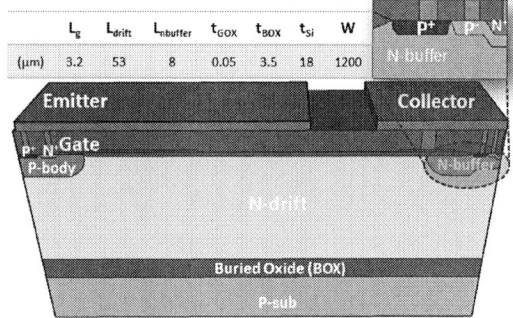

	L_g	L_{drift}	$L_{nbuffer}$	t_{GOX}	t_{BOX}	t_{si}	W
(μm)	3.2	53	8	0.05	3.5	18	1200

Fig. 2: The schematic of the lateral IGBT (LIGBT) used in this study. The LIBGT were fabricated on the SOI substrate with channel direction of <110>/(001). The device dimensions are shown in the table. Both uniaxial tensile strain and uniaxial compressive strain are applied along the channel direction in this study.

In this paper, the strains with a positive number represent the uniaxial tensile strains and the strains with the negative number represent the uniaxial compressive strains. The on-state characteristics of the LIGBT under the uniaxial strain parallel to the channel direction are examined. Fig. 3 shows the I_C-V_{CE} curves of LIGBT under uniaxial compressive and tensile strains at V_g of 5 V, 10 V and 15 V. It is shown that the uniaxial tensile strain will increase I_C and the uniaxial compressive strain will decrease I_C. Fig. 4 shows the I_C-V_g curves of LIGBT under uniaxial compressive and tensile strains at V_{CE} of 3 V. It is shown that the uniaxial tensile strain will increase I_C and the uniaxial compressive strain will decrease I_C. The threshold voltage does not change. Thus, the variation of I_C is attributed to the strain modulated carrier mobility. Fig. 5 illustrates the ΔI_C ($=I_{C\_strain}/ I_{C0}-1$) versus V_{CE} curves under uniaxial compressive and tensile strains at V_g of 5 V, 10 V and 15 V. ΔI_C is strongly correlated with both V_{CE} and V_g. ΔI_C has the maximum value in the linear region and decreases with increasing V_{CE}. The maximum value of ΔI_C (ΔI_{C_linear}) increases with increasing V_g and the minimum value of ΔI_C ($\Delta I_{C\_saturation}$) decreases with increasing V_g. Fig. 6 shows the ΔI_{C_linear} and $\Delta I_{C\_saturation}$ versus the amount of uniaxial strain curves at V_g of 5 V and 10 V. The 22% I_C enhancement is achieved at V_g of 10 V in the linear region.

The piezoresistance (π-) coefficients represents the resistance change ($\Delta R=|R_{strain}/R_0|-1$) under strain ($\pi=\Delta R/Strain$). It is used to represent the strain effect. Fig. 7 shows the piezoresistance (π-) coefficients of LIGBT in the liner region (π_linear) and saturation region ($\pi\_saturation$) at V_g of 5 V and 10 V. The π_linear (-101×10^{-11} Pa^{-1}) is more than three times of the nMOSFET and nDMOS (-32×10^{-11} Pa^{-1}). This is because of the carriers' bipolar transportation. Thus, the uniaxial strain is very effective in LIGBT. Fig. 8 illustrates the he BV characteristics of LIGBT under the uniaxial compressive and tensile strains. Neglectable V_{bd} degradation is observed. The leakage current is modulated slightly by the strain. This result proves the down-shift of the R_{on} versus BV curve under the uniaxial tensile strain, which can help to break through the Si-limit.

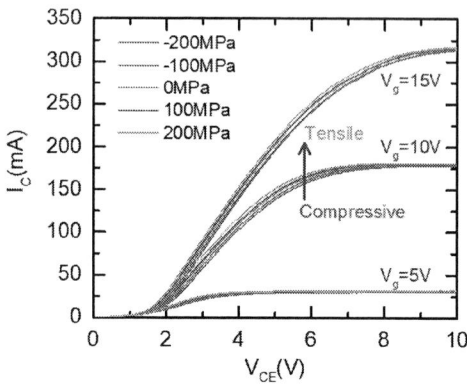

Fig. 3: The I_C-V_{CE} curves of LIGBT under uniaxial compressive and tensile strains at V_g of 5 V, 10 V and 15 V. It is shown that the uniaxial tensile strain will increase I_C and the uniaxial compressive strain will decrease I_C.

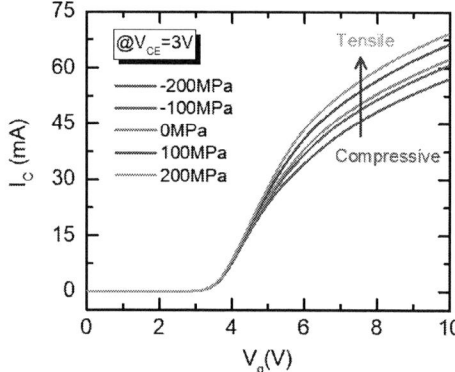

Fig. 4: The I_C-V_g curves of LIGBT under uniaxial compressive and tensile strains at V_{CE} of 3 V. It is shown that the uniaxial tensile strain will increase I_C and the uniaxial compressive strain will decrease I_C. The threshold voltage does not change.

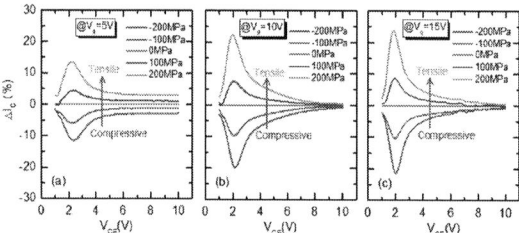

Fig. 5: The ΔI_C ($=I_{C\_strain}/ I_{C0}-1$) versus V_{CE} curves under uniaxial compressive and tensile strains at V_g of (a) 5 V, (b) 10 V and (c) 15 V. ΔI_C is strongly correlated with both V_{CE} and V_g. ΔI_C has the maximum value in the linear region and decreases with increasing V_{CE}. The maximum value of ΔI_C (ΔI_{C_linear}) increases with increasing V_g and the minimum value of ΔI_C ($\Delta I_{C\_saturation}$) decreases with increasing V_g.

978-1-7281-0582-6/19 $31.00 © 2019 IEEE

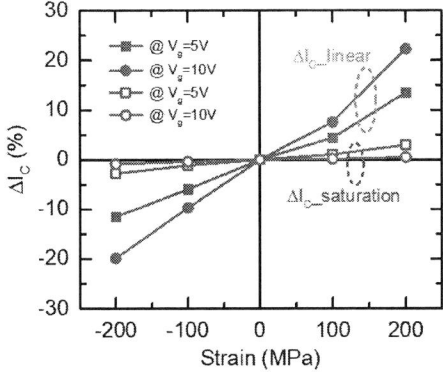

Fig. 6: The $\Delta I_C\_$linear and $\Delta I_C\_$saturation versus the amount of uniaxial strain curves at V_g of 5 V and 10 V. The 22% I_C enhancement is achieved at V_g of 10 V in the linear region.

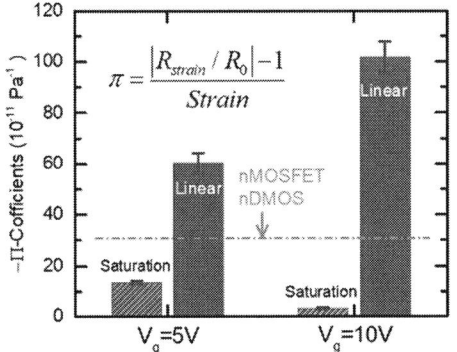

$$\pi = \frac{\left| R_{strain} / R_0 \right| - 1}{Strain}$$

Fig. 7: The piezoresistance (π-) coefficients of LIGBT in the liner region ($\pi\_$linear) and saturation region ($\pi\_$saturation) at V_g of 5 V and 10 V. The $\pi\_$linear (-101×10^{-11} Pa^{-1}) is more than three times of the nMOSFET and nDMOS (-32×10^{-11} Pa^{-1}).

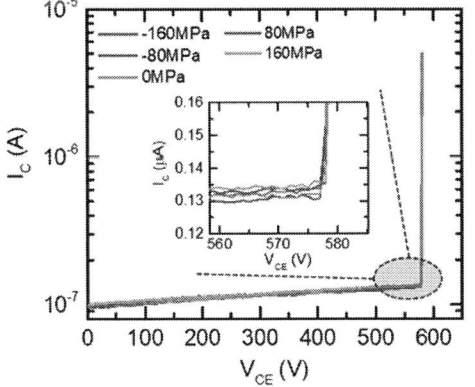

Fig. 8: The BV characteristics of LIGBT under the uniaxial compressive and tensile strains. Neglectable V_{bd} degradation is observed. The leakage current is modulated slightly by the strain.

The turn-off characteristics of the LIGTB is critical as the switching device. The turn-off waveforms of the LIGBT is measured by the circuit shown in the inset of Fig. 9. Fig. 9 shows the turn-off waveforms of the LIGBT under the uniaxial compressive and tensile strains. The uniaxial tensile strain will enhance turn-off time (t_{off}) and the uniaxial compressive strain will decrease t_{off}. Fig. 10 illustrates the t_{off} and turn-off loose (E_{off}) of the LIGBT versus the amount of uniaxial strain curves. Both t_{off} and E_{off} increases with the uniaxial tensile strain and decreases with the uniaxial compressive strain. Fig. 11 shows the trade-off between on-state voltage at I_C of 100 mA (V_{on}) and E_{off} of the LIGBT under the uniaxial compressive and tensile strains. The uniaxial tensile strain will decrease the V_{on} while increase the E_{off}. The uniaxial compressive strain is the opposite.

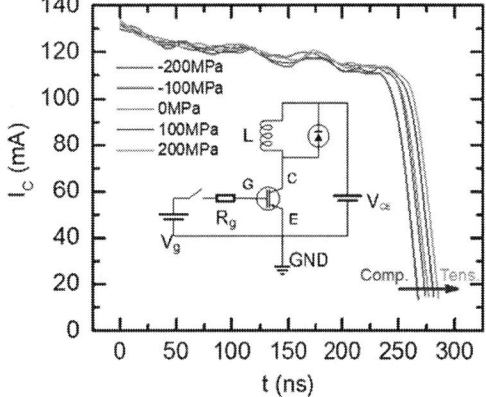

Fig. 9: Turn-off waveforms of the LIGBT under the uniaxial compressive and tensile strains. The uniaxial tensile strain will enhance t_{off} and the uniaxial compressive strain will decrease t_{off}.

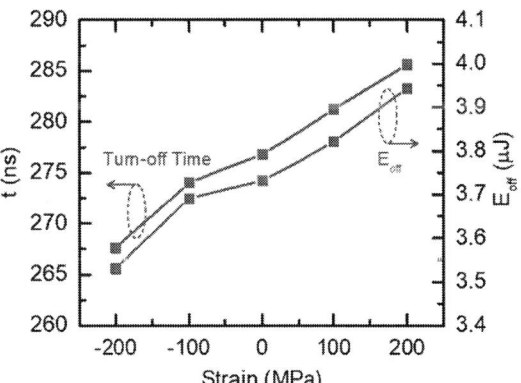

Fig. 10: The t_{off} and E_{off} of the LIGBT versus the amount of uniaxial strain curves. Both t_{off} and E_{off} increases with the uniaxial tensile strain and decreases with the uniaxial compressive strain.

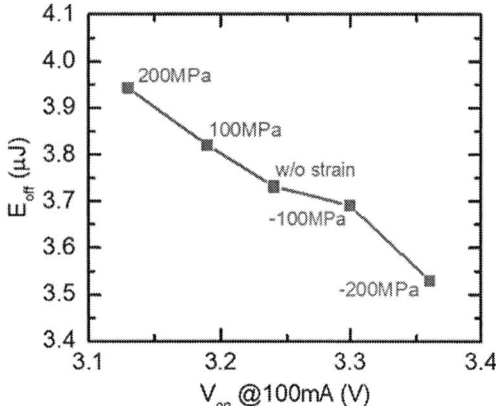

Fig. 11: The trade-off between V_{on} and E_{off} of the LIGBT under the uniaxial compressive and tensile strains. The uniaxial tensile strain will decrease the V_{on} while increase the E_{off}. The uniaxial compressive strain is the opposite.

III. CONCLUSIONS

In this work, we investigate the electrical properties of lateral LIGBT under the uniaxial mechanical strain. The on-state characteristics and turn-off characteristics of LIGBT under uniaxial strains parallel to the channel direction are studied extensively. The uniaxial tensile strain will increase the collector I_C and the uniaxial compressive strain will decrease I_C, which is because the modulation of carriers' mobility. The breakdown voltage and threshold voltage do not change. Both the V_g and V_{CE} play important roles on the strain effects. The piezoresistance (π-) coefficients of LIGBT are evaluated. The π-coefficient of LIGBT is more than three times of nMOSFET and nDMOS because of the carriers' bipolar transportation. The uniaxial tensile strain will decrease the V_{on} while increase the t_{off} and E_{off}. The uniaxial compressive strain is the opposite.

ACKNOWLEDGMENT

The work is supported by National Natural Science Foundation of China (61704025, 61674030), Natural Science Foundation of Jiangsu Province (BK20181140, BK20160691).

REFERENCES

[1] D. Disney, T. Letavic, T. Trajkovic, T. Terashima, and A. Nakagawa, "High-voltage integrated circuits: history, state of the art, and future prospects," IEEE Transactions on Electron Devices, vol. PP, pp. 1-15, 2017.

[2] A. Nakagawa, H. Funaki, Y. Yamaguchi, and F. Suzuki, "Improvement in lateral IGBT design for 500 V 3 A one chip inverter ICs," in International Symposium on Power Semiconductor Devices & Ics, 1999.

[3] N. Iwamuro and T. Laska, "IGBT history, state-of-the-art, and future prospects," IEEE Transactions on Electron Devices, vol. 64, pp. 741-752, 2017.

[4] Z. Long, Z. Jing, W. Sun, C. Meng, M. Zhao, X. Huang, J. Chen, Y. Qian, and L. Shi, "Novel snapback-free reverse-conducting SOI-LIGBT with dual embedded diodes," IEEE Transactions on Electron Devices, vol. PP, pp. 1-6, 2017.

[5] Z. Long, Z. Jing, W. Sun, M. Zhao, J. Chen, X. Huang, D. Ding, C. Jian, and L. Shi, "A U-shaped channel SOI-LIGBT with dual trenches," IEEE Transactions on Electron Devices, vol. PP, pp. 1-5, 2017.

[6] S. E. Thompson, M. Armstrong, C. Auth, and S. Cea, "A logic nanotechnology featuring strained-silicon," Electron Device Letters IEEE, vol. 25, pp. 191-193, 2004.

[7] S. Takagi, T. Irisawa, T. Tezuka, T. Numata, S. Nakaharai, N. Hirashita, Y. Moriyama, K. Usuda, E. Toyoda, S. Dissanayake, M. Shichijo, R. Nakane, S. Sugahara, M. Takenaka, and N. Sugiyama, "Carrier-transport-enhanced channel CMOS for improved power consumption and performance," Ieee Transactions on Electron Devices, vol. 55, pp. 21-39, Jan 2008.

[8] S. E. Thompson, S. Suthram, Y. Sun, and G. Sun, "Future of strained Si/semiconductors in nanoscale MOSFETs," in International Electron Devices Meeting, 2006, pp. 1-4.

[9] S. Thompson, N. Anand, M. Armstrong, C. Auth, B. Arcot, M. Alavi, P. Bai, J. Bielefeld, R. Bigwood, and J. Brandenburg, A 90 nm logic technology featuring 50 nm strained silicon channel transistors, 7 layers of Cu interconnects, low k ILD, and 1 /spl mu/m/sup 2/ SRAM cell, 2002.

[10] S. E. Thompson, G. Sun, Y. S. Choi, and T. Nishida, "Uniaxial-process-induced strained-Si: extending the CMOS roadmap," IEEE Transactions on Electron Devices, vol. 53, pp. 1010-1020, 2006.

[11] C. Auth, A. Aliyarukunju, M. Asoro, D. Bergstrom, V. Bhagwat, J. Birdsall, N. Bisnik, M. Buehler, V. Chikarmane, and G. Ding, "A 10nm high performance and low-power CMOS technology featuring 3 rd generation FinFET transistors, self-aligned quad patterning, contact over active gate and cobalt local interconnects," in IEEE International Electron Devices Meeting, 2017, pp. 29.1.1-29.1.4.

[12] P. Moens, et al., "Stress-induced mobility enhancement for integrated power transistors", IEDM, 2007, pp: 877-880.

[13] M. Miyamoto, et al., "Low-on-resistance strain-controlled LDMOS transistors for 0.25-μm power ICs", ISPSD, 2011, pp: 168-171.

[14] W. Wu, et al., "Comprehensive investigation on mechanical strain induced performance boosts in LDMOS", ISPSD, 2018, pp: 60-63.

Proceedings of the 31st International Symposium on Power Semiconductor Devices & ICs
May 19-23, 2019, Shanghai, China

Circuit Dependent Plasma Charging Effect Robustness in 0.16 um BCD Technology Platform

Michele Basso
Smart Power Technology R&D
ST Microelectronics
Agrate Brianza, Italy
michele.basso@st.com

Damiano Riccardi
Smart Power Technology R&D
ST Microelectronics
Agrate Brianza, Italy
damiano.riccardi@st.com

Antonino Martino
Smart Power Technology R&D
ST Microelectronics
Agrate Brianza, Italy
antonino.martino@st.com

Paola Galbiati
Smart Power Technology R&D
ST Microelectronics
Agrate Brianza, Italy
paola.galbiati@st.com

Simone Bertaiola
Smart Power Technology R&D
ST Microelectronics
Agrate Brianza, Italy
simone.bertaiola@st.com

Abstract— **BCD and High Voltage technologies are addressing an increasingly high number of disparate and peculiar applications. Although different applicative scopes, a very common circuit configuration requires a large device (power MOS) connected to a smaller one (sense-FET, copy MOS) acting as its sensor [1]. This configuration can be very "unbalanced" in terms of metal area and number of vias insisting on the power stage with respect to the sense-FET one. In this paper, the effect of different circuit topologies on n-channel HV MOSFET robustness against plasma charging is studied. Particular focus was given to the circuital block indicated above, showing that in this configuration the small HV MOS is the most critical element. By tuning the diffusion process and by using large antennas, plasma damage was amplified and its effect applied on a number of realistic circuits. These different blocks allowed us to identify the factors leading to plasma damage and to provide a qualitative model that explains the plasma-related damage propagation. Design rules created ad-hoc should be implemented to avoid the occurrence of this circuit-dependent plasma effect.**

Keywords— *plasma, charging, damage, PID, circuit robustness, HV MOS, DMOS, BCD, power, sense-FET.*

I. INTRODUCTION

Plasma process-Induced Damage (PID) is a threat in modern VLSI and ULSI technologies leading to yield losses and reliability concerns [2]. The role of plasma damage with device scaling has been widely studied. A number of significant results have been reported targeting the charging effects from different process conditions [3] [4], gate oxide thickness impact [5] [6], antenna topologies and fast plasma detection [7]. However, studies focused on damage dependence with respect to the connections of the device under analysis [8] [9] are relatively recent and usually do not target HV platforms and devices.

In this work, we thoroughly analyze the behavior of a number of circuits subjected to plasma charging. Even if these circuits are designed to be fully compliant with well-known antenna rules specific for gate routing (Cumulated Antenna Ratio – CAR - and Partial Antenna Ratio - PAR) they show a statistically distributed abnormally high gate leakage current. This affects overall yield. This paper demonstrates that this behavior is due to plasma damage phenomenon. We succeed in measuring the impact of different connections and

topologies of elementary devices on the final failure rate and we provide a detailed failure model that explains these effects. Furthermore, this model provides suitable design guidelines to prevent plasma damage occurrence.

II. EXPERIMENTAL PROCEDURE

The Devices Under Test (DUTs) were diffused using a 0.16 um dual flavor poly gate BCD technology platform with Aluminum interconnects. The nominal gate oxide thickness is 70Å. The circuit block under characterization is composed of a power HV MOS with a shared gate contact with its sensor MOS. The HV devices are both n-channel MOS rated for applications up to 40 V (Fig. 1). It is worth to mention that this type of HV MOS embeds an integrated diode between drain and substrate with a breakdown voltage higher than 70 V. As suggested in [10], gate current measurement is performed to extract the failure rate. The circuit design allowed to measure the separate contributions of the two devices. OBIRCH was used to confirm the failure and detect its exact position.

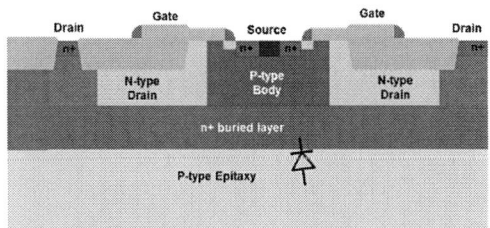

Fig. 1. Cross Section of the HV MOS under study. The component has a body shorted to source configuration and external drain. The diode between the n+ buried layer (isolation) and the p-type epitaxial layer is shown.

III. EXPERIMENTAL RESULTS

A. Reference Circuit Evaluation

The first characterized circuit represents the reference of the study. Two identical 40 V HV MOS have been used as power (MOS1, total width on silicon: 48.8 mm) and sense-FET (MOS2, width: 13.8 um) in a common gate configuration, as shown in Fig. 2a. Gate leakage lower than 100 pA is measured at Vgs = 3.3V (Fig. 2b).

978-1-7281-0582-6/19 $31.00 © 2019 IEEE
399

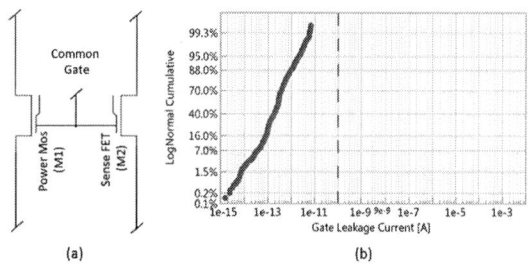

Fig. 2. (a) Basic Circuit block of a power MOS (MOS1) and its twin Sense-FET (MOS2) with common gate. (b) The baseline structure shows no fails.

B. Diode Leakage at Sense-FET side

With respect to the reference situation, in real circuits, the Sense-FET is connected to other devices (CMOS, HV MOS, diodes, resistors). Diodes of different area have been inserted between the source of MOS2 and ground (Fig. 3a). By increasing the diode area, it is possible to tune its leakage current, allowing a more effective connection to ground. The electrical measurements together with subsequent OBIRCH analysis assesses that the failing HV MOS is MOS2 (Fig. 3 and Fig. 4). The failure rate is highly impacted by diode area (from zero to 100% fails) suggesting that a stronger ground connection on MOS2 silicon is responsible for the device failure: it is reasonable to suppose that the gate terminal during process steps is biased to a potential high enough to provoke damage at gate oxide level.

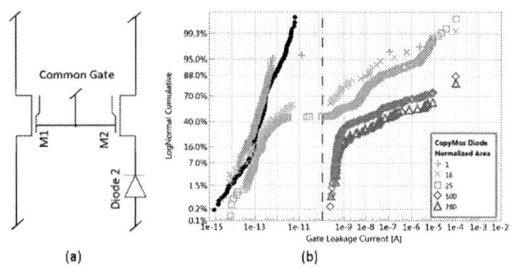

Fig. 3. (a) Configuration with a diode connected to sense-FET source. (b) Failure rate of this structure with different diode areas. The failure rate strongly increases when diode area becomes bigger (black dots represent the structure without diode).

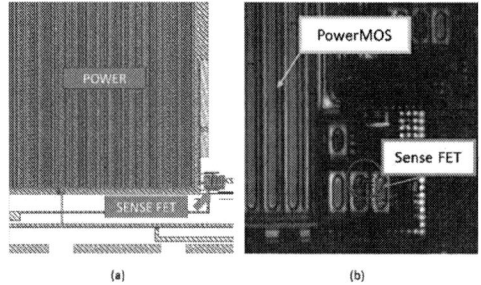

Fig. 4 – (a) layout view of the power and Sense-FET system. (b) OBIRCH images showing the position of the failure: emission found on Sense-FET.

C. Vias multiplicity on Power Source and Drain

In this case, the same circuit configuration with diode on MOS2 was studied, showing the dependence of failure rate with respect to the number of vias on power MOS source affecting the plasma current. We succeed in producing an

increasingly high failure rate (up to 100%) with a high enough number of vias. The correlation between vias number of power source/drain and Sense-FET fail suggests that the plasma current onto the power device affects the voltage drop on MOS2 gate oxide. No differences have been measured unbalancing the vias number (power side) on drain instead of source: the failure rate trend does not change. To enhance plasma charging, worst case trials were performed at IMD filling and vias barrier deposition level (sputtering, thickness, temperature and bias): see Fig. 5.

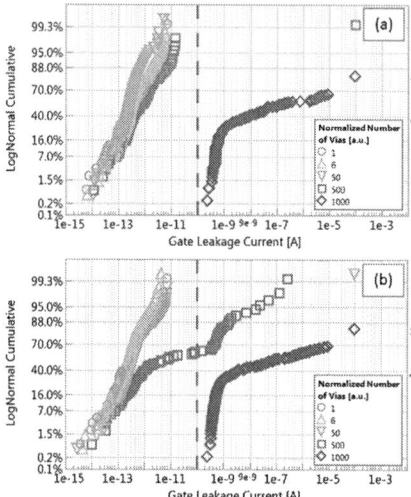

Fig. 5. Failure rate with different vias number for different process split. With "500" vias a.u. the failure rate is negligible in (a) process but becomes very significant (60%) in worst-case split (b).

D. Diode insertion at power side

Starting from the previous circuit (with a diode on MOS2) and adding a diode on power source side (Fig. 6) the number of failures collapses to zero. The diodes used have the same properties with respect to the one connected to the Sense-FET. The leakage path provided by additional diode on power MOS ties to ground the source and consequently the power and sense-FET common gate.

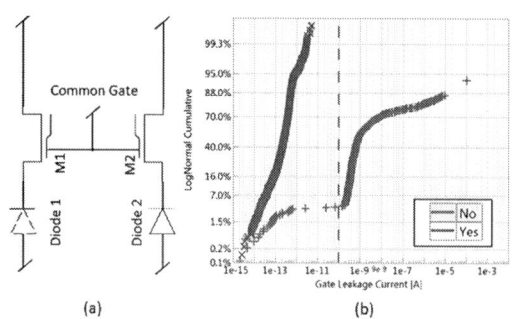

Fig. 6. (a) Circuital block with diodes both at MOS1 source (red dotted diode) and at MOS2 source. (b) Failure Rate collapses when a diode is inserted on power MOS source (blue line).

IV. FAILURE MODEL DESCRIPTION

Typical design rules for plasma damage management focus on gate routing, and the gate protection is achieved with a diode in case of high cumulative antenna ratios. Even if the analysed configuration is compliant with CAR and PAR rules, it has been demonstrated that the high current during the plasma event on source (drain) of the power MOS plays the most significant role in the damage of the sense-FET. Moreover, the source and drain of the devices have been proven interchangeable: the circuital schematic can be simplified (Fig. 7). The two HV MOSFET behave simply like MOS capacitors.

When high plasma current (Ip1) insists on power MOS source (drain), the leakage of the intrinsic vertical diode (buried layer vs. substrate) is not capable of maintaining a ground connection and it induces a bias in the silicon region as seen in Fig. 8 . It is interesting to point out that this effect is more significant when the plasma process is cold and so more harmful with advanced technology lithographic nodes [11]. If no clamp (i.e. Diode 1) is present, the node potential can increase (nominally up to the vertical junction breakdown that, as specified in the introduction, can be of several tens of volt in a smart power platform).

Thanks to the huge gate capacitance of the power stage (C1), the common gate is coupled with power MOS source and drain. If the source and drain of the Sense-FET can float (MOS2 has the same bulk diode of the power MOS) they can match the whole system bias and no deterioration occurs (Fig. 2).

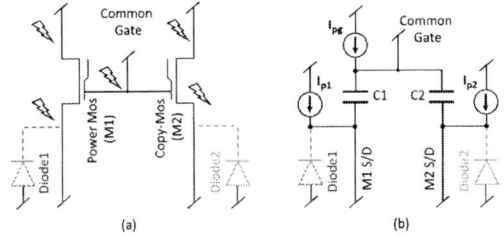

(a)   (b)

Fig. 7. (a) Circuit used in the experiments. Flashes represent the plasma charging occurring during process flow. Dotted lines exemplify that the connection to additional diode can be present or not. (b) Simplified circuit for plasma damage model of Power MOS and Sense-FET. C1 and C2 represent the gate oxide related capacitance of MOS1 and MOS2, respectively. The drain to substrate diode is shown in black. Ip1, Ip2 and Ipg are the current generators used to reproduce the plasma effect.

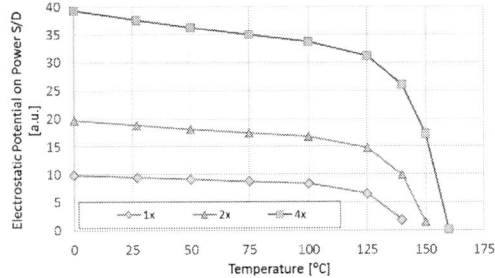

Fig. 8. Electrostatic Potential (source and drain terminals, substrate grounded) dependence in case of three different constant currents (1x, 2x and 4x) flowing inside the drain to substrate diode, at different temperatures. The lowest the temperature the highest is the voltage drop between the power source and drain and the substrate: this is due to the lower leakage of the substrate diode.

When an additional diode (Diode 2) is inserted at MOS 2 source (drain) side, if its leakage current is higher than the one of the intrinsic diode, it ties this node to ground and the system behaves like a capacitive divider: the voltage drop on the gate oxide of the Sense-FET can cause serious damage (Fig. 3). It is worth to highlight that this phenomenon can arise if and only if MOS1 and MOS2 common gate is already connected when plasma process occurs: it has been experimentally observed that a gate metal bridge between the two components is an effective way to avoid plasma-related bias propagation.

V. SIMULATION

Electrical simulation was performed to support the proposed model. The circuit under investigation (with and without diode at Sense-FET side, Fig. 9) is implemented in a commercial spice tool, using the technology database. To simplify the circuit and allow a more robust convergence of the simulation, a *voltage generator* is used to represent the plasma current flowing in the bulk diode. This voltage supply is function of the number of vias as seen in Fig. 5. Only the plasma current of the power MOS has been taken into account. The other nodes are floating: they are connected to a current generator set with current equal to zero. In fact, in standard POWER IC circuits the number of vias and the antenna ratio of sensing structures is largely smaller than the power ones. Different power (MOS1) to Sense-FET (MOS2) width ratios have been simulated with and without the presence of the diode on MOS2. The voltage drop on the gate-source of MOS2 follows the model previously described, as shown in Fig. 10. When no diode is connected, the gate to source voltage of MOS2 is drastically reduced.

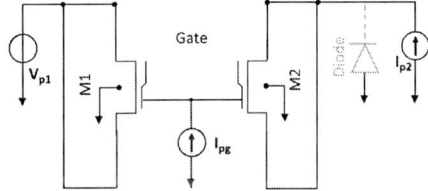

Fig. 9. Circuit used for the Spice simulation. The Vp1 is the equivalent voltage of the Ip1 plasma current flowing in the drain to substrate diode of the power MOS. Ipg and Ip2 are set to zero current. Diode (in green) can be either present or not and with different areas.

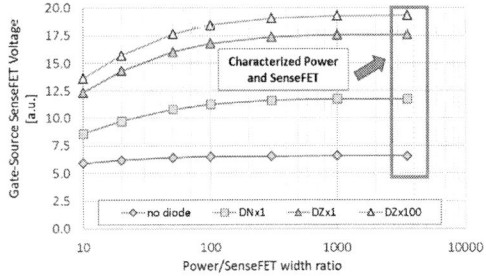

Fig. 10. Voltage drop across the gate oxide of the Sense-FET in different configurations (simulations). The geometrical properties of the system power MOS and Sense-FET characterized in this paper are highlighted. The voltage supply has been set to 20 a.u. and depends on plasma current. DN diode has a reverse-bias leakage lower than DZ. When no diode is present at MOS2 side, the voltage drop is negligible. Gate to Source voltage drop on Sense-FET depends on the area (i.e. total leakage current) of the inserted diodes.

Moreover, by increasing the power transistor width and/or the area of the diode placed at the sense-FET side, the gate to source voltage drop of MOS2 strongly increases: it can be high enough to create a damage in the oxide, while the electrical field on gate oxide of MOS1 remains negligible.

VI. SUMMARY

In the present paper, a thorough analysis of the behavior under DC plasma of a very common circuit used in smart power applications has been provided. The standard protection approach is to limit the gate voltage increase towards ground. We showed that the plasma charging on gate is not the only source of plasma damage in a common smart power circuit. In these platforms, the well pockets are capable of withstanding several tens (hundreds) of volts and the coupling (due to capacitance and/or other parasitic effects) of the gate with these pockets during wafer processing can lead to a failure. In the case under study, current sense with a common gate between a power device and a small transistor, a high voltage drop occurs on the gate oxide of the small FET.

This failure mechanism represents a novelty and therefore classical plasma rules and control methods [12] are not enough to prevent this type of damage. Appropriate protection techniques, such as bridging, must be used to avoid potential issues during wafer diffusion. Simple antenna protection diode (commonly used in most designs) has been proven to be not always adequate and even harmful in few configurations, if its connection is not carefully studied.

ACKNOWLEDGMENT

The authors gratefully acknowledge Giuseppe Croce, Vincenzo D'Urzo and Simone Alba for the helpful and productive technical discussion.

REFERENCES

[1] D.A. Grant and J. Gowar, " Power MOSFETs: theory and applications" John Wiley & Sons, Ed. 1989.

[2] Xiaoyu Li et al, " Effect of plasma poly etch on effective channel length and hot carrier reliability in submicron transistors" in IEEE Electron Device Letters, vol. 15, no. 4, pp. 140-141, 1994.

[3] S. Fang and J. P. McVittie., " Oxide damage from plasma charging: breakdown mechanism and oxide quality" in IEEE Transactions on Electron Devices, vol. 41, no. 6, pp. 1034-1039, 1994.

[4] T. Gu et al., "Impact of polysilicon dry etching on 0.5 μm NMOS transistor performance: the presence of both plasma bombardment damage and plasma charging damage" in IEEE Electron Device Letters, vol. 15, no. 2, pp. 48-50, 1994.

[5] S. Krishnan, A. Amerasekera, S. Rangan and S. Aur, "Antenna Device Reliability for ULSI Processing" in International Electron Devices Meeting 1998. Technical Digest, pp. 601–604, 1998.

[6] D. Park and C. Hu, "Plasma Charging Damage on Ultrathin Gate Oxides" in IEEE electron device letters, vol. 19, n. 1, 1998.

[7] D. Beckmeier and A. Martin, "Plasma process induced damage detection by fast wafer level reliability monitoring for automotive applications" in Elsevier Microelectronics Reliability, vol.64 pp.189-193, 2016.

[8] W. Lin and G. Sery, "Role of Source/Drain Junction on Plasma Induced Gate Charging Damage in N MOSFET" in 6th International Symposium on Plasma Process-induced Damage, pp.112-116, American Vacuum Society, 2001.

[9] W. Lin, "A multiple-terminal gate charging model" in IEEE Electron Device Letters, vol. 24, no. 8, pp. 521-523, 2003.

[10] C. T. Gabriel and J. L. Educato, "Application Of Damage Measurement Techniques To A Study Of Antenna Structure Charging" in 2nd International Symposium on Plasma Process-Induced Damage, pp. 91-94, 1997.

[11] Z. Wang, A. Scarpa, S.M. Smits, C. Salm and F.G. Kuper "Temperature effect on antenna protection strategy for plasma-process induced charging damage" in 7th International symposium of Plasma Process-Induced Damage, pp. 134-137, 2002.

[12] C. T. Gabriel and E. de Muizon, "Quantifying a simple antenna design rule" in 5th International Symposium on Plasma Process-Induced Damage, pp. 153-156, 2000.

Proceedings of the 31st International Symposium on Power Semiconductor Devices & ICs
May 19-23, 2019, Shanghai, China

A Novel Self-Regulated Potential SOI LIGBT With Low ON-State Voltage and Turn-off Loss

Yun Xia, Wanjun Chen*, Wuhao Gao, Bin Qiao, Chao Liu, Yijun Shi, Yajie Xin, Fangzhou Wang, Yu Shi,
Ruize Sun, Qi Zhou, Zhaoji Li, and Bo Zhang

State Key Laboratory of Electronic Thin Films and Integrated Devices, University of Electronic Science and
Technology of China, Chengdu610054, China
(e-mail: wjchen@uestc.edu.cn)

Abstract—**In this paper, a novel Self-Regulated Potential SOI-LIGBT (SRP-LIGBT) with low on-state voltage (V_{on}) and turn-off loss (E_{off}) is proposed and demonstrated by TCAD simulation. By employing two NMOS structures to regulate the potential of P body and N+ source, respectively, the proposed device can operates under latch up mode at low anode voltage and operates under saturation mode at higher anode voltage. Thus stronger conductivity modulation and more uniform carriers distribution in the drift region will be achieved, which brings the proposed device with a low V_{on} of 1.22V, showing 34.76% reduction when compared with conventional LIGBT (Con-LIGBT) at the same E_{off}. Moreover, there is no sacrifice in the complexity or cost of process.**

Keywords—*LIGBT, SOI, On-state Voltage, Turn-off Loss.*

I. INTRODUCTION

Silicon-On-Insulator Lateral insulated gate bipolar transistor (SOI-LIGBT) is a key device for high voltage monolithic ICs due to its high input impedance, low power dissipation and excellent isolation [1-5]. In order to reduce the power consumption and shrink the chip size, SOI-LIGBTs with low on-state voltage (V_{on}) and turn-off loss (E_{off}) are preferred. Several structures with low V_{on} and E_{off} such as trench barrier structure [6-7] and carriers store layer structure [8-9] have been reported. However, these structures with deep trench or carriers store layer increase the fabrication complexity and the cost.

In this work, a Self-Regulated Potential SOI-LIGBT (SRP-LIGBT) is proposed and demonstrated by TCAD simulation. By employing two extra n-channel MOSFET (NMOS) structures to regulate the potential of P body and N+ source of the conventional LIGBT (Con-LIGBT) respectively, the proposed device can operates under latch up mode at low anode voltage and operates under saturation mode at higher anode voltage, thus inducing stronger conductivity modulation and more uniform carriers distribution in the drift region. Compared with the Con-LIGBT, the proposed structure achieves a better V_{on}-E_{off} tradeoff performance without raising the complexity or cost of process. V_{on} of the proposed one (1.22V) is 34.76% lower than Con-LIGBT (1.87V) under turn-off loss (E_{off}) of 3.5mJ/cm².

This work was supported in part by the National Natural Science Foundation of China (No. 51877030), the Sichuan Youth Science and Technology Foundation (No. 2017JQ0020), the Major Science and Technology Special Projects in Guangdong under Grant (No. 2017B010112003) and the Fundamental Research Funds for the Central Universities (No. ZYGX2016Z006).

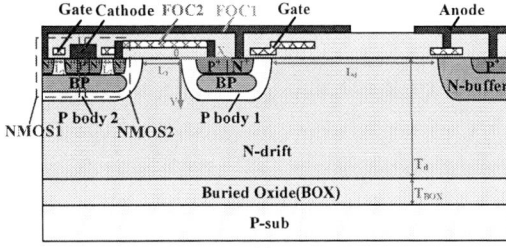

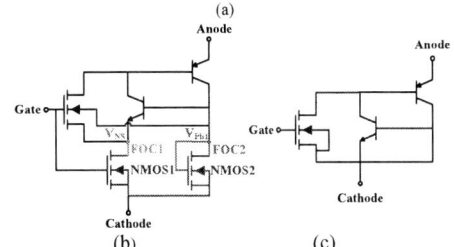

Fig. 1. (a) Cross section view of the proposed SRP-LIGBT. Simplified equivalent circuit of (b) SRP-LIGBT and (c) Con-LIGBT.

II. DEVICE STRUCTURE AND MECHANISM

Fig.1 shows the schematic view of the proposed SRP-LIGBT, as well as the simplified equivalent circuits of SRP-LIGBT and Con-LIGBT. Compared with the Con-LIGBT, the SRP-LIGBT features a different cathode zone which introduces NMOS1 and NMOS2 connected to the N+ source and P body of the Con-LIGBT, respectively. The gate electrode of NMOS1 is connected to the gate electrode above P body 1, and the drain electrode of NMOS1 is connected to the N+ source through floating ohmic contact 1(FOC1). The gate electrode and the drain electrode of the NMOS2 are shorted by floating ohmic contact 2 (FOC2) which is also contacted to the P body 1. NMOS2 can automatically turn on with the increase of anode voltage (V_A), consequently, an improved V_{on}-E_{off} tradeoff relationship is achieved.

When the device turns on, most of the holes injected from the P+ anode are gathered by P body 1, which increases the potential of P body 1(V_{Pb1}). Once the potential difference of P body1 and N+ source (ΔV_{PN}, $\Delta V_{PN}=V_{Pb1}-V_{NS}$) is greater than 0.7V, the P body1/N+ source junction will turn on and latch up issue occurs, which reduces the on resistance dramatically.

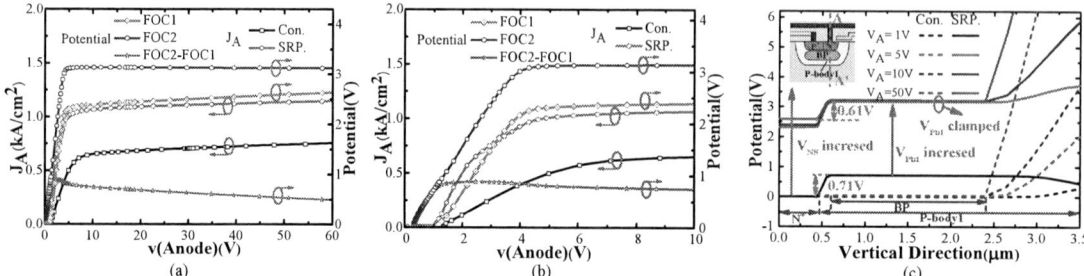

Fig.2. (a) The forward I-V characteristics and the equivalent potential of FOC1 and FOC2. (b)Zoomed-in forward I-V characteristics and the equivalent potential of FOC1 and FOC2 at low V_A. (c) The vertical potential distribution along line AA' (V_G=10V) with varying V_A.

Table I Key Device Parameters

Parameters	Definitions	Values
T_d	The thickness of the SOI layer	18μm
T_{BOX}	The thickness of the BOX	3μm
L_d	The length of the drift	47μm
N_{drift}	The concentration of the drift	8.3E14 cm^{-3}
N_{pbody}	The concentration of the P body	7.8E16 cm^{-3}
N_{BP}	The buried p layer (BP) concentration	3E18 cm^{-3}
N_{p+}	All the P+ concentration	5E19 cm^{-3}
N_{N+}	All the N+ concentration	2E20 cm^{-3}

Since V_{Pb1} is increased with the increase of V_A, when it exceeds the threshold voltage of NMOS2, NMOS2 will conduct automatically and the holes in the P body 1 will flow into the cathode electrode through NMOS2, then V_{Pb1} will be clamped by NMOS2. In the meantime, with the increase of conducting current, the potential of N+ source (V_{NS}) is increased because of the on resistance of NMOS1, resulting in ΔV_{PN} less than 0.7V, hence the SRP-LIGBT drops out of the latch up mode and falls into saturation mode.

III. RESULTS AND DISSCUSSION

Simulations are carried out by MEDICI. Some optimized key device parameters are listed in Table I and carriers lifetime of 10μs is set.

Fig.2 (a) and (b) show the forward I-V characteristics of the two LIGBT structures and the equivalent potential of FOC1 and FOC2.The potential difference value (V(FOC2-FOC1)) of FOC2 and FOC1 is approximate to the potential difference value of P body1 and N+ source (i.e., ΔV_{PN}). It is clear that V(FOC2-FOC1) first rises greater than 0.7V then descends less than 0.7V, which shows the proposed LIGBT is first latching up then saturating. Fig.2(c) shows the vertical potential distribution along line AA' with varying V_A. The V_{NS} is increased with the augment of V_A steadily while V_{Pb1} is first increased with V_A then clamped by NMOS2, which is consistent with the variation trend of V(FOC2-FOC1). Therefore, the proposed LIGBT achieves stronger conductivity modulation in the drift due to latch up issue occurred at low V_A, which helps to achieve a lower on-state voltage drop.

Fig.3 (a) and (b) show the current path of the SRP-LIGBT when V_A= 1V and V_A=50V, respectively, which indicates that the proposed device is under latch up mode when V_A=1V and

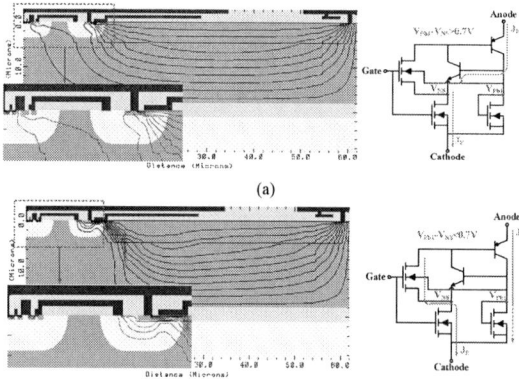

(a)

(b)

Fig.3. Current path of the SRP-LIGBT when (a) V_G= 10V, V_A= 1V (b) V_G= 10V, V_A=50V.

under saturation mode when V_A=50V. As shown in Fig.3 (a), the NMOS2 does not turn on yet and the P body 1/N+ source junction has turned on, so that the conductance modulation in the drift region on the P body 1 side is enhanced dramatically. As shown in Fig.3 (b), the NMOS2 has turned on, the P body 1/N+ source junction has turned off and the holes current in the drift flows to the cathode electrode through NMOS2, which indicate that the device is under saturation mode. Fig.3 (a) and (b) show that NMOS2 can automatically turn on when anode voltage is high enough, which turns the device from latch up mode to saturation mode.

Note that the I-V characteristic of the proposed LIGBT is closely related to the conduction of both NMOS1 and NMOS2, hence related variables (NMOS1 channel length L_1, NMOS2 channel length L_2, distance between P body 1 and P body 2 L_3, gate oxide thickness T_{ox}) should be carefully designed. NMOS2 should be designed to keep the holes accumulated in P body 1 at low V_A to enhance conductivity modulation and discharge the holes at higher V_A to clamp V_{Pb1}; NMOS1's on resistance should be large enough to turn the device from latch up mode to saturation mode while keep low on state-voltage. Fig.4 (a) shows that P body 1 and P body 2 should keep a long enough distance to avoid too much holes flow to the cathode electrode through P body 2 at low V_A, otherwise a snapback phenomenon would occur due to the turn on delay of P body 1/N+ source junction. Fig.4 (b) shows the thinner the T_{ox} is, the lower the V_{on}.

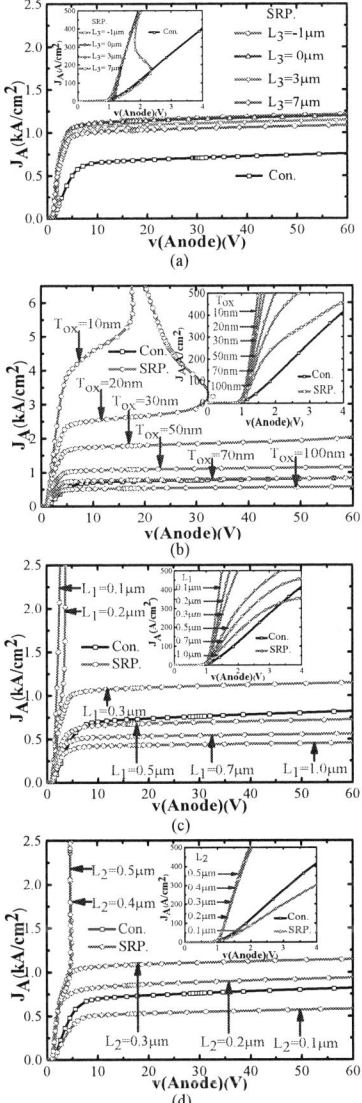

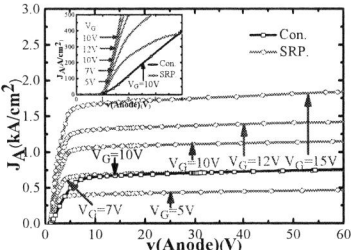

Fig.5. The forward I-V characteristics of the Con-LIGBT (V_G=10V) and the SRP-LIGBT with varying V_G.

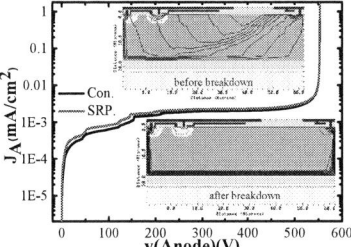

Fig.6. The forward BV characteristics of the Con-LIGBT and the SRP-LIGBT.

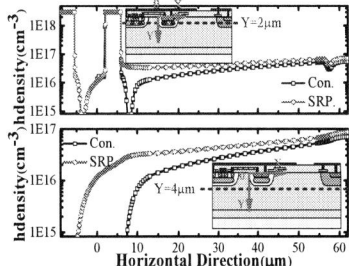

Fig.7. The hole distribution along Y=2μm and Y=4μm when J_A=100A/cm^2 and V_G=10V.

Fig.4. The forward I-V characteristics of the Con-LIGBT (T_{ox}=50nm) and SRP-LIGBT with (a) varying L_3 (V_G=10V, L_1=0.3μm, L_2=0.3μm, T_{ox}=50nm), minus means P body 1 and P body 2 is partially overlapped. (b) varying T_{ox} (V_G=10V, L_1=0.3μm, L_2=0.3μm, L_3=3μm). (c) varying L_1 (V_G=10V, T_{ox}=50nm, L_2=0.3μm, L_3=3μm). (d) varying L_2 (V_G=10V, T_{ox}=50nm, L_1=0.3μm, L_3=3μm).

However, a too thin gate oxide would induce too stronger conductance modulation, then the saturation current would be too large to withstand. Fig.4 (c) shows that the shorter the L_1 is, the lower the V_{on} and the saturation current, but if L_1 is too short, the device will be hard to turn from latch up mode to saturation mode. Fig.4 (d) shows that a too short L_2 would enlarge the V_{on} because holes cannot be fully blocked at low V_A owing to the

leakage current of NMOS2 induced by the too short channel. In conclusion, T_{ox}=50nm, L_1=0.3μm, L_2=0.3μm, L_3=3μm is a proper choice.

Fig.5 shows the I-V characteristics of the Con-LIGBT and the SRP-LIGBT with varying gate voltage (V_G). The result reveals that even with a much smaller gate voltage, the proposed structure achieves a lower V_{on} when compared with the conventional one, therefore the power consumption of gate drive could be reduced.

Fig.6 depicts that the forward blocking characteristics of the both structures. The proposed structure has a slight increase in leakage current due to a slight increase in V_{Pb1} induced by the conduction of NMOS2 during forward blocking.

Fig.7 exhibits the hole distribution of the both structures when J_A=100A/cm^2, more uniform carriers distribution of the SRP-LIGBT is achieved, and more hole carriers around the P body 1 are observed, which results in an ultra-low on-state voltage and a better tradeoff performance.

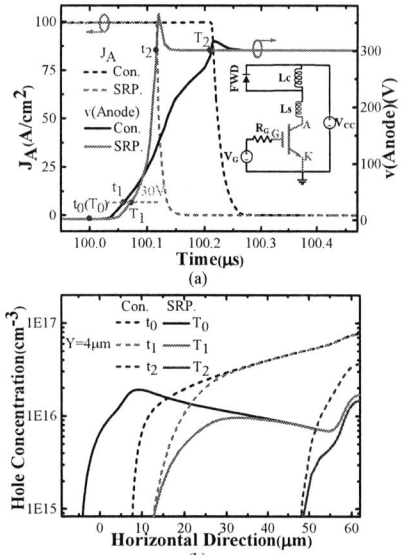

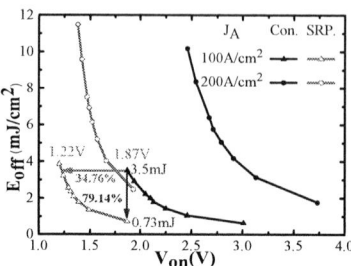

Fig.9. Tradeoff relationship between V_{on} and E_{off} of the Con-LIGBT and SRP-LIGBT.

Fig.8. (a) Turn-off waves under inductive load, V_{on} of both structures are adjusted to the same as 1.87V by change the doping concentration of P+ anode. (b) Hole distribution in the drift during turn-off. Simulated turn-off density is J_A=100A/cm^2, bus voltage is V_{CC}=300V and load inductance is L_c=1mH, stray inductance is L_S=10nH, R_G=5Ω, cell area is 0.3cm^2 and V_G is from 10 to 0V.

The turn-off waveforms under inductive load of two LIGBT structures are shown in Fig.8 (a), V_{on} of both structures are adjusted to the same as 1.87V at 100A/cm^2 by change the doping concentration of P+ anode. The proposed structure exhibits a faster turn-off due to an improved carriers distribution in the drift region. Fig.8 (b) shows the hole distribution in the drift of two structures at different times ($t_0 \sim t_2 / T_0 \sim T_2$) during turn off. Owing to stronger conductivity modulation in the cathode side, high carrier density at the cathode side and low carrier density at the anode side are realized, resulting in a faster extraction of holes in the drift during turn off, thus E_{off} is decreased.

Fig.9 comparatively depicts the tradeoff relationship between V_{on} and E_{off} of the two structures at J_A of 100A/cm^2 and 200A/cm^2. An improved trade-off performance of the SRP-LIGBT is achieved, in which E_{off} of the proposed one (0.73mJ/cm^2) decreases 79.14% compared with the Con-LIGBT (3.5mJ/cm^2) at the same V_{on} of 1.87V under J_A =100A/cm^2。

IV. CONCLUSION

A novel self-regulated potential SOI-LIGBT is proposed and investigated in this work. With an improved carrier distribution in drift region, the proposed structure can achieve a significantly improvement in V_{on}-E_{off} tradeoff relationship. The simulation results indicate that, at the same E_{off}, V_{on} of the proposed one is 34.76% lower than the conventional one and its E_{off} is 79.14% lower than the other at the same V_{on}.

REFERENCES

[1] D. Disney, T. Letavic, T. Trajkovic, T. Terashima and A. Nakagawa, "High-Voltage Integrated Circuits: History, State of the Art, and Future Prospects," in IEEE Transactions on Electron Devices, vol. 64, no. 3, pp. 659-673

[2] K. Hara et al., "600V single chip inverter IC with new SOI technology," 2014 IEEE 26th International Symposium on Power Semiconductor Devices & IC's (ISPSD), Waikoloa, HI, 2014, pp. 418-421.

[3] A. Nakagawa, H. Funaki, Y. Yamaguchi and F. Suzuki, "Improvement in lateral IGBT design for 500 V 3 A one chip inverter ICs," 11th International Symposium on Power Semiconductor Devices and ICs. ISPSD'99 Proceedings (Cat. No.99CH36312), Toronto, Ont., Canada, 1999, pp. 321-324.

[4] K. Sakurai, D. Maeda and H. Hasegawa, "Three-Input Type Single-Chip Inverter IC including a Function to Generate Six Signals and Dead Time," 2008 20th International Symposium on Power Semiconductor Devices and IC's, Orlando, FL, 2008, pp. 323-326.

[5] W. Sun et al., "A Novel Silicon-on-Insulator Lateral Insulated-Gate Bipolar Transistor With Dual Trenches for Three-Phase Single Chip Inverter ICs," in IEEE Electron Device Letters, vol. 36, no. 7, pp. 693-695, July 2015.

[6] T. Matsudai, M. Kitagawa and A. Nakagawa, "A trench-gate injection enhanced lateral IEGT on SOI," Proceedings of International Symposium on Power Semiconductor Devices and IC's: ISPSD '95, Yokohama, Japan, 1995, pp. 141-145.

[7] L. Zhang et al., "Low-Loss SOI-LIGBT With Dual Deep-Oxide Trenches," in IEEE Transactions on Electron Devices, vol. 64, no. 8, pp. 3282-3286, Aug. 2017.

[8] Shigeki, Akio, Youichi, Satoshi and Norihito, "Carrier-storage effect and extraction-enhanced lateral IGBT (E^{2}LIGBT): A super-high speed and low on-state voltage LIGBT superior to LDMOSFET," 2012 24th International Symposium on Power Semiconductor Devices and ICs, Bruges, 2012, pp. 393-396.

[9] J. Sakano et al., "Large current capability 270V lateral IGBT with multi-emitter," 2010 22nd International Symposium on Power Semiconductor Devices & IC's (ISPSD), Hiroshima, 2010, pp. 83-86.

978-1-7281-0582-6/19 $31.00 © 2019 IEEE

Proceedings of the 31st International Symposium on Power Semiconductor Devices & ICs
May 19-23, 2019, Shanghai, China

Performance and Reliability Co-design of LDMOS-SCR for Self-Protected High Voltage Applications On-Chip

Nagothu Karmel Kranthi[1], B. Sampath Kumar[1], Akram Salman[2], Gianluca Boselli[2] and Mayank Shrivastava[1]

[1]Department of ESE, Indian Institute of Science, Bangalore, Karnataka, India;
[2]Texas Instruments Inc, Dallas, USA. E-mail: mayank@iisc.ac.in

Abstract— **In this work we address turn-on vulnerability of conventional LDMOS-SCR devices under standard circuit operation window. This behavior is correlated with early ESD / SoA failure and power-to-fail scalability issue in HV LDMOS-SCR devices. The 3D TCAD is used to Develop physical insights into the performance and reliability limiters of LDMOS-SCR device. Different engineered designs are proposed to mitigate turn-on vulnerability and ESD power to fail scalability, while keeping channel performance and hot carrier degradation unaffected.**

Index Terms—**Electrostatic Discharge, Laterally Double Diffused MOS (LDMOS), Silicon Controlled rectifier (SCR).**

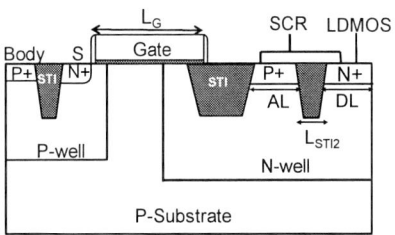

Fig. 1: Cross-sectional View of conventional LDMOS-SCR. To extract intrinsic LDMOS characteristics of LDMOS-SCR design, P+ (in N-Well) terminal was left floating.

I. INTRODUCTION

LDMOS devices are used to implement high voltage (HV) circuit functionalities in System on Chips (SoCs), Power SoCs (PwrSoC) and automotive ICs. Given the exposure of these HV applications to outside world and extreme conditions, LDMOS devices in these functionalities are often prone to early ESD damage. This is known to be attributed to space charge modulation induced current filament formation [1-2]. An embedded parasitic SCR in LDMOS (LDMOS-SCR) is proposed to have higher ESD robustness, which is a potential contender for self-protected HV applications [3-4]. However, it's failure under longer time electrostatic discharge – common in automotive environments – is still a major concern [5]. Besides, these devices can have functional SCR turn-on. These two issues hinder its self-protection capability and circuit functionality, respectively. This work for the first time attempts to provide physical insights into these issues. Moreover, this work provides physics-based device design to address these keeping channel performance and hot carrier reliability unaffected.

II. FUNCTIONAL TURN-ON ISSUES IN CONVENTIONAL DESIGN

The conventional LDMOS-SCR designs as depicted in Fig.1, when used as MOS switch, found to have SCR turn-on at high gate voltages (Fig. 2a & 2b) as signified by jump in drain current in I_D-V_D and loss of gate control in I_D-V_G characteristics. In presence of early SCR turn-on, the I-V characteristics drift away from the intrinsic LDMOS characteristics (Fig. 2c). The intrinsic LDMOS characteristics are obtained by not making a contact to the P+ island in the N-well. The SCR turn on in the device functional region is attributed to accumulation of channel injected electrons in the N-well region below P+ contact, before they reach N+ drain terminal. This results in early SCR action (Fig. 3), Which hinders device's usability as a switch or amplifier. A flipped configuration is proposed to address the SCR turn-on in the

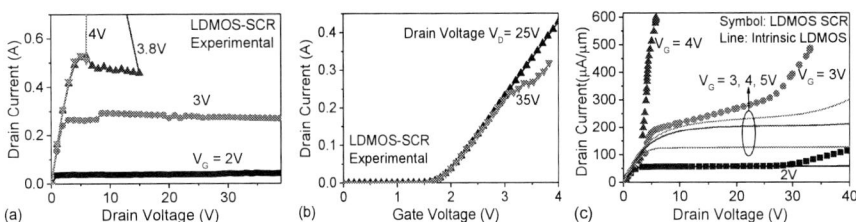

Fig. 2 (a) Measured DC I_D-V_D characteristics of LDMOS-SCR depicting an early SCR turn-on during the functional region. (b) Measured I_D-V_G characteristics show fall in drain current with increase in drain voltage, which is unlikely in intrinsic LDMOS. This is attributed to partial loss of gate's control once SCR is turned-on. (c) Simulated DC I_D-V_D characteristics of LMDOS-SCR depicting presence of strong SCR turn-on and its comparison with I_D-V_D characteristics of intrinsic LDMOS device.

978-1-7281-0582-6/19 $31.00 © 2019 IEEE

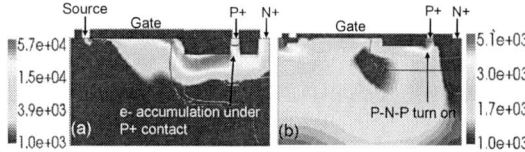

Fig. 3 Electron and hole current densities extracted at V_{GS}= 3V and V_{DS}=30V. Flooding of the gate induced electrons is observed under the P+ (in N-Well) contact, turns-on the parasitic SCR in the functional region.

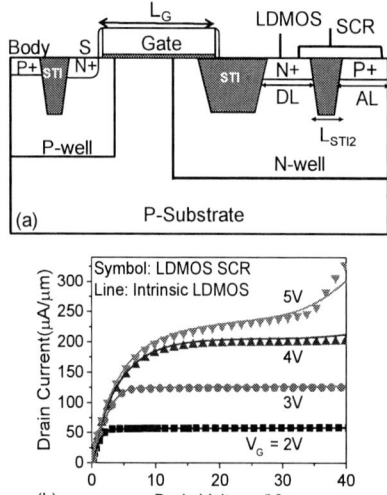

Fig. 4 (a) Cross-sectional view of flipped LDMOS-SCR device which suppresses early SCR action. Position of the N+ Drain and P+ contacts in the N-well are swapped in flipped device. (b) DC I_D-V_D characteristic of the LDMOS-SCR compared with intrinsic LDMOS characteristics confirms absence of SCR action in the functional region.

functional region. Position of the N+ and P+ diffusions in N-well are interchanged (Fig. 4). Proposed flipped configuration found to recover the MOS action by suppressing the PNP turn-on (Fig. 5). The electrons in the N-well are collected efficiently by N+, before they accumulate underneath the P+. However, such a weakening of the PNP action results in severe power scalability issue, as depicted in Fig.6.

III. POWER SCALABILITY PROBLEM

LDMOS-SCR devices, when used in automotive applications, can experience longer ESD discharges than the qualified 100 ns duration pulses [5]. The duration of these discharge depends on RC lines for which the high voltage I/O are connected. LDMOS-SCR is stressed with pulses of different duration, using a high impedance load Transmission Line Pulse (TLP) system. As depicted in Fig.6, LDMOS-SCR survives the snapback and fail at high current when stressed with pulse widths less than 100 ns. For longer pulse times (>100ns), device was found to fail at the verge of snapback (increase in leakage is depicted in fig.6(b)). As a result, the

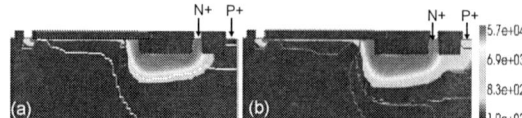

Fig. 5: (a) Electron current density across flipped LDMOS-SCR device extracted at (a) V_D=30 V and V_D=40 V for maximum gate voltage (5V). Gate induced electrons in the N-well are effectively collected by the N+ Drain contact before they accumulate under the P+ contact at higher drain voltage. This phenomenon shifted the onset of SCR turn-on in flipped device beyond device's operating I-V window, mentioned parameters are the knob to tune the performance and voltage rating of the device.

failure current and failure power do not follow expected power law curve as depicted in Fig.7. The device failure is also found to be specific to a window between the snapback and holding current, above which device survives failure. 3D TCAD simulations are used to understand the unique failure physics near the snapback region. The transient lattice temperature plotted for different injected current reveals (Fig. 8) unique device behavior. At low current, device temperature increases linearly with time. At currents near the snapback region, lattice temperature increases above a critical value (Fig. 8), which leads to device failure due to filament induced hotspot formation as depicted in Fig.9. It is observed that majority of the current is collected at N+ contact in the N-well, hotspot is observed underneath the N+ in N-well.

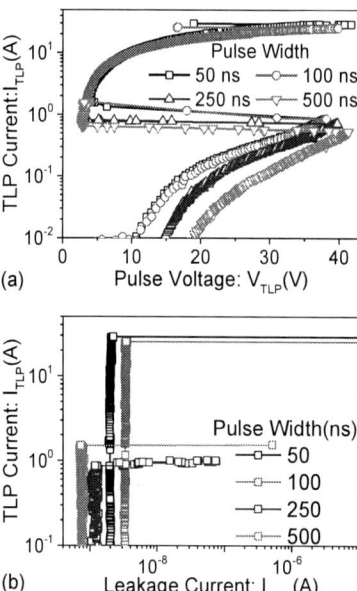

Fig. 6: (a) Measured TLP I-V characteristics of LDMOS-SCR device for different stress pulse widths in grounded gate configuration (b) Leakage current measured after each pulse. Device stressed beyond 100ns duration are observed to fail during the voltage snapback region. The observed failure is found to be specific to window of current near the snapback region.

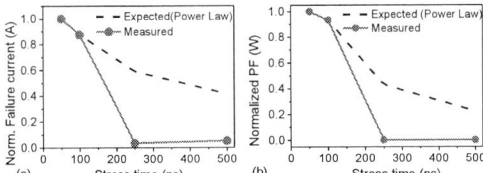

Fig. 7: (a) Normalized Failure current (Norm. with I_{Fail} @ 50ns) and (b) Norm. power to fail extracted at different stress times. The LDMOS-SCR device has shown a sudden fall in failure current from 25A at 100 ns to 1.5 A for 250 ns, depicting severe power-to-fail scalability issues. The expected failure current scaling according to the power law behavior is plotted in the same graph.

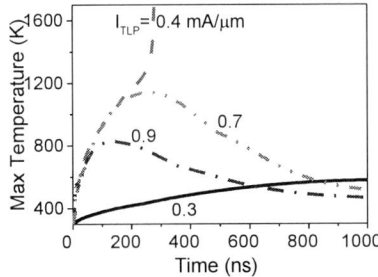

Fig. 8: Maximum lattice temperature as a function of stress time in flipped LDMOS-SCR device for different injected TLP currents, in the low current snapback region. For smaller currents lattice temperature increases linearly with time, for medium current levels, the lattice temperature increases exponentially leading to peak temperature above failure temperature. However, with increasing TLP current the temperature falls to lower values after reaching a maximum. A unique device failure is visible in a window of currents.

With further increase in injected current, the magnitude of peak reduces and time at which the peak appeared shift to lower time scales (Fig.8). This is attributed to faster turn-on of the parasitic SCR (Fig.10) at higher injected currents. This allowed current spreading, mitigated device heating and allowed device to survive failure at high currents (Fig.10). Majority of the current is observed to be collected by P+ in the N-well, shows the SCR turn on (Fig.10). Different physical events are summarized in fig.11.

IV. DESIGN SOLUTIONS

Two physics-based design solutions are proposed to modulate the P-N-P and or SCR turn on time. (1) Reducing N+ drain diffusion length DL. This weakens the LDMOS action, increases electron accumulation in N-well and leads to faster SCR turn on (Fig.12). Faster SCR turn-on reduces the lattice temperature near the snapback causes safe snapback. (2) Increasing AL too lowers the SCR turn-on time, which is due to increase in emitter area (Fig. 13). These two designs yielded power law behavior or power-to-fail scalability (Fig. 12c & 13c). While increasing AL doesn't affect channel performance and hot carrier reliability, DL reduction was found to improve channel performance (Fig.14) and increase Hot carrier degradation (Fig.15).

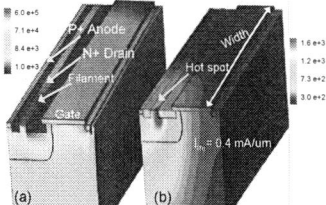

Fig. 9: (a) Current density (A/cm²) and (b) lattice temperature (K) for an injected current density of 0.4 mA/μm, extracted at 250ns. Filament induced hotspot causes device failure. The Parasitic SCR is not triggered in this case. Majority of the current being collected by N+ drain contact, not by P+ contact, depicting absence of SCR action.

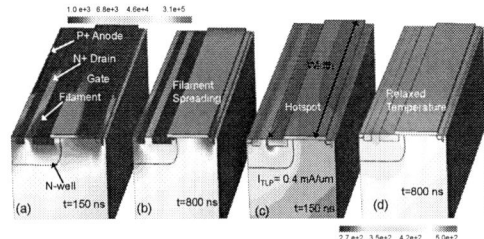

Fig. 10: (a) – (b) Current density (A/cm²) and (c) – (d) lattice temperature (K) extracted at different times: (a,c) 150ns & (b,d) 800ns, for an injected current of 0.9 mA/μm. (a,c) Absence of SCR turn-on (before device failure) leads to filament driven hot spot formation and device failure. (b,d) Fast PNP turn-on at higher currents enables SCR turn-on before LDMOS driven filament failure. This allows filament to spread and avoids early failure.

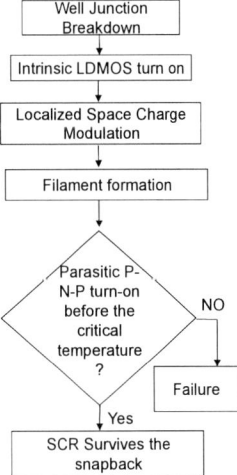

Fig. 11: various physical events responsible for poor power scalability in LDMOS-SCR.

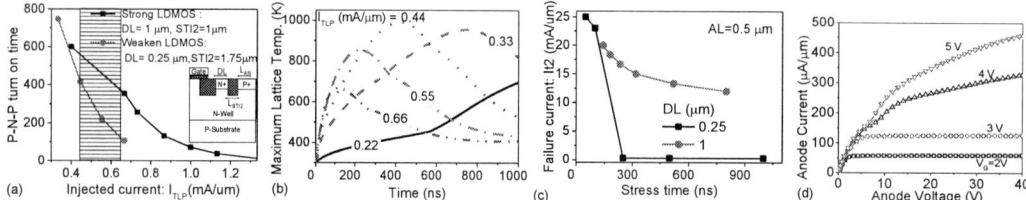

Fig. 12 (a) Comparison of SCR's _P-N-P turn-on time in LDMOS-SCR device in flipped configuration for different DL. By lowering DL, to reduce collection of electrons from N-Well, SCR turn-on gets strengthened. This is because of accumulation of electron in N-Well, which trigger SCR's parasitic PNP faster. (b) Temperature profile for different injection currents when DL was reduced. (c) Power scalability recovered with improved SCR action. (d) Output characteristics slightly deviate from the intrinsic LDMOS.

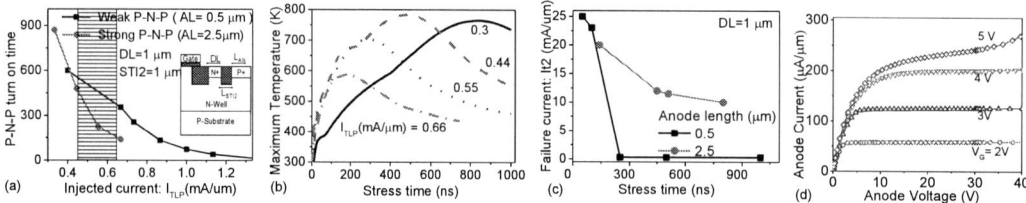

Fig. 13: (a) Comparison of SCR's P-N-P turn-on time in flipped LDMOS-SCR device configuration for different SCR strengths while keeping intrinsic LDMOS intact. By increasing P+ length (AL) SCR and PNP turn-on was strengthened. (b) Temperature profile for different stress currents when intrinsic PNP strength was enhanced by increasing AL. (c) Power scalability recovered when SCR's parasitic PNP turn-on was improved by increasing AL. (d) Output characteristics of device with increase AL.

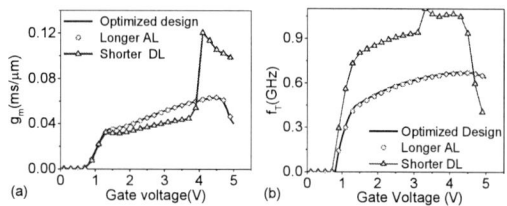

Fig. 14: (a) Transconductance, (b) cut-off frequency and (c) Gate to drain Capacitance of three different designs.

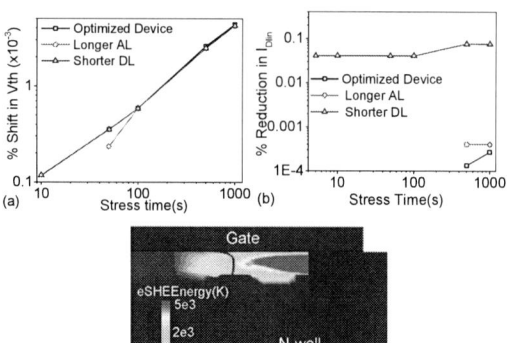

Fig. 15: Comparison of threshold voltage and linear drain current shift with stress duration for three different designs under stress. Shorter DL device found to has higher degradation in Idlin. (c) Electron Hot carrier energy.

V. Conclusion

The conventional LDMOS-SCR devices are found to be vulnerable for SCR turn on in the functional region attributed to high electron density under the P+ region in N-well. The engineered flipped device configuration can mitigate the SCR turn on because of weak P-N-P action. However, such design suffers from Power scalability issues when TLP stressed with long duration pulses. The inherent LDMOS forms filament due to the non-uniform space charge modulation near the snapback region and causes device failure. When stressed with higher current, the SCR turns on before the filament temperature reaches critical value, causes filament spreading and hotspot reduction. The proposed design solutions, where increasing AL and reducing DL have yielded improved power scalability in flipped designs. A reduction in DL causes more electron accumulation in N-well contributes to faster SCR turn on. Increasing AL implies larger emitter area of P-N-P that leads to improved SCR turn on time .

REFERENCES

[1] G. Boselli, V. Vassilev and C. Duvvury, "Drain Extended NMOS High Current Behavior and ESD Protection Strategy for HV Applications in Sub-100nm CMOS Technologies," *2007 IEEE International Reliability Physics Symposium Proceedings. 45th Annual*, Phoenix, AZ, 2007, pp. 342-347.

[2] M. Shrivastava and H. Gossner, "A Review on the ESD Robustness of Drain-Extended MOS Devices," in *IEEE Transactions on Device and Materials Reliability*, vol. 12, no. 4, pp. 615-625, Dec. 2012.doi: 10.1109/TDMR.2012.2220358.

[3] S. Pendharkar, R. Teggatz, J. Devore, J. Carpenter, T. Efland and C. -. Tsai, "SCR-LDMOS. A novel LDMOS device with ESD robustness," 12th International Symposium on Power Semiconductor Devices & ICs. Proceedings (Cat. No.00CH37094), Toulouse, France, 2000, pp. 341-344.

[4] A. Griffoni, S. -. Chen, S. Thijs, D. Linten, M. Scholz and G. Groeseneken, "Charged device model (CDM) ESD challenges for laterally diffused nMOS (nLDMOS) silicon controlled rectifier (SCR) devices for high-voltage applications in standard low-voltage CMOS technology," *2010 International Electron Devices Meeting*, San Francisco, CA, 2010, pp. 35.5.1-35.5.4.

[5] G. Boselli, A. Salman, J. Brodsky and H. Kunz, "The relevance of long-duration TLP stress on system level ESD design," Electrical Overstress/Electrostatic Discharge Symposium Proceedings 2010, Reno, NV, 2010, pp. 1-1

978-1-7281-0582-6/19 $31.00 © 2019 IEEE   410

Proceedings of the 31st International Symposium on Power Semiconductor Devices & ICs
May 19-23, 2019, Shanghai, China

Revealing the Positive Bias Temperature Instability in Normally-OFF AlGaN/GaN MIS-HFETs by Constant-Capacitance DLTS

Sen Huang, Xinhua Wang, Xinyu Liu, Xuanwu Kang, Jie Fan,
Shuo Yang, Haibo Yin, Ke Wei, Yingkui Zheng, Xiaolei Wang,
Wenwu Wang, Jingyuan Shi
Institute of Microelectronics of Chinese Academy of Sciences (CAS) and University of CAS
Beijing, China
Email: huangsen@ime.ac.cn; wangxinhua@ime.ac.cn;
xyliu@ime.ac.cn

Hongwei Gao, Qian Sun
Suzhou Institute of Nano-Tech and Nano-Bionics, CAS
Suzhou, China

Kevin J. Chen
Hong Kong University of Science and Technology, Clear Water Bay, Kowloon
Hong Kong

Abstract—Effects of oxide/III-nitride interface and oxide states on the threshold voltage instability (V_{TH}-instability) of normally-OFF Al_2O_3/AlGaN/GaN MIS-HFETS, i.e., metal-insulator-semiconductor heterojunction field-effect transistors, were revealed by constant-capacitance deep level transient spectroscopy (CC-DLTS). It is confirmed that a technique of *in-situ* remote plasma pretreatments could effectively suppress the D_{it} with level depth (E_C-E_T) larger than 0.4 eV down to below 1.3×10^{12} $cm^{-2}eV^{-1}$, in spite of the presence of a discrete level with E_C-E_T and capture cross section (σ_n) being 0.33 eV and 4.0×10^{-15} cm^2 respectively. However, electron charging of oxide states occurs when the MIS-HFETs are pulsed by a high positive gate bias (e.g., > 8 V), as confirmed by a reduced tunnel constant d_0 of 0.79 nm. High electric field induced tunnel filling of gate oxide states could be an assignable cause for the positive bias temperature instability in normally-OFF III-nitride MIS-HFETs.

Keywords—GaN, normally-OFF, MIS-HFETs, threshold voltage instability, interface states, oxide states, constant-capacitance DLTS, tunnel filling

I. INTRODUCTION

Normally-OFF AlGaN/GaN metal-insulator-semiconductor heterojunction field-effect transistors (MIS-HFETs), are highly preferred for high-voltage power switches due to their ultralow gate leakage and wide gate swing as compared with Schottky or PN-junction based HFETs [1]. However, the insertion of gate dielectrics creates an additional dielectric/III-nitride interface, accommodating not only a high density of interface states but also some deep bulk/border states in the gate dielectric itself [2],[3]. The dynamic capture/emission processes of these interface/bulk states could lead to threshold voltage (V_{TH}) instability issues of PBTI (positive bias temperature instability) and NBTI (negative bias temperature instability), which are regarded as a principal technology hurdle for commercialization of insulated-gate GaN-based power devices [4]. Therefore, it is worthwhile to investigate the capture/emission transient of

This work was supported in part by the Key Research Program of Frontier Sciences, Chinese Academy of Sciences (CAS) (No. QYZDB-SSW-JSC012), in part by Natural Science Foundation of China (No. 61822407, 61534007, 61527816, 11634002, 61334002, and 61631021), in part by the National Key R&D Program of China (No. 2016YFB0400105, 2017YFB0403000), and in part by the Opening Project of Key Laboratory of Microelectronic Devices & Integrated Technology, Institute of Microelectronics, CAS.

gate-dielectric/III-nitride interface (D_{it}) and bulk/border states, to shed light on the physical mechanism of V_{TH}-instability in GaN-based MIS-HFETs.

The dielectric/III-nitride interface have been intensively studied by pulsed transfer, capacitance-voltage, conductance and deep level transient spectroscopy (DLTS) techniques [5]. However, in-situ capture of V_{TH} shift as well as mapping the spatial distribution of interface/bulk states is difficult, due to the presence of several hetero-interfaces like Al_2O_3/AlGaN and AlGaN/GaN. Sophisticated model are required for D_{it} extraction. In this work, constant-capacitance DLTS (CC-DLTS) was adopted to investigate the V_{TH}-instability in normally-OFF Al_2O_3/AlGaN/GaN MIS-HFETs under PBTI conditions. By virtue of a fixed space-charge-region (SCR) associated with the CC-DLTS technique, V_{TH} transient were successfully captured. A wide mapping of the Al_2O_3/AlGaN interface with $0.02 < E_C$-$E_T < 0.91$ eV, was also realized. It is confirmed that remote plasma pretreatments (RPP) in ALD could effectively suppress deep interface states. Significant filling of states in the Al_2O_3 gate dielectric occurs at high gate bias (e.g., > 8 V), leading to PBTI in normally-OFF III-nitride MIS-HFETs.

II. LOW V_{TH}-HYSTERESIS NORMALLY-OFF Al_2O_3/ALGAN/GAN MIS-HFETS FABRICATED ON UTB-ALGAN/GAN HETEROSTRUCTURE

An ultrathin-barrier (UTB) AlGaN(<6 nm)/GaN heterostructure featuring a natural pinched-off 2-D electron gas (2DEG) channel, was utilized for fabrication of the normally-OFF MIS-HFETs [6]. The heterostructure was grown by metal-organic chemical vapor deposition

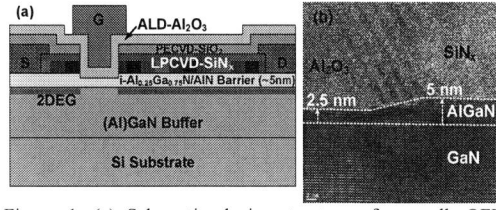

Figure 1. (a) Schematic device structure of normally-OFF Al_2O_3/AlGaN/GaN MIS-HFETs fabricated on ultrathin-barrier AlGaN/GaN heterostructures grown in Si substrate. (b) TEM cross sectional view of the device's gate corner.

978-1-7281-0582-6/19 $31.00 © 2019 IEEE

(MOCVD) on 4-inch Si (111) substrate. The fabrication process for the MIS-HFETs was similar as that reported in [6], as shown in Fig. 1(a). Note that after fluorine-based dry etching to open the gate region, the AlGaN barrier was further thinned by three cycles of plasma-oxidation/HCl-dipping digital recess. Then the wafer was transferred into the chamber of an Oxford instruments PEALD system, and *in-situ* RPP featuring NH$_3$-Ar-N$_2$ plasma applied to the sample in sequence at a substrate temperature of 300 °C. After RPP, a 30-nm Al$_2$O$_3$ layer was deposited *in situ*, followed by a post-deposition annealing (PDA) at 500 °C in oxygen ambient [3]. High-resolution transmission-electron spectroscopy (TEM) revealed that about 2.5-nm AlGaN barrier remained in the gate region (Fig. 1(b)).

The 2DEG sheet resistance (R_{2DEG}) of the MIS-HFETs, as determined by transfer length method (TLM), was ~286 Ω/□, indicating efficient recovery of 2DEG in the UTB-AlGaN/GaN heterostructure with low-pressure chemical vapor deposited (LPCVD) SiN$_x$ passivation. Fig. 2 shows dc transfer and breakdown characteristics of the normally-OFF MIS-HFETs. A saturated output current I_{Dmax} of 626 mA/mm was obtained at 25 °C at gate voltage V_{GS} of 12 V. The ON/OFF current ratio reached 10^{10} at drain bias of 10 V. The gate leakage current was well suppressed to below 5×10^{-8} mA/mm at $V_{GS} < 10$ V. An OFF-state breakdown voltage of 635 V was achieved at a drain current criterion of 10 μA/mm with substrate grounded.

Temperature dependent dc characteristics of the MIS-HFETs were plotted in Fig. 3 and 4. The I_{Dmax} decreased from 730 to 463 mA/mm as the substrate temperature was increased from -50 to 150 °C, and the extracted ON-resistance increased monotonically from 7.3 to 18.8 Ω·mm (Fig. 3(b)). Stable ohmic contact resistance R_{C_Ohmic} (~0.89 Ω·mm) is realized over the entire temperature range. The access R_{access} and gate channel $R_{channel}$ resistance can be calculated as follows:

$$R_{ON} = R_{channel} + 2R_{C\_Ohmic} + R_{access}, R_{access} = R_{2DEG}(L_{GS} + L_{GD})/W_G \quad (1)$$

Both of them exhibits similar increasing tendency with temperature, which is likely due to optical phonon scattering of 2DEG at high temperatures. Clockwise hysteresis is observed during double-mode transfer measurements at V_{DS} = 1 V (Fig. 4(a)). The extracted V_{TH} varies slightly between 0.5 and 0.7 V, while its hysteresis increases from 0.22 to 0.54 V as the temperature is increased from -50 to 100 °C

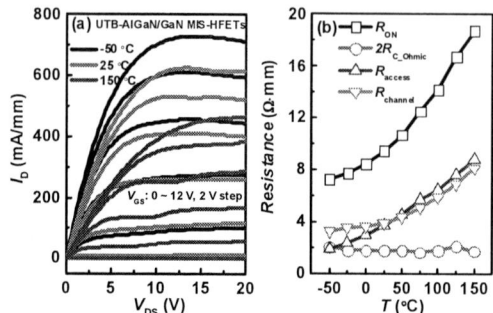

Figure 3. (a) dc *I-V* of the fabricated normally-OFF Al$_2$O$_3$/AlGaN/GaN MIS-HFETs measured at various temperatures. (b) Evolution of the extracted R_{ON}, $2R_{C\_Ohmic}$, R_{access}, and $R_{channel}$ of the MIS-HFETs with temperature.

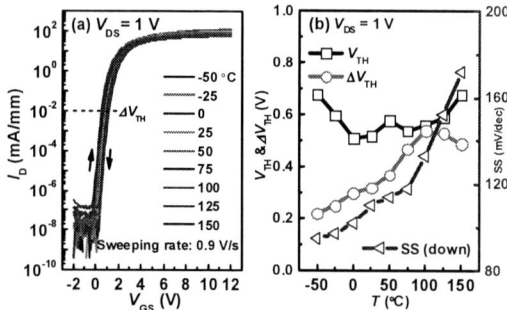

Figure 4. (a) Transfer characteristics of the fabricated MIS-HFETs at various temperatures. (b) Evolution of the extracted V_{TH}, ΔV_{TH}, and subthreshold slope SS with temperature.

and then dropped a little towards 150 °C (Fig. 4(b)). It indicates that some deeper interface/bulk states response at high temperatures, which is also reflected by the increased subthreshold slop (SS). Nevertheless, RPP in PEALD could be a compelling technique for suppression of interface states in III-nitride MIS devices [3].

III. INVESTIGATION OF THE PHYSICAL MECHANISM OF PBTI IN THE AL$_2$O$_3$/ALGAN/GAN MIS-HFETS BY CC-DLTS

To investigate the physical origin of the V_{TH}-instability in the normally-OFF MIS-HFETs, CC-DLTS were adopted for characterization of *large-area* MIS diodes featuring similar structure as the MIS-HFETs. Fig. 5(a) shows *C-V* characteristics of the MIS diode at -50, 25 and 150 °C, along with simulated curves. The increase of accumulation capacitance (C_{ACC}) with increasing temperature suggests activation of more deep levels. The CC-DLTS measurements were carried out in a PhysTech GmbH FT1030 high energy-resolution analysis (HERA) DLTS system, which was equipped with a Lakeshore low-temperature stage (10–400 K). The emission of interface/oxide states is monitored by a voltage transient (shift) $\Delta V_G(t)$, as shown in Fig. 5(b) and (c). It is generated by a feedback loop operating between the capacitance meter and the bias voltage source to maintain a constant capacitance C_R, and thus a fixed space-charge region (SCR). The MIS diodes were continually pulsed into accumulation by a gate bias U_P (pulse width t_P) to fill deep levels. Starting from t_0, a transient period of t_W will be

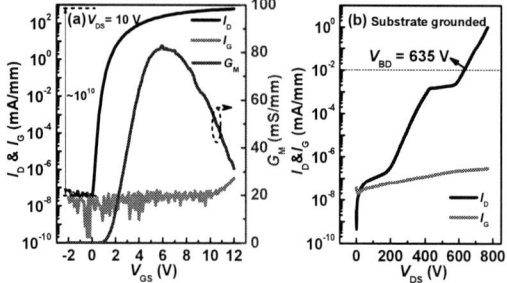

Figure 2. dc *I-V* characteristics of the fabricated normally-OFF Al$_2$O$_3$/AlGaN/GaN MIS-HFETs ($L_{GS}/L_G/L_{GD}/W_G$ of 3/2/10/100 μm). (a) dc transfer characteristics at V_{DS} = 10 V. (b) OFF-state breakdown characteristics measured with substrate grounded.

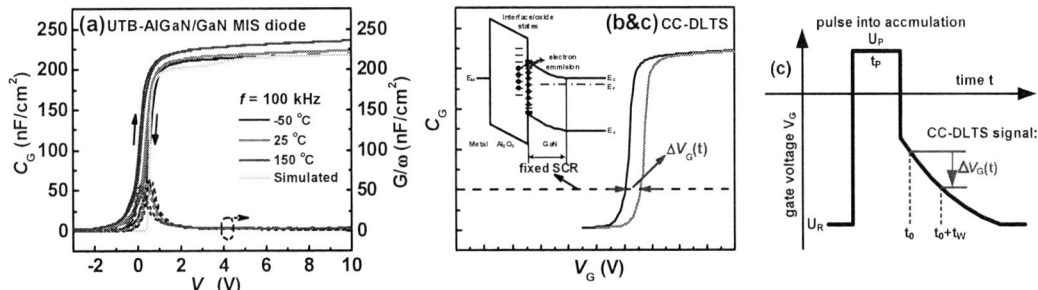

Figure 5. (a) Measured and simulated C-V curves of a fabricated Al_2O_3/AlGaN/GaN MIS diode at different temperatures. (b)&(c) Schematic band diagram (fixed space charge region), capacitance-voltage, and gate-voltage waveforms in CC-DLTS measurements of MIS diodes.

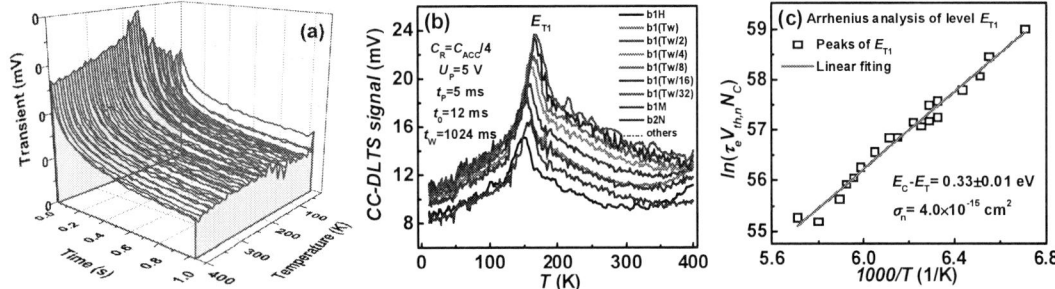

Figure 6. (a) Transients mapping of $\Delta V_G(t)$ in CC-DLTS measurement of the fabricated Al_2O_3/AlGaN/GaN MIS diode at a temperature range of 10-400 K. (b) CC-DLTS signals extracted from different coefficients of Fourier Transformation during a temperature scan. (c) Arrhenius plot of $ln(\tau_e V_{th,n} N_C)$ for the detected level E_{T1} in (b).

recorded and analyzed by Fourier Transformation (FT) method [7], [8].

Fig. 6(a) illustrates a 3D mapping of $\Delta V_G(t)$ for the MIS diode from 10 to 400 K (U_P: 5 V, t_P: 5 ms). C_R was set to be a quarter of C_{ACC}. A distinct peak appeared at about 165 K. By using FT analysis, CC-DLTS signals extracted from different FT coefficients, corresponding to different rate windows in conventional DLTS, are plotted in Fig. 6(b). A discrete deep level named as E_{T1} appears, and its level depth E_C-E_T and σ_n are determined to be 0.33±0.01 eV and 4.0 × 10^{-15} cm^2 by Arrhenius analysis shown in Fig. 6(c). It probably originated from remnant nitrogen vacancy on AlGaN surface [9], and optimization of the RPP is ongoing.

Evolution of density of detected states N_{ss} with t_P, was measured at various U_P to explore the capture behavior of interface/oxide states, using conventional DLTS (Fig. 7(a)). At U_P of 2 V, the measured density of state increased slowly with t_P and saturated at ~380 μs. When U_P was increased to 5 and 11 V, it increased sharply at first at $t_P < 6.4$ μs, and then got slowdown until saturation. The former might be a capturing behavior of interface states, while the latter is likely due to filling of oxide states, as electron tunneling is required [10]. The density of interface (N_{it}) and oxide (N_{ox}) states that being charged by electrons during capture measurements, can be given by [11],

$$N_{it}(t_P) = \int_{E_{FR}}^{E_{FP}} N_{ss}(E)(1-\exp(-t_p / \tau_c(E)))dE, \quad \tau_c(E) = \frac{1}{v_{th,n}\sigma_n(E)n_0} \quad (2)$$

$$N_{ox}(t_P) = \int_{0}^{d_{ox}} N_T(x)(1-\exp(-t_p / \tau_c(x)))dx \approx N_T d_0 \ln(t_p / \tau_c) \quad (3)$$

$$d_0 = \frac{d_{ox}}{\ln(t_{Ps} / \tau_c)}$$

where $\tau_c(E)$ and $\tau_c(x)$ are the capture time constant for interface and oxide states, as the latter may also depend on the distance between oxide states and oxide/semiconductor interface (x), $\sigma_n(E)$ is the capture cross section, and d_0 is the tunnel constant, E_{FR} and E_{FP} are Fermi level corresponding to gate bias of U_R and U_P, N_{ss} is the density of interface states in cm^{-2}eV^{-1}, N_T is the bulk density of oxide states in cm^{-3}, d_{ox} is the thickness of gate oxide, and t_{Ps} is the saturated filling pulse width.

The capture process of the interface states needs further investigation due to the lack of definite $\sigma_n(E)$ throughout the Al_2O_3/(Al)GaN interface. While for the oxide states, it can be observed that the a section of the capture curves with t_P larger than 6.4 μs, can be well fitted by equation (3) following a logarithmic time law, excepting for the curve measured at $U_P = 11$ V. The extracted d_0 decreases from 3.58 to 0.79 nm as U_P increases from 2 to 8 V, and N_T of detected oxide states is in the order of 10^{16} cm^{-3}. Therefore, a high gate bias can effectively reduce d_0 and thus accelerate the tunnel filling of oxide states.

The distribution of the interface/oxide states N_{ss} in the MOS diodes was successfully mapped by the CC-DLTS, with both U_P and t_P varies (Figure 7(b)). Note that an energy-independent σ_n of 4.0 × 10^{-15} cm^2 is assumed across the forbidden gap to comply with level E_{T1}. The measured N_{ss}

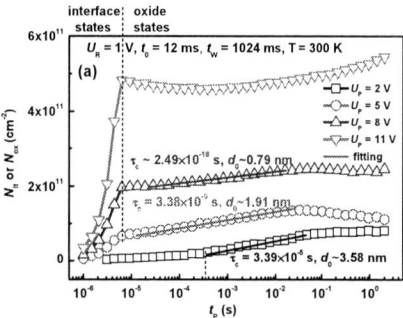

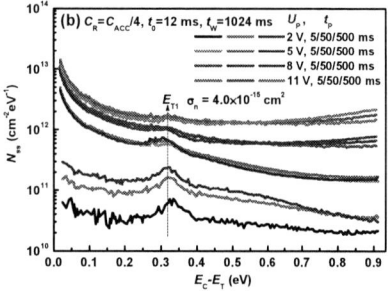

Figure 7. (a) Evolution of detected interface/oxide states (N_{it}/N_{ox}) with filling pulse width t_P measured by isothermal capture transient (T = 300 K). The capturing process boundary between interface and oxide states are marked by dashed lines. (b) Distribution of interface/oxide states N_{ss} measured at different U_P and t_P by CC-DLTS.

increases remarkably as t_P is increased from 5 to 500 ms at U_P of 2 V. While at U_P above 5 V, it exhibits minor changes with t_P, implying that t_P of 5 ms is enough for filling most of the interface/oxide states. The deep states with $E_C-E_T > 0.4$ eV keeps increase to a density of 1.3×10^{12} cm^{-2}eV^{-1} as U_P reaches 11 V, which is probably caused by tunnel filling of the oxide states. Such phenomenon is consistent with the increasing tendency of N_{it}/N_{ox} with U_P shown in Figure 7(a).

Based on the simulated C-V curve in Fig. 5(a), energy band diagrams of the MIS-HFETs at different $V_G(U_P)$, were simulated to shed light on the capture/emission mechanism (Fig. 8). Most of the deep interface states ($E_C-E_T < 0.9$ eV), oxide states E_{ox} and the 2DEG channel are not filled at $V_G = 0$ V, suggesting good normally-OFF behavior. As the filling pulse U_P is increased to 8 V (Fig. 8(b)), most of the interface states are charged to facilitate 2DEG spill over to the Al$_2$O$_3$/AlGaN interface. Moreover, the tunneling distance

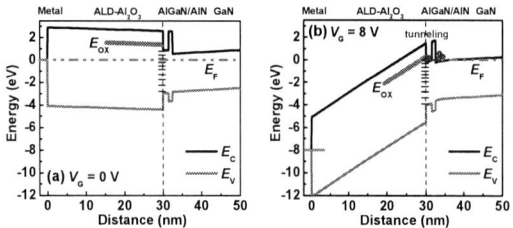

Fig. 8. Simulated energy band diagrams of the Al$_2$O$_3$/AlGaN/GaN MIS-HFETs at V_G of (a) 0 and (b) 8 V respectively. Oxide states E_{ox} in the ALD-Al$_2$O$_3$ gate dielectric are also indicated.

between 2DEG channel or even the electron-spilled interface and the oxide states E_{ox} are effectively reduced to 0.79 nm owing to strong band bending. Therefore, more electrons can be injected into E_{ox}, giving rise to a higher density of deep levels. Fermi level at $U_P = 11$ V may reach to level E_{ox}, resulting in early saturation of N_{ss} at 6.4 µs in Fig. 7(a).

IV. CONCLUSION

The capture/emission transient behavior of oxide/III-nitride interface and oxide states, and their effects on the V_{TH}-instability of normally-OFF Al$_2$O$_3$/AlGaN/GaN MIS-HFETs were investigated by CC-DLTS. It is demonstrated that *in-situ* RPP in PEALD is an effective method to suppress the states at Al$_2$O$_3$/(Al)GaN interface, while high gate bias induced tunnel filling of oxide states becomes an assignable cause for PBTI in normally-OFF III-nitride MIS-HFETs.

REFERENCES

[1] K. J. Chen, O. Haberlen, A. Lidow, C. lin Tsai, T. Ueda, Y. Uemoto, and Y. Wu, "GaN-on-Si Power Technology: Devices and Applications," *IEEE Trans. Electron Devices*, vol. 64, no. 3, pp. 779–795, Mar. 2017.

[2] S. Huang, S. Yang, J. Roberts, and K. J. Chen, "Threshold Voltage Instability in Al$_2$O$_3$/GaN/AlGaN/GaN Metal–Insulator–Semiconductor High-Electron Mobility Transistors," *Jpn. J. Appl. Phys.*, vol. 50, no. 11, p. 110202, Oct. 2011.

[3] S. Yang, Z. Tang, K.-Y. Wong, Y.-S. Lin, Y. Lu, S. Huang, and K. J. Chen, "Mapping of interface traps in high-performance Al$_2$O$_3$/AlGaN/GaN MIS-heterostructures using frequency- and temperature-dependent C-V techniques," in *2013 IEEE International Electron Devices Meeting*, 2013, p. 6.3.1-6.3.4.

[4] S. Zafar, A. Kumar, E. Gusev, and E. Cartier, "Threshold voltage instabilities in high-k gate dielectric stacks," *IEEE Trans. Device Mater. Reliab.*, vol. 5, no. 1, pp. 45–64, Mar. 2005.

[5] R. Engel-Herbert, Y. Hwang, and S. Stemmer, "Comparison of methods to quantify interface trap densities at dielectric/III-V semiconductor interfaces," *J. Appl. Phys.*, vol. 108, no. 12, p. 124101, Dec. 2010..

[6] S. Huang, X. Liu, X. Wang, X. Kang, J. Zhang, J. Fan, J. Shi, K. Wei, Y. Zheng, H. Gao, Q. Sun, M. Wang, B. Shen, and K. J. Chen, "Ultrathin-Barrier AlGaN/GaN Heterostructure: A Recess-Free Technology for Manufacturing High-Performance GaN-on-Si Power Devices," *IEEE Trans. Electron Devices*, vol. 65, no. 1, pp. 207–214, Jan. 2018.

[7] X. Liu, X. Wang, Y. Zhang, K. Wei, Y. Zheng, X. Kang, H. Jiang, J. Li, W. Wang, X. Wu, X. Wang, and S. Huang, "Insight into the Near-Conduction Band States at the Crystallized Interface between GaN and SiN x Grown by Low-Pressure Chemical Vapor Deposition," *ACS Appl. Mater. Interfaces*, vol. 10, no. 25, pp. 21721–21729, Jun. 2018.

[8] *FT 1030 HERA-DLTS Theory Manual*, http://www.phystech.de/products/dlts/dlts.htm (accessed Nov 21, 2016).

[9] S. Kim, Y. Hori, W.-C. Ma, D. Kikuta, T. Narita, H. Iguchi, T. Uesugi, T. Kachi, and T. Hashizume, "Interface Properties of Al$_2$O$_3$/n-GaN Structures with Inductively Coupled Plasma Etching of GaN Surfaces," *Jpn. J. Appl. Phys.*, vol. 51, no. 6 PART 1, p. 060201, May 2012.

[10] Sieghard Weiss, "Semiconductor Investigations with the DLTFS (Deep-Level Transient Fourier Spectroscopy)." Ph. D. Thesis, University of the Country of Hessen, Kassel, 1991.

[11] P. van Staa, H. Rombach, and R. Kassing, "Time-dependent response of interface states in indium phosphide metal-insulator-semiconductor capacitors investigated with constant-capacitance deep-level transient spectroscopy," *J. Appl. Phys.*, vol. 54, no. 7, pp. 4014–4021, Jul. 1983.

Proceedings of the 31st International Symposium on Power Semiconductor Devices & ICs
May 19-23, 2019, Shanghai, China

Demonstration of Electron/Hole Injections in the Gate of *p*-GaN/AlGaN/GaN Power Transistors and Their Effect on Device Dynamic Performance

Xi Tang[1,2,3], Baikui Li[1,*], Jun Zhang[2], Hui Li[3], Jisheng Han[2], Nam-Trung Nguyen[2], Sima Dimitrijev[2], and Jiannong Wang[3]

Email: libk@szu.edu.cn s.dimitrijev@griffith.edu.au phjwang@ust.hk

[1]College of Physics and Optoelectronics, Shenzhen University, Shenzhen, China
[2]Queensland Micro- and Nanotechnology Centre, Nathan, QLD 4111, Australia
[3]Department of Physics, The Hong Kong University of Science and Technology, Hong Kong, China

Abstract— **In this work, we demonstrated the electron/hole injections in the gate of *p*-GaN/AlGaN/GaN power transistors at forward gate bias by capturing the electroluminescence (EL) emission from the gate region. The EL included both visible and ultraviolet (UV) luminescence. The dynamic switching tests were carried out to investigate the effect of electron and hole injections on device performances. The electron injection and trapping at the forward gate bias caused positive threshold voltage shift, while the injection of holes into the GaN channel induced the emission of the UV light and the resulting leakage current increase in the device.**

Keywords—p-GaN, HEMTs, electron and hole injections, electroluminescence, threshold voltage shift, leakage current increase.

I. INTRODUCTION

A *p*-doped GaN layer on top of the AlGaN barrier can be employed in AlGaN/GaN power high-electron-mobility transistors (HEMTs) to deplete the two-dimensional electron gas (2DEG) at zero gate bias [1], [2]. These normally-OFF *p*-GaN/AlGaN/GaN HEMTs are now widely adopted by device manufacturers and research institutes due to the strong control over the gate region, less parasitic inductance, and a superior $R_{ON} \cdot Q_G$ figure of merit compared to other approaches to realizing the normally-OFF operation [3]. Also, the *p*-GaN gate technology features a good balance between performance and cost.

In recent years, the *p*-GaN gate related reliability issues are attracting more research interest and many studies have been performed on this topic. Among these reliability issues, the device performances such as the threshold voltage (V_{TH}) shift [4]–[10] and leakage current increase [11], [12] have attracted the most attention. The *p*-GaN/AlGaN/GaN heterostructure is a direct-bandgap hetero- PIN junction, in which the electron and hole injection processes are normal. In this work, we systematically investigated the electron and hole injections in the gate region and their effects on the device performances of a normally-OFF *p*-GaN/AlGaN/GaN power HEMT.

II. DEVICE FABRICATION

The *p*-GaN/AlGaN/GaN HEMT fabricated in this work is

This work was supported by the National Natural Science Foundation of China under Grant 61604098, the Shenzhen Science and Technology Innovation Commission under Grant JCYJ20170412110137562, and the Research Grants Council of the Hong Kong SAR under Grants C7036-17W-1 and C6013-16E.

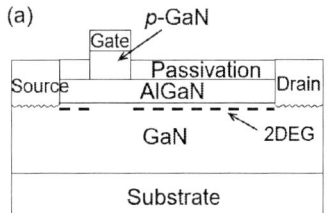

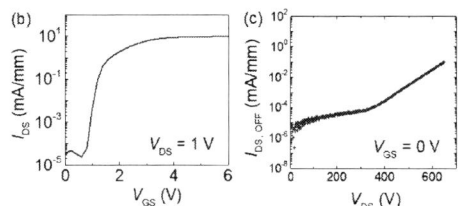

Fig. 1. (a) Schematic cross-section view of a *p*-GaN/AlGaN/GaN HEMT. (b) Transfer and (c) OFF-state leakage characteristics.

schematically illustrated in Fig. 1(a). The *p*-GaN/AlGaN/GaN heterostructure consists of a 70-nm *p*-type GaN, a 15-nm $Al_{0.25}Ga_{0.75}N$ barrier and a 4-μm GaN layer, grown by metalorganic vapour phase epitaxy (MOCVD) on a 4-inch *p*-type Si (111) substrate. The fabrication process was started with the ohmic contact formation. After defining the ohmic region, the *p*-GaN layer was etched by low-power chlorine-based inductively coupled plasma reactive ion etching (ICP-RIE). Then, a Ti/Al/Ni/Au (20/150/50/80 nm) metal stack was deposited in the ohmic region and annealed. After that, a metal stack of Ni/Cr was deposited in the gate region for (i) the gate metal contact and (ii) the patterning mask for the following *p*-GaN etching in the device access region. A plasma-enhanced chemical vapor deposition (PECVD) SiN_x (50 nm) was employed as the passivation layer, followed by the device active region isolation with fluorine ion implantation [13]. The pad electrode was formed by Ni/Au (20/200 nm) metal stack. Finally, the device was annealed at 400 °C for 10 min in nitrogen ambiance.

The static current-voltage (*I-V*) characteristics of the *p*-GaN/AlGaN/GaN HEMT are shown in Fig. 1(b) and 1(c). The characteristics were carried out on HEMTs with $L_{GS} = 2$ μm, $(W/L)_G = 300/2$ μm, and $L_{GD} = 10$ μm. The HEMT features a static threshold voltage (V_{TH}) of approximately 1 V at $I_{DS} = 10$ mA/mm. The breakdown voltage of the HEMT at $V_{GS} = 0$ V is over 600 V with the Si substrate grounded.

978-1-7281-0582-6/19 $31.00 © 2019 IEEE

III. ELECTRON/HOLE INJECTIONS AND ELECTROLUMINESCENCE EMISSION

The electron and hole injection processes in the *p*-GaN/AlGaN/GaN HEMTs were clarified by analyzing

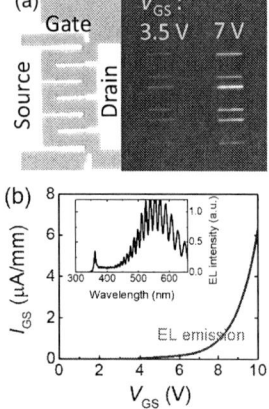

Fig. 2. (a) The bright-field photograph of the *p*-GaN/AlGaN/GaN HEMT and the EL image in the dark field. Light emission from the gate region was captured at forward gate bias (V_{GS} = 3.5 and 7 V, respectively) using a microscope camera. Here, the picture is in monochromatic. (b) Gate *I-V* characteristics of the *p*-GaN/AlGaN/GaN HEMT at room temperature. The insert shows the EL spectrum at the gate bias of 8 V.

electroluminescence (EL) emission. Here, a test device (DUT) was fabricated, which employed a semitransparent Schottky metal stack of Ni/Cr (5/6 nm) as the gate metal, for the purpose of picturing the EL emission from the gate region. The EL emission in the DUT was captured with a Nikon DS-Qi2 monochrome camera mounted on a Nikon Eclipse Ti2-E microscope as shown in Fig. 2(a). The light emission from the gate region at the forward gate bias was obtained in the dark field with its EL intensity getting stronger as the gate bias increased. Figure 2(b) shows the gate *I-V* characteristics of the DUT. A turn-on voltage of approximately 7 V can be identified from the gate linear *I-V* curve, which indicates the full turn-on of the gate hetero-PIN junction and the holes are continuously injected into the channel [1], [8]. The EL spectrum at V_{GS} = 8 V was measured using an Ocean Optics charge-coupled device (CCD) spectrometer as shown in the insert of Fig. 2(b). The EL spectrum contains both visible luminescence and GaN band-edge ultraviolet luminescence (UV). The visible emission was detected at a lower bias range while the UV component was observed when the gate-bias voltage was larger than 7 V. It can be deduced from the EL emission that, under forward gate bias, the electrons/holes are first injected into the *p*-GaN layer at low bias, generating the visible luminescence, and then the holes are injected into the GaN channel layer at higher bias, generating the UV emission [1], [14]. Although the EL spectrum varies with different epilayer structures, the obtained EL emission and spectrum demonstrate the electron and hole injection processes in this *p*-GaN/AlGaN/GaN heterostructure induced by forward gate bias.

To schematically illustrate the aforementioned injection processes, we draw the band diagram of the *p*-GaN/AlGaN/GaN heterostructure in Fig. 3. At forward gate

bias, the electrons from the 2DEG are injected into the *p*-GaN through the AlGaN barrier, as well as the holes start to drift in the *p*-GaN layer towards the *p*-GaN/AlGaN interface. The injected electrons can recombine with holes in the *p*-GaN through a donor-to-acceptor transition, generating the visible emissions [15], [16]. When V_{GS} further increases, the holes arriving at the *p*-GaN/AlGaN heterojunction interface will be injected across the AlGaN barrier into the GaN channel region, and recombine with the 2DEG electrons through a band-to-band transition, generating the GaN band-edge UV emission [1], [14], [17]. The generated UV emission needs to go through the *p*-GaN layer to be received

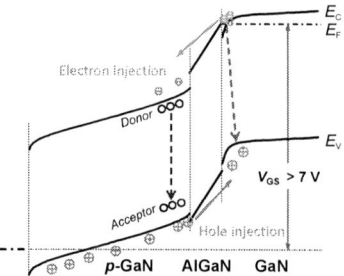

Fig. 3. Energy band diagram of the *p*-GaN/AlGaN/GaN heterostructure at forward biases larger than 7 V. The processes of electron/hole injections (solid arrow) and recombination (dashed arrow) are schematically illustrated.

by the spectrometer where the band-edge absorption was quite severe [18], [19], therefore, the intensity of the UV component revealed in Fig. 2(b) was underestimated. The analysis on possible origins and related energy levels of these donors/acceptors, which has been widely discussed in either *p*-type or unintentionally doped GaN [20], [21], is beyond the scope of this work.

IV. EFFECTS ON DEVICE DYNAMIC PERFORMANCES

A. Threshold Voltage Instability

In this section, the dynamic characteristics were carried out to investigate the effect of electron/hole injections on the *p*-GaN/AlGaN/GaN HEMT using an Agilent B1505A Power Device Analyzer/Curve Tracer. Figure 4(a) illustrates the waveforms of the pulsed *I-V* measurement in the dynamic characteristics. Within one cycle, the HEMT was first stressed at a gate bias and then switched to the current measurement point. The pulse width (time for the measurement) was 50 μs, and the pulse period (one switching cycle) was 5 ms. The small duty cycle of 1% assured that the self-heating effect was negligible in the measurement [22].

Figure 4(b) summarizes the measured dynamic transfer curves at different gate-bias voltages ($V_{GS, \text{stress}}$). The dynamic transfer curves shift towards the positive direction at forward gate-bias voltages. The positive shift of V_{TH} is due to the injection and trapping of electrons into the *p*-GaN layer [7], [8]. At forward gate bias, the electrons are sourced from the 2DEG channel across the AlGaN barrier as shown in Fig. 3. The trapped electrons cannot escape immediately when the gate bias is removed and the *p*-GaN layer becomes negatively charged, which leads to the positive shift of V_{TH}. No current collapse was observed in the dynamic transfer curves when the gate/channel is fully turned on due to the low drain bias applied ($V_{DS} = 1$ V) [6], [23].

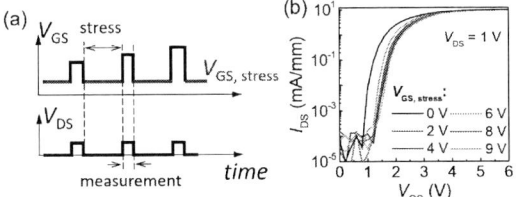

Fig. 4. (a) Pulsed I-V waveforms for the measurement of dynamic transfer characteristics. The magnitude of the gate-bias voltage ($V_{GS, stress}$) can be tuned. (b) Dynamic transfer characteristics measured at different $V_{GS, stress}$.

The V_{TH} was extracted from the dynamic transfer curves and plotted as a function of forward gate-bias voltages as shown in Fig. 5. From the result, the V_{TH} first positively shifted as the gate bias increased, but the shifting speed "decreased" when the bias voltage is larger than 3 V. Moreover, when the bias reached 7 V, the V_{TH} even became negatively shifted as gate bias further increased. The movement of holes and their recombination with electrons caused this unique V_{TH} shifting behavior.

Trapped electrons can be de-trapped through a hole compensation process [24] or an optical pumping process [25], [26]. At forward gate bias, the holes in the p-GaN layer start to drift towards the p-GaN/AlGaN interface. As a result, the holes can compensate and partially neutralize the negatively-charged trapping region in the p-GaN layer. At the same time, the electron/hole recombination happens and generates light emission, but the energy (i.e., visible emissions) is not large enough to pump the trapped electrons. So in this stage, the hole compensation is the dominated process and results in the "decrease" of the V_{TH} shifting speed. As the gate bias further increases, more holes arriving at the p-GaN/AlGaN interface are injected into the GaN

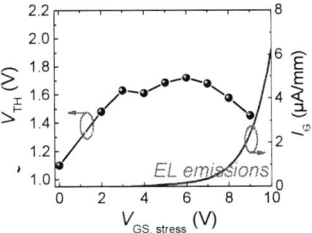

Fig. 5. The V_{TH} as a function of forward gate-bias voltages and the gate I-V of the p-GaN/AlGaN/GaN HEMT. The V_{TH} is extracted from the dynamic transfer curves at $I_{DS} = 10$ mA/mm.

channel, recombine with the 2DEG electrons and generate GaN band-edge UV emission. Due to the large photon energy, the UV light can effectively pump the trapped electrons in the p-GaN layer. As a result, the electron trapping is effectively suppressed and the V_{TH} is negatively shifted as the gate bias further increases.

B. Leakage Current Increase

In this section, an increase of OFF-state leakage current ($I_{DS, OFF}$) in p-GaN/AlGaN/GaN HEMTs, induced by forward gate bias, was observed and studied. In the test, the dynamic behavior of $I_{DS, OFF}$ was recorded right after an ON-state forward gate bias ($V_{GS, ON}$). The applied bias waveforms are schematically shown in Fig. 6(a). With a 10-V/1-s ON-state gate bias, as shown in Fig. 6(b), the $I_{DS, OFF}$ increases by approximately four orders of magnitude and takes 20 s to recover to its equilibrium level in dark at room temperature. Meanwhile, the gate leakage current ($I_{GS, OFF}$) shows no obvious change which indicates that the self-heating effect is negligible [27]. Figure 6(b) shows the dynamic behavior of $I_{DS, OFF}$ after various $V_{GS, ON}$. A higher forward gate bias leads to a larger leakage current. We proposed that the initial increase of OFF-state leakage current and its subsequent decay with time are due to persistent photoconductivity (PPC) effects in GaN, induced by the EL emission during the forward gate bias [11], [12].

Figure 7(a) plots the $I_{DS, OFF}$ as a function of ON-state forward bias durations. The $I_{DS, OFF}$ increases with longer bias durations since the buildup of the PPC is a slow process due to the small capture cross-section of the donor-complex (DX) centers [28], [29]. From the trend as illustrated by the red solid line in Fig. 7(a), it can be deduced that with an ON-state forward gate bias of 10 µs, the $I_{DS, OFF}$ increase can be neglected. In practical switching applications, the HEMT is turned ON/OFF periodically. Therefore, the accumulation effect of $I_{DS, OFF}$ increase needs to be explored. By driving the HEMT with a designed waveform shown in the insert of Fig. 7(b), the $I_{DS, OFF}$ after each pulse is recorded. As shown in Fig. 7(b), the magnitude of $I_{DS, OFF}$ increases with the continuous gate bias pulses but eventually saturates to a

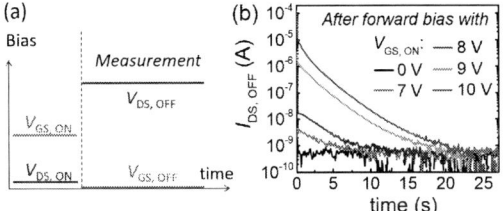

Fig. 6. (a) I-V waveforms for the ON-state and OFF-state bias conditions in the test. The ON-to-OFF switching time is approximately 10 ms. (b) The drain leakage current $I_{DS, OFF}$ after ON-to-OFF switching, measured at OFF-state with $V_{GS, OFF} = 0$ V, $V_{DS, OFF} = 20$ V, following different gate-bias voltages at ON-state with the gate bias duration of 1 s.

certain level. This saturation phenomenon is expected since the EL emission and the buildup of PPC are discrete with limited durations, while the decay process or recovery of PPC is continuous in the entire switching operation. This is important for practical power switching applications: (i) the $I_{DS, OFF}$ increase induced by forward gate bias will not keep increasing but will saturate at a certain level at which the buildup and decay processes of the underlying PPC effect are balanced; (ii) the $I_{DS, OFF}$ increase is not a random process, but one that will eventually become orderly and stable. Therefore, a proper gate driver is expected to maximize the p-GaN/AlGaN/GaN transistor's performance while maintaining low OFF-state leakage current and avoiding potential device degradation.

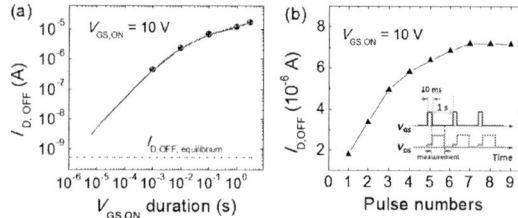

Fig. 7. (a) OFF-state ($V_{DS,\,OFF} = 20$ V, $V_{GS,\,OFF} = 0$ V) leakage $I_{DS,\,OFF}$ increase as a function of ON-state gate bias durations. The solid red line is a visual guide. (b) $I_{DS,\,OFF}$ increase as a function of continuous forward gate-bias pulses. The pulse waveform is illustrated in the figure insert.

V. Conclusions

In this work, we verified the electron/hole injections in the gate of p-GaN/AlGaN/GaN power transistors. The EL emission from the gate region was observed at forward gate bias including both visible and UV emissions. We also investigated the effect of electron and hole injections on device dynamic performances, such as the threshold voltage and leakage current.

References

[1] Y. Uemoto *et al.*, 'Gate Injection Transistor (GIT) - A Normally-Off AlGaN/GaN Power Transistor Using Conductivity Modulation', *IEEE Trans. Electron Devices*, vol. 54, no. 12, pp. 3393–3399, Dec. 2007.

[2] D. Marcon, Y. N. Saripalli, and S. Decoutere, '200mm GaN-on-Si epitaxy and e-mode device technology', in *2015 IEEE International Electron Devices Meeting (IEDM)*, Washington, DC, USA, 2015, pp. 16.2.1-16.2.4.

[3] K. J. Chen *et al.*, 'GaN-on-Si Power Technology: Devices and Applications', *IEEE Trans. Electron Devices*, vol. 64, no. 3, pp. 779–795, Mar. 2017.

[4] L. Sayadi *et al.*, 'Charge Injection in Normally-Off p-GaN Gate AlGaN/GaN-on-Si HFETs', in *2018 48th European Solid-State Device Research Conference (ESSDERC)*, Dresden, 2018, pp. 18–21.

[5] J. He, G. Tang, and K. J. Chen, 'V_{TH} Instability of p-GaN Gate HEMTs under Static and Dynamic Gate Stress', *IEEE Electron Device Lett.*, vol. 39, no. 10, pp. 1576–1579, Oct. 2018.

[6] J. Wei *et al.*, 'Charge Storage Mechanism of Drain Induced Dynamic Threshold Voltage Shift in p-GaN Gate HEMTs', *IEEE Electron Device Lett.*, early access, 2019.

[7] A. N. Tallarico, S. Stoffels, N. Posthuma, S. Decoutere, E. Sangiorgi, and C. Fiegna, 'Threshold Voltage Instability in GaN HEMTs with p-type Gate: Mg Doping Compensation', *IEEE Electron Device Lett.*, early access, 2019.

[8] X. Tang, B. Li, H. A. Moghadam, P. Tanner, J. Han, and S. Dimitrijev, 'Mechanism of Threshold Voltage Shift in p-GaN Gate AlGaN/GaN Transistors', *IEEE Electron Device Lett.*, vol. 39, no. 8, pp. 1145–1148, Aug. 2018.

[9] M. Ruzzarin *et al.*, 'Degradation Mechanisms of GaN HEMTs With p-Type Gate Under Forward Gate Bias Overstress', *IEEE Trans. Electron Devices*, vol. 65, no. 7, pp. 2778–2783, Jul. 2018.

[10] Y. Shi *et al.*, 'Bidirectional threshold voltage shift and gate leakage in 650 V p-GaN AlGaN/GaN HEMTs: The role of electron-trapping and hole-injection', in *2018 IEEE 30th International Symposium on Power Semiconductor Devices and ICs (ISPSD)*, Chicago, IL, 2018, pp. 96–99.

[11] Y. Wang *et al.*, 'Dynamic OFF-State Current (Dynamic I_{OFF}) in p-GaN Gate HEMTs With an Ohmic Gate Contact', *IEEE Electron Device Lett.*, vol. 39, no. 9, pp. 1366–1369, Sep. 2018.

[12] X. Tang, B. Li, H. A. Moghadam, P. Tanner, J. Han, and S. Dimitrijev, 'Effect of Hole-Injection on Leakage Degradation in a p-GaN Gate AlGaN/GaN Power Transistor', *IEEE Electron Device Lett.*, vol. 39, no. 8, pp. 1203–1206, Aug. 2018.

[13] Z. Tang *et al.*, '600-V Normally Off/AlGaN/GaN MIS-HEMT With Large Gate Swing and Low Current Collapse', *IEEE Electron Device Lett.*, vol. 34, no. 11, pp. 1373–1375, Nov. 2013.

[14] B. Li, X. Tang, J. Wang, and K. J. Chen, 'P-doping-free III-nitride high electron mobility light-emitting diodes and transistors', *Appl. Phys. Lett.*, vol. 105, no. 3, p. 032105, Jul. 2014.

[15] U. Kaufmann *et al.*, 'Nature of the 2.8 eV photoluminescence band in Mg doped GaN', *Appl. Phys. Lett.*, vol. 72, no. 11, pp. 1326–1328, Mar. 1998.

[16] B. Z. Qu *et al.*, 'Photoluminescence of Mg-doped GaN grown by metalorganic chemical vapor deposition', *J. Vac. Sci. Technol. Vac. Surf. Films*, vol. 21, no. 4, pp. 838–841, Jul. 2003.

[17] B. Li *et al.*, 'Schottky-on-heterojunction optoelectronic functional devices realized on AlGaN/GaN-on-Si platform', in *2014 IEEE International Electron Devices Meeting (IEDM)*, 2014, pp. 11.4.1-11.4.4.

[18] X. Tang, B. Li, Z. Zhang, G. Tang, J. Wei, and K. J. Chen, 'Characterization of Static and Dynamic Behaviors in AlGaN/GaN-on-Si Power Transistors With Photonic-Ohmic Drain', *IEEE Trans. Electron Devices*, vol. 63, no. 7, pp. 2831–2837, Jul. 2016.

[19] X. Tang, Z. Zhang, J. Wei, B. Li, J. Wang, and K. J. Chen, 'Photon emission and current-collapse suppression of AlGaN/GaN field-effect transistors with photonic–ohmic drain at high temperatures', *Appl. Phys. Express*, vol. 11, no. 7, p. 071003, Jul. 2018.

[20] Q. Yan, A. Janotti, M. Scheffler, and C. G. Van de Walle, 'Role of nitrogen vacancies in the luminescence of Mg-doped GaN', *Appl. Phys. Lett.*, vol. 100, no. 14, p. 142110, Apr. 2012.

[21] M. A. Reshchikov, P. Ghimire, and D. O. Demchenko, 'Magnesium acceptor in gallium nitride. I. Photoluminescence from Mg-doped GaN', *Phys. Rev. B*, vol. 97, no. 20, May 2018.

[22] H. Wang, J. Wei, R. Xie, C. Liu, G. Tang, and K. J. Chen, 'Maximizing the Performance of 650-V p-GaN Gate HEMTs: Dynamic RON Characterization and Circuit Design Considerations', *IEEE Trans. Power Electron.*, vol. 32, no. 7, pp. 5539–5549, Jul. 2017.

[23] P. Moens *et al.*, 'On the impact of carbon-doping on the dynamic Ron and off-state leakage current of 650V GaN power devices', in *2015 IEEE 27th International Symposium on Power Semiconductor Devices IC's (ISPSD)*, 2015, pp. 37–40.

[24] S. Kaneko *et al.*, 'Current-collapse-free operations up to 850 V by GaN-GIT utilizing hole injection from drain', in *2015 IEEE 27th International Symposium on Power Semiconductor Devices & IC's (ISPSD)*, 2015, pp. 41–44.

[25] X. Tang *et al.*, 'III-Nitride transistors with photonic-ohmic drain for enhanced dynamic performances', in *2015 IEEE International Electron Devices Meeting (IEDM)*, 2015, pp. 35–3.

[26] B. Li, X. Tang, and K. J. Chen, 'Optical pumping of deep traps in AlGaN/GaN-on-Si HEMTs using an on-chip Schottky-on-heterojunction light-emitting diode', *Appl. Phys. Lett.*, vol. 106, no. 9, p. 093505, Mar. 2015.

[27] X. Tang *et al.*, 'Mechanism of leakage current increase in p-GaN gate AlGaN/GaN power devices induced by ON-state gate bias', *Jpn. J. Appl. Phys.*, vol. 57, no. 12, p. 124101, Dec. 2018.

[28] D. V. Lang, R. A. Logan, and M. Jaros, 'Trapping characteristics and a donor-complex (DX) model for the persistent-photoconductivity trapping center in Te-doped $Al_xGa_{1-x}As$', *Phys. Rev. B*, vol. 19, no. 2, pp. 1015–1030, Jan. 1979.

[29] M. T. Hirsch, J. A. Wolk, W. Walukiewicz, and E. E. Haller, 'Persistent photoconductivity in n-type GaN', *Appl. Phys. Lett.*, vol. 71, no. 8, pp. 1098–1100, Aug. 1997.

Proceedings of the 31st International Symposium on Power Semiconductor Devices & ICs
May 19-23, 2019, Shanghai, China

Trading Off between Threshold Voltage and Subthreshold Slope in AlGaN/GaN HEMTs with a p-GaN Gate

Benoit Bakeroot*, Steve Stoffels[†], Niels Posthuma[†], Dirk Wellekens[†], Stefaan Decoutere[†]

*CMST, imec & Ghent University, Technologiepark 126, 9000 Ghent, Belgium, benoit.bakeroot@ugent.be

[†]imec, Kapeldreef 75, 3001 Leuven, Belgium

Abstract—The measured values of the threshold voltage of AlGaN/GaN high-electron-mobility transistors (HEMTs) with a p-GaN gate are generally more positive than what is expected from a classical HEMT. The transfer characteristics exhibit subthreshold slopes which are higher compared to the standard 60 mV per decade at room temperature. The higher threshold voltage values and subthreshold slopes are related to the specific structure of the p-GaN gate, consisting of two back-to-back diodes. The dominating diode—either the metal to p-GaN Schottky diode or the p-GaN/AlGaN barrier/GaN channel diode—dictates how much gate current flows, determines the subthreshold behavior, and, thus also the threshold voltage. The trade-off between the threshold voltage and the subthreshold slope is discussed revealing the intricate dynamic relation between those two device metrics and the gate leakage current.

I. INTRODUCTION

Enhancement-mode (E-mode) operation for high-voltage devices is of paramount importance in high-voltage, high-power electronics. Many circuits in power electronics require E-mode (or normally-off) devices as these ensure fail-safe behavior and eliminate the need for complex gate drive circuitry.

The quest for E-mode devices in the AlGaN/GaN material system, where strong polarization charges at the heterointerfaces automatically yield depletion-mode (D-mode) high-electron-mobility transistors (HEMTs)—without special precautions for the gate structure—seems to crystallize in a solution based on p-type (Al)GaN as part of the gate structure, witness the fact that many companies [1]–[3] have come to the market with these type of devices.

The threshold voltage V_T of AlGaN/GaN HEMTs with a p-GaN gate have been calculated analytically [4]. It was shown that E-mode operation ($V_T > 0$) is possible for a limited range of AlGaN barrier thicknesses and mole fractions and gets more positive with thinner and less Al-rich AlGaN barriers. However, the barrier's thickness nor mole fraction (Al percentage) can't be too low, since (gate) leakage current through the barrier will increase to unacceptably large values, rendering the device useless.

Yet, GaN HEMTs with a p-GaN gate and with a reasonably thick and Al-rich AlGaN barrier have been reported to suffer from V_T instability [5]–[7], and from gate leakage current [8], [9]. Threshold voltage instability is caused by leakage current through the AlGaN barrier and trapping of charges in the barrier or at its interfaces; whereas the gate leakage current

is also determined by the Schottky metal-to-p-GaN contact. In this paper, the correlation between the barrier leakage current and the metal-p-GaN leakage current on the one hand and the threshold voltage and subthreshold slope on the other hand is discussed. A direct, dynamic trade-off between the V_T and the subthreshold slope is observed, depending on the level of leakage current through one of the parts of the p-GaN gate.

II. THE DEVICE UNDER STUDY

The reference process of the device under study has been described in [10], [11]: a thin TiN layer is deposited and etched prior to the selective p-GaN to AlGaN dry etch. The result of such a gate process as simulated with Technology Computer Aided Design (TCAD) software [12] is depicted in Fig. 1. The p-GaN gate consists of two back-to-back diodes: the first one is the Schottky metal-to-p-GaN diode (i.e., the Schottky diode) and the second one is the p-GaN/AlGaN/GaN heterostructure (i.e., the PIN diode). The simulated band diagram features the Schottky contact to the p-GaN (note the band bending in the p-GaN next to the metal contact) as well as a two-dimensional hole gas (2DHG) at the p-GaN/AlGaN barrier interface.

The p-GaN HEMT device under study has a barrier thickness and a mole fraction of 12.5 nm and 25%, respectively; which, according to the calculations [4] yields a V_T of 0.70 V (Fig. 2). Fig. 3 shows a typical transfer characteristic of a p-GaN HEMT for the reference process (process A) compared with a quasi-zero gate leakage current (case 1 in Table I) TCAD simulated p-GaN HEMT. The pinch-off voltage (defined at $I_D = 1\,\mu A/mm$) is ~ 0.7 V, in accordance with the analytical calculation and the numerical simulation. Yet, the V_T at maximum transconductance g_m is higher, at ~ 2.8 V. One clearly observes that the subthreshold slope diverts from the TCAD simulated one (typically ~ 60 mV/dec for a quasi-zero gate leakage current device), and that the turn-on also happens more moderate than in the simulated device (i.e., the transition from diffusion-limited to drift current, around $V_{GS} \approx 1$ V).

The measured I_D-V_{GS} and I_G-V_{GS} as shown in Fig. 3 are a result of a process optimization aiming at lower gate currents. Fig. 4 displays typical I_D-V_{GS} and I_G-V_{GS} curves before (B) and after (A) process optimization, showing the trade-off between a better subthreshold slope and a less positive V_T.

978-1-7281-0582-6/19 $31.00 © 2019 IEEE

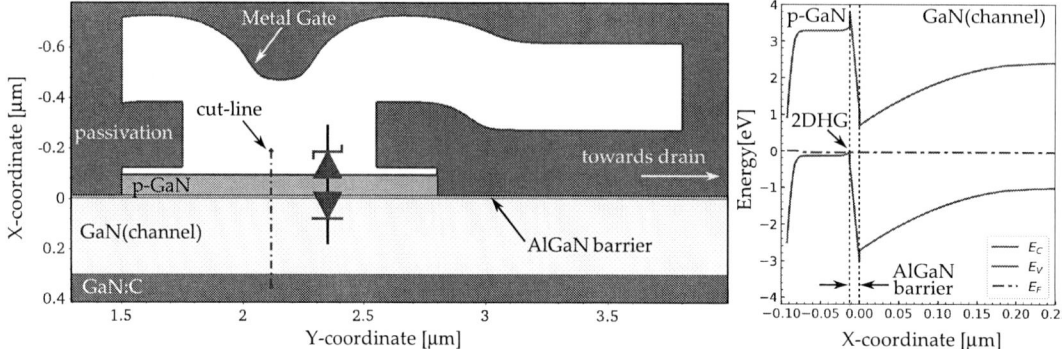

Fig. 1. Detailed cross-section showing the p-GaN gate of an AlGaN/GaN HEMT: the metal itself is left blank and contacts the p-GaN layer. The two back-to-back diodes are also depicted schematically (left). Band diagram at equilibrium (right) along the cut-line as indicated on the left figure. Note that a 2DHG at the p-GaN/AlGaN interface is present as indicated on the figure.

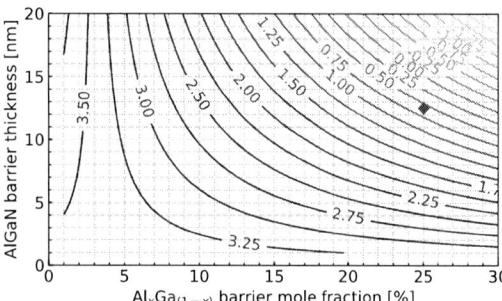

Fig. 2. Calculated [4] threshold voltage (in Volt) of the p-GaN gate AlGaN/GaN HEMT as a function of mole fraction and thickness of the AlGaN barrier. The diamond symbol represents our AlGaN barrier reference.

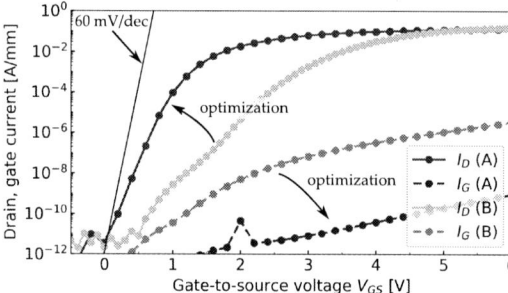

Fig. 4. Measured transfer characteristics and gate leakage current before (B) and after (A) process optimization at $V_{DS} = 1.0\,\text{V}$.

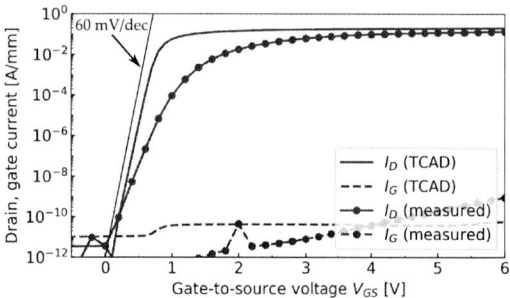

Fig. 3. Measured (after process optimization, referred to as process A) and TCAD simulated (for a quasi-zero leakage device: case 1 in Table1) transfer characteristics and gate leakage current at $V_{DS} = 1.0\,\text{V}$. The theoretical subthreshold I-V with a subthreshold slope of $60\,\text{mV/dec}$ is also shown.

III. SOME THEORETICAL BACKGROUND

In order to understand the operation of the p-GaN HEMT, and especially the role of the back-to-back diodes in the gate structure, some of the results presented in our work [4] are

repeated here. Upon close inspection of the band diagram at equilibrium (Fig. 1), one realizes that the gate voltage is divided between the Schottky diode and the barrier/channel diode. As long as the doping level in the p-GaN is high enough and/or the gate voltage is not too high (which is the case for the p-GaN gate and the gate biases discussed in this paper), the voltage drop in the p-GaN due to the Schottky diode (in reverse bias for positive gate voltages) will be limited to a few tens of nanometer inside the p-GaN. Hence, there is a quasi-neutral region inside the p-GaN where the electric field is zero for all gate biases under consideration. This allows us to apply Gauss' law on the Schottky diode and on the PIN diode separately, yielding the following equations:

$$\sigma_m = \sigma_{p1} + \sigma_s, \tag{1a}$$

$$\sigma_e = \sigma_h - \sigma_b - \sigma_4, \tag{1b}$$

where σ_m, σ_{p1}, σ_s, σ_e, σ_h, σ_b, and σ_4 are the *absolute* values of the charge densities: in the metal, the p-GaN polarization charge, the space-charge in the depletion layer of the Schottky diode, the two-dimensional electron gas (2DEG) and 2DHG charge density, the charge in the AlGaN barrier due to ionized

Mg, and the charge at the channel/back-barrier interface, respectively. We will assume that σ_b and σ_4 are constant for the remainder of the paper. Although both quantities can change (e.g., the ionized Mg can trap holes, altering σ_b) and can be the cause of V_T instabilities, this assumption will not compromise the main conclusions of this paper and is only made to simplify the discussion. Thus, *the 2DEG density can only increase if the 2DHG density increases*:

$$\Delta\sigma_e = \Delta\sigma_h \qquad (2)$$

or, *the ON-state can only be attained if there is a sufficient supply of holes to the 2DHG with increasing gate bias*.

IV. FOUR CASES OF GATE LEAKAGE

The discrepancy between the simulated and measured subthreshold slope (Fig. 3) and the optimization of the subthreshold slope and gate leakage current (Fig. 4) can be understood with this important finding in hand. Four separate cases can be discerned as summarized in Table I:

1) Low Schottky diode leakage and low PIN diode leakage

This condition has been shown in Fig. 3 and is repeated in Fig. 5 in the top row. The *I-V* plot in the left most column shows that the gate current is low (i.e., a displacement current as these are transient simulations at $1\,\mathrm{V/s}$). The tunneling model at the Schottky contact has been switched off for this case. As a result, the only possible source of holes is from within the p-GaN itself: the Schottky diode depletion layer width increases between $V_{GS} = 0$ and $1\,\mathrm{V}$ and is able to supply a sufficient amount of holes so as to assure a 2DHG increase. However, once the PIN diode gets (slightly) forward biased, this limited supply of holes will no longer suffice. The band diagrams at $V_{GS} = 2, 4,$ and $6\,\mathrm{V}$ reveal what happens at higher gate bias: the AlGaN energy barrier is reducing since holes and electrons are accumulating on either side of the barrier and are compensating the polarization charges. As a result, more holes (and electrons) get emitted thermionically into the barrier. This is a self-limiting choking effect: the ON-state current remains almost constant as the 2DEG is no longer increasing because the supply of holes is no longer sufficient. Since the voltage drop in the AlGaN barrier is choked, any further increase of gate voltage will be taken up by the reverse biased Schottky diode.

2) Low Schottky diode leakage and high PIN diode leakage

If the holes, that get displaced from within the Schottky diode depletion layer towards the 2DHG, leak away through the AlGaN barrier, then these holes do not contribute to the 2DHG increase. Therefore, based on (2), the 2DEG density can not increase neither and the ON-state current is choked at low forward gate bias as is visible in Fig. 5 (second row). A similar effect can occur even when the Schottky diode is leaky: when the PIN diode leaks even more than the combined hole current from within the p-GaN and from the Schottky contact, the 2DHG will not increase and 2DEG modulation is not possible neither.

TABLE I
FOUR CASES OF GATE LEAKAGE STUDIED WITH TCAD SIMULATIONS

Level of leakage current	Schottky	P-i-N Diode
Case 1	LOW	LOW
Case 2	LOW	HIGH
Case 3	HIGH	LOW
Case 4	HIGH	HIGH

3) High Schottky diode leakage and low PIN diode leakage

If the Schottky diode leaks more than the PIN diode, then no problems occur during subthreshold as the hole supply is sufficient. The 2DEG modulation can even be stronger than in case 1. Yet, eventually the PIN diode gets forward biased and the gate current increases (see Fig. 5, third row). The level of gate leakage current is determined by the leaky Schottky contact.

4) High Schottky diode leakage and high PIN diode leakage

If both diodes leak, then the 2DEG modulation depends on how well the 2DHG is able to increase with increasing gate bias. It is not the total gate current that is decisive, but the relative amount of leakage current through the Schottky diode compared to the PIN diode: as long as the hole supply to the 2DHG is larger than the leakage through the PIN diode, the ON-state current increases. However, the subthreshold slope and the transconductance are deteriorated due to the more moderate 2DEG modulation (see Fig. 5, fourth row).

This last case is what happens in reality (Fig. 3) and what can be controlled (Fig. 4): by optimizing the process conditions, one can reduce the overall gate leakage current and at the same time enhance the subthreshold behavior.

V. CONCLUSION

The Schottky/p-GaN and p-GaN/AlGaN barrier/GaN channel diode are both leaky to some extent in reality. The shape of the transfer characteristic is governed by the relative amount of both diode leakage currents. Variations in electrical parameters (V_T, S, g_m, $I_{D,ON}$ ...) stem from the ease or difficulty at which the 2DHG at the p-GaN/AlGaN barrier interface is built up and may vary depending on the gate voltage. The performance of the p-GaN HEMT is not hampered by the level of gate leakage current but by the loss of gate control on the 2DHG variation.

REFERENCES

[1] E. A. Jones, F. F. Wang, and D. Costinett, "Review of Commercial GaN Power Devices and GaN-Based Converter Design Challenges," *IEEE Journal of Emerging and Selected Topics in Power Electronics*, vol. 4, no. 3, SI, pp. 707–719, Sep. 2016, IEEE Workshop on Wide-Bandgap Power Devices and Applications, Knoxville, TN, Oct., 2014.

[2] K. J. Chen, O. Haeberlen, A. Lidow, C. L. Tsai, T. Ueda, Y. Uemoto, and Y. Wu, "GaN-on-Si Power Technology: Devices and Applications," *IEEE Transactions on Electron Devices*, vol. 64, no. 3, pp. 779–795, Mar. 2017.

[3] H. Amano *et al.*, "The 2018 GaN power electronics roadmap," *Journal of Physics D-Applied Physics*, vol. 51, no. 16, Apr. 2018.

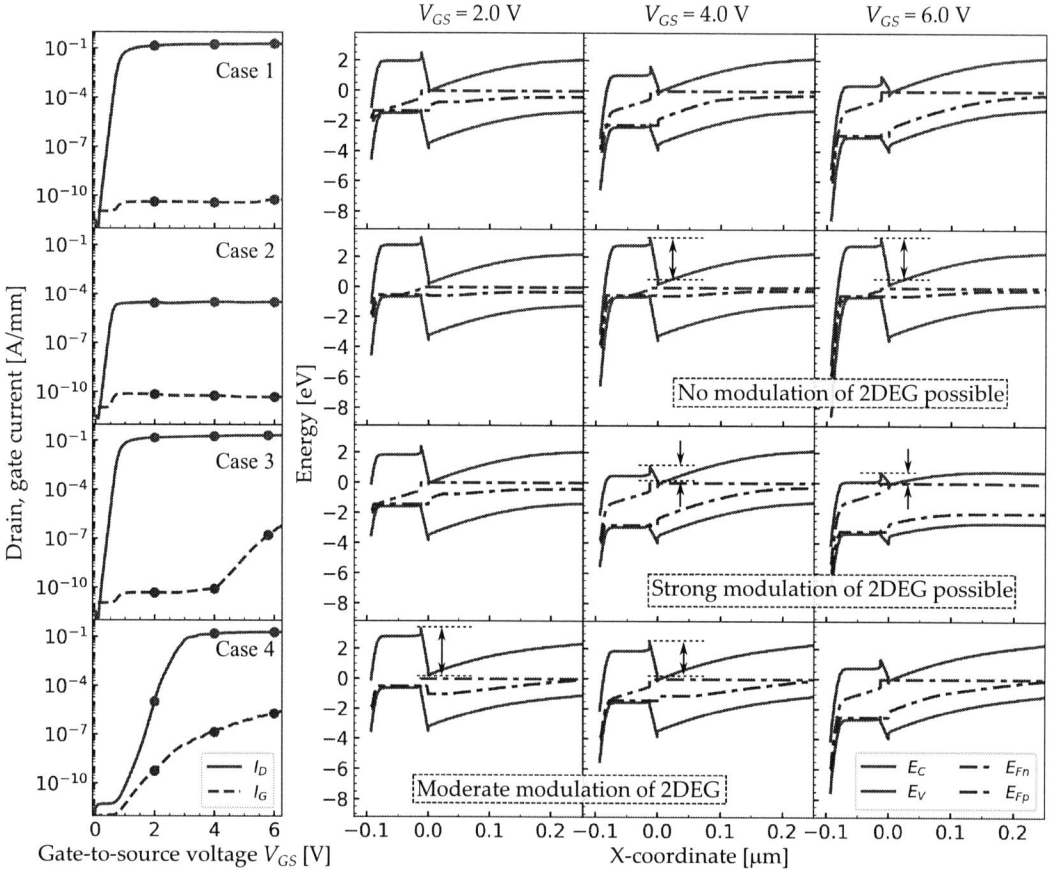

Fig. 5. Simulated transfer characteristics (left most column) and simulated band diagrams for all 4 cases as given in Table I. Each row corresponds to the case as indicated on the left most column. The band diagrams at $V_{GS} = 2$ V, 4 V, and 6 V show what is going on inside the device. Case 1 exhibits a strong modulation of the 2DEG in the subthreshold region, and less strong modulation once the device is in the ON-state due to a lack of hole supply. Case 2 shows that no transistor action is obtained when the barrier leaks considerably more than the Schottky contact—here the holes leak away through the barrier and are not supplied by the metal contact yielding a strong modulation of the Schottky diode, but no modulation of the PIN diode, and, thus, no modulation of the 2DEG. Case 3 shows that with enough hole supply, the 2DEG can be modulated even stronger than in case 1 (note the smaller energy drop in the barrier in case 3 compared to case 1). Case 4 demonstrates a realistic situation where both the Schottky and the PIN diode leak: depending on the level of leakage current through each of the diodes in a certain voltage range, the gate voltage will be shared differently between both diodes.

[4] B. Bakeroot, A. Stockman, N. Posthuma, S. Stoffels, and S. Decoutere, "Analytical Model for the Threshold Voltage of p-(Al)GaN High-Electron-Mobility Transistors," *IEEE Transactions on Electron Devices*, vol. 65, no. 1, pp. 79–86, Jan. 2018.

[5] A. N. Tallarico, S. Stoffels, N. Posthuma, P. Magnone, D. Marcon, S. Decoutere, E. Sangiorgi, and C. Fiegna, "PBTI in GaN-HEMTs With p-Type Gate: Role of the Aluminum Content on Delta V-TH and Underlying Degradation Mechanisms," *IEEE Transactions on Electron Devices*, vol. 65, no. 1, pp. 38–44, Jan. 2018.

[6] L. Sayadi, G. Iannaccone, S. Sicre, O. Haeberlen, and G. Curatola, "Threshold Voltage Instability in p-GaN Gate AlGaN/GaN HFETs," *IEEE Transactions on Electron Devices*, vol. 65, no. 6, pp. 2454–2460, Jun. 2018.

[7] X. Tang, B. Li, H. A. Moghadam, P. Tanner, J. Han, and S. Dimitrijev, "Mechanism of Threshold Voltage Shift in p-GaN Gate AlGaN/GaN Transistors," *IEEE Electron Device Letters*, vol. 39, no. 8, pp. 1145–1148, Aug. 2018.

[8] N. Xu *et al.*, "Gate leakage mechanisms in normally off p-GaN/AlGaN/GaN high electron mobility transistors," *Applied Physics Letters*, vol. 113, no. 15, Oct. 2018.

[9] A. Stockman, F. Masin, M. Meneghini, E. Zanoni, G. Meneghesso, B. Bakeroot, and P. Moens, "Gate Conduction Mechanisms and Lifetime Modeling of p-Gate AlGaN/GaN High-Electron-Mobility Transistors," *IEEE Transactions on Electron Devices*, vol. 65, no. 12, pp. 5365–5372, Dec. 2018.

[10] N. E. Posthuma, S. You, S. Stoffels, D. Wellekens, H. Liang, M. Zhao, and S. Decoutere, "Gate architecture design for enhancement mode p-GaN gate HEMTs for 200 and 650 V applications," in *30th International Symposium on Power Semiconductor Devices and ICs (ISPSD)*, 2018, pp. 188–191.

[11] N. E. Posthuma *et al.*, "An Industry-Ready 200 mm p-GaN E-mode GaN-on- Si power Technology," in *30th International Symposium on Power Semiconductor Devices and ICs (ISPSD)*, 2018, pp. 284–287.

[12] *TCAD Documentation*, Synopsys, June 2018.

Proceedings of the 31st International Symposium on Power Semiconductor Devices & ICs
May 19-23, 2019, Shanghai, China

Observation of self-recoverable gate degradation in p-GaN AlGaN/GaN HEMTs after long-term forward gate stress: The trapping & detrapping dynamics of hole/electron

Yuanyuan Shi, Qi Zhou †, Wei Xiong, Xi Liu, Xin Ming, Zhaoji Li, Wanjun Chen, and Bo Zhang
State Key Laboratory of Electronic Thin Films and Integrated Devices
University of Electronic Science and Technology of China Chengdu, P.R. China
† Email address: zhouqi@uestc.edu.cn

Abstract—The gate reliability of normally off p-GaN gate AlGaN/GaN HEMTs and its underlying physical mechanism were investigated in a long-term recovery period after long-term forward gate stress. For the first time, a nondestructive gate degradation featuring a first degradation and then followed by a self-recovery dynamic was observed after a rigorous long-term (13-53 ks) forward gate stress. To elucidate the gate degradation kinetics, the static gate current, transfer characteristic and transient gate current were characterized before and after the gate stress as well as in the recovery intervals. During the stress, hole-injection together with trap-generation and hole-trapping concurred in AlGaN barrier is observed. Subsequently, electrons attracted by the trapped holes are persistently trapped in the AlGaN barrier, which leads to an electron energy barrier lowering and a continuing gate current increase in the recovery intervals. The hole-injection, hole-trapping, electron-trapping and hole/electron -detrapping take place in sequence during the stress and recovery process are reinforced by the inhomogeneous shift in VTH. The observed unique gate degradation dynamics and its underlying mechanism are of great value for p-GaN gate optimization to achieve improved gate reliability.

Keywords—*p-GaN gate, AlGaN/GaN HEMT, threshold voltage instability; gate reliability; gate leakage; long-term gate stress; long-term recovery period, electron trapping; hole injection*

I. INTRODUCTION

The p-GaN/AlGaN/GaN HEMTs currently appear to be the mainstream commercialized normally-off GaN transistors for a new generation power electronics achieving high speed, high power efficiency [1-3]. Owing to the ultra-high operation frequency (e.g. up to ~10 MHz) of GaN power circuits, unintentional gate voltage overshoot and noise stem from the circuitry parasitic could be introduced during the switch-on transient, which is fatal for p-GaN transistor due to its small reliable forward gate swing (e.g. <7 V) [4-8]. Thus, it is essential to investigate the ruggedness of p-GaN gate in positive gate bias and understand the gate degradation kinetics.

Traditionally the gate reliability of p-GaN HEMTs has been extensively investigated by comparing device characteristics before and rightly after gate stress [9-13]. However, it is time-consuming for the traps to exchange charge via the III-N barrier in GaN HEMTs. Hence Lagger *et al.* investigated the long-term V_{TH} shift in GaN MISHEMTs over a broad range of

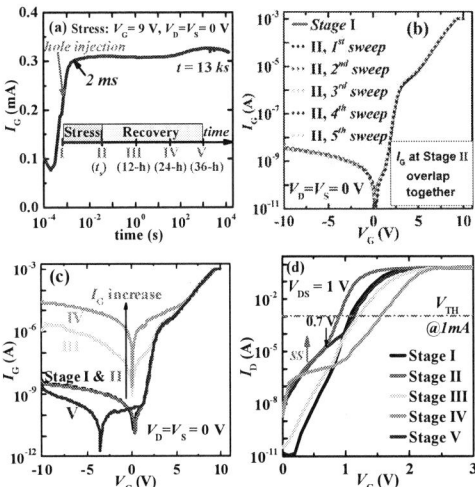

Fig. 1 (a) The typical I_G of DUT monitored during the forward stress with $V_G = 9$ V. (b) The I_G measured in fresh device and right after gate stress for several continuous tests. (c) The I_G-V_G characteristics of the DUT at Stage I-V. (d) Transfer curves of the DUT at Stage I-V.

stress and recovery times [14]. Moreover, considering the lower hole mobility in GaN material, it may need more time to study the reliability issues related to hole-trapping/detrapping in p-GaN HEMTs [3]. However, literatures on the p-GaN gate degradation in a long-term recovery period after long-term gate stress have not been reported extensively yet.

In this work, the commercial 650 V rating p-GaN HEMTs were investigated in a long-term recovery period after long-term forward gate stress. For the first time, a sophisticated non-destructive gate degradation featuring a first degradation and then followed by a self-recovery process without gate failure is observed after a rigorous long-term (13-53 ks). The device characterization including the static gate current, the transfer characterization and the gate current transient were conducted before and after stress as well as at different recovery intervals after stress to capture the degradation dynamics. A physical model relating to the trapping/ detrapping of hole and electron was proposed to responsible for the observed self-recoverable gate degradation.

This work was supported in part by the National Natural Science Foundation of China under Grant 61674024, and in part by the Assembly Pre-research Project under Grant JZX2017-1643/Y537, and in part by the National Science and Technology Major Project under Grant 2013ZX02308-005. (*Corresponding Author. Qi Zhou*)

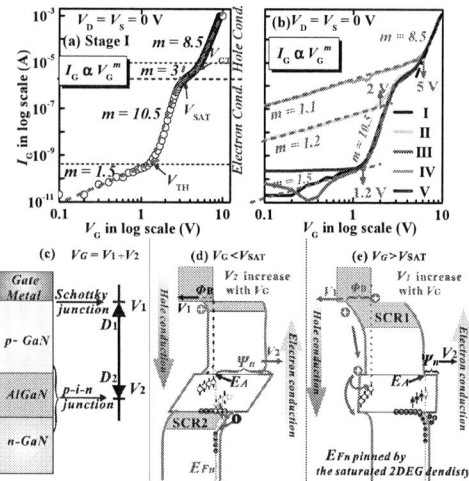

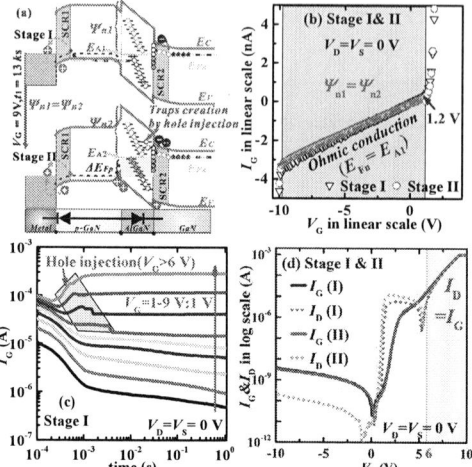

Fig. 2 log I_G-log V_G of the DUT (a) at Stage-I and (b) Stage-I-V. (c) Equivalent circuit of p-GaN gate stack. Schematic energy diagram of p-GaN gate stack (d) for $V_G < V_{SAT}$ and (d) for $V_G > V_{SAT}$.

Fig. 3. (a) Schematic energy diagram of p-GaN gate stack at Stage-I and II. (b) I_G-V_G of the DUT at Stage I & II for $V_G < V_{TH}$. (c) Time-resolved I_G at Stage-I. (d) I_D & I_G-V_G of the DUT at Stage I & II.

II. GATE STRESS AND CHARACTRIZATION METHOD

The p-GaN gate HEMT (DUT) used in the experiments is the state-of-the-art 650 V rating commercial devices. The DUT was stressed at $V_G = 9$ V while $V_S = V_D = 0$ V, prior to which the fresh device was characterized for reference. The measurement strategy is shown in the inset of Fig. 1 (a). The device characteristics including the static gate current, the transfer characteristic and the gate current transient were respectively measured right after the stress and at different intervals after the stress (i.e. 12-, 24-, and 36-hour) to capture the gate degradation kinetics. Then, another stress run with a much longer forward stress of 53 kilo second also with $V_G = 9$ V was further applied to the DUT.

III. RESULT AND DISCUSSION

During the forward gate stress, as shown in Fig.1 (a), the I_G exhibits an abrupt increase revealing massive hole injection and then reaches to a steady state. The steady I_G maintained for up to 13 kilo seconds without gate breakdown, which is quite different from the reported observation at a very comparable V_G[10-11]. The I_G measured for several times right after stress (@ Stage-II) exhibit negligible variation compared with that in the fresh device as shown in Fig. 1 (b). However, it is compelling to observe in Fig. 1 (c) that the I_G significantly increased at Stage III & IV and eventually recovered back to the initial level at Stage V during the recovery duration, while the gate still features channel modulation functionality without failure after the rigorous long-term forward stress.

The detailed I_G characterization was further carried out to reveal such self-recoverable gate degradation mechanisms. The I_G-V_G at different stages plotted in log-scale are shown in Fig. 2 (a) and 2 (b). As reported in our previous works [15-16]

by fitting with $I_G \propto V_G^m$, the varied exponential coefficient m indicates different gate current transport mechanisms. As illustrated in Fig. 2 (c), the p-GaN gate stack is analogous to two back-to-back connected diodes, namely, the metal/p-GaN Schottky junction diode D_1 and the p-GaN/AlGaN/GaN heterojunction diode D_2. At Stage-I shown in Fig. 2 (a), the I_G is dominated by Ohmic conduction with the $m \approx 1$ for $V_G < V_{TH}$. In this region, the I_G can be described by thermally activated electrons over the gate electron energy barrier ψ_n, which is the energy difference between the AlGaN conduction band (E_C^{AlGaN}) and the electron quasi-Fermi level (E_{Fn}), as shown in Fig. 2 (d). For $V_{TH} < V_G < V_{SAT}$ (2DEG saturation voltage), the pronounced electron injection overwhelms thermal electron-ionization, which moves the E_{Fn} moves to E_C^{AlGaN} and leads to a reducing ψ_n. Hence, a steep increase in I_G (m=10.5) governed by space-charge-limited-current (SCLC) is observed. For $V_G > V_{SAT}$, the E_{Fn} is pinned by the saturated 2DEG density, resulting in a saturated voltage drop across D_2. Particularly for $V_G > V_{GT}$, effective hole injection was triggered by the reverse turn-on of the Schottky junction D_1, leading to the second steep increase in I_G with $m = 8.5$.

Driven by high electric-field on forward gate stress, holes can be effectively injected and trapped in the AlGaN barrier as well as traps created by hot holes [17], which leads to a negative V_{TH} shift and an degraded sub threshold slope at Stage-II as shown in Fig. 1 (d). Effective hole-trapping shifts the hole quasi-Fermi level E_{Fp} towards the valence band E_V^{AlGaN} without varying the E_{Fn} as shown in Fig. 3 (a). Consequently, the electron barrier at Stage-II is identical to that at Stage-I, i.e. $\psi_{n1} = \psi_{n2}$ as illustrated in Fig. 3 (a). Thus, as shown in Fig. 3 (b), the I_G at the Ohmic conduction region is identical to that of Stage I. Limited by the same electron barrier, the I_G transient exhibits a slump at low gate bias with $V_G < 6$ V as plotted in Fig. 3 (c). Effective hole-injection is

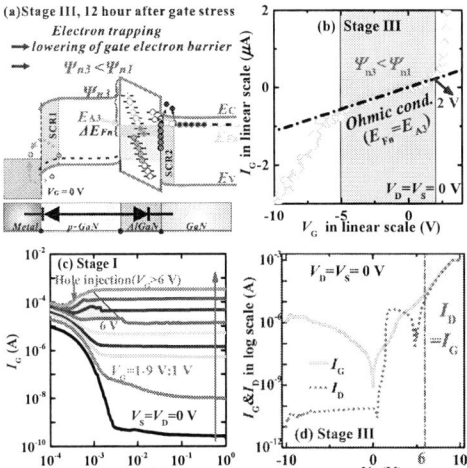

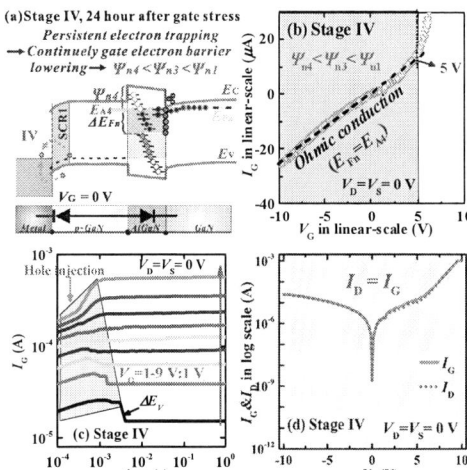

Fig. 4 (a) Schematic energy band diagram of the p-GaN gate at Stage III ($V_G = 0$ V). (b) The I_G at Stage III with an extended Ohmic conduction region to $V_G = 2$ V. (c) Time resolved I_G characteristics the DUT at Stage-III. (d) I_D & I_G-V_G of the DUT at Stage-III.

Fig. 5 (a) Schematic energy band diagram of the p-GaN gate at Stage IV ($V_G = 0$ V). (b) The I_G at Stage IV with an extended Ohmic conduction region to $V_G = 5$ V. (c) Time resolved I_G characteristics the DUT at Stage-IV. (d) I_D & I_G-V_G of the DUT measured at Stage-IV.

captured for $V_G > 6$ V, which induces the increase in I_G-time curves. On the other hand, the identical I_D and I_G for $V_G > 6$ V in Fig. 4(d) reinforces the turn-on of the p-GaN gate where the channel leakage current is sustained by the gate leakage.

Driven by the positively charged holes, electrons have been persistently trapping in the AlGaN barrier in the 24-hour after the stress, resulting in the successive V_{TH} shift from Stage II to Stage IV as shown in Fig.1 (d). More importantly, such an electron trapping process results in the energy barrier lowering, i.e. $\psi_{n4} < \psi_{n3} < \psi_{n1}$ as illustrated in Fig. 4 (a) and 5 (a) respectively. More electrons can climb over shallower gate electron barrier, which leads to a higher I_G in the Ohmic conduction region. Hence as shown in Fig. 4(a), the I_G at Stage III is 3 orders of magnitude higher than that of Stage-I &II. The I_G at Stage-IV is ten times higher than I_G at Stage-III. Meanwhile, the Ohmic conduction region extends to 2 V at Stage III as shown in Fig. 4 (b) and 5 V at Stage IV in Fig. 5(b) when the degraded electron barrier ψ_{n3} and ψ_{n4} are further reduced by the applied gate voltage. Massive hole-injection is once again the reason for the rise in the I_G transient and the identical I_G and I_D for $V_G > 6$ V as shown in Fig. 4(c) and 4(d). However, the ψ_{n4} is reduced to such a low level that electrons move freely over the diode D_2, which leads to an identical I_G and I_D for all applied V_G from -10 V to 10 V in Fig. 5 (d). Thus the applied V_G is solely sustained by the Schottky diode D_1, which effectively triggered hole-injection at low V_G in Fig. 5 (c).

In a further recovery duration (36-h after stress), the release of the trapped electrons and holes enhances the gate electron energy barrier ψ_{n5} back to the initial state ψ_{n1} as illustrated in Fig. 6 (a). Hence as shown in Fig. 6 (b), the I_G at Stage V also reduced back to the comparable level as that in the fresh device. Moreover, the V_{TH} at Stage-V returns back to the level

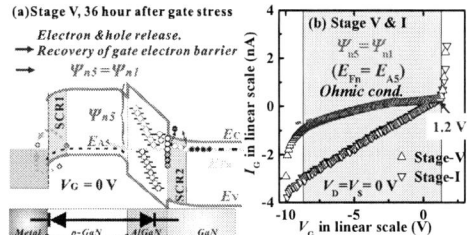

Fig. 6 (a) Schematic energy band diagram of the p-GaN gate at Stage-V with $V_G = 0$ V. (b) I_G-V_G of the DUT at Stage V & I in the ohmic conduction region with $V_G < 1.2$ V.

of the fresh device excepting the degraded sub threshold slope induced by the generated traps during the gate stress as shown in Fig. 1 (d).

After the recovery of the initial gate electron energy barrier at Stage-V, the DUT was further stressed again for another 53 ks as plotted in Fig. 7 (a). The monitored I_G during the 2nd gate stress exhibits again no obvious gate current degradation. However, the gate current at Stage-VI increases up to ~10 μA at low gate bias as shown in Fig. 7 (e), owing to the gate electron barrier lowering by electron-trapping and severe hole-injection and trap generation during the 2nd gate stress. Even though, the I_G can recover back to the initial level at Stage-VII after another 36-hour rest and the gate still feature channel modulation capability. However, the original low I_G can be only measured in a single sweep at Stage VII and then the I_G significantly increased by up to 4-order in the following I_G - V_G sweep as shown in Fig. 7 (e) due to severe electron trapping in the severe degraded AlGaN barrier.

978-1-7281-0582-6/19 $31.00 © 2019 IEEE

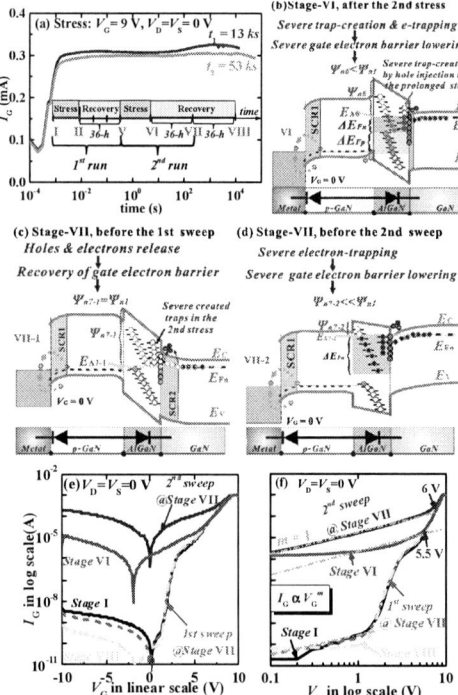

Fig. 7. (a) I_G of the DUT monitored during the 1st and 2nd forward stress with $V_G = 9$ V. Schematic energy band diagram of the p-GaN gate with $V_G = 0$ V (b) at Stage-VI , (c) before the 1st sweep at Stage-VII and (d)2nd sweep at Stage-VII. (e) I_G-V_G of the DUT after the 2nd gate stress.

IV. CONCLUSION

We report a nondestructive gate degradation dynamic of the 650 V commercial normally-off p-GaN/AlGaN/GaN HEMTs in the long-term recovery period after a long-term positive gate stress. The p-GaN gate starts to exhibit a degradation with significant gate leakage increase and then the gate leakage recovers to the initial level that comparable to that measured in the fresh device. Even though the gate shows an accelerate degradation after positive stress with longer stress time, it also features gate control functionality without permanent failure. The underlying carrier transport mechanisms were revealed by the voltage and time dependent gate current. The hole-injection together with trap-states generation take place in the forward gate stress duration and followed by electron-trapping as well as the eventually detrapping of holes and electrons is found to be responsible for the observed sophisticated gate degradation dynamics. Such a carrier transport kinetics in the p-GaN gate region is validated by the V_{TH} shift during the device characterization duration. The nondestructive and self-recoverable gate degradation and the underlying mechanism is beneficial for understanding the ruggedness of p-GaN gate.

REFERENCES

[1] K.J. Chen, O. Haberlen, A. Lidow, C. L. Tsai, T. Ueda, Y. Uemoto, and Y. Wu, "GaN-on-Si Power Technology: Devices and Applications," *IEEE Trans. Electron Devices*, vol. 64, no. 3, pp. 779-795

[2] D. Marcon, Y. N. Saripalli, and S. Decoutere, "200mm GaN-on-Si epitaxy and e-mode device technology," in *IEEE International Electron Devices Meeting (IEDM)*, 2015, p. 16.2.1-16.2.4

[3] Y. Uemoto, M. Hikita, H. Ueno, H. Matsuo, M. Ishida, M. Yanagihara, T. Ueda, T. Tanaka, and D. Ueda, "Gate injection transistor (GIT) - A normally-off AlGaN/GaN power transistor using conductivity modulation," *IEEE Trans. Electron Devices*, vol. 54, no. 12, pp. 3393–3399, 2007

[4] EPC. EPC2050 Datasheet. Accessed: Apr. 2018. [Online]. Available: http://epc-co.com/epc

[5] Panasonic. PGA26E19BA Datasheet. Accessed: Mar. 2015. [Online]. Available: http://www.mouser.hk

[6] GaN Systems. GS66504B Datasheet. Accessed: Dec. 2015. [Online]. Available: https://gansystems.com/

[7] Infineon. IGT60R070D1 Datasheet. Accessed: Oct. 2018. [Online]. Available: https://www.infineon.com/

[8] T. L. Wu, D. Marcon, S. You, N. Posthuma, B. Bakeroot, S. Stoffels, M. V. Hove, G. Groeseneken, and S. Decoutere, "Forward bias gate breakdown mechanism in enhancement-mode p-GaN gate AlGaN/GaN high-electron mobility transistors," *IEEE Electron Device Lett.*, vol. 36, no. 10, pp. 1001–1003, Oct. 2015.

[9] I. Rossetto, M. Meneghini, O. Hilt, E. Bahat-Treidel, C. De Santi, S. Dalcanale, J. Wuerfl, E. Zanoni, and G. Meneghesso, "Time-Dependent Failure of GaN-on-Si Power HEMTs with p-GaN Gate," *IEEE Trans. Electron Devices*, vol. 63, no. 6, pp. 2334–2339, Jun. 2016.

[10] M. Ťapajna, O. Hilt, E. Bahat-Treidel, J. Würfl, and J. Kuzmik, "Gate reliability investigation in normally-off p-type-GaN Cap/AlGaN/GaN HEMTs under forward bias stress," *IEEE Electron Device Lett.*, vol. 37, no. 4, pp. 385–388, Apr. 2016.

[11] A. Stockman, F. Masin, M. Meneghini, E. Zanoni, G. Meneghesso, B. Bakeroot, and P. Moens, "Gate Conduction Mechanisms and Lifetime Modeling of p-Gate AlGaN/GaN High-Electron-Mobility Transistors," *IEEE Trans. Electron Devices*, vol. 65, no. 12, pp. 5365-5371, Dec. 2018

[12] M. Ruzzarin, M. Meneghini, A. Barbato, V. Padovan, O. Haeberlen, M. Silvestri, T. Detzel, G. Meneghesso, and E. Zanoni, "Degradation Mechanisms of GaN HEMTs with p-Type Gate under Forward Gate Bias Overstress," *IEEE Trans. Electron Devices*, vol. 65, no. 7, pp. 2778–2783, 2018

[13] A. N. Tallarico, S. Stoffels, N. Posthuma, P. Magnone, D. Marcon, S. Decoutere, E. Sangiorgi, and C. Fiegna, "PBTI in GaN-HEMTs with p-Type gate : role of the Aluminum content on V_{TH} and underlying degradation mechanisms," *IEEE Trans. Electron Devices*, vol. 65, no. 1, pp. 38–44, Jan. 2018

[14] P. Lagger, M. Reiner, D. Pogany, and C. Ostermaier, "Comprehensive study of the complex dynamics of forward bias-induced threshold voltage drifts in GaN based MIS-HEMTs by stress/recovery experiments," *IEEE Trans. Electron Devices*, vol. 61, no. 4, pp. 1022–1030, Apr. 2014.

[15] Y. Shi, Q. Zhou, Q. Cheng, P. Wei, L. Zhu, D. Wei, A. Zhang, W. Chen, and B. Zhang, "Bidirectional threshold voltage shift and gate leakage in 650 V p-GaN AlGaN/GaN HEMTs: The role of electron-trapping and hole-injection," in *Proc. 30th ISPSD*, May 2018, pp. 96–99.

[16] Y. Shi, Q. Zhou, Q. Cheng, P. Wei, L. Zhu, D. Wei, A. Zhang, W. Chen, and B. Zhang, "Carrier Transport Mechanisms Underlying the Bidirectional VTH Shift in p-GaN Gate HEMTs Under Forward Gate Stress," *IEEE Trans. Electron Devices*, vol. 66, no. 2, pp. 876–882, Feb. 2019.

[17] M. Hua, J. Wei, Q. Bao, Z. Zhang, Z. Zheng, and K. J. Chen, "Dependence of VTH Stability on Gate-Bias Under Reverse-Bias Stress in E-mode GaN MIS-FET," *IEEE Electron Device Lett.*, vol. 39, no. 3, pp. 413–416, Mar. 2018.

978-1-7281-0582-6/19 $31.00 © 2019 IEEE

Proceedings of the 31st International Symposium on Power Semiconductor Devices & ICs
May 19-23, 2019, Shanghai, China

Recess-Free Normally-off GaN MIS-HEMT Fabricated on Ultra-Thin-Barrier AlGaN/GaN Heterostructure

line 1: Ping-Cheng Han
line 2: *International college of Semiconductor Technology*
line 3: *National Chiao Tung University*
line 4: Hsinchu, Taiwan R.O.C.
line 5: han9747ahc@gmail.com

line 1: Zong-Zheng Yan
line 2: *Department of Materials Science & Engineering*
line 3: *National Chiao Tung University*
line 4: Hsinchu, Taiwan R.O.C.
line 5: buoy077123@gmail.com

line 1: Chia-Hsun Wu
line 2: *Department of Materials Science & Engineering*
line 3: *National Chiao Tung University*
line 4: Hsinchu, Taiwan R.O.C.
line 5: s906713.mse97@gmail.com

line 1: Edward Yi Chang
line 2: *Department of Materials Science & Engineering*
line 3: *National Chiao Tung University*
line 4: Hsinchu, Taiwan R.O.C.
line 5: edc@mail.nctu.edu.tw

line 1: Yu-Hsuan Ho
line 2: *Institute of Lighting and Energy Photonics*
line 3: *National Chiao Tung University*
line 4: Hsinchu, Taiwan R.O.C.
line 5: sandyho16@gmail.com

Abstract—**In this work, a recess-free thin AlGaN barrier metal-insulator-semiconductor high electron mobility transistor (MIS-HEMT) with high threshold voltage of 3.19V, high maximum drain current of 716 mA/mm and high breakdown voltage of 906V is demonstrated for the first time. Three different thin barrier structures were compared for device performance. In addition, long-term reliability measurements were carried out to investigate the interface quality of the devices with and without gate recess. The recess-free device exhibits a smaller V_{th} hysteresis and a more stable performance than those of gate recessed devices. Overall, the recess-free thin barrier AlGaN/GaN MIS-HEMT demonstrates promising performance for power switching applications.**

Keywords—AlGaN, HEMT, normally-off, thin barrier, recess-free

I. INTRODUCTION

GaN high electron mobility transistors (HEMT) show high device performance for power switching applications due to the high breakdown field and low on-resistance (R_{ON}). Owing to the false turn-on issue and the requirements for fail-safe operation for electrical vehicles, a high threshold voltage (V_{th}) normally-off GaN HEMT is desired. Gate recess technology is one of the most commonly used technology for normally-off HEMT. However, the dry etching process generates interface states which effect process controllability. Ultra-thin barrier HEMT device which exhibits well-controlled V_{th} uniformity [1] has high potential for integrated GaN circuit applications [2]. But it suffers from low output current if without proper passivation. SiN passivation with nitrogen plasma pretreatment was utilized to reduce the sheet resistance (R_{sheet}) and increase the maximum drain current ($I_{DS,max}$). On the other hand, the use of ferroelectric charge trap gate stack [3] results in a positive V_{th} shift after the initialization process. Therefore, by combining the SiN passivation and the ferroelectric charge trap gate stack, recess-free thin barrier device can obtain a high V_{th} with a high $I_{DS,max}$. Different barrier compositions were compared for the optimized device performance. Furthermore, to verify the long-term reliability, on-state stress test and positive bias temperature instability (PBTI) measurement were performed for devices with and without gate recess.

II. DEVICE FABRICATION

The metal-insulator-semiconductor high electron mobility transistor (MIS-HEMT) structure is plotted in Fig. 1. The devices started with mesa isolation using Cl_2-based dry etching by inductively coupled plasma reactive ion etching (ICP-RIE) machine. Ti/Al/Ni/Au alloy was annealed under 800℃ for 60 seconds as ohmic contact, SiN_x deposition with nitrogen plasma pre-treatment using plasma-enhanced chemical vapor deposition (PECVD) machine was carried out as passivation layer. The gate window was opened by using CF_4-based ICP-RIE dry etching. Devices with different gate recess depth were fabricated by adjusting the etching time. The ferroelectric charge trap gate stack was then deposited by using plasma enhanced atomic layer deposition (PEALD) machine. The gate stack used was the same as that reported in the literature [3]. Post deposition annealing was carried out at 400℃ for 10 minutes in nitrogen ambient. Ni/Au was deposited as gate metal. The gate length (L_G), gate-source spacing (L_{GS}) and gate-drain spacing (L_{GD}) were 4µm, 2µm and 14µm, respectively. Agilent B1505A power device analyzer was used for I-V characterization.

III. RESULTS AND DISCUSSION

Fig. 2 shows the increase in output current for thin barrier devices after SiN passivation. SiN passivation with nitrogen plasma pretreatment was utilized to reduce the R_{sheet} and therefore increase the $I_{DS,max}$.

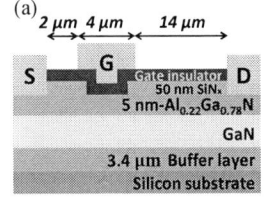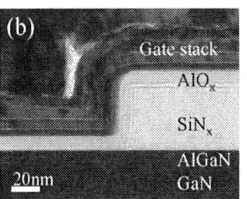

Fig. 1. (a) Schematic cross section of the recess-free thin barrier AlGaN/GaN MIS-HEMT. (b) Cross-sectional TEM in the gate edge region.

978-1-7281-0582-6/19 $31.00 © 2019 IEEE

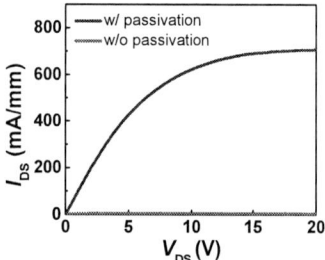

Fig. 2. Output characteristics of the thin barrier devices with and without passivation.

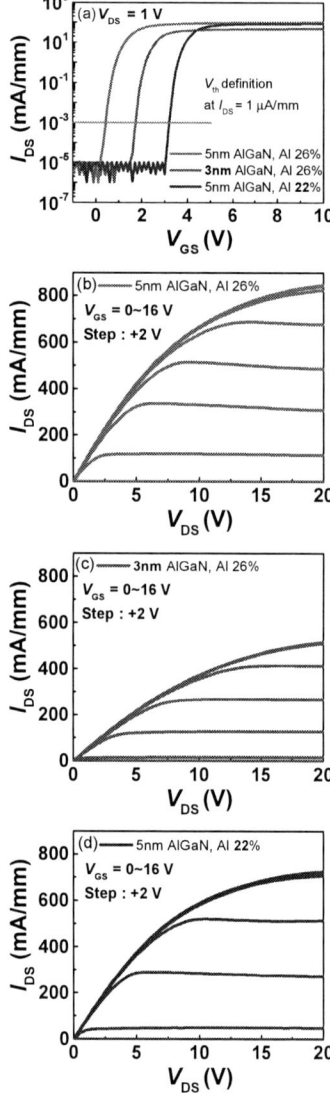

Fig. 3. (a) Transfer characteristics of devices with different barrier structures. Output characteristics of devices with (b) 5nm AlGaN barrier and 26% Al concentration, (c) 3nm AlGaN barrier and 26% Al concentration and (d) 5nm AlGaN barrier and 22% Al concentration.

The ferroelectric charge trap gate stack was utilized to shift the V_{th} positively. After a 16V pulse being applied on the gate electrode, the electrons were stored in the gate stack, resulting in a normally-off device.

Fig. 3 shows the DC characteristics of the devices with three different barrier structures. The thinning of the barrier layer leads to a significant degradation in $I_{DS,max}$, while reducing the Al concentration results in a higher V_{th} (Fig. 4). Table I shows the summary of device performance and the results from Hall measurement. The Hall mobility and carrier concentration were extracted from the sheet resistance under gate region (R_g), which was measured from the samples with ferroelectric gate stack directly deposited on the AlGaN surface (Fig. 5). The barrier structure with higher drain current exhibits a smaller Hall resistance. Besides, the V_{th} value is inversely related to the product of carrier concentration and mobility. V_{th} is defined as the gate voltage at which the drain current reaches 1μA/mm with 1V drain biasing. Among the three different barrier structures, the sample with 5nm AlGaN barrier and 22% Al concentration exhibits the highest V_{th} of 3.19V and a moderate $I_{DS,max}$ of 716 mA/mm, which is then chosen as the most suitable barrier structure in this study. The device performance compared to other reports is plotted in Fig. 6. The breakdown voltage is 906V and the dynamic R_{ON} increases to 1.71 times at the quiescent bias of 600V (Fig. 7).

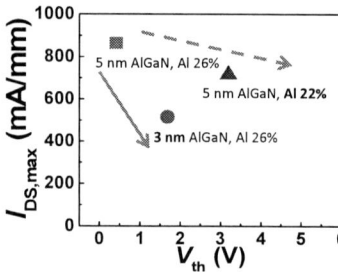

Fig. 4. Summary of $I_{DS,max}$ versus V_{th} of devices with different barrier structures.

TABLE I. SUMMARY OF DEVICE PERFORMANCE AND THE RESULTS OF HALL MEASUREMENT IN DEVICES WITH DIFFERENT BARRIER STRUCTURES.

AlGaN barrier (thickness, Al concentration)	5nm, Al 26%	**3nm**, Al 26%	5nm, Al **22%**
R_{ON} (Ω·mm)	10.60	21.59	12.25
$I_{DS,max}$ (mA/mm)	864	516	716
V_{th} (V)	0.41	1.68	3.19
R_{SD} (Ω/sq)	529	1226	491
R_g (Ω/sq)	365	443	591
Hall mobility (cm²/V sec.)	1600	1550	1640
Carrier Conc. (cm⁻²)	1.01×10^{13}	9.07×10^{12}	7.05×10^{12}

978-1-7281-0582-6/19 $31.00 © 2019 IEEE

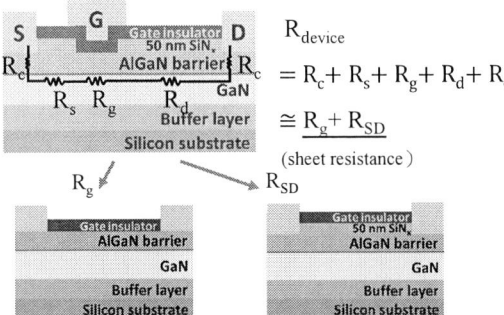

$$R_{device} = R_c + R_s + R_g + R_d + R_c$$
$$\cong \underline{R_g + R_{SD}}$$
(sheet resistance)

Fig. 5. Schematic of the sheet resistances in MIS-HEMT devices for Hall measurement. The sheet resistance of gate-source and gate-drain access region (R_{SD}) was measured from the samples with SiN passivation and gate stack, while the sheet resistance under gate region (R_g) was measured from the samples with gate stack deposited on AlGaN surface. Hall mobility and carrier concentration were extracted from R_g.

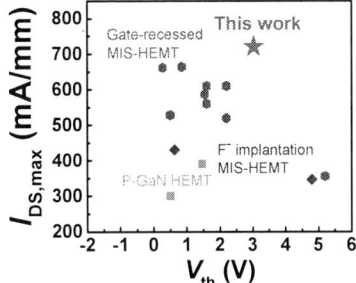

Fig. 6. Benchmark plot of $I_{DS,max}$ versus V_{th} for published in literature normally-off GaN HEMTs [4-16].

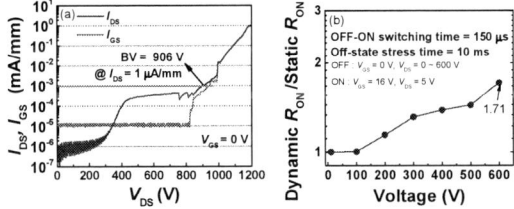

Fig. 7. (a) Off-state breakdown characteristics. (b) Ratio between dynamic R_{ON} and static R_{ON} at different off-state biases.

The DC characteristics of the devices with and without gate recess were shown in Fig. 8 and Fig. 9. As the recess depth increases, the V_{th} value increases but the $I_{DS,max}$ reduces accordingly. For the gate recessed devices, etching damage on the barrier formed due to the longer etching time. The generated interface states result in the widening of V_{th} hysteresis. On the other hand, the small V_{th} hysteresis in the recess-free device indicates that dry etching damage during gate window opening after SiN passivation was minimized. In addition, a tight distribution of V_{th} of 0.09V was achieved in the recess-free devices (Fig. 10).

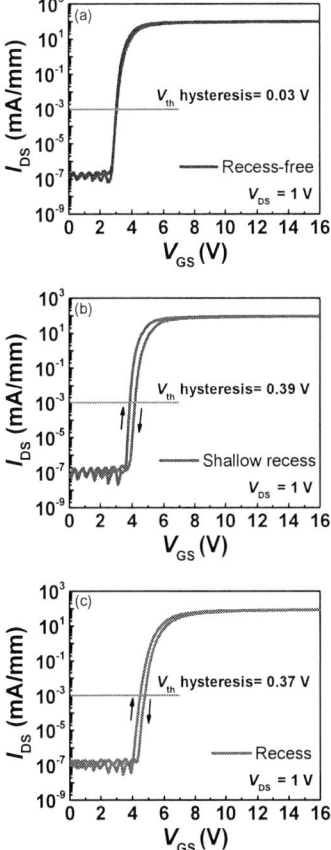

Fig. 8. Transfer characteristics of (a) recess-free (b) shallow-recessed (c) recessed thin barrier devices.

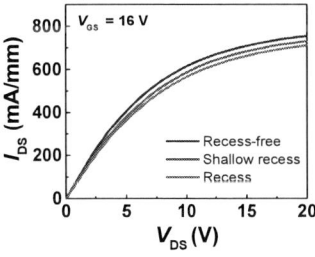

Fig. 9. Output characteristics of devices with different recess depth.

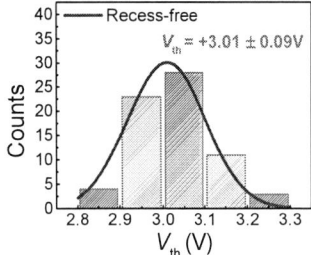

Fig. 10. V_{th} uniformity of the recess-free devices.

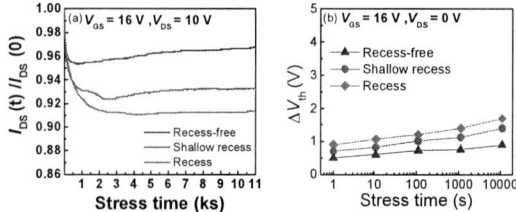

Fig. 11. (a) Ratio of the measured output current (I_{DS}) during the on-state stress test. (b) Stress time evolution of V_{th} shift.

Fig. 11 shows the on-state stress test and the PBTI results. The gate recess process generated interface states. During the stress measurement, some of the electrons were trapped at the AlGaN/dielectric interface, leading to a positive shift of V_{th} and the degradation of output current. With V_{GS} of 16V and V_{DS} of 10V bias, only 4% degradation in output current was observed for the recess-free device after on-state stress. Besides, the recess-free device exhibited the smallest V_{th} shift of 0.89V compared to the gate recessed devices after PBTI measurement, revealing that the interface quality was preserved.

IV. CONCLUSION

The recess-free normally-off GaN MIS-HEMT with thin barrier exhibits excellent device performance and stable properties. The ultra-thin-barrier structure with ferroelectric charge trap gate stack results in a high V_{th} of 3.19V and a small hysteresis. The SiN passivation with nitrogen plasma pretreatment leads to a high $I_{DS,max}$ of 716 mA/mm. Overall, the recess-free thin barrier AlGaN/GaN MIS-HEMT is a promising candidate for power switching applications.

ACKNOWLEDGMENT

The authors would like to thank Dr. K.-C. Lin, Dr. J.-H. Liu and Dr. Y.-L. Hsiao (Elite Advanced Laser Corporation) for their valuable discussions and technical support on the epitaxial structure design.

REFERENCES

[1] S. Huang, X. Liu, X. Wang, X. Kang, J. Zhang, J. Fan, *et al.*, "Ultrathin-Barrier AlGaN/GaN Heterostructure: A Recess-Free Technology for Manufacturing High-Performance GaN-on-Si Power Devices," *IEEE Transactions on Electron Devices*, vol. 65, pp. 207-214, Jan. 2018.

[2] X. Kang, X. Wang, S. Huang, J. Zhang, J. Fan, S. Yang, *et al.*, "Recess-free AlGaN/GaN lateral Schottky barrier controlled Schottky rectifier with low turn-on voltage and high reverse blocking," in *2018 IEEE 30th International Symposium on Power Semiconductor Devices and ICs (ISPSD)*, pp. 280-283, May 2018.

[3] C. Wu, S. Liu, C. Huang, Y. Chiu, P. Han, P. Chang, *et al.*, "High V_{th} enhancement mode GaN power devices with high $I_{D,max}$ using hybrid ferroelectric charge trap gate stack," in *VLSI Technology, 2017 Symposium on*, pp. T60-T61, June 2017.

[4] Y. Wang, Y. C. Liang, G. S. Samudra, H. Huang, B. Huang, S. Huang, *et al.*, "6.5 V High Threshold Voltage AlGaN/GaN Power Metal-Insulator-Semiconductor High Electron Mobility Transistor Using Multilayer Fluorinated Gate Stack," *IEEE Electron Device Letters*, vol. 36, pp. 381-383, April 2015.

[5] Q. Zhou, L. Liu, A. Zhang, B. Chen, Y. Jin, Y. Shi, *et al.*, "7.6 V Threshold Voltage High-Performance Normally-Off Al2O3/GaN MOSFET Achieved by Interface Charge Engineering," *IEEE Electron Device Letters*, vol. 37, pp. 165-168, Feb. 2016.

[6] O. Hilt, R. Zhytnytska, J. Böcker, E. Bahat-Treidel, F. Brunner, A. Knauer, *et al.*, "70 mΩ/600 V normally-off GaN transistors on SiC and Si substrates," in *2015 IEEE 27th International Symposium on Power Semiconductor Devices & IC's (ISPSD)*, pp. 237-240, May 2015.

[7] Z. Tang, Q. Jiang, Y. Lu, S. Huang, S. Yang, X. Tang, *et al.*, "600-V Normally Off SiNx/AlGaN/GaN MIS-HEMT With Large Gate Swing and Low Current Collapse," *IEEE Electron Device Letters*, vol. 34, pp. 1373-1375, Nov. 2013.

[8] M. Wang, Y. Wang, C. Zhang, B. Xie, C. P. Wen, J. Wang, *et al.*, "900 V/1.6 mΩ · cm² Normally Off Al2O3/GaN MOSFET on Silicon Substrate," *IEEE Transactions on Electron Devices*, vol. 61, pp. 2035-2040, June 2014.

[9] Q. Hu, S. Li, T. Li, X. Wang, X. Li, and Y. Wu, "Channel Engineering of Normally-OFF AlGaN/GaN MOS-HEMTs by Atomic Layer Etching and High- κ Dielectric," *IEEE Electron Device Letters*, vol. 39, pp. 1377-1380, Sept. 2018.

[10] M. Hua, J. Wei, Q. Bao, Z. Zhang, Z. Zheng, and K. J. Chen, "Dependence of V_{TH} Stability on Gate-Bias Under Reverse-Bias Stress in E-mode GaN MIS-FET," *IEEE Electron Device Letters*, vol. 39, pp. 413-416, March 2018.

[11] K. J. Chen, O. Häberlen, A. Lidow, C. l. Tsai, T. Ueda, Y. Uemoto, *et al.*, "GaN-on-Si Power Technology: Devices and Applications," *IEEE Transactions on Electron Devices*, vol. 64, pp. 779-795, March 2017.

[12] Y. Wang, M. Wang, B. Xie, C. P. Wen, J. Wang, Y. Hao, *et al.*, "High-Performance Normally-Off Al2O3/GaN MOSFET Using a Wet Etching-Based Gate Recess Technique," *IEEE Electron Device Letters*, vol. 34, pp. 1370-1372, Nov. 2013.

[13] H. Handa, S. Ujita, D. Shibata, R. Kajitani, N. Shiozaki, M. Ogawa, *et al.*, "High-speed switching and current-collapse-free operation by GaN gate injection transistors with thick GaN buffer on bulk GaN substrates," in *2016 IEEE International Electron Devices Meeting (IEDM)*, pp. 10.3.1-10.3.4, Dec. 2016.

[14] M. Hua, Z. Zhang, J. Wei, J. Lei, G. Tang, K. Fu, *et al.*, "Integration of LPCVD-SiNx gate dielectric with recessed-gate E-mode GaN MIS-FETs: Toward high performance, high stability and long TDDB lifetime," in *2016 IEEE International Electron Devices Meeting (IEDM)*, pp. 10.4.1-10.4.4, Dec. 2016.

[15] Y. Shi, S. Huang, Q. Bao, X. Wang, K. Wei, H. Jiang, *et al.*, "Normally OFF GaN-on-Si MIS-HEMTs Fabricated With LPCVD-SiNx Passivation and High-Temperature Gate Recess," *IEEE Transactions on Electron Devices*, vol. 63, pp. 614-619, Feb. 2016.

[16] S. Huang, X. Liu, X. Wang, X. Kang, J. Zhang, J. Fan, *et al.*, "Ultrathin-Barrier AlGaN/GaN Heterostructure: A Recess-Free Technology for Manufacturing High-Performance GaN-on-Si Power Devices," *IEEE Transactions on Electron Devices*, vol. 65, pp. 207-214, Jan. 2018.

Proceedings of the 31st International Symposium on Power Semiconductor Devices & ICs
May 19-23, 2019, Shanghai, China

Over Kilovolt GaN Vertical Super-Junction Trench MOSFET: Approach for Device Design and Optimization

Peng Huang, Qi Zhou†, Kuangli Chen, Xiaoqi Han, Dong Wei, Yuanyuan Shi, Wanjun Chen, and Bo Zhang
State Key Laboratory of Electronic Thin Films and Integrated Device
University of Electronic Science and Technology of China (UESTC), Chengdu, P.R. China
†Email address: zhouqi@uestc.edu.cn

Abstract—In this work, the fundamental concept of P/N-junction for voltage sustaining was introduced in GaN vertical device to push the *BV* of the device beyond kilovolts. Accordingly, a novel GaN vertical super-junction (SJ) trench MOSFET was proposed and studied by TCAD simulation. In order to realize the typical gradient doping transition stack of $P^+/P/N^+$ in conventional Si-SJ structure, the P-GaN/UID-GaN/N^+-GaN stack, where the top P-GaN and bottom N^+-GaN act as the field-stop (FS) layer, is employed avoiding using the hardly achievable high quality P^+-GaN. In order to mitigate the premature punch-through in the top P-GaN/N-drift junction due to the possible under design of the P-GaN thickness or the avalanche breakdown triggered by the high *E*-field due to the possible under design of the UID-GaN thickness, the device design scheme by obtaining a uniform *E*-field distribution at the P-GaN/UID-GaN/N^+-GaN to N-drift region interface is proposed. By optimizing the thickness ratio of P-GaN to UID-GaN, the balance between the top punch-through and bottom avalanche can be obtained, which enables achieving the maximum overall *BV* of the GaN SJ-MOS with a given total thickness of N-GaN drift region. The proposed device structure and the methodology of device design are of great interests for GaN vertical device to achieve kilovolts *BV* for ultra-high power electronics.

Keywords—Gallium Nitride, MOSFET, high breakdown, vertical, Super-Junction, Electric Field Modulation, avalanche breakdown, punch-through breakdown

I. INTRODUCTION

GaN vertical power devices are considered to be suitable for kilowatt rating (e.g. ⩾1.2 kV/100 A) power electronics owing to its great potential of simultaneously delivering high current and high breakdown voltage (*BV*) without excessive wafer size consumption. Besides, in GaN vertical device, the peak electric-field (*E*-field) is normally buried in GaN bulk, which effectively eliminates the surface trapping effect and the corresponding dynamic R_{on} degradation. Hence, GaN vertical devices have attracted tremendous attention in very recent years[1]-[10].

Nevertheless, the normally reported *BV* of GaN vertical devices are still less than 2 kV that is even inferior to its lateral counterpart[11]-[14]. However, the carbon doping in transition/buffer layer is essential for lateral GaN device to obtain off-state blocking capability, which is detrimental to dynamic R_{on} and could be even severe at kilovolts rating applications that must be avoid in GaN vertical devices.[15][16] Hence, in order to push the *BV* to kilovolt rating or even higher, the fundamental concept of P/N-junction for voltage sustaining should be employed in GaN vertical devices.

This work was supported in part by the National Natural Science Foundation of China under Grant 61674024, in part by the Assembly Pre-Research Project under Grant JZX2017-1643/Y537, in part by the National Science and Technology Major Project under Grant 2013ZX02308-005. (*Corresponding Author: Qi Zhou.*)

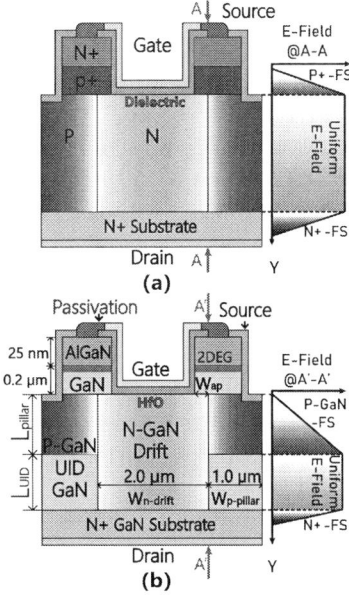

Fig. 1. (a) Conventional Si SJ trench MOSFET and the ideal *E*-field distribution. (b) Schematic cross section of the proposed vertical GaN Super-Junction trench MOSFET with expected *E*-field distribution.

Besides pursuing high *BV*, the R_{on} is required to be minimized for improved power efficiency. Because the super-junction technology has been approved capable of achieving the optimized trade-off between *BV* and R_{on}, it is of great interests to explore the adaption of super-junction for GaN power device in view of device design and optimization.

In this work, an over kilovolt GaN vertical super-junction trench MOSFET was proposed and studied by TCAD-simulation. By taking into account the availability of P-GaN and N-GaN, the P^+ layer acting as the FS layer in conventional Si SJ-MOS is removed while an unintentional doped (UID) GaN layer was inserted between the P-GaN and N^+-GaN substrate, which composes the P/UID/N^+-GaN stack with P/N^+ act as the FS layers that is analogous to the typical $P^+/N/N^+$ structure in Si SJ device. The device design consideration and optimization strategy of GaN vertical SJ MOSFET is investigated for over kilovolt power applications, which is of great value for experimental reference.

II. DEVICE STRUCTURE AND OPERATION MECHANISM

Fig. 1 shows the device structures of conventional Si super-junction and the proposed GaN super-junction (SJ) trench MOSFET. The nitride passivation layer is placed on AlGaN barrier layer.Owing to the lack of high quality P^+ GaN (e.g. low-effective P-type doping & low hole mobility),

978-1-7281-0582-6/19 $31.00 © 2019 IEEE

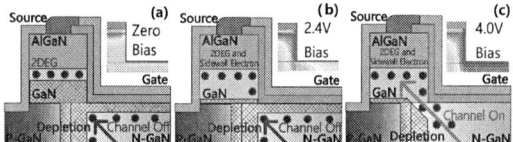

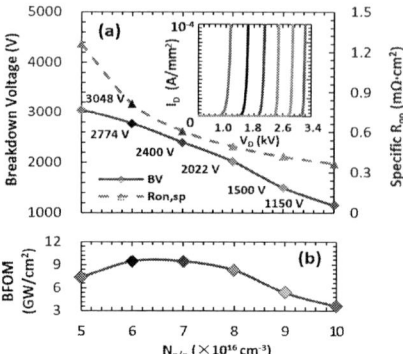

Fig. 2. Simulation results for electron density distribution and schematic cross sections of (a) fully depleted channel region at V_{gs}=0 V, (b) channel region with electrons accumulated at the sidewall at V_{gs}=2.4 V and (c) on-state channel region at V_{gs}=4 V.

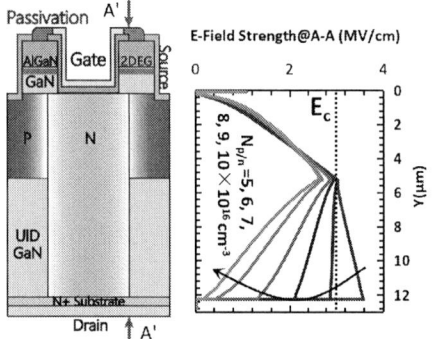

Fig. 3. E-field distribution along A-A as a function of effective N/P doping concentration $N_{p/n}$.

Fig. 4 (a) The relationship of BV, $R_{on,sp}$ and doping concentration with various $N_{p/n}$. (b) The $BFOM$ of the proposed device with various $N_{p/n}$.

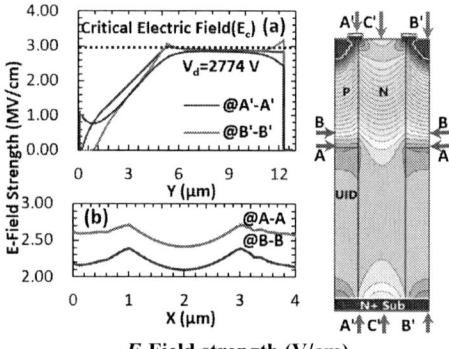

Fig. 5. (a) The absolute of E-field strength distribution at V_{gs}=0 V and V_{ds}=2774 V in vertical direction (b) The absolute of E-field strength distribution at V_{gs}=0 V and V_{ds}=2774 V in lateral direction.

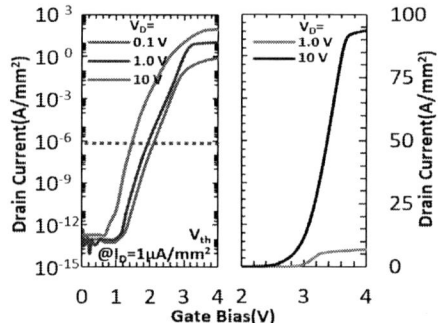

Fig. 6. Transfer characteristics of the proposed device at V_{ds}=0.1, 1.0, 10 V in (a) log-scale and (b) linear-scale. The device features a high on/off current ratio of over ~12-order.

the P$^+$ body as a FS layer is replaced by the AlGaN/GaN heterojunction, which can take the advantage of high density and high mobility 2DEG favoring R_{on} reduction. Meanwhile, an unintentional doped (UID) GaN layer with the width of 1 μm is inserted between the P-pillar and N$^+$ GaN substrate. In this manner, a gradient doping can be realized by the P-GaN/UID-GaN/N$^+$-GaN that is similar to that in the Si SJ by P$^+$/P/N$^+$ structure. Hence, the P-GaN in the proposed device performs as the FS layer while the lateral depletion between P-GaN and N-drift pillar with the width of 2 μm is also capable of sustaining the off-state voltage. In addition, the P-GaN may deplete the above GaN (in AlGaN/GaN) that facilitates pinch-off the gated side-wall channel, which is beneficial for overall leakage lowering.

Fig. 2 illustrates the turn-on dynamic of the proposed device. The current conduction channel of the device consists of the gated side-wall channel and the aperture channel between the P-GaN and the gate which is turned on in sequence with the increased gate bias. In order to achieve the high BV, an uniform E-field distribution both in lateral and vertical in the device is required. Hence, the effective doping concentration of P-GaN and N-GaN drift region ($N_{p/n}$) in the simulation is defaulted to be identical for an uniform E-field distribution in lateral.

Further optimizing $N_{p/n}$, the E-field profile in vertical direction in UID-GaN can be modulated and turned from a trapezoid-like into a triangle-like distribution with the increased $N_{p/n}$ as shown in Fig. 3. It can be seen that the slope of E-field in UID-GaN region of the proposed device with L_{pillar}=5 μm and L_{UID}=7 μm reduces along with increased $N_{p/n}$, and therefore, a BV reduction is shown for its triangle E-field, while $N_{p/n}$=6×10^{16} cm^{-3} can provide an optimized uniform E-field in UID-GaN region. The relationship of $BFOM$, BV, $R_{on,sp}$ and doping concentration with various $N_{p/n}$ is presented in Fig. 4. The $R_{on,sp}$ and BV of the device reduce along with the increased $N_{p/n}$, while a peak of $BFOM$=BV^2/$R_{on,sp}$ for the device is obtained from the device within $N_{p/n}$=6~7×10^{16} cm^{-3}.

Hence, referring Fig. 3, a uniform E-field in UID-GaN is obtained for $N_{p/n}$=6×10^{16} cm^{-3} which simultaneously yields the maximum Baliga's figure-of-merit ($BFOM$) of 9.46 GW/cm^2.

III. RESULTS AND DISCUSSION

The E-field distribution in the device with L_{pillar}=5 μm, L_{UID}=7 μm and $N_{p/n}$=6×10^{16} cm^{-3} at V_D=2774 V is shown in Fig. 5. It can be seen that, with the optimized $N_{p/n}$, the E-field in lateral direction exhibits a uniform distribution. In vertical direction, the E-field achieves a uniform plateau in UID-GaN

978-1-7281-0582-6/19 $31.00 © 2019 IEEE

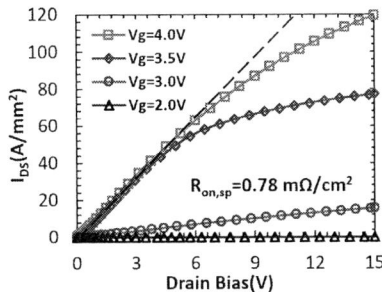

Fig. 7. The output characteristic of the proposed device with L_{pillar}=5 μm and L_{UID}=7 μm, and $N_{p/n}$=6E+16 cm⁻³. A $R_{on,sp}$ of 0.78 mΩ·cm² is obtained.

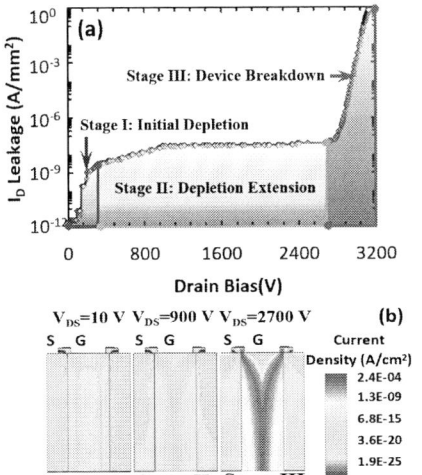

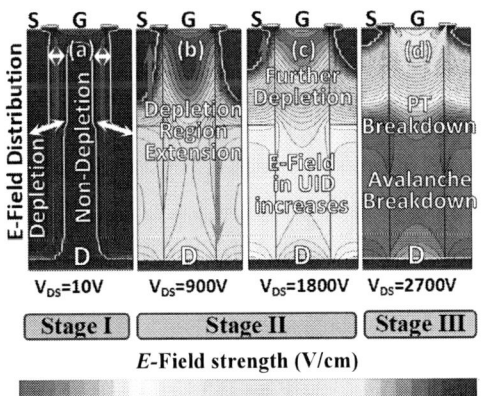

Fig. 8. (a)Simulated breakdown I-V of proposed GaN SJ MOSFET. (b)The leakage current density distribution at V_{ds}=10, 900, 2700V.

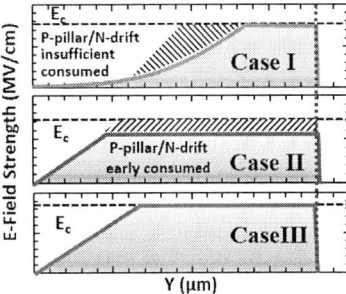

Fig. 10. Schematic E-field distribution in vertical direction along A-A at breakdown for three types of breakdown mechanisms of the proposed device for a given overall GaN thickness (L_{pillar}+L_{UID}).

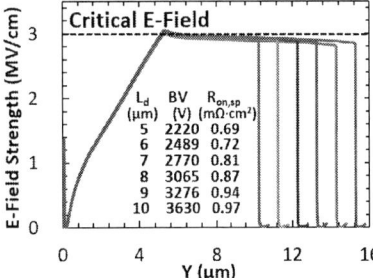

Fig. 11. E-field strength distribution of the proposed device with a given L_{pillar}=5 μm, L_{UID}=5~10 μm and $N_{p/n}$=6×10¹⁶ cm⁻³.

Fig. 9. E-field strength distribution at V_{gs}=0V and various V_{ds} is presented in (a) to (d).

region, while it significantly reduced to the minimum in the pillar and N⁺-GaN region (i.e. field-stop). Accordingly, the transfer and output curves of the device are shown in Fig. 6 and 7, respectively. In Fig. 6, it can be observed that threshold voltage shifts towards negative X axis until 1 V along with increased V_{gs}, while a $R_{on,sp}$ of 0.78 mΩ·cm² can be obtained in Fig. 7, respectively.

The off-state drain leakage shown in Fig. 8 can be separated into 3 stages while a BV of 2774 V can be obtained

from Fig. 8(a). The depletion region simultaneously extends from the UID-GaN-to-N-drift junction in the bottom part and the P-GaN-to-N-drift junction in the top part with the increased V_{DS} as shown in Fig. 9, respectively: 1) In stage I, the depletion region extends from P/N junction and UID/N junction with the increased V_{ds}; 2) In stage II, E-field in UID-GaN region increases with increased V_{gs}; 3) In stage III, the depletion region extends to source and the punch-through could be induced, while high E-field could trigger avalanche breakdown in UID-GaN region. In this manner, the device breakdown is governed by two competing mechanisms as shown in Fig. 10: 1) In the bottom part, an impact ionization could be triggered while the E-field increases higher than the critical E-field E_C (i.e. 3 MV/cm); 2) In the top part, a punch-through (PT) could be induced together with the depletion region extends to the source.

Fig. 11 shows the E-field strength distribution for a given L_{pillar} with varied L_{UID}. The BV can increase from 2.2 to 3.6 kV while the L_{UID} increases from 5 to 10 μm unless the PT was not triggered for the given L_{pillar}. It can be seen that thicker UID-GaN region and N-Gan drift could sustain high E-field in a larger region. Hence, with the increased thickness of UID-GaN region, BV and $R_{on,sp}$ increase. Therefore, besides the $N_{p/n}$ optimization for uniform E-field distribution, it is essential to delicately design the thickness ratio of P-pillar to UID-GaN (L_{pillar}/L_{UID}) to achieve the identical PT and avalanche breakdown respectively in the top (pillar-region) and bottom part (UID-region) for the achievable maximum BV with the minimum over design of the overall thickness of the vertical SJ device. Fig. 12 shows the device breakdown dependence on L_{pillar}/L_{UID} ratio as well as the optimization strategy. For a given overall N-drift thickness (i.e. 12 μm), the breakdown of the device turned from the avalanche limited by UID-region into the PT limited by pillar-region with the decrease of L_{pillar}/L_{UID} ratio.

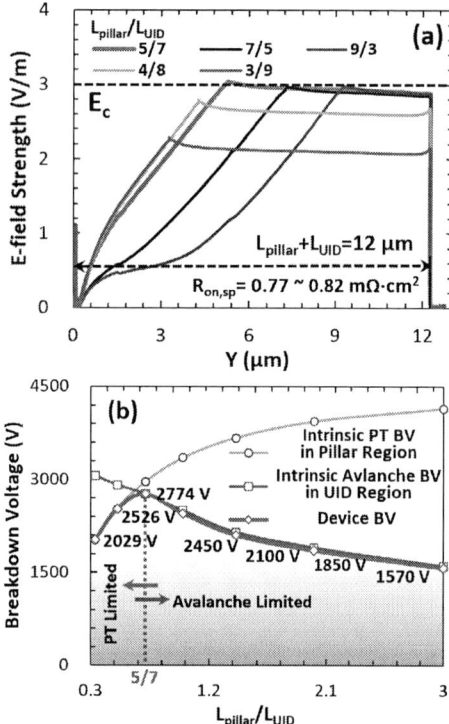

Fig. 12. (a) Vertical E-field strength distribution of the proposed device with $N_{p/n}=6\times10^{16}$ cm^{-3}, various L_{pillar}/L_{UID} ratio and a given overall thickness of 12.775 μm ($L_{pillar}+L_{UID}=12$ μm). (b) Relationship of breakdown voltage, PT limitation, avalanche limitation and L_{pillar}/L_{UID} ratio.

The optimized device features a L_{pillar}/L_{UID} of 5/7 and a $R_{on,sp}$ of 0.78 mΩ·cm^2 while the breakdown of the device is 2774 V that is simultaneously triggered by both of the PT in pillar-region and avalanche breakdown in UID-region. It can be seen that the vertical E-field strength distribution of the device is ideal in P-pillar region and UID-GaN region as Fig. 1 and Fig. 11 shown.

IV. CONCLUSION

In this work, a GaN vertical super-junction trench MOSFET was proposed, designed and optimized. By taking into account the unavailability of high quality P$^+$-GaN, the P-GaN/UID-GaN/N$^+$-GaN stack is used in the proposed device to replace the typical P$^+$/N/N$^+$ structure in conventional Si SJ-MOS to obtain the gradient doping transition for E-field modulation. Based on the proposed device optimization scheme, for the device with a 12-μm N-GaN drift region, the optimized effective doping concentration of N-GaN drift region and P-GaN is found to be $N_{p/n}=6\times10^{16}$ cm^3 and the thickness ratio of P-GaN to UID-GaN is determined to be 5/7. Accordingly, the optimized trade-off between the BV and $R_{on,sp}$ is obtained that enables the device delivers a BV of 2774 V and $R_{on,sp}$ of 0.78 mΩ·cm^2 and yields a high Baliga's figure-of-merit ($BFOM$) of 9.46 GW/cm^2. The obtained

device performance reveals that the junction concept is of great potential and even essential for GaN device targeting over kilovolt applications.

REFERENCES

[1] Ruopu Zhu, Qi Zhou, Hong Tao *et al.* , "A Split Gate Vertical GaN Power Transistor with Intrinsic Reverse Conduction Capability and Low Gate Charge", at *International Symposium on Power Semiconductor Devices and ICs (ISPSD)*, Chicago, IL, USA, May 13-17, 2018.

[2] Dong Ji , Anchal Agarwal, Haoran Li *et al.* , "880 V/2.7 m·cm^2 MIS Gate Trench CAVET on Bulk GaN Substrates", *IEEE Electron Device Letters*, vol. 39, no. 6, pp. 863–865, Jun. 2018.

[3] Dong Ji, Anchal Agarwal, Wenwen Li *et al.* , "Demonstration of GaN Current Aperture Vertical Electron Transistors With Aperture Region Formed by Ion Implantation", *IEEE Transactions on Electron Devices*, vol. 65, no. 2, pp. 483–487, Feb. 2018.

[4] Srabanti Chowdhury, Man Hoi Wong, Brian L. Swenson *et al.* , "CAVET on Bulk GaN Substrates Achieved With MBE-Regrown AlGaN/GaN Layers to Suppress Dispersion", *IEEE Electron Device Letters*, vol. 39, no. 6, pp. 863–865, Jun. 2018.

[5] Chirag Gupta, Cory Lund, Silvia H. Chan *et al.* , "In Situ Oxide, GaN Interlayer-Based Vertical Trench MOS FET (OG-FET) on Bulk GaN substrates ", *IEEE Electron Device Letters*, vol. 38, no. 3, pp. 353–355, Mar. 2017.

[6] Ray Li, Yu Cao, Mary Chen *et al.* , "600 V/1.7 Normally-Off GaN Vertical Trench Metal – Oxide – Semiconductor Field-Effect Transistor", *IEEE Electron Device Letters*, vol. 37, no. 11, pp. 1466–1469, Nov. 2016.

[7] Min Sun, Yuhao Zhang, Xiang Gao *et al.* . "High-Performance GaN Vertical Fin Power Transistors on Bulk GaN Substrates", *IEEE Electron Device Letters*, vol. 38, no. 4, pp. 509–512, Apr. 2017.

[8] Isik C. Kizilyalli, Andrew P. Edwards, Ozgur Aktas *et al.* , "Vertical Power p-n Diodes Based on Bulk GaN", *IEEE Transactions on Electron Devices*, vol. 62, no. 2, pp. 414–422, Feb. 2015.

[9] Zhongda Li and T. Paul Chow. "Design and Simulation of 5–20-kV GaN Enhancement-Mode Vertical Superjunction HEMT", *IEEE Transactions on Electron Devices*, vol. 60, no. 10, pp. 3230–3237, Oct. 2013.

[10] Yuhao Zhang, Mengyang Yuan, Nadim Chowdhury *et al.* , "720-V/0.35-m·cm^2 Fully Vertical GaN-on-Si Power Diodes by Selective Removal of Si Substrates and Buffer Layers", *IEEE Electron Device Letters*, vol. 39, no. 5, pp. 715–718, May. 2018.

[11] Kevin J. Chen, Oliver Häberlen, Alex Lidow *et al.* , "GaN-on-Si Power Technology: Devices and Applications", *IEEE Transactions on Electron Devices*, vol. 64, no. 3, pp. 779–794, Mar. 2017.

[12] Jiacheng Lei, Jin Wei, Gaofei Tang *et al.* , "Reverse-Blocking AlGaN/GaN Normally-Off MISHEMT with Double-Recessed Gated Schottky Drain", at *International Symposium on Power Semiconductor Devices and ICs (ISPSD)*, Chicago, IL, USA, May 13-17, 2018.

[13] Puneet Srivastava, Jo Das, Domenica Visalli *et al.* , "Record Breakdown Voltage (2200 V) of GaN DHFETs on Si With 2-μm Buffer Thickness by Local Substrate Removal", *IEEE Electron Device Letters*, vol. 32, no. 1, pp. 30–32, Jan. 2011.

[14] N. Herbecq, I. Roch-Jeune, A. Linge *et al.* , "GaN-on-silicon high electron mobility transistors with blocking voltage of 3 kV", *Electronics Letters*, vol. 51, no. 19, pp. 1532–1534, Sept. 2015.

[15] M. J. Uren *et al.*, "Punch-through in short-channel AlGaN/GaN HFETs," *IEEE Trans. Electron Devices*, vol. 53, no. 2, pp. 395–398, Feb. 2006.

[16] O. Hilt, E. Bahat-Treidel, E. Cho, S. Singwald, *et al.*, "Impact of buffer composition on the dynamic on-state resistance of high-voltage AlGaN/GaN HFETs," in *Proc. 24th Int. Symp. Power Semiconductor Devices ICs*, 2012, pp. 345–348.

Proceedings of the 31st International Symposium on Power Semiconductor Devices & ICs
May 19-23, 2019, Shanghai, China

Identifying the Location of Hole-Induced Gate Degradation in LPCVD-SiN$_x$/GaN MIS-FETs under High Reverse-Bias Stress

Zheyang Zheng[1], Mengyuan Hua[2], Jin Wei[1], Zhaofu Zhang[1] and Kevin J. Chen[1]

[1]Department of Electronic and Computer Engineering, The Hong Kong University of Science and Technology, Hong Kong

[2]Department of Electrical and Electronic Engineering, Southern University of Science and Technology, Shenzhen, P. R. China

Phone: +852-2358-8535, Email: zzhengah@connect.ust.hk, eekjchen@ust.hk

Abstract— **When SiN$_x$/GaN MIS-FETs are under high reverse-bias stress, holes can be generated by impact ionization, leak to gate electrode through the gate dielectric and generate defects that induce V_{TH} instability. In this work, we identify the location of such degradation by separately probing V_{TH} at source-side and drain-side of the MIS channel. It is revealed that the hole-induced gate degradation under the reverse-bias stress is more uniformly distributed along the gate with a less negative V_{GS}. As a result, the stability of device under high reverse-bias stress is enhanced. To further suppress the degradation, holes should be prevented from either going through or accumulating under the gate dielectric.**

Keywords— LPCVD SiN$_x$/GaN MIS-FET, Hole-induced degradation, Hole-barrier-free, Location identification, E-field uniformity

I. INTRODUCTION

GaN metal-insulator-semiconductor field-effect transistors (MIS-FETs) with the AlGaN barrier at the gate region fully-recessed are highly attractive for normally-off operation in power switch applications, for the large gate voltage swing, low gate leakage and good threshold voltage (V_{TH}) controllability [1]. It has been reported that MIS-FETs with SiN$_x$ prepared by low-pressure chemical vapor deposition (LPCVD-SiN$_x$) as gate dielectric could deliver prolonged time-dependent dielectric breakdown (TDDB) lifetime owing to the enhanced dielectric quality [2].

When the MIS-FET is biased at OFF-state with large V_{DS} (i.e. high reverse-bias), holes can be generated by impact ionization initiated by electrons injected into the high electric-field (E-field) region and result in various reliability issues [3]–[5]. In MIS-FETs with Al$_2$O$_3$ as gate dielectric (which presents a barrier to holes from GaN), premature breakdown has been observed at the source-side gate corner where there is a potential minimum for holes to accumulate [3], because the increment of positive charges could enhance the local E-field in the dielectric. A different hole-induced degradation mechanism has been revealed in SiN$_x$/GaN MIS-FETs as SiN$_x$ exhibits a type-II alignment with GaN (i.e. without a hole-barrier to GaN) and allows holes to pass through, which could consequently generate defects in the dielectric. When the MIS-FET is turned on, such defects could capture electrons to result in positive V_{TH} shift, which has been used as an indicator of gate dielectric degradation [4]–[6].

With identical gate-to-drain (V_{GD}) stress that generates almost identical high E-field region at the drain-side gate region, negative gate-to-source (V_{GS}) bias at OFF-state could accelerate the hole-induced degradation process in LPCVD-SiN$_x$/GaN MIS-FETs [6]. We expected that it is related to the location dependence of gate degradation along the channel as the E-field is highly non-uniform under high reverse-bias stress, which has not been experimentally confirmed yet.

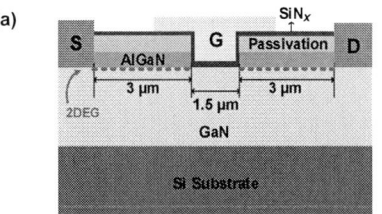

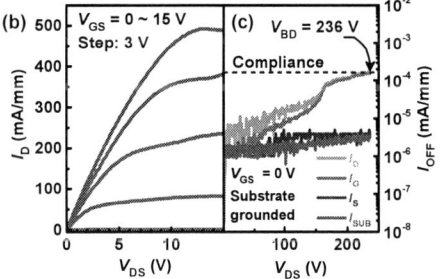

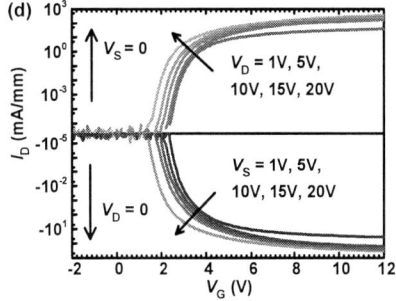

Fig. 1: (a) Schematics of the device cross-section, (b) output, (c) OFF-state I-V and (d) transfer characteristics of the SiN$_x$/GaN MIS-FET with $L_{GS}/L_G/L_{GD} = 3/1.5/3$ μm used in this work. (d) shows that the device is symmetric.

This work is supported by Hong Kong Innovation Technology Fund under ITS/412/17FP.

978-1-7281-0582-6/19 $31.00 © 2019 IEEE

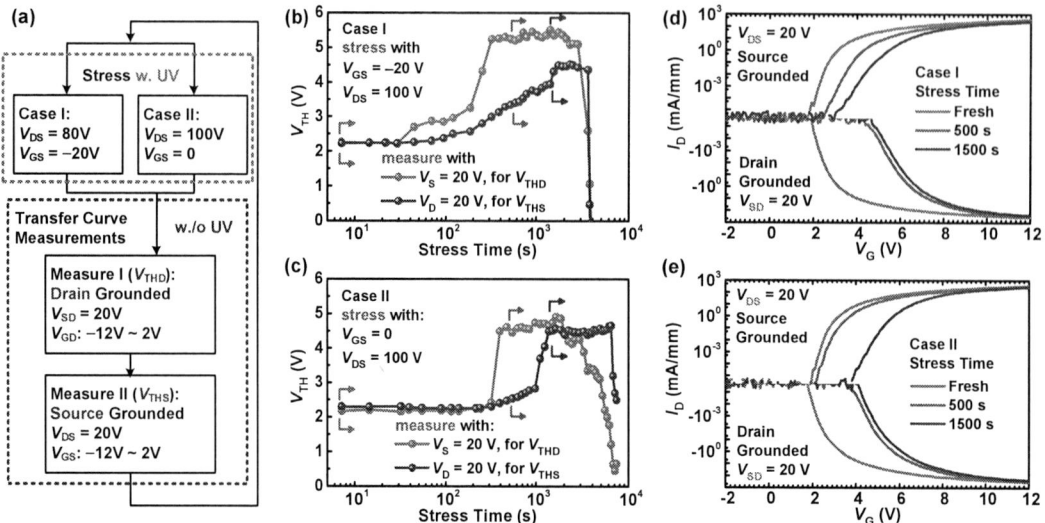

Fig. 2: (a) Flow chart of a 'stress-measure' cycle. Substrates are always grounded. In the stress stage, V_{GD} are all 100 V. Measured V_{TH} (V_G at $I_D = 1$ µA/mm) vs. reverse-bias stress time: (b) Case I with $V_{GS} = -20$ V stress; (c) Case II with $V_{GS} = 0$ stress. (d) and (e): complete down-sweep transfer curves of 3 sampling points marked in (b) and (c), respectively.

In this work, we present a characterization and analysis approach to identify the locations of hole-induced degradation in LPCVD-SiN$_x$/GaN E-mode MIS-FETs. The location dependence is investigated by selectively probing V_{TH} of different sides of the gate region and discussed from the perspective of the transportation of holes in the highly non-uniform E-fields.

II. DEVICE CHARACTERIZATION

A. Device Structure and Static Characterizations

LPCVD-SiN$_x$/GaN E-mode MIS-FETs with identical gate-to-source and gate-to-drain spacing ($L_{GS}/L_G/L_{GD} = 3/1.5/3$ µm) were employed in this work, as shown in Fig. 1(a). The devices fabricated with the same process reported in [2] were designed to be symmetrical to eliminate the influence of source and drain access regions when monitoring V_{TH} shift at different positions of the channel. At the gate region, the AlGaN barrier is fully removed so that the channel features a metal-SiN$_x$-GaN MIS structure. Fig. 1(b) shows the output characteristics of the device with an ON-resistance (R_{ON}) of 16.5 Ω·mm. In OFF-state, the leakage current could be limited within 10^{-4} mA/mm under 200 V stress, as shown in Fig. 1(c). The transfer characteristics showed in Fig. 1(d) were measured with different V_{DS} and V_{SD} by grounding source and drain terminals, respectively. The fresh device shows good symmetry in its electrical performance, and the induced asymmetry behaviors observed during high reverse-bias stress could serve as an indication of distribution of hole-induced degradations.

B. Reverse-Bias Stress and Location-Dependent V_{TH} Measurements

The devices were subject to a sequence of 'stress-measurement' cycles (with the substrate terminals always grounded) to identify the location of hole-induced degradations under high reverse-bias stress. Fig. 2(a) shows the flow chart of one typical cycle. In the stress stage, an identical V_{GD} of 100 V was used in two different stress cases to generate similar lateral E-field that pull the holes toward the gate. Different V_{GS} was applied in Case I ($V_{GS} = -20$ V) and Case II ($V_{GS} = 0$) to have different E-field distributions near the channel. Since different drain-to-source voltage (V_{DS}) in Case I and Case II could probably induce different leakage current that initiates the impact ionization at the high E-field region, ultraviolet (UV) illumination was applied to generate a large number of electron-hole pairs in both stress cases so that the influence of V_{DS} could be ignored. In the transfer-curve measurement stage (without UV illumination), by biasing the device to saturation mode ($|V_{DS}| = 20$ V) and exchanging the roles of the source and drain terminals, the channel's V_{TH} on the drain or source sides (V_{THD} or V_{THS}) can be probed separately [7]. For instance, when a large V_{DS} (20 V) is applied, there is always a depletion region on the drain side of the channel. Therefore, the V_{TH} extracted from the transfer curve is the V_{TH} on the source side, i.e. V_{THS}. Similarly, when a large V_{SD} is applied, the source side gate region is always depleted and V_{THD} could be measured. In this work, the V_{TH} was extracted with the criteria of $I_D = 1$ µA/mm.

By separately measuring the shifts in V_{THS} and V_{THD} after high reverse-bias stress, the location of the hole-induced gate degradation can be identified. Since similar number of holes would be generated under the two stress cases, different gate degradation behavior should be induced by the different hole transport paths under the gate.

III. RESULTS AND DISCUSSION

A. V_{TH} Instability in Two Stress Cases

Fig 2(b) and (c) show the V_{THD} and V_{THS} monitored during two different high reverse-bias stress cases by periodically interrupting the stress to measure the transfer characteristics. Stressed with $V_{GS} = -20$ V and $V_{DS} = 80$ V (i.e. Case I, $V_{GD} =$

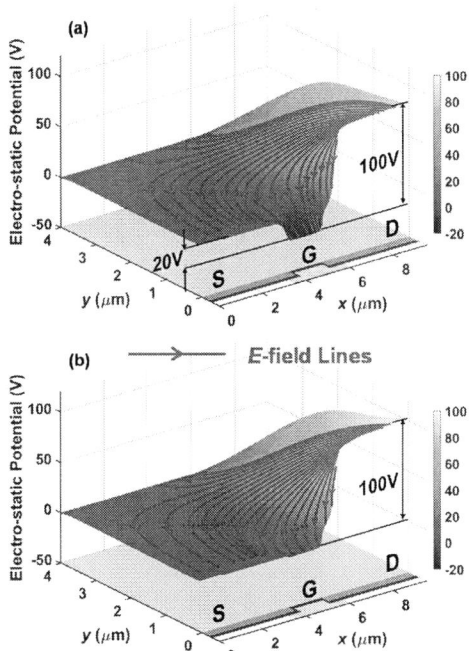

Fig. 3: Simulated electro-static potential of the GaN part under two different stress cases. (a) $V_{GS} = -20V$; (b) $V_{GS} = 0$. V_{GD} are identical. Red curves are E-field lines. After generated by UV, holes will roughly drift following the E-field lines.

100 V), V_{THD} exhibits much stronger and faster positive-shift than V_{THS}. On the contrary, when the device is stressed with $V_{GS} = 0$ and $V_{GD} = V_{DS} = 100$ V (Case II), V_{THD} and V_{THS} exhibit similar shifting behaviors but with different timing. Fig. 2(d) and (e) plot complete transfer curves measured at three typical sampling points during stress Case I and Case II, respectively. In Case I, the device gets asymmetric during the stress, which means that degradations mainly distribute at the drain-side of the gate region while the source side suffer from much less damages. In Case II, the device becomes asymmetric at the beginning but gets symmetric again with longer stress time. It is also noted that significant V_{THD} occurs with less stress time in Case I than in Case II. These results are consistent with [6] and indicate that to enhance the V_{TH} stability, the negative gate bias should be confined within a pre-identified small range when GaN MIS-FETs are employed as power switches.

B. Insight of Hole-Induced Degradation Process

After holes are generated by impact ionization or UV illumination, they could get accelerated by E-field in GaN and transport roughly following the E-field lines. In both stress cases, since the UV illumination and V_{GD} are identical, the number of generated holes should be similar. Thus, the discrepancies observed in two stress cases are probably originated from the transportation of holes, which could be modulated by different V_{GS}. We conducted numerical simulations with TCAD tools to plot the E-field distributions under two stress cases. Fig. 3 shows the distributions of electro-static potentials and E-field lines in GaN in a 2-dimensional way. In Case I, the E-field is more concentrated at the drain-side gate region with a negative V_{GS} of -20 V [Fig.

3(a)]. Therefore, more holes would transport to the drain corner of the channel. When stressed with $V_{GS} = 0$ (Case II), E-field is more uniformly distributed from drain side to source side [Fig. 3(b)] and the number of holes that transport to the drain side could be reduced.

As SiN_x/GaN features a type-II band alignment (i.e. SiN_x presents no hole-barrier to GaN), holes that arrive at the SiN_x/GaN interface could easily drift into SiN_x layer under the strong vertical E-field in the gate region, especially at the drain-side, as shown in Fig. 4(b). Fig. 4(a) compares the lateral and vertical E-fields near the 2DEG channel. Under the MIS-gate, the vertical E-field is always stronger than the lateral E-field. Therefore, holes tend to enter the SiN_x without longitudinal drifting along the channel. The region with more E-field lines terminated could have more energetic holes injected into the SiN_x layer. Correspondingly, more defects that could capture electrons when the device is switched on (i.e. electron traps) are generated when holes transport in the dielectric. As a result, the V_{THD} shift in Case I is the most significant and occurs earliest.

As for holes that reach the AlGaN/GaN interface at the drain side access region, most of them would be blocked by the hole-barrier of AlGaN [Fig. 4(c)] and drift toward the gate region under the lateral E-field [Fig. 4(a)]. These holes also contribute to the drain-side gate dielectric degradation. Therefore, even the E-field distribution is more uniform in Case II, V_{THD} still exhibits positive shift earlier than V_{THS}.

With prolonged stress under either Case I or Case II conditions, negative V_{TH} shift is observed [Fig. 2 (b, c)]. It is probably due to the formation of effective leakage path in the gate dielectric, which could be evidenced by the increase of I_G. Fig. 5(a) shows three typical I_G-V_G curves monitored during

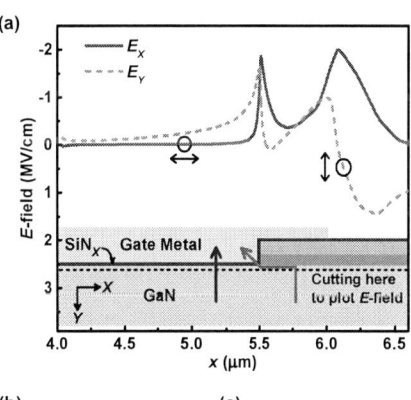

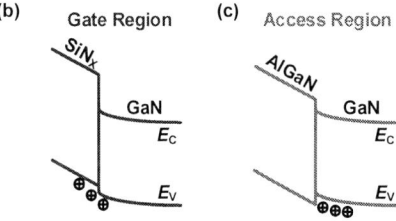

Fig. 4: (a) Simulated E-field distribution along the channel in Case II. (It is similar in Case I.) Vertical E-field (E_Y) dominates in the channel region. In the access region lateral E-field (E_X) is also significant. (b) Schematics of band alignment at the gate region (hole-barrier-free) and (c) the access region (with hole-barrier).

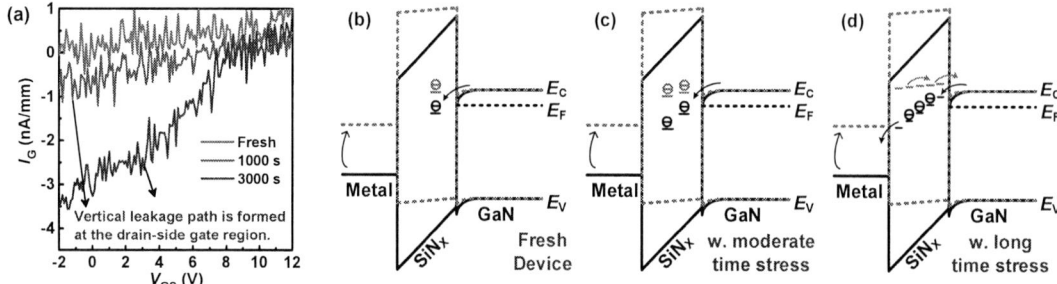

Fig. 5: (a) The Gate leakage monitored with $(V_S, V_D) = (0, 20V)$ during stress Case I. In Case II, similar leakage behavior can be observed. (b) – (d) Schematics of band bending near the channel during transfer curve measurements. Black/solid lines: the device is fully turned on, and some electrons are trapped in SiN_x. Red/dash lines: V_G is swept down, and the device enters the subthreshold regime.

transfer curve measurements after stress. A significant increase of I_G was observed with long time stress. Fig. 5(b-d) illustrates three stages of V_{TH} shift. For a fresh device, only few electron traps exist in the SiN_x. With moderate time stress, electron traps are gradually generated by holes transporting through the dielectric. During the transfer curve measurement, electrons captured into the dielectric cannot be de-trapped instantaneously and result in a negative V_{TH} shift. As the stress continues, trap density keeps on increasing and electrons could leak out via adjacent trap states. Therefore, traps in the dielectric cannot hold electrons anymore and V_{TH} exhibits negative shift.

IV. CONCLUSIONS

Location of hole-induced degradation in LPCVD-SiN_x/GaN MIS-FETs under high reverse-bias stress have been identified by selectively measuring V_{THD} and V_{THS} during the stress. Since SiN_x presents no hole-barrier to GaN, more severe degradation locates at the drain-side gate region, which is different from hole-induced degradation in devices with Al_2O_3 as gate dielectric. The spatial distribution of degradation is determined by the E-field distribution in depleted GaN under different bias conditions. With a less negative V_{GS} bias during the stress, hole-induced degradation is more uniformly distributed, and the V_{TH} stability of MIS-FETs could be maintained for a longer time. As results of such degradations, V_{TH} exhibit positive shift with moderate stress time but negative shift with prolonged stress. Details of the negative V_{TH} shift need to be further investigated. Since both kinds of dielectric, i.e. with or without hole-barriers to GaN, could be accelerated to degrade by holes generated under high-reverse

bias, these holes should be prevented from accumulating under or going through gate dielectric. A promising solution is to deploy hole blocking layer at the channel, e.g. GaO_xN_{1-x} [8].

REFERENCES

[1] K. J. Chen *et al.*, "GaN-on-Si Power Technology: Devices and Applications," *IEEE Transactions on Electron Devices*, vol. 64, no. 3, pp. 779–795, Mar. 2017.

[2] M. Hua *et al.*, "Integration of LPCVD-SiN$_x$ gate dielectric with recessed-gate E-mode GaN MIS-FETs: Toward high performance, high stability and long TDDB lifetime," in *2016 IEEE International Electron Devices Meeting (IEDM)*, 2016, pp. 10.4.1-10.4.4.

[3] S. R. Bahl, M. V. Hove, X. Kang, D. Marcon, M. Zahid, and S. Decoutere, "New source-side breakdown mechanism in AlGaN/GaN insulated-gate HEMTs," in *2013 25th International Symposium on Power Semiconductor Devices IC's (ISPSD)*, 2013, pp. 419–422.

[4] M. Hua *et al.*, "Reverse-bias stability and reliability of hole-barrier-free E-mode LPCVD-SiN$_x$/GaN MIS-FETs," in *2017 IEEE International Electron Devices Meeting (IEDM)*, 2017, pp. 33.2.1-33.2.4.

[5] M. Hua *et al.*, "Hole-Induced Threshold Voltage Shift Under Reverse-Bias Stress in E-Mode GaN MIS-FET," *IEEE Transactions on Electron Devices*, vol. 65, no. 9, pp. 3831–3838, Sep. 2018.

[6] M. Hua, J. Wei, Q. Bao, Z. Zhang, Z. Zheng, and K. J. Chen, "Dependence of V_{TH} Stability on Gate-Bias Under Reverse-Bias Stress in E-mode GaN MIS-FET," *IEEE Electron Device Letters*, vol. 39, no. 3, pp. 413–416, Mar. 2018.

[7] R. B. Fair and R. C. Sun, "Threshold-voltage instability in MOSFET's due to channel hot-hole emission," *IEEE Transactions on Electron Devices*, vol. 28, no. 1, pp. 83–94, 1981.

[8] M. Hua *et al.*, "Suppressed Hole-Induced Degradation in E-mode GaN MIS-FETs with Crystalline GaO$_x$N$_{1-x}$ Channel," in *2018 IEEE International Electron Devices Meeting (IEDM)*, 2018, pp. 30.3.1-30.3.4.

Proceedings of the 31st International Symposium on Power Semiconductor Devices & ICs
May 19-23, 2019, Shanghai, China

Surge Current Capability of GaN E-HEMTs in Reverse Conduction Mode

Yinxiang Liu, Shaowen Han, Shu Yang*, Kuang Sheng
College of Electrical Engineering, Zhejiang University, Hangzhou, China
*E-mail: eesyang@zju.edu.cn

Abstract—**In this paper, we investigate the surge current capability of two types of commercial GaN enhancement-mode HEMTs (E-HEMTs) in the reverse conduction mode, and reveal the impacts of the p-GaN technology and gate-to-source voltage (V_{GS}). The surge current capability of GaN E-HEMT can be enhanced with ohmic p-GaN contact and positive V_{GS}, owing to higher efficiency of hole injection and channel modulation. To our best knowledge, it is the first report to study the surge current capability and to reveal its underlying mechanisms in lateral GaN power devices.**

Keywords—*GaN, hole injection, normally-off, p-GaN, surge current*

I. INTRODUCTION

GaN-based high-electron-mobility transistors (HEMTs) can deliver low power losses and high switching frequency for power electronic applications [1, 2]. Owing to its unique symmetric structure, the lateral GaN enhancement-mode-HEMTs (E-HEMTs), without a p-n body diode, can still conduct current (i.e. source-to-drain current I_{SD}) in the reverse direction with zero reverse recovery. Compared with the GaN E-HEMT anti-paralleled with an external diode to conduct current in the reverse direction, using GaN E-HEMT for both forward- and reverse-conduction (Fig. 1) enables lower parasitic capacitance and lower current overshoot/oscillation during turning on, leading to reduced overall switching losses (Tab I.) [3, 4]. Therefore, GaN E-HEMT is highly desirable in bidirectional DC/DC converter [5] and DC/RF inter-conversion systems [6, 7].

GaN E-HEMTs would undergo a high surge current in the reverse conduction mode under the circumstance of current overshoot/oscillation, which could possibly lead to thermal runaway. However, investigation on the surge current capability of GaN E-HEMTs in the reverse conduction mode is still lacking to date.

In this work, the surge current capabilities of two types of commercial GaN E-HEMTs in the 3rd quadrant are evaluated, and the underlying mechanisms of p-GaN technology and gate-to-source voltage (V_{GS}) in influencing the surge current capability have been investigated. Surge current capability of the GaN E-HEMT can be enhanced by positive V_{GS} and ohmic-type p-GaN gate, which can yield more effective hole injection and higher channel modulation efficiency.

II. CHARACTERIZATION PLATFORM

A. Test Circuit

As shown in Fig. 2, the surge current characterizations are based on an LC resonator with a pulse width of

$$W = \pi \cdot (L \cdot C)^{1/2} \qquad (1)$$

The capacitor is firstly charged up by the DC power supply voltage V_{in} with the switch turned on. After turning off the switch, the Si controlled rectifier (SCR) is triggered while the LC resonance can generate a half sine current pulse to go through the device under test (DUT). The current flowing through the DUT was sensed by a coaxial

Table I. Comparison of different schemes of the power switch in bidirectional converters [3, 4]

Schemes	GaN E-HEMT anti-paralleled with an external diode	GaN E-HEMT only
Parasitic capacitance	High	Low
Current overshoot/Current oscillation (during turn-on)	High	Low
Turn-on loss	High	Low
Turn-off loss	Low	High
Total power losses	High	Low

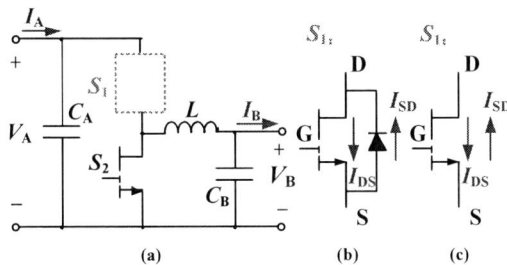

Fig. 1. (a) Schematic of a bidirectional DC/DC converter with different schemes of the upper switch S_1 Illustration of the upper switch S_1 using (b) GaN E-HEMT anti-paralleled with an external diode, and (c) GaN E-HEMT only.

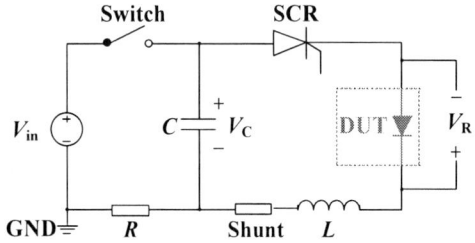

Fig. 2. Schematic circuit for the surge current characterizations. After each measurement, the leakage current of the DUT under a reverse bias (V_R) was monitored.

current shunt. After each measurement, the leakage current under a reverse bias (V_R) was monitored to determine whether the DUT degrades during the previous surge current test. As shown in Fig. 3(a), a half sine surge current pulse with a width of 10-ms was used in this work, and the peak current (I_{peak}) can be varied by adjusting the capacitor voltage (V_C), according to [8]

$$I_{peak} = V_C \cdot (C / L)^{1/2} \qquad (2)$$

B. Characteristics of DUT

In this work, the surge current capabilities of two types of the state-of-the-art commercial GaN E-HEMTs with different p-GaN technologies are investigated. Fig. 3(b) illustrates the operation mode of the GaN E-HEMTs in the 3rd quadrant. The key parameters of the two types of GaN E-HEMTs [9, 10] with similar voltage/current ratings but different p-GaN technologies are given in Tab II. The *I-V* characteristics of the GaN E-HEMTs in the 1st and 3rd quadrants were measured by the curve tracer Tektronix 371. Fig. 4(a) shows the I_{DS}-V_{DS} characteristics in the 1st quadrant and the R_{DS_ON} extraction for devices A and B. The I_{SD}-V_{SD} characteristics in the 3rd quadrant with gate and source terminals electrically shorted (i.e., $V_{GS} = 0$ V)

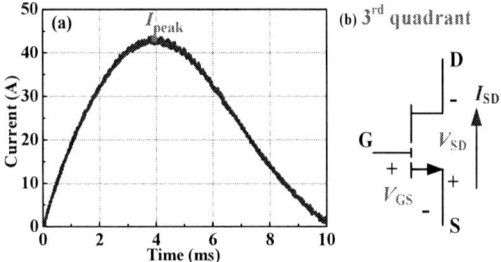

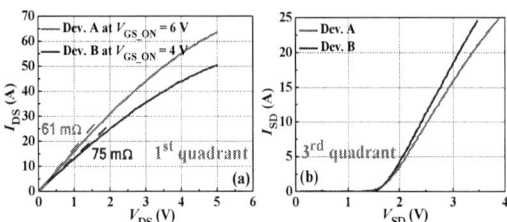

Fig. 3. (a) Waveform of the half sine surge current pulse with a width of 10-ms. (b) Illustration of the operation mode of the GaN E-HEMT in the 3rd quadrant.

Table II. Key parameters of DUT [9, 10]

Parameters	Device A	Device B
Voltage Rating	650 V	600 V
Current Rating	30 A	26 A
R_{DS_ON} *	61 mΩ	75 mΩ
p-GaN Contact	Schottky	Ohmic

* The R_{DS_ON} values are extracted from the output curves at the recommended on-state gate bias (V_{GS_ON}).

Fig. 4. (a) I_{DS}-V_{DS} characteristics of the two types of GaN E-HEMTs in the 1st quadrant under the recommended V_{GS_ON}. (b) I_{SD}-V_{SD} characteristics in the 3rd quadrant with gate and source terminals electrically shorted.

show that the ohmic-gate GaN E-HEMT exhibits a slightly higher current and lower resistance, as shown in Fig. 4(b).

III. MESUREMENT RESULTS OF SURGE CURRENT

A. Impact of The P-GaN Technology

Surge current of the GaN E-HEMTs in the 3rd quadrant was evaluated with gradually increased I_{peak}. Fig. 5 and Fig. 6 show the voltage waveforms and the corresponding I_{SD}-V_{SD} curves of the GaN E-HEMTs with different p-GaN technologies in the 3rd quadrant with $V_{GS} = 0$ V. The GaN

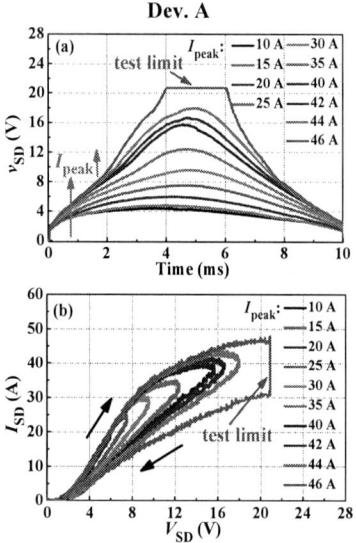

Fig. 5. (a) Voltage waveforms $v(t)$ and (b) the corresponding I_{SD}-V_{SD} curves of Device A in the 3rd quadrant with gate and source shorted. Device A can withstand a maximum surge current of 44 A.

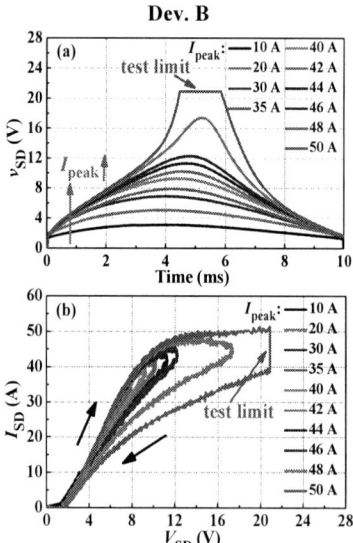

Fig. 6. (a) Voltage waveforms $v(t)$ and (b) the corresponding I_{SD}-V_{SD} curves of Device B in the 3rd quadrant with gate and source shorted. Device B can withstand a maximum surge current of 48A.

978-1-7281-0582-6/19 $31.00 © 2019 IEEE

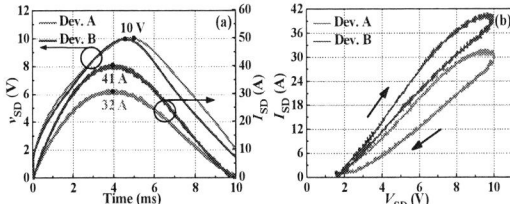

Fig. 7. (a) Voltage and current waveforms, and (b) the corresponding I_{SD}-V_{SD} characteristics of devices A and B in the 3rd quadrant with the same V_{max} of 10 V and V_{GS} of 0 V.

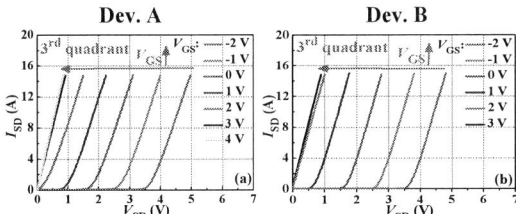

Fig. 8. I_{SD}-V_{SD} characteristics of devices (a) A and (b) B in the 3rd quadrant with V_{GS} varying from -2 V to 4 V, which were measured by the curve tracer Tektronix 371.

E-HEMT with an ohmic p-GaN contact can withstand a higher surge current of 48 A, compared with 44 A in that with a Schottky p-GaN contact. At a relatively small I_{peak}, the maximum voltage (V_{max}) occurs almost simultaneously with I_{peak}. On the other hand, at a higher I_{peak}, V_{max} lags behind the occurrence of I_{peak}, possibly caused by the considerable thermal generation.

Fig. 7 compares the voltage and current waveforms, as well as the corresponding I_{SD}-V_{SD} characteristics of the two types of GaN E-HEMTs with different p-GaN technologies in the 3rd quadrant. At the same V_{max} of 10 V, E-HEMT with the ohmic p-GaN contact is capable of withstanding higher surge current while maintaining a lower R_{ON}.

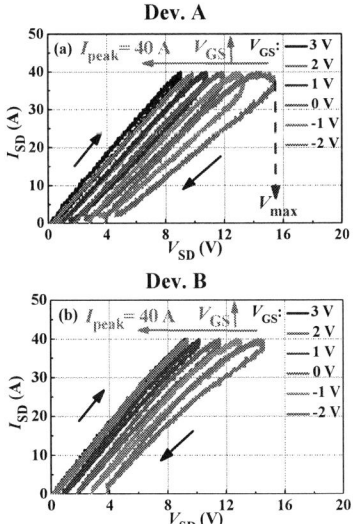

Fig. 9. I_{SD}-V_{SD} characteristics of devices (a) A and (b) B in the 3rd quadrant with the same I_{peak} of 40 A and V_{GS} varying from -2 V to 3 V.

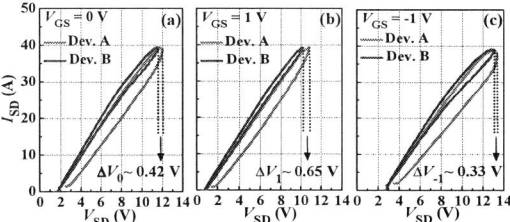

Fig. 10. I_{SD}-V_{SD} characteristics with (a) V_{GS} = 0 V, (b) V_{GS} = 1 V and (c) V_{GS} = -1 V, as extracted from Fig. 9. The differences in V_{max} between devices A and B under various V_{GS} are denoted as ΔV_0, ΔV_1, ΔV_{-1}, respectively.

B. Imapct of V_{GS}

The impact of positive/negative V_{GS} on the surge current capability was also investigated. Fig. 8 shows the I_{SD}-V_{SD} characteristics of the GaN E-HEMTs in the 3rd quadrant with V_{GS} varying from -2 V to 4 V. A positive V_{GS} can reduce the R_{ON} of both GaN E-HEMTs in the 3rd quadrant by enhancing the channel conductivity, which saturates at V_{GS} of ~3 V and ~2 V for E-HEMTs with Schottky and ohmic p-GaN contact, respectively.

Similarly, in the surge current characterizations, V_{max} corresponding to the same I_{peak} of 40 A in both E-HEMTs decreases at higher V_{GS} (Fig. 9), suggesting that a positive V_{GS} can enhance the surge current capability.

As shown in Fig. 10, V_{max} in the E-HEMT with an ohmic p-GaN contact is lower than that with a Schottky contact. In addition, the difference in V_{max} (ΔV) between the two E-HEMTs is more significant at a larger V_{GS}. It suggests that the surge current capability of the E-HEMT with an ohmic p-GaN contact is more susceptible to V_{GS}, as the ohmic p-GaN contact can facilitate an effective modulation of hole injection and channel conduction.

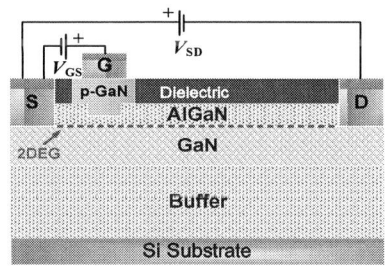

Fig. 11. Schematic cross section of GaN E-HEMT in the 3rd quadrant.

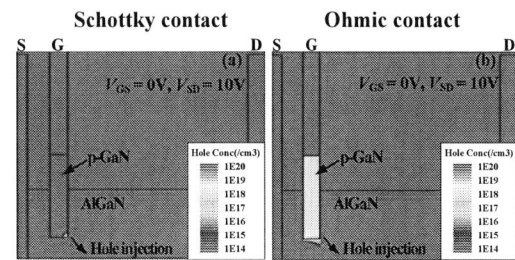

Fig. 12. Hole concentration distribution of E-HEMTs with different p-GaN contact in the 3rd quadrant under V_{GS} = 0 V and high V_{SD} = 10 V. (a) Schottky contact. (b) Ohmic contact.

978-1-7281-0582-6/19 $31.00 © 2019 IEEE   441

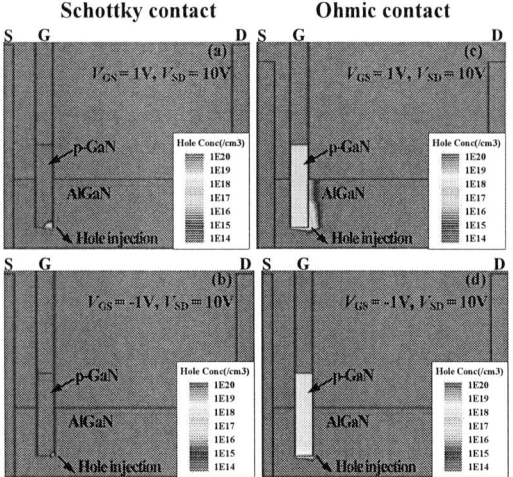

Fig. 13. Hole concentration distribution of E-HEMTs with different p-GaN contacts in the 3rd quadrant under different V_{GS} and a high V_{SD} of 10 V. (a) Schottky contact with V_{GS} = 1 V. (b) Schottky contact with V_{GS} = -1 V. (c) Ohmic contact with V_{GS} = 1 V. (d) Ohmic contact with V_{GS} = -1 V.

IV. MECHANISM ANALYSIS

The mechanisms of p-GaN contact and V_{GS} in influencing the surge current capability were investigated using TCAD simulations with device parameters described in [11] (Fig. 11).

At a considerably large V_{SD}, the positive gate-to-channel bias can induce hole injection into the III-nitride underneath the gate electrode, enabling the channel conductivity modulation and surge current capability [12, 13]. As shown in Fig. 12, the E-HEMT with an ohmic p-GaN contact can deliver higher hole injection efficiency and consequently superior surge current capability at V_{GS} of 0 V.

Fig. 13 shows the 2D hole concentration distribution of the GaN E-HEMTs with different p-GaN contacts in the 3rd quadrant under different V_{GS}. Higher V_{GS} can further enhance the hole injection and channel conductivity, leading to a higher surge current. Furthermore, the ohmic p-GaN contact enables a higher modulation efficiency of V_{GS} over the hole injection and channel conduction, leading to a higher susceptibility of surge current capability to V_{GS}.

V. CONCLUSIONS

The surge current capabilities of the state-of-the-art commercial GaN E-HEMTs with different types of p-GaN technologies in the 3rd quadrant are characterized and investigated in this work. The LC resonator was used to generate half sine surge current pulses, whereby the surge current of E-HEMTs can be evaluated. The impacts of different p-GaN technologies and V_{GS} on the surge current capability have been analyzed and revealed. It is found that the ohmic-type p-GaN gate and a positive V_{GS} can further enhance the hole injection and channel modulation

efficiency, leading to a higher surge current. Moreover, with the ohmic p-GaN contact, V_{GS} can impose a more effective modulation to the hole injection and channel conduction.

The analysis in this work suggests that the efficiency of hole injection and channel modulation can be increased by a low-resistivity p-GaN contact, positive gate voltage and/or relatively low gate capacitance to further improve the surge current capability of GaN E-HEMTs.

ACKNOWLEDGMENT

This work was supported in part by the National Key Research and Development Program of China under Grant 2018YFB0104601, in part by the National Natural Science Foundation of China under Grant 51807175 and in part by the Fundamental Research Funds for the Central Universities.

REFERENCES

[1] K. J. Chen, O. Haberlen, A. Lidow, C. L. Tsai, T. Ueda, et al. "GaN-on-Si power technology: Devices and applications," *IEEE Trans. Electron Devices*, vol. 64, no.3, pp. 779-794, Mar. 2017.

[2] E. A. Jones, F. Wang, and B. Ozpineci, "Application-based review of GaN HFETs," in *Workshop Wide Bandgap Power Devices Appl. (WiPDA)*, 2014, pp. 24-29.

[3] H. Qin, Y. Zhang, D. Wang, D. Fu, and C. Zhao, "Evaluating self-commutated reverse conduction characterization of enhancement-mode GaN HFET," in *Proc. IEEE Int. Exhib. Conf. for Power Electr. Intell. Motion, Renew. Energy Energy Manag. (PCIM-Asia)*, 2017, pp. 68-73.

[4] J. Lu, H. Bai, S. Averitt, D. Chen, and J. Styles, "An e-mode GaN HEMTs based three-level bidirectional DC/DC converter used in robert bosch DC-grid system," in *Workshop Wide Bandgap Power Devices Appl. (WiPDA)*, 2016, pp. 334-340.

[5] S. Kloetzer, U. Muter, S. Fahlbusch, and K. F. Hoffmann, "Compact diode-less bidirectional GaN based buck converter for mobile DC-DC applications," in *Proc. IEEE Int. Exhib. Conf. for Power Electr. Intell. Motion, Renew. Energy Energy Manag. (PCIM-Europe)*, 2017, pp. 1143-1150.

[6] T. Yasui, R. Ishikawa and K. Honjo, "GaN HEMT DC I-V device model for accurate RF rectifier simulation," *IEEE Microw. Wireless Compon. Lett.*, vol. 27, no. 10, pp. 930-932, Oct. 2017.

[7] R. Ishikawa, and K. Honjo, "High-efficiency DC-to-RF/RF-to-DC interconversion switching module at C-band," in *Proc. 45th European Microwave Conference (EuMC)*, 2015, pp. 295-298.

[8] S. Fichtner, S. Frankeser, J. Lutz, R. Rupp, T. Basler, et al., "Ruggedness of 1200V SiC MPS diodes," *Microelectron. Reliab.*, vol.55, no.9-10, pp. 1677-1681, Jun. 2015.

[9] "Enhancement-mode gallium nitride high electron mobility transisitor," *GS66508T GaN Systems Datasheet*, 2018. [Online] Available: http://www.gansystems.com.

[10] "Enhancement-mode gallium nitride high electron mobility transisitor," *PGA26E07BA Panasonic Datasheet*, 2017. [Online] Available: http://www.panasonic.com.

[11] K. Tanaka, T. Morita, H. Umeda, S. Kaneko, M. Kuroda, et al., "Suppression of current collapse by hole injection from drain in a normally-off GaN-based hybrid-drain-embedded gate injection transistor," *Appl. Phys. Lett.*, vol. 107, no. 16, pp. 163502-1-163502-4, Oct. 2015.

[12] T. Wu, D. Marcon, S. You, N. Posthuma, B. Bakeroot, et al., "Forward bias gate breakdown mechanism in enhancement-mode p-GaN gate AlGaN/GaN high-electron mobility transistors," *IEEE Electron Device Lett.*, vol. 36, no. 10, pp. 1001-1003, Oct. 2015.

[13] Y. Uemoto, M. Hikita, H. Ueno, H. Matsuo, H. Ishida, et al., "Gate injection transistor (GIT)-a normally-off AlGaN/GaN power transistor using conductivity modulation," *IEEE Trans. Electron Devices*, vol. 54, no.12, pp. 3393-3399, Dec. 2007.

Proceedings of the 31st International Symposium on Power Semiconductor Devices & ICs
May 19-23, 2019, Shanghai, China

Impact of Carrier Accumulation on the Transient Behavior of p-Gate GaN HEMTs

Thorsten Oeder, Martin Pfost
Chair of Energy Conversion, TU Dortmund
Emil-Figge-Straße 68, Dortmund, Germany - email: thorsten.oeder@tu-dortmund.de

Abstract—**A measurement methodology is introduced to identify the characteristics of carrier accumulation in p-gate GaN HEMTs. To analyse the influence of the gate- and drain-biasing independently, the DUT is operated in a half-bridge configuration. Thereby, the gate can be charged prior to the onset of the drain-current. The impact can be observed in the transient drain-current behavior. Two commercially available 650 V-class devices are evaluated. To achieve enhancement-mode operation one of the devices utilizes a layer of p-AlGaN, while the other uses p-GaN. For both devices the time-, voltage- and current-dependencies of carrier accumulation are investigated.**

I. INTRODUCTION

Semiconductor devices based on wide bandgap materials are a significant milestone for power electronic applications. The high electron mobility transistor (HEMT) forms a two-dimensional electron gas (2DEG) based on an AlGaN/GaN heteroepitaxy structure and is a promising solution for future power electronic applications. Due to the introduction of the p-doped gate structures [1], that provides enhancement-mode operation, the p-gate GaN HEMT became competitive GaN HEMT challenges more established technologies, like the SiC-MOSFET. In recent studies on the short-circuit robustness of p-gate GaN HEMTs, the electrical failure is attributed to an accumulation of holes under the gate [2], [3]. So far, the accumulation has been proven to exist through gate leakage measurements and numerical device simulations [4]–[7].

In this work a novel measurement methodology is introduced to identify time-, voltage and current-dependencies of trapped or accumulated carriers. For this purpose, the transient drain-current response is investigated as a function of various gate-biasing conditions.

II. MEASUREMENT SETUP

Two commercially available 650 V-class devices are evaluated. The enhancement-mode operation in "GaN A" is realized using a layer of p-doped AlGaN. As a result, the gate stack forms a PIN-junction between the gate and the 2DEG. In case of "GaN B", a layer of p-doped GaN is used, which leads to a Schottky-gate structure. The devices under test (DUT) are listed in Tab. I, their cross-section is depicted in Fig. 1. A schematic of the measurement setup is shown in Fig. 2. The setup is optimized for low inductance and uses a custom digital high-speed gate driver. The bias voltages for the on- and off-state can be adjusted independently with $V_{GS,on}$ and $V_{GS,off}$. The DUT is operated as the LS-switch in a half-bridge, with a Si-based MOSFET ($R_{DS,on} \approx 17\,m\Omega$) on the high-side.

TABLE I
OVERVIEW OF THE DEVICES UNDER INVESTIGATION

	Gate structure	$R_{DS,on}$ [Ω]	V_{th} [V]
GaN A	p-AlGaN	150	1.3
GaN B	p-GaN	120	1.3

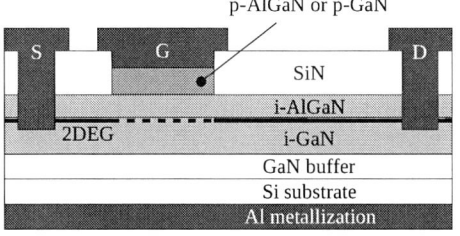

Fig. 1. Cross-section of the HEMT structure.

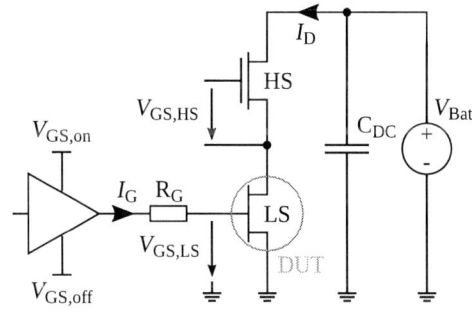

Fig. 2. Setup used for pulsed measurements.

III. TIME-DEPENDENCY

Due to driving the DUT in a half-bridge, the impact of its gate and the drain path can be decoupled by being able to turn them on and off independently. Thereby, the gate can be charged prior to the onset of the drain-current. At $V_{GS,on} = 1.7\,V$, the DUT is operated just above its threshold voltage. Hence, even at $V_{DS} = 2\,V$ the device is in saturation and the current is

978-1-7281-0582-6/19 $31.00 © 2019 IEEE

determined by the channel. Furthermore, no hot carriers and self-heating have to be considered. The amount of free carriers in the channel determines the conductivity of the channel. Since trapping and accumulation are time dependent phenomena, their impact can be observed in the transient drain-current behavior $I_D(t)$. To identify this time-dependency, the DUT gate is turned on for the precharge time t_{pre} before applying V_{DS}, as exemplified in Fig. 3.

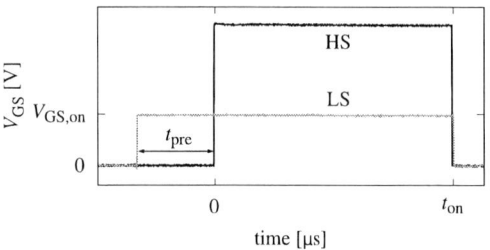

Fig. 3. Gate driver waveforms exemplifying the precharge time t_{pre} on the LS (the DUT) prior to the HS.

The impact of t_{pre} on $I_D(t)$ is shown in Fig. 4. For the purpose of maintaining a comparable sample rate across several orders of magnitude, the depicted waveforms are constructed from several measurements. For the uncharged state, with no precharge time ($t_{pre} = 0$ s), GaN A has a rather constant $I_D(t)$ after the settling processes. On the other hand, GaN B significantly drops off after around $t = 20$ ms. With $t_{pre} = 10$ s an increase occurs for GaN A, while a significant decrease is present for GaN B. However, the drain-currents with a t_{pre} of 0 s and 10 s converge for both devices to a certain steady state.

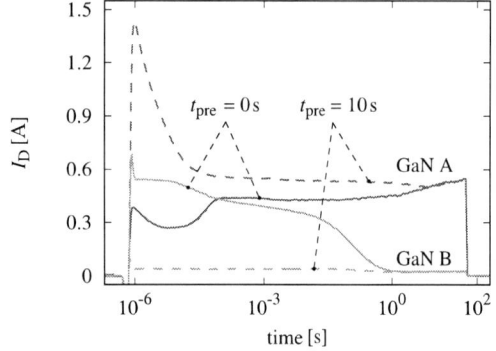

Fig. 4. Transient drain-current behavior $I_D(t)$ on a time-logarithmic scale, comparing the behavior with two precharge times t_{pre} of 0 s and 10 s. The two different devices GaN A and B are operated at $V_{DS} = 2$ V and $V_{GS,on} = 1.7$ V.

The major impact of the time-dependency can be observed in the initial current peak $I_{D,peak}$. Therefore, the variations of $I_{D,peak}$ for a sweep of t_{pre} are depicted in Fig. 5. In this context, the variation is determined as the discrepancy of $I_{D,peak}$ for a

given t_{pre}, normalized to the $I_{D,peak}$ of an uncharged state with $t_{pre} = 0$ s. Therefore, the $\Delta I_{D,peak}$ is calculated as:

$$\Delta I_{D,peak}(t_{pre}) = \frac{I_{D,peak}(t_{pre})}{I_{D,peak}\big|_{t_{pre}=0\,s}}$$

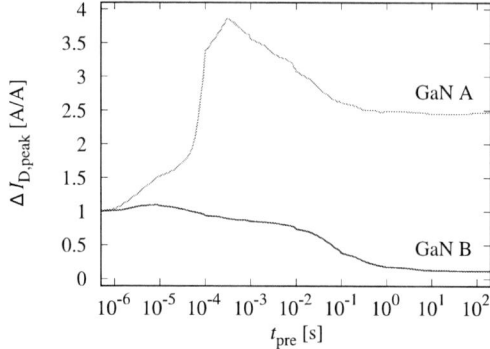

Fig. 5. Variation of the drain-current peak value $\Delta I_{D,peak}$ for a sweep of the precharge time t_{pre}, normalized to the uncharged state of $t_{pre} = 0$ s.

With GaN A, a significant increase for $t_{pre} \leq 500\,\mu s$ is present, which decreases until $t_{pre} \leq 100$ ms and remains constant thereafter. With GaN B, a roughly linear decrease is present until $t_{pre} \leq 1$ s, which remains constant, too.

The t_{pre}-dependent behavior of $I_D(t)$ and the variation of $I_{D,peak}$ substantiates the presence of a bias-dependent interference of the free carriers in the channel. Fast changing processes such as $\Delta I_{D,peak}$ with $t_{pre} \leq 500\,\mu s$ may indicate an accumulation. On the other hand, slow changing processes and the observed saturation are likely to be a trapping phenomenon.

IV. VOLTAGE-DEPENDENCY

To investigate a voltage-dependency, the time-dependency of the observed effects need to be negligible. The $I_D(t)$ of both devices are reaching a steady state within an on-time of at least $t \geq 10$ s. Therefore, the DUTs are precharged with their on-state bias for $t_{pre} = 10$ s. Afterwards, the DUTs are either negatively or positively biased prior to the turn on of V_{DS}, as shown in Fig. 6.

The duration $t_{bias} = 100\,\mu s$ is determined empirically, ensuring comparability especially for high values of the positive and negative precharge voltages $V_{GS,pre}$. The on-state voltages $V_{GS,on}$ are chosen to be 1.7 V for GAN A and 2.3 V for GAN B, in order to accomplish a similarly steady state of the DUTs.

The impact of the negative and positive $V_{GS,pre}$ are depicted in Fig. 7 and Fig. 8, respectively. Here, the variations of $I_D(t)$ with $V_{GS,pre}$ are compared to the behavior of the devices precharged to their on-state ($V_{GS,pre} = V_{GS,on}$ for $t_{pre} = 10$ s). Applying a negative $V_{GS,pre}$ on GaN A leads to an increase of $I_D(t)$, while GaN B remains nearly independent. With a positive $V_{GS,pre}$ an even further increase of $I_{D,peak}$ is present for GaN A, whereas GaN B appears to remain fully depleted during the early stages up to $t \leq 100\,\text{ms}$. Furthermore, for GaN A $V_{\text{GS,pre}}$ is limited to $+5\,\text{V}$, since the current through the PIN gate stack reaches very high values. However, the drain-currents for all variations

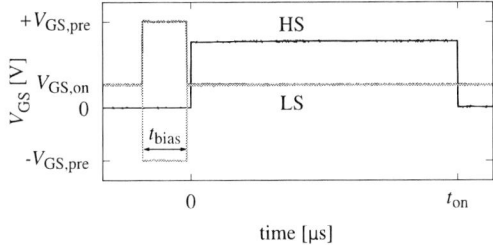

Fig. 6. Gate driver waveforms exemplifying the applied precharge voltage $V_{\text{GS,pre}}$ on the LS (the DUT) prior to the turn on of the HS. The *red* line corresponds to a positive and the *green* line to a negative $V_{\text{GS,pre}}$.

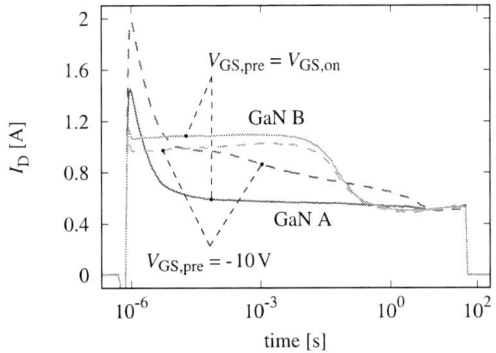

Fig. 7. Transient drain-current behavior $I_{\text{D}}(t)$ on a time-logarithmic scale. Here, the negative precharge voltage $V_{\text{GS,pre}}$ of $-10\,\text{V}$ is compared with the on-state precharged behavior of $V_{\text{GS,on}}$. The two different devices are operated at $V_{\text{DS}} = 2\,\text{V}$ with GaN A at $V_{\text{GS,on}} = 1.7\,\text{V}$ and GaN B at $V_{\text{GS,on}} = 2.3\,\text{V}$.

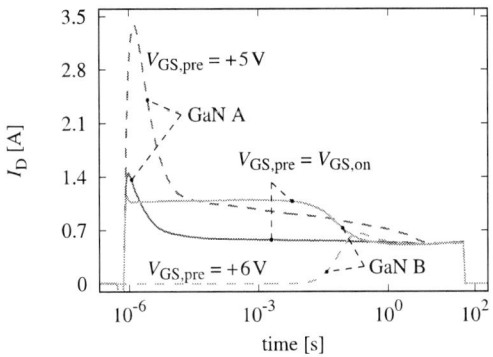

Fig. 8. Transient drain-current behavior $I_{\text{D}}(t)$. Here, the positive precharge voltage $V_{\text{GS,pre}}$ of $+5\,\text{V}$ (GaN A) and $+6\,\text{V}$ (GaN B) is compared with the on-state precharged behavior of $V_{\text{GS,on}}$. The two different devices are operated at $V_{\text{DS}} = 2\,\text{V}$ with GaN A at $V_{\text{GS,on}} = 1.7\,\text{V}$ and GaN B at $V_{\text{GS,on}} = 2.3\,\text{V}$.

of $V_{\text{GS,pre}}$ converge in a steady state for both devices, similarly to the observations made in Sec. III.

Again, the major impact of the voltage-dependency can be observed in $I_{\text{D,peak}}$. The variations for a sweep of $V_{\text{GS,pre}}$ are depicted in Fig. 9. The steady state of both devices is reached independently of $V_{\text{GS,pre}}$ within at least $t = 10\,\text{s}$. Thus, the $\Delta I_{\text{D,peak}}$ is calculated as:

$$\Delta I_{\text{D,peak}}(V_{\text{GS,pre}}) = \frac{I_{\text{D,peak}}(V_{\text{GS,pre}})}{I_{\text{D}}(t = 10\,\text{s})\big|_{V_{\text{GS,pre}}}}$$

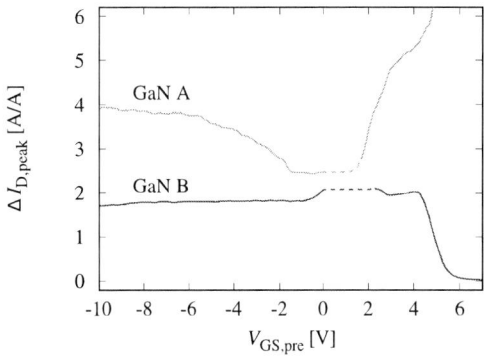

Fig. 9. Variation of the drain-current peak value $\Delta I_{\text{D,peak}}$ for a sweep of the precharge time t_{pre}, normalized to the uncharged state of $t_{\text{pre}} = 0\,\text{s}$.

For GaN A, the $I_{\text{D,peak}}$ variation increases considerably and continues to saturate for negative $V_{\text{GS,pre}}$. On the other hand, the decrease of GaN B appears to be negligible. For GaN A a significant and linear increase of $I_{\text{D,peak}}$ is present for positive $V_{\text{GS,pre}}$. Then again, GaN B remains roughly independent up until $V_{\text{GS,pre}} \approx +4\,\text{V}$ where the current and consequently $\Delta I_{\text{D,peak}}$ drops to zero.

In the range of the linear increase of GaN A, hole injection occurs during the biasing phase through the PIN gate stack. However, accumulated carriers may cause the temporary increase of channel conductivity.

V. CURRENT-DEPENDENCY

The influence of increasing currents need to be considered, to verify the significance of the observed time- and voltage-dependencies. For comparability of the varying on-state voltages up to $V_{\text{GS,on}} = 3.5\,\text{V}$, it is crucial to keep the DUTs in saturation in all operational points. Therefore, the drain-source voltage is increased to $V_{\text{DS}} = 7\,\text{V}$ for both devices.

The impact of the current-dependency is investigated by applying several precharge conditions to a given operating point, as presented in Fig. 10 for GaN A and Fig. 11 for GaN B. To verify the *time-dependency*, a precharged state ($t_{\text{pre}} = 10\,\text{s}$) with $V_{\text{GS,pre}} = V_{\text{GS,on}}$ is used. Therefore, the *uncharged* state ($t_{\text{pre}} = 0\,\text{s}$) is used as a reference, as presented in Sec. III. For the *voltage-dependency* only the positive $V_{\text{GS,pre}}$ is considered, since the negative $V_{\text{GS,pre}}$ appears to be negligible for higher

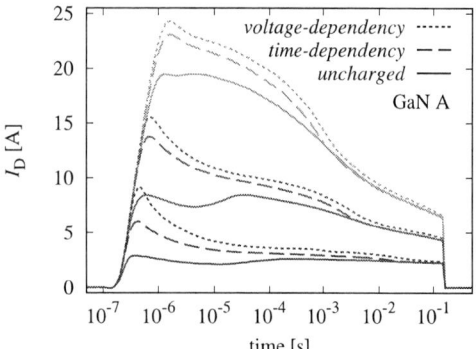

Fig. 10. Transient drain-current behavior $I_D(t)$ of GaN A. The operational points correspond to $V_{GS,on} = 2\,V$ (*blue*), $V_{GS,on} = 2.5\,V$ (*red*) and $V_{GS,on} = 3.5\,V$ (*green*). Furthermore, the *uncharged* state correlates to $t_{pre} = 0\,s$, the *time-dependency* to $t_{pre} = 10\,s$ with $V_{GS,pre} = V_{GS,pre}$ and the *voltage-dependency* to $t_{pre} = 10\,s$ with $V_{GS,pre} = 4\,V$ for $t_{bias} = 100\,\mu s$.

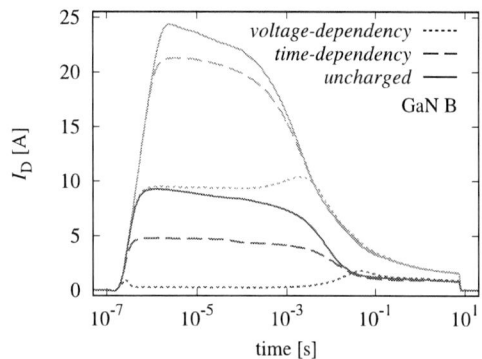

Fig. 11. Transient drain-current behavior $I_D(t)$ of GaN B. The operational points correspond to $V_{GS,on} = 2.5\,V$ (*red*) and $V_{GS,on} = 3.5\,V$ (*green*). Furthermore, the *uncharged* state correlates to $t_{pre} = 0\,s$, the *time-dependency* to $t_{pre} = 10\,s$ with $V_{GS,pre} = V_{GS,pre}$ and the *voltage-dependency* to $t_{pre} = 10\,s$ with $V_{GS,pre} = 6\,V$ for $t_{bias} = 100\,\mu s$.

currents. The $V_{GS,pre}$ is $+4\,V$ for GaN A and $+6\,V$ for GaN B, with a duration of $t_{bias} = 100\,\mu s$. Here, the precharged state (*time-dependency*) is used as a reference, similar to Sec. IV. The *time-dependency* appears to be relevant for GaN A and B, although its impact tends to saturate for increasing currents. A similar behavior is present with the *voltage-dependency* for GaN A, but its influence is less significant. On the other hand, the impact of the *voltage-dependency* is substantial for GaN B, although the device does not remain fully depleted in the early

stages anymore. However, all observed effects converge over time, but the convergence tends to shift to earlier points in time for increasing currents.

The ratio carriers influencing the channel conductivity, compared to the free carriers drifting through, is lower for higher currents. Furthermore, the lifetime of accumulated and/or trapped carriers reduces with the amount of free carriers due to neutralization. This is in accordance with the time shift of the convergence for increasing currents.

VI. CONCLUSIONS

Both devices show a significant influence regarding the charging time of the gate. Here, processes with high- and low time-constants have been observed. A considerable impact of the gate voltage is present for the device with the p-AlGaN gate, which could be caused by hole-injection with subsequent carrier accumulation. For the p-GaN device, enhanced voltages can cause a depletion during the early states after the turn on. Furthermore, the time- and voltage-dependencies appear to be relevant up very high currents. However, all observed effects tend to convergence, wherefore they are likely to be caused by trapping and accumulation phenomena.

REFERENCES

[1] H. Okita, M. Hikita, A. Nishio, T. Sato, K. Matsunaga, H. Matsuo, M. Mannoh, and Y. Uemoto, "Through recessed and regrowth gate technology for realizing process stability of gan-gits," in *2016 28th International Symposium on Power Semiconductor Devices and ICs (ISPSD)*, 2016, pp. 23–26.

[2] X. Huang, D. Y. Lee, V. Bondarenko, A. Baker, D. C. Sheridan, A. Q. Huang, and B. J. Baliga, "Experimental study of 650v algan/gan hemt short-circuit safe operating area (scsoa)," in *2014 IEEE 26th International Symposium on Power Semiconductor Devices IC's (ISPSD)*, 2014, pp. 273–276.

[3] T. Oeder, A. Castellazzi, and M. Pfost, "Electrical and thermal failure modes of 600v p-gate gan hemts," *Microelectronics Reliability*, vol. 76-77, pp. 321 –326, 2017. [Online]. Available: http://www.sciencedirect.com/science/article/pii/S0026271417302342.

[4] Y. Shi, Q. Zhou, Q. Cheng, P. Wei, L. Zhu, D. Wei, A. Zhang, W. Chen, and B. Zhang, "Bidirectional threshold voltage shift and gate leakage in 650 v p-gan algan/gan hemts: The role of electron-trapping and hole-injection," in *2018 IEEE 30th International Symposium on Power Semiconductor Devices and ICs (ISPSD)*, 2018, pp. 96–99.

[5] H. Chonan, T. Ide, X.-Q. Shen, and M. Shimizu, "Effect of hole injection in algan/gan hemt with git structure by numerical simulation," *physica status solidi c*, vol. 9, no. 34, pp. 847–850, eprint: https://onlinelibrary.wiley.com/doi/pdf/10.1002/pssc.201100330. [Online]. Available: https://onlinelibrary.wiley.com/doi/abs/10.1002/pssc.201100330.

[6] A. Stockman, E. Canato, A. Tajalli, M. Meneghini, G. Meneghesso, E. Zanoni, P. Moens, and B. Bakeroot, "On the origin of the leakage current in p-gate algan/gan hemts," in *2018 IEEE International Reliability Physics Symposium (IRPS)*, 2018, 4B.5–1–4B.5–4.

[7] X. Tang, B. Li, H. A. Moghadam, P. Tanner, J. Han, and S. Dimitrijev, "Effect of hole-injection on leakage degradation in a p-gan gate algan/gan power transistor," *IEEE Electron Device Letters*, vol. 39, no. 8, pp. 1203–1206, 2018.

Proceedings of the 31st International Symposium on Power Semiconductor Devices & ICs
May 19-23, 2019, Shanghai, China

Integrated GaN MIS-HEMT with Multi-Channel Heterojunction SBD Structures

Sheng Li, Siyang Liu, Chi Zhang, Jiaxing Wei,
Long Zhang, Weifeng Sun*
National ASIC System Engineering Research Center
Southeast University
Nanjing, China
*E-mail: swffrog@seu.edu.cn

Youhua Zhu, Tingting Zhang, Dongsheng
Wang, Yinxia Sun, Yiheng Li, Tinggang Zhu
CorEnergy Semiconductor Co., LTD
Zhangjiagang, China

Abstract—**In this paper, the integrated GaN metal-insulator-semiconductor high electron mobility transistor (MIS-HEMT) with multi-channel heterojunction Schottky barrier diode (SBD) structures are proposed. By growing HEMT upon multi-channel heterojunction SBD with a polarization effect-free AlN insertion layer between them, the reverse conduction voltage (V_{R_F}) are decreased by 41% compared with traditional MIS-HEMT device. Meanwhile, the blocking characteristic is excellent and the forward conduction performances are maintained. On the other hand, the proposed device needs a relatively simple manufacturing process and the wafer area efficiency is relatively high.**

Keywords—reverse conduction, integrated structures, multi-channel

I. INTRODUCTION

GaN-based power devices become expected to renovate the traditional power systems to improve the conversion efficiency [1][2]. More and more GaN devices are applied to power electronic systems with bridge circuit topology or LLC resonant circuit topology (seen in Fig. 1)[3]-[5]. In these applications, it is inevitable for GaN devices to work at reverse conduction state. However, the reverse voltage drop (V_{R_F}) on the GaN devices with traditional structures is modulated by the gate bias and is usually even high, which brings higher energy loss and lower efficiency [6][7]. The above shortcomings become the obstacles in front of the developments of GaN devices.

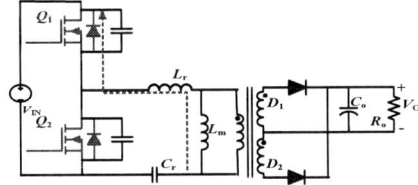

Fig. 1 LLC resonant circuit topology using GaN devices. The red line indicates the current path at freewheeling state.

The method that places anti-paralleled Schottky barrier diodes (SBDs) and high electron mobility transistor (HEMTs) into one chip is valid to reduce the high V_{R_F}. However, extra SBDs bring parasitic capacitances and inductances [8]. To minimize the parasitic values, an integrating way is required. Among previous approaches, the method that places a Schottky contact in the gate-to-drain region is compact [9][10], but the off-state leakage current is high. Also, an interdigitated metal-insulator-semiconductor

(MIS) HEMT and SBD structure is proposed [11], but the wafer area efficiency is limited. In this work, the integrated GaN based MIS-HEMT with multi-channel heterojunction SBD structures are proposed to avoid the shortcomings. The investigation results show that the proposed devices exhibit low V_{R_F} and excellent blocking characteristics. Meanwhile, the forward conduction performances of the devices are maintained.

II. DEVICE STRUCTURE AND OPERATING PRINCIPLES

Fig. 2 shows the cross-section diagram of the proposed integrated MIS-HEMT with heterojunction SBD structures. The HEMT is grown upon heterojunction SBDs with a polarization effect-free AlN insertion layer between them. To maintain the electrical performances of the HEMT structure, the GaN buffer layer under the channel of HEMT (D_{ch1}) is relatively thick, which may make the polarization effect at SBD channel weaker and reduce the electron concentration. The polarization effect-free AlN layer is inserted between HEMT and SBD to maintain electron concentration in SBD channel. Once even lower V_{R_F} is required, multi-channel SBD structures can be applied to the device to achieve even lower reverse on-state resistance (R_{on_R}). To make the contacts, deep recesses should be made to access the barrier layer of SBDs. The drain contact is ohmic type, the gate contact is Schottky type, while the source contact is Schottky/ohmic mixed type. In this way, SBDs are integrated with the HEMT upon the same substrate efficiently.

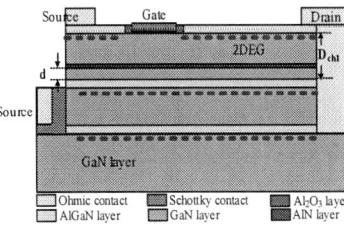

Fig. 2 The cross-section diagram of the integrated MIS-HEMT with heterojunction SBD structures

When applying the devices into the power systems, the basic operating principles are the same with traditional GaN MIS-HEMT devices, however, the reverse conduction capability is improved. At reverse conduction state, the gate to source voltage (V_{gs}) is usually 0V to −3V, while drain to source is reverse biased (V_{ds}<0). Under these bias conditions, three reverse current channels (Fig. 3(a)) are formed, resulting

978-1-7281-0582-6/19 $31.00 © 2019 IEEE

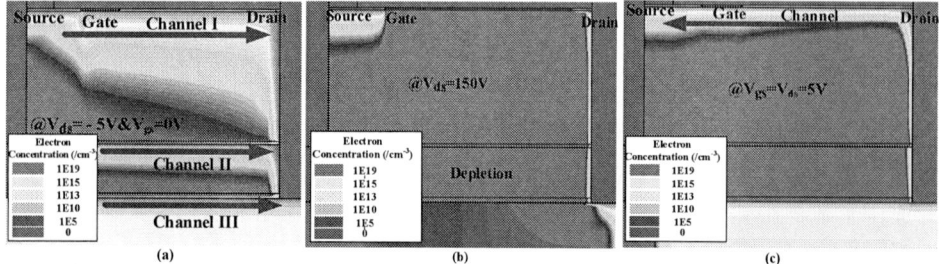

Fig. 3 The proposed device at different working state. (a) The electrons distribution of the device at reverse conduction state. (b) The electrons distribution at blocking state. (c) The electrons distribution of the device at forward conduction state.

in a lower V_{R_F} (decreased by 41% at 0V gate bias compared with original MIS-HEMT). At blocking state, the V_{gs} is still 0V to −3V but the V_{ds} is highly forward biased. Under these bias conditions, the Schottky gate and Schottky source help each other deplete the electrons (Fig. 3(b)), improving the breakdown voltage (BV, increased by 58%). At forward conduction state, the V_{gs} and V_{ds} are both forward biased, under these bias conditions, the forward conduction performances of the device are maintained since the forward current channel stays the same (Fig .3(c)).

III. ELECTRICAL PERFORMANCE OPTIMIZATION

The electrical performances of the proposed device together with the influences of layer depth and Schottky metal work-function have been investigated. Fig. 4 shows the reverse conduction performances with different layer depth and work-function. The reverse conduction capability of proposed devices are much better than traditional MIS-

HEMT devices. At reverse conduction state, the reverse turn-on voltage (V_{R_T}) of the device is determined by that of SBD, moreover, the R_{on_R} are much smaller than traditional device since the current paths are more. To achieve better performances, the structure parameters should be optimized. As seen in Fig. 4(a), when the metal work-function is confirmed, larger D_{ch1} and smaller d make lower V_{R_F}. These phenomena indicate that the layer depth should be appropriate to make sure that the polarization charge density at HEMT and SBD channels is maintained. The results presented in Fig. 4(b) indicate that lower metal work-function of Schottky contact makes the reverse turn-on voltage smaller when the layer depth is confirmed, for the SBDs are easier to be turned-on with lower metal work-function.

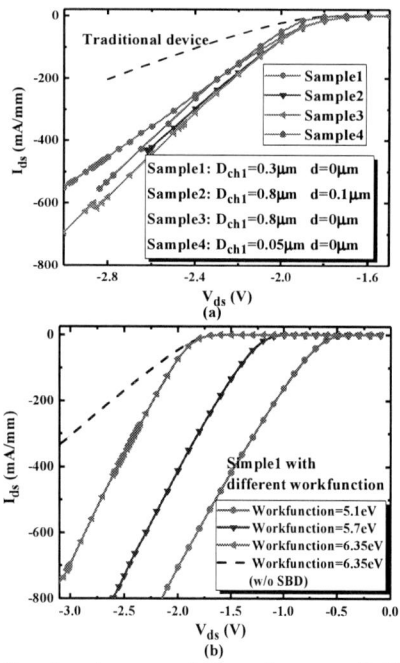

Fig. 4 The enhanced reverse conduction performances with different layer depth and Schottky metal work-function. (a)The reverse conduction performances of samples with the same Schottky metal work-function but different layer depth. (b) The reverse conduction performances with the same layer depth but different Schottky metal work-function.

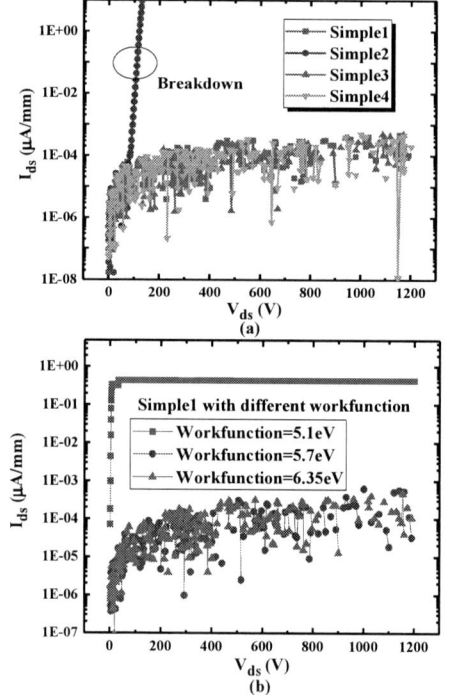

Fig. 5 The blocking characteristics of proposed devices (a) The blocking characteristics of the devices with different layer depth. (b) The blocking characteristics of sample1 with different work-function of the Shcottky metal.

Fig. 5 presents the blocking characteristics of different samples. As shown, sample2 (D_{ch1}=0.8μm, d=0.1μm) performs an extremely low BV. To find the mechanisms of

978-1-7281-0582-6/19 $31.00 © 2019 IEEE   448

this failure, more investigations have been accomplished. From Fig. 6(a) and (b), the drain to source leakage current mainly flows through the HEMT channel due to the high potential in the top GaN layer. The electron concentrations in the three AlGaN/GaN heterojunctions are extracted to understand this phenomenon. It can be seen from Fig. 6(c) that the larger D_{ch1} and d bring more electrons in channel II,

conducting higher potential from drain to the top GaN layer. In this way, the BV of sample2 is reduced. Moreover, lower gate metal work-function also makes larger leakage current of the HEMT due to its weaker depletion capability on electrons in HEMT channel, as seen in Fig. 6(d). Thus, the position of insertion layer and metal work-function should be appropriate when designing the proposed devices.

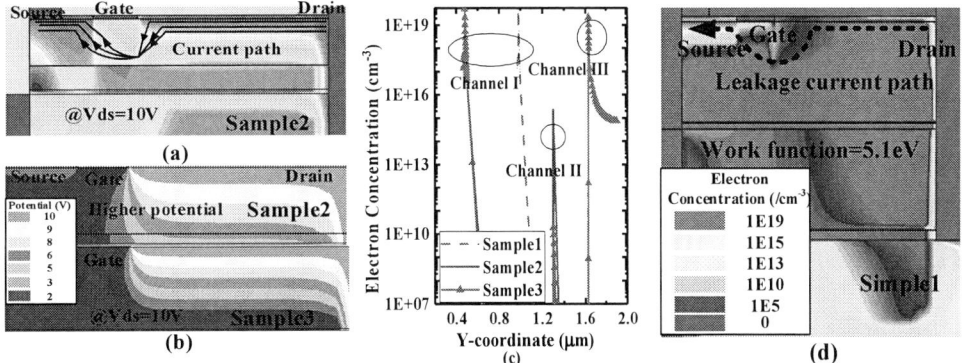

Fig. 6 Analyses on the devices under blocking state. (a) The leakage current path of simple2. (b) The potential distribution under the channels of simple2 and simple3. (c) The extracted electrons concentrations in channels at equilibrium. (d) The leakage current path of simple1 with gate metal work-function equaling to 5.1eV.

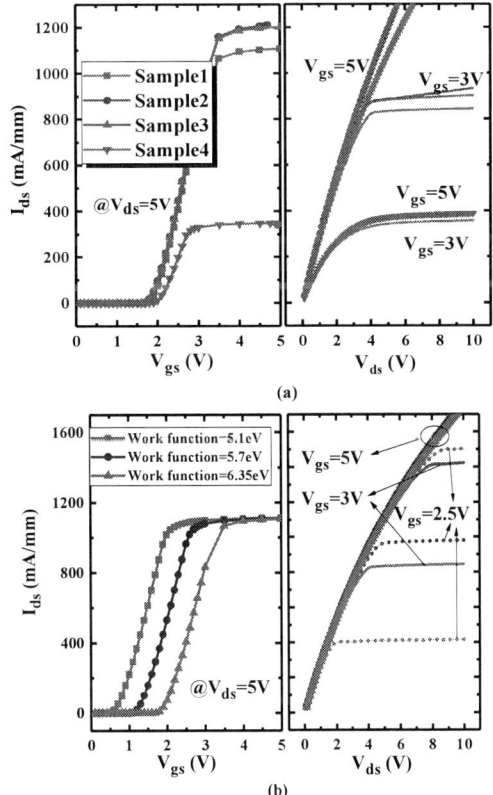

Fig. 7 The forward conduction performances of the proposed devices. (a) Influences of D_{ch1} and d on transfer and I-V characteristics. (b) Influences of Schottky metal work-function on transfer and I-V characteristics.

Also, the forward conduction performances of the proposed devices are presented in Fig. 7. The results indicate that D_{ch1} cannot impact the threshold voltage (V_{th}), however, smaller D_{ch1} can bring forward on-state resistance (R_{on}) increase even at high V_{gs}. These phenomena result from the narrowing of channel in HEMT structure, indicating that relatively thicker channel depth of HEMT is required to maintain the forward conduction performances. As for the influences of metal work-function, the V_{th} increases with the work-function, however, the R_{on} cannot be changed when the V_{gs} is high. Thus, thicker channel layer of HEMT and appropriate work-function are the key points to maintain the forward conduction performances.

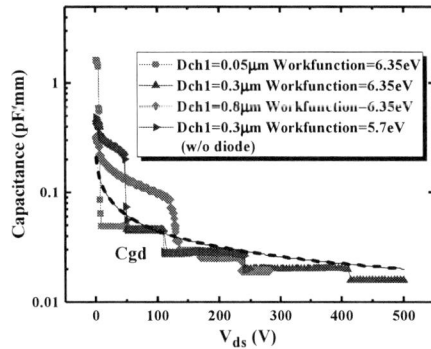

Fig. 8 Influences of D_{ch1} and Schottky metal work-function on capacitance characteristics

Reverse capacitance (C_{gd}) is an important parameter when estimating the switching performances of power devices. Fig. 8 presents the C_{gd} characteristics of different samples. The results show that the larger D_{ch1} make the reverse capacitance of the device larger at low V_{ds}, but the influences can be neglected at high V_{ds} under the same D_{ch1}. In conclusion, the C_{gd} of the proposed device is maintained at high V_{ds}.

978-1-7281-0582-6/19 $31.00 © 2019 IEEE

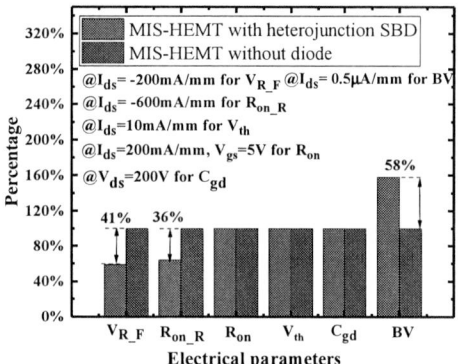

Fig. 9 Comparisons of the electrical parameters between the MIS-HEMT without diode (D_{ch1}=0.3mm) and MIS-HEMT with heterojunction SBD (D_{ch1}=0.3mm, d=0). The work-function for gate metal is 6.35eV, and the work-function for Schottky source metal is 5.1eV.

After optimizing the structure parameters, the comparisons of the electrical parameters between the optimal proposed device and original MIS-HEMT are presented in Fig. 9. The integrated MIS-HEMT with heterojunction SBD structure exhibits enhanced reverse conduction characteristic and blocking characteristic without changing forward conduction characteristic, performed as BV increased from 545V to 860V, V_{R_F} decreased from 2.6V to 1.55V, and R_{on_R} decreased from 2.5Ω·mm to 1.61Ω·mm.

IV. CONCLUSIONS

In this paper, the integrated GaN MIS-HEMT with multi-channel heterojunction SBD structures are proposed. By growing HEMT upon heterojunction SBDs with a polarization effect-free AlN insertion layer between them, the reverse conduction characteristic is enhanced. After optimizing the layer depth and metal work-function of proposed devices, the results show that the V_{R_F} is decreased by 41%, R_{on_R} is decreased by 36% and BV is increased by 58% compared with the traditional MIS-HEMT, meanwhile, the forward conduction performances of the device are maintained.

ACKNOWLEDGMENT

This work was supported by the National Natural Science Foundation of China (61604038), the Foundation of State Key Laboratory of Wide-bandgap Semiconductor Power Electronics Devices (2017KF003), the Key R&D Plan of Jiangsu Province (BE2018082), and the Fundamental Research Funds for the Central Universities.

REFERENCES

[1] B. Liu, R. Ren, E. A. Jones, F. Wang, D. Costinett and Z. Zhang, "A Modulation Compensation Scheme to Reduce Input Current Distortion in GaN-Based High Switching Frequency Three-Phase Three-Level Vienna-Type Rectifiers," in *IEEE Transactions on Power Electronics*, vol. 33, no. 1, pp. 283-298, Jan. 2018.

[2] C. Yao *et al.*, "Adaptive Constant Power Control of MHz GaN-Based AC/DC Converters for Low Power Applications," in *IEEE Transactions on Industry Applications*, vol. 54, no. 3, pp. 2525-2533, May-June 2018.

[3] A. Hariya *et al.*, "Circuit Design Techniques for Reducing the Effects of Magnetic Flux on GaN-HEMTs in 5-MHz 100-W High Power-Density LLC Resonant DC-DC Converters," in *IEEE Transactions on Power Electronics*, vol. 32, no. 8, pp. 5953-5963, Aug. 2017.

[4] M. H. Ahmed, M. A. de Rooij and J. Wang, "High-Power Density, 900-W LLC Converters for Servers Using GaN FETs: Toward Greater Efficiency and Power Density in 48 V to 6V/12 V Converters," in *IEEE Power Electronics Magazine*, vol. 6, no. 1, pp. 40-47, March 2019.

[5] T. Ou *et al.*, "A Novel Transformer Structure Used in a 1.4 MHz LLC Resonant Converter with GaNFETs," *2018 IEEE International Power Electronics and Application Conference and Exposition (PEAC)*, Shenzhen, 2018, pp. 1-5.

[6] T. Sun, X. Ren, Q. Chen, Z. Zhang and X. Ruan, "Reliability and efficiency improvement in LLC resonant converter by adopting GaN transistor," *2015 IEEE Applied Power Electronics Conference and Exposition (APEC)*, Charlotte, NC, 2015, pp. 2459-2463.

[7] S. Park and J. Rivas-Davila, "Power loss of GaN transistor reverse diodes in a high frequency high voltage resonant rectifier," *2017 IEEE Applied Power Electronics Conference and Exposition (APEC)*, Tampa, FL, 2017, pp. 1942-1945.

[8] R. Zhu *et al.*, "A split gate vertical GaN power transistor with intrinsic reverse conduction capability and low gate charge," *2018 IEEE 30th International Symposium on Power Semiconductor Devices and ICs (ISPSD)*, Chicago, IL, 2018, pp. 212-215.

[9] R. Reiner *et al.*, "Integrated reverse-diodes for GaN-HEMT structures," *2015 IEEE 27th International Symposium on Power Semiconductor Devices & IC's (ISPSD)*, Hong Kong, 2015, pp. 45-48.

[10] Park B. R. *et al.*, "Normally-Off AlGaN/GaN-on-Si Power Switching Device with Embedded Schottky Barrier Diode". Applied Physics Express, 2013, 6(3):031001.

[11] J. Lei *et al.*, "An interdigitated GaN MIS-HEMT/SBD normally-off power switching device with low ON-resistance and low reverse conduction loss," *2017 IEEE International Electron Devices Meeting (IEDM)*, San Francisco, CA, 2017, pp. 25.2.1-25.2.4.

Proceedings of the 31st International Symposium on Power Semiconductor Devices & ICs
May 19-23, 2019, Shanghai, China

Damage accumulation in GaN GITs exposed to repetitive short-circuit

F. D'Aniello, A. Fayyaz, A. Castellazzi
Power Electronics, Machines and Control (PEMC) Group,
University of Nottingham, Nottingham, NG7 2RD, UK
Email: asad.fayyaz@nottingham.ac.uk

T. Oeder, M. Pfost
Chair of Energy Conversion, TU Dortmund, Emil-Figge-
Straße 68, Dortmund 44227, Germany

Abstract — **Short-circuit withstand capability is one of the key elements when it comes to the selection of semiconductor power devices in different mainstream power electronics applications (i.e industrial, e-mobility etc). Alongside the single pulse SC robustness, aging of power devices is also of concern when subjected to repetitive SC stress since devices may experience such transient events relatively quite frequently during their lifetime. This paper investigates the effect of repetitive short-circuit robustness on 600 V rated commercially available GaN gate injection transistors (GITs). In particular, monitors of device degradation as stress accumulates on the device under test (DUT) have been reiterated in this paper. Moreover, the effect of SC energy and the applied drain-source bias (V_{DS}) during SC have also been investigated. Lastly, de-capsulation of the degraded devices were also performed for microscopic analysis.**

Keywords — short-circuit; gallium nitride; GaN HEMT; GIT; transient robustness; repetitive stress; aging

I. INTRODUCTION

Gallium Nitride (GaN), a wide bandgap (WBG) semiconductor material, is well-known and widely discussed for making power devices due its superior material properties when compared to the traditional silicon (Si) technology [1]. In early stages of any new device technology, research efforts are needed to have an in-depth device characterization for robustness and reliability prior to its mainstream implementation in different strategic and commercial power application domains, by no means limited to, such as renewable energies and railway traction.

GaN gate injection transistors (GITs) have been recently shown to exhibit excellent single pulse short-circuit (SC) withstand capability [2, 3]. In [3], two different SC failure mechanisms have been identified namely as electrical and thermal failure. The failure modes have been characterized by the peak drain current ($I_{D(pk)}$) and the applied drain-source bias (V_{DS}). Devices tend to withstand SC times well in excess of the nominal SC withstand capability of 10 μs prior to thermal failure. However, higher V_{DS} (≥ 450 V) could also trigger an almost instantaneous electrical failure with no SC withstand capability due to really high presence of electric field inside the structure as discussed in more detail in [3, 4]. With careful gate drive design (as discussed in section II) to avoid electrical failure, such characteristics makes them a very valuable candidate technology option for drive applications

(i.e., industrial, e-mobility), in which, next to single pulse robustness, the ability to withstand repetitive short-circuit (SC) conditions with contained degradation of the device characteristics is also critical. Recently, monitors of device aging as a result of repetitive stress application have been identified and presented [4]. In this work, the study is extended with the aim of quantifying the amount of accumulated damage and degradation based on the single-pulse energy and the corresponding semiconductor chip temperature excursion. Moreover, the impact of electric field for the same applied energy is also investigated for a thorough analysis.

II. GATE DRIVE AND DEVICE UNDER TEST (DUT)

Results presented here are on the lateral 600 V rating gate injection GaN AlGaN/GaN HEMT transistor (Device under test (DUT): PGA26E07BA) [5]. Further details about device physics and structure can be found in [4]. The typical gate drive circuit for this particular type of GaN transistor is presented in Fig. 1. The additional branch containing capacitance (C_{RC}) and resistance (R_{RC}) connected in parallel to gate resistance (R_G) is needed to ensure fast and low dissipative switching by supplying the in-rush gate current at switching instances. These values for three components namely C_{RC}, R_{RC} and R_G are carefully chosen as per the desired application requirements.

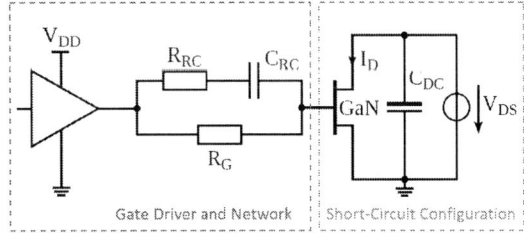

Fig. 1: Typical gate drive circuit for driving GaN GIT transistors

The corresponding transient V_{GS} waveform profile for this type of transistor is also illustrated in Fig. 2. The $V_{GS(pk)}$ is mainly influenced by the values of R_{RC} and C_{RC}. The determination of negative V_{GS} at turn off going to zero is also

978-1-7281-0582-6/19 $31.00 © 2019 IEEE

dependent on R_{RC} and C_{RC} values. However, additional circuitry could also be implemented for more versatile gate drive and quick return of negative V_{GS} to zero [6]. Lastly, the $V_{GS(sat)}$ value is controlled by the R_G value. Therefore, R_{RC}, C_{RC} and R_G values help optimize the switching and/or on-state performance as required. Here during turn-on switching transient for SC, the slope of the drain current (I_D) is determined by the gate drive switching loop, the applied drain source bias (V_{DS}) and peak gate-source voltage ($V_{GS(pk)}$).

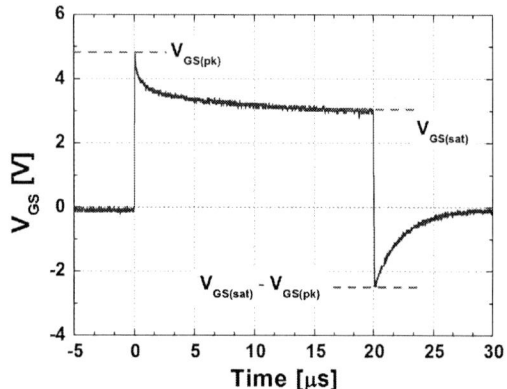

Fig. 2: Typical V_{GS} waveform for GaN transistors (DUT)

III. REPETITIVE STRESS AGING

Here, the aim of the aging tests was to investigate the effect of cumulative SC energy ($E_{SC(CM)}$) in terms of functional and structural evolution of the DUT. In order to have an electro-thermally coupled analysis, the experimental results were also used to input into a detailed functional and structural 3D electro-thermal device model (as detailed in [7]) to obtain the maximum junction temperature ($T_{J(max)}$) estimates during the SC conditions discussed here.

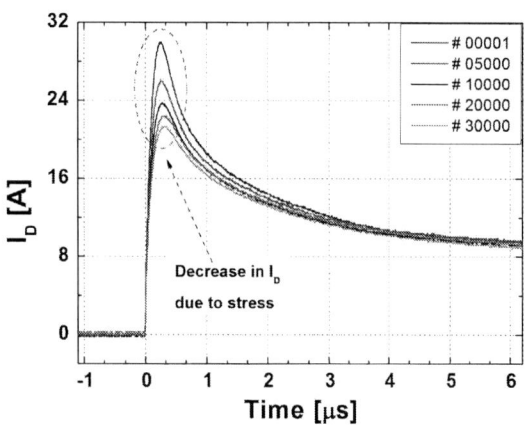

Fig. 3: *Decrease* in I_D as the stress accumulated; $V_{DC} = 400$ V; $t_{SC} = 150$ µs; $T_{CASE} = 150$ °C

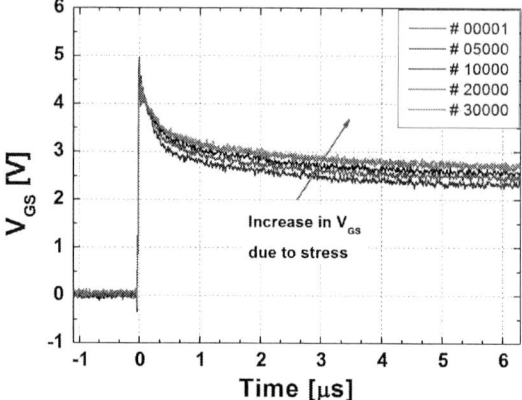

Fig. 4: *Increase* in V_{GS} as the stress accumulated; $V_{DC} = 400$ V; $t_{SC} = 150$ µs; $T_{CASE} = 150$ °C

The SC results presented in Fig. 3 and 4 were carried out at $V_{DS} = 400$ V, $t_{SC} = 150$ µs and $T_{CASE} = 150$ °C. Fig. 3 shows the zoom-in of the I_D waveforms for selective repetitive SC stress accumulation up to a total of 30,000 pulses to manifest degradation. Here, the SC energy (E_{SC}) equated to 350 mJ. In order to ensure no build-up of temperature within the DUT, a delay of 1 second was chosen between pulses to allow enough time for the device temperature to regulate back to T_{CASE} prior to the next SC pulse. The maximum change in peak drain current ($\Delta I_{D(pk)}$) in Fig. 3 was approximately 8.5 A. Here, a prominent effect was the decrease of $I_{D(pk)}$ as the DUT was repetitively subjected to electro-thermal stress (i.e. high junction temperatures (T_J)) leading to damage being accumulated on the DUT. In the first 5000 pulses, there is already a significant decrease in $I_{D(pk)}$ and the change gets smaller and smaller as $I_{D(pk)}$ keeps decreasing before it finally saturates. Subsequently, Fig. 4 shows the increase in V_{GS} for I_D waveforms presented in Fig. 3.

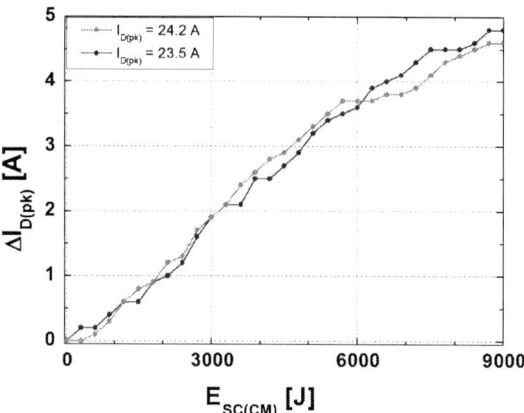

Fig. 5: $\Delta I_{D(pk)}$ versus $E_{SC(CM)}$ for two devices with similar $I_{D(pk)}$ with $E_{SC} = 300$ mJ

978-1-7281-0582-6/19 $31.00 © 2019 IEEE

In order to quantify the device degradation, relevant test parameters were changed i.e. t_{SC} and V_{DS} while other parameters including the gate driver variables which were kept same. Prior to starting any detailed analysis, it was crucial to ensure the repeatability and accuracy of the experiments. Therefore, reference tests were performed with $V_{DS} = 400$ V and $E_{SC} = 300$ mJ on two different DUTs where the starting $I_{D(pk)}$ and E_{SC} was kept the same in order to observe $\Delta I_{D(pk)}$ for devices with similar parameters. Results for $\Delta I_{D(pk)}$ versus $E_{SC(CM)}$ have shown a good match with $\Delta I_{D(pk)}$ of approximately 4.5 A for both devices as illustrated in Fig. 5.

Fig. 6 shows the plot of $\Delta I_{D(pk)}$ versus cumulative SC energy ($E_{SC(CM)}$) for two different SC energies; $E_{SC} = 300$ mJ and 350 mJ. The $\Delta I_{D(pk)}$ for $E_{SC} = 300$ mJ and 350 mJ was 8.8 A and 4.6 A respectively. The $\Delta I_{D(pk)}$ is bigger for $E_{SC} = 350$ mJ which clearly indicates that ΔE_{SC} of as small as 50 mJ is substantial to give rise to quite different $T_{J(max)}$ in the device. Here, it is to be noted that the different E_{SC} and T_J as presented in Fig. 6 and 7 actually happened as a result of variations in device parameters (most probably threshold voltage (V_{TH}) and/or drain-source on-state resistance ($R_{DS(on)}$)). Recent studies investigating the impact of device parameters on the robustness of SiC power MOSFETs are presented in [8, 9]. In order to confirm this hypothesis, a device model was used for T_J simulations and the corresponding profiles for the abovementioned stress test conditions are presented. Fig. 7 shows the simulated T_J profiles for both test conditions; $E_{SC} = 300$ mJ and 350 mJ, where $T_{J(max)}$ was approximately 855 K and 940 K respectively. Quantitatively, also from Fig. 7, it is obvious that the used test conditions gave rise to a ΔT_J of approximately 85 K which clearly explains the different levels of degradation observed (i.e. $\Delta I_{D(pk)}$) as a result of different damage accumulated on the devices (i.e. $E_{SC(CM)}$). Moreover, as also evident in Fig. 7, the temperature increases rapidly during the beginning of the switching pulse and then the increase in T_J is not so high during steady state. Therefore, t_{SC} and SC power (P_{SC}) level was appropriately selected to ensure T_J of interest to trigger device degradation.

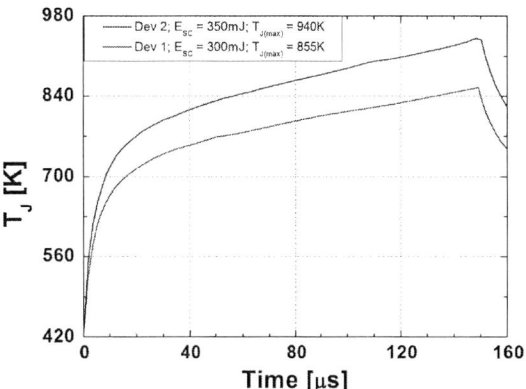

Fig. 7: Simulated T_J for different E_{SC} but same V_{DS} (400 V); $E_{SC} = 300$ mJ and 350 mJ

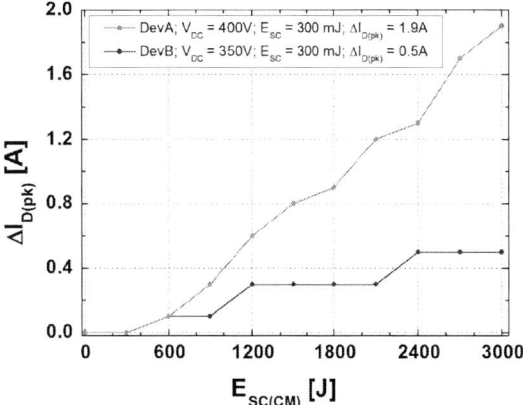

Fig. 8: $\Delta I_{D(pk)}$ versus $E_{SC(CM)}$ for $V_{DS} = 350$ V and 400 V respectively

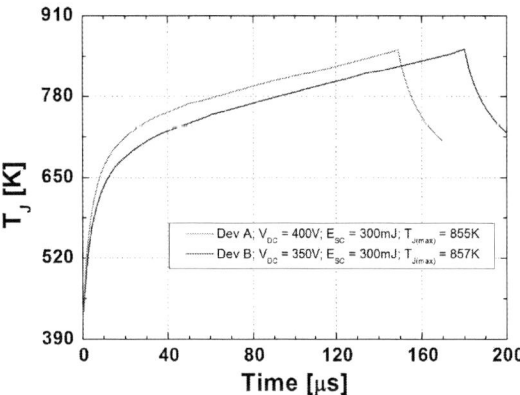

Fig. 9: Simulated T_J for different V_{DC} but same E_{SC} (300 mJ); $V_{DC} = 350$ V and 400 V

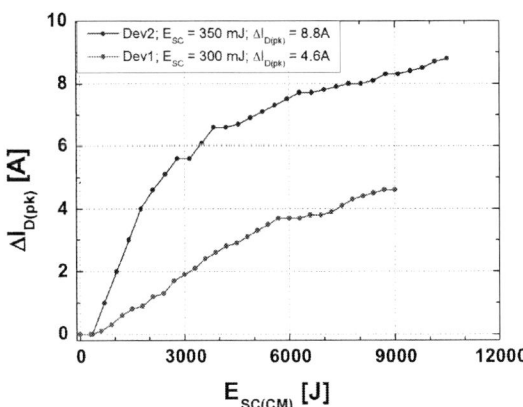

Fig. 6: $\Delta I_{D(pk)}$ versus $E_{SC(CM)}$ for $E_{SC} = 300$ mJ and 350 mJ respectively

Moreover, tests were also carried out to understand the effect of V_{DS} bias on degradation. Fig. 8 shows the $\Delta I_{D(pk)}$ versus $E_{SC(CM)}$ for two different applied drain-source bias; V_{DS} = 350 V and 400 V while keeping E_{SC} the same. The $\Delta I_{D(pk)}$ after 3000 pulses for V_{DS} = 350 V and 400 V was 1.9 A and 0.5 A respectively. Even though E_{SC} is kept the same (E_{SC} = 300 mJ), lower V_{DS} resulted in relatively less degradation as shown in Fig. 8 which shows that ΔV_{DS} of as small as 50 V is substantial to give rise to quite different electric field in the device. Fig. 9 shows the simulated T_J profiles for both test conditions; V_{DS} = 350 V and 400 V, where $T_{J(max)}$ was approximately 855 K and 957 K respectively. An important result is presented in Fig. 9 which confirms that DUTs tested with different V_{DS} had approximately same $T_{J(max)}$ which clearly indicates that a difference in V_{DS} (i.e. electric field) and peak short-circuit power ($P_{SC(pk)}$) also plays a role in degradation.

A recent study included in [11] showed the formation of a crack with aluminum extrusion during power cycling of Si integrated circuits. Hypothesis formed here is that the observed degradation during repetitive SC is because of a similar kind of phenomena. The degradation mechanism during aging test resembles that of the single pulse thermal type of failure as discussed in [4], but here, the evolution of crack formation and aluminum extrusion happens gradually.

Lastly, Fig. 10 shows the opened failed device. Here, parts of the device metallization may have been attacked by the acid during the etching process; however, parts of the device which is burnt out and left a crater as circled in Fig. 10 (b) is clearly due to stress accumulation on the device.

IV. CONCLUSION

GaN GITs have been investigated for their repetitive SC robustness. GaN transistors discussed here exhibit extreme robustness during SC. Results presented have quantified the change in $I_{D(pk)}$ as a function of cumulative stress (i.e. $E_{SC(CM)}$) under different thermal (i.e. dependant on E_{SC}) and applied electric field (i.e. dependant on V_{DS}) stress conditions. It is shown that both the thermal and electric field stresses contribute to the degradation of the DUT. The extreme stress applied to the device resulted in apparent damage to the bare die device as presented within the microscopic analysis in previous section.

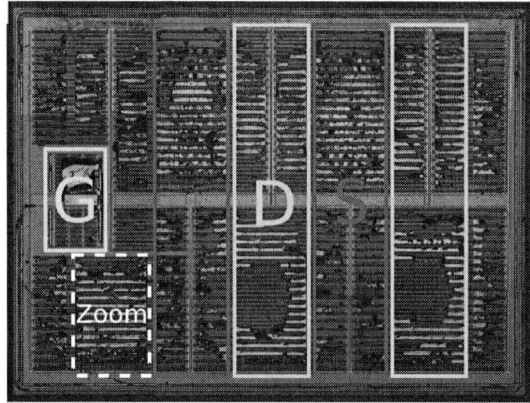

(a) – Complete Device

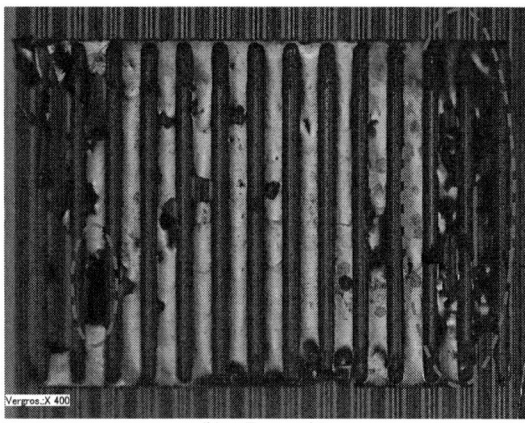

(b) – Zoomed In

Fig. 10: Microscopic top view of the opened failed devices

REFERENCES

[1] Lidow, Alex, Johan Strydom, Michael De Rooij, and David Reusch. *GaN transistors for efficient power conversion*. John Wiley & Sons, 2014.

[2] Oeder, Thorsten, Alberto Castellazzi, and Martin Pfost. "Experimental study of the short-circuit performance for a 600V normally-off p-gate GaN HEMT." In *Power Semiconductor Devices and IC's (ISPSD), 2017 29th International Symposium on*, pp. 211-214. IEEE, 2017.

[3] Oeder, Thorsten, Alberto Castellazzi, and Martin Pfost. "Electrical and thermal failure modes of 600 V p-gate GaN HEMTs." *Microelectronics Reliability* 76 (2017): 321-326.

[4] Castellazzi, Alberto, Asad Fayyaz, Siwei Zhu, Thorsten Oeder, and Martin Pfost. "Single pulse short-circuit robustness and repetitive stress aging of GaN GITs." In *Reliability Physics Symposium (IRPS), 2018 IEEE International*, pp. 4E-1. IEEE, 2018.

[5] https://www.mouser.com/ds/2/315/pga26e07ba-product-standards-1112741.pdf

[6] Wu, Han, Asad Fayyaz, and Alberto Castellazzi. "P-gate GaN HEMT gate-driver design for joint optimization of switching performance, freewheeling conduction and short-circuit robustness." In *2018 IEEE 30th International Symposium on Power Semiconductor Devices and ICs (ISPSD)*, pp. 232-235. IEEE, 2018.

[7] Mocanu, Manuela, Christian Unger, Martin Pfost, Patrick Waltereit, and Richard Reiner. "Thermal stability and failure mechanism of schottky gate algan/gan hemts." *IEEE Transactions on Electron Devices* 64, no. 3 (2017): 848-855.

[8] Castellazzi, Alberto, Asad Fayyaz, and Rainer Kraus. "SiC MOSFET device parameter spread and ruggedness of parallel multichip structures." In *Materials Science Forum*, vol. 924, pp. 811-817. Trans Tech Publications, 2018.

[9] Fayyaz, A., B. Asllani, A. Castellazzi, M. Riccio, and A. Irace. "Avalanche ruggedness of parallel SiC power MOSFETs." *Microelectronics Reliability* 88 (2018): 666-670.

[10] Tanaka, Kenichiro, et al. "Reliability of hybrid-drain-embedded gate injection transistor." *Reliability Physics Symposium (IRPS), 2017 IEEE International*. IEEE, 2017.

[11] Jacob, Peter, and Giovanni Nicoletti. "Failure causes generating aluminium protrusion/extrusion." *Microelectronics Reliability* 53, no. 9-11 (2013): 1553-1557.

Proceedings of the 31st International Symposium on Power Semiconductor Devices & ICs
May 19-23, 2019, Shanghai, China

Soft-Switching Losses in GaN and SiC Power Transistors Based on New Calorimetric Measurements

Julian Weimer and Ingmar Kallfass

Institute of Robust Power Semiconductor Systems, University of Stuttgart, Germany. Email: julian.weimer@ilh.uni-stuttgart.de

Abstract—The use of modern GaN and SiC power semiconductors in soft-switching power electronic applications makes it possible to minimize switching losses and to achieve a significant increase in power density. However, the loss mechanisms in soft-switching operation of power semiconductors are poorly understood and electrical measurements of these losses are affected by a large error. Therefore calorimetric measuring methods are becoming state of the art for determining soft-switching losses. This paper for the first time presents empirical power dissipation data for high voltage GaN and SiC power semiconductors of low power class at application-oriented switching frequencies for zero voltage switching (ZVS) operation. The losses were measured using a transient calorimetric measurement setup with the goal of determining also low power dissipation with high accuracy while maintaining the application-oriented commutation cell layout.

Keywords—**Soft-Switching Losses, ZVS, Calorimetric Measurement, High Voltage GaN/SiC Power Transistor.**

I. INTRODUCTION

In order to achieve a high power density for a $150\,\mathrm{W}$ ac/dc charger, a switching frequency of up to $1\,\mathrm{MHz}$ should be reached to minimize the size of the passive components. The very low power dissipation budget due to passive cooling requires soft-switching and the use of modern wide bandgap semiconductors. To reduce losses, the choice of a suitable semiconductor depends on a good knowledge of the expected switching and conduction losses for different current and voltage loads.

Unlike conduction losses, soft-switching losses cannot be predicted based on data sheet values. Although ideally no switching losses should occur in the transistors for soft-switching operation, C_{OSS}-losses can be observed for charging and discharging the output capacitance [1]. Only recently the origin of the C_{OSS}-losses for GaN-on-Si HEMTS could be determined [2]. A prediction of the soft-switching losses, however, remains difficult and is influenced not only by the semiconductors but also by packaging and layout due to parasitic capacitances. Electrical measurements of the small switching energies are affected by a too large error because of switching slopes in the ns range [3].

Therefore calorimetric measuring methods are used for determining soft-switching losses [2]–[8]. High measurement speeds and high accuracy can be achieved by shifting the measurement from the thermally steady state to the heating phase.

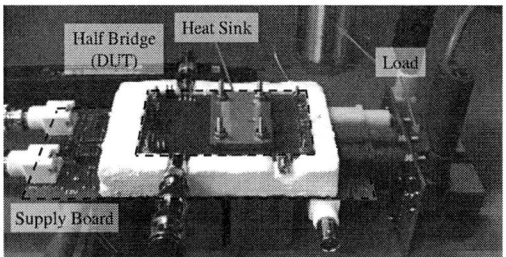

Fig. 1. Measurement setup with open thermal insulation consisting of supply board PCB, half bridge PCB, heat sink and load.

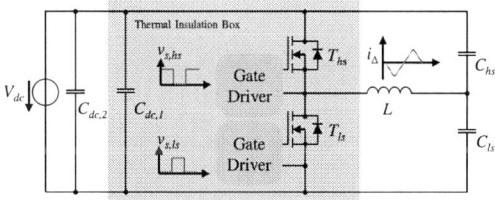

Fig. 2. Schematic of the calorimetric measurement setup, consisting of half bridge and capacitive-inductive load.

However, in current approaches a simplified thermal model is used which is no longer valid for very low semiconductor losses for which large heat sinks achieve too little temperature change. By utilizing smaller heat sinks, positioning the temperature measurement close to the semiconductors and calibrating the calorimetric measurement system to the corresponding time dependent thermal impedance, even very small losses can be determined with high accuracy. Furthermore, only minor changes are made to the layout of the commutation cell, so that the measured losses allow predictions for the target application. Fig. 1 shows the measurement setup with the corresponding Schematic in Fig. 2. The three examined half bridges with similar layout but different power transistors are shown in Fig. 3.

978-1-7281-0582-6/19 $31.00 © 2019 IEEE

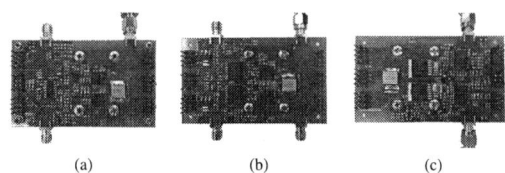

(a)   (b)   (c)

Fig. 3. Half bridge DUTs: PGA26E19BA GaN HEMT (a), LMG3410 GaN Power Stage (b), C3M0120090J SiC MOSFET (c).

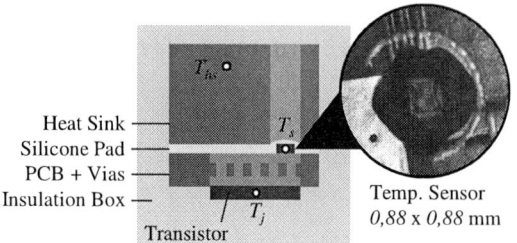

Fig. 4. Temperature sensor fixed with SMD adhesive on the transistor heat dissipation pad and cross-section of the measurement setup indicating the temperature sensor position.

II. Transient Calorimettric Measurement

A. Electrical Measurement Setup

All DUTs were designed as comparable half bridges which can be used in the target application. The half bridges were each realized as single PCBs, that are connected to the supply board via elevated connectors (see Fig. 3). For calorimetric power dissipation measurement, the semiconductors must switch over several seconds to minutes at defined switching currents with soft-switching transitions. In addition, the losses must be distributed symmetrically between the high-side (HS) and the low-side (LS) transistor. The topology with an inductive capacitive load from Fig. 2 has proven to be suitable for this [2], [6]–[8]. During operation with a 50 % duty cycle, the triangular current I_Δ is generated across the load. The soft-switching transitions always occur at the triangular peak currents and by selecting the load inductance, the same transistor switching and effective currents can be achieved for different switching frequencies: $L = \frac{V_{dc}}{8 f_{sw} I_\Delta}$. By exchanging the inductance L, an operation at different switching frequencies with constant conduction losses can be realized. The load inductances used were designed as air coils to avoid non-linearity due to saturating core material.

To minimize the parasitic inductance of the commutation cell, a ceramic capacitor $C_{dc,1}$ is located on each half bridge PCB near the switching power transistors. Together with the transistor-near placement of the gate drivers, optimal switching operations can be ensured. The supply board provides the galvanically separated high side and low side driver voltages and stabilizes the DC link voltage with the capacitance $C_{dc,2}$ which consists of six film capacitors with $5\,\mu\mathrm{F}$ each. The switching signals are generated with variable frequency, dead time and duration by means of a frequency generator.

B. Thermal Measurement Setup

For the calorimetric measurement of transistor losses, a temperature measurement independent of other power dissipation sources is required. Due to the division into half bridge board and supply board and the use of a thermal insulation box, the losses of the half bridge are thermally separated from the losses in supply and load. The LMT70A analog IC from Texas Instruments is used as the temperature sensor. In addition, the temperature signal in the measuring range from $20\,^\circ\mathrm{C}$ to $50\,^\circ\mathrm{C}$ is amplified to $0\,\mathrm{V}$ to $5\,\mathrm{V}$ via an op-amp circuit. In order to measure a sufficient temperature swing even for low

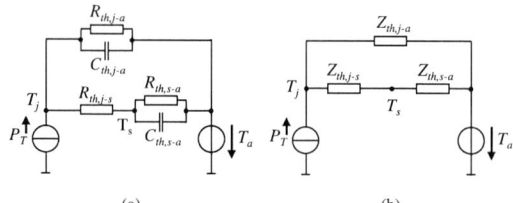

(a)   (b)

Fig. 5. Simplified (a) and applied (b) thermal Model.

transistor losses, the temperature sensor is placed on the heat sink side of the heat dissipation pad of a power transistor in contrast to the usual positioning at the heat sink. Fig. 4 shows the temperature sensor position and the schematic structure in cross-section. The placement on the heat sink side ensures only minor interference with the commutation cell layout. The advantage of this good thermal connection goes hand in hand with increased electromagnetic compatibility (EMC) problems. The sensor and the amplifier circuit are therefore supplied by battery and the sensor voltage measurement is galvanically isolated with the help of a differential probe. This also makes it possible to measure the sensor temperature on the drain potential of the SiC MOSFETs which are cooled via the drain connector. An aluminum block connected via a silicone pad was selected as the heat sink, whose heat capacity was designed according $C_{th} = \frac{P_T \Delta T_{hs}}{\Delta t}$ for a temperature swing of $\Delta T_{hs} = 10\,\mathrm{K}$ after an average transistor power loss of $P_T = 2\,\mathrm{W}$ over $\Delta t = 200\,\mathrm{s}$. Due to the positioning close to the transistor, a temperature swing of $\Delta T_s = 10\,\mathrm{K}$ is already achieved for $P_T = 1.5\,\mathrm{W}$.

C. Calibration Measurement

To calibrate the calorimetric measurement setup, several known transistor conduction loss steps are carried out and the temperature step responses of the sensor are recorded. For this purpose, a constant current is applied via the turned-on HS and LS transistors. The average losses can be calculated electrically via the calibration transistor current $I_{d,C}$ and the voltage drop across the two semiconductors $V_{ds,C}$ using

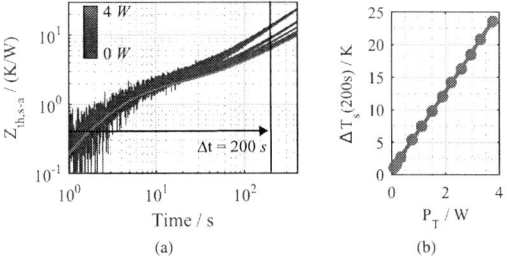

Fig. 6. Time-dependent thermal impedance measurement for various power dissipations and derived calibration curve.

$P_T = P_{T,hs} + P_{T,ls} = V_{ds,C} I_{d,C}$. Based on this calibration measurements, unknown losses in switching operation can then be calculated by evaluating the temperature step response. Previous transient calorimetric soft-switching measurements are based on a simplified thermal model as shown in Fig. 5a [2], [6]–[8]. Assuming a large thermal capacitance $C_{th,s-a}$, a linear increase of the sensor voltage can be expected after a settling time of several seconds. Since a comparably small heat sink must be used for the low power dissipation in this measurement setup, the dynamic transient response can no longer be neglected. The thermal behavior must therefore be recorded as time-dependent such as in the calorimetric driver and dead time optimization from [9]. By measuring the sensor voltage using an oscilloscope, a high time resolution is ensured, which enables the thermal impedance $Z_{th,s-a}$ to be measured. Fig. 5b shows the underlying thermal equivalent circuit diagram. The thermal impedance $Z_{th,s-a}$ measured for half bridge (c) for power losses from 0 W to 4 W can be seen in Fig. 6a. Based on the thermal step responses, the calibration characteristic $\Delta T_s = Z_{th,s-a}(\Delta t) P_T$ can be derived for any Δt. Fig. 6b shows the calibration characteristic curve for $\Delta t = 200\,\mathrm{s}$ of half bridge (c). Due to the precise measurement of the thermal impedance, measurement times well below $200\,\mathrm{s}$ can be achieved. For a sufficient temperature swing, the thermal capacity of the heat sink may have to be adjusted. For power dissipations smaller than 1 W, non-linearities were observed in the thermal impedance measurement. Since a linear thermal system is expected the non-linearity probably results within the temperature measurement. However, since this non-linearity is included in the calibration characteristic, the influence on the results can be minimized.

D. Power Loss Measurement

To determine the frequency-dependent soft-switching losses, the total transistor losses for switched triangular currents of different frequencies with equal peak currents are measured. Since a temperature measurement during operation is not possible due to EMC, the switching takes place over the pre-defined duration Δt without temperature measurement. A few milliseconds after switching operation, the sensor temperature T_s settles and can be measured. Together with the

ambient temperature T_a recorded for $t < 0\,\mathrm{s}$, the temperature swing ΔT_s is determined. Using the calibration characteristic $\Delta T_s(P_T)$ the total losses are calculated. Fig. 7a shows the frequency-dependent total losses for different DUTs, DC link voltages and switching currents. A linear dependency of the total losses on the frequency is shown in each case. Since the frequency generator is limited by 100×10^6 pulses in the maximum duration, no switching frequencies above $500\,\mathrm{kHz}$ were considered. The measuring accuracy was determined approximately with an error of $\pm 0.05\,\mathrm{K}$ for the temperature sensor, a relative error of $2\,\%$ for the differential probe and a relative error of $1\,\%$ for the current probe (see Fig. 7a).

E. Parasitic Heat Sources

A large parasitic heat source within the insulation box is the driver circuit. In order to calculate its influence for the loss measurements, the temperature swing ΔT_s for switched operation without load was determined frequency dependent. Moreover, electrical measurements of the driver losses showed only a small influence of the load. Before determining the transistor loss, the parasitic influence of the driver circuit can be taken into account for the measured switching frequency by subtracting the temperature swing of the driver circuit without load upfront. The parasitic influence of the DC link capacitance $C_{dc,2}$ is neglected due to the relatively low losses.

F. Soft-Switching Energy

The soft-switching energy can be determined by linear approximation of the frequency-dependent switching losses. Since the division into conductive and switching losses, however, does not provide an ideal line through the origin, the switching energies were determined from the slope of the linear approximation of the total losses (see Fig. 7b). Dead times during switching operation were set to the duration of the soft recharging of the output capacitance based on the electrically measured switching transitions. Corresponding losses can therefore be neglected. Even in real operation of the target application, the dead time can not be significantly further optimized. Remaining losses are thus added to the switching losses and are included in the switching energy calculation. The conductive losses can be calculated with the effective triangular current and the forward resistance according to $P_T = 2 R_{ds,on} I_{\Delta,rms}^2$ the current and temperature dependent forward resistance $R_{ds,on}$ is determined by the current and voltage measurements of the calibration measurement for different temperatures and currents and can be approximated for pulsed operation. Due to the small temperature swings of $\Delta T_{s,max} = 30\,°\mathrm{C}$, however, a change of less than $3\,\%$ relative to the minimum measured $R_{ds,on}$ remains. Since only small conduction losses occur in comparison to the total losses, the influence of the current and voltage dependency off $R_{ds,on}$ on the calorimetric measurement of the soft-switching energies of these DUTs is small.

III. Transistor Switching Engergy Benchmarking

Fig. 7b shows the switching energies approximated from the calorimetric measurements for soft-switching operations

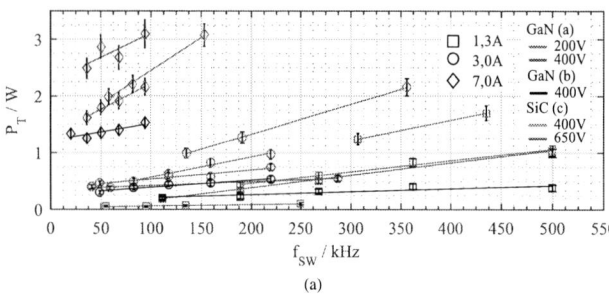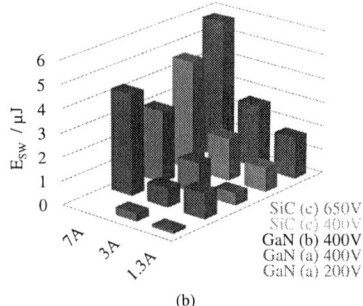

Fig. 7. Total transistor losses determined by calorimetric measurement for various DUTs and operating points (a). Derived switching energies for the half bridge for different DUTs and operating points (b).

(ZVS) for the three DUTs for different switching currents and different DC link voltages. The switching energies include the turn-on and turn-off energies for both the HS and LS transistors. Compared to the results from [6] for larger semiconductors with larger output capacitance, the switching losses are reduced further. In addition, these results also suggest an exponential relationship between switching current and switching energies. In addition to the strong current dependence, a high dependence on the DC link voltage is also visible. The advantage of the increased blocking voltage of SiC can be used for higher DC link voltages, but is purchased with significantly increase in losses. Overall, the switching energies for the soft-switching of the GaN half bridges investigated are lower than for the analyzed SiC half bridge. For the GaN half bridge with integrated driver the lowest losses could be determined despite increased output capacitance and lower $R_{ds,on}$. The increased switching losses at 400 V and 1.3 A for the GaN (a) semiconductors, which were determined at higher frequencies compared to the 3 A and 7 A values, are particularly noticeable. Based on these results, it can be assumed that by measuring switching energies at different switching frequencies, the effect of the dynamic on-resistance, in particular of GaN power transistors can be shown. Despite similar operating points for the GaN (b) semiconductors, this behavior does not occur with this transistor. In principle, however, changes in total losses due to dynamic on-resistance should be visible. A more detailed investigation in this regard requires more measurement points over a larger frequency spectrum.

IV. CONCLUSION

Calorimetric power loss measurements were performed to compare the soft-switching losses of various GaN and SiC high voltage power transistors for the target application of a 150 W ac/dc charger. Due to the low transistor losses of this power class, the thermal measurement setup and previous calorimetric measurement methods had to be adjusted. By placing the temperature sensor close to the transistor and calibrating it to the thermal impedance of the power elec-

tronic system, the switching energies for the DUTs could be determined for different operating points. The parasitic influence of the driver circuit was taken into account to allow for a application-oriented analysis without large half bridge layout adjustments being necessary. The calorimetric switching energy results suggest that besides the comparison of the transistors for the target application also a further investigation of the dynamic $R_{ds,on}$ effect might be possible.

REFERENCES

[1] K. Surakitbovorn and J. R. Davila, "Evaluation of gan transistor losses at mhz frequencies in soft switching converters," in *2017 IEEE 18th Workshop on Control and Modeling for Power Electronics (COMPEL)*. IEEE, 2017, pp. 1–6.

[2] M. Guacci, M. Heller, D. Neumayr, D. Bortis, J. W. Kolar, G. Deboy, C. Ostermaier, and O. Häberlen, "On the origin of the coss-losses in soft-switching gan-on-si power hemts," *IEEE Journal of Emerging and Selected Topics in Power Electronics*, 2018.

[3] J. A. Anderson, C. Gammeter, L. Schrittwieser, and J. W. Kolar, "Accurate calorimetric switching loss measurement for 900 v 10 mΩ sic mosfets," *IEEE Transactions on Power Electronics*, vol. 32, no. 12, pp. 8963–8968, 2017.

[4] H. Li, X. Li, Z. Zhang, J. Wang, L. Liu, and S. Bala, "A simple calorimetric technique for high-efficiency gan inverter transistor loss measurement," in *2017 IEEE 5th Workshop on Wide Bandgap Power Devices and Applications (WiPDA)*. IEEE, 2017, pp. 251–256.

[5] S. Tiwari, J. K. Langelid, O.-M. Midtgård, and T. M. Undeland, "Soft switching loss measurements of a 1.2 kv sic mosfet module by both electrical and calorimetric methods for high frequency applications," in *2017 19th European Conference on Power Electronics and Applications (EPE'17 ECCE Europe)*. IEEE, 2017, pp. 1–10.

[6] D. Neumayr, M. Guacci, D. Bortis, and J. W. Kolar, "New calorimetric power transistor soft-switching loss measurement based on accurate temperature rise monitoring," in *2017 29th International Symposium on Power Semiconductor Devices and IC's (ISPSD)*. IEEE, 2017, pp. 447–450.

[7] D. Rothmund, D. Bortis, and J. W. Kolar, "Accurate transient calorimetric measurement of soft-switching losses of 10-kv sic mosfets and diodes," *IEEE Transactions on Power Electronics*, vol. 33, no. 6, pp. 5240–5250, 2018.

[8] A. Anurag, S. Acharya, Y. Prabowo, G. Gohil, H. Kassa, and S. Bhattacharya, "An accurate calorimetrie method for measurement of switching losses in silicon carbide (sic) mosfets," in *2018 IEEE Applied Power Electronics Conference and Exposition (APEC)*. IEEE, 2018, pp. 1695–1700.

[9] L. Hoffmann, C. Gautier, S. Lefebvre, and F. Costa, "Optimization of the driver of gan power transistors through measurement of their thermal behavior," *IEEE Transactions on Power Electronics*, vol. 29, no. 5, pp. 2359–2366, 2014.

978-1-7281-0582-6/19 $31.00 © 2019 IEEE

Proceedings of the 31st International Symposium on Power Semiconductor Devices & ICs
May 19-23, 2019, Shanghai, China

Charge-Modulated Schottky Barrier Lowering Effect in GaN Double-Channel Lateral Power SBDs with Gated Anode

Jiacheng Lei, Jin Wei, Gaofei Tang, Zhaofu Zhang, Qingkai Qian, Mengyuan Hua,
Zheyang Zheng, Yuru Wang, and Kevin J. Chen

Department of Electronic and Computer Engineering
The Hong Kong University of Science and Technology
Hong Kong SAR, CHINA
E-mail: jleiaa@connect.ust.hk, eekjchen@ust.hk

Abstract—The charge-modulated Schottky barrier lowering (SBL) effect is revealed and studied on an AlGaN/GaN lateral SBD with gated anode. The fabricated SBD features a metal-2DEG Schottky contact and a leakage suppression MIS field plate to provide both low turn-on voltage (V_T) and low reverse leakage current (I_R). It is revealed that the SBDs with higher carrier density in the channel under the MIS field plate (MIS-channel) exhibits lower V_T, higher I_R and lower Schottky barrier height (φ_B, which is extracted from the forward I-V curves based on the thermionic emission model). The φ_B of the metal-GaN junction can be lowered by the image charge in the metal electrode. A higher carrier density in the MIS-channel yields a narrower Schottky barrier, which benefits a more prominent image force induced SBL effect and leads to a lower φ_B.

Keywords—*AlGaN/GaN, double-channel, double-recessed, image charge, metal-2DEG anode, Schottky barrier diode, Schottky barrier lowering effect.*

I. INTRODUCTION

GaN based power devices are promising for next generation power switching applications, since these devices could provide low loss and high power density [1]. As two terminal power devices, the AlGaN/GaN based lateral rectifiers (including the Schottky barrier diodes and field-effect rectifiers) have drawn much attention [2]–[7]. In the fabrication of the power Schottky barrier diode (SBD), major efforts have been made to deliver a SBD with both low turn-on voltage (V_T) and low reverse leakage current (I_R). A commonly used approach is by employing a planar anode with a MIS field plate deployed on partially recessed AlGaN barrier [4], [6]. The partially recessed AlGaN barrier benefits the carrier-transport so that a reduced V_T is ensured; while the MIS field plate can pinch-off the channel underneath and shield the Schottky contact from strong electric field [8], leading to a suppressed I_R. Another widely employed anode architecture is the metal-2DEG anode [3], [5], in which the Schottky metal contacts 2DEG directly from the exposed GaN sidewall. This absence of an AlGaN barrier between the Schottky metal and GaN delivers a significant benefit in

This work is supported by Hong Kong Innovation and Technology Fund under grant ITS/234/16.

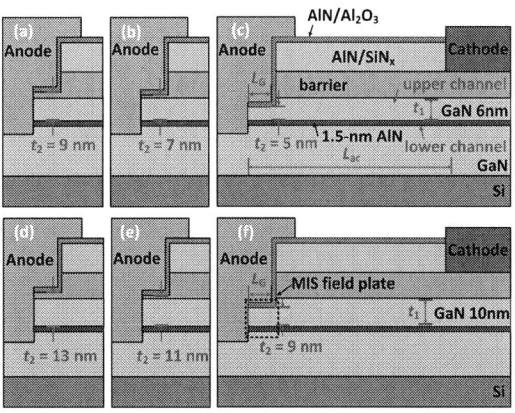

Fig. 1: Schematic cross-section of the GaN double-channel SBDs. The SBD (a) (b) and (c) are fabricated on a heterojunction with 6-nm upper GaN channel layer, while the SBD (d) (e) and (f) are with 10-nm upper channel. The SBDs feature multiple cycles of digital-etch conducted on the MIS field plate region, and the thickness of the remaining barrier/GaN upper channel layer (t_2) is as shown.

reduced V_T.

Recently, an AlGaN/GaN double-channel lateral power SBD with gated anode was demonstrated [7]. The device features a double-recess process so that a metal-2DEG Schottky contact and a leakage suppression MIS field plate are achieved simultaneously (Fig. 1 (c)), leading to low V_T, low I_R and high breakdown voltage.

In this work, it is experimentally revealed that the Schottky barrier height (φ_B, extracted from the forward I-V based on a thermionic emission model) of the AlGaN/GaN double-channel metal-2DEG SBD exhibits a strong dependence on the thickness of the remaining barrier layer at the MIS field plate region. A thicker remaining AlGaN barrier features a stronger polarization effect than a thinner barrier, and introduces more carriers in the channel under the MIS field plate (MIS-channel), which facilitates a more prominent image force induced Schottky barrier lowering (SBL) effect [9]–[12] and leads to a lower φ_B. In the design

978-1-7281-0582-6/19 $31.00 © 2019 IEEE   459

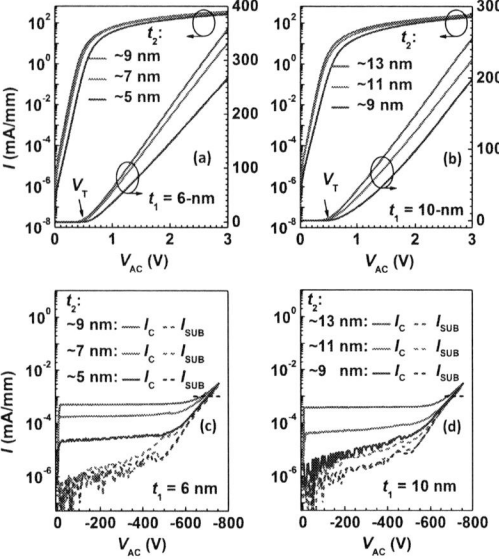

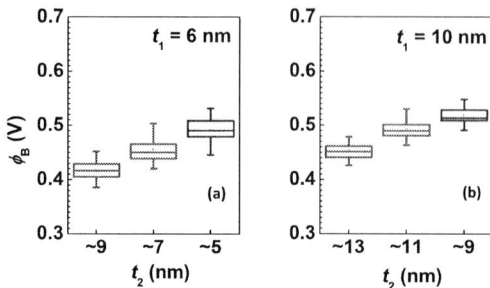

Fig. 3: The statistics of the barrier height are illustrated in (a) and (b). ~20 devices are measured for each type of SBDs. A deeper recess depth (i.e. lower t_2) exhibits a higher barrier height.

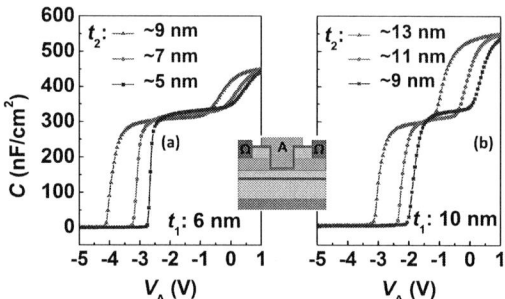

Fig. 4: The representative C-V characteristics of the MIS field plate with (a) 6-nm and (b) 10-nm upper GaN channel. The C-V curves are measured on MIS-diodes. With the t_2 decreasing, the C-V curves exhibit a positive shift (i.e. the increasing of the V_p). The recess on barrier layer results in a reduced carrier density.

Fig. 2: The representative forward I-V characteristics of the double-channel SBDs with (a) 6-nm and (b) 10-nm upper GaN channel. The high voltage reverse I-V curves are shown in (c) and (d). The anode-to-cathode distance (L_{AC}) and length of MIS field plate (L_G) is 15 µm and 1.5 µm, respectively. With the reduction of t_2, the V_T increases and the reverse leakage current decreases.

and fabrication of the AlGaN/GaN lateral SBDs with the metal-2DEG anode, a desired Schottky barrier height can be achieved by adjusting the charge density in the adjacent MIS-channel through AlGaN barrier recess-depth control.

II. EXPERIMENT

The SBDs were fabricated on a GaN-on-Si AlGaN/GaN double-channel heterojunction grown by MOCVD. The epi-structure consists of a barrier layer (including 3-nm GaN cap, 17-nm AlGaN and 1.5-nm AlN), a 6-nm/10-nm upper GaN channel layer, a 1.5-nm AlN-insertion-layer and a 4-µm GaN buffer/transition layer. The upper channel and lower channels is below the barrier layer and AlN-insertion-layer, respectively. The heterostructure with upper GaN channel layer of 6 nm/10 nm exhibits a total 2DEG density of $8.6 \times 10^{12}/8.9 \times 10^{12}$ cm^{-2}, and an overall carrier mobility of 2080/1811 cm^2/V·s, which are determined by Hall measurement [13]–[15]. The SBDs were fabricated with the similar process in reference [7].

Device fabrication started with AlN/SiN$_x$ passivation stack deposition in sequence by PEALD/PECVD (plasma enhanced atomic layer deposition/plasma enhanced chemical vapor deposition) [16]. After the removal of passivation stack at the cathode region, the Ti/Al/Ni/Au ohmic contacts were patterned and formed. Subsequently, a planar isolation was performed using F$^-$ ion implantation. The anode formation features two recess steps to form the MIS field plate and metal-2DEG Schottky contact:

(1) Formation of leakage suppression MIS field plate: After the anode region passivation removal (anode-window-1), the 1st recess (shallow recess) was performed using multiple cycles of digital-etch [17], [18]. The recess stopped at the

partially recessed barrier (or upper GaN channel layer). Before the deposition of PEALD-grown Al$_2$O$_3$ dielectric to form the MIS-structure, a PEALD-grown AlN-interfacial-layer (~1 nm) was deposited so that a normally-on MIS-channel was ensured.

(2) The fabrication of the metal-2DEG Schottky contact: After another anode-window (anode-window-2) was defined inside the anode-window-1, the exposed AlN/Al$_2$O$_3$ was removed by diluted KOH. Then a dry etch (2nd recess) employing ICP-RIE (inductively coupled plasma reactive ion etching) using BCl$_3$/Cl$_2$ mixture plasma was performed to expose the sidewall of the GaN channel. Subsequently, the Ni/Au Schottky metal was evaporated and patterned, contacting the GaN channel from the sidewall and forming the metal-2DEG anode. Finally, contacting pads were formed.

Six types of devices were fabricated and their schematic cross-sections are shown in Fig. 1. The recess depth is determined by AFM (Atomic Force Microscope) during device fabrication, and the distance from the MIS-dielectric to the AlN-insertion-layer is determined accordingly.

III. DEVICE CHARACTERIZATION AND DISCUSSION

The I-V characteristics of the GaN double-channel SBDs are as shown in Fig. 2. The devices feature an anode-to-cathode distance and MIS field plate length of 15 µm and 1.5 µm, respectively. All of the SBDs show similar

978-1-7281-0582-6/19 $31.00 © 2019 IEEE

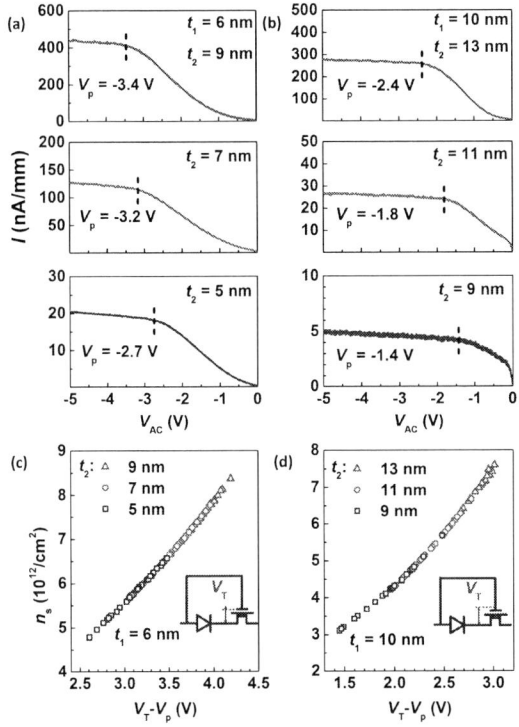

Fig. 5: The representative reverse I-V characteristics of the SBDs with (a) 6-nm and (b) 10-nm upper GaN channel. The V_p of the MIS field plate is extracted from the reverse I-V curves. The plots of the V_T-V_p against n_s of a large amount of SBDs are shown in (c) and (d). The plots are obtained by integrating the C-V from V_p to V_T.

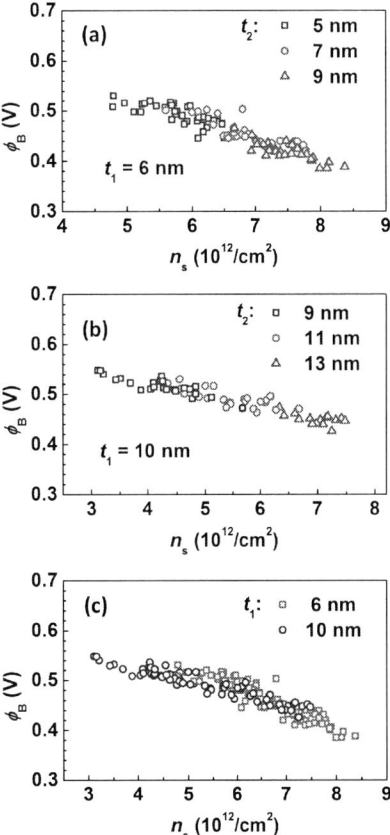

Fig. 6: The φ_B-n_s plot of the SBDs fabricated on heterojunction with (a) 6-nm and (b) 10-nm upper GaN channel. (c) φ_B-n_s plot of the two samples. A higher carrier density in MIS-channel exhibits a lower Schottky barrier height.

breakdown voltage (BV) of >650 V at a drain current of 1 μA/mm with a backside-to-anode connection [19], and the substrate leakage dominates the drain current at breakdown. Among the devices fabricated on the heterojunction with the same upper GaN channel thickness (i.e. with the same t_1), a thicker remaining barrier/upper GaN channel (i.e. larger t_2) results in a lower turn-on voltage (V_T) and a higher reverse leakage current (I_R). The statistical results in Fig. 3 exhibits that a thicker remaining barrier also results in a lower Schottky barrier height (φ_B) that is extracted using thermionic emission model.

The carrier density (n_s) under the MIS field plate should be extracted so that the n_s dependent Schottky barrier height can be studied. The MIS field plate adjacent to the Schottky contact is characterized by the C-V measurement on a MIS-diode with a radius of 80 μm (Fig. 4). A deeper etch depth of the 1st recess leads to a positive shift on the C-V curves, indicating a less negative threshold voltage and a lower carrier density in the MIS-channel. When the SBD is at reverse OFF-state, the carriers under the MIS field plate are depleted once the reverse bias reaches the pinch-off voltage of the MIS-channel (V_p). When the SBD is at forward ON-state, the voltage between the Schottky metal and MIS-channel is the turn-on voltage (V_T) of the SBD. Thus, the n_s in the MIS-channel can be calculated by integrating the C-V curves from V_p to V_T. The V_p is extracted from the reverse I-V curve as illustrated in Fig. 5 (a) and (b). Fig. 5 (c) and (d)

plot the MIS-channel carrier density of an amount of fabricated SBDs. The corresponding φ_B of these SBDs are also extracted from the forward I-V curves. As illustrated by the φ_B-n_s plot in Fig. 6, a higher n_s in the MIS-channel leads to a lower φ_B.

This carrier modulated Schottky barrier lowering (SBL) effect is depicted in the Fig. 7. The schematic band diagram under the MIS field plate and at the Schottky junction of a SBD with a higher MIS-channel carrier density are illustrated in Fig. 7 (a) and (b); while the schematic band diagram of a SBD with a lower MIS-channel carrier density are shown in Fig. 7 (c) and (d). For simplicity, the dielectric (ALD-grown Al_2O_3 and AlN) and remaining barrier (AlGaN and AlN-insertion-layer) are illustrated as two layers. The Schottky barrier height at the metal-2DEG junction can be affected by the image charge in the metal electrode as the electrons approach the interface of the Schottky contact. This image charge induced potential is described by $\varphi=q/(16\pi\varepsilon x)$, where ε is the dielectric constant of the GaN channel and x is the distance from the Schottky contact interface. With electrons approaching the metal-GaN interface, the image charge induced potential drops rapidly, which results in a

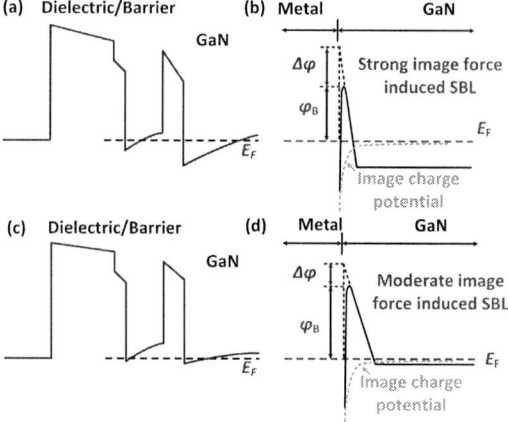

Fig. 7: Schematic band diagram at (a) the MIS field plate and (b) the metal-2DEG junction with a higher carrier density under the field plate. (c) (d) Schematic band diagram with a lower carrier density. More carriers in MIS-channel lead to a narrower barrier, facilitating the image force induced barrier lowing effect across the Schottky junction.

significant reduction on the conduction band of the GaN channel. A higher carrier density in the MIS-channel results in a narrower initial Schottky barrier (i.e. the Schottky barrier without considering the image charge), facilitating the image force induced Schottky barrier lowering (SBL) effect.

IV. CONCLUSIONS

The charge-modulated Schottky barrier lowering Effect in GaN double-channel SBDs with metal-2DEG anode is revealed and studied. The conduction band of the GaN channel can be lowered by the image charge at the metal-GaN interface, leading to a reduced Schottky barrier height at forward bias. Compared to the SBD with a lower carrier density in the MIS-channel, the SBD with more carriers in the MIS-channel exhibits a narrower initial Schottky barrier that results in a more prominent image charge induced barrier lowering effect at the junction. In designing and fabricating of the AlGaN/GaN lateral metal-2DEG SBDs, a desired Schottky barrier height can be obtained by adjusting the charge density in the channel under the leakage suppression MIS field plate.

REFERENCES

[1] K. J. Chen, O. Häberlen, A. Lidow, C. l Tsai, T. Ueda, Y. Uemoto, Y. Wu, "GaN-on-Si Power Technology: Devices and Applications," *IEEE Trans. Electron Devices,* vol. 64, no. 3, pp. 779–795, Mar. 2017.

[2] Wanjun Chen, King-Yuen Wong, Wei Huang, and Kevin J. Chen, "High-performance AlGaN/GaN lateral field-effect rectifiers compatible with high electron mobility transistors," *Appl. Phys. Lett.,* vol. 92, no. 25, p. 253501, Jun. 2008.

[3] E. Bahat-Treidel, O. Hilt, R. Zhytnytska, A. Wentzel, C. Meliani, J. Wurfl, G. Trankle, "Fast-Switching GaN-Based Lateral Power Schottky Barrier Diodes With Low Onset Voltage and Strong Reverse Blocking," *IEEE Electron Device Lett.,* vol. 33, no. 3, pp. 357–359, Mar. 2012.

[4] S. Lenci, B. De Jaeger, L. Carbonell, J. Hu, G. Mannaert, , D. Wellekens, S. You, B. Bakeroot, S. Decoutere, "Au-Free AlGaN/GaN Power Diode on 8-in Si Substrate With Gated Edge Termination," *IEEE Electron Device Lett.,* vol. 34, no. 8, pp. 1035–1037, Aug. 2013.

[5] M. Zhu, B. Song, M. Qi, Z. Hu, K. Nomoto, X. Yan, Y. Cao, W. Johnson, E. Kohn, D. Jena, H. G. Xing, "1.9-kV AlGaN/GaN Lateral Schottky Barrier Diodes on Silicon," *IEEE Electron Device Lett.,* vol. 36, no. 4, pp. 375–377, Apr. 2015.

[6] J. Hu, S. Stoffels, S. Lenci, B. Bakeroot, B. De Jaeger, M. Van Hove, N. Ronchi, R. Venegas, H. Liang, M. Zhao, G. Groeseneken, S. Decoutere, "Performance Optimization of Au-Free Lateral AlGaN/GaN Schottky Barrier Diode With Gated Edge Termination on 200-mm Silicon Substrate," *IEEE Trans. Electron Devices,* vol. 63, no. 3, pp. 997–1004, Mar. 2016.

[7] J. Lei, J. Wei, G. Tang, Z. Zhang, Q. Qian, Z. Zheng, M. Hua, K. J. Chen, "650-V Double-Channel Lateral Schottky Barrier Diode With Dual-Recess Gated Anode," *IEEE Electron Device Lett.,* vol. 39, no. 2, pp. 260–263, Feb. 2018.

[8] J. Ma, D. C. Zanuz, and E. Matioli, "Field Plate Design for Low Leakage Current in Lateral GaN Power Schottky Diodes: Role of the Pinch-off Voltage," *IEEE Electron Device Lett.,* vol. 38, no. 9, pp. 1298–1301, Sep. 2017.

[9] L. Yuan, H. Chen, and K. J. Chen, "Normally Off AlGaN/GaN Metal–2DEG Tunnel-Junction Field-Effect Transistors," *IEEE Electron Device Lett.,* vol. 32, no. 3, pp. 303–305, Mar. 2011.

[10] L. Yuan, H. Chen, Q. Zhou, C. Zhou, and K. J. Chen, "A novel normally-off GaN power tunnel junction FET," in *Proc. IEEE Int. Symp. Power Semiconductor Device ICs (ISPSD),* May. 2011, pp. 276–279.

[11] L. Yuan, H. Chen, Q. Zhou, C. Zhou, and K. J. Chen, "Gate-Induced Schottky Barrier Lowering Effect in AlGaN/GaN Metal–2DEG Tunnel Junction Field Effect Transistor," *IEEE Electron Device Lett.,* vol. 32, no. 9, pp. 1221–1223, Sep. 2011.

[12] Y. Yao, Jian Zhong, Yue Zheng, Fan Yang, Yiqiang Ni, Zhiyuan He, Zhen Shen, Guilin Zhou, Shuo Wang, Jincheng Zhang, Jin Li, Deqiu Zhou, Zhisheng Wu, Baijun Zhang, Yang Liu, "Current transport mechanism of AlGaN/GaN Schottky barrier diode with fully recessed Schottky anode," *Jpn. J. Appl. Phys.,* vol. 54, no. 1, p. 011001, Jan. 2015.

[13] J. Wei, S. Liu, B. Li, X. Tang, Y. Lu, C. Liu, M. Hua, Z. Zhang, G. Tang, K. J. Chen, "Enhancement-mode GaN double-channel MOS-HEMT with low on-resistance and robust gate recess," in *IEDM Tech. Dig.,* Dec. 2015, pp. 9.4.1-9.4.4.

[14] J. Wei, S. Liu, B. Li, X. Tang, Y. Lu, C. Liu, M. Hua, Z. Zhang, G. Tang, K. J. Chen, "Low On-Resistance Normally-Off GaN Double-Channel Metal-Oxide-Semiconductor High-Electron-Mobility Transistor," *IEEE Electron Device Lett.,* vol. 36, no. 12, pp. 1287–1290, Dec. 2015.

[15] J. Wei, J. Lei, X. Tang, B. Li, S. Liu, and K. J. Chen, "Channel-to-Channel Coupling in Normally-Off GaN Double-Channel MOS-HEMT," *IEEE Electron Device Lett.,* vol. 39, no. 1, pp. 59–62, Jan. 2018.

[16] Z. Tang, S. Huang, Q. Jiang, S. Liu, C. Liu, and K. J. Chen, "High-Voltage (600-V) Low-Leakage Low-Current-Collapse AlGaN/GaN HEMTs with AlN/SiNx Passivation," *IEEE Electron Device Lett.,* vol. 34, no. 3, pp. 366–368, Mar. 2013.

[17] Y. Wang, M. Wang, B. Xie, C. P. Wen, J. Wang, Y. Hao, W. Wu, K. J. Chen, B. Shen, "High-Performance Normally-Off Al2O3/GaN MOSFET Using a Wet Etching-Based Gate Recess Technique," *IEEE Electron Device Lett.,* vol. 34, no. 11, pp. 1370–1372, Nov. 2013.

[18] S. Liu, S. Yang, Z. Tang, Q. Jiang, C. Liu, M. Wang, K. J. Chen, "Al2O3/AlN/GaN MOS-Channel-HEMTs With an AlN Interfacial Layer," *IEEE Electron Device Lett.,* vol. 35, no. 7, pp. 723–725, Jul. 2014.

[19] J. A. Croon, G. A. M. Hurkx, J. J. T. M. Donkers, and J. Šonský, "Impact of the backside potential on the current collapse of GaN SBDs and HEMTs," in *Proc. IEEE Int. Symp. Power Semiconductor Device ICs (ISPSD),* May 2015, pp. 365–368.

978-1-7281-0582-6/19 $31.00 © 2019 IEEE

Proceedings of the 31st International Symposium on Power Semiconductor Devices & ICs
May 19-23, 2019, Shanghai, China

Characterization of Dynamic I_{OFF} in Schottky-Type p-GaN Gate HEMTs

Yuru Wang, Jin Wei, Song Yang, Jiacheng Lei, Mengyuan Hua, and Kevin J. Chen

Deparement of Electronic and Computer Engineering
The Hong Kong University of Science and Technology
Hong Kong SAR, CHINA
E-mail: ywangfk@connect.ust.hk

Abstract—In this work, systematic characterization of dynamic OFF-state leakage current (I_{OFF}) in Schottky-type p-GaN gate high-electron-mobility transistors (HEMTS) is presented based on pulsed I-V measurement and consecutive switching measurement. The high I_{OFF} under dynamic pulse mode without hole injection is found to be a result of the reduced voltage blocking capabilities (both lateral and vertical) with weaker trapping effect in the buffer, and the dynamic I_{OFF} induced by ON-state hole injection is attributed to further increased lateral conductivity through the buffer from source to drain. Under continuous ON/OFF switching operation, saturation of dynamic I_{OFF} is observed due to a balanced trapping/de-trapping process of buffer traps. Higher temperature is found to be beneficial to the reduction of the dynamic I_{OFF} induced by ON-state hole injection, and a sufficiently large negative OFF-state gate bias ($V_{GS,OFF}$) of -3 V is shown to completely eliminate the dynamic I_{OFF} induced by hole injection to minimize the OFF-state power consumption in practical power switching applications.

Keywords—*Schottky-type, p-GaN gate HEMTs, dynamic I_{OFF}, characterization, ON-state hole injection.*

I. INTRODUCTION

GaN based high-electron mobility transistors (HEMTs), with a p-GaN gate structure have become the first single-chip enhance-mode (E-mode) GaN power devices to be commercialized, with a good balance among performance, reliability, and cost [1]-[3]. There are two approaches to realize the p-GaN gate HEMTs. One is to adopt an Ohmic contact to the p-GaN gate [4]. This type of p-GaN gate HEMT has been developed under the concept of GIT (gate injection transistor) and has shown impressive performance [5] and reliability [6]. This device is current driving [7], and features larger gate driver power consumption and compromising speed. The other approach is to use a Schottky gate contact to p-GaN [8]. The Schottky-type p-GaN gate HEMT is a voltage driven device and features the virtues of reduced gate leakage current and enlarged gate swing [8]-[10], but has to pay the price of a dynamic V_{TH} due to the floating nature of the p-GaN layer [11], [12].

Apart from the distinct gate characteristics, the corresponding dynamic behaviors of GaN-on-Si power HEMTs under switching operations have been the major focus. Extensive investigation has been carried out to understand and reduce the current collapse induced dynamic ON-resistance (R_{ON}) [5], [13]-[15]. There are less studies on other application-relevant dynamic behaviors such as the dynamic behavior of the OFF-state leakage current (I_{OFF}). Systematic characterization of dynamic I_{OFF} can provide an

This work was supported by the Hong Kong Innovation and Technology Fund under Grant ITS/412/17FP.

accurate estimation on I_{ON}/I_{OFF} ratio and OFF-state power consumption in high-speed switching applications for GaN power devices.

Recently, we reported the dynamic I_{OFF} behaviors in the p-GaN gate injection transistors (GITs) [16]. It is found that the I_{OFF} under dynamic pulse-mode operation is many orders of magnitude higher than the I_{OFF} obtained in the quasi-static measurement, and dynamic I_{OFF} is further enhanced with ON-state hole injection through the gate. The dynamic I_{OFF} behavior is explained by the dynamic trapping/de-trapping processes of the traps in the buffer. However, the corresponding dynamic I_{OFF} characteristics and especially the behaviors under continuous switching operation are still lacking for the commercially available Schottky-type p-GaN gate HEMTs.

In this work, a comprehensive characterization and evaluation of dynamic I_{OFF} in the Schottky-type p-GaN gate HEMTs was carried out with fast pulsed I-V measurement and consecutive switching measurement. The detailed dynamic I_{OFF}-V_{DS} characteristics with and without hole injections are illustrated to clarify the underlying physical mechanisms of the high dynamic I_{OFF}. The corresponding behaviors under continuous switching operation with different switching conditions are analyzed to identify suitable gate driving conditions. It is found that applying negative turn-off gate bias is an effective approach to suppressing the high dynamic I_{OFF} and lower the OFF-state power consumption in practical switching applications.

II. DYNAMIC I_{OFF}-V_{DS} CHARACTERISTICS

The device used in this work is commercially available 650-V/7.5-A Schottky-type p-GaN gate HEMTs [17]. From

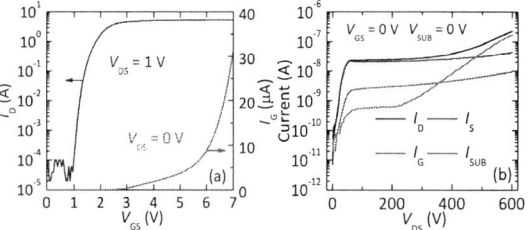

Fig. 1. Static I-V characteristics of a Schottky-type p-GaN gate HEMT: (a) Transfer (V_{DS} = 1 V) and I_G-V_{GS} (V_{DS} = 0 V) characteristics; (b) Quasi-static I_{OFF}-V_{DS} characteristics with a gate bias of 0 V. The Si substrate is grounded.

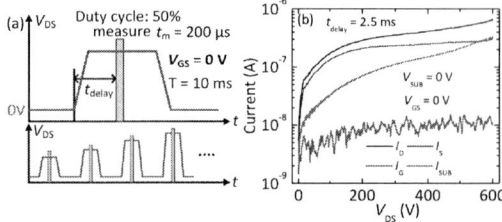

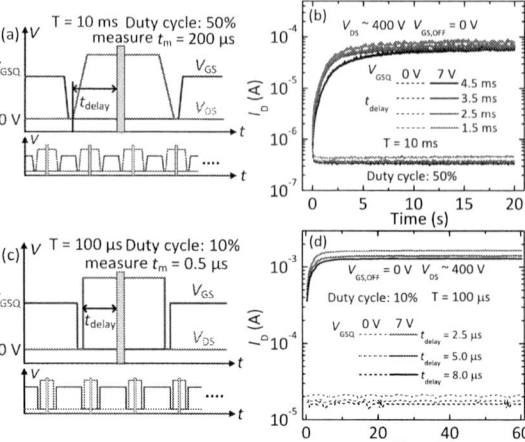

Fig. 2. (a) Pulsed V_{DS} waveforms for the measurement of dynamic I_{OFF} without gate hole injection (V_{GSQ} = 0 V) by Agilent B1505A. (b) Dynamic I_{OFF}-V_{DS} characteristics with t_{delay} = 2.5 ms and a gate bias of 0 V. The Si substrate is grounded.

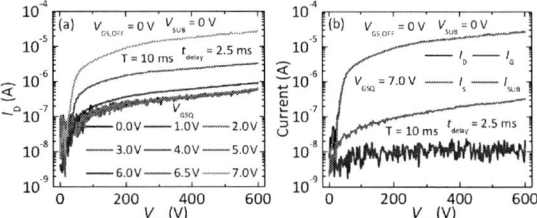

Fig. 3. (a) Dynamic I_{OFF} measured using Agilent B1505A with t_{delay} = 2.5 ms, period T = 10 ms, $V_{GS,OFF}$ = 0 V, V_{SUB} = 0 V, and different V_{GSQ} under soft-switching waveforms. (b) Dynamic I_{OFF}-V_{DS} characteristics with gate hole injection with t_{delay} = 2.5 ms, T = 10 ms, and V_{GSQ} = 7 V.

the static I-V characteristics (Fig. 1), the device features a threshold voltage (V_{TH}) of 1.2 V (at V_{DS} = 1 V, I_D = 10 mA), and a static R_{ON} around 185 mΩ (at V_{GS} = 7 V, V_{DS} = 1 V, yielding an ON-state current of 5.4 A). The forward ON-state I_G is ~ 31 μA at V_{GS} of 7 V and the quasi-static I_{OFF} is ~ 40 nA at V_{DS} of 400 V.

The dynamic I_{OFF}-V_{DS} characteristics were measured using a power semiconductor device parameter analyzer Agilent B1505A. The measurements were conducted with a pulsed sweeping V_{DS} with and without gate hole injections (determined by the quiescent gate bias, V_{GSQ}), and the dynamic I_{OFF} was measured during each pulse of V_{DS} with a preset measurement delay time, t_{delay}. Differing from the measurement in the slow-ramping quasi-static mode (the slow ramping rate allows sufficient time for the empty traps in the buffer to capture electrons and the energy barrier in the leakage path is raised to block the leakage current in the OFF-state), the I_{OFF} under fast dynamic pulse mode is measured in each pulse before the buffer trap filling process can be completed. The overall negative space charge in the buffer is less and the barrier height in the leakage path (both lateral and vertical) is lower than that in the quasi-static mode, resulting in reduced voltage blocking capabilities in the buffer and larger dynamic I_{OFF} [16]. As a result, dynamic I_{OFF} even without hole injection at V_{DS} = 400 V dramatically increases to 0.4 μA (i.e., an order of magnitude increase from the static I_{OFF} of 40 nA) with a t_{delay} of 2.5 ms, and the drastic change in I_{OFF} is not coming from I_G but originates from increase in both the lateral drain-to-source leakage current and vertical drain-to-substrate leakage current, as shown in the Fig. 2.

To investigate the influence of ON-state hole injection on dynamic I_{OFF}, the measurement started with ON-OFF soft-switching cycles using Agilent B1505A. During each switching cycle, the device was first stressed in the ON-state with a gate bias of V_{GSQ}, and then switched to the OFF-state for dynamic I_{OFF} measurement with V_{DS} varying from 0 V up

to 600 V. As shown in the Fig. 3(a), with a large V_{GSQ} (> 5 V), the further increase of dynamic I_{OFF} induced by ON-state hole injection is observed in the Schottky-type p-GaN gate HEMTs. With a sufficiently large gate bias, holes can be injected from the p-GaN layer to the 2DEG channel. These holes could either recombine with electrons in the channel to emit photons of 3.4 eV and pump electrons from the buffer traps [18], or further transport under the influence of the vertical E-field to compensate the buffer traps with electrons [16]. Both processes could lead to reduced negative space charge under the gate in the buffer layer compared to the case without hole injection. As the device is switched back to the OFF-state, the hole injection is turned off immediately while the ionized buffer traps would persist for a period of time before recapturing electrons [16], [19]. As a result, the energy barrier in the lateral leakage path is lowered than that when hole injection is absent, leading to the increased lateral buffer conductivity from source to drain and larger drain-to-source leakage current. The increase in I_{OFF} induced by hole injection is from 0.4 μA to 20 μA (~50 times) at V_{DS} of 400 V with a V_{GSQ} = 7 V. The further increased dynamic I_{OFF} induced by ON-state hole injection is not coming from the I_G and I_{SUB}, and is dominated by the lateral drain-to-source leakage current, as shown in the Fig. 3(b).

III. DYNAMIC I_{OFF} UNDER CONTINUOUS SWITCHING OPERATION

Aiming at evaluating the dynamic I_{OFF} under practical switching operations, the time-dependent continuous soft-switching waveforms at V_{DS} of 400 V were measured using Agilent B1505A for slow dynamic measurement (Fig. 4(a)) and high-speed AMCAD pulse I-V system for fast dynamic measurement (Fig. 4(c)). Similar phenomenon could be observed in both slow and fast dynamic measurements. Under continuous soft-switching operation, dynamic I_{OFF} increases in the beginning, and then saturates at a certain level as a result of the balanced trapping/de-trapping process of buffer traps, shown in the Fig. 4(b) and Fig. 4(d). With a sufficiently large forward gate bias (V_{GSQ} = 7 V), the

Fig. 4. Continuous soft-switching waveforms for the measurement of dynamic I_{OFF} by (a) Agilent B1505A with slow pulse and (c) AMCAD pulse I-V system with fast pulse. Time evolution of dynamic I_{OFF} under corresponding continuous soft-switching measured by (b) Agilent B1505A and (d) AMCAD pulse I-V system with V_{DS} ~ 400 V, V_{GSQ} = 0/7 V, $V_{GS,OFF}$ = 0 V and different t_{delay}.

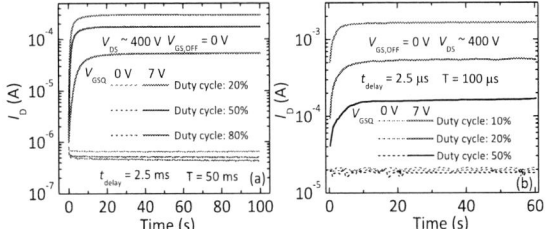

Fig. 5. Dynamic I_{OFF} under (a) slow and (b) fast continuous soft-switching with $V_{DS} \sim 400$ V, $V_{GS,OFF} = 0$ V, $V_{GSQ} = 0/7$ V and different OFF-state duty cycle.

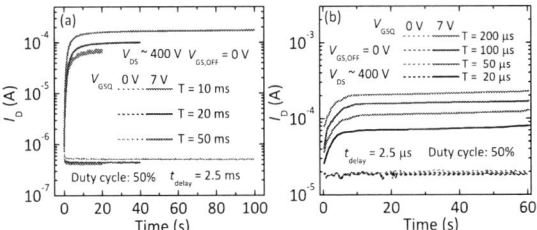

Fig. 6. Dynamic I_{OFF} under (a) slow and (b) fast continuous soft-switching with $V_{DS} \sim 400$ V, $V_{GS,OFF} = 0$ V, $V_{GSQ} = 0/7$ V and an OFF-state duty cycle of 50% with different period (frequency).

negative space charge in the buffer layer is reduced by the ON-state hole injection, leading to a lowered energy barrier in the leakage path and larger dynamic I_{OFF}. The continuous decrease of negative space charge in each cycle leads to the increase of dynamic I_{OFF}. After certain number of switching cycles, a balanced process between ionizing electrons from the buffer traps in the ON-state and recapturing electrons of the ionized buffer traps in the OFF-state is achieved, leading to a constant negative space charge in the buffer and a saturation of dynamic I_{OFF} with a certain t_{delay}. With a much shorter t_{delay} ($\sim \mu$s), the dynamic I_{OFF} is increased further by more than one order of magnitude as less electrons can be effectively recaptured by the buffer traps during each switching cycle.

Fig. 5(a) and (b) show the dynamic I_{OFF} under slow and fast continuous soft-switching with $V_{DS} \sim 400$ V, $V_{GS,OFF} = 0$ V, $V_{GSQ} = 0/7$ V and different OFF-state duty cycle. A smaller OFF-state duty cycle leads to a longer ON-state duration time in each cycle with stronger hole injection, resulting in the higher dynamic I_{OFF}. With the increase of pulse period while keeping the OFF-state duty cycle and t_{delay} unchanged, the ON-state duration time in each cycle also increases, resulting in stronger ON-state hole injection and higher dynamic I_{OFF}, as shown in the Fig. 6. With $t_{delay} = 2.5$ μs, T = 100 μs and an OFF-state duty cycle of 10%, dynamic I_{OFF} with hole injection ($V_{GSQ} = 7$ V) at $V_{DS} \sim 400$ V dramatically increases to about 2 mA (i.e., a 50,000 times increase from the static I_{OFF} of 40 nA), and the OFF-state power loss is estimated to reach $\sim 10\%$ of the ON-state conduction loss.

The temperature-dependent dynamic I_{OFF} with and without hole injections are plotted in the Fig. 7. Without hole injection ($V_{GSQ} = 0$ V), an increase in temperature could further enhance electron emission over the energy barrier in the leakage path, and leads to the increase of dynamic I_{OFF}

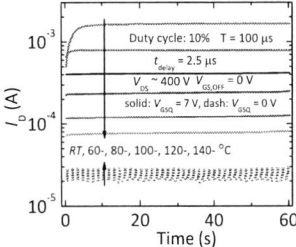

Fig. 7. Temperature dependence of dynamic I_{OFF} under fast continuous soft-switching with $V_{DS} \sim 400$ V, $V_{GS,OFF} = 0$ V, T = 100 μs, $t_{delay} = 2.5$ μs, $V_{GSQ} = 0/7$ V and an OFF-state duty cycle of 10%.

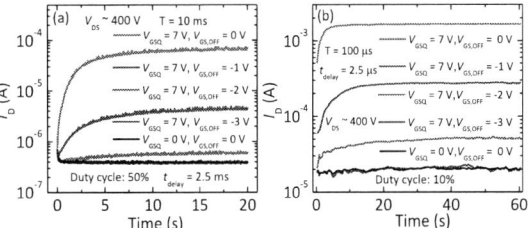

Fig. 8. Dynamic I_{OFF} under (a) slow and (b) fast continuous soft-switching with $V_{DS} \sim 400$ V, $V_{GSQ} = 0/7$ V and different OFF-state gate bias, $V_{GS,OFF}$.

[20]. With hole injection ($V_{GSQ} = 7$ V), the dynamic I_{OFF} significantly decreases with higher temperature. It is explained that there are more electrons available during the trapping process (which occurs at OFF-state) at higher temperatures because of stronger thermionic emission. Thus, the recovery of the "lost" electrons in the buffer layer induced by ON-state hole injection becomes faster at higher temperatures. As a result, for the power switching devices operating at high voltage/high current state, the higher junction temperature in the Schottky type p-GaN gate HEMTs is beneficial to the reduction of dynamic I_{OFF} induced by the ON-state hole injection into the buffer.

The impacts of OFF-state gate bias ($V_{GS,OFF}$) on dynamic I_{OFF} are shown in the Fig. 8. With a negative $V_{GS,OFF}$, the increased dynamic I_{OFF} induced by ON-state hole injection can be effectively suppressed, and with a sufficiently large negative $V_{GS,OFF}$ of -3 V, a two orders of magnitude reduction in dynamic I_{OFF} is obtained and the part of increased dynamic I_{OFF} induced by ON-state hole injection is completely eliminated.

To gain physical insights, 2-D numerical device simulations were performed using the Synopsys TCAD tool. Fig. 9 shows the simulated hole density distribution profile with different ON-state gate bias (V_{GSQ}) in the Schottky-type p-GaN gate HEMTs. With $V_{GSQ} > 5$ V in the ON-state, the holes can be injected from the p-GaN layer to the 2DEG channel. The holes could either recombine with electrons in the channel to emit photons of 3.4 eV and pump electrons from the buffer traps, or further pass the channel to compensate the electrons in buffer traps, leading to the increased lateral conductivity from source to drain and larger dynamic I_{OFF}. A rather shallow penetration depth for the injected holes is obtained with $V_{GSQ} = 7$ V, as shown in the Fig. 9(c). So, with a sufficient negative $V_{GS,OFF}$ (e.g. -3 V), the energy barrier under the gate is raised sufficiently so that it can effectively offset the energy barrier lowering effect induced by ON-state hole injection. Consequently, the energy barrier under the gate can be restored to block the lateral leakage, resulting in lower dynamic I_{OFF}, as shown in

978-1-7281-0582-6/19 $31.00 © 2019 IEEE   465

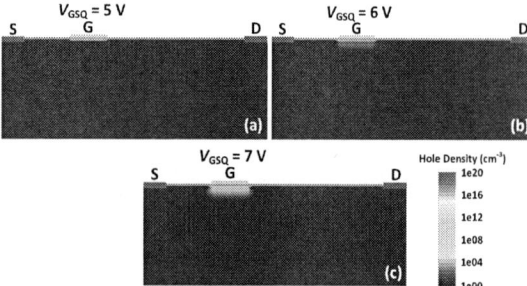

Fig. 9. Simulated hole density distribution profile with an ON-state gate bias (V_{GSQ}) of (a) 5 V, (b) 6 V and (c) 7 V in the Schottky-type p-GaN gate HEMTs.

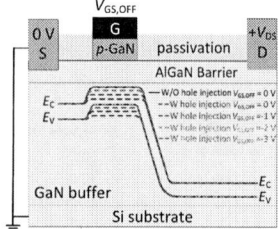

Fig. 10. Schematic cross-section and band diagrams along channel of the Schottky-type p-GaN gate HEMTs at different OFF-state gate bias, $V_{GS,OFF}$ with and without hole injections.

the Fig. 10. For the Schottky-type p-GaN gate HEMTs, a sufficiently large negative $V_{GS,OFF}$ ($\leq$ -3 V) is recommended to the gate driver turn-off voltage, which can effectively suppress the high dynamic I_{OFF} and lower the OFF-state power consumption in practical power switching applications.

IV. CONCLUSION

Dynamic I_{OFF}-V_{DS} characteristics and the behaviors under continuous switching operations in Schottky type p-GaN gate HEMTs have been investigated. It is found that the higher dynamic I_{OFF} even without hole injection is a result of the reduced lateral and vertical voltage blocking capabilities with weaker trapping effect in the buffer. With a large gate bias (> 5 V), the ON-state injected holes further increase the dynamic I_{OFF} due to the increased lateral buffer conductivity from source to drain. Under continuous switching operation, saturation of dynamic I_{OFF} is observed due to a balanced trapping/de-trapping process of buffer traps. The existing high junction temperature in practical switching applications is found to be beneficial to suppress the dynamic I_{OFF} induced by ON-state hole injection. To totally eliminate the dynamic I_{OFF} induced by hole injection, a sufficiently large negative $V_{GS,OFF}$ ($\leq$ -3 V) can be applied to the gate driver turn-off voltage.

REFERENCES

[1] K. J. Chen, O. Häberlen, A. Lidow, C. Tsai, T. Ueda, Y. Uemoto, and Y. Wu, "GaN-on-Si power technology: devices and applications," *IEEE Trans. Electron Devices*, vol. 64, no. 3, pp. 779–795, Mar. 2017.

[2] M. Ishida, T. Ueda, T. Tanaka, and D. Ueda, "GaN on Si technologies for power switching devices," *IEEE Trans. Electron Devices*, vol. 60, no. 10, pp. 3053–3059, Oct. 2013.

[3] C. Tsai, Y. Wang, M. Kwan, P. Chen, F. Yao, S. Liu, J. Yu, C. Yeh, R. Su, W. Wang, W. Yang, K. Wong, Y. Lin, M. Lin, H. Wu, C.

Chen, C. Yu, C. Wu, M. Chang, J. You, T. Huang, S. Wang, L. Tsai, C. Chern, H.Tuan, and A. Kalnitsky, "Smart GaN platform: performance & challenges," in *IEDM Tech. Dig.*, Dec. 2017, pp. 33.1.1-33.1.4.

[4] Y. Uemoto, M. Hikita, H. Ueno, H. Matsuo, H. Ishida, M. Yanagihara, T. Ueda, T. Tanaka, and D. Ueda, "Gate injection transistor (GIT): A normally-off AlGaN/GaN power transistor using conductivity modulation," *IEEE Trans. Electron Devices*, vol. 54, no. 12, pp. 3393–3399, Dec. 2007.

[5] S. Kaneko, M. Kuroda, M. Yanagihara, A. Ikoshi, H. Okita, T. Morita, K. Tanaka, M. Hikita, Y. Uemoto, S. Takahashi, and T. Ueda, "Current-collapse-free operations up to 850 V by GaN-GIT utilizing hole injection from drain," in *Proc. IEEE Int. Symp. Power Semiconductor Device ICs (ISPSD)*, May 2015, pp. 41–44.

[6] K. Tanaka, T. Morita, M. Ishida, T. Hatsuda, T. Ueda, K. Yokoyama, A. Ikoshi, M. Hikita, M. Toki, M. Yanagihara, and Y. Uemoto, "Reliability of hybrid-drain-embedded gate injection transistor," in *Proc. IEEE Int. Rel. Phys. Symp.*, Apr. 2017, pp. 4B-2.1–4B-2.10.

[7] A. Cai, H. A. Carrera, S. B. How, and S. Liter, "Gate driver IC for GaN GIT for high slew rate and cross conduction protection," in *Proc. PCIM Europe*, May 2017, pp. 1537-1544.

[8] I. Hwang, J. Kim, H. Choi, H. Choi, J. Lee, K. Kim, J. Park, J. Lee, J. Ha, J. Oh, J. Shin, and U. Chung, "p-GaN gate HEMTs with tungsten gate metal for high threshold voltage and low gate current," *IEEE Electron Device Lett.*, vol. 34, no. 2, pp. 202–204, Feb. 2013.

[9] H. Wang, J. Wei, R. Xie, C. Liu, G. Tang, and K. J. Chen, "Maximizing the performance of 650-V p-GaN HEMTs: dynamic R_{ON} characterization and circuit design considerations," *IEEE Trans. Power Electronics*, vol. 32, no. 7, pp. 5539–5549, July 2017.

[10] M. Meneghini, O. Hilt, J. Wuerfl, and G. Menehesso, "Technology and reliability of normally-off GaN HEMTs with p-type gate," *Energies*, vol. 10, no. 2, pp. 1–15, Jan. 2017.

[11] J. He, G. Tang, and K. J. Chen, "V_{TH} instability of p-GaN gate HEMTs under static and dynamic gate stress," *IEEE Electron Device Lett.*, vol. 39, no. 10, pp. 1576-1579, Oct. 2018.

[12] J. Wei, R. Xie, H. Xu, H. Wang, Y. Wang, M. Hua, K. Zhong, G. Tang, J He, M. Zhang, and K. J. Chen, "Charge storage mechanism of drain induced dynamic threshold voltage shift in p-GaN gate HEMTs," *IEEE Electron Device Lett.*, Early Access.

[13] S. Huang, Q. Jiang, S. Yang, C. Zhou, and K. J. Chen, "Effective passivation of AlGaN/GaN HEMTs by ALD-grown AlN thin film," *IEEE Electron Device Lett.*, vol. 33, no. 4, pp. 516–518, Apr. 2012.

[14] S. Karmalkar, and U. K. Mishra, "Enhancement of breakdown voltage in AlGaN/GaN high electron mobility transistors using a field plate," *IEEE Trans. Electron Devices*, vol. 48, no. 8, pp. 1515–1521, Aug. 2001.

[15] M. J. Uren, S. Karboyan, I. Chatterjee, A. Pooth, P. Moens, A. Banerjee, and M. Kuball, ""Leaky dielectric" mode for the suppression of dynamic R_{ON} in carbon-doped AlGaN/GaN HEMTs", *IEEE Trans. Electron Devices*, vol. 64, no. 7, pp. 2826–2834, July 2017.

[16] Y. Wang, M. Hua, G. Tang, J. Lei, Z. Zheng, J. Wei, and K. J. Chen, "Dynamic OFF-state current (Dynamic I_{OFF}) in p-GaN gate HEMTs with an ohmic gate contact," *IEEE Electron Device Lett.*, vol. 39, no. 9, pp. 1366–1369, Sep. 2018.

[17] GaN Systems, GS66502B Datasheet. ([Online]. Available: http://gansystems.com/wp-content/uploads/2018/04/GS66502B-DS-Rev-180420.pdf).

[18] B. Li, X. Tang, and K. J. Chen "Optical pumping of deep traps in AlGaN/GaN-on-Si HEMTs using an on-chip Schottky-on-heterojunction light-emitting diode," *Appl. Phys. Lett.*, vol. 106, no. 9, pp. 093505-1-093505-4, Mar. 2015.

[19] B. Li, Q. Jiang, S. Liu, C. Liu, and K. J. Chen, "Degradation of transient OFF-state leakage current in AlGaN/GaN HEMTs induced by ON-state gate overdrive," *Phys. Status Solidi C*, vol. 11, no. 3-4, pp. 928–931, Feb. 2014.

[20] C. Zhou, Q. Jiang, S. Huang, and K. J. Chen, "Vertical leakage/Breakdown mechanisms in AlGaN/GaN-on-Si devices," *IEEE Electron Device Lett.*, vol. 33, no. 8, pp. 1132–1134, Aug. 2012.

Proceedings of the 31st International Symposium on Power Semiconductor Devices & ICs
May 19-23, 2019, Shanghai, China

Effects of Substrate Termination on Reverse-bias Stress Reliability of Normally-off Lateral GaN-on-Si MIS-FETs

Mengyuan Hua[1,2], Song Yang[2], Zheyang Zheng[2], Jin Wei[2], Zhaofu Zhang[2], and Kevin J. Chen[2]

[1]Department of Electrical and Electronic Engineering, Southern University of Science and Technology, Shenzhen, China
[2] The HKUST Shenzhen Research Institute, The Hong Kong University of Science and Technology, Shenzhen, China
E-mail address: huamy@sustech.edu.cn, eekjchen@ust.hk

Abstract—In this work, reliability characterization under reverse-bias (i.e. off-state with high V_{DS}) stress was conducted on the E-mode GaN MIS-FETs with different substrate terminations. The MIS-FETs with a floating substrate exhibit stronger V_{TH} instability than those with a grounded substrate. A non-monotonic dependence of V_{TH} shifts on the positive substrate bias was also observed. The underlying mechanisms are proposed to be the impacts of positive substrate bias on holes drift during long-term reverse-bias stress.

Keywords—GaN, MIS-FET, reverse-bias stress, reliability, substrate termination

I. INTRODUCTION

With GaN power devices being commercialized progressively, reliability and stability issues such as time-dependent dielectric breakdown (TDDB), bias-temperature instability (BTI), reverse-bias stress instability, are becoming the focus of many on-going investigations [1]–[3]. Long TDDB lifetime of GaN devices with a metal-insulator-semiconductor (MIS) gate can be achieved with a suitable dielectric film, such as SiN_x prepared by low-pressure chemical vapor deposition (LPCVD) [4], [5]. Owing to the high process temperature at around 800 °C, LPCVD-SiN_x is of high film quality and low defects density, which has been under intense investigation in GaN-based power devices as both passivation layer [6]–[8] and gate dielectric [9]–[12]. With interface protection layers, the LPCVD-SiN_x can be integrated with fully recessed-gate structure (i.e. with the barrier layer completely removed in the gate region) to achieve high-performance Enhancement-mode MIS-FET with long gate operation lifetime [5]. The interlayers are essential to prevent the etched-GaN surface from degradation during high temperature LPCVD process [5], [13], which benefits a sharp SiN_x/GaN interface with low border/ interface trap density. Consequently, small BTI can been obtained in the E-mode GaN MIS-FETs simultaneously with long TDDB lifetime [14].

Device instability under reverse-bias stress (i.e. stress at off-state with high V_{DS}) remains one of the most critical reliability issues for the GaN MIS-FETs. Recently, in the E-mode LPCVD-SiN_x MIS-FET, an obvious dependence of threshold voltage (V_{TH}) shifts on the negative gate bias was

This work was supported in part by the Shenzhen Science and Technology Innovation Commission under Grant JCYJ20160229205511222 and in part by the Ministry of Science and Technology of the People's Republic of China under Grant 2017YFB0403002.

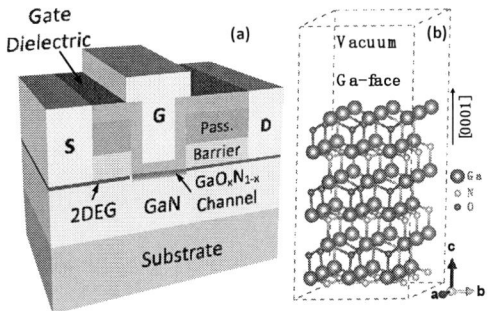

Fig. 1. (a) Schematic cross-sectional view of the MIS-FETs with LPCVD-SiN_x gate dielectric and GaO_xN_{1-x} (simplified as GaON) channel. (b) Schematic atomic structure of the crystalline GaON. The position of small atoms changes with N replaced by O every other layer.

observed [15] and the underlying mechanism was revealed to be hole-induced gate-dielectric degradation. Under reverse-bias stress, impact ionization could occur in the high electric field region at gate edge near the drain side [16], resulting in electron-hole pairs generation and the subsequent hole-drift toward the gate [17]. The holes passing through the gate dielectric will lead to a generation of new defects in the dielectric film during a long-term stress, which could cause non-recoverable positive V_{TH} shifts and devastating time-dependent breakdown [18]. This hole-induced degradation mechanism reveals an important indication that negative gate bias is preferably confined within a suitable range in high-power switching applications of GaN E-mode MIS-FET for a stable V_{TH} and longer operation lifetime.

Although the hole-induced degradation mechanism has been identified, the impacts of substrate termination on reverse-bias stress reliability still have not been considered. Different from the vertical power devices, the substrate of lateral GaN-on-Si power devices could serve as an independent termination port, which is typically shorted to the source. However, floating substrate (FS) is suggested by some works in view of dynamic performance and voltage rating [19]. In addition, in the monolithic integration on GaN-on-Si technology platform, all the GaN devices share the same substrate potential (e.g. half-bridge integration [20], [21]). Therefore, it is necessary to consider the impacts of substrate termination on the device reliability, especially

978-1-7281-0582-6/19 $31.00 © 2019 IEEE 467

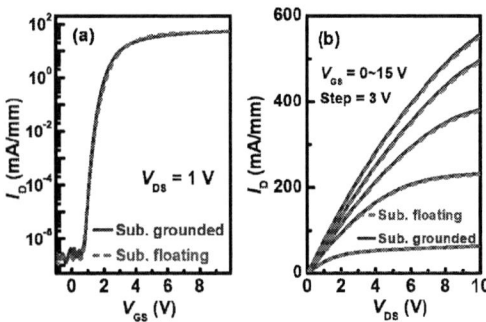

Fig. 2. (a) Transfer and (b) output characteristics of the E-mode MIS-FET with substrate grounded and floating. The MIS-FETs measured in this work are without field-plate structures, and have device dimension of $L_{GS}/L_G/L_{GD}$=2/1.5/15 μm.

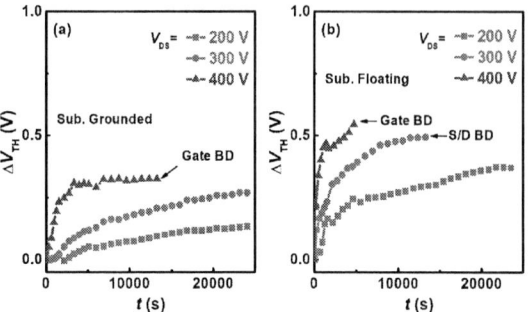

Fig. 4. Threshold voltage shifts of LPCVD-SiN$_x$ MIS-FETs during reverse-bias stress with V_{GS} of 0 V, V_{DG} of 200 V, 300 V, 400 V and substrate (a) grounded and (b) floating, respectively. V_{TH} is defined as V_{GS} at $I_D = 1$ μA/mm.

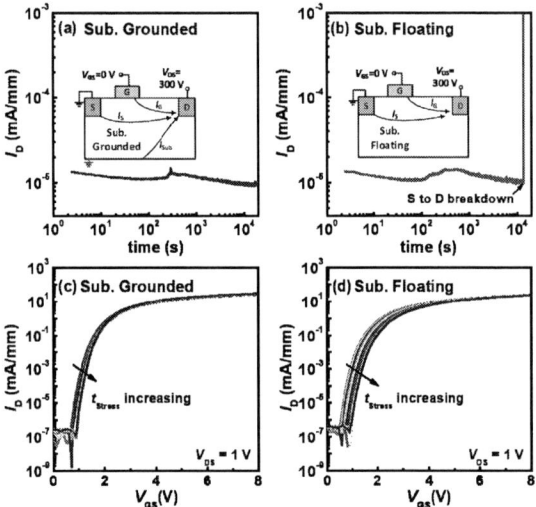

Fig. 3. Transfer characteristics of the E-mode MIS-FETs measured during the long-term reverse-bias stress with V_{GS} of 0 V, V_{DS} of 300 V and substrate (a) grounded and (b) floating, respectively.

under reverse-bias stress, yet detailed characterizations and understanding of the underlying mechanisms are lacking. In this work, reverse-bias stress tests were conducted with different substrate terminations (i.e. floating, grounded and positive V_{sub}). The effects of substrate terminations on V_{TH} stability and device reliability were observed, and the dominant mechanism was found to be related with the hole-induced degradation for the E-mode GaN-on-Si MIS-FETs.

II. DEVICE CHARACTERIZATION

A. Device Fabrication and Static Performance

The structure of MIS-FETs characterized in this work is shown in Fig. 1 (a), which has a LPCVD-SiN$_x$ gate dielectric and crystalline GaO$_x$N$_{1-x}$ (GaON) channel for enhanced device stability and reliability. The GaON was formed with periodic displacements of N/O columns in every other Ga-N layer (Fig. 1 (b)) [22], which has a larger bandgap (~4.1 eV) than the GaN. The valence band offset between GaON and the surrounding GaN creates a hole-blocking ring around the

gate dielectric, preventing holes from flowing to the gate dielectric and therefore mitigating the hole-induced degradation [23].

The E-mode MIS-FETs with GaO$_x$N$_{1-x}$ (GaON) channel and were fabricated on a GaN-on-Si wafer with device dimensions of $L_{GS}/L_G/L_{GD} = 2/1.5/15$ μm. The fabrication commenced with a passivation layer deposition, followed by gate window opening. After gate recess etching, with stronger oxidation of the exposed GaN surface (than that previously developed for the formation of a thin interfacial layer [24]) and subsequent high-temperature *in-situ* annealing, a crystalline GaO$_x$N$_{1-x}$ channel will be formed. In particular, the exposed-GaN surface was oxidized in an ICP chamber with an O$_2$ flow of 40 sccm and plate/coil power of 10/10 W for 10 minutes. Afterward, the sample was *in-situ* annealed in the LPCVD chamber at 780 °C in NH$_3$ ambiance to form the crystalline GaON layer prior to SiN$_x$ deposition. The crystalline GaON will serve as a channel layer in the MIS-FETs structure. The remaining process steps, including the source/drain ohmic contact formation, planar device isolation, and gate/pad metal definitions, are the same as those reported in ref. [13].

The E-mode MIS-FET has a large positive threshold voltage (V_{TH}~1.3 V @ $I_D = 1$ μA/mm), a small threshold voltage hysteresis ($\Delta V_{TH} \sim 0.15$ V), a small subthreshold swing SS (~ 90 mV/dec) (Fig. 2 (a)) and small R_{ON} (~12 Ω·mm) (Fig. 2 (b)). In addition, the transfer and output characteristics are slightly affected by substrate grounded or floating. These devices are without field-plate structure, and thus will be subjected to harsh electrical stress in terms of the electric field strength near the gate edge. Each reverse-bias stress test under different bias conditions was conducted on fresh devices to avoid the effect of pre-stress.

B. Reverse-Bias Stress with GS and FS

The reverse-bias stress tests were conducted on the E-mode MIS-FETs with grounded substrate (GS) and floating substrate (FS) to investigate the impacts of substrate terminations on the threshold voltage instability. During the tests, the MIS-FETs were stressed with a constant V_{DG} of 300 V and a V_{GS} of 0 V, while transfer characteristics were monitored with V_{sub} of 0 V and V_{DS} of 1 V by interrupting the stress periodically [5]. The drain leakage current during the test with GS and FS was shown in Fig. 3 (a) and (b), respectively. For the MIS-FET with FS, device breakdown

978-1-7281-0582-6/19 $31.00 © 2019 IEEE

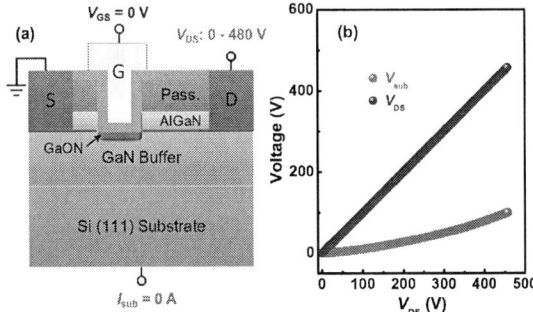

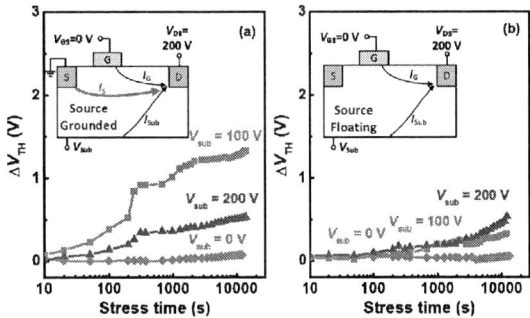

Fig. 5. (a) Test set-up for measuring the substrate voltage induced by floating substrate. (b) OFF-state substrate potential (V_{sub}) of MIS-FETs with substrate floating as a function of drain bias (V_{DS}).

Fig. 7. Threshold voltage shifts during reverse-bias stress with a constant V_{DS} of 200 V and source (a) grounded and (b) floating, respectively.

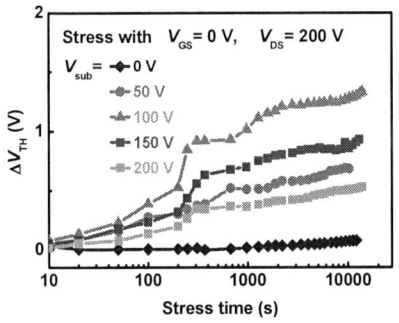

Fig. 6. Effects of substrate voltage (V_{sub}) on threshold voltage shifts during reverse-bias stress with V_{GS} = 0 V, V_{DS} = 200 V.

happens after stress for ~1.2×10^4 s, which is caused by the sudden increase of source-to-drain leakage current while I_G remains at a small value. The MIS-FET with FS also shows larger transfer curves dispersion (Fig. 3 (d)) than the MIS-FET with GS (Fig. 3 (a)).

Fig. 4 (c) and (d) show the threshold voltage shifts (ΔV_{TH}) during the reverse-bias stress with V_{GS} of 0 V, various V_{DS} of 200 V, 300 V, 400 V, and GS and FS, respectively. With higher V_{DS}, MIS-FETs with both GS and FS show larger ΔV_{TH}, which could be caused by the larger electric-field at gate edge near the drain side and the subsequent enhanced both electrons trapping in the gate region and hole-induced degradation. The ΔV_{TH} of MIS-FET with a GS is smaller than 0.3 V during the long-term stress with various V_{DS} (Fig. 2 (c)), while the ΔV_{TH} of MIS-FET with a FS becomes obviously larger, and earlier hard breakdown happens with V_{DS} of 400 V and 300 V (Fig. 2(d)).

C. Reverse-Bias Stress with Positive V_{sub}

The difference between FS and GS terminations is that a positive voltage will be induced at the substrate with a FS because of the asymmetric vertical leakage *I-V* characteristics [19], [25]. The substrate voltage induced by FS was measured by maintaining V_{GS} of 0 V, I_{sub} of 0 A and sweeping V_{DS} from 0 V up to 480 V, as shown in Fig. 5 (a). A V_{sub} of 85 V is induced with a V_{DS} of 400 V (Fig. 5 (b)). The positive V_{sub} could be responsible for the accelerated V_{TH} shifts and shorter time-to-breakdown.

To further understand the impacts of positive V_{sub} on reverse-bias stress reliability, tests with V_{DS} of 200 V, V_{GS} of 0 V and various positive V_{sub} were conducted. During long-term reverse-bias stress, the ΔV_{TH} increases monotonically with longer stress time, but shows a non-monotonic dependence on V_{sub} (Fig. 6). With V_{sub} increased from 0 V to 100 V, the ΔV_{TH} after an identical stress time is larger with a higher V_{sub}; while with V_{sub} increased from 100 V to 200 V, the ΔV_{TH} after a same stress time turns out to be smaller with a higher V_{sub}. This observation indicates that certain mechanisms responsible for degradation in the gate region is suppressed by higher V_{sub} (> 100 V).

III. MECHANISM AND DISCUSSION

To identify the origins of the gate region degradation, reverse-bias stress tests were conducted with source grounded (Fig. 7 (a)) and floating (Fig. 7 (b)) with the same V_{DS} and V_{DG}, respectively. In reverse-bias stress with a floating source, the electron injection from the source side was eliminated, while the peak value of electric field near gate edge could be slightly larger compared with the situation with a grounded source. During all the reverse-bias stress tests with different V_{sub}, the ΔV_{TH} of MIS-FET with a floating source is much smaller than that of MIS-FET with a grounded source. It indicates that the dominant reason of the large ΔV_{TH} with different V_{sub} is the electron injection from the source side to the high field region, rather than the electric-field induced degradation. According to the previous reports, the source injected electrons damage the gate dielectric in the way of hole-induced degradation [15].

Based on the above characterizations and analysis, a physical model of V_{sub}-affected hole-induced gate degradation is proposed. On one hand, more holes will be pushed to the gate side as they are blocked from flowing to substrate with a positive V_{sub}. The holes passing through the gate dielectric will accelerate the defects generation in the dielectric film, and therefore larger ΔV_{TH} was observed with higher V_{sub} (e.g. 100 V). But without source injection, this effect is greatly suppressed (Fig. 7 (b)). On the other hand, positive V_{sub} will reduce the I_S to the drain side, as part of source leakage will be diverted into the substrate following the electrical potential (Fig. 8). Therefore, the subsequent hot-electron and e$^-$-h$^+$ pairs generation will be reduced, resulting in suppressed hole-induced degradation. This mechanism is responsible for the reduced ΔV_{TH} with V_{sub} larger than 100 V (Fig. 6), as well as the smaller difference

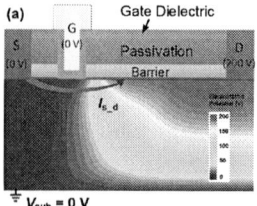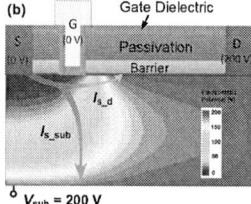

Fig. 8. Simulated electrostatic potential distribution with V_{DS} of 200 V and (a) substrate grounded and (b) V_{sub} = 200 V. The leakage from source to drain could be reduced with positive V_{sub}.

between reverse-bias stress with source floating and grounded at V_{sub} of 200 V (Fig. 7).

IV. CONCLUSION

In this work, the impacts of substrate terminations on reverse-bias stress instability were investigated in the E-mode GaN MIS-FETs. The V_{TH} shifts have non-monotonic dependence on the positive substrate bias. For moderate V_{sub}, hole-induced degradation will be enhanced and results in accelerated V_{TH} shifts and shorter t_{BD}. While with higher V_{sub}, electrons injection from source to the drain side will be reduced, which helps to mitigate the hole-induced degradation. But its still worse than that with substrate grounded. For a good V_{TH} stability, it is not suggested to leave the substrate floating or positively biased.

REFERENCES

[1] K. J. Chen, O. Haberlen, A. Lidow, C. Tsai, T. Ueda, Y. Uemoto, and Y. Wu, 'GaN-on-Si Power Technology: Devices and Applications', *IEEE Trans. Electron Devices*, vol. 64, no. 3, pp. 779–795, Mar. 2017.

[2] C. Ostermaier, P. Lagger, M. Reiner, and D. Pogany, 'Review of bias-temperature instabilities at the III-N/dielectric interface', *Microelectron. Reliab.*, vol. 82, pp. 62–83, Mar. 2018.

[3] G. Meneghesso, M. Meneghini, I. Rossetto, D. Bisi, S. Stoffel, M. Hove, S. Decoutere, and E. Zanoni, 'Reliability and parasitic issues in GaN-based power HEMTs: a review', *Semicond. Sci. Technol.*, vol. 31, no. 9, p. 093004, 2016.

[4] M. Hua *et al.*, 'Characterization of Leakage and Reliability of SiN$_x$ Gate Dielectric by Low-Pressure Chemical Vapor Deposition for GaN-based MIS-HEMTs', *IEEE Trans. Electron Devices*, vol. 62, no. 10, pp. 3215–3222, Oct. 2015.

[5] M. Hua, C. Liu, S. Yang, S. Liu, K. Fu, Z. Dong, Y. Cai, B. Zhang, and K. J. Chen, 'Integration of LPCVD-SiN$_x$ gate dielectric with recessed-gate E-mode GaN MIS-FETs: Toward high performance, high stability and long TDDB lifetime', in *2016 IEEE International Electron Devices Meeting (IEDM)*, 2016, pp. 10.4.1-10.4.4.

[6] S. Huang, X. Liu, X. Wang, X. Kang, J. Zhang, Q. Bao, K. Wei, Y. Zheng, C. Zhao, H. Gao, Q. Sun, Z. Zhang, and K. J. Chen, 'High Uniformity Normally-OFF GaN MIS-HEMTs Fabricated on Ultra-Thin-Barrier AlGaN/GaN Heterostructure', *IEEE Electron Device Lett.*, vol. PP, no. 99, pp. 1–1, 2016.

[7] Y. Shi, S. Huang, Q. Bao, X. Wang, K. Wei, H. Jiang, J. Li, C. Zhao, S. Li, Y. Zhou, H. Gao, Q. Sun, H. Yang, J. Zhang, W. Chen, Q. Zhou, B. Zhang, and X. Liu, 'Normally OFF GaN-on-Si MIS-HEMTs Fabricated With LPCVD-SiN$_x$ Passivation and High-Temperature Gate Recess', *IEEE Trans. Electron Devices*, vol. 63, no. 2, pp. 614–619, Feb. 2016.

[8] X. Wang, S. Huang, Y. Zheng, K. Wei, X. Chen, G. Liu, T. Yuan, W. Luo, L. Pang, H. Jiang, J. Li, C. Zhao, H. Zhang, and X.Liu, 'Robust SiN$_x$/AlGaN Interface in GaN HEMTs Passivated by Thick LPCVD-Grown SiN$_x$ Layer', *IEEE Electron Device Lett.*, vol. 36, no. 7, pp. 666–668, Jul. 2015.

[9] M. Hua, C. Liu, S. Yang, S. Liu, K. Fu, Z. Dong, Y. Cai, B. Zhang, and K. J. Chen, 'GaN-Based Metal-Insulator-Semiconductor High-Electron-Mobility Transistors Using Low-Pressure Chemical Vapor Deposition SiN$_x$ as Gate Dielectric', *IEEE Electron Device Lett.*, vol. 36, no. 5, pp. 448–450, May 2015.

[10] Z. Zhang, G. Yu, X. Zhang, X. Deng, S. Li, Y. Fan, S. Sun, L. Song, S. Tan, D. Wu, W. Li, W. Huang, K. Fu, Y. Cai, Q. Sun, amd B. Zhang, 'Studies on High-Voltage GaN-on-Si MIS-HEMTs Using LPCVD-SiN$_x$ as Gate Dielectric and Passivation Layer', *IEEE Trans. Electron Devices*, vol. PP, no. 99, pp. 1–8, 2016.

[11] S. A. Jauss, K. Hallaceli, S. Mansfeld, S. Schwaiger, W. Daves, and O. Ambacher, 'Reliability Analysis of LPCVD SiN Gate Dielectric for AlGaN/GaN MIS-HEMTs', *IEEE Trans. Electron Devices*, vol. 64, no. 5, pp. 2298–2305, 2017.

[12] Z. Zhang, G. Yu, X. Zhang, S. Tan, D. Wu, K. Fu, W. Huang, Y. Cai, and B. Zhang, '16.8 A/600 V AlGaN/GaN MIS-HEMTs employing LPCVD-Si$_3$N$_4$ as gate insulator', *Electron. Lett.*, vol. 51, no. 15, pp. 1201–1203, 2015.

[13] M. Hua, J. Wei, G. Tang, Z. Zhang, Q. Qian, X. Cai, N. Wang, and K. J. Chen, 'Normally-Off LPCVD-SiN$_x$/GaN MIS-FET With Crystalline Oxidation Interlayer', *IEEE Electron Device Lett.*, vol. 38, no. 7, pp. 929–932, Jul. 2017.

[14] M. Hua, Q. Qian, J. Wei, Z. Zhang, G. Tang, and K. J. Chen, 'Bias Temperature Instability of Normally-Off GaN MIS-FET with Low-Pressure Chemical Vapor Deposition SiN$_x$ Gate Dielectric', *Phys. Status Solidi A*, vol. 215, no. 10, p. 1700641, 2018.

[15] M. Hua, J. Wei, Q. Bao, Z. Zhang, Z. Zheng, and K. J. Chen, 'Dependence of V_{th} Stability on Gate-Bias Under Reverse-Bias Stress in E-mode GaN MIS-FET', *IEEE Electron Device Lett.*, vol. 39, no. 3, pp. 413–416, Mar. 2018.

[16] G. Meneghesso, G. Verzellesi, F. Danesin,F. Rampazzo, F. Zanon, A. Tazzoli, M. Meneghini, and E. Zanoni, 'Reliability of GaN High-Electron-Mobility Transistors: State of the Art and Perspectives', *IEEE Trans. Device Mater. Reliab.*, vol. 8, no. 2, pp. 332–343, Jun. 2008.

[17] M. Hua, J. Wei, Q. Bao, Z. Zhang, Z. Zheng, J. He, and K. J. Chen, 'Hole-Induced Threshold Voltage Shift Under Reverse-Bias Stress in E-Mode GaN MIS-FET', *IEEE Trans. Electron Devices*, vol. 65, no. 9, pp. 3831–3838, Sep. 2018.

[18] M. Hua, J. Wei, Q. Bao, J. He, Z. Zhang, Z. Zheng, J. Lei, and K. J. Chen, 'Reverse-Bias Stability and Reliability of Hole-Barrier-Free E-mode LPCVD-SiN$_x$/GaN MIS-FETs', presented at the 2017 IEEE International Electron Devices Meeting (IEDM), San Francisco, USA, 2017, pp. 33.2.1-33.2.2.

[19] G. Tang, J. Wei, Z. Zhang,X. Tang, M. Hua, H. Wang, and K. J. Chen, 'Dynamic R_{ON} of GaN-on-Si Lateral Power Devices with a Floating Substrate Termination', *IEEE Electron Device Lett.*, vol. 38, no. 7, pp. 937–940, Jul. 2017.

[20] C. Tsai *et al.*, 'Smart GaN platform: Performance and challenges', in *2017 IEEE International Electron Devices Meeting (IEDM)*, 2017, pp. 33.1.1-33.1.4.

[21] Y. Uemoto *et al.*, 'GaN monolithic inverter IC using normally-off gate injection transistors with planar isolation on Si substrate', in *2009 IEEE International Electron Devices Meeting (IEDM)*, 2009, pp. 1–4.

[22] X. Cai, M. Hua, Z. Zhang, S. Yang, Z. Zheng, Y. Cai, K. J. Chen, and N. Wang, 'Atomic-scale identification of crystalline GaON nanophase for enhanced GaN MIS-FET channel', *Appl. Phys. Lett.*, vol. 114, no. 5, p. 053109, Feb. 2019.

[23] M. Hua, X. Cai, S. Yang, Z. Zhang, Z. Zheng, J. Wei, N. Wang, and K. J. Chen, 'Suppressed Hole-Induced Degradation in E-mode GaN MIS-FETs with Crystalline GaO$_x$N$_{1-x}$ Channel', in *2018 IEEE International Electron Devices Meeting (IEDM)*, 2018, pp. 30.3.1-30.3.4.

[24] M. Hua, Z. Zhang, Q. Qian, J. Wei, Q. Bao, G. Tang, and K. J. Chen, 'High-performance fully-recessed enhancement-mode GaN MIS-FETs with crystalline oxide interlayer', in *2017 29th International Symposium on Power Semiconductor Devices and IC's (ISPSD)*, 2017, pp. 89–92.

[25] S. Yang, C. Zhou, S. Han, J. Wei, K. Sheng, and K. J. Chen, 'Impact of Substrate Bias Polarity on Buffer-Related Current Collapse in AlGaN/GaN-on-Si Power Devices', *IEEE Trans. Electron Devices*, vol. 64, no. 12, pp. 5048–5056, Dec. 2017.

Proceedings of the 31st International Symposium on Power Semiconductor Devices & ICs
May 19-23, 2019, Shanghai, China

Novel 2000 V Normally-off MOS-HEMTs using AlN/GaN Superlattice Channel

line 1: 1st Ming Xiao
line 2: *School of Microelectronics*
line 3: *Xidian University*
line 4: Xi'an, China
line 5: xm427@126.com,
mxiao@vt.edu

line 1: 4th Hong Zhou
line 2: *School of Microelectronics*
line 3: *Xidian University*
line 4: Xi'an, China
line 5: hongzhou@xidian.edu.cn

line 1: 7th Yue Hao
line 2: *School of Microelectronics*
line 3: *Xidian University*
line 4: Xi'an, China
line 5: yhao@xidian.edu.cn

line 1: 2nd Weihang Zhang
line 2: *School of Microelectronics*
line 3: *Xidian University*
line 4: Xi'an, China
line 5: whzhang@stu.xidian.edu.cn

line 1: 5th Kui Dang
line 2: *School of Microelectronics*
line 3: *Xidian University*
line 4: Xi'an, China
line 5: 1103476535@qq.com

line 1: 3rd Yuhao Zhang
line 2: *Center for Power Electronics Systems*
line 3: *Virginia Polytechnic Institute and State University*
line 4: Blacksburg, USA
line 5: yhzhang@vt.edu

line 1: 6th Jincheng Zhang
line 2: *School of Microelectronics*
line 3: *Xidian University*
line 4: Xi'an, China
line 5: jchzhang@xidian.edu.cn

Abstract—**We demonstrate for the first time a GaN-based metal oxide semiconductor high electron mobility transistor (MOS-HEMT) with AlN/GaN superlattice (SL) channels. This new channel structure allows for superior voltage blocking capabilities and thermal stability than conventional GaN channels, as well as higher electron mobility than AlGaN channels. State-of-the-art static and dynamic performance has been achieved in this new MOS-HEMT, including a breakdown voltage over 2000 V, a high ON current density of 768 mA/mm, a threshold voltage (V_{TH}) of 1.0 V, a specific on-resistance (R_{ON}) of 7.7 m$\Omega \cdot$cm^2, thermal stability up to 225 ºC and good dynamic R_{ON}. These results show the great potential of our novel AlN/GaN-SL-channel MOS-HEMTs for high-voltage and high temperature power switching applications**

Keywords—superlattices, normally-off, metal oxide semiconductor high-electron mobility transistor (MOS-HEMT), gallium nitride, aluminum gallium nitride, high breakdown voltage

I. INTRODUCTION

GaN-based HEMTs have shown great potential for applications in microwave power amplifier and power conversion applications due to their fast switching speed and low switching loss [1, 2]. Recently, they have achieved initial commercialization up to 650 V, but encountered great challenges in further scaling up breakdown voltages (*BV*) and improving device reliability at high temperatures. AlGaN channel devices have been proposed as a potential solution for these challenges. Due to the larger bandgap and better thermal stability of AlGaN over GaN, AlGaN channel devices have great potential advantages than conventional GaN channel devices for high-voltage and high-temperature applications [3, 4]. Recently, the AlGaN channel MOSFETs and high performance HEMTs have been reported [5, 6].

Despite the excellent performance achieved in AlGaN channel HEMTs, the poor crystal quality and low electrical properties have become the key bottlenecks to further improve restrict the performance improvement and application of AlGaN channel HEMT devices [?] Especially, the current density of AlGaN channel HEMTs is

limited to the low channel electronic mobility owing to strong alloy disorder scattering of AlGaN channels and the low electronic density due to poor heterostructure quality. Although optimization of the AlGaN layer can increase electronic density, improvement of the electron mobility is difficult, since it cannot decrease alloy disorder scattering.

In this work, we propose a novel HEMT structure using the AlN/GaN superlattice (SL) channels. This new channel structure could allow for a significantly higher electron mobility compared to AlGaN channels, due to the reduction of alloy disorder scattering, while maintaining the good voltage-blocking capabilities and thermal stability of AlGaN channels. Based on this new channel structure, we demonstrate 2000 V normally-off MOS-HEMTs with the gate recess structure.

II. DEVICE DESIGN AND FABRICATION

The schematic of our AlN/GaN SL channel MOS-HEMTs is shown in Fig. 1. The epitaxial structure consists of a 1 nm GaN cap-layer, a 12 nm Al$_{0.37}$Ga$_{0.63}$N barrier, a 2 nm AlN space-layer and a 1.7 μm composite buffer layer, all grown by low-pressure metal-organic chemical vapor deposition on SiC substrate. The 1.7 μm buffer layer is composed of four parts, including 130 nm AlN/GaN SLs (30

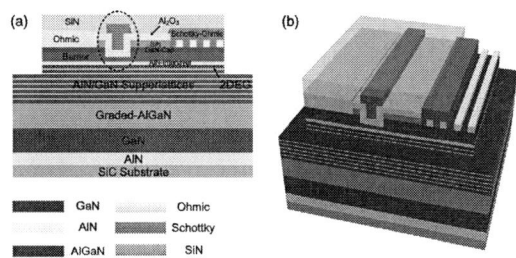

Fig. 1. (a) Cross-sectional and (b) side-view three-dimensional schematics of the device structure with single gate finger. The gate length (L_G) is 1 μm and three gate-drain distance (L_{GD}) are 5 μm, 10 μm and 20 μm, respectively. The drain used of the Ohmic-Schottky hybrid structure

978-1-7281-0582-6/19 $31.00 © 2019 IEEE   471

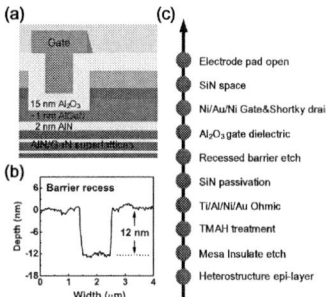

Fig. 2. (a) The structure diagram under gate. (b) AFM measurement of the barrier recess formed by low power ICP-RIE etching. (c) Key process steps to fabricate the AlN/GaN SL MOS-HEMTs

cycles) layer, 70 nm $Al_{0.12}Ga_{0.88}N$ layer, 500 nm graded-AlGaN and a 1 μm GaN layer. Room-temperature Hall measurements revealed a sheet electron density of 6.1×10^{12} cm^{-2} and a high electron mobility of 1179 $cm^2/V\cdot s$. As far as we know, this is the highest mobility in all reported AlGaN channel HEMTs. As shown in Fig. 1, an Ohmic/Schottky hybrid structure was used in the drain electrode, with a stripe width of 2 μm. The structure under gate consists of 15 nm Al_2O_3, ~1 nm AlGaN and 2 nm AlN layer. All the devices in this work have the L_{GS} = 1.5 μm, L_G = 1 μm and L_{GD}=5, 10 and 20 μm.

The device fabrication (Fig. 2) started with the formation of mesa isolation obtained using Cl_2/BCl_3 plasma dry etching in an inductively coupled plasma (ICP) reactive ion etching (RIE) system, followed by a hot TMAH treatment to reduce leakage current [7, 8]. Then, the Ohmic contacts were formed by a Ti/Al/Ni/Au metal stack, followed by rapid thermal annealing at 840 °C for 40 s in a N_2 atmosphere. The Ohmic contact resistance and the sheet resistance extracted from the transmission line measurements were 0.5 Ω·mm and 886 Ω/sq, respectively. Next, a 60 nm plasma enhanced chemical vapor deposition Si_3N_4 layer was deposited, and a dry etching in a RIE system was used to pattern a 1 μm gate-trench region. Barrier recess of 12 nm depth was then made by low-power ICP-RIE etching (Fig. 2(b)). A 15 nm Al_2O_3 dielectric layer was then deposited by plasma enhanced atomic layer deposition followed by an evaporated gate metal (Ni/Au/Ni) layer. Finally, a 200 nm Si_3N_4 space layer

was deposited followed by the electrode pad open by RIE.

III. RESULTS AND DISCUSSION

Firstly, an appropriate buffer layer structure is important to avoid the production of mismatch dislocation and parasitic electron channel. Fig. 3(a) shows the cross-sectional scanning transmission electron microscopy (STEM) images of the entire epitaxial layer. Good crystal quality was observed in the composite buffer layer, where the GaN layer was grown to reduce the dislocation density and the graded AlGaN was grown in the intermediate to adjust the epitaxial stress. The AlN/GaN SLs, $Al_{0.12}Ga_{0.88}N$ layer and graded AlGaN are lattice-matched to GaN layer which was proved by reciprocal space mapping. This lattice-matching avoided the production of mismatch dislocation by suppressing strain relaxation of AlGaN. As shown in Fig. 3(a), no production of new dislocation was observed in the composite buffer layer. Fig. 3(b) shows the HR-STEM image of AlGaN barrier and AlN/GaN SLs. Perfect superlattices cycle structures and good heterostructure interface quality are exhibited. Because the two dimensional electron gas (2DEG) are concentrated in a super thin area near the interface, the first GaN single layer near barrier serves as the main electron channel, which results in a high electron mobility of 1179 $cm^2/V\cdot s$. This mobility is significantly higher than the one (807 $cm^2/V\cdot s$) obtained in our control AlGaN channel sample with a similar Al component [6]. This demonstrates the effectiveness of our SL structure in reducing the alloy disorder scattering.

AlN-GaN-SL material has a single band-gap due to the super-short cycle. In this case, the small inset image in Fig. 4(a) shows a comparison of the calculated band diagram of the novel AlGaN/AlN-GaN-SL heterostructure and the

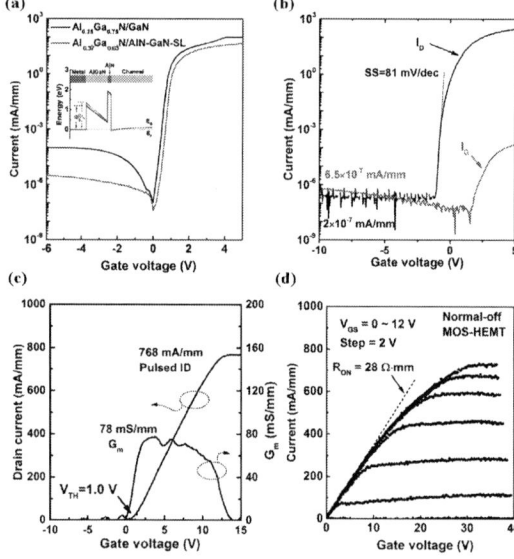

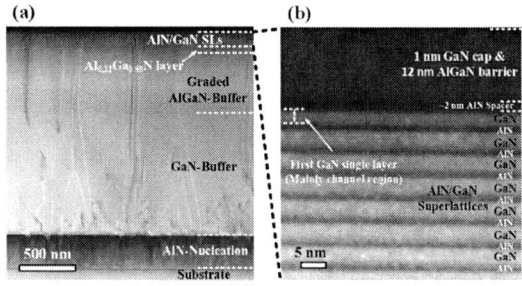

Fig. 3. (a) Cross-sectional STEM of the barrier layer and AlN/GaN SL channel layer, demonstrating the buffer layer consist of the 130-nm-AlN/GaN SL, 70-nm $Al_{0.12}Ga_{0.88}N$ layer, 500-nm-graded-AlGaN and 1 μm GaN layer. (b) Cross-sectional HR-STEM of the barrier layer and multicycle AlN/GaN SL channel layer. The channel is mostly located at the first GaN single layer of the SLs. The single cycle thickness is 4.3 nm, including 0.5 nm AlN and 3.8 nm GaN.

Fig. 4. (a) Gate current and calculated band diagram of $Al_{0.25}Ga_{0.75}N$/GaN channel and $Al_{0.37}Ga_{0.63}N$/AlN-GaN SL channel at gate bias of 0 V. The potential at AlGaN/AlN-GaN-SL channel is lifted up higher owing to the larger band-gap of channel material than that of conventional GaN channel (b) DC (V_{DS}=10 V) transfer curve and gate leakage current curve, (c) Pulsed (V_{DS}=40 V) transfer curve and G_m curve, (d) Pulsed output characteristics of the AlN/GaN SL channels HEMT with L_{GD}=20 μm.

conventional AlGaN/GaN heterostructure. As shown, the potential at heterojunction is lifted up in our structure, due to the larger bandgap of AlN-GaN SLs than GaN. This band lift-up could contribute to a move positive V_{TH} and a reduced gate leakage current. It can be approved by the Schottky gate leakage current curves shown in Fig. 4(a). The Schottky gate leakage current of AlN-GaN SLs channel HEMT structure is lower than that of GaN channel HEMT structure.

To illustrate the electrical characteristics of AlN/GaN SL MOSHEMTs, Fig. 4(b), (c) and (d) show the DC transfer, pulsed transfer and pulsed output characteristics of AlN/GaN SL MOS-HEMT with L_{GD}=20 μm, respectively. A threshold voltage of 1.0 V and a large drain current of up to 768 mA/mm were obtained. The off-state current, gate leakage current, peak transconductance (G_m), subthreshold swing (SS) and I_{ON}/I_{OFF} are 2×10^{-7} mA/mm, 6.5×10^{-7} mA/mm, 78 mS/mm, 81 mV/dec and 3.8×10^{9}, respectively. An R_{ON} of 28 $\Omega\cdot$mm was demonstrated in the devices with L_{GD}=20 μm.

Furthermore, Fig. 5(a) shows the representative off-state characteristics of a AlN/GaN SL channel MOS-HEMT. A BV over 2000 V was demonstrated with low drain leakage current (~0.1 μA/mm) at high drain bias. It should be noted that no field plate structures are used in our devices. Therefore, the device BV could be further increased by introducing the field plate and other field engineering structures. Fig. 5(b) shows the representative transfer characteristics at 25 °C to 225 °C of our device and a conventional MOS-HEMTs with GaN channels. As shown,

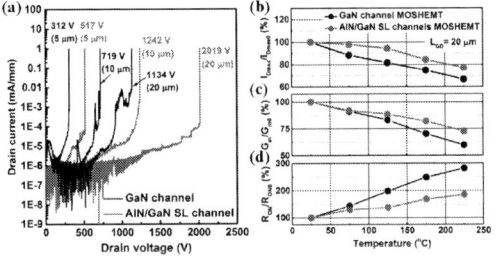

Fig. 5. (a) Off-state I-V characteristics of AlN/GaN SL and GaN channel MOS-HEMTs with L_{GD}=5, 10, 20 μm. A breakdown voltage over 2000 V for L_{GD}=20 μm AlN/GaN SL MOS-HEMT. Relative (b) I_{Dmax}, (c) G_m and (d) R_{ON} of GaN channel and the AlN/GaN SL MOS-HEMT with L_{GD}=20 μm changed with temperature from 25 °C to 225 °C.

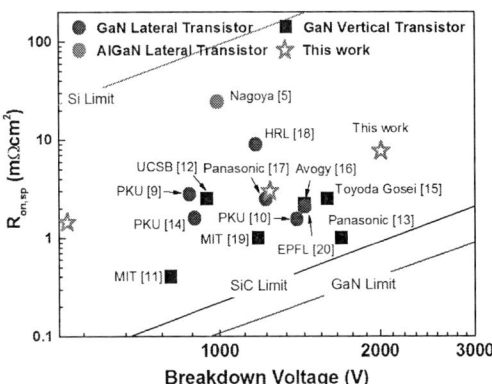

Fig. 6. Specific R_{ON} v.s. BV of the AlN/GaN SL MOS-HEMT, benchmarked with other normally-off GaN and AlGaN transistors. This work represents the R_{ON} normalized with the total active area.

whether it is in I_{Dmax} and G_m, or in R_{ON}, the AlN/GaN SLs channel MOS-HEMTs demonstrate better thermal stability than GaN channel MOS-HEMTs.

In order to more intuitively compare the characteristics of different GaN Normally-off devices, Fig. 6 benchmarks the R_{ON} v.s. BV for our AlN/GaN SL channel devices with othernormally-off lateral and vertical GaN and AlGaN transistors. With a BV of 2000 V and a R_{ON} of 7.7 m$\Omega\cdot$cm^2, our device demonstrated a Baliga's figure of merit up to 520 MW/cm^2, which is among the best in all AlGaN power devices. Table I summarizes the key device metrics of our AlN/GaN SL MOS-HEMTs, including R_{ON}, BV, V_{TH}, on-state drain-to-source saturation current density ($I_{ON,SAT}$) and off-state leakage current density (I_{OFF}) at 1200 V, benchmarked with other reported normally-off GaN and AlGaN transistors. As show, our AlN/GaN SL MOS-HEMTs with L_{GD}=20 μm show the highest BV and the second lowest I_{OFF} at 1200 V.

IV. CONCLUSION

This work demonstrates the first MOS-HEMT using a AlN/GaN SL channel. A BV over 2000 V, high electron mobility, a V_{TH} of 1.0 V, a R_{ON} of 7.7 m$\Omega\cdot$cm^2, high ON current density and good thermal stability were demonstrated. The AlN/GaN SL structure dramatically

TABLE I. SUMMARY AND BENCHMARK OF DEVICE STRUCTURES AND KEY DEVICE METRICS FOR THE GaN AND ALGaN TRANSISTORS (VERTICAL AND LATERAL) WITH HIGH CURRENT RATINGS REPORTED.

		Device Structure		R_{on} (m$\Omega\cdot$cm^2)	BV (V)	V_{TH} (V)	$I_{ON,SAT}$ (A/cm^2)	I_{OFF}@1200 (A/cm^2)
L	AlGaN	**This work**	L_{GD}=5μm	**1.5**	**517**	**1.0**	**6400**	---
			L_{GD}=10μm	**3.0**	**1242**	**1.0**	**4570**	---
			L_{GD}=20μm	**7.7**	**2019**	**1.0**	**2792**	**1.2E-5**
		AlGaN channel MOSFET		24.5	993	2.4	0.09 A/mm	---
V	GaN	Gate Injector Transistor		2.3	1250	3	0.4 A/mm	10^{-7} A/mm
		Gate-recess HEMT		9	1200	0.64	255	0.005
		FinFET		1.0	1200	1	25000	<10^{-4}
		CAVET		2.2	1500	0.5	1500	0.02
		Trench CAVET		1.0	1700	2.5	3500	0.01
		Trench MOSFET		2.7	1600	2	~1100	<10^{-8}

Note: 'L' represent 'Lateral' and 'V' present 'Vertical'

improved the electron mobility of AlGaN channel device, while maintaining the advantages in blocking voltage and thermal stability over conventional GaN channels. This performance demonstrates the great potential of AlN/GaN SL channel MOS-HEMTs for high-power and high temperature application..

ACKNOWLEDGMENT

This work was supported by the National Key Research and Development Project under Grant 2016YFB0400105.

REFERENCES

[1] Y. Wu, M. Jacob-Mitos, M. L. Moore, and S. Heikman, "A 97.8% Efficient GaN HEMT Boost Converter With 300-W Output Power at 1 MHz," IEEE Electron Device Letters, vol. 29, pp. 824-826, Aug 2008.

[2] W. Saito et al., "A 120-W Boost Converter Operation Using a High-Voltage GaN-HEMT," IEEE Electron Device Letters, vol. 29, pp. 8-10, Jan 2007.

[3] T. Nanjo et al., "AlGaN Channel HEMT with Extremely High Breakdown Voltage," IEEE Transactions on Electron Devices, vol. 60, pp. 1046-1053, Mar 2013.

[4] T. Nanjo et al., "First Operation of AlGaN Channel High Electron Mobility Transistors," Applied Physics Express, vol. 1, pp. 155-162, Dec 2008.

[5] J. J. Freedsman, T. Hamada, M. Miyoshi, and T. Egawa, "Al$_2$O$_3$/AlGaN Channel Normally-Off MOSFET on Silicon With High Breakdown Voltage," IEEE Electron Device Letters, vol. 38, pp. 497-500, Feb 2017.

[6] M. Xiao et al., "High Performance Al0.10Ga0.90N Channel HEMTs," (in English), IEEE Electron Device Letters, Article vol. 39, pp. 1149-1151, Aug 2018.

[7] Q. Zhou et al., "Schottky-Contact Technology in InAlN/GaN HEMTs for Breakdown Voltage Improvement," IEEE Transactions on Electron Devices, Article vol. 60, pp. 1075-1081, Mar 2013.

[8] Y. Zhang et al., "Origin and Control of OFF-State Leakage Current in GaN-on-Si Vertical Diodes," IEEE Transactions on Electron Devices, vol. 62, pp. 2155-2161, Jul 2015.

[9] M. Tao et al., "Characterization of 880 V Normally Off GaN MOSHEMT on Silicon Substrate Fabricated With a Plasma-Free, Self-Terminated Gate Recess Process," IEEE Transactions on Electron Devices, vol. 65, pp. 1453-1457, Apr 2018.

[10] J. Gao et al., "Gate-Recessed Normally OFF GaN MOSHEMT With High-Temperature Oxidation/Wet Etching Using LPCVD Si3N4 as the Mask," IEEE Transactions on Electron Devices, vol. 65, pp. 1728-1733, May 2018.

[11] M. Sun, Y. Zhang, X. Gao, and T. Palacios, "High-Performance GaN Vertical Fin Power Transistors on Bulk GaN Substrates," IEEE Electron Device Letters, vol. 38, pp. 509-512, Apr 2017.

[12] C. Gupta et al., "In Situ Oxide, GaN Interlayer-Based Vertical Trench MOSFET (OG-FET) on Bulk GaN substrates," IEEE Electron Device Letters, vol. 38, pp. 353-355, Mar 2017.

[13] D. Shibata et al., "1.7 kV / 1.0 m Omega cm(2) Normally-off Vertical GaN Transistor on GaN substrate with Regrown p-GaN/AlGaN/GaN Semipolar Gate Structure," in 2016 IEEE International Electron Devices Meeting [IEEE International Electron Devices Meeting, 2016].

[14] M. J. Wang et al., "900 V/1.6 m Omega . cm(2) Normally Off Al2O3/GaN MOSFET on Silicon Substrate," IEEE Transactions on Electron Devices, vol. 61, pp. 2035-2040, Jun 2014.

[15] T. Oka, Y. Ueno, T. Ina, and K. Hasegawa, "Vertical GaN-based trench metal oxide semiconductor field-effect transistors on a free-standing GaN substrate with blocking voltage of 1.6 kV," Applied Physics Express, vol. 7, p. 021002, Feb 2014.

[16] H. Nie et al., "1.5-kV and 2.2-m Omega-cm(2) Vertical GaN Transistors on Bulk-GaN Substrates," IEEE Electron Device Letters, vol. 35, pp. 939-941, Sep 2014.

[17] M. Ishida, T. Ueda, T. Tanaka, and D. Ueda, "GaN on Si Technologies for Power Switching Devices," IEEE Transactions on Electron Devices, vol. 60, pp. 3053-3059, Oct 2013.

[18] R. Chu et al., "1200-V Normally Off GaN-on-Si Field-Effect Transistors With Low Dynamic ON-Resistance," IEEE Electron Device Letters, vol. 32, pp. 632-634, May 2011.

[19] Y. Zhang et al., "1200 V GaN Vertical Fin Power Field-Effect Transistors," in 2017 IEEE International Electron Devices Meeting, Jan 2017 [IEEE International Electron Devices Meeting, 2017].

[20] J. Ma and E. Matioli, "High Performance Tri-Gate GaN Power MOSHEMTs on Silicon Substrate," IEEE Electron Device Letters, vol. 38, pp. 367-370, Jan 2017.

Proceedings of the 31st International Symposium on Power Semiconductor Devices & ICs
May 19-23, 2019, Shanghai, China

Analysis the complex tradeoff among E_{on}-V_{CEsat}-SCSOA and EMI noise through the single chip evaluation method

Koichi Nishi, Tetsuo Takahashi and Atsushi Narazaki
Power Device Works, Mitsubishi Electric Corporation
1-1-1 Imajyuku-Higashi, Nishi-ku, Fukuoka, 819-0192, Japan
Phone: +81-92-805-3332 E-mail: Nishi.Koichi@cw.MitsubishiElectric.co.jp

Abstract— As a compensation for the superior V_{CEsat}, Carrier Stored Trench gate Bipolar Transistor (CSTBT™) has the more complex tradeoff among E_{on}-V_{CEsat}-SCSOA and EMI noise compared to IGBT. EMI noise evaluation is known to need special facilities such as a shield room and antenna detector, which deteriorates an evaluation efficiency and consistency. We found that EMI noise can be indirectly but simply and accurately evaluated by using the surge current under small current turn-on of single-chip IGBTs without any EMI shield facilities. By using this method, maintaining the CSTBT's better tradeoff among E_{on}-V_{CEsat}-SCSOA, we have succeeded to reduce the EMI noise by decreasing Carrier Stored Layer doping concentration and/or gate-collector capacitance while adjusting channel density.

Keywords—IGBT, CSTBT, EMI noise, dv/dt, di/dt, Current surge, Turn-on switching loss, Short Circuit SOA

I. INTRODUCTION

For the progress of power electronics, it is necessary to enhance the performance of Insulated Gate Bipolar Transistors (IGBTs). In the power electronics application fields, our CSTBT™ has been playing an important role as one of the best total loss IGBT structures, but also has some weak points to be improved as mentioned below. Both on-state forward voltage (V_{CEsat}) and turn-on switching loss (E_{on}) can be lowered due to the extra n-layer called Carrier Stored layer (CS-layer), especially for Intelligent Power Module application (IPM) with protection circuit [1]. One of the weak points of CSTBT™ is relatively narrow Short Circuit Safe Operation Area (SCSOA) based upon the high saturation current. While SCSOA properties of CSTBT™ has been improved by cell pattern optimization and/or advanced thin wafer process for the latest CSTBT™, the tradeoff among E_{on}-V_{CEsat}-SCSOA still exists [2-4]. Another weak point is electromagnetic interference (EMI) noise through the high dv/dt rate during the turn-on period [5]. Owing to this complex tradeoff among E_{on}-V_{CEsat}-SCSOA and EMI noise, CSTBT™ has been facing to a dilemma of improving both V_{CEsat} and EMI noise simultaneously without any sacrifice.

On the other hand, an EMI noise evaluation of power modules needs complex facilities such as a shield room, a barrack setting of an evaluation circuit and an antenna detector. From the viewpoints of not only an evaluation efficiency but also consistency of all the tradeoff among E_{on}-V_{CEsat}-SCSOA and EMI noise, single chip evaluation methods are very effective in the research and development phase, in which it is necessary to evaluate various IGBT chips with various kinds of structural parameters and/or combinations of various IGBT and diode chips. For this reason, we need a simple method to evaluate EMI noise by using single-chip IGBT.

II. APPROACH AND SIMULATION

A. Approach

The EMI noise is caused by the high dv/dt under small current turn-on condition [6-7], and also related to the common-mode current (I_{common}) [8]. We studied the relationships between I_{common}, dv/dt, di/dt and IGBT's waveform under small current (below 1A) turn-on condition by using the device simulation. Figure 1 is the equivalent circuit under an inductive load. A ground stray capacitance is added to the standard full bridge circuit. The IGBT is 15A rated Trench IGBT (T-IGBT) or CSTBT™. I_{common} is calculated as the summation of I_1 and I_2.

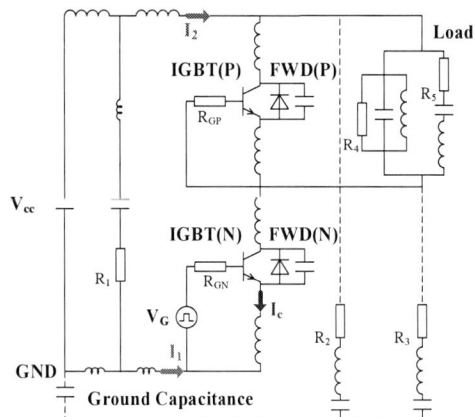

Fig. 1 Equivalent circuit in the device simulation with 15A rated T-IGBT or CSTBT™

978-1-7281-0582-6/19 $31.00 © 2019 IEEE

B. Simulation Results

Figure 2 (a) and (b) show the simulated I_{common} and I_C waveforms under the N-side small current turn-on condition, respectively. We found that I_{common} has strong correlation with the surge current (I_{surge}). Figure 3 (a) and (b) show the relationships between I_{surge} and (a) dv/dt(max) and (b) di/dt(max), respectively. Considering the higher I_{surge} of CSTBT^{TM} than T-IGBT at the same dv/dt(max) and/or di/dt(max) condition (point A), I_{surge} under low current turn-on can be a good indicator of EMI noise instead of dv/dt(max) and di/dt(max).

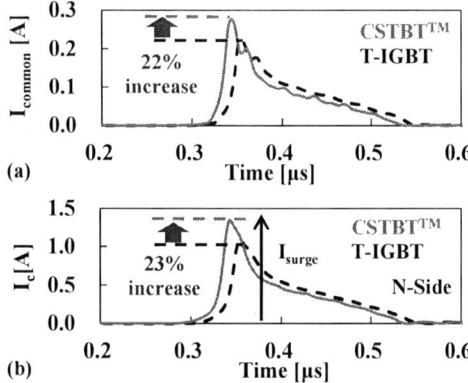

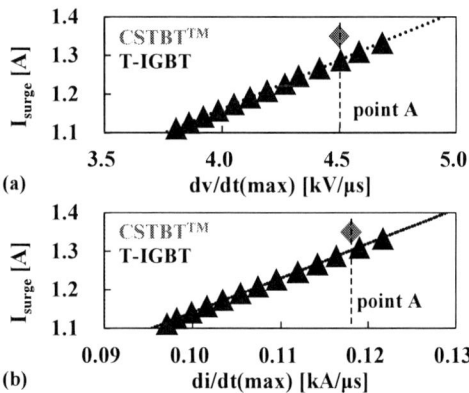

Fig. 2 Simulated I_{common} and I_{surge} of T-IGBT and CSTBT^{TM} at small current turn-on (T_j=25°C, V_{cc}=300V, V_{GE}=15/0V).

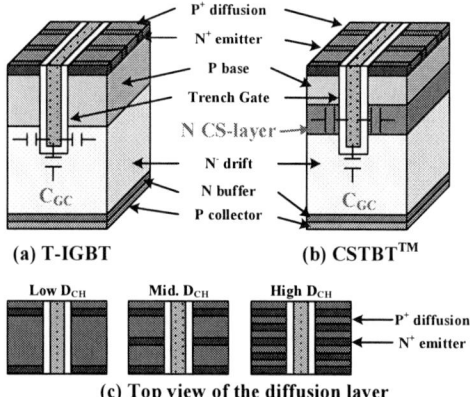

Fig. 3 Simulated relationships between I_{surge} and (a) dv/dt(max) and (b) di/dt(max) at small current turn-on.

III. EXPERIMENTAL RESULTS

The two sets of experiments were carried out. In the first step, we evaluated EMI noise of power modules mounted with various 600V/15A rated CSTBT^{TM} chips with the different CS-layer dose, channel density (D_{CH}) and gate-collector capacitance (C_{GC}), as shown in figure 4. Here, D_{CH} is controlled by altering the N$^+$ emitter – P$^+$ diffusion stripe ratio, and C_{GC} is also controlled by altering the trench depth. The single-chip assembly was fabricated to evaluate the turn-on characteristics such as dv/dt, di/dt and I_{surge}. To find a good indicator of EMI noise by using single-chip assembly, the correlation between EMI noise of power modules and turn-on characteristics of single-chip assembly were compared.

In the second step, the tradeoff among E_{on}-V_{CEsat}-SCSOA and EMI noise of CSTBT^{TM} was analyzed by using the single-chip assembly. As a reference, we also fabricated T-IGBT single-chip assembly.

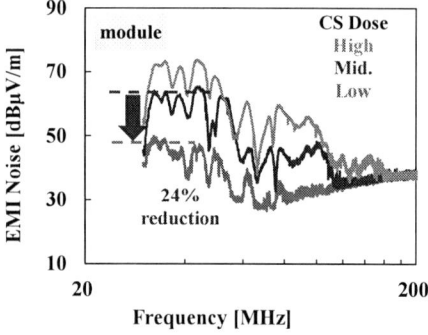

Fig. 4 Schematic 3D views of (a) T-IGBT, (b) CSTBT^{TM} and (c) top views of N$^+$ emitter arrangement

A. EMI noise and Small current turn-on

Figure 5 shows EMI noises of power modules mounted with different CS dose CSTBT^{TM}, measured in a shield room. EMI noise is reduced in the entire frequency range and its peak value is reduced 24% by lowering CS dose.

Fig. 5 Experimental results of the relationship between EMI noise of power module and CS dose.

978-1-7281-0582-6/19 $31.00 © 2019 IEEE

Figure 6 (a) and (b) show the experimental results of small current and 15A turn-on waveforms of the single-chip CSTBT™ with different CS dose, measured without any EMI shield facilities, respectively. By lowering CS dose, the small current turn-on waveforms become softer (low dv/dt, di/dt and I$_{surge}$). Considering the sensitivities (responsiveness) of 15A turn-on I$_{surge}$, small current turn-on dv/dt, di/dt or I$_{surge}$ are good candidates for the indicator of EMI noise.

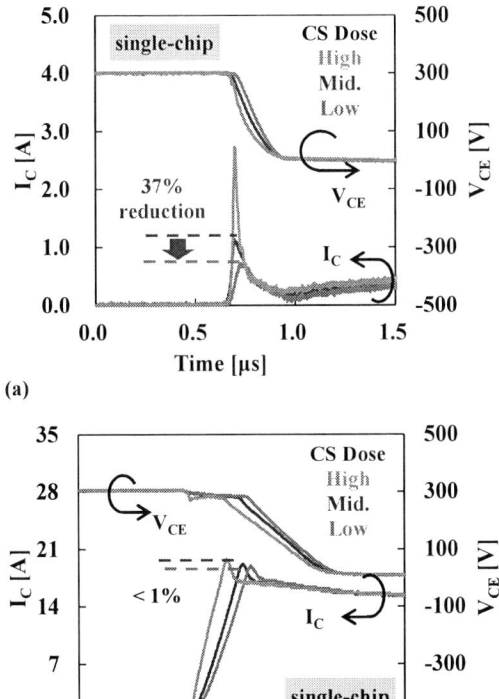

(a)

(b)

Fig. 6 Experimental results of the relationship between (a) small current turn-on I$_{surge}$ and (b) 15A turn-on I$_{surge}$ of single-chip CSTBT™ and CS dose.

Figure 7 shows the relationship between EMI noise (peak value) of power modules and small current turn-on dv/dt(max) of single-chip CSTBT™ with different CS dose, D$_{CH}$ and C$_{GC}$. While EMI noise can be reduced by lowering CS dose, D$_{CH}$ and/or C$_{GC}$, decrease of dv/dt(max) limits at around 2kV/us. This means there is no correlation between EMI noise and dv/dt(max) at low dv/dt region. Figure 8 shows the relationship between EMI noise and small current turn-on di/dt(max). Similar to dv/dt(max), there is no correlation between EMI noise and di/dt(max) at low di/dt region. On the other hand, as shown in figure 9, the correlation between EMI noise and small current turn-on I$_{surge}$ was fairly maintained almost up to the graph origin of

0A I$_{surge}$. Considering these results, I$_{surge}$ at low current turn-on is the better indicator of EMI noise than dv/dt and/or di/dt.

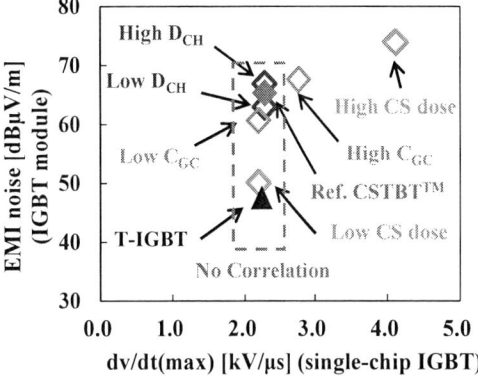

Fig. 7 Relationship between EMI noise and dv/dt(max).

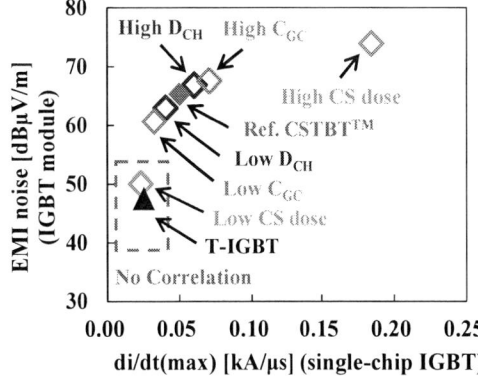

Fig. 8 Relationship between EMI noise and di/dt(max).

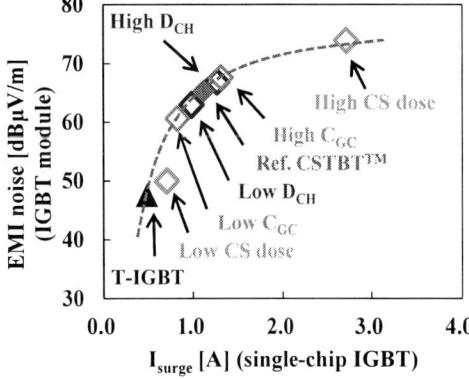

Fig. 9 Relationship between EMI noise and I$_{surge}$.

B. Analysis of the E_{on}-V_{CEsat}-SCSOA-EMI noise tradeoff

By using small current turn-on I_{surge} as the indicator of EMI noise, we analyzed the complex tradeoff among E_{on}-V_{CEsat}-SCSOA and EMI noise (I_{surge}). Figure 10 shows the tradeoff between small current turn-on I_{surge} and 15A turn-on E_{on} with three different gate resistances (R_G). The tradeoff is improved by lowering CS dose, which is also translated to C_{GC} reduction. Figure 11 shows the tradeoff relationships between V_{CEsat} and I_{surge} under constant E_{on} condition controlled by R_G. Commonly known, the capacitance related parameters such as CS dose and C_{GC} affects both EMI noise (I_{surge}) and V_{CEsat}, which results in the strong V_{CEsat}-I_{surge} tradeoff. Compared to the strong dependence on CS dose and C_{GC}, the dependence of I_{surge} on D_{CH} is weak, which results in V_{CEsat} reduction by D_{CH} increase under constant E_{on} condition. Figure 12 shows the tradeoff relationships between V_{CEsat} and SCSOA endurance time (t_w). While capacitance related parameters are known to be not effective to control t_w, the dependence of tw on D_{CH} is strong. Then, we can control EMI noise (I_{surge}) and t_w independently, by using the capacitance related parameters and D_{CH}, and optimize the total characteristics of CSTBT^{TM}.

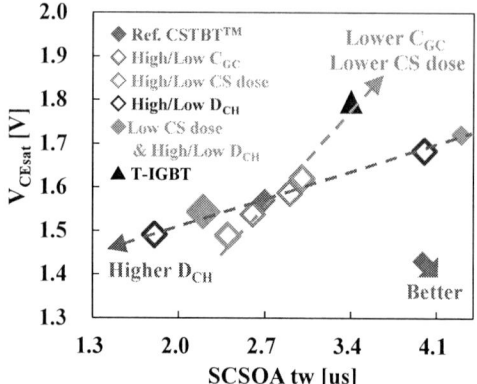

Fig. 12 Tradeoff between V_{CEsat} and SCSOA/tw.

IV. CONCLUSION

We found that EMI noise could be simply evaluated by using the low current turn-on I_{surge} of single-chip IGBTs without any EMI shield facilities. Through this method, maintaining the CSTBT^{TM}'s better tradeoff among E_{on}-V_{CEsat}-SCSOA, we could improve the EMI noise by reducing CS dose and/or C_{GC} while adjusting D_{CH}.

ACKNOWLEDGMENT

The authors would like to thank Dr. T. Minato, Mr. A. Furukawa, Mr. Y. Asai and Mr. T. Tadakuma for the support throughout this work.

REFERENCES

[1] H. Takahashi, H. Haruguchi, H. Hagino and T. Yamada, "Carrier Stored Trench-Gate Bipolar Transistor (CSTBT) –A Novel Power Device for High Voltage Application-", *Proc. ISPSD1996*, pp. 349-352, 1996.

[2] T. Takahashi, Y. Tomomatsu and K. Sato, "CSTBT^{TM}(III) as the next generation IGBT", *Proc. ISPSD2008*, pp.72-75, 2008.

[3] R. Kamibaba, K. Konishi, Y. Fukada, A. Narazaki and M. Tarutani, "Next Generation 650V CSTBT^{TM} with improved SOA fabricated by an Advanced Thin Wafer Technology", *Proc. ISPSD2015*, pp. 29-32, 2015.

[4] K. Suzuki, K. Nishi, M. Kaneda and A. Furukawa, "N-buffer design optimization for Short Circuit SOA ruggedness in 1200V class IGBT", *Proc. ISPSD2018*, pp. 128-131, 2018.

[5] K. Konishi, R. Kamibaba, M. Umeyama, A. Narazaki, T. Takahashi, A. Furukawa and M. Tarutani, "Experimental Demonstration of the Active Trench Layout Tuned 1200V CSTBT^{TM} for Lower dV/dt Surge and Turn-on Switching Loss", *Proc. ISPSD2016*, pp. 363-366, 2016.

[6] G. Busatto, C. Abbate, F. Iannuzzo, L. Fratelli, B. Cascone, G. Giannini, "EMI Characterisation of high power IGBT modules for Traction Application", PESC 2005, pp.2180-2186, 2005.

[7] S. Momota, M. Otsuki, K. Ishii, H. Takubo and Y. Seki, "Analysis on the Low Current Turn-on Behavior of IGBT Module", *Proc. ISPSD2000*, pp. 359-362, 2000.

[8] H. Bishnoi, P. Mattavelli, R. Burgos and D. Boroyevich, "EMI Behavioral Models of DC-Fed Three-Phase Motor Drive Systems", *IEEE Trans. on Power Electronics, vol.29, No.9*, pp.4633-4645, 2014

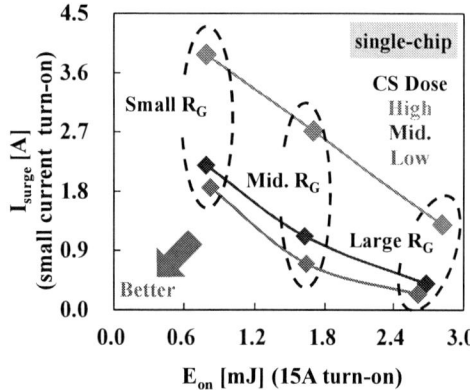

Fig. 10 Tradeoff between I_{surge} and E_{on}, controlled by R_G.

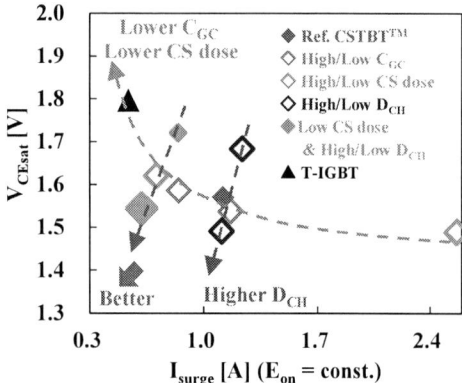

Fig. 11 Tradeoff between V_{CEsat} and I_{surge}.

Proceedings of the 31st International Symposium on Power Semiconductor Devices & ICs
May 19-23, 2019, Shanghai, China

An Ultra-Low Qrr Cell-Distributed Schottky Contacts SJ-MOSFET with Integrated Isolated NMOS

Shaohong Li, Ajiang Li, Tian Tian, Jing Zhu, Long Zhang, Weifeng Sun*, Guichuang Zhu, Yanqin Zou, Xin Tong, Yangyang Lu, Jiaxing Wei and Ran Ye

National ASIC System Engineering Research Center, Southeast University, Nanjing, China

*swffrog@seu.edu.cn

Abstract—An ultra-low reverse recovery charge (Q_{rr}) Cell-Distributed Schottky Contacts (CDSC) Superjunction MOSFET (SJ-MOSFET) is proposed in this paper. The proposed structure features that an isolated lateral NMOS is integrated in p-body region of SJ device. The conduction of the SJ-VDMOS intrinsic diode can be effectively suppressed by the reversed body diode of the integrated NMOS, which almost eliminates hole current in the voltage-sustaining layer during freewheeling period. Combining the CDSC with the integrated NMOS, an excellent reverse recovery characteristic of body diode in SJ-MOSFET is realized. The simulated results show that the proposed SJ-MOSFET can achieve a 97% and 91.8% lower Q_{rr} compared with the conventional and the CDSC SJ-MOSFETs at the same I_F of 3.2A. Moreover, the Figure-of-Merit (FOM = $R_{on,sp} \cdot Q_{rr}$) of the proposed structure is only 29.3mΩ·nC, which is much superior to that of the conventional and the CDSC SJ-MOSFETs.

Keywords—Superjunction MOSFETs; cell-distributed Schottky contacts; isolated integrated NMOS; reverse recovery charge

I. INTRODUCTION

For the applications requiring low power loss and high switching frequencies, SJ-MOSFETs are the most attractive candidates because of their low specific on-state resistance ($R_{on,sp}$). Unfortunately, they usually suffer from the adverse reverse recovery (RR) issue of their intrinsic diodes [1]. SJ-MOSFET with CDSC is a good approach [2], because it can be fully compatible to the conventional manufacturing process and achieve the improved RR characteristics. However, in the case of high freewheeling current (I_F) level, the portion of freewheeling current passing through Schottky contacts would be dramatically reduced in SJ-MOSFETs with paralleled Schottky diodes [3-4]. This is due to the body diodes of these SJ-MOSFETs that play a leading role in conducting the current, resulting in a significant reduction in the improvement of the RR.

In this paper, an ultra-low reverse recovery charge (Q_{rr}) CDSC SJ-MOSFETs with integrated isolated NMOS is proposed and investigated by TCAD simulations. By

This work was supported in part by the National Natural Science Foundation of China under Grant 61804026, Grant 61874026, Grant 61674030 and Grant 61504025, in part by the Postdoctoral Research Funding of Jiangsu Province under Grant 2018K001A, and in part by the National key research and development plan under Grant 2017YFB0402900.

employing the CDSC and integrated NMOS, the excellent RR characteristics of intrinsic diode at high freewheeling current can be obtained.

TABLE I. Key design parameters of the SJ-MOSFETs used in Sentaurus TCAD simulations.

Parameters	Description	Conv.	CDSC	Prop.
N_{Pb}	Doping concentration of the P-body region (cm^{-3})	2×10^{16}	2×10^{16}	2×10^{16}
N_{Pp}	Doping concentration of the P-pillar region (cm^{-3})	7.5×10^{15}	7.5×10^{15}	7.5×10^{15}
N_N	Doping concentration of the N-drift region (cm^{-3})	1×10^{15}	1×10^{15}	1×10^{15}
L_{CP}	Length of the cell pitch (μm)	20.2	20.2	20.2
L_{SC}	Length of the Schottky contact (μm)	/	2~8	2~8
T_{epi}	Thickness of the epitaxial layer (μm)	47	47	47
W_{SC}	Work function of the Schottky contact (eV)	/	4.7	4.7
K	The ratio of Lsc to Lcp (%)	/	0~39.6	0~39.6

II. DEVICE STRUCTURE AND MECHNISM

Fig. 1(a) and (b) show the cross section views of the conventional SJ-MOSFET and the cell-distributed Schottky contacts (CDSC) SJ-MOSFET. The proposed SJ-MOSFET features that an isolated lateral NMOS structure is integrated in the body region of the SJ device, which can been seen in Fig. 1(d). Fig. 1(c) is a partial enlarged view of Fig. 1(d). Table I lists the key structural parameters of the devices used in the Sentaurus TCAD simulations. The length of Schottky contact and cell pitch are L_{SC} and L_{CP}, respectively. The Schottky contact area ratio is defined as K ($K = L_{SC}/L_{CP}$).

Fig. 2(a) and (c) show the on-state hole and electron current flowlines distributions of body diodes in the three devices at $I_F = 3.2A$, K = 19.8%. In conventional and CDSC structures, the hole current mainly comes from p-body region. However, in proposed structure, the on-state current is almost entirely composed of electron current, which can be seen in Fig. 2(b) and (d), the hole current density along line Y =

978-1-7281-0582-6/19 $31.00 © 2019 IEEE

20.2µm of the proposed structure is much lower than that of the conventional and CDSC structures, and the electron current density of the proposed structure is the highest one. Consequently, the excess carrier in drift region can be almost eliminated, which contributes to the ultra-low body-diode Q_{rr} in proposed SJ-MOSFET.

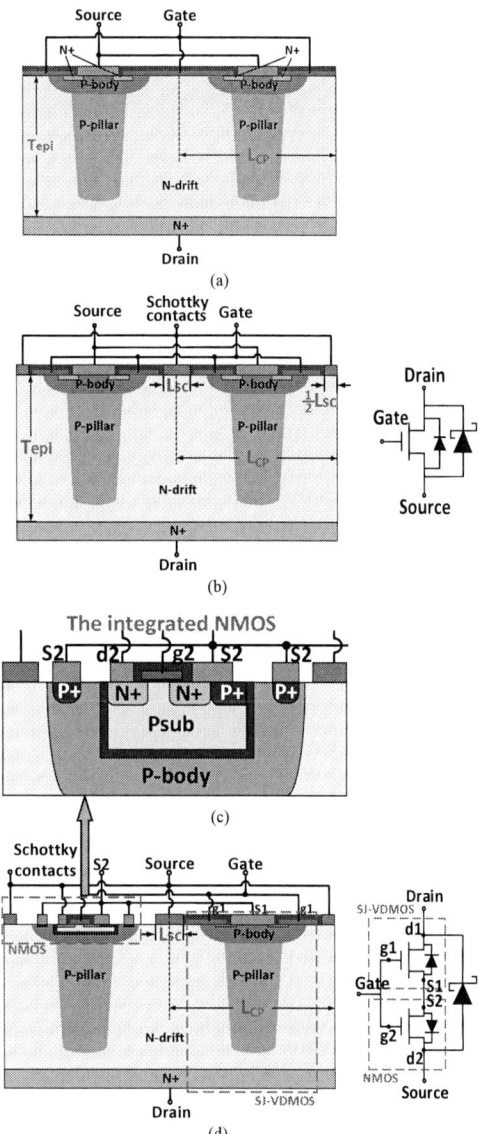

Fig. 1. The cross section views of (a) the conventional SJ-MOSFET, (b) the cell-distributed Schottky contacts (CDSC) SJ-MOSFET, (c) the enlarged view of the integrated NMOS in proposed structure and (d) the proposed SJ-MOSFET.

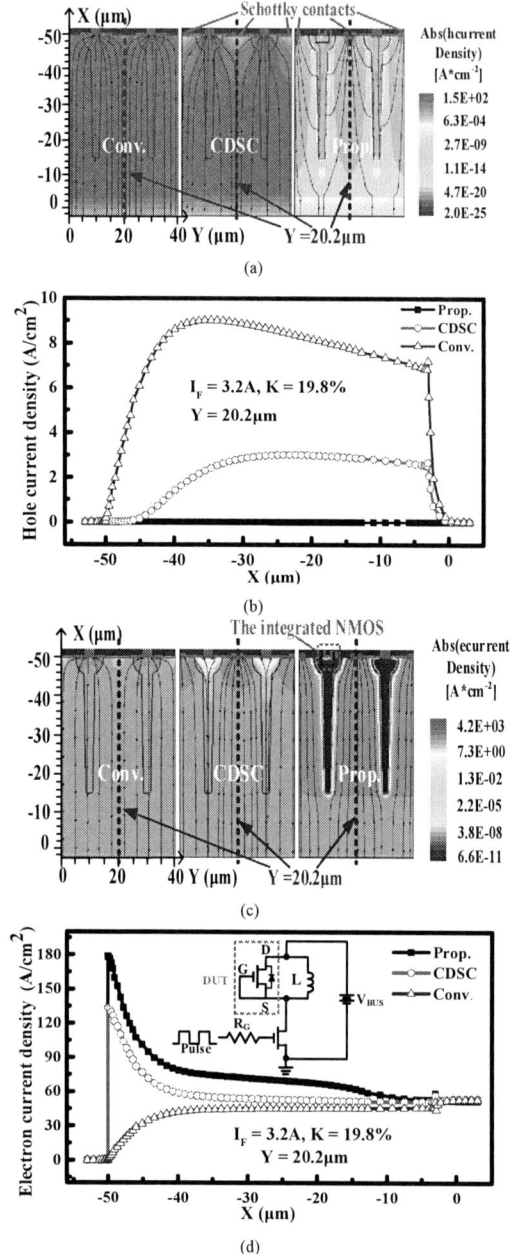

Fig. 2. The on-state current flowlines distributions of the conventional, the CDSC and the proposed structures at I_F = 3.2A, K = 19.8%. (a) The distribution of hole current flowlines. (b) Hole current density profile along Y = 20.2µm of the three structures. (c) The distribution of electron current flowlines. (d) Electron current density profile along Y = 20.2µm of the three structures.

978-1-7281-0582-6/19 $31.00 © 2019 IEEE

III. RESULTS AND DISSCUSSIONS

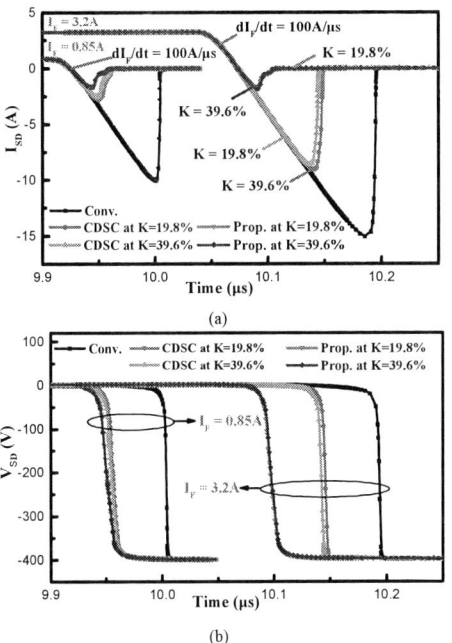

(a)

(b)

Fig. 3. Body-diode reverse recovery waveforms of the conventional, the CDSC and the proposed SJ-MOSFETs, (a) current waveforms (b) voltage waveforms.

Fig. 3(a) and (b) show the reverse-recovery characteristics of the body diodes of the three SJ-MOSFETs at different freewheeling current (I_F) level. At I_F = 0.85A, K = 19.8%, the RR of the CDSC and proposed structure both have a great improvement. Compared with the CDSC Structure, the proposed structure can achieve the superior RR at high I_F. At I_F = 3.2A, K = 19.8%, the body-diode Q_{rr} of the proposed structure is significantly reduced by 97.1% compared with the conventional structure, while the CDSC structure is only reduced by 63.2%. In CDSC structure, body-diode Q_{rr} decreases with the increase of K, however, in proposed structure, body-diode Q_{rr} hardly changes with the increase of K. The schematic of the reverse recovery test circuit is inserted in Fig. 2(d).

Fig. 4(a) and (b) show the electrical characteristics of the three devices. Both leakage current and on-resistance of the CDSC and proposed devices increase with the increase of K. The breakdown voltage of the structures with Schottky contacts would not degrade as the N-drift region can be rapidly pinched off by adjacent p-pillars, which provides electrostatic-shielding to the Schottky contacts. The leakage current of the structures with Schottky contacts nearly two orders of magnitude larger than that of the conventional structure at K = 39.6%. The on-resistance of the proposed structure is nearly two times of that of the CDSC structure, because the integrated NMOS in proposed structure occupies

half of the conductive channel. It is worth noting that the on-resistance of the CDSC structure only slightly increases at K = 9.9%. However, the FOM (FOM = $R_{on,sp} \cdot Q_{rr}$) of proposed structure is 94.9% and 86.3% lower than that of the conventional and CDSC structures at I_F = 3.2A, K = 24.7%, respectively, which can be revealed in Fig. 5.

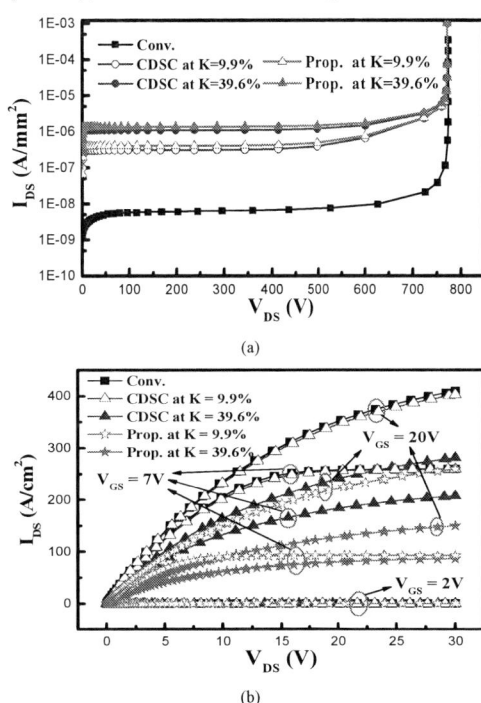

(a)

(b)

Fig. 4. Electrical characteristics of the conventional, the CDSC and the proposed SJ-MOSFETs, (a) reverse I-V characteristics (b) forward I-V characteristics.

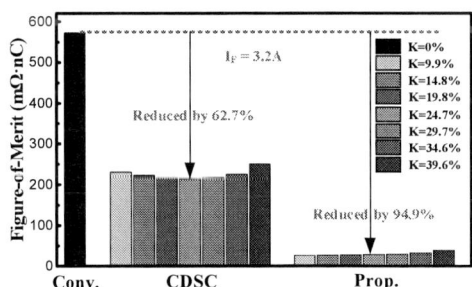

Fig. 5. The relationship between Figure-of-Merit and K among the three devices at I_F = 3.2A.

Fig. 6 shows the trade-off curve between the Q_{rr} of the body-diode and the specific on-resistance ($R_{on,sp}$) for the trench filling SJ-MOSFETs at I_F = 3.2A in this study. For the CDSC and the proposed structures, the larger K is good for Q_{rr}, but is not conducive to $R_{on,sp}$. Compared with the CDSC structure, the proposed structure can achieve the superior Q_{rr}. When K = 29.7%, the Q_{rr} of the CDSC structure is ten times higher than that of the proposed structure. In addition, the Q_{rr} has almost no change within the given K range for the proposed SJ-MOSFET.

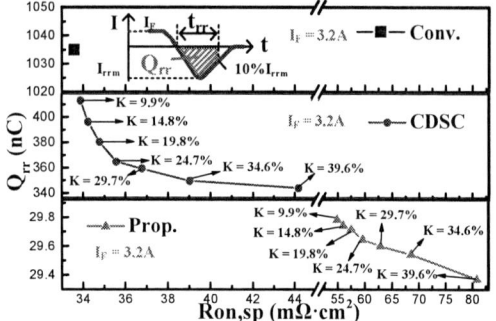

Fig. 6. The trade-off curve between the body-diode Q_{rr} and the $R_{on,sp}$ for the three SJ-MOSFETs studied in this paper.

IV. CONCLUSION

SJ-MOSFET with CDSC and integrated isolated NMOS is proposed and investigated by simulations in this paper. Combining the CDSC with the integrated NMOS, an excellent RR characteristics of body diode in SJ-MOSFET is realized. The proposed SJ-MOSFET can achieve a 97% and 91.8% lower Q_{rr} compared with the conventional and the CDSC SJ-MOSFETs at the same I_F of 3.2A.

REFERENCES

[1] R. Matsui et al., "Modeling of reverse recovery effect for embedded diode in SJ MOSFET," in Proc. IEEE EDSSC, Jun. 2015, pp. 375–378.

[2] Xu CHENG et al., "Improving the CoolMOS Body-Diode Switching Performance with Integrated Schottky Contacts," in Proc. 15th ISPSD, April 2003, pp. 304-307.

[3] Zhi Lin et al., "Low-Reverse Recovery Charge Superjunction MOSFET With a p-Type Schottky Body Diode," IEEE Electron Device Letters., vol. 38, pp. 1059–1062, 2017.

[4] KRISHNA SHENA et al., "Monolithically Integrated Power MOSFET and Schottky Diode with Improved Reverse Recovery Characteristics," in IEEE Trans Electron Devices, vol.37, pp.1167-1169, 1990.

Proceedings of the 31st International Symposium on Power Semiconductor Devices & ICs
May 19-23, 2019, Shanghai, China

1.6 kV Vertical Ga_2O_3 FinFETs With Source-Connected Field Plates and Normally-off Operation

Zongyang Hu[1], Kazuki Nomoto[1], Wenshen Li[1], Riena Jinno[1,2], Tohru Nakamura[3], Debdeep Jena[1,4,5] and Huili (Grace) Xing[1,4,5]

[1] School of Electrical and Computer Engineering, Cornell University, Ithaca, NY 14853, USA.
[2] Graduate School of Electric Science Engineering, Kyoto University, Kyoto, 606-8501, Japan.
[3] Center of Micro-Nano Technology, Hosei University, Koganei, Tokyo 184-0003, Japan.
[4] Department of Materials Science and Engineering, Cornell University, Ithaca, NY 14853, USA.
[5] Kavli Institute at Cornell for Nanoscale Science, Cornell University, Ithaca, NY 14853, USA.
Email: zh249@cornell.edu

Abstract—High performance Ga_2O_3 vertical FinFETs with a breakdown voltage of 1.6 kV, a drain current density of 600 A/cm^2 have been demonstrated in this work. Fin-shaped channels with sub-micron widths lead to a high threshold voltage of 4 V, and also provide strong RESURF effects to reduce the drain leakage current and increase the breakdown voltage. A low leakage current of lower than 10^{-3} A/cm^2 is maintained until the hard breakdown of the transistor. In order to sustain high voltage, a source-connected field plate (FP) supported by a dielectric layer is implemented at the outer edge of the gate pad as the edge termination, which enabled a breakdown voltage almost 2 times as high as those without FP. Device simulation shows that the highest electric field peak appears at the FP edge, which suggests further improvement of the breakdown voltage is possible by optimizing the FP design or implementing additional edge terminations.

Keywords—Vertical transistor, Ga_2O_3, FinFET, normally-off, breakdown voltage, field plate

I. INTRODUCTION

In recent years, Ga_2O_3 has been identified as one of the most important semiconductors for power applications. In its most stable crystal structure, monoclinic β-Ga_2O_3 has an ultra wide band gap of up to 4.9 eV, a high estimated breakdown electric field of 8 MV/cm [1] and a decent electron mobility of up to 250 cm^2/Vs [2], enabling high voltage, high current and stable device operations even under harsh environments. One of the most attractive aspects of Ga_2O_3 for both research and applications is the availability of large-area low dislocation density (~10^2 cm^{-2}) bulk Ga_2O_3 substrates through melt growth methods [3] [4]. This serves as the fundation for high quality epitaxial layers with electron mobility up to 150 cm^2/Vs [5] and > 10 μm thick n-type layers with controllable doping concentrations [6], which are excellent building blocks for power device development.

Similar to many other wide band gap semiconductors, Ga_2O_3 power devices are limited to unipolar conduction due to the lack of high conductivity p-type Ga_2O_3. To this end, MOS-based structures provide a RESURF (Reduced Surface Field) effect that enhances the breakdown voltages (*BV*), therefore Field plate (FP) Schottky Barrier Diodes (SBD) [7] and lateral MOSFETs [8] have shown high *BV* in excess of 1 kV. Furthermore, we have proposed and demonstrated vertical FinFETs [9] [10] and vertical trench SBDs [11] [12] with stronger RESURF effects near the channel region similar to those in p-n super junctions, and achieved record-high performance in vertical Ga_2O_3 devices well surpassing

the Si unipolar limit. These vertical Ga_2O_3 power devices directly benefit from the high quality bulk substrates and Halide Vapor Phaze Epitaxy (HVPE), and have the potential to deliver higher power than the lateral power devices. In our recent development of the vertical FinFETs, both experiment and simulation suggest that electric field crowding near the edge of the gate pads could limit the *BV* of the transistor [13]. In this new generation of vertical FinFETs, we have designed a source-connected FP with perfect compatibility with our existing process flow. This modification significantly improves the *BV* of the transistors to > 1.6 kV, compared to *BV* ~870 V for those without FP on the same sample. FP together with a higher charge concentration in the drift layer also leads to a slightly reduced on-resistance thus improved Baliga's figure of merit (BFOM).

II. EXPERIMENT

The sample used in this work is a 10 μm n$^-$-Ga_2O_3 grown by HVPE method on a (001) Ga_2O_3 substrate. The process flow is decribed as the following: The source contact Cr/Pt is deposited on a layer of n$^+$-Ga_2O_3 formed by Si ion implantation on the top surface of the wafer. The channel of the transistor is defined by electron beam lithogratphy, and formed by dry etching in an ICP-RIE system and subsequent acid/base wet chemical treatments to remove the plasma damages. From the SEM images taken in **Fig. 1**, a typical fin geometry is about 1.3 μm tall and 480-560 nm wide from the source-end to the drain-end. The bottleneck structure shown

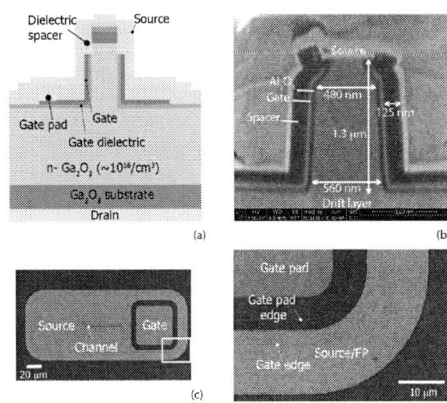

Fig. 1: (a) Schematic structure of a vertical Ga_2O_3 (001) FinFET with source-connected field plates, (b) an SEM cross section image, (c) SEM top-view image and (d) zoomed in image of the field plate region.

978-1-7281-0582-6/19 $31.00 © 2019 IEEE

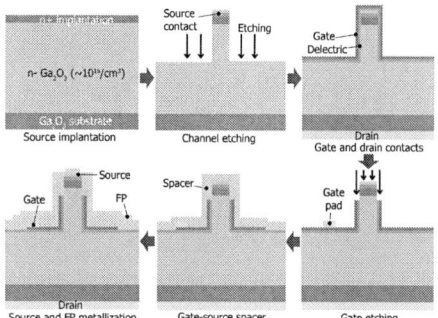

Fig. 2: Device fabrication process flow of vertical Ga_2O_3 FinFETs.

in Fig. 1(b) is likely a result of the wet chemical reaction. The drain ohmic contact is then metallized, followed by conformal depositions of a thin gate dielectric (Al_2O_3 by Atomic Layer Deposition (ALD)) and a gate metal (Cr) layer to form MOS structures on both sidewalls of the fin channel. The gate-source spacing is defined in the gate etchback step which includes a critical photoresist based planarization process [14]. A thick layer of ALD Al_2O_3 is used to realize the source-gate isolation, as well as to support the field plate extension. Finally, the source pad is conformally deposited in a sputtering system, forming the source-connected field plate at the same time. Relevant processing steps are schematically shown in **Fig. 2**. Compared to previous generations of Ga_2O_3 FinFETs [9] [10], the new device design features a reduced gate area and an extended source pad outside the gate edge. This allows the implementation of the FP without complicating the processing steps. This is essential in improving device yield and facilitating low-cost fabrication. The finished Ga_2O_3 FinFET (**Fig. 1**) has a 30-nm gate dielectric (ALD Al_2O_3), a 125-nm (h_{fp}) ALD Al_2O_3 source-gate spacer and a source metal with a 10-μm FP extension (L_{fp}) outside of the gate edges.

III. RESULTS

High frequency C-V measurements are taken on vertical MOS capacitors on etched (001) surface, and the net charge concentration in the n$^-$-Ga_2O_3 is calculated as about 1.2×10^{16} cm^{-3} (**Fig. 3**). DC I-V characteristics of a transistor with a fin width of ~0.5 μm (same channel geometries as the SEM image in Fig. 1(b)) show a maximum drain current density of 600 A/cm^2, a differential on-resistance of 5.5 mΩcm^2 at

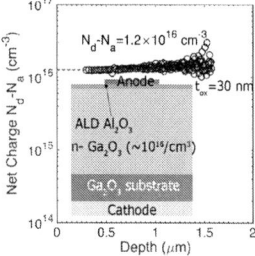

Fig. 3: Charge distribution in the drift layer extracted from high frequency (1 MHz) C-V measurements.

This work was in part supported by AFOSR FA9550-17-1-0048, NSF DMREF 1534303 and AFOSR FA9550-18-1-0529.

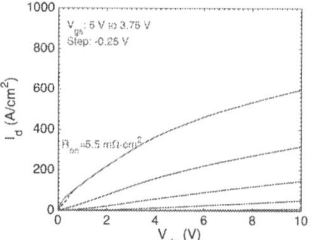

Fig. 4: DC I_d-V_{ds} characteristics of vertical Ga_2O_3 FinFETs with 0.5 μm fin width.

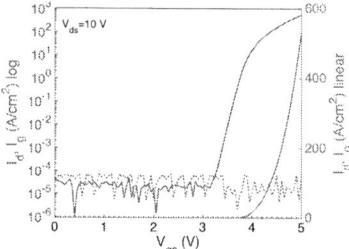

Fig. 5: DC I_d, I_g-V_{gs} characteristics of vertical Ga_2O_3 FinFETs with 0.5 μm fin width in log/linear scales.

V_{gs}=5 V (**Fig. 4**) and a threshold voltage V_{th}~3.8 V (**Fig. 5**) at V_{ds}=10 V. The current density and on-resistance are both normalized using the footprint of the vertical channel. The gate leakage current is lower than the limit of our measurement system for all applied voltages below the BV. The drain current on/off ratio is about 10^8, with its off state leakage at the same level as the gate leakage thanks to the strong RESURF effect provided by the fin channel. The unusually high V_{th} may be explained by a sheet of negative interface charge on the order of 2-5×10^{12} cm^{-2} on the sidewalls between the Al_2O_3 gate dielectric and the Ga_2O_3 channel [13]. Since this sheet charge concentration is significantly higher than the total donor concentration in the channel (3×10^{11} cm^{-2}), it reverses the electric field in the gate dielectric layer and causes V_{th} to be insensitive to the channel width.

2D simulations are performed to guide the design of the device for high BVs. It can be shown by both electrostatic analysis and numerical simulation that the electric field at the bottom of the fin channel is reduced significantly with a narrower fin channel width (W_{ch}). For example, for W_{ch}~0.5 μm channels in the FinFETs in this work and parameters used in **Fig. 6**, the electric field peak value at the bottom corner of the channel is 3.05 MV/cm, while peak value underneath the central area of the gate pad is 3 MV/cm. In comparison, the peak value at the edge of the gate pad is 9 MV/cm without any RESURF effects (all values are taken at a depth of 0.1 μm below the etched Ga_2O_3 surface). The stark difference is explained as the following: a gate without an edge termination has a lateral depletion region outside the edge, leading to severe electric field crowding typically described by the cylindrical junction model [15]. Due to the symmetry of the double-sided gate structure, the lateral depletion width near the bottom of the fin channel is limited to W_{ch}/2. This effectively removes majority of the space charges outside of the gate edge that would have caused the

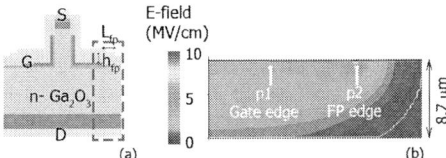

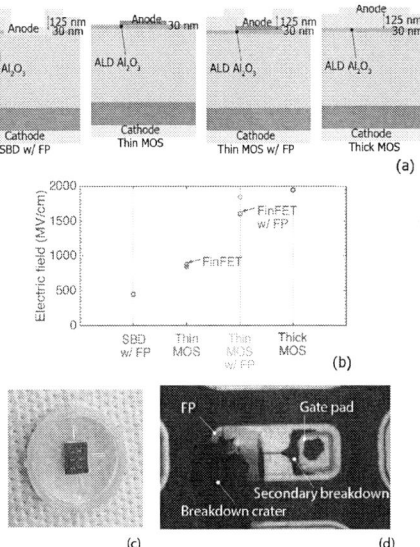

Fig. 6: Vertical Ga_2O_3 FinFET structure with FP. (b) Simulation of off-state electric field distribution at V_{ds}=1600 V showing electric field peaks at the gate (p1) and FP (p2) edges. L_{fp}=10 μm, h_{fp}=0.125 μm, N_d=1.2×10^{16} cm^{-3}.

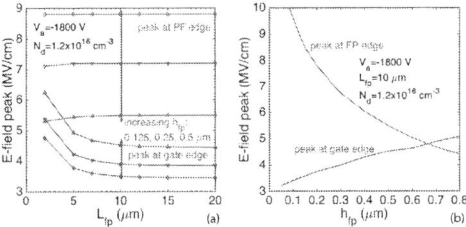

Fig. 7: Simulation of electric field peaks at the gate and FP edge as functions of (a) FP length (L_{fp}) and (b) FP height (h_{fp}). E-field values are taken at a depth of 0.1 μm inside Ga_2O_3. Other parameters are the same as those in Fig. 6.

electric field crowding. This effect is fundamentally very similar to the deep mesa edge termination [16], but with added benefits of easier experimental implementaion and the fact that it applies to unipolar power devices.

The electric field distribution near the gate edge with the addition of the FP show two peaks at the gate edge (p1) and FP edge (p2) respectively (**Fig. 6**). Two main parameters are considered for their impact on electric field peaks: the thickness of the supporting dielectric h_{fp} and FP extension ouside of the gate L_{fp}. It is discovered that for the voltage range considered, a FP extension L_{fp} of 10 μm or longer is able to suppress the electric field peak p1 for any thickness value of the h_{fp}, while the electric field at p2 is not significantly affected by L_{fp} (**Fig. 7(a)**). For a L_{fp} value of 10 μm, the tradeoff between p1 and p2 values and their heavy dependence on h_{fp} is clearly seen in **Fig. 7(b)**, where the optimal value of h_{fp} can be determined when the 2 field peaks have comparable magnitude.

Experimentally, a dielectric thickness h_{fp} of 0.125 μm is used due to the practical limitation of the device processes. Vertical diodes with various edge terminations (**Fig. 8(a)**) are measured as a reference to the transistor results, and the highest BV of each structure is shown in **Fig. 8(b)**. BV of the thick (155 nm dielectric) MOS diode is measured at 1950 V, which is more than twice of the BV measured on thin (30 nm dielectric) MOS diodes without FP. This is because the increase of the dielectric thickness has reduced the peak electric field at the surface of Ga_2O_3 at the same reverse voltage. The breakdown mechanism in dielectric materials [17] suggests that its BV will be limited by the electric field near the Al_2O_3/Ga_2O_3 interface, therefore much higher reverse voltage can be sustained by the thick MOS. The Schottky barrier diodes (SBD) with FP fabricated on the same sample have much lower BVs most likely due to plasma etch damage since all diodes are fabricated on dry-etched Ga_2O_3 surfaces. From the electric field simulation, the electric field peak at the gate edge (p1) of the thin MOS is

Fig. 8: Reverse breakdown voltages of Ga_2O_3 SBD and MOS structures: (a) schematic cross sections of diodes with different edge terminations, (b) measured highest reverse BVs of each structure, (c) fabricated Ga_2O_3 vertical FinFET sample size 10mm×7.5mm and (d) an optical image of a typical FinFET after breakdown.

significantly reduced by the FP. The thin MOS diode with FP has a BV of 1840 V, similar to those measured in thick MOS diodes. Off-state (V_{gs}=0 V) breakdown measurements of vertical FinFETs with and without FP show BVs of 1605 V and 876 V respectively (**Fig. 9**), which are similar to the results on vertical MOS diodes. The gate and drain leakage currents stay lower than 10^{-3} A/cm^2 until dielectric breakdown. Multiple FinFETs with fin channel widths ranging from 0.4 μm to 0.6 μm have been tested, and their BVs are largely independent of fin widths, indicating that breakdown is not dominated by the fin channel region. Both experimental data and simulation suggest that the breakdown is likely dominated by the dielectric breakdown at the edge of the FP (p2), which is then substantiated by the visual evidence that the most material damage after the destructive breakdown appears at the edge of the devices (**Fig. 8(d)**). In order to further improve the breakdown voltage of the transistor, additional edge termination may be implemented such as the multiple field plates, resistive ion implantation and floating guard rings.

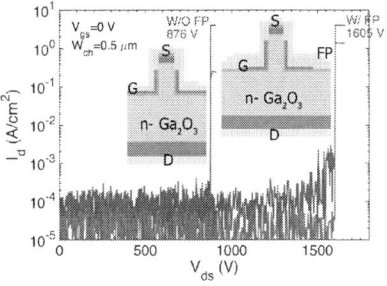

Fig. 9: Off-state (V_{gs}=0 V) I_d, I_g-V_{ds} and BV measured on vertical Ga_2O_3 FinFETs with and without FP.

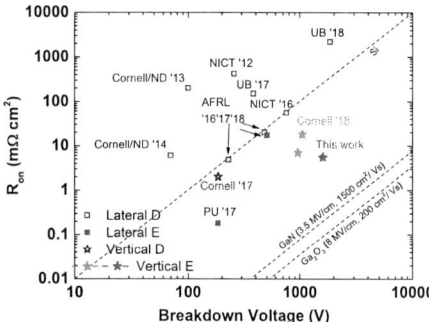

Fig. 10: On resistance and breakdown voltage benchmark of state-of-the-art Ga₂O₃ lateral and vertical power transistors [18]-[26]. The data

On-resistance and *BV* benchmark for state-of-the-art lateral and vertical Ga₂O₃ transistors is summarized in **Fig. 10**. Compared to previous generation of vertical FinFETs with a lower charge concentration of 10^{15} cm^{-3} in the drift layer [10], the FP FinFETs in this work has slightly lower R_{on} and much higher *BV*, thus significantly improved BFOM comparable to the state-of-the-art lateral Ga₂O₃ FETs. Furthermore, the vertical FinFET in this work combine a superior electrostatic design based on RESURF principles and a novel fabrication process flow, which makes it an attractive power transistor concept that can be applied to any wide band gap semiconductors.

ACKNOWLEDGMENT

This work was performed in part at Cornell NanoScale Facility, an NNCI member supported by NSF Grant No. ECCS-1542081.

REFERENCES

[1] M. Higashiwaki, A. Kuramata, H. Murakami and Y. Kumagai, "State-of-the-art technologies of gallium oxide power devices", J. Phys. D: Appl. Phys. Vol. 50, pp. 333002-1-12, July 2017.

[2] N. Ma, N. Tanen, A. Verma, Z. Guo, T. Luo, H. G. Xing and D. Jena, "Intrinsic electron mobility limits in β-Ga₂O₃", Appl. Phys. Lett., Vol. 109, pp. 212101-1-5, Nov. 2016.

[3] T. Oishi, Y. Koga, K. Harada and M. Kasu, "High-mobility β-Ga₂O₃ (-201) single crystals grown by edge-defined film-fed growth method and their Schottky barrier diodes with Ni contact", Appl. Phys. Express, Vol. 8, pp. 031101-1-3, Feb. 2015.

[4] E. G. Villora, K. Shimamura, Y. Yoshikawa, K. Aoki and No. Ichinose, "Large-size β-Ga₂O₃ single crystals and wafers", J. Crystal Growth, Vol. 270, pp. 420-426, Aug. 2004.

[5] M. Baldini, M. Albrecht, A. Fiedler, K. Irmscher, R. Schewski and G. Wagner, "Si- and Sn-Doped Homoepitaxial β -Ga₂O₃ Layers Grown by MOVPE on (010)-Oriented Substrates", ECS J. Solid State Sci. Tech., Vol. 6, No. 2, pp. Q3040-Q3044, Oct. 2016.

[6] H. Murakami, K. Nomura, K. Goto, K. Sasaki, K. Kawara, Q. T. Theiu et al, "Homoepitaxial growth of β-Ga₂O₃ layers by halide vapor phase epitaxy", Appl. Phys. Express, Vol. 8, pp. 015503-1-4, Dec. 2014.

[7] K. Konishi, K. Goto, H. Murakami, Y. Kumagai, A. Kuramata, S. Yamakoshi and M. Higashiwaki, "1-kV vertical Ga₂O₃ field-plated Schottky barrier diodes", Appl. Phys. Lett., Vol. 110, pp. 103506-1-4, Mar. 2017.

[8] K. Zeng, A. Vaidya and U. Singisetti, "1.85 kV Breakdown Voltage in Lateral Field-Plated Ga₂O₃ MOSFETs", IEEE Electron Device Lett., Vol. 39, No. 9, pp. 1385-1388, Sept. 2018.

[9] Z. Hu, K. Nomoto, W. Li, L. J. Zhang, J.-H. Shin, N. Tanen, T. Nakamura, D. Jena and H. G. Xing, "Vertical fin Ga₂O₃ power field

effect transistors with on/off ratio > 10^{9}", Device Research Conf. (DRC), pp. 1-3, June 2017.

[10] Z. Hu, K. Nomoto, W. Li, N. Tanen, K. Sasaki, A. Kuramata, D. Jena and H. G. Xing, "Enhancement-mode Ga₂O₃ Vertical Transistors With Breakdown Voltage > 1kV", IEEE Electron Device Lett., Vol. 39, No. 6, pp. 869-872, June 2018.

[11] W. Li, Z. Hu, K. Nomoto, Z. Zhang, J.-Y. Hsu, Q. T. Thieu, K. Sasaki, A. Kuramata, D. Jena and H. Xing, "1230 V β-Ga₂O₃ trench Schottky barrier diodes with an ultra-low leakage current of < 1 μA/cm²", Appl. Phys. Lett., Vol. 113, pp. 202101-1-5, Nov. 2018.

[12] W. Li, Z. Hu, K. Nomoto, R. Jinno, Z. Zhang, T. Q. Tu, K. Sasaki, A. Kuramata, D. Jena and H. Xing, "2.44 kV Ga₂O₃ vertical trench Schottky barrier diodes with very low reverse leakage current", IEEE IEDM digest, pp. 8.5.1-8.5.4, Dec. 2018.

[13] Z. Hu, K. Nomoto, W. Li, Z. Zhang, N. Tanen, Q. T. Thieu, K. Sasaki, A. Kuramata, T. Nakamura, D. Jena and H. Xing, "Breakdown mechanism in 1 kA/cm² and 960 V E-mode β-Ga₂O₃ vertical transistors", Appl. Phys. Lett., Vol. 113, pp. 122103-1-5, Sept. 2018.

[14] Y. Zhang, M. Sun, D. Piedra, J. Hu, Z. Liu, Y. Lin, X. Gao, K. Shepard and T. Palacios, "1200 V GaN Vertical Fin Power Field-Effect Transistors", IEEE IEDM digest, pp. 9.2.1-9.2.4, Dec. 2017.

[15] B. J. Baliga, "Fundamentals of Power Semiconductor Devices", Chapter 3, New York, NY, USA: Springer-Verlag, 2008.

[16] H. Fukushima, S. Usami, M. Ogura, Y. Ando, A. Tanaka, M. Deki et al, "Vertical GaN p-n diode with deeply etched mesa and the capability of avalanche breakdown", Appl. Phys. Express, Vol. 12, pp. 026502-1-4, Feb. 2019.

[17] J. W. McPherson, "Time dependent dielectric breakdown physics – Models resisited", Micro. Reliability, Vol. 52, pp. 1753-1760, July, 2012.

[18] W. S. Hwang, A. Verma, H. Peelaers, V. Protasenko, S. Rouvimov, H. G. Xing, A. Seabaugh, W. Haensch, C. Van de Walle, Z. Galazka, M. Albrecht, R. Fornari and D. Jena, "High-voltage field effect transistors with wide-bandgap β-Ga₂O₃ nanomembranes", Appl. Phys. Lett., Vol. 104, pp. 203111-1-5, May 2014.

[19] W. S. Hwang, A. Verma, V. Protasenko, S. Rouvimov, H. G. Xing, A. Seabaugh, W. Haensch, C. Van de Walle, Z. Galazka, M. Albrecht, R. Fornari and D. Jena, "Nanomembrane β-Ga₂O₃ High-Voltage Field Effect Transistors", Device Research Conf. (DRC), pp. 207-208, June 2013.

[20] M. Higashiwaki, K. Sasaki, A. Kuramata, T. Masui and S. Yamakoshi, "Gallium oxide (Ga₂O₃) metal-semiconductor field-effect transistors on single-crystal β-Ga₂O₃ (010) substrates", Appl. Phys. Lett., Vol. 100, pp. 013504-1-3, Jan. 2012.

[21] M. H. Wong, K. Sasaki, A. Kuramata, S. Yamakoshi and M. Higashiwaki, "Field-Plated Ga₂O₃ MOSFETs With a Breakdown Voltage of Over 750 V", IEEE Electron Device Lett., Vol. 37, No. 2, pp. 212-215, Feb. 2016.

[22] A. J. Green, K. D. Chabak, E. R. Heller, R. C. Fitch, M. Baldini, A. Fiedler, K. Irmscher, G. Wagner, Z. Galazka, S. E. Tetlak, A. Crespo, K. Leedy and G. H. Jessen, "3.8 MV/cm breakdown strength of MOVPE-Grown Sn-Doped β-Ga₂O₃ MOSFETs", IEEE Electron Device Lett., Vol. 37, No. 7, pp. 902-905, July 2016.

[23] N. Moser, J. McCandless, A. Crespo, K. Leedy, A. Green, A. Neal, S. Mou, E. Ahmadi, J. Speck, K. Chabak, N. Peixoto and G. Jessen, "Ge-Doped β-Ga₂O₃ MOSFETs", IEEE Electron Device Lett., Vol. 38, No. 6, pp. 775-778, June 2017.

[24] K. D. Chabak, J. P. McCandless, N. A. Moser, A. J. Green, K. Mahalingam, A. Crespo, N. Hendricks, B. M. Howe, S. E. Tetlak, K. Leedy, R. C. Fitch, D. Wakimoto, K. Sasaki, A. Kuramata and G. H. Jessen, "Recessed-Gate Enhancement-Mode β -Ga₂O₃ MOSFETs", IEEE Electron Device Lett., Vol. 39, No. 1, pp. 67-70, Jan. 2018.

[25] K. Zeng, J. S. Wallace, C. Heimburger, K. Sasaki, A. Kuramata, T. Masui, J. A. Gardella Jr. and U. Singisetti, "Ga₂O₃ MOSFETs using Spin-On-Glass Source/Drain Doping Technology", IEEE Electron Device Lett., Vol. 38, No. 4, pp. 513-516, Apr. 2017.

[26] H. Zhou, M. Si, S. Alghamdi, G. Qiu, L. Yang and P. D. Ye, "High-Performance Depletion/Enhancement-Mode β-Ga₂O₃ on Insulator (GOOI) Field-Effect-Transistors With Record Drain Currents of 600/450 mA/mm", IEEE Electron Device Lett., Vol. 38, No. 1, pp. 103-106, Jan. 2017.

978-1-7281-0582-6/19 $31.00 © 2019 IEEE

Proceedings of the 31st International Symposium on Power Semiconductor Devices & ICs
May 19-23, 2019, Shanghai, China

Influence of Carrier Lifetime on Silicon Carbide Power Devices for Pulsed Power Application

Kun Zhou*, Yingxing Cui, Lianghui Li, Yunfei Gu, Lin Zhang, Shuairong Deng, Zhiqiang Li, and Juntao Li
Microsystem & Terahertz Research Center, China Academy of Engineering Physics, Chengdu/China
Institute of Electronic Engineering, China Academy of Engineering Physics, Mianyang/China
zhoukun@mtrc.ac.cn

Abstract—In this paper, we investigate the influence of carrier lifetime on SiC GTO and PiN devices for pulsed power application. Both the device characteristics and the circuit dynamic performance in pulsed discharging system are studied based on simulation and experiments. The influence of τ_{HL} on the static and dynamic discharging characteristics in both single- and multi-stage pulsed power network (PFN) are analyzed by extensive mix-mode simulation considering thermoelectric effect within SiC devices. Simulation results show that the SiC GTO is more sensitive to carrier lifetime and induces the power dissipation 10X higher than that of SiC PiN during pulsed discharging. The discharging capability of SiC GTO decreases for $\tau_{HL}<0.4\mu s$. The heat generates in the drift region and tends to shift towards the top anode region in long pulse width discharging. The simulation is demonstrated by experimental data of our fabricated SiC GTOs.

Keywords—silicon carbide (SiC), power device, gate turn-off (GTO) thyristor, carrier lifetime, pulsed power application

I. INTRODUCTION

Silicon carbide (SiC) power devices are attractive for power electronics and pulsed power applications[1]. Compared with silicon, the SiC material exhibits 10× higher critical electric field, higher operation temperature, and much higher thermal conductivity. Hence the SiC power devices are more favorable for high voltage and high power applications. For pulsed power application, the switching power device has to support high voltage of several kilovolts in the blocking state and conduct the current up to several kiloampere, releasing the energy stored in the capacitor. As the bipolar devices, the SiC GTO and PiN are favorable due to both the large current handling capability and high blocking voltage. In particular, the SiC GTO is more suitable for pulsed power application than the voltage controlled devices, e.g. the power MOSFET and IGBT, since the on-state current characteristic of SiC GTO is mostly close to that of PiN diode and has no limit of the saturation current that is typical for power MOSFET and IGBT. Extensive attention is given to SiC GTO for pulsed power application[2,3].

For bipolar devices, the high level carrier lifetime (τ_{HL}) is critical for conductivity modulation effect in the drift region and thus has significant impact on the on-state performance. For Si material, the τ_{HL} is tens of microseconds or even higher and needs to be reduced to improve the switching speed for Si IGBTs. Differently, the SiC material still suffers from a low carrier lifetime due to the defects e.g. $Z_{1/2}$ and EH_{67}, limiting the conductivity modulation effect and thus the conduction performance of SiC bipolar devices. At

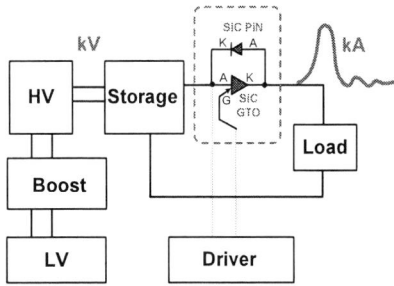

Fig. 1. Schematic topology of the pulsed power system. The SiC GTO is used as high voltage switch and the SiC PiN diode acts as the freewheeling diode.

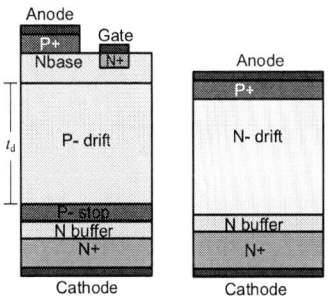

Fig. 2. Schematic cross-section views of SiC GTO and PiN devices.

present, most efforts are focused on carrier lifetime enhancement in SiC as well as the influence of carrier lifetime on the static performance from the prospective of power electronics l[4-7]. Few attentions are given on how the carrier lifetime influences the thermoelectric conduction behaviors of SiC switching devices especially in pulsed power application to our best knowledge.

In this paper, we present the impact of carrier lifetime on the device characteristics and circuit dynamic performance in pulsed discharging system by simulation and experiments. The influence of τ_{HL} on the device static characteristics and dynamic discharging performance is evaluated in the single- and multi-stage pulsed power network (PFN) by extensive mix-mode simulation considering the heating effect within SiC devices. The simulation is demonstrated by experimental data of our fabricated SiC GTOs. Both quantitative and physical insight are provided in this work for evaluating the impact of carrier lifetime on SiC bipolar devices for pulsed power application.

978-1-7281-0582-6/19 $31.00 © 2019 IEEE

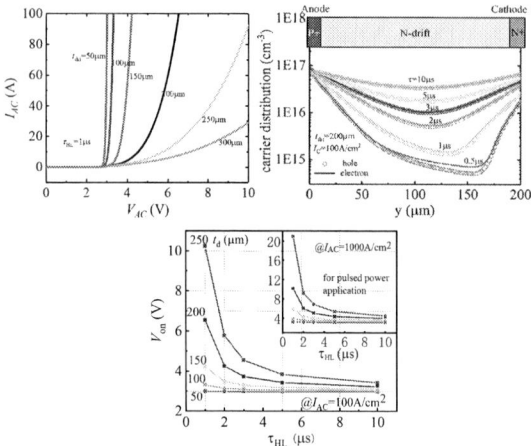

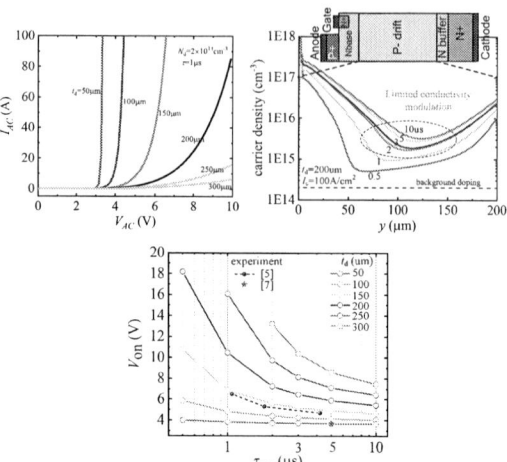

Fig. 3. (a) I-V characteristics of SiC PiN at τ_{HL}=1μs. (b) Carrier distribution in SiC PiN for different τ_{HL} values. (c) V_{on} versus τ_{HL} for SiC PiN as the function of t_d.

Fig. 4. (a) I-V characteristics of SiC GTO at τ_{HL}=1μs. (b) Carrier distribution in SiC GTO for different τ_{HL} values. (c) V_{on} versus τ_{HL} for SiC GTO at different t_d. Overall, the improvement of V_{on} by increasing τ_{HL} is not so dramatic for SiC GTO as that for SiC PiN. Note that the simulation agrees well with experiments.

II. CIRCUIT AND DEVICE STRUCTURES

The topology of pulsed power system is shown in Fig. 1. The low voltage input is converted to high voltage by the booster module charging the capacitor. In the RLC discharging circuit, the SiC GTO acts as the switching device and SiC PiN is anti-parallel connected as the freewheeling diode in order to protect the SiC GTO considering its low reverse breakdown voltage (*BV*). Fig. 2 shows the device structures and key parameters of the SiC GTO and PiN devices.

III. RESULTS AND DISCUSSION

In this section, the influence of carrier lifetime on both static and dynamic characteristics of SiC GTO and PiN devices is investigated. The electron carrier lifetime τ_n is assumed to be five times of hole carrier lifetime τ_p. The ambipolar lifetime for high level injection is $\tau_n+\tau_p$. The dependence of carrier lifetime on temperature and doping concentration is taken into account.

A. Static Characteristics

Fig. 3 shows the on-state characteristics of SiC PiN with different drift region thickness (t_d). For a constant carrier lifetime of τ_{HL}=1μs, the current capability of SiC PiN decreases with the increasing t_d, as shown in Fig. 3(a). The SiC PiN for the high *BV* of 7kV~35kV requires the t_d in the range of 50μm~300μm. The V_{on} (@100A/cm²) of SiC PiN rises up to over 10V for $t_d \geq 250$μm due to the degraded conductivity modulation effect. Fig. 3(b) shows the carrier distribution in the drift region of SiC PiN at on-state current of 100A/cm². For t_d=200μm and τ_{HL}=1μs, the carrier density in the drift region reduces to 10^{15}cm⁻³. Increasing the carrier lifetime effectively improves conductivity modulation and thus the carrier density is increased to over 10^{16}cm⁻³ for τ_{HL}>3μs. Fig. 3(c) shows the influence of τ_{HL} on the V_{on} at 100A/cm² and 1000A/cm² for SiC PiN with different t_d. It is observed that the V_{on} at 100A/cm² of SiC PiN with various t_d values can be reduced to below 6V for τ_{HL}=2~4μs. Such V_{on}

is sufficiently low for power electronic applications, e.g. the inverter or DC-DC converter. For pulsed power application where the involved SiC PiN is to conduct much higher current density up to 10^3A/cm², the carrier lifetime of τ_{HL}>5μs is needed to obtain a low V_{on}, as shown in the inset of Fig. 3(c).

Fig. 4 shows the impact of τ_{HL} on the conduction characteristics of SiC GTO. Although similar as SiC PiN, the SiC GTO shows weakened conductivity modulation effect and higher V_{on} than that of SiC PiN. For the SiC GTO, the effect of τ_{HL} on conduction performance is closely related to the gate driving current (I_G). For the SiC p-GTO, excess holes in the drift region are provided via the topside PNP bipolar transistor. For a given I_G, the hole current injected into the drift region is determined by the gain (β_{PNP}), i.e. $I_G \times \beta_{PNP}$. The indirect injection of holes leads to the limited modulation conductivity in the drift region of SiC GTO. Therefore the influence of the increasing carrier lifetime on the conduction characteristics for SiC GTO is not as significant as that for SiC PiN diodes, if the I_G is small. As shown in Fig. 4(b), for I_G=100mA, the carrier density in the middle of the drift region remains below 10^{16}cm⁻³ although the is τ_{HL} increased up to 10μs. Fig. 4(c) shows the dependence of V_{on} on τ_{HL} as a function of t_d. It is observed that the V_{on} of SiC p-GTO significantly reduces with the increasing τ_{HL} in the range of 1~5μs and shows limited reduction for τ_{HL}>5μs. Different from the SiC PiN, the SiC GTO still exhibits the increasing V_{on} as the t_d, even with a high τ_{HL} of 10μs. Hence, both the high τ_{HL} and I_G are necessary for SiC GTO used in pulsed discharging application. Fig. 4(c) also shows that our simulation results agree well with the reported experimental data.

B. Dynamic Characteristics

Fig. 5(a) shows the simplified single-stage pulsed discharging topology including the charging and discharging circuits. Here, the storage capacitor (C_{st}) is 2μF, the load resistor (R) is 60mΩ, the stray inductance (Ls) is 100nH, the supply voltage (V_{CC}) is 1kV. For the SiC GTO and PiN, the

978-1-7281-0582-6/19 $31.00 © 2019 IEEE

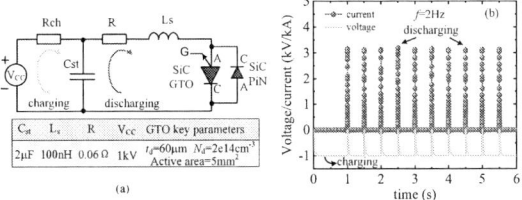

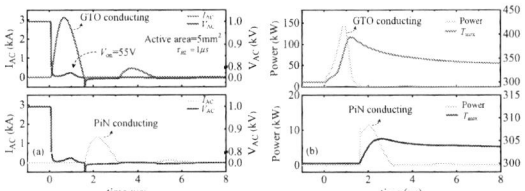

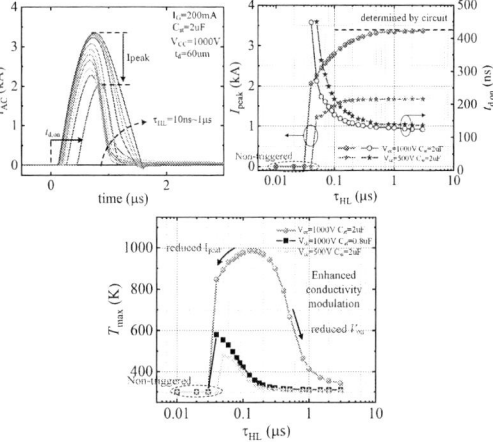

Fig. 5. (a) Single-stage pulsed discharging circuit. (b) Simulated continuous charging-discharging waveforms with f=2Hz. Discharging pulsed width is 1.6μs and I_{peak} of SiC GTO reaches 3.1kA.

Fig. 6. I-V discharging waveforms (a) and transient power (b) of SiC GTO and PiN with τ_{HL}=1μs. The SiC GTO endures large I_{AC} and power dissipation up to 150kW. The thermal resistance (R_{th}) of 0.05cm²K/W is used in simulation. The rise of T_{max} in SiC PiN is not severe.

Fig. 7. (a) Influence of τ_{HL} on SiC GTO current. The τ_{HL} greatly influences the SiC GTO discharging performance. The current waveform shrinks leading to reduced discharging energy if τ_{HL}<0.5μs. (b) Dependence of I_{peak} and t_{don} on τ_{HL} at different discharging conditions. The high level discharging is more sensitive to τ_{HL}. (c) T_{max} versus τ_{HL}. Note that there exists a maximum T_{max} in 50ns<τ_{HL}<0.3μs, due to both the large I_{peak} and high V_{on}.

t_d=60μm and the active area of 5mm² are used. Both the electrical and thermal behaviors of the device are taken into account in the mix-mode TCAD simulation. Fig. 5(b) shows the I-V waveforms for continuous discharging with the frequency of f=2Hz. With the SiC GTO biased at the blocking state, the power supply charges the C_{st} to a high voltage. When a triggering I_G is applied, the SiC GTO is switched-on and conducts large current discharging the energy stored in C_{st}, determined as $0.5 \times C_{st} \times V_{CC}^2$. The peak current ($I_{peak}$) reaches 3kA for V_{CC}=1000V and C_{st}=2μF.

Fig. 6 shows the electrical and thermal characteristics of SiC GTO and PiN in pulsed charging transient. For τ_{HL}=1μs, the SiC GTO conducts the I_{peak} of 3.1kA (60kA/cm²) and the V_{on} reaches 55V, as shown in Fig. 7(a). The large I_{peak} and V_{on} lead to the peak power of around 150kW, as shown in Fig. 6(b). Although the enduring time is only 1.6μs, such high power density of 30kW/mm² still yields significant heat dissipation inside the SiC GTO, leading to the maximum temperature (T_{max}) of 394K. The heat is generated initially at the middle drift region due to the relatively lower conductivity modulation effect. The SiC PiN conducts the current of 1.2kA after the discharging half-cycle of SiC GTO. Since the lower current and stronger conductivity modulation effect, the SiC PiN delivers the power dissipation 90% lower than that of SiC GTO and thus the heat generation is not severe, as shown in Fig. 6(b).

The discharging performance of the system is closely related to the switching devices. Fig. 7 shows the influence of carrier lifetime on the current waveforms of SiC GTO. As observed in Fig. 7(a), the τ_{HL} is critical for the discharging capability of SiC GTO. The current waveform shrinks leading to reduced discharging energy if τ_{HL}<0.5μs, since the energy is consumed by the increased on-resistance of the devices. The I_{peak} significantly reduces as the decreasing τ_{HL} in the range of τ_{HL}≤100ns, as a result of the degraded conductivity modulation effect. In addition, the turn-on delay time also increases, as shown in Fig. 7(b). For τ_{HL}≤40ns the SiC GTO cannot be turned on. Although the I_{peak}=3.1kA can

be achieved in the range of τ_{HL}>0.1μs, the T_{max} of SiC GTO exhibits a maximum >800K in the middle range of τ_{HL}=50ns~0.3μs, as shown in Fig. 7(c). This is attributed to both the large I_{peak} and V_{on} in this τ_{HL} range. Such case should be considered if the SiC devices are subjected to neutron irradiation which induces the τ_{HL} degradation and the severe heating may induce failure at metal contact or package. Fig. 7(c) shows that higher τ_{HL} value is required for high energy discharging with large C_{st}-V_{CC} conditions. The τ_{HL}>1μs is beneficial for the SiC GTO to remain the T_{max}<400K.

We have also evaluated the conduction performance of SiC GTO as well as the influence of τ_{HL} for wide pulse discharging in the 10-stage pulse-forming-network (PFN). Fig. 8(a) shows the circuit topology and key device parameters. The current waveforms of SiC GTO with different τ_{HL} values are given in Fig. 8(b). In the 10-stage PFN system, the SiC GTO conducts I_{AC}=2.45kA for 100μs long pulse width at V_{cc}=2.7kV. For such large current and wide pulse conduction, the SiC GTO endures high transient power consumption and heat dissipation. Hence it is necessary to reduce the on-resistance of the SiC GTO by carrier lifetime enhancement. As shown in Fig. 8(b), the device suffers thermal runaway and the current collapses for τ_{HL}≤0.2μs. Only τ_{HL}≥0.3μs satisfies 2.45kA large current discharging. At I_{peak}=2.45kA, the T_{max} within the SiC GTO can be remained below 400K for τ_{HL}≥0.4μs. Fig. 8(c) shows the dependence of T_{max} on τ_{HL} for different current levels. At raised current levels, higher τ_{HL}≥2μs is necessary to remain T_{max}<400K for the SiC GTO. As also revealed in Fig.8(c), the hot point arises in the middle drift region during the discharging process, as shown in ② of Fig. 8(b) and (c). After discharging, the heating-up stops and the T_{max} reaches the maximum platform. The hot point shifts to the anode region subjected to the largest current density. This would be risk for reliability of the metal contact. Careful consideration should be given to the anode current uniformity during discharging from perspectives of both the anode pattern design and the wire bonding.

978-1-7281-0582-6/19 $31.00 © 2019 IEEE

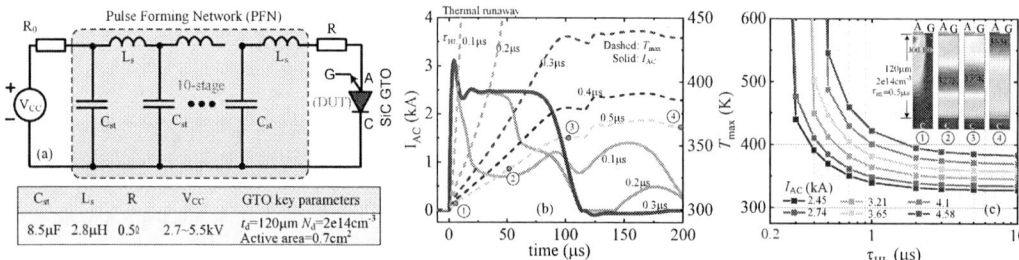

Fig. 8. (a) 10-stage PFN system for long pulse width discharging and circuit-device parameters. (b) I_{AC} and T_{max} of SiC GTO with τ_{HL}=0.1~0.5μs at the discharging condition of V_{cc}=2.7kV. Only τ_{HL}≥0.3μs satisfies 2.45kA current. (c) Dependence of T_{max} on τ_{HL} for SiC GTO at different current levels. As revealed, the heat generation occurs in the drift region during the discharging process (see ②) and shifts to anode region (see ④) with largest current density.

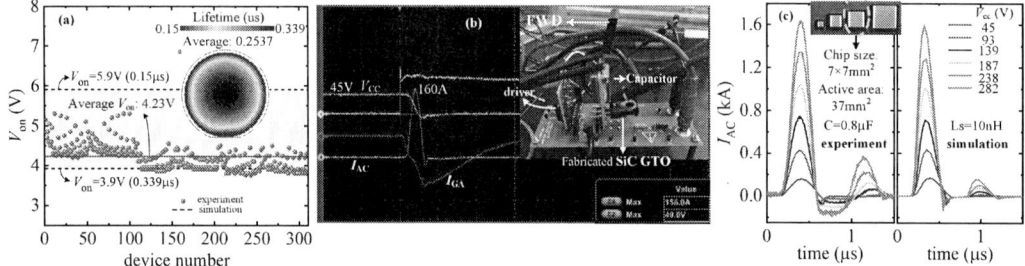

Fig. 9. (a) V_{on} distribution of our massive fabricated SiC GTOs. Carrier lifetime measured with u-PCD distributes in 0.15~0.339μs in 4-inch 4H-SiC wafer. Measured V_{on} distributes in 3.9~5.5V in accordance with the simulated range (dashed lines). (b) Measured discharging waveforms at V_{cc}=286V and C=0.8μF. (c) Simulated and measured pulsed I_{AC}=160~**1600**A at different V_{cc} using packaged 7×7mm² SiC GTOs. Good agreement is obtained.

C. Experimental Results

We have also fabricated SiC GTOs with t_d=60μm. The tested 4-inch SiC epi-wafer exhibits a mean carrier lifetime of 0.25μs and the V_{on} of massive SiC GTOs distributes in 3.9~5.5V as shown in Fig. 9(a). The distribution range agrees well with our static simulation. The pulsed discharging experiment is performed using packaged 7×7mm² SiC GTO. Fig. 9(b) shows the devices on board and the measured discharging waveforms at V_{cc}=286V and C=0.8μF. The wide range discharging of I_{peak}=160~1600A at different V_{cc} is as shown in Fig. (c). It is should be noted that the dynamic simulation agrees well with experimental data

IV. CONCLUSION

In this work, we present the impact of carrier lifetime on SiC GTO and PiN devices for pulsed power application. Both the device characteristics and circuit dynamic performance in pulsed discharging system are investigated based on simulation and experiments. The τ_{HL} effect on the device static characteristics and dynamic discharging performance in both single- and multi-stage PFN by mix-mode simulation considering thermoelectric effect within SiC devices. The results revealed that the τ_{HL}>2μs contributes to a sufficient low V_{on}<6V of SiC PiN for t_d≤250μm. However, the SiC GTO is more sensitive to carrier lifetime and withstands the power dissipation 10X higher than that of SiC PiN during pulsed discharging. The I_{peak} of SiC GTO tends to decrease for τ_{HL}<0.4μs in the single-stage discharging. For long pulse width discharging in 10-stage PFN, the heat generates in the drift region and tends to shift towards the top anode region. The simulation is verified by experimental data of our fabricated SiC GTOs.

ACKNOWLEDGMENT

This work is supported by Science Challenge Project, TZ2018003-1.

REFERENCES

[1] X. She, A. Q. Huang, Ó. Lucía and B. Ozpineci, "Review of Silicon Carbide Power Devices and Their Applications," in *IEEE Transactions on Industrial Electronics*, vol. 64, no. 10, pp. 8193-8205, Oct. 2017.

[2] L. Cheng *et al.*, "20 kV, 2 cm², 4H-SiC gate turn-off thyristors for advanced pulsed power applications," *2013 19th IEEE Pulsed Power Conference (PPC)*, San Francisco, CA, 2013, pp. 1-4.

[3] A. Agarwal *et al.*, "9 kV, 1 cm×1 cm SiC super gto technology development for pulse power," *2009 IEEE Pulsed Power Conference*, Washington, DC, 2009, pp. 264-269.

[4] N. Kaji, H. Niwa, J. Suda and T. Kimoto, "Ultrahigh-Voltage SiC p-i-n Diodes With Improved Forward Characteristics," in *IEEE Transactions on Electron Devices*, vol. 62, no. 2, pp. 374-381, Feb. 2015.

[5] S. Ryu, D.J. Lichtenwalner, M. O'Loughlin, et al. "Blocking Performance Improvements for 4H-SiC P-GTO Thyristors with Carrier Lifetime Enhancement Processes," Materials Science Forum. Vol. 924, pp 633-636. 2018.

[6] S. Ryu, D.J. Lichtenwalner, E. Van Brunt, et al. "Impact of Carrier Lifetime Enhancement Using High Temperature Oxidation on 15 kV 4H-SiC P-GTO Thyristor," Materials Science Forum. Vol. 897, pp 587-590. 2017.

[7] L. Cheng *et al.*, "Advanced silicon carbide gate turn-off thyristor for energy conversion and power grid applications," *2012 IEEE Energy Conversion Congress and Exposition (ECCE)*, Raleigh, NC, 2012, pp. 2249-2252.

Proceedings of the 31st International Symposium on Power Semiconductor Devices & ICs
May 19-23, 2019, Shanghai, China

An 8.5kV Sacrificial Bypass Thyristor with Unprecedented Rupture Resilience

Tobias Wikström
ABB Switzerland Ltd., Semiconductors
Lenzburg, Switzerland
Email: tobias.wikstroem@ch.abb.com

Bjørn Ødegård and Remo Baumann
ABB Switzerland Ltd., FACTS
Turgi, Switzerland
Email: bjoern.oedegard@ch.abb.com, remo.baumann@ch.abb.com

Abstract—**The paper details the development of a device intended for use as a sacrificial cell bypass switch in cascaded multi-level topologies with serial redundancy of cells of up to 4.6 kV (DC). The design is based on an 8.5 kV phase-control thyristor for industrial use. Maintaining rupture resilience constituted the main challenge. It ultimately triggered surges of more than 210 MA^{2}s without rupturing. Secondary challenges include a guaranteed surge *in*capability and ensuring a long-term stable short-circuit failure mode.**

I. INTRODUCTION

Modular Multi-level Converter topologies (MMC) can offer serial redundancy if a reliable means of discharging a faulty cell's energy storage and shorting its terminals is supplied. The stored energy in some high power applications is large enough to put the installation at risk when discharging due to the risk of semiconductor housing rupture, external arcing, capacitor explosion or electro-mechanical rupture of electrical connections. In many existing power systems, semiconductor explosions are accepted in fault cases because they cannot be avoided. Instead, explosions are contained and the cell terminals are shorted using a mechanical switch.

This work describes the development and application of a semiconductor-based bypass switch. This is not a new concept [1], [2], [3] but to the authors' knowledge, this publication marks the first time that all requirements for complete non-rupture conformance of the system have been addressed and fulfilled.

A. Intended application

In spite of having remarkable closing speed and non-rupture resilience, it is important to note that the device is *one* of the parts needed to achieve non-rupture conformance of the whole cell. The following is needed from the rest of the cell.

1) The active power semiconductor devices—switches and diodes—must retain non-rupture capability to some degree.
2) The control system must detect faults and trigger the bypass on a time-scale that doesn't allow other circuit elements to rupture.

Figure 1 illustrates a voltage-source MMC cell layout applicable for this device ("Bypass"). The choke ($\geq 3\mu$H) provides both surge-current moderation and slows the surge current transient. The cell controller (CC) detects faults by monitoring

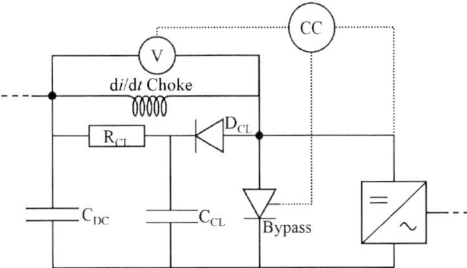

Fig. 1. The circuit diagram of a choked, voltage-source MMC cell of otherwise general architecture. The cell energy stored in C_{DC} can exceed 100 kJ. For a sense of the destructive potential, this energy is comparable to dropping the bypass thyristor from the top of Mount Everest.

the choke voltage. The bypass control unit (not drawn) monitors the cell voltage and can trigger the bypass autonomously if the voltage exceeds safe limits; an ultimate safety measure in case of, for example, a defective cell controller. The inverting circuit (DC/AC conversion symbol) is typically a half- or full-bridge, like in this work. More details on system ruggedness considerations can be found in [4].

B. Non-rupture criterion

Thyristor housing rupture is a concept standardised by IEC [5], and the ambition was to follow the standard as closely as possible. Most importantly, the housing technology for containing an unmitigated[1] edge arc at the targeted energy levels does not yet exist, assuming a reasonable device size. Spontaneous semiconductor failures *must* be mitigated, or explosions will inevitably occur. Two other aspects of the standard were disregarded. Firstly, IEC's test conditions were created with fused circuits in mind, which is why an application-specific fuse-free test circuit was developed. Secondly, the standard lists either loss of hermeticity after, or external detection of plasma during the surge as the rupture criterion, independent of the appearance of small cracks. As this work progressed, it became clear that it suffices to visually inspect the device after the surge, given that any optical

---

[1]The word "unmitigated" is used for arcs that are allowed burn in the device's internal air until the energy source is depleted without opening a parallel current path by triggering.

changes to the outside—e.g. ceramic hairline cracks or flange discoloration—are classified as ruptures.

II. FINAL DESIGN AND PERFORMANCE

A. Wafer design

The wafer design is straightforward and based on an existing thyristor wafer type, exclusively for non-technical reasons. There are two challenges for the wafer: limiting the maximal surge-current capability to an acceptable level and steering spontaneous faults away from the wafer edge. The first challenge was met, while catastrophic edge faults could had to be mitigated by trigger assistance that guided the fault away from the edge.

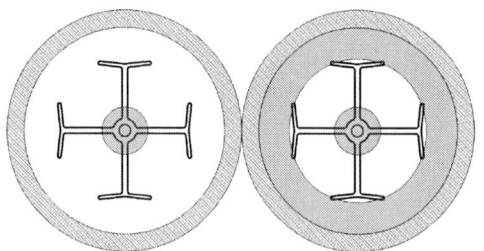

Fig. 2. Electron irradiation mask designs (light gray), overlaid with the wafer gate structure (black outline) and the silicone rubber (red striped fill). The left sketch shows the final design of the electron irradiation mask. To the right, the design of a two-part mask that was investigated but rejected, is shown.

The taken approach was to constrict the current to a small central portion of the wafer using a high-dose electron irradiation on the rest of the active area, shown in the left half of fig. 2. The on-state voltage increased to around 3 V at 100 A after irradiation. An unirradiated wafer would conduct thousands of Amperes at the same voltage. This irradiation led to a reasonably low maximal ITSM capability, needed for safe shorting of faulty cells without fully charged capacitors. It is an approximate 20 kA current pulse train shown in fig. 3. In the example, the device survives the first period (blocking the negative half-wave), but fails in the second positive half-wave at around 1.6 ms, seen as the sudden drop of the on-state voltage. The on-state voltage before triggering is incorrect (due to a measurement technicality) and should be disregarded. The test rig was set to pass three full periods through the DUT, which is why the oscillation ends at 3.3 ms.

The relatively rapid success of the approach led to limited investigation of other ideas and approaches. One approach that *was* investigated was to mask the electron irradiation also at the periphery of the active area, shown to the right of fig. 2. The motivating idea, loosely based on [6], was that the long lifetime region could self-trigger with help from the filament and move the arc into the active area. If this variant had any effect on the non-rupture capability of unmitigated edge-arcs, test resolution was too coarse to observe it, and it was abandoned because it increased the ITSM capability significantly.

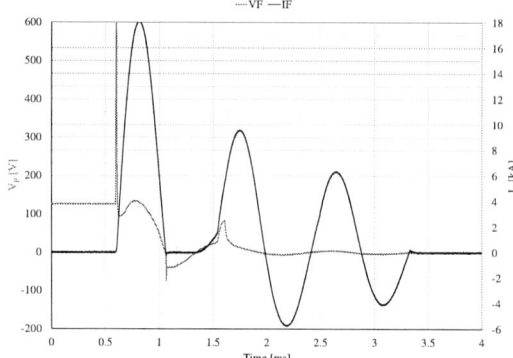

Fig. 3. Waveforms from a test of the minimal surge required to fail the bypass device.

The bypass device is operated at zero duty cycle under a DC-voltage that can reach 4.6kV. Avoiding numerous cosmic-ray failures forced the implementation of a wafer with 8.5kV blocking capability. Being a non-punch-through voltage design, such a wafer is quite thick—around 1.5 mm—and increases the on-state voltage of an electric arc, whose electrodes are separated by the wafer thickness. A designated bypass wafer design could be made using a punch-through inhibiting buffer thanks to the absence of reverse voltages in the intended application. This would allow for a thinner wafer to be used, maintaining cosmic-ray stability but with significantly lower energy absorption during a surge.

B. Housing design

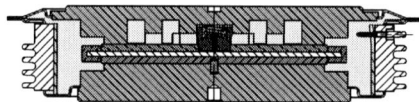

Fig. 4. Cross-sectional illustration of the assembled final housing.

Foregoing intermediate results, it was found that the most important aspect to achieve triggered non-rupture conformance at the highest energy levels was to make space inside the cathode pole-piece; an expansion volume that both decreases the gas pressure and improves heat transfer from the plasma to the cathode pole-piece. Figure 4 shows how the expansion volume (cyan) is included in the lid (upper orange region). The volume available for expansion in the original housing is indicated in dark blue colour—a minuscule region inside the gate contact—for visual comparison. Further improvements indicated in fig. 4 are: The ceramic walls are lined with a (green) silicone rubber strip, to protect the wall from cracking due plasma that may escape the pole-piece in spite of the expansion volume. The cathode sealing flange is somewhat protected by a labyrinth seal between the lid and the ceramic wall. Finally, the external cathode centring bore was made shallower (the original bore is identical to that in the anode pole-piece), because a central arc could easily melt through the original weak-spot between the bore and gate contact seat. The design of fig. 4 passed all case non-rupture tests that the test rig could safely produce. The limitation corresponds to $i^2t \approx 215\text{MA}^2\text{s}$ and $\hat{i}_{\max} \approx 360$ kA, depending on the DUT impedance. The capability exceeds requirements, which is why it was not extended to provoke actual ruptures of the final design. The waveforms in fig. 5 is an example of a surge at the limit of the capability of the equipment.

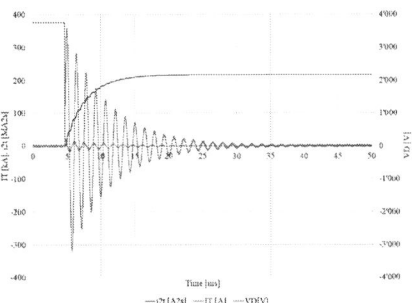

Fig. 5. Thyristor current and voltage wave-forms from the toughest surge the test rig could produce. $\hat{i}_{\max} = 363$kA, $i^2t_{\max} = 217\text{MA}^2\text{s}$.

Figure 6 shows a 3D model of the developed lid to the left. In this illustration, the plasma expansion trenches are blue and connection trenches are green. The yellow area indicates the seal intended to keep the plasma in the expansion volume. The gate cable trench (red) breaches this seal, unfortunately, by necessity. A photography of the inside of a prototype lid after a maximally tough surge is shown to the right of fig. 6. Apart from noting that the hermetic seal was intact after the surge, there are a few observations to be made. First, the arc's pressure still sufficed to push the gate cable out of its trench. The amount of escaped plasma (and its temperature) did obviously not suffice to rupture the ceramic wall or the flanges, and is only visible as soot stains on the pole-piece at the mouth of the cable trench. It is also noteworthy that the labyrinth seal keeps the plasma from penetrating to the cathode flange where no soot stains are visible. However, since the anode flange is not protected by such a seal, the rupture capability (discoloration of the outside of the anode flange) could be quite close to the tested conditions. The insert in the lower right corner shows the inside of the anode flange, blackened by soot around the mouth of the gate trench.

III. HOUSING DESIGN—ABANDONED CONCEPTS

The failed experiments are at least as interesting as the ultimate successful design. Figure 7 conveys an idea of how far from non-rupture conformance a device with damaged edge is, without any housing modifications or trigger-assist.

Attempts to reinforce the hermetic seal to allow unmitigated edge arcs all proved fruitless. The plasma is penetrant and

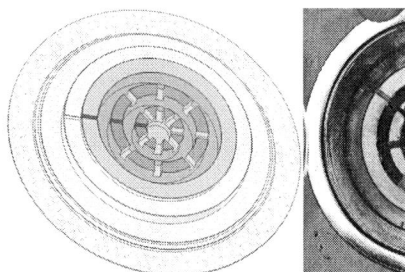

Fig. 6. A 3D illustration of the final lid (left). To the right, a photograph of the inside of a prototype lid after a maximally energetic surge. The insert shows cathode Moly-disc on top of the wafer and the rest of the anode-side assembly.

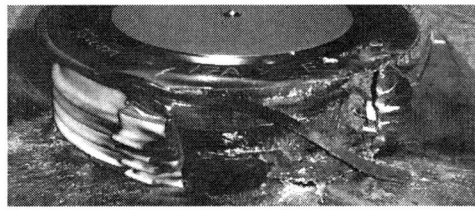

Fig. 7. One of the first non-rupture tests performed on a 102 mm housing with an unmitigated edge failure. $\hat{i}_{\max} = 199$ kA and $i^2t = 37\text{MA}^2\text{s}$.

aggressive, finding its way past obstacles to a point where the shielding effort becomes too expensive or difficult to manufacture. As a result, the unmitigated edge arc requirement was abandoned for the housing, and trigger assistance was designed into the control circuit. The arc is reliably moved from the periphery to the centre of the device if triggered quickly enough. However, even with a central arc, non-rupture conformance proved non-trivial or even difficult. The reason is that there is almost no volume available for plasma expansion in the original housing. The original available volume is a subset of the final housing cross-section, indicated in fig. 4 as blue areas around inside the gate contact. Lacking expansion volume, the plasma creates the following paths to either the internal or external atmosphere:

- Primarily, it flows through the channel made for routing the gate lead through the pole-piece, erratically pushing the gate lead out of the channel, which widens the path. All attempts to close the channel, using resins and metals, failed.
- The original housing lid had a deep external centring bore in the cathode pole-piece. The arc easily melted the weak spot between the bore and the gate contact seat, creating a path to the outside. By decreasing the depth of the centring bore, the path was closed.
- A few observations suggest that the plasma pressure trumped the contact force (40kN), resulting in a temporary separation of the pole-pieces that would increase contact resistance and the thyristor's absorbed share of

the stored energy.

Three housing sizes (102, 120 and 150 mm flange outer diameters) were fitted for the selected wafer size ($\oslash \approx 75$ mm). The expansion trenches were only tested with the 120 mm housing. There is tendency towards better performance with increasing housing size which suggests that an arbitrarily large housing would meet expectations at some point. This tendency is illustrated in fig. 8, showing how a triggered surge causes an explosion in the 102 mm housing in the upper half, compared to the slight cracking of the ceramic wall of a 150 mm housing, barely visible on the outside ceramic wall. While both surge events lead to ejection of the gate cable and were classified as ruptures, the plasma obviously loses some of its destructive potential on its way to the wall of the larger housing. The bottom picture shows the coiled-up gate cable resulting from being pushed out of the trench by the explosion.

Fig. 8. Post-surge photographs of a 102 mm housing (top), compared to a 150 mm housing (bottom). The bottom picture is taken after the housing was opened, at the inner ceramic wall in the direction of the gate cable trench.

Early housing rupture testing often yielded erratic results, changing dramatically with the incorporation of plasma expansion trenches for which no single rupture was observed. Part of the explanation is surely that testing with high resolution is laborious and expensive. A device can only be tested once, regardless of whether it passes or fails. If conditions are far away from the capability in any direction, much detail is lost. But it must also be admitted that plasma behaviour is complex and that there probably were influential parameters that were not controlled in the experiments.

IV. RELIABILITY

Device requirements in operation are trivial in comparison to other devices. Thanks to the low active losses, the device assumes approximately the ambient temperature in operation, leading to trivial load- and temperature cycling requirements.

After triggering, its short-circuit failure mode (SCFM) remain low-ohmic until the next planned maintenance. To verify the stability, SCFM testing was performed at maximal foreseen phase current, 1300 A_{RMS}, for over a year. A range of different variants were tested, both hermetic and punctured, with and without the wafer sandwiched between 0.5 mm thick Aluminium foils. While all tests passed the test electrically with monotonous on-state voltages varying between 0.1 and 1.75 V, the appearance after the test varied substantially. The use of Aluminium foils had a tremendous impact on the post-SCFM appearance. Two extremes are shown in fig. 9. Interestingly, these devices display similar electrical behaviour in spite of the standard-assembly wafer melting completely. However, the use of foils could not be motivated in the final design, since it did not improve electrical SCFM performance measurably.

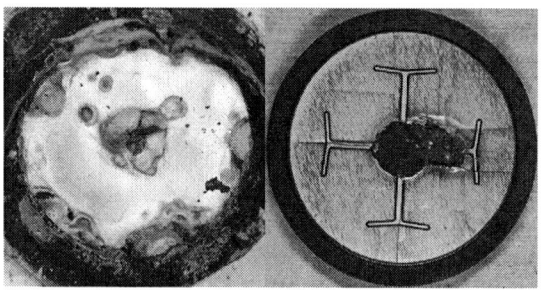

Fig. 9. Two wafers after one year in SCFM at 1300 A_{RMS}, both wtih punctured, non-hermetic, housings. The wafer to the right was sandwiched between two 0.5 mm thick Aluminum sheets. The wafer to the left had a standard assembly, the only identifiable remains being a bit of red rubber in the lower left corner.

V. CONCLUSION

A semiconductor-based bypass device for use in Modular Multilevel Converter topologies with a cell voltage of up to 4.6kV DC is proposed. The device will not rupture up to or exceeding 363 kA or 217 MA^2s, as long as it is triggered during the fault. The key to achieving such high non-rupture ratings is to provide volume inside the pole-piece for the byproducts of the electric arc to expand and cool. After a fault, the device displays a stable short circuit for more than a year under 1300 A_{RMS} at a voltage drop below 1.75 V_{RMS}.

REFERENCES

[1] Patent application WO 2019/007532 A1, N. Stahlut, Jan. 10, 2019.
[2] German patent office publication Nr. DE10323220A1, M. Glinka, Dec. 12, 2004.
[3] US Patent Nr. 4.399.452, K. Nakashima, Y. Araki, Y. Igarashi and S. Kojima, Aug. 16, 1983.
[4] "Rugged MMC converter cell for high power applications", Ødegård, Weiss, Baumann and Wikström, Proc. EPE 2016
[5] IEC international standard 60747-6, Section 6.3.6, "Peak case non-rupture current". http://www.iec.ch
[6] Europeant patent application Nr. EP1098371A3. K. Yoshikawa, June 25, 2003.

Proceedings of the 31st International Symposium on Power Semiconductor Devices & ICs
May 19-23, 2019, Shanghai, China

Power cycling capability of silicon low-voltage MOSFETs under different operation conditions

Christian Schwabe,
Chair of Power Electronics and EMC
Chemnitz University of Technology
Chemnitz, Germany
christian.schwabe@etit.tu-chemnitz.de

Peter Seidel
Chair of Power Electronics and EMC
Chemnitz University of Technology
Chemnitz, Germany
peter.seidel@etit.tu-chemnitz.de

Josef Lutz
Chair of Power Electronics and EMC
Chemnitz University of Technology
Chemnitz, Germany
josef.lutz@etit.tu-chemnitz.de

Abstract—Silicon low-voltage MOSFETs with a breakdown voltage lower than 80 V were tested for lifetime estimation under different operation conditions. The on-state resistance for these devices is very low; therefore, a special power cycling method with switching and conduction losses is used to be able to reach high temperature swings. These results are compared with the power cycling capability of inverse diodes and MOSFETs with reduced gate voltage under standard test conditions. The results with switching losses indicate a significantly higher lifetime compared to the other operation conditions. A simulation was performed to investigate the behavior. The temperature and current density distribution differ extremely due to different temperature coefficients. This leads to lower thermal-mechanical stress for the devices in switched mode and therefore to a higher lifetime as well.

Keywords—power cycling, switching losses, MOSFET, thermal stress, lifetime estimation

I. INTRODUCTION

Power MOSFETs have a wide range of application in power electronic systems, especially in fast switching operations such as DC-DC converters. One advantage of the silicon devices with low breakdown voltage are their low conduction losses. In accelerated power cycling tests, the low conduction losses pose a challenge because it is usually not possible to reach high temperature swings while applying the nominal current and nominal gate voltage at the same time. The standard power cycling method predominantly used for these devices is the operation of the inverse diode because of the additional voltage drop of the pn-junction.

II. TESTING SETUP

In standard power cycling tests for power semiconductors as specified in standards [1], only conduction losses in on-state mode are used to heat up the devices under test. The tested device in the following paper has an on-state resistance $R_{DS,on}$ of 1.35 milli Ohms. Therefore, it is not possible to reach a high temperature swing without exceeding the rated current. To compensate this disadvantage, a special test bench was developed to provide both conduction losses and switching losses [2]. This test bench allows to stay in the current limitation and still reach high temperature swings, respectively a decent testing time. A simplified schematic of phase 1 is shown in Fig. 1. It is possible to test four devices at once in switching mode and two reference devices under standard power cycling conditions with conduction losses only. The possibility of deploying reference devices is very useful, as it allows direct comparison of the power cycling

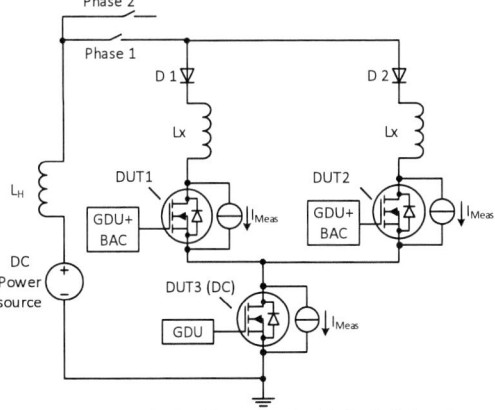

Fig. 1 schematic circuit of the test bench with detailed phase 1

capability with different methods, because all ambient conditions like load current, cooling and on/off-times are exactly the same. Every devices under test comes with a separate gate drive unit (GDU), while the switched devices are equipped with an additional boosted active clamping circuit (BAC).

The load current is switched actively between device 1 and 2 with a high frequency of several kilohertz. While turn-off of the load current by the MOSFETs, the stray inductance L_x causes a voltage overshoot. This would expose the MOSFET to an avalanche breakdown. For several low-voltage MOSFETs, this behavior is tolerated by the manufacturer; however, due to the unknown influence of this effect to the power cycling capability, an avalanche breakdown should be avoided. Therefore, the BAC circuit is used. A Zener-diode in combination with an amplifier stage limits the voltage to approx. 80% of the rated breakdown voltage. With this effect it is possible to generate suitable turn-off losses. The temperature swing of a single turn-off event should be lower than 1 Kelvin to safely stay in the elastic region, to prevent accelerated aging by the switching operation. The test bench is able to switch with a maximum frequency of 100 kHz, even though this frequency was not necessary for the tests. When the heating phase is finished, the current is switched to phase 2 to allow a cool down time for phase 1, as it is standard in power cycling tests. Hence, temperature swings of 100 Kelvin are possible for low voltage devices without exceeding the rated current.

978-1-7281-0582-6/19 $31.00 © 2019 IEEE

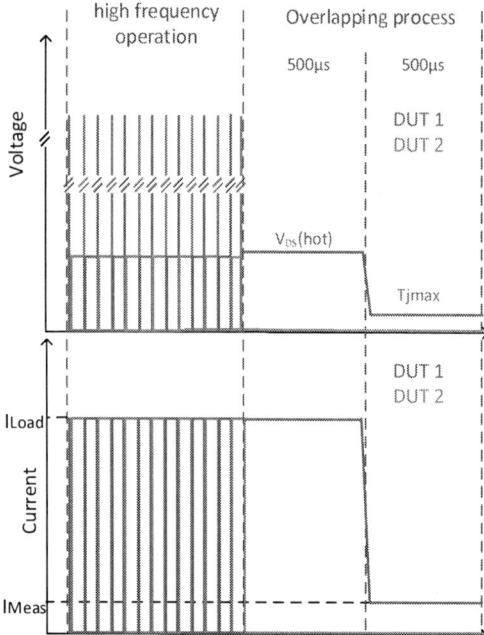

Fig. 2 example for voltage and current measurement for hot state [5]

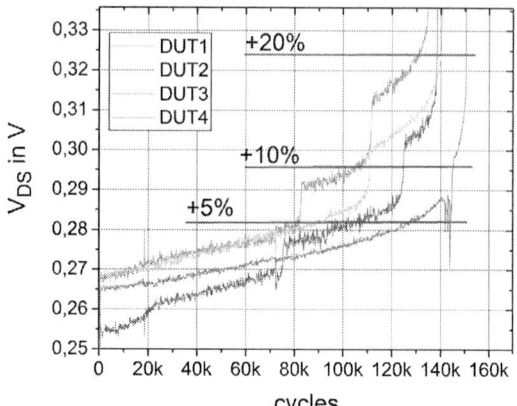

Fig. 3 drain source voltage development during power cycling

III. MEASUREMENT PROCEDURE

During the test, all relevant parameters are tracked, such as drain-source voltage drop (hot and cold), chip temperature, temperature swing and thermal resistance. To be able to measure the temperature of the chip, the $V_{CE}(T)$-method is used [3]. This is standard for power cycling tests. The MOSFET has no active pn-junction in on-state, so the inverse body diode is used for this purpose. This method is suitable for MOSFETs, especially SiC devices [4]. A high accuracy temperature measurement is very important to achieve meaningful results [5]. In addition, the measurement accuracy was investigated with an ANSYS simulation (see chapter VI).

In Fig. 2, the measurement pattern is displayed. All relevant measurements are performed in the so called overlapping phase. The most important measurement of the T_{jmax} temperature is an average value for the chip temperature measured between 50µs and 350µs after turn off. The accuracy of this method is ± 1 Kelvin.

IV. FAILURE CRITERA

There are three possible main criteria indicating end-of-life (EOL): an increase of the drain source voltage of +5%, a thermal resistance increase of +20% or a loss of function. The failure criterion for 5% drain source voltage increase is not very suitable for the low-voltage MOSFETs. At first, the voltage drop at nominal current is only in the range of some 100 millivolts, hence an increase of 5% is so small that it might be affected by noise. A second point can be seen in Fig. 3. The drain source voltage drop for the low voltage MOSFET increases slowly from 0 to approx. 120,000 cycles, already reaching the 5% increase. However, the graph indicates that after 140,000 cycles the device is in fact failing and the drain source voltage is increasing rapidly.

Hence, to be able to evaluate the test results, a failure criterion of 20% drain source voltage increase for the MOSFET devices is used. Some authors even use a 100% increase as failure criterion [7].

V. EXPERIMENTAL RESULTS

For all tests, a low voltage MOSFET from Infineon IRFP4004 in a molded TO-247 package is used [7]. Two tests (A&B) were performed with three different operation conditions. In Test A MOSFETs with additional switching losses (1) are compared to the inverse body diode (2). In Test B, again some MOSFETs in switched mode are compared to MOSFETs with reduced gate voltage and only conduction losses. Table 1 gives a short overview of the testing parameters, while ΔT is the temperature swing, T_{jmax} the maximum junction temperature, T_{inlet} the coolant temperature of the heat exchanger, I_{load} the load current and f_{switch} the switching frequency.

Table 1 Detailed test settings for the different tests

Parameter	Test settings	
	Test A	Test B
ΔT	75 K	60 K
T_{jmax}	150 °C	150 °C
T_{inlet}	42 °C	65 °C
$t_{on} = t_{off}$	2 s	2 s
I_{load}	127 A	106 A
f_{switch}	8,5 kHz	8,5 kHz

Results of Test A are detailed in Fig. 4. The results can be divided into two groups. The first group, failing at approx. 70,000 cycles, are the inverse diodes under standard power cycling conditions. The second group are the MOSFETs with 75 % switching losses. They have a higher scattering and fail at approx. 135,000 cycles. A similar behavior can be seen in Test B. The MOSFETs with low gate voltage (5.5-6V separately adjusted for each device) fail at 110,000 cycles in average, while the MOSFETs with switching losses fail after 430,000 cycles. This is a huge difference between the switched devices and the devices heated up with conduction losses. In both tests, it becomes visible that the switched MOSFETs last significantly longer in power cycling compared to the reference devices.

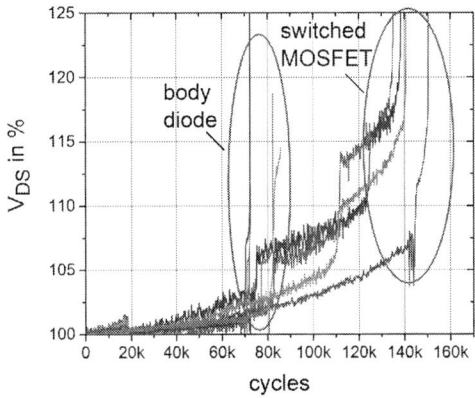

Fig. 4 results of power cycling of Test A

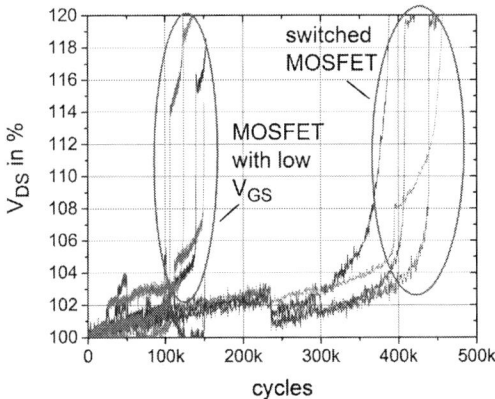

Fig. 5 results of power cycling of Test B

The root cause for this behavior is assumed to be the different temperature characteristics for each operation mode. The MOSFET during on-state has a positive temperature coefficient; in contrast, the inverse diode has a strong negative temperature coefficient (-1.3 mV/K). The MOSFET with the low gate voltage has an even stronger temperature coefficient (-7.5 mV/K) than the inverse diode. These characteristics might be an indicator for different current and temperature distributions during the test, which are leading to different lifetime estimations.

VI. SIMULATION

A three dimensional (3D) simulation was developed in ANSYS (version 17). It is a transient direct linked electro-thermal simulation. The ambient conditions were set as close as possible to the real test bench settings, nevertheless some simplifications have to be made. The water flow in the heat sink is simplified by a constant convection on the bottom side of the heat sink. Moreover, the connecting cables and the mounting screw are not included. Fig. 6 and Table 2 show an overview of the model in cross-section and the used materials. Detailed material information is given in [8]. In the first simulation, the measurement error for the MOSFET is investigated. The temperature deviation between ($T_{jmax}(t=0)$) and the measurement average between 50μs and 350μs is smaller than 1 Kelvin.

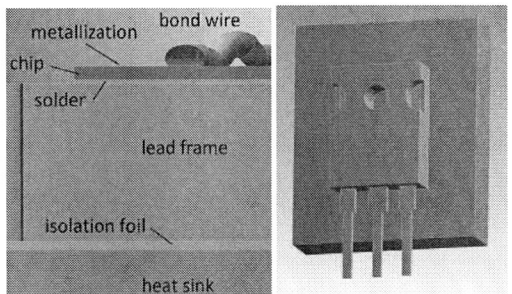

Fig. 6 cross-section of simulation model with one example bond wire and top-side view of the whole model

Table 2 Material overview for the simulation model

Name	Thickness	Material
Bond wire	Ø 500 μm	Aluminium
Metalization	4 μm	
Chip total:	200 μm	Silicon
• Active area (MOSFET)	20 μm	Silicon
• pn-junction (diode)	20 μm	Silicon
• Chip substrate	180 μm	Silicon
Solder	50 μm	Sn96.5Ag3.5
Lead frame	2.3 mm	Copper
Isolation foil	200 μm	Kunze CG-20
Housing	5 mm	Mold

This result lies in an acceptable range for the measurement error [9] in accordance with the results in [5]. The MOSFET is modelled with a resistive behavior which causes 25% of the losses and an internal heat generation in the chip, which is representing the switching losses and amounts the remaining 75%. The diode conduction losses are modelled with $P = I_{Load} \cdot (V_{F0} + R_{diff} \cdot I_{Load})$, R_{diff} is extracted from measurement and P is used as input for the next calculation loop. The method is described in detail in [8].

The comparison of the temperature distribution indicates a clear trend in Fig. 7. Both chips have approx. the same average temperature, which is 151.2°C for the MOSFET mode (top) and 149.9 °C for the inverse diode (bottom).

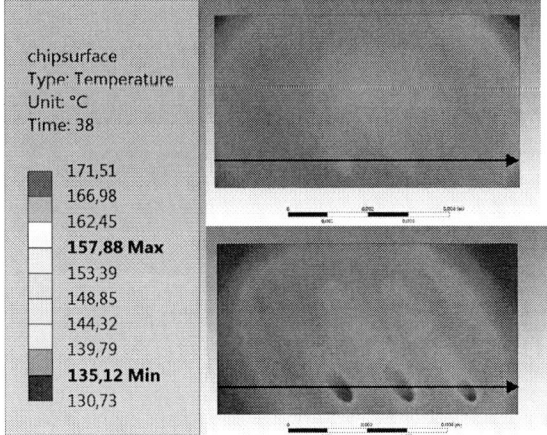

Fig. 7 temperature distribution on the chip surface of the switched MOSFET (top) and the diode mode (bottom) for Tjmax

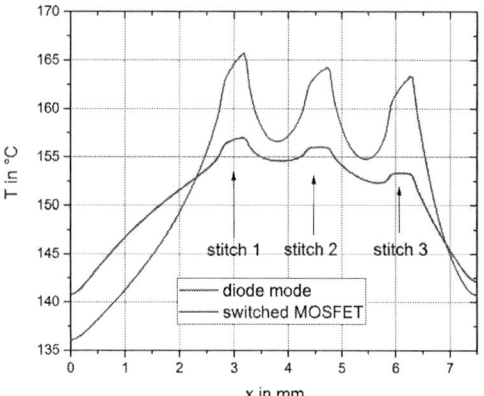

Fig. 8 temperature path results for switched MOSFET and inverse body diode

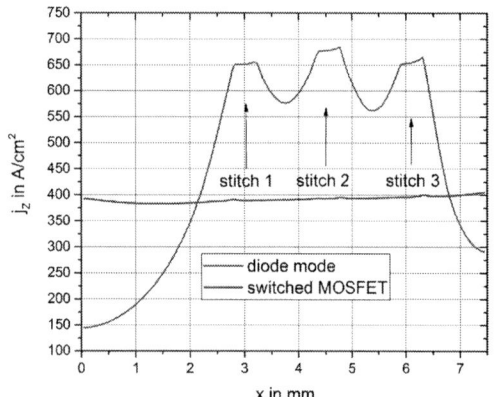

Fig. 9 current crowding on the chip surface along the evaluation path during hot state Tjmax

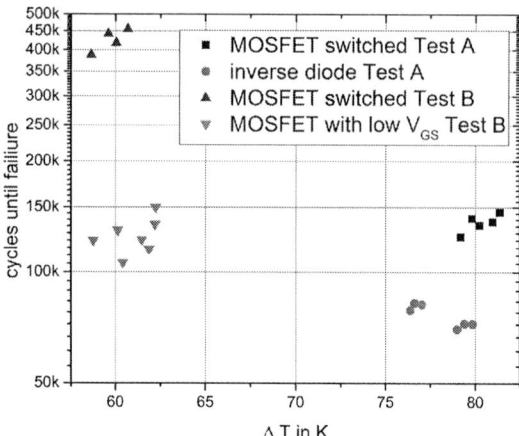

Fig. 10 power cycling results depending on the temperature swing ΔT

The simulation shows that the temperature and current distribution is not even over the whole chip surface, as the devices with a negative temperature coefficient current crowding around the first bond stitches cause higher stress at the bond wire interconnection. The different factors for the two groups compared to the lifetime of the switched devices can be explained with regard to the different negative temperature coefficients. The measurement results indicate a higher negative coefficient for the MOSFET with low gate voltage (-7.5mV/K) in contrast to the inverse diode (-1.3mV/K). The power cycling tests presented a clear trend that low voltage MOSFETs under operation conditions have a longer lifetime than testing of the inverse diode would predict.

However, the temperature distribution along the first bond stitches is very different. A path along the first bond stitches (cf. Fig. 7 arrow) is presented in Fig. 8 and Fig. 9 to clarify the differences. The simulation results show the root cause of the different lifetimes during power cycling. Due to the negative temperature coefficient of the inverse diode during the test, the current is crowded towards the bond stitches. Hence, the temperature on the bond feet is increasing which is accelerating bond wire fatigue. This results in a shorter lifetime expectation in contrast to the MOSFET mode, where the current distribution is quite homogeneous due to the positive temperature coefficient.

VII. CONCLUSION

Fig. 10 shows an overview of the power cycling capability of the low-voltage MOSFETs under different operation conditions. All devices have the same T_{jmax} and time settings, while the current and inlet temperature are different. The main outcome here is that the MOSFETs under switching condition have approx. twice the lifetime than the reverse diodes under the same thermal stress conditions. The MOSFET in switched mode has an approx. 3.5 to 4 times longer lifetime expectation compared to the MOSFET with low gate voltage and conduction losses only.

VIII. REFERENCES

[1] IEC 61709; Electric components - Reliability - Reference conditions for failure rates and stress models for conversion; version 3.0; 2017

[2] P. Seidel, C. Herold, J. Lutz, C. Schwabe and R. Warsitz, "Power cycling test with power generated by an adjustable part of switching losses," *2017 19th European Conference on Power Electronics and Applications (EPE'17 ECCE Europe)*, Warsaw, 2017, pp. P.1-P.10.

[3] R. Schmidt and U. Scheuermann, "Using the chip as a temperature sensor — The influence of steep lateral temperature gradients on the Vce(T)-measurement," *2009 13th European Conference on Power Electronics and Applications*, Barcelona, 2009, pp. 1-9.

[4] F. Hoffmann, Y. Auth, N. Kaminski, Evaluation of the VSD-Method for Temperature Estimation During Power Cycling of SiC-MOSFETs, *14th ISPS Interantional Seminar on Power Semiconductors*, Prague 2018, pp. 117-123

[5] G. Zeng, L. Borucki, O. Wenzel, O. Schilling and J. Lutz, "First Results of Development of a Lifetime Model for Transfer Molded Discrete Power Devices," *PCIM Europe 2018*; Nuremberg, Germany, 2018, pp. 1-8.

[6] S. Russo, R. Letor, O. Viscuso, L. Torrisi and G. Vitali. "Fast thermal fatigue on top metal layer of power devices." *Microelectronics Reliability 42 (2002)*, 1617-1622.

[7] https://www.infineon.com/dgdl/irfp4004pbf.pdf?fileId=5546d462533600a40153562905f41ffc, datasheet, date: 05.03.2019

[8] C. Schwabe, P. Seidel, J. Lutz, Simulation of current crowding in inverse diodes of low-voltage Si MOSEFTs at power cycling, *21st Conference on Power Electronics and Applications EPE 2019*, in press

[9] C. Herold, J. Franke, R. Bhojani, A. Schleicher, J. Lutz, Requirements in power cycling for precise lifetime estimation, *Microelectronics Reliability 58 (2015)*, 82-89.

Proceedings of the 31st International Symposium on Power Semiconductor Devices & ICs
May 19-23, 2019, Shanghai, China

Die-attach on Copper by Pressureless Silver Sintering in Formic Acid

Meiyu Wang
School of Materials Science and Engineering
Tianjin University
Tianjin, China
meiyuwang@tju.edu.cn

Yijing Xie
School of Materials Science and Engineering
Tianjin University
Tianjin, China
13920189330@163.com

Yunhui Mei
(Corresponding author)
School of Materials Science and Engineering
Tianjin University
Tianjin, China
yunhui@tju.edu.cn

Xin Li
School of Materials Science and Engineering
Tianjin University
Tianjin, China
xinli@tju.edu.cn

Guo-Quan Lu
School of Materials Science and Engineering of Tianjin University
Department of Materials Science and Engineering, and the Bradley Department of Electrical and Computer Engineering of Virginia Tech
Blacksburg, VA, USA
gqlu@vt.edu

Abstract—Sintered-silver die-attach has emerged as an excellent lead-free and reliable solution for high-temperature packaging. However, the most widely reported process involves pressure-assisted silver sintering on direct-bond-copper (DBC) substrates that are surface-finished with silver or gold plating, which adds to the cost of implementing this die-attach technology. One way to lower the cost is to die-attach on uncoated copper by pressure-less silver sintering. However, a challenge for this process is to prevent copper oxidation without significantly slowing the kinetics of silver sintering found in air. In this study, we explored the process of die-attach on copper by pressure-less silver sintering in a mixture atmosphere of formic-acid gas and air. The effects of oxygen fraction in the sintering atmosphere on the performance of an IGBT device bonded on an uncoated DBC substrate were examined. Strong, oxygen-free bond-line with die-shear strength over 30 MPa and excellent thermal and electrical performance of the packaged IGBT were obtained. The results in this study show the feasibility of a low-cost solution for implementing the silver sintering technology in manufacturing power semiconductor devices and/or modules.

Keywords—bonding by silver sintering, pressure-less silver sintering on uncoated copper, IGBT packaging, sinter in formic acid

I. INTRODUCTION

The development of wide band-gap (WBG) semiconductor devices targeting high-power density and high-temperature applications is driving adoption of the low-temperature silver sintering die-attach technology for manufacturing power semiconductor discrete devices and modules [1-3]. Compared with the traditional die-attach by soldering, the sintered-silver die-attach has over five times higher thermal and electrical conductivities and at least two times higher reliability [4, 5]. Furthermore, the sintering technology has a low processing temperature below 300°C that is several hundreds of degrees below the bond melting temperature of 960°C [6, 7]. These features make the technology an excellent lead-free, reliable, and low-cost solution for packaging WBG power electronics.

Thus far, the most widely reported process with the sintering die-attach is pressure-assisted silver sintering on direct-bond copper (DBC) substrates that are surface-finished with silver or gold, which adds to the cost of implementing this die-attach technology [8, 9]. To lower the cost, it is significant to process die-attach on uncoated copper by pressure-less silver sintering. However, a challenge for this process is that the copper is prone to oxidation at sintering temperature, and copper oxides prohibit strong bonding strength [10-13]. Therefore, in order to obtain a strong sintered silver die-attach on copper surface finish, it is critical to prevent copper oxidation during sintering.

In this study, in order to prevent copper oxidation without totally eliminating the oxygen effect on silver sintering, the die-attach on copper was pressure-less silver-sintered in a mixture atmosphere of formic acid and air with oxygen fraction varied from 50×10^3 ppm to 200×10^3 ppm. The bonding process was applied to packaging an IGBT device with the semiconductor chips bonded on uncoated DBC substrates. The IGBT packages were characterized for die-shear strength, bond-line microstructure, and thermal and electrical performance. The feasibility of a low-cost silver sintering technology for IGBT packaging was demonstrated by pressure-less silver sintering on uncoated copper in a mixture atmosphere of formic-acid and air containing 100×10^3 ppm oxygen.

II. SILVER PASTE MATERIAL

A silver paste designed for pressure-less sintering was acquired from NBE Technologies, LLC. Fig. 1(a) shows the scanning electron microscope (SEM) image of the bonding material, which consists of mostly micron-size silver particles. Fig. 1(b) shows its thermogravimetric analysis (TGA) and differential scanning calorimetry (DSC) test results when heated in nitrogen. The weight loss and endothermic transition started immediately upon heating as a result of evaporation of solvents in the paste. After much of the solvents evaporated at temperatures higher than 170°C, an exothermic reaction took place accompanied by a further loss in weight. This is mostly attributed to the decomposition of the organic elements, such as binders and dispersants, covering the silver particles [14]; and, it is possible that part of the heat release came from

978-1-7281-0582-6/19 $31.00 © 2019 IEEE

reduction of surface area by the sintering of silver particles. It is pointed out that after the exothermic reaction, there is no further weight loss beyond 300°C, suggesting a complete removal of organics.

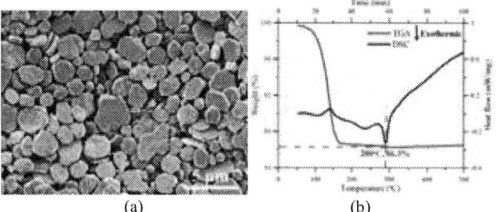

Fig. 1 (a) SEM image and (b) TGA and DSC test results of the silver paste.

Die-attach samples were fabricated by bonding "dummy" and functional IGBT chips on uncoated DBC substrate (Curamik Rogers). The "dummy" chips were from a silver-coated silicon wafer cut to 6.5 mm × 6.5 mm × 0.15 mm (thick) and were used for testing die-shear strength. The bonded IGBT chips (Infineon, IGC15T65QE) of 3.92 mm × 3.88 mm × 0.07 mm (thick) were for testing the saturation voltage $V_{CE(sat.)}$ and transient thermal impedance (Z_{th}) after packaging. The bond-line thickness for all attached chips was at 50 μm prior to heating. Afterwards, the attached samples were pressure-less sintered using different temperature-time-atmosphere sintering profiles.

III. SINTERING IN CONTROLLED ATMOSPHERE

A. Sintering in air

The widely practiced pressure-assisted silver sintering process is generally done in air because (1) silver does not oxidize at the sintering temperature, and (2) there are studies [10, 11]suggesting that oxygen and or hydroxyl groups in the sintering atmosphere promote silver sintering. Therefore, the pressure-less silver sintering on DBC copper surface was firstly tested in air. Sintering temperature of 280oC was selected based on the above TGA and DSC test results. Shown in Fig. 2 is SEM micrographs of a cross-sectional view and a plane-view of the bond-line sintered in air. It is clear that all the sintered-silver particles bonded together through neck-formation. The organics were effectively removed by burning in the presence of oxygen in air. One would expect the sintered bond-line to have a high cohesive strength. However, the DBC copper surface was extensively oxidized in air, which would suggest poor adhesive strength at the bonding interface. Therefore, to prevent copper oxidation, the die-attach needs to be sintered in a reducing atmosphere.

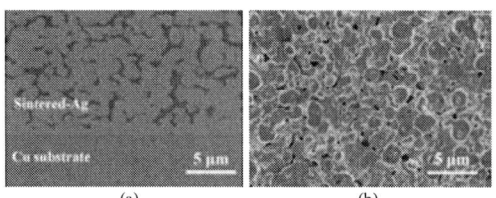

Fig. 2 SEM micrographs of bond-line sintered in air (a) cross-sectional view and (b) plane-view.

B. Sintering in formic acid

Formic acid (HCOOH) is widely used in solder-reflow equipment to prevent oxidation and remove oxides on PCB board resulting in high quality solder joints on assembled boards. This practice is based on the thermal decomposition of formic acid into H_2 and CO [15, 16] of both are reducing agents of copper. We believe that H_2 produced by the decomposition also would help removing the organic dispersant and binder in the silver paste prior to the silver sintering [12]. This is the reason why we selected formic acid for silver sintering die-attach on copper surface. Fig. 3 shows the SEM micrographs of the silver paste sintered in formic acid gas at temperatures in range of 220°C and 310°C. The sintered density increased with sintering temperature. This is not surprising because at higher temperatures the decomposition of organic dispersant and binder in the paste is expected to be faster and more complete leading to faster sintering kinetics of silver [7]. The sintered densities at 280°C and 310°C are similar although the grain size in the 310°C sintered microstructure is larger. We expect the two microstructures to have similar bond-line cohesive strengths because of their similar densities. Of the two temperatures, the lower one would be preferred because processing at a lower temperature reduces the rate of copper oxidation.

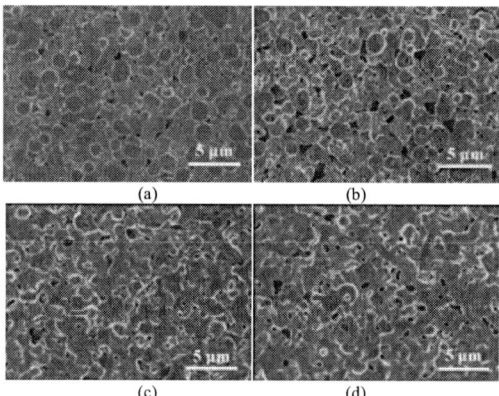

Fig. 3 SEM micrographs of pressure-less sintered-silver in formic acid at (a) 220°C, (b) 250°C, (c) 280°C, and (d) 310°C.

Fig. 4 shows the respective SEM images of a cross-sectional view and plane-view of the sintered silver in formic acid at 280°C. No delamination was observed in the cross-section of the bond-line. However, compared with the sintered-silver in air as shown in Fig. 2, the neck-formation of the sintered-silver in formic acid was to a lesser extent. Fig. 5 shows a comparison of the sintered-silver density, the wire bonding force on sintered copper surface, and the die-shear strength measured from a sample sintered in air versus a sample sintered in formic acid. The density and die-shear strength of the die-attach sintered in air was higher than that sintered in formic acid, while the wire bonding force is on the contrary. Specifically, the die-attach sintered in air had 66% higher die-shear strength and 3.4% higher sintered bond-line density than that sintered in formic acid. This is likely that the oxygen in air promoted organic dispersant and binder decomposition and increased the sintering kinetics of silver leading to higher density and bonding strength. On the contrary, the ultrasonic bonding force of aluminum wires on

DBC copper surface sintered in air is 51% lower than that sintered in formic acid. This is because the oxygen in air lead to substantial oxidation of the copper surface, thus causing the wire bonding force to drop. The H_2 and CO decomposed from formic acid could reduce copper oxides, thus strong wire bonding force were achieved by strong metal-metal metallic bonds [15, 16].

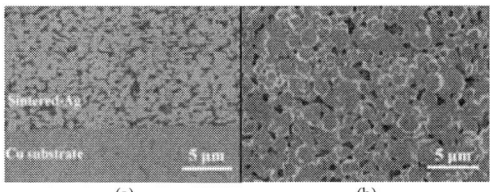

(a)   (b)

Fig. 4 SEM micrographs of bond-line sintered in formic acid (a) cross-sectional view and (b) plane-view.

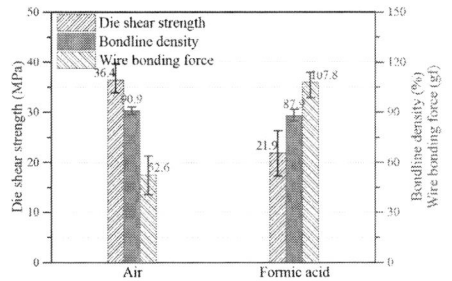

Fig. 5 Comparison of die-shear strength, bond-line density, and wire bonding force of silver sintering on copper in air and in formic acid atmosphere.

C. Sintering in mixture of formic acid and air

To prevent copper oxidation without totally eliminating the oxygen effect on silver sintering, we processed silver sintering die-attach in a mixture atmosphere of formic acid and air. Fig. 6 shows the heating profile. To avoid formation of voids in the bond-line caused by rapid pumping out of solvents, a drying process was used to slowly evaporate the solvents in the paste before vacuum pumping at 200°C. Then the sintering was done in a mixture atmosphere of formic acid and air with oxygen fraction varied from 50×10^3 ppm to 200×10^3 ppm (the oxygen level in air). The effects of oxygen fraction on IGBT package's mechanical, thermal, and electrical performance are discussed next. Fig. 7 shows the die shear strength, wire bonding force, transient thermal impedance (Z_{th}), and saturate voltage $V_{CE(sat.)}$ of die-attach sintered on copper in the mixture atmosphere of formic acid and air with oxygen fraction varied from 50×10^3 ppm to 200×10^3 ppm. When the oxygen fraction increases, the die-shear strength, Z_{th}, and $V_{CE(sat.)}$ increase, due to higher cohesive strength, thermal and electrical conductivity resulted from higher densification sintering of the bond-line, whereas the wire bonding force decreases with the increasing oxygen fraction, due to the increasing copper oxidation on the surface.

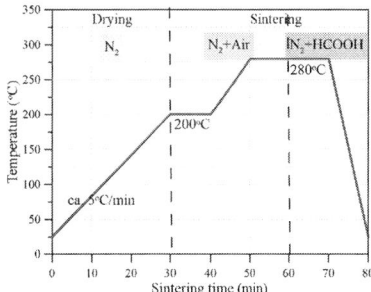

Fig. 6 Pressure-less silver sintering profile in the mixture atmosphere of formic acid and air.

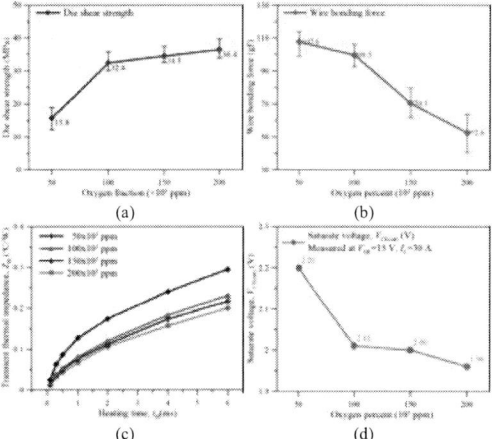

(a)   (b)

(c)   (d)

Fig. 7 (a) Die-shear strength, (b) wire bonding force, (c) transient thermal impedance (Z_{th}), and (d) saturate voltage $V_{CE(sat.)}$ of die-attach sintered on copper in formic acid and air with oxygen fraction varied from 50×10^3 ppm to 200×10^3 ppm.

There is no obvious difference in the mechanical, thermal, and electrical performance of the sintered silver in oxygen fraction above 100×10^3 ppm. Therefore, the microstructures of the sintered silver in oxygen fraction of 50×10^3 ppm and 100×10^3 ppm was compared, as shown in Fig. 8. The silver particles almost keep separate when sintering in oxygen fraction of 50×10^3 ppm, while that are neck-formed together when sintered in oxygen fraction of 100×10^3 ppm. The results demonstrate that the higher oxygen fraction promotes organic dispersant and binder decomposition and sintering kinetics of silver particles. Fig. 9 shows the microstructure of the fracture surface on the DBC substrate after die-shear test. When sintering in oxygen fraction of 50×10^3 ppm, the de-bonding mostly occurs on the copper surface, indicating a weak bonding strength. When increasing the oxygen fraction to 100×10^3 ppm, the fracture is a cohesive failure, indicating a strong bonding strength. Fig. 10 shows the SEM microstructure and EDS chemistry analysis of the cross-sectional sintered silver bond-line. The SEM image shows well-bonded interface without delamination, and the EDS result indicates no copper oxidation at the bonding interface. Overall, the die-attach sintered in the mixture atmosphere of formic acid and air with oxygen fraction of 100×10^3 ppm has the best combination of mechanical, thermal, and electrical performance, which could achieve robust pressure-less silver sintering on copper.

978-1-7281-0582-6/19 $31.00 © 2019 IEEE   501

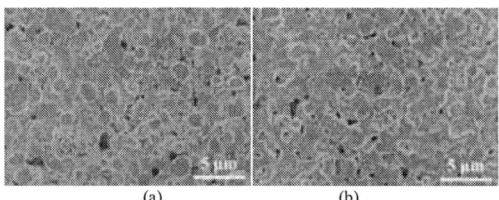

(a)   (b)

Fig. 8 SEM microstructures of silver sintered in the mixture atmosphere of formic acid and air with oxygen fraction of: (a)50 $\times10^3$ ppm, and (b) 100×10^3 ppm.

(a)   (b)

Fig. 9 Fracture surface after die-shear test of die-attach sintered in the mixture atmosphere of formic acid and air with oxygen fraction of: (a)50 $\times10^3$ ppm, and (b) 100×10^3 ppm.

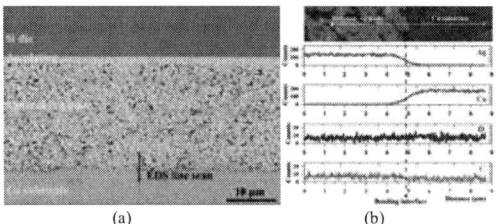

(a)   (b)

Fig. 10 (a) SEM microstructure and (b) EDS chemistry of sintered silver bond-line in the mixture atmosphere of formic acid and air with oxygen fraction of 100×10^3 ppm.

IV. CONCLUSIONS

A low-cost silver sintering die-attach process for packaging power devices or modules was developed in this study. The process involves pressure-less silver sintering on uncoated direct-bond-copper substrates. To prevent copper oxidation, the sintering process was carried out in a mixture atmosphere of formic acid and air. Oxidation-free bond-line with die-shear strength over 30 MPa was obtained by processing at 280°C in formic acid with oxygen fraction of 100×10^3 ppm. Packages of IGBT devices made by the low-cost die-attach process had excellent mechanical, thermal, and electrical performance.

ACKNOWLEDGMENT

This work was supported by the National Natural Science Foundation of China (No. 51877147), the Science Challenge Project (No. TZ2018003), and the Tianjin Municipal Natural Science Foundation (No.17JCYBJC19200).

REFERENCES

[1] R. Khazaka, L. Mendizabal, and D. Henry, "Review on joint shear strength of nano-silver paste and its long-term high temperature reliability," Journal of electronic materials, vol. 43, no. 7, pp. 2459-2466, 2014.

[2] V. R. Manikam and K. Y. Cheong, "Die attach materials for high temperature applications: A review," IEEE Transactions on Components, Packaging and Manufacturing Technology, vol. 1, no. 4, pp. 457-478, 2011.

[3] R. Khazaka, L. Mendizabal, D. Henry, and R. Hanna, "Survey of high-temperature reliability of power electronics packaging components," IEEE Transactions on power Electronics, vol. 30, no. 5, pp. 2456-2464, 2015.

[4] K. S. Siow and Y. Lin, "Identifying the development state of sintered silver (Ag) as a bonding material in the microelectronic packaging via a patent landscape study," Journal of Electronic Packaging, vol. 138, no. 2, p. 020804, 2016.

[5] S. Fu, Y. Mei, X. Li, C. Ma, and G.-Q. Lu, "Reliability Evaluation of Multichip Phase-Leg IGBT Modules Using Pressureless Sintering of Nanosilver Paste by Power Cycling Tests," IEEE Transactions on Power Electronics, vol. 32, no. 8, pp. 6049-6058, 2017.

[6] K. S. Siow, "Are sintered silver joints ready for use as interconnect material in microelectronic packaging?," Journal of electronic materials, vol. 43, no. 4, pp. 947-961, 2014.

[7] G.-Q. Lu, M. Wang, Y. Mei, and X. Li, "Advanced Die-attach by Metal-powder Sintering: The Science and Practice," CIPS 2018, 10th International Conference on Integrated Power Electronics Systems, VDE, 2018.

[8] M. Wang, Y. Mei, R. Burgos, D. Boroyevich, and G.-Q. Lu, "Effect of substrate surface finish on bonding strength of pressure-less sintered silver die-attach," presented at the 2018 International Conference on Electronics Packaging and iMAPS All Asia Conference (ICEP-IAAC), 2018.

[9] C. Chen, K. Suganuma, T. Iwashige, K. Sugiura, and K. Tsuruta, "High-temperature reliability of sintered microporous Ag on electroplated Ag, Au, and sputtered Ag metallization substrates," Journal of Materials Science: Materials in Electronics, vol. 29, no. 3, pp. 1785-1797, 2018.

[10] J. Li, X. Li, L. Wang, Y.-H. Mei, and G.-Q. Lu, "A novel multiscale silver paste for die bonding on bare copper by low-temperature pressure-free sintering in air," Materials & Design, vol. 140, pp. 64-72, 2018.

[11] Z. Zhang et al., "Low-temperature and pressureless sinter joining of Cu with micron/submicron Ag particle paste in air," Journal of Alloys and Compounds, vol. 780, pp. 435-442, 2019.

[12] H. Zheng, D. Berry, K. D. Ngo, and G.-Q. Lu, "Chip-bonding on copper by pressureless sintering of nanosilver paste under controlled atmosphere," IEEE Transactions on Components, Packaging and Manufacturing Technology, vol. 4, no. 3, pp. 377-384, 2014.

[13] A. Masson, C. Buttay, H. Morel, C. Raynaud, S. Hascoët, and L. Grémillard, "High-temperature die-attaches for SiC power devices," in Power Electronics and Applications (EPE 2011), Proceedings of the 2011-14th European Conference on, 2011, pp. 1-10: IEEE.

[14] S. Wang, M. Li, H. Ji, and C. Wang, "Rapid pressureless low-temperature sintering of Ag nanoparticles for high-power density electronic packaging," Scripta Materialia, vol. 69, no. 11, pp. 789-792, 2013.

[15] H. Yan, Y.-H. Mei, X. Li, C. Ma, and G.-Q. Lu, "A Multichip Phase-Leg IGBT Module Using Nanosilver Paste by Pressureless Sintering in Formic Acid Atmosphere," IEEE Transactions on Electron Devices, vol. 65, no. 10, pp. 4499-4505, 2018.

[16] M. F. Kuehnel, D. W. Wakerley, K. L. Orchard, and E. Reisner, "Photocatalytic Formic Acid Conversion on CdS Nanocrystals with Controllable Selectivity for H2 or CO," Angewandte Chemie, vol. 127, no. 33, pp. 9763-9767, 2015.

Proceedings of the 31st International Symposium on Power Semiconductor Devices & ICs
May 19-23, 2019, Shanghai, China

Mutual inductance influence to switching speed and TDR measurements for separating self- and mutual inductances in the package

H. Iida, K. Hasegawa, I. Omura
Department of Biological Function Engineering
Kyushu Institute of Technology, Kitakyushu, 8080196 Japan
iida.hirotaka884@mail.kyutech.jp

Abstract—Parasitic inductances in power semiconductor packages affect the device switching speeds. The self-inductance of the source terminal has been considered to be the main factor that limits the switching speed, but mutual inductances among the three terminals also influence the speed. This paper proposes a method to measure the source self-inductance and the mutual inductances in the package using time-domain reflectometry (TDR) and time-domain transmissometry (TDT) that reveals the limitation of the switching speed. The measurement results agreed well with those of Q3D simulation.

Keywords—mutual inductance, switching speed, power semiconductor package, TDR, TDT

I. INTRODUCTION

As power semiconductor devices attain higher switching frequencies and higher current capabilities, the parasitic inductances in these devices are having increasingly strong effects on their switching characteristics [1-5]. The self-inductance of the source terminal mainly influences the switching speed in larger devices, while mutual inductance mainly affects the speed in smaller devices.

Mutual inductances among the gate, drain and source terminals affect the switching speed of the device package. Figure 1 shows the current density and the magnetic field of a TO-247 package. The source current i_S increases and decreases rapidly during the turn-on and turn-off periods, respectively. The source self-inductance L_S and the magnetic field add an induced voltage to the gate-source voltage V_{GS} in proportion to the slew rate of i_S. The effect of the magnetic field is changed by both the direction and the switching speed of the current. The magnetic effect and the source inductance limit the switching speed of power semiconductor packages, but the relationships between the two inductances and the switching speed have not been investigated theoretically to date.

Two points must be addressed to evaluate the effects of the parasitic inductances on the switching speed. The first involves clarification of the effects of the self- and mutual inductances from a theoretical viewpoint and acquisition of the corresponding equations. The second point is that the self- and mutual inductances among the three terminals must be measured separately.

This paper analyzes the limitations of the switching speed of a power device package based on consideration of the source self-inductance and the three mutual inductances and then introduces a measurement method using time domain

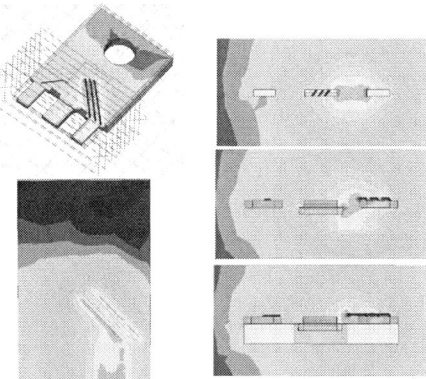

Fig. 1. Current density and magnetic field of TO-247 package.

reflectometry (TDR) and time domain transmissometry (TDT) that gives these four parameters separately.

II. SWITCHING SPEED LIMIT

The mutual inductances in a power semiconductor package influence its switching speed. The source self-inductance and the mutual inductances among the three terminals add an induced voltage to the gate-source voltage V_{GS} during the turn-off period. V_{GS} is expressed using the following equation in the equivalent circuit shown in Fig. 1.

$$V_{gg} - V_{GS} = L_G C_{GS} \frac{d^2 V_{GS}}{dt^2}$$
$$+\{(L_S + M_{GD} + M_{GS} + M_{DS})gm + R_G C_{GS}\}\frac{dV_{GS}}{dt} \quad (1)$$

From this equation, the authors defined a time constant composed of $(L_S+M_{GD}+M_{GS}+M_{DS})gm+R_G C_{GS}$ as the "M-factor". When the mutual conductance gm is sufficiently high, the maximum value of the slew rate of the drain current I_D is given by the following equation:

$$\left[\frac{dI_D}{dt}\right]_{max} \cong \frac{(V_{gg} - V_{th})}{L_S + M_{GD} + M_{GS} + M_{DS}} \quad (2)$$

The switching speed limit can be estimated based on the M-factor and $[dI_D/dt]_{max}$, and its value is dependent on the package structure. Based on these equations, the limit of the switching speed of the package is determined by measuring $L_S+M_{GD}+M_{GS}+M_{DS}$.

978-1-7281-0582-6/19 $31.00 © 2019 IEEE

503

III. SIMULATION

A. SPICE Simulation

In this work, a SPICE simulation was used to confirm the influence of the M-factor and $[dI_D/dt]_{max}$. We measured the parasitic inductances in some example packages and these inductance values were input into the SPICE circuit model for a double pulse test. The switching speeds determined using equations (1) and (2) were then compared with the speeds given by SPICE simulation of the inductance model.

B. Q3D Simulation

Quasi-3D (Q3D) simulations are used to confirm the accuracy of the parasitic inductance measurements. Fig. 2 shows a TO-247 package and the model of the package used for the Q3D simulation. The gate terminal and the source terminal are connected to the front side of the chip through bonding wires. In this case, the left side is the gate, the center is the drain, and the right side is the source.

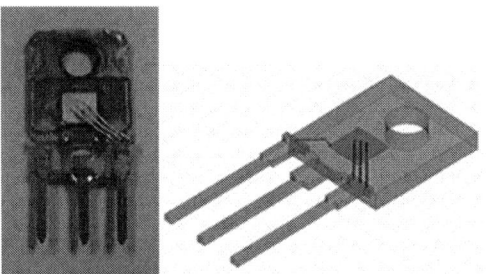

Fig. 2. TO-247 package and model used for the Q3D simulation.

IV. MEASUREMENT

A. Three-Port Measurement

This method measures six inductances from the three terminals. The authors have introduced a "neutral point" in the package [1] that separates the three terminals into six inductances. This neutral point is placed at the center of the metal-oxide-semiconductor field-effect transistor (MOSFET) chip in the package and is grounded. Fig. 3 shows the proposed TDR and TDT measurement setup. Two of the three terminals were connected to a digital oscilloscope through 50 Ω microstrip lines and the other is terminated at 50 Ω. The neutral point is connected to the ground on the back side of the strip line. The digital oscilloscope provided a high-speed step voltage for input into one of the terminals. The reflected and transmitted waves, denoted by e_r and e_t, respectively, were then observed via the oscilloscope. The following equations give the self-inductance L and the mutual inductance M [5].

$$L = \frac{z_0}{2e_i} \int_0^\infty (e_i + e_r)\, dt \qquad (3)$$

$$M = \frac{z_0}{2e_i} \int_0^\infty e_t\, dt \qquad (4)$$

The self-inductances and mutual inductances of each terminal are calculated from the reflected and transmitted waves.

B. Two-Port Measurement

Another measurement method using TDR and TDT [6] was also used in this paper. The value of $L_S + M_{GD} + M_{GS} + M_{DS}$ can be measured directly without a "neutral point" when using this method. Fig. 4 shows the two-port measurement setup. This method grounds the source and connects the other two terminals to the digital oscilloscope through the 50 Ω microstrip line. The digital oscilloscope then inputs a high-

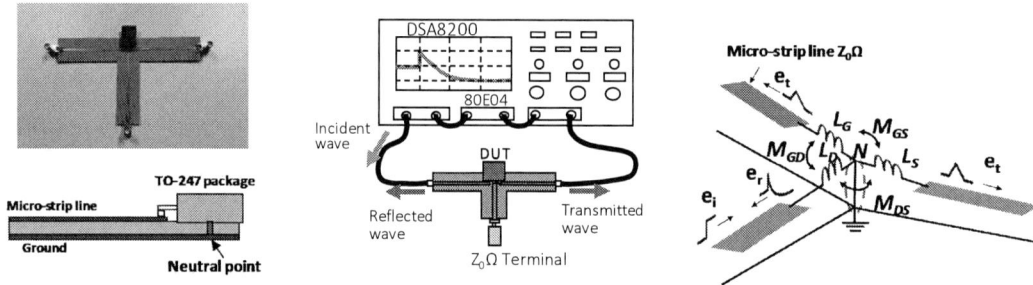

Fig. 3. Self- and mutual inductance measurement setup (DSA8200 oscilloscope with 80E04 TDR sampling module and microstrip line) and reflected/transmitted waves on the strip lines.

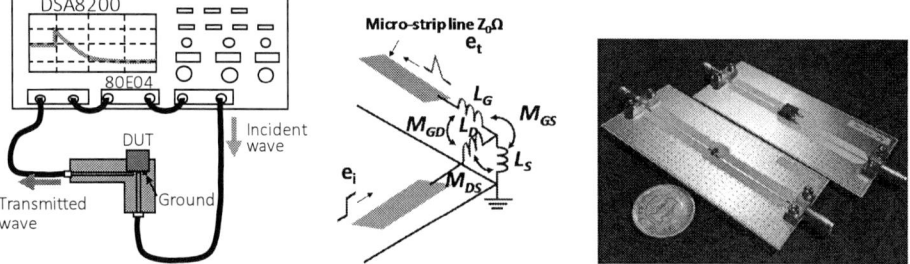

Fig. 4. Two-port measurement setup (DSA 8200 with 80E04 module and microstrip line) and reflected/transmitted waves on the strip lines. Photograph shows an example of microstrip line for the two-port setup for surface-mount packages.

speed step voltage to the drain terminal and measures the transmitted voltage. The following equation gives $L_S+M_{GD}+M_{GS}+M_{DS}$:

$$L_S + M_{GD} + M_{GS} + M_{DS} = \frac{Z_0}{2e_i} \int_0^\infty e_t \, dt \qquad (5)$$

$L_S+M_{GD}+M_{GS}+M_{DS}$ are calculated from the transmitted wave.

C. Results and Influence of Mutual Inductance

The measurement results were evaluated via SPICE and Q3D simulations. Fig. 5 shows an example of the reflected and transmitted waveforms from the strip line connected to the drain that were measured using the three-port measurement method. Fig. 6 shows a comparison between the results of the experiments and the simulations. The error was less than 1 nH. SPICE circuit simulations and double pulse testing confirmed the influence of the mutual inductance. Figs. 7 and 8 show the simulated results with value of measured inductances when using low- and high-voltage MOSFETs, respectively. These figures show the turn-on period waveforms, which were compared in the cases of zero inductance, self-inductance only and both self- and mutual inductances. V_{DS} falls most rapidly in the case without inductance, and falls most slowly in the case with self-inductance. The turn-on speed with the self- and mutual

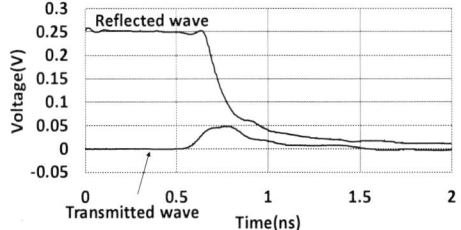

Fig. 5. Example of measured reflected waveform for strip line connected to drain.

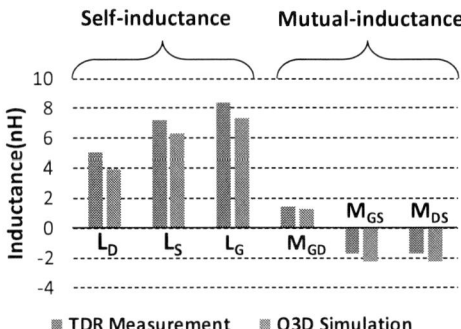

Fig. 6. Comparison of experiments and simulations.

inductances is faster than that with self-inductance only. The mutual inductances increased the switching speed during the turn-on period. Measurement results are provided for several packages in this work. Table 1 shows the results for $L_S+M_{GD}+M_{GS}+M_{DS}$ for five packages. The two measurement methods gave different values and the two-port measurement results were closer to those of the Q3D simulations than the three-port measurements. The error is caused by the conductor that connects the package to the ground and the capacitance of the chip. Fig. 9 shows the SPICE simulation results for the turn-on time of the TO-247 package as a function of the M-factor for both LV-MOSFETs and HV-MOSFETs. Fig. 10 shows the limits of the drain current slew rate.

TABLE 1. Results for $L_S+M_{GD}+M_{GS}+M_{DS}$

	3 ports measurement $L_S+M_{GD}+M_{GS}+M_{DS}$	2 ports measurement $L_S+M_{GD}+M_{GS}+M_{DS}$	
	Measurement(nH)	Measurement (nH)	Q3D (nH)
TO-247 1000V MOSFET (FQH8N100C)	5.2	5.3	4.0
TO-220 120V MOSFET (TK32E12N1,S1X)	9.7	3.4	3.5
DPAK 250V MOSFET (IRFR224PBF)	5.51	2.7	3.4
SOT-223 60V MOSFET (NVf3055L108T1G)		2.9	2.5
SOT-23 60V MOSFET (2N7002ET1G)		0.94	0.85

V. CONCLUSION

This paper has clarified the source self-inductance and the mutual inductances among three terminals to give switching speed limits for power semiconductor device packages. Measurement methods were proposed and their accuracy was confirmed using Q3D simulations and SPICE simulations.

REFERENCES

[1] K. Hasegawa, K. Wada, and I. Omura, "Mutual Inductance Measurement for Power Device Package Using Time Domain Reflectometry," IEEE ECCE, pp. 1-6, 2010.

[2] . S. Hashino and T. Shimizu, "Separation measurement of parasitic impedance on a power electronics circuit board using TDR" IEEE ECCE, pp. 2700-2705, 2010.

[3] Z. chen, D. Boroyevich, and R. Burgos, "Experimental Parametric Study of the Parasitic Inductance Influence on MOSFET Switching Characteristics," IEEE IPEC, pp. 164-169, 2010

[4] Y. Xiao, H. Shah, T. P. Chow and R. J. Gutmann, "Analytical Modeling and Experimental Evaluation of Interconnect Parasitic Inductance on MOSFET Switching Characteristics," IEEE APEC, pp. 516-521, 2004

[5] TDA systems, "TDR Technique for Characterization and Modeling of Electronic Packaging," Application note PKGM-0301, pp. 1-16, 2001

[6] K. Aikawa, T. Shiida, R. Matsumoto, K. Umetani, E. Hiraki, "Measurement of the Common Source Inductance of Typical Switching Device Packages," IEEE IFECC 2017 – ECCE Asia, pp. 1172-1177, 2017

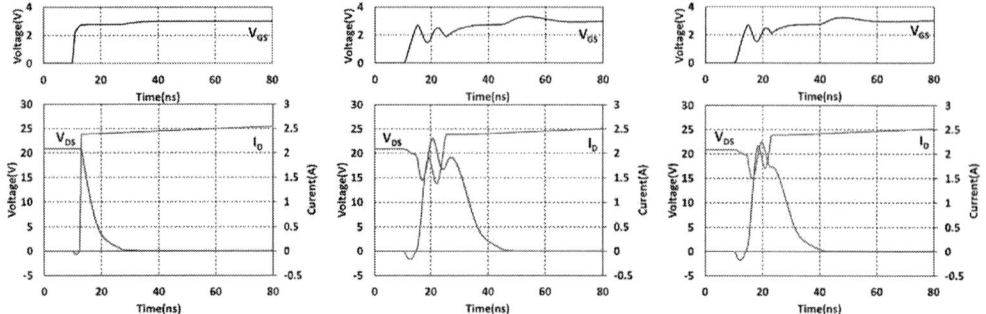

(a) Waveform without stray inductance. (b) Waveform with self–inductance. (c) Waveform with self and mutual inductances.

Fig. 7. Simulated results with various measured inductances (low-voltage MOSFET).

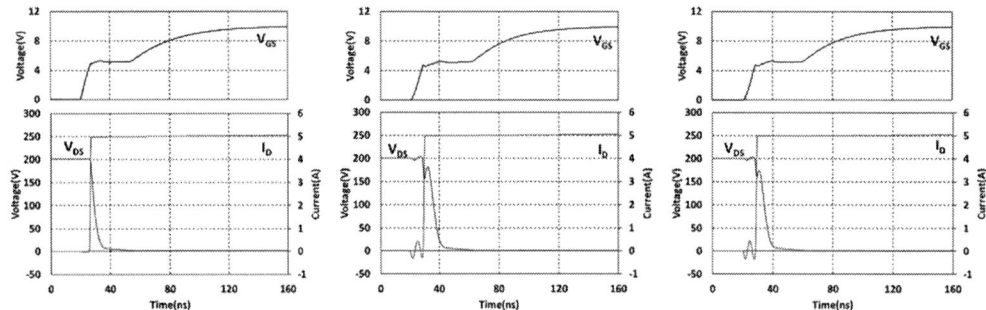

(a) Waveform without stray inductance. (b) Waveform with self- inductance. (c) Waveform with self and mutual inductances.

Fig. 8. Simulated results with various measured inductances (high-voltage MOSFET).

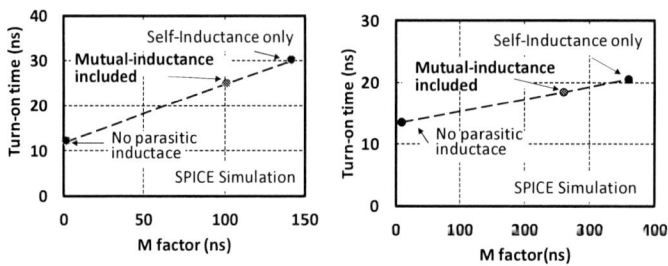

Fig. 9. Turn-on time SPICE simulation results as functions of M-factor $=(L_S+M_{GD}+M_{GS}+M_{DS})gm+R_GC_{GS}$ for LV-MOSFET and HV-MOSFET.

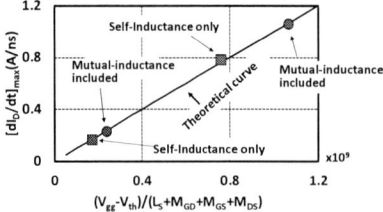

Fig. 10. Limits of drain current slew rate (blue: LV-MOSFET; red: HV-MOSFET).

Proceedings of the 31st International Symposium on Power Semiconductor Devices & ICs
May 19-23, 2019, Shanghai, China

Experiments of a Novel low on-resistance LDMOS with 3-D Floating Vertical Field Plate

Guangsheng Zhang[1], Wentong Zhang[1, 2, †], Junqing He[2], Xuhan Zhu[2], Sen Zhang[1], Jingchuan Zhao[1], Zhili Zhang[1], Ming Qiao[2], Xin Zhou[2], Zhaoji Li[2] and Bo Zhang[2]

[1]Technology Development Department, CSMC Technologies Corporation, Wuxi, P. R. China
[2]State Key Laboratory of Electronic Thin Films and Integrated Devices, University of Electronic Science and Technology of China, Chengdu, P. R. China, †zhwt@uestc.edu.cn

Abstract—**A novel lateral double-diffused metal–oxide semiconductor (LDMOS) with three-dimensional floating vertical field plate (3-D F-VFP) is proposed in this paper. The 3-D F-VFP LDMOS features a discrete pillar-type VFP array through the N-type drift region and the VFP pillars with the same distance from the source are connected to each other to realize the equal potential. In the off-state, the potential along the drift region are pinned by the series of equal-potential F-VFP rings and a new full-region depletion mode is introduced into the bulk of the device. Therefore, the highly doped drift region is depleted by the 3-D F-VFP, establishing a self-adaptive charge balance inside the drift region. In the on-state, the current flows through the gaps among all the F-VFP pillars, preventing the long current path and reducing the on resistance R_{on}. A 3-D F-VFP LDMOS is experimentally implemented, which obtains a breakdown voltage V_B of 630 V and a R_{on} of 344.8 Ω compared with 550 V and 722.5 Ω of the device without the F-VFP. The measured saturation current of the 3-D F-VFP LDMOS is more than 4.5 times of that of the device without F-VFP.**

Keywords—floating vertical field plate (F-VFP), LDMOS, charge balance, on resistance R_{on}, breakdown voltage V_B

I. INTRODUCTION

The lateral double-diffused metal–oxide semiconductor (LDMOS), as the heart of the smart power ICs, shows the wide applications in the fields of automotive electronics, communication, industrial and consumer market, and so on. To reduce the on resistance R_{on} of the LDMOS, the charge balanced voltage sustaining layers are introduced into the devices [1]-[6]. The charge balance can be realized by two typical methods: (1) using the PN junctions, e. g., the lateral super junction devices [1]-[4]; (2) using the MIS structure, e. g., the field plate is used to deplete the full drift region by covering all the drift regions [5]-[6] of the thin drift layer devices. For the conventional lateral field plate, its influence is only located at the surface of the device, which causes a weak depletion effect in the deep drift region. To solve this problem, the vertical field plate (VFP) is proposed for the thick drift layer devices [7]. Because the VFP introduces the bulk depletion effect to achieve both the low R_{on} and high breakdown voltage V_B, the similar mechanism has been expanded into other devices [8]-[12]. However, the VFP device needs the wide trench and thick oxide layer to sustain the high V_B, resulting in the process difficulties.

In this paper, a novel three-dimensional floating vertical field plate (3-D F-VFP) concept is proposed and experimentally demonstrated to reduce R_{on} by a new full-region depletion mode. Section II gives the 3-D F-VFP concept and mechanism of the F-VFP LDMOS. The design

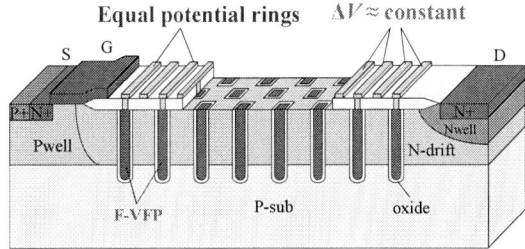

Fig. 1: Structure of the proposed 3-D F-VFP LDMOS. The new structure features the 3-D F-VFP inserted through the drift region and all the F-VFP pillars with the same distance from the source are connected to each other to provide equal-potential rings along the full drift region.

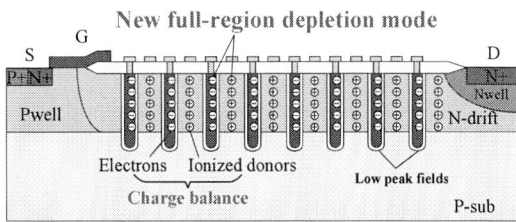

Fig. 2: Mechanism of the 3-D F-VFP LDMOS. From the 2-D cross-sectional view along the center of the F-VFP trench in Fig. 1, the new charge balance between the ionized positive charges in the drift region and the negative charges in the F-VFP is realized in the new device. The F-VFP introduces a full-region depletion mode into the drift region.

and process are discussed in Section III and the experiments of the F-VFP LDMOS are discussed in Section IV.

II. STRUCTURE AND MECHANISM

A. Structure of the F-VFP LDMOS

The structure of the 3-D F-VFP LDMOS is shown in Fig. 1. The F-VFP features a discrete pillar-type VFP array through the N-type drift region and inserted into the p-sub, which prevents the peak electric fields near the bottom of the trench. The VFP pillars with the same distance from the source are connected to each other by the surface metal rings. The metal rings make sure the equal potential in each of the VFP arrays.

In the off-state, the high voltage biased at drain is sustained by the equal potential VFP arrays and the potential along the drift region shows an almost linearly increased distribution. The F-VFP arrays provide a series of equidistant equal-potential planes. Therefore, the potential along the drift region of the F-VFP LDMOS is pinned by the series of equal-potential F-VFPs

978-1-7281-0582-6/19 $31.00 © 2019 IEEE

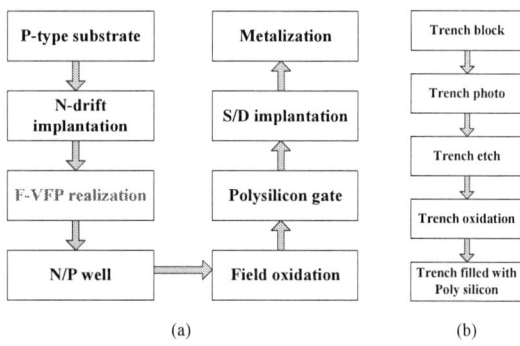

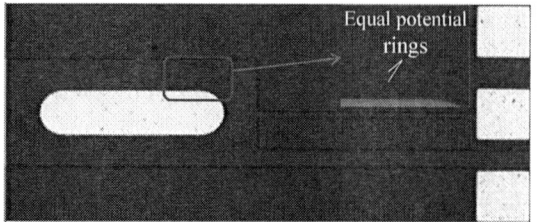

Fig. 5: Micro-photo of the proposed F-VFP LDMOS. The equal-potential rings are shown in the insert.

TABLE I.   KEY PARAMETERS IN EXPERIMENTS

Key parameter values	Values
Rectangular trench width	2 μm
Distance between trenches	3 μm
Depth of trenches	11.5 μm
Drift length	40 μm
Number of F-VFP Rings	15

(a)   (b)

Fig. 3: The key process flow of the 3-D F-VFP LDMOS. (a) the F-VFP LDMOS shows a similar process flow as that of the conventional LDMOS except for the F-VFP realization; (b) the process steps of the 3-D F-VFP, which is realized by the etch, oxidation and poly filling of the trench pillars and connected to each other in the process of metalization.

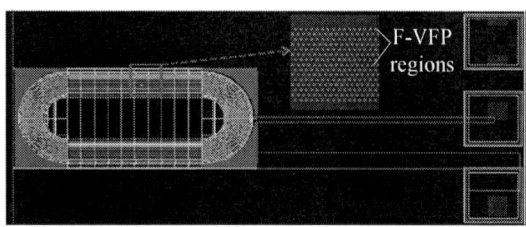

Fig. 4: Layout of the proposed F-VFP LDMOS. The insert gives the metal rings and the corresponding F-VFP regions with rectangular layouts.

B. Mechanism

Unlike the conventional LDMOS, the potential distribution in the bulk of the F-VFP LDMOS is determined by the F-VFP. A new full-region depletion mode is introduced into the bulk of the device, as shown in Fig. 2. The equidistant equal-potential planes are formed in the full drift region and the potential difference ΔV between every two adjacent planes is almost a constant:

$$\Delta V \approx V_{\mathrm{B}} / N \qquad (1)$$

in which N is the array number in the drift region.

The segmented drift regions of the 3-D F-VFP LDMOS are depleted by the adjacent F-VFP plane. The self-adaptive charge balance is realized between the ionized positive charges in the drift region and the negative charges in the F-VFP. So the higher doping concentration can be realized in the 3-D F-VFP LDMOS by the new full-region depletion mode. On the other hand, the discrete pillar-type VFP structure also provides the current path along the gaps among all the F-VFP pillars. Therefore, both the optimized V_B and low R_{on} can be realized by the 3-D F-VFP.

III.   DESIGN AND PROCESS

The main difference between the process of the new device and the standard process of the conventional LDMOS is the realization of the F-VFP. Fig. 3(a) gives the process flow of the 3-D F-VFP LDMOS. The F-VFP is fabricated after the N-drift implantation and followed by the N/P well and field oxidation. At last, the equal-potential metal rings are obtained with the same metalization process.

The F-VFP realization is the key process with detailed steps in Fig.3(b). After the etching of the rectangular deep trenches, the oxide layer of the F-VFP is obtained by the trench oxidation with a thickness of 0.4 μm. After that, the trench is filled with a highly doped polysilicon. The high temperature of the trench oxidation is also used in the driving process of the N-drift. In our device, the F-VFP is fabricated through the drift region and inserted into the substrate to realize the best full-region depletion. The current in the on-state can follow along the gaps of the F-VFP pillars.

Fig. 4 gives the racetrack-shaped layout of the 3-D F-VFP LDMOS. The insert gives the F-VFP regions. The discrete trench pillar arrays are etched in the entire drift region. Other layout details are the same as those of the conventional LDMOS. The process of the F-VFP is also suitable for the isolation structure with a closed layout around the boundary of the device.

IV.   EXPERIMENTS

Based on the process flow in Fig. 3, the 3-D F-VFP LDMOS was fabricated based on the 0.5 μm CSMC process platform. The micro-photo of the 3-D F-VFP LDMOS is shown in Fig. 5. The insert gives the equal-potential metal rings above the drift region. With these metal rings, a series of potentials with the equal ΔV are realized along the F-VFP arrays. The bulk potential of the drift region is thus clamped and maintains the same potential as that of the local F-VFP array, ensuring the almost linearly increased floating potentials from the source to the drain. Detailed parameter values are illustrated in Table I.

Fig 6(a) gives the SEM of the 3-D F-VFP LDMOS. The bulk floating field plate and the surface equal potential metal rings are marked in the figure. The F-VFP arrays are fabricated deep into the bulk with a number of 15 and only 8 arrays are shown in Fig. 4 because of the staggered layout. It is found from the SEM that the equidistant F-VFP arrays are realized in the entire drift region with a length of 40 μm. More details about the F-VFP are given in Fig. 6(b). The

978-1-7281-0582-6/19 $31.00 © 2019 IEEE

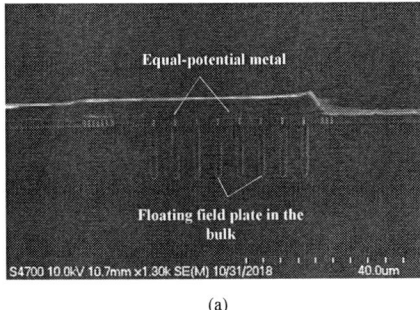

(a)

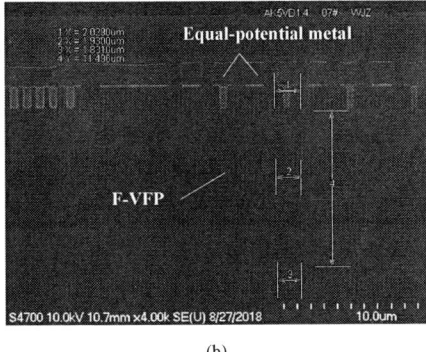

(b)

Fig. 6: The SEM of the 3-D F-VFP LDMOS. (a) the full device with a drift length of 40 μm; (b) the local region of the F-VFP with a width of 2 μm and a depth of 11.5 μm. Key parameter values are shown in Table I.

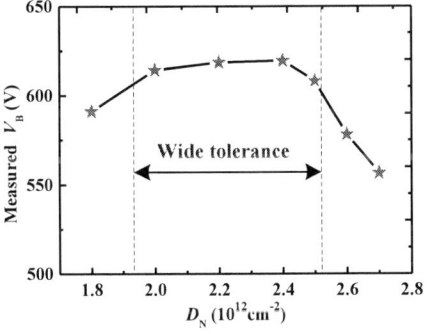

Fig. 7: Measured off-state V_B of the 3-D F-VFP LDMOS. The V_B values sustain almost a constant value in the wide doping range from 2 to 2.5 × 10^{12} cm^{-2} because of the self-adaptive charge balance.

width and depth of the rectangular F-VFP pillars are 2 μm and 11.5 μm, respectively. The distance between two F-VFPs is 3 μm. The equal-potential metal rings are realized on the surface of the drift region and connected with the F-VFP by the via hole structure.

The 3-D F-VFP LDMOS is fabricated with different doping dose D_N of the drift region. The corresponding measured off-state V_B is shown in Fig. 7. The 3-D F-VFP LDMOS has a wide tolerance with D_N. V_B is almost 630 V with D_N from 2 to 2.4 × 10^{12} cm^{-2} and above 550 V with D_N from 1.8 to 2.7 × 10^{12} cm^{-2}. The 15 equidistant F-VFP arrays along the drift region act as the parallel plate capacitive

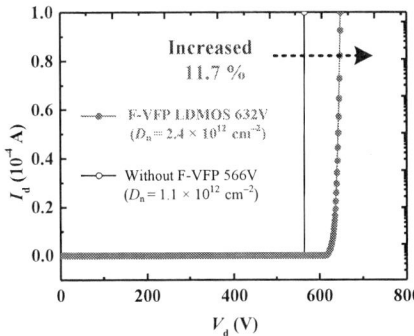

Fig. 8: Measured off-state breakdown curves of the 3-D F-VFP LDMOS and the device without the F-VFP. The maximum V_B of the device without the F-VFP is obtained with a doping dose of 1.1 × 10^{12} cm^{-2}. V_B of the new device is increased from 566 V of the device without F-VFP to 632 V.

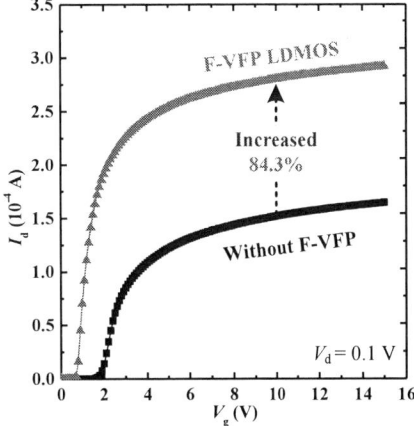

Fig. 9: The measured $I_d - V_g$ curves with the drain voltate V_d = 0.1 V. The current of the 3D F-VFP LDMOS is increased by 84.3% from that of the device without F-VFP.

voltage divider for the applied drain voltage, resulting in an almost linearly increased potential distribution in the arrays. Because the potential in the bulk of the drift is pinned by the local F-VFP pillar, the wide tolerance is realized by the self-adaptive charge balance between the ionized donors in the drift region and the electrons in the F-VFP.

The measured off-state breakdown lines of the 3-D F-VFP LDMOS and the device without the VFP are compared in Fig. 8. The maximum V_B of the new device is increased from 566 V to 632 V. The optimized D_N is also increased from 1.1 to 2.4 × 10^{12} cm^{-2} by the full region depletion mode. Therefore, the potential difference ΔV between every two adjacent F-VFPs is 42 V and the average electric field along the drift region is 15.8 V/μm.

Fig. 9 compares the measured I_d - V_g curves of the 3-D F-VFP LDMOS and the LDMOS without F-VFP under the same condition of V_d = 0.1 V. Owing to the high doping concentration of the 3-D F-VFP LDMOS, its I_d features a significant increase of 84.3% compared with that of the device without F-VFP.

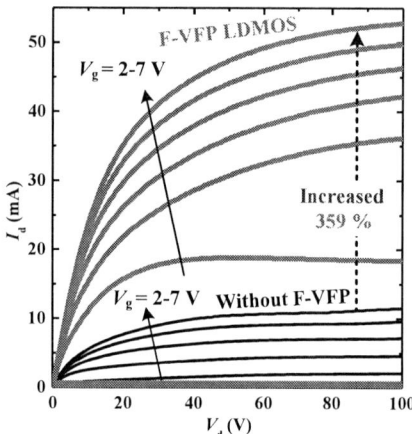

Fig. 10: Measured $I_d - V_d$ curves with V_g from 2 to 7 V. The significant current increasement is observed in the 3D F-VFP LDMOS, which is more than 4.5 times of the current of the device without the F-VFP.

The influence of the F-VFP on the output characteristic curves with the gate voltage V_g from 2 to 7 V is shown in Fig. 10. The maximum saturation current of the proposed 3-D F-VFP LDMOS is more than 4.5 times of that of the device without F-VFP. The resistance R_{on} of the 3-D F-VFP LDMOS is reduced from 722.5 Ω of the device without the F-VFP to·344.8 Ω at a gate voltage V_g of 7 V. The high current capacity of the new device is explained as follows: firstly, the full-region depletion mode introduced by the F-VFP allows the high doping concentration of the drift region. Secondly, the potential in the bulk of the drift region is still pinned by the F-VFP arrays and the parasitic JFET effect between the N-drift and the P-sub is weakened.

V. CONCLUSION

In this paper, the 3-D F-VFP concept is proposed for the first time to introduce a new full-region depletion mode in the lateral LDMOS devices. The F-VFP depletes the highly doped drift region and establishes a self-adaptive charge balance inside the device. The 3-D F-VFP LDMOS was fabricated based on the standard 0.5 μm CSMC process platform. The measured results of the new device exhibit a reduced R_{on} from 722.5 Ω of the conventional LDMOS to 344.8 Ω with a V_B from 566V to 632 V, which also shows an increased saturation current up to 4.5 times compared with that of the device without F-VFP.

ACKNOWLEDGMENT

This work was supported in part by the National Natural Science Foundation of China (grants 61704020, 61674027), the Natural Science Foundation of Guangdong Province under Grant 2018A030310675 and Grant 2016A030311022, the China Postdoctoral Science Foundation (grant 2018M643447), and the Fundamental Research Funds for the Central Universities (grant ZYGX2017KYQD159).

REFERENCES

[1] Bo Zhang, Wentong Zhang, Zehong Li, Ming Qiao, and Zhaoji Li, "Equivalent substrate model for lateral super junction device," IEEE Trans. Electron Devices, vol. 61, pp. 525-532, Feb 2014.

[2] Wentong Zhang, Song Pu, Chunlan Lai, Li Ye, Shikang Cheng, and Sen Zhang, "Non-full depletion mode and its experimental realization of the lateral superjunction," ISPSD 2018, pp.475-478, Chicago, Americ.

[3] Wentong Zhang, Bo Zhang, Ming Qiao, Zehong Li, Xiaorong Luo, and Zhaoji Li, "Optimization of lateral superjunction based on the minimum specific ON-resistance," IEEE Trans. Electron Devices, vol. 63, pp. 1984-1990, May 2016.

[4] Bo Zhang, Wentong Zhang, Ming Qiao, Zhenya Zhan, and Zhaoji Li "Concept and design of super junction devices," Journal of Semiconductors, Vol. 39, pp. 5-16, Feb 2018.

[5] Wentong Zhang, Zhenya Zhan, Yang Yu, Shikang Cheng, Yan Gu, and Sen Zhang, "Novel superjunction LDMOS (> 950 V) with a thin layer SOI," IEEE Electron Device Lett, vol. 38, pp.1555-1558, Nov 2017.

[6] Kenji Hara, Tomoko Kakegawa, Shinichiro Wada,Tomoyuki Utsumi, and Tetsuo Oda, "Low on-resistance high voltage thin layer SOI LDMOS transistors with stepped field plates," ISPSD 2017, pp. 307-310, Sapporo, Japan.

[7] Wentong Zhang, Bo Zhang, Ming Qiao, Lijuan Wu, Kun Mao, Zhaoji Li, and Qiao M, "A novel vertical field plate lateral device with ultralow specific on-resistance," IEEE Trans. Electron Devices, vol. 61, pp. 518-524, Feb 2014.

[8] Wentao Yang, Xianda Zhou, Chao Xiao, Hao Feng, Yong Liu, and Xiangming Fang, "Optimization of trench sidewall for low leakage current of the sloped field plate trench edge termination," ISPSD 2018, pp. 172-175, Chicago, Americ.

[9] Wentong Zhang, Li Ye, Dong Fang, Ming Qiao, Kui Xiao, Boyong He, "Model and Experiments of Small-Size Vertical Devices With Field Plate," IEEE Trans. Electron Devices, vol. 66, pp. 1416-1421, March 2019.

[10] Ying Wang, Zhi-Yuan Li, Yue Hao, Xin Luo, Jun-Peng Fang, and Ya-chao Ma, "Evaluation by Simulation of AlGaN/GaN Schottky Barrier Diode (SBD) With Anode-Via Vertical Field Plate Structure," IEEE Trans. Electron Devices, vol. 65, pp. 2552-2557, June 2018.

[11] Lijuan Wu, Wentong Zhang, Qin Shi, Pengfei Cai, and Hangcheng He, "Trench SOI LDMOS with vertical field plate," Electronics Letters, vol. 50, pp. 1982-1984, December 2014.

[12] Chao Xia, Xinhong Cheng, Zhongjian Wang, Dawei Xu, Duo Cao, and Li Zheng, "Improvement of SOI trench LDMOS performance with double vertical metal field plate," IEEE Trans. Electron Devices, vol. 61, pp. 3477-3482, Oct 2014.

978-1-7281-0582-6/19 $31.00 © 2019 IEEE

Proceedings of the 31st International Symposium on Power Semiconductor Devices & ICs
May 19-23, 2019, Shanghai, China

A Laterally Monolithic-Integrated Multi-Cascode for Applications with 600V and more based on 20V-FINFETs in 90nm Technology

Rolf Weis, Marko Lemke, Marco Müller, Ralf Rudolf, Knut Stahrenberg,
Thomas Bertrams, Martin Bartels, Ahmed Mahmoud*, Nicolas Nagel*
Rolf.Weis@infineon.com, Infineon Technologies Dresden GmbH&Co.KG, Dresden/Germany,
*Infineon Technologies AG, München/Germany

Abstract—Over decades the usage of the charge compensation principle for power devices has improved the performance and efficiency of typical applications in almost all voltage classes like power supplies, converters and drives. Different types of charge compensation concepts were applied to various device structures like the superjunction compensation, the field plate compensation and the dynamic compensation in form of bipolar electron-hole plasma. For each voltage class, unique technologies have been developed to adjust and optimize drift zone length, doping concentrations and avalanche robustness, leading to a large variety of different vertical and lateral power device technologies. In this paper, laterally monolithic-integrated multi-transistor cascodes are presented as high-voltage MOSFET transistors for a large range of breakdown voltages from 20V to multiple hundred volts.

Keywords — HV-MOSFET, LV-MOSFET, FINFET, Cascode, Multi-Transistor Cascode, Integrated PowerTechnology

I. NEW CONCEPT FOR HVMOS: THE MULTI-CASCODE

The concept of the multi-transistor cascode is shown in Figure 1. It consists of multiple depletion FETs (DFETs) connected in series to one enhancement FET (EFET), with which the cascode can be controlled like a conventional MOSFET [1]. Each DFET D_n of the cascode has its gate terminal connected to the source of the neighboring transistor D_{n-1} on its source side. Therefore the drain-source voltage Vds of this neighboring transistor is the negative gate-source voltage Vgs of the regarded DFET and therefore it will control its conductivity with $V_{GS}(D_n)=-V_{DS}(D_{n-1})$. The cascode is in on-state, if the EFET is turned-on. If a current is flowing thru the cascode, for low-ohmic devices there is only a small voltage drop Vds over each of the transistors and therefore Vgs is almost zero with almost no impact on the conductivity on all DFETs. If the EFET is turned-off, the EFET starts to pick-up voltage on its drain-source path. This Vds starts to decrease the conductivity of the neighboring DFET on the drain side until this DFET also starts to block and the drain-source path picks up voltage too. The next DFET starts to block and so on. Finally the cascode is switched-off and an applied voltage is split over multiple transistors of the cascode. Each transistor in the cascode is protected by an avalanche or zener diode in

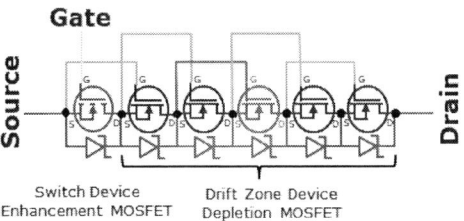

Fig. 1: High-Voltage MOSFET with external pads for source, gate and drain composed of multiple low-voltage MOSFETs in series arranged as a multi-cascode configuration.

parallel to its source-drain path, so it stays within its safe operating area (SOA). Basically, this protection can also be realized by the intrinsic avalanche capability of the used MOSFETs itself. This voltage-limiting element also protects the gate-source path of the neighboring DFET on the drain side. If the EFET is turned-on again, the drain-source voltage drops, turning-on the next DFET and so on, until all transistors are in on-state.

II. POWER DEVICE FORMULAS FOR THE CASCODE

For monolithic integration, new lateral 20V-enhancement and 20V-depletion FINFET transistors were developed as illustrated in Figure 2. They were fabricated with a single 90nm-technology and are protected on input and output path by laterally-integrated 20V-avalanche diodes not shown. The breakdown voltage V^C_{BVDSS} and the specific resistance $(R_{DSon}*A)^C$ of a cascode with MOSFETs are determined by the properties of the single low-voltage transistors. The breakdown voltage of a single EFET or DFET stage in the cascode is V^S_{BVDSS}. It is given by the breakdown voltage of the avalanche diode in parallel and the specific resistance of it is $(R_{DSon}*A)^S$. For a cascode with N transistor-diode pairs in series the cascode breakdown voltage V^C_{BVDSS} is

$$V^C_{BVDSS}=N*V^S_{BVDSS} \qquad (1)$$

and

$$(R_{DSon}*A)^C=(R_{DSon}*A)^S*N^2$$

$$=(R_{DSon}*A)^S*(V^C_{BVDSS}/V^S_{BVDSS})^2 \qquad (2)$$

978-1-7281-0582-6/19 $31.00 © 2019 IEEE

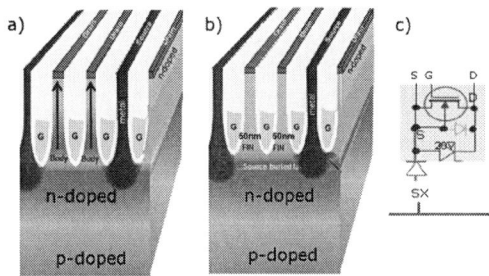

Fig. 2: FINFETs with vertical current flow and more than 25V blocking capability were used in the cascode. The FINFETs have source, drain and gate contacts accessible on the silicon surface for wiring purposes. A vertical metal contact was used on one hand to contact a buried layer with source potential and to deliver a source and body contact at the bottom of the device and on the other hand as field plate for the drift region extending from the channel area up to the surface. Enhancement FINFET a) and depletion FINFET b) differ only in channel doping c) DFET schematics with full functionality.

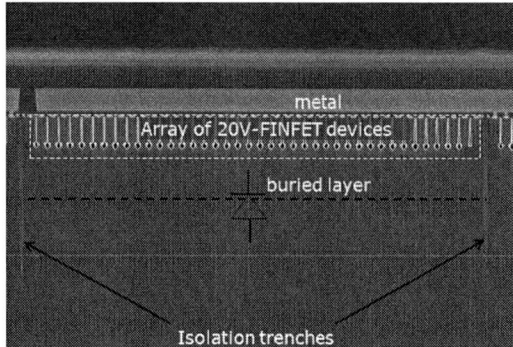

Fig. 3: High-Voltage MOSFET cascode manufactured in 90nm technology: Arrays of FINFETs are isolated vertically with an n-doped buried layer resulting in a diode in blocking state towards the p-substrate on the lowest potential and laterally array against array by deep oxide isolation trenches.

III. NEW FINFETs AS BASIC ELEMENTS FOR THE CASCODE

According to Eq. (2), single transistor devices with low $(R_{DSon}*A)^S$ are the base for achieving low $(R_{DSon}*A)^C$ for the cascode. A key requirement for the functionality of such a cascode is, that one single transistor device in the chain controls the next device with its drain-source voltage. Therefore for all depletion FETs the maximum input voltage must be equal to the maximum output voltage $|V_{DSmax}|$ $=|-V_{GSmax}|$. This condition is well known for logic devices, but not for power devices with around 20V operating voltage. The structure of a FINFET with two gates arranged around a channel area like shown in Fig. 2 allows to combine thick gate oxides with a very good channel control by the fully-depletion of a thin silicon fin area. Especially, the threshold voltage V_{TH}

of such a FINFET can be easily tuned from normally-off to normally-on by just changing the doping concentration and type in the channel area from p-doping to n-doping.. With the arrangement of the channel and the drift zone of the FINFET in the vertical direction, enhancement and depletion FETs can be easily manufactured by choosing the doping profile accordingly in the vertical depth direction. In order to achieve a low Miller-Capacitance C_{GD} for fast switching, the source contact of the transistor was chosen to be connected to a n-doped buried layer at the lower end. Accordingly the drain contact was arranged towards the wafer surface with a small footprint. With this vertical arrangement, an additional local metal contact to the buried source layer could be used as field plate for the drift zone to achieve lowest $R_{DSon}*A$. With using two of these metal contacts to surround two fins with shortened drain, the FINFETs do not require any additional edge construction. A periodical arrangement in an array is easily possible.

Fig. 3 shows an integrated part of a cascode realized in 90nm technology. Multiple FINFETs were connected in parallel to such an array. Isolation trenches separate the different voltage stages in the cascode. As the voltage in a cascode changes only from array to array at these isolation trench borders, the geometrical length of the driftzone for blocking an overall voltage of the cascode is not a function of blocking voltage anymore. Instead it can be chosen independently by designing the width and length of the arrays in the cascode. This is giving an additional freedom in realizing the geometry of a certain R^C_{DSon} and V^C_{BVDSS} of an MOSFET-cascode, which is not available for MOSFETs based on the compensation principle.

IV. ELECTRICAL RESULTS

In Fig. 4, the output characteristics of the two types of FINFET transistors, an enhancement and a depletion transistor required for a MOSFET-multi-transistor cascode are shown. The two FINFET-types have a breakdown voltage larger than 25V, which is higher than the breakdown voltage $V^S_{BVDSS}=20V$ of an avalanche diode arranged in parallel, which is used to protect the input and output path of all MOSFETs from over-voltages according to Fig.1.

With Eq. (1), different cascode breakdown voltages V^C_{BVDSS} can be obtained as multiples of 20V by connecting N transistor-diode pairs in series. In Fig. 5 the output characteristics of 5- and 8-transistor cascodes demonstrate the easy scalability of the concept to different breakdown voltages. Breakdown voltages up to more than 800V were achieved with this concept on one wafer with one manufacturing process flow.

Low leakage can be achieved until all N avalanche diodes are achieving their breakdown voltage V^S_{BVDSS} at the same time. At that point, when $V^C_{BVDSS}=N*V^S_{BVDSS}$ is reached, the cascode starts to show a significant increase of drain current due to the cascode overall breakdown.

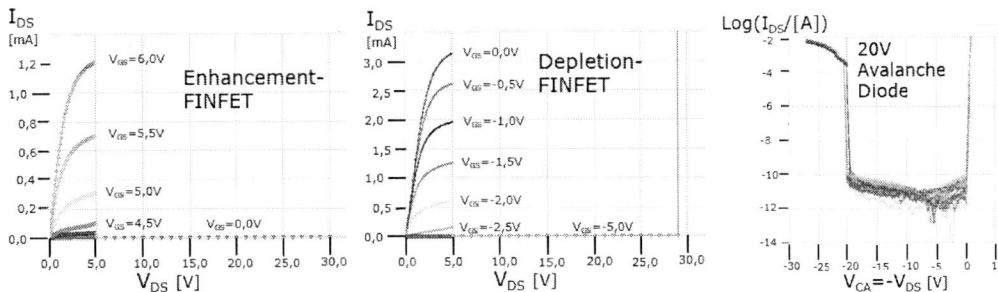

Fig. 4: The output characteristics of the used enhancement and depletion FINFET transistors are shown. The V_{TH} of the cascodes can be changed by the V_{TH} of the enhancement FET only. Separate laterally monolithic integrated avalanche diodes were used to limit the source-drain voltage and the gate-source voltage of the FINFETs according to Fig. 1 to a maximum of 20V. The characteristic of these diodes is shown from blocking to conducting voltages V_{CA} Cathode to Anode. As diodes are in parallel to FINFETs, $-V_{CA}$ equals applied V_{DS} of the FINFETs and therefore avalanche always occurs in the diodes first for protection of the FINFETs from over-voltages.

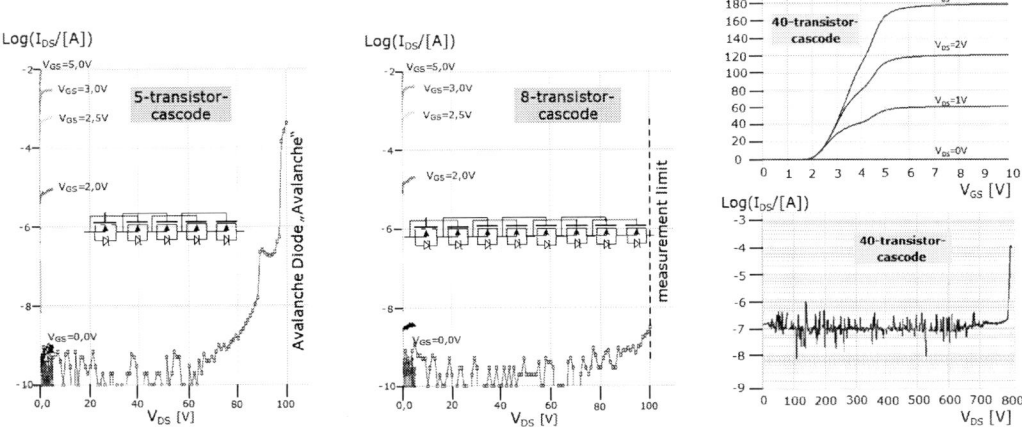

Fig. 5: Using these low-voltage 20V-FINFETs in multi-transistor cascode configuration, different breakdown voltages in steps of 20V can be realized like demonstrated up to 800V. Increasing the number N of transistor-diode pairs results in increased breakdown voltage of the cascode while turn-on behavior is still dominated by the enhancement FINFET (see 5- vs. 8-transistors) and its V_{TH} (see 40-transistor device). In cascode breakdown all avalanche diodes reach their breakdown voltage together leading to a drain-source current increase. The external drain-source-leakage stays low up to that over-all cascode breakdown voltage point.

As visible in Fig. 4, the avalanche diode breakdown characteristic is designed to have an increasing breakdown voltage with increasing current. This avoids thermal destruction by current filamentation and leads to a uniform distribution of the avalanche energy over all laterally distributed avalanche diodes in the cascode.

Fig. 6 shows the temperature dependence of the depletion FINFETs for its typical operation conditions in the cascode in an I_{DS} verses $V_{DS}=-V_{GS}$ chart. For low voltage increases in V_{DS}, the transistor current I_{DS} still rises until the depletion by the control voltage $|-V_{GS}|$ starts to decrease the current. Blocking is reached with low leakage below the negative V_{TH} of a normally on-transistor. The leakage current I_{DS} stays low

up to $V^S_{BVDSS}=20V$ limited in the cascode by the external avalanche diode in parallel even though the drain-gate voltage reaches 40V at that point. Only less than 20% increase in R^S_{DSon} is observed between 40°C and 125°C for the FINFET.

In Fig. 7 the superior low-temperature dependence of the cascode as effect of the high doping concentration in the single low-voltage FINFET is shown. Less than 20% resistance increase from 25°C to 130°C is demonstrated in comparison to approximately a factor of two known for 600V-MOSFETs based on the compensation principle like CoolMOS with superjunction. In addition, a high transconductance can be observed with a linear dependency of

978-1-7281-0582-6/19 $31.00 © 2019 IEEE

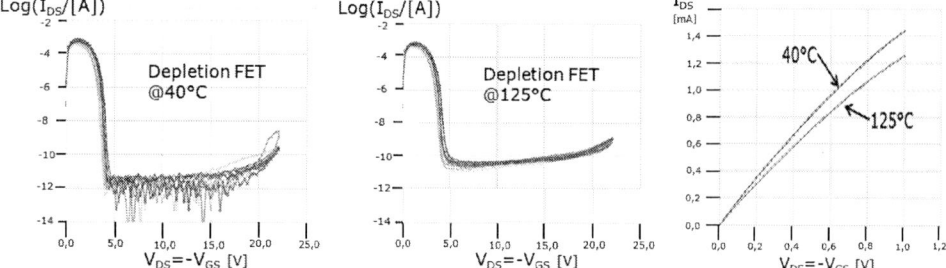

Fig. 6: Single depletion FINFET devices show less than 20% R_{DSon}-increase from 40°C to 125°C.This is due to the high doping concentration with already limited temperature dependence based on the degraded mobility.

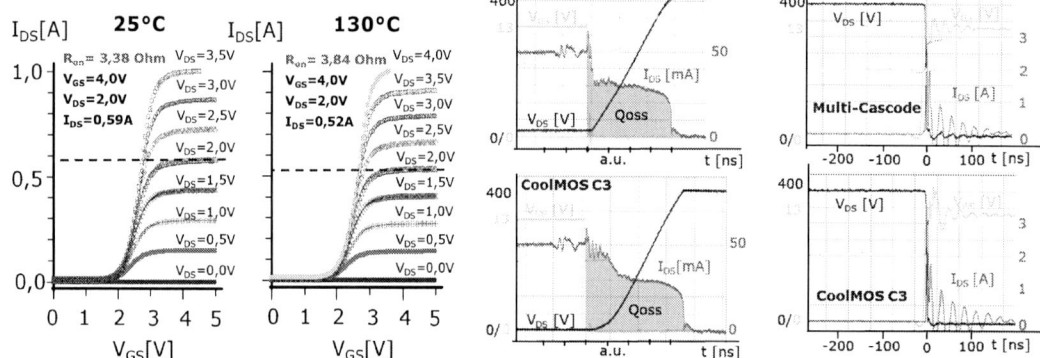

Fig. 7: The low temperature-dependence of the single FINFETs is also transferred to the cascode, i.e. a high-voltage cascode device with 3,38 Ohm as shown has less than 20% R_{DSON}-increase from 25°C to 130°C instead of a typical factor 2 observed for a typical 600V compensation MOSFET.

Fig. 8: Inductive switching of high-ohmic devices against a SIC diode reveal a linear voltage switch-off characteristic of multi-cascodes and lower Q_{oss} compared to a CoolMOS C3 due to a capacitive switching mechanism The turn-on behavior is similar to that of a CoolMOS C3.

the drain current I_{DS} versus the input voltage V_{GS} in the saturation case. In Fig. 8 inductive switching was performed at 400V with a 33-transistor cascode. Similar switching losses were achieved compared to a CoolMOS C3, a superjunction MOSFET. Whereas there are almost no differences in switch-on behavior, the multi-transistor cascode shows an almost linear switching slope for switch-off with similar rise time like the superjunction device under similar switching conditions. It looks like, there is no delay for the multi-transistor cascode before picking up the voltage like observed for CoolMOS transistors. This delay is caused by discharging the drift zone of the superjunction device at low voltages. Obviously, this is not necessary in case of a multi-transistor cascode, giving a hint to the mechanism of capacitive switching with a lower Q_{oss} compared to a vertical, compensation based CoolMOS C3 device. C3 was chosen for this comparison, as it is an older technology generation having a larger superjunction pitch, therefore lower doping concentration with higher mobility and therefore an even lower Q_{oss} than recent generations like C7.

V. SUMMARY

It was demonstrated, that the concept of multi-transistor cascodes with one technology for a series of laterally-arranged low-voltage transistors leads to competitive high-voltage MOSFETs. Various breakdown voltages can be realized by chaining these single devices on one wafer. Different and also superior device properties compared to MOSFETs based on the compensation principle were measured.

ACKNOWLEDGMENT

We like to thank the Fab of Infineon Technologies Dresden for the development and manufacturing of the devices. This work was co-financed by tax money based on the budget agreed upon by the members of the Saxon State Parliament.

REFERENCES

[1] R. Weis, F. Hirler, M. Stecher, A. Willmeroth, G. Deboy, and M. Feldtkeller, "Semiconductor Device Arrangement with a First Semiconductor Device and with a Plurality of Second Semiconductor Devices," oct ober 2011. Patent online available under https://patents.google.com/patent/US20120175635A1/

Proceedings of the 31st International Symposium on Power Semiconductor Devices & ICs
May 19-23, 2019, Shanghai, China

Cu Double Side Plating Technology for High Performance and Reliable Si Power Devices

Hitoshi Kobayashi, Tatsuya Ohguro, Tetsuya Kai, Takako Motai, Masaaki Ogawa, Mie Matsuo, Kenichi Oohashi, Shinsuke Kozumi,
Yoshiharu Takada, Hideharu Kojima, Shingo Masuko, Naoki Yonezawa, Akira Komatsu, Tatsuya Nishiwaki, Takuma Hara,
Mari Takahashi, Akira Ezaki, Kenichi Ohtsuka, Seiji Inumiya and Kyoichi Suguro
Discrete Semiconductor Division, Toshiba Electronic Devices & Storage Corporation, Tokyo, Japan
Email: jin.kobayashi@toshiba.co.jp

Abstract— **In this work, we have developed vertical Si power MOSFETs with high performance and high reliability by using Cu double side plating technology. 20 μm thick Cu plating layers are formed on both sides of devices with 50 μm thick Si substrate. In this structure, even though Si substrate is thinner, Safety Operating Area (SOA) is wider and the warpage of chip is smaller thanks to front and back side Cu plating layers.**

Keywords—Cu, Double side, plating, warpage, SOA, Ron

I. INTRODUCTION

In order to realize lower on-resistance (Ron) of vertical Si power MOSFETs [1,2], thinner Si substrate is desirable to reduce the Si substrate resistance [3,4]. Safety operating area (SOA) [5-9], however, degrades due to lower heat content and warpage of chip becomes larger. So, it was required to introduce a novel technology to achieve both lower Ron and wider SOA with small warpage of chips. In this paper, Cu double side plating process with thinner Si substrate is introduced as a novel technology.

II. SAMPLE FABRICATION

As shown in Fig. 1 we have developed Cu double side plating structure [10,11]. This consists of 20 μm thick Cu plating layers on both sides of devices with 50 μm thick Si substrate.

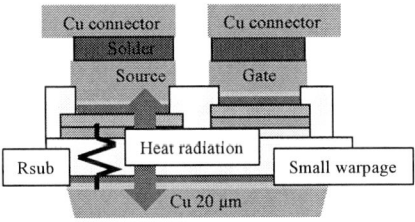

Fig. 1 The effect of vertical Si power MOSFETs with thin Si substrate and Cu double side plating.

Fig. 2 shows the estimation of wafer warpage for various Si substrate and Cu plate thickness. In the case of double side plating with the same Cu thickness, the warpage becomes significantly small. Then we have considered simultaneous both sides Cu plating to prevent wafer warpage caused by thick Cu film stress. And also as means of suppression of thinner wafer warpage, backside grinding for wafer thinning is provided only inner side of Si wafer and the rim region of Si wafer remains thick. This structure enables simultaneous both sides plating because there're no support materials on silicon surfaces for maintaining wafer strength. Thus

simultaneous both side Cu plating on thinner substrate is accomplished.

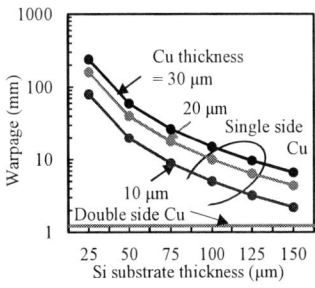

Fig. 2 The estimation of wafer warpage for various Si substrate and Cu plate thickness.

● MOSFETs formation for power device
● Backside grinding for wafer thinning
● Patterning for front side Cu plate
● Cu plating on both sides of Si wafer
● Patterning and etching off back side Cu film on dicing line
● Dicing and assembly

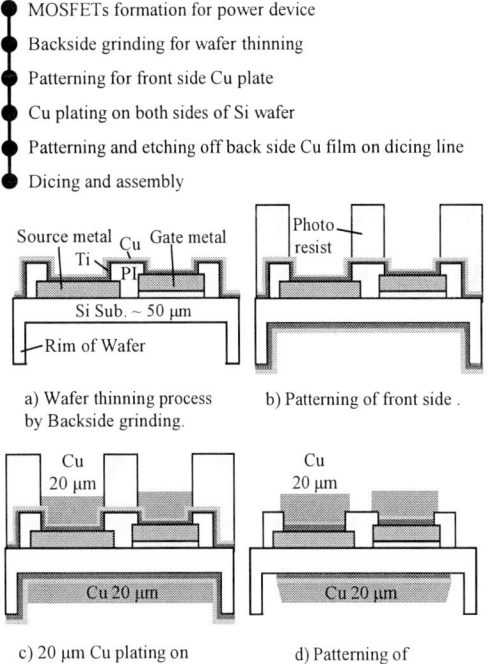

a) Wafer thinning process by Backside grinding.

b) Patterning of front side .

c) 20 μm Cu plating on both sides of Si wafer.

d) Patterning of back side Cu film.

Fig. 3 Process flow of Cu double side plating for vertical Si power MOSFETs.

Fig. 3 shows process flow of Cu double side plating. After formation of MOSFETs, wafer thinning by backside grinding

978-1-7281-0582-6/19 $31.00 © 2019 IEEE   515

down to 50 μm is processed. Then, lithography process on front side is performed for Cu electroplating. It is ideal to process both sides patterning before simultaneous electroplating, but such a both sides lithography is difficult from the point of handling. Therefore front side patterning is processed at first. After that simultaneous both sides Cu plating is done by double side plating equipment. The thickness of both plates is about 20 μm. After that back side patterning is processed in order to get rid of Cu plates in dicing region.

To implement above process flow, both sides lithography and electroplating are the key points. Features of those processes in this fabrication are shown as below.

A. Requirements for patterning process

As stated above, back side surface of the thinning wafer has a rim structure. So front side lithography must be processed without touching with the rim as shown in Fig. 3-b). Therefore stage structure of exposure equipment must be compatible to rim part. Also in case of back side patterning it is necessary to align with front side patterning. And exposure equipment must be able to see front side alignment mark facing exposure stage because there are no alignment mark on back side of Si wafer. And also thick Cu plating has already been formed on back side surface. Therefore it is difficult to see front side alignment mark by infrared from upward of exposure equipment. So we'd have to see front side alignment mark facing exposure stage from downward of stage. An adopted exposure machine in this fabrication is MEMS Stepper NES2W-i06 (Nikon Engineering). This machine meets above requirements.

B. Requirements for Cu plating process

Another process point for this fabrication is double side electroplating. To suppress wafer warpage due to Cu film stress it is necessary to be able to process simultaneous both sides plating and also it is desirable to be able to adjust plating thickness on each side of wafer surface. Thereby we can control wafer warpage due to different surface coverage of Cu plates on each side. And also electroplating equipment is required to be compatible to thinner wafer, for example gentle handling, little damage wafer drying like IPA drying (Maragoni drying). An adopted Cu plating equipment in this fabrication is Multiplate (Atotech). This equipment meets above requirements.

Fig. 4 shows the cross-sectional view of the device with double side Cu plating fabricated by our introduced process

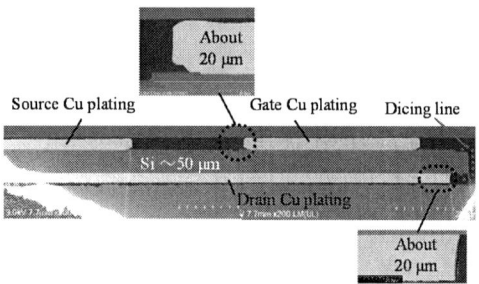

Fig. 4 The cross-sectional view of Si power MOS device with double side Cu plating. The thickness of both front and back is about 20 μm.

flow. As shown Fig. 4 about 20 μm thickness Cu plates are formed on both sides of thinner wafer which is about 50 μm thickness.

III. RESULTS AND DISCUSSION

A. Warpage suppression of the chip

Fig. 5-a) shows temperature dependences of chip warpage with single side and double side Cu plating, comparing to a chip with conventional metal. In single side Cu plating and conventional metal cases, the warpage becomes larger toward convex upward with increasing temperature from 50 to 360 degrees and that becomes smaller toward convex downward with decreasing from 360 to 70 degrees because the stress and thermal expansion are not equivalent. On the other hand, the warpage is extremely small and no dependence of temperature in double side Cu plates case thanks to good stress balance. Fig. 5-b) shows the temperature dependence of warpage when the front side and back side Cu plate thickness are 20 and 10 μm, 10 and 20 μm. In the case of different Cu plate thickness for front and back sides, the warpage has dependence on temperature which is almost same as single side Cu plate case. These results show the controllability of Cu plate thickness is important to suppress the warpage during thermal process.

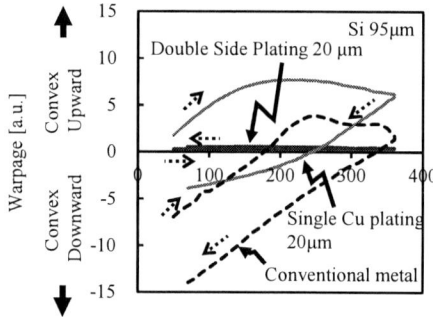

Fig. 5-a) Temperature dependence of warpage with single side and double side Cu plating.

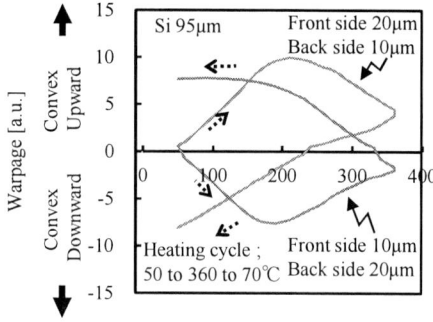

Fig. 5-b) Temperature dependence of warpage when the front side and back side are 20 and 10 μm, 10 and 20 μm.

Fig. 5-c) shows the X-ray observation of chip at 350 and 150 degrees for single Cu plate and double side Cu plating. The larger voids in solder layer are observed in single Cu plate case because of the chip with larger warpage. It is important to reduce the void for wider SOA because the voids are smaller thermal conductivity (0.02 W/m•k). In the double side Cu plates structure, the voids become smaller thanks to smaller warpage.

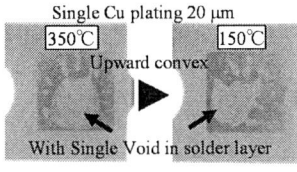

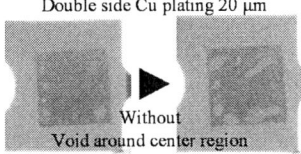

Fig. 5-c) X-ray observation of chip at 350 and 150 degrees for single Cu plate and double side Cu plating.

B. Superior on-resistance (Ron)

Fig. 6 shows the resistance ratio of some elements related to Ron of vertical 40 V Si power MOSFETs. The thinning of Si substrate has a great role for lower Ron because the ratio of Si substrate is about 25 %. The Ron is decreased by 10 % by thinning Si substrate from 95 μm to 50 μm as shown in Fig. 7. Superior Ron in Si 50 μm and Cu 20 μm is obtained comparing to that in Si 95 μm and no Cu plate. This result is good agreement with our estimation as shown in Fig. 7.

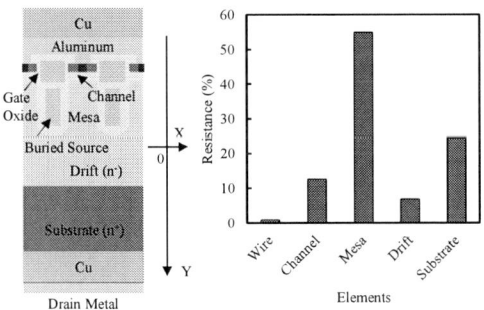

Fig. 6 The resistance ratio of some elements related to on-resistance (Ron) of vertical 40V Si power MOSFETs with buried source.

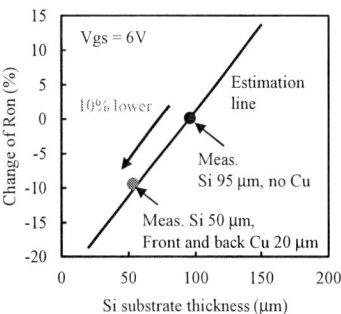

Fig. 7 The comparison of on-resistance between Si power MOSFETs with Si 95 μm and no Cu, and Si 50 μm and 20 μm Cu plate on both sides. The estimation line is calculated by the ratio as shown in Fig. 6.

C. Improvement of Safety operating area (SOA)

As shown in Fig. 8, in order to evaluate SOA, the change of forward-bias at drain before and after power applying is observed when Ids is a fixed value [12]. The operating time (Tpulse) is 100 μs. The Ids is controlled by gate bias.

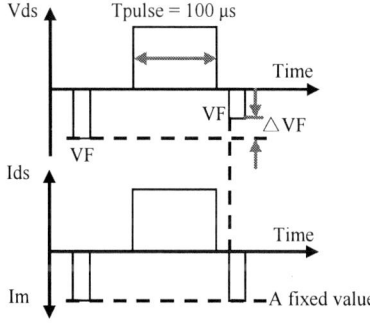

Fig. 8 Procedure of SOA evaluation.

Fig. 9-a) shows the simulation result of time dependence of the maximum temperature and Fig. 9-b) shows temperature profile in the device with and without double side Cu film. In this simulation, Id and Tmax are calculated at a fixed Vg. Id increases with temperature because the Vth decreases. That of active area with double side Cu plating becomes lower compared with no Cu plate structure after power applying because thicker Cu plate has larger heat content. Accordingly, SOA has been improved as below. Fig. 10-a) shows the definition of SOA in our experiments. The SOA is defined as Id when a fixed value in ΔVF becomes larger than the reference line. Fig. 10-b) shows the ratio of SOA (ΔId/Id1) of devices with and without double side Cu film. That becomes larger with increasing of Vds. As a result, 100 % larger SOA at Vds = 32 V is observed thanks to 20 μm Cu plate with larger heat content.

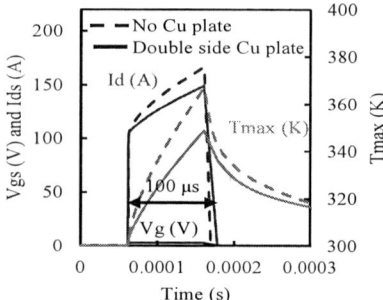

a) Time dependence of the maximum temperature.

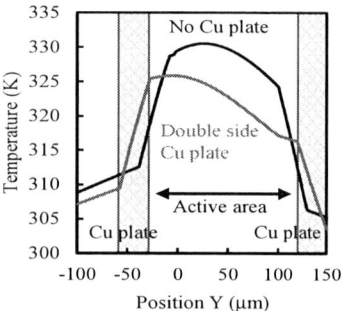

b) Temperature profile in the device.

Fig. 9 The simulation results of temperature with and without double side Cu film.

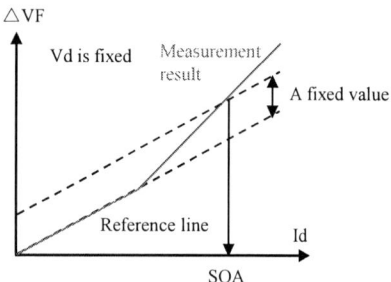

a) Definition of SOA in our experiments.

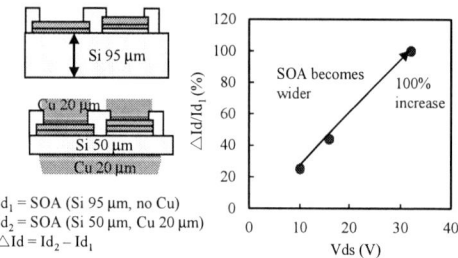

Id_1 = SOA (Si 95 μm, no Cu)
Id_2 = SOA (Si 50 μm, Cu 20 μm)
$\triangle Id = Id_2 - Id_1$

b) The results of SOA with and without double side Cu film.

Fig. 10 Definition and the results of SOA.

IV. CONCLUSION

The double side Cu plating technology is introduced to Si power MOSFET with thin Si substrate for the first time. This structure has lower on-resistance (Ron) by decrease of substrate resistance, wider safety operating area (SOA) by larger heat content of thicker Cu plate. Additionally larger warpage and larger void in solder layer can be suppressed by double side Cu film.

ACKNOWLEDGMENT

The authors gratefully acknowledge the contributions of Nikkon engineering for sample fabrication.

REFERENCES

[1] B. J. Baliga, "Power semiconductor devices having improved highfrequency switching and breakdown characteristics", United States Patent, No. 5,998,833(1998).

[2] M. A. Gajda, S.W. Hodgkiss, L.A. Mounfield, N.T. Irwin, G.E.J. Koops and R. van Dalen,", "Industrialisation of Resurf Stepped Oxide Technology for Power Transistors", Proceedings of ISPSD, pp. 109-112, 2006.

[3] J. N. Burghartz, "Ultra-thin chip technology and applications", pp. 321-326. Springer.

[4] Q. Wang, M. Li, Y. Sokolov, A. Black, H. Yilmaz, J. V. Mancelita, and R. Nanatad, "Power trench MOSFET devices on metal substrate," IEEE Trans. ED, vol.29, No.9, pp.1040-1042, 2008.

[5] P. Hower, C-Y. Tsai, S. Merchant, T. Efland, S. Pendharkar, R. Steinhoff, and J. Brodsky, "Avalanche-induced thermal instability in Ldmos transistors," Proc. ISPSD, pp.153-156, 2001.

[6] Y. S. Chung, T. Willett, V. Macary, S. Merchant, and B. Baird, "Energy capability of power devices with Cu layer integration," Proceedings of ISPSD, pp. 63-66, 1999.

[7] G. V. den bosch, T. Webers, E. Driessens, B. Elattari, D. Wojciechowski, P. Gassoe, P. Moens, G. Groeseneken, "Design and characterization of a post-processed copper heat sink for smart power drivers," Proc. ICMTS, pp. 27-31, 2005.

[8] C. Hu and M-H. Chi, "Second Breakdown of Vertical Power MOSFET's," IEEE Trans. ED., vol. 29, No.8, pp.1287-1293, 1982.

[9] W-Y. Chen and M-D. Ker, "Characterization of SOA in Time Domain and the Improvement Techniques for Using in High-Voltage Integrated Circuits" IEEE Trans. DMR., vol. 12, No.2, pp.382-390 2012.

[10] C. Melvin and B. Roelfs, "Simultaneous front and back side Cu metallization on power chips," Proc. ASMC, pp. 189-191, 2017.

[11] M. R. Marks, Z. Hassan and K. Y. Cheong, "Effect of nanosecond laser dicing on the mechanical strength and fracture mechanism of ultrathin Si dies with Cu stabilization layer," IEEE Trans. CPMT, vol. 5, No.12, pp.1885-1897, 2015.

[12] P. Moens G. V. den bosch, "Characterization of Total Safe Operating Area of Lateral DMOS Transistors" IEEE Trans. DMR, vol. 6, No.3, pp.349-357, 200

THE 32ND INTERNATIONAL SYMPOSIUM ON POWER SEMICONDUCTOR DEVICES AND ICS

ISPSD is the premier forum for technical discussion in all areas of power semiconductor devices and power integrated circuits. ISPSD2020 will be held in the city of Vienna, described as Europe's cultural capital, UNESCO world heritage site and a metropolis with unique charm, vibrancy and flair.

TOPICS OF INTEREST INCLUDE BUT ARE NOT LIMITED TO:

High Voltage Power Devices (HV): High voltage silicon based discrete devices (>200V) such as super junction MOSFETs, IGBTs, thyristors, GTOs and pn-diodes

Low Voltage Power Devices and Power IC Technology (LVT): Low voltage silicon based discrete power devices ($\leq$ 200V) and power devices for power ICs of all voltage ranges

Power IC Design (ICD): Circuit design and demonstration using power IC technology platform

GaN and Other Compound Materials (GaN): GaN and other compound material (e.g. AIN, Ga_2O_3, GaAs) based power devices, technology and integration, materials, and processing

SiC and Other Materials (SiC): SiC and other material (e.g. diamond) based power devices, technology and integration, materials, and processing

Module and Package Technologies (PK): Module and package technology for discrete power devices and power ICs

ABSTRACT SUBMISSION:

Abstract Submission Deadline: November 8, 2019
www.ispsd2020.com

Author Notification: January 20, 2020

Late News Submission (limited acceptance): February 28, 2020

Final Manuscript Submission Deadline: March 13, 2020

Submission Requirements:
A PDF abstract should be submiced through the website Including a single page text summary in English (500 words maximum) and up to two additional pages of supporting figures.

GENERAL CHAIR:
Dr. Oliver Häberlen, Infineon Technologies Austria
Email: oliver.haeberlen@infineon.com

TECHNICAL PROGRAM CHAIR:
Prof. Nando Kaminski, University of Bremen
Email: nando.kaminski@uni-bremen.de

TECHNICALLY CO-SPONSORED BY

1ST ANNOUNCEMENT AND CALL FOR PAPERS   www.ispsd2020.com

AUTHOR INDEX

A

Abe, Seiya	339
Adachi, Kohei	39
Agarwal, Aditi	159
Aiba, Ruito	23, 167
Aida, Kikuo	91, 99
Alatise, Olayiwola	207
Alexandrov, Peter	191
Alfieri, Giovanni	103
Allerstam, F.	163
Ambacher, Oliver	111
Andenna, Maxi	47
Antoniou, Marina	175
Arai, Daisuke	335
Arango, Y.	103, 211
Asaba, S.	139
Asada, Takeshi	335
Azam, Misbah	95

B

Bakeroot, Benoit	287, 419
Baliga, B. Jayant	159
Bartels, Martin	511
Basso, Michele	399
Bat-Ochir, Bat-Otgon	251
Bau, Plinio	75
Baumann, Remo	491
Bayarkhuu, Battuvshin	251
Bellini, M.	103, 211, 215
Bertaiola, Simone	399
Bertrams, Thomas	511
Bhalla, Anup	191
Bianda, E.	103, 211, 215
Bittner, R.	235
Boettcher, N.	351
Bolotnikov, Alexander	179
Borghese, A.	215
Bort, Lorenz	219
Boselli, Gianluca	407
Breglio, G.	215, 315
Buetow, S.	235
Buitrago, Elizabeth	47
Burani, N.	235

C

Canato, Eleonora 287
Cao, Han 247
Cao, Lei 227
Cao, Shilin 387
Castellazzi, A. 451
Chang, Edward Yi 427
Chang, J.Y. 239
Chang, T.C. 239
Chang, Wen-Che 299
Chen, Kevin J. **199, 263, 275, 291, 295, 411, 435, 459, 463, 467**
Chen, Kuangli 431
Chen, Nan 331
Chen, P.W. 67
Chen, Wanjun **271, 331, 347, 403, 423, 431**
Cheng, Chu Yao 83
Cheng, Chuanyi 279
Chiu, H.-C. 67
Cho, Inwook 391
Choi, Junghun 391
Chu, Kuo-Ting **135, 255**
Constant, A. 163
Corvasce, C. **47, 55**
Cougo, Bernardo 75
Cousineau, Marc 75
Cui, Xiang 355
Cui, Yingxin 487

D

Dang, Kui 471
D'Aniello, F. 451
Decoutere, Stefaan 419
Deng, Erping 243
Deng, Gaoqiang 359
Deng, Shuairong 487
Deng, Xiaochuan 231
Dengel, Gabriel 95
Deviny, I. 323
Diaz Reigosa, P. 55
Dick, Jan Frederik 343
Dimitrijev, Sima 415
Duan, Baoxing 379
Dugarjav, Bayasgalan 251
Dymond, Harry C.P. 79

E

Ebihara, Yasuhiro 35
Eikyu, Katsumi 371
Elahipanah, Hossein 115
Elsayed, Ahmed 343

Eon, David	151
Erlbacher, T.	351
Ezaki, Akira	515

F

Fan, Diao	359
Fan, Jie	411
Fang, Jian	123
Fayyaz, A.	451
Feng, Hao	143
Feng, Xu-dong	127
Flumian, Didier	75
Franchi, J.	163
Fujishima, Naoto	143
Fujita, Ryuusei	259
Fukatsu, S.	139
Fuks, Adam	17
Fukui, Munetoshi	43, 311, 339
Furukawa, Kazuyoshi	43, 311
Furukawa, M.	139

G

Galbiati, Paola	399
Gao, Hongwei	411
Gao, Wei	51
Gao, Wuhao	403
Gao, Zijian	227
Gerfer, Alexander	7
Ghandi, Reza	179
Grossner, Ulrike	219
Gu, Miaosong	355
Gu, Yan	279, 387
Gu, Yunfei	487
Guo, Qing	155, 227

H

Han, Jisheng	415
Han, Kijeong	159
Han, Ping-Cheng	427
Han, Shaowen	63, 439
Han, W.K.	239
Han, Xiaoqi	431
Hao, Yue	471
Hara, Takuma	515
Harada, Shinsuke	23, 31, 167
Hasegawa, K.	251, 339, 503
He, Jiabei	291, 295
He, Junqing	507
He, Simon	67
Heckel, T.	351

Herzer, R. 235
Hino, Shiro 27
Hiramoto, Toshiro 43, 311, 339
Hirasawa, Wataru 335
Ho, Yu-Hsuan 427
Hoffmann, Felix 147
Hollis, Simon J. 79
Honda, Masaaki 335
Hong, Seunghyun .... WITHDRAWN 283
Horita, Masahiro 59
Hoshii, Takuya 43, 311
Hossain, Zia 383
Hsu, Fu-Jen 135, 255
Hu, Li 127
Hu, Zongyang 483
Hua, Mengyuan 291, 295, 435, 459, 463, 467
Huang, Alex Q. 231
Huang, Boning 1
Huang, Peng 431
Huang, Sen 411
Huang, Yongzhang 243
Hung, Chien-Chung 135, 255
Hutchings, J. 323

I

Iannuzzo, F. 55
Ichimura, Aicko 35
Ichinoseki, Kentaro 91, 99
Iida, H. 503
Iijima, R. 139
Iizuka, Koji 375
Ina, Tsutomu 303
Inumiya, Seiji 515
Ipposhi, Takashi 375
Irace, A. 215
Irace, Andrea 315
Ishigaki, Takashi 259
Ito, T. 139
Itou, Kazuo 43, 311, 339
Iwai, Hiroshi 43, 311
Iwamuro, Noriyuki 23, 167, 187

J

Jadav, Sumit 267
Jeliazkov, Stoyan 267
Jena, Debdeep 483
Ji, Shiyang 39
Jiang, Huaping 199
Jiang, Pengkai 119
Jiang, Qian 347
Jiang, Qimeng 1

Jin, Rui 355
Jinno, Riena 483

K

Kachi, Tetsu 59
Kachi, Tsuyoshi 371
Kai, Tetsuya 515
Kakarla, Bhagyalakshmi 219
Kakushima, Kuniyuki 43, 311
Kallfass, Ingmar 111, 455
Kalnitsky, Alex 299
Kaminski, Nando 147
Kanamori, Taiga 23, 167
Kanechika, Masakazu 59
Kang, H. 319
Kang, Xuanwu 411
Kang, Yuhui 247
Kao, K.S. 239
Kargarrazi, Saleh 115
Karmel Kranthi, Nagothu 407
Kato, Hiroaki 99
Kawada, Yasuyuki 39
Kawaguchi, Yusuke 91, 99
Kawahara, Koutarou 27
Kennerly, Stacey 179
Kim, Daehoon 391
Kim, Jaeeuk 391
Kimoto, Tsunenobu 59
Kimura, Koji 363
Kinoshita, Koyo 259
Kinoshita, Tomoko 363
Knoll, L. 103, 211, 215
Kobayashi, Hitoshi 515
Kobayashi, Kenya 99
Kobayashi, Yusuke 23, 31
Kojima, Hideharu 91, 515
Komatsu, Akira 515
Komatsu, Kanako 363
Kong, C. 323
Konishi, Kumiko 259
Koseki, Kunio 171
Kosugi, Ryoji 39
Kozumi, Shinsuke 515
Kranz, L. 103, 211
Kujath, M. 235
Kumar, B. Sampath 407
Kumazawa, Teruaki 31
Kundu, Aritra 107
Kwan, Man-Ho 299
Kyogoku, Shinya 31

L

Lansbergen, Gabriel Petrus	299
Lee, Chwan-Ying	135, 255
Lee, Jaehee	391
Lee, John	67
Lee, Junho WITHDRAWN	283
Lee, Kuemju	391
Lee, Lurng-Shehng	135, 255
Lei, Jiacheng	459, 463
Lemke, Marko	511
Leng, Yahui	83
Letellier, Juliette	151
Li, Ajiang	479
Li, Ankang	387
Li, Anqi	243
Li, Baikui	199, 415
Li, Donghua	51
Li, Hui	415
Li, Juntao	487
Li, Lei	247
Li, Lianghui	487
Li, Ming	123
Li, Seiya	67
Li, Shaohong	479
Li, Sheng	447
Li, Wenshen	483
Li, Xin	499
Li, Xu	231
Li, Xuan	231
Li, Xuebao	355
Li, Xueqing	191
Li, Yiheng	447
Li, Yongkai	63
Li, Zehong	51
Li, Zhaoji	51, 331, 347, 367, 403, 423, 507
Li, Zhiqiang	487
Liang, Y.C.	271
Liao, Henry	67
Lilienfeld, David	179
Lin, H.H.	239
Lin, Ming-Cheng	299
Lin, Y.T.	239
Ling, Rongxun	119
Liu, Chao	331, 347, 403
Liu, Dawei	79
Liu, Pengku	231
Liu, Siyang	395, 447
Liu, Xi	423
Liu, Xinyu	411
Liu, Yinxiang	63, 439
Liu, Yong	143

Long, Hu	155
Lophitis, Neophytos	175
Losee, Peter	191
Lu, Guo-Quan	499
Lu, Yangyang	279, 479
Lu, Yu Shen	83
Lu, Yuliang	1
Luo, H.	323
Luo, Ping	119
Luo, Xiaorong	359, 367
Luo, Yunzhong	123
Lutz, Josef	495
Lynch, Justin	223

M

Ma, D. Brian	87
Ma, Jie	387
Ma, Jun	71
Maeda, Takuya	59
Mahmoud, Ahmed	511
Maresca, L.	215, 315
Martino, Antonino	399
Masante, Cédric	151
Masuko, Shingo	515
Matioli, Elison	71
Matsudai, Tomoko	43, 311
Matsuo, Mie	515
Matsuoka, Fumitomo	363
Mawby, Philip	207
Mei, Yunhui	499
Meijer, Maurice	17
Meneghesso, Gaudenzio	287
Meneghini, Matteo	287
Mesemanolis, Athanasios	47
Mihaila, A.	103, 147, 211, 215
Ming, Xin	127, 423
Mitani, Shuhei	35
Mitsui, Yohei	27
Miura, Naruhisa	27
Mizoguchi, Takeshi	307
Mochizuki, Kazuhiro	39
Moench, Stefan	111
Moens, P.	163, 287
Mori, Takahiro	375
Mori, Yuki	259
Morikawa, Takahiro	259
Morimoto, Tadao	31
Motai, Takako	515
Mullapudi, Raghuram	383
Müller, Marco	511
Mulpuri, Vamsi	267

Muneta, Iriya 311
Murata, Tatsunori 259

N

Nagamatsu, Takahito 363
Nagel, Nicolas 511
Nakabayashi, Y. 139
Nakajima, Akira 311
Nakamura, Tohru 483
Nakashima, Junichi 27
Narazaki, Atsushi 475
Narita, Tetsuo 59
Nava, Melvin 191
Nela, Luca 71
Ng, Wai Tung 83
Nguyen, Nam-Trung 415
Ngwendson, L. 323
Ning, Puqi 247
Ning, Runtao 107
Ninomiya, Tamotsu 339
Nishi, Koichi 475
Nishiguchi, Toshifumi 99
Nishii, Junya 303
Nishiwaki, Tatsuya 91, 99, 515
Nishizawa, Shin-ichi 43, 311
Nobe, Tsuguo 11
Noborio, Masato 35
Nomoto, Kazuki 483
Numasawa, Yohichiroh 43

O

Oasa, Kohei 91, 99
Oda, Tetsuo 259
Ødegård, Bjørn 491
Oeder, T. 443, 451
Ogawa, Masaaki 515
Ogura, Atsushi 43
Ohashi, Hiromichi 43, 311
Ohguro, Tatsuya 91, 515
Ohno, Tetsuya 99
Ohtsuka, Kenichi 515
Oka, Tohru 303
Okawa, Masataka 23, 167
Okumura, Hajime 39
Omura, I. 43, 251, 339, 503
Onozawa, Yuichi 143
Oohashi, Kenichi 515
Ortiz Gonzalez, Jose 207
Ouyang, Dongfa 367

P

Papadopoulos, C. 47, 55, 211
Park, Chanho 95
Pelaic, K. 351
Peng, Cheng 355
Peng, Xin 51
Pernot, Julien 151
Pfost, M. 195, 443, 451
Popescu, Dan 327
Posthuma, Niels 419
Prasmusinto, A. 103, 211

Q

Qian, Qingkai 459
Qiao, Bin 347, 403
Qiao, Ming 507
Qin, Yao 127
Quay, Rüdiger 111

R

Rahimo, M. 47, 55
Ravisekhar, Raju 179
Reiner, Richard 111
Ren, Bin 243
Ren, Min 51
Ren, Na 183, 203
Riccardi, Damiano 399
Riccio, M. 215, 315
Richardeau, Frederic 75
Romano, G. 211, 215
Rouger, Nicolas 75, 151
Rudolf, Ralf 511

S

Saddiqui, I. 323
Saito, Takashi 371
Saito, Wataru 43, 307
Sakaguchi, Shoichi 363
Sakamoto, Toshihiro 363
Sakiyama, Yoko 307
Salman, Akram 407
Saraya, Takuya 43, 311, 339
Satoh, Katsumi 43, 311
Sayama, Hirokazu 375
Schulze, Joerg 343
Schwabe, Christian 495
Segawa, Satoshi 39
Seidel, Peter 495
Senesky, Debbie 115
Shan, Zhou Rong   WITHDRAWN   283

Shao, Bin	17
Shen, Z. John	107
Sheng, Kuang	63, 155, 183, 203, 227, 439
Shi, Jingyuan	411
Shi, Qi	331
Shi, Yijun	347, 403
Shi, Yu	403
Shi, Yuanyuan	423, 431
Shi, Yue	131
Shibib, Ayman	95
Shigyo, Naoyuki	43, 311
Shima, Akio	259
Shimizu, T.	139
Shimomura, Saya	99
Shioda, Saori	363
Shrivastava, Mayank	407
Sin, Johnny K.O.	143
Singh, Ranbir	267
Soler, Victor	147
Song, Kinam WITHDRAWN	283
Song, Wei	1
Song, Yingxin	51
Spehar, James	17
Stahrenberg, Knut	511
Stark, Bernard H.	79
Stockman, A.	163, 287
Stoffels, Steve	419
Suda, Jun	59
Sudario, Frank	191
Sudo, Masaki	339
Suguro, Kyoichi	515
Sumaoang, Deborah	191
Sun, Defu	51
Sun, Jiahui	263
Sun, Licheng	379
Sun, Qian	411
Sun, Ruize	271, 403
Sun, Tao	359, 367
Sun, Weifeng	279, 387, 395, 447, 479
Sun, Yinxia	447
Sundaresan, Siddarth	267
Sung, Woongje	223
Surdi, Harshad	383
Suzuki, Noriaki	335
Suzuki, Shinichi	43, 311, 339
Suzuki, T.	139

T

Takada, Yoshiharu	91, 515
Takahashi, Keita	363
Takahashi, Mari	515

Takahashi, Tetsuo ... 475
Takakura, Toshihiko ... 43, 311, 339
Takei, Manabu ... 31
Takeuchi, Kiyoshi ... 43, 311, 339
Takeuchi, Yuichi ... 35
Tanaka, Yasunori ... 171
Tang, Gaofei ... 275, 459
Tang, Xi ... 415
Tang, Xinling ... 355
Terrill, Kyle ... 95
Tian, Tian ... 479
Tiwari, Amit K. ... 175
Tominaga, Takaaki ... 27
Tomohisa, Shingo ... 27
Tong, Xin ... 479
Tong, Zikang ... 115
Treiber, Maximilian ... 327
Tsai, Chun-Lin ... 299
Tsibizov, Alexander ... 219
Tsukamoto, Naoto ... 307
Tsukuda, Masanori ... 43, 251, 339
Tsuruta, Kazuhiro ... 35
Tsutsui, Kazuo ... 43, 311
Tuan, Hsiao-Chin ... 299
Tzeng, C.M. ... 239

U

Udrea, F. ... 175, 319
Ueda, Hiroyuki ... 59
Uehara, Junichi ... 35
Ueno, Yukihisa ... 303
Uesugi, Tsutomu ... 59
Unger, Christian ... 195
Urata, Akihiro ... 363

V

van Beek, Joost ... 17
Vauclair, Marc ... 17
Vinnac, Sebastien ... 75
Vobecky, Jan ... 47

W

Wakabayashi, Hitoshi ... 43, 311
Wakimoto, Setsuko ... 143
Waltereit, Patrick ... 111
Wang, Dongsheng ... 447
Wang, Fangzhou ... 331, 403
Wang, Hanxing ... 291
Wang, Jianjing ... 79
Wang, Jiannong ... 415

Wang, Kang L. 183
Wang, Meiyu 499
Wang, Wenwu 411
Wang, Xiaolei 411
Wang, Xinhua 411
Wang, Y. 323, 379
Wang, Yanhao 243
Wang, Yaohui 395
Wang, Yuan 331
Wang, Yunkun 131
Wang, Yuru 263, 291, 459, 463
Wang, Zhuo 127
Watanabe, Masahiro 43, 311
Watanabe, Yuji 335
Wei, Dong 431
Wei, Jiaxing 447, 479
Wei, Jin 199, 263, 275, 291, 295, 435, 459, 463, 467
Wei, Ke 411
Weimer, Julian 455
Weis, Rolf 511
Wellekens, Dirk 419
Wikström, Tobias 491
Wirths, S. 103, 211, 215
Wong, Roy K.-Y. 67
Wu, Cheng-Pao 299
Wu, Chia-Hsun 427
Wu, Haw-Yun 299
Wu, Jiupeng 183, 203
Wu, Macro 67
Wu, Wangran 395
Wu, Y.B. 67
Wu, Yucao 119

X

Xia, Yun 347, 403
Xiao, Ming 471
Xie, Andy 67
Xie, Qiang 1
Xie, Ruiliang 291
Xie, Yijing 499
Xin, Yajie 403
Xing, Huili 483
Xiong, Wei 423
Xu, Han 275, 291
Xu, Hongyi 183
Xu, Xiaorui 331, 347
Xue, Peng 315

Y

Yamaguchi, Takeshi 335
Yamaji, Mizue 335

Yamamoto, Masayuki ... 171
Yamashiro, Yusuke ... 31
Yan, Ding ... 279
Yan, Zong-Zheng ... 427
Yang, Bowei ... 279
Yang, Chao ... 367
Yang, Guangan ... 395
Yang, Liping ... 231
Yang, Shu ... 63, 439
Yang, Shuo ... 411
Yang, Song ... 295, 463, 467
Yang, Wentao ... 143
Yang, Yang ... 51
Yang, Yintang ... 379
Yang, Zhuo ... 387
Yano, Hiroshi ... 23, 167, 187
Yao, Kailun ... 187
Ye, Ran ... 479
Yen, Cheng-Tyng ... 135, 255
Yeo, Yee-Chia ... 271
Yin, Haibo ... 411
Yonemura, Koji ... 363
Yonezawa, Naoki ... 515
Yonezawa, Yoshiyuki ... 39
Yu, Jingshu ... 83
Yu, Jiun-Lei ... 299
Yu, Siyuan ... 279
Yuan, Tianshu ... 247
Yuan, Yandong ... 131
Yun, Nick ... 223

Z

Zetterling, Carl-Mikael ... 115
Zhang, A.B. ... 67
Zhang, Anbang ... 367
Zhang, Bo ... 51, 123, 127, 131, 231, 271, 331, 347, 359, 367, 403, 423, 431, 507
Zhang, Chi ... 447
Zhang, Erli ... 123
Zhang, Guangsheng ... 507
Zhang, J.H. ... 67
Zhang, Jeff ... 67
Zhang, Jincheng ... 471
Zhang, Jinping ... 51
Zhang, Jun ... 415
Zhang, Kenan ... 331
Zhang, Lin ... 487
Zhang, Long ... 387, 395, 447, 479
Zhang, Martin ... 67
Zhang, Meng ... 199, 291
Zhang, Sen ... 279, 387, 507
Zhang, Tingting ... 447

Zhang, Wei Jia	83
Zhang, Weihang	471
Zhang, Wentong	507
Zhang, Xuan	127
Zhang, Yuhao	471
Zhang, Yunwu	279, 387
Zhang, Zhaofu	435, 459, 467
Zhang, Zhili	507
Zhang, Zhi-wen	127
Zhao, Cezhou	271
Zhao, Jingchuan	507
Zhao, Thomas	67
Zhao, Yishang	51
Zhao, Zhibin	243, 355
Zheng, Yingkui	411
Zheng, Zheyang	199, 263, 435, 459, 467
Zhong, Kailun	291, 295
Zhou, C.	67
Zhou, Hong	471
Zhou, Kun	487
Zhou, Qi	127, 331, 347, 403, 423, 431
Zhou, Qijun	347
Zhou, Xiao	119
Zhou, Xin	507
Zhou, Yuanfeng	107
Zhou, Zekun	131
Zhu, C.	323
Zhu, Guichuang	479
Zhu, Jing	279, 387, 395, 479
Zhu, Kuncun	51
Zhu, Mike	191
Zhu, Minghua	71
Zhu, Tinggang	447
Zhu, Xuhan	507
Zhu, Youhua	447
Ziemann, Thomas	219
Zou, Peng	1
Zou, Y.B.	67
Zou, Yanqin	387, 479

IEEE
445 Hoes Lane
Piscataway, NJ 08854-4141

ISBN 978-1-7281-0582-6

2010 Twenty-Fifth Annual IEEE Applied Power Electronics Conference and Exposition (APEC 2010)

Palm Springs, California, USA
21-25 February 2010

IEEE Catalog Number: CFP10APE-POD
ISBN: 978-1-42444-782-4